FIELD THEORY OF GUIDED WAVES

Second Edition

ROBERT E. COLLIN

CASE WESTERN RESERVE UNIVERSITY

IEEE Antennas and Propagation Society, *Sponsor*

The Institute of Electrical
and Electronics Engineers, Inc.
New York

Oxford University Press
Oxford

IEEE PRESS
445 Hoes Lane, PO Box 1331
Piscataway, NJ 08855-1331

Printed in the United States of America

10   9   8   7   6   5

This is a copublication of the IEEE Press in association with Oxford University Press.
IEEE ISBN 0-87942-237-8
IEEE Order Number: PC2568

Oxford University Press, Oxford OX2 6DP
Oxford is a trade mark of Oxford University Press
OUP ISBN 0 19 859213 2

Library of Congress Cataloging-in-Publication Data

Collin, Robert E.
  Field theory of guided waves / R. E. Collin. — 2nd ed.
    p.   cm.
  Includes bibliographical references and index.
  ISBN 0-87942-237-8
  1. Electromagnetic theory.   2. Wave guides.   3. Field theory
(Physics)   I. Title.
QC670.C55   1991
530.1'41—dc20

Contents

3   **Transverse Electromagnetic Waves**   173

4   **Transmission Lines**   247

      5.1   General Properties of Cylindrical Waveguides   330
      5.2   Orthogonal Properties of the Modes   333
      5.3   Power, Energy, and Attenuation   337
      5.4   The Rectangular Waveguide   349
      5.5   Circular Cylindrical Waveguides   354
      5.6   Green's Functions   356
      5.7   Analogy with Transmission Lines   367
      5.8   The Tangent Method for the Experimental Determina-
            tion of the Equivalent-Circuit Parameters   373
      5.9   Electromagnetic Cavities   377
      5.10 Cavity with Lossy Walls   387
      5.11 Variational Formulation for Cavity Eigenvalues   395
      5.12 Cavity Perturbation Theory   400
            References and Bibliography   402
            Problems   404

6   **Inhomogeneously Filled Waveguides and
      Dielectric Resonators**   **411**

      6.1   Dielectric-Slab–Loaded Rectangular Guides   411
      6.2   The Rayleigh–Ritz Method   419
      6.3   A dielectric Step Discontinuity   430
      6.4   Ferrite Slabs in Rectangular Guides   433
      6.5   Dielectric Waveguides   441
      6.6   Dielectric Resonators   459
            References and Bibliography   467
            Problems   470

7   **Excitation of Waveguides and Cavities**   **471**

      7.1   The Probe Antenna   471
      7.2   The Loop Antenna   483
      7.3   Coupling by Small Apertures   499
      7.4   Cavity Coupling by Small Apertures   523
      7.5   General Remarks on Aperture Coupling   531
      7.6   Transients in Waveguides   533
            References and Bibliography   537
            Problems   539

8   **Variational Methods for Waveguide
      Discontinuities**   **547**

      8.1   Outline of Variational Methods   547
      8.2   Capacitive Diaphragm   569

Preface

When the IEEE Press expressed interest in reprinting the original edition of *Field Theory of Guided Waves* I was, of course, delighted. However, I felt that some revision of the original book would greatly enhance its value. The original edition was published in 1960, and since that time the field of applied electromagnetics has advanced on several fronts, and a variety of new problems have come into prominence. There was a clear need to include some of these advances in a revised edition. We agreed that a modest revision would be undertaken. As the revision proceeded it became clear that space limitations would not allow in-depth treatment of many of the newer developments. Even with this constraint, the revised edition contains approximately 40% new material, considerably more than was originally envisioned.

The constraints I placed on myself in carrying out the revision were to use as much of the original material as possible without rewriting and to limit the amount of new material to what I felt was most urgently needed in support of current research activities. The unfortunate consequence of such a decision is that one is committed to using the old notation and retaining the original development of many topics. Because one's preferred approach to the development of a particular topic or theory changes with time, the result is not always optimum. I have made a concerted effort to blend new material with old material such that the overall presentation forms a coherent overall treatment. I hope the reader will find that this goal has been achieved to a satisfactory level.

The main focus of the revised edition is essentially the same as in the original: A theoretical treatment of wave-guiding structures and related phenomena along with the development of analytical methods for the solution of important engineering problems. Perhaps the greatest development in electromagnetics research in the past three decades is that of numerical analysis and solutions of complex problems on computers. A significant portion of current research is numerically oriented, to the extent that one sometimes is led to believe that analytical methods are of secondary importance. It is my firm conviction that successful numerical work depends critically on analytical techniques, not only for robust problem formulation but also as a necessity to develop physical understanding of complex electromagnetic phenomena. Thus in the revised edition the analytical approach is stressed with very little reference to numerical methods. Numerical methods are very important, and are treated in depth in the recent IEEE Press book *Numerical Methods for Passive Microwave and Millimeter Wave Structures* by R. Sorrentino. Thus there was no need to include a treatment of numerical methods.

Chapter 1 is a review of basic electromagnetic theory and includes a discussion of boundary conditions, new material on field behavior in source regions, field behavior at the edge of a conducting wedge, and added material on field singularities at a dielectric edge or corner. The original material on field equivalence principles has been improved, and Babinet's principle is developed in a more general way. A development of expressions for the electric and magnetic energy densities in dispersive media has also been added. Also included in this chapter is the standard theory for vector, scalar, and Hertzian potential functions.

Chapter 2 is essentially all new and gives a broad and comprehensive account of scalar and dyadic Green's functions. This extensive chapter covers Green's functions for the Sturm-Liouville equation; alternative representations for Green's functions; synthesis of multidimensional Green's functions from characteristic one-dimensional Green's functions; the eigenfunction expansion of dyadic Green's functions in rectangular, cylindrical, and spherical coordinates; and the conversion of the eigenfunction expansions into representations in terms of modes. The chapter concludes with a development of the electric and magnetic field integral equations for scattering, a discussion of uniqueness, and the use of dyadic Green's functions with boundary values interpreted in terms of equivalent surface sources. Scattering by a conducting sphere is included as an example to illustrate the occurrence of resonances when the electric field or magnetic field integral equations are used to solve the scattering problem.

Chapter 3 deals with plane waves in homogeneous isotropic, anisotropic, and ferrite media. The transmission line theory for propagation through a multilayered medium is developed. The new material added to this chapter is on dyadic Green's functions for layered media, which is important for many current problems associated with planar transmission lines and microstrip patch resonators. Also new is the discussion on group, signal, and phase velocities and a short treatment of dipole radiation in an anisotropic dielectric medium.

The theory of transverse electromagnetic (TEM) transmission lines is developed in Chapter 4. Variational methods and conformal mapping methods for determining the line capacitance and characteristic impedance are covered. A considerable amount of new material on microstrip and coupled microstrip lines has been added. The spectral domain Galerkin method for microstrip lines is developed in detail. In addition, the potential theory for planar transmission lines is developed. Conformal mapping techniques are described that enable the dominant part of the Green's functions to be diagonalized over the microstrip, both for the single line and the coupled line. This technique is an alternative to Lewin's singular integral equation techniques and leads to efficient and robust formulations for line parameter evaluation on a computer.

The theory of uniform metallic waveguides is developed in the fifth chapter. In addition to the material in the original edition, new material on eigenfunction expansions and mode representations of dyadic Green's functions for waveguides has been added. Specific results for rectangular and circular waveguides are given. I have also taken this opportunity to include the theory of electromagnetic cavities and dyadic Green's functions for cavities. Perturbation theory for a cavity containing a small dielectric or magnetic obstacle is also developed and provides the basis for a well-known technique to measure the complex permittivity of materials.

In the original edition, Chapter 6 covered the topics of dielectric and ferrite slabs in rectangular waveguides and variational methods for calculating the propagation constants. This material has been retained in the revised edition as well. The subjects of dielectric waveguides and resonators have been of great interest in recent years and should have received an in-depth treatment. However, we chose to limit the discussion on these topics in the interest of space. Thus we only provide an introductory treatment of variational methods that form the basis for the finite element method, a discussion of the boundary element method, and an introduction to dielectric resonators.

Chapter 7 treats a number of topics related to the excitation of waveguides by probes and loops, aperture coupling of waveguides, and aperture coupling of waveguides and cavities. A more complete theory of the basic waveguide probe problem is given along with a more careful consideration of the limitations of the variational formulation for the probe impedance. The original small-aperture theory formulated by Bethe had one major shortcoming, which was that it did not give a solution for the radiation conductance of the aperture. As a consequence, the results of the theory could be interpreted in terms of an equivalent circuit only by invoking other considerations. For coupling between dissimilar regions, it was often difficult to construct a meaningful physical equivalent circuit for the coupling problem. We have overcome this deficiency by adding a radiation reaction term to the aperture polarizing field. The resultant theory is now fully internally self-consistent and leads directly to physically meaningful equivalent circuits for the coupling. It also enables one to treat the problems of aperture coupling between waveguides and cavities, again yielding physical equivalent circuits. This improved small-aperture theory is presented along with a number of examples that illustrate how it is applied in practice.

After the revision of Chapter 7, it became evident that new material could be added to the remaining chapters only at the expense of some of the old material. However, I found that relatively little original material could be eliminated since much of it was still essential as background material for any new topics that might be introduced. Consequently I chose to keep Chapters 8 through 12 essentially unchanged, with some minor exceptions.

Chapter 8 contains classical material on variational methods for waveguide discontinuities and serves to illustrate a number of special techniques useful for rectangular waveguide discontinuities. These methods are readily extended to other waveguides. It was my original intention to expand the number of examples, but because of space limitations, only a short treatment of the inductive post and a brief, general discussion of scattering from obstacles in a waveguide were added. There is an abundance of papers on waveguide discontinuities, as well as several books on the subject, so the reader will have no difficulty in finding examples to study and review.

Chapter 9 on periodic structures has been left unchanged. It covers the fundamentals in sufficient depth that the extension of the theory and its application to specific structures should follow quite readily.

An additional example has been added to Chapter 10 on integral transform and function-theoretic techniques. This example is that of a bifurcated parallel-plate waveguide with a dielectric slab. This particular example provides useful physical insight into the basic properties of a wide microstrip line and follows quite closely the theory developed by El-Sherbiny. In particular, it illustrates the crucial importance of edge conditions in order to obtain a unique solution. It also illustrates that the LSE and LSM modes in a microstrip line are coupled through the edge conditions in accordance with the theory given by Omar and Shünemann.

Chapter 11 on surface waveguides and Chapter 12 on artificial dielectrics are unchanged from the original edition. I had considered deleting Chapter 12 and expanding Chapter 11, but decided against it on the basis that a number of people had expressed the hope that a discussion of artificial dielectrics would remain in the revised edition.

During the years I used the original book for graduate courses, I generally found that graduate students were not very familiar with the use of Fourier and Laplace transforms in the complex plane. I would normally provide the students with supplementary material on this topic. In the revised edition this material has been added to the Mathematical Appendix. In addition, a more complete account is given of the steepest descent method for the asymptotic evaluation of radiation integrals when a simple pole lies close to the saddle point. The relationship between the steepest descent method and the method of stationary phase is also discussed. Some general results for the factorization problem for Wiener–Hoff integral equations has also been added since this was very sketchy in the original edition. Finally, for the convenience of the reader, a collection of useful relationships for Bessel functions, Legendre functions, and formulas from vector analysis have been added.

The reader is encouraged to examine the problems at the end of each chapter since many of these contain additional specific results that are not given in the text. For example, Problems 1.17 and 2.35 provide a somewhat different proof of the uniqueness of the solution for the external scattering problem, which does not require any assumption about a finite loss for the medium.

There does not seem to be any magical way to eliminate typographical errors or even errors in derived equations. I hope that there are not too many and will appreciate if readers will bring errors that they discover to my attention so that future printings can be corrected.

Most of my knowledge of electromagnetics has been gained from study of the work of other people, and there are a great many. To these people I am indebted and regret that it is not possible to cite their work except in a very limited way. The literature is so voluminous that any thought of compiling a meaningful list of important contributions is quickly put aside. However, I would like to specifically acknowledge a number of helpful discussions on integral equations as presented in Chapter 2 with Dr. Maurice Sancer.

This revised edition would not have appeared except for Reed Crone who recommended reprinting the original edition or publishing a revised edition. I appreciate his efforts in initiating the revision.

The material for the revised edition was typed by Sue Sava. In addition to her expertise in technical typing she also provided helpful editorial changes when my grammar went astray.

The editorial staff at the IEEE Press, in particular Randi Scholnick and Anne Reifsnyder, have been of great assistance. I have marveled at their insight in sorting out what I was trying to say even when I said it poorly. I would also like to thank Barbara Palumbo for providing thorough and useful indexes for this edition.

The last acknowledgment goes to my wife Kathleen, who provided the encouragement and willingness to forgo many other activities so that the revision could be completed in a timely manner.

ROBERT E. COLLIN

1
Basic Electromagnetic Theory

The early history of guided waves and waveguides dates back to around the end of the nineteenth century. In 1897 Lord Rayleigh published an analysis of electromagnetic-wave propagation in dielectric-filled rectangular and circular conducting tubes, or waveguides as they are now called. This was followed by an analysis of wave propagation along dielectric cylinders by Hondros and Debye in 1910. Several other workers contributed in both theory and experiment to our knowledge of waveguides during the same period. However, the true birth of the waveguide as a useful device for transmission of electromagnetic waves did not come about before 1936. At that time, Carson, Mead, Schelkunoff, and Southworth from the Bell Telephone Laboratories published an extensive account of analytical and experimental work on waveguides that had been under way since 1933. The result of other work was published the same year by Barrow.

The period from 1936 to the early 1940s saw a steady growth in both the theoretical and experimental work on waveguides as practical communication elements. However, even though this early growth was very important, our present state of knowledge of guided electromagnetic waves owes much to the tremendous efforts of a multitude of physicists, mathematicians, and engineers who worked in the field during World War II. The desire for radars operating at microwave frequencies put high priority on work related to waveguides, waveguide devices, antennas, etc. A variety of devices were developed to replace conventional low-frequency lumped-circuit elements. Along with these devices, methods of analysis were also being developed. The formulation and solution of intricate boundary-value problems became of prime importance.

The postwar years have seen continued development and refinement in both theory and techniques. Our efforts in this book are devoted to a presentation of the major theoretical results and mathematical techniques, obtained to date, that underlie the theory of guided waves. The treatment places the electromagnetic field in the foreground, and for this reason we begin with a survey of that portion of classical electromagnetic field theory that is pertinent to those devices and problems under examination. Very few of the boundary-value problems that will be dealt with can be solved exactly. However, provided the problem is properly formulated, very accurate approximate solutions can be found, since the availability of high-speed computers has virtually removed the limitations on our ability to evaluate lengthy series solutions and complex integrals. The formulation of a problem and its reduction to a suitable form for numerical evaluation require a good understanding of electromagnetic field phenomena, including the behavior of the field in source regions and at edges and corners. The accuracy and efficiency with which numerical results can be obtained depend on how well a problem has been reduced analytically to a form suitable for numerical evaluation. Thus even though computers are available to remove the drudgery of obtaining numerical answers, the results can be unsatisfactory because of round-off errors, numerical instability, or the need for extensive computational effort. It is therefore just as important today as in the past to carry out the analytical solutions as far as possible so as to obtain final expressions or a system of equations that can be numerically evaluated in an efficient manner. In addition, one gains considerable insight through analysis with regard to the basic phenomena involved.

Analytical techniques rely on knowledge of a number of mathematical topics. For the convenience of the reader the topics that are frequently needed throughout the book have been summarized in the Mathematical Appendix. The reader is also assumed to have a basic knowledge of electromagnetic theory, waveguides, and radiation. (See, for example, [1.1].[1])

1.1. Maxwell's Equations

Provided we restrict ourselves to the macroscopic domain, the electromagnetic field is governed by the classical field equations of Maxwell. In derivative form these are

$$\nabla \times \mathcal{E} = -\frac{\partial \mathcal{B}}{\partial t} \tag{1a}$$

$$\nabla \times \mathcal{H} = \frac{\partial \mathcal{D}}{\partial t} + \mathcal{J} \tag{1b}$$

$$\nabla \cdot \mathcal{D} = \rho \tag{1c}$$

$$\nabla \cdot \mathcal{B} = 0 \tag{1d}$$

$$\nabla \cdot \mathcal{J} = -\frac{\partial \rho}{\partial t}. \tag{1e}$$

The above equations are not all independent since, if the divergence of (1a) is taken, we get the result that

$$\nabla \cdot \frac{\partial \mathcal{B}}{\partial t} = 0$$

because the divergence of the curl of any vector is identically zero. By interchanging the order of the time and space derivatives and integrating with respect to time, (1d) is obtained, provided the constant of integration is taken as zero. In a similar fashion (1c) follows from (1b) with the assistance of (1e). Equation (1a) is just the differential form of Faraday's law of induction, since the curl of a vector is the line integral of this vector around an infinitesimal contour divided by the area of the surface bounded by the contour. Equation (1b) is a generalization of Ampère's circuital law (also referred to as the law of Biot–Savart) by the addition of the term $\partial \mathcal{D}/\partial t$, which is called the displacement current density. Equation (1c) is the differential form of Gauss' law, while (1d) expresses the fact that the magnetic lines of flux form a system of closed loops and nowhere terminate on "magnetic" charge. Finally, (1e) is just the equation of conservation of charge; the amount of charge diverging away per second from an infinitesimal volume element must equal the time rate of decrease of the charge contained within.

At this time it is convenient to restrict ourselves to fields that vary with time according to the complex exponential function $e^{j\omega t}$, where ω is the radian frequency. There is little loss in generality in using such a time function since any physically realizable time variation can be decomposed into a spectrum of such functions by means of a Fourier integral, i.e.,

$$f(t) = \int_{-\infty}^{\infty} g(\omega)e^{j\omega t}\,d\omega$$

[1]Bracketed numbers are keyed to references given at the end of each chapter.

and

$$g(\omega) = \frac{1}{2\pi} \int_{-\infty}^{\infty} f(t)e^{-j\omega t}\,dt.$$

With the above assumed time variation all time derivatives may be replaced by $j\omega$. We will not include the factor $e^{j\omega t}$ explicitly as this factor always occurs as a common factor in all terms. The field vectors are now represented by boldface roman type and are complex functions of the space coordinates only.

The basic field equations may be expressed in equivalent integral form also. An application of Stokes' theorem to (1a) and (1b) gives

$$\iint_{S} \nabla \times \mathbf{E}\cdot d\mathbf{S} = \oint_{C} \mathbf{E}\cdot d\mathbf{l} = -j\omega \iint_{S} \mathbf{B}\cdot d\mathbf{S} \qquad (2a)$$

$$\oint_{C} \mathbf{H}\cdot d\mathbf{l} = \iint_{S} (j\omega \mathbf{D} + \mathbf{J})\cdot d\mathbf{S}. \qquad (2b)$$

Using the divergence theorem, the remaining three equations become

$$\oiint_{S} \mathbf{D}\cdot d\mathbf{S} = \iiint_{V} \rho\,dV \qquad (2c)$$

$$\oiint_{S} \mathbf{B}\cdot d\mathbf{S} = 0 \qquad (2d)$$

$$\oiint_{S} \mathbf{J}\cdot d\mathbf{S} = -j\omega \iiint_{V} \rho\,dV. \qquad (2e)$$

In (2a) and (2b) the surface S is an open surface bounded by a closed contour C, while in (2c) through (2e) the surface S is a closed surface with an interior volume V. The vector element of area $d\mathbf{S}$ is directed outward. This latter description of the field equations is particularly useful in deriving the boundary conditions to be applied to the field vectors at a boundary between two mediums having different electrical properties.

Equations (1) contain considerably more information than we might anticipate from a cursory glance at them. A good deal of insight into the meaning of these equations can be obtained by decomposing each field vector into the sum of an irrotational or lamellar part and a rotational (solenoidal) part. The lamellar part has an identically vanishing curl, while the rotational part always has a zero divergence. A vector field is completely specified only when both the lamellar and solenoidal parts are given.[2] Let the subscript l denote the lamellar part and the subscript r denote the rotational or solenoidal part. For an arbitrary vector field $\mathbf{C}$ we have

$$\mathbf{C} = \mathbf{C}_l + \mathbf{C}_r \qquad (3a)$$

where

$$\nabla \times \mathbf{C}_l \equiv 0 \qquad (3b)$$

[2]See Section A.1e.

and

$$\nabla \cdot \mathbf{C}_r \equiv 0. \tag{3c}$$

When the field vectors are decomposed into their basic parts and the results of (3b) and (3c) are used, the field equations (1) become

$$\nabla \times \mathbf{E}_r = -j\omega \mathbf{B}_r \tag{4a}$$

$$\nabla \times \mathbf{H}_r = j\omega(\mathbf{D}_l + \mathbf{D}_r) + \mathbf{J}_l + \mathbf{J}_r \tag{4b}$$

$$\nabla \cdot \mathbf{D}_l = \rho \tag{4c}$$

$$\nabla \cdot \mathbf{B}_r = 0 \tag{4d}$$

$$\nabla \cdot \mathbf{J}_l = -j\omega\rho. \tag{4e}$$

From (4c) and (4e) we find that the divergence of the lamellar part of the displacement current is related to the divergence of the lamellar part of the conduction current. The rotational part of the conduction current has zero divergence and hence exists in the form of closed current loops. It does not terminate on a distribution of charge. On the other hand, the lamellar part $\mathbf{J}_l$ is like a fragment of $\mathbf{J}_r$ with the ends terminating in a collection of charge. The lamellar part of the displacement current provides the continuation of $\mathbf{J}_l$, so that the total lamellar current $\mathbf{J}_l + j\omega\mathbf{D}_l$ forms the equivalent of a rotational current. Equations (4c) and (4e) show at once that

$$\nabla \cdot \mathbf{J}_l + j\omega\nabla \cdot \mathbf{D}_l = 0. \tag{5}$$

In view of this latter result, we are tempted to integrate (5) to obtain $\mathbf{J}_l + j\omega\mathbf{D}_l = 0$ and to conclude that (4b) is equivalent to

$$\nabla \times \mathbf{H}_r = j\omega\mathbf{D}_r + \mathbf{J}_r$$

or, in words, that the rotational part of $\mathbf{H}$ is determined only by the rotational part of the total current (displacement plus conduction). This line of reasoning is incorrect since the integration of (5) gives, in general,

$$\mathbf{J}_l + j\omega\mathbf{D}_l = \mathbf{C}_r$$

where $\mathbf{C}_r$ is a rotational vector. By comparison with (4b), we see that $\mathbf{C}_r$ is the difference between $\nabla \times \mathbf{H}_r$ and $j\omega\mathbf{D}_r + \mathbf{J}_r$. It is the total current consisting of both the lamellar part and the rotational part that determines $\nabla \times \mathbf{H}_r$ and not vice versa. Furthermore, since the total lamellar part of the current, that is, $\mathbf{J}_l + j\omega\mathbf{D}_l$, is a rotational current in its behavior, there is no reason to favor the separate rotational parts $\mathbf{J}_r$ and $j\omega\mathbf{D}_r$ over it.

Equations (4a) and (4d) are more or less self-explanatory since $\mathbf{B}$ does not have a lamellar part.

We can also conclude from the above discussion that the lamellar part of $\mathbf{D}$ will, in general, not be zero outside the source region where $\mathbf{J}$ and ρ are zero.

At this point we make note of other terms that are commonly used to identify the lamellar and rotational parts of a vector field. The lamellar part that has zero curl is also called the longitudinal part and the solenoidal or rotational part that has zero divergence is called the transverse part. This terminology has its origin in the Fourier integral representation of the field as a spectrum of plane waves. In this representation the divergence of the field takes the form $-j\mathbf{k}\cdot\mathbf{E(k)}$ and the curl takes the form $-j\mathbf{k}\times\mathbf{E(k)}$ where $\mathbf{E(k)}$ is the three-dimensional Fourier spatial transform of $\mathbf{E(r)}$ and $\mathbf{k}$ is the transform variable, i.e.,

$$\mathbf{E(r)} = \frac{1}{(2\pi)^3} \iiint\limits_{-\infty}^{\infty} \mathbf{E(k)}e^{-j\mathbf{k}\cdot\mathbf{r}}\, d\mathbf{k}.$$

The scalar product $\mathbf{k}\cdot\mathbf{E(k)}$ selects the component (longitudinal part) of the plane wave spectrum in the direction of the propagation vector, while the cross product $\mathbf{k}\times\mathbf{E(k)}$ selects the transverse component that is perpendicular to $\mathbf{k}$.

1.2. Relation between Field Intensity Vectors and Flux Density Vectors

The field equations cannot be solved before the relationship between $\mathbf{B}$ and $\mathbf{H}$ and that between $\mathbf{E}$ and $\mathbf{D}$ are known. In a vacuum or free space the relationship is a simple proportion, i.e.,

$$\mathbf{B} = \mu_0\mathbf{H} \qquad (6a)$$

$$\mathbf{D} = \epsilon_0\mathbf{E} \qquad (6b)$$

where $\mu_0 = 4\pi \times 10^{-7}$ henry per meter and $\epsilon_0 = (36\pi)^{-1} \times 10^{-9}$ farad per meter. The constant μ_0 is called the magnetic permeability of vacuum, and ϵ_0 is the electric permittivity of vacuum. For material bodies the relationship is generally much more complicated with μ_0 and ϵ_0 replaced by tensors of rank 2, or dyadics. Letting a bar above the quantity signify a dyadic, we have

$$\mathbf{B} = \bar{\mu}\cdot\mathbf{H} \qquad (7a)$$

$$\mathbf{D} = \bar{\epsilon}\cdot\mathbf{E} \qquad (7b)$$

where

$$\bar{\mu} = \mu_{xx}\mathbf{a}_x\mathbf{a}_x + \mu_{xy}\mathbf{a}_x\mathbf{a}_y + \mu_{xz}\mathbf{a}_x\mathbf{a}_z + \mu_{yx}\mathbf{a}_y\mathbf{a}_x + \mu_{yy}\mathbf{a}_y\mathbf{a}_y + \mu_{yz}\mathbf{a}_y\mathbf{a}_z$$
$$+ \mu_{zx}\mathbf{a}_z\mathbf{a}_x + \mu_{zy}\mathbf{a}_z\mathbf{a}_y + \mu_{zz}\mathbf{a}_z\mathbf{a}_z$$

with a similar definition for $\bar{\epsilon}$. The unit vectors along the coordinate axis are represented by $\mathbf{a}_x$, etc. The dot product of a dyadic and a vector or of two dyadics is defined as the vector or dyadic that results from taking the dot product of the unit vectors that appear on either side of the dot. Dyadic algebra has a one-to-one correspondence with matrix algebra with the advantage of a somewhat more condensed notation. In general, the dot product is not commutative, i.e.,

$$\bar{\mathbf{A}}\cdot\bar{\mathbf{C}} \neq \bar{\mathbf{C}}\cdot\bar{\mathbf{A}}$$

unless the product, $\bar{\mathbf{A}}$, and $\bar{\mathbf{C}}$ are all symmetric, so that $A_{ij} = A_{ji}$, etc. With the definition of the dot product established, we find that the x component of (7a) is $B_x = \mu_{xx}H_x + \mu_{xy}H_y + \mu_{xz}H_z$, with similar expressions existing for the y and z components. Thus, each component of **B** is related to all three components of **H**. The coefficients μ_{ij} may also be functions of **H** so that the relationship is not a linear one. With a suitable rotation of axes to the principal axes u, v, w, the dyadic $\bar{\mu}$ becomes a diagonal dyadic of the form

$$\bar{\mu} = \mu_{uu}\mathbf{a}_u\mathbf{a}_u + \mu_{vv}\mathbf{a}_v\mathbf{a}_v + \mu_{ww}\mathbf{a}_w\mathbf{a}_w.$$

Although we introduced a somewhat involved relationship above, in practice we find that for many materials $\bar{\mu}$ and ϵ reduce to simple scalar quantities and, furthermore, are essentially independent of the field strength. When μ and ϵ are constant, the divergence equations for **D** and **B** give

$$\nabla \cdot \mathbf{D} = \rho \qquad \text{or} \qquad \nabla \cdot \mathbf{E} = \rho/\epsilon \tag{8a}$$

$$\nabla \cdot \mathbf{B} = 0 \qquad \text{or} \qquad \nabla \cdot \mathbf{H} = 0 \tag{8b}$$

and **H** is seen to be a solenoidal field also. When μ and ϵ vary with position, we have

$$\nabla \cdot \mathbf{D} = \nabla \cdot \epsilon \mathbf{E} = \epsilon \nabla \cdot \mathbf{E} + \mathbf{E} \cdot \nabla \epsilon = \rho$$

$$\text{or}$$

$$\nabla \cdot \mathbf{E} = \rho/\epsilon - \mathbf{E} \cdot \nabla \ln \epsilon \tag{9a}$$

and similarly

$$\nabla \cdot \mathbf{H} = -\mathbf{H} \cdot \nabla \ln \mu \tag{9b}$$

since $\nabla \ln \epsilon = (1/\epsilon)\nabla \epsilon$ etc.

When the material under consideration has finite losses, μ and ϵ are complex with negative imaginary parts.

The physical meaning of Eqs. (9) will require a consideration of the properties of matter that give rise to relationships of the form expressed by (7) with μ and ϵ scalar functions of position.

When an electric field is applied to a dielectric material, the electron orbits of the various atoms and molecules involved become perturbed, resulting in a dipole polarization **P** per unit volume. Some materials have a permanent dipole polarization also, but, since these elementary dipoles are randomly oriented, there is no net resultant field in the absence of an applied external field. The application of an external field tends to align the dipoles with the field, resulting in a decrease in the electric field intensity in the material. The displacement flux vector **D** is defined by

$$\mathbf{D} = \epsilon_0 \mathbf{E} + \mathbf{P} \tag{10}$$

in the interior of a dielectric material. In (10), **E** is the net field intensity in the dielectric, i.e., the vector sum of the applied field and the field arising from the dipole polarization. For many materials **P** is collinear with the applied field and also proportional to the applied field and hence proportional to $\mathbf{E}$ also. Thus we have

$$\mathbf{P} = \chi_e \epsilon_0 \mathbf{E} \tag{11}$$

where χ_e is called the electric susceptibility of the material and is a dimensionless quantity. For the scalar case we have

$$\mathbf{D} = \epsilon_0 (\chi_e + 1)\mathbf{E} = \epsilon \mathbf{E}$$

and so

$$\epsilon = (1 + \chi_e)\epsilon_0. \tag{12}$$

The relative dielectric constant is $\kappa = \epsilon/\epsilon_0$ and is equal to $1 + \chi_e$. Equation (11) is not universally applicable, since $\mathbf{P}$ will be in the direction of the applied field only in materials having a high degree of symmetry in their crystal structure. In general, $\bar{\chi}_e$ is a dyadic quantity, and hence ϵ becomes the dyadic quantity $(\bar{\mathbf{I}} + \bar{\chi}_e)\epsilon_0$, where $\bar{\mathbf{I}}$ is the unit dyadic.

The situation for materials having a permeability different from μ_0 is somewhat analogous, with $\mathbf{H}$ being defined by the relation

$$\mathbf{H} = \mathbf{B}/\mu_0 - \mathbf{M} \tag{13}$$

where $\mathbf{M}$ is the magnetic dipole polarization per unit volume. Logically we should now take $\mathbf{M}$ proportional to $\mathbf{B}$ for those materials for which such a proportion exists, but convention has adopted

$$\mathbf{M} = \chi_m \mathbf{H} \tag{14}$$

where χ_m is the magnetic susceptibility of the medium. Equation (14) leads at once to the following result for μ:

$$\mu = (1 + \chi_m)\mu_0. \tag{15}$$

As in the dielectric case, $\mathbf{M}$ is not always in the direction of $\mathbf{B}$, and hence χ_m and μ are dyadic quantities in general.

Materials for which μ and ϵ are scalar constants are referred to as homogeneous isotropic materials. When μ and ϵ vary with position, the material is no longer homogeneous but is still isotropic. When μ and ϵ are dyadics, the material is said to be anisotropic. In the latter case the material may again be homogeneous or nonhomogeneous, depending on whether or not $\bar{\mu}$ and $\bar{\epsilon}$ vary with position. A more general class of materials has the property that $\mathbf{B}$ and $\mathbf{D}$ are linearly related to both $\mathbf{E}$ and $\mathbf{H}$. These are called bi-isotropic or bi-anisotropic materials according to whether or not the constitutive parameters are scalars or dyadics [1.46].

In material bodies with μ and ϵ scalar functions of position, Maxwell's curl equations may be written as

$$\nabla \times \mathbf{E} = -j\omega\mu_0 \mathbf{H} - j\omega\mu_0\chi_m \mathbf{H} \tag{16a}$$

$$\nabla \times \mathbf{H} = j\omega\epsilon_0 \mathbf{E} + j\omega\epsilon_0\chi_e \mathbf{E} + \mathbf{J}. \tag{16b}$$

In (16b) the term $j\omega\epsilon_0\chi_e\mathbf{E}$ may be regarded as an equivalent polarization current $\mathbf{J}_e$. Similarly in (16a) it is profitable at times to consider the term $j\omega\chi_m\mathbf{H}$ as the equivalent of a magnetic polarization current $\mathbf{J}_m$.

Introducing the polarization vectors $\mathbf{P}$ and $\mathbf{M}$ into the divergence equations gives

$$\nabla\cdot\mathbf{D} = \epsilon_0\nabla\cdot\mathbf{E} + \nabla\cdot\mathbf{P} = \rho \tag{17a}$$

$$\nabla\cdot\mathbf{B} = \mu_0\nabla\cdot\mathbf{H} + \mu_0\nabla\cdot\mathbf{M} = 0. \tag{17b}$$

The additional terms $\nabla\cdot\mathbf{P}$ and $\nabla\cdot\mathbf{M}$ may be interpreted as defining an electric-polarization charge density $-\rho_e$ and a magnetic-polarization charge density $-\rho_m$, respectively. The latter definition is a mathematical one and should not be interpreted as showing the existence of a physical magnetic charge. Introducing the parameters χ_e and χ_m, we obtain the following equations giving the polarization charge densities:

$$\rho_e = -\nabla\cdot\mathbf{P} = -\epsilon_0\nabla\cdot\chi_e\mathbf{E}$$

$$= -\epsilon_0\chi_e\nabla\cdot\mathbf{E} - \epsilon_0\mathbf{E}\cdot\nabla\chi_e \tag{18a}$$

$$\rho_m = -\nabla\cdot\mathbf{M} = -\nabla\cdot\chi_m\mathbf{H}$$

$$= -\chi_m\nabla\cdot\mathbf{H} - \mathbf{H}\cdot\nabla\chi_m. \tag{18b}$$

When the susceptibilities are dyadic functions of position, we may obviously make a similar interpretation of the additional terms that arise in Maxwell's equations. In the analysis of scattering by material bodies such an interpretation is often convenient since it leads directly to an integral equation for the scattered field in terms of the polarization currents.

At high frequencies, the inertia of the atomic system causes the polarization vectors $\mathbf{P}$ and $\mathbf{M}$ to lag behind the applied fields. As a result, ϵ and μ must be represented by complex quantities in order to account for this difference in time phase. The real parts will be designated by ϵ', μ' and the imaginary parts by ϵ'', μ''; thus

$$\epsilon = \epsilon' - j\epsilon''$$

$$\mu = \mu' - j\mu''.$$

For many materials ϵ'' and μ'' are negligible, i.e., for materials with small losses.

For many material bodies the conduction current, which flows as a result of the existence of a field $\mathbf{E}$, is directly proportional to $\mathbf{E}$, so that we may put

$$\mathbf{J} = \sigma\mathbf{E} \tag{19}$$

where σ is a parameter of the medium known as the conductivity. Equation (19) is not true in general, of course, since σ may depend on both the magnitude and the direction of $\mathbf{E}$, and hence is a nonlinear dyadic quantity, as for example in semiconductors at a rectifying boundary. When (19) holds, the curl equation for $\mathbf{H}$ becomes

$$\nabla\times\mathbf{H} = (j\omega\epsilon + \sigma)\mathbf{E} = j\omega\epsilon'\left(1 - j\frac{\epsilon''}{\epsilon'} - j\frac{\sigma}{\omega\epsilon'}\right)\mathbf{E}$$

$$= j\omega\epsilon'(1 - j\tan\delta)\mathbf{E}. \tag{20}$$

The effective dielectric permittivity is now complex, even if ϵ'' is zero, and is given by

$$\epsilon_e = \epsilon'(1 - j \tan \delta) \tag{21}$$

where $\tan \delta$ is the loss tangent for the material.

If a material body has an appreciable conductivity, the density ρ of free charge in the interior may be taken as zero. Any initial free-charge density ρ_0 decays exponentially to zero in an extremely short time. The existence of a charge distribution ρ implies the existence of a resultant electric field $\mathbf{E}_0$ and hence a current $\mathbf{J}_0 = \sigma \mathbf{E}_0$. Now $\nabla \cdot \mathbf{J}_0 = -\partial \rho/\partial t$ and $\nabla \cdot \epsilon \mathbf{E}_0 = \rho$. When ϵ and σ are constants, we find, upon eliminating $\mathbf{E}_0$ and $\mathbf{J}_0$, that

$$\frac{\partial \rho}{\partial t} = -\frac{\sigma}{\epsilon} \rho$$

and hence

$$\rho = \rho_0 e^{-(\sigma/\epsilon)t}. \tag{22}$$

For metals the decay time $\tau = \epsilon/\sigma$ is of the order of 10^{-18} second, so that for frequencies well beyond the microwave range the decay time is extremely short compared with the period of the impressed fields.

The above result means that in conducting bodies we may take $\nabla \cdot \mathbf{D}$ equal to zero. If ϵ and σ are constants, the divergence of $\mathbf{E}$ and $\mathbf{J}$ will also be zero. In practice, we often simplify our boundary conditions at a conducting surface by assuming that σ is infinite. It is then found that the conduction current flows on the surface of the conducting boundary. Both lamellar and solenoidal surface currents may exist. The lamellar current terminates in a surface charge distribution. The normal component of the displacement vector $\mathbf{D}$ also terminates in this charge distribution and provides the continuation of $\mathbf{J}$ into the region external to the conducting surface. The perfectly conducting surface is a mathematical idealization, and we find at such a surface a discontinuity in the field vectors that does not correspond to physical reality. The nature of the fields at a conducting surface as σ approaches infinity will be considered in more detail in a later section.

The constitutive relations given by (7) and (19) are strictly valid only in the frequency domain for linear media. These relations are analogous to those that relate voltage and current through impedance and admittance functions in network theory. The complex phasors that represent the fields at the radian frequency ω can be interpreted as the Fourier transforms of the fields in the time domain. In the time domain we would then have, for example,

$$\mathfrak{D}(\mathbf{r}, t) = \frac{1}{2\pi} \int_{-\infty}^{\infty} \epsilon(\omega) \mathbf{E}(\mathbf{r}, \omega) e^{j\omega t} \, d\omega$$

which can be expressed as a convolution integral

$$\mathfrak{D}(\mathbf{r}, t) = \int_{-\infty}^{\infty} \epsilon(\tau) \mathcal{E}(\mathbf{r}, t - \tau) \, d\tau$$

where $\epsilon(\tau)$ is an integral-differential operator whose Fourier transform is $\epsilon(\omega)$. The conductivity σ is not a constant for all values of ω. For very large values of ω it is a function of ω and the simple exponential decay of charge as described by (22) is fundamentally incorrect. The conductivity σ can be assumed to be constant up to frequencies corresponding to infrared radiation. The prediction of very rapid decay of free charge in a metal is also correct but the decay law is more complex than that given by (22). For nonhomogeneous media the constitutive parameters (operators) are functions of the spatial coordinates and possibly also derivatives with respect to the spatial coordinates. In the latter case the medium is said to exhibit spatial dispersion.

1.3. ELECTROMAGNETIC ENERGY AND POWER FLOW

The time-average amount of energy stored in the electric field that exists in a volume V is given by

$$W_e = \frac{1}{4} \operatorname{Re} \iiint_V \mathbf{D} \cdot \mathbf{E}^* \, dV. \tag{23a}$$

The time-average energy stored in the magnetic field is

$$W_m = \frac{1}{4} \operatorname{Re} \iiint_V \mathbf{B} \cdot \mathbf{H}^* \, dV. \tag{23b}$$

The asterisk (*) signifies the complex conjugate value, and the additional factor $\frac{1}{2}$, which makes the numerical coefficient $\frac{1}{4}$, arises because of the averaging over one period in time. We are using the convention that $\mathbf{D}$, etc., represents the peak value of the complex field vector. The imaginary parts of the integrals in (23) represent power loss in the material. The interpretation of (23) as representing stored energy needs qualification. The concept of stored energy as a thermodynamic state function is clear only for media without dissipation in which case stored energy can be related to the work done on the system. Even for dissipationless media Eqs. (23) fail to be correct when the constitutive parameters ϵ and μ depend on ω. Thus in media with dispersion Eqs. (23) need to be modified; we will present this modification later.

The time-average complex power flow across a surface S is given by the integral of the complex Poynting vector over that surface,

$$P = \frac{1}{2} \iint_S \mathbf{E} \times \mathbf{H}^* \cdot d\mathbf{S}. \tag{24}$$

The real power is given by the real part of P while the imaginary part represents energy stored in the electric and magnetic fields. The integral of the complex Poynting vector over a closed surface yields a result that may be regarded as a fundamental theorem in the theory of equivalent circuits for waveguide discontinuities. It is also of fundamental importance in the problem of evaluating the input impedance of an antenna. The integral of $\mathbf{E} \times \mathbf{H}^*$ over a closed surface S gives

$$\frac{1}{2} \oiint_S \mathbf{E} \times \mathbf{H}^* \cdot d\mathbf{S} = -\frac{1}{2} \iiint_V \nabla \cdot \mathbf{E} \times \mathbf{H}^* \, dV$$

if the vector element of area is taken as directed into the volume V. Now $\nabla \cdot \mathbf{E} \times \mathbf{H}^* = \mathbf{H}^* \cdot \nabla \times \mathbf{E} - \mathbf{E} \cdot \nabla \times \mathbf{H}^*$. The curl of $\mathbf{E}$ and $\mathbf{H}^*$ may be replaced by $-j\omega\mu\mathbf{H}$ and

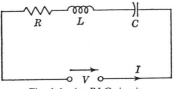

Fig. 1.1. An *RLC* circuit.

$-j\omega\epsilon^*\mathbf{E}^* + \sigma\mathbf{E}^*$ from Maxwell's equations. Thus we get

$$\frac{1}{2}\oint_S \mathbf{E}\times\mathbf{H}^*\cdot d\mathbf{S} = \frac{1}{2}\iiint_V [j\omega(\mathbf{B}\cdot\mathbf{H}^* - \mathbf{D}^*\cdot\mathbf{E}) + \sigma\mathbf{E}\cdot\mathbf{E}^*]\,dV$$

$$= 2j\omega(W_m - W_e) + P_L. \tag{25}$$

It is in the interpretation of (25) that the expressions (23) for the energy stored in the field are obtained. Since ordinary materials are passive, it follows from (25) that ϵ and μ have negative imaginary parts in order to represent a loss of energy. The power loss P_L in V is the sum of the conduction loss and $j\omega$ times the imaginary parts of (23). The imaginary part of the integral of the complex Poynting vector over the closed surface S gives a quantity proportional to the difference between the time-average energy stored in the magnetic field and that stored in the electric field. The right-hand side of (25), when the fields are properly normalized, can be related to or essentially used to define a suitable input impedance if the complex power flows into the volume V through a waveguide or along a transmission line. We shall have several occasions to make use of this result later.

Consider a simple *RLC* circuit as illustrated in Fig. 1.1. The input impedance to this circuit is $Z = R + j\omega L + 1/j\omega C$. The complex time-average power flow into the circuit is given by

$$P = \frac{1}{2}VI^* = \frac{1}{2}ZII^* = \frac{1}{2}j\omega\left(LII^* - \frac{II^*}{\omega^2 C}\right) + \frac{1}{2}II^*R.$$

The time-average reactive energy stored in the magnetic field around the inductor is $\frac{1}{4}LII^*$, while the time-average electric energy stored in the capacitor is $(C/4)(I/\omega C)(I^*/\omega C)$, since the peak voltage across C is $I/\omega C$. We therefore obtain the result that

$$P = \frac{1}{2}ZII^* = 2j\omega(W_m - W_e) + P_L \tag{26}$$

which at least suggests the usefulness of (25) in impedance calculations. In (26) the average energy dissipated in R per second is $P_L = \frac{1}{2}RII^*$.

When there is an impressed current source $\mathbf{J}$ in addition to the current $\sigma\mathbf{E}$ in the medium it is instructive to rewrite (25) in the form

$$-\frac{1}{2}\iiint_V \mathbf{J}^*\cdot\mathbf{E}\,dV = \frac{1}{2}\iiint_V [j\omega(\mathbf{B}\cdot\mathbf{H}^* - \mathbf{D}^*\cdot\mathbf{E}) + \sigma\mathbf{E}\cdot\mathbf{E}^*]\,dV + \frac{1}{2}\oint_S \mathbf{E}\times\mathbf{H}^*\cdot d\mathbf{S}$$

$$\tag{27}$$

where the vector element of area is now directed out of the volume V. The real part of the left-hand side is now interpreted as representing the work done by the impressed current source against the radiation reaction field **E** and accounts for the power loss in the medium and the transport of power across the surface S.

A significant property of the energy functions W_e and W_m is that they are positive functions. For the electrostatic field an important theorem known as Thomson's theorem states that the charges that reside on conducting bodies and that give rise to the electric field **E** will distribute themselves in such a way that the energy function W_e is minimized. Since we shall use this result to obtain a variational expression for the characteristic impedance of a transmission line, a proof of this property of the electrostatic field will be given below.

Consider a system of conductors $S_1, S_2, \ldots, S_n$, as illustrated in Fig. 1.2, that are held at potentials $\Phi_1, \Phi_2, \ldots, \Phi_n$. The electric field between the various S_i may be obtained from the negative gradient of a suitable scalar potential function Φ, where Φ reduces to Φ_i on S_i, with $i = 1, 2, \ldots, n$. The energy stored in the electric field is given by

$$W_e = \frac{\epsilon}{2} \iiint_V \nabla\Phi \cdot \nabla\Phi \, dV.$$

Let the charges be displaced from their equilibrium positions by a small amount but still keep the potentials of the conductors constant. The potential distribution around the conductors will change to $\Phi + \delta\Phi$, where $\delta\Phi$ is a small change in Φ. The change in Φ, that is, $\delta\Phi$, is of course a function of the space coordinates but will vanish on the conductors S_i since Φ_i is maintained constant. The change in the energy function W_e is

$$\delta W_e = \frac{\epsilon}{2} \iiint_V \nabla(\Phi + \delta\Phi) \cdot \nabla(\Phi + \delta\Phi) \, dV - \frac{\epsilon}{2} \iiint_V \nabla\Phi \cdot \nabla\Phi \, dV$$

$$= \frac{\epsilon}{2} \iiint_V 2\nabla\Phi \cdot \nabla\delta\Phi \, dV + \frac{\epsilon}{2} \iiint_V \nabla\delta\Phi \cdot \nabla\delta\Phi \, dV.$$

The first term may be shown to vanish as follows. Let $u\nabla v$ be a vector function; using the divergence theorem, we get

$$\iiint_V \nabla \cdot u\nabla v \, dV = \iiint_V (\nabla u \cdot \nabla v + u\nabla^2 v) \, dV$$

$$= \oiint_S u\nabla v \cdot d\mathbf{S} = \oiint_S u\frac{\partial v}{\partial n} \, dS$$

where **n** is the outward normal to the surface S that surrounds the volume V. The above result is known as Green's first identity. If we let $\Phi = v$ and $\delta\Phi = u$, and recall that Φ is a solution of Laplace's equation $\nabla^2\Phi = 0$, in the region surrounding the conductors we obtain the result

$$\iiint_V \nabla\Phi \cdot \nabla\delta\Phi \, dV = \oiint_S \delta\Phi \frac{\partial\Phi}{\partial n} \, dS.$$

The surface S is chosen as the surface of the conductors, plus the surface of an infinite sphere, plus the surface along suitable cuts joining the various conductors and the infinite sphere. The integral along the cuts does not contribute anything since the integral is taken along the cuts

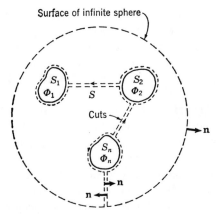

Fig. 1.2. Illustration of cuts to make surface S simply connected.

twice but with the normal $\mathbf{n}$ oppositely directed. Also the potential Φ vanishes at infinity at least as rapidly as r^{-1}, and so the integral over the surface of the infinite sphere also vanishes. On the conductor surfaces $\delta\Phi$ vanishes so that the complete surface integral vanishes. The change in the energy function W_e is, therefore, of second order, i.e.,

$$\delta W_e = \frac{\epsilon}{2} \iiint_V \nabla\delta\Phi\cdot\nabla\delta\Phi\,dV. \tag{28}$$

The first-order change is zero, and hence W_e is a stationary function for the equilibrium condition. Since δW_e is a positive quantity, the true value of W_e is a minimum since any change from equilibrium increases the energy function W_e.

A Variational Theorem

In order to obtain expressions for stored energy in dispersive media we must first develop a variational expression that can be employed to evaluate the time-average stored energy in terms of the work done in building up the field very slowly from an initial value of zero. For this purpose we need valid expressions for fields that have a time dependence $e^{j\omega t + \alpha t}$ where α is very small. We assume that the phasors $\mathbf{E}(\mathbf{r}, \omega)$, $\mathbf{H}(\mathbf{r}, \omega)$ are analytic functions of ω and can therefore be expanded in Taylor series to obtain $\mathbf{E}(\mathbf{r}, \omega - j\alpha)$, $\mathbf{H}(\mathbf{r}, \omega - j\alpha)$. The medium is assumed to be characterized by a dyadic permittivity $\epsilon(\omega)$ and a dyadic permeability $\bar{\mu}(\omega)$ with the components ϵ_{ij} and μ_{ij} being complex in general. In the complex Poynting vector theorem given by (25) the quantities $\mathbf{B}\cdot\mathbf{H}^*$ and $\mathbf{D}^*\cdot\mathbf{E}$ become $\mathbf{H}^*\cdot\bar{\mu}\cdot\mathbf{H}$ and $\mathbf{E}\cdot\epsilon^*\cdot\mathbf{E}^*$. We now introduce the hermitian and anti-hermitian parts of $\bar{\mu}$ and ϵ as follows:

$$\epsilon' = \frac{1}{2}(\epsilon + \epsilon_t^*) \qquad \bar{\mu}' = \frac{1}{2}(\bar{\mu} + \bar{\mu}_t^*)$$

$$-j\epsilon'' = \frac{1}{2}(\epsilon - \epsilon_t^*) \qquad -j\bar{\mu}'' = \frac{1}{2}(\bar{\mu} - \bar{\mu}_t^*)$$

where ϵ_t^* and $\bar{\mu}_t^*$ are the complex conjugates of the transposed dyadics, e.g., ϵ_{ij} is replaced by ϵ_{ji}^*. The hermitian parts satisfy the relations

$$(\epsilon')_t^* = \frac{1}{2}(\epsilon_t^* + \epsilon) = \epsilon' \qquad (\bar{\mu}')_t^* = \bar{\mu}'$$

while for the anti-hermitian parts $(-j\epsilon'')_t^* = j\epsilon''$ and $(-j\bar{\mu}'')_t^* = j\bar{\mu}''$. By using these relationships we find that $\mathbf{H}^* \cdot \bar{\mu}' \cdot \mathbf{H}$ and $\mathbf{E} \cdot (\bar{\epsilon}' \cdot \mathbf{E})^*$ are real quantities and that $\mathbf{H}^* \cdot (-j\bar{\mu}'' \cdot \mathbf{H})$ and $\mathbf{E} \cdot (-j\bar{\epsilon}'' \cdot \mathbf{E})^*$ are imaginary. It thus follows that the anti-hermitian parts of $\bar{\epsilon}$ and $\bar{\mu}$ give rise to power dissipation in the medium and that lossless media are characterized by hermitian dyadic permittivities and permeabilities. To show that $\mathbf{H}^* \cdot \bar{\mu}' \cdot \mathbf{H}$ is real we note that this quantity also equals $\mathbf{H} \cdot \bar{\mu}_t' \cdot \mathbf{H}^* = \mathbf{H} \cdot (\bar{\mu}' \cdot \mathbf{H})^* = (\mathbf{H}^* \cdot \bar{\mu}' \cdot \mathbf{H})^*$ because $\bar{\mu}'$ is hermitian. In a similar way we have $\mathbf{H}^* \cdot (-j\bar{\mu}'' \cdot \mathbf{H}) = \mathbf{H} \cdot (-j\bar{\mu}_t'' \cdot \mathbf{H}^*) = \mathbf{H} \cdot (j\bar{\mu}'' \cdot \mathbf{H})^* = -[\mathbf{H}^* \cdot (-j\bar{\mu}'' \cdot \mathbf{H})]^*$ and hence is imaginary. Since we are going to relate stored energy to the work done in establishing the field we need to assume that $\bar{\epsilon}''$ and $\bar{\mu}''$ are zero, i.e., $\bar{\epsilon}$ and $\bar{\mu}$ are hermitian.

The variational theorem developed below is useful not only for evaluating stored energy but also for deriving expressions for the velocity of energy transport, Foster's reactance theorem, and other similar relations. It will be used on several other occasions in later chapters for this purpose. The complex conjugates of Maxwell's curl equations are

$$\nabla \times \mathbf{E}^* = j\omega\bar{\mu}^* \cdot \mathbf{H}^* \qquad \nabla \times \mathbf{H}^* = -j\omega\bar{\epsilon}^* \cdot \mathbf{E}^*.$$

A derivative with respect to ω gives

$$\nabla \times \frac{\partial \mathbf{E}^*}{\partial \omega} = j\omega\bar{\mu}^* \cdot \frac{\partial \mathbf{H}^*}{\partial \omega} + j\frac{\partial \omega\bar{\mu}^*}{\partial \omega} \cdot \mathbf{H}^*$$

$$\nabla \times \frac{\partial \mathbf{H}^*}{\partial \omega} = -j\omega\bar{\epsilon}^* \cdot \frac{\partial \mathbf{E}^*}{\partial \omega} - j\frac{\partial \omega\bar{\epsilon}^*}{\partial \omega} \cdot \mathbf{E}^*.$$

Consider next

$$\nabla \cdot \left(\mathbf{E} \times \frac{\partial \mathbf{H}^*}{\partial \omega} + \frac{\partial \mathbf{E}^*}{\partial \omega} \times \mathbf{H} \right) = \frac{\partial \mathbf{H}^*}{\partial \omega} \cdot \nabla \times \mathbf{E} - \mathbf{E} \cdot \nabla \times \frac{\partial \mathbf{H}^*}{\partial \omega}$$

$$+ \mathbf{H} \cdot \nabla \times \frac{\partial \mathbf{E}^*}{\partial \omega} - \frac{\partial \mathbf{E}^*}{\partial \omega} \cdot \nabla \times \mathbf{H} = -j\omega \frac{\partial \mathbf{H}^*}{\partial \omega} \cdot \bar{\mu} \cdot \mathbf{H}$$

$$+ j\omega\mathbf{H} \cdot \bar{\mu}^* \cdot \frac{\partial \mathbf{H}^*}{\partial \omega} + j\omega\mathbf{E} \cdot \bar{\epsilon}^* \cdot \frac{\partial \mathbf{E}^*}{\partial \omega} + j\mathbf{E} \cdot \frac{\partial \omega\bar{\epsilon}^*}{\partial \omega} \cdot \mathbf{E}^*$$

$$+ j\mathbf{H} \cdot \frac{\partial \omega\bar{\mu}^*}{\partial \omega} \cdot \mathbf{H}^* - j\omega \frac{\partial \mathbf{E}^*}{\partial \omega} \cdot \bar{\epsilon} \cdot \mathbf{E}.$$

We now make use of the assumed hermitian property of the constitutive parameters which then shows that most of the terms in the above expression cancel and we thus obtain the desired *variational theorem*

$$-\nabla \cdot \left(\mathbf{E}^* \times \frac{\partial \mathbf{H}}{\partial \omega} + \frac{\partial \mathbf{E}}{\partial \omega} \times \mathbf{H}^* \right) = j\mathbf{H}^* \cdot \frac{\partial \omega\bar{\mu}}{\partial \omega} \cdot \mathbf{H} + j\mathbf{E}^* \cdot \frac{\partial \omega\bar{\epsilon}}{\partial \omega} \cdot \mathbf{E} \qquad (29)$$

where the complex conjugate of the original expression has been taken.

Derivation of Expression for Stored Energy [1.10]–[1.14]

When we assume that the medium is free of loss, the stored energy at any time t can be evaluated in terms of the work done to establish the field from an initial value of zero at $t = -\infty$. Consider a system of source currents with time dependence $\cos \omega t$ and contained within a finite volume V_0. In the region V outside V_0 a field is produced that is given by

$$\mathcal{E}(\mathbf{r},\, t) = \operatorname{Re} \mathbf{E}(\mathbf{r},\, \omega)e^{j\omega t} = \frac{\mathbf{E}}{2}e^{j\omega t} + \frac{\mathbf{E}^*}{2}e^{-j\omega t}$$

$$\mathcal{H}(\mathbf{r},\, t) = \operatorname{Re} \mathbf{H}(\mathbf{r},\, \omega)e^{j\omega t} = \frac{\mathbf{H}}{2}e^{j\omega t} + \frac{\mathbf{H}^*}{2}e^{-j\omega t}.$$

The fields $\mathbf{E}(\mathbf{r}, \omega), \mathbf{H}(\mathbf{r}, \omega)$ are solutions to the problem where the current source has a time dependence $e^{j\omega t}$ while $\mathbf{E}^*, \mathbf{H}^*$ are the solutions for a current source with time dependence of $e^{-j\omega t}$. If the current source is assumed to have a dependence on t of the form $e^{\alpha t} \cos \omega t, -\infty \le t < \infty$ and $\alpha \ll \omega$, then this is equivalent to two problems with time dependences of $e^{\alpha t + j\omega t}$ and $e^{\alpha t - j\omega t}$. The solutions for $\mathbf{E}$ and $\mathbf{H}$ may be obtained by using the first two terms of a Taylor series expansion about ω when $\alpha \ll \omega$. Thus

$$\mathbf{E}(\mathbf{r},\, \omega - j\alpha) = \mathbf{E}(\mathbf{r},\, \omega) - j\alpha \frac{\partial \mathbf{E}(\mathbf{r},\, \omega)}{\partial \omega}$$

$$\mathbf{H}(\mathbf{r},\, \omega - j\alpha) = \mathbf{H}(\mathbf{r},\, \omega) - j\alpha \frac{\partial \mathbf{H}(\mathbf{r},\, \omega)}{\partial \omega}$$

are the solutions for the time dependence $e^{\alpha t + j\omega t}$. The solutions for the conjugate time dependence are given by the complex conjugates of the above solutions. The physical fields are given by

$$\mathcal{E}(\mathbf{r},\, t) = \frac{1}{2}[\mathbf{E}(\mathbf{r},\, \omega - j\alpha)e^{\alpha t + j\omega t} + \mathbf{E}^*(\mathbf{r},\, \omega - j\alpha)e^{\alpha t - j\omega t}]$$

$$= \frac{1}{2}[\mathbf{E}(\mathbf{r},\, \omega)e^{\alpha t + j\omega t} + \mathbf{E}^*(\mathbf{r},\, \omega)e^{\alpha t - j\omega t}]$$

$$- \frac{j\alpha}{2}\left[\frac{\partial \mathbf{E}(\mathbf{r},\, \omega)}{\partial \omega}e^{\alpha t + j\omega t} - \frac{\partial \mathbf{E}^*(\mathbf{r},\, \omega)}{\partial \omega}e^{\alpha t - j\omega t}\right]$$

and a similar expression for $\mathcal{H}(\mathbf{r}, t)$.

The energy supplied per unit volume outside the source region is given by

$$\int_{-\infty}^{t} -\nabla \cdot \mathcal{E}(\mathbf{r},\, t) \times \mathcal{H}(\mathbf{r},\, t)\, dt.$$

This expression comes from interpreting the Poynting vector $\mathcal{E} \times \mathcal{H}$ as giving the instantaneous rate of energy flow. That is, the rate of energy flow into an arbitrary volume V is

$$\oiint_S \mathcal{E} \times \mathcal{H} \cdot (-d\mathbf{S}) = \iiint_V -\nabla \cdot \mathcal{E} \times \mathcal{H}\, dV$$

and hence $-\nabla \cdot \mathcal{E} \times \mathcal{H}$ is the rate of energy flow into a unit volume. When we substitute the expressions for $\mathcal{E}$ and $\mathcal{H}$ given earlier we obtain

$$\frac{1}{4} \int_{-\infty}^{t} - \nabla \cdot \left[(\mathbf{E} \times \mathbf{H}^* + \mathbf{E}^* \times \mathbf{H}) e^{2\alpha t} + j\alpha \left(-\frac{\partial \mathbf{E}}{\partial \omega} \times \mathbf{H}^* \right. \right.$$

$$\left. \left. + \frac{\partial \mathbf{E}^*}{\partial \omega} \times \mathbf{H} + \mathbf{E} \times \frac{\partial \mathbf{H}^*}{\partial \omega} - \mathbf{E}^* \times \frac{\partial \mathbf{H}}{\partial \omega} \right) e^{2\alpha t} \right] dt$$

$$+ \text{ terms multiplied by } e^{\pm 2j\omega t} + \text{ terms multiplied by } \alpha^2.$$

Note that $\mathbf{E}, \mathbf{H}, \partial \mathbf{E}/\partial \omega, \partial \mathbf{H}/\partial \omega$ are evaluated at ω, i.e., for $\alpha = 0$. In a loss-free medium the expansion of $\nabla \cdot (\mathbf{E} \times \mathbf{H}^* + \mathbf{E}^* \times \mathbf{H})$ can be easily shown to be zero upon using Maxwell's equations. Alternatively, one can infer from the complex Poynting vector theorem given by (25) that

$$\text{Re} \oiint_{S} \mathbf{E} \times \mathbf{H}^* \cdot (-d\mathbf{S}) = 0$$

since there is no power dissipation in the medium in V that is bounded by the surface S. We are going to eventually average over one period in time and let α tend to zero and then the terms multiplied by $e^{\pm 2j\omega t}$ and α^2 will vanish. Hence we will not consider these terms any further. Thus up to time t the work done to establish the field on a per unit volume basis is

$$U(t) = -\int_{-\infty}^{t} \nabla \cdot \mathcal{E} \times \mathcal{H} \, dt = -\frac{j\alpha}{4} \nabla \cdot \left(-\frac{\partial \mathbf{E}}{\partial \omega} \times \mathbf{H}^* \right.$$

$$\left. + \frac{\partial \mathbf{E}^*}{\partial \omega} \times \mathbf{H} + \mathbf{E} \times \frac{\partial \mathbf{H}^*}{\partial \omega} - \mathbf{E}^* \times \frac{\partial \mathbf{H}}{\partial \omega} \right) \frac{e^{2\alpha t}}{2\alpha}.$$

When we average over one period and let α tend to zero we obtain the average work done and hence the average stored energy, i.e.,

$$\frac{1}{T} \int_{0}^{T} U(t) \, dt = U_e + U_m = -\frac{j}{8} \nabla \cdot \left(-\frac{\partial \mathbf{E}}{\partial \omega} \times \mathbf{H}^* + \frac{\partial \mathbf{E}^*}{\partial \omega} \times \mathbf{H} + \mathbf{E} \times \frac{\partial \mathbf{H}^*}{\partial \omega} - \mathbf{E}^* \times \frac{\partial \mathbf{H}}{\partial \omega} \right).$$

$$(30a)$$

By using the variational theorem given by (29) and the assumed hermitian nature of ϵ and $\bar{\mu}$ we readily find that

$$U_e + U_m = \frac{1}{4} \left(\mathbf{H}^* \cdot \frac{\partial \omega \bar{\mu}}{\partial \omega} \cdot \mathbf{H} + \mathbf{E}^* \cdot \frac{\partial \omega \epsilon}{\partial \omega} \cdot \mathbf{E} \right). \qquad (30b)$$

We are thus led to interpret the average stored energy in a lossless medium with dispersion (ϵ and $\bar{\mu}$ dependent on ω) as given by

$$U_e = \frac{1}{4} \mathbf{E}^* \cdot \frac{\partial \omega \epsilon}{\partial \omega} \cdot \mathbf{E} \qquad (31a)$$

$$U_m = \frac{1}{4} \mathbf{H}^* \cdot \frac{\partial \omega \bar{\mu}}{\partial \omega} \cdot \mathbf{H} \qquad (31b)$$

on a per unit volume basis. When the medium does not have dispersion, ϵ and $\bar{\mu}$ are not functions of ω and Eqs. (31) reduce to the usual expressions as given by (23). At this point we note that the real and imaginary parts of the components ϵ_{ij} and μ_{ij} of ϵ and $\bar{\mu}$ are related by the Kröning–Kramers relations [1.5], [1.10]. Thus an ideal lossless medium does not exist although there may be frequency bands where the loss is very small. The derivation of (31) required that the medium be lossless so the work done could be equated to the stored energy. It was necessary to assume that α was very small so that the field could be expanded in a Taylor series and would be essentially periodic. The terms in $e^{\pm 2j\omega t}$ represent energy flowing into and out of a unit volume in a periodic manner and do not contribute to the average stored energy. The average stored energy is only associated with the creation of the field itself.

1.4. Boundary Conditions and Field Behavior in Source Regions [1.36]–[1.39]

In the solution of Maxwell's field equations in a region of space that is divided into two or more parts that form the boundaries across which the electrical properties of the medium, that is, μ and ϵ, change discontinuously, it is necessary to know the relationship between the field components on adjacent sides of the boundaries. This is required in order to properly continue the solution from one region into the next, such that the whole final solution will indeed be a solution of Maxwell's equations. These boundary conditions are readily derived using the integral form of the field equations.

Consider a boundary surface S separating a region with parameters ϵ_1, μ_1 on one side and ϵ_2, μ_2 on the adjacent side. The surface normal $\mathbf{n}$ is assumed to be directed from medium 2 into medium 1. We construct a small contour C, which runs parallel to the surface S and an infinitesimal distance on either side as in Fig. 1.3. Also we construct a small closed surface S_0, which consists of a "pillbox" with faces of equal area, parallel to the surface S and an infinitesimal distance on either side. The line integral of the electric field around the contour C is equal to the negative time rate of change of the total magnetic flux through C. In the limit as h tends to zero the total flux through C vanishes since $\mathbf{B}$ is a bounded function. Thus

$$\lim_{h \to 0} \oint_C \mathbf{E} \cdot d\mathbf{l} = 0 = (E_{1t} - E_{2t})l$$

or

$$E_{1t} = E_{2t} \tag{32}$$

where the subscript t signifies the component of $\mathbf{E}$ tangential to the surface S, and l is the length of the contour. The negative sign preceding E_{2t} arises because the line integral is in the opposite sense in medium 2 as compared with its direction in medium 1. Furthermore, the contour C is chosen so small that $\mathbf{E}$ may be considered constant along the contour.

The line integral of $\mathbf{H}$ around the same contour C yields a similar result, provided the displacement flux $\mathbf{D}$ and the current density $\mathbf{J}$ are bounded functions. Thus we find that

$$H_{1t} = H_{2t}. \tag{33}$$

Equations (32) and (33) state that the tangential components of $\mathbf{E}$ and $\mathbf{H}$ are continuous across a boundary for which μ and ϵ change discontinuously. A discontinuous change in μ and ϵ is again a mathematical idealization of what really consists of a rapid but finite rate of change of parameters.

18   FIELD THEORY OF GUIDED WAVES

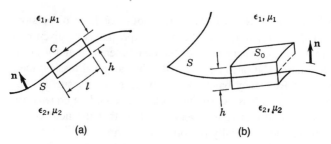

Fig. 1.3. Contour C and surface S_0 for deriving boundary conditions.

The integral of the outward flux through the closed surface S_0 reduces in the limit as h approaches zero to an integral over the end faces only. Provided there is no surface charge on S, the integral of **D** through the surface S_0 gives

$$\lim_{h\to 0}\oiint_S \mathbf{D}\cdot d\mathbf{S} = 0 = (D_{1n} - D_{2n})S_0$$

or

$$D_{1n} = D_{2n}. \tag{34}$$

Similarly we find that

$$B_{1n} = B_{2n} \tag{35}$$

where the subscript n signifies the normal component. It is seen that the normal components of the flux vectors are continuous across the discontinuity boundary S.

The above relations are not completely general since they are based on the assumptions that the current density **J** is bounded and that a surface charge density does not exist on the boundary S. If a surface density of charge ρ_s does exist, then the integral of the normal outward component of **D** over S_0 is equal to the total charge contained within S_0. Under these conditions the boundary condition on **D** becomes

$$\mathbf{n}\cdot(\mathbf{D}_1 - \mathbf{D}_2) = \rho_s. \tag{36}$$

The field vectors remain finite on S, but a situation, for which the current density **J** becomes infinite in such a manner that

$$\lim_{h\to 0} h\mathbf{J} = \mathbf{J}_s$$

a surface distribution of current, arises at the boundary of a perfect conductor for which the conductivity σ is infinite. In this case the boundary condition on the tangential magnetic field is modified. Let τ be the unit tangent to the contour C in medium 1. The positive normal $\mathbf{n}_1$ to the plane surface bounded by C is then given by $\mathbf{n}_1 = \mathbf{n}\times\tau$. The normal current density through C is $\mathbf{n}_1\cdot\mathbf{J}_s$. The line integral of **H** around C now gives

$$(\mathbf{H}_1 - \mathbf{H}_2)\cdot\tau = \mathbf{n}_1\cdot\mathbf{J}_s = \mathbf{n}\times\tau\cdot\mathbf{J}_s.$$

Rearranging gives

$$(\mathbf{H}_1 - \mathbf{H}_2)\cdot\boldsymbol{\tau} = -\boldsymbol{\tau}\cdot\mathbf{n} \times \mathbf{J}_s$$

and, since the orientation of C and hence $\boldsymbol{\tau}$ is arbitrary, we must have

$$\mathbf{H}_1 - \mathbf{H}_2 = -\mathbf{n} \times \mathbf{J}_s$$

which becomes, when both sides are vector-multiplied by $\mathbf{n}$,

$$\mathbf{n} \times (\mathbf{H}_1 - \mathbf{H}_2) = \mathbf{J}_s. \tag{37}$$

If medium 2 is the perfect conductor, then the fields in medium 2 are zero, so that

$$\mathbf{n} \times \mathbf{H}_1 = \mathbf{J}_s. \tag{38}$$

At a perfectly conducting boundary the boundary conditions on the field vectors may be specified as follows:

$$\mathbf{n} \times \mathbf{H} = \mathbf{J}_s \tag{39a}$$

$$\mathbf{n} \times \mathbf{E} = 0 \tag{39b}$$

$$\mathbf{n}\cdot\mathbf{B} = 0 \tag{39c}$$

$$\mathbf{n}\cdot\mathbf{D} = \rho_s \tag{39d}$$

where the normal $\mathbf{n}$ points outward from the conductor surface.

At a discontinuity surface it is sufficient to match the tangential field components only, since, if the fields satisfy Maxwell's equations, this automatically makes the normal components of the flux vectors satisfy the correct boundary conditions. The vector differential operator ∇ may be written as the sum of two operators as follows:

$$\nabla = \nabla_t + \nabla_n$$

where ∇_t is that portion of ∇ which represents differentiation with respect to the coordinates on the discontinuity surface S, and ∇_n is the operator that differentiates with respect to the coordinate normal to S. The curl equation for the electric field now becomes

$$\nabla_t \times (\mathbf{E}_t + \mathbf{E}_n) + \nabla_n \times (\mathbf{E}_t + \mathbf{E}_n) = -j\omega\mu(\mathbf{H}_t + \mathbf{H}_n).$$

The normal component of the left-hand side is $\nabla_t \times \mathbf{E}_t$ since $\nabla_n \times \mathbf{E}_n$ is zero, and the remaining two terms lie in the surface S. If $\mathbf{E}_t$ is continuous across S, then the derivative of $\mathbf{E}_t$ with respect to a coordinate on the surface S is also continuous across S. Hence the normal component of $\mathbf{B}$ is continuous across S. A similar result applies for $\nabla_t \times \mathbf{H}_t$ and the normal component of $\mathbf{D}$. If $\mathbf{H}_t$ is discontinuous across S because of a surface current density $\mathbf{J}_s$, then $\mathbf{n}\cdot\mathbf{D}$ is discontinuous by an amount equal to the surface density of charge ρ_s. Furthermore, the surface current density and charge density satisfy the surface continuity equation.

In the solution of certain diffraction problems it is convenient to introduce fictitious current sheets of density $\mathbf{J}_s$. Since the tangential magnetic field is not necessarily zero on either side, the correct boundary condition on **H** at such a current sheet is that given by (37). The tangential electric field is continuous across such a sheet, but the normal component of **D** is discontinuous. The surface density of charge is given by the continuity equation as

$$\rho_s = (j/\omega)\nabla\cdot\mathbf{J}_s$$

and $\mathbf{n}\cdot\mathbf{D}$ is discontinuous by this amount.

For the same reasons, it is convenient at times to introduce fictitious magnetic current sheets across which the tangential electric field and normal magnetic flux vectors are discontinuous [1.2], [1.30]. If a fictitious magnetic current $\mathbf{J}_m$ is introduced, the curl equation for **E** is written as

$$\nabla \times \mathbf{E} = -j\omega\mu\mathbf{H} - \mathbf{J}_m. \tag{40}$$

The divergence of the left-hand side is identically zero, and since $\nabla\cdot\mathbf{J}_m$ is not necessarily taken as zero, we must introduce a fictitious magnetic charge density as well. From (40) we get

$$\nabla\cdot\mathbf{J}_m = -j\omega\nabla\cdot\mathbf{B} = -j\omega\rho_m. \tag{41}$$

An analysis similar to that carried out for the discontinuity in the tangential magnetic field yields the following boundary condition on the electric field at a magnetic current sheet:

$$\mathbf{n} \times (\mathbf{E}_1 - \mathbf{E}_2) = -\mathbf{J}_m. \tag{42}$$

The negative sign arises by virtue of the way $\mathbf{J}_m$ was introduced into the curl equation for **E**. The normal component of **B** is discontinuous according to the relation

$$\mathbf{n}\cdot(\mathbf{B}_1 - \mathbf{B}_2) = \rho_m \tag{43}$$

where ρ_m is determined by (41). At a magnetic current sheet, the tangential components of **H** and the normal component of **D** are continuous across the sheet. To avoid introducing new symbols, $\mathbf{J}_m$ and ρ_m in (42) and (43) are used to represent a surface density of magnetic current and charge, respectively. It should be emphasized again that such current sheets are introduced only to facilitate the analysis and do not represent a physical reality. Also, the magnetic current and charge introduced here do not correspond to those associated with the magnetic polarization of material bodies that was introduced in Section 1.2.

If the fictitious current sheets that we introduced do not form closed surfaces and the current densities normal to the edge are not zero, we must include a distribution of charge around the edge of the current sheet to take account of the discontinuity of the normal current at the edge [1.30]. Let C be the boundary of the open surface S on which a current sheet of density $\mathbf{J}_s$ exists. Let τ be the unit tangent to C, and **n** the normal to the surface S, as in Fig. 1.4. The unit normal to the boundary curve C that is perpendicular to **n** and directed out from S is given by

$$\mathbf{n}_1 = \tau \times \mathbf{n}.$$

The charge flowing toward the boundary C per second is $\mathbf{n}_1\cdot\mathbf{J}_s$ per unit length of C. Replacing

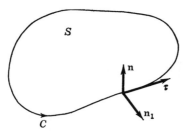

Fig. 1.4. Coordinate system at boundary of open surface S.

$\mathbf{J}_s$ from (37) by the discontinuity in $\mathbf{H}$ gives

$$j\omega\sigma_e = (\boldsymbol{\tau} \times \mathbf{n})\cdot\mathbf{n} \times (\mathbf{H}_1 - \mathbf{H}_2)$$
$$= -\mathbf{n} \times (\boldsymbol{\tau} \times \mathbf{n})\cdot(\mathbf{H}_1 - \mathbf{H}_2)$$
$$= -\boldsymbol{\tau}\cdot(\mathbf{H}_1 - \mathbf{H}_2) \tag{44}$$

where $j\omega\sigma_e$ is the time rate of increase of electric charge density per unit length along the boundary C. A similar analysis gives the following expression for the time rate of increase of magnetic charge along the boundary C of an open surface containing a magnetic surface current:

$$j\omega\sigma_m = \boldsymbol{\tau}\cdot(\mathbf{E}_1 - \mathbf{E}_2). \tag{45}$$

These line charges are necessary in order for the continuity equations relating charge and current to hold at the edge of the sheet.

A somewhat more complex situation is the behavior of the field across a surface on which we have a sheet of normally directed electric current filaments $\mathbf{J}_n$ as shown in Fig. 1.5(a). At the ends of the current filaments charge layers of density $\rho_s = \pm J_n/j\omega$ exist and constitute a double layer of charge [1.36]. The normal component of $\mathbf{D}$ terminates on the negative charge below the surface and emanates with equal strength from the positive charge layer above the surface. Thus $\mathbf{n}\cdot\mathbf{D}$ is continuous across the surface as Gauss' law clearly shows since no net charge is contained within a small "pillbox" erected about the surface as in Fig. 1.3(b). The normal component of $\mathbf{B}$ is also continuous across this current sheet since there is no equivalent magnetic charge present. We will show that, in general, the tangential electric field will undergo a discontinuous change across the postulated sheet of normally directed current. The electric field between the charge layers is made up from the locally produced field and that arising from remote sources. We will assume that the surface S is sufficiently smooth that a small section can be considered a plane surface which we take to be located in the xz plane as shown in Fig. 1.5(c) and (d). The locally produced electric field has a value of $\rho_s(x, z)/\epsilon_0$, and is directed in the negative y direction. We now consider a line integral of the electric field around a small contour with sides parallel to the x axis and lying on adjacent sides of the double layer of charge and with ends at x and $x + \Delta x$. Since the magnetic flux through the area enclosed by the contour is of order $h\Delta x$ and will vanish as h and Δx are made to approach zero, the line integral gives, to first order,

$$(-E_{n1} + E_{n2})h + (E_{t2} - E_{t1})\Delta x = 0.$$

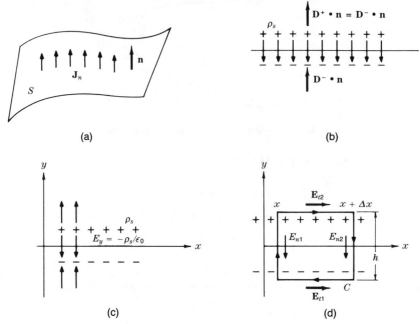

Fig. 1.5. (a) A sheet of normally directed current filaments. (b) The associated double layer of charge. (c) The behavior of the normal component of the electric field across the double layer of charge. (d) Illustration of contour C.

This can be expressed in the form

$$\epsilon_0(E_{t2} - E_{t1})\Delta x = -\left[-\rho_s(x, z) + \rho_s(x, z) + \frac{\partial \rho_s}{\partial x}\Delta x\right] h$$

$$= -\frac{\partial \rho_s}{\partial x}\Delta xh = \frac{-1}{j\omega}\frac{\partial J_n}{\partial x}\Delta xh$$

or as

$$j\omega\epsilon_0(E_{x2} - E_{x1}) = -\frac{\partial J_y}{\partial x}h.$$

A similar line integral along a contour parallel to the z axis would give

$$j\omega\epsilon_0(E_{z2} - E_{z1}) = -\frac{\partial J_y}{\partial z}h.$$

These two relations may be combined and expressed in vector form as

$$j\omega\epsilon_0\mathbf{a}_y \times (\mathbf{E}_2 - \mathbf{E}_1) = h\nabla \times \mathbf{J}_n$$

since $\mathbf{J}_n$ has only a component along y. We now let the current density increase without limit but such that the product $h\mathbf{J}_n$ remains constant so as to obtain a surface density of normal current $\mathbf{J}_{ns}$ amperes per meter [compare with the surface current $\mathbf{J}_s$ that is tangential to the surface in (37)]. We have thus established that across a surface layer of normally directed electric current of density $\mathbf{J}_{ns}$ the tangential electric field undergoes a discontinuous change given by the boundary condition

$$\mathbf{n} \times (\mathbf{E}_1 - \mathbf{E}_2) = \frac{1}{j\omega\epsilon_0} \nabla \times \mathbf{J}_{ns} \tag{46}$$

where the unit normal $\mathbf{n}$ points into region 1. The normally directed electric current produces a discontinuity equivalent to that produced by a sheet of tangential magnetic current of density $\mathbf{J}_m$ given by

$$\mathbf{J}_m = -\frac{1}{j\omega\epsilon_0} \nabla \times \mathbf{J}_{ns} \tag{47}$$

as (42) shows.

Inside the source region the normal electric field has the value $\rho_s/\epsilon_0 = J_n/j\omega\epsilon_0 = J_n h/j\omega\epsilon_0 h$. In the limit as h approaches zero this can be expressed in the form

$$E_y = -\frac{J_{ys}\delta(y)}{j\omega\epsilon_0}$$

since

$$-\int_{-h/2}^{h/2} E_y dy = \frac{J_y h}{j\omega\epsilon_0} = \frac{J_{ys}}{j\omega\epsilon_0}.$$

The quantity J_{ns}/h represents a pulse of height J_{ns}/h and width h. The area of this pulse is J_{ns}. Hence in the limit as h approaches zero we can represent this pulse by the Dirac delta symbolic function $\delta(y)$.

A similar derivation would show that a surface layer of normally directed magnetic current of density $\mathbf{J}_{mn}$ would produce a discontinuity in the tangential magnetic field across the surface according to the relation

$$\mathbf{n} \times (\mathbf{H}_1 - \mathbf{H}_2) = \frac{1}{j\omega\mu_0} \nabla \times \mathbf{J}_{mn}. \tag{48}$$

This surface layer of normally directed magnetic current produces a discontinuous change in $\mathbf{H}$ equivalent to that produced by a tangential electric current sheet of density

$$\mathbf{J}_s = \frac{1}{j\omega\mu_0} \nabla \times \mathbf{J}_{mn}. \tag{49}$$

The relations derived above are very useful in determining the fields produced by localized point sources, which can often be expressed as sheets of current through expansion of the point source in an appropriate two-dimensional Fourier series (or integral).

1.5. FIELD SINGULARITIES AT EDGES [1.15]–[1.25]

We now turn to a consideration of the behavior of the field vectors at the edge of a conducting wedge of internal angle ϕ, as illustrated in Fig. 1.6. The solutions to diffraction problems are not always unique unless the behavior of the fields at the edges of infinite conducting bodies

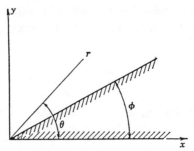

Fig. 1.6. Conducting wedge.

is specified [1.15], [1.19]. In particular, it is found that some of the field components become infinite as the edge is approached. However, the order of singularity allowed is such that the energy stored in the vicinity of the diffracting edge is finite. The situation illustrated in Fig. 1.6 is not the most general by any means, but will serve our purpose as far as the remainder of this volume is concerned. A more complete discussion may be found in the references listed at the end of this chapter.

In the vicinity of the edge the fields can be expressed as a power series in r. It will be assumed that this series will have a dominant term r^α where α can be negative. However, α must be restricted such that the energy stored in the field remains finite.

The sum of the electric and magnetic energy in a small region surrounding the edge is proportional to

$$\int\!\!\!\int\!\!\!\int_0^r (\epsilon \mathbf{E}\cdot\mathbf{E}^* + \mu \mathbf{H}\cdot\mathbf{H}^*) r \, d\theta \, dr \, dz. \tag{50}$$

Since each term is positive, the integral of each term must remain finite. Thus each term must increase no faster than $r^{2\alpha}, \alpha > -1$, as r approaches zero. Hence each field component must increase no faster than r^α as r approaches zero. The minimum allowed value is, therefore, $\alpha > -1$. The singular behavior of the electromagnetic field near an edge is the same as that for static and quasi-static fields since when r is very small propagation effects are not important because the singular behavior is a local phenomenon. In other words, the spatial derivatives of the fields are much larger than the time derivatives in Maxwell's equations so the latter may be neglected.

The two-dimensional problem involving a conducting wedge is rather simple because the field can be decomposed into waves that are transverse electric (TE, $E_z \equiv 0$) and transverse magnetic (TM, $H_z \equiv 0$) with respect to z. Furthermore, the z dependence can be chosen as e^{-jwz} where w can be interpreted as a Fourier spectral variable if the fields are Fourier transformed with respect to z. The TE solutions can be obtained from a solution for H_z, while the TM solutions can be obtained from the solution for E_z. The solutions for E_z and H_z in cylindrical coordinates are given in terms of products of Bessel functions and trigonometric functions of θ. Thus a solution for the axial fields is of the form

$$\begin{bmatrix} J_\nu(\gamma r) \\ \text{or} \\ H_\nu^2(\gamma r) \end{bmatrix} \begin{bmatrix} \sin \nu\theta \\ \text{or} \\ \cos \nu\theta \end{bmatrix}$$

where $\gamma = \sqrt{k_0^2 - w^2}$ and ν is, in general, not an integer because θ does not cover the full range from 0 to 2π. The Hankel function $H_\nu^2(\gamma r)$ of order ν and of the second kind represents an outward propagating cylindrical wave. Since there is no source at $r = 0$ this function should not be present on physical grounds. From a mathematical point of view the Hankel function must be excluded since for r very small it behaves like $r^{-\nu}$ and this would result in some field components having an $r^{-\nu-1}$ behavior. But $\alpha > -1$ so such a behavior is too singular.

An appropriate solution for E_z is

$$E_z = J_\nu(\gamma r) \sin \nu(\theta - \phi)$$

since $E_z = 0$ when $\theta = \phi$. The parameter ν is determined by the condition that $E_z = 0$ at $\theta = 2\pi$ also. Thus we must have $\sin \nu(2\pi - \phi) = 0$ so the allowed values of ν are

$$\nu = \frac{n\pi}{2\pi - \phi}, \qquad n = 1, 2, \ldots . \tag{51}$$

The smallest value of ν is $\pi/(2\pi - \phi)$. For a wedge with zero internal angle, i.e., a half plane, $\nu = 1/2$, while for a $90°$ corner $\nu = \pi/(2\pi - \pi/2) = 2/3$. When $\phi = 180°, \nu = 1$. Maxwell's equations[3] show that E_r is proportional to $\partial E_z/\partial r$, while E_θ is proportional to $\partial E_z/r\,\partial\theta$. We thus conclude that E_z remains finite at the edge but E_r and E_θ become infinite like $r^\alpha = r^{\nu-1}$ since $J_\nu(\gamma r)$ varies like r^ν for r approaching zero. For a half plane E_r and E_θ are singular with an edge behavior like $r^{-1/2}$. From Maxwell's equations one finds that H_r and H_θ are proportional to E_θ and E_r, respectively, so these field components have the same edge singularity.

In order that the tangential electric field derived from H_z will vanish on the surface of the perfectly conducting wedge, H_z must satisfy the boundary conditions

$$\frac{\partial H_z}{\partial \theta} = 0; \qquad \theta = \phi, 2\pi.$$

Hence an appropriate solution for H_z is

$$H_z = J_\nu(\gamma r) \cos \nu(\theta - \phi).$$

The boundary condition at $\theta = 2\pi$ gives $\sin \nu(\theta - \phi) = 0$ so the allowed values of ν are again given by (51). However, since H_z depends on $\cos \nu(\theta - \phi)$ the value of $\nu = 0$ also yields a nonzero solution, but this solution will have a radial dependence given by $J_0(\gamma r)$. Since $dJ_0(\gamma r)/dr = -\gamma J_1(\gamma r)$ and vanishes at $r = 0$, both H_r and E_θ derived from H_z with $\nu = 0$ remain finite at the edge.

We can summarize the above results by stating that near the edge of a perfectly conducting wedge of internal angle less than π the normal components of **E** and **H** become singular, but the components of **E** and **H** tangent to the edge remain finite. The charge density ρ_s on the surface is proportional to the normal component of **E** and hence also exhibits a singular behavior. Since $\mathbf{J}_s = \mathbf{n} \times \mathbf{H}$, the current flowing toward the edge is proportional to H_z and will vanish at the edge (an exception is the $\nu = 0$ case for which the current flows around the corner but remains finite). The current tangential to the edge is proportional to H_r and thus has a singular behavior like that of the charge density.

---

[3]Expressions for the transverse field components in terms of the axial components are given in Chapter 5.

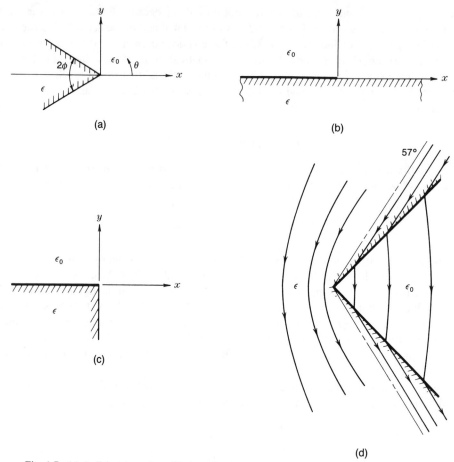

Fig. 1.7. (a) A dielectric wedge. (b) A conducting half plane on a dielectric medium. (c) A 90°
dielectric corner and a conducting half plane. (d) Field behavior in a dielectric wedge for $\kappa = 10$.

Singular Behavior at a Dielectric Corner [1.20]–[1.24]

The electric field can exhibit a singular behavior at a sharp corner involving a dielectric
wedge as shown in Fig. 1.7(a). The presence of dielectric media around a conducting wedge
such as illustrated in Fig. 1.7(b) and (c) will also modify the value of α that governs the
singular behavior of the field. In order to obtain expressions for the minimum allowed value
of α we will take advantage of the fact that the field behavior near the edge under dynamic
conditions is the same as that for the static field.

For the configuration shown in Fig. 1.7(a) let ψ be the scalar potential that satisfies Laplace's
equation $\nabla^2 \psi$ and is symmetrical about the xz plane. In cylindrical coordinates a valid solution
for ψ that is finite at $r = 0$ is (no dependence on z is assumed)

$$\psi = \begin{cases} C_1 r^\nu \cos \nu\theta, & |\theta| \leq \pi - \phi \\ C_2 r^\nu \cos \nu(\pi - \theta), & \pi - \phi \leq \theta \leq \pi + \phi. \end{cases}$$

At $\theta = \pm(\pi - \phi)$ the potential ψ must be continuous along with D_θ. These two boundary conditions require that

$$C_1 \cos \nu(\pi - \phi) = C_2 \cos \nu\phi$$

$$C_1 \sin \nu(\pi - \phi) = -C_2 \kappa \sin \nu\phi.$$

When we divide the second equation by the first we obtain

$$-\kappa \sin \nu\phi \cos \nu(\pi - \phi) = \cos \nu\phi \sin \nu(\pi - \phi).$$

By introducing $\beta = \pi/2 - \phi$ as an angle and expanding the trigonometric functions the following expression can be obtained:

$$\sin \nu(\pi - 2\phi) = \frac{\kappa + 1}{\kappa - 1} \sin \nu\pi. \tag{52}$$

For a 90° corner (52) simplifies to

$$\sin(\nu\pi/2) = [(\kappa + 1)/(\kappa - 1)]2 \sin(\nu\pi/2) \cos(\nu\pi/2)$$

or

$$\nu = \frac{2}{\pi} \cos^{-1} \frac{\kappa - 1}{2(1 + \kappa)}. \tag{53}$$

The smallest value of $\nu > 0$ determines the order of the singularity. For $\kappa = 2, \nu = 0.8934$ while for large values of κ the limiting value of ν is $\frac{2}{3}$. Since E_r and E_θ are given by $-\nabla\psi$ the normal electric field components have an $r^{-1/3}$ singularity for large values of κ. This is the same singularity as that for a conducting wedge with a 90° internal angle. For small values of κ the singularity is much weaker.

It is easily shown that for an antisymmetrical potential proportional to $\sin \nu\theta$ the eigenvalue equation is the same as (52) with a minus sign. This case corresponds to a dielectric wedge on a ground plane and the field becomes singular when $\phi > \pi/2$. This behavior is quite different from that of a conducting wedge that has a nonsingular field for $\phi \geq \pi/2$. Even when κ approaches infinity the dielectric wedge has a singular field; e.g., for $\phi = 3\pi/4$, the limiting value of ν is $\frac{2}{3}$. The field inside the dielectric is not zero when κ becomes infinite so the external field behavior is not required to be the same as that for a conducting wedge for which the scalar potential must be constant along the surface. A sketch of the field lines for an unsymmetrical potential field with $\kappa = 10$ is shown in Fig. 1.7(d).

For the half plane on a dielectric half space a suitable solution for ψ is

$$\psi = \begin{cases} C_1 r^\nu \sin \nu(\pi - \theta), & 0 \leq \theta \leq \pi \\ C_2 r^\nu \sin \nu(\pi + \theta), & -\pi \leq \theta \leq 0. \end{cases}$$

The boundary conditions at $\theta = 0$ require that $C_1 = C_2$ and $C_1 \cos \nu\pi = -\kappa C_2 \cos \nu\pi$. For a nontrivial solution $\nu = \frac{1}{2}$ and the field singularity is the same as that for a conducting half plane.

For the configuration shown in Fig. 1.7(c) solutions for ψ are

$$\psi = \begin{cases} C_1 r^\nu \sin \nu(\pi - \theta), & -\pi/2 \leq \theta \leq \pi \\ C_2 r^\nu \sin \nu(\pi + \theta), & -\pi \leq \theta \leq -\pi/2. \end{cases}$$

The boundary conditions at $\theta = -\pi/2$ require that $C_1 \sin(3\nu\pi/2) = C_2 \sin(\nu\pi/2)$ and $-C_1 \cos(3\nu\pi/2) = C_2 \kappa \cos(\nu\pi/2)$. These equations may be solved for ν to give

$$\nu = \frac{1}{\pi} \cos^{-1} \frac{1 - \kappa}{2(1 + \kappa)}. \tag{54}$$

The presence of the dielectric makes the field less singular than that for a conducting half plane since ν given by (54) is greater than $\frac{1}{2}$. For large values of κ the limiting value of ν is $\frac{2}{3}$.

In the case of a dielectric wedge without a conductor the magnetic field does not exhibit a singular behavior because the dielectric does not affect the static magnetic field. If the dielectric is replaced by a magnetizable medium, then the magnetic field can become singular. The nature of this singularity can be established by considering a scalar potential for $\mathbf{H}$ and results similar to those given above will be obtained.

A discussion of the field singularity near a conducting conical tip can be found in the literature [1.23].

Knowledge of the behavior of the field near a corner involving conducting and dielectric materials is very useful in formulating approximate solutions to boundary-value problems. When approximate current and charge distributions on conducting surfaces are specified in such a manner that the correct edge behavior is preserved, the numerical evaluation of the field converges more rapidly and the solution is more accurate than approximate solutions where the edge conditions are not satisfied.

1.6. THE WAVE EQUATION

The electric and magnetic field vectors $\mathbf{E}$ and $\mathbf{H}$ are solutions of the inhomogeneous vector wave equations. These vector wave equations will be derived here, but first we must make a distinction between the impressed currents and charges that are the sources of the field and the currents and charges that arise because of the presence of the field in a medium having finite conductivity. The latter current is proportional to the electric field and is given by $\sigma \mathbf{E}$. This conduction current may readily be accounted for by replacing the permittivity ϵ by the complex permittivity $\epsilon'(1 - j \tan \delta)$ as given by (21) in Section 1.2. The density of free charge, apart from that associated with the impressed currents, may be assumed to be zero. Thus the charge density ρ and the current density $\mathbf{J}$ appearing in the field equations will be taken as the impressed charges and currents. Any other currents will be accounted for by the complex electric permittivity, although we shall still write simply ϵ for this permittivity.

The wave equation for $\mathbf{E}$ is readily obtained by taking the curl of the curl of $\mathbf{E}$ and substituting for the curl of $\mathbf{H}$ on the right-hand side. Referring back to the field equations (1), it is readily seen that

$$\nabla \times \nabla \times \mathbf{E} = -j\omega\mu\nabla \times \mathbf{H} = \omega^2\mu\epsilon\mathbf{E} - j\omega\mu\mathbf{J}$$

where it is assumed that μ and ϵ are scalar constants. Using the vector identity $\nabla \times \nabla \times \mathbf{E} =$

$\nabla\nabla\cdot\mathbf{E} - \nabla^2\mathbf{E}$ and replacing $\nabla\cdot\mathbf{E}$ by ρ/ϵ gives the desired result

$$\nabla^2\mathbf{E} + k^2\mathbf{E} = j\omega\mu\mathbf{J} + \frac{\nabla\rho}{\epsilon} \qquad (55a)$$

where $k = \omega(\mu\epsilon)^{1/2}$ and will be called the wavenumber. The wavenumber is equal to ω divided by the velocity of propagation of electromagnetic waves in a medium with parameters μ and ϵ when ϵ is real. It is also equal to $2\pi/\lambda$, where λ is the wavelength of plane waves in the same medium. The impressed charge density is related to the impressed current density by the equation of continuity so that (55a) may be rewritten as

$$\nabla^2\mathbf{E} + k^2\mathbf{E} = j\omega\mu\mathbf{J} - \frac{\nabla\nabla\cdot\mathbf{J}}{j\omega\epsilon}. \qquad (55b)$$

In a similar fashion we find that $\mathbf{H}$ is a solution of the equation

$$\nabla^2\mathbf{H} + k^2\mathbf{H} = -\nabla \times \mathbf{J}. \qquad (55c)$$

The magnetic field is determined by the rotational part of the current density only, while the electric field is determined by both the rotational and lamellar parts of $\mathbf{J}$. This is as it should be, since $\mathbf{E}$ has both rotational and lamellar parts while $\mathbf{H}$ is a pure solenoidal field when μ is constant. The operator ∇^2 may, in rectangular coordinates, be interpreted as the Laplacian operator operating on the individual rectangular components of each field vector. In a general curvilinear coordinate system this operator is replaced by the grad div − curl curl operator since

$$\nabla^2 \equiv \nabla\nabla\cdot - \nabla \times \nabla \times .$$

The wave equations given above are not valid when μ and ϵ are scalar functions of position or dyadics. However, we will postpone the consideration of the wave equation in such cases until we take up the problem of propagation in inhomogeneous and anisotropic media. In a source-free region both the electric and magnetic fields satisfy the homogeneous equation

$$\nabla^2\mathbf{E} + k^2\mathbf{E} = 0 \qquad (56)$$

and similarly for $\mathbf{H}$. Equation (56) is also known as the vector Helmholtz equation.

The derivation of (55a) shows that $\mathbf{E}$ is also described by the equation

$$\nabla \times \nabla \times \mathbf{E} - k^2\mathbf{E} = -j\omega\mu\mathbf{J}. \qquad (57)$$

This equation is commonly called the vector wave equation to distinguish it from the vector Helmholtz equation. However, the latter is also a wave equation describing the propagation of steady-state sinusoidally varying fields.

Since the current density vector enters into the inhomogeneous wave equations in a rather complicated way, the integration of these equations is usually performed by the introduction of auxiliary potential functions that serve to simplify the mathematical analysis. These auxiliary potential functions may or may not represent clearly definable physical entities (especially in the absence of sources), and so we prefer to adopt the viewpoint that these potentials are just useful mathematical functions from which the electromagnetic field may be derived. We shall discuss some of these potential functions in the following section.

1.7. Auxiliary Potential Functions

The first auxiliary potential functions we will consider are the vector and scalar potential functions $\mathbf{A}$ and Φ, which are just extensions of the magnetostatic vector potential and the electrostatic scalar potential. These functions are here functions of time and vary with time according to the function $e^{j\omega t}$, which, however, for convenience will be deleted. Again we will assume that μ and ϵ are constants.

The flux vector $\mathbf{B}$ is always solenoidal and may, therefore, be derived from the curl of a suitable vector potential function $\mathbf{A}$ as follows:

$$\mathbf{B} = \nabla \times \mathbf{A} \qquad (58)$$

since $\nabla \cdot \nabla \times \mathbf{A} = 0$, this makes $\nabla \cdot \mathbf{B} = 0$ also. Thus one of Maxwell's equations is satisfied identically. The vector potential $\mathbf{A}$ may have both a solenoidal and a lamellar part. At this stage of the analysis the lamellar part is entirely arbitrary since $\nabla \times \mathbf{A}_l = 0$. Substituting (58) into the curl equation for $\mathbf{E}$ gives

$$\nabla \times (\mathbf{E} + j\omega \mathbf{A}) = 0.$$

Since $\nabla \times \nabla \Phi = 0$, the above result may be integrated to give

$$\mathbf{E} = -j\omega \mathbf{A} - \nabla \Phi \qquad (59)$$

where Φ is a scalar function of position and is called the scalar potential. So far we have two of Maxwell's field equations satisfied, and it remains to find the relation between Φ and $\mathbf{A}$ and the condition on Φ and $\mathbf{A}$ so that the two remaining equations $\nabla \cdot \mathbf{D} = \rho$ and $\nabla \times \mathbf{H} = j\omega \epsilon \mathbf{E} + \mathbf{J}$ are satisfied. The curl equation for $\mathbf{H}$ gives

$$\nabla \times \mu \mathbf{H} = \nabla \times \nabla \times \mathbf{A} = \nabla \nabla \cdot \mathbf{A} - \nabla^2 \mathbf{A} = j\omega \epsilon \mu \mathbf{E} + \mu \mathbf{J}$$
$$= k^2 \mathbf{A} - j\omega \epsilon \mu \nabla \Phi + \mu J. \qquad (60)$$

Since Φ and the lamellar part of $\mathbf{A}$ are as yet arbitrary, we are at liberty to choose a relationship between them. With a view toward simplifying the equation satisfied by the vector potential $\mathbf{A}$, i.e., (60), we choose

$$\nabla \cdot \mathbf{A} = -j\omega \epsilon \mu \Phi. \qquad (61)$$

This particular choice is known as the Lorentz condition and is by no means the only one possible. Using (61), we find that (60) reduces to

$$\nabla^2 \mathbf{A} + k^2 \mathbf{A} = -\mu \mathbf{J}. \qquad (62)$$

In order that the divergence equation for $\mathbf{D}$ shall be satisfied it is necessary that

$$\nabla \cdot \mathbf{E} = \frac{\rho}{\epsilon} = -j\omega \nabla \cdot \mathbf{A} - \nabla^2 \Phi \qquad \text{from (59).}$$

Using the Lorentz condition to eliminate $\nabla \cdot \mathbf{A}$ gives the following equation to be satisfied by the scalar potential Φ:

$$\nabla^2 \Phi + k^2 \Phi = \frac{-\rho}{\epsilon}. \tag{63}$$

The vector potential must be a solution of the inhomogeneous vector wave equation (62), while the scalar potential must be a solution of the scalar inhomogeneous wave equation (63). Clearly some advantage has been gained inasmuch as the wave equations for $\mathbf{A}$ and Φ are simpler than those for $\mathbf{E}$ and $\mathbf{H}$. These potentials are not unique since we may add to them any solution of the homogeneous wave equations provided the new potentials still obey the Lorentz condition. Consider a new vector potential $\mathbf{A}' = \mathbf{A} - \nabla \psi$, where ψ is an arbitrary scalar function satisfying the scalar equation

$$\nabla^2 \psi + k^2 \psi = 0.$$

This transformation leaves the magnetic field $\mathbf{B}$ unchanged since $\nabla \times \nabla \psi \equiv 0$. In order that the electric field, as given by (59), shall remain invariant under this transformation, it is necessary that the scalar potential Φ be transformed to a new scalar potential Φ' according to the relation $\Phi' = \Phi + j\omega\psi$. Thus we have

$$\mathbf{E} = -j\omega\mathbf{A}' - \nabla\Phi' = -j\omega\mathbf{A} + j\omega\nabla\psi - \nabla\Phi - j\omega\nabla\psi = -j\omega\mathbf{A} - \nabla\phi.$$

The transformation condition on Φ' is just the condition that the new potentials shall satisfy the Lorentz condition.[4] The above transformation to the new potentials $\mathbf{A}'$ and Φ' is known as a gauge transformation, and the property of the invariance of the field vectors under such a transformation is known as gauge invariance.

Using the Lorentz condition, the fields may be written in terms of the vector potential alone as follows:

$$\mathbf{B} = \nabla \times \mathbf{A} \tag{64a}$$

$$\mathbf{E} = -j\omega\mathbf{A} + \frac{\nabla\nabla\cdot\mathbf{A}}{j\omega\epsilon\mu} \tag{64b}$$

where $\mathbf{A}$ is a solution of (62), and Φ is a solution of (63) as well as being related to $\mathbf{A}$ by (61). We may include in $\mathbf{A}$ any arbitrary solution to the homogeneous scalar Helmholtz equation, that is, $-\nabla\psi$, and, provided the new scalar potential is taken as $\Phi + j\omega\psi$, we will always derive the same electromagnetic field by means of the operations given in (58) and (59). Consequently, ψ may be chosen in such a manner that, if possible, a simplification in the solution of Maxwell's equations is obtained. For example, ψ may be chosen so that $\nabla \cdot \mathbf{A}' = 0$. Then we find that

$$\mathbf{B} = \nabla \times \mathbf{A}' \tag{65a}$$

$$\mathbf{E} = -j\omega\mathbf{A}' - \nabla\Phi' \tag{65b}$$

---

[4]In the particular gauge where $\nabla \cdot \mathbf{A}$ is zero, the Lorentz condition can no longer be applied since, in general, both $\nabla \cdot \mathbf{A}$ and Φ cannot be equal to zero.

and the fields are explicitly separated into their solenoidal and lamellar parts. The vector potential is now a solution of

$$\nabla^2 \mathbf{A}' + k^2 \mathbf{A}' = -\mu \mathbf{J} + j\omega\epsilon\mu\nabla\Phi' \tag{66a}$$

and the scalar potential is a solution of

$$\nabla^2 \Phi' = -\frac{\rho}{\epsilon}. \tag{66b}$$

Thus the lamellar part of the electric field is determined by a scalar potential that is a solution of Poisson's equation with the exception that both Φ' and ρ vary with time. Since current and charge are related by the continuity equation, that is, $\nabla \cdot \mathbf{J} = -j\omega\rho$, we find that the divergence of the right-hand side of (66a) vanishes, in view of the relation between Φ' and ρ expressed by (66b). The term $j\omega\epsilon\mu\nabla\Phi'$ may be regarded as canceling the lamellar part of $\mathbf{J}$ in (66a), and hence permits the solution for a vector potential having zero divergence, i.e., no lamellar part.

In the region outside of that containing the sources $\nabla \cdot \mathbf{E} = 0$, but this does not mean that the lamellar part of the electric field will be identically zero since the solution of (66) will give a nonzero scalar function Φ' outside the source region. Since $j\omega\epsilon\mathbf{E} = \nabla \times \mathbf{H}$ outside the source region the electric field can be expressed in terms of the curl of another vector function. However, this does not imply that $\nabla \cdot \mathbf{E} = 0$ everywhere; it merely requires that $\nabla^2 \Phi' = 0$ outside the source region.

The arbitrariness in the vector potential is only in the lamellar part. In a source-free region it is usually more convenient to find a vector potential function with a nonzero divergence, and the fields are then determined by (64). Since $\mathbf{E}$ has zero divergence in a source-free region, it is quite apparent that $\mathbf{E}$ could equally well be placed equal to the curl of a vector potential function. The details for this case will now be considered under the subject of Hertzian potentials.

In a homogeneous isotropic source-free region $\nabla \cdot \mathbf{E} = 0$, and so $\mathbf{E}$ may be derived from the curl of an auxiliary vector potential function which we will call a magnetic type of Hertzian vector potential. The reason for the terminology "magnetic" will be explained later. Thus we take

$$\mathbf{E} = -j\omega\mu\nabla \times \mathbf{\Pi}_h \tag{67}$$

where $\mathbf{\Pi}_h$ is the magnetic Hertzian potential. The curl equation for $\mathbf{H}$ gives

$$\nabla \times \mathbf{H} = k^2 \nabla \times \mathbf{\Pi}_h$$

so that

$$\mathbf{H} = k^2 \mathbf{\Pi}_h + \nabla\Phi \tag{68}$$

where Φ is as yet an arbitrary scalar function. Substituting in the curl equation for $\mathbf{E}$ gives

$$\nabla \times \nabla \times \mathbf{\Pi}_h = \nabla\nabla \cdot \mathbf{\Pi}_h - \nabla^2 \mathbf{\Pi}_h = k^2 \mathbf{\Pi}_h + \nabla\Phi. \tag{69a}$$

Since both Φ and $\nabla \cdot \mathbf{\Pi}_h$ are still arbitrary, we make them satisfy a Lorentz type of condition; i.e., put $\nabla \cdot \mathbf{\Pi}_h = \Phi$, so that the equation for $\mathbf{\Pi}_h$ becomes

$$\nabla^2 \mathbf{\Pi}_h + k^2 \mathbf{\Pi}_h = 0. \tag{69b}$$

Since $\nabla \cdot \mathbf{B} = 0$, we find, by taking the divergence of (68) and remembering that μ is a constant, that Φ is a solution of

$$\nabla^2 \Phi + k^2 \Phi = 0. \tag{70}$$

The fields are now given by the relations

$$\mathbf{E} = -j\omega\mu \nabla \times \mathbf{\Pi}_h \tag{71a}$$

$$\mathbf{H} = k^2 \mathbf{\Pi}_h + \nabla\nabla \cdot \mathbf{\Pi}_h$$

$$= \nabla \times \nabla \times \mathbf{\Pi}_h \tag{71b}$$

with $\mathbf{\Pi}_h$ a solution of the vector Helmholtz equation (69b). In practice, we do not need to determine the scalar function Φ explicitly since it is equal to $\nabla \cdot \mathbf{\Pi}_h$ and has been eliminated from the equation determining the magnetic field $\mathbf{H}$.

In a source-free region where $\mu \neq \mu_0$ and $\epsilon = \epsilon_0$, the Hertzian potential $\mathbf{\Pi}_h$ will be shown to be the solution of the inhomogeneous vector Helmholtz equation with the source function being the magnetic polarization $\mathbf{M}$ per unit volume. It is for this reason that $\mathbf{\Pi}_h$ is called a magnetic Hertzian potential. When the magnetic polarization is taken into account explicitly, we rewrite (67) as

$$\mathbf{E} = -j\omega\mu_0 \nabla \times \mathbf{\Pi}_h. \tag{72}$$

Equation (68) now becomes

$$\mathbf{H} = k_0^2 \mathbf{\Pi}_h + \nabla \Phi \tag{73}$$

where $k_0^2 = \omega^2 \mu_0 \epsilon_0$. In place of (69a), we get

$$\nabla\nabla \cdot \mathbf{\Pi}_h - \nabla^2 \mathbf{\Pi}_h = k_0^2 \mathbf{\Pi}_h + \nabla\Phi + \mathbf{M}$$

since $\mathbf{B} = \mu_0(\mathbf{H} + \mathbf{M})$. Again, equating Φ to $\nabla \cdot \mathbf{\Pi}_h$, we get the final result

$$\nabla^2 \mathbf{\Pi}_h + k_0^2 \mathbf{\Pi}_h = -\mathbf{M}. \tag{74}$$

Usually it is more convenient to absorb $\mathbf{M}$ into the parameter μ. An exception arises when we wish to determine the field scattered from a magnetic body, in which case (74) can, at times, be fruitfully applied to set up an integral equation for the scattered field. This approach depends on our ability to solve the resultant integral equation, by either rigorous means or a perturbation technique, if it is to be of any value to us.

The electric Hertzian potential $\mathbf{\Pi}_e$ is introduced in somewhat the same manner as the vector potential $\mathbf{A}$. Thus let

$$\mathbf{H} = j\omega\epsilon \nabla \times \mathbf{\Pi}_e. \tag{75}$$

Carrying through an analysis along lines that by now are no doubt quite familiar, we find that Π_e is a solution of

$$\nabla^2 \Pi_e + k^2 \Pi_e = 0 \tag{76}$$

and that the electric field is given by

$$\mathbf{E} = k^2 \Pi_e + \nabla \nabla \cdot \Pi_e = \nabla \times \nabla \times \Pi_e. \tag{77}$$

In a source-free medium where $\mu = \mu_0$ and $\epsilon \neq \epsilon_0$, the source for Π_e is readily shown to be the dielectric polarization $\mathbf{P}$ per unit volume. If ϵ is replaced by ϵ_0 in (75) and the analysis carried out again, we find that Π_e is a solution of

$$\nabla^2 \Pi_e + k_0^2 \Pi_e = -\frac{\mathbf{P}}{\epsilon_0} \tag{78}$$

and $\mathbf{E}$ is now given by

$$\mathbf{E} = k_0^2 \Pi_e + \nabla \nabla \cdot \Pi_e$$
$$= \nabla \times \nabla \times \Pi_e - \frac{\mathbf{P}}{\epsilon_0}. \tag{79}$$

In the study of wave propagation in cylindrical waveguides we will find that the transverse electric modes may be derived from a Hertzian potential Π_h having a single component along the axis of propagation, while the transverse magnetic modes may be derived from the electric Hertzian potential Π_e with a single component directed along the axis of propagation. In other words, the modes with a longitudinal electric field component (E modes) will be derived from Π_e, while the modes with a longitudinal magnetic field component (H modes) will be derived from Π_h. In a similar manner, the analysis of the modes of propagation in an inhomogeneously filled rectangular waveguide is facilitated by the use of Hertzian vector potential functions.

When ϵ and μ vary with position, the auxiliary potential functions discussed above satisfy more complicated partial differential equations. It is difficult to give a general formulation, so that we are forced to say that, in practice, it is best to treat each case separately in order to take advantage of any simplification that may be possible.

1.8. Some Field Equivalence Principles

The term "field equivalence principle" will be interpreted in a fairly broad sense to include any principle that permits the solution of a given field boundary-value problem by replacing it by another equivalent problem for which we either know the solution or can find the solution with less difficulty. The definition also includes the artifice of replacing the actual sources by a system of virtual sources without affecting the field in a certain region exterior to that containing the sources. The classical method of images used for the solution of potential problems comes under our definition of a field equivalence principle. The following five topics will be considered in turn:

- Love's field equivalence theorem,
- Schelkunoff's field equivalence principles,
- Babinet's principle,

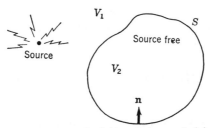

Fig. 1.8. Illustration for field equivalence principle.

- the Schwarz reflection principle,
- method of images.

Love's Field Equivalence Theorem [1.9], [1.26]–[1.30]

Let a closed surface S separate a homogeneous isotropic medium into two regions V_1 and V_2 as in Fig. 1.8. All the sources are assumed to be located in region V_1 so that both V_2 and S are source-free. The theorem states that the field in V_2 produced by the given sources in V_1 is the same as that produced by a system of virtual sources on the surface S. Furthermore, if the field produced by the original sources is $\mathbf{E}, \mathbf{H}$, then the virtual sources on S consist of an electric current sheet of density $\mathbf{n} \times \mathbf{H}$ and a magnetic current sheet of density $\mathbf{E} \times \mathbf{n}$, where the normal $\mathbf{n}$ points from V_1 into V_2. In the region external to S, that is, in V_1, these virtual sources produce a null field. The theorem is readily proved by an application of the uniqueness theorem for electromagnetic fields. We will prove the uniqueness theorem first.

Let $\mathbf{E}_1, \mathbf{H}_1$ be an electromagnetic field in a region V arising from a system of sources external to V. Let S be a closed surface enclosing V and excluding sources. Let $\mathbf{E}_2, \mathbf{H}_2$ be another possible field in the region V. Since both sets of fields satisfy Maxwell's equations, the difference field $\mathbf{E}_2 - \mathbf{E}_1$, $\mathbf{H}_2 - \mathbf{H}_1$ does also. If this difference field is used in the integral of the complex Poynting vector over S, as given by (25) in Section 1.3, we get

$$\oint_S (\mathbf{E}_2 - \mathbf{E}_1) \times (\mathbf{H}_2 - \mathbf{H}_1)^* \cdot d\mathbf{S} = j\omega \left(\iiint_V \mu |\mathbf{H}_2 - \mathbf{H}_1|^2 \, dV \right.$$

$$\left. - \iiint_V \epsilon^* |\mathbf{E}_2 - \mathbf{E}_1|^2 \, dV \right) + \iiint_V \sigma |\mathbf{E}_2 - \mathbf{E}_1|^2 \, dV.$$

If either the tangential components of $\mathbf{E}_2$ and $\mathbf{E}_1$ or the tangential components of $\mathbf{H}_2$ and $\mathbf{H}_1$ are equal over the surface S, then either $(\mathbf{E}_2 - \mathbf{E}_1) \times \mathbf{n}$ or $(\mathbf{H}_2 - \mathbf{H}_1) \times \mathbf{n}$ vanishes and the surface integral vanishes. Thus the real and imaginary parts of the right-hand side above must vanish also. This is possible only if $\mathbf{E}_2 - \mathbf{E}_1 = 0$ so that the real part vanishes, and if $\mathbf{H}_2 - \mathbf{H}_1 = 0$ so that the remaining part of the imaginary term vanishes. Thus, provided either the tangential component of $\mathbf{E}$ or the tangential component of $\mathbf{H}$ satisfies the required values on the boundary S, the field interior to the surface S is unique.

The proof for uniqueness becomes invalid when the conductivity σ is zero because in this case there may exist nonzero values of $|\mathbf{H}_2 - \mathbf{H}_1|$ and $|\mathbf{E}_2 - \mathbf{E}_1|$ that still allow the difference

$$\iiint_V [(\mu' - j\mu'')|\mathbf{H}_2 - \mathbf{H}_1|^2 - (\epsilon' + j\epsilon'')|\mathbf{E}_2 - \mathbf{E}_1|^2] \, dV$$

to equal zero if also $\mu'' = \epsilon'' = 0$. A resonant mode in a cavity bounded by the surface S has an electric field $\mathbf{E}$ that satisfies the boundary condition $\mathbf{n} \times \mathbf{E} = 0$ on S and in the volume V the average stored electric and magnetic energies are equal so that

$$\iiint\limits_{V} \mu' \mathbf{H} \cdot \mathbf{H}^* \, dV = \iiint\limits_{V} \epsilon' \mathbf{E} \cdot \mathbf{E}^* \, dV.$$

Similar resonant modes that satisfy the boundary condition $\mathbf{n} \times \mathbf{H} = 0$ on S exist and also have equal stored electric and magnetic energies. These resonant mode solutions exist only for certain discrete values of ω corresponding to the resonant frequencies. Resonant modes satisfying the boundary condition $\mathbf{n} \times \mathbf{E} = 0$ or $\mathbf{n} \times \mathbf{H} = 0$ also exist when ϵ'' and μ'' are nonzero but these solutions only occur for discrete complex values of ω. Thus, with the exception of these resonant mode solutions that occur at certain discrete values of ω, we can state that the field in V is uniquely determined by a knowledge of either the tangential electric field or the tangential magnetic field on the closed surface S.

To prove the equivalence theorem, let $\mathbf{E}_1, \mathbf{H}_1$ be the field $\mathbf{E}, \mathbf{H}$ in V_2 and a null field in V_1. The field $\mathbf{E}_1, \mathbf{H}_1$ as defined is discontinuous across the surface S by the amount $\mathbf{n} \times \mathbf{H}_1 = \mathbf{n} \times \mathbf{H}$ and $\mathbf{E}_1 \times \mathbf{n} = \mathbf{E} \times \mathbf{n}$. Since this field $\mathbf{E}_1, \mathbf{H}_1$ satisfies the correct boundary conditions at the electric and magnetic current sheets on S, it follows from the uniqueness theorem that the field $\mathbf{E}_1, \mathbf{H}_1$ as defined is the unique field generated by the virtual sources on S.

The field in V_2 is readily found from the vector potential functions, which, in turn, are determined by the virtual sources on S. In addition to the usual vector potential function $\mathbf{A}_e$ arising from an electric current distribution, we must also introduce a vector potential function $\mathbf{A}_m$ which has as its source the magnetic current distribution. The fields obtained from $\mathbf{A}_e$ are given by

$$\mathbf{B} = \nabla \times \mathbf{A}_e \tag{80a}$$

$$\mathbf{E} = -j\omega \mathbf{A}_e + \frac{\nabla \nabla \cdot \mathbf{A}_e}{j\omega\epsilon\mu}. \tag{80b}$$

An analysis similar to that used for obtaining (80) shows that the fields obtained from a vector potential function $\mathbf{A}_m$ are given by

$$\epsilon \mathbf{E} = -\nabla \times \mathbf{A}_m \tag{81a}$$

$$\mathbf{H} = -j\omega \mathbf{A}_m + \frac{\nabla \nabla \cdot \mathbf{A}_m}{j\omega\epsilon\mu}. \tag{81b}$$

The potential $\mathbf{A}_m$ is closely related to the Hertzian potential Π_h, as may be seen from a comparison of (81a) and (67). The total field is the sum of those given by (80) and (81). The vector potentials are given in terms of the virtual sources as follows:

$$\mathbf{A}_e(\mathbf{r}) = \oiint\limits_{S} \frac{\mu \mathbf{n} \times \mathbf{H}(\mathbf{r}_0)}{4\pi R} e^{-jkR} \, dS_0 \tag{82a}$$

$$\mathbf{A}_m(\mathbf{r}) = \oiint\limits_{S} \frac{\epsilon \mathbf{E}(\mathbf{r}_0) \times \mathbf{n}}{4\pi R} e^{-jkR} \, dS_0 \tag{82b}$$

where $R = |r - r_0|$ is the distance from a source point $\mathbf{r}_0$ on S to the observation point $\mathbf{r}$ and dS_0 is a differential element of surface area on S.

Equations (82) are derived in Section 1.9. If part of the surface S consists of a perfectly conducting boundary S_e, the tangential electric field vanishes on this part of S, and hence the only virtual source (in this case actually a real source) on S_e is an electric current sheet. The counterpart of a perfectly conducting surface or electric wall is called a magnetic wall and is defined as a surface over which the tangential magnetic field vanishes.

Regions V_1 and V_2 in Fig. 1.8 can be interchanged. Thus the sources can be contained in a bounded volume V_1 and the volume V_2 can extend to infinity. The field equivalence principle given above applies equally well to this situation. However, in this case the field must satisfy a radiation condition at infinity, i.e., must correspond to outward propagating waves generated by the equivalent sources on S. When the boundary conditions on S are satisfied along with the radiation condition at infinity the solution for the exterior problem is unique. A proof for the uniqueness is developed in Problem 2.35.

Schelkunoff's Field Equivalence Principles [1.9], [1.27], [1.28]

Two modifications of Love's field equivalence theorem were introduced by Schelkunoff. The surface sources $\mathbf{n} \times \mathbf{H}$ and $\mathbf{E} \times \mathbf{n}$ in Love's field equivalence theorem radiate into a free space environment and produce a null field in V_1. We can, therefore, introduce a perfectly conducting surface coinciding with S but an infinitesimal distance away from the sources; i.e., the current sources are placed adjacent to a perfectly conducting surface. This problem, namely, the one of equivalent electric currents $\mathbf{J}_e = \mathbf{n} \times \mathbf{H}$ and magnetic currents $\mathbf{J}_m = -\mathbf{n} \times \mathbf{E}$ placed on a perfect electric conducting surface, is fully equivalent to the one represented by Love's equivalence theorem as far as calculating the electromagnetic field in region V_2. We now consider the partial fields produced by the currents $\mathbf{J}_e$ and $\mathbf{J}_m$ acting separately.

It can be shown that an electric current element tangent to and adjacent to a perfect electric conducting surface does not produce any radiated field (see Problem 1.9). The current $\mathbf{J}_e$ results in an equal and opposite current density induced on the conducting surface. Hence the total effective current is zero and the resultant partial field is also zero. Thus the field in V_2 can be determined in terms of the equivalent magnetic currents $\mathbf{E} \times \mathbf{n}$ on S. These currents, however, no longer radiate in a free space environment. The field may be determined from (82b) using (81) but must be subject to the boundary condition $\mathbf{n} \times \mathbf{E} = 0$ for observation points on the interior side (V_1 side) of S. This requires the addition of a solution to the homogeneous Maxwell's equations to the particular solution obtained from (81). For convenience we will let S^- denote the V_1 side of the surface S. Since (82b) will determine a primary field $\mathbf{E}_p$ that has the property that its tangential value $\mathbf{E}_p \times \mathbf{n}$ undergoes a discontinuous change equal to $\mathbf{E} \times \mathbf{n}$ as the observation point crosses the magnetic current sheet, the field assumes a value $\mathbf{E}_p^- \times \mathbf{n}$, that is generally not zero, on the perfectly conducting surface S^-. Thus a solution to the homogeneous Maxwell's equations, i.e., a scattered field $\mathbf{E}_s$, must be added to the primary field such that $\mathbf{E}_s \times \mathbf{n} = -\mathbf{E}_p^- \times \mathbf{n}$ on S^-. The total field will then satisfy the boundary condition $\mathbf{E} \times \mathbf{n} = 0$ on S^- and the required discontinuity condition $\mathbf{E}_p^+ \times \mathbf{n} - \mathbf{E}_p^- \times \mathbf{n} = \mathbf{E} \times \mathbf{n}$ since $\mathbf{E}_s \times \mathbf{n}$ is continuous across the magnetic current sheet. The scattered field in V_2 is the same as the free space field produced by the equivalent source $\mathbf{n} \times \mathbf{H}$ on S that is part of Love's field equivalence theorem. Thus in practice this field equivalence principle of Schelkunoff does not give any computational advantage but does provide an alternative way of formulating a boundary-value problem. Note that the scattered field found from $\mathbf{n} \times \mathbf{H}$ is determined by regarding $\mathbf{n} \times \mathbf{H}$ as radiating into a free space environment so this is not a contradiction of

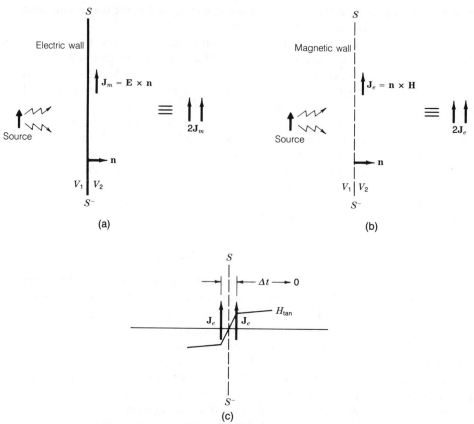

Fig. 1.9. (a, b) Illustration of Schelkunoff's field equivalence principles for a plane surface S. (c) Behavior of the tangential magnetic field across the current sheet of strength $2\mathbf{J}_e$. The separation Δt is implied to vanish.

the statement that tangential electric currents in front of an electric conducting surface do not radiate. By introducing an appropriate dyadic Green's function (see Chapter 2) the total electric field can be given directly in terms of $\mathbf{J}_m$ without adding an additional scattered field.

The other field equivalence principle of Schelkunoff is the dual of the one just described in that a perfect magnetic conducting surface (a fictitious surface on which the boundary condition $\mathbf{n} \times \mathbf{H} = 0$ holds) is placed adjacent to the surface sources. In this case the magnetic currents do not radiate and the primary field is caused only by the electric currents $\mathbf{n} \times \mathbf{H}$ on S. However, a scattered field $\mathbf{H}_s$ must be added to cancel $\mathbf{n} \times \mathbf{H}_p^-$ on S^- so that the total field will satisfy the boundary condition $\mathbf{n} \times \mathbf{H} = 0$ on S^-.

There is one practical situation where Schelkunoff's field equivalence principles have an advantage over Love's theorem. When the surface S is an infinite plane surface as shown in Fig. 1.9, the boundary condition $\mathbf{n} \times \mathbf{E} = 0$ on S^- can be satisfied by replacing the magnetic current sheet backed by a conducting electric wall by the original magnetic current sheet and its image. This simply doubles the strength of the magnetic current sheet to a value $2\,\mathbf{E} \times \mathbf{n}$ but allows the field to be found in V_2 from this new current sheet considered to be radiating in a free space environment. Similarly, when S is an infinite plane surface the current $\mathbf{n} \times \mathbf{H}$ in front of a magnetic conducting wall can be replaced by twice its value and viewed as radiating into a free space environment. The image of $\mathbf{n} \times \mathbf{H}$ and the symmetry of the problem make the resultant radiated magnetic field have a tangential part that is an odd function about the normal to S and hence $\mathbf{n} \times \mathbf{H}$ goes to zero at the center of the current sheet, which coincides with S^-.

In Chapter 2 the above field equivalence principles will emerge as a natural part of the theory of Green's functions.

Babinet's Principle [1.30]–[1.32], [1.35]

Maxwell's equations in a source-free region are

$$\nabla \times \mathbf{E} = -j\omega\mu\mathbf{H} \quad \nabla \times \mathbf{H} = j\omega\epsilon\mathbf{E} \quad \nabla\cdot\mathbf{D} = 0 \quad \nabla\cdot\mathbf{B} = 0.$$

A high degree of symmetry is exhibited in the expressions for $\mathbf{E}$ and $\mathbf{H}$. If a field $\mathbf{E}_1, \mathbf{H}_1$ satisfies the above equations, then a second field $\mathbf{E}_2, \mathbf{H}_2$, obtained from $\mathbf{E}_1, \mathbf{H}_1$ by the transformations

$$\mathbf{E}_2 = \mp \left(\frac{\mu}{\epsilon}\right)^{1/2} \mathbf{H}_1 \tag{83a}$$

$$\mathbf{H}_2 = \pm \left(\frac{\epsilon}{\mu}\right)^{1/2} \mathbf{E}_1 \tag{83b}$$

is also a solution to the source-free field equations. The above equations are a statement of the duality property of the electromagnetic field and constitute the basis of Babinet's principle. We must, of course, consider also the effect of the transformations (83) on the boundary conditions to be satisfied by the new field. Clearly, since the roles of $\mathbf{E}$ and $\mathbf{H}$ are interchanged, we must interchange all electric walls for magnetic walls and magnetic walls for electric walls. At a boundary where the conditions on $\mathbf{E}_1, \mathbf{H}_1$ were that the tangential components be continuous across a surface with ϵ and μ equal to ϵ_1, μ_1 on one side and ϵ_2, μ_2 on the other side, the new boundary conditions on $\mathbf{E}_2$ and $\mathbf{H}_2$ are still the continuity of the tangential components, and this changes the relative amplitudes of the fields on the two sides of the discontinuity surface. This is permissible since $\mathbf{E}_2$ and $\mathbf{H}_2$ in (83) can be multiplied by an arbitrary constant and still remain solutions of the field equations. When a parallel-polarized wave is incident on a plane–dielectric interface, we obtain a reflected and a transmitted wave. The solution for the case of a perpendicular-polarized incident wave is readily deduced from the solution of the parallel-polarized case by transforming the incident, reflected, and transmitted waves according to (83) and matching these new field components at the interface. That the field $\mathbf{E}_2, \mathbf{H}_2$ is indeed a solution to the field equations is readily demonstrated. For example,

$$\nabla \times \mathbf{E}_2 = \left(\frac{\mu}{\epsilon}\right)^{1/2} \nabla \times \mathbf{H}_1 = j\omega\epsilon \left(\frac{\mu}{\epsilon}\right)^{1/2} \mathbf{E}_1 = -j\omega\mu\mathbf{H}_2$$

and similarly for the other field equations.

As an example, we will apply the duality principle to show the equivalence between the problem of diffraction by an aperture in a perfectly conducting infinite-plane screen and the complementary problem of diffraction by an infinitely thin, perfectly conducting disk having the same shape as the aperture opening in the first problem. Let the screen be located in the xy plane, and denote the aperture surface by S_m and the conducting screen surface by S_e as in Fig. 1.10. The complementary problem is also illustrated in the same figure.

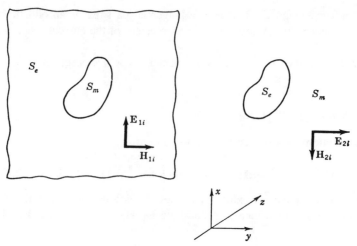

Fig. 1.10. Illustration of Babinet's principle.

In order to use the transformations (83), we must reduce the problem to one where the aperture surface S_m corresponds to a magnetic wall. Consider first the case of unsymmetrical excitation with the incident fields chosen as follows:

$$\mathbf{E}_i = \mathbf{E}_{it}(x, y, z) + \mathbf{E}_{iz}(x, y, z)$$
$$\mathbf{H}_i = \mathbf{H}_{it}(x, y, z) + \mathbf{H}_{iz}(x, y, z) \qquad z < 0 \qquad (84a)$$

$$\mathbf{E}_i = -\mathbf{E}_{it}(x, y, -z) + \mathbf{E}_{iz}(x, y, -z)$$
$$\mathbf{H}_i = \mathbf{H}_{it}(x, y, -z) - \mathbf{H}_{iz}(x, y, -z). \qquad z > 0 \qquad (84b)$$

From the unsymmetrical nature of the incident electric field and the symmetry of the screen about the $z = 0$ plane, we conclude that the transverse components of the scattered electric field in the region $z < 0$ are the negative of those in the region $z > 0$. The transverse part of the scattered electric field is therefore an odd function of z. Since the divergence of $\mathbf{E}$ is zero, we conclude that the z component of the scattered electric field is an even function of z because

$$\nabla \cdot \mathbf{E}_s = \nabla_t \cdot \mathbf{E}_{st} + \frac{\partial E_{sz}}{\partial z} = 0$$

and the relation can hold on both sides of the screen only if $\partial E_{sz}/\partial z$ is an odd function of z, and hence E_{sz} is an even function of z. The subscript t signifies the transverse, that is, x and y, components while the subscript s denotes the scattered field. From Maxwell's equation for the curl of $\mathbf{E}$ it is found that $-j\omega\mu\mathbf{H}_t = \nabla_t \times \mathbf{E}_z + \mathbf{a}_z \times \partial\mathbf{E}_t/\partial z$, and hence, if the transverse part of the scattered electric field is an odd function of z, the transverse part of the scattered magnetic field must be an even function of z. The scattered field may be represented as follows in view of the above considerations:

$$\mathbf{E}_s^o = \mathbf{E}_{st}^o(x, y, z) + \mathbf{E}_{sz}^o(x, y, z)$$
$$\mathbf{H}_s^o = \mathbf{H}_{st}^o(x, y, z) + \mathbf{H}_{sz}^o(x, y, z) \qquad z < 0 \qquad (85a)$$

$$\mathbf{E}_s^o = -\mathbf{E}_{st}^o(x, y, -z) + \mathbf{E}_{sz}^o(x, y, -z)$$
$$\mathbf{H}_s^o = \mathbf{H}_{st}^o(x, y, -z) - \mathbf{H}_{sz}^o(x, y, -z). \qquad z > 0 \qquad (85b)$$

The superscript identifies the scattered field as that due to an incident field having a transverse electric field that is an odd function of z. The negative sign in front of z for $z > 0$ is required since z is of opposite signs on the two sides of the screen.

In the aperture surface the transverse part of the electric field must be continuous and so from (84) and (85) we obtain

$$\mathbf{E}_{it}(x, y, 0) + \mathbf{E}_{st}^o(x, y, 0) = -\mathbf{E}_{it}(x, y, 0) - \mathbf{E}_{st}^o(x, y, 0)$$

which is a possible result only if both sides are equal to zero. For this case we then get the result that the electric field tangential to the aperture vanishes in the aperture, and hence the aperture surface can be replaced by an electric wall. The problem is now simply that of reflection from an infinite perfectly conducting plane, and hence the scattered field is given by the continuation of the incident fields into the opposite half spaces; thus

$$\begin{aligned}
\mathbf{E}_s^o &= -\mathbf{E}_{it}(x, y, -z) + \mathbf{E}_{iz}(x, y, -z) \\
\mathbf{H}_s^o &= \mathbf{H}_{it}(x, y, -z) - \mathbf{H}_{iz}(x, y, -z)
\end{aligned} \qquad z < 0 \qquad (86a)$$

$$\begin{aligned}
\mathbf{E}_s^o &= \mathbf{E}_{it}(x, y, z) + \mathbf{E}_{iz}(x, y, z) \\
\mathbf{H}_s^o &= \mathbf{H}_{it}(x, y, z) + \mathbf{H}_{iz}(x, y, z).
\end{aligned} \qquad z > 0 \qquad (86b)$$

Consider next the case of symmetrical excitation with incident fields given by

$$\begin{aligned}
\mathbf{E}_i &= \mathbf{E}_{it}(x, y, z) + \mathbf{E}_{iz}(x, y, z) \\
\mathbf{H}_i &= \mathbf{H}_{it}(x, y, z) + \mathbf{H}_{iz}(x, y, z)
\end{aligned} \qquad z < 0 \qquad (87a)$$

$$\begin{aligned}
\mathbf{E}_i &= \mathbf{E}_{it}(x, y, -z) - \mathbf{E}_{iz}(x, y, -z) \\
\mathbf{H}_i &= -\mathbf{H}_{it}(x, y, -z) + \mathbf{H}_{iz}(x, y, -z).
\end{aligned} \qquad z > 0 \qquad (87b)$$

A consideration of symmetry conditions this time leads us to conclude that the transverse part of the scattered electric field is now an even function of z with the transverse part of the scattered magnetic field an odd function of z. Thus the scattered fields can be represented as follows:

$$\begin{aligned}
\mathbf{E}_s^e &= \mathbf{E}_{st}^e(x, y, z) + \mathbf{E}_{sz}^e(x, y, z) \\
\mathbf{H}_s^e &= \mathbf{H}_{st}^e(x, y, z) + \mathbf{H}_{sz}^e(x, y, z)
\end{aligned} \qquad z < 0 \qquad (88a)$$

$$\begin{aligned}
\mathbf{E}_s^e &= \mathbf{E}_{st}^e(x, y, -z) - \mathbf{E}_{sz}^e(x, y, -z) \\
\mathbf{H}_{sz}^e &= -\mathbf{H}_{st}^e(x, y, -z) + \mathbf{H}_{sz}^e(x, y, -z).
\end{aligned} \qquad z > 0 \qquad (88b)$$

Since the transverse magnetic field is continuous across the aperture surface we must have

$$\mathbf{H}_{it}(x, y, 0) + \mathbf{H}_{st}^e(x, y, 0) = -\mathbf{H}_{it}(x, y, 0) - \mathbf{H}_{st}^e(x, y, 0)$$

which is possible only if both sides vanish. In this case we may replace the aperture surface by a magnetic wall since the tangential magnetic field vanishes in the aperture. The scattered field, however, cannot be determined by elementary means since we have reflection from a nonhomogeneous screen in this case. The solution to the diffraction problem with a wave incident from $z < 0$ is readily obtained from a superposition of the above two solutions. This solution is given by

$$\mathbf{E}_i = 2[\mathbf{E}_{it} + \mathbf{E}_{iz}]$$
$$\qquad\qquad\qquad\qquad z < 0 \qquad\qquad\qquad\qquad (89a)$$
$$\mathbf{H}_i = 2[\mathbf{H}_{it} + \mathbf{H}_{iz}]$$

$$\mathbf{E}_i = \mathbf{H}_i = 0 \qquad\qquad z > 0 \qquad\qquad\qquad\qquad\qquad (89b)$$

$$\mathbf{E}_s = -\mathbf{E}_{it}(x, y, -z) + \mathbf{E}_{iz}(x, y, -z) + \mathbf{E}^e_{st}(x, y, z) + \mathbf{E}^e_{sz}(x, y, z)$$
$$\qquad\qquad\qquad\qquad\qquad\qquad\qquad\qquad z < 0 \quad (89c)$$
$$\mathbf{H}_s = \mathbf{H}_{it}(x, y, -z) - \mathbf{H}_{iz}(x, y, -z) + \mathbf{H}^e_{st}(x, y, z) + \mathbf{H}^e_{sz}(x, y, z)$$

$$\mathbf{E}_s = \mathbf{E}_{it}(x, y, z) + \mathbf{E}_{iz}(x, y, z) + \mathbf{E}^e_{st}(x, y, -z) - \mathbf{E}^e_{sz}(x, y, -z)$$
$$\qquad\qquad\qquad\qquad\qquad\qquad\qquad\qquad z > 0 \quad (89d)$$
$$\mathbf{H}_s = \mathbf{H}_{it}(x, y, z) + \mathbf{H}_{iz}(x, y, z) - \mathbf{H}^e_{st}(x, y, -z) + \mathbf{H}^e_{sz}(x, y, -z).$$

Note the interesting feature that the total tangential magnetic field in the aperture is simply the incident magnetic field since $\mathbf{H}^e_{st}$ is an odd function of z.

We may handle the complementary problem in the same way. With unsymmetrical excitation we get again simply the solution for reflection from a perfectly conducting infinite plane. For symmetrical excitation the solution is readily obtained using the transformations (83), since the new boundary conditions are simply an interchange of magnetic and electric walls. The solution for symmetrical excitation is thus

$$\mathbf{E}'_i = Z\mathbf{H}_{it}(x, y, z) + Z\mathbf{H}_{iz}(x, y, z)$$
$$\qquad\qquad\qquad\qquad\qquad z < 0 \qquad\qquad\qquad (90a)$$
$$\mathbf{H}'_i = -Y\mathbf{E}_{it}(x, y, z) - Y\mathbf{E}_{iz}(x, y, z)$$

$$\mathbf{E}'_i = Z\mathbf{H}_{it}(x, y, -z) - Z\mathbf{H}_{iz}(x, y, -z)$$
$$\qquad\qquad\qquad\qquad\qquad z > 0 \qquad\qquad\qquad (90b)$$
$$\mathbf{H}'_i = Y\mathbf{E}_{it}(x, y, -z) - Y\mathbf{E}_{iz}(x, y, -z)$$

$$\mathbf{E}'_s = Z\mathbf{H}^e_{st}(x, y, z) + Z\mathbf{H}^e_{sz}(x, y, z)$$
$$\qquad\qquad\qquad\qquad\qquad z < 0 \qquad\qquad\qquad (90c)$$
$$\mathbf{H}'_s = -Y\mathbf{E}^e_{st}(x, y, z) - Y\mathbf{E}^e_{sz}(x, y, z)$$

$$\mathbf{E}'_s = Z\mathbf{H}^e_{st}(x, y, -z) - Z\mathbf{H}^e_{sz}(x, y, -z)$$
$$\qquad\qquad\qquad\qquad\qquad z > 0 \qquad\qquad\qquad (90d)$$
$$\mathbf{H}'_s = Y\mathbf{E}^e_{st}(x, y, -z) - Y\mathbf{E}^e_{sz}(x, y, -z)$$

where $Z = 1/Y = (\mu/\epsilon)^{1/2}$.

For $z < 0$ the lower signs in the transformations (83) are used, while for $z > 0$ the upper signs are used to obtain (90) from (87) and (88). This is necessary in order to obtain a solution that is continuous across the open portion of the screen. We cannot have a discontinuous solution since the magnetic wall does not have physical existence. It is only by choosing special forms of excitation for the two complementary problems that they can be reduced to the duals of each other in the half space $z < 0$. The solution must be continued analytically into the region $z > 0$ bearing in mind the nonexistence of physical magnetic walls.

Superimposing a simple solution of the problem with unsymmetrical excitation with an amplitude sufficient to cancel the incident wave for $z > 0$ yields the following solution to the problem of diffraction by a disk:

$$\mathbf{E}'_i = 2Z\mathbf{H}_i$$
$$\qquad\qquad\qquad z > 0 \qquad\qquad\qquad\qquad (91a)$$
$$\mathbf{H}'_i = -2Y\mathbf{E}_i$$

$$\mathbf{E}_i' = \mathbf{H}_i' = 0 \qquad z < 0 \tag{91b}$$

$$\mathbf{E}_s' = -Z\mathbf{H}_{it}(x, y, -z) + Z\mathbf{H}_{iz}(x, y, -z) + Z\mathbf{H}_{st}^e(x, y, z)$$

$$+ Z\mathbf{H}_{sz}^e(x, y, z)$$

$$\mathbf{H}_s' = -Y\mathbf{E}_{it}(x, y, -z) + Y\mathbf{E}_{iz}(x, y, -z) - Y\mathbf{E}_{st}^e(x, y, z) \qquad z < 0 \tag{91c}$$

$$- Y\mathbf{E}_{sz}^e(x, y, z)$$

$$\mathbf{E}_s' = Z\mathbf{H}_{it}(x, y, z) + Z\mathbf{H}_{iz}(x, y, z) + Z\mathbf{H}_{st}^e(x, y, -z)$$

$$- Z\mathbf{H}_{sz}^e(x, y, -z)$$

$$\mathbf{H}_s' = -Y\mathbf{E}_{it}(x, y, z) - Y\mathbf{E}_{iz}(x, y, z) + Y\mathbf{E}_{st}^e(x, y, -z) \qquad z > 0 \tag{91d}$$

$$- Y\mathbf{E}_{sz}^e(x, y, -z).$$

The duality principle thus allows us to readily obtain the solution to diffraction by a disk from the solution to the complementary problem of diffraction by an aperture or vice versa.

From an examination of the scattered field for the two problems the following relations are found to hold:

$$\mathbf{E}_s + Z\mathbf{H}_s' = -2\mathbf{E}_i(x, y, -z)$$
$$\mathbf{H}_s - Y\mathbf{E}_s' = 2\mathbf{H}_i(x, y, -z) \qquad z < 0 \tag{92a}$$

$$\mathbf{E}_s - Z\mathbf{H}_s' = 2\mathbf{E}_i(x, y, z)$$
$$\mathbf{H}_s + Y\mathbf{E}_s' = 2\mathbf{H}_i(x, y, z). \qquad z > 0 \tag{92b}$$

Note that $2\mathbf{E}_i$ and $2\mathbf{H}_i$ are simply the total incident field from $z < 0$ propagated into the region $z > 0$. Thus (92b) states that a superposition of the scattered fields for the two complementary problems equals the incident field in $z > 0$. In optics when the incident wave is a randomly polarized quasi-monochromatic field, the diffraction patterns for the two complementary problems have the shadow regions and lit regions interchanged so that their superposition is a uniformly lighted region corresponding to that produced by the incident field.

A generalization of Babinet's principle to a resistive screen has been given by Senior [1.35]. Equations (92) constitute the mathematical statement of Babinet's principle for complementary screens, which is a result of the dual property of the electromagnetic field. It should be noted that the duality principle is often referred to as Babinet's principle.

The Schwarz Reflection Principle [1.33]

The Schwarz reflection principle states that the analytic continuation of a function $f(x, y)$ across a surface S, say the plane $y = 0$, on which $f(x, 0)$ vanishes is given by an odd reflection of f in the plane $y = 0$. The analytic continuation of a function $f(x, y)$ for which $\partial f/\partial y$ vanishes on the plane $y = 0$ is given by an even reflection of f in the plane $y = 0$. It should be noted at this time that we made use of this principle several times in the discussion of Babinet's principle. This principle has been used to demonstrate the equivalence between the problem of a semi-infinite bifurcation of a waveguide and the problem of a semidiaphragm in a waveguide. Since we shall consider these problems in a later section, we will postpone further comment of the reflection principle until then.

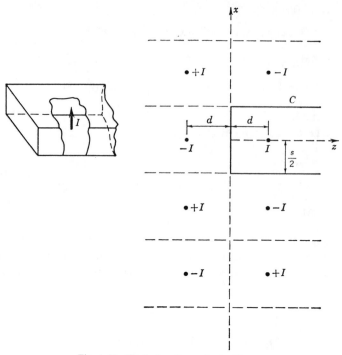

Fig. 1.11. Illustration for method of images.

Method of Images

The method of images is closely akin to the reflection principle except that here we are concerned with replacing a source within a given boundary by a system of sources, which are images of the original source in the boundary. The underlying ideas are best explained by considering a particular application. We choose the particular problem of a linear current element in a rectangular waveguide as illustrated in Fig. 1.11. The field radiated by the linear current element is such that the tangential electric field vanishes on the boundary C. The method of images shows that the field radiated by the source I in the presence of the boundary C is the same as that radiated by the source and the image sources I located at $x = \pm 2n's, z = d; x = \pm(2n+1)s, z = -d$; and image sources $-I$ at $x = \pm(2n+1)s, z = d; x = \pm 2ns, z = -d$; where n is any integer from zero to infinity, and the prime on n in the first term means exclusion of the value $n = 0$, since this corresponds to the original source. The proof follows at once by symmetry considerations. The field radiated by the real source and all the virtual sources is such that it vanishes on the boundary C.

1.9. Integration of the Inhomogeneous Helmholtz Equation

In this section we consider the integration of the inhomogeneous Helmholtz equation having the form

$$\nabla^2 \mathbf{A} + k^2 \mathbf{A} = -\mu \mathbf{J}. \tag{93}$$

As a preliminary step we prove the following theorem:

$$I = \iiint\limits_{V} \mathbf{P}(x_0, y_0, z_0) \nabla^2 \frac{1}{R} \, dV_0 = \begin{cases} 0 & (x, y, z) \text{ not in } V \\ -4\pi \mathbf{P}(x, y, z) & (x, y, z) \text{ in } V \end{cases} \tag{94}$$

where $\mathbf{P}$ is an arbitrary vector function that is continuous at the point (x, y, z) and R is the distance from the point (x, y, z) to the point (x_0, y_0, z_0); that is, $R^2 = (x - x_0)^2 + (y - y_0)^2 + (z - z_0)^2$. Direct differentiation shows that $\nabla^2(1/R) = 0$ at all points except $R = 0$, where it has a singularity. The first part of the theorem follows at once since $R \neq 0$ if (x, y, z) is not in V. For the second part of the theorem we need be concerned only about the value of the integral in the vicinity of the point $R = 0$. We surround the point (x, y, z) by a small sphere of volume V_0 and surface S_0. We choose V_0 so small that $\mathbf{P}$ may be considered constant throughout, and in particular equal to its value $\mathbf{P}(x, y, z)$ at the center. This is always possible in view of the postulated continuity of $\mathbf{P}$ at the point (x, y, z). We may now take $\mathbf{P}$ outside the integral sign and, using the result that $\nabla^2(1/R) = \nabla_0^2(1/R)$, where ∇_0 indicates differentiation with respect to the x_0, y_0, z_0 coordinates, we get

$$I = \mathbf{P}(x, y, z) \iiint\limits_{V_0} \nabla_0 \cdot \nabla_0 \frac{1}{R} \, dV_0 = \mathbf{P}(x, y, z) \oiint\limits_{S_0} \nabla_0 \frac{1}{R} \cdot d\mathbf{S}_0$$

upon using the divergence theorem. Now $\nabla_0(1/R) = \mathbf{R}/R^3$, and

$$d\mathbf{S}_0 = -R\mathbf{R} \, d\Omega$$

where $d\Omega$ is an element of solid angle, and $\mathbf{R}$ points from the x_0, y_0, z_0 coordinates to the x, y, z coordinates. Using this result shows at once that $I = -4\pi \mathbf{P}(x, y, z)$ and proves the second part of the theorem, or does it?

The proof can be criticized on the basis that the divergence theorem is being used in connection with a function $\nabla(1/R)$ that becomes singular at $\mathbf{r} = \mathbf{r}_0$ and hence might not be valid. We will, therefore, give an alternative proof that is more rigorous.

The function $1/4\pi R$ is the scalar potential produced by a point source of unit strength and located at the point $\mathbf{r}_0$. A symbolic way to represent a point source of unit strength is the use of Dirac's three-dimensional delta function; thus

$$\text{Unit point source} = \delta(x - x_0)\delta(y - y_0)\delta(z - z_0) = \delta(\mathbf{r} - \mathbf{r}_0) \tag{95}$$

with the last form being a condensed notation. The three-dimensional delta function is defined as having the operational property that $\delta(\mathbf{r} - \mathbf{r}_0) = 0$ for $\mathbf{r} \neq \mathbf{r}_0$ and for any vector function $\mathbf{P}(\mathbf{r}_0)$ that is continuous at $\mathbf{r} = \mathbf{r}_0$:

$$\iiint\limits_{V} \mathbf{P}(\mathbf{r})\delta(\mathbf{r} - \mathbf{r}_0) \, dV = \begin{cases} \mathbf{P}(\mathbf{r}_0) & \mathbf{r}_0 \text{ in } V \\ 0 & \mathbf{r}_0 \text{ not in } V. \end{cases} \tag{96}$$

We will have much more to say about the delta function and the different ways to represent it in Chapter 2, but for now we will adopt a simple model consisting of a uniform source density contained in a small spherical volume of radius a. In order that the total source strength will be unity the source density must be $3/4\pi a^3$. Thus we choose

$$\delta(\mathbf{r} - \mathbf{r}_0) = \lim_{a \to 0} \frac{3}{4\pi a^3}, \qquad |\mathbf{r} - \mathbf{r}_0| \le a. \tag{97}$$

The delta function is not a true mathematical function that exhibits continuity or differentiability so it is called a symbolic function.

In a formal way we write

$$\nabla^2 \frac{1}{4\pi R} = -\delta(\mathbf{r} - \mathbf{r}_0). \tag{98}$$

The solution $1/4\pi R$, as we will show, is the limit as $a \to 0$ of the function $g(R, a)$ that is a solution of the problem

$$\nabla^2 g(R, a) = \begin{cases} -3/4\pi a^3, & R \le a \\ 0, & R > a \end{cases} \tag{99}$$

where g and $\partial g/\partial R$ are to be made continuous at $R = a$. The latter two conditions are sufficient to allow the divergence theorem to be applied. Thus for points $R < a$, taking into account that g depends only on R because of the spherical symmetry involved, we obtain

$$\iiint_V \nabla^2 g \, dV = \oiint_S \nabla g \cdot d\mathbf{S} = 4\pi R^2 \frac{\partial g}{\partial R} = -\iiint_{\Delta V} \frac{3}{4\pi a^3} \, dV$$

$$= -\frac{R^3}{a^3}$$

where ΔV is a spherical volume with radius less than a. Hence we obtain $\partial g/\partial R = -R/4\pi a^3$ for $R < a$. If we choose ΔV such that $R > a$ on S we obtain $\partial g/\partial R = -1/4\pi R^2$. We now integrate these equations and set $g = 0$ at $R = \infty$ and apply the boundary conditions, stated earlier, at $R = a$ to get

$$g(R, a) = \begin{cases} \dfrac{a^2 - R^2}{8\pi a^3} + \dfrac{1}{4\pi a}, & R \le a \\[2ex] \dfrac{1}{4\pi R}, & R \ge a. \end{cases} \tag{100}$$

As $a \to 0$ the potential function g tends to infinity within the source region. Outside the source region it equals $1/4\pi R$ which then becomes the solution for the potential outside a point source. This is the solution to (98).

If we now return to (94) and use $\nabla^2 g$ in place of $\nabla^2(1/4\pi R)$ we can easily prove that when a is sufficiently small so that $\mathbf{P}(\mathbf{r}_0)$ can be replaced by $\mathbf{P}(\mathbf{r})$ within the small sphere of radius a that $(\mathbf{r}, \mathbf{r}_0$ are relabeled)

$$\lim_{a \to 0} \iiint_{V_0} \mathbf{P}(\mathbf{r}_0) \nabla^2 g \, dV_0 = \lim_{a \to 0} \mathbf{P}(\mathbf{r}) \iiint_{V_0} \nabla^2 g \, dV_0 = -\mathbf{P}(\mathbf{r})$$

upon using either the defining equation (99) for g or the divergence theorem. The integral equals -1 if V_0 encloses the sphere of radius a centered on $\mathbf{r}$ and hence is independent of a. This allows us to choose a as small as we like so the limit exists as $a \to 0$.

What we have thus established is that by a suitable limiting process we can regard the function $\nabla^2(-1/4\pi R)$ as a unit point source and represent it symbolically in terms of the delta function (a symbolic function). Obviously $\nabla^2(-1/4\pi R)$ is an ordinary function that happens to equal zero for all $R \neq 0$. At $R = 0$ it is undefined but when regarded as the limit of the function $\nabla^2 g$ it has the operational property defined by (96). We have now established the result (94) in a reasonably rigorous manner so we return to the problem of integrating the inhomogeneous Helmholtz equation.

Consider the following result:

$$\nabla^2 \frac{e^{-jkR}}{R} = e^{-jkR} \nabla^2 \frac{1}{R} - \frac{k^2}{R} e^{-jkR}$$

which may be readily derived by direct differentiation. Rearranging terms gives the useful result

$$\nabla^2 \frac{e^{-jkR}}{R} + k^2 \frac{e^{-jkR}}{R} = e^{-jkR} \nabla^2 \frac{1}{R}. \tag{101}$$

Now $\nabla^2(1/R) = 0$ except at $R = 0$, so (101) shows at once that (e^{-jkR}/R) is a solution to the scalar Helmholtz equation with a source function $e^{-jkR}\nabla^2(1/R)$ or $\nabla^2(1/R)$, since the factor e^{-jkR} may be dropped because it equals unity at $R = 0$, the only point where $\nabla^2(1/R)$ differs from zero. The source function $-\nabla^2(1/4\pi R)$ has all the properties of a unit source or a delta function by virtue of its properties as expressed by (94). For an arbitrary source distribution the solution follows by superposition; for example, for a source function $-\rho/\epsilon$ the solution is

$$\iiint_V \frac{\rho}{\epsilon} \frac{e^{-jkR}}{4\pi R} \, dV_0.$$

The vector inhomogeneous Helmholtz equation is readily integrated in a similar fashion to yield

$$\mathbf{A} = \frac{\mu}{4\pi} \iiint_V \frac{\mathbf{J} e^{-jkR}}{R} \, dV_0 \tag{102}$$

That this is the solution of (43) is readily demonstrated as follows:

$$\nabla^2 \mathbf{A} + k^2 \mathbf{A} = \frac{\mu}{4\pi} \left[\iiint_V \mathbf{J}(x_0, \dots) \nabla^2 \frac{e^{-jkR}}{R} \, dV_0 + k^2 \iiint_V \mathbf{J} \frac{e^{-jkR}}{R} \, dV_0 \right]$$

where the ∇^2 operator could be brought inside the integral sign since it affects only the x, y, z coordinates, while the integration is over the x_0, y_0, z_0 coordinates. Using (101), we find that the right-hand side reduces to

$$\frac{\mu}{4\pi} \iiint_V \mathbf{J}(x_0, \dots) \nabla^2 \frac{1}{R} \, dV_0$$

which from (94) equals $-\mu\mathbf{J}(x, y, z)$. If the current source is a surface or line distribution, then the solution is, of course, given by a surface or a line integral.

We now take up the problem of determining the vector and scalar potentials from an electric current sheet distributed over an open surface S and possibly having a line distribution of charge on the boundary C. If the normal current density on the boundary C is zero, there are no line charges on the boundary, and then the vector and scalar potentials are solutions of (62) and (63) given in Section 1.7. These solutions are simply

$$\mathbf{A} = \frac{\mu}{4\pi} \iint_S \frac{\mathbf{J}_s(x_0, \ldots)}{R} e^{-jkR} \, dS_0 \tag{103a}$$

$$\Phi = \frac{1}{4\pi\epsilon} \iint_S \frac{\rho_s(x_0, \ldots)}{R} e^{-jkR} \, dS_0 \tag{103b}$$

where ρ_s is given by $-\nabla\cdot\mathbf{J}_s/j\omega$. The electric and magnetic fields are given by

$$\mathbf{B} = \nabla \times \mathbf{A} \tag{104a}$$

$$\mathbf{E} = -j\omega\mathbf{A} - \nabla\Phi = -j\omega\mathbf{A} + \frac{\nabla\nabla\cdot\mathbf{A}}{j\omega\epsilon\mu} \tag{104b}$$

with $\mathbf{A}$ and Φ obeying the Lorentz condition since $\mathbf{J}_s$ and ρ_s satisfy the continuity equation. When a distribution of line charge σ_e is located on the boundary C, the new scalar potential is given in [1.29]:

$$\Phi_1 = \frac{1}{4\pi\epsilon} \iint_S \frac{\rho_s(x_0, \ldots)}{R} e^{-jkR} \, dS_0 + \frac{1}{4\pi\epsilon} \oint_C \sigma_e \frac{e^{-jkR}}{R} \, dl_0 \tag{105}$$

where $j\omega\sigma_e = J_n$, and J_n is the normal component of the current at the boundary C. The equation for the electric field must now be changed to

$$\mathbf{E} = -j\omega\mathbf{A} - \nabla\Phi_1 = -j\omega\mathbf{A} + \frac{\nabla\nabla\cdot\mathbf{A}}{j\omega\epsilon\mu}. \tag{106}$$

The validity of (105) and (106) depends on $\mathbf{A}$ and Φ_1 obeying the Lorentz condition $\nabla\cdot\mathbf{A} + j\omega\epsilon\mu\Phi_1 = 0$. If the line charge σ_e were not placed on the boundary C, then $\mathbf{A}$ and Φ_1 would not satisfy the Lorentz condition, and the integral for $\mathbf{A}$ would not be given by (103a). It is not surprising to find that inclusion of a line charge makes $\mathbf{A}$ and Φ_1 satisfy the Lorentz condition since the Lorentz condition is equivalent to the condition that current and charge should obey a continuity equation. We now prove that $\nabla\cdot\mathbf{A} = -j\omega\epsilon\mu\Phi_1$ with Φ_1 given by (105). We have

$$\nabla\cdot\mathbf{A} = \frac{\mu}{4\pi} \iint_S \nabla\cdot\mathbf{J}_s(x_0, \ldots) \frac{e^{-jkR}}{R} \, dS_0.$$

Now

$$\nabla\cdot\frac{\mathbf{J}(x_0, \ldots)}{R} e^{-jkR} = \mathbf{J}\cdot\nabla\frac{e^{-jkR}}{R} = -\mathbf{J}\cdot\nabla_0\frac{e^{-jkR}}{R}$$

$$= -\nabla_0\cdot\frac{\mathbf{J}(x_0, \ldots)}{R} e^{-jkR} + \frac{e^{-jkR}}{R}\nabla_0\cdot\mathbf{J}$$

where ∇_0 indicates differentiation with respect to the x_0, y_0, z_0 coordinates. Inserting this result into the integral, we get

$$\nabla \cdot \mathbf{A} = \frac{\mu}{4\pi} \left(\iint_S \frac{e^{-jkR}}{R} \nabla_0 \cdot \mathbf{J}\, dS_0 - \iint_S \nabla_0 \cdot \frac{\mathbf{J}}{R} e^{-jkR}\, dS_0 \right).$$

Now $\nabla_0 \cdot \mathbf{J} = -j\omega \rho_s$, and the second integral may be converted to a line integral around C by using the two-dimensional form of the divergence theorem. Thus we finally get

$$\nabla \cdot \mathbf{A} = -j\omega\mu\epsilon \left(\iint_S \frac{\rho_s e^{-jkR}}{4\pi\epsilon R}\, dS_0 + \frac{1}{j\omega\epsilon} \int_C J_n \frac{e^{-jkR}}{4\pi R}\, dl_0 \right)$$
$$= -j\omega\epsilon\mu \Phi_1$$

upon comparison with (105) and recognizing that the line-charge density σ_e is given by $J_n/j\omega$, with J_n being the normal outward-directed current on the boundary C. The significance of this result is that since $\mathbf{A}$ and Φ_1 satisfy the Lorentz condition, the field may be determined from the vector potential alone. In general, this requires a line charge σ_e on the boundary C, but this in no way enters into the evaluation of $\mathbf{A}$ as given by (103a).

The evaluation of the potentials from a magnetic current sheet proceeds in a similar manner. The normal component of magnetic current density at the boundary must be terminated in a magnetic line-charge distribution. This line-charge distribution must be taken into account in evaluating the scalar potential Φ_m.

1.10. LORENTZ RECIPROCITY THEOREM

One of the most useful theorems in electromagnetic field theory is the Lorentz reciprocity theorem. We will derive this theorem for the particular case where $\bar{\mu}$ and ϵ are symmetrical constant tensors of rank 2, or dyadics.

Let $\mathbf{E}_1, \mathbf{H}_1$ be the field generated, in the volume V bounded by a closed surface S, by a volume distribution of electrical current $\mathbf{J}_{e1}$ and magnetic current $\mathbf{J}_{m1}$. Magnetic currents do not exist physically, but we will include them for generality because of their use in various field equivalence principles. Let $\mathbf{E}_2, \mathbf{H}_2$ be the field generated by a second system of sources $\mathbf{J}_{e2}$ and $\mathbf{J}_{m2}$. The fields are solutions of Maxwell's equations:

$$\nabla \times \mathbf{E}_1 = -j\omega\bar{\mu} \cdot \mathbf{H}_1 - \mathbf{J}_{m1}$$
$$\nabla \times \mathbf{H}_1 = j\omega\epsilon \cdot \mathbf{E}_1 + \mathbf{J}_{e1}$$
$$\nabla \times \mathbf{E}_2 = -j\omega\bar{\mu} \cdot \mathbf{H}_2 - \mathbf{J}_{m2}$$
$$\nabla \times \mathbf{H}_2 = j\omega\epsilon \cdot \mathbf{E}_2 + \mathbf{J}_{e2}.$$

Expanding the quantity $\nabla \cdot (\mathbf{E}_1 \times \mathbf{H}_2 - \mathbf{E}_2 \times \mathbf{H}_1)$ gives

$$\nabla \cdot (\mathbf{E}_1 \times \mathbf{H}_2 - \mathbf{E}_2 \times \mathbf{H}_1) = \mathbf{H}_2 \cdot \nabla \times \mathbf{E}_1 - \mathbf{E}_1 \cdot \nabla \times \mathbf{H}_2 - \mathbf{H}_1 \cdot \nabla \times \mathbf{E}_2 + \mathbf{E}_2 \cdot \nabla \times \mathbf{H}_1.$$

Substituting for the curl of the field vectors from Maxwell's equations and noting that $\bar{\mu} \cdot \mathbf{H} =$

$\mathbf{H} \cdot \bar{\mu}, \epsilon \cdot \mathbf{E} = \mathbf{E} \cdot \epsilon$ because of the symmetry of the dyadics, we get

$$\nabla \cdot (\mathbf{E}_1 \times \mathbf{H}_2 - \mathbf{E}_2 \times \mathbf{H}_1) = -\mathbf{H}_2 \cdot \mathbf{J}_{m1} - \mathbf{E}_1 \cdot \mathbf{J}_{e2} + \mathbf{H}_1 \cdot \mathbf{J}_{m2} + \mathbf{E}_2 \cdot \mathbf{J}_{e1}. \qquad (107)$$

All the other terms in the right-hand side of the expansion of the divergence cancel. Integrating (107) over the volume V and converting the volume integral of the divergence to a surface integral give

$$\oiint_S (\mathbf{E}_1 \times \mathbf{H}_2 - \mathbf{E}_2 \times \mathbf{H}_1) \cdot d\mathbf{S} = \iiint_V (\mathbf{H}_1 \cdot \mathbf{J}_{m2} - \mathbf{E}_1 \cdot \mathbf{J}_{e2} - \mathbf{H}_2 \cdot \mathbf{J}_{m1} + \mathbf{E}_2 \cdot \mathbf{J}_{e1}) \, dV. \qquad (108)$$

Equation (108) is the general form of the Lorentz reciprocity theorem for media characterized by symmetrical tensor permeability and permittivity. Various special forms of the theorem may be obtained by considering a source-free volume, a volume bounded by a perfectly conducting surface, no magnetic current sources, etc.

As an example, consider a volume V that is bounded by a perfectly conducting surface S and in which two linear current elements $\mathbf{J}_{e1}, \mathbf{J}_{e2}$ exist. From (97) we get

$$\mathbf{E}_1 \cdot \mathbf{J}_{e2} = \mathbf{E}_2 \cdot \mathbf{J}_{e1}$$

since $\mathbf{n} \times \mathbf{E}_1 = \mathbf{n} \times \mathbf{E}_2 = 0$ on S, and the surface integral vanishes. The above equation states that a current element $\mathbf{J}_{e1}$ generates a field $\mathbf{E}_1$ with a component along $\mathbf{J}_{e2}$ that is equal to the component of $\mathbf{E}_2$ along $\mathbf{J}_{e1}$, where $\mathbf{E}_2$ is the field generated by $\mathbf{J}_{e2}$. This is the Rayleigh–Carson form of the reciprocity theorem.

The Lorentz reciprocity theorem is useful for deriving reciprocal properties of wave-guiding devices and antennas, for deriving orthogonality properties of waveguide modes, for constructing the field radiated from current in waveguides, and so forth. The theorem will be used on several occasions throughout the remainder of this book.

The reciprocity theorem is readily extended to the case of sources and fields in media characterized by nonsymmetrical permittivity and permeability tensors. It is only necessary to choose the field $\mathbf{E}_2, \mathbf{H}_2$ as the field in a medium characterized by the transposed tensors ϵ_t and $\bar{\mu}_t$ (Problem 1.15). Reciprocity theorems are closely related to the self-adjointness property, or lack thereof, of the differential equation operator that the field is a solution of.

REFERENCES AND BIBLIOGRAPHY

Electromagnetic Theory

[1.1]   S. Ramo, J. Whinnery, and T. Van Duzer, *Fields and Waves in Communication Electronics*.   New York, NY: Wiley, 1984.
[1.2]   J. A. Stratton, *Electromagnetic Theory*.   New York, NY: McGraw-Hill Book Company, Inc., 1941.
[1.3]   W. R. Smythe, *Static and Dynamic Electricity*, 2nd ed.   New York, NY: McGraw-Hill Book Company, Inc., 1950.
[1.4]   W. K. H. Panofsky and M. Phillips, *Classical Electricity and Magnetism*.   Reading, MA: Addison-Wesley Publishing Company, 1955.
[1.5]   L. D. Landau and E. M. Lifshitz, *Electrodynamics of Continuous Media*.   Reading: MA: Addison-Wesley Publishing Company, 1960.
[1.6]   D. S. Jones, *The Theory of Electromagnetism*.   New York, NY: The Macmillan Company, 1964.
[1.7]   J. D. Jackson, *Classical Electrodynamics*.   New York, NY: Wiley, 1952.
[1.8]   J. Van Bladel, *Electromagnetic Fields*. Reprinted by Hemisphere Publishing Corp., Washington/New York/London, 1985.

[1.9]   R. F. Harrington, *Time Harmonic Electromagnetic Fields*.   New York, NY: McGraw-Hill Book Company, Inc., 1961.

Energy in Dispersive Media and Kröning–Kramers Relations

[1.10]   D. E. Kerr, *Propagation of Short Radio Waves*.   New York, NY: McGraw-Hill Book Company, Inc., 1951, ch. 8.

[1.11]   B. S. Gouray, "Dispersion relations for tensor media and their applications to ferrites," *J. Appl. Phys.*, vol. 28, pp. 283–288, 1957.

[1.12]   A. Tonning, "Energy density in continuous electromagnetic media," *IRE Trans. Antennas Propagat.*, vol. AP-8, pp. 428–434, 1960.

[1.13]   V. L. Ginzburg, *Propagation of Electromagnetic Waves in Plasmas*.   New York, NY: Gordon and Breach Science Publishers, Inc., 1961, sects. 22 and 24. (Translated from the 1960 Russian edition.)

[1.14]   L. B. Felsen and N. Marcuvitz, *Radiation and Scattering of Waves*.   Englewood Cliffs, NJ: Prentice-Hall, 1973, ch. 1.

Boundary Conditions at an Edge

[1.15]   A. E. Heins and S. Silver, "The edge conditions and field representation theorems in the theory of electromagnetic diffraction," *Proc. Cambridge Phil. Soc.*, vol. 51, pp. 149–161, 1955.

[1.16]   C. J. Bouwkamp, "A note on singularities at sharp edges in electromagnetic theory," *Physica*, vol. 12, pp. 467–474, Oct. 1946.

[1.17]   J. Meixner, "The edge condition in the theory of electromagnetic waves at perfectly conducting plane screens," *Ann. Physik*, vol. 6, pp. 2–9, Sept. 1949.

[1.18]   J. Meixner, "The behavior of electromagnetic fields at edges," New York Univ. Inst. Math. Sci. Research Rep. EM-72, Dec. 1954.

[1.19]   D. S. Jones, "Note on diffraction by an edge," *Quart. J. Mech. Appl. Math.*, vol. 3, pp. 420–434, Dec. 1950.

[1.20]   R. A. Hurd, "The edge condition in electromagnetics," *IEEE Trans. Antennas Propagat.*, vol. AP-24, pp. 70–73, 1976.

[1.21]   J. Meixner, "The behavior of electromagnetic fields at edges," *IEEE Trans. Antennas Propagat.*, vol. AP-20, pp. 442–446, 1972.

[1.22]   M. S. Bobrovnikov and V. P. Zamaraeva, "On the singularity of the field near a dielectric wedge," *Sov. Phys. J.*, vol. 16, pp. 1230–1232, 1973.

[1.23]   G. H. Brooke and M. M. Z. Kharadly, "Field behavior near anisotropic and multidielectric edges," *IEEE Trans. Antennas Propagat.*, vol. AP-25, pp. 571–575, 1977.

[1.24]   J. Bach Anderson and V. V. Solodukhov, "Field behavior near a dielectric wedge," *IEEE Trans. Antennas Propagat.*, vol. AP-26, pp. 598–602, 1978.

[1.25]   R. De Smedt and J. G. Van Bladel, "Field singularities at the tip of a metallic cone of arbitrary cross section," *IEEE Trans. Antennas Propagat.*, vol. AP-34, pp. 865–870, 1986.

Field Equivalence Principles

[1.26]   A. E. H. Love, "The integration of the equations of propagation of electric waves," *Phil. Trans. Roy. Soc. London, Ser. A*, vol. 197, pp. 1–45, 1901.

[1.27]   S. A. Schelkunoff, "Some equivalence theorems of electromagnetics and their application to radiation problems," *Bell Syst. Tech. J.*, vol. 15, pp. 92–112, 1936.

[1.28]   S. A. Schelkunoff, "Kirchhoff's formula, its vector analogue and other field equivalence theorems," *Commun. Pure Appl. Math.*, vol. 4, pp. 43–59, 1951.

[1.29]   J. A. Stratton and L. J. Chu, "Diffraction theory of electromagnetic waves," *Phys. Rev.*, vol. 56, pp. 99–107, 1939.

[1.30]   B. B. Baker and E. T. Copson, *The Mathematical Theory of Huygens' Principle*, 2nd ed.   New York, NY: Oxford University Press, 1950.

[1.31]   H. G. Booker, "Slot aerials and their relationships to complementary wire aerials," *J. IEE (London)*, vol. 93, part IIIA, pp. 620–626, 1946.

[1.32]   E. T. Copson, "An integral equation method for solving plane diffraction problems," *Proc. Roy. Soc. (London), Ser. A*, vol. 186, pp. 100–118, 1946.

[1.33]   S. N. Karp and W. E. Williams, "Equivalence relations in diffraction theory," *Proc. Cambridge Phil. Soc.*, vol. 53, pp. 683–690, 1957.

[1.34]   N. Marcuvitz and J. Schwinger, "On the representation of the electric and magnetic fields produced by currents and discontinuities in waveguides," *J. Appl. Phys.*, vol. 22, pp. 806–819, June 1951.

[1.35] T. B. A. Senior, "Some extensions of Babinet's principle in electromagnetic theory," *IEEE Trans. Antennas Propagat.*, vol. AP-25, pp. 417–420, 1977.

See also [1.9] for a good general treatment of field equivalence principles of the Love–Schelkunoff kind.

Field Behavior in Source Regions and Potential Theory

[1.36] C. T. Tai, "Equivalent layers of surface charge, current sheet, and polarization in the eigenfunction expansion of Green's functions in electromagnetic theory," *IEEE Trans. Antennas Propagat.*, vol. AP-29, pp. 733–739, 1981.

[1.37] P. H. Pathak, "On the eigenfunction expansion of electromagnetic dyadic Green's functions," *IEEE Trans. Antennas Propagat.*, vol. AP-31, pp. 837–846, 1983.

[1.38] O. D. Kellogg, *Foundations of Potential Theory*. Berlin: Springer-Verlag, 1929.

[1.39] A. R. Panicali, "On the boundary conditions at surface current distributions described by the first derivative of Dirac's impulse," *IEEE Trans. Antennas Propagat.*, vol. AP-25, pp. 901–903, 1977.

[1.40] P. M. Morse and H. Feshbach, *Methods of Theoretical Physics*, part I. New York, NY: McGraw-Hill Book Company, Inc., 1953. (See also [1.2]–[1.7].)

Reciprocity Principles

[1.41] R. F. Harrington and A. T. Villeneuve, "Reciprocal relations for gyrotropic media," *IRE Trans. Microwave Theory Tech.*, vol. MTT-6, pp. 308–310, July 1958.

[1.42] Lord Rayleigh, *The Theory of Sound*, vol. I, pp. 98, 150–157, and vol. II, p. 145. New York, NY: The Macmillan Company, 1877 (vol. I), 1878 (vol. II). New York, NY: Dover Publications, 1945.

[1.43] J. R. Carson, "A generalization of the reciprocal theorem," *Bell Syst. Tech. J.*, vol. 3, pp. 393–399, July 1924.

[1.44] J. R. Carson, "Reciprocal theorems in radio communication," *Proc. IRE*, vol. 17, pp. 952–956, June 1929.

[1.45] W. J. Welch, "Reciprocity theorems for electromagnetic fields whose time dependence is arbitrary," *IRE Trans. Antennas Propagat.*, vol. AP-8, pp. 68–73, 1960.

[1.46] J. A. Kong, "Theorems of bianisotropic media," *Proc. IEEE*, vol. 60, pp. 1036–1046, 1972. This paper discusses reciprocity for bianisotropic media as well as other properties.

Problems

1.1. Consider an ionized gas with N free electrons per unit volume. Under the action of a time harmonic electric field the equation of motion for the electron velocity $\mathbf{v}$ is $j\omega m\mathbf{v} = -e\mathbf{E}$ where m is the mass and $-e$ is the charge of the electron. The induced current is $-Ne\mathbf{v} = \mathbf{J}$. Use this in Maxwell's $\nabla \times \mathbf{H}$ equation and show that the effective dielectric constant for the plasma medium is $1 - \omega_p^2/\omega^2$ where $\omega_p^2 = Ne^2/\epsilon_0 m$. Show that the average stored energy per unit volume is given by the sum of the free space energy and the kinetic energy, i.e., $\frac{1}{4}\epsilon_0\mathbf{E}\cdot\mathbf{E}^* + \frac{1}{4}Nm\mathbf{v}\cdot\mathbf{v}^*$.

1.2. Carry out the details in the derivation of (52).

1.3. The differential equation for the electric field lines in cylindrical coordinates is $E_r/dr = E_\theta/r\,d\theta$. For the dielectric wedge shown in Fig. 1.7(d) show that the field lines are described by $r = (D_1/\cos \nu\theta)^{1/\nu}$ in the free space region and by $r = [D_2/\cos \nu(\pi - \theta)]^{1/\nu}$ in the dielectric region, where D_1 and D_2 are constants. Show that the field lines in the dielectric approach the $\theta = 57°$ radial line when $\kappa = 10$ and $2\phi = 3\pi/2$.

1.4. Consider a current $\mathbf{J} = I_0 \cos \theta \mathbf{a}_r e^{-jwz}$ located on the surface of a cylinder with radius a. Show that the discontinuity in the tangential components of the electric field is given by $E_z^+ - E_z^- = (wZ_0/k_0)I_0\cos \Theta e^{-jwz}, E_\theta^+ - E_\theta^- = -j(Z_0/k_0 a)I_0\sin \Theta^{-jwz}$. Show that an equivalent magnetic current sheet is

$$\mathbf{J}_m = jk_0^{-1}Z_0(-jw\mathbf{a}_\theta \cos \theta + a^{-1}\mathbf{a}_z \sin \theta)I_0 e^{-jwz}.$$

1.5. The E_{nm} modes in a rectangular waveguide of cross-sectional dimensions a by b can be derived from the vector potential function $\mathbf{A} = \mathbf{a}_z \sin(n\pi x/a) \sin(m\pi y/b)e^{\pm\Gamma_{nm}z}$, where $\Gamma_{nm}^2 = n^2\pi^2/a^2 + m^2\pi^2/b^2 - k^2$. Show that $\mathbf{A}$ is a solution of the vector Helmholtz equation. Use the Lorentz relation to find the scalar potential Φ associated with $\mathbf{A}$. Find a gauge transformation that will give a new set of potentials $\mathbf{A}'$ and Φ' such that $\nabla\cdot\mathbf{A}' = 0$. Find also the solenoidal and lamellar parts of the electric field.

1.6. A magnetic current sheet of density $\mathbf{J}_m = \mathbf{a}_x \sin x \sin y$ is located on the surface defined by $z = 0$ and the four lines $x = 0, x = 1$, and $y = 0, y = 1$. Find the discontinuity in the tangential electric field and normal magnetic flux across the magnetic current sheet. Also determine the density of magnetic line charge that must be placed on the boundary in order that the magnetic current and charge shall obey the continuity equation at the boundary. Set up the integral expressions for the vector and scalar potentials $\mathbf{A}_m$ and Φ_m.

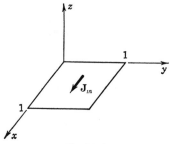

Fig. P1.6.

1.7. Consider a uniform line current $\mathbf{a}_x J_0$ along the x axis and extending from $-1 \leq x \leq 1$. Set up the expressions for the vector and scalar potentials in integral form. The scalar potential arises from the two point charges $+J_0/j\omega$ at $x = 1$ and $-J_0/j\omega$ at $x = -1$. Show that $\nabla \cdot \mathbf{A} = -j\omega\mu\epsilon\Phi$. The inclusion of the two point charges is the one-dimensional form of a line charge along a surface boundary. The electric field is thus seen to be derivable from the vector potential $\mathbf{A}$ alone.

1.8. Consider a system of sources radiating a field $\mathbf{E}_0$, $\mathbf{H}_0$ into a region in which a volume V consisting of inhomogeneous material exists. Let all sources lie external to the surface S which bounds V. In the absence of the inhomogeneous material in V, let the sources radiate a field $\mathbf{E}_i$, $\mathbf{H}_i$. In the presence of the material in V, the total field radiated or transmitted into V is $\mathbf{E}_t$, $\mathbf{H}_t$, while external to V the field is the sum of the incident field and a reflected field, that is, $\mathbf{E}_i + \mathbf{E}_r$, $\mathbf{H}_i + \mathbf{H}_r$. Show that the scattered field $\mathbf{E}_t$, $\mathbf{H}_t$ and $\mathbf{E}_r$, $\mathbf{H}_r$ can be derived from a system of virtual sources on the boundary S, in particular, from an electric current sheet $\mathbf{n} \times \mathbf{H}_i$ and a magnetic current sheet $\mathbf{E}_i \times \mathbf{n}$, where $\mathbf{n}$ is the inward normal to S. This is Schelkunoff's induction theorem [S. A. Schelkunoff, "Some equivalence theorems of electromagnetics and their application to radiation problems," *Bell Syst. Tech. J*, vol. 15, pp. 92–112, 1936]. When the inhomogeneous material in V is removed, the reflected field vanishes, and the induction theorem becomes Love's equivalence theorem.

1.9. Use the Lorentz reciprocity theorem to show that a tangential electric current or a normal magnetic current adjacent to a perfectly conducting surface does not radiate an electromagnetic field. Will a normal electric current or a tangential magnetic current adjacent to a perfectly conducting surface radiate a field?

HINT: Consider the field components, at the conductor surface, radiated by arbitrary currents remote from the surface.

1.10. Derive the inhomogeneous Helmholtz equations satisfied by the vector potential functions $\mathbf{A}_e$ and $\mathbf{A}_m$ when the electric and magnetic polarizations $\mathbf{P}$ and $\mathbf{M}$ of a material body are considered as the source functions. The results will be similar to that expressed by (74) and (78).

1.11. Consider an infinite-plane perfectly conducting screen containing an aperture. For the case of odd excitation show that the normal electric field and tangential magnetic field in the aperture are equal to twice those associated with one of the incident waves. For the case of even excitation show that the normal electric field and tangential magnetic field in the aperture are zero. Next superimpose the two cases and show that, for a field incident on only one side of the screen, the normal electric field and tangential magnetic field in the aperture are equal to the corresponding field components of the incident wave.

1.12. Consider a current $\mathbf{J}(\mathbf{r}_0)$ in a volume V_0. From the integral for the vector potential and using (80b) show that very far from the source region to terms of order $1/r$ the electric field $\mathbf{E}(\mathbf{r})$ is given by $-j\omega(\mathbf{A} - A_r\mathbf{a}_r)$ with

$$\mathbf{A} = (\mu_0/4\pi r)e^{-jk_0r} \iiint \mathbf{J}(\mathbf{r}_0)e^{jk_0\mathbf{a}_r \cdot \mathbf{r}_0}\, dV.$$

Use $R \approx r$ in the denominator and $R = |\mathbf{r} - \mathbf{r}_0| \approx r - \mathbf{a}_r \cdot \mathbf{r}_0$, as obtained from a binomial expansion, in the exponential term.

1.13. Consider a current $\mathbf{J}_1$ in V_1 and a current $\mathbf{J}_2$ in V_2. Show that at infinity $(\mathbf{E}_1 \times \mathbf{H}_2 - \mathbf{E}_2 \times \mathbf{H}_1) \cdot \mathbf{a}_r = 0$ to terms of order $1/r^2$.

HINT: See Problem 1.12 and express the magnetic field in terms of the electric field.

1.14. Consider sources $\mathbf{J}_{e1}$ and $\mathbf{J}_{m1}$ that produce a field $\mathbf{E}_1$, $\mathbf{H}_1$ and sources $\mathbf{J}_{e2}$ and $\mathbf{J}_{m2}$ that produce a field $\mathbf{E}_2$, $\mathbf{H}_2$. Show that

$$\iiint (\mathbf{J}_{e1} \cdot \mathbf{E}_2 - \mathbf{J}_{m1} \cdot \mathbf{H}_2)\, dV = \iiint (\mathbf{J}_{e2} \cdot \mathbf{E}_1 - \mathbf{J}_{m2} \cdot \mathbf{H}_1)\, dV.$$

1.15. Consider a medium characterized by a tensor permeability $\bar{\mu}$ and a tensor permittivity ϵ. Let sources $\mathbf{J}_{e1}$, $\mathbf{J}_{m1}$ radiate a field $\mathbf{E}_1$, $\mathbf{H}_1$ in this medium. Let sources $\mathbf{J}_{e2}$, $\mathbf{J}_{m2}$ radiate a field $\mathbf{E}_2$, $\mathbf{H}_2$ into a medium characterized by the transposed parameters $\bar{\mu}_t$ and $\bar{\epsilon}_t$. Show that the reciprocity statement given in Problem 1.14 will be true.

1.16. Let current elements $\mathbf{J}_1$ and $\mathbf{J}_2$ radiate fields $\mathbf{E}_1$, $\mathbf{H}_1$ and $\mathbf{E}_2$, $\mathbf{H}_2$ in unbounded space. The fields are subjected to the radiation condition (see Section 2.12)

$$\lim_{r \to \infty} r(\nabla \times \mathbf{E} + jk_0\mathbf{a}_r \times \mathbf{E}) = 0 \qquad \text{or} \qquad \lim_{r \to \infty} (-j\omega\mu_0\mathbf{H} + jk_0\mathbf{a}_r \times \mathbf{E}) = 0.$$

Show that

$$\lim_{r \to \infty} \oiint_S (\mathbf{E}_1 \times \mathbf{H}_2 - \mathbf{E}_2 \times \mathbf{H}_1) \cdot \mathbf{a}_r \, dS = 0$$

by using the radiation condition to eliminate the magnetic fields. Thus show that in unbounded free space $\mathbf{E}_1 \cdot \mathbf{J}_2 = \mathbf{E}_2 \cdot \mathbf{J}_1$.

1.17. The results of Problem 1.16 may be used to establish the following theorem that is useful in proving the uniqueness of the external scattering problem. Let $\mathbf{E}$, $\mathbf{H}$ be a source-free solution of Maxwell's equations in all of space and which also satisfies the radiation condition. Let $\mathbf{E}'$, $\mathbf{H}'$ be a solution to Maxwell's equations due to an arbitrary current element $\mathbf{J}(\mathbf{r}')$ and which also satisfies the radiation condition. Show that this implies $\mathbf{J}(\mathbf{r}') \cdot \mathbf{E}(\mathbf{r}') = 0$ for all $\mathbf{r}'$. Hence a solution to Maxwell's equations in all space that is source free and satisfies the radiation condition must be a null field. Since $\mathbf{J}(\mathbf{r}')$ is arbitrary and nonzero, $\mathbf{E}$ must be zero.

2
Green's Functions

In 1828 George Green wrote an essay entitled "On the application of mathematical analysis to the theories of electricity and magnetism" in which he developed a method for obtaining solutions to Poisson's equation in potential theory. The method, which makes use of a potential function that is the potential from a point or line source of unit strength, has been expanded to deal with a large number of different partial differential equations. The technique always makes use of a function that is the field from a source of unit strength; this function is called the Green's function for the problem. In the early applications of Green's method no mathematical expression was introduced to represent a unit source. The Green's function was defined instead through its singular behavior in the vicinity of the unit-strength source.

In 1927 the delta function was introduced by Dirac as the limit of a sequence of functions, i.e.,

$$\delta(x - x') = \lim_{n \to \infty} U_n(x - x')$$

where $U_n(x - x')$ can be, for example,

$$\frac{n}{\sqrt{\pi}} e^{-n^2(x-x')^2} \qquad \text{or} \qquad \frac{\sin n(x - x')}{\pi(x - x')}.$$

These functions have the property that

$$\lim_{n \to \infty} \int f(x') U_n(x - x') \, dx' = f(x).$$

The area under the curve of $U_n(x - x')$ is unity for all n. It is now common practice to use the delta function to represent a unit source in one dimension and to use a product of delta functions to represent a unit source in a multidimensional space. The functions $U_n(x - x')$ are legitimate functions, but once the limit has been taken the result is no longer a function in the classical sense and $\delta(x - x')$ is called, for this reason, a symbolic function. It can be defined for most purposes by the operational property given by (96) in Chapter 1. The delta function did not come into widespread use before around 1940. Even Baker and Copson in their 1950 book [1.30] did not make use of the delta function. In 1950 Laurent Schwartz developed the distribution theory which provided a rigorous nonoperational method of handling entities such as the delta function. A short introduction to distribution theory is given in Section 2.21.

The first two sections of this chapter present the general Green's function method for solving Poisson's equation. This is done to provide the reader with an introduction to the subject. A number of the following sections take up the problem of how to construct Green's functions for differential equations of the Sturm–Liouville type. The construction of Green's functions for multidimensional problems and techniques for obtaining alternative representations are then given. The chapter includes a treatment of dyadic Green's functions which are needed for the vector problem. Integral equations for scattering are also discussed. In Section 2.20 we present a brief discussion of procedures that must be followed to construct Green's functions for non-self-adjoint differential equations.

The field point is described in terms of the coordinates x, y, z or the position vector **r**. The source point is described in terms of the coordinates x_0, y_0, z_0 or x', y', z' with corresponding position vector $\mathbf{r}_0$ or $\mathbf{r}'$. In general, we use the x_0, y_0, z_0 notation when a number of differentiations with respect to x_0, y_0, z_0 are involved. The distance $|\mathbf{r} - \mathbf{r}_0|$ is represented by the symbol R.

2.1. GREEN'S FUNCTIONS FOR POISSON'S EQUATION

The scalar potential Φ in electrostatics is a solution of Poisson's equation

$$\nabla^2 \Phi = -\frac{\rho}{\epsilon} \tag{1}$$

where ρ is the density of the charge distribution which gives rise to the scalar potential Φ, and ϵ is assumed constant. In the absence of any boundaries the solution is given by the integral

$$\Phi(x, y, z) = \iiint\limits_V \frac{\rho(x_0, y_0, z_0)}{4\pi\epsilon R} \, dV_0 \tag{2}$$

where $R = [(x - x_0)^2 + (y - y_0)^2 + (z - z_0)^2]^{1/2}$ and is the distance from the source point to the point in space where the potential is evaluated. A source of unit strength located at the point (x_0, y_0, z_0) can be conveniently represented by a three-dimensional delta function $\delta(x - x_0)\delta(y - y_0)\delta(z - z_0)$, where the delta function is zero at all points except $x_0 = x$, $y_0 = y$, $z_0 = z$.

If $\mathbf{P}(x_0, y_0, z_0)$ is an arbitrary vector function which is continuous at (x, y, z), then [see (96) in Chapter 1]

$$\iiint\limits_V \mathbf{P}(x_0, \ldots)\delta(x - x_0)\delta(y - y_0)\delta(z - z_0) \, dV_0$$

$$= \begin{cases} 0, & (x, y, z) \text{ not in } V \\ \mathbf{P}(x, y, z), & (x, y, z) \text{ in } V. \end{cases} \tag{3}$$

A similar result holds for a scalar function. Our discussion of the solution of the inhomogeneous Helmholtz equation in Section 1.9 showed that $-\nabla^2(4\pi R)^{-1}$ has the properties given by (3) above. The three-dimensional solution of the following inhomogeneous differential equation defines the free-space three-dimensional Green's function for Poisson's equation:

$$\nabla^2 G(x, y, z | x_0, y_0, z_0) = -\delta(x - x_0)\delta(y - y_0)\delta(z - z_0). \tag{4}$$

The Green's function G is a function of both the point of observation or field point and the source point. From the known properties of the function $(4\pi R)^{-1}$ we see at once that this is the solution of (4), and hence

$$G(x, y, z | x_0, y_0, z_0) = 1/4\pi R. \tag{5}$$

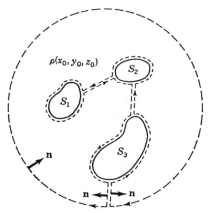

Fig. 2.1. Illustration for Green's theorem.

The Green's function is symmetrical in the variables x_0, y_0, z_0 and x, y, z. This property will be found to be true in most cases that we will encounter, and is equivalent to the field obeying a reciprocity principle. Once the solution for a unit source is known, the solution for any given arbitrary source distribution follows at once by superposition, i.e.,

$$\Phi(x, \ldots) = \frac{1}{\epsilon} \iiint\limits_V G(x, y, z | x_0, y_0, z_0) \rho(x_0, y_0, z_0) \, dV_0 \qquad (6)$$

which, with the aid of (5), is certainly recognizable as the standard solution of Poisson's equation.

In addition to the notation used above, it will be convenient, in what follows, to use the following abbreviated notation as well:

$$\Phi(\mathbf{r}) = \frac{1}{\epsilon} \iiint\limits_V G(\mathbf{r}, \mathbf{r}_0) \rho(\mathbf{r}_0) \, dV_0.$$

When G is symmetrical, $G(\mathbf{r}, \mathbf{r}_0) = G(\mathbf{r}_0, \mathbf{r})$.

Let Φ and ψ be two scalar functions of position which are continuous with continuous partial derivatives. Let the region of space V contain conducting bodies with surfaces $S_1, S_2, \ldots, S_n$, as illustrated in Fig. 2.1. By introducing suitable cuts, the region of interest can be made simply connected. Let S be a closed surface consisting of the surfaces $S_1, \ldots, S_n$, the surfaces along the cuts, and the surface of an infinite sphere which surrounds all the conducting bodies.

Applying Green's second identity[1] to the functions Φ and ψ gives

$$\iiint\limits_V (\Phi \nabla_0^2 \psi - \psi \nabla_0^2 \Phi) \, dV_0 = \oiint\limits_S \left(\psi \frac{\partial \Phi}{\partial n} - \Phi \frac{\partial \psi}{\partial n} \right) dS_0$$

where $\mathbf{n}$ is the normal to S and is directed into the volume V. We now let Φ be the required solution of Poisson's equation, and let ψ be the Green's function as given by (5). Replacing

[1] Section A.1c.

$\nabla_0^2 \Phi$ by $-\rho/\epsilon$ and $\nabla_0^2 G$ by the negative delta function, we find that

$$-\iiint_V \Phi(\mathbf{r}_0)\delta(\mathbf{r} - \mathbf{r}_0)\,dV_0 + \frac{1}{\epsilon}\iiint_V \rho(\mathbf{r}_0)G(\mathbf{r}, \mathbf{r}_0)\,dV_0$$

$$= \oiint_S \frac{\partial \Phi(\mathbf{r}_0)}{\partial n}G(\mathbf{r}, \mathbf{r}_0)\,dS_0 - \oiint_S \Phi(\mathbf{r}_0)\frac{\partial G(\mathbf{r}, \mathbf{r}_0)}{\partial n}\,dS_0 \qquad (7a)$$

where

$$\frac{\partial G}{\partial n} = (\nabla_0 G)\cdot\mathbf{n} \qquad \frac{\partial \Phi}{\partial n} = (\nabla_0 \Phi)\cdot\mathbf{n}$$

and ∇_0 designates differentiation with respect to x_0, y_0, z_0. It is important to note that, if x_0, y_0, z_0 are used as the variables of integration in Green's identity, then all derivatives of functions in the integrands must also be with respect to x_0, y_0, z_0. Using the property of the delta function as given by (3) shows that the required solution for Φ is

$$\Phi(\mathbf{r}) = \frac{1}{\epsilon}\iiint_V \rho(\mathbf{r}_0)G(\mathbf{r}, \mathbf{r}_0)\,dV_0 + \oiint_S \left(\Phi\frac{\partial G}{\partial n} - G\frac{\partial \Phi}{\partial n}\right)dS_0. \qquad (7b)$$

The surface integrals along the cuts do not contribute since they are traversed twice in the opposite sense and with the normal $\mathbf{n}$ oppositely directed. The three terms contributing to Φ are the volume distribution of charge ρ, a surface distribution of charge $-\epsilon(\partial\Phi/\partial n)$, and a surface distribution of electric dipoles of dipole moment $\epsilon\Phi$ per unit area [2.1].

In a two-dimensional space the Green's function is defined by the differential equation

$$\frac{\partial^2 G}{\partial x^2} + \frac{\partial^2 G}{\partial y^2} = -\delta(x - x_0)\delta(y - y_0). \qquad (8a)$$

The source function in this case is an infinite line source of unit density. The solution for G is readily obtained by rewriting (8a) in cylindrical coordinates r, θ as follows:

$$\frac{1}{r}\frac{\partial}{\partial r}r\frac{\partial G}{\partial r} + \frac{1}{r^2}\frac{\partial^2 G}{\partial \theta^2} = -\delta(\mathbf{r}) \qquad (8b)$$

where the origin for the r, θ coordinate system has been chosen coincident with the line charge, i.e., at x_0, y_0. From symmetry considerations we see that G is not a function of θ, and hence depends only on the radial distance r from the line source. If we surround the line source by a cylindrical surface S as illustrated in Fig. 2.2 and integrate (8b) throughout the volume enclosed by a unit length of S, we get

$$\iiint_V \nabla\cdot\nabla G\,dV = \oiint_S \nabla G\cdot d\mathbf{S} = -\iiint_V \delta(\mathbf{r})\,dV = -1$$

by virtue of the properties of the delta function. Since G is a function of r only, we get $(\partial G/\partial r)2\pi r = -1$, and a further integration gives the desired result, which is

$$G = -\frac{\ln r}{2\pi}. \qquad (9a)$$

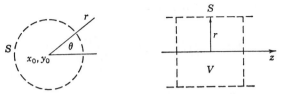

Fig. 2.2. Surface S surrounding a line source.

Transforming back to our x, y coordinate system, we have

$$G(x, y | x_0, y_0) = -\frac{\ln [(x - x_0)^2 + (y - y_0)^2]^{1/2}}{2\pi}. \tag{9b}$$

2.2. Modified Green's Functions

The solution for Φ given by (7b) is not too useful in practice since usually we know Φ on part of the boundary and $\partial \Phi / \partial n$ on the remainder, but rarely do we know both on the complete surface S. However, by a suitable modification of our Green's function it is possible, in principle at least, to obtain a solution for Φ from a knowledge of either Φ or $\partial \Phi / \partial n$ on the boundary. If the required boundary conditions specify the value of Φ on the complete boundary, the potential Φ is said to satisfy Dirichlet boundary conditions. If the normal derivative $\partial \Phi / \partial n$ is specified on the whole boundary, the corresponding boundary conditions are referred to as Neumann boundary conditions. Finally, if Φ is specified on part of the boundary and $\partial \Phi / \partial n$ on the remainder, we have mixed boundary conditions.

If Φ is known on the complete boundary, we see from an examination of the solution (7b) that the offending term involving $\partial \Phi / \partial n$ can be eliminated by making the Green's function vanish on the complete boundary, i.e., on the surface S. Thus we define the Green's function of the first kind as a solution to the equation

$$\nabla^2 G_1(\mathbf{r}, \mathbf{r}_0) = -\delta(\mathbf{r} - \mathbf{r}_0) \tag{10a}$$

and obeying the boundary condition

$$G_1 = 0 \qquad \text{on } S. \tag{10b}$$

The solution for the potential Φ is thus given by

$$\Phi(\mathbf{r}) = \iiint_V \frac{\rho}{\epsilon} G_1 \, dV_0 + \oiint_S \Phi \frac{\partial G_1}{\partial n} \, dS_0. \tag{11}$$

When $\partial \Phi / \partial n$ is given everywhere on S, we modify our Green's function to make $\partial G / \partial n$ vanish on the boundary. The Green's function of the second kind is defined as a solution of

$$\nabla^2 G_2(\mathbf{r}, \mathbf{r}_0) = -\delta(\mathbf{r} - \mathbf{r}_0) \tag{12a}$$

subject to the boundary condition

$$\frac{\partial G_2}{\partial n} = 0 \qquad \text{on } S. \tag{12b}$$

The solution for Φ in this case is given by

$$\Phi(\mathbf{r}) = \iiint_V \frac{\rho}{\epsilon} G_2 \, dV_0 - \oiint_S \frac{\partial \Phi}{\partial n} G_2 \, dS_0. \tag{13}$$

When we have mixed boundary conditions, it is desirable, if possible, to find a Green's function which vanishes on that part of S for which we know Φ, and which has a normal derivative equal to zero over that part of S for which we know $\partial \Phi / \partial n$. From (11) and (13) we see that the Green's functions are useful for solving boundary-value problems, when the volume distribution of charge ρ is zero, as well. Needless to say, quite often in practice it is about as difficult to find a solution for the Green's functions as it is to solve the original boundary-value problem. The use of Green's functions converts a differential equation together with the boundary conditions to an integral equation, which in many cases is more readily attacked by approximate techniques.

We will now show that the Green's function is symmetric in the variables x, y, z and x_0, y_0, z_0, that is,

$$G(\mathbf{r}, \mathbf{r}_0) = G(\mathbf{r}_0, \mathbf{r}). \tag{14}$$

This property might have been anticipated since the delta function is symmetrical, i.e., $\delta(\mathbf{r} - \mathbf{r}_0) = \delta(\mathbf{r}_0 - \mathbf{r})$. This reciprocity principle states that the potential at (x, y, z) from a unit charge at (x_0, y_0, z_0) is the same as the potential at (x_0, y_0, z_0) due to a unit charge at (x, y, z). Let $G_1(\mathbf{r}, \mathbf{r}_{01})$ be the Green's function of the first kind for a point charge at (x_{01}, y_{01}, z_{01}), and let $G_1(\mathbf{r}, \mathbf{r}_{02})$ be the Green's function for a point charge at (x_{02}, y_{02}, z_{02}). If these two functions are used in Green's second identity, we get

$$\iiint_V [G_1(\mathbf{r}, \mathbf{r}_{01}) \nabla^2 G_1(\mathbf{r}, \mathbf{r}_{02}) - G_1(\mathbf{r}, \mathbf{r}_{02}) \nabla^2 G_1(\mathbf{r}, \mathbf{r}_{01})] \, dV$$

$$= -\oiint_S \left[G_1(\mathbf{r}, \mathbf{r}_{01}) \frac{\partial G_1(\mathbf{r}, \mathbf{r}_{02})}{\partial n} - G_1(\mathbf{r}, \mathbf{r}_{02}) \frac{\partial G_1(\mathbf{r}, \mathbf{r}_{01})}{\partial n} \right] dS. \tag{15}$$

Since G_1 vanishes on S and

$$\nabla^2 G_1(\mathbf{r}, \mathbf{r}_{01}) = -\delta(\mathbf{r} - \mathbf{r}_{01})$$

$$\nabla^2 G_1(\mathbf{r}, \mathbf{r}_{02}) = -\delta(\mathbf{r} - \mathbf{r}_{02})$$

we obtain the following result from the volume integral:

$$G_1(\mathbf{r}_{02}, \mathbf{r}_{01}) = G_1(\mathbf{r}_{01}, \mathbf{r}_{02}).$$

Since (x_{01}, y_{01}, z_{01}) and (x_{02}, y_{02}, z_{02}) are arbitrary points, the stated reciprocity principle is proved. A similar proof holds for the Green's function of the second kind since $\partial G_2 / \partial n$ equals zero in this case and the surface integral again vanishes.

2.3. STURM–LIOUVILLE EQUATION

Many of the boundary-value problems encountered in connection with the study of guided electromagnetic waves lead to the Sturm–Liouville equation. In this section we will review the basic properties of the Sturm–Liouville system but without detailed derivation.

The Sturm–Liouville equation is of the form

$$\frac{d}{dx}p(x)\frac{d\psi(x)}{dx} + [q(x) + \lambda\sigma(x)]\psi(x) = 0 \tag{16}$$

where λ is a separation constant. In practice both p and σ are usually positive and continuous functions of x. We will consider the properties of this equation when the interval in x is finite, say $0 \leq x \leq a$.

The solutions to (16) are fixed only if certain boundary conditions are first specified. The three common boundary conditions are $\psi = 0$, $d\psi/dx = 0$ and $\psi + K(d\psi/dx) = 0$ at $x = 0, a$, where K is a suitable constant. Once a set of boundary conditions has been specified (one condition at $x = 0$ and one at $x = a$) then one finds an infinite set of solutions ψ_n to (16) corresponding to particular values of λ, say λ_n. The functions ψ_n are called eigenfunctions and the λ_n are called eigenvalues. In all cases, the ψ_n form an orthogonal set of functions over the interval $0 \leq x \leq a$. The orthogonality of the functions may be proved as follows: Multiplying the equation for ψ_n by ψ_m and vice versa, subtracting and integrating gives

$$\int_0^a \left(\psi_n \frac{d}{dx}p\frac{d\psi_m}{dx} - \psi_m \frac{d}{dx}p\frac{d\psi_n}{dx}\right) dx = \int_0^a (\lambda_n - \lambda_m)\sigma\psi_n\psi_m \, dx$$

since the term involving q vanishes. Integrating the left-hand side by parts gives

$$\left(\psi_n \frac{d\psi_m}{dx} - \psi_m \frac{d\psi_n}{dx}\right)p \bigg|_0^a - \int_0^a p \left(\frac{d\psi_m}{dx}\frac{d\psi_n}{dx} - \frac{d\psi_n}{dx}\frac{d\psi_m}{dx}\right) dx$$

$$= p \left(\psi_n \frac{d\psi_m}{dx} - \psi_m \frac{d\psi_n}{dx}\right)\bigg|_0^a = (\lambda_n - \lambda_m)\int_0^a \sigma\psi_n\psi_m \, dx.$$

When ψ_n and ψ_m satisfy the same boundary conditions, of the type given earlier, the integrated terms vanish. For example, if $\psi_i + K(d\psi_i/dx) = 0$ at $x = 0, a$, with $i = n, m$, then we can write

$$\psi_n \frac{d\psi_m}{dx} - \psi_m \frac{d\psi_n}{dx} = \left(\psi_n + K\frac{d\psi_n}{dx}\right)\frac{d\psi_m}{dx} - \left(\psi_m + K\frac{d\psi_m}{dx}\right)\frac{d\psi_n}{dx}$$

which clearly vanishes at $x = 0, a$. When $\lambda_n \neq \lambda_m$ we conclude that

$$\int_0^a \sigma\psi_n\psi_m = 0 \tag{17}$$

i.e., the ψ_n form an orthogonal set with respect to the weighting function $\sigma(x)$. If $\lambda_n = \lambda_m$ we have a degeneracy but suitable combinations of ψ_n and ψ_m can be chosen to yield an orthogonal set of functions.

For example, if ψ_1 and ψ_2 are eigenfunctions with degenerate eigenvalues $\lambda_1 = \lambda_2$ and $\int_0^a \sigma\psi_1\psi_2 \, dx \neq 0$, then we can choose new eigenfunctions $\bar{\psi}_1 = \psi_1$ and $\bar{\psi}_2 = \psi_1 + C\psi_2$

and force these to be orthogonal. This means that the constant C is chosen such that $\int_0^a \sigma(\psi_1^2 + C\psi_1\psi_2)\, dx = 0$. The new eigenfunctions are now orthogonal but have the common eigenvalue λ_1. This procedure may be extended to the case where there are N degenerate eigenvalues with corresponding eigenfunctions. The procedure just described is an example of the Gram–Schmidt orthogonalization procedure.

We will now assume that the ψ_n have been normalized so that they form an orthonormal set, i.e.,

$$\int_0^a \sigma\psi_n\psi_m\, dx = \begin{cases} 1, & n = m \\ 0, & n \neq m. \end{cases} \tag{18}$$

The eigenfunctions form a complete set and may be used to expand an arbitrary piecewise continuous function $f(x)$ into a Fourier-type series. Thus let

$$f(x) = \sum_{n=1}^{\infty} a_n\psi_n(x). \tag{19}$$

To find the unknown coefficients multiply both sides by $\sigma\psi_m$, integrate over 0 to a, and use (18); thus

$$\int_0^a \sigma f\psi_m\, dx = \sum_{n=1}^{\infty} a_n \int_0^a \sigma\psi_n\psi_m\, dx = a_m. \tag{20}$$

Our main interest in the orthogonality property of the eigenfunctions is for the purpose of finding the coefficients in Fourier series-type expansions.

The notion of completeness for the space of functions ψ_n defined on the interval $0 \leq x \leq a$ involves the following: Let $f(x)$ be a piecewise continuous function on $0 \leq x \leq a$ that is quadratically integrable with $\sigma(x)$ as a weighting function, i.e.,

$$\int_0^a |f(x)|^2\, \sigma(x)\, dx < \infty.$$

We assume that $\sigma(x)$ is always positive.

Consider now the approximation

$$\sum_{n=1}^{N} C_n\psi_n(x), \qquad C_n = \int_0^a \sigma(x)f(x)\psi_n(x)\, dx$$

to $f(x)$. If the limit as $N \to \infty$ of the integrated square of the error tends to zero, then the ψ_n form a complete set. Completeness thus implies that

$$\lim_{N\to\infty} \int_0^a \left| f(x) - \sum_{n=1}^{N} C_n\psi_n(x) \right|^2 \sigma(x)\, dx = 0. \tag{21}$$

This equation becomes, in the limit,

$$\int_0^a |f(x)|^2\, \sigma(x)\, dx = \sum_{n=1}^{\infty} |C_n|^2 \tag{22}$$

which is Parseval's theorem. The space of functions ψ_n is said to be complete when (21) is true. The functions ψ_n then span the space or domain of functions that the Sturm–Liouville differential operator operates on. The functions are said to form a basis for this domain, which means that they may be used to expand functions in the domain into Fourier-like series. It is important to note that the functions ψ_n are not a basis for expanding functions that do not lie in the domain of the Sturm–Liouville differential operator. For example, they cannot be used to give an expansion of the function defined on the interval $0 \le x \le b$ with $b > a$.

A method of showing that the eigenfunctions for the Sturm–Liouville differential operator form a complete set is given in Chapter 6 and is based on an associated variational principle for the eigenvalues. The eigenfunctions form an orthogonal set when the differential operator is self-adjoint. When the operator is not self-adjoint it is necessary to find the eigenfunctions of the adjoint operator in order to have an orthogonality principle that can be used to develop eigenfunction expansions. Non-self-adjoint operators are discussed in Section 2.20.

The normalized eigenfunctions correspond to unit vectors in a linear vector function space. When the eigenfunctions are orthogonal they correspond to orthogonal unit vectors. The symmetrical scalar product

$$\langle f, \psi_n \rangle = \int_0^a f(x)\psi_n(x)\, dx \tag{23}$$

is analogous to finding the projection of the vector $f(x)$ onto the unit vector $\psi_n(x)$. When the operator is not self-adjoint the eigenfunctions are not orthogonal and we then need to introduce a reciprocal set of unit vectors in order to find the components of f. The normalized eigenfunctions of the adjoint operator correspond to the reciprocal set of unit vectors in a nonorthogonal system.

2.4. GREEN'S FUNCTION $G(x, x')$

The Green's function is the response of a linear system to a point source of unit strength. A source of unit strength at the position x' can be conveniently represented by the Dirac delta function $\delta(x - x')$. From a heuristic viewpoint we can think of $\delta(x - x')$ as a narrow rectangular pulse of width Δx and height $1/\Delta x$, and thus of unit area, in the limit as $\Delta x \to 0$ as shown in Fig. 2.3(a). The Green's function $G(x, x')$ is the solution of the Sturm–Liouville equation when the source term or forcing function is a point source of unit strength. The Green's function satisfies the equation,

$$\frac{d}{dx} p \frac{dG}{dx} + (q + \lambda\sigma)G = -\delta(x - x'). \tag{24}$$

The Green's function must also satisfy appropriate boundary conditions at $x = 0, a$. We will discuss two basic solutions to (24).

Method I

We may expand G in terms of the eigenfunctions ψ_n, with eigenvalues λ_n, that satisfy the same boundary conditions as G. Thus let $G(x, x') = \sum_{n=1}^{\infty} a_n \psi_n(x)$. Substituting in the equation for G gives $\sum_{n=1}^{\infty} a_n(\lambda - \lambda_n)\sigma\psi_n = -\delta(x - x')$ since ψ_n is a solution of (16) for

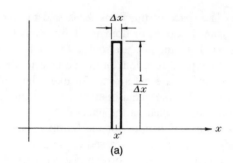

(a)

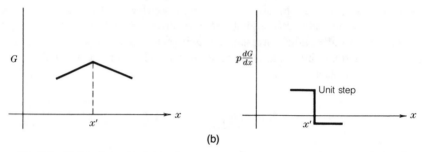

(b)

Fig. 2.3. (a) One-dimensional delta function. (b) Behavior of G and its derivative near x'.

$\lambda = \lambda_n$. If we now multiply both sides by ψ_m, integrate over 0 to a, and use (18) we obtain

$$\sum_{n=1}^{\infty} a_n(\lambda - \lambda_n) \int_0^a \sigma \psi_n \psi_m \, dx = a_m(\lambda - \lambda_m) = -\int_0^a \psi_m \delta(x - x') \, dx = -\psi_m(x')$$

upon using (3) to evaluate the last integral. Using this solution for a_m we obtain

$$G(x, x') = -\sum_{n=1}^{\infty} \frac{\psi_n(x)\psi_n(x')}{\lambda - \lambda_n}. \tag{25}$$

Note that G has poles at the points $\lambda = \lambda_n$. We will make use of this property later on.

Method II

We note that at all points $x \neq x'$ the Green's function is a solution of the homogeneous equation $(d/dx)p(dG/dx)+(q+\lambda\sigma)G = 0$. At x' G must be continuous but $p(dG/dx)$ must be discontinuous by a unit amount [see Fig. 2.3(b)]. The second derivative $(d/dx)p(dG/dx)$ then yields a delta function singularity. If G was not continuous at x' the resulting singularity would be of too high an order.

Let $A\Phi_1(x)$ be a solution of the homogeneous equation that satisfies the boundary condition at $x = 0$. Similarly, let $B\Phi_2(x)$ be a solution, linearly independent of Φ_1, that satisfies the boundary condition at $x = a$. Thus let $G = A\Phi_1(x)$, $x \leq x'$, and $G = B\Phi_2(x)$, $x \geq x'$. Continuity of G at $x = x'$ requires that $A\Phi_1(x') = B\Phi_2(x')$. If we integrate the equation for G from $x' - \tau$ to $x' + \tau$ and let $\tau \to 0$ and note that $\lim_{\tau \to 0} \int_{x'-\tau}^{x'+\tau}(q + \lambda\sigma)G \, dx = 0$ since

q, σ, and G are all continuous at x', then we obtain

$$\lim_{\tau \to 0} \int_{x'-\tau}^{x'+\tau} \frac{d}{dx} p \frac{dG}{dx} \, dx = p \left. \frac{dG}{dx} \right|_{x'_-}^{x'_+} = -\int_{x'-\tau}^{x'+\tau} \delta(x - x') \, dx = -1.$$

In other words, $p \, dG/dx$ evaluated between adjacent sides of the point x' must undergo a step change as shown in Fig. 2.3(b). Hence we have

$$\left. \frac{dG}{dx} \right|_{x'_-}^{x'_+} = -\frac{1}{p(x')} = B\Phi_2'(x') - A\Phi_1'(x')$$

where Φ' stands for $d\Phi/dx$. The solution for A and B may be found from the above two conditions and is $A = -\Phi_2(x')/p(x')W(x')$, $b = -\Phi_1(x')/p(x')W(x')$ where the Wronskian determinant $W(x') = \Phi_1(x')\Phi_2'(x') - \Phi_1'(x')\Phi_2(x')$ does not equal zero since Φ_1 and Φ_2 are linearly independent solutions.

Our solution for the Green's function is thus

$$G(x, x') = \begin{cases} -\dfrac{\Phi_1(x)\Phi_2(x')}{p(x')W(x')}, & 0 \le x \le x' \\[2mm] -\dfrac{\Phi_1(x')\Phi_2(x)}{p(x')W(x')}, & x' \le x \le a. \end{cases} \tag{26}$$

The Wronskian determinant W will be a function of the parameter λ and will vanish whenever $\lambda = \lambda_n$ so G will have the poles that are explicitly exhibited by the first solution (25). Although (25) and (26) are different in form they are equal.

It is convenient to define the following:

$$x_> \equiv \text{the greater of } x \text{ or } x'$$

$$x_< \equiv \text{the lesser of } x \text{ or } x'.$$

We can then express the solution for $G(x, x')$ in the condensed form

$$G(x, x') = -\frac{\Phi_1(x_<)\Phi_2(x_>)}{p(x')W(x')} \tag{27}$$

which includes both forms given by (26).

A useful property of the Sturm–Liouville equation is that the product of $p(x)W(x)$ is a constant. We can prove this result as follows: We form the product of $\Phi_1(x)$ with the equation for $\Phi_2(x)$ and subtract from this the corresponding result with Φ_1 and Φ_2 interchanged; thus

$$\Phi_1 \frac{d}{dx} p \frac{d\Phi_2}{dx} + (q + \lambda\sigma)\Phi_1\Phi_2 - \Phi_2 \frac{d}{dx} p \frac{d\Phi_1}{dx} - (q + \lambda\sigma)\Phi_1\Phi_2$$

$$= \Phi_1 \frac{d}{dx} p \frac{d\Phi_2}{dx} - \Phi_2 \frac{d}{dx} p \frac{d\Phi_1}{dx}$$

$$= \frac{d}{dx} \left(\Phi_1 p \frac{d\Phi_2}{dx} - \Phi_2 p \frac{d\Phi_1}{dx} \right) = 0.$$

From the last result we conclude that

$$p\left(\Phi_1 \frac{d\Phi_2}{dx} - \Phi_2 \frac{d\Phi_1}{dx}\right) = p(x)W(x) = C \tag{28}$$

where the constant C is a function of λ.

Example

We wish to find the eigenfunctions and eigenvalues for the equation $d^2\psi/dx^2 + \lambda\psi = 0$ with boundary conditions $\psi = 0$ at $x = 0, a$. This is a special case with $p = \sigma = 1$ and $q = 0$. A general solution for ψ is $A \sin \sqrt{\lambda}x + B \cos \sqrt{\lambda}x$. To make $\psi = 0$ at $x = 0$ we must choose $B = 0$. To make ψ vanish at $x = a$ we must have $\sin \sqrt{\lambda}a = 0$. Thus $\sqrt{\lambda}a = n\pi$, $n = 1, 2, 3, \ldots$ or $\lambda_n = (n\pi/a)^2$. Hence the eigenfunctions are $\sin n\pi x/a$. Since $\int_0^a \sin^2(n\pi x/a)\,dx = a/2$ the normalized eigenfunctions are $\sqrt{2/a} \sin(n\pi x/a) = \psi_n$.

We now wish to solve $d^2G/dx^2 + \lambda G = -\delta(x - x')$. According to Method I we let $G = \sum_{n=1}^{\infty} a_n \sqrt{2/a} \sin(n\pi x/a)$. By substituting in the equation for G we find $\sum_{n=1}^{\infty} a_n(\lambda - n^2\pi^2/a^2)\sqrt{2/a} \sin(n\pi x/a) = -\delta(x - x')$. We now multiply both sides by $\sqrt{2/a} \sin(m\pi x/a)$, integrate from 0 to a, and obtain

$$a_m\left(\lambda - \frac{m^2\pi^2}{a^2}\right) = -\sqrt{\frac{2}{a}}\int_0^a \delta(x - x') \sin\frac{m\pi x}{a}\,dx = -\sqrt{\frac{2}{a}}\sin\frac{m\pi x'}{a}.$$

Thus

$$G(x, x') = -\sum_{n=1}^{\infty} \frac{\sqrt{\frac{2}{a}}\sin\frac{n\pi x}{a}\sqrt{\frac{2}{a}}\sin\frac{n\pi x'}{a}}{\lambda - n^2\pi^2/a^2}. \tag{29a}$$

We may also use Method II. We let $\Phi_1 = \sin \sqrt{\lambda}x$ and $\Phi_2 = \sin \sqrt{\lambda}(a - x)$. Note that Φ_1 and Φ_2 are linearly independent and satisfy the boundary conditions at $x = 0, a$, respectively. The Wronskian determinant $W(x') = \Phi_1(x')\Phi_2'(x') - \Phi_1'(x')\Phi_2(x') = (\sin \sqrt{\lambda}x')[-\sqrt{\lambda} \cos \sqrt{\lambda}(a - x')] - \sqrt{\lambda} \cos \sqrt{\lambda}x' \sin \sqrt{\lambda}(a - x') = -\sqrt{\lambda} \sin \sqrt{\lambda}a$. From (27) we get

$$G(x, x') = \frac{\sin \sqrt{\lambda}x_< \sin \sqrt{\lambda}(a - x_>)}{\sqrt{\lambda} \sin \sqrt{\lambda}a}. \tag{29b}$$

Note that $\sin \sqrt{\lambda}a = 0$ whenever $\lambda = (n\pi/a)^2$ so that this second solution for G does have the poles at $\lambda = n^2\pi^2/a^2$ that are shown explicitly in the first solution. A Fourier series expansion of the second solution would yield the first solution.

2.5. Solution of Boundary-Value Problems

Let it be required to find a solution to the equation

$$\frac{d}{dx}p\frac{d\psi}{dx} + (q + \lambda\sigma)\psi = -f(x) \tag{30}$$

with either ψ, $d\psi/dx$, or $\psi + K(d\psi/dx)$ specified at $x = 0, a$. In (30) $f(x)$ is a distributed forcing function. Let $G(x, x')$ be a Green's function that is a solution of

$$\frac{d}{dx}p\frac{dG}{dx} + (q + \lambda\sigma)G = -\delta(x - x').$$

(31)

We now replace x by x' for convenience, multiply the equation for ψ by G and the equation for G by ψ, subtract, and integrate to obtain

$$\int_0^a \left[G(x', x)\frac{d}{dx'}p(x')\frac{d\psi(x')}{dx'} - \psi(x')\frac{d}{dx'}p(x')\frac{dG(x', x)}{dx'} \right] dx'$$

$$= \int_0^a -f(x')G(x', x)\,dx' + \int_0^a \psi(x')\delta(x' - x)\,dx'$$

since the terms involving the factor $q + \lambda\sigma$ cancel. After an integration by parts we obtain

$$\psi(x) = \int_0^a G(x', x)f(x')\,dx' + \left[G(x', x)p(x')\frac{d\psi(x')}{dx'} \right.$$

$$\left. - \psi(x')p(x')\frac{dG(x', x)}{dx'} \right]\Big|_0^a.$$

(32)

If ψ satisfies one of the boundary conditions $\psi = 0$, $d\psi/dx' = 0$, $\psi + K(d\psi/dx') = 0$, at each end $x = 0$ or a, and we choose G to satisfy the same boundary conditions, then the boundary terms vanish and the solution for ψ is simply

$$\psi(x) = \int_0^a G(x', x)f(x')\,dx'.$$

(33)

This solution is a superposition of the response from each differential element of force $f\,dx'$ with G being the response of the system due to a point source of unit strength. It is the ability to represent the solution to a linear system driven by an arbitrary forcing function $f(x)$ by a superposition integral such as (33) that makes the theory of Green's functions of great importance in practice.

Sometimes the boundary conditions on ψ may be of the form $K_1\psi + K_2(d\psi/dx) = K_3$ (K_1 or K_2 might be zero and the K_i may be different at each end). In this case we choose the boundary conditions on G in such a way that we can evaluate the boundary terms in (32) in terms of known quantities. If ψ is given on the boundary ($K_2 = 0$) we choose $G = 0$ on the boundary; if $d\psi/dx$ is given ($K_1 = 0$) we choose $dG/dx = 0$; and finally if $K_1\psi + K_2(d\psi/dx)$ is given we choose $K_1G + K_2(dG/dx) = 0$ on the boundary. We will illustrate the last case explicitly. The boundary terms in (32) may be written as

$$p(x')\left[\frac{G}{K_2}\left(K_1\psi + K_2\frac{d\psi}{dx'} \right) - \frac{\psi}{K_2}\left(K_1G + K_2\frac{dG}{dx'} \right) \right]\Big|_0^a$$

which by virtue of the boundary conditions on ψ and G becomes the known quantity $(K_3/K_2)p(x')G(x', x)|_0^a$.

At this point we digress in order to introduce some mathematical concepts that will be more

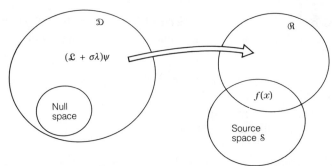

Fig. 2.4. Mapping of functions from the domain $\mathfrak{D}$ of an operator into its range space $\mathfrak{R}$. The source function $f(x)$ from the source space $\mathfrak{S}$ should lie in the range space for a solution of the differential equation to exist. Functions in the null space are not mapped into the range space.

important in connection with dyadic Green's functions but which can be illustrated in a simple setting here.

For convenience we repeat (30) in the form

$$\mathfrak{L}\psi + \lambda\sigma\psi = -f(x) \tag{34}$$

where $\mathfrak{L}$ denotes the operator $(d/dx)p(x)(d/dx)+q(x)$. This operator operates on functions ψ that are in the space of functions called the domain $\mathfrak{D}$ of the operator. This space is the space of functions that are twice differentiable, are quadratically integrable as defined in Section 2.3, exist on the interval $0 \leq x \leq a$, and satisfy the boundary conditions at $x = 0, a$. The boundary conditions are an important part of the specification of the domain of the operator. The operator $\mathfrak{L}+\sigma\lambda$ can be viewed as operating on functions ψ in the domain $\mathfrak{D}$ and producing new functions that lie in a space of functions called the range space $\mathfrak{R}$ of the operator. This concept is illustrated in Fig. 2.4. It should be obvious that (34) does not have a solution if the source function $f(x)$ is not in the range space $\mathfrak{R}$ of the operator. Hence a necessary condition for (34) to have a solution is that the source $f(x)$ be in the range space, i.e., $f(x) \in \mathfrak{R}$ where $\in$ means that $f(x)$ is in $\mathfrak{R}$.

The solution for ψ in (34) when $f(x) \in \mathfrak{R}$ is unique only if there are no solutions except the trivial one $u = 0$ to the homogeneous equation [solutions to the homogeneous equation are not related to the source $f(x)$]

$$\mathfrak{L}u + \sigma\lambda u = 0.$$

Any nonzero solution u to this homogeneous equation is said to be in the null space of the operator and is not mapped into the range space. For $\lambda = \lambda_n$, an eigenvalue, $u = \psi_n$ is a nonzero solution and is in the null space of $(\mathfrak{L} + \sigma\lambda_n)$. Hence for a unique solution to (34) λ must not be equal to an eigenvalue. All values of λ that correspond to the discrete eigenvalues λ_n make up the point spectrum of the operator. All values of λ for which (34) has a bounded (finite) solution belong to the resolvent set. Later on we will encounter problems where λ has a continuous range of values forming a continuous spectrum. The continuous spectrum is not part of the resolvent set. When λ is in the resolvent set (34) can be solved (resolved).

The completeness of the eigenfunction set enables us to replace the operator $\mathfrak{L}+\sigma\lambda$ operating on ψ by

$$(\mathfrak{L} + \sigma\lambda)\psi = \sum_n \sigma(\lambda - \lambda_n)C_n\psi_n$$

where

$$C_n = \int_0^a \sigma\psi(x')\psi_n(x')\,dx'.$$

We can represent the operator $\mathcal{L} + \sigma\lambda$ as

$$\mathcal{L} + \sigma\lambda = \sum_n (\lambda - \lambda_n)\sigma\psi_n(x)\psi_n(x')$$

where the action of the operator on a function $\psi(x')$ is obtained by multiplying by $\psi(x')$ and integrating over x'. This representation of the operator is called the spectral representation. If we define the scalar product $\langle\sigma\psi_n, \psi\rangle$ to be

$$\langle\sigma\psi_n(x'), \psi(x')\rangle = \int_0^a \sigma(x')\psi_n(x')\psi(x')\,dx'$$

then symbolically we can write

$$\langle\mathcal{L} + \sigma\lambda, \psi\rangle = \sum_n (\lambda - \lambda_n)\psi_n(x)\langle\sigma\psi_n, \psi\rangle$$

as describing the action of the operator on ψ.

When we substitute the series expansion for ψ into (34) we obtain

$$\sum_n C_n(\lambda - \lambda_n)\sigma\psi_n(x) = -f(x).$$

By means of the standard procedure

$$C_n = \frac{-1}{\lambda - \lambda_n}\int_0^a f(x')\psi_n(x')\,dx'$$

and

$$\psi(x) = -\sum_{n=1}^{\infty}\frac{\psi_n(x)}{\lambda - \lambda_n}\int_0^a f(x')\psi_n(x')\,dx'. \tag{35a}$$

We now rewrite this solution in the form

$$\psi(x) = \int_0^a f(x')\left[-\sum_{n=1}^{\infty}\frac{\psi_n(x')\psi_n(x)}{\lambda - \lambda_n}\right]dx' \tag{35b}$$

which allows us to identify the Green's function operator as

$$G(x', x) = -\sum_{n=1}^{\infty}\frac{\psi_n(x')\psi_n(x)}{\lambda - \lambda_n}. \tag{36}$$

This is the spectral representation of the inverse operator $(\mathcal{L} + \sigma\lambda)^{-1}$. It is irrelevant whether or not this series converges since it should be viewed as an operator that is to be used in a term by term integration so as to generate the series expansion for the solution ψ as in (35a). For one-dimensional problems the Green's function series does converge to an ordinary function, but in two- and three-dimensional problems the Green's function series is not convergent when the field point and source point coincide. For this reason it is not meaningful to always consider summing the series, which cannot be done in general. The concept of the Green's function as an operator as used in (35a) is, however, valid.

In the space of functions $\mathfrak{D}$ we have, for a function $h(x)$,

$$h(x) = \sum_{n=1}^{\infty} a_n \psi_n(x), \qquad a_n = \int_0^a h(x')\sigma(x')\psi_n(x')\,dx'$$

and hence

$$h(x) = \sum_{n=1}^{\infty} \psi_n(x) \int_0^a h(x')\sigma(x')\psi_n(x')\,dx'$$
$$= \int_0^a h(x') \left[\sum_{n=1}^{\infty} \sigma(x')\psi_n(x')\psi_n(x) \right] dx'. \qquad (37)$$

The series in (37) is seen to have the same operational property as the symbolic function $\delta(x - x')$. Hence the series, which is not a convergent one for all x, x', is a representation for $\delta(x - x')$ in the space $\mathfrak{D}$, i.e.,

$$\delta(x - x') = \sum_{n=1}^{\infty} \sigma(x')\psi_n(x')\psi_n(x). \qquad (38)$$

This representation for $\delta(x - x')$ is also to be viewed as an operator, not as a function, and be used in a term by term integration to yield $h(x)$ as in (37). It is important to note that this eigenfunction representation for $\delta(x - x')$ is valid only for functions in the domain $\mathfrak{D}$ for which the eigenfunctions form a complete set of functions, i.e., span the domain $\mathfrak{D}$. With these qualifications the relationship exhibited in (38) is said to be the completeness relation for the eigenfunction set.

Consider the equation $(\mathfrak{L} + \sigma\lambda_k)\psi = -f$ for a given specified eigenvalue λ_k. The eigenfunction ψ_k is in the null space of the operator $\mathfrak{L} + \sigma\lambda_k$ since $(\mathfrak{L} + \sigma\lambda_k)\psi_k = 0$. Clearly the space of functions in the range space of $\mathfrak{L} + \sigma\lambda_k$ does not contain the function ψ_k. Consequently, the equation has a solution only if the source function f is orthogonal to ψ_k. We see that the range space is orthogonal to the null space for a self-adjoint operator.

Not all differential operators are self-adjoint operators. Non-self-adjoint operators are described in Section 2.20. In this section we will give an example of a non-self-adjoint operator and its implication. Consider a differential operator of the form

$$\mathfrak{L}\psi = \frac{d^2\psi}{dx^2} - j\frac{d\psi}{dx} + \lambda\psi = 0$$

with boundary conditions $\psi = 0$ at $x = 0, a$. Let this equation be multiplied by a function Φ and integrate by parts so as to transform the differential operators onto Φ; thus the symmetric product

$$\langle \Phi, \mathfrak{L}\psi \rangle = \int_0^a \Phi\frac{d^2\psi}{dx^2}\,dx - j\int_0^a \Phi\frac{d\psi}{dx}\,dx + \lambda\int_0^a \Phi\psi\,dx = 0$$

becomes

$$\Phi\frac{d\psi}{dx}\bigg|_0^a - \psi\frac{d\Phi}{dx}\bigg|_0^a + \int_0^a \psi\frac{d^2\Phi}{dx^2}\,dx - j\Phi\psi\bigg|_0^a$$

$$+ j\int_0^a \psi\frac{d\Phi}{dx}\,dx + \lambda\int_0^a \Phi\psi\,dx = 0.$$

If we wish to make the integral of the product $\Phi\psi$ equal to zero, then Φ must be a solution of

$$\mathcal{L}_a\psi = \frac{d^2\Phi}{dx^2} + j\frac{d\Phi}{dx} + \mu\Phi = 0, \qquad \mu \neq \lambda$$

and in order to make the integrated terms vanish $\Phi = 0$ at $x = 0, a$. The new differential operator shown above is called the adjoint operator. It differs from the original operator by having j instead of $-j$ as a multiplier in front of the first derivative. With the conditions above imposed, we write

$$\langle \Phi, \mathcal{L}\psi \rangle = \langle \mathcal{L}_a\Phi, \psi \rangle \tag{39}$$

which is the formal definition for the adjoint operator. If an operator is not self-adjoint, then it is the eigenfunctions, say Φ_n, of the adjoint operator that are orthogonal to the eigenfunctions, say ψ_m, of the original operator. A different expansion procedure is required to obtain a spectral expansion of the solution for the inhomogeneous equation. The details for the modifications required are described in Section 2.20.

When the eigenfunctions are complex we can choose to define the scalar product in the following way:

$$\langle \Phi, \psi \rangle = \int_0^a \Phi\psi^* \,dx$$

where ψ^* is the complex conjugate of ψ. If we choose this as our scalar product then a similar analysis would show that $\mathcal{L}_a = \mathcal{L}$ so that the operator is now self-adjoint. Thus the choice of scalar product dictates in many instances whether or not an operator is self-adjoint. Sometimes $\mathcal{L}_a = \mathcal{L}$ but the boundary conditions for Φ are different from those for ψ, in which case the operator is classified as non-self-adjoint. An operator is not fully defined until the boundary conditions are also given since this determines the domain of the operator. Generally, when we have complex eigenfunctions it is preferable to use the scalar product involving $\Phi\psi^*$. The mathematical theory for differential and integral operators that involve complex functions is based on the use of the scalar product involving $\Phi\psi^*$ because this allows the concept of length (norm) to be defined as a positive real quantity, i.e., the length of a function Φ would be given by

$$\|\Phi\| = \langle \Phi, \Phi \rangle^{1/2} = \left[\int_0^a \Phi\Phi^* \,dx\right]^{1/2}$$

which is analogous to the length of a Euclidean vector $\mathbf{A}$ with real components. The latter is given by

$$|\mathbf{A}| = (\mathbf{A}\cdot\mathbf{A})^{1/2} = (A_x^2 + A_y^2 + A_z^2)^{1/2}$$

and is always a real positive quantity. These concepts are described more completely in Section 2.20.

2.6. MULTIDIMENSIONAL GREEN'S FUNCTIONS AND ALTERNATIVE REPRESENTATIONS [2.10], [2.13], [2.14]

We have noted that there is more than one representation for a one-dimensional Green's function. Also we have noted that the Green's function has poles for certain values of the separation constant λ. These poles are called the spectrum of the Green's function. For finite-range problems the spectrum is discrete, but for infinite-range problems the spectrum is continuous and the Green's function will have branch point singularities instead of poles.

In two and three dimensions there are many different possible representations for the Green's functions. For this reason we wish to present a unified approach to multidimensional Green's functions.

First we will present a theorem that is useful in constructing two- and three-dimensional Green's functions from Green's functions for one-dimensional problems.

Theorem I:

$$\frac{1}{2\pi j} \oint_C G(x, x', \lambda) \, d\lambda = -\frac{\delta(x - x')}{\sigma(x')} \tag{40}$$

where C is a closed contour in the complex λ plane that encloses all the singularities of $G(x, x', \lambda)$.

Proof: Use the form

$$G(x, x', \lambda) = -\sum_{n=1}^{\infty} \frac{\psi_n(x)\psi_n(x')}{\lambda - \lambda_n}$$

and Cauchy's theorem to give

$$-\frac{1}{2\pi j} \sum_{n=1}^{\infty} \oint_C \frac{\psi_n(x)\psi_n(x')}{\lambda - \lambda_n} \, d\lambda = -\sum_{n=1}^{\infty} \psi_n(x)\psi_n(x').$$

To complete the proof we show that this equals the right-hand side of (40) by expanding $\delta(x - x')$ in terms of a series in the ψ_n. Let $\delta(x - x') = \sum_{n=1}^{\infty} C_n \psi_n(x)$. We multiply both sides by $\sigma(x)\psi_m(x)$ and use (18) to obtain $C_m = \int_0^a \delta(x - x')\sigma(x)\psi_m(x) \, dx = \sigma(x')\psi_m(x')$. Hence

$$\frac{\delta(x - x')}{\sigma(x')} = \sum_{n=1}^{\infty} \psi_n(x)\psi_n(x') \tag{41}$$

which completes the proof.

Example

Consider $d^2G/dx^2 + \lambda G = -\delta(x - x')$, $G = 0$ at $x = 0, a$. By using Method II we readily find that

$$G(x, x', \lambda) = \frac{\sin \sqrt{\lambda} x_< \sin \sqrt{\lambda}(a - x_>)}{\sqrt{\lambda} \sin \sqrt{\lambda} a}.$$

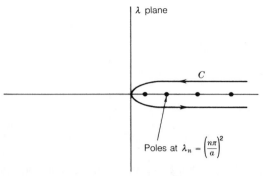

Fig. 2.5. Contour C enclosing the poles of the Green's function.

Since G is an even function of $\sqrt{\lambda}$ it has no branch points; its only singularities are poles at $\sqrt{\lambda}a = n\pi$ or $\lambda = (n\pi/a)^2$, $n = 1, 2, 3, \ldots$. Theorem I gives

$$\frac{1}{2\pi j} \oint_C G \, d\lambda = -\delta(x - x')$$

where C is the contour shown in Fig. 2.5. Near $\lambda_n = (n\pi/a)^2$ we have $\sin \sqrt{\lambda}a = \sin a(\sqrt{\lambda} - \sqrt{\lambda_n} + \sqrt{\lambda_n}) = \sin[(\sqrt{\lambda} - \sqrt{\lambda_n})a + n\pi] = \cos n\pi \sin(\sqrt{\lambda} - \sqrt{\lambda_n})a \rightarrow (\cos n\pi)(\sqrt{\lambda} - \sqrt{\lambda_n})a = \cos n\pi[(\lambda - \lambda_n)/(\sqrt{\lambda} + \sqrt{\lambda_n})]a$. Thus the residue from the term $1/(\sqrt{\lambda} \sin \sqrt{\lambda}a)$ at the nth pole is $(2\sqrt{\lambda_n})/(a\sqrt{\lambda_n} \cos n\pi) = (-1)^n(2/a)$. The residue expansion of the integral may now readily be found:

$$\frac{1}{2\pi j} \oint_C G d\lambda = \sum_{n=1}^{\infty} (-1)^n \frac{2}{a} \sin \frac{n\pi x_<}{a} \sin \frac{n\pi}{a}(a - x_>)$$

$$= -\sum_{n=1}^{\infty} \frac{2}{a} \sin \frac{n\pi x}{a} \sin \frac{n\pi x'}{a}$$

since

$$\sin \frac{n\pi}{a}(a - x_>) = -\cos n\pi \sin \frac{n\pi x_>}{a}.$$

The latter series is easily verified to be the expansion of $-\delta(x - x')$ in conformity with Theorem I.

We will use the above theorem to show that multidimensional Green's functions can be constructed in the form of contour integrals taken over the spectrum of a product of associated one-dimensional Green's functions. The theory will be developed by means of suitable examples.

Green's Function for Two-Dimensional Laplace's Equation

Consider

$$\frac{\partial^2 G}{\partial x^2} + \frac{\partial^2 G}{\partial y^2} = -\delta(x - x')\delta(y - y'), \qquad G = 0 \text{ at } x = 0, a; \; y = 0, b. \tag{42}$$

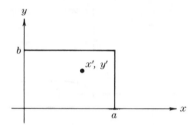

Fig. 2.6. Line charge along z.

Apart from a factor $1/\epsilon_0$ the function G represents the electric potential from a z-directed line charge inside a conducting rectangular tube as shown in Fig. 2.6.

Let $\mathcal{L} = \partial^2/\partial x^2 + \partial^2/\partial y^2$ and $\mathcal{L}_x = \partial^2/\partial x^2$, $\mathcal{L}_y = \partial^2/\partial y^2$. The equation $\mathcal{L}\psi = 0$ separates into two one-dimensional equations $\mathcal{L}_x\psi_x(x) + \lambda_x\psi_x = 0$ and $\mathcal{L}_y\psi_y(y) + \lambda_y\psi_y = 0$ with $\lambda_x + \lambda_y = 0$ provided we assume a solution in product form, i.e., $\psi(x, y) = \psi_x(x)\psi_y(y)$. Corresponding to each of the separated equations there are associated one-dimensional Green's function problems of the form

$$\frac{d^2G_x}{dx^2} + \lambda_x G_x = -\delta(x - x'), \qquad G_x = 0 \text{ at } x = 0, a$$

$$\frac{d^2G_y}{dy^2} + \lambda_y G_y = -\delta(y - y'), \qquad G_y = 0 \text{ at } y = 0, b. \qquad (43)$$

We will now show that the solution to the original problem (42) can be expressed in terms of the solutions to the simpler one-dimensional problems (43). Specifically, we will show that

$$G(x, x', y, y') = \frac{-1}{2\pi j} \oint_{C_x} G_x(x, x', \lambda_x)G_y(y, y', -\lambda_x)\,d\lambda_x$$

$$= \frac{-1}{2\pi j} \oint_{C_y} G_x(x, x', -\lambda_y)G_y(y, y', \lambda_y)\,d\lambda_y \qquad (44)$$

where C_x encloses the singularities of G_x and excludes those of G_y and similarly C_y encloses only the singularities of G_y. The formula (44) synthesizes an appropriate two-dimensional Green's function from the associated one-dimensional Green's functions.

To prove (44) we only need to show that it is a solution of (42). Clearly G satisfies the correct boundary conditions in view of the conditions imposed on G_x and G_y at the boundaries. We also have

$$\mathcal{L}G = (\mathcal{L}_x + \lambda_x + \mathcal{L}_y + \lambda_y)G$$

$$= -\frac{1}{2\pi j} \oint_{C_x} (\mathcal{L}_x + \lambda_x + \mathcal{L}_y + \lambda_y)G_x(x, x', \lambda_x)G_y(y, y', \lambda_y)\,d\lambda_x$$

$$= -\frac{1}{2\pi j} \oint_{C_x} [-\delta(x - x')G_y(y, y', -\lambda_x) - G_x(x, x', \lambda_x)\delta(y - y')]\,d\lambda_x$$

upon using (43) to replace $(\mathcal{L}_x + \lambda_x)G_x$ by $-\delta(x - x')$ and $(\mathcal{L}_y + \lambda_y)G_y$ by $-\delta(y - y')$. Since C_x excludes the singularities of G_y we obtain $\mathcal{L}G = (1/2\pi j)\oint_{C_x} G_x(x, x', \lambda_x)\,d\lambda_x\,\delta(y -$

$y') = -\delta(x - x')\delta(y - y')$ upon using Theorem I (note that $\sigma = 1$ in this problem). In other words, the contour integral of G_y is zero because no singularities are enclosed by virtue of the manner in which C_x was chosen. A similar proof holds if we use the form involving integration around the contour C_y in (44).

We will now illustrate the above procedure by constructing G directly and then show that the same solution is obtained from using (44) and solutions to (43).

The normalized eigenfunctions of the equation $d^2G_x/dx^2 + \lambda_x G_x = 0$ which vanish at $x = 0, a$, are $\sqrt{2/a}\,\sin(n\pi x/a)$, $n = 1, 2\ldots$, and $\lambda_x = (n\pi/a)^2$. These form a complete set so we may assume that

$$G(x, x', y, y') = \sum_{n=1}^{\infty} a_n(y)\Phi_n(x), \qquad \Phi_n = \sqrt{\frac{2}{a}}\,\sin\frac{n\pi x}{a}.$$

When we substitute into the equation for G, i.e., into (42), we obtain

$$\sum_{n=1}^{\infty} \left[\frac{d^2 a_n(y)}{dy^2} - \left(\frac{n\pi}{a}\right)^2 a_n(y)\right]\Phi_n(x) = -\delta(x - x')\delta(y - y').$$

We now multiply both sides by $\Phi_m(x)$ and integrate to get

$$\frac{d^2 a_m}{dy^2} - \left(\frac{m\pi}{a}\right)^2 a_m = -\Phi_m(x')\delta(y - y')$$

since the functions Φ_n are orthogonal. We can readily solve this equation for $a_m(y)$ according to Method II given earlier. We find that

$$a_m(y) = \frac{\Phi_m(x')\sinh\dfrac{m\pi}{a}y_< \sinh\dfrac{m\pi}{a}(b - y_>)}{\dfrac{m\pi}{a}\sinh\dfrac{m\pi b}{a}}$$

upon choosing $\Phi_1 = \sinh(m\pi y/b)$, $\Phi_2 = \sinh[(m\pi/a)(b - y)]$. Our solution for G is thus

$$G = \sum_{n=1}^{\infty}\frac{2}{n\pi}\frac{\sin\dfrac{n\pi x}{a}\,\sin\dfrac{n\pi x'}{a}\,\sinh\dfrac{n\pi}{a}y_< \sinh\dfrac{n\pi}{a}(b - y_>)}{\sinh\dfrac{n\pi b}{a}}. \qquad (45)$$

We could equally well have made the first expansion with respect to y and we would then have found that

$$G = \sum_{n=1}^{\infty}\frac{2}{n\pi}\frac{\sin\dfrac{n\pi y}{b}\,\sin\dfrac{n\pi y'}{b}\,\sinh\dfrac{n\pi}{b}x_< \sinh\dfrac{n\pi}{b}(a - x_>)}{\sinh\dfrac{n\pi a}{b}}. \qquad (46)$$

We can find yet another form for G by assuming that it can be expanded as a double Fourier

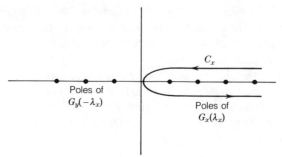

Fig. 2.7. The proper contour C_x.

series. Thus let

$$G = \sum_{n=1}^{\infty}\sum_{m=1}^{\infty} C_{nm} \sin \frac{n\pi x}{a} \sin \frac{m\pi y}{b}.$$

If we substitute this solution into (42) we obtain

$$-\sum_{n=1}^{\infty}\sum_{m=1}^{\infty} C_{nm} \left[\left(\frac{n\pi}{a}\right)^2 + \left(\frac{m\pi}{b}\right)^2 \right] \sin \frac{n\pi x}{a} \sin \frac{m\pi y}{b} = -\delta(x - x')\delta(y - y').$$

The coefficient C_{nm} may be found by multiplying both sides by $\sin(n\pi x/a) \sin(m\pi y/b)$ and integrating over x and y. The final result is

$$G = \sum_{n=1}^{\infty}\sum_{m=1}^{\infty} \frac{4}{ab} \frac{\sin \dfrac{n\pi x}{a} \sin \dfrac{n\pi x'}{a} \sin \dfrac{m\pi y}{b} \sin \dfrac{m\pi y'}{b}}{\left(\dfrac{n\pi}{a}\right)^2 + \left(\dfrac{m\pi}{b}\right)^2}. \qquad (47)$$

We will now show how the different solutions (45)–(47) are all contained in the general formulation (44). We first need the solutions to the one-dimensional Green's function problems given in (43). By using Method II these are readily found in the forms

$$G_x = \frac{\sin \sqrt{\lambda_x}\, x_< \, \sin \sqrt{\lambda_x}(a - x_>)}{\sqrt{\lambda_x} \, \sin \sqrt{\lambda_x}\, a} \qquad (48a)$$

$$G_y = \frac{\sin \sqrt{\lambda_y}\, y_< \, \sin \sqrt{\lambda_y}(b - y_>)}{\sqrt{\lambda_y} \, \sin \sqrt{\lambda_y}\, b}. \qquad (48b)$$

We note that $G_x(x, x', \lambda_x)$ has poles at $\lambda_x = (n\pi/a)^2$, $n = 1, 2, \ldots$, and that $G_y(y, y', -\lambda_x)$ has poles when $\sin \sqrt{-\lambda_x}\, b = 0$ or at $\lambda_x = -(n\pi/b)^2$. Thus we choose the contour C_x as shown in Fig. 2.7. According to (44) we now have

$$G = -\frac{1}{2\pi j} \oint_{C_x} G_x(\lambda_x) G_y(-\lambda_x)\, d\lambda_x.$$

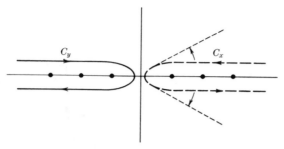

Fig. 2.8. Deformation of contour C_x into C_y.

The residue expansion[2] of this integral in terms of the residues at the poles of $G_x(\lambda_x)$ gives the following solution for G:

$$G = -\sum_{n=1}^{\infty} \frac{\sin \dfrac{n\pi}{a} x_< \sin \dfrac{n\pi}{a}(a - x_>) \sin j\dfrac{n\pi}{a} y_< \sin j\dfrac{n\pi}{a}(b - y_>)}{j \dfrac{n\pi}{a} \dfrac{a}{2} \cos n\pi \sin j\dfrac{n\pi b}{a}}$$

$$= \sum_{n=1}^{\infty} \frac{2 \sin \dfrac{n\pi x}{a} \sin \dfrac{n\pi x'}{a} \sinh \dfrac{n\pi}{a} y_< \sinh \dfrac{n\pi}{a}(b - y_>)}{n\pi \sinh \dfrac{n\pi b}{a}}$$

which is the same as (45).

Since the integral over any portion of a circle with infinite radius is zero, the contour C_x may be deformed into the contour C_y as shown in Fig. 2.8. If this is done the integral may be evaluated in terms of the residues at the poles of $G_y(-\lambda_x)$ which occur at $\lambda_x = -(n\pi/b)^2$. This evaluation would give the same solution as (46).

If we construct the solution for G_y according to Method I we would obtain [see (29a)]

$$G_y(y, y', \lambda_y) = -\sum_{m=1}^{\infty} \frac{\dfrac{2}{b} \sin \dfrac{m\pi y}{b} \sin \dfrac{m\pi y'}{b}}{\lambda_y - (m\pi/b)^2}. \tag{49}$$

If we use (48a) for G_x and (49) for G_y in the formula (44) we would obtain the Green's function solution given by (47) provided that we integrate around a contour C_x enclosing the singularities of $G_x(x, x', \lambda_x)$. We leave the details as Problem 2.6 to be carried out by the reader.

The eigenfunctions $\psi_{nm}(x, y)$ of the two-dimensional Laplacian operator are solutions of

$$\left(\frac{\partial^2}{\partial x^2} + \frac{\partial^2}{\partial y^2}\right) \psi_{nm} + \lambda_{nm}\psi_{nm} = 0 \tag{50}$$

---

[2]The residues may be found by evaluating the derivative of the denominator with respect to λ_x at the poles.

Fig. 2.9. A line source in a rectangular waveguide.

where $\psi_{nm} = 0$ on the boundary shown in Fig. 2.6. The normalized eigenfunctions are

$$\sqrt{\frac{4}{ab}} \sin \frac{n\pi x}{a} \sin \frac{m\pi y}{b}$$

and $\lambda_{nm} = (n\pi/a)^2 + (m\pi/b)^2$. When these eigenfunctions are used to solve (42) the solution is that given by (47). Thus (47) represents the complete eigenfunction expansion for the Green's function. The other solutions given by (45) and (46) are eigenfunction expansions with respect to x or y and closed-form solutions with respect to y or x, respectively.

2.7. GREEN'S FUNCTION FOR A LINE SOURCE IN A RECTANGULAR WAVEGUIDE

Figure 2.9 shows a uniform current filament directed along y, at x', z', in a rectangular waveguide. The current has a time dependence $e^{j\omega t}$ and is not a function of y. This current source will excite TE$_{n0}$ modes only. The vector potential is a solution of $(\mathbf{A} = A_y \mathbf{a}_y = \psi \mathbf{a}_y)$

$$\left(\frac{\partial^2}{\partial x^2} + \frac{\partial^2}{\partial z^2} + k_0^2\right)\psi = -\mu_0 I_g \delta(x - x')\delta(z - z') \tag{51}$$

with $\psi = 0$ at $x = 0, a$ and ψ representing outward-propagating waves at $|z|$ approaching infinity. If we solve the Green's function problem

$$\left(\frac{\partial^2}{\partial x^2} + \frac{\partial^2}{\partial z^2} + k_0^2\right)G = -\delta(x - x')\delta(z - z'), \qquad G = 0; \ x = 0, a \tag{52}$$

then $\psi = \mu_0 I_g G$ and $E_y = -j\omega\mu_0 I_g G$. In (51) and (52) $k_0^2 = \omega^2 \mu_0 \epsilon_0 = \omega^2/c^2$.

Consider now the homogeneous equation $(\partial^2/\partial x^2 + \partial^2/\partial z^2 + k_0^2)G = 0$. If we assume $G(x, x', z, z') = G_x(x, x')G_z(z, z')$ we obtain

$$\frac{1}{G_x}\frac{\partial^2 G_x}{\partial x^2} + \frac{1}{G_z}\frac{\partial^2 G_z}{\partial z^2} + k_0^2 = 0$$

so we must have

$$\frac{\partial^2 G_x}{\partial x^2} + \lambda_x G_x = 0, \qquad \frac{\partial^2 G_z}{\partial z^2} + \lambda_z G_z = 0$$

and

$$\lambda_x + \lambda_z = k_0^2. \tag{53}$$

From these we can identify the associated one-dimensional Green's function problems to be

$$\frac{\partial^2 G_x}{\partial x^2} + \lambda_x G_x = -\delta(x - x'), \qquad G_x = 0; \; x = 0, a \tag{54a}$$

$$\frac{\partial^2 G_z}{\partial z^2} + \lambda_z G_z = -\delta(z - z'). \tag{54b}$$

G_z is an outward-propagating wave as $|z| \to \infty$.

We may readily solve (54a) by Method I or II. If we use Method I we obtain [see (29a)]

$$G_x = -\sum_{n=1}^{\infty} \frac{2}{a} \frac{\sin \dfrac{n\pi x}{a} \sin \dfrac{n\pi x'}{a}}{\lambda_x - n^2 \pi^2 / a^2}.$$

Since the range in z is infinite we use Fourier transforms to solve (54b). We will define $\hat{G}_z(\beta)$ to be the transform of $G_z(z)$, i.e.,

$$\hat{G}_z(\beta) = \int_{-\infty}^{\infty} G_z(z) e^{j\beta z} \, dz. \tag{55a}$$

Then

$$G_z(z) = \frac{1}{2\pi} \int_{-\infty}^{\infty} \hat{G}_z(\beta) e^{-j\beta z} \, d\beta. \tag{55b}$$

If we take the Fourier transform of (54b) and note that the transform of $\partial^2 G_z / \partial z^2$ is $-\beta^2 \hat{G}_z$ (obtained through two integrations by parts) we find that

$$\hat{G}_z = -\frac{e^{j\beta z'}}{\lambda_z - \beta^2} = \frac{e^{j\beta z'}}{(\beta - \sqrt{\lambda_z})(\beta + \sqrt{\lambda_z})}. \tag{56}$$

We wish to consider λ_z as a complex variable and so we will arbitrarily choose that branch of the two-valued function $\sqrt{\lambda_z}$ which has Imag. $\sqrt{\lambda_z} < 0$. In the complex β plane $\hat{G}_z$ then has two poles located as shown in Fig. 2.10. The solution for G_z is given by

$$G_z = \frac{1}{2\pi} \int_{-\infty}^{\infty} \frac{e^{-j\beta(z-z')}}{(\beta - \sqrt{\lambda_z})(\beta + \sqrt{\lambda_z})} \, d\beta.$$

When $z > z'$ the factor $e^{-j\beta(z-z')}$ becomes exponentially small (we write β as $\beta' + j\beta''$) in the lower half plane [proportional to $e^{\beta''(z-z')}$ with $\beta'' < 0$]. Hence we can close the contour by a semicircle of infinite radius in the lower half plane and evaluate the integral in terms of the residue at the pole at $\beta = \sqrt{\lambda_z}$. For $z < z'$ we can close the contour in the upper half plane and evaluate the integral in terms of the residue at $-\sqrt{\lambda_z}$. Note that there is no contribution to the integral from the semicircles and that a negative sign enters when the contour is traversed in a clockwise sense. We are thus led to the following solution for G_z:

$$G_z = \begin{cases} -\dfrac{j}{2\sqrt{\lambda_z}} e^{-j\sqrt{\lambda_z}(z-z')}, & z > z' \\[3mm] -\dfrac{j}{2\sqrt{\lambda_z}} e^{j\sqrt{\lambda_z}(z-z')}, & z < z' \end{cases} \tag{57}$$

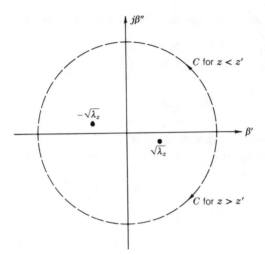

Fig. 2.10. Pole locations in the complex β plane.

which may be also expressed as

$$G_z = -\frac{j}{2\sqrt{\lambda_z}}e^{-j\sqrt{\lambda_z}|z-z'|} \qquad \text{all } z. \tag{58}$$

Note that with Imag. $\sqrt{\lambda_z} < 0$ chosen as the branch of the function $\sqrt{\lambda_z}$ that G_z will tend toward zero as $|z| \to \infty$ for complex λ_z. Also note that G_z does not have any poles but instead has a branch point at $\lambda_z = 0$. If we wish to always remain on the branch for which $\sqrt{\lambda_z}$ has a negative imaginary part, then the angle of λ_z in the complex plane must be restricted to -2π to 0. This may be accomplished by placing a branch line (or cut) along the positive real axis but an infinitesimal distance above it as shown in Fig. 2.11. This forces us to measure the angle of λ_z from the positive real axis in a clockwise or negative sense and hence restricts the angle to the range 0 to -2π. The spectrum of G_z is the continuous range of values of λ_z along the branch cut (singular line, which if crossed changes the sign of $\sqrt{\lambda_z}$). This is in accord with our earlier statement that for infinite-interval problems the Green's function has a continuous spectrum.

We now apply (44) to obtain G in the form

$$G(x, x', z, z') = -\frac{1}{2\pi j}\oint_{C_x} G_x(x, x', \lambda_x)G_z(z, z', k_0^2 - \lambda_x)\,d\lambda_x$$

$$= -\frac{1}{2\pi j}\oint_{C_z} G_x(x, x', k_0^2 - \lambda_z)G_z(z, z', \lambda_z)\,d\lambda_z. \tag{59}$$

Note that when we integrate over λ_x we use the condition (53) to express λ_z in terms of λ_x and vice versa. In the λ_x plane the poles of G_x are at $\lambda_x = (n\pi/a)^2$ and the branch line for G_z becomes the line along which Imag. $\sqrt{k_0^2 - \lambda_x} = 0$. This is the line extending from $\lambda_x = k_0^2$ (branch point) along the negative axis. The contour C_x is chosen to enclose the poles but not the branch point at k_0^2 as shown in Fig. 2.12(a). On the other hand, if we choose to integrate over λ_z, then the contour C_z is chosen to enclose the branch cut for G_z but such that the poles of G_x, which now occur when $k_0^2 - \lambda_z = (n\pi/a)^2$ or $\lambda_z = k_0^2 - (n\pi/a)^2$, are

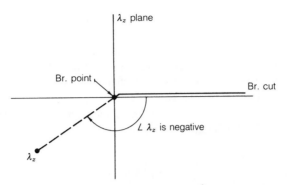

Fig. 2.11. Branch cut for $G_z(z, z', \lambda_z)$.

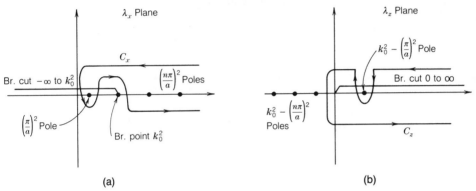

(a)   (b)

Fig. 2.12. Proper choice of the contours C_x and C_z.

excluded [Fig. 2.12(b)]. If only the TE_{10} mode propagates, then $k_0 > \pi/a$ but $k_0 < n\pi/a$ for $n > 1$. The contour C_z can be deformed to cross the branch cut and thus exclude the pole at $\lambda_z = k_0^2 - (\pi/2)^2$ without changing the value of the integral, as long as the branch point is enclosed.

Either of the two prescriptions given by (59) may be used for determining G. If we use the first one then the integral may be evaluated in terms of the residues at the poles. We have

$$G = -\frac{1}{2\pi j} \oint_{C_x} \frac{je^{-j\sqrt{k_0^2 - \lambda_x}|z-z'|}}{a\sqrt{k_0^2 - \lambda_x}} \sum_{n=1}^{\infty} \frac{\sin \dfrac{n\pi x}{a} \sin \dfrac{n\pi x'}{a}}{\lambda_x - (n\pi/a)^2} d\lambda_x$$

$$= \frac{1}{a}\sum_{n=1}^{\infty} \frac{1}{\Gamma_n} \sin \frac{n\pi x}{a} \sin \frac{n\pi x'}{a} e^{-\Gamma_n|z-z'|} \qquad (60)$$

where $\Gamma_n = \sqrt{(n\pi/a)^2 - k_0^2}$ and Imag. $\Gamma_n > 0$. If $k_0 < n\pi/a$ for $n > 1$, then the solution consists of the propagating dominant mode for which $\Gamma_1 = j\sqrt{k_0^2 - \pi^2/a^2}$ plus an infinite number of evanescent or nonpropagating TE_{n0} modes. These modes exist on both sides of the source point at z'. If $x' = a/2$, then only the modes for $n = 1, 3, 5, \ldots$, etc. are excited.

If the prescription involving the contour C_z in (59) is used, then the solution is expressed simply as a branch cut integral which may be interpreted as a continuous sum (integral) over the continuous spectrum of G_z in place of the sum over the discrete spectrum of G_x as exhibited in (60).

We could equally well have chosen the branch Imag. $\sqrt{\lambda_z} > 0$ and would have arrived at the same solution. The reader may find it of interest to make this choice and to show that (60) is still obtained for the final result. If we make the change of variable $w^2 = k_0^2 - \lambda_x$ the branch cut integral in (60) becomes an inverse Fourier transform.

At this point it will be instructive to show the relationship of this problem to the concepts introduced in Section 2.5. The equation of interest is

$$\left(\frac{\partial^2}{\partial x^2} + \frac{\partial^2}{\partial z^2} + k_0^2\right)\psi(x, z) = -f(x, z) \tag{61}$$

with $\psi = 0$ at $x = 0, a$; ψ bounded at $z = \pm\infty$; and $f(x, z)$ being a source function. We wish to solve this equation using the eigenfunctions of

$$\left(\frac{\partial^2}{\partial x^2} + \frac{\partial^2}{\partial z^2} + \lambda\right)\psi = 0.$$

Since $-\infty \leq z \leq \infty$ we first Fourier transform this equation with respect to z to obtain

$$\left(\frac{\partial^2}{\partial x^2} + \lambda - w^2\right)\int_{-\infty}^{\infty}\psi(x, z)e^{jwz}\,dz = 0$$

or

$$\left(\frac{\partial^2}{\partial x^2} + \lambda - w^2\right)\hat{\psi}(x, w) = 0.$$

For the x dependence we choose $\sin n\pi x/a$ and we then find that a solution exists if

$$\lambda = w^2 + \left(\frac{n\pi}{a}\right)^2. \tag{62}$$

Corresponding to this eigenvalue is the unnormalized eigenfunction

$$\psi_{nw}(x, z) = e^{-jwz}\sin\frac{n\pi x}{a}.$$

The normalization factor is obtained from the integral

$$\int_{-\infty}^{\infty}\int_0^a \psi_{nw}(x, z)\psi_{nw'}^*(x, z)\,dx\,dz = \frac{a}{2}\int_{-\infty}^{\infty}e^{-j(w-w')z}\,dz$$

$$= 2\pi\frac{a}{2}\delta(w - w') \tag{63}$$

upon using the Fourier transform property

$$\int_{-\infty}^{\infty}e^{j\gamma x}\,d\gamma = 2\pi\delta(x).$$

This normalization procedure is analogous to setting the integral of the product of the one-dimensional eigenfunction $\sin(n\pi x/a)$ with itself equal to unity. The normalization factor and orthogonality properties are expressed through the integral

$$\int_{-\infty}^{\infty}\int_{0}^{a} \sin\frac{n\pi x}{a} \sin\frac{m\pi x}{a} e^{j(w-w')z}\,dx\,dz$$

$$= \pi a\delta_{nm}\delta(w-w') \tag{64}$$

where the Kronecker delta $\delta_{nm} = 0$ if $n \neq m$ and $\delta_{nn} = 1$. When a continuous spectrum is involved the normalization–orthogonality integral results in a Dirac delta function. The normalized eigenfunctions are seen to be

$$\frac{1}{\sqrt{\pi a}} \sin\frac{n\pi x}{a} e^{-jwz}. \tag{65}$$

We can expand the solution of (61) in terms of these as follows: Let

$$\psi(x,z) = \sum_{m=1}^{\infty}\int_{-\infty}^{\infty} C_m(w') \sin\frac{m\pi x}{a} e^{-jw'z}\frac{dw'}{\sqrt{\pi a}}$$

which is equivalent to solving (61) using a combination of Fourier series and a Fourier transform. The sum over the continuous spectrum of eigenfunctions is an integral over w'. We now substitute into (61) to obtain

$$\sum_{m=1}^{\infty}\int_{-\infty}^{\infty}\frac{1}{\sqrt{\pi a}} C_m(w')[k_0^2 - \lambda] \sin\frac{m\pi x}{a} e^{-jw'z}\,dw' = -f(x,z)$$

where $\lambda = w'^2 + (m\pi/a)^2$. In order to find $C_n(w)$ we multiply by the complex conjugated eigenfunction $(1/\sqrt{\pi a}) \sin(n\pi x/a)e^{jwz}$, integrate over x and z, and use (64) to obtain

$$\sum_{m=1}^{\infty}\int_{-\infty}^{\infty}(k_0^2 - \lambda)C_m(w')\delta_{nm}\delta(w-w')\,dw'$$

$$= (k_0^2 - \lambda)C_n(w)$$

$$= -\frac{1}{\sqrt{\pi a}}\int_{-\infty}^{\infty}\int_{0}^{a} f(x,z) \sin\frac{n\pi x}{a} e^{jwz}\,dx\,dz.$$

The expansion coefficients are now known and thus the solution for ψ is readily constructed:

$$\psi(x,z) = \sum_{n=1}^{\infty}\int_{-\infty}^{\infty}\frac{1}{\pi a(\lambda - k_0^2)} \sin\frac{n\pi x}{a} e^{-jwz}$$

$$\times \int_{-\infty}^{\infty}\int_{0}^{a} f(x',z') \sin\frac{n\pi x'}{a} e^{jwz'}\,dx'\,dz'\,dw. \tag{66}$$

We can rewrite this solution in a form that will enable us to identify the Green's function operator. We have

$$\psi(x, z) = \int_{-\infty}^{\infty} \int_0^a f(x', z')$$

$$\times \left[\sum_{n=1}^{\infty} \int_{-\infty}^{\infty} \frac{\sin \dfrac{n\pi x'}{a} \sin \dfrac{n\pi x}{a}}{\pi a(\lambda - k_0^2)} e^{-jw(z-z')} \, dw \right] dx' \, dz' \qquad (67)$$

which shows that

$$G(x', z'; x, z) = \sum_{n=1}^{\infty} \int_{-\infty}^{\infty} \frac{\sin \dfrac{n\pi x'}{a} \sin \dfrac{n\pi x}{a}}{\pi a(\lambda - k_0^2)} e^{-jw(z-z')} \, dw \qquad (68)$$

where $\lambda = w^2 + (n\pi/a)^2$. This Green's function is expressed as a complete eigenfunction expansion in terms of the eigenfunctions of the operator given after (61). Its use is according to the sequence of operations shown in (66) so as to generate the eigenfunction expansion of $\psi(x, z)$. If the integral over w in (68) is carried out, then the eigenfunction expansion of the Green's function gets converted over to the expansion in terms of the waveguide modes as given by (60). Note that the expansion of G in terms of modes is not the same as the eigenfunction expansion even though they are equivalent. The integrand in (68) has poles of order one at $w = \pm j\Gamma_n$ so the mode expansion is readily found from an evaluation of the integral in terms of the residues at the poles.

When the order of integration over x', z' and the integration over w and summation over n are interchanged from that shown in (66) the interchange must, in general, be justified with regard to its validity. When the integration over w leads to well-defined functions, which it does for the present problem, it can be carried out first.

The series that occurs in the mode expansion of G is convergent for all values of $x \neq x'$, even if $z = z'$ so that the exponential decay of the individual terms vanishes. For large values of n we have $\Gamma_n \approx n\pi/a$ so the behavior of G is governed by the dominant series

$$S = \sum_{n=1}^{\infty} \frac{1}{n\pi} \sin \frac{n\pi x}{a} \sin \frac{n\pi x'}{a} e^{-n\pi|z-z'|/a}$$

$$= \sum_{n=1}^{\infty} \frac{1}{2n\pi} \left[\cos \frac{n\pi(x - x')}{a} - \cos \frac{n\pi(x + x')}{a} \right] e^{-n\pi|z-z'|/a}$$

$$= \operatorname{Re} \sum_{n=1}^{\infty} \frac{1}{2n\pi} [e^{jn\pi(x-x')/a} - e^{jn\pi(x+x')/a}] e^{-n\pi|z-z'|/a}.$$

The sum of series of this type is given in the Mathematical Appendix. Thus we find that G behaves like the function S given by

$$S = -\frac{1}{4\pi} \ln \left[\frac{\cosh^2 \dfrac{\pi(z - z')}{2a} - \cos^2 \dfrac{\pi(x - x')}{2a}}{\cosh^2 \dfrac{\pi(z - z')}{2a} - \cos^2 \dfrac{\pi(x + x')}{2a}} \right]. \qquad (69)$$

When $z - z'$ and $x - x'$ are both very small we readily find that S has the limiting form

$$S \sim -\frac{1}{4\pi} \ln \left[\frac{(x-x')^2 + (z-z')^2}{(z-z')^2 + 2\left(\dfrac{2a}{\pi}\right)^2 \sin^2 \dfrac{\pi(x+x')}{2a}} \right] \tag{70}$$

which shows that G has the expected logarithmic singularity near the line source. For an infinitely long line source in free space the static vector potential is given by (I is the current in the line source)

$$A_y = -\frac{\mu_0 I}{4\pi} \ln [(x-x')^2 + (z-z')^2]$$

which is similar to the expression in (70).

In practice it is often useful to extract the dominant part (singular part) of the Green's function as a closed-form expression since the remainder is a rapidly converging series; i.e., we express G in the form

$$G = S + (G - S)$$

where $(G - S)$ converges rapidly because the terms in S become essentially equal to those in G for large n. The singular behavior of the Green's function is the same as that for static conditions ($k_0 = 0$).

We can use the eigenfunctions of the differential operator in (61) to also obtain a representation of the delta function operator $\delta(x - x')\delta(z - z')$ that is valid for functions in the domain of the operator. A function $h(x, z)$ in the domain of the operator can be expanded as follows:

$$h(x, z) = \sum_{n=1}^{\infty} \int_{-\infty}^{\infty} A_n(w) \frac{1}{\sqrt{\pi a}} \sin \frac{n\pi x}{a} e^{-jwz} \, dw.$$

By following the standard procedure we find that

$$\sqrt{\pi a} A_n(w) = \int_{-\infty}^{\infty} \int_0^a h(x', z') \sin \frac{n\pi x'}{a} e^{jwz'} \, dx' \, dz'$$

and hence

$$h(x, z) = \sum_{n=1}^{\infty} \int_{-\infty}^{\infty} \frac{1}{\pi a} \sin \frac{n\pi x}{a} e^{-jwz}$$

$$\times \int_{-\infty}^{\infty} \int_0^a h(x', z') \sin \frac{n\pi x'}{a} e^{jwz'} \, dx' \, dz' \, dw.$$

We now rewrite this last expression as

$$h(x, z) = \int_{-\infty}^{\infty} \int_0^a h(x', z')$$

$$\times \left[\sum_{n=1}^{\infty} \int_{-\infty}^{\infty} \frac{1}{\pi a} \sin \frac{n\pi x'}{a} \sin \frac{n\pi x}{a} e^{-jw(z-z')} \, dw \right] dx' \, dz' \tag{71}$$

from which we can identify the delta function operator as

$$\delta(x - x')\delta(z - z') = \sum_{n=1}^{\infty} \int_{-\infty}^{\infty} \frac{1}{\pi a} \sin \frac{n\pi x'}{a} \sin \frac{n\pi x}{a} e^{-jw(z-z')} \, dw. \tag{72}$$

This result can also be obtained by carrying out a formal expansion of the two-dimensional delta function in terms of a Fourier series in x and a Fourier integral in z.

From the results given above it can be correctly inferred that the Green's function operator and the delta function operator or unit source function can be represented in terms of the eigenfunctions of the differential operator in a multidimensional space. When one or more of the coordinate variables extends over an infinite interval (or semi-infinite interval) the related eigenvalue spectrum is a continuous one rather than a discrete one. For a continuous spectrum the sum over eigenfunctions becomes an integral and the normalization integral results in a delta function replacing the Kronecker delta δ_{nm}. In rectangular coordinates the continuous spectrum is described by a Fourier transform. In cylindrical and spherical coordinates we will find that the continuous spectrum is described in terms of the Hankel transforms.

2.8. THREE-DIMENSIONAL GREEN'S FUNCTIONS

The theory presented in the previous sections may be applied to three-dimensional problems also. We will illustrate the case of the scalar Helmholtz equation in rectangular coordinates. We have

$$(\nabla^2 + k_0^2)G = -\delta(x - x')\delta(y - y')\delta(z - z'). \tag{73}$$

The equation $(\nabla^2 + k_0^2)\psi = 0$ separates into three one-dimensional equations of the form $\partial^2 \psi_x / \partial x^2 + \lambda_x \psi_x = 0$, $\partial^2 \psi_y / \partial y^2 + \lambda_y \psi_y = 0$, $\partial^2 \psi_z / \partial z^2 + \lambda_z \psi_z = 0$ with

$$\lambda_x + \lambda_y + \lambda_z = k_0^2. \tag{74}$$

Consequently, the associated one-dimensional Green's function problems are

$$\frac{d^2 G_x}{dx^2} + \lambda_x G_x = -\delta(x - x') \tag{75a}$$

$$\frac{d^2 G_y}{dy^2} + \lambda_y G_y = -\delta(y - y') \tag{75b}$$

$$\frac{d^2 G_z}{dz^2} + \lambda_z G_z = -\delta(z - z') \tag{75c}$$

along with appropriate boundary conditions.

When (75a)–(75c) have been solved then the solution to (73) is given by

$$\begin{aligned}
G(x, y, z, x', y', z') &= \left(\frac{-1}{2\pi j}\right)^2 \oint_{C_x} \oint_{C_y} G_x(\lambda_x) G_y(\lambda_y) G_z(k_0^2 - \lambda_x - \lambda_y) \, d\lambda_x \, d\lambda_y \\
&= \left(\frac{-1}{2\pi j}\right)^2 \oint_{C_x} \int_{C_z} G_x(\lambda_x) G_y(k_0^2 - \lambda_x - \lambda_z) G_z(\lambda_z) \, d\lambda_x \, d\lambda_z \\
&= \left(\frac{-1}{2\pi j}\right)^2 \oint_{C_y} \oint_{C_z} G_x(k_0^2 - \lambda_y - \lambda_z) G_y(\lambda_y) G_z(\lambda_z) \, d\lambda_y \, d\lambda_z.
\end{aligned} \tag{76}$$

Note that there is an integration over two of the λ_i and that the third λ is expressed in terms of the two being integrated over by means of the condition (74) on the separation constants. The contours C_i enclose only the singularities of the G_i with $i = x, y, z$. The proof is readily developed by noting that

$$
(\nabla^2 + k_0^2)G = \left[\left(\frac{\partial^2}{\partial x^2} + \lambda_x\right) + \left(\frac{\partial^2}{\partial y^2} + \lambda_y\right) + \left(\frac{\partial^2}{\partial z^2} + k_0^2 - \lambda_x - \lambda_y\right)\right]G
$$

$$
= \left(\frac{-1}{2\pi j}\right)^2 \oint_{C_x}\oint_{C_y} [-G_y(\lambda_y)G_z(k_0^2 - \lambda_x - \lambda_y)\delta(x - x')
$$

$$
- G_x(\lambda_x)G_z(k_0^2 - \lambda_x - \lambda_y)\delta(y - y') - G_x(\lambda_x)G_y(\lambda_y)\delta(z - z')]\,d\lambda_x\,d\lambda_y.
$$

The integral around C_x of the first term vanishes while the integral around C_y of the second term vanishes because no singularities of G_z are included. The integral of the third term gives $-\delta(x - x')\delta(y - y')\delta(z - z')$ by application of Theorem I twice.

The above technique is applicable in other coordinate systems also. Some applications are given in papers by Marcuvitz [2.13] and Felsen [2.14]. The method can be applied in a very effective way to obtain the Green's function in the form of a very rapidly converging series for the problem of diffraction by a very large cylinder or sphere [2.13]–[2.15].

2.9. Green's Function as a Multiple-Reflected Wave Series

The Green's function can be constructed in terms of the free-space waves radiated by the source and with these waves undergoing multiple reflections at the boundaries. This technique is useful in a number of different problems for the purpose of generating a series representation of the Green's function that may be more rapidly converging. The method of images is a special case of the method of multiple-reflected waves. We will illustrate it by considering a uniform line source that is parallel to the z axis and located between two concentric cylinders of radii a and b as illustrated in Fig. 2.13. The mathematical statement of the problem is

$$
\frac{1}{r}\frac{\partial}{\partial r}r\frac{\partial \psi}{\partial r} + \frac{1}{r^2}\frac{\partial^2 \psi}{\partial \theta^2} + k_0^2\psi = -\frac{\delta(r - r')}{r'}\delta(\theta), \qquad \psi = 0 \text{ at } r = a, b. \tag{77}
$$

We can expand $\psi(r, \theta)$ into a Fourier series in θ; thus

$$
\psi(r, \theta) = \sum_{n=0}^{\infty} f_n(r)\cos n\theta. \tag{78}
$$

We now substitute this expansion into (77), multiply by $\cos n\theta$, integrate over θ from 0 to 2π, and then obtain

$$
\frac{1}{r}\frac{\partial}{\partial r}r\frac{\partial f_n}{\partial r} - \frac{n^2}{r^2}f_n + k_0^2 f_n = -\frac{\epsilon_{0n}}{2\pi r'}\delta(r - r') \tag{79}
$$

where $\epsilon_{0n} = 1$ if $n = 0$ and equals 2 for $n > 0$. In this equation the source function can be viewed as a uniform cylindrical sheet of current at r'. When $r \neq r'$ the homogeneous equation is Bessel's differential equation for which the solutions are Bessel functions of order n, i.e., $J_n(k_0 r)$ and $Y_n(k_0 r)$ or combinations of these functions in the form of Hankel functions

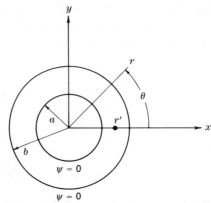

Fig. 2.13. A line source radiating between two cylinders.

$H_n^1(k_0r)$ and $H_n^2(k_0r)$ of the first and second kinds where

$$H_n^1(k_0r) = J_n(k_0r) + jY_n(k_0r)$$
$$H_n^2(k_0r) = J_n(k_0r) - jY_n(k_0r).$$

When the argument k_0r is large the asymptotic forms for the Hankel functions are

$$H_n^1(k_0r) \sim \sqrt{\frac{2}{\pi k_0 r}} e^{-j(2n+1)\pi/4} e^{jk_0r}$$

$$H_n^2(k_0r) \sim \sqrt{\frac{2}{\pi k_0 r}} e^{j(2n+1)\pi/4} e^{-jk_0r}.$$

These asymptotic forms show that the Hankel function of the first kind represents an inward-propagating cylindrical wave, while the Hankel function of the second kind represents an outward-propagating cylindrical wave. Bessel's differential equation is a Sturm–Liouville equation with $p(r) = r$, $q(r) = -n^2/r$, and $\sigma(r) = r$, and k_0^2 can be regarded as the eigenvalue parameter λ.

Initially we will ignore the boundaries at $r = a, b$. We can then choose

$$f_n(r) = \begin{cases} A_n H_n^1(k_0r), & r < r' \\ B_n H_n^2(k_0r), & r > r' \end{cases}$$

as the solution for $f_n(r)$. At $r = r'$ the solution must be continuous so

$$A_n H_n^1(k_0r') = B_n H_n^2(k_0r').$$

We can integrate the differential equation about a vanishingly small interval centered on r' to obtain

$$r \frac{df_n}{dr}\bigg|_{r'}^{r'_+} = -\frac{\epsilon_{0n}}{2\pi} = k_0r' \left(B_n \frac{dH_n^2}{dk_0r} - A_n \frac{dH_n^1}{dk_0r} \right)\bigg|_{r'}$$

which fixes the step change in the first derivative across the point r'. These two source-related boundary conditions allow us to find the amplitude coefficients A_n and B_n. The solutions for A_n and B_n involve the Wronskian determinant

$$W = H_n^2 \frac{dH_n^1}{dk_0 r} - H_n^1 \frac{dH_n^2}{dk_0 r}$$

evaluated at $r = r'$. We can use the property that $p(r)W(r)$ is a constant and then use the asymptotic expressions given earlier to evaluate the constant since $W(r')$ is readily found from the asymptotic forms. Since $p(r) = r$ we find that

$$W(r') = \frac{4j}{\pi k_0 r'}.$$

After solving for A_n and B_n we obtain

$$f_n(r) = -\frac{j\epsilon_{0n}}{8} H_n^1(k_0 r_<) H_n^2(k_0 r_>). \tag{80}$$

We now take the effect of the boundaries into account in the following way. The inward-propagating wave, apart from a multiplying factor, is $H_n^1(k_0 r)$ and is reflected at $r = a$ to produce an outward-propagating wave $H_n^2(k_0 r)$. At $r = a$ the reflected wave must cancel the incident wave so the reflection coefficient Γ_a is given by the condition

$$H_n^1(k_0 a) + \Gamma_a H_n^2(k_0 a) = 0.$$

Thus the reflected wave is

$$\Gamma_a H_n^2(k_0 r) = -\frac{H_n^1(k_0 a)}{H_n^2(k_0 a)} H_n^2(k_0 r).$$

This reflected outward-propagating wave is in turn reflected at $r = b$ and generates a new inward-propagating wave $\Gamma_b H_n^1(k_0 r)$ where

$$\Gamma_b = -H_n^2(k_0 b)/H_n^1(k_0 b).$$

This new wave is again reflected at $r = a$. The sum of all of the multiple-reflected waves generated by the primary inward-propagating wave is

$$\Gamma_a H_n^2(k_0 r) + \Gamma_a \Gamma_b H_n^1(k_0 r) + \Gamma_a^2 \Gamma_b H_n^2(k_0 r) + \Gamma_a^2 \Gamma_b^2 H_n^1(k_0 r) + \cdots$$

$$= \sum_{m=1}^{\infty} [\Gamma_a^m \Gamma_b^{m-1} H_n^2(k_0 r) + \Gamma_a^m \Gamma_b^m H_n^1(k_0 r)]. \tag{81}$$

The initially outward-propagating primary wave $H_n^2(k_0 r)$ is reflected at $r = b$ and subsequently undergoes multiple reflections thus creating the wave series

$$\Gamma_b H_n^1(k_0 r) + \Gamma_a \Gamma_b H_n^2(k_0 r) + \Gamma_a \Gamma_b^2 H_n^1(k_0 r) + \cdots$$

$$= \sum_{m=1}^{\infty} [\Gamma_b^m \Gamma_a^{m-1} H_n^1(k_0 r) + \Gamma_a^m \Gamma_b^m H_n^2(k_0 r)]. \tag{82}$$

The solution for the function $f_n(r)$ is a superposition of the primary solution given by (80) and the two wave series (81) and (82) multiplied respectively by the factors $-j\epsilon_{0n} H_n^2(k_0 r')/8$ and $-j\epsilon_{0n} H_n^1(k_0 r')/8$. Hence a solution for f_n is

$$f_n(r) = -\frac{j\epsilon_{0n}}{8} H_n^1(k_0 r_<) H_n^2(k_0 r_>)$$

$$-\frac{j\epsilon_{0n}}{8} \sum_{m=1}^{\infty} (\Gamma_a \Gamma_b)^{m-1} \{ [H_n^2(k_0 r) + \Gamma_b H_n^1(k_0 r)] \Gamma_a H_n^2(k_0 r') + [H_n^1(k_0 r) + \Gamma_a H_n^2(k_0 r)] \Gamma_b H_n^1(k_0 r') \}$$

We can use this series representation in (79) to complete the solution for the Green's function.

The method just described is quite general and can be applied to problems with other boundary conditions by using the appropriate expressions for the reflection coefficients. If the method is applied to the problem of a line source in a rectangular waveguide as analyzed earlier, the x-dependent part of the Green's function is the image series.

For the rectangular waveguide problem the Green's function can be expressed as an inverse Fourier transform in the form

$$G(x, x'; z, z') = \frac{1}{2\pi} \int_{-\infty}^{\infty} f(x, x', w) e^{-jw(z-z')} \, dw \tag{84}$$

where f is a solution of

$$\left(\frac{d^2}{dx^2} + \gamma^2 \right) f = -\delta(x - x') \tag{85}$$

with $\gamma^2 = k_0^2 - w^2$ and $f = 0$ at $x = 0, a$. One solution for this Green's function is given by (60) and also by (68). If we ignore the boundary conditions at $x = 0, a$, then the primary field that is a solution of (85) can be found using Method II and is

$$f = -\frac{j}{2\gamma} e^{-j\gamma|x-x'|}. \tag{86}$$

The primary field $e^{j\gamma(x-x')}$ incident on the reflecting boundary at $x = 0$ produces the multiple-reflected wave series

$$e^{-j\gamma x'} [\Gamma e^{-j\gamma x} + \Gamma \Gamma_a e^{j\gamma x} + \Gamma^2 \Gamma_a e^{-j\gamma x} + \Gamma^2 \Gamma_a^2 e^{j\gamma x} + \cdots]$$

$$= \sum_{n=1}^{\infty} (\Gamma^n \Gamma_a^{n-1} e^{-j\gamma x} + \Gamma^n \Gamma_a^n e^{j\gamma x}) e^{-j\gamma x'} \tag{87}$$

where $\Gamma = -1$ and $\Gamma_a = -e^{-2j\gamma a}$. In a similar way the primary wave $e^{-j\gamma(x-x')}$ incident on the boundary at $x = a$ produces the multiple-reflected wave series

$$e^{j\gamma x'} [\Gamma_a e^{j\gamma x} + \Gamma \Gamma_a e^{-j\gamma x} + \Gamma \Gamma_a^2 e^{j\gamma x} + \cdots]$$

$$= \sum_{n=1}^{\infty} (\Gamma_a^n \Gamma^{n-1} e^{j\gamma x} + \Gamma_a^n \Gamma^n e^{-j\gamma x}) e^{j\gamma x'}. \tag{88}$$

By combining the two wave series and the primary field we find, after simplification, that the solution for f is

$$f(x, x', w) = -\frac{j}{2\gamma}\left[e^{-j\gamma|x-x'|} + \sum_{n=-\infty}^{\infty}{}' e^{-j\gamma|x-x'-2na|} - \sum_{n=-\infty}^{\infty} e^{-j\gamma|x+x'+2na|}\right] \quad (89)$$

where the prime on the summation sign signifies that the $n = 0$ term is omitted. This series can be used in (84) to obtain the Green's function G. The series represents waves generated by the images of the line source.

The integral in (84) over w can be evaluated by using the Fourier transform relation

$$\int_{-\infty}^{\infty} \frac{e^{-jwu-j\gamma v}}{\gamma} \, dw = \pi H_0^2(k_0\sqrt{u^2+v^2}). \quad (90)$$

By using this result the Green's function can be expressed as a series of cylindrical waves radiated from the original line source and its images in the two side walls of the waveguide. This representation is

$$G = -\frac{j}{4}H_0^2(k_0\sqrt{(x-x')^2+(z-z')^2}) + \frac{j}{4}H_0^2(k_0\sqrt{(x+x')^2+(z-z')^2})$$

$$-\frac{j}{4}\sum_{n=-\infty}^{\infty}{}'[H_0^2(k_0\sqrt{(x-x'-2na)^2+(z-z')^2}) - H_0^2(k_0\sqrt{(x+x'+2na)^2+(z-z')^2})].$$
$$(91)$$

The expansion of G in terms of waveguide modes can be obtained by using the Poisson summation formula to transform the series in (91) into series in terms of the modes (see the Mathematical Appendix).

It should be apparent by now that there are many equivalent expressions for the Green's function for a given problem. Which form is best depends on the physical nature of the problem to be solved. As a general rule, if one series representation is very slowly converging an alternative series representation usually converges more rapidly. For example, in (91) only the first term is singular at $x = x'$, $z = z'$ and since $H_0^2 \sim -(j/\pi)\ln[(x-x')^2+(z-z')^2]$ for small arguments the singular part of the Green's function is known explicitly.

2.10. Free-Space Green's Dyadic Function

The vector potential $\mathbf{A}$ is a solution of the inhomogeneous vector Helmholtz equation

$$\nabla^2\mathbf{A} + k^2\mathbf{A} = -\mu\mathbf{J}. \quad (92)$$

In Section 1.9 it was shown that a particular solution is

$$\mathbf{A}(\mathbf{r}) = \frac{\mu}{4\pi}\iiint_V \mathbf{J}(\mathbf{r}_1)\frac{e^{-jk|\mathbf{r}-\mathbf{r}_1|}}{|\mathbf{r}-\mathbf{r}_1|}\,dV_1. \quad (93)$$

If $\mu\mathbf{J}$ is replaced by a unit vector source $(\mathbf{a}_x+\mathbf{a}_y+\mathbf{a}_z)\delta(\mathbf{r}_1-\mathbf{r}_0)$, we find at once by substituting into (93) that the solution is

$$(\mathbf{a}_x + \mathbf{a}_y + \mathbf{a}_z)\frac{e^{-jk|\mathbf{r}-\mathbf{r}_0|}}{4\pi|\mathbf{r}-\mathbf{r}_0|} \quad (94)$$

and this may be considered as a vector Green's function. For a general current distribution $\mathbf{J}$, the solution would be obtained by a superposition integral provided we make the rule that the x component of the current J_x is to be associated with the unit vector $\mathbf{a}_x$ in the Green's function, and similarly for J_y and J_z. This rule is readily taken account of by introducing the dyadic Green's function, which is defined as a solution of the equation

$$\nabla^2 \bar{\mathbf{G}}(\mathbf{r}, \mathbf{r}_0) + k^2 \bar{\mathbf{G}} = -\bar{\mathbf{I}}\delta(\mathbf{r} - \mathbf{r}_0) \tag{95}$$

where $\bar{\mathbf{I}}$ is the unit dyadic or idemfactor $\mathbf{a}_x\mathbf{a}_x + \mathbf{a}_y\mathbf{a}_y + \mathbf{a}_z\mathbf{a}_z$. Since multiplying the source by $\bar{\mathbf{I}}$ simply multiplies the solution by $\bar{\mathbf{I}}$ also, we have the result

$$\bar{\mathbf{G}}(\mathbf{r}, \mathbf{r}_0) = \bar{\mathbf{I}}\frac{e^{-jk|\mathbf{r}-\mathbf{r}_0|}}{4\pi|\mathbf{r} - \mathbf{r}_0|}. \tag{96}$$

Each component of the current gives rise to a vector field, and hence the general linear relation between the current and the field is a dyadic relation. Therefore, for vector fields the Green's function is a dyadic quantity. Equation (96) is the free-space dyadic Green's function for the vector Helmholtz equation. For an arbitrary current distribution the solution for the vector potential is

$$\mathbf{A}(\mathbf{r}) = \mu\iiint\limits_V \bar{\mathbf{G}}(\mathbf{r}, \mathbf{r}_0)\cdot\mathbf{J}(\mathbf{r}_0)\,dV_0 \tag{97}$$

since the scalar product of $\mathbf{J}$ with $\bar{\mathbf{G}}$ associates J_x with $\mathbf{a}_x$, etc. The above dyadic Green's function is a symmetrical dyadic so that $\mathbf{J}\cdot\bar{\mathbf{G}} = \bar{\mathbf{G}}\cdot\mathbf{J}$. It is also symmetrical in the variables $\mathbf{r}$ and $\mathbf{r}_0$. The use of dyadic Green's functions for the vector Helmholtz equation results in a considerable advantage in notation.

2.11. MODIFIED DYADIC GREEN'S FUNCTIONS

In the presence of boundaries on which the electromagnetic field must satisfy prescribed boundary conditions, the free-space dyadic Green's function suffers from the same shortcomings that the free-space Green's function for Poisson's equation does in the solution of electrostatic-potential boundary-value problems. The dyadic Green's functions are, therefore, modified so as to facilitate the solution of the given boundary-value problem. The sort of modification we should make is best determined by an examination of the type of solution we get by using the free-space dyadic Green's function.

Consider a region of space V bounded by a closed surface S, within which a current distribution $\mathbf{J}$ exists. Let $\mathbf{n}$ be the unit inward normal to S as illustrated in Fig. 2.14. Consider the following dyadic quantities:

$$\nabla_0\cdot[\mathbf{A} \times (\nabla_0 \times \bar{\mathbf{B}})] \tag{98a}$$

$$\nabla_0\cdot[(\nabla_0 \times \mathbf{A}) \times \bar{\mathbf{B}}] \tag{98b}$$

$$\nabla_0\cdot[(\nabla_0\cdot\mathbf{A})\bar{\mathbf{B}}] \tag{98c}$$

$$\nabla_0\cdot[\mathbf{A}(\nabla_0\cdot\bar{\mathbf{B}})] \tag{98d}$$

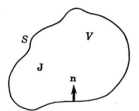

Fig. 2.14. Region V with source function $\mathbf{J}$.

where $\bar{\mathbf{B}}$ is a symmetric dyadic. By expanding the above expressions, we shall be able to obtain a combined vector and dyadic equivalent of Green's second identity. The dyadic $\bar{\mathbf{B}}$ will be replaced by the dyadic Green's function later on. Although $\bar{\mathbf{B}}$ is a symmetric dyadic, $\nabla_0 \times \bar{\mathbf{B}}$ is not symmetric, and special care must be taken in expanding the above expressions, and also in the order in which the terms are written. Perhaps the easiest way to proceed is to write the dyadic $\bar{\mathbf{B}}$ as the sum of three vectors associated with the unit vectors $\mathbf{a}_x$, $\mathbf{a}_y$, $\mathbf{a}_z$, that is,[3]

$$\bar{\mathbf{B}} = \mathbf{B}_1 \mathbf{a}_x + \mathbf{B}_2 \mathbf{a}_y + \mathbf{B}_3 \mathbf{a}_z.$$

Thus, for (98a) we have

$$\nabla_0 \cdot [\mathbf{A} \times (\nabla_0 \times \bar{\mathbf{B}})] = \nabla_0 \cdot (\mathbf{A} \times \nabla_0 \times \mathbf{B}_1) \mathbf{a}_x$$
$$+ \nabla_0 \cdot (\mathbf{A} \times \nabla_0 \times \mathbf{B}_2) \mathbf{a}_y + \nabla_0 \cdot (\mathbf{A} \times \nabla_0 \times \mathbf{B}_3) \mathbf{a}_z.$$

Each term may now be expanded according to the rules of vector calculus to obtain

$$\nabla_0 \cdot (\mathbf{A} \times \nabla_0 \times \mathbf{B}_1) = (\nabla_0 \times \mathbf{A}) \cdot (\nabla_0 \times \mathbf{B}_1) - (\nabla_0 \times \nabla_0 \times \mathbf{B}_1) \cdot \mathbf{A}$$
$$= (\nabla_0 \times \mathbf{A}) \cdot (\nabla_0 \times \mathbf{B}_1) - \mathbf{A} \cdot \nabla_0 \nabla_0 \cdot \mathbf{B}_1 + \mathbf{A} \cdot \nabla_0^2 \mathbf{B}_1 \qquad (99)$$

and similarly for the other two terms. By arranging the terms in the expansion so that the vector $\mathbf{B}_1$ always appears on the right-hand side of each term, we may again associate $\mathbf{B}_1$ with $\mathbf{a}_x$, and hence we can combine the end results for all three associated vectors into a dyadic again. Thus the expansion of (98a) gives

$$\nabla_0 \cdot [\mathbf{A} \times (\nabla_0 \times \bar{\mathbf{B}})] = (\nabla_0 \times \mathbf{A}) \cdot (\nabla_0 \times \bar{\mathbf{B}}) - \mathbf{A} \cdot (\nabla_0 \nabla_0 \cdot \bar{\mathbf{B}}) + \mathbf{A} \cdot (\nabla_0^2 \bar{\mathbf{B}}). \qquad (100a)$$

Expanding (98b)–(98d) in a similar fashion, we obtain

$$\nabla_0 \cdot [(\nabla_0 \times \mathbf{A}) \times \bar{\mathbf{B}}] = (\nabla_0 \nabla_0 \cdot \mathbf{A}) \cdot \bar{\mathbf{B}} - (\nabla_0^2 \mathbf{A}) \cdot \bar{\mathbf{B}} - (\nabla_0 \times \mathbf{A}) \cdot (\nabla_0 \times \bar{\mathbf{B}}) \qquad (100b)$$

$$\nabla_0 \cdot [(\nabla_0 \cdot \mathbf{A}) \bar{\mathbf{B}}] = (\nabla_0 \cdot \mathbf{A})(\nabla_0 \cdot \bar{\mathbf{B}}) + (\nabla_0 \nabla_0 \cdot \mathbf{A}) \cdot \bar{\mathbf{B}} \qquad (100c)$$

$$\nabla_0 \cdot [\mathbf{A}(\nabla_0 \cdot \bar{\mathbf{B}})] = (\nabla_0 \cdot \mathbf{A})(\nabla_0 \cdot \bar{\mathbf{B}}) + \mathbf{A} \cdot (\nabla_0 \nabla_0 \cdot \bar{\mathbf{B}}). \qquad (100d)$$

Adding (100a), (100b), and (100d) and subtracting (100c) from the sum gives

$$\nabla_0 \cdot [\mathbf{A} \times (\nabla_0 \times \bar{\mathbf{B}}) + (\nabla_0 \times \mathbf{A}) \times \bar{\mathbf{B}} + \mathbf{A}(\nabla_0 \cdot \bar{\mathbf{B}}) - (\nabla_0 \cdot \mathbf{A}) \bar{\mathbf{B}}]$$
$$= \mathbf{A} \cdot (\nabla_0^2 \bar{\mathbf{B}}) - (\nabla_0^2 \mathbf{A}) \cdot \bar{\mathbf{B}} \qquad (101)$$

---

[3] Section A.2.

which may now be used in the divergence theorem to give us the equivalent of Green's second identity. Integrating (101) over the volume V and converting the volume integral of the divergence to a surface integral gives

$$\iiint_V [\mathbf{A} \cdot \nabla_0^2 \bar{\mathbf{B}} - (\nabla_0^2 \mathbf{A}) \cdot \bar{\mathbf{B}}] \, dV_0$$

$$= -\oiint_S \mathbf{n} \cdot [\mathbf{A} \times (\nabla_0 \times \bar{\mathbf{B}})$$

$$+ (\nabla_0 \times \mathbf{A}) \times \bar{\mathbf{B}} + \mathbf{A}(\nabla_0 \cdot \bar{\mathbf{B}}) - (\nabla_0 \cdot \mathbf{A})\bar{\mathbf{B}}] \, dS_0 \qquad (102a)$$

when $\mathbf{n}$ is the unit inward normal to S. A term such as $\nabla_0^2 \mathbf{A} \cdot \bar{\mathbf{B}}$ may be written as $\bar{\mathbf{B}} \cdot \nabla_0^2 \mathbf{A}$ since $\bar{\mathbf{B}}$ is a symmetric dyadic. The first term in the surface integral in (102a) may be written as

$$(\mathbf{n} \times \mathbf{A}) \cdot (\nabla_0 \times \bar{\mathbf{B}}),$$

the second term as

$$[\mathbf{n} \times (\nabla_0 \times \mathbf{A})] \cdot \bar{\mathbf{B}},$$

and the third and fourth terms as

$$\mathbf{n} \cdot \mathbf{A} \nabla_0 \cdot \bar{\mathbf{B}} \qquad \mathbf{n} \cdot \bar{\mathbf{B}} \nabla_0 \cdot \mathbf{A}.$$

Hence, we obtain the following final result, which is the basic starting point for the application of dyadic Green's functions to the solution of the vector Helmholtz equation:

$$\iiint_V [(\nabla_0^2 \mathbf{A}) \cdot \bar{\mathbf{B}} - \mathbf{A} \cdot \nabla_0^2 \bar{\mathbf{B}}] \, dV_0$$

$$= \oiint_S [(\mathbf{n} \times \mathbf{A}) \cdot (\nabla_0 \times \bar{\mathbf{B}})$$

$$+ (\mathbf{n} \times \nabla_0 \times \mathbf{A}) \cdot \bar{\mathbf{B}} + \mathbf{n} \cdot \mathbf{A} \nabla_0 \cdot \bar{\mathbf{B}} - \mathbf{n} \cdot \bar{\mathbf{B}} \nabla_0 \cdot \mathbf{A}] \, dS_0. \qquad (102b)$$

This result is also valid when $\bar{\mathbf{B}}$ is not a symmetric dyadic.

If the dyadic $\bar{\mathbf{B}}$ is now replaced by a dyadic Green's function which is a solution of (95) and $\mathbf{A}$ is the vector potential satisfying (92), then $\nabla_0^2 \bar{\mathbf{B}}$ and $\nabla_0^2 \mathbf{A}$ may be replaced by $-k^2 \bar{\mathbf{B}} - \bar{\mathbf{I}} \delta(\mathbf{r} - \mathbf{r}_0)$ and $-k^2 \mathbf{A} - \mu \mathbf{J}$, respectively, in (102b). The terms in k^2 cancel, and, by using the property of the three-dimensional delta function, we obtain the following solution for the vector potential $\mathbf{A}$:

$$\mathbf{A}(\mathbf{r}) = \iiint_V \mu \mathbf{J}(\mathbf{r}_0) \cdot \bar{\mathbf{G}}(\mathbf{r}_0, \mathbf{r}) \, dV_0 + \oiint_S [(\mathbf{n} \times \mathbf{A}) \cdot (\nabla_0 \times \bar{\mathbf{G}})$$

$$+ (\mathbf{n} \times \nabla_0 \times \mathbf{A}) \cdot \bar{\mathbf{G}} + \mathbf{n} \cdot \mathbf{A} \nabla_0 \cdot \bar{\mathbf{G}} - \mathbf{n} \cdot \bar{\mathbf{G}} \nabla_0 \cdot \mathbf{A}] \, dS_0. \qquad (103)$$

The electric field is given by

$$\mathbf{E} = -j\omega\mathbf{A} + \frac{\nabla\nabla\cdot\mathbf{A}}{j\omega\epsilon\mu} = -j\omega\mathbf{A} - \nabla\Phi \tag{104a}$$

while the magnetic field is given by

$$\mu\mathbf{H} = \nabla \times \mathbf{A}. \tag{104b}$$

The scalar potential Φ is related to the vector potential $\mathbf{A}$ by the Lorentz condition

$$-j\omega\epsilon\mu\Phi = \nabla\cdot\mathbf{A}. \tag{105}$$

If the boundary conditions specify that the tangential component of $\mathbf{E}$ vanishes on S, then $\mathbf{n} \times \mathbf{E} = 0$ on S, and hence from (104a) we see that we must have $\mathbf{n} \times \mathbf{A} = 0$ on S, $\mathbf{n} \times \nabla\Phi = 0$ on S. The condition $\mathbf{n} \times \nabla\Phi = 0$ on S means that the derivatives of Φ on the surface S are zero, and hence Φ is constant, which may be taken as zero, on the surface S. If Φ vanishes on S, then $\nabla_0\cdot\mathbf{A} = 0$ on S also by virtue of (105).

An examination of (103) shows that in this case the first and last terms in the surface integral vanish. If we are able to modify our Green's function so that $\mathbf{n} \times \bar{\mathbf{G}}$ and $\nabla_0\cdot\bar{\mathbf{G}}$ vanish on the surface S, then the whole surface integral vanishes, and the solution for $\mathbf{A}$ is given by the volume integral of $\mu\mathbf{J}\cdot\bar{\mathbf{G}}$ only.

Sometimes it is more convenient to deal directly with the equation

$$\nabla \times \nabla \times \mathbf{E} - k^2\mathbf{E} = -j\omega\mu\mathbf{J} \tag{106}$$

which the electric field $\mathbf{E}$ satisfies. In this case the Green's function is taken as a solution of the corresponding equation

$$\nabla \times \nabla \times \bar{\mathbf{G}} - k^2\bar{\mathbf{G}} = \bar{\mathbf{I}}\delta(\mathbf{r} - \mathbf{r}_0). \tag{107}$$

The appropriate Green's identity to use in this case is obtained by adding (100a) and (100b) and converting the volume integral of the divergence to a surface integral again to get

$$\iiint\limits_V [(\nabla_0 \times \nabla_0 \times \mathbf{A})\cdot\bar{\mathbf{B}} - \mathbf{A}\cdot(\nabla_0 \times \nabla_0 \times \bar{\mathbf{B}})]\,dV_0$$

$$= -\oiint\limits_S \{[\mathbf{n} \times (\nabla_0 \times \mathbf{A})]\cdot\bar{\mathbf{B}} + (\mathbf{n} \times \mathbf{A})\cdot(\nabla_0 \times \bar{\mathbf{B}})\}\,dS_0. \tag{108}$$

Replacing $\mathbf{A}$ by $\mathbf{E}$ and $\bar{\mathbf{B}}$ by $\bar{\mathbf{G}}$ and using (106) and (107), the following solution for $\mathbf{E}$ is obtained:

$$\mathbf{E}(\mathbf{r}) = -j\omega\mu \iiint\limits_V \mathbf{J}(\mathbf{r}_0)\cdot\bar{\mathbf{G}}(\mathbf{r}_0, \mathbf{r})\,dV_0$$

$$+ \oiint\limits_S [(\mathbf{n} \times \nabla_0 \times \mathbf{E})\cdot\bar{\mathbf{G}} + (\mathbf{n} \times \mathbf{E})\cdot\nabla_0 \times \bar{\mathbf{G}}]\,dS_0. \tag{109}$$

Since $\nabla \times \mathbf{E} = -j\omega\mu\mathbf{H}$, the first term in the surface integral may be interpreted as a contribution from a surface current of density $\mathbf{n} \times \mathbf{H}$ on S. The second term may be considered a contribution from a magnetic current sheet of density $\mathbf{E} \times \mathbf{n}$. If the tangential components of $\mathbf{E}$, that is, $\mathbf{n} \times \mathbf{E}$, are known or specified on S, then, if possible, we should modify our Green's function so that $\mathbf{n} \times \bar{\mathbf{G}}$ vanishes on S. Thus only the second term, the one which we know, contributes to the surface integral. In practice, the particular boundary values we are given dictate the choice of which Green's function to use.

2.12. Solution for Electric Field Dyadic Green's Function

In this section we will derive a solution for the electric field dyadic Green's function in free space. This Green's function is a solution of (107) and will now be represented by the symbol $\bar{\mathbf{G}}_e$ to emphasize that it is the dyadic Green's function for (106). One way to obtain the solution is to make use of the relationship between the electric field and the vector potential in the Lorentz gauge. Since

$$\mathbf{E} = -j\omega\mathbf{A} - j(\omega/k_0^2)\nabla\nabla\cdot\mathbf{A}$$

we readily see that for a current element $\mathbf{J}(\mathbf{r}_0)$ and using the vector potential Green's dyadic given by (96) that

$$\begin{aligned}
\mathbf{E}(\mathbf{r}) &= -j\omega\mu_0\mathbf{J}(\mathbf{r}_0)\cdot\bar{\mathbf{G}}_e(\mathbf{r}_0, \mathbf{r}) \\
&= -j\omega\mu_0\left[\mathbf{J}(\mathbf{r}_0)\cdot\bar{\mathbf{G}}(\mathbf{r}_0, \mathbf{r}) + \frac{1}{k_0^2}\nabla\nabla\cdot\bar{\mathbf{G}}(\mathbf{r}, \mathbf{r}_0)\cdot\mathbf{J}(\mathbf{r}_0)\right] \\
&= -j\omega\mu_0\mathbf{J}(\mathbf{r}_0)\cdot\left[\bar{\mathbf{I}}\frac{e^{-jk_0R}}{4\pi R} + \frac{1}{k_0^2}\nabla\nabla\frac{e^{-jk_0R}}{4\pi R}\right]
\end{aligned}$$

where $R = |\mathbf{r} - \mathbf{r}_0|$. Note that the operation $\nabla\nabla\cdot\bar{\mathbf{G}}$ is equal to $\nabla\nabla\cdot\bar{\mathbf{I}} = \nabla\nabla$ operating on the scalar Green's function. From the above relation we find that

$$\bar{\mathbf{G}}_e(\mathbf{r}_0, \mathbf{r}) = \left(\bar{\mathbf{I}} + \frac{1}{k_0^2}\nabla\nabla\right)\frac{e^{-jk_0R}}{4\pi R}. \tag{110}$$

Another way to derive (110) is a generalization of the method used in Section 1.9 to obtain a solution for the vector potential. The scalar Green's function $g(\mathbf{r}_0, \mathbf{r}) = e^{-jk_0R}/4\pi R$ is a solution of the equation

$$(\nabla^2 + k_0^2)g = -\delta(\mathbf{r} - \mathbf{r}_0) \tag{111}$$

so the left-hand side can be viewed as a representation for the delta function. The divergence of (107) gives $-k_0^2\nabla\cdot\bar{\mathbf{G}}_e = \nabla\cdot\bar{\mathbf{I}}\delta(\mathbf{r} - \mathbf{r}_0) = \nabla\delta(\mathbf{r} - \mathbf{r}_0)$. If we expand the $\nabla \times \nabla \times$ operator and use the above result we obtain

$$\nabla\nabla\cdot\bar{\mathbf{G}}_e - \nabla^2\bar{\mathbf{G}}_e - k_0^2\bar{\mathbf{G}}_e = \bar{\mathbf{I}}\delta(\mathbf{r} - \mathbf{r}_0)$$

or equivalently

$$(\nabla^2 + k_0^2)\bar{\mathbf{G}}_e = -\left(\bar{\mathbf{I}} + \frac{1}{k_0^2}\nabla\nabla\right)\delta(\mathbf{r} - \mathbf{r}_0).$$

We now use (111) to replace the unit source function; thus

$$(\nabla^2 + k_0^2)\bar{\mathbf{G}}_e = \left(\bar{\mathbf{I}} + \frac{1}{k_0^2}\nabla\nabla\right)(\nabla^2 + k_0^2)g$$

or

$$(\nabla^2 + k_0^2)\left[\bar{\mathbf{G}}_e - \left(\bar{\mathbf{I}} + \frac{1}{k_0^2}\nabla\nabla\right)g\right] = 0.$$

A particular solution to this equation is

$$\bar{\mathbf{G}}_e = \left(\bar{\mathbf{I}} + \frac{1}{k_0^2}\nabla\nabla\right)g. \tag{112}$$

If necessary, solutions to the homogeneous equation can be added to take care of boundary conditions. In free space $\bar{\mathbf{G}}_e$ and g both satisfy the radiation condition at infinity; i.e., they represent outward-propagating waves so only the particular solution is needed. Hence we arrive at the solution given earlier by (110).

The above method of derivation was first used by Levine and Schwinger [2.16]. We can apply it also to the case where g is represented in terms of cylindrical waves instead of spherical waves. Consider a nonuniform line source $I(z)$ located along the z axis and the scalar field which is a solution of

$$\frac{1}{r}\frac{\partial}{\partial r}r\frac{\partial\psi}{\partial r} + \frac{\partial^2\psi}{\partial z^2} + k_0^2\psi = -\frac{I(z)\delta(r)}{2\pi r} \tag{113}$$

and represents outward-propagating waves. By means of a Fourier transform with respect to z we obtain ($\gamma^2 = k_0^2 - w^2$)

$$\frac{1}{r}\frac{\partial}{\partial r}r\frac{\partial\hat{\psi}(r, w)}{\partial r} + \gamma^2\hat{\psi}(r, w) = -\frac{\delta(r)}{2\pi r}\hat{I}(w).$$

For $r > 0$ the solution to Bessel's equation that represents outward-propagating cylindrical waves is $\hat{\psi}(r, w) = CH_0^2(\gamma r)$ where C is a constant to be determined. If we apply the divergence theorem in two-dimensional form to the equation

$$(\nabla_t^2 + \gamma^2)\hat{\psi} = -\hat{I}\frac{\delta(r)}{2\pi r}$$

we obtain, for a circular disk of radius a,

$$\int_0^a \int_0^{2\pi} (\nabla_t \cdot \nabla_t\hat{\psi} + \gamma^2\hat{\psi})r\,d\theta\,dr$$

$$= -\frac{\hat{I}}{2\pi}\int_0^a \int_0^{2\pi} \delta(r)\,d\theta\,dr$$

$$= \int_0^{2\pi} \nabla_t\hat{\psi}\cdot\mathbf{a}_r a\,d\theta + \int_0^a \int_0^{2\pi} \gamma^2\hat{\psi}r\,d\theta\,dr = -\hat{I}.$$

When we make the radius a vanishingly small, the integral of $\gamma^2\hat{\psi}$ vanishes since for r very small $H_0^2(\gamma r) \sim -j(2/\pi)\ln r$. We also can use $\nabla_t CH_0^2(\gamma r) = -j(2/\pi)C\mathbf{a}_r/r$ for r very small and we then obtain $j4C = \hat{I}$. The solution for ψ is now seen to be

$$\psi = \frac{1}{2\pi}\int_{-\infty}^{\infty} -\frac{j\hat{I}(w)}{4}H_0^2(\gamma r)e^{-jwz}\,dw$$

$$= \frac{1}{2\pi}\int_{-\infty}^{\infty} I(z')\left[\int_{-\infty}^{\infty}\frac{-j}{4}H_0^2(r\sqrt{k_0^2-w^2})e^{-jw(z-z')}\,dw\right]dz'$$

from which we can identify the scalar Green's function, expressed in cylindrical coordinates, as

$$g = \frac{1}{2\pi}\int_{-\infty}^{\infty}\frac{-j}{4}H_0^2(r\sqrt{k_0^2-w^2})e^{-jw(z-z')}\,dw.$$

If the unit source is located at x', y' then we replace $r = (x^2+y^2)^{1/2}$ by $[(x-x')^2+(y-y')^2]^{1/2}$ to obtain

$$g(\mathbf{r}, z|\mathbf{r}', z') = \frac{-j}{8\pi}\int_{-\infty}^{\infty} H_0^2(|\mathbf{r}-\mathbf{r}'|\sqrt{k_0^2-w^2})e^{-jw(z-z')}\,dw \qquad (114)$$

where $\mathbf{r} = x\mathbf{a}_x + y\mathbf{a}_y$ and $\mathbf{r}' = x'\mathbf{a}_x + y'\mathbf{a}_y$. We can apply the operation shown in (112) to this function so as to obtain the dyadic Green's function for the electric field as a function of cylindrical coordinates.

The solutions for the dyadic Green's functions are the solutions outside the unit dyadic point source. These solutions are the limit of the solutions for a uniform-density source contained, for example, within a small spherical volume of radius a as the volume shrinks to zero but the total source strength is kept equal to unity. The limiting procedure is that described in Section 1.9. It is important to keep this limiting aspect in mind in order to obtain a correct interpretation of the electric field solution given by (109).

For currents contained in a finite volume V in free space the surface integral in (109) is taken over a sphere of infinite radius. When $\mathbf{r}$ is much larger than all $\mathbf{r}_0$ in V the asymptotic form of the radiated electric field is

$$\mathbf{E}(\mathbf{r}) \sim \mathbf{f}(\theta, \phi)\frac{e^{-jk_0r}}{4\pi r}$$

where $\mathbf{f}$ is a vector function of the polar angle θ and azimuth angle ϕ and has $\mathbf{a}_\theta$ and $\mathbf{a}_\phi$ components only, i.e., is an outward-propagating spherical TEM wave. We can approximate $R = |\mathbf{r} - \mathbf{r}_0|$ by $r - \mathbf{a}_r\cdot\mathbf{r}_0$ upon using the binomial expansion and we then find that the asymptotic form for the dyadic Green's function is

$$\bar{G}_e \sim (\mathbf{a}_\theta\mathbf{a}_\theta + \mathbf{a}_\phi\mathbf{a}_\phi)\frac{e^{-jk_0r+jk_0\mathbf{a}_r\cdot\mathbf{r}_0}}{4\pi r}$$

at infinity. For an outward-propagating spherical wave at infinity the magnetic field is related to the electric field by the equation $Z_0\mathbf{H} = \mathbf{a}_r \times \mathbf{E}$. Since $\nabla \times \mathbf{E} = -j\omega\mu_0\mathbf{H} = -jk_0Z_0\mathbf{H}$ the radiation condition at infinity can be stated in the form

$$\lim_{r\to\infty} r(\nabla \times \mathbf{E} + jk_0\mathbf{a}_r \times \mathbf{E}) = 0 \qquad (115a)$$

and for the Green's dyadic function

$$\lim_{r \to \infty} r(\nabla \times \bar{\mathbf{G}}_e + jk_0 \mathbf{a}_r \times \bar{\mathbf{G}}_e) = 0. \tag{115b}$$

These relations are valid in any isotropic medium as well provided k_0 is replaced by k. With these radiation conditions the reader can readily verify that to terms of order $1/r^2$ the integrand in the surface integral in (109) vanishes. Hence in free space the solution for the electric field is given by

$$\mathbf{E}(\mathbf{r}) = -j\omega\mu_0 \iiint\limits_V \mathbf{J}(\mathbf{r}_0) \cdot \bar{\mathbf{G}}_e(\mathbf{r}_0, \mathbf{r}) \, dV_0. \tag{116}$$

If the asymptotic form (115) is used for $\bar{\mathbf{G}}_e$ the electric field clearly has the form given earlier for $r \gg r_0$.

For observation points $\mathbf{r}$ that are outside V there is no difficulty in evaluating the integral in (116) since $\bar{\mathbf{G}}_e$ is not singular within V. However, if $\mathbf{r}$ is located in V then $\mathbf{r}$ and $\mathbf{r}_0$ can coincide and $\bar{\mathbf{G}}_e$ will become singular. The most singular part of $\bar{\mathbf{G}}_e$ is the term $\nabla\nabla(e^{-jk_0R}/4\pi k_0^2 R)$. For $|\mathbf{r} - \mathbf{r}_0|$ very small the exponential function can be replaced by unity so the dominant singular part is

$$\frac{1}{4\pi k_0^2} \nabla\nabla \frac{1}{R} = \frac{1}{4\pi k_0^2} \nabla_0 \nabla_0 \frac{1}{R}.$$

One way to overcome the difficulty in integrating this singular term is to reformulate the problem using the finite unit source discussed in Section 1.9. Thus we consider a scalar Green's function that is a solution of

$$(\nabla_0^2 + k_0^2)g = \begin{cases} -\dfrac{3}{4\pi a^3}, & R = |\mathbf{r} - \mathbf{r}_0| \le a \\ 0, & R = |\mathbf{r} - \mathbf{r}_0| > a \end{cases}$$

where a is the radius of a small spherical volume V_0 centered on $\mathbf{r}$. Outside the source region the solution for g is still given by $Ae^{-jk_0R}/4\pi R$ since this is a solution of the homogeneous equation. In this solution A is a constant to be determined. The Green's function g that satisfies the above equation is finite everywhere so in the region $R \le a$ we can choose a power series

$$g = \sum_{n=0}^{\infty} C_n R^n$$

for the solution of the Helmholtz equation in spherical coordinates,

$$\frac{1}{R^2} \frac{\partial}{\partial R} R^2 \frac{\partial g}{\partial R} + k_0^2 g = -\frac{3}{4\pi a^3}, \qquad R \le a.$$

There is no dependence on the angles θ and ϕ because of the spherical symmetry. When we substitute the series expansion into this equation we obtain

$$\sum_{n=0}^{\infty} C_n n(n+1) R^{n-2} + k_0^2 \sum_{n=0}^{\infty} C_n R^n = -\frac{3}{4\pi a^3}.$$

By equating the coefficients of like powers of R it is readily found that for the particular solution $C_0 = C_1 = C_3 = C_5 = \cdots = 0$

$$C_2 = -\frac{1}{8\pi a^3}, \qquad C_n = -\frac{k_0^2}{n(n+1)} C_{n-2}, \qquad n = 4, 6, \ldots .$$

When the series is written out it can be recognized as the solution

$$g_1 = \frac{3}{4\pi a^3 k_0^2} \left[\frac{\sin k_0 R}{k_0 R} - 1 \right], \qquad R \le a.$$

A solution to the homogeneous equation that is finite at $R = 0$ is

$$g_0 = B \frac{\sin k_0 R}{k_0 R}, \qquad R \le a.$$

We now make g and $\partial g / \partial R$ continuous at $R = a$. This allows the two constants A and B to be determined. The final solution obtained for g is

$$g = \begin{cases} \dfrac{3}{4\pi a^3 k_0^2} \left(\dfrac{\sin k_0 R}{k_0 R} - 1 \right) + \dfrac{3 \sin k_0 R}{4\pi k_0^3 a^3 R} [(1 + jk_0 a)e^{-jk_0 a} - 1], & R \le a \\ \\ \dfrac{3}{k_0^3 a^3} (\sin k_0 a - k_0 a \cos k_0 a) \dfrac{e^{-jk_0 R}}{4\pi R}, & R \ge a. \end{cases} \tag{117}$$

If we choose $k_0 a$ to be very small, then $k_0 R$ is also small in the volume V_0 and we can approximate g as follows:

$$g \approx \begin{cases} \left(\dfrac{3}{8\pi a} - \dfrac{R^2}{8\pi a^3} \right), & R \le a \\ \\ \dfrac{e^{-jk_0 R}}{4\pi R}, & R \ge a. \end{cases} \tag{118}$$

This approximation is obtained by using the small argument approximation for the trigonometric functions. The approximation for g in $R \le a$ is the static field approximation used in Section 1.9.

In (116) we now split the volume into a small spherical volume V_0 plus the remainder $V - V_0$. The integral over $V - V_0$ can be carried out since $\bar{\mathbf{G}}_e$ is not singular in this region. We can choose V_0 so small that $\mathbf{J}(\mathbf{r}_0)$ can be approximated by its value at the center. In addition, the only term that will remain as we make V_0 shrink to zero is the term $\nabla_0 \nabla_0 g$ and we evaluate the integral of this term using the approximation given in (118). We now have

$$- j\omega\mu_0 \mathbf{J}(\mathbf{r}) \cdot \iiint\limits_{V_0} \nabla_0 \nabla_0 \left(\frac{-R^2}{k_0^2 8\pi a^3} \right) dV_0 = j\omega\mu_0 \mathbf{J}(\mathbf{r}) \cdot \oiint\limits_{S_0} \frac{\mathbf{a}_R \nabla_0 R^2}{k_0^2 8\pi a^3} \bigg|_{R=a} dS_0$$

upon converting the volume integral of the gradient to a surface integral using a vector identity (see the Mathematical Appendix). At this point we note that $\nabla_0 R^2 = 2R\mathbf{a}_R$ and then by expressing $\mathbf{a}_R \mathbf{a}_R$ in rectangular coordinates we readily find that the integral gives a finite result independent of the radius of the small sphere. Hence we find that the contribution to the electric field from the currents in V_0 is given by

$$\lim_{a \to 0} - j\omega\mu_0 \iiint\limits_{V_0} \mathbf{J}(\mathbf{r}_0) \cdot \bar{\mathbf{G}}_e(\mathbf{r}_0, r) \, dV_0 = j\omega\mu_0 \frac{\bar{\mathbf{I}} \cdot \mathbf{J}(r)}{3k_0^2}. \tag{119}$$

At the end of Section 2.15 we will show that the above result can also be obtained by using the eigenfunction expansion of the dyadic Green's function.

In classical potential theory where a scalar Green's function behaving like $1/R$ is involved the integral in (109) can be successfully defined as the limit of an improper integral. This approach has also been applied to (109) [2.20], [2.21], but as shown by the author [2.17] it is not an entirely satisfactory method.

The classical approach is to surround the singular point $\mathbf{r}$ by a small volume V_0, apply Green's theorem to the region $V - V_0$, and take the limit as V_0 tends to zero. For the scalar Helmholtz equation one term obtained from the limiting procedure is the scalar field at the singular point. If we apply (109) with the point $\mathbf{r}$ excluded we obtain

$$- j\omega\mu_0 \lim_{V_0 \to 0} \iiint_{V-V_0} \mathbf{J}(\mathbf{r}_0) \cdot \bar{\mathbf{G}}_e(\mathbf{r}_0, \mathbf{r}) \, dV_0$$

$$+ \lim_{V_0 \to 0} \oiint_{S_0} [(\mathbf{n} \times \nabla_0 \times \mathbf{E}) \cdot \bar{\mathbf{G}}_e + (\mathbf{n} \times \mathbf{E}) \cdot \nabla_0 \times \bar{\mathbf{G}}_e] \, dS_0 = 0. \tag{120}$$

It can be shown that both limits depend on the shape of the excluded volume [2.17], [2.21]. For a spherical volume the limit as V_0 goes to zero gives

$$- \frac{2}{3}\mathbf{E}(\mathbf{r}) - \frac{1}{3k_0^2} \nabla \times \nabla \times \mathbf{E}(\mathbf{r})$$

$$- j\omega\mu_0 \lim_{V_0 \to 0} \iiint_{V-V_0} \mathbf{J} \cdot \bar{\mathbf{G}}_e \, dV_0 = 0 \tag{121}$$

which is not an explicit solution for the electric field [2.17]. However, if we assume that the electric field in (121) does obey the original equation (106) then we obtain

$$\mathbf{E}(\mathbf{r}) = j\omega\mu_0 \frac{\bar{\mathbf{I}} \cdot \mathbf{J}(\mathbf{r})}{3k_0^2} - j\omega\mu_0 \lim_{V_0 \to 0} \iiint_{V-V_0} \mathbf{J} \cdot \bar{\mathbf{G}}_e \, dV_0 \tag{122}$$

which is the same result as that given by (119) when the contribution from the volume $V - V_0$ is added. Since the principal-volume approach does not yield a direct solution for the electric field the dyadic Green's function does not serve as the inverse operator for the electric field differential equation (106). For this reason we do not regard the principal-volume procedure as the appropriate procedure for obtaining an inverse operator that will solve (106). The limiting procedure using a finite unit source as given earlier does permit the integration to be carried out over the whole volume containing $\mathbf{J}$ even when the field point is located in that volume. As we will see later, the eigenfunction expansion of $\bar{\mathbf{G}}_e$ also enables the integral to be carried out.

Thus (116) can be evaluated and clearly $\bar{\mathbf{G}}_e$ then functions as an inverse operator that solves (106), whereas the principal-volume approach results in a new differential equation for $\mathbf{E}(\mathbf{r})$ as given by (121).

The use of the principal-volume result (122) to evaluate the electric field also presents difficulties that often mitigate against its use. In a cavity or a waveguide the satisfaction of the boundary conditions is most easily accomplished by using an eigenfunction or mode expansion of the dyadic Green's function. The use of (122) then considerably complicates the solution procedure since the eigenfunctions are not orthogonal over the volume $V - V_0$. As a consequence, infinite series occur that must be handled carefully in the limiting process since some of these series become divergent. When an eigenfunction expansion of the dyadic Green's function is used there is no need to introduce an exclusion volume since no singular integrals occur.

2.13. RECIPROCITY RELATION FOR DYADIC GREEN'S FUNCTIONS

In free space, the dyadic Green's function for the vector Helmholtz equation is a symmetrical dyadic. In the presence of boundaries on which $\bar{\mathbf{G}}$ satisfies prescribed boundary conditions, the dyadic Green's function is not a symmetrical dyadic in general. The nonsymmetry is brought about by the boundary conditions that $\bar{\mathbf{G}}$ must satisfy. Any number of examples may be constructed to demonstrate the validity of the above statements. With reference to Fig. 2.15, let $\bar{\mathbf{G}}$ be the Green's function for the half space $z > 0$ and subject to the boundary condition $\mathbf{n} \times \bar{\mathbf{G}} = 0$ on the perfectly conducting plane at $z = 0$. The Green's dyadic is a function of the coordinates of the field point, the coordinates x_0, y_0, z_0 of the source point, and the coordinates x_0, y_0, $-z_0$ of the image source. It may be constructed from the free-space Green's dyadic by using image theory. However, we will not require the analytical form of $\bar{\mathbf{G}}$ to demonstrate the nonsymmetry of the dyadic. A current element $\mathbf{a}_x J_x$ produces an electric field with a z component given by $J_x G_{xz}$, while a current element $\mathbf{a}_z J_z$ produces a field with an x component given by $J_z G_{zx}$. If $G_{xz} = G_{zx}$, that is, if $\bar{\mathbf{G}}$ is a symmetric dyadic, these components of the field would be equal when $J_x = J_z$. For the problem under consideration, the current $\mathbf{a}_x J_x$ produces a field with a nonzero z component, while $\mathbf{a}_z J_z$ produces a field with an x component which vanishes on the conducting plane. Therefore, $G_{xz} \neq G_{zx}$, and the Green's dyadic is not a symmetrical dyadic in general.

Even though $\bar{\mathbf{G}}$ is not a symmetric dyadic in general, a reciprocity principle does hold for $\bar{\mathbf{G}}$. Let $\bar{\mathbf{G}}(\mathbf{r}_1, \mathbf{r})$ be the Green's dyadic such that a current $\mathbf{J}_1(\mathbf{r}_1)$ at $\mathbf{r}_1$ produces an electric field $\mathbf{E}_1(\mathbf{r}_2) = \mathbf{J}_1(\mathbf{r}_1) \cdot \bar{\mathbf{G}}(\mathbf{r}_1, \mathbf{r}_2)$ at the point $\mathbf{r}_2$. Similarly, a current element $\mathbf{J}_2(\mathbf{r}_2)$ produces a field $\mathbf{E}_2(\mathbf{r}_1) = \mathbf{J}_2(\mathbf{r}_2) \cdot \bar{\mathbf{G}}(\mathbf{r}_2, \mathbf{r}_1)$ at the point $\mathbf{r}_1$. According to the Lorentz reciprocity theorem (Section 1.10 and assuming that the surface integral is zero),

$$\mathbf{E}_1(\mathbf{r}_2) \cdot \mathbf{J}_2(\mathbf{r}_2) = \mathbf{E}_2(\mathbf{r}_1) \cdot \mathbf{J}_1(\mathbf{r}_1)$$

and hence

$$\mathbf{J}_1(\mathbf{r}_1) \cdot \bar{\mathbf{G}}(\mathbf{r}_1, \mathbf{r}_2) \cdot \mathbf{J}_2(\mathbf{r}_2) = \mathbf{J}_2(\mathbf{r}_2) \cdot \bar{\mathbf{G}}(\mathbf{r}_2, \mathbf{r}_1) \cdot \mathbf{J}_1(\mathbf{r}_1). \tag{123}$$

Since $\mathbf{J}_1$ and $\mathbf{J}_2$ are arbitrary, we must have for an arbitrary current element

$$\mathbf{J}(\mathbf{r}_0) \cdot \bar{\mathbf{G}}(\mathbf{r}_0, \mathbf{r}) = \bar{\mathbf{G}}(\mathbf{r}, \mathbf{r}_0) \cdot \mathbf{J}(\mathbf{r}_0) = \mathbf{J}(\mathbf{r}_0) \cdot \bar{\mathbf{G}}_t(\mathbf{r}, \mathbf{r}_0) \tag{124}$$

where $\bar{\mathbf{G}}_t$ is the transposed dyadic. Thus the Green's dyadic with the variables $\mathbf{r}$ and $\mathbf{r}_0$ interchanged is equal to the transpose of the original dyadic. If $\mathbf{J}_1 = \mathbf{a}_x$ and $\mathbf{J}_2 = \mathbf{a}_y$, we obtain from (123) the result $G_{xy}(\mathbf{r}_1, \mathbf{r}_2) = G_{yx}(\mathbf{r}_2, \mathbf{r}_1)$, which is a special case of (124). A formal proof of (124) may be obtained by using the vector form of Green's identity instead of the Lorentz reciprocity theorem. The details are left as a problem at the end of the chapter.

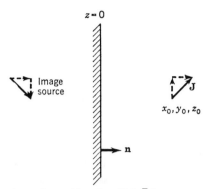

Fig. 2.15. Boundary-value problem for which $\bar{\mathbf{G}}$ is not a symmetric dyadic.

2.14. Eigenfunction Expansions of Dyadic Green's Functions

In order to find the spectral representation of the differential operator in the vector Helmholtz equation and the vector wave equation we need to construct the vector eigenfunctions for the two equations

$$(\nabla^2 + \lambda)\psi = 0 \tag{125a}$$

$$(\nabla \times \nabla \times - \lambda)\psi = 0 \tag{125b}$$

where ψ is a vector eigenfunction and λ is the eigenvalue parameter. The vector eigenfunctions can be separated into solenoidal or transverse eigenfunctions $\mathbf{F}$ that have the properties

$$\nabla \cdot \mathbf{F} \equiv 0, \qquad \nabla \times \mathbf{F} \neq 0$$

and irrotational or longitudinal eigenfunctions $\mathbf{L}$ with the properties

$$\nabla \times \mathbf{L} \equiv 0, \qquad \nabla \cdot \mathbf{L} \neq 0.$$

If we expand the curl curl operator in (125b) to get $\nabla\nabla \cdot - \nabla^2$, then since $\nabla \cdot \mathbf{F} = 0$ we see that

$$(\nabla^2 + \lambda)\mathbf{F} = 0$$

so the solenoidal functions can be eigenfunctions of both operators shown in (125). If we substitute the longitudinal functions $\mathbf{L}$ into (125b) we see that because $\nabla \times \mathbf{L} = 0$ these nonzero functions lie in the null space of the curl curl operator and the eigenvalue λ must be zero. This is an important point that we will return to when we consider the eigenfunction expansion for the electric field dyadic Green's function.

Our ability to construct vector eigenfunctions is limited to a relatively small number of coordinate systems. For spherical (and conical) coordinate systems Hansen [2.23] showed that the solenoidal functions could be obtained from the scalar functions of the scalar Helmholtz equation

$$(\nabla^2 + \lambda)\psi = 0.$$

The **F** functions consist of two sets called the **M** and **N** functions which are given by

$$\mathbf{M} = \nabla \times (\mathbf{a}_r r\psi) \tag{126a}$$

$$\sqrt{\lambda}\mathbf{N} = \nabla \times \nabla \times (\mathbf{a}_r r\psi). \tag{126b}$$

The longitudinal functions can be generated from the gradient of ψ, i.e.,

$$\mathbf{L} = \nabla\psi. \tag{126c}$$

The reader can readily verify that in spherical coordinates the **M** and **N** functions satisfy both equations in (125) and that the **L** functions satisfy (125a). For the free-space problem the same scalar function ψ generates all three sets of vector functions. In cavity and waveguide problems the scalar functions that generate the **M**, **N**, and **L** functions are generally different because they are required to satisfy different boundary conditions.

In cylindrical coordinate systems such as circular, elliptical, parabolic, and rectangular cylindrical coordinates the vector functions can be found from the operations

$$\mathbf{M} = \nabla \times \mathbf{a}_z\psi = -\mathbf{a}_z \times \nabla\psi \tag{127a}$$

$$\sqrt{\lambda}\mathbf{N} = \nabla \times \nabla \times \mathbf{a}_z\psi = \nabla\frac{\partial\psi}{\partial z} - \mathbf{a}_z\nabla^2\psi \tag{127b}$$

$$\mathbf{L} = \nabla\psi \tag{127c}$$

where again ψ is a solution of the scalar Helmholtz equation. In rectangular coordinates the unit vector $\mathbf{a}_z$ can be replaced by $\mathbf{a}_x$ or $\mathbf{a}_y$ also.

Vector Eigenfunctions in Spherical Coordinates

The scalar Helmholtz equation in spherical coordinates is

$$\frac{1}{r^2}\frac{\partial}{\partial r}r^2\frac{\partial\psi}{\partial r} + \frac{1}{r^2\sin\theta}\frac{\partial}{\partial\theta}\sin\theta\frac{\partial\psi}{\partial\theta} + \frac{1}{r^2\sin^2\theta}\frac{\partial^2\psi}{\partial\phi^2} + k^2\psi = 0 \tag{128}$$

where θ is the polar angle relative to the z axis and ϕ is the azimuth angle measured from the x axis. We have also replaced λ by k^2 as the eigenvalue parameter. If we assume a product solution of the form $\psi = f(r)g(\theta)h(\phi)$, then the Helmholtz equation separates into the following three ordinary differential equations:

$$\frac{d^2h}{d\phi^2} + \lambda_\phi h = 0 \tag{129a}$$

$$\frac{d}{d\theta}\sin\theta\frac{dg}{d\theta} - \frac{\lambda_\phi}{\sin\theta}g + \lambda_r\sin\theta g = 0 \tag{129b}$$

$$\frac{d}{dr}r^2\frac{df}{dr} - \lambda_r f + k^2r^2f = 0 \tag{129c}$$

where λ_ϕ and λ_r are separation constants. The above equations are all of the Sturm–Liouville type. For the free-space problem the angle ϕ covers the whole interval 0 to 2π so $h(\phi)$ must be periodic in ϕ and hence $\lambda_\phi = m^2$ where m is an integer 0, 1, 2, 3, ... The solutions for h are $\cos m\phi$ and $\sin m\phi$.

The equation for $g(\theta)$ is Legendre's equation. There are two independent solutions and both are singular on the z axis unless λ_r is chosen equal to $n(n+1)$ where n is an integer. In this case one solution is finite everywhere and is called the associated Legendre polynomial designated by the symbol $P_n^m(\cos \theta)$. When $m = 0$ the solutions are called Legendre polynomials. These solutions can be obtained using Rodrigues' formula

$$P_n^m(u) = \frac{(1 - u^2)^{m/2}}{2^n n!} \frac{d^{n+m}}{du^{n+m}} (u^2 - 1)^n. \tag{130}$$

For $m > n$ the solutions are zero. The orthogonality of the different solutions for different values of n is with respect to $\sin \theta$ as a weighting factor. For the same n but different m the solutions are orthogonal with respect to $1/\sin \theta$ as a weighting factor. These orthogonality relations follow directly from the properties of the Sturm–Liouville equation. A number of useful formulas, including recurrence relations, are summarized in the Mathematical Appendix.

The equation for the function $f(r)$ is a modified form of Bessel's differential equation. The solutions are called spherical Bessel functions and are related to the Bessel cylinder functions as follows:

$$j_n(kr) = \sqrt{\frac{\pi}{2kr}} J_{n+1/2}(kr) \tag{131a}$$

$$y_n(kr) = \sqrt{\frac{\pi}{2kr}} Y_{n+1/2}(kr) \tag{131b}$$

$$h_n^1(kr) = j_n(kr) + j y_n(kr) \tag{131c}$$

$$h_n^2(kr) = j_n(kr) - j y_n(kr). \tag{131d}$$

The basic properties and recurrence relations for the spherical Bessel functions are also summarized in the Mathematical Appendix.

In rectangular coordinates the continuous spectrum of eigenfunctions can be represented by a Fourier integral. In spherical coordinates the radial eigenfunction can be chosen as the $j_n(kr)$ function and its completeness properties are expressed through the Hankel transforms

$$f(r) = \sqrt{\frac{2}{\pi}} \int_0^\infty F(k) j_n(kr) k^2 \, dk \tag{132a}$$

$$F(k) = \sqrt{\frac{2}{\pi}} \int_0^\infty f(r) j_n(kr) r^2 \, dr \tag{132b}$$

and the completeness relations

$$\frac{2}{\pi} \int_0^\infty j_n(kr) j_n(k'r) r^2 \, dr = \frac{\delta(k - k')}{k^2} \tag{133a}$$

$$\frac{2}{\pi} \int_0^\infty j_n(kr) j_n(kr') k^2 \, dk = \frac{\delta(r - r')}{r^2}. \tag{133b}$$

In view of these relations the eigenfunctions for the scalar Helmholtz equation, for free space, can be seen to be

$$\psi_{nm}^{e,o} = P_n^m(\cos\theta)j_n(kr)\begin{array}{c}\cos\\\sin\end{array}m\phi \qquad (134)$$

where e stands for the even function $\cos m\phi$ and o identifies the odd function $\sin m\phi$. These functions remain finite at the origin and vanish at infinity.

From the scalar function given in (134) we can now construct the vector eigenfunctions using the operations shown in (126). Thus we have

$$\mathbf{L}_{nm}^{e,o} = \nabla\psi_{nm}^{e,o} = P_n^m\begin{array}{c}\cos\\\sin\end{array}m\phi\, k\frac{dj_n}{d(kr)}\mathbf{a}_r$$

$$+ j_n\begin{array}{c}\cos\\\sin\end{array}m\phi\frac{1}{r}\frac{dP_n^m}{d\theta}\mathbf{a}_\theta$$

$$\mp\frac{m}{r\sin\theta}P_n^m j_n\begin{array}{c}\sin\\\cos\end{array}m\phi\mathbf{a}_\phi \qquad (135a)$$

$$\mathbf{M}_{nm}^{e,o} = \nabla\times kr\mathbf{a}_r\psi_{nm}^{e,o} = \mp\frac{mk}{\sin\theta}P_n^m j_n\begin{array}{c}\sin\\\cos\end{array}m\phi\mathbf{a}_\theta$$

$$- kj_n\begin{array}{c}\cos\\\sin\end{array}m\phi\frac{dP_n^m}{d\theta}\mathbf{a}_\phi \qquad (135b)$$

$$\mathbf{N}_{nm}^{e,o} = \frac{1}{k}\nabla\times\nabla\times kr\mathbf{a}_r\psi_{nm}^{e,o} = \frac{1}{k}\nabla\times\mathbf{M}_{nm}^{e,o}$$

$$= \frac{n(n+1)}{r}P_n^m j_n\begin{array}{c}\cos\\\sin\end{array}m\phi\mathbf{a}_r + \frac{dP_n^m}{d\theta}\begin{array}{c}\cos\\\sin\end{array}m\phi\frac{1}{r}\frac{d(krj_n)}{d(kr)}\mathbf{a}_\theta$$

$$\mp\frac{m}{\sin\theta}P_n^m\begin{array}{c}\sin\\\cos\end{array}m\phi\frac{1}{r}\frac{d(krj_n)}{d(kr)}\mathbf{a}_\phi. \qquad (135c)$$

Note that we have included a factor k in the definition of the solenoidal functions.

The vector eigenfunctions are orthogonal among themselves as well as with respect to each other when integrated over all values of r, θ, and ϕ. Consider the product $\mathbf{L}\cdot\mathbf{M} = k\nabla\psi\cdot\nabla\times(r\psi\mathbf{a}_r) = k\nabla\cdot[\psi\nabla\times(r\psi\mathbf{a}_r)] - \psi\nabla\cdot\nabla\times(r\psi\mathbf{a}_r)$. Hence we have, upon using the divergence theorem,

$$\iiint\limits_V \mathbf{L}\cdot\mathbf{M}\, dV = \oiint\limits_S \psi\mathbf{M}\cdot\mathbf{a}_r\, dS = 0$$

because $\mathbf{M}$ does not have a radial component. In a similar way we find that

$$\iiint\limits_V \mathbf{L}\cdot\mathbf{N}\, dV = \oiint\limits_S \psi\mathbf{N}\cdot\mathbf{a}_r\, dS = 0$$

because ψ decreases like r^{-1}, $\mathbf{N} \cdot \mathbf{a}_r$ decreases like r^{-2}, and the surface area of a sphere increases only as r^2 so that for the surface of the sphere with a radius that becomes infinite the surface integral vanishes. We have thus shown that the $\mathbf{L}$ functions are orthogonal to the $\mathbf{M}$ and $\mathbf{N}$ functions.

Consider next the relation $\nabla \cdot (kr\psi \mathbf{a}_r) \times \mathbf{N} = \nabla \times (kr\psi \mathbf{a}_r) \cdot \mathbf{N} - kr\psi \mathbf{a}_r \cdot \nabla \times \mathbf{N} = \mathbf{M} \cdot \mathbf{N}$ since $\nabla \times \mathbf{N} = k\mathbf{M}$ and does not have a radial component. We now use the divergence theorem to get

$$\iiint_V \mathbf{M} \cdot \mathbf{N} \, dV = \oiint_S kr\psi \mathbf{a}_r \times \mathbf{N} \cdot \mathbf{a}_r \, dS = 0$$

because $\mathbf{a}_r \times \mathbf{N} \cdot \mathbf{a}_r = (\mathbf{a}_r \times \mathbf{a}_r) \cdot \mathbf{N} = 0$. Consequently, we find that the $\mathbf{M}$ and $\mathbf{N}$ functions are also orthogonal. Note that from (135c) we have $k\nabla \times \mathbf{N} = \nabla \times \nabla \times \mathbf{M} = k^2 \mathbf{M}$ which is the relationship we used above.

The orthogonality of the vector eigenfunctions among themselves for pairs of values nmk and $n'm'k'$ follows in a straightforward way from the orthogonality properties of the P_n^m, $\cos m\phi$, $\sin m\phi$, and the $j_n(kr)$ functions, the latter being expressed by (133a).

The normalization integrals for the $\mathbf{M}$ and $\mathbf{N}$ functions are the same. We can prove this by direct evaluation. The normalization integral for the $\mathbf{N}$ functions is not straightforward so we will consider it in detail and follow the procedure given by Tai [2.12].

By using (135c) and integrating over ϕ we readily find that

$$\iiint_V \mathbf{N}_{nm}^{e,o}(\theta, \phi, kr) \cdot \mathbf{N}_{nm}^{e,o}(\theta, \phi, k'r) \, dV$$

$$= \frac{2\pi}{\epsilon_{0m}} \int_0^\pi \int_0^\infty \left\{ [n(n+1)P_n^m]^2 j_n(kr) j_n(k'r) + \left[\left(\frac{dP_n^m}{d\theta} \right)^2 + \frac{(mP_n^m)^2}{\sin^2 \theta} \right] \right.$$

$$\times \left. \left[\frac{dr j_n(kr)}{dr} \frac{dr j_n(k'r)}{dr} \right] \right\} \sin \theta \, dr \, d\theta.$$

We now use the known integrals

$$\int_0^\pi [P_n^m(\cos \theta)]^2 \sin \theta \, d\theta = \frac{2}{2n+1} \frac{(n+m)!}{(n-m)!} \tag{136a}$$

$$\int_0^\pi \left[\left(\frac{dP_n^m}{d\theta} \right)^2 + \frac{(mP_n^m)^2}{\sin^2 \theta} \right] \sin \theta \, d\theta = \frac{2n(n+1)}{2n+1} \frac{(n+m)!}{(n-m)!} \tag{136b}$$

to obtain

$$\frac{2\pi}{\epsilon_{0m}} \frac{2n(n+1)}{2n+1} \frac{(n+m)!}{(n-m)!} \int_0^\infty \left\{ n(n+1) j_n(kr) j_n(k'r) \right.$$

$$\left. + \frac{dr j_n(kr)}{dr} \frac{dr j_n(k'r)}{dr} \right\} dr.$$

In order to proceed further we simplify the integrand by using the following two relations for the spherical Bessel functions:

$$\frac{d}{du}[u j_n(u)] = \frac{u}{2n+1}[(n+1)j_{n-1} - n j_{n+1}] \tag{137a}$$

$$j_n(u) = \frac{u}{2n+1}(j_{n-1} + j_{n+1}). \tag{137b}$$

The use of these relations enables us to obtain

$$\frac{d}{dr} r j_n(kr) \frac{d}{dr} r j_n(k'r)$$

$$= \frac{kk'r^2}{2n+1}[(n+1)j_{n-1}(kr)j_{n-1}(k'r) + n j_{n+1}(kr)j_{n+1}(k'r)] - n(n+1)j_n(kr)j_n(k'r)$$

by eliminating the cross-product terms $j_{n-1}j_{n+1}$ through use of the expansion of $j_n(kr)j_n(k'r)$ and (137b). By employing the above relation the integrand reduces to

$$kk'[(n+1)j_{n-1}(kr)j_{n-1}(k'r) + n j_{n+1}(kr)j_{n+1}(k'r)]r^2/(2n+1).$$

The integration may be completed by using (133a) to give the final result, which is

$$\iiint\limits_{V} \mathbf{N}_{nm}^{e,o}(\theta, \phi, kr) \cdot \mathbf{N}_{nm}^{e,o}(\theta, \phi, k'r)\, dV = \frac{\pi^2 2n(n+1)(n+m)!}{\epsilon_{0m}(2n+1)(n-m)!}\delta(k-k'). \tag{138a}$$

The other two normalization integrals are straightforward to obtain and are

$$\iiint\limits_{V} \mathbf{M}_{nm}^{e,o}(\theta, \phi, kr) \cdot \mathbf{M}_{nm}^{e,o}(\theta, \phi, k'r)\, dV = \frac{\pi^2 2n(n+1)(n+m)!}{\epsilon_{0m}(2n+1)(n-m)!}\delta(k-k'). \tag{138b}$$

$$\iiint\limits_{V} \mathbf{L}_{nm}^{e,o}(\theta, \phi, kr) \cdot \mathbf{L}_{nm}^{e,o}(\theta, \phi, k'r)\, dV = \frac{2\pi^2(n+m)!}{\epsilon_{0m}(2n+1)(n-m)!}\delta(k-k'). \tag{138c}$$

In evaluating the normalization integral (138c) the following relationship obtained from (137) was used:

$$\frac{u\, dj_n(u)}{du} = \frac{u}{2n+1}[n j_{n-1}(u) - (n+1)j_{n+1}(u)].$$

For later use we will define the normalization factor Q_{nm} by the relation

$$Q_{nm} = \frac{2\pi^2(n+m)!}{\epsilon_{0m}(2n+1)(n-m)!}. \tag{139}$$

Now that the complete set of vector eigenfunctions in spherical coordinates has been obtained we are in a position to consider the eigenfunction expansion of the solutions for the vector Helmholtz and vector wave equations. This topic is taken up next for the vector and scalar potentials in the Coulomb gauge. The results are then used to obtain the complete eigenfunction expansion of the electric field.

Eigenfunction Expansion of Vector and Scalar Potentials: Coulomb Gauge

In Section 1.7 it was shown that in the Coulomb gauge where $\nabla \cdot \mathbf{A} \equiv 0$ the vector and scalar potentials are solutions of the equations

$$\nabla \times \nabla \times \mathbf{A} - k_0^2 \mathbf{A} = \mu_0 \mathbf{J}(\mathbf{r}) - j\omega\epsilon_0\mu_0 \nabla \Phi(\mathbf{r}) = \mathbf{S}(\mathbf{r}) \qquad (140a)$$

$$\nabla^2 \Phi = -\frac{\rho(\mathbf{r})}{\epsilon_0}. \qquad (140b)$$

The source function $\mathbf{S}(\mathbf{r})$ for $\mathbf{A}(\mathbf{r})$ has zero divergence since $\nabla \cdot \mathbf{J} = -j\omega\rho$ so that $\nabla \cdot \mathbf{S} = -j\omega\mu_0\rho - j\omega\mu_0\epsilon_0 \nabla^2 \Phi = 0$ when Φ satisfies (140b). The $\mathbf{L}$ functions lie in the null space of the curl curl operator but since both $\mathbf{A}$ and $\mathbf{S}$ are orthogonal to this space of functions the solution of (140a) can be constructed using only the $\mathbf{M}$ and $\mathbf{N}$ functions. We assume that the current and charge are contained in a bounded volume V_0 in free space. We will order the eigenfunction labels n, m, e, and o in such a way that a single subscript j will denote a particular combination. We can easily show that the projection of the source function $\mathbf{S}$ onto the space of $\mathbf{L}_j$ functions is zero by showing that

$$\iiint_V \mathbf{L}_j \cdot \mathbf{S}\,dV = 0$$

for all j. We note that $\mathbf{L}_j \cdot \mathbf{S} = \nabla \psi_j \cdot \mathbf{S} = \nabla \cdot (\psi_j \mathbf{S}) - \psi_j \nabla \cdot \mathbf{S} = \nabla \cdot (\psi_j \mathbf{S})$. Hence by using the divergence theorem we have

$$\iiint_V \mathbf{L}_j \cdot \mathbf{S}\,dV = \iint_S \psi_j \mathbf{S} \cdot \mathbf{n}\,dS$$

where S is the surface at infinity. On S_0, the boundary of the source volume, we have $\mathbf{S} \cdot \mathbf{n} = \mu_0 \mathbf{J} \cdot \mathbf{n} - j\omega\epsilon_0\mu_0 \mathbf{n} \cdot \nabla \Phi$. If $\mathbf{n} \cdot \mathbf{J}$ is not zero on S_0, then a compensating layer of surface charge with density

$$\rho_s = -\epsilon_0 \nabla \Phi \cdot \mathbf{n}\big|_-^+ = \frac{\mathbf{n} \cdot \mathbf{J}}{j\omega}$$

exists on S_0 where $\nabla \Phi \cdot \mathbf{n}\big|_-^+$ is the discontinuity in the potential gradient across the charge layer. The two discontinuities, namely, that in $\mathbf{n} \cdot \mathbf{J}$ and that in $\mathbf{n} \cdot \nabla \Phi$, cancel so $\mathbf{n} \cdot \mathbf{S}$ is continuous across S_0. For this reason we could apply the divergence theorem to the total volume V. Since ψ_j and $\mathbf{S}$ vanish at infinity faster than r^{-2} the surface integral vanishes thus proving that $\mathbf{S}$ has no projection on the functions $\mathbf{L}_j$.

We can let $\mathbf{A}$ be expanded as follows:

$$\mathbf{A}(\mathbf{r}) = \sum_j \int_0^\infty [B_j(k)\mathbf{M}_j(\theta, \phi, kr) + C_j(k)\mathbf{N}_j(\theta, \phi, kr)]\,dk.$$

We will also express $\mathbf{M}_j(\theta, \phi, kr)$ in the form $\mathbf{M}_j(\mathbf{r}, k)$ and similarly for the $\mathbf{N}_j$ function.

When we substitute this expression into (140a) we obtain

$$\sum_j \int_0^\infty (k^2 - k_0^2)[B_j \mathbf{M}_j + C_j \mathbf{N}_j]\, dk = \mathbf{S}.$$

We can get a solution for $B_j(k)$ by scalar multiplying both sides by $\mathbf{M}_j(\theta, \phi, k'r)$ and integrating over V. By virtue of the orthogonal properties of the vector eigenfunctions and using the normalization integral (138b) we obtain

$$(k'^2 - k_0^2)B_j(k')n(n+1)Q_{nm} = \iiint_V \mathbf{S}(\mathbf{r}_0)\cdot\mathbf{M}_j(\mathbf{r}_0, k')\, dV_0$$

since the volume integral, as (138b) shows, gives a delta function factor $\delta(k - k')$ and the integral over k then selects the value $k = k'$. The coefficient $C_j(k')$ is obtained in a similar way. We easily find that

$$\mathbf{A}(\mathbf{r}) = \sum_j \int_0^\infty \frac{1}{n(n+1)Q_{nm}(k^2 - k_0^2)} \left[\mathbf{M}_j(\mathbf{r}, k) \iiint_V \mathbf{M}_j(\mathbf{r}_0, k)\cdot\mathbf{S}(\mathbf{r}_0)\, dV_0 \right.$$

$$\left. + \mathbf{N}_j(\mathbf{r}, k)\cdot \iiint_V \mathbf{N}_j(\mathbf{r}_0, k)\cdot\mathbf{S}(\mathbf{r}_0)\, dV_0 \right] dk \qquad (141)$$

which is the eigenfunction expansion of the solution to (140a). The solution for $\mathbf{A}(\mathbf{r})$ can be rewritten in the form

$$\mathbf{A}(\mathbf{r}) = \iiint_V \mathbf{S}(\mathbf{r}_0) \cdot \left\{ \sum_j \int_0^\infty \frac{dk}{n(n+1)Q_{nm}(k^2 - k_0^2)} \left[\mathbf{M}_j(\mathbf{r}_0, k)\mathbf{M}_j(\mathbf{r}, k) \right.\right.$$

$$\left.\left. + \mathbf{N}_j(\mathbf{r}_0, k)\mathbf{N}_j(\mathbf{r}, k) \right] \right\} dV_0. \qquad (142)$$

From this expression we can identify the free-space dyadic Green's function for the vector potential in the Coulomb gauge; thus

$$\bar{\mathbf{G}}(\mathbf{r}_0, \mathbf{r}) = \sum_j \int_0^\infty \frac{dk}{n(n+1)Q_{nm}(k^2 - k_0^2)}[\mathbf{M}_j(\mathbf{r}_0, k)\mathbf{M}_j(\mathbf{r}, k) + \mathbf{N}_j(\mathbf{r}_0, k)\mathbf{N}_j(\mathbf{r}, k)]. \qquad (143)$$

We emphasize that the proper interpretation of this eigenfunction expansion of the dyadic Green's function is that it is to be used in a term by term integration so as to yield the same solution as that given by (141).

For functions $\mathbf{F}$ that have $\nabla\cdot\mathbf{F} \equiv 0$ the vector eigenfunctions $\mathbf{M}_j$ and $\mathbf{N}_j$ form a complete set in the domain of the operator. Consequently, in this space of functions the unit source function has the representation

$$\mathbf{I}\delta(\mathbf{r} - \mathbf{r}_0) = \sum_j \int_0^\infty \frac{dk}{n(n+1)Q_{nm}}[\mathbf{M}_j(\mathbf{r}_0, k)\mathbf{M}_j(\mathbf{r}, k) + \mathbf{N}_j(\mathbf{r}_0, k)\mathbf{N}_j(\mathbf{r}, k)]. \qquad (144)$$

This result can be obtained by a formal expansion of the left-hand side or by beginning with the eigenfunction expansion of $\mathbf{F}$ which is

$$\mathbf{F}(\mathbf{r}) = \sum_j \int_0^\infty \frac{dk}{n(n+1)Q_{nm}} \left[\mathbf{M}_j(\mathbf{r}, k) \iiint_V \mathbf{M}_j(\mathbf{r}_0, k) \cdot \mathbf{F}(\mathbf{r}_0) \, dV_0 \right.$$

$$\left. + \mathbf{N}_j(\mathbf{r}, k) \iiint_V \mathbf{N}_j(\mathbf{r}_0, k) \cdot \mathbf{F}(\mathbf{r}_0) \, dV_0 \right] \tag{145}$$

and rewriting this in a form that will lead to the identification of the delta function operator.

Note that even though the current $\mathbf{J}$ and charge ρ are zero outside V_0 the source function $\mathbf{S}$ has a nonzero term proportional to $\nabla\Phi$ outside V_0. However, $\nabla\Phi$ and the lamellar part $\mathbf{J}_l$ of $\mathbf{J}$ have a zero projection on the functions $\mathbf{M}_j$ and $\mathbf{N}_j$. Consider, for example,

$$\iiint_V \nabla\Phi \cdot \mathbf{M}_j \, dV = \iiint_V [\nabla \cdot (\Phi \mathbf{M}_j) - \Phi \nabla \cdot \mathbf{M}_j] \, dV.$$

Now Φ is continuous across the surface S_0 and $\nabla \cdot \mathbf{M}_j = 0$ so the divergence theorem can be applied to give

$$\oiint_S \Phi \mathbf{M}_j \cdot \mathbf{n} \, dS.$$

This integral vanishes when S is a spherical surface since $\mathbf{M}_j \cdot \mathbf{a}_r = 0$. A similar result holds for the $\mathbf{N}_j$ functions because $\Phi \mathbf{N}_j \cdot \mathbf{a}_r$ varies like r^{-3} at infinity. It now follows that $\mathbf{S}$ can be replaced by $\mu_0 \mathbf{J}$ in the solution given by (141) or (142).

The solution for the scalar potential can be expanded in terms of the scalar eigenfunctions of the Laplacian operator. This eigenfunction set is the set $\psi_j = \psi_{nm}^{e,o}$. Thus we let

$$\Phi(\mathbf{r}) = \sum_j \int_0^\infty D_j(k)\psi_j(\mathbf{r}, k) \, dk.$$

The normalization integral for the ψ_j functions is

$$\iiint_V \psi_j(\mathbf{r}, k)\psi_j(\mathbf{r}, k') \, dV = \frac{2\pi^2(n+m)!}{\epsilon_{0m}k^2(2n+1)(n-m)!}\delta(k-k'). \tag{146}$$

By following the standard procedure we obtain

$$\Phi(\mathbf{r}) = \sum_j \int_0^\infty \frac{\epsilon_{0m}(2n+1)(n-m)!}{2\pi^2(n+m)!}\psi_j(\mathbf{r}, k) \iiint_V \frac{\rho(\mathbf{r}_0)}{\epsilon_0}\psi_j(\mathbf{r}_0, k) \, dV_0 \, dk. \tag{147}$$

We now use $\nabla \cdot \mathbf{J} = -j\omega\rho$, $\mathbf{n} \cdot \mathbf{J} = j\omega\rho_s$ on S_0, and $\nabla \cdot (\psi_j \mathbf{J}) = \psi_j \nabla \cdot \mathbf{J} + \nabla\psi_j \cdot \mathbf{J}$ to express the volume integral in the form

$$\iiint_V \frac{\nabla_0 \cdot \mathbf{J}}{-j\omega\epsilon_0}\psi_j \, dV_0 = \iiint_{V_0} \frac{\nabla_0 \psi_j \cdot \mathbf{J}}{j\omega\epsilon_0} \, dV_0 - \oiint_{S_0} \frac{\psi_j \mathbf{J} \cdot \mathbf{n}}{j\omega\epsilon_0} \, dS_0. \tag{148}$$

If $\mathbf{n} \cdot \mathbf{J}$ does not equal zero on S_0 then the volume integral in (147) must also include the surface integral

$$\oiint_{S_0} \frac{\rho_s(\mathbf{r}_0)}{\epsilon_0} \psi_j(\mathbf{r}_0, k) \, dS_0$$

where $j\omega\rho_s = \mathbf{n} \cdot \mathbf{J}$ and ρ_s is the charge density on S_0. This integral needs to be added to (148); when this is done we obtain

$$\iiint_V \frac{\nabla_0 \cdot \mathbf{J}}{-j\omega\epsilon_0} \psi_j \, dV_0 + \oiint_{S_0} \frac{\rho_s(\mathbf{r}_0)}{\epsilon_0} \psi_j \, dS_0 = \iiint_V \frac{\mathbf{L}_j(\mathbf{r}_0, k) \cdot \mathbf{J}(\mathbf{r}_0)}{j\omega\epsilon_0} \, dV_0. \qquad (149)$$

We can use this result to express the solution for Φ in the form

$$\Phi(\mathbf{r}) = \frac{1}{j\omega\epsilon_0} \sum_j \int_0^\infty \frac{2\pi^2(n+m)!}{\epsilon_{0m}(2n+1)(n-m)!} \psi_j(\mathbf{r}, k) \iiint_V \mathbf{L}_j(\mathbf{r}_0, k)$$

$$\cdot \mathbf{J}(\mathbf{r}_0) \, dV_0 \, dk. \qquad (150)$$

The electric field is given by $\mathbf{E}(\mathbf{r}) = -j\omega\mathbf{A}(\mathbf{r}) - \nabla\Phi(\mathbf{r})$. By using the solutions obtained for the potentials we can easily write down the eigenfunction expansion for the electric field. We will do this in a form that will allow us to identify the Green's function operator $\bar{\mathbf{G}}_e$ for the electric field. From (142) and (150) we obtain

$$\mathbf{E}(\mathbf{r}) = -j\omega\mu_0 \iiint_{V_0} \mathbf{J}(\mathbf{r}_0) \cdot \left\{ \sum_j \int_0^\infty \frac{dk}{n(n+1)Q_{nm}(k^2 - k_0^2)} \left[\mathbf{M}_j(\mathbf{r}_0, k)\mathbf{M}_j(\mathbf{r}, k) \right. \right.$$

$$\left. \left. + \ \mathbf{N}_j(\mathbf{r}_0, k)\mathbf{N}_j(\mathbf{r}, k) - n(n+1)\frac{k^2 - k_0^2}{k_0^2} \mathbf{L}_j(\mathbf{r}_0, k)\mathbf{L}_j(\mathbf{r}, k) \right] \right\} dV_0. \qquad (151)$$

Clearly the electric field dyadic Green's function operator is

$$\bar{\mathbf{G}}_e(\mathbf{r}_0, \mathbf{r}) = \sum_j \int_0^\infty \frac{dk}{Q_{nm}} \left\{ \frac{1}{n(n+1)(k^2 - k_0^2)} [\mathbf{M}_j(\mathbf{r}_0, k)\mathbf{M}_j(\mathbf{r}, k) \right.$$

$$\left. + \ \mathbf{N}_j(\mathbf{r}_0, k)\mathbf{N}_j(\mathbf{r}, k)] - \frac{1}{k_0^2}\mathbf{L}_j(\mathbf{r}_0, k)\mathbf{L}_j(\mathbf{r}, k) \right\}. \qquad (152)$$

The expansion of the electric field requires both the transverse and longitudinal vector eigenfunctions and hence the dyadic Green's function must also have both sets of eigenfunctions in its expansion. Even though $\nabla \cdot \mathbf{E} = 0$ outside the source region the scalar potential Φ is not zero and hence $\mathbf{E}$ has a lamellar part given by $-\nabla\Phi$ outside the source region. We point out once more that (152) is to be used in a term by term integration to yield the eigenfunction expansion of the electric field.

If we want to obtain an eigenfunction expansion of the electric field directly we consider the equation

$$\nabla \times \nabla \times \mathbf{E} - k_0^2 \mathbf{E} = -j\omega\mu_0 \mathbf{J}. \qquad (153)$$

Since the $\mathbf{L}_j$ functions lie in the null space of the curl curl operator and hence are not mapped into the range space of the operator, (153) does not have a solution in terms of the eigenfunctions of the differential operator unless $\mathbf{J}$ lies in the range space. But, in general, $\nabla \cdot \mathbf{J} \neq 0$ so $\mathbf{J}$ is not always in the range space of the operator. The range space only contains functions with nonzero curl. The part of $\mathbf{J}$ that does not lie in the range space $\mathfrak{R}$ lies in the orthogonal complement $\mathfrak{R}_\perp$ of the range space. This is the space containing the $\mathbf{L}_j$ functions and coincides with the null space of the curl curl operator since the latter is a self-adjoint operator. We can overcome the difficulty in solving (153) by the simple expedient of finding the projection of both sides on the space of $\mathbf{L}_j$ functions and removing this part from the equation. Thus we split $\mathbf{E}$ into its longitudinal and transverse parts $\mathbf{E}_l + \mathbf{E}_t$ and let

$$\mathbf{E}_l = \sum_j \int_0^\infty C_j(k) \mathbf{L}_j(\mathbf{r}, k) \, dk.$$

Similarly, we let the longitudinal part of $\mathbf{J}$ be represented by

$$\mathbf{J}_l = \sum_j \int_0^\infty D_j(k) \mathbf{L}_j(\mathbf{r}, k) \, dk.$$

From (153) we see that

$$\nabla \times \nabla \times \mathbf{E}_t - k_0^2 \mathbf{E}_t - k_0^2 \mathbf{E}_l = -j\omega\mu_0 \mathbf{J}_t - j\omega\mu_0 \mathbf{J}_l$$

and thus we must have $k_0^2 C_j = j\omega\mu_0 D_j$. By following the procedure that by now should be clear we readily find that

$$\mathbf{E}_l(\mathbf{r}) = j\omega\mu_0 \sum_j \int_0^\infty \frac{dk}{Q_{nm}k_0^2} \iiint_V \mathbf{J}(\mathbf{r}_0) \cdot \mathbf{L}_j(\mathbf{r}_0, k) \mathbf{L}_j(\mathbf{r}, k) \, dV_0.$$

When we remove the longitudinal part from (153) the equation that remains can be solved in terms of the $\mathbf{M}_j$ and $\mathbf{N}_j$ functions which are the eigenfunctions of the curl curl operator. The result is, of course, the same solution as that given earlier by (151).

Another way to approach (153) is to take the divergence of both sides to obtain

$$-k_0^2 \nabla \cdot \mathbf{E} = -j\omega\mu_0 \nabla \cdot \mathbf{J}.$$

Now let $\mathbf{E}_l = -\nabla\Phi$ and use the continuity equation to get

$$\nabla^2 \Phi = -\frac{j\omega\mu_0}{k_0^2} \nabla \cdot \mathbf{J} = -\frac{\rho}{\epsilon_0}.$$

The solution for the scalar potential can be constructed in terms of the ψ_j functions and simply returns us to our earlier solution (150). Note that even though $\mathbf{J}$ may be zero outside a finite volume V_0 the separate parts $\mathbf{J}_l$ and $\mathbf{J}_t$ are generally not zero outside V_0. However, outside V_0 the transverse part $\mathbf{J}_t$ cancels the longitudinal part $\mathbf{J}_l$.

The eigenfunction expansion of the scalar Green's function for the Helmholtz equation

$$(\nabla^2 + k_0^2)g = -\delta(\mathbf{r} - \mathbf{r}_0) \tag{154}$$

can be identified from the solution (147) for $\Phi(\mathbf{r})$ by inserting a factor $k^2 - k_0^2$ into the denominator and keeping the k^2 factor in the normalization constant. The solution for g is

$$g(\mathbf{r}, \mathbf{r}_0) = \frac{e^{-jk_0R}}{4\pi R}$$

$$= \sum_j \int_0^\infty \frac{\epsilon_{0m}(2n+1)(n-m)!}{2\pi^2(n+m)!} \frac{k^2 \psi_j(\mathbf{r}, k)\psi_j(\mathbf{r}_0, k)}{k^2 - k_0^2} \, dk. \tag{155}$$

For the scalar Green's function j includes the $n = m = 0$ terms.

2.15. Expansion of the Electric Field in Spherical Modes [2.18]

In this section we will show that the eigenfunction expansion (151) for the electric field can be converted into an expansion in terms of spherical standing waves and outward-propagating spherical waves. The mode expansion outside the source region involves only $\mathbf{M}$ and $\mathbf{N}$ functions that are derived from the scalar wave functions

$$\psi_j = P_n^m(\cos\theta) \frac{\cos}{\sin} m\phi\, j_n(k_0 r)$$

$$\psi_j^+ = P_n^m(\cos\theta) \frac{\cos}{\sin} m\phi\, h_n^2(k_0 r).$$

The $\mathbf{N}$ function spectrum of eigenfunctions contributes a zero frequency or "static-like" set of modes that cancels the longitudinal eigenfunction contributions outside the source region. Inside the source region the cancellation is not complete but the remainder can be expressed as a delta function contribution. The mode expansion is obtained by evaluating the integral over k in the eigenfunction expansion of the dyadic Green's function given by (152).

We will consider the mode expansion of the part of (143) involving the $\mathbf{N}_j$ functions first. If we examine (135c) for the $\mathbf{N}_j$ functions we see that the basic integral over k that has to be evaluated is

$$I = \int_0^\infty \frac{j_n(kr)j_n(kr_0)}{(k^2 - k_0^2)} \, dk. \tag{156}$$

Actually, various derivatives with respect to r and r_0 are also involved. According to the sequence of operations specified in (141) the derivatives should be taken before the integration over k is carried out. However, we can keep the derivative operations outside the integral provided due care is taken in differentiation of the resultant functions of r and r_0, some of which have discontinuous derivatives at the point $r = r_0$.

In order to evaluate the integral we convert it to a contour integral by using the substitution

$$j_n(kr) = \tfrac{1}{2}[h_n^1(kr) + h_n^2(kr)]$$

when $r > r_0$. We note that when kr is very small $j_n(kr)$ behaves like $(2kr)^n n!/(2n+1)!$ while $h_n^{1,2}(kr)$ behaves like $\mp j2(2n)!/n!(2kr)^{n+1}$. Thus the product $j_n(kr_0)h_n^{1,2}(kr)$ is asymptotic to $\mp jr_0^n/(2n+1)r^{n+1}k$ and has a first-order pole at $k = 0$. In the sum $h_n^1 + h_n^2$ the pole term is canceled and indeed the integrand in (156) is regular at $k = 0$. We will return to this point shortly.

The circuit relations

$$h_n^1(-kr) = e^{-jn\pi} h_n^2(kr), \qquad j_n(-kr) = e^{jn\pi} j_n(kr)$$

can be used to replace the integral of $j_n(kr_0)h_n^1(kr)$ from η to infinity by an integral from minus infinity to $-\eta$, i.e.,

$$\int_\eta^\infty \frac{j_n(kr_0)h_n^1(kr)}{k^2 - k_0^2} \, dk = \int_{-\eta}^{-\infty} \frac{j_n(-kr_0)h_n^1(-kr)(-dk)}{k^2 - k_0^2}$$

$$= \int_{-\infty}^{-\eta} \frac{j_n(kr_0)h_n^2(kr)}{k^2 - k_0^2} \, dk.$$

Later on we will take the limit as η approaches zero. When we split the integral of $j_n(kr_0)[h_n^1(kr) + h_n^2(kr)]$ in the manner indicated we must take the resultant integral as a Cauchy principal-value integral in order to obtain the required pole cancellation at $k = 0$. With these considerations the integral in (156) can be replaced by

$$I = \frac{1}{2} P \int_{-\infty}^\infty \frac{j_n(kr_0)h_n^2(kr)}{k^2 - k_0^2} \, dk$$

$$= \lim_{\eta \to 0} \int_{-\infty}^{-\eta} \frac{j_n(kr_0)h_n^2(kr)}{2(k^2 - k_0^2)} \, dk + \lim_{\eta \to 0} \int_\eta^\infty \frac{j_n(kr_0)h_n^2(kr)}{2(k^2 - k_0^2)} \, dk. \qquad (157)$$

We can convert the principal-value integral to a contour integral along either of the two contours shown in Fig. 2.16, but we must then add or subtract, respectively, $2\pi j$ times one-half of the residue at the pole $k = 0$. This last term is $\pm \pi j(jr_0^n/(2n + 1)r^{n+1}) = \mp \pi r_0^n/(2n + 1)r^{n+1}$. We can initially consider the medium to have a small loss so that $k_0 = k_0' - jk_0''$. Thus the pole at $k = k_0$ lies above the contour. The solution must be an analytic function of the loss parameters so consequently when we reduce the loss to zero the contour must be indented above the pole at k_0 and below the pole at $-k_0$ as shown in Fig. 2.16. We cannot allow the pole to move across the contour since this will create a discontinuity in the function equal to $\pm 2\pi j$ times the residue at the pole.

When kr is very large the spherical Bessel functions have the following asymptotic behavior:

$$j_n(kr) \sim \frac{1}{kr} \cos\left(kr - \frac{n+1}{2}\pi\right)$$

$$h_n^2(kr) \sim \frac{1}{kr} j^{n+1} e^{-jkr}.$$

Consequently, in the lower half of the complex k plane the product $j_n(kr_0)h_n^2(kr)$ becomes exponentially small for $r > r_0$. We can, therefore, close the contour by a semicircle in the lower half plane; when the radius of this circle becomes infinite we do not get any contribution to the integral from this part of the contour. We can now evaluate the contour integral using residue theory. We get residue contributions from the two poles at $k = 0$ and $k = k_0$ when we use the contour shown in Fig. 2.16(a). With this choice of contour we must add to the

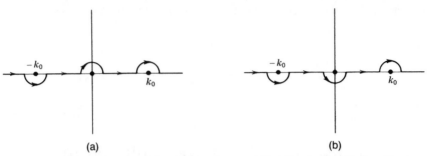

Fig. 2.16. Contour for converting the eigenfunction expansion into a mode expansion. The two alternative semi-circle contours about the point at the origin converts the principal value integral to a contour integral.

contour integral one-half of the pole contribution at $k = 0$. Thus we find that the integral I has the value

$$I(r, r_0) = -2\pi j \frac{j_n(k_0 r_0) h_n^2(k_0 r)}{4 k_0} - \frac{\pi r_0^n}{2 k_0^2 (2n + 1) r^{n+1}}, \qquad r \geq r_0. \qquad (158)$$

The result is valid for $r = r_0$ also since the contribution to the integral from the semicircle is still zero when $r = r_0$ because the integrand decreases like k^{-4} while the circumference of the circle increases only like $|k|$.

When $r < r_0$ we use the substitution $2 j_n(k r_0) = h_n^1(k r_0) + h_n^2(k r_0)$. A similar analysis then shows that the value of the integral is given by (158) but with the roles of r and r_0 interchanged. For all values of r, r_0 we have

$$I = -\frac{\pi j}{2 k_0} j_n(k_0 r_<) h_n^2(k_0 r_>) - \frac{\pi r_<^n}{2 k_0^2 (2n + 1) r_>^{n+1}}. \qquad (159)$$

The contribution to $\bar{\mathbf{G}}_e$ from the $\mathbf{N}_j \mathbf{N}_j$ terms in (152) can now be expressed in the form

$$\bar{\mathbf{G}}_{e1} = \sum_j \frac{1}{Q_{nm} n(n + 1)} \left[\nabla_0 \times \nabla_0 \times \mathbf{a}_{r_0} r_0 P_n^m(\cos \theta_0) \begin{matrix} \cos \\ \sin \end{matrix} m\phi_0 \right]$$

$$\times \left[\nabla \times \nabla \times \mathbf{a}_r r P_n^m(\cos \theta) \begin{matrix} \cos \\ \sin \end{matrix} m\phi \right] I(r_0, r) \qquad (160)$$

where $I(r_0, r)$ is given by (159). For convenience we will define the following functions:

$$\mathbf{N}_j(\mathbf{r}) = \nabla \times \nabla \times \mathbf{a}_r r j_n(k_0 r) P_n^m(\cos \theta) \begin{matrix} \cos \\ \sin \end{matrix} m\phi \qquad (161a)$$

$$\mathbf{N}_j^+(\mathbf{r}) = \nabla \times \nabla \times \mathbf{a}_r r h_n^2(k_0 r) P_n^m(\cos \theta) \begin{matrix} \cos \\ \sin \end{matrix} m\phi \qquad (161b)$$

$$\mathbf{N}_{j\sigma}(\mathbf{r}) = \begin{cases} \mathbf{N}_j(\mathbf{r}), & r < r_0 \\ \mathbf{N}_j^+(\mathbf{r}), & r > r_0 \end{cases} \qquad (161c)$$

$$\mathbf{N}_{j\sigma}(\mathbf{r}_0) = \begin{cases} \mathbf{N}_j(\mathbf{r}_0), & r_0 < r \\ \mathbf{N}_j^+(\mathbf{r}_0), & r_0 > r. \end{cases} \tag{161d}$$

The subscript σ is an indicator that tells which of $\mathbf{r}$, $\mathbf{r}_0$ is to be used according to the rules given in (161c) and (161d).

We also need symbols for the static-like modes generated from the other term in $I(r_0, r)$. For these we will use the symbols $\mathbf{N}_{j0}(\mathbf{r})$, $\mathbf{N}_{j0}^+(\mathbf{r})$, and $\mathbf{N}_{j\sigma 0}(\mathbf{r})$. The function $\mathbf{N}_{j0}(\mathbf{r})$ is the same as that in (161a) but with $j_n(k_0 r)$ replaced by r^n. The function $\mathbf{N}_{j0}^+(\mathbf{r})$ is the same as that in (161b) but with $h_n^2(k_0 r)$ replaced by r^{-n-1}. The σ indicator is applied according to the rules already given.

The function $I(r_0, r)$ is continuous at $r = r_0$, but the derivative with respect to r or r_0 has a step change across the point $r = r_0$. The operations in (161) require the derivative operation

$$\frac{d}{dr_0}\left[r_0 \frac{d}{dr} r I(r_0, r)\right] = I + r\frac{dI}{dr} + r_0 \frac{dI}{dr_0} + r r_0 \frac{d^2 I}{dr_0\, dr}$$

in order to evaluate the $\mathbf{a}_{\theta_0}\mathbf{a}_\theta$ and $\mathbf{a}_{\phi_0}\mathbf{a}_\phi$ contributions to $\bar{\mathbf{G}}_{e1}$. This operation produces delta function terms. The delta function terms come from the second derivative of I with respect to r_0 and r. The step change in dI/dr is

$$\left.\frac{dI}{dr}\right|_{r_0^-}^{r_0^+} = -\frac{j\pi}{2}[h_n'^2(k_0 r_0)j_n(k_0 r_0) - h_n^2(k_0 r_0)j_n'(k_0 r_0)] + \frac{\pi}{2k_0^2 r_0^2}$$

where the prime means a derivative with respect to $k_0 r_0$. By using the Wronskian relation

$$W = h_n'^2(u)j_n(u) - h_n^2(u)j_n'(u) = -j/u^2$$

we obtain

$$\left.\frac{dI}{dr}\right|_{r_0^-}^{r_0^+} = -\frac{\pi}{2k_0^2 r_0^2} + \frac{\pi}{2k_0^2 r_0^2}$$

so the two step changes are equal in value but of opposite sign. When dI/dr is regarded as a function of r, the two terms undergo the above step changes as r crosses the point r_0 from the region $r < r_0$ to the region $r > r_0$. But when dI/dr_0 is viewed as a function of r_0, the step changes are the negative of the above amounts as the point r_0 crosses the point r from the region $r_0 < r$ to the region $r_0 > r$. Clearly, the delta functions generated from the second derivative cancel. If the derivative operations are retained in the integrand, then the integrand can be rewritten in a form that will give the same cancellation of terms.

With all of the above definitions and canceling delta function terms taken into account we can express $\bar{\mathbf{G}}_{e1}$ in the concise form

$$\bar{\mathbf{G}}_{e1} = \sum_j \left[\frac{-j\pi}{2k_0 Q_{nm} n(n+1)}\mathbf{N}_{j\sigma}(\mathbf{r}_0)\mathbf{N}_{j\sigma}(\mathbf{r}) - \frac{\pi}{2k_0^2 Q_{nm} n(n+1)(2n+1)}\mathbf{N}_{j\sigma 0}(\mathbf{r}_0)\mathbf{N}_{j\sigma 0}(\mathbf{r})\right].$$

$$\tag{162}$$

For the $\mathbf{L}_j$ function contribution to the Green's dyadic function the integral to be evaluated is

$$I = \int_0^\infty \frac{j_n(kr_0)j_n(kr)}{-k_0^2} \, dk.$$

By means of the same substitutions as used before, the integral can be transformed into the same form as in (157) except that the denominator term is simply $-k_0^2$. The only pole present is that at $k = 0$ so the value of the integral is given by the second term only in (159), i.e.,

$$I(r_0, r) = -\frac{\pi r_<^n}{2k_0^2(2n+1)r_>^{n+1}}.$$

The contribution to $\bar{\mathbf{G}}_e$ from the $\mathbf{L}_j\mathbf{L}_j$ terms is

$$\bar{\mathbf{G}}_{e2} = -\sum_j \frac{\pi}{2k_0^2(2n+1)Q_{nm}} \nabla_0 \nabla \left[P_n^m(\cos\theta_0)P_n^m(\cos\theta) \begin{matrix}\cos\\\sin\end{matrix} m\phi_0 \begin{matrix}\cos\\\sin\end{matrix} m\phi \right] \frac{r_<^n}{r_>^{n+1}}.$$

By evaluating the gradients we find that

$$\bar{\mathbf{G}}_{e2} = \sum_j \frac{\pi}{2k_0^2 Q_{nm}n(n+1)(2n+1)} \mathbf{N}_{j\sigma0}(\mathbf{r}_0)\mathbf{N}_{j\sigma0}(\mathbf{r})$$

$$-\frac{1}{k_0^2}\sum_j \frac{\pi}{2Q_{nm}} P_n^m(\cos\theta_0)P_n^m(\cos\theta) \begin{matrix}\cos\\\sin\end{matrix} m\phi_0 \begin{matrix}\cos\\\sin\end{matrix} m\phi \frac{\delta(r-r_0)}{r_0^2}\mathbf{a}_r\mathbf{a}_r. \quad (163)$$

The derivative of a unit step function, which gives a delta function, is sometimes called a generalized derivative. The sum of the series is equal to $\delta(\theta - \theta_0)\delta(\phi - \phi_0)/\sin\theta$. This can be verified by expanding the delta functions in terms of the Legendre functions and the trigonometric functions. By comparing (163) and (162) we see that the $\mathbf{L}_j\mathbf{L}_j$ terms cancel the static-like modes that are a part of the $\mathbf{N}_j$ spectrum of modes outside the source region. Inside the source region the cancellation is not complete and the residual part can be expressed by the delta function term

$$-\frac{\mathbf{a}_r\mathbf{a}_r}{k_0^2}\frac{\delta(r-r_0)}{r^2}\frac{\delta(\theta-\theta_0)}{\sin\theta}\delta(\phi-\phi_0) = -\frac{\mathbf{a}_r\mathbf{a}_r}{k_0^2}\delta(\mathbf{r}-\mathbf{r}_0). \quad (164)$$

The last contribution to $\bar{\mathbf{G}}_e$ is that from the $\mathbf{M}_j\mathbf{M}_j$ terms. The integral over k that must be evaluated for these terms is

$$I = \int_0^\infty \frac{k^2 j_n(kr_0)j_n(kr)}{k^2 - k_0^2} \, dk.$$

Since the numerator has a factor k^2, there is no pole at $k = 0$. The evaluation of the integral is carried out using the same procedure as before and it has the value

$$I = -\frac{j\pi k_0}{2} j_n(k_0 r_<)h_n^2(k_0 r_>).$$

At this point we introduce the functions

$$\mathbf{M}_j(\mathbf{r}) = \nabla \times \mathbf{a}_r k_0 r P_n^m(\cos\theta) \,\genfrac{}{}{0pt}{}{\cos}{\sin}\, m\phi \, j_n(k_0 r) \qquad (165a)$$

$$\mathbf{M}_j^+(\mathbf{r}) = \nabla \times \mathbf{a}_r k_0 r P_n^m(\cos\theta) \,\genfrac{}{}{0pt}{}{\cos}{\sin}\, m\phi \, h_n^2(k_0 r) \qquad (165b)$$

$$\mathbf{M}_{j\sigma}(\mathbf{r}) = \begin{cases} \mathbf{M}_j(\mathbf{r}), & r < r_0 \\ \mathbf{M}_j^+(\mathbf{r}), & r > r_0 \end{cases} \qquad (165c)$$

with similar definitions when $\mathbf{r}$ and $\mathbf{r}_0$ are interchanged. The mode expansion for the electric field dyadic Green's function in free space can now be expressed in the form

$$\bar{\mathbf{G}}_e(\mathbf{r}_0, \mathbf{r}) = -\sum_j \frac{j\pi}{2k_0 n(n+1)Q_{nm}} [\mathbf{M}_{j\sigma}(\mathbf{r}_0)\mathbf{M}_{j\sigma}(\mathbf{r}) + \mathbf{N}_{j\sigma}(\mathbf{r}_0)\mathbf{N}_{j\sigma}(\mathbf{r})] - \frac{\mathbf{a}_r\mathbf{a}_r}{k_0^2}\delta(\mathbf{r} - \mathbf{r}_0).$$

$$(166)$$

This mode expansion for the dyadic Green's function was first obtained by Tai [2.24].

The eigenfunction expansion of $\bar{\mathbf{G}}_e$ as given by (152) clearly shows that the dyadic Green's function has a nonzero lamellar part outside the source region. The mode expansion given by (166) on the other hand involves only transverse wave functions outside the source region. However, these transverse wave functions have components that are not all continuous with continuous derivatives in the source region, and therefore do not represent purely transverse or solenoidal fields. Consequently, $\bar{\mathbf{G}}_e$ as given by (166) should not be interpreted to mean that $\bar{\mathbf{G}}_e$ does not have a lamellar part outside the source region.

The mode expansion for $\bar{\mathbf{G}}_e$ can be effectively employed to evaluate the electric field in the interior of a small spherical volume V_0 of radius a for a current density $\mathbf{J}(\mathbf{r}_0)$ that we will assume can be considered as constant and in the z direction. The source point $\mathbf{r}_0$ and the field point $\mathbf{r}$ both lie within the small volume V_0 but are otherwise arbitrary points. In view of the symmetry about the z axis, all of the terms $m > 0$ in $\mathbf{a}_z \cdot \bar{\mathbf{G}}_e$ when integrated over the angle ϕ give zero. Consequently, the $\mathbf{M}_j$ functions do not contribute to the field. By referring to (135c) it can be seen that the scalar product $\mathbf{a}_z \cdot \mathbf{N}_j$ gives rise to two integrals over θ_0 that are given by

$$\int_0^\pi P_n(\cos\theta_0)\cos\theta_0 \sin\theta_0 \, d\theta_0$$

$$-\int_0^\pi \frac{dP_n(\cos\theta_0)}{d\theta_0} \sin^2\theta_0 \, d\theta_0.$$

Since $\cos\theta_0 = P_1(\cos\theta_0)$ the first integral is zero if $n \neq 1$ because of the orthogonal property of the Legendre polynomials. For $n = 1$ the value of the first integral is, from (136a), $\frac{2}{3}$. The second integral can be integrated by parts to give

$$-P_n(\cos\theta_0)\sin^2\theta_0\big|_0^\pi + 2\int_0^\pi P_n \cos\theta_0 \sin\theta_0 \, d\theta_0.$$

The integrated term vanishes and the integral that remains has a nonzero value of $\frac{4}{3}$ only for $n = 1$. In view of these results we can conclude that only the $\mathbf{N}_j$ mode for $m = 0$, $n = 1$ is needed to describe the electric field. This result is solely dependent on the fact that the current is constant and in the z direction and is true for any spherical volume V_0.

Inside the volume V_0 the electric field is given by

$$-j\omega\mu_0 J \left(\frac{-3j}{8\pi k_0} \right) \iiint\limits_{V_0} \mathbf{a}_z \cdot \mathbf{N}_{1\sigma}(\mathbf{r}_0)\mathbf{N}_{1\sigma}(\mathbf{r})\,dV_0$$

$$+ j\omega\mu_0 J \frac{\mathbf{a}_z \cdot \mathbf{a}_r \mathbf{a}_r}{k_0^2} \iiint\limits_{V_0} \delta(\mathbf{r} - \mathbf{r}_0)\,dV_0$$

upon using (151) and (166). After performing the integration over θ_0 and ϕ_0 we obtain

$$\mathbf{E}(\mathbf{r}) = \frac{jZ_0 J \cos\theta}{k_0}\mathbf{a}_r - Z_0 J \left\{ \mathbf{N}_1^+(\mathbf{r})\int_0^r \left[r_0 j_1(k_0 r_0) + r_0 \frac{d}{dr_0}r_0 j_1(k_0 r_0) \right] dr_0 \right.$$

$$\left. + \mathbf{N}_1(\mathbf{r})\int_r^a \left[r_0 h_1^2(k_0 r_0) + r_0 \frac{d}{dr_0}r_0 h_1^2(k_0 r_0) \right] dr_0 \right\}.$$

The spherical Bessel functions of order 1 are given by

$$j_1(k_0 r_0) = \frac{\sin k_0 r_0}{(k_0 r_0)^2} - \frac{\cos k_0 r_0}{k_0 r_0}$$

$$h_1^2(k_0 r_0) = j\frac{e^{-jk_0 r_0}}{(k_0 r_0)^2} - \frac{e^{-jk_0 r_0}}{k_0 r_0}.$$

From these expressions we find that the two integrands reduce to

$$r_0 \sin k_0 r_0 \qquad \text{and} \qquad jr_0 e^{-jk_0 r_0}.$$

The integration over r_0 is easily done to give

$$\mathbf{E}(\mathbf{r}) = \frac{jZ_0 J \cos\theta}{k_0}\mathbf{a}_r - \frac{Z_0 J}{k_0^2}[(\sin k_0 r - k_0 r \cos k_0 r)\mathbf{N}_1^+(\mathbf{r})$$

$$+ ((j - k_0 a)e^{-jk_0 a} - (j - k_0 r)e^{-jk_0 r})\mathbf{N}_1(\mathbf{r})], \qquad r < a \qquad (167)$$

where $\mathbf{N}_1^+$ and $\mathbf{N}_1$ are given by

$$\mathbf{N}_1 = \frac{2}{r}\cos\theta\, j_1(k_0 r)\mathbf{a}_r - \frac{1}{r}\sin\theta\frac{dr j_1(k_0 r)}{dr}\mathbf{a}_\theta$$

$$\mathbf{N}_1^+ = \frac{2}{r}\cos\theta\, h_1^2(k_0 r)\mathbf{a}_r - \frac{1}{r}\sin\theta\frac{dr h_1^2(k_0 r)}{dr}\mathbf{a}_\theta.$$

When $k_0 r$ is small $\sin k_0 r - k_0 r \cos k_0 r \approx (k_0 r)^3/3$ so the term involving $\mathbf{N}_1^+$ remains finite at $r = 0$. If we set $r = 0$ then the electric field at the center of the spherical volume is found to be equal to

$$\mathbf{E}(0) = \frac{jZ_0J}{3k_0}\mathbf{a}_z - \frac{jZ_0J}{3k_0}(k_0a)^2\mathbf{a}_z \tag{168}$$

when k_0a is also small. The limit of (168) as the radius a tends to zero is thus a constant and finite value as found earlier and given by (119). Outside the sphere the electric field is given by

$$\mathbf{E}(\mathbf{r}) = -\frac{Z_0J}{k_0^2}(\sin k_0a - k_0a \cos k_0a)\mathbf{N}_1^+(\mathbf{r}). \tag{169}$$

For k_0a very small (169) reduces to $-(Z_0k_0I/4\pi)\mathbf{N}_1^+(\mathbf{r})$ where $I = 4\pi a^3J/3$. For suitable values of k_0a, which are apparent from (169), the radiated field vanishes.

2.16. DYADIC GREEN'S FUNCTION EXPANSION IN CYLINDRICAL COORDINATES

In cylindrical coordinates r, ϕ, z the solutions to the scalar Helmholtz equation that are finite at the origin are

$$\psi_n^{e,o} = J_n(kr)e^{-jwz} \begin{matrix} \cos \\ \sin \end{matrix} n\phi \tag{170}$$

where

$$\nabla^2\psi_n + \lambda\psi_n = 0 \tag{171}$$

and $k^2 = \lambda - w^2$. For this problem there are two continuous eigenvalue parameters k and w since the interval in z is infinite and that for r is semi-infinite. A general solution to the scalar Helmholtz equation is of the form

$$\int_{-\infty}^{\infty}\int_0^{\infty}\sum_{n=0}^{\infty}[C_n(w,k)\cos n\phi + D_n(w,k)\sin n\phi]J_n(kr)e^{-jwz}\,dk\,dw$$

where C_n and D_n are expansion coefficients depending on w and k. The orthogonality and normalization properties of the eigenfunctions e^{-jwz} are expressed by the Fourier transform property

$$\int_{-\infty}^{\infty}e^{-jwz+jw'z}\,dz = 2\pi\delta(w - w'). \tag{172}$$

The corresponding property for the eigenfunctions $J_n(kr)$ is given by the Hankel transform relations for the cylinder Bessel functions:

$$\int_0^{\infty}F(k)J_n(kr)k\,dk = f(r) \tag{173a}$$

$$\int_0^{\infty}f(r)J_n(kr)r\,dr = F(k) \tag{173b}$$

$$\int_0^\infty J_n(kr)J_n(k'r)r\,dr = \frac{\delta(k-k')}{k} \tag{173c}$$

$$\int_0^\infty J_n(kr)J_n(kr')k\,dk = \frac{\delta(r-r')}{r}. \tag{173d}$$

The vector eigenfunctions are given by (j stands for a particular n, e, o combination)

$$\mathbf{L}_j = \nabla\psi_j \tag{174a}$$

$$\mathbf{M}_j = \nabla \times \mathbf{a}_z\psi_j = -\mathbf{a}_z \times \nabla\psi_j \tag{174b}$$

$$p\mathbf{N}_j = \nabla \times \nabla \times \mathbf{a}_z\psi_j = \nabla\nabla\cdot\mathbf{a}_z\psi_j - \mathbf{a}_z\nabla^2\psi_j$$
$$= p^2\mathbf{a}_z\psi_j - jw\nabla\psi_j = \mathbf{a}_z p^2\psi_j - jw\mathbf{L}_j \tag{174c}$$

where $p^2 = \lambda$.

In order to derive the normalization integrals the following two Bessel function differential and recurrence relations are used:

$$\frac{dJ_n(u)}{du} = \frac{1}{2}[J_{n+1}(u) - J_{n-1}(u)] \tag{175a}$$

$$J_n(u) = \frac{u}{2n}[J_{n+1}(u) + J_{n-1}(u)]. \tag{175b}$$

The explicit forms for the vector eigenfunctions are

$$\mathbf{L}_j = \mathbf{a}_r\frac{dJ_n(kr)}{dr}e^{-jwz}\genfrac{}{}{0pt}{}{\cos}{\sin}n\phi \mp \mathbf{a}_\phi\frac{n}{r}J_n(kr)e^{-jwz}\genfrac{}{}{0pt}{}{\sin}{\cos}n\phi$$

$$- \mathbf{a}_z jwJ_n(kr)e^{-jwz}\genfrac{}{}{0pt}{}{\cos}{\sin}n\phi \tag{176a}$$

$$\mathbf{M}_j = -\mathbf{a}_\phi\frac{dJ_n(kr)}{dr}e^{-jwz}\genfrac{}{}{0pt}{}{\cos}{\sin}n\phi \mp \mathbf{a}_r\frac{n}{r}J_n(kr)e^{-jwz}\genfrac{}{}{0pt}{}{\sin}{\cos}n\phi \tag{176b}$$

$$\mathbf{N}_j = -\mathbf{a}_r\frac{jw}{p}\frac{dJ_n(kr)}{dr}e^{-jwz}\genfrac{}{}{0pt}{}{\cos}{\sin}n\phi \pm \mathbf{a}_\phi\frac{jw}{p}\frac{n}{r}J_n(kr)$$

$$\times e^{-jwz}\genfrac{}{}{0pt}{}{\sin}{\cos}n\phi + \mathbf{a}_z\frac{k^2}{p}J_n(kr)e^{-jwz}\genfrac{}{}{0pt}{}{\cos}{\sin}n\phi. \tag{176c}$$

Normalization

In this section we will use ψ_j with k, w and ψ_j^* with k', w' as the eigenvalue parameters. By using (175) it is readily found that the scalar product $\mathbf{L}_j\cdot\mathbf{L}_j^*$, after integration over ϕ, equals

$$\int_0^{2\pi} \mathbf{L}_j\cdot\mathbf{L}_j^*\,d\phi = \left\{\frac{kk'}{2}[J_{n+1}(kr)J_{n+1}(k'r) + J_{n-1}(kr)J_{n-1}(k'r)]\right.$$

$$\left. + ww'J_n(kr)J_n(k'r)\right\}(2\pi/\epsilon_{0n})e^{-j(w-w')z}.$$

The integral over all values of r and z, upon using (172) and (173c), is given by

$$\int_{-\infty}^{\infty}\int_{0}^{\infty}\int_{0}^{2\pi} \mathbf{L}_j \cdot \mathbf{L}_j^* r \, d\phi \, dr \, dz = \frac{4\pi^2}{\epsilon_{0n}}(kk' + ww')\delta(w - w')\frac{\delta(k - k')}{k'}$$

$$= \frac{4\pi^2}{\epsilon_{0n}}\frac{k^2 + w^2}{k}\delta(w - w')\delta(k - k'). \qquad (177a)$$

Next consider $\mathbf{M}_j \cdot \mathbf{M}_j^*$ and carry out the integration over ϕ to obtain

$$\int_{0}^{2\pi} \mathbf{M}_j \cdot \mathbf{M}_j^* \, d\phi = \frac{2\pi}{\epsilon_{0n}}\left[kk'J_n'(kr)J_n'(k'r) + \frac{n^2}{r^2}J_n(kr)J_n(k'r)\right]e^{-j(w - w')z}.$$

By using (175) to simplify the integrand, the result is readily integrated over r and z to yield

$$\iiint_{V} \mathbf{M}_j \cdot \mathbf{M}_j^* \, d\mathbf{r} = \frac{4\pi^2}{\epsilon_{0n}}k\delta(w - w')\delta(k - k'). \qquad (177b)$$

In a very similar way one obtains

$$\iiint_{V} \mathbf{N}_j \cdot \mathbf{N}_j^* \, d\mathbf{r} = \frac{4\pi^2}{\epsilon_{0n}}k\delta(w - w')\delta(k - k'). \qquad (177c)$$

Orthogonality

A volume integral of $\mathbf{N}_n \cdot \mathbf{M}_{n'}^*$, $\mathbf{N}_n \cdot \mathbf{L}_{n'}^*$, or $\mathbf{M}_n \cdot \mathbf{L}_{n'}^*$ is clearly zero if $n \neq n'$ and $w \neq w'$ because of the orthogonal property of the cos $n\phi$ and sin $n\phi$ functions and the Fourier integral relation (172). Hence, it suffices to consider only the case $n' = n$ and $w' = w$, $k \neq k'$.

First consider

$$\int_{0}^{2\pi}\int_{-\infty}^{\infty} \mathbf{L}_j \cdot \mathbf{N}_j^* \, dz \, d\phi = \frac{4\pi^2}{\epsilon_{0n}}\left(\frac{-jw}{p'}\right)\left[k'^2 J_n(kr)J_n(k'r)\right.$$

$$\left. - kk'J_n'(kr)J_n'(k'r) - \frac{n^2}{r^2}J_n(kr)J_n(k'r)\right]\delta(w - w').$$

When we compare the second term on the right with the normalization integral for $\mathbf{M}_j \cdot \mathbf{M}_j^*$ and also use (173c) we see that the integral over r gives a factor

$$(k'^2 - k^2)\frac{\delta(k - k')}{k} = 0$$

so $\mathbf{L}_j$ and $\mathbf{N}_j^*$ are orthogonal.

After integrating over ϕ and z the other two scalar products, namely, $\mathbf{L}_j \cdot \mathbf{M}_j^*$ and $\mathbf{N}_j \cdot \mathbf{M}_j^*$, are both found to have the factor

$$[kJ_n'(kr)J_n(k'r) + k'J_n(kr)J_n'(k'r)]r^{-1}.$$

The use of (175) allows this factor to be expressed as

$$\frac{kk'}{2n}[J_{n+1}(kr)J_{n+1}(k'r) - J_{n-1}(kr)J_{n-1}(k'r)]$$

for which the integral over r equals zero. Thus the orthogonality of the vector eigenfunction set is established.

Dyadic Green's Function

The dyadic Green's function $\bar{\mathbf{G}}_e$ is a solution of the equation

$$(\nabla \times \nabla \times - k_0^2)\bar{\mathbf{G}}_e = \bar{\mathbf{I}}\delta(\mathbf{r} - \mathbf{r}'). \tag{178}$$

The fact that the $\mathbf{L}_j$ functions are in the null space of the curl curl operator does not cause any difficulty if we simply include the $\mathbf{L}_j$ functions in the expansion of $\bar{\mathbf{G}}_e$ since this will automatically make the projection of $-k_0^2\bar{\mathbf{G}}_e$ and $\bar{\mathbf{I}}\delta(\mathbf{r} - \mathbf{r}')$ onto this space of functions equal. Hence we let

$$\bar{\mathbf{G}}_e(\mathbf{r}, \mathbf{r}') = \sum_j \int_0^\infty [\mathbf{M}_j \mathbf{A}_j(k, w) + \mathbf{N}_j \mathbf{B}_j(k, w) + \mathbf{L}_j \mathbf{C}_j(k, w)]\, dk$$

where $\mathbf{A}_j$, $\mathbf{B}_j$, and $\mathbf{C}_j$ are vector expansion coefficients. It is a straightforward matter to use the orthogonality and normalization properties of the eigenfunctions to solve for the expansion coefficients after the assumed solution for $\bar{\mathbf{G}}_e$ is substituted into (178). We will omit the details and only give the final result which is

$$\bar{\mathbf{G}}_e(\mathbf{r}, \mathbf{r}') = \frac{1}{4\pi^2}\sum_{n=0}^\infty \epsilon_{0n}$$

$$\times \int_0^\infty \int_{-\infty}^\infty \left[\frac{\mathbf{M}_j(k, w, \mathbf{r})\mathbf{M}_j^*(k, w, \mathbf{r}') + \mathbf{N}_j(k, w, \mathbf{r})\mathbf{N}_j^*(k, w, \mathbf{r}')}{k(k^2 + w^2 - k_0^2)}\right.$$

$$\left. - \frac{k}{k_0^2}\frac{\mathbf{L}_j(k, w, \mathbf{r})\mathbf{L}_j^*(k, w, \mathbf{r}')}{(k^2 + w^2)}\right]\, dw\, dk. \tag{179}$$

The vector eigenfunctions are given explicitly by (176).

The eigenfunction expansion of $\bar{\mathbf{G}}_e$ can be converted into a mode expansion. We can carry out the integral over w to obtain a mode expansion along z while leaving a spectral integral over k. Alternatively, we can carry out the integration over k and then be left with an inverse Fourier transform to be carried out to obtain the explicit dependence on z. Unfortunately, in either case it is only possible to evaluate the first integral using residue theory. We will evaluate the integral over k and leave it to the reader to evaluate the alternative solution obtained by integrating over w first.

The integral over k will be converted to an integral from $-\infty$ to ∞ by using

$$J_n(kr) = \tfrac{1}{2}[H_n^1(kr) + H_n^2(kr)] \tag{180a}$$

$$H_n^1(-kr) = -e^{-jn\pi}H_n^2(kr) \tag{180b}$$

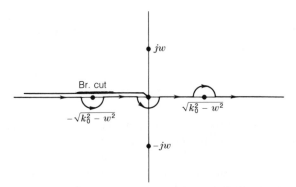

Fig. 2.17. Contour for obtaining the mode expansion.

$$J_n(-kr) = e^{jn\pi} J_n(kr). \tag{180c}$$

The integrand in (179) does not have any singularities at $k = 0$ since the $\mathbf{M}_j$ and $\mathbf{N}_j$ functions vanish for $k = 0$. When the Hankel functions are introduced the integral from $-\infty$ to ∞ must be chosen as a principal-value integral so as to avoid introducing any pole singularity at $k = 0$. The Y_n function which is part of the Hankel function has a part that contains a logarithmic factor $\ln kr$ and thus a branch point at the origin is introduced along with the Hankel functions. The branch point, however, does not cause any difficulty in the evaluation of the integrals. The Bessel function circuit relations given in (180) are valid when the phase angle of kr is between $-\pi$ and π. A branch cut along the negative real axis is introduced and the contour of integration is chosen as in Fig. 2.17.

A typical integral to be evaluated is of the form

$$\int_0^\infty f(k)kJ_n(kr)J_n(kr')\,dk$$

where $f(k)$ is an even function of k. For $r > r'$ we use (180a) to replace $J_n(kr)$ and then upon using (180b) and (180c) we have

$$\lim_{\eta \to 0} \int_\eta^\infty f(k)kJ_n(kr)J_n(kr')\,dk$$

$$= P\int_{-\infty}^\infty \frac{1}{2}f(k)kH_n^2(kr)J_n(kr')\,dk$$

by following the same steps used to obtain (157) for the mode expansion in spherical coordinates. The derivative operations with respect to r and r' can be brought outside the integral and applied later with due care being taken in differentiating functions that have a discontinuous first derivative when $r = r'$. The principal-value integral can be expressed as a contour integral along the path shown in Fig. 2.17 provided πj times the residue at any pole at $k = 0$ is subtracted. The contour can be closed by a semicircle in the lower half of the complex k plane since the asymptotic behavior of the Bessel functions makes the product $J_n(kr')H_n^2(kr)$ behave like $(1/k)e^{-jk(r-r')}$ which is exponentially decaying in the lower half plane when $r > r'$. The integral over the semicircle of infinite radius does not contribute as long as the conditions for Jordan's lemma hold. This requires that when $r > r'$ the integrand behaves like

$|k|^\tau$ with $\tau < 0$, apart from the exponential factor. When $r = r'$ the integrand must decrease at least as fast as $|k|^{-1+\tau}$ with $\tau < 0$. The only term for which these conditions are violated is the $\mathbf{a}_r \mathbf{a}_{r'}$ component of the $\mathbf{L}_j \mathbf{L}_j^*$ contribution. For this term the integrand is of the form

$$\frac{k^3}{k^2 + w^2} J_n'(kr) J_n'(kr')$$

and cannot be evaluated using residue theory. However, when the derivatives with respect to kr and kr' are brought outside the integral, a k^2 factor is eliminated and residue theory can be used. The resultant function of r and r' has a discontinuous derivative at $r = r'$ so the result of applying the operator $d^2/dr\,dr'$ is a delta function term in addition to the usual terms that arise. An alternative procedure is to express $k^2/(k^2 + w^2)$ as $1 - w^2/(k^2 + w^2)$. The integrand for the $\mathbf{a}_r \mathbf{a}_{r'}$ part of the $\mathbf{L}_j \mathbf{L}_j^*$ contribution is

$$\frac{k}{k^2 + w^2} \frac{dJ_n(kr)}{dr} \frac{dJ_n(kr')}{dr'}.$$

We can use (175) and the splitting of $k^2/(k^2 + w^2)$ just noted to rewrite the integrand in the form

$$\frac{-n^2 k J_n(kr) J_n(kr')}{(k^2 + w^2) rr'} - \frac{kw^2}{2(k^2 + w^2)} [J_{n+1}(kr) J_{n+1}(kr')$$

$$+ J_{n-1}(kr) J_{n-1}(kr')] + \frac{k}{2} [J_{n+1}(kr) J_{n+1}(kr')$$

$$+ J_{n-1}(kr) J_{n-1}(kr')].$$

The integral over k of the last term gives $\delta(r - r')/r$ as reference to (173d) shows. The remaining terms can be integrated using standard residue theory after the replacement of one of the Bessel functions by a Hankel function. In general, we can justify the formal procedure of differentiating discontinuous functions to obtain delta functions by extracting the delta function contributions directly from the integrals by rewriting the integrands in such a form that (173d) can be used. Of course (173d) should not be viewed as a classical function; it is an eigenfunction representation of the delta function operator.

When $r < r'$ the Bessel function $J_n(kr')$ is expressed in terms of the Hankel functions and a similar evaluation is carried out. Note that the conjugation operation shown in (179) applies only to the jw factor. The complex conjugate of the Hankel functions is not taken since these are replacements for the real J_n functions.

For the $\mathbf{M}_j \mathbf{M}_j^*$ contribution in (179) the denominator is $k(k^2 + w^2 - k_0^2)$ so poles at $k = 0$ and $k = (k_0^2 - w^2)^{1/2}$ occur. We will show that the residue contribution at the $k = 0$ pole is canceled by a corresponding contribution from the $\mathbf{N}_j \mathbf{N}_j^*$ terms. For the $\mathbf{N}_j \mathbf{N}_j^*$ terms the denominator is $k(k^2 + w^2)(k^2 + w^2 - k_0^2)$ because of the extra factor $1/p^2$ that occurs. The residue at the pole $k = -jw$ in the lower half plane will be shown to be canceled by a corresponding contribution from the $\mathbf{L}_j \mathbf{L}_j^*$ term. In fact, these residue terms completely cancel the $\mathbf{L}_j \mathbf{L}_j^*$ contributions outside, as well as inside, the source region. The only remaining contribution from the $\mathbf{L}_j \mathbf{L}_j^*$ terms is a delta function term coming from the $\mathbf{a}_r \mathbf{a}_{r'}$ part as described earlier. Thus, apart from the delta function term, the mode expansion is made up solely from the residue contributions at the pole $k = (k_0^2 - w^2)^{1/2}$ that is present in the integral for the $\mathbf{M}_j$ and $\mathbf{N}_j$ contributions.

At this point we must verify the residue cancellations described above and also evaluate the residual delta function term that is left over as part of the contribution from the longitudinal eigenfunctions. In order to evaluate the singular behavior at $k = 0$, in the various integrals we have to deal with, the following small argument approximations for the Bessel functions are used:

$$J_n(kr') \sim \frac{1}{n!} \left(\frac{kr'}{2} \right)^n - \frac{1}{(n+1)!} \left(\frac{kr'}{2} \right)^{n+2} \tag{181a}$$

$$H_n^2(kr) \sim \left[1 - j\frac{2}{\pi} \left(\gamma + \ln \frac{kr}{2} \right) \right] J_n(kr) + \frac{j}{\pi}(n-1)! \left(\frac{2}{kr} \right)^n \tag{181b}$$

$$H_0^2(kr) \sim -j\frac{2}{\pi} \left(\gamma + \ln \frac{kr}{2} \right) \tag{181c}$$

where $\gamma = 0.5772$ is Euler's constant. From these relations it is clear that the product $J_n(kr')H_n^2(kr)$ is finite at $k = 0$ and $\ln(kr/2)$ is multiplied by a k^n factor. Only the $n = 0$ term exhibits a logarithmic singularity. We will use the notation that $\mathbf{L}_j^+$, $\mathbf{M}_j^+$, and $\mathbf{N}_j^+$ denote the functions given in (176) but with $J_n(kr)$ replaced by $H_n^2(kr)$. The $\mathbf{a}_z$ component of the $\mathbf{N}_j$ functions is multiplied by k^2 so this component vanishes at $k = 0$ and does not have a residue contribution. Apart from the z- and ϕ-dependent factors, the parts of the $\mathbf{M}_j^+\mathbf{M}_j^*$ and $\mathbf{N}_j^+\mathbf{N}_j^*$ terms that contribute nonzero residues at $k = 0$ are (note that these are the factors multiplying the same $\sin n\phi$ or $\cos n\phi$ terms)

$$\left[-\mathbf{a}_\phi \frac{dH_n^2(kr)}{dr} \pm \mathbf{a}_r \frac{n}{r} H_n^2(kr) \right] \left[-\mathbf{a}_{\phi'} \frac{dJ_n(kr')}{dr'} \pm \mathbf{a}_{r'} \frac{n}{r'} J_n(kr') \right]$$

and

$$\frac{w^2}{p^2} \left[\mp \mathbf{a}_\phi \frac{n}{r} H_n^2(kr) + \mathbf{a}_r \frac{dH_n^2(kr)}{dr} \right] \left[\mp \mathbf{a}_{\phi'} \frac{n}{r'} J_n(kr') + \mathbf{a}_{r'} \frac{dJ_n(kr')}{dr'} \right].$$

Upon using the small argument approximations these expressions become

$$\frac{n}{\pi} \frac{r'^{n-1}}{r^{n+1}} (\mathbf{a}_\phi \pm \mathbf{a}_r)(\mathbf{a}_{\phi'} \mp \mathbf{a}_{r'})$$

$$\frac{w^2}{p^2} \frac{n}{\pi} \frac{r'^{n-1}}{r^{n+1}} (\pm \mathbf{a}_\phi + \mathbf{a}_r)(\mp \mathbf{a}_{\phi'} + \mathbf{a}_{r'}).$$

At the pole $k = 0$ the factor w^2/p^2 equals unity since $p^2 = k^2 + w^2$. Thus, as is readily apparent, the above two factors will cancel so the residues at the $k = 0$ pole cancel.

Next we consider the residues at the pole $k = -jw$ that occurs in connection with the $\mathbf{L}_j^+\mathbf{L}_j^*$ and $\mathbf{N}_j^+\mathbf{N}_j^*$ terms. The integrand for the $\mathbf{N}_j^+\mathbf{N}_j^*$ term is proportional to the expression

$$[\mathbf{a}_z(k^2 + w^2)H_n^2(kr) - jw\hat{\mathbf{L}}_j^+][\mathbf{a}_z(k^2 + w^2)J_n(kr') + jw\hat{\mathbf{L}}_j^*]$$

as can be seen by referring to (174c). The caret above the $\mathbf{L}_j$ functions signifies that the common functions of ϕ, ϕ', z, and z' are not included. The residue contributions come only from the part

$$\frac{w^2 \hat{\mathbf{L}}_j^+ \hat{\mathbf{L}}_j^*}{k(k^2 + w^2)(k^2 + w^2 - k_0^2)}$$

for the $\mathbf{N}_j$ functions. For the $\mathbf{L}_j$ functions the residues come from the factor

$$-\frac{k}{k_0^2} \frac{\hat{\mathbf{L}}_j^+ \hat{\mathbf{L}}_j^*}{k^2 + w^2}.$$

When $k = -jw$ the two multiplying factors are w^2/jwk_0^2 and jw/k_0^2, respectively, and hence the residue contributions are of opposite sign and cancel.

As noted earlier, the integral of the $\mathbf{a}_r \mathbf{a}_{r'}$ part of the $\mathbf{L}_j^+ \mathbf{L}_j^*$ contribution has to be evaluated either by extracting the delta function part or before the operation $d^2/dr\,dr'$ is carried out. This is not necessary for the $\mathbf{N}_j^+ \mathbf{N}_j^*$ terms. The result is that we obtain a delta function contribution which is given by the second derivative at $r = r'$, i.e.,

$$\mathbf{a}_r \mathbf{a}_{r'} \sum_n \frac{j\epsilon_{0n}}{8\pi k_0^2} \int_{-\infty}^{\infty} e^{-jw(z-z')} \begin{matrix} \cos \\ \sin \end{matrix} n\phi \begin{matrix} \cos \\ \sin \end{matrix} n\phi' \left. \frac{d^2}{dr_> dr_<} H_n^2(-jwr_>) J_n(-jwr_<) \right|_{r=r'} dw.$$

The step change in the first derivative with respect to r at $r = r'$ is

$$J_n(-jwr') \left. \frac{dH_n^2(-jwr)}{dr} \right|_{r'} - H_n^2(-jwr') \left. \frac{dJ_n(-jwr)}{dr} \right|_{r'} = -\frac{2j}{\pi r'}$$

upon using the Wronskian relationship for the Bessel functions involved. The step change, when the derivative is viewed as a function of r', is the negative of this so the delta function term that is produced is $(2j/\pi r')\delta(r - r')$. We now note that the integral over w can be carried out to give $2\pi\delta(z - z')$; hence we get

$$-\mathbf{a}_r \mathbf{a}_{r'} \delta(z - z') \frac{\delta(r - r')}{k_0^2 r} \sum_{n=0}^{\infty} \frac{\epsilon_{0n}}{2\pi} (\cos n\phi \cos n\phi' + \sin n\phi \sin n\phi')$$

$$= -\frac{1}{k_0^2} \frac{\delta(r - r')}{r} \delta(\phi - \phi') \delta(z - z') \mathbf{a}_r \mathbf{a}_r = -\frac{\mathbf{a}_r \mathbf{a}_r}{k_0^2} \delta(\mathbf{r} - \mathbf{r}')$$

since the series is the expansion of $\delta(\phi - \phi')$.

The final expression for the mode expansion of the electric field dyadic Green's function in cylindrical coordinates is

$$\bar{\mathbf{G}}_e(\mathbf{r}, \mathbf{r}') = \int_{-\infty}^{\infty} \sum_j \frac{j\epsilon_{0n}}{8\pi} [\mathbf{M}_{j\sigma}(\mathbf{r})\mathbf{M}_{j\sigma}(\mathbf{r}') + \mathbf{N}_{j\sigma}(\mathbf{r})\mathbf{N}_{j\sigma}(\mathbf{r}')] \frac{dw}{w^2 - k_0^2} - \frac{\mathbf{a}_r \mathbf{a}_r}{k_0^2} \delta(\mathbf{r} - \mathbf{r}') \quad (182)$$

where the mode functions are defined by

$$\mathbf{M}_j(\mathbf{r}) = \nabla \times \mathbf{a}_z J_n(\gamma r) e^{-jwz} \begin{matrix} \cos \\ \sin \end{matrix} n\phi \quad (183a)$$

$$\mathbf{M}_j^+(\mathbf{r}) = \nabla \times \mathbf{a}_z H_n^2(\gamma r) e^{-jwz} \frac{\cos}{\sin} n\phi \tag{183b}$$

$$k_0 \mathbf{N}_j(\mathbf{r}) = \nabla \times \nabla \times \mathbf{a}_z J_n(\gamma r) e^{-jwz} \frac{\cos}{\sin} n\phi \tag{183c}$$

$$k_0 \mathbf{N}_j^+(r) = \nabla \times \nabla \times \mathbf{a}_z H_n^2(\gamma r) e^{-jwz} \frac{\cos}{\sin} n\phi \tag{183d}$$

$$\mathbf{M}_{j\sigma}(\mathbf{r}) = \begin{cases} \mathbf{M}_j(\mathbf{r}), & r < r' \\ \mathbf{M}_j^+(\mathbf{r}), & r > r' \end{cases} \tag{183e}$$

$$\mathbf{N}_{j\sigma}(\mathbf{r}) = \begin{cases} \mathbf{N}_j(\mathbf{r}), & r < r' \\ \mathbf{N}_j^+(\mathbf{r}), & r > r' \end{cases} \tag{183f}$$

with $\gamma = (k_0^2 - w^2)^{1/2}$. The functions of $\mathbf{r}'$ are defined the same way except that e^{-jwz} is replaced by $e^{jwz'}$. The σ indicator is also applied the same way but with the roles of $\mathbf{r}$ and $\mathbf{r}'$ interchanged. The mode expansion including the delta function term given in (182) was also derived first by Tai [2.24].

If the integral over w is carried out first then there will be residue terms at $w = (k_0^2 - k^2)^{1/2}$ for the $\mathbf{M}_j \mathbf{M}_j^*$ and $\mathbf{N}_j \mathbf{N}_j^*$ terms and these produce a mode expansion along z. There are residues from poles at $w = -jk$ for the $\mathbf{L}_j \mathbf{L}_j^*$ and $\mathbf{N}_j \mathbf{N}_j^*$ terms also. However, these residue terms cancel with the exception of a delta function term that arises from the $\mathbf{a}_z \mathbf{a}_z$ part of the $\mathbf{L}_j \mathbf{L}_j^*$ contribution. In evaluating the integrals over w the $-jw$ factors that appear in the expressions for the $\mathbf{M}_j$, $\mathbf{N}_j$, and $\mathbf{L}_j$ functions in (176) are brought outside the integral as a d/dz operator so as to allow evaluation of the integrals using residue theory. The result of such an evaluation gives the following alternative mode expansion for the dyadic Green's function:

$$\bar{\mathbf{G}}_e(\mathbf{r}, \mathbf{r}') = -\frac{j}{4\pi} \sum_j \int_0^\infty \frac{\mathbf{M}_{j\sigma}(\mathbf{r})\mathbf{M}_{j\sigma}(\mathbf{r}') + \mathbf{N}_{j\sigma}(\mathbf{r})\mathbf{N}_{j\sigma}(\mathbf{r}')}{k\sqrt{k_0^2 - k^2}} dk - \frac{\mathbf{a}_z \mathbf{a}_z}{k_0^2} \delta(\mathbf{r} - \mathbf{r}') \tag{184}$$

where now the mode functions are given by

$$\mathbf{M}_j^+(\mathbf{r}) = \nabla \times \mathbf{a}_z J_n(kr) e^{-j\gamma z} \frac{\cos}{\sin} n\phi \tag{185a}$$

$$\mathbf{M}_j^-(\mathbf{r}) = \nabla \times \mathbf{a}_z J_n(kr) e^{j\gamma z} \frac{\cos}{\sin} n\phi \tag{185b}$$

$$\mathbf{M}_{j\sigma}(\mathbf{r}) = \begin{cases} \mathbf{M}_j^+(\mathbf{r}), & z > z' \\ \mathbf{M}_j^-(\mathbf{r}), & z < z' \end{cases} \tag{185c}$$

$$k_0 \mathbf{N}_j^+(\mathbf{r}) = \nabla \times \nabla \times \mathbf{a}_z J_n(kr) e^{-j\gamma z} \frac{\cos}{\sin} n\phi \tag{185d}$$

$$k_0 \mathbf{N}_j^-(r) = \nabla \times \nabla \times \mathbf{a}_z J_n(kr) e^{j\gamma z} \frac{\cos}{\sin} n\phi \tag{185e}$$

$$k_0 \mathbf{N}_{j\sigma}(\mathbf{r}) = \begin{cases} \mathbf{N}_j^+(\mathbf{r}), & z > z' \\ \mathbf{N}_j^-(\mathbf{r}), & z < z' \end{cases} \tag{185f}$$

where $\gamma = (k_0^2 - k^2)^{1/2}$.

The functions of $\mathbf{r}'$ are defined the same way but with the roles of $\mathbf{r}$ and $\mathbf{r}'$ interchanged. The σ indicator specifies the functions with superscript $+$ or $-$ in accordance with $z' > z$ or $z' < z$, respectively.

2.17. ALTERNATIVE REPRESENTATIONS FOR DYADIC GREEN'S FUNCTIONS

For certain types of scattering and diffraction problems it is useful to have other representations for the dyadic Green's functions that may exhibit a more rapid convergence than the series forms given in the two previous sections. This is very much the case for scattering and diffraction by very large cylinders and spheres for which the standard mode expansions converge very slowly.

One way to obtain an alternative representation is to synthesize an alternative form for the scalar Green's function g that is a solution of the Helmholtz equation and obtain the Green's dyadic function by means of the operations shown in (112). We will illustrate the method in connection with cylindrical coordinates.

Associated with the three-dimensional scalar Green's function problem

$$(\nabla^2 + k_0^2)g = -\delta(\mathbf{r} - \mathbf{r}') \tag{186}$$

are three one-dimensional characteristic Green's function problems which can be identified from the separated partial differential equation. These one-dimensional Green's function problems are

$$\frac{d^2 g_\phi}{d\phi^2} + \lambda_\phi g_\phi = -\delta(\phi - \phi') \tag{187a}$$

$$\frac{d^2 g_z}{dz^2} + \lambda_z g_z = -\delta(z - z') \tag{187b}$$

$$\frac{1}{r}\frac{d}{dr}r\frac{dg_r}{dr} + \lambda_r g_r - \frac{\nu^2}{r^2}g_r = -\frac{\delta(r - r')}{r} \tag{187c}$$

where $\lambda_\phi = \nu^2$, $\lambda_r = k_0^2 - \lambda_z$, in order that $g_r(r)$, $g_\phi(\phi)$, $g_z(z)$ satisfy the homogeneous Helmholtz equation when $\mathbf{r} \neq \mathbf{r}'$.

For a specific problem to solve we choose the scattering of the field from an infinitely long cylinder of radius a as illustrated in Fig. 2.18. On the cylinder surface we will impose Dirichlet boundary conditions which require $g = 0$ at $r = a$. In addition, g must represent outward-propagating waves at infinity.

The function g must be periodic in ϕ with a period of 2π. A useful solution can be obtained by considering ϕ as a rectangular coordinate that varies from $-\infty$ to ∞. The periodicity requirement can be met by placing unit sources at $\phi = \phi' + 2n\pi$, $n = \pm 1, \pm 2, \dots$. For the single source at ϕ' we can solve (187a) by choosing

$$g_\phi = \begin{cases} C_1 e^{-j\nu\phi}, & \phi > \phi' \\ C_2 e^{j\nu\phi}, & \phi < \phi' \end{cases}$$

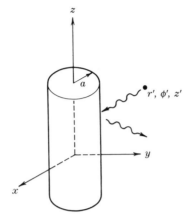

Fig. 2.18. A cylindrical scattering problem.

where $\nu = \sqrt{\lambda_\phi}$. At $\phi = \phi'$ we require g_ϕ to be continuous and in addition

$$\left. \frac{dg_\phi}{d\phi} \right|_{\phi'_-}^{\phi'_+} = -1.$$

By imposing these source boundary conditions we find that a solution for g_ϕ is

$$g_\phi = -\frac{j}{2\nu} e^{-j\nu|\phi-\phi'|}.$$

This wave solution is called a creeping wave since it represents a field that propagates or creeps around the cylinder. We can now superimpose the solutions from all the other sources to obtain the Green's function that has the required periodicity. Thus a valid Green's function is

$$g_\phi = -\frac{j}{2\nu} \sum_{n=-\infty}^{\infty} e^{-j\nu|\phi-\phi'-2n\pi|}. \tag{188}$$

A similar solution exists for g_z so we choose

$$g_z = -\frac{j}{2\sqrt{\lambda_z}} e^{-j\sqrt{\lambda_z}|z-z'|}. \tag{189}$$

For the function g_r a suitable solution is

$$G_r = \begin{cases} C_1[H_\nu^2(\sqrt{\lambda_r}r)J_\nu(\sqrt{\lambda_r}a) - H_\nu^2(\sqrt{\lambda_r}a)J_\nu(\sqrt{\lambda_r}r)], & a \le r \le r' \\ C_2H_\nu^2(\sqrt{\lambda_r}r), & r \ge r'. \end{cases}$$

This solution has the desired properties of vanishing at $r = a$ and corresponding to outward-propagating waves for large r. We require g_r to be continuous at $r = r'$ and also

$$\left. r\frac{dg_r}{dr} \right|_{r'_-}^{r'_+} = -1.$$

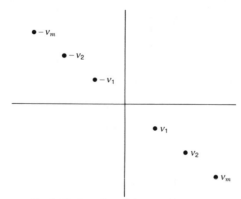

Fig. 2.19. Location of the complex roots ν_m.

When we impose the source boundary conditions and use the Wronskian relation

$$J_\nu(\sqrt{\lambda_r}r')\frac{dH_\nu^2(\sqrt{\lambda_r}r')}{dr'} - H_\nu^2(\sqrt{\lambda_r}r')\frac{dJ_\nu(\sqrt{\lambda_r}r')}{dr'} = -\frac{2j}{\pi r'}$$

we readily find that

$$g_r = \frac{j\pi}{2}\frac{H_\nu^2(\sqrt{\lambda_r}r_>)}{H_\nu^2(\sqrt{\lambda_r}a)}[H_\nu^2(\sqrt{\lambda_r}r_<)J_\nu(\sqrt{\lambda_r}a) - H_\nu^2(\sqrt{\lambda_r}a)J_\nu(\sqrt{\lambda_r}r_<)]. \qquad (190)$$

The spectra associated with g_ϕ and g_z arise from the branch points at the origin. In order that the wave solutions do not grow exponentially the branches are chosen such that

$$\text{Imag.}\ \sqrt{\lambda_\phi} = \text{Imag.}\ \nu \leq 0 \qquad (191a)$$

$$\text{Imag.}\ \sqrt{\lambda_z} \leq 0. \qquad (191b)$$

The function $H_\nu^2(\sqrt{\lambda_r}a)$ has an infinite number of zeroes for complex values of ν. Hence g_r has a discrete spectrum. The zeroes are located in the second and fourth quadrants of the complex ν plane as shown in Fig. 2.19. The two sets of zeroes are the negatives of each other. In the λ_ϕ plane only the zeroes with Imag. $\nu < 0$ lie on the Riemann sheet of interest in view of the manner in which the desired branch was specified by (191a).

According to the synthesis method presented in Section 2.8 the three-dimensional Green's function is given by

$$g(\mathbf{r}, \mathbf{r}') = \left(\frac{-1}{2\pi j}\right)^2 \oint_{C_\phi} \oint_{C_z} g_\phi(\lambda_\phi)g_r(-\lambda_\phi, k_0^2 - \lambda_z)g_z(\lambda_z)\,d\lambda_\phi\,d\lambda_z \qquad (192)$$

where we have chosen to integrate around the branch cuts in the λ_ϕ and λ_z planes. These cuts are placed along, but just above, the real axis so that the angle of λ_ϕ and λ_z ranges from 0 to -2π. This ensures that the conditions given by (191) will hold. It is convenient to change variables according to

$$w^2 = \lambda_z, \qquad d\lambda_z = 2w\,dw$$

$$\nu^2 = \lambda_\phi, \qquad d\lambda_\phi = 2\nu\,d\nu.$$

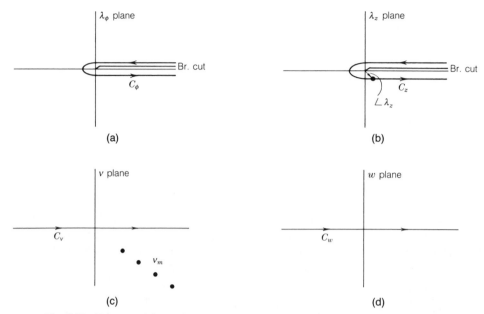

Fig. 2.20. Contours of integration for synthesizing the three-dimensional Green's function. (a) Branch cut and contour in λ_ϕ plane. (b) Branch cut and contour in λ_z plane. (c) Location of poles in ν plane. (d) The contour C_z mapped into the w plane.

The new contours now become the whole real axis in the w and ν planes. The various contours are shown in Fig. 2.20.

We can express g in the following form after the above change of variables has been made:

$$g(\mathbf{r}, \mathbf{r}') = \frac{1}{4\pi^2} \sum_{n=-\infty}^{\infty} \int_{-\infty}^{\infty} \int_{-\infty}^{\infty} g_r(-\nu^2, k_0^2 - w^2) e^{-j\nu|\phi-\phi'-2n\pi|}$$

$$\times e^{-jw|z-z'|} \, dw \, d\nu. \tag{193}$$

For a line source the solution is obtained by integrating over z' from $-\infty$ to ∞ to give $2\pi\delta(w)$. The integral over w then selects the value $w = 0$. We can close the contour by a semicircle in the lower half of the complex ν plane and evaluate the integral over ν in terms of the residues at the poles $\nu = \nu_m = \sigma_m - j\eta_m$. If we denote the residue associated with $H_\nu^2(\sqrt{k_0^2 - w^2}a)$ by

$$R_{\nu_m} = \frac{1}{\left.\dfrac{d}{d\nu} H_\nu^2\left(\sqrt{k_0^2 - w^2}a\right)\right|_{\nu_m}}$$

our solution for g can be expressed as

$$g(\mathbf{r}, \mathbf{r}') = \frac{1}{2\pi j} \sum_{n=-\infty}^{\infty} \sum_{m=1}^{\infty} e^{-(j\sigma_m + \eta_m)|\phi-\phi'-2n\pi|} \frac{j\pi}{2} R_{\nu_m} H_{\nu_m}^2(\gamma r_>)$$

$$\times [H_{\nu_m}^2(\gamma r_<) J_{\nu_m}(\gamma a) - H_{\nu_m}^2(\gamma a) J_{\nu_m}(\gamma r_<)] e^{-jw|z-z'|} \, dw \tag{194}$$

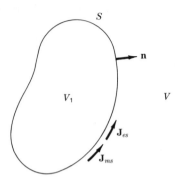

Fig. 2.21. Illustration for a field equivalence principle.

where $\gamma = (k_0^2 - w^2)^{1/2}$. The significant advantage obtained with this solution is that when γa is large the imaginary parts η_m of ν_m are large so that only a few terms in the sum over n and m need be retained because of the rapid decay of the ϕ function. The evaluation of the roots and the residues is quite complex and the reader is referred to the literature for the details [2.15], [2.25].

We can construct a dyadic Green's function from (194) but this function will, in general, not satisfy the boundary conditions $\mathbf{n} \times \bar{\mathbf{G}}_e = 0$ on the cylinder surface. However, if we have a z-directed electric current element the function

$$\bar{\mathbf{G}}_e(\mathbf{r}, \mathbf{r}') \cdot \mathbf{a}_z = \mathbf{a}_z g + \frac{1}{k_0^2} \nabla \frac{\partial g}{\partial z} \tag{195}$$

will have zero tangential components on the cylinder surface and the problem of scattering of electromagnetic waves, radiated by a z-directed electric current element, from a perfectly conducting cylinder is solved. For the more general case it would be necessary to add solutions to the homogeneous vector wave equation to take care of boundary conditions.

2.18. Dyadic Green's Functions and Field Equivalence Principles

Let the sources for an electromagnetic field be contained in a volume V_1 bounded by a smooth closed surface S as shown in Fig. 2.21. The unit normal directed out from V_1 is $\mathbf{n}$. The electric field outside S is a solution of the source-free equation

$$\nabla \times \nabla \times \mathbf{E} - k_0^2 \mathbf{E} = 0. \tag{196}$$

Consider also a Green's dyadic function that is a solution of

$$\nabla \times \nabla \times \bar{\mathbf{G}}(\mathbf{r}, \mathbf{r}_0) - k_0^2 \bar{\mathbf{G}} = \bar{\mathbf{I}} \delta(\mathbf{r} - \mathbf{r}_0) \tag{197}$$

and satisfies a radiation condition at infinity. We now form the scalar products

$$\mathbf{E} \cdot \nabla \times \nabla \times \bar{\mathbf{G}} - \nabla \times \nabla \times \mathbf{E} \cdot \bar{\mathbf{G}} = \mathbf{E}(\mathbf{r}) \delta(\mathbf{r} - \mathbf{r}_0).$$

The left-hand side can be written as

$$-\nabla \cdot [\mathbf{E} \times \nabla \times \bar{\mathbf{G}} + \nabla \times \mathbf{E} \times \bar{\mathbf{G}}]$$

as direct expansion of the divergence will verify. We now integrate over the volume V outside S and use the divergence theorem to obtain

$$\oiint_S (\mathbf{n} \times \mathbf{E} \cdot \nabla \times \bar{\mathbf{G}} + \mathbf{n} \times \nabla \times \mathbf{E} \cdot \bar{\mathbf{G}}) \, dS = \begin{cases} \mathbf{E}(\mathbf{r}_0), & \mathbf{r}_0 \in V \\ 0, & \mathbf{r}_0 \in V_1 \end{cases} \tag{198}$$

where the integral over a spherical surface at infinity has been set to zero since it vanishes because both $\bar{\mathbf{G}}$ and $\mathbf{E}$ satisfy a radiation condition. From Maxwell's equations $\nabla \times \mathbf{E} = -j\omega\mu_0\mathbf{H}$, so for $\mathbf{r}_0$ in V

$$\mathbf{E}(\mathbf{r}_0) = \oiint_S \mathbf{n} \times \mathbf{E} \cdot \nabla \times \bar{\mathbf{G}} \, dS - j\omega\mu_0 \oiint_S \mathbf{n} \times \mathbf{H} \cdot \bar{\mathbf{G}} \, dS \tag{199}$$

which gives the field in V in terms of boundary values on S. This is the mathematical statement of Love's field equivalence theorem discussed in Section 1.8. The boundary values can be interpreted as equivalent surface currents $\mathbf{J}_{ms} = -\mathbf{n} \times \mathbf{E}$ and $\mathbf{J}_{es} = \mathbf{n} \times \mathbf{H}$. The dyadic Green's function can be chosen as the one for unbounded free space.

A similar solution for the magnetic field can be constructed since $\mathbf{H}$ also satisfies (196). Hence

$$\mathbf{H}(\mathbf{r}_0) = \oiint_S \mathbf{n} \times \mathbf{H} \cdot \nabla \times \bar{\mathbf{G}} \, dS + j\omega\epsilon_0 \oiint_S \mathbf{n} \times \mathbf{E} \cdot \bar{\mathbf{G}} \, dS. \tag{200}$$

The solutions given by (199) and (200) have the expected property that the functions of $\mathbf{r}_0$ given by the integrals involving $\nabla \times \bar{\mathbf{G}}$ have tangential components that undergo a step change equal to the boundary value $\mathbf{n} \times \mathbf{E}$ or $\mathbf{n} \times \mathbf{H}$ as the observation point $\mathbf{r}_0$ passes through the surface S from the exterior side V to the interior side V_1. We can readily prove this property.

Let S_0 be a very small circular disk that is a part of S as shown in Fig. 2.22. The surface integral in (199) is split into an integral over S_0 and an integral over $S - S_0$. The latter integral is a continuous function of $\mathbf{r}_0$ since $\mathbf{r} \neq \mathbf{r}_0$ and hence does not have a discontinuous behavior across S. For the integral over S_0 we choose S_0 so small that $\mathbf{n} \times \mathbf{E}$ can be considered to be constant on S_0. If we use the free-space dyadic Green's function given by (112), then

$$\nabla \times \bar{\mathbf{G}} = \nabla \times \bar{\mathbf{I}} \frac{e^{-jk_0 R}}{4\pi R} \approx \frac{1}{4\pi} \nabla \times \frac{\bar{\mathbf{I}}}{R}$$

since the exponential factor can be approximated by unity over the small disk and in the region immediately surrounding it. The integral of interest is thus approximated as follows:

$$\frac{1}{4\pi} \mathbf{n} \times \mathbf{E} \cdot \iint_{S_0} \nabla \times \frac{\bar{\mathbf{I}}}{R} \, dS = \frac{1}{4\pi} \mathbf{n} \times \mathbf{E} \cdot \iint_{S_0} \nabla \frac{1}{R} \, dS \times \bar{\mathbf{I}}.$$

From Fig. 2.22 we see that $\nabla(1/R)$ can be expressed as

$$\nabla \frac{1}{R} = -\frac{\mathbf{a}_R}{R^2} = \frac{\mathbf{n}}{R^2} \cos \theta - \frac{\mathbf{a}_r}{R^2} \sin \theta$$

where $\mathbf{a}_R = (\mathbf{r} - \mathbf{r}_0)/R$.

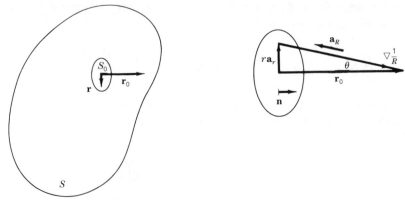

Fig. 2.22. A small circular area S_0 on S.

From symmetry considerations the integral of the component along $\mathbf{a}_r$ over the circular disk is zero. The remaining integral

$$\mathbf{n} \iint_{S_0} \frac{\cos \theta}{R^2} \, dS = \mathbf{n} \iint_{S_0} d\Omega = \mathbf{n}\Omega$$

where Ω is the solid angle subtended by the disk at the observation point. Our final result for the integral is thus

$$\mathbf{n} \times \mathbf{E} \cdot \mathbf{n} \times \bar{\mathbf{I}} \frac{\Omega}{4\pi} = (\mathbf{n} \times \mathbf{E}) \times \mathbf{n} \frac{\Omega}{4\pi} = -\mathbf{n} \times (\mathbf{n} \times \mathbf{E}) \frac{\Omega}{4\pi}. \tag{201}$$

The solid angle Ω increases up to a value of 2π as $\mathbf{r}_0$ approaches S and then jumps to -2π as $\mathbf{r}_0$ passes through S. Since $\mathbf{n} \times (\mathbf{n} \times \mathbf{E}) = (\mathbf{n} \cdot \mathbf{E})\mathbf{n} - (\mathbf{n} \cdot \mathbf{n})\mathbf{E} = -\mathbf{E}_{\text{tan}}$, we see that (201) shows that the tangential value of $\mathbf{E}$ given by

$$\lim_{S_0 \to 0} \iint_{S_0} \mathbf{n} \times \mathbf{E} \cdot \nabla \times \bar{\mathbf{G}} \, dS$$

approaches a value equal to one-half of the boundary value obtained from $\mathbf{n} \times \mathbf{E}$ as $\mathbf{r}_0$ approaches S from the exterior region. The other half of the total tangential electric field comes from the sources on $S - S_0$. Note from (201) that the integral does not contribute a normal component when S_0 is a circular disk.

Let the z axis be oriented along $\mathbf{n}$ from the center of the circular disk assumed to have a radius a. We now consider

$$\iint_{S_0} \mathbf{n} \times \mathbf{H} \cdot \bar{\mathbf{G}} \, dS \approx \mathbf{n} \times \mathbf{H} \cdot \iint_{S_0} \left(\bar{\mathbf{I}} + \frac{1}{k_0^2} \nabla\nabla \right) \frac{1}{4\pi R} \, dS$$

where we have again approximated $e^{-jk_0 R}$ by one. If $\mathbf{r}_0$ lies on the z axis at z_0 and ρ is a radial coordinate on the disk, then the first integral is given by

$$\int_0^{2\pi} \int_0^a \frac{\rho \, d\rho \, d\phi}{\sqrt{\rho^2 + z_0^2}} = 2\pi[\sqrt{a^2 + z_0^2} - |z_0|].$$

This function is continuous as the point z_0 passes through the surface and vanishes as a approaches zero.

In order to evaluate the second integral we will orient the x axis parallel to $\mathbf{n} \times \mathbf{H} = \mathbf{J}_{es}$ and first find the x component of the contribution to the field. This is given by (the integral is evaluated first and then z_0 can be set equal to zero)

$$-\frac{j\omega\mu_0}{4\pi k_0^2} J_{es} \int_0^{2\pi} \int_0^a \frac{\partial^2}{\partial x^2} \frac{1}{\sqrt{\rho^2 + z_0^2}} \rho \, d\rho \, d\phi$$

where $\rho^2 = x^2 + y^2$. The integral can be expressed as

$$-\frac{j\omega\mu_0}{4\pi k_0^2} J_{es} 4 \int_0^a \int_0^{\sqrt{a^2 - y^2}} \frac{\partial^2}{\partial x^2} \frac{1}{\sqrt{x^2 + y^2 + z_0^2}} \, dx \, dy$$

$$= -\frac{j\omega\mu_0}{4\pi k_0^2} J_{es} 4 \int_0^a \frac{\partial}{\partial x} \frac{1}{\sqrt{x^2 + y^2 + z_0^2}} \Bigg|_0^{\sqrt{a^2 - y^2}} \, dy$$

$$= \frac{j\omega\mu_0}{4\pi k_0^2} J_{es} 4 \int_0^a \frac{\sqrt{a^2 - y^2}}{(a^2 + z_0^2)^{3/2}} \, dy$$

$$= \frac{j\omega\mu_0 J_{es}}{2\pi k_0^2 (a^2 + z_0^2)^{3/2}} \left[y\sqrt{a^2 - y^2} + a^2 \sin^{-1} \frac{y}{a} \right]\Bigg|_0^a$$

$$= \frac{j\omega\mu_0}{4 k_0^2} J_{es} \frac{a^2}{(a^2 + z_0^2)^{3/2}}.$$

This contribution is also a continuous function of z_0 and vanishes when a approaches zero for all finite values of z_0. For $z_0 = 0$ the tangential electric field increases proportional to a^{-1}. This singular behavior is an artificial one due to the line charge $\rho_\ell = \boldsymbol{\tau} \cdot \mathbf{J}_{es}/j\omega$ on the boundary C on S_0 when the current disk is isolated. The integral over $S - S_0$ produces a compensating singularity. For numerical evaluation the patch size would be kept finite. From a theoretical point of view the integral can be evaluated by an integration by parts to show explicitly the line charge contribution that produces the singular behavior, as we will show later. If we excluded the line charge contribution from both the integral over S_0 and that over $S - S_0$ both integrals would have much better convergence properties.

By means of a similar analysis, we find that the contribution to the tangential component of the electric field that is perpendicular to $\mathbf{J}_{es}$ is zero.

The normal component of the electric field at the center of the disk cannot be found using the assumption that the current $\mathbf{J}_{es}$ on the disk is constant since this is equivalent to assuming that the charge density given by $j\omega\rho_s = -\nabla_s \cdot \mathbf{J}_{es}$ is zero. In order to evaluate the normal component we consider (the limit as S_0 approaches zero is implied)

$$-\frac{j\omega\mu_0}{4\pi k_0^2} \mathbf{n} \cdot \iint_{S_0} \left(\nabla\nabla \frac{1}{R} \right) \cdot \mathbf{J}_{es} \, dS$$

$$= \frac{j\omega\mu_0}{4\pi k_0^2} \mathbf{n} \cdot \iint_{S_0} \left(\nabla_0 \nabla \frac{1}{R} \right) \cdot \mathbf{J}_{es} \, dS$$

$$= \frac{j\omega\mu_0}{4\pi k_0^2} \mathbf{n} \cdot \iint_{S_0} \nabla_0 \left[\nabla \cdot \left(\frac{\mathbf{J}_{es}}{R} \right) - \frac{1}{R} \nabla \cdot \mathbf{J}_{es} \right] dS.$$

When S_0 approaches zero we can replace $\nabla \cdot \mathbf{J}_{es}$ by $-j\omega\rho_s$ outside the integral. Hence we obtain, upon using the divergence theorem,

$$\frac{j\omega\mu_0}{4\pi k_0^2}\oint_C \boldsymbol{\tau}\cdot\mathbf{J}_{es}\,\mathbf{n}(\mathbf{r}_0)\cdot\nabla_0\frac{1}{R}\,d\ell - \frac{\rho_s}{4\pi\epsilon_0}\mathbf{n}\cdot\iint_{S_0}\nabla_0\frac{1}{R}\,dS$$

where $\boldsymbol{\tau}$ is a unit outward normal to the boundary of S_0 and lies on S.

We can choose $\mathbf{r}_0$ to coincide with the center of S_0. On S_0 we can expand $\mathbf{J}_{es}(\mathbf{r})$ in a Taylor series about $\mathbf{r}_0$. The dominant contribution to the contour integral comes from the $\mathbf{J}_{es}(\mathbf{r}_0)$ term. For this term $\boldsymbol{\tau}\cdot\mathbf{J}_{es}(\mathbf{r}_0) = |\mathbf{J}_{es}(\mathbf{r}_0)|\cos\phi$ where ϕ is the angle between $\boldsymbol{\tau}$ and the constant vector $\mathbf{J}_{es}(\mathbf{r}_0)$. When $\mathbf{r}_0$ lies on S_0 we have $\mathbf{n}(\mathbf{r}_0)\cdot\nabla_0(1/R) = \cos\theta/R^2$ where θ is the angle between $\mathbf{n}(\mathbf{r}_0)$ and $\mathbf{a}_R$. On C the distance R is never zero and since R is a constant along the contour C of the circular disk the contour integral is zero because the integral of $\cos\phi$ over the range 0 to 2π is zero. The second integral equals $\rho_s(\mathbf{r}_0)\Omega/4\pi\epsilon_0$ which is the expected value since for $z_0 = 0$ this makes $\mathbf{n}\cdot\mathbf{E} = \rho_s/2\epsilon_0$ at the center of the disk. There is an equal and opposite directed electric field normal to the disk on the interior side. The sources on $S - S_0$ produce a field that will cancel the interior field, which, as (198) shows, is zero.

In principle we can construct dyadic Green's functions that satisfy one of the following boundary conditions on S:

$$n \times \bar{\mathbf{G}} = 0 \qquad \text{on } S$$

$$\mathbf{n} \times \nabla \times \bar{\mathbf{G}} = 0 \qquad \text{on } S.$$

In this instance the solutions for $\mathbf{E}(\mathbf{r}_0)$ reduce to

$$\mathbf{E}(\mathbf{r}_0) = \oiint_S \mathbf{n} \times \mathbf{E}\cdot\nabla \times \bar{\mathbf{G}}\,dS \tag{202a}$$

and

$$\mathbf{E}(\mathbf{r}_0) = -j\omega\mu_0 \oiint_S \mathbf{n} \times \mathbf{H}\cdot\bar{\mathbf{G}}\,dS. \tag{202b}$$

These two solutions correspond to Schelkunoff's field equivalence principles.

The two Green's functions appearing in (202) are functions of $\mathbf{r}$ and $\mathbf{r}_0$ but cannot be continued as a function of $\mathbf{r}_0$ into the region V_1. When boundary conditions are imposed on a Green's function its region of validity becomes restricted to the region for which it was constructed. Nevertheless, the function of $\mathbf{r}_0$ in (202a) will exhibit the same discontinuous behavior in its tangential value in that as $\mathbf{r}_0$ passes through the surface S the tangential value given by $\mathbf{n} \times \mathbf{E}$ on the exterior side is reduced to zero on the interior side. This property stems from the boundary condition $\mathbf{n} \times \bar{\mathbf{G}} = 0$ on S. The integral in (202b) has a discontinuous curl. The discontinuous behavior exhibited by (202a) and (202b) is consistent with the interpretation of $-\mathbf{n} \times \mathbf{E}$ in (202a) being an equivalent magnetic surface current on a perfect electric conducting surface and that of $\mathbf{n} \times \mathbf{H}$ in (202b) being an equivalent electric current on a perfect magnetic conducting surface.

In practice there are relatively few problems for which dyadic Green's functions satisfying either of the two above boundary conditions can be constructed. For those problems for which the Green's functions can be constructed the discontinuous behavior described above can be readily verified.

2.19. Integral Equations for Scattering

The scattering of an incident electromagnetic field by an obstacle may be formulated as an integral equation to be solved numerically rather than by solving the partial differential equation with boundary conditions. In order to formulate the basic integral equations for scattering, consider a current source $\mathbf{J}_a$ outside a perfectly conducting surface S (see Fig. 2.21). In place of (199) one readily finds that the total electric field is given by

$$\mathbf{E}(\mathbf{r}_0) = \oiint_S \mathbf{n} \times \mathbf{E} \cdot \nabla \times \bar{\mathbf{G}} \, dS - j\omega\mu_0 \oiint_S \mathbf{n} \times \mathbf{H} \cdot \bar{\mathbf{G}} \, dS - j\omega\mu_0 \iiint_V \mathbf{J}_a \cdot \bar{\mathbf{G}} \, dS.$$

The last integral gives the incident electric field $\mathbf{E}_i$ in the absence of the obstacle when $\bar{\mathbf{G}}$ is the free-space dyadic Green's function. The induced current $\mathbf{J}_{es}$ on S is given by $\mathbf{n} \times \mathbf{H}$. The integral equation for the unknown current $\mathbf{J}_{es}$ is obtained by setting the tangential component of the total electric field equal to zero on the perfectly conducting surface S; thus

$$\mathbf{n} \times \mathbf{E}(\mathbf{r}_0) = j\omega\mu_0 \oiint_S \mathbf{n} \times \mathbf{H} \cdot \bar{\mathbf{G}} \times \mathbf{n}(\mathbf{r}_0) \, dS + \mathbf{n} \times \mathbf{E}_i = 0$$

or equivalently

$$j\omega\mu_0 \oiint_S \mathbf{J}_{es}(\mathbf{r}) \cdot \bar{\mathbf{G}}(\mathbf{r}, \mathbf{r}_0) \times \mathbf{n}(\mathbf{r}_0) \, dS = -\mathbf{n}(\mathbf{r}_0) \times \mathbf{E}_i(\mathbf{r}_0), \qquad \mathbf{r}_0 \text{ on } S \qquad (203)$$

which is the electric field integral equation (EFIE).[4] This integral equation has a unique solution provided there are no current distributions on S for which the integral on the left gives a value of zero. In order that a solution exists the source function $-\mathbf{n} \times \mathbf{E}_i$ must be orthogonal to the null space of the integral operator; i.e., it must lie completely in the range space of the operator. Unfortunately, at discrete values of k_0 corresponding to the internal resonant frequencies of the cavity formed by S the currents $\mathbf{J}_{es}$ that correspond to those associated with the cavity resonant modes are in the null space of the integral operator. When the scattering object is large in terms of wavelength the resonant frequencies are closely spaced so the numerical solution of (203) is not feasible without some modification of the integral equation. A suitable augmentation will be described later.

The proof that the current on S that corresponds to a resonant cavity mode is in the null space of the operator is straightforward. The application of Green's theorem in the region V_1 shows that

$$\mathbf{E}(\mathbf{r}_0) = j\omega\mu_0 \oiint_S \mathbf{J}_{es} \cdot \bar{\mathbf{G}} \, dS - \oiint_S \mathbf{n} \times \mathbf{E} \cdot \nabla \times \bar{\mathbf{G}} \, dS, \qquad \mathbf{r}_0 \in V_1$$

where $\mathbf{n}$ is still directed out from V_1.

The eigenvalue equation for $\mathbf{J}_{es}$ for a resonant mode is obtained by equating $\mathbf{n} \times \mathbf{E}$ to zero

---

[4]The integral must be evaluated using an appropriate limiting procedure as described in the previous section.

on S. This gives

$$j\omega\mu_0 \oiint_S \mathbf{J}_{es}(\mathbf{r}) \cdot \bar{\mathbf{G}}(\mathbf{r}, \mathbf{r}_0) \times \mathbf{n}(\mathbf{r}_0) \, dS = 0, \qquad \mathbf{r}_0 \text{ on } S.$$

But this equation is the same as the homogeneous equation obtained from (203) when $\mathbf{n} \times \mathbf{E}_i$ is set equal to zero. Hence, resonant mode surface currents are in the null space of the integral operator. The eigenvalue equation has solutions at the resonant wavenumbers for the cavity.

An eigenmode current $\mathbf{J}_{es}$ does not produce a radiated field outside S since it gives $\mathbf{n} \times \mathbf{E} = 0$ on S. We can show that this current is orthogonal to the incident field $\mathbf{E}_i$. For this purpose we need to choose an appropriate scalar product. We will choose as the scalar product the integral of $\mathbf{E}_i \cdot \mathbf{J}_{es}$ over the surface S for the simple reason that the results we wish to establish later can then be obtained without introducing an adjoint integral operator. The same results can be obtained by using $\mathbf{E}_i \cdot \mathbf{J}_{es}^*$ in the integral over S.

From a physical point of view, the resonant cavity mode current distribution on S supports the nonzero interior cavity resonant mode field. This field has a zero tangential electric field on the closed surface S, a finite tangential magnetic field on the interior side of S, and a zero field on the exterior side of S.

When $\mathbf{J}_a$ is the source for the incident field outside S the reciprocity theorem gives

$$\iiint_V \mathbf{E}_i \cdot \mathbf{J}_{es} \, dV = \iiint_V \mathbf{E} \cdot \mathbf{J}_a \, dV = 0$$

where $\mathbf{E}$, the field radiated by $\mathbf{J}_{es}$, is zero outside S. The volume integral on the left reduces to the surface integral

$$\oiint_S \mathbf{E}_i \cdot \mathbf{J}_{es} \, dS = 0$$

and involves only the tangential part of $\mathbf{E}_i$. Since the integral is zero we conclude that $\mathbf{E}_{i,\text{tan}}$ is orthogonal to the eigenmode current. For this reason a solution to the integral equation exists. In principle, the solution can be made unique by removing the eigenmode currents from any current distribution $\mathbf{J}_{es}$ on S. In practice, it is not easy to accomplish this goal since neither the eigenfrequencies nor the eigenmode currents are known.

An approximation numerical solution of the EFIE (203) can be achieved using the method of moments [2.26]. The current $\mathbf{J}_{es}$ may be expanded in terms of a set of vector basis functions defined on S, i.e.,

$$\mathbf{J}_{es}(\mathbf{r}) = \sum_{n=1}^{N} I_n \mathbf{J}_n(\mathbf{r})$$

where the I_n are unknown amplitudes and the $\mathbf{J}_n(\mathbf{r})$ are basis functions in the domain of the operator. These should be part of a complete set of functions on S. When this expansion is used in (203) and the integration over $\mathbf{r}$ is carried out we obtain

$$j\omega\mu_0 \sum_{n=1}^{N} I_n \mathbf{G}_n(\mathbf{r}_0) \approx -\mathbf{n}(\mathbf{r}_0) \times \mathbf{E}_i(\mathbf{r}_0) \tag{204}$$

where

$$\mathbf{G}_n(\mathbf{r}_0) = \oiint_S \mathbf{J}_n(\mathbf{r}) \cdot \bar{\mathbf{G}}(\mathbf{r}, \mathbf{r}_0) \times \mathbf{n}(\mathbf{r}_0) \, dS.$$

In evaluating the $\mathbf{G}_n(\mathbf{r}_0)$ an appropriate limiting procedure must be used when the two points $\mathbf{r}$ and $\mathbf{r}_0$ can coincide. For example, a small circular disk area can be isolated in the manner described earlier so that the integral is split into an integral over S_0 and one over $S - S_0$. When only a finite number of basis functions are used the integral equation can hold in an approximate sense only; this is reflected in (204).

A system of N algebraic equations for determining the I_n can be obtained by introducing N weighting functions $\mathbf{W}_m(\mathbf{r}_0)$ that are part of a complete set in the range space of the operator, and equating the N integrated weights of the two sides of (204) to each other. Thus

$$j\omega\mu_0 \sum_{n=1}^{N} I_n G_{nm} = S_m, \qquad M = 1, 2, \ldots, N \tag{205}$$

where

$$G_{nm} = \oiint_S \mathbf{G}_n(\mathbf{r}_0) \cdot \mathbf{W}_m(\mathbf{r}_0) \, dS_0$$

$$S_m = -\oiint_S \mathbf{n}(\mathbf{r}_0) \times \mathbf{E}_i(\mathbf{r}_0) \cdot \mathbf{W}_m(\mathbf{r}_0) \, dS_0.$$

The reciprocity theorem shows that $\mathbf{J}_n(\mathbf{r}) \cdot \bar{\mathbf{G}}(\mathbf{r}, \mathbf{r}_0) \cdot \mathbf{J}_m(\mathbf{r}_0) = \mathbf{J}_m(\mathbf{r}_0) \cdot \bar{\mathbf{G}}(\mathbf{r}_0, \mathbf{r}) \cdot \mathbf{J}_n(\mathbf{r})$ and since the free-space dyadic Green's function is symmetrical in $\mathbf{r}$, $\mathbf{r}_0$, i.e., $\bar{\mathbf{G}}(\mathbf{r}, \mathbf{r}_0) = \bar{\mathbf{G}}(\mathbf{r}_0, \mathbf{r})$, we see that if we choose $\mathbf{W}_m = \mathbf{J}_m$ the matrix with elements G_{nm} will be symmetrical. The above system of equations can be solved provided the matrix with elements G_{nm} is not singular. However, when k_0 is equal to an interior cavity resonant wavenumber the matrix has a zero (or near zero, for finite N) determinant because an eigenvector exists at each resonant frequency and the eigenvalue of the operator is zero. Even if k_0 is only close to a resonant wavenumber the inversion of (205) is numerically difficult because the matrix is ill conditioned, i.e., almost singular. A solution can be obtained using generalized eigenvectors [2.9] but it is preferable to find some way to avoid a singular matrix altogether.

An integral equation involving the magnetic field can also be formulated. For a source current $\mathbf{J}_a(\mathbf{r})$ the magnetic field is a solution of

$$\nabla \times \nabla \times \mathbf{H} - k_0^2 \mathbf{H} = \nabla \times \mathbf{J}_a.$$

This equation can be solved using the same free-space dyadic Green's function. The solution is given by

$$\mathbf{H}(\mathbf{r}_0) = \oiint_S \mathbf{n} \times \mathbf{H} \cdot \nabla \times \bar{\mathbf{G}} \, dS + j\omega\epsilon_0 \oiint_S \mathbf{n} \times \mathbf{E} \cdot \bar{\mathbf{G}} \, dS + \iiint_V \nabla \times \mathbf{J}_a \cdot \bar{\mathbf{G}} \, dS.$$

On the surface S we have $\mathbf{n} \times \mathbf{E} = 0$, $\mathbf{n} \times \mathbf{H} = \mathbf{J}_{es}$, so upon replacing the last integral by the incident magnetic field $\mathbf{H}_i$ we obtain

$$\mathbf{H}(\mathbf{r}_0) = \oint\!\!\!\oint_S \mathbf{J}_{es} \cdot \nabla \times \bar{\mathbf{G}}\, dS + \mathbf{H}_i. \tag{206}$$

The surface integral gives the scattered magnetic field $\mathbf{H}_s$.

In the formulation of the electric field integral equation (203) the observation point $\mathbf{r}_0$ could be placed on the surface S since the tangential component of the integral on the left is continuous across S. In (206) the tangential component of the integral is a discontinuous function of $\mathbf{r}_0$ and changes discontinuously as $\mathbf{r}_0$ crosses the surface S [see (201) for a similar case]. For the magnetic field we have

$$\lim_{\mathbf{r}_0 \to S^+} \mathbf{n} \times (\mathbf{H}_s + \mathbf{H}_i) = \mathbf{J}_{es}(\mathbf{r}_0) \tag{207}$$

where S^+ denotes the exterior side of S.

If the integral in (206) is split into an integral over $S - S_0$ and an integral over S_0 where S_0 is a small circular disk, then

$$-\mathbf{n} \times \mathbf{H}(\mathbf{r}_0) = \lim_{S_0 \to 0}\left[\iint_{S-S_0} \mathbf{J}_{es} \cdot \nabla \times \bar{\mathbf{G}}\, dS \times \mathbf{n}\right] + \lim_{\substack{S_0 \to 0 \\ \mathbf{r}_0 \to S^+}} \iint_{S_0} \mathbf{J}_{es} \cdot \nabla \times \bar{\mathbf{G}}\, dS \times \mathbf{n} - \mathbf{n} \times \mathbf{H}_i.$$

The first integral is called the principal-value integral even though the term principal value as used here is not the same as the Cauchy principal value of an integral of a complex function.[5] The second integral has the value $(\mathbf{H} \times \mathbf{n})/2 = -\mathbf{J}_{es}/2$ as may be deduced from (201) which was obtained for a similar integral. Since $\mathbf{n} \times \mathbf{H}(\mathbf{r}_0) = \mathbf{J}_{es}$ we finally obtain

$$\frac{\mathbf{J}_{es}(\mathbf{r}_0)}{2} + \lim_{S_0 \to 0}\left[\iint_{S-S_0} \mathbf{J}_{es}(\mathbf{r}) \cdot \nabla \times \bar{\mathbf{G}}(\mathbf{r}, \mathbf{r}_0)\, dS \times \mathbf{n}(\mathbf{r}_0)\right] = \mathbf{n}(\mathbf{r}_0) \times \mathbf{H}_i(\mathbf{r}_0), \qquad \mathbf{r}_0 \text{ on } S \tag{208a}$$

which is Maue's magnetic field integral equation (MFIE).

If the derivation of (201) is examined it can be seen that

$$\lim_{z_0 \to 0} \mathbf{n} \iint_{S_0} \frac{\cos\theta}{R^2}\, dS = \lim_{z_0 \to 0} \mathbf{n} \iint_{S_0} d\Omega = 2\pi\mathbf{n}$$

is independent of the shape of S_0. Thus the factor of $\frac{1}{2}$ multiplying $\mathbf{J}_{es}(\mathbf{r}_0)$ in (208a), which stems from this integral, is independent of the shape of S_0.

If the unit normal $\mathbf{n}(\mathbf{r}_0)$ is brought inside the integral then the integrand behaves like $(|\mathbf{r} - \mathbf{r}_0|)^{-1}$ as $\mathbf{r}_0$ approaches $\mathbf{r}$ and it is not necessary to introduce a limiting procedure. A common

---

[5]Sancer has pointed out that the use of the phrase "principal value" implies that the integral converges only if S_0 is a symmetric area centered on the point $\mathbf{r}_0$ such that in the integration negative and positive diverging contributions cancel. In the MFIE canceling contributions are not required to obtain a converging improper integral [2.2]. The integral, without the limiting designation, is often written $\iint$ as a shorthand notation to mean that it is evaluated in a "principal-value sense."

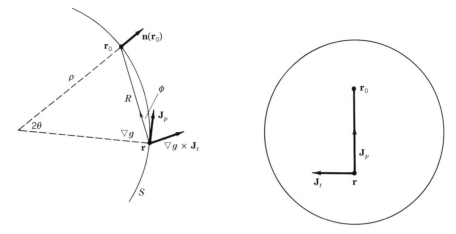

Fig. 2.23. Illustration for the magnetic field integral equation.

form of the MFIE often quoted in the literature is

$$\frac{\mathbf{J}_{es}(\mathbf{r}_0)}{2} + \oiint\limits_S [\mathbf{J}_{es}(\mathbf{r}) \times \nabla g(\mathbf{r}, \mathbf{r}_0)] \times \mathbf{n}(\mathbf{r}_0)\, dS = \mathbf{n}(\mathbf{r}_0) \times \mathbf{H}_i(\mathbf{r}_0) \qquad (208b)$$

where g is the free-space scalar Green's function. The validity of (208b) is readily established. We will not give a detailed proof but will show in a geometrical way how the proof can be constructed.

With reference to Fig. 2.23 we note that ∇g is a vector along the line joining the two points $\mathbf{r}$ and $\mathbf{r}_0$, which lie on the surface S. We now pass a plane through the surface S which cuts the surface along the curve joining the points $\mathbf{r}$ and $\mathbf{r}_0$. The current $\mathbf{J}_{es}$ may be expressed as a component $\mathbf{J}_p$ along the curve and a component $\mathbf{J}_t$ perpendicular to $\mathbf{J}_p$. As the point $\mathbf{r}_0$ approaches $\mathbf{r}$ the angle ϕ approaches θ and the cross product $\mathbf{J}_p \times \nabla g$ has a magnitude that approaches

$$|\nabla g| J_p \sin \theta = \frac{1}{4\pi |\mathbf{r} - \mathbf{r}_0|^2} J_p \frac{|\mathbf{r} - \mathbf{r}_0|}{2\rho}$$

where ρ is the radius of curvature of the surface, in the chosen plane, at the point $\mathbf{r}$. The other vector component $\mathbf{J}_t \times \nabla g$ approaches $\mathbf{n}(\mathbf{r})$ in direction as $\mathbf{r}_0$ approaches $\mathbf{r}$ such that $|(\mathbf{J}_t \times \nabla g) \times \mathbf{n}(\mathbf{r}_0)|$ is also proportional to $\sin \theta$. Thus the integrand in (208b) has only the integrable singularity $|\mathbf{r} - \mathbf{r}_0|^{-1}$ as $\mathbf{r}_0$ approaches $\mathbf{r}$ and a limiting formulation is not needed when $\mathbf{n}(\mathbf{r}_0)$ is brought under the integral sign.[6] If the cross product with $\mathbf{n}(\mathbf{r}_0)$ is not taken first, then $\mathbf{J}_t \times \nabla g$ will have a $|\mathbf{r} - \mathbf{r}_0|^{-2}$ singularity.

The magnetic field integral equation is a statement of a necessary boundary condition. In general, it is not a sufficient condition that will guarantee a unique solution for the scattered magnetic field. At those frequencies corresponding to the resonant frequencies of the internal cavity modes, the solution is not unique. An example of the failure of the MFIE to provide a unique solution at a resonant frequency will be given later for the case of scattering by a perfectly conducting sphere.

---

[6]For this reason the integral operator in (208b) is a compact operator (see [2.3]).

The introduction of a limiting procedure in (208a) is only for the purpose of providing a method of evaluating the surface integral. An equivalent expression for the magnetic field integral equation that comes directly from (207) is

$$\mathbf{J}_{es} + \lim_{\mathbf{r}_0 \to S^+} \oiint_S \mathbf{J}_{es} \cdot \nabla \times \bar{\mathbf{G}}\, dS \times \mathbf{n} = \mathbf{n} \times \mathbf{H}_i.$$

If the homogeneous equation, obtained by setting $\mathbf{H}_i = 0$, has a nonzero solution it follows that the integral equation does not have a unique solution.

Equation (208), which is a Fredholm integral equation of the second kind since $\mathbf{J}_{es}$ also occurs outside the integral, has solutions that make the left-hand side vanish when k_0 is equal to any one of the resonant wavenumbers for the internal cavity bounded by S. A proof for this is developed below.

For the interior problem Green's theorem gives

$$\mathbf{H}(\mathbf{r}_0) = -\oiint_S \mathbf{n} \times \mathbf{H} \cdot \nabla \times \bar{\mathbf{G}}\, dS - j\omega\epsilon_0 \oiint_S \mathbf{n} \times \mathbf{E} \cdot \bar{\mathbf{G}}\, dS, \qquad \mathbf{r}_0 \in V_1.$$

For a cavity with a perfectly conducting surface $\mathbf{n} \times \mathbf{E} = 0$ on S. We again introduce a limiting procedure and let $\mathbf{r}_0$ approach S^-, i.e., S from the interior side. If we then use the result in (201) and the fact that now $\mathbf{n} \times \mathbf{H} = -\mathbf{J}_{es}$, since $\mathbf{n}$ is still directed out of V_1, we obtain

$$\frac{\mathbf{J}_{es}(\mathbf{r}_0)}{2} - \lim_{S_0 \to 0} \left[\iint_{S-S_0} \mathbf{J}_{es}(\mathbf{r}) \cdot \nabla \times \bar{\mathbf{G}}(\mathbf{r}, \mathbf{r}_0)\, dS \times \mathbf{n}(\mathbf{r}_0) \right] = 0 \qquad (209)$$

for the interior eigenvalue problem. The eigenmode current that satisfies this equation is not the solution of the homogeneous equation obtained from (208a) even though the integral operators in the two equations are the same. However, as we will show, the two solutions are closely related.

In order to conveniently relate the two solutions we will recast the interior and exterior eigenvalue problems in a different but equivalent form. In place of $\mathbf{J}_{es}$ in (209) write $-\mathbf{n} \times \mathbf{H}$ and bring $\mathbf{n}(\mathbf{r}_0)$ inside the integral; thus

$$\oiint_S \mathbf{H} \cdot [\mathbf{n}(\mathbf{r}) \times \nabla \times \bar{\mathbf{G}}(\mathbf{r}, \mathbf{r}_0) \times \mathbf{n}(\mathbf{r}_0)]\, dS - \frac{1}{2}\mathbf{H}(\mathbf{r}_0) \cdot [\mathbf{n}(\mathbf{r}_0) \times \bar{\mathbf{I}}] = 0 \qquad (210a)$$

where we have used $\mathbf{n} \times \mathbf{H} = -\mathbf{H} \times \mathbf{n} = -\mathbf{H} \cdot \bar{\mathbf{I}} \times \mathbf{n} = -\mathbf{H} \cdot (\mathbf{n} \times \bar{\mathbf{I}})$. The homogeneous equation for the exterior problem can be written in a similar form, namely,

$$\oiint_S \mathbf{H} \cdot [\mathbf{n}(\mathbf{r}) \times \nabla \times \bar{\mathbf{G}} \times \mathbf{n}(\mathbf{r}_0)]\, dS + \frac{1}{2}\mathbf{H}(\mathbf{r}_0) \cdot [\mathbf{n}(\mathbf{r}_0) \times \bar{\mathbf{I}}] = 0. \qquad (210b)$$

We can show that the scalar product $\mathbf{H}_1(\mathbf{r}) \cdot [\mathbf{n}(\mathbf{r}) \times \nabla \times \bar{\mathbf{G}}(\mathbf{r}, \mathbf{r}_0) \times \mathbf{n}(\mathbf{r}_0)] \cdot \mathbf{H}_2(\mathbf{r}_0)$ is symmetric for any two vectors $\mathbf{H}_1$ and $\mathbf{H}_2$ that are tangential to S. Let $\mathbf{J}_1(\mathbf{r}) = \mathbf{n}(\mathbf{r}) \times \mathbf{H}_1(\mathbf{r})$ and $\mathbf{J}_2(\mathbf{r}_0) = \mathbf{n}(\mathbf{r}_0) \times \mathbf{H}_2(\mathbf{r}_0)$. The scalar product becomes $-\mathbf{J}_1(\mathbf{r}) \cdot \nabla \times \bar{\mathbf{G}} \cdot \mathbf{J}_2(\mathbf{r}_0) = -\mathbf{J}_1(\mathbf{r}) \cdot (\nabla g \times \bar{\mathbf{I}}) \cdot \mathbf{J}_2(\mathbf{r}_0) = -\mathbf{J}_1(\mathbf{r}) \cdot \nabla g \times \mathbf{J}_2(\mathbf{r}_0) = \mathbf{J}_2(\mathbf{r}_0) \cdot \nabla g \times \mathbf{J}_1(\mathbf{r})$ where we have used

$\nabla \times \bar{\mathbf{G}} = \nabla \times \bar{\mathbf{I}}g = \nabla g \times \bar{\mathbf{I}}$ and g is the scalar Green's function

$$g(\mathbf{r}, \mathbf{r}_0) = \frac{e^{-jk_0|\mathbf{r}-\mathbf{r}_0|}}{4\pi|\mathbf{r} - \mathbf{r}_0|}.$$

If we interchange the labels on $\mathbf{r}$ and $\mathbf{r}_0$ and note that $\nabla g = -\nabla_0 g$, then we obtain

$$\mathbf{J}_2(\mathbf{r})\cdot\nabla_0 g \times \mathbf{J}_1(\mathbf{r}_0) = -\mathbf{J}_2(\mathbf{r})\cdot\nabla g \times \mathbf{J}_1(\mathbf{r}_0)$$

$$= -\mathbf{J}_2(\mathbf{r})\cdot\nabla \times \bar{\mathbf{G}}\cdot\mathbf{J}_1(\mathbf{r}_0)$$

$$= \mathbf{H}_2(\mathbf{r})\cdot[\mathbf{n}(\mathbf{r}) \times \nabla \times \bar{\mathbf{G}} \times \mathbf{n}(\mathbf{r}_0)]\cdot\mathbf{H}_1(\mathbf{r}_0)$$

which proves that the scalar product is symmetric.

The other scalar product $\mathbf{H}_1\cdot(\mathbf{n} \times \bar{\mathbf{I}})\cdot\mathbf{H}_2$ is antisymmetric since $\mathbf{H}_1\cdot(\mathbf{n} \times \bar{\mathbf{I}})\cdot\mathbf{H}_2 = \mathbf{H}_1\cdot\mathbf{n} \times \mathbf{H}_2 = \mathbf{H}_2\cdot\mathbf{H}_1 \times \mathbf{n} = -\mathbf{H}_2\cdot\mathbf{n} \times \mathbf{H}_1 = -\mathbf{H}_2\cdot(\mathbf{n} \times \bar{\mathbf{I}})\cdot\mathbf{H}_1$. Note that $\mathbf{n} \times \bar{\mathbf{I}}$ is a rotation operator. It rotates a vector by $90°$.

Let the solution for $\mathbf{H}$ in (210a) be expanded in terms of a complete set of vector basis functions $\mathbf{J}_n(\mathbf{r})$ defined on S; thus

$$\mathbf{H}(\mathbf{r}) = \sum_n I_n^i \mathbf{J}_n(\mathbf{r}).$$

We use this expansion in (210a), integrate over $\mathbf{r}$, then scalar multiply by $\mathbf{J}_m(\mathbf{r}_0)$ and integrate over $\mathbf{r}_0$. This leads to the following system of equations:

$$\sum_n I_n^i \left(G_{nm} - \frac{1}{2}T_{nm} \right) = 0, \qquad m = 1, 2, \ldots \qquad (211a)$$

where

$$G_{nm} = \oiint_S \oiint_S \mathbf{J}_n(\mathbf{r})\cdot[\mathbf{n}(\mathbf{r}) \times \nabla \times \bar{\mathbf{G}} \times \mathbf{n}(\mathbf{r}_0)]\cdot\mathbf{J}_m(\mathbf{r}_0)\, dS\, dS_0$$

$$T_{nm} = \oiint_S \mathbf{J}_n(\mathbf{r}_0)\cdot[\mathbf{n}(\mathbf{r}_0) \times \bar{\mathbf{I}}]\cdot\mathbf{J}_m(\mathbf{r}_0)\, dS_0.$$

Note that $G_{nm} = G_{mn}$, $T_{mn} = -T_{nm}$, and $T_{nn} = 0$ because of the symmetrical and antisymmetrical properties of the scalar products. In (211a) the I_n^i are the unknown amplitudes.

For the solution to (210b) we assume an expansion of the form

$$\mathbf{H}(\mathbf{r}) = \sum_n I_n^e \mathbf{J}_n(\mathbf{r}).$$

By a similar procedure we obtain the system of equations

$$\sum_n I_n^e \left(G_{nm} + \frac{1}{2}T_{nm} \right) = 0, \qquad m = 1, 2, \ldots. \qquad (211b)$$

For a numerical solution only a finite number of basis functions can be used.

Solutions for I_n^i and I_n^e exist only if the determinant of the system of equations vanishes.

The transpose of the matrix in (211b) has elements

$$G_{mn} + \frac{1}{2}T_{mn} = G_{nm} - \frac{1}{2}T_{nm}$$

which are the elements of the matrix in (211a). Both systems have the same determinant. Hence, if there is an eigenvector for the interior problem there will also be an eigenvector for the exterior problem. We can express (211a) and (211b) in matrix form as

$$\left[G - \frac{1}{2}T\right][I^i] = 0 \tag{212a}$$

$$\left[G + \frac{1}{2}T\right][I^e] = 0 = [I^e]_t \left[G - \frac{1}{2}T\right]. \tag{212b}$$

By subtracting the two equations we get

$$[G]\{[I^i] - [I^e]\} - \frac{1}{2}[T]\{[I^i] + [I^e]\} = 0.$$

If we assume that $[I^i] = [I^e]$, then since $[T]$ is a nonzero matrix we would get $[I^i] = [I^e] = 0$. Hence we conclude that $[I^i] \neq [I^e]$. We see that if $[I^i]$ is a right eigenvector of $[G - \frac{1}{2}T]$ then $[I^e]$ is a left eigenvector. Thus the interior and exterior eigenvalue problems have solutions at the resonant wavenumbers for the internal cavity but the two current distributions on S are not the same. This result is a very important one since it allows us to combine the electric field and magnetic field integral equations into a combined field integral equation that does not have any nonzero solutions in the null space.

Yaghjian has given a physical description of the current which is a solution of the homogeneous MFIE [2.29]. We will present the same results from a somewhat different approach. Let $\mathbf{F}_n(\mathbf{r})$ be a solution inside S of $\nabla \times \nabla \times \mathbf{F}_n - k_0^2\mathbf{F}_n = 0$ with the boundary condition $\mathbf{n} \times \mathbf{F}_n = 0$ on S. Thus $\mathbf{F}_n$ corresponds to the electric field of an interior cavity resonant mode. The corresponding magnetic field and current on S are

$$\mathbf{H} = \frac{j}{k_0 Z_0}\nabla \times \mathbf{F}_n$$

$$\mathbf{J}_{es} = \mathbf{J}_{en} = -\mathbf{n} \times \mathbf{H} = -\frac{j}{k_0 Z_0}\mathbf{n} \times \nabla \times \mathbf{F}_n$$

where the unit normal $\mathbf{n}$ is directed outward from S. By duality, a solution of Maxwell's equations with $\mathbf{H} = \mathbf{F}_n$ also exists and represents the resonant mode in a cavity with perfectly conducting magnetic walls. The electric field of this mode has a tangential value on S given by

$$\mathbf{n} \times \mathbf{E} = -\frac{j}{k_0 Y_0}\mathbf{n} \times \nabla \times \mathbf{H} = -\frac{j}{k_0 Y_0}\mathbf{n} \times \nabla \times \mathbf{F}_n = \frac{\mu_0}{\epsilon_0}\mathbf{J}_{en}.$$

The tangential component of $\mathbf{H}$ is zero on S. We can construct a unique solution for the electric field outside S such that this electric field has tangential components that match those of the interior electric field on S. From this electric field we can find the magnetic field using

Maxwell's equations. We will call the external field $\mathbf{E}_s$, $\mathbf{H}_s$. In order to support the external magnetic field a current $\mathbf{n} \times \mathbf{H}_s$ must be placed on S. This current is also given by

$$\mathbf{J}_{es} = \mathbf{n} \times \mathbf{H}_s = \frac{j}{k_0 Z_0} \mathbf{n} \times \nabla \times \mathbf{E}_s$$

and is different from $\mathbf{J}_{en}$. We will now show that the current $\mathbf{J}_{es}$ as specified is a solution of the homogeneous MFIE.

By means of an application of Green's theorem, the magnetic field exterior to S is given in terms of boundary values on S by (200) and is

$$\mathbf{H}_s = \oiint_S \mathbf{n} \times \mathbf{H}_s \cdot \nabla \times \bar{\mathbf{G}} \, dS + jk_0 Y_0 \oiint_S \mathbf{n} \times \mathbf{E}_s \cdot \bar{\mathbf{G}} \, dS.$$

On the surface S we have

$$\mathbf{n} \times \mathbf{H}_s + \lim_{\mathbf{r}_0 \to S^+} \oiint_S \mathbf{n} \times \mathbf{H}_s \cdot \nabla \times \bar{\mathbf{G}} \, dS \times \mathbf{n}(\mathbf{r}_0) = -jk_0 Y_0 \oiint_S \mathbf{n} \times \mathbf{E}_s \cdot \bar{\mathbf{G}} \, dS \times \mathbf{n}(\mathbf{r}_0).$$

The integral on the right side of this equation equals zero because on S we have $\mathbf{n} \times \mathbf{E}_s = (\mu_0/\epsilon_0)\mathbf{J}_{en}$ so the integral gives the tangential value of the electric field radiated by the resonant cavity mode current $\mathbf{J}_{en}$ and the latter is zero. If we now replace $\mathbf{n} \times \mathbf{H}_s$ by $\mathbf{J}_{es}$ we obtain the homogeneous MFIE.

In order to be a valid solution of Maxwell's equations, the solution of the homogeneous MFIE would require that an actual physical current $\mathbf{J}_{es}$ be placed on S. In a scattering problem such a current distribution does not exist on S. Hence the solution for the homogeneous MFIE is spurious. It corresponds to an undamped external oscillation; such undamped oscillations can only be maintained by an impressed source since energy is lost by radiation. In addition, this homogeneous solution has an internal cavity electromagnetic field associated with it.

As we have noted, the solutions to the homogeneous EFIE and MFIE represent different current distributions on the surface of the scatterer, but at the same frequency. Neither current distribution can satisfy the homogeneous equation that results from the combination of the two field integral equations.

The combined field integral equation is, upon combining (203) and (208) [2.27]:

$$-jk_0 Z_0 \oiint_S \mathbf{J}_{es} \cdot \bar{\mathbf{G}} \times \mathbf{n}(\mathbf{r}_0) \, dS + \frac{\gamma \mathbf{J}_{es}(\mathbf{r}_0)}{2} + \gamma \oiint_S \mathbf{J}_{es} \cdot \nabla \times \bar{\mathbf{G}} \times \mathbf{n}(\mathbf{r}_0) \, dS$$

$$= \mathbf{n}(\mathbf{r}_0) \times \mathbf{E}_i(\mathbf{r}_0) + \gamma \mathbf{n}(\mathbf{r}_0) \times \mathbf{H}_i(\mathbf{r}_0) \tag{213}$$

where γ is an arbitrary constant. Since $|Z_0 \mathbf{H}_i|$ is equal to $|\mathbf{E}_i|$ for an incident plane wave, γ is commonly chosen equal to Z_0.

Another procedure that can be used to obtain an integral equation that does not have a resonant mode solution gives the combined source integral equation [2.28]. In this formulation the scattered electric field is expressed as a field produced by a system of electric and magnetic currents on S. This integral equation is obtained from Love's field equivalence theorem as stated by (199). The combined source integral equation is

$$\mathbf{n} \times \mathbf{E}_s = -\mathbf{n} \times \mathbf{E}_i = \lim_{\mathbf{r}_0 \to S^+} \mathbf{n} \times \oiint_S \mathbf{J}_{ms} \cdot \nabla \times \bar{\mathbf{G}} \, dS - jk_0 Z_0 \mathbf{n} \times \oiint_S \mathbf{J}_{es} \cdot \bar{\mathbf{G}} \, dS$$

or

$$\frac{\mathbf{J}_{ms}}{2} + \lim_{S_0 \to 0} \mathbf{n} \times \iint_{S - S_0} \mathbf{J}_{ms} \cdot \nabla \times \bar{\mathbf{G}} \, dS - jk_0 Z_0 \mathbf{n} \times \oiint_S \mathbf{J}_{es} \cdot \bar{\mathbf{G}} \, dS = -\mathbf{n} \times \mathbf{E}_i \qquad \text{on } S.$$

$$(214)$$

In obtaining (214) we replaced $\mathbf{n} \times \mathbf{E}$ by $-\mathbf{J}_{ms}$ and $\mathbf{n} \times \mathbf{H}$ by $\mathbf{J}_{es}$ in (199). Mautz and Harrington chose $\mathbf{J}_{ms} = Z_0 \mathbf{n} \times \mathbf{J}_{es}$ and then showed that the homogeneous equation obtained from (214) does not have a solution [2.27]. The two sources in (214) are equivalent sources only. The actual current induced on S has to be determined by finding the total tangential magnetic field at the surface S.

Yaghjian has reviewed the various integral equation formulations and the reader is referred to [2.29] for more details on methods that can be used to circumvent the problem associated with the cavity resonances. Yaghjian has also shown that the electric field and magnetic field integral equations can be made unique by imposing a condition on the normal components of the field at the surface S [2.29]. This latter method has the advantage of keeping the resultant integral equations relatively simple, whereas the combined field and combined source equations are quite complex. The resonant mode currents can also be eliminated by requiring that the total field inside S vanish at a number of selected points. A review of the literature using this approach has also been given by Yaghjian [2.29].

The solution of the integral equations can be developed in terms of the eigenvectors of the matrix operators shown in (205) and (211b). These eigenvectors are called the characteristic modes and can be used as a basis to expand the current or tangential magnetic field respectively on the surface S. A treatment of the characteristic modes associated with scattering can be found in [2.30].

The magnetic field integral equation cannot be used to solve the scattering problem involving an open surface where the surface consists of an infinitely thin perfectly conducting metal surface. The reason is that the incident magnetic field has a tangential component with equal values on the two sides of the surface and hence the boundary condition $\mathbf{n} \times \mathbf{H}_s^+ - \mathbf{n} \times \mathbf{H}_s^- = \mathbf{J}_{es}$ does not involve $\mathbf{H}_i$.

In recent years a large amount of work has been done on transient or pulsed electromagnetic field scattering from perfectly conducting obstacles using Baum's singularity expansion method (SEM). An extensive bibliography of the relevant literature is given in a recent volume of the *Electromagnetics Journal* [2.4]. A conducting body supports an infinite number of external natural damped oscillations corresponding to complex resonant frequencies. The scattered field may be described in terms of these natural modes [2.31]. The spurious solution that is obtained for the homogeneous MFIE with steady-state sinusoidal excitation is not part of the spectrum of natural modes because it is an undamped solution. Undamped natural oscillations exterior to a scattering body do not exist because the oscillating modes must lose energy by radiation. In the exterior transient problem the pure imaginary $j\omega$ axis poles do not contribute when the Laplace transform solution is inverted so the spurious modes are eliminated.[7] Sancer

---

[7]In actuality the $j\omega$ axis poles in the eigenfunction expansion are canceled by numerator mode coupling coefficients that vanish at the same frequencies.

et al. have shown that current distributions on the scatterer that arise at the interior cavity resonant frequencies are not excited [2.32]. Thus the transient scattering problem is a better posed problem than that of scattering by a time harmonic incident field.

Example: Scattering from a Perfectly Conducting Sphere

The problem of scattering of an incident field by a perfectly conducting sphere of radius a can be solved exactly. Thus this problem provides an instructive example of the application of the EFIE and MFIE and insight into the effects of the homogeneous integral equation solutions on the determination of unique solutions for the scattered fields.

We will assume that the source for the incident field is located at a distance $r > R > a$ away from the sphere. In the region $r \leq R$ the incident field can then be expanded in terms of the $\mathbf{M}_{nm}$ and $\mathbf{N}_{nm}$ functions given by (135) with $k = k_0$. We will assume that the incident field can be represented by a finite number of modes; thus[8]

$$\mathbf{E}_i = \sum_{n=1}^{N} \sum_{m=0}^{M} (A_{nm}\mathbf{M}_{nm}^o + B_{nm}\mathbf{N}_{nm}^e) \tag{215a}$$

$$\mathbf{H}_i = \sum_{n=1}^{N} \sum_{m=0}^{N} jY_0(A_{nm}\mathbf{N}_{nm}^o + B_{nm}\mathbf{M}_{nm}^e). \tag{215b}$$

The magnetic field is obtained by using $\nabla \times \mathbf{E} = -j\omega\mu_0\mathbf{H}$, $\nabla \times \mathbf{M}_{nm} = k_0\mathbf{N}_{nm}$, and $\nabla \times \mathbf{N}_{nm} = k_0\mathbf{M}_{nm}$. For simplicity we choose the $\cos m\phi$ and $\sin m\phi$ functions in the expressions for $\mathbf{M}_{nm}$ and $\mathbf{N}_{nm}$ given by (135) so that $E_{i\theta}$ is an even function about $\phi = 0$. For the scattering problem we also need the outward-propagating modes $\mathbf{M}_{nm}^{o,+}$ and $\mathbf{N}_{nm}^{e,+}$. These mode functions are also given by (135) with $k = k_0$ and $j_n(k_0 r)$ replaced by $h_n^2(k_0 r)$.

The classical method to solve the scattering problem is to let the scattered field be described in terms of a series of outward-propagating modes in the form

$$\mathbf{E}_s = \sum_{n=1}^{N} \sum_{m=0}^{M} (C_{nm}\mathbf{M}_{nm}^{o,+} + D_{nm}\mathbf{N}_{nm}^{e,+}). \tag{216}$$

The coefficients C_{nm}, D_{nm} are found by requiring that $\mathbf{a}_r \times \mathbf{E}_s = -\mathbf{a}_r \times \mathbf{E}_i$ at $r = a$. This gives, in view of the orthogonal property of the modes,

$$C_{nm}\mathbf{M}_{nm}^{o,+} \times \mathbf{a}_r = -A_{nm}\mathbf{M}_{nm}^o \times \mathbf{a}_r \qquad \text{at } r = a$$

$$D_{nm}\mathbf{N}_{nm}^{e,+} \times \mathbf{a}_r = -B_{nm}\mathbf{N}_{nm}^e \times \mathbf{a}_r \qquad \text{at } r = a. \tag{217}$$

We will now assume that $j_1(k_0 a) = 0$ so that $\mathbf{M}_{11}^e$, $\mathbf{M}_{10}^e$, and $\mathbf{M}_{11}^o$ represent degenerate resonant modes inside the sphere. Thus, part of the incident field will have an electric field with a zero tangential component on the surface of the sphere. The sphere will not scatter the $\mathbf{M}_{11}^{o,+}$ mode as is quite clear from the solutions for C_{nm} and D_{nm}; i.e., $C_{11} = 0$ because the $\mathbf{M}_{nm}^o$ functions have a $j_n(k_0 r)$ radial dependence and $j_1(k_0 a) = 0$.

In order to implement the EFIE we use the dyadic Green's function given by (166) and note

---

[8]For an incident plane wave an infinite number of terms is needed in the expansion.

that the field point $\mathbf{r}_0$ approaches the spherical surface from the exterior. Thus the EFIE is

$$
\mathbf{E}_i \times \mathbf{a}_{r_0} = \sum_{n=1}^{\infty} \sum_{m=0}^{n} \oint_S \mathbf{J}_{es}(\mathbf{r}) \cdot \left\{ \frac{\pi Z_0}{2n(n+1)Q_{nm}} [\mathbf{M}_{nm}^{o}(\mathbf{r})\mathbf{M}_{nm}^{o,+}(\mathbf{r}_0) + \mathbf{M}_{nm}^{e}(\mathbf{r})\mathbf{M}_{nm}^{e,+}(\mathbf{r}_0) \right.
$$

$$
\left. + \mathbf{N}_{nm}^{e}(\mathbf{r})\mathbf{N}_{nm}^{e,+}(\mathbf{r}_0) + \mathbf{N}_{nm}^{o}(\mathbf{r})\mathbf{N}_{nm}^{o,+}(\mathbf{r}_0)] \right\} \times \mathbf{a}_{r_0} \, dS \tag{218}
$$

where $r = r_0 = a$.

At this point we will introduce the following surface vector functions:

$$
\mathbf{S}_{nm}^{e,o} = \mp \frac{mP_n^m}{\sin\theta} \frac{\sin}{\cos} m\phi \mathbf{a}_\theta - \frac{dP_n^m}{d\theta} \frac{\cos}{\sin} m\phi \mathbf{a}_\phi \tag{219a}
$$

$$
\mathbf{T}_{nm}^{e,o} = \frac{dP_n^m}{d\theta} \frac{\cos}{\sin} m\phi \mathbf{a}_\theta \mp \frac{mP_n^m}{\sin\theta} \frac{\sin}{\cos} m\phi \mathbf{a}_\phi. \tag{219b}
$$

Note that $\mathbf{S}_{nm}^{e,o} = \mathbf{T}_{nm}^{e,o} \times \mathbf{a}_r$ and $\mathbf{T}_{nm}^{e,o} = -\mathbf{S}_{nm}^{e,o} \times \mathbf{a}_r$. These vector functions are all mutually orthogonal with respect to integration over the surface of a sphere. They have the same normalization values which can be found using (136b); thus

$$
\int_0^{2\pi} \int_0^{\pi} \mathbf{T}_{nm}^{e,o} \cdot \mathbf{T}_{nm}^{e,o} \sin\theta \, d\theta \, d\phi = \int_0^{2\pi} \int_0^{\pi} \mathbf{S}_{nm}^{e,o} \cdot \mathbf{S}_{nm}^{e,o} \sin\theta \, d\theta \, d\phi
$$

$$
= \frac{2n(n+1)(n+m)!}{(2n+1)(n-m)!} \frac{2\pi}{\epsilon_{om}} = \frac{2n(n+1)}{\pi} Q_{nm} \tag{219c}
$$

where Q_{nm} is given by (139). These vector functions may be used as a basis set to expand the current on the sphere. Note that the $\mathbf{M}_{nm}$ functions are equal to $\mathbf{S}_{nm}$ multiplied by a function of r and that the θ and ϕ components of $\mathbf{N}_{nm}$ are given by $\mathbf{T}_{nm}$ multiplied by a radial function. The vector functions given by (219) are the characteristic modes for a sphere. Note that the $\mathbf{S}_{10}^{o}$ and $\mathbf{T}_{10}^{o}$ modes do not exist.

The current on the sphere can be expanded in the form

$$
\mathbf{J}_{es} = \sum_{n,m} (I_{nm}^{o} \mathbf{S}_{nm}^{o} + I_{nm}^{e} \mathbf{S}_{nm}^{e} + J_{nm}^{o} \mathbf{T}_{nm}^{o} + J_{nm}^{e} \mathbf{T}_{nm}^{e}). \tag{220}
$$

The $I_{nm}^{o,e}$ and $J_{nm}^{e,o}$ are unknown amplitude constants. When we use this expansion in (218) and then scalar multiply (218) by the basis functions used to expand $\mathbf{J}_{es}$ it is clear that the matrix multiplying the I_{nm} and J_{nm} will be diagonal and the matrix elements corresponding to the I_{11}^{o}, I_{10}^{e}, and I_{11}^{e} terms will be zero. Thus the determinant of the matrix is zero. However, the term $\mathbf{E}_i \times \mathbf{a}_{r_0}$ when scalar multiplied by $\mathbf{S}_{11}^{o}$ and integrated over S is also zero.

When the expansion (220) is used in the EFIE (218) the following system of equations is obtained:

$$
(k_0 a j_n) A_{nm} = a Z_0 I_{nm}^{o} (k_0 a j_n)(k_0 a h_n^2)
$$

$$
(k_0 a j_n)' B_{nm} = a Z_0 J_{nm}^{e} (k_0 a j_n)'(k_0 a h_n^2)'
$$

$$0 = aZ_0I^e_{nm}(k_0aj_n)(k_0ah^2_n)$$

$$0 = aZ_0J^o_{nm}(k_0aj_n)'(k_0ah^2_n)'$$

$$m = 0, 1, 2, \cdots, n; \qquad n = 1, 2, \cdots, \infty.$$

From these equations we readily see that I^e_{10} and I^e_{11} are unspecified when $j_1(k_0a) = 0$. Thus the currents $I^e_{10}S^e_{10}$ and $I^e_{11}S^e_{11}$ are solutions of the homogeneous EFIE.[9] All the other I^e_{nm} and J^o_{nm} coefficients are zero (except at the resonant frequencies where $j_n(k_0a)$ or $[k_0aj_n]'$ vanish). The solutions for I^o_{nm} and J^e_{nm} for $n = 1, \ldots, N$, $m = 1, \ldots, M$, are

$$I^o_{nm} = \frac{A_{nm}}{aZ_0(k_0ah^2_n)} \qquad J^e_{nm} = \frac{B_m}{aZ_0(k_0ah^2_n)'}$$

provided we cancel the common factors on both sides of the equations for I^o_{nm} and J^e_{nm}. All of the other I^o_{nm} and J^e_{nm} are zero. If we do not cancel the common factor $j_1(k_0a)$ then the equation for I^o_{11} is indeterminant. Consequently, we do not obtain a unique solution for I^o_{11}; this is consistent with the fact that S^o_{11} is also a solution of the homogeneous EFIE. Apart from the lack of uniqueness for the solutions for I^o_{11}, I^e_{10}, and I^e_{11} the results agree with those obtained by the classical method. The current modes S^o_{11}, S^e_{10}, and S^e_{11} do not radiate because they do not couple to the mode spectrum outside the sphere since the M^o_{11}, M^e_{10}, and M^e_{11} factors in the Green's function are zero at $r = a$.

If the surface integrals were evaluated on a computer, then because of the finite precision of the computer the matrix element and the coupling of the incident field to the S^o_{11} current mode would be small but not zero. As a consequence, the $M^{o,+}_{11}$ mode would appear as part of the scattered field. If the current were expanded in terms of a set of basis functions that do not lead to a diagonal matrix we would have an ill-conditioned (almost singular) matrix and would not be able to easily identify the current that is a solution of the homogeneous integral equation. The same difficulties would occur if $j_1(k_0a)$ were very small but not equal to zero.

From an analytical point of view the EFIE gives the correct solution for the sphere problem, but the solution is not unique because the homogeneous integral equation has a nontrivial solution.

If we examine the expression for the incident magnetic field as given by (215) we find that the modes $jY_0B_{10}M^e_{10}$ and $jY_0B_{11}M^e_{11}$ equal zero at $r = a$ because $j_1(k_0a) = 0$. From the classical solution to the scattering problem as given by (217) we see that the scattered magnetic field modes $jY_0D_{10}M^{e,+}_{10}$ and $jY_0D_{11}M^{e,+}_{11}$ do not have zero amplitudes. Consequently, we can anticipate that it might be difficult to determine the scattered magnetic field modes from the MFIE.

The MFIE based on (207) can be expressed in the form

$$\mathbf{a}_{r_0} \times \mathbf{H}_i = \mathbf{J}_{es} + \sum_{n=1}^{\infty}\sum_{m=0}^{n} \oiint_S \mathbf{J}_{es} \cdot \left[\frac{-j\pi}{2n(n+1)Q_{nm}} \right] [\mathbf{N}^e_{nm}(\mathbf{r})\mathbf{M}^{e,+}_{nm}(\mathbf{r}_0) + \mathbf{N}^o_{nm}(\mathbf{r})\mathbf{M}^{o,+}_{nm}(\mathbf{r}_0)$$

$$+ \mathbf{M}^e_{nm}(\mathbf{r})\mathbf{N}^{e,+}_{nm}(\mathbf{r}_0) + \mathbf{M}^o_{nm}(\mathbf{r})\mathbf{N}^{o,+}_{nm}(\mathbf{r}_0)] \, dS \times \mathbf{a}_{r_0}. \tag{221}$$

A complete expansion for the current is given by (220). We can simplify (221) by introducing

[9]The origin for ϕ can always be chosen so that the incident field does not have the $\mathbf{M}^e_{11}$ mode (Problem 2.33).

the surface vector functions and using (219). Thus we find that (221) can be expressed as

$$\sum_{n=1}^{N}\sum_{m=0}^{M}[-jY_0(k_0aj_n)'A_{nm}S_{nm}^o + jY_0(k_0aj_n)B_{nm}T_{nm}^e]$$

$$= a\mathbf{J}_{es} - ja\sum_{n=1}^{\infty}\sum_{m=0}^{n}[(k_0aj_n)(k_0ah_n^2)'(I_{nm}^e\mathbf{S}_{nm}^e + I_{nm}^o\mathbf{S}_{nm}^o)$$

$$- (k_0aj_n)'(k_0ah_n^2)(J_{nm}^e\mathbf{T}_{nm}^e + J_{nm}^o\mathbf{T}_{nm}^o)]$$

where the prime denotes a derivative with respect to k_0a. We now scalar multiply these equations by the various basis functions and integrate over θ and ϕ. This procedure yields the equations

$$-jY_0(k_0aj_n)'A_{nm} = aI_{nm}^o - ja(k_0aj_n)(k_0ah_n^2)'I_{nm}^o$$

$$jY_0(k_0aj_n)B_{nm} = aJ_{nm}^e + ja(k_0aj_n)'(k_0ah_n^2)J_{nm}^e$$

$$0 = aI_{nm}^e - ja(k_0aj_n)(k_0ah_n^2)'I_{nm}^e$$

$$0 = aJ_{nm}^o + ja(k_0aj_n)'(k_0ah_n^2)J_{nm}^o$$

$$m = 0, 1, 2, \cdots, n; \qquad n = 1, 2, \cdots, \infty.$$

We make note of the following Wronskian relationship:

$$\frac{d[aj_1(k_0a)]}{da}h_1^2(k_0a) - \frac{[dah_1^2(k_0a)]}{da}j_1(k_0a) = \frac{d[aj_1(k_0a)]}{da}h_1^2(k_0a) = \frac{j}{k_0a}. \quad (222)$$

The formal solutions for I_{nm}^o and J_{nm}^e are

$$I_{nm}^o = \frac{-jY_0A_{nm}(k_0aj_n)'}{a[1 - j(k_0aj_n)(k_0ah_n^2)']} = \frac{A_{nm}}{aZ_0(k_0ah_n^2)}$$

$$J_{nm}^e = \frac{jY_0B_{nm}(k_0aj_n)}{a[1 + j(k_0aj_n)'(k_0ah_n^2)]} = \frac{B_{nm}}{aZ_0(k_0ah_n^2)'}$$

upon using (222). These solutions are the same as those obtained using the EFIE. The solutions remain finite even at the resonant frequencies. However, if we had not used the Wronskian relation (222) the solutions for J_{nm}^e would have taken the indeterminate form of zero divided by zero and would have required a limiting procedure for evaluation. From a numerical point of view indeterminant forms cannot be easily evaluated on a computer.

The solutions to the homogeneous MFIE are the current modes $J_{11}^o\mathbf{T}_{11}^o$, $J_{10}^e\mathbf{T}_{10}^e$, and $J_{11}^e\mathbf{T}_{11}^e$ at the frequency where $j_1(k_0a) = 0$. These solutions are different from those for the homogeneous EFIE at the same resonant frequency. The coefficient $aI_{11}^o = -jY_0(k_0aj_1)'A_{11}$ and the current $I_{11}^o\mathbf{S}_{11}^o$ supports the incident $jY_0A_{11}\mathbf{N}_{11}^o = jY_0A_{11}(k_0aj_1)'a^{-1}\mathbf{T}_{11}^o$ mode. All of the coefficients I_{nm}^o, J_{nm}^e for n up to N and m up to M can be solved for, but the solutions for

J_{10}^e and J_{11}^e are not unique because the equations for them have a zero factor on both sides.

It is seen that the MFIE also gives a correct solution for the sphere problem but the solution is not unique at the resonant frequencies. Furthermore, the solutions to the homogeneous MFIE give rise to spurious radiated modes since the $\mathbf{N}_{11}^o$, $\mathbf{N}_{10}^e$, and $\mathbf{N}_{11}^e$ mode functions are not zero at $j_1(k_0 a) = 0$.

As a final comment we point out that if the scattered magnetic field inside the sphere is desired then in the Green's dyadic function the modal functions representing outward-propagating waves will appear on the left. By using the Wronskian determinant one can then readily show that

$$\oiint_S \mathbf{J}_{es} \cdot \nabla \times \bar{\mathbf{G}} \, dS \times \mathbf{a}_r$$

has the correct discontinuous behavior as the observation point $\mathbf{r}_0$ passes through S from the exterior side to the interior side. Of course, this is not a surprising outcome since the correct discontinuous behavior is a built-in feature of the eigenfunction expansion of the dyadic Green's function. One also readily finds that for the interior homogeneous MFIE the solution is $I_{11}^e \mathbf{S}_{11}^e + I_{11}^o \mathbf{S}_{11}^o + I_{10}^e \mathbf{S}_{10}^e$ which is the current associated with the $\mathbf{M}_{11}^e$, $\mathbf{M}_{10}^e$, and $\mathbf{M}_{11}^o$ electric field modes.

2.20. NON-SELF-ADJOINT SYSTEMS

Let $\mathcal{L}$ denote a differential operator and consider the equation

$$\mathcal{L}\psi + \lambda\psi = f(x). \tag{223}$$

The variable is x where $0 \leq x \leq a$. The domain $\mathcal{D}$ of the operator $\mathcal{L}$ is the domain of functions ψ that satisfy certain continuity conditions and specified boundary conditions. We can define a scalar product in more than one way—for example,

$$\langle \phi, \psi \rangle = \int_0^a \phi(x)\psi(x) \, dx \tag{224a}$$

or

$$\langle \phi, \psi \rangle = \int_0^a \phi(x)\psi(x)\sigma(x) \, dx \tag{224b}$$

or

$$\langle \phi, \psi \rangle = \int_0^a \phi(x)\psi^*(x) \, dx \tag{224c}$$

etc. The definition (224c) is used in quantum mechanics (Hilbert space), while (224b) is used if (223) has the function σ as a factor with λ.

Once the scalar product has been defined the adjoint operator $\mathcal{L}_a$ is defined to be that operator which satisfies

$$\langle \phi, \mathcal{L}\psi \rangle = \langle \mathcal{L}_a \phi, \psi \rangle. \tag{225}$$

If $\mathcal{L} = \mathcal{L}_a$ and the domain $\mathcal{D}_a$ of $\mathcal{L}_a$ coincides with the domain $\mathcal{D}$ of $\mathcal{L}$, then $\mathcal{L}$ is called a self-adjoint operator. Sometimes $\mathcal{D}$ and $\mathcal{D}_a$ differ even though $\mathcal{L} = \mathcal{L}_a$ and in this case $\mathcal{L}$ is only formally self-adjoint. In practice $\mathcal{L}_a$ is determined by integration by parts to transfer differentiations on ψ to differentiations on ϕ.

Example

$$\frac{d^2\psi}{dx^2} + \lambda\psi = f, \quad \psi + j\frac{d\psi}{dx} = 0, \quad x = 0, a.$$

ψ may be expressed in terms of the eigenfunctions of $d^2\psi/dx^2 + \lambda\psi = 0$. It is easily verified that $\psi_n = \sin(n\pi x/a) - j(n\pi/a)\cos(n\pi x/a)$. If we choose (224a) for the scalar product then

$$\int_0^a \psi_n\psi_m \, dx = 0, \quad n \neq m.$$

The domain of $\mathcal{L}$ is the domain of functions that satisfy $\psi + j(d\psi/dx) = 0$ at $x = 0, a$.
   With (224a) as the scalar product we have

$$\langle \phi, \mathcal{L}\psi \rangle = \int_0^a \phi\frac{d^2\psi}{dx^2} \, dx = \phi\frac{d\psi}{dx}\bigg|_0^a - \int_0^a \frac{d\phi}{dx}\frac{d\psi}{dx} \, dx$$

$$= \left(\phi\frac{d\psi}{dx} - \frac{d\phi}{dx}\psi\right)\bigg|_0^a + \int_0^a \psi\frac{d^2\phi}{dx^2} \, dx = \langle \mathcal{L}_a\phi, \psi \rangle.$$

The integrated terms vanish if $\phi + j(d\phi/dx) = 0$ at $x = 0, a$. Hence $\mathcal{L}_a = \mathcal{L}$; since $\mathcal{D} = \mathcal{D}_a$ also, the operator is self-adjoint.
   However, if we choose (224c) as the scalar product then

$$\langle \phi, \mathcal{L}\psi \rangle = \left(\phi\frac{d\psi^*}{dx} - \frac{d\phi}{dx}\psi^*\right)\bigg|_0^a + \int_0^a \psi^*\frac{d^2\phi}{dx^2} \, dx.$$

The boundary terms which can be written as

$$\left[\left(\phi - j\frac{d\phi}{dx}\right)\frac{d\psi^*}{dx} - \left(\psi^* - j\frac{d\psi^*}{dx}\right)\frac{d\phi}{dx}\right]\bigg|_0^a$$

vanish if $\phi - j(d\phi/dx) = 0$ at $x = 0, a$. In this case $\mathcal{L}_a = \mathcal{L}$ but $\mathcal{D}_a \neq \mathcal{D}$ so the operator is not self-adjoint. Note that the adjointness properties of an operator are dependent on the choice of scalar product. When an operator is not self-adjoint the eigenfunctions are not orthogonal with respect to the scalar product used.
   With the scalar product (224a) $\mathcal{L}$ is self-adjoint and $\int_0^a \psi_n\psi_m \, dx = 0$, $n \neq m$. With the scalar product (224c) $\mathcal{L}$ is not self-adjoint and this implies $\int_0^a \psi_n\psi_m^* \, dx \neq 0$ for $n \neq m$. In a non-self-adjoint system the appropriate orthogonality principle is instead

$$\int_0^a \phi_n\psi_m^* \, dx = 0, \quad n \neq m. \tag{226}$$

We may show this as follows: Let ϕ_n, μ_n be the eigenfunctions and eigenvalues of $\mathcal{L}_a$; then $\mathcal{L}_a\phi_n + \mu_n\phi_n = 0$. Since $\mathcal{L}\psi_m + \lambda_m\psi_m = 0$ also, we have $\langle\phi_n, \mathcal{L}\psi_m\rangle = -\lambda_m^*\langle\phi_n, \psi_m\rangle = \langle\mathcal{L}_a\phi_n, \psi_m\rangle = -\mu_n\langle\phi_n, \psi_m\rangle$ if (224c) is used for the scalar product [if (224a) is used, replace λ_m^* by λ_m]. If $\lambda_m^* \neq \mu_n$, then $\langle\phi_n, \psi_m\rangle = 0$.

The eigenvalues of the adjoint operator are related to those of the original operator. We have $\langle(\mathcal{L}_a + \mu_n)\phi_n, \psi\rangle = 0$ for all ψ in $\mathcal{D}$. But this also equals $\langle\mathcal{L}_a\phi_n, \psi\rangle + \mu_n\langle\phi_n, \psi\rangle = \langle\phi_n, \mathcal{L}\psi\rangle + \langle\phi_n, \mu_n^*\psi\rangle = \langle\phi_n, (\mathcal{L} + \mu_n^*)\psi\rangle = 0$. Since ϕ_n is not zero $(\mathcal{L} + \mu_n^*)\psi = 0$ so μ_n^* is an eigenvalue of $\mathcal{L}$ when (224c) is used for the scalar product. When (224a) is used μ_n is an eigenvalue of both $\mathcal{L}_a$ and $\mathcal{L}$. With proper indexing $\mu_n = \lambda_n^*$ (or $\mu_n = \lambda_n$).

In our example if we use (224c) for the scalar product then $\phi_n = \psi_n^*$ and the orthogonality relation is $\langle\psi_n, \phi_m\rangle = 0$, $n \neq m$. But this equals $\int_0^a \psi_n\phi_m^* \, dx = \int_0^a \psi_n\psi_m \, dx = 0$ in agreement with the results obtained if (224a) is used for the scalar product.

For the Sturm–Liouville equation we choose $\mathcal{L} = (1/\sigma)(d/dx)p(d/dx) + q/\sigma$ so the equation becomes $\mathcal{L}\psi + \lambda\psi = f/\sigma$. We now use (224b) for the scalar product. For $\langle\phi, \mathcal{L}\psi\rangle$ we get

$$\int_0^a \sigma\phi\left[\frac{1}{\sigma}\frac{d}{dx}p\frac{d\psi}{dx} + \frac{q}{\sigma}\psi\right] dx = \left[p\phi\frac{d\psi}{dx} - p\psi\frac{d\phi}{dx}\right]\Big|_0^a + \int_0^a \sigma\psi\left[\frac{1}{\sigma}\frac{d}{dx}p\frac{d\phi}{dx} + \frac{q}{\sigma}\phi\right] dx$$

upon integrating by parts twice. If $K_1\psi + K_2\,d\psi/dx = 0$ on the boundary then the integrated terms vanish if $K_1\phi + K_2\,d\phi/dx = 0$ on the boundary. Hence $\mathcal{L} = \mathcal{L}_a$ and $\mathcal{D} = \mathcal{D}_a$ so the Sturm–Liouville equation is self-adjoint.

Many people, outside the field of quantum mechanics, use (224a) for the scalar product. We can use this for the Sturm–Liouville equation also if we choose $(d/dx)p(d/dx) + q$ as the operator $\mathcal{L}$. The orthogonality condition is then expressed as $\langle\phi_n, \sigma\phi_m\rangle = 0$, $n \neq m$. In the next section on Green's function we will follow this convention.

Green's Function

Consider the Green's function for a self-adjoint system with

$$\mathcal{L}G(x, x') + \lambda G = -\delta(x - x'). \tag{227}$$

We then have

$$\langle G(x, x_1'), \mathcal{L}G(x, x_2')\rangle - \langle G(x, x_2'), \mathcal{L}G(x, x_1')\rangle$$
$$+ \lambda\langle G(x, x_1'), G(x, x_2')\rangle - \lambda\langle G(x, x_2'), G(x, x_1')\rangle$$
$$= -\langle G(x, x_1'), \delta(x - x_2')\rangle + \langle G(x, x_2'), \delta(x - x_1')\rangle.$$

But $\langle G(x, x_1'), \mathcal{L}G(x, x_2')\rangle = \langle\mathcal{L}G(x, x_1'), G(x, x_2')\rangle$ for a self-adjoint system so the left-hand side vanishes. Thus $\langle G(x, x_1'), \delta(x - x_2')\rangle = \langle G(x, x_2'), \delta(x - x_1')\rangle$ or

$$G(x_2', x_1') = G(x_1', x_2') \tag{228}$$

i.e., the Green's function is symmetrical in x_1', x_2'.

For a non-self-adjoint system we consider (227) along with

$$\mathcal{L}_aG_a(x, x') + \lambda G_a = -\delta(x - x'). \tag{229}$$

We can now write

$$\langle G(x, x_1'), \mathcal{L}_a G_a(x, x_2') \rangle + \lambda \langle G, G_a \rangle - \langle G_a(x, x_2'), \mathcal{L}G(x, x_1') \rangle - \lambda \langle G_a, G \rangle$$
$$= -\langle G(x, x_1'), \delta(x - x_2') \rangle + \langle G_a(x, x_2'), \delta(x - x_1') \rangle = 0$$

since $\langle G, \mathcal{L}_a G_a \rangle = \langle \mathcal{L}G, G_a \rangle = \langle G_a, \mathcal{L}G \rangle$. Hence

$$G(x_2', x_1') = G_a(x_1', x_2').   \tag{230}$$

For a non-self-adjoint system the Green's function is not symmetric.

Consider now a *linear* system

$$\mathcal{L}\psi + \lambda\psi = f.   \tag{231}$$

If we solve (227) and make G satisfy the same boundary conditions as ψ does, then by *superposition* the solution for ψ is

$$\psi(x) = -\int_0^a G(x, x')f(x')\,dx'   \tag{232}$$

since $G(x, x')$ is the field at x due to a unit source at x'. From (230) we see that this solution can also be written as

$$\psi(x) = -\int_0^a G_a(x', x)f(x')\,dx'.   \tag{233}$$

If ψ_n, λ_n are the eigenfunctions and eigenvalues of $\mathcal{L}$, and ϕ_n, λ_n, those of $\mathcal{L}_a$, then

$$G(x, x') = -\sum_n \frac{\psi_n(x)\phi_n(x')}{\lambda - \lambda_n}   \tag{234a}$$

and

$$G_a(x, x') = -\sum_n \frac{\psi_n(x')\phi_n(x)}{\lambda - \lambda_n}.   \tag{234b}$$

For the purpose of obtaining eigenfunction expansions and expressions for Green's functions the choice of scalar (inner) product to use is often dictated by what is most convenient for the problem at hand. For many problems in electromagnetic theory the use of (224a) as the scalar product makes it unnecessary to introduce an adjoint operator. The standard reciprocity theorems are based on (224a) as the scalar product, e.g.,

$$\iiint_V \mathbf{E}_1 \cdot \mathbf{J}_2\,dV = \iiint_V \mathbf{E}_2 \cdot \mathbf{J}_1\,dV.$$

On the other hand, if we wish to investigate convergence properties of series expansions that involve complex functions the scalar product shown in (224c) is used. In a linear vector space of complex functions the concept of length or norm of a function is a positive quantity. Thus if $\psi_n(x)$ is a complex eigenfunction and we want its length to be unity we normalize this function by dividing it by the scalar quantity

$$\sqrt{\langle \psi_n, \psi_n \rangle} = \left[\int_0^a \psi_n(x)\psi_n^*(x)\,dx \right]^{1/2}.$$

A complete linear vector space of complex functions with the metric (measure of distance) defined by (224c) as the scalar product is called a Hilbert space. In this space the distance between two functions ψ and Φ is given by

$$(\|\psi - \Phi\|)^{1/2} = \langle \psi - \Phi, \psi - \Phi \rangle^{1/2} = \left[\int_0^a (\psi - \Phi)(\psi - \Phi)^* \, dx \right]^{1/2}$$

when the functions are defined on the interval $0 \leq x \leq a$. If ψ is a sequence of functions ψ_N, which might be generated by a Fourier series,

$$\psi_N = \sum_{n=1}^N C_n \Phi_n$$

then we say that ψ_N converges to Φ as N tends to infinity if the distance between ψ_N and Φ vanishes as N tends to infinity.

The use of the scalar product (224c) is important if we want to make use of the well-developed mathematical theory of linear operators in a Hilbert space [2.9], [2.11].

The condition for a system such as (231) to have a solution is that the source function be in the range of the operator. For a unique solution the equation should not have any solutions for the homogeneous equation obtained by setting f equal to zero. If $\mathcal{L}$ is not a self-adjoint operator the solvability condition can also be stated as the requirement that f be orthogonal to the functions in the null space of the adjoint operator [2.11].

2.21. DISTRIBUTION THEORY

In this section we will give a brief introduction to distribution theory for the purpose of showing how this theory gives a mathematical structure within which the delta function can be treated in a rigorous manner.

In distribution theory a function $f(x)$ is characterized by a functional mapping. As a function, $f(x)$ is characterized by the scalar numbers $y = f(x)$ for each assigned value of x. In the Fourier integral of $f(x)$, i.e.,

$$F(w) = \int_{-\infty}^{\infty} e^{jwx} f(x)\,dx \tag{235}$$

the function $f(x)$ is characterized by the scalars $F(w)$ for each value of w. This is an example of a functional mapping and $f(x)$ is viewed as a functional on the space of e^{jwx} functions.

Consider now a space of continuous functions $\phi(\omega, x)$ with continuous derivatives of all orders and with bounded support [all $\phi(\omega, x)$ vanish for $|x|$ greater than some finite number]. The members of the set of functions $\phi(\omega, x)$ are identified by the parameter ω. The $\phi(\omega, x)$ are called testing functions. An example of a testing function that vanishes for $|x| \geq 1$ is

$$\phi(x) = \begin{cases} \exp \dfrac{1}{x^2 - 1}, & |x| < 1 \\ 0, & |x| \geq 1. \end{cases} \tag{236}$$

This function has continuous derivatives of all orders and they all vanish at $x = \pm 1$.

If $f(x)$ is an integrable function it can be characterized in terms of the distribution $F(\omega)$ defined by

$$F(\omega) = \int_{-\infty}^{\infty} \phi(\omega, x) f(x) \, dx. \tag{237}$$

It is common practice to simply write $f(x)$ for the distribution $F(\omega)$, it being understood that $f(x)$ viewed as a distribution is given by (237). We will deviate from this practice and use a subscript d to denote the distribution given by (237), i.e., $f_d(x) = F(\omega)$. The functional mapping given by (237) is unique. All functions that have or generate the same distribution are identical with the exception of possible values assigned at a number of discrete points. For example, the two functions

$$f_1(x) = x^2, \qquad 0 \le x \le 1$$

$$f_2(x) = \begin{cases} x^2, & 0 \le x < \frac{1}{2}, \ \frac{1}{2} < x \le 1 \\ 1, & x = \frac{1}{2} \end{cases}$$

have the same distribution. For physical problems functions such as f_2 do not arise. The distribution $f_d(x)$ is not a function of x. The notation is merely a reminder that the distribution was generated from a function $f(x)$ of x. The definition (237) characterizes $f(x)$ as a functional on the space of all testing functions. The derivative of a distribution $f_d(x)$ is defined as

$$f_d'(x) = \int_{-\infty}^{\infty} \phi(\omega, x) \frac{df}{dx} \, dx = f(x)\phi(\omega, x)|_{-\infty}^{\infty} - \int_{-\infty}^{\infty} f(x) \frac{d\phi(\omega, x)}{dx} \, dx$$

$$= - \int_{-\infty}^{\infty} f(x) \frac{d\phi(\omega, x)}{dx} \, dx. \tag{238}$$

The integrated part vanishes since ϕ has a bounded support.

Since $\phi(\omega, x)$ has a bounded support the Heaviside unit step function $h(x)$ given by

$$h(x) = \begin{cases} 0, & x < 0 \\ 1, & x > 0 \end{cases}$$

can be characterized as a distribution

$$h_d(x) = \int_{-\infty}^{\infty} h(x)\phi(\omega, x) \, dx = \int_0^{\infty} \phi \, dx. \tag{239a}$$

By means of the definition (238) we can also define the derivative of $h(x)$ in a distributional sense; i.e., we write

$$h_d' = \frac{dh_d(x)}{dx} = - \int_{-\infty}^{\infty} h(x) \frac{d\phi(\omega, x)}{dx} \, dx = - \int_0^{\infty} \frac{d\phi}{dx} \, dx = \phi(\omega, 0). \tag{239b}$$

The derivative does not exist in the classical sense but does exist as a distribution since $-d\phi/dx$ is also a testing function, as is any derivative of ϕ. In (239b) a knowledge of all $d\phi(\omega, x)/dx$ defines the functional h_d'.

The delta function distribution $\delta_d(x - x')$ is defined as follows (the integral is only a formal intermediate step):

$$\delta_d(x - x') = \int_{-\infty}^{\infty} \delta(x - x')\phi(\omega, x)\, dx = \phi(\omega, x'). \qquad (240)$$

Consequently, the distribution $\delta_d(x) = \phi(\omega, 0)$. But this is equal to (239b) so in a distributional sense the derivative of the Heaviside unit step distribution is the delta function distribution.

Consider the potential $\psi(r)$ from a unit point charge. It is given by $1/4\pi r$ and is a solution of Laplace's equation for $r > 0$

$$\frac{\partial}{\partial r} r^2 \frac{\partial}{\partial r} \frac{1}{4\pi r} = 0, \qquad r \neq 0.$$

The derivative of $1/r$ is not defined at $r = 0$ in a classical sense. From a distributional point of view

$$\begin{aligned}
\left(\frac{\partial}{\partial r} r^2 \frac{\partial}{\partial r} \frac{1}{4\pi r}\right)_d &= -\int_0^{\infty} \left(r^2 \frac{\partial}{\partial r} \frac{1}{4\pi r}\right) \frac{\partial \phi}{\partial r}\, dr \\
&= -\left(r^2 \frac{\partial \phi}{\partial r}\right) \frac{1}{4\pi r}\bigg|_0^{\infty} + \int_0^{\infty} \frac{1}{4\pi r} \frac{\partial}{\partial r} r^2 \frac{\partial \phi}{\partial r}\, dr \\
&= \int_0^{\infty} \frac{1}{4\pi r} \frac{\partial}{\partial r} r^2 \frac{\partial \phi}{\partial r}\, dr.
\end{aligned}$$

We can rewrite this integral in the form

$$I = \lim_{a \to 0} \int_a^{\infty} \frac{1}{4\pi r} \frac{\partial}{\partial r} r^2 \frac{\partial \phi}{\partial r}\, dr$$

since the derivatives of ϕ exist and are bounded at $r = 0$. Thus the limit exists. We now integrate by parts to reverse our equation back to the original form, i.e.,

$$\begin{aligned}
I &= \lim_{a \to 0} \left[\frac{r}{4\pi} \frac{\partial \phi}{\partial r}\bigg|_a^{\infty} - \int_a^{\infty} r^2 \frac{\partial \phi}{\partial r} \frac{\partial}{\partial r}\left(\frac{1}{4\pi r}\right) dr\right] \\
&= \lim_{a \to 0} \left[-\phi r^2 \frac{\partial}{\partial r}\left(\frac{1}{4\pi r}\right)\bigg|_a^{\infty} + \int_a^{\infty} \phi \frac{\partial}{\partial r} r^2 \frac{\partial}{\partial r}\left(\frac{1}{4\pi r}\right) dr\right].
\end{aligned}$$

The first integrated part vanishes as $a \to 0$ and for $r = \infty$. The last integral vanishes since for all finite r, $(\partial/\partial r)r^2(\partial\psi/\partial r) = 0$. Also $\partial(1/r)/\partial r = -1/r^2$ so

$$I = \lim_{a \to 0} \left[\frac{\phi(\omega, r)}{4\pi}\bigg|_a^{\infty}\right] = -\frac{\phi(\omega, 0)}{4\pi} = -\frac{\delta_d(r)}{4\pi}$$

by comparison with (240). Hence in a distributional sense

$$\frac{\partial}{\partial r} r^2 \frac{\partial \psi}{\partial r} = -\frac{\delta(r)}{4\pi}.$$

The potential ψ is now viewed as a distribution. The source for ψ_d can be regarded as a delta function distribution. The factor of 4π in the denominator arises because we used a one-dimensional form of Laplace's equation. The total source strength is unity since

$$\int_0^\pi \int_0^{2\pi} \int_0^\infty \frac{\delta(r)}{4\pi r^2} r^2 \sin\theta \, dr \, d\phi \, d\theta = 1.$$

The above examples clearly show that distribution theory does provide a mathematical framework in which entities such as the delta function and the derivative of a step function can be defined in a rigorous manner.

When we represent $\delta(x - x')$ in terms of an infinite series of eigenfunctions that series should be viewed as a distribution. The series may be differentiated and integrated term by term in a distributional sense.

Consider the space of testing functions with support $0 \le x \le a$. The distribution

$$\delta_d(x - x') = \sum_{n=1}^\infty \int_0^a \frac{2}{a} \sin\frac{n\pi x'}{a} \sin\frac{n\pi x}{a} \phi(\omega, x) \, dx \tag{241}$$

is a distribution depending on x' as a parameter and represents the delta function as a distributional series. A function $f(x)$ has the distribution

$$f_d(x) = \int_0^a f(x')\phi(\omega, x') \, dx' \tag{242a}$$

when $f(0) = f(a) = 0$. We can also express this in the form

$$\int_0^a f(x') \sum_{n=1}^\infty \frac{2}{a} \sin\frac{n\pi x'}{a} \int_0^a \phi(\omega, x) \sin\frac{n\pi x}{a} \, dx \, dx' \tag{242b}$$

upon introducing a Fourier series representation for $\phi(\omega, x')$, which always exists. By comparison with (241) we see that the distribution $f_d(x)$ is also given by

$$f_d(x) = \int_0^a f(x')\delta_d(x - x') \, dx' \tag{242c}$$

where $\delta_d(x - x')$ is regarded as a distribution. We can also write

$$f_d(x) = \int_0^a \phi(\omega, x) \left[\sum_{n=1}^\infty \frac{2}{a} \sin\frac{n\pi x}{a} \int_0^a f(x') \sin\frac{n\pi x'}{a} \, dx' \right] dx \tag{242d}$$

which is clearly the distributional definition of the Fourier series representation of $f(x)$. Note that in (242a) to (242d) $f_d(x)$ as a distribution is not a function of x. The convention of representing the function and its distribution by the same notation $f(x)$ requires careful attention to the context. To assist the reader in this regard we have attached the subscript d to the distribution.

As a final example consider the Fourier series for the triangle function

$$f(x) = \begin{cases} 2x/a, & 0 \le x \le a/2 \\ 2(1 - x/a), & a/2 \le x \le a \end{cases}$$

which is

$$f(x) = \sum_{n=1}^{\infty} \frac{8}{(n\pi)^2} \sin \frac{n\pi}{2} \sin \frac{n\pi x}{a}. \tag{243}$$

This series is uniformly convergent. The derivative of $f(x)$ is the piecewise continuous function $2/a$ for $0 \le x < a/2$ and $-2/a$ for $a/2 < x \le a$. The second and higher order derivatives of $f(x)$ do not exist at $x = a/2$ in the classical sense. In a distributional sense derivatives of $f(x)$ of any order have a meaning. The distribution for $d^2 f/dx^2$ is

$$f_d''(x) = \int_0^a \sum_{n=1}^{\infty} -\frac{8}{a^2} \sin \frac{n\pi}{2} \sin \frac{n\pi x}{a} \phi(\omega, x)\, dx = -\frac{4}{a}\delta_d(x - a/2) \tag{244}$$

by comparison with (241). We note that the step discontinuity at $x = a/2$ is $-4/a$ so formally we would express $f''(x)$ as $-4/a$ multiplied by the delta function. The distributional representation of the third derivative is

$$-\frac{4}{a}\frac{d\delta_d(x - a/2)}{dx} = f_d'''(x) = \int_0^a \sum_{n=1}^{\infty} \frac{8}{a^2} \sin \frac{n\pi}{2} \sin \frac{n\pi x}{a} \frac{d\phi}{dx}\, dx$$

$$= -\int_0^a \sum_{n=1}^{\infty} \frac{8n\pi}{a^3} \sin \frac{n\pi}{2} \cos \frac{n\pi x}{a} \phi(\omega, x)\, dx \tag{245}$$

with the latter obtained by an integration by parts. This example shows that distribution theory gives a meaning to highly divergent series when these are viewed as distributions.

It is not necessary to know the explicit form of the testing functions $\phi(\omega, x)$. It is sufficient to know that they exist in order to apply the theory.

The above discussion of distribution theory is only a sketchy outline and is presented mainly for the purpose of showing in a heuristic way how distribution theory deals with delta functions and divergent series that occur in boundary-value problems and the eigenfunction representation of Green's functions. The reader should consult the references for a rigorous development of the theory. For most problems of engineering importance one can use the operational approach followed in this chapter with the assurance that a rigorous approach exists that justifies the formal procedures.

REFERENCES AND BIBLIOGRAPHY

[2.1] J. A. Stratton, *Electromagnetic Theory*. New York, NY: McGraw-Hill Book Company, Inc., 1941, ch. 3.
[2.2] M. I. Sancer, private communication. (See also: M. I. Sancer, "The magnetic field integral equation singularity and its numerical consequences," Tech. Dig., URSI National Radio Science Meeting, Boulder, CO, Jan. 1981.)
[2.3] D. S. Jones, *Methods in Electromagnetic Wave Propagation*. Oxford: Clarendon Press, 1979, sect. 6.16.
[2.4] *Electromagnetics*, vol. 1, no. 4, 1981.

General Mathematical Theory

[2.5] P. M. Morse and H. Feshbach, *Methods of Theoretical Physics*, parts I and II. New York, NY: McGraw-Hill Book Company, Inc., 1953.
[2.6] A. G. Webster, *Partial Differential Equations of Mathematical Physics*. New York, NY: Hafner Publishing Company, 1950.

[2.7]   R. Courant and D. Hilbert, *Methods of Mathematical Physics*, vol. 1, English ed.   New York, NY: Interscience Publishers, Inc., 1953.

[2.8]   A. Sommerfeld, *Partial Differential Equations in Physics*.   New York, NY: Academic Press, Inc., 1949.

[2.9]   B. Friedman, *Principles and Techniques of Applied Mathematics*.   New York, NY: John Wiley & Sons, Inc., 1956.

[2.10] E. C. Titchmarsh, *Eigenfunction Expansions Associated with Second Order Differential Equations*. Oxford: Clarendon Press, 1946.

[2.11] I. Stakgold, *Green's Functions and Boundary Value Problems*.   New York, NY: John Wiley & Sons, Inc., 1979.

[2.12] C. T. Tai, *Dyadic Green's Functions in Electromagnetic Theory*.   Scranton, PA: Intext, 1971.

Special Topics

[2.13] N. Marcuvitz, "Field representations in spherically stratified regions," *Commun. Pure Appl. Math.*, vol. 4, pp. 263–315, sect. 5, Aug. 1951.

[2.14] L. Felson, "Alternative field representations in regions bounded by spheres, cones, and planes," *IRE Trans. Antennas Propagat.*, vol. AP-5, pp. 109–121, Jan. 1957.

[2.15] S. Sensiper, "Cylindrical radio waves," *IRE Trans. Antennas Propagat.*, vol. AP-5, pp. 56–70, 1957.

[2.16] H. Levine and J. Schwinger, "On the theory of electromagnetic wave diffraction by an aperture," *Commun. Pure Appl. Math.*, vol. 3, pp. 355–391, 1950.

[2.17] R. E. Collin, "The dyadic Green's function as an inverse operator," *Radio Sci.*, vol. 21, pp. 883–890, 1986.

[2.18] R. E. Collin, "Dyadic Green's function expansions in spherical coordinates," *Electromagnetics J.*, vol. 6, pp. 183–207, 1986.

[2.19] L. Wilson Pearson, "On the spectral expansion of the electric and magnetic dyadic Green's functions in cylindrical coordinates," *Radio Sci.*, vol. 18, pp. 166–174, 1983.

[2.20] J. Van Bladel, "Some remarks on Green's dyadic for infinite space," *IEEE Trans. Antennas Propagat.*, vol. AP-9, pp. 563–566, 1961.

[2.21] A. D. Yaghjian, "Electric dyadic Green's functions in source regions," *Proc. IEEE*, vol. 68, pp. 248–263, 1980. (See also: A. D. Yaghjian, "Maxwellian and cavity fields within continuous sources," *Amer. J. Phys.*, vol. 53, pp. 859–863, 1985.)

[2.22] W. A. Johnson, Q. Howard, and D. G. Dudley, "On the irrotational component of the electric Green's dyadic," *Radio Sci.*, vol. 14, pp. 961–967, 1979.

[2.23] W. W. Hansen, "A new type of expansion in radiation problems," *Phys. Rev.*, vol. 47, pp. 139–143, 1935.

[2.24] C. T. Tai, "Singular terms in the eigenfunction expansion of the electric dyadic Green's functions," Tech. Rep. RL-750, Rad. Lab., The University of Michigan, Mar. 1980.

[2.25] J. R. Wait, *Electromagnetic Radiation from Cylindrical Structures*.   New York, NY: Pergamon Press, 1959.

[2.26] R. F. Harrington, *Field Computation by Moment Methods*.   Malabar, FL: Kreiger Publishing Co., Inc., 1968.

[2.27] J. R. Mautz and R. F. Harrington, "*H*-field, *E*-field, and combined field solutions for conducting bodies of revolution," *A.E.U.* (Germany), vol. 32, pp. 157–164, 1978.

[2.28] J. R. Mautz and R. F. Harrington, "A combined-source solution for radiation from a perfectly conducting body," *IEEE Trans. Antennas Propagat.*, vol. AP-27, pp. 445–454, 1979.

[2.29] A. D. Yaghjian, "Augmented electric and magnetic field integral equations," *Radio Sci.*, vol. 16, pp. 987–1001, 1981.

[2.30] R. F. Harrington and J. R. Mautz, "Theory of characteristic modes for conducting bodies," *IEEE Trans. Antennas Propagat.*, vol. AP-19, pp. 622–628, 1971.

[2.31] L. Marin, "Natural-mode representation of transient scattered fields," *IEEE Trans. Antennas Propagat.*, vol. AP-21, pp. 809–818, 1973.

[2.32] M. Sancer, A. D. Varvatsis, and S. Siegel, "Pseudosymmetric eigenmode expansion for the magnetic field integral equation and SEM consequences," Interaction Note 355, Air Force Weapons Laboratory, Kirtland Air Force Base, NM, 87117, 1978.

[2.33] J. Van Bladel, *Singularities of the Electromagnetic Field*.   London: Oxford University Press. To be published.

PROBLEMS

2.1. Make a Fourier series expansion of (29b) in terms of the eigenfunctions $\sqrt{2/a}\,\sin(n\pi x/a)$ and show that (29a) is the result obtained.

2.2. Consider the equation $d^2\psi/dx^2 + \lambda\psi = 0$ with boundary conditions $\psi + 2\,d\psi/dx = 0$ at $x = 0$ and $\psi = 0$ at $x = a$. Show that the eigenvalues are determined by the transcendental equation $\tan\sqrt{\lambda}a = 2\sqrt{\lambda}$ and that the eigenfunctions (unnormalized) are $\sin\sqrt{\lambda_n}x - 2\sqrt{\lambda_n}\cos\sqrt{\lambda_n}x$, with λ_n the nth eigenvalue.

2.3. Consider the lossless transmission-line circuit illustrated in Fig. P2.3. The line has inductance L and capacitance C per meter. At the end $z = a$ the line is terminated in a reactance jX, while at $z = 0$ it is short-circuited. At $z = z'$ the line is excited by a time harmonic current generator with output $I_g e^{j\omega t}$. Because of the current source at z' the current on the line undergoes a step change I_g at z' as shown in the figure. Consequently, $\partial I/\partial z$ has a delta function change $I_g\delta(z - z')$ at z'. The voltage is continuous at z' but its slope, proportional to I, has a step change as shown. The equations describing the line are $\partial V/\partial z = -j\omega LI$, $\partial I/\partial z = -j\omega CV + I_g\delta(z - z')$. Show that $\partial^2 V/\partial z^2 + \omega^2 LCV = -j\omega LI_g\delta(z - z')$. Find two solutions for V (Methods I and II). Note that at $z = 0$, $V = 0$ and at $z = a$, $V = jXI = -(X/\omega L)(\partial V/\partial z)$ or $V + (X/\omega L)(\partial V/\partial z) = 0$. Note that the poles determine the resonant frequencies.

Answers:

Method I.

$$V = \sum_{n=1}^{\infty} \frac{j\omega LI_g \sin k_n z_< \sin k_n z_>}{(\lambda_n - k_0^2)\left(\dfrac{a}{2} - \dfrac{\sin 2k_n a}{4k_n}\right)}$$

where $\lambda_n = k_n^2$ and $\tan k_n a = -k_n X/\omega L$.

Method II.

$$V = jZ_c I_g \frac{\sin k_0 z_< [XY_c \cos k_0(a - z_>) + \sin k_0(a - z_>)]}{\sin k_0 a + XY_c \cos k_0 a}$$

where $k_0^2 = \omega^2 LC$ and $Y_c^2 = C/L$.

Resonant frequencies are given by $\tan k_n a = -XY_c$ where $k_n = \omega_n\sqrt{LC}$.

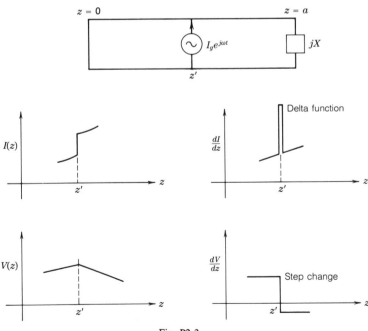

Fig. P2.3.

2.4. For the above problem let $\lambda = \omega^2 LC$. The boundary condition is then $(\tan\sqrt{\lambda}a)/\sqrt{\lambda} = -X/\omega L = -XY_c/\sqrt{\lambda}$ at $z = a$ and the eigenvalue is now part of the boundary condition. This is no longer a standard boundary-value problem. A solution can be obtained by extending the definition of the operator and that of its domain. The procedure to follow is given below.

Let $\mathbf{U}$ be the two-element column matrix

$$\mathbf{U} = \begin{bmatrix} \Psi(z) \\ \Phi(z) \end{bmatrix}$$

and define $\mathfrak{L}\mathbf{U}$ by

$$\mathfrak{L}\mathbf{U} = \begin{bmatrix} d\Phi(z)/dz \\ -d\Psi(z)/dz \end{bmatrix}.$$

Show that the system

$$\frac{d^2\Psi(z)}{dz^2} + k_0^2\Psi(z) = 0; \qquad \Psi(0) = 0, \quad \frac{d\Psi(a)}{dz} = -\frac{k_0 Z_c \Psi(a)}{X}$$

is equivalent to

$$\mathfrak{L}\mathbf{U} = \begin{bmatrix} d\Phi/dz \\ -d\Psi/dz \end{bmatrix} = k_0 \begin{bmatrix} \Psi \\ \Phi \end{bmatrix}$$

with boundary conditions $\Psi(a) = XY_c\Phi(a)$ and $\Psi(0) = d\Phi(0)/dz = 0$.

Consider the eigenvalue problem

$$\mathfrak{L}\mathbf{U}_n = k_n \begin{bmatrix} \Psi_n \\ \Phi_n \end{bmatrix}.$$

Show that for two solutions $\mathbf{U}_n$ and $\mathbf{U}_m$

$$\langle \mathbf{U}_n, \mathbf{U}_m \rangle = \int_0^a (\Psi_n\Psi_m + \Phi_n\Phi_m)\, dz = 0$$

so the eigenfunctions are orthogonal. Choose $\Psi_n = \sin k_n z$ and $\Phi_n = -\cos k_n z$. You will need to use $\tan k_n a = -XY_c$.

Show that $\langle \mathbf{U}_n, \mathbf{U}_n \rangle = a$ which is the normalization constant.

Show that if $\mathbf{V}$ satisfies the same boundary conditions as $\mathbf{U}$, $\langle \mathbf{V}, \mathfrak{L}\mathbf{U} \rangle = \langle \mathbf{U}, \mathfrak{L}\mathbf{V} \rangle$ so the operator $\mathfrak{L}$ is self-adjoint.

Show that

$$\frac{d^2 V}{dz^2} + k_0^2 V = -j\omega L I_g \delta(z - z')$$

is equivalent to

$$\mathfrak{L} \begin{bmatrix} V \\ \Phi \end{bmatrix} - k_0 \begin{bmatrix} V \\ \Phi \end{bmatrix} = \begin{bmatrix} jZ_c I_g \delta(z - z') \\ 0 \end{bmatrix}.$$

Let

$$\begin{bmatrix} V \\ \Phi \end{bmatrix} = \sum C_n \begin{bmatrix} \sin k_n z \\ -\cos k_n z \end{bmatrix}$$

and use the orthogonality of the eigenfunctions to show that a third solution for $V(z)$ is

$$V(z) = -\sum_{n=1}^{\infty} \frac{jZ_c I_g \, \sin k_n z_< \, \sin k_n z_>}{(\omega - \omega_n)\sqrt{LC}a}.$$

Note that this solution involves different eigenvalues and eigenfunctions than the solutions for Problem 2.3. (See B. Friedman, *Principles and Techniques of Applied Mathematics*. New York, NY: John Wiley & Sons, Inc., 1956, p. 205.)

2.5. Verify the equivalence of the solutions obtained in Problems 2.3 and 2.4. Evaluate the following integral using residue theory

$$\frac{1}{2\pi j}\oint_C \frac{jZ_c I_g \omega \sqrt{LC} \, \sin wz_< \left[\dfrac{wX}{\omega L}\cos w(a - z_>) + \sin w(a - z_>)\right]}{(w^2 - \omega^2 LC)\left(\sin wa + \dfrac{wX}{\omega L}\cos wa\right)}\, dw$$

and show that the residue series from the poles at $w = \pm \sqrt{LC}\omega$ gives the Method II solution in Problem 2.3 and the residue series from the poles at $\tan w_n a = -w_n X/\omega L$, i.e., $w = \pm \sqrt{\lambda n}$, gives the negative of the Method I solution. The residues can be found by evaluating the derivative of the denominator at the poles. You will need to use the eigenvalue equation to simplify the results. Show that on circle C with infinite radius, with $w = u + jv$ the integrand behaves like $\{\exp[-|v|(z_> - z_<)]\}/w$ and hence the integral is zero and solutions 1 and 2 in Problem 2.3 are equivalent. Consider next the integral

$$\frac{1}{2\pi j}\oint_C \frac{jZ_c I_g \, \sin wz_< [XY_c \cos w(a - z_>) + \sin w(a - z_>)]}{(w - \omega\sqrt{LC})(\sin wa + XY_c \cos wa)}\, dw$$

and show that the residue evaluation demonstrates that solutions 1 and 3 are equivalent. Thus all three solutions are equivalent.

2.6. Use (48a) and (49) in (44) and integrate around a contour C_x enclosing the poles of G_x and show that (47) is the solution obtained for G.

2.7. Consider an infinitely long transmission line excited by a current source $I_g e^{j\omega t}$ at z' (see Fig. P2.7). Solve $d^2 V/dz^2 + kk_0^2 V = -j\omega LI_g\delta(z - z')$ by means of a Fourier transform.

HINT: Assume that there is a small shunt conductance so that $k_0^2 = \omega^2 LC(1 - jG/\omega C)$ and $k_0 = k_0' - jk_0''$. This will displace the poles away from the real axis. Note that for $z > z'$ the inversion contour can be closed in the lower half plane, and for $z < z'$ it can be closed in the upper half plane. Thus the inverse transform can be evaluated in terms of residues.

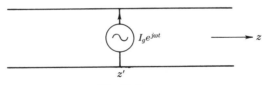

$I_g e^{j\omega t}$

z'

Fig. P2.7.

2.8. Consider an axial line source inside a rectangular pipe as shown in Fig. P2.8. Find the Green's function which satisfies

$$\frac{\partial^2 G}{\partial x^2} + \frac{\partial^2 G}{\partial y^2} = -\delta(x - x')\delta(y - y'), \qquad G = 0 \text{ on boundary}$$

in terms of a double Fourier series of eigenfunctions for the equation

$$\left(\frac{\partial^2}{\partial x^2} + \frac{\partial^2}{\partial y^2} + \lambda_{nm}\right)\psi_{nm} = 0, \qquad \psi_{nm} = 0 \text{ on boundary}.$$

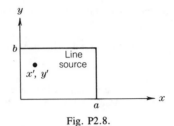

Fig. P2.8.

2.9. Find G for Problem 2.8 in terms of an eigenfunction expansion along x but as a closed form in the y variable, i.e., in the form

$$G = \sum_{n=1}^{\infty} f_n(y) \sin \frac{n\pi x}{a}.$$

2.10. Find the Green's function for Problem 2.8 in terms of a contour integral of $G_x(\lambda_x)G_y(\lambda_y)$ where

$$\frac{\partial^2 G_x}{\partial x^2} + \lambda_x G_x = -\delta(x - x')$$

$$\frac{\partial^2 G_y}{\partial y^2} + \lambda_y G_y = -\delta(y - y').$$

Verify your result with those obtained in Problems 2.8 and 2.9.

2.11. Determine the Green's function of the first kind ($G = 0$ at $r = a$) for Poisson's equation for a line source parallel and outside a conducting cylinder of radius a. The source is at $r' > a$. Use (a) the method of images and (b) expansion in a suitable set of eigenfunctions.

2.12. For the line source in a rectangular waveguide use Method II to show that an equivalent solution for G_x is

$$G_x = \frac{\sin \sqrt{\lambda_x} x_< \sin \sqrt{\lambda_x}(a - x_>)}{\sqrt{\lambda_x} \sin \sqrt{\lambda_x} a}.$$

Use this solution in (59) and make the variable change $\lambda_z = w^2$ to show that

$$G = -\frac{1}{2\pi} \int_{-\infty}^{\infty} \frac{\sin \gamma x_< \sin \gamma(a - x_>)}{\gamma \sin \gamma a} e^{-jw|z-z'|} \, dw$$

where $\gamma = \sqrt{k_0^2 - w^2}$. Evaluate the integral by closing the contour in the lower half plane and using residue theory. Note that the contour runs above the pole at $w = \gamma$ and below the pole at $w = -\gamma$.

2.13. By using Laplace transforms, find the Green's function for the following differential equation:

$$L \left(\frac{d^2 G}{dt^2} + \frac{R}{L} \frac{dG}{dt} + \frac{1}{LC} G \right) = \delta(t - t')$$

$$G = \frac{dG}{dt} = 0 \qquad \text{at } t = 0.$$

This Green's function represents the charge flowing in a series RLC circuit with a unit voltage impulse input at time $t = t'$.

Answer:

$$G(t|t') = \frac{1}{\omega L} e^{-\alpha(t+t')} \sin \omega(t - t'), \qquad t > t'$$

where

$$\alpha = \frac{R}{2L}$$

$$\omega = (LC)^{-1/2}\left(1 - \frac{R^2 C}{4L}\right)^{1/2}.$$

2.14. For Problem 2.13 show that

$$\int_{-\infty}^{t} e^{j\omega_0 t'} G(t|t')\,dt'$$

gives the usual steady-state solution for the charge in the circuit due to an impressed voltage source $e^{j\omega_0 t}$.

2.15. Let $\bar{G}(\mathbf{r}|\mathbf{r}_0)$ be a solution of $\nabla \times \nabla \times \bar{G} - k^2\bar{G} = \bar{I}\delta(\mathbf{r} - \mathbf{r}_0)$ in a volume V. On the surface S surrounding V let $\bar{G}$ satisfy one of the boundary conditions, $\mathbf{n} \times \bar{G}$ or $\mathbf{n} \times \nabla \times \bar{G} = 0$. Use the vector form of (108) to show that

$$\mathbf{J}_1(\mathbf{r}_1)\cdot\bar{G}(\mathbf{r}_1, \mathbf{r}_2)\cdot\mathbf{J}_2(\mathbf{r}_2) = \mathbf{J}_2(\mathbf{r}_2)\cdot\bar{G}(\mathbf{r}_2, \mathbf{r}_1)\cdot\mathbf{J}_1(\mathbf{r}_1).$$

HINT: Replace the dyadic $\bar{B}$ in (108) by a vector function $\mathbf{B}$. Next choose for $\mathbf{A}$ and $\mathbf{B}$ the following:

$$\mathbf{A} = \bar{G}(\mathbf{r}, \mathbf{r}_1)\cdot\mathbf{J}_1(\mathbf{r}_1)$$

$$\mathbf{B} = \bar{G}(\mathbf{r}, \mathbf{r}_2)\cdot\mathbf{J}_2(\mathbf{r}_2).$$

2.16. Derive the normalization integrals (138b), (138c).

2.17. Derive the solution for the scalar Green's function g given by (155).

2.18. Evaluate the integral over k in (155) and show that

$$g = \frac{e^{-jk_0 R}}{4\pi R} = -jk_0 \sum_{n=0}^{\infty}\sum_{m=0}^{n} \frac{\epsilon_{0m}(2n+1)(n-m)!}{4\pi(n+m)!} P_n^m(\cos\theta)P_n^m(\cos\theta')\cos m(\phi - \phi')j_n(k_0 r_<)h_n^2(k_0 r_>).$$

2.19. Show that the Green's function for Poisson's equation in free space, which is a solution of

$$\nabla^2 g = -\delta(\mathbf{r} - \mathbf{r}') = -\frac{\delta(r - r')\delta(\theta - \theta')\delta(\phi - \phi')}{r^2 \sin\theta},$$

is given by

$$g = \frac{1}{4\pi R} = \sum_{n=0}^{\infty}\sum_{m=0}^{n}\int_0^{\infty} \frac{\epsilon_{0m}(2n+1)(n-m)!}{2\pi^2(n+m)!} P_n^m(\cos\theta)P_n^m(\cos\theta')\cos m(\phi - \phi')j_n(kr)j_n(kr')\,dk.$$

HINT: Expand the solution in terms of the eigenfunctions of the Laplacian operator.

2.20. Evaluate the integral over k in Problem 2.19 and show that another solution is

$$g = \frac{1}{4\pi R} = \sum_{n=0}^{\infty}\sum_{m=0}^{n} \frac{\epsilon_{0m}(n-m)!}{4\pi(n+m)!} P_n^m(\cos\theta)P_n^m(\cos\theta')\cos m(\phi - \phi')\frac{r_<^n}{r_>^{n+1}}.$$

Take the limit of the solution in Problem 2.18 as k_0 approaches zero and show that the same result is obtained.

2.21. Find a solution for the Green's function in Problem 2.19 using the following method: Expand g in the form

$$g = \sum_{n=0}^{\infty}\sum_{m=0}^{n} f_n(r)P_n^m(\cos\theta)\cos m(\phi - \phi').$$

Substitute this into the differential equation and make use of the orthogonal properties of the P_n^m and $\cos m(\phi - \phi')$ functions to show that

$$\frac{1}{r^2}\frac{d}{dr}r^2\frac{df_n}{dr} - \frac{n(n+1)}{r^2}f_n = -\frac{\epsilon_{0m}(2n+1)(n-m)!}{4\pi(n+m)!}P_n^m(\cos\theta)\frac{\delta(r-r')}{r^2}.$$

Use Method I and a Hankel transform to find a solution for f_n; i.e., let

$$f_n(r) = \int_0^\infty F_n(\lambda)j_n(\lambda r)\lambda^2\,d\lambda.$$

Also use Method II and let

$$f_n(r) = \begin{cases} C_n\dfrac{r^n}{(r')^{n+1}}, & r \le r' \\[2ex] C_n\dfrac{(r')^n}{r^{n+1}}, & r \ge r' \end{cases}$$

to find a solution. Verify your solutions by comparison with those given in Problems 2.19 and 2.20.

2.22. In cylindrical coordinates show that a solution of

$$(\nabla^2 + k_0^2)g = -\frac{\delta(r-r')\delta(\phi-\phi')\delta(z-z')}{r}$$

for the free-space scalar Green's function is

$$g = \sum_{n=0}^{\infty}\int_{-\infty}^{\infty}\int_0^\infty \frac{\epsilon_{0n}}{4\pi^2}\cos n(\phi-\phi')e^{-jw(z-z')}\frac{kJ_n(kr)J_n(kr')}{k^2+w^2-k_0^2}\,dk\,dw.$$

Evaluate the integral over k and show that

$$g = -j\sum_{n=0}^{\infty}\frac{\epsilon_{0n}}{8\pi}\int_{-\infty}^{\infty}e^{-jw(z-z')}J_n(\gamma r_<)H_n^2(\gamma r_>)\cos n(\phi-\phi')\,dw$$

where $\gamma = \sqrt{k_0^2 - w^2}$. Show that when $r' = 0$ the solution reduces to

$$g = \frac{e^{-jk_0R}}{4\pi R} = -\frac{j}{8\pi}\int_{-\infty}^{\infty}H_0^2(\gamma r)e^{-jw(z-z')}\,dw$$

where $R = \sqrt{r^2 + (z-z')^2}$. Note that a useful inverse Fourier transform has been evaluated.

2.23. For Problem 2.22 expand g in the form

$$g = \sum_{n=0}^{\infty}\int_{-\infty}^{\infty}f_n(r)\cos n(\phi-\phi')e^{-jwz}\,dw$$

and by Fourier analysis show that

$$\frac{1}{r}\frac{d}{dr}r\frac{df_n}{dr} - \frac{n^2}{r^2}f_n + \gamma^2 f_n = -\frac{\epsilon_{0n}}{4\pi^2}e^{jwz'}\frac{\delta(r-r')}{r}.$$

Assume that $f_n = C_n J_n(\gamma r_<)H_n^2(\gamma r_>)$ and use the Wronskian relation $W = J_n H_n^{2\prime} - J_n' H_n^2 = -2j/\pi\gamma r'$ to show that $C_n = -j(\epsilon_{0n}/8\pi)e^{jwz'}$ and that the solution given in Problem 2.22 is obtained.

2.24. Consider the scalar Green's function within a sphere of radius a that is a solution of

$$\frac{d}{dr} r^2 \frac{dg}{dr} - n(n+1)g + k_0^2 r^2 g = -\delta(r - r'), \qquad g(a) = 0.$$

Show that two solutions for g are

$$g = \sum_{i=1}^{\infty} \frac{2j_n(k_i r_<)j_n(k_i r_>)}{a^3 [j_n'(k_i a)]^2 (k_0^2 - k_i^2)} = -k_0 \frac{j_n(k_0 r_<)[j_n(k_0 a)y_n(k_0 r_>) - y_n(k_0 a)j_n(k_0 r_>)]}{j_n(k_0 a)}$$

where k_i are the roots of $j_n(ka) = 0$. The Wronskian determinant is $W = j_n(u)y_n'(u) - j_n'(u)y_n(u) = 1/u^2$. Show that when k_0 approaches zero the limiting solution is

$$g = \frac{r_<^n}{(2n+1)a^n} \left[\frac{a^n}{r_>^{n+1}} - \frac{r_>^n}{a^{n+1}} \right].$$

The series that represents this function can be found by putting $k_0 = 0$ in the above series.

2.25. For Problem 2.24 let a become infinite and use Hankel transforms to show that g is given by

$$g = \frac{2}{\pi} \int_0^{\infty} \frac{j_n(kr)j_n(kr')}{k^2 - k_0^2} k^2 \, dk.$$

Evaluate the integral over k to show that

$$g = -jk_0 j_n(k_0 r_<)h_n^2(k_0 r_>).$$

2.26. Show that when the boundary condition in Problem 2.24 is changed to $d(rg)/dr = 0$ at $r = a$ the two solutions for g are

$$g = \sum_{i=0}^{\infty} \frac{2j_n(l_i r_<)j_n(l_i r_>)}{a^2 j_n^2(l_i a) \left[1 - \frac{n(n+1)}{l_i^2 a^2} \right] (k_0^2 - l_i^2)} = -k_0 \frac{j_n(k_0 r_<)[y_n(k_0 r_>)(k_0 a j_n(k_0 a))' - j_n(k_0 r_>)(k_0 a y_n(k_0 a))']}{[k_0 a j_n(k_0 a)]'}$$

where l_i are the roots of $d[rj_n(kr)]/dr = 0$ at $r = a$. Show that the limiting form as k_0 tends to zero is

$$g = \frac{r_<^n}{(n+1)(2n+1)} \left[\frac{(n+1)a^n}{r_>^{n+1}} + \frac{nr_>^n}{a^{n+1}} \right].$$

Derive a series representation for this function.

2.27. Show that the two solutions for g given in Problem 2.24 have the same residues at $k_0 = k_i$.

HINT: Expand $j_n(\sqrt{k_0^2}a)$ in a Taylor series in k_0^2 about the point $k_0^2 = k_i^2$. Use the Wronskian determinant to evaluate $y_n(k_i a)$ to obtain $y_n = -(k_i^2 a^2 j_n')^{-1}$. Show that g vanishes when k_0 becomes infinite. Note that two functions with the same pole singularities in the complex plane and that vanish at infinity must be identical. Hence the two solutions represent the same function.

2.28. A point source at $r' > a$ radiates in the presence of a sphere of radius a. Find the solution for the scalar Green's function that satisfies the Dirichlet boundary condition $g = 0$ at $r = a$.

HINT: Use the mode solution in Problem 2.18 and superimpose outward-propagating waves that cancel the incident field at $r = a$.

2.29. In the solution given in Problem 2.18 let r' tend to infinity, use the asymptotic value of $h_n^2(k_0 r)$, and derive the following expansion for a plane wave:

$$e^{jk_0 r \cdot a_{r_0}} = \sum_{n=0}^{\infty} \sum_{m=0}^{n} j^n \frac{\epsilon_{0m}(2n+1)(n-m)!}{(n+m)!} P_n^m(\cos\theta) P_n^m(\cos\theta') \cos m(\phi - \phi') j_n(k_0 r)$$

for r finite.

Hint: Express $e^{-jk_0R}/4\pi R$ in the form

$$\frac{e^{-jk_0|\mathbf{r}-\mathbf{r}_0|}}{4\pi|\mathbf{r}-\mathbf{r}_0|} \sim \frac{e^{jk_0r_0}}{4\pi r_0}e^{jk_0\mathbf{r}\cdot\mathbf{a}_{r_0}}.$$

If the point source is located on the z axis only the terms for $m = 0$ are needed.

2.30. A dyadic Green's function for the electric field may be obtained from (155) by applying the operator shown in (112). Comment on this solution with respect to the decomposition of the dyadic Green's function into transverse spherical TE and TM waves, longitudinal waves, and orthogonality properties.

2.31. Take a three-dimensional Fourier transform of

$$(\nabla \times \nabla \times - k_0^2)\bar{\mathbf{G}} = \delta(\mathbf{r} - \mathbf{r}')\bar{\mathbf{I}}$$

to obtain

$$[(k^2 - k_0^2)\bar{\mathbf{I}} - \mathbf{k}\mathbf{k}]\cdot\hat{\bar{\mathbf{G}}} = e^{j\mathbf{k}\cdot\mathbf{r}'}\bar{\mathbf{I}}.$$

To invert the dyadic assume that

$$\bar{\mathbf{D}}^{-1} = [(k^2 - k_0^2)\bar{\mathbf{I}} - \mathbf{k}\mathbf{k}]^{-1} = A\mathbf{k}\mathbf{k} + B\bar{\mathbf{I}}$$

and use $\bar{\mathbf{D}}\cdot(A\mathbf{k}\mathbf{k} + B\bar{\mathbf{I}}) = \bar{\mathbf{I}}$ to find A and B. Thus show that

$$\bar{\mathbf{G}}(\mathbf{r}, \mathbf{r}') = \frac{1}{8\pi^3}\iiint\limits_{-\infty}^{\infty}\frac{k_0^2\bar{\mathbf{I}} - \mathbf{k}\mathbf{k}}{k_0^2(k^2 - k_0^2)}e^{-j\mathbf{k}\cdot(\mathbf{r}-\mathbf{r}')}\,d\mathbf{k}.$$

In order to evaluate the integral, change to spherical coordinates in $\mathbf{k}$ space. Choose the polar axis to lie along $\mathbf{r} - \mathbf{r}'$. Then

$$\bar{\mathbf{G}} = \frac{1}{8\pi^3}\int_0^\infty\int_0^{2\pi}\int_0^\pi\frac{k_0^2\bar{\mathbf{I}} - \mathbf{k}\mathbf{k}}{k_0^2(k^2 - k_0^2)}e^{-jkR\cos\theta}k^2\,\sin\theta\,d\theta\,d\phi\,dk.$$

The integration over θ and ϕ is readily carried out after bringing $\mathbf{k}\mathbf{k}$ outside the integral as the operator $-\nabla\nabla$. The final integral over k can be expressed as an integral from $-\infty$ to ∞ and evaluated by residues. Complete the details of this evaluation and show that the solution given by (112) is obtained.

The solution for $\bar{\mathbf{G}}$ given above is an eigenfunction expansion in terms of plane waves. It can be split into longitudinal and transverse waves. The longitudinal waves can be obtained by taking a scalar product with $\mathbf{k}/k$. Thus show that

$$\bar{\mathbf{G}} = \frac{1}{8\pi^3}\iiint\limits_{-\infty}^{+\infty}\left[\frac{k^2\mathbf{I} - \mathbf{k}\mathbf{k}}{k^2(k^2 - k_0^2)} - \frac{\mathbf{k}\mathbf{k}}{k_0^2k^2}\right]e^{-j\mathbf{k}\cdot(\mathbf{r}-\mathbf{r}')}\,d\mathbf{k}.$$

The first integral is well behaved and can be evaluated by bringing $k^2\mathbf{I} - \mathbf{k}\mathbf{k}$ outside the integral sign as the operator $(\nabla\nabla\cdot - \nabla^2)\mathbf{I}$. Show that the contribution to the Green's dyadic is

$$\frac{1}{k_0^2}\nabla\times\nabla\times\mathbf{I}\frac{e^{-jk_0R}}{4\pi R}.$$

If in the second term $\mathbf{k}\mathbf{k}$ is brought outside the integral sign as the operator $-\nabla\nabla$ show that the contribution is

$$\frac{1}{k_0^2}\nabla\nabla\frac{1}{R}.$$

However, the second integral is not a convergent one so replacing $\mathbf{kk}$ by $-\nabla\nabla$ cannot be justified. What this means is simply that the eigenfunction solution for the Green's dyadic "function" cannot generally be summed to give a true function. However, the above solution can be used in the usual way by scalar multiplying with the source function $\mathbf{J}(\mathbf{r}')$ and integrating over $\mathbf{r}'$ before completing the integration over $\mathbf{k}$.

A mode expansion in terms of waves propagating along z can also be obtained by integrating over k_z first. Show that in the expansion of $\bar{\mathbf{G}}$ the longitudinal modes are canceled by contributions from the transverse modes apart from a residual delta function term $-\mathbf{a}_z\mathbf{a}_z\delta(\mathbf{r}-\mathbf{r}')/k_0^2$. Show that the remaining transverse mode contribution is given by

$$\frac{-j}{8\pi^2}\int\!\!\!\int_{-\infty}^{+\infty}\frac{[\mathbf{I}k_0^2-(\mathbf{a}_xk_x+\mathbf{a}_yk_y+\mathbf{a}_zk_z\,\mathrm{sg}\,(z-z'))^2]}{k_0^2k_z}e^{-j\mathbf{k}_t\cdot(\mathbf{r}-\mathbf{r}')-jk_z|z-z'|}\,d\mathbf{k}_t$$

where $\mathbf{k}_t=\mathbf{a}_xk_x+\mathbf{a}_yk_y$ and $k_z=\sqrt{k_0^2-k_t^2}$.

Note that the longitudinal modes have the factor $e^{-k_t|z-z'|}$ in the integrand and are nonpropagating modes along z. An often overlooked point in connection with the above Fourier transform solution for $\bar{\mathbf{G}}$ is the fact that $\bar{\mathbf{G}}$ in the form

$$\left(\mathbf{I}+\frac{1}{k_0^2}\nabla\nabla\right)\frac{e^{-jk_0R}}{4\pi R}$$

cannot be Fourier transformed if the differentiations are carried out because of the $1/R^3$ singularity. Hence the solution given above should be interpreted as the plane wave eigenfunction expansion of the Green's dyadic operator and integration over the source coordinates is to be performed first before inverting the Fourier transform.

2.32. Consider the one-dimensional Green's function problem for the interior of a cylinder

$$\frac{1}{r}\frac{d}{dr}r\frac{dg}{dr}+k^2g=\frac{\delta(r-r')}{r},\qquad\frac{dg}{dr}=0\text{ at }r=a.$$

Show that two solutions are

$$g=\sum_{m=1}^{\infty}\frac{2J_0(q_mr/a)J_0(q_mr'/a)}{a^2J_0^2(q_m)[k^2-(q_m/a)^2]}+\frac{2}{a^2k^2}=\frac{\pi}{2J_1(ka)}J_0(kr_<)[J_1(ka)Y_0(kr_>)-J_0(kr_>)Y_1(ka)]$$

where q_m are the roots of $dJ_0(kr)/dr=0$ at $r=a$. Note that because of the Neumann boundary condition at $r=a$ there is an $m=0$ eigenfunction with a zero eigenvalue.

2.33. Let the incident electric field given by (215a) contain the additional mode $C_{11}\mathbf{M}_{11}^e$. Show that the following current is induced on the sphere (assume that the common factor $j_1(k_0a)$ is canceled): $(A_{11}\mathbf{S}_{11}^o+C_{11}\mathbf{S}_{11}^e)/[aZ_0(k_0ah_1^2)]$. Let $A_{11}=\sqrt{A_{11}^2+C_{11}^2}\cos\alpha$ and $C_{11}=\sqrt{A_{11}^2+C_{11}^2}\sin\alpha$ and show that the induced current is proportional to $\sqrt{A_{11}^2+C_{11}^2}\mathbf{S}_{11}^o(\phi+\alpha)$. Show that $\mathbf{S}_{11}^e(\phi+\alpha)$ is an orthogonal current distribution and a solution of the homogeneous integral equation in the limit as $j_1(k_0a)$ approaches zero.

2.34. Consider the radiation condition given by (115) and the solution (109) for the electric field where S is chosen as a sphere of infinite radius. Use (115) to eliminate $\lim_{r\to\infty}r\nabla\times\mathbf{E}$ and $\lim_{r\to\infty}r\nabla\times\bar{\mathbf{G}}$ and show that the integrand in (109) vanishes when the radiation condition holds.

2.35. Let $\mathbf{E}$, $\mathbf{H}$ be the difference field in V_1 as discussed in Section 1.8 where the sources for the field are contained in V_2. Then on S and the sphere of infinite radius

$$\oint_S\mathbf{E}\times\mathbf{H}\cdot\mathbf{n}\,dS+\lim_{r\to\infty}\oint_{S_\infty}\mathbf{E}\times\mathbf{H}\cdot\mathbf{n}\,dS=0$$

since $\mathbf{n}\times(\mathbf{E}_1-\mathbf{E}_2)=\mathbf{n}\times\mathbf{E}=0$ and $\mathbf{n}\times(\mathbf{H}_1-\mathbf{H}_2)=\mathbf{n}\times\mathbf{H}=0$ on S and the difference field satisfies the radiation condition.

Consider a field $\mathbf{E}'$, $\mathbf{H}'$ that is a solution of Maxwell's equations due to an arbitrary source $\mathbf{J}$ in V_1 and which satisfies the radiation condition. Then

$$\oint_S(\mathbf{E}\times\mathbf{H}'-\mathbf{E}'\times\mathbf{H})\cdot\mathbf{n}\,dS=0$$

and according to an extension of the results of Problem 1.17 $\mathbf{J}(\mathbf{r}') \cdot \mathbf{E}(\mathbf{r}') = 0$ for all $\mathbf{r}'$ in V_1. Hence $\mathbf{E} = 0$ in V_1 and therefore the two solutions $\mathbf{E}_1$ and $\mathbf{E}_2$ are equal throughout V_1. Maxwell's equations then require that $\mathbf{H}_1 = \mathbf{H}_2$ and hence the solution to the external radiation problem is unique if the radiation condition is satisfied and $\mathbf{n} \times \mathbf{E}_1$ and $\mathbf{n} \times \mathbf{H}_1$ are given on the surface S. If only $\mathbf{n} \times \mathbf{E}_1$ is given on S then $\mathbf{E}'$, $\mathbf{H}'$ can be chosen so that $\mathbf{n} \times \mathbf{E}' = 0$ on S and the reciprocity theorem as used above will again show that $\mathbf{E}$ is zero throughout V_1 because the surface integrals over S and the sphere at infinity are zero.

For scattering from a perfectly conducting object the two candidate solutions $\mathbf{E}_1$ and $\mathbf{E}_2$ are required to satisfy the boundary condition $\mathbf{n} \times \mathbf{E}_1 = \mathbf{n} \times \mathbf{E}_2 = -\mathbf{n} \times \mathbf{E}_i$ where $\mathbf{E}_i$ is the incident field. Thus on S we have $\mathbf{n} \times (\mathbf{E}_1 - \mathbf{E}_2) = \mathbf{n} \times \mathbf{E} = 0$ but $\mathbf{n} \times \mathbf{H}$ is left unspecified. As noted above, the solution is unique. When the scattering problem is solved using either the EFIE or the MFIE the lack of uniqueness is a fault of the method of solution and is not due to an intrinsic nonuniqueness in the problem. (A much longer and detailed proof of uniqueness based on the properties of the spherical vector wave functions may be found in: C. Müller, *Mathematical Theory of Electromagnetic Waves.* New York, NY: Springer-Verlag, 1969.)

3
Transverse Electromagnetic Waves

A plane transverse electromagnetic (TEM) wave is a wave for which the electric and magnetic field vectors lie in a plane that is transverse or perpendicular to the axis of propagation. This is the principal type of wave which exists on an ideal transmission line. There may be other types such as spherical and cylindrical transverse electromagnetic waves as well as plane TEM waves. Far away from radiating structures the field is essentially a TEM wave. In the discussion to follow a time factor $e^{j\omega t}$ will be assumed.

3.1. PLANE TEM WAVES

We will refer our wave to an xyz rectangular coordinate frame and take the z axis as the direction along which the wave propagates. Maxwell's curl equations are

$$\nabla \times \mathbf{E}_t = -j\omega\mu\mathbf{H}_t \tag{1a}$$

$$\nabla \times \mathbf{H}_t = j\omega\epsilon\mathbf{E}_t \tag{1b}$$

where we have attached a subscript t to the field vectors to indicate that they have *transverse*, that is, x and y, components only. The medium is assumed to be lossless, isotropic, and homogeneous with electrical parameters μ, ϵ. The vector operator ∇ may be split into two parts, a transverse part

$$\nabla_t = \mathbf{a}_x\frac{\partial}{\partial x} + \mathbf{a}_y\frac{\partial}{\partial y}$$

and a longitudinal part

$$\nabla_z = \mathbf{a}_z\frac{\partial}{\partial z}.$$

Now $\nabla_t \times \mathbf{E}_t$ is a vector perpendicular to the transverse plane, and similarly with $\nabla_t \times \mathbf{H}_t$. Since the fields are purely transverse, Eqs. (1) reduce to

$$\mathbf{a}_z \times \frac{\partial\mathbf{E}_t}{\partial z} = -j\omega\mu\mathbf{H}_t \tag{2a}$$

$$\mathbf{a}_z \times \frac{\partial\mathbf{H}_t}{\partial z} = j\omega\epsilon\mathbf{E}_t \tag{2b}$$

$$\nabla_t \times \mathbf{E}_t = 0 \tag{2c}$$

$$\nabla_t \times \mathbf{H}_t = 0. \tag{2d}$$

The general theory of vector fields[1] shows that, when the curl of a vector is identically zero, it may be derived from the gradient of a suitable scalar function since the curl of the gradient is identically zero. Thus (2c) and (2d) will be automatically satisfied if we take

$$\mathbf{E}_t = g_1(z)\nabla_t\Phi \tag{3a}$$

$$\mathbf{H}_t = g_2(z)\nabla_t\psi \tag{3b}$$

where g_1, g_2, Φ, and ψ are scalar functions to be determined. In (3), Φ and ψ are functions of the transverse coordinates x and y only, since it is only the transverse curl which vanishes. The form of solution given in (3) is permissible because the vector Helmholtz equation is separable in rectangular coordinates. Since we are considering a source-free medium, the divergence of the field vectors is zero. From (3) it is then seen that both Φ and ψ are solutions of Laplace's equation in the transverse plane; that is, $\nabla_t^2\Phi = 0$, and similarly for ψ.

The scalar functions Φ and ψ are not independent since $\mathbf{E}_t$ and $\mathbf{H}_t$ are related by (2a) and (2b). Differentiating (2a) with respect to z and substituting from (2b) after cross multiplying both equations by $\mathbf{a}_z$ gives

$$\mathbf{a}_z \times \left(\mathbf{a}_z \times \frac{\partial^2\mathbf{E}_t}{\partial z^2}\right) - \omega^2\mu\epsilon\mathbf{E}_t = 0. \tag{4}$$

Expanding this result gives

$$\frac{\partial^2\mathbf{E}_t}{\partial z^2} + k^2\mathbf{E}_t = 0 \tag{5}$$

where $k^2 = \omega^2\mu\epsilon$. Equation (5) follows from (4) since $\mathbf{a}_z\cdot(\partial^2\mathbf{E}_t/\partial z^2) = 0$ and $\mathbf{a}_z\cdot\mathbf{a}_z = 1$. In a similar fashion it may be shown that $\mathbf{H}_t$ satisfies the same type of equation. The solution to (5) gives the function $g_1(z)$ as $A^\pm e^{\mp jkz}$, where A^+ and A^- are suitable amplitude coefficients. Since a time function $e^{j\omega t}$ was assumed, $A^+ e^{-jkz}$ represents propagation along the positive z direction, and $A^- e^{jkz}$ represents propagation along the negative z direction. The propagation constant k is equal to

$$\omega(\mu\epsilon)^{1/2} = \frac{\omega}{v_c} = \frac{2\pi f}{v_c} = \frac{2\pi}{\lambda}$$

where v_c is the velocity of light in a medium with parameters μ, ϵ, and λ is the wavelength corresponding to the frequency f in the same medium. The solution for the electric field may be written as

$$\mathbf{E}_t = A^\pm\nabla_t\Phi(x, y)e^{\mp jkz}. \tag{6}$$

From (2a) the solution for $\mathbf{H}_t$ is found to be

$$\mathbf{H}_t = \pm\left(\frac{\epsilon}{\mu}\right)^{1/2}\mathbf{a}_z \times \nabla_t\Phi A^\pm e^{\mp jkz}. \tag{7}$$

[1] Section A.1e.

From (7) the appropriate solution for $g_2(z)$ and ψ in terms of $g_1(z)$ and Φ may be found. However, we will not require explicit expressions for these functions since both $\mathbf{E}_t$ and $\mathbf{H}_t$ are uniquely defined by $g_1(z)$ and Φ alone. Equation (7) may be rewritten as

$$\mathbf{H}_t = \pm \left(\frac{\epsilon}{\mu}\right)^{1/2} \mathbf{a}_z \times \mathbf{E}_t \tag{8}$$

or

$$H_x = \mp \left(\frac{\epsilon}{\mu}\right)^{1/2} E_y$$

$$H_y = \pm \left(\frac{\epsilon}{\mu}\right)^{1/2} E_x$$

The factor $(\mu/\epsilon)^{1/2}$ has the dimensions of ohms and is called the characteristic or intrinsic impedance of the medium. It is numerically equal to the ratio of the transverse electric field to the mutually perpendicular transverse magnetic field in a plane TEM wave. The reciprocal quantity $(\epsilon/\mu)^{1/2}$ is called the intrinsic admittance of the medium. In free space $Z = (\mu_0/\epsilon_0)^{1/2} = 120\pi = 377$ ohms. The ratio of the transverse electric field to the mutually perpendicular transverse magnetic field is called the wave impedance of the wave. For plane TEM waves, the wave impedance is equal to the intrinsic impedance of the medium in which the wave propagates.

The scalar function ψ is related to Φ as follows:

$$\nabla_t \psi = \pm Y \mathbf{a}_z \times \nabla_t \Phi$$

where $Y = (\epsilon/\mu)^{1/2}$, and hence ψ is a function whose gradient is everywhere perpendicular to the gradient of Φ. The family of curves $\Phi = $ constant is mutually orthogonal to the family $\psi = $ constant. Thus the electric and magnetic field vectors are also mutually perpendicular. The transverse magnetic field is consequently related to the transverse electric field by a vector equation such as (7). For this reason it is convenient to introduce a dyadic admittance and impedance which we shall take as follows:

$$\bar{\mathbf{Y}} = Y(\mathbf{a}_y \mathbf{a}_x - \mathbf{a}_x \mathbf{a}_y) \tag{9a}$$

$$\bar{\mathbf{Z}} = Z(\mathbf{a}_x \mathbf{a}_y - \mathbf{a}_y \mathbf{a}_x) \tag{9b}$$

where the bar on top signifies a dyadic quantity. We may now rewrite (8) as

$$\mathbf{H}_t = \pm \bar{\mathbf{Y}} \cdot \mathbf{E}_t \tag{10a}$$

and also

$$\mathbf{E}_t = \pm \bar{\mathbf{Z}} \cdot \mathbf{H}_t. \tag{10b}$$

By definition, the dot product of a dyadic with a vector is the vector formed by dotting the unit vectors immediately adjacent to and on either side of the dot.[2] The product is not, in general,

---

[2] Section A.2.

commutative, that is, $\bar{Y} \cdot E_t \neq E_t \cdot \bar{Y}$. In the above case $\bar{Y} \cdot E_t = -E_t \cdot \bar{Y}$, but this is not true in general. We have chosen the signs in (10) so that the positive sign applies for propagation in the positive z direction. The dyadic may be thought of as an operator that changes a vector into another vector. In the application above, the dyadic $a_x a_y - a_y a_x$ is a rotation operator which rotates a vector through 90° in a clockwise direction. This operation is required in order to relate the magnetic field components to the mutually orthogonal electric field components.

The simplest solution to $\nabla_t^2 \Phi = 0$ is $\nabla_t \Phi = $ a constant vector, which may be chosen as a vector along the x axis. This leads to the solution for a plane, uniform, linearly polarized TEM wave. This solution is as follows:

$$E_t = A^{\pm} a_x e^{\mp jkz} \tag{11a}$$

$$H_t = \pm A^{\pm} a_y Y e^{\mp jkz}. \tag{11b}$$

The time-average power flow along the z axis per unit area in the xy plane is given by the dot product of the complex Poynting vector with the unit vector a_z:

$$P = \tfrac{1}{2}(E \times H^* \cdot a_z) = \tfrac{1}{2}(Y|A^{\pm}|^2). \tag{12}$$

The above wave is called a linearly polarized wave. The plane of polarization is defined as the plane that contains the E vector and the axis of propagation, in the above case the xz plane. A more general type of TEM wave is one composed of two mutually perpendicular, linearly polarized waves differing in time phase. The resultant wave is called an elliptically polarized wave for reasons to be discussed below.

Consider two mutually perpendicular, linearly polarized waves with a time phase of θ:

$$E_x = \mathrm{Re}\, E_1 e^{-jkz+j\omega t} \tag{13a}$$

$$E_y = \mathrm{Re}\, E_2 e^{-jkz+j\omega t+j\theta} \tag{13b}$$

where Re denotes the real part. The imaginary part is another possible solution. In a transverse plane, say $z = 0$, we have

$$E_x = E_1 \cos \omega t \tag{14a}$$

$$E_y = E_2 \cos(\omega t + \theta) = E_2(\cos \omega t \cos \theta - \sin \omega t \sin \theta). \tag{14b}$$

In the xy plane, the total instantaneous electric field is

$$E = a_x E_1 \cos \omega t + a_y E_2(\cos \omega t \cos \theta - \sin \omega t \sin \theta). \tag{15}$$

Since E_x and E_y are functions of time, we may express E_x as a function of E_y by eliminating the time t between them. From (14a) we get

$$\frac{E_x}{E_1} = \cos \omega t \qquad \left(1 - \frac{E_x^2}{E_1^2}\right)^{1/2} = \sin \omega t.$$

Substituting in (14b), we get

$$\frac{E_y}{E_2} - \cos \theta \frac{E_x}{E_1} = -\sin \theta \left(1 - \frac{E_x^2}{E_1^2}\right)^{1/2}.$$

Squaring both sides gives

$$\left(\frac{E_x}{E_1}\right)^2 + \left(\frac{E_y}{E_2}\right)^2 - 2\frac{E_x E_y}{E_1 E_2}\cos\theta = \sin^2\theta. \tag{16}$$

This is the equation of an ellipse, and thus, in general, the combination of two linearly polarized waves differing in time phase results in an elliptically polarized wave, i.e., a wave whose electric vector describes the locus of an ellipse in a transverse plane. The electric vector rotates around the axis of propagation at an average angular rate ω, and varies in absolute magnitude at the same time so as to trace out an ellipse. If, when the wave is viewed along the direction of propagation, the electric vector rotates clockwise, the wave is called right elliptically polarized. If the sense of rotation is anticlockwise, it is called left elliptically polarized.

When the time phase θ is 0, π, 2π, etc., Eqs. (14) show that the resultant wave is linearly polarized, i.e.,

$$\mathbf{E} = (\mathbf{a}_x E_1 \pm \mathbf{a}_y E_2)\cos\omega t$$

or

$$|\mathbf{E}| = (E_1^2 + E_2^2)^{1/2} \qquad \angle\mathbf{E} = \alpha = \tan^{-1}\frac{\pm E_2}{E_1} \tag{17}$$

where α is the angle that the $\mathbf{E}$ vector makes with the x axis. When $E_1 = E_2$ and $\theta = \pi/2 + n\pi$, (16) reduces to

$$E_x^2 + E_y^2 = E_1^2 \tag{18}$$

which is the equation of a circle. Under these conditions the wave is circular-polarized. For $\theta = \pi/2$, $\pi/2 + 2\pi$, etc., it is left-circular-polarized. For $\theta = 3\pi/2$, $3\pi/2 + 2\pi$, etc., it is right-circular-polarized. The direction of rotation is easily obtained as follows. The angle α which the electric vector makes with the x axis is $\tan^{-1}(E_y/E_x) = \tan^{-1}[\cos(\omega t + \theta)/\cos\omega t]$. When $\theta = \pi/2$, $\alpha = -\tan^{-1}\tan\omega t = -\omega t$, and the wave is left-circular-polarized. When $\theta = 3\pi/2$, $\alpha = \omega t$, the electric vector rotates clockwise, and the wave is called right-circular-polarized. For circular polarization the rate of rotation is uniform and equal to ω. For an elliptical-polarized wave, $\alpha = \tan^{-1}[E_2\cos(\omega t + \theta)/E_1\cos\omega t]$, and the angular rate of rotation of the $\mathbf{E}$ vector is nonuniform. When $E_1 \neq E_2$ but $\theta = \pi/2 + n\pi$, we again obtain an elliptical-polarized wave since (16) becomes

$$\left(\frac{E_x}{E_1}\right)^2 + \left(\frac{E_y}{E_2}\right)^2 = 1. \tag{19}$$

This ellipse has major and minor axes defined by E_1 and E_2, and these axes are parallel to the x and y axes. By referring $\mathbf{E}$ to a new coordinate frame uvz obtained by rotating the xyz frame about the z axis through an angle β in a counterclockwise direction, (16) can be reduced to the form of (19). The angle of rotation β is found to be

$$\beta = \frac{1}{2}\tan^{-1}\left(2E_1 E_2 \frac{\cos\theta}{E_2^2 - E_1^2}\right) \tag{20a}$$

and (16) reduces to

$$E_u^2 \left(\frac{\cos^2 \beta}{E_1^2} + \frac{\sin^2 \beta}{E_2^2} + \frac{\sin 2\beta \cos \theta}{E_1 E_2} \right)$$

$$+ E_v^2 \left(\frac{\cos^2 \beta}{E_2^2} + \frac{\sin^2 \beta}{E_1^2} - \frac{\sin 2\beta \cos \theta}{E_1 E_2} \right) = \sin^2 \theta. \tag{20b}$$

3.2. TEM WAVES IN ORTHOGONAL CURVILINEAR COORDINATE SYSTEMS

In this section we will investigate the solution for transverse electromagnetic waves in orthogonal curvilinear coordinate systems. Let u_1, u_2, u_3 form a right-handed curvilinear coordinate frame with unit vectors $\mathbf{a}_1$, $\mathbf{a}_2$, $\mathbf{a}_3$ along the coordinate curves and with scale factors h_1, h_2, and h_3. We will attempt to find solutions for $\mathbf{E}$ and $\mathbf{H}$ which have no components along the u_3 coordinate axis. This latter axis is thus the axis along which the wave propagates. In order to proceed without an undue amount of mathematical difficulties, we shall impose certain restrictions on the type of coordinate system we will deal with. These restrictions will be explained during the development of the theory. Our basic approach will be based on a generalization of the results presented in the previous section.

Let us attempt to find a solution for $\mathbf{E}_t$ of the form

$$\mathbf{E}_t = \nabla_t \Phi(u_1, u_2) g(u_3) \tag{21}$$

where the subscript t indicates the transverse components. The operator ∇_t is given by $(\mathbf{a}_1/h_1)(\partial/\partial u_1) + (\mathbf{a}_2/h_2)(\partial/\partial u_2)$ in orthogonal curvilinear coordinates.[3] The electric field $\mathbf{E}$ is a solution of the vector Helmholtz equation

$$\nabla \times \nabla \times \mathbf{E} - k^2 \mathbf{E} = 0. \tag{22}$$

A solution of the form (21) is only possible in those coordinate systems for which the vector Helmholtz equation is separable in the form of (21). Since $\nabla \cdot \mathbf{E}_t = 0$, we must have

$$\nabla \cdot \nabla_t \Phi g = g \nabla_t^2 \Phi = \frac{g}{h_1 h_2 h_3} \left(\frac{\partial}{\partial u_1} \frac{h_2 h_3}{h_1} \frac{\partial \Phi}{\partial u_1} + \frac{\partial}{\partial u_2} \frac{h_1 h_3}{h_2} \frac{\partial \Phi}{\partial u_2} \right) = 0 \tag{23}$$

so Φ is a solution of Laplace's equation in the $u_1 u_2$ curvilinear plane. Using the expression for the curl in curvilinear coordinates, we obtain[4]

$$\nabla \times \mathbf{E}_t = \frac{-\mathbf{a}_1}{h_2 h_3} \frac{\partial \Phi}{\partial u_2} \frac{\partial g}{\partial u_3} + \frac{\mathbf{a}_2}{h_1 h_3} \frac{\partial \Phi}{\partial u_1} \frac{\partial g}{\partial u_3} + \frac{\mathbf{a}_3}{h_1 h_2} \left(\frac{\partial^2 \Phi}{\partial u_1 \partial u_2} - \frac{\partial^2 \Phi}{\partial u_2 \partial u_1} \right) g. \tag{24}$$

An examination of (24) shows that the component along the u_3 axis vanishes, and hence (24) determines $-j\omega\mu \mathbf{H}_t$ as a vector function with no component along the u_3 axis. Taking the

---

[3] Section A.1d.
[4] Section A.1e.

curl of (24) and substituting in (22) gives

$$\frac{\mathbf{a}_1}{h_2 h_3} \frac{\partial \Phi}{\partial u_1} \frac{\partial}{\partial u_3} \left(\frac{h_2}{h_1 h_3} \frac{\partial g}{\partial u_3} \right) + \frac{\mathbf{a}_2}{h_1 h_3} \frac{\partial \Phi}{\partial u_2} \frac{\partial}{\partial u_3} \left(\frac{h_1}{h_2 h_3} \frac{\partial g}{\partial u_3} \right)$$

$$- \frac{\mathbf{a}_3}{h_1 h_2} \frac{\partial g}{\partial u_3} \left(\frac{\partial}{\partial u_1} \frac{h_2 h_3}{h_1 h_3^2} \frac{\partial \Phi}{\partial u_1} + \frac{\partial}{\partial u_2} \frac{h_1 h_3}{h_2 h_3^2} \frac{\partial \Phi}{\partial u_2} \right)$$

$$+ k^2 g \left(\frac{\mathbf{a}_1}{h_1} \frac{\partial \Phi}{\partial u_1} + \frac{\mathbf{a}_2}{h_2} \frac{\partial \Phi}{\partial u_2} \right) = 0. \tag{25}$$

Since the term multiplied by k^2, that is, $\mathbf{E}_t$, has no $\mathbf{a}_3$ component, the component along $\mathbf{a}_3$ in (25) must vanish. An examination of this term shows that, apart from the h_3^2 factor, it is proportional to $\nabla_t^2 \Phi$. Hence, provided h_3 is not a function of the u_1 and u_2 coordinates, we may bring this factor outside the differentiation operators, and the whole term vanishes, provided $\nabla_t^2 \Phi = 0$. Equating separately the components along $\mathbf{a}_1$ and $\mathbf{a}_2$ to zero gives

$$\frac{\partial \Phi}{\partial u_1} \left(\frac{\partial}{\partial u_3} \frac{h_2}{h_1 h_3} \frac{\partial g}{\partial u_3} + \frac{h_2 h_3}{h_1} k^2 g \right) = 0 \tag{26a}$$

$$\frac{\partial \Phi}{\partial u_2} \left(\frac{\partial}{\partial u_3} \frac{h_1}{h_2 h_3} \frac{\partial g}{\partial u_3} + \frac{h_1 h_3}{h_2} k^2 g \right) = 0. \tag{26b}$$

Since g cannot satisfy two different differential equations simultaneously, (26a) and (26b) must reduce to the same form. This will be the case provided h_1/h_2 is not a function of u_3. In this case Eqs. (26) become

$$\frac{1}{h_3} \frac{\partial}{\partial u_3} \frac{1}{h_3} \frac{\partial g}{\partial u_3} + k^2 g = 0. \tag{27}$$

This is not the only condition for which a solution for g may be found. Φ may be a function of only one of the transverse coordinates u_1 or u_2. If Φ is not a function of u_2, $\partial \Phi / \partial u_2$ equals zero, and g is then determined by (26a) only. In this case there is no restriction on h_1/h_2. Similarly, when Φ is a function of u_2 only, the propagation function g is determined by (26b) only. If Φ is a function of one coordinate only, then, since $\nabla_t^2 \Phi = 0$, the only permissible solution for $\nabla_t \Phi$ is a constant vector. This solution, therefore, always gives a homogeneous TEM wave, i.e., a wave whose properties do not vary with the transverse coordinates. A solution for a nonhomogeneous TEM wave exists only in those coordinate systems for which h_1/h_2 is not a function of u_3. The orthogonal curvilinear coordinate systems for which this latter condition is true are

1. rectangular coordinates,
2. all cylindrical coordinates with propagation along the z axis,
3. spherical coordinates with propagation along the radial axis, and
4. conical coordinates with propagation along the radial axis.

The conical coordinates have spheres and cones of elliptical cross section as constant coordinate surfaces. The coordinate systems for which an axis, say along u_3, can be found, and whose scale factor h_3 is a function of u_3 only, are those listed above and, in addition, the circular cylindrical coordinates with propagation along the radial axis.

In a cylindrical coordinate system θ, z, r, the scale factors are $h_1 = r$, $h_2 = 1$, $h_3 = 1$. If we take Φ as a function of θ only, then $\mathbf{E}_t$ has a θ component and $\mathbf{H}_t$ a z component only. For one solution the propagation function g must satisfy the equation

$$\frac{d}{dr}\frac{1}{r}\frac{dg}{dr} + \frac{1}{r}k^2 g = 0 \tag{28a}$$

as obtained from (26a). When Φ is chosen as a function of z only, $\mathbf{E}_t$ has a z component and $\mathbf{H}_t$ a θ component only. The function g is now a solution of

$$\frac{d}{dr}r\frac{dg}{dr} + rk^2 g = 0. \tag{28b}$$

We will consider this latter solution first. If the substitution $kr = x$ is made in (28b), this equation becomes

$$\frac{d^2 g}{dx^2} + \frac{1}{x}\frac{dg}{dx} + g = 0$$

which is the standard form for Bessel's equation of order 0. [3.1], [3.2]. The solutions to this equation are linear combinations of the Bessel and Neumann functions of order 0 as follows:

$$g = AJ_0(kr) + BY_0(kr) \tag{29}$$

where A and B are suitable amplitude coefficients, and x has been replaced by kr. When $|kr|$ is very large, the following asymptotic forms are valid:

$$J_0(kr) \sim \left(\frac{2}{\pi kr}\right)^{1/2} \cos\left(kr - \frac{\pi}{4}\right) \tag{30a}$$

$$Y_0(kr) \sim \left(\frac{2}{\pi kr}\right)^{1/2} \sin\left(kr - \frac{\pi}{4}\right). \tag{30b}$$

If we want to make g represent an outward-propagating wave, the coefficient B must be chosen as $-jA$, and the asymptotic form of (29) becomes

$$g \sim A\left(\frac{2}{\pi kr}\right)^{1/2} e^{-j(kr - \pi/4)}, \qquad |kr| \gg 1. \tag{31}$$

The particular combinations of the functions J_0 and Y_0 such as $J_0 \pm jY_0$ are called Hankel functions of the first and second kind, respectively; i.e,

$$H_0^1 = J_0 + jY_0 \tag{32a}$$

$$H_0^2 = J_0 - jY_0. \tag{32b}$$

The Hankel function of the first kind represents an inward-propagating wave, while the Hankel function of the second kind represents an outward-propagating wave.

The solution for our TEM wave is

$$E_z = AH_0^2(kr) \tag{33a}$$

$$H_\theta = \frac{1}{j\omega\mu}\frac{\partial E_z}{\partial r} = -\frac{kA}{j\omega\mu}H_1^2(kr) \tag{33b}$$

where H_1^2 is the Hankel function of the second kind and order 1. The asymptotic form of this function is

$$H_1^2 \sim \left(\frac{2}{\pi kr}\right)^{1/2} e^{-j(kr-3\pi/4)}$$

and hence, for $|kr|$ large, we get

$$\frac{-E_z}{H_\theta} \approx \left(\frac{\mu}{\epsilon}\right)^{1/2}.$$

Far away from the origin the wave impedance of a cylindrical TEM wave is the same as that of a plane TEM wave.

We return now to (26a) and make the substitutions $g = rG$, $x = kr$. The equation for G becomes

$$\frac{d^2G}{dx^2} + \frac{1}{x}\frac{dG}{dx} + \left(1 - \frac{1}{x^2}\right)G = 0$$

which is Bessel's equation of order 1. Appropriate solutions for g in this case are the Hankel functions of order 1 multiplied by r, that is,

$$g = \begin{cases} rA_1H_1^1(kr) \\ rA_2H_1^2(kr) \end{cases}. \tag{34}$$

This leads to a similar type of TEM wave but with the roles of $\mathbf{E}_t$ and $\mathbf{H}_t$ interchanged. The solution for $\mathbf{E}_t$ is of the form

$$E_\theta \propto H_1^i(kr), \qquad i = 1, 2, \ldots$$

since $\nabla_t\Phi(\theta) = (\mathbf{a}_\theta/r)(\partial\Phi/\partial\theta)$, and hence a factor $1/r$ occurs which cancels the factor r in (34). The rest of the details for this solution are left for the reader as one of the problems at the end of this chapter.

3.3. REFLECTION AND TRANSMISSION AT A DISCONTINUITY INTERFACE

Let the half space $z > 0$ be filled with a lossless homogeneous dielectric with a relative dielectric constant $\kappa = \epsilon/\epsilon_0$, and let the half-space $z < 0$ be free space as in Fig. 3.1. From the region $z < 0$, let a TEM wave be incident on the interface. When this wave strikes the discontinuity interface, a part is transmitted into the region $z > 0$. Let the reflected and transmitted waves be

$$\mathbf{E}_R = \mathbf{e}_R e^{jk_0z} \tag{35a}$$

$$\mathbf{E}_T = \mathbf{e}_T e^{-jkz} \tag{35b}$$

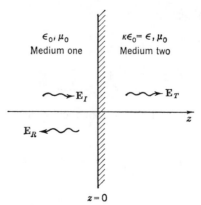

Fig. 3.1. A discontinuity interface.

while the incident wave is

$$\mathbf{E}_I = \mathbf{e}_I e^{-jk_0 z} \tag{35c}$$

where $k_0 = \omega(\mu_0\epsilon_0)^{1/2} = 2\pi/\lambda_0$, $k = \omega(\mu_0\epsilon)^{1/2} = \kappa^{1/2}k_0 = 2\pi/\lambda$. The transverse magnetic field components are

$$\mathbf{h}_I = \bar{\mathbf{Y}}_0 \cdot \mathbf{e}_I \tag{36a}$$

$$\mathbf{h}_R = -\bar{\mathbf{Y}}_0 \cdot \mathbf{e}_R \tag{36b}$$

$$\mathbf{h}_T = \bar{\mathbf{Y}} \cdot \mathbf{e}_T \tag{36c}$$

where

$$\bar{\mathbf{Y}}_0 = (\epsilon_0/\mu_0)^{1/2}(\mathbf{a}_y \mathbf{a}_x - \mathbf{a}_x \mathbf{a}_y)$$
$$\bar{\mathbf{Y}} = \kappa^{1/2}\bar{\mathbf{Y}}_0.$$

The quantities Y_0, Y and Z_0, Z are the wave admittances and impedances, respectively. The vectors $\mathbf{e}$ and $\mathbf{h}$ are used to represent the transverse fields without the propagation factor $e^{\pm jkz}$. At the discontinuity interface, the boundary conditions require that the tangential field components be continuous. The discontinuity interface is uniform in the xy plane, and, consequently, there will be no waves excited other than those given by (35a) and (35b). The conditions for the continuity of the tangential electric field is

$$\mathbf{e}_I + \mathbf{e}_R = \mathbf{e}_T \tag{37a}$$

while

$$\mathbf{h}_I + \mathbf{h}_R = \mathbf{h}_T \tag{37b}$$

is the required condition for continuity of the transverse magnetic field components. Using (36), we may write (37b) as follows:

$$\bar{\mathbf{Y}}_0 \cdot (\mathbf{e}_I - \mathbf{e}_R) = \bar{\mathbf{Y}} \cdot \mathbf{e}_T$$

or

$$\bar{\mathbf{Y}}_0 \cdot (\mathbf{e}_I - \mathbf{e}_R - \kappa^{1/2} \mathbf{e}_T) = 0. \tag{37c}$$

In order for Eqs. (37) to hold for all values of x and y, it is necessary for all the fields to have the same functional form in the xy plane. The reflection coefficient R_i is defined as the ratio of reflected to incident electric field wave amplitudes, and the transmission coefficient T_{ij} as the ratio of transmitted to incident wave amplitudes; thus

$$\mathbf{e}_R = R_1 \mathbf{e}_I \qquad \text{at } z = 0 \tag{38a}$$

$$T_{12} \mathbf{e}_I = \mathbf{e}_T \qquad \text{at } z = 0. \tag{38b}$$

With these definitions in mind we may write (37a) and (37c) as follows:

$$1 + R_1 = T_{12} \tag{39a}$$

$$1 - R_1 = \kappa^{1/2} T_{12} \tag{39b}$$

or

$$Y_0(1 - R_1) = YT_{12}$$

since $\bar{\mathbf{Y}}_0 = Y_0(\mathbf{a}_y \mathbf{a}_x - \mathbf{a}_x \mathbf{a}_y)$ and $\bar{\mathbf{Y}} = Y(\mathbf{a}_y \mathbf{a}_x - \mathbf{a}_x \mathbf{a}_y)$. Solving Eqs. (39) for R_1 and T_{12} gives

$$R_1 = \frac{Y_0 - Y}{Y_0 + Y} = \frac{Z - Z_0}{Z + Z_0} = \frac{1 - \kappa^{1/2}}{1 + \kappa^{1/2}} \tag{40a}$$

$$T_{12} = \frac{2Y_0}{Y_0 + Y} = \frac{2Z}{Z + Z_0} = \frac{2}{1 + \kappa^{1/2}}. \tag{40b}$$

We use the subscript 1 to indicate that R_1 is the reflection coefficient for a wave incident in region 1. Similarly, T_{12} indicates transmission from medium 1 into medium 2. For a wave incident in region 2 we find that

$$R_2 = \frac{Y - Y_0}{Y + Y_0} = -R_1 \tag{41a}$$

$$T_{21} = 1 + R_2 = \frac{2Y}{Y_0 + Y}. \tag{41b}$$

The input admittance and impedance (normalized with respect to the characteristic admittance and impedance of medium 1) looking into region 2 at the plane $z = 0$ are Y/Y_0 and Z/Z_0, respectively, and are given by the following relations, derivable from (40a):

$$\frac{Y}{Y_0} = \frac{1 - R_1}{1 + R_1} = \frac{Z_0}{Z}. \tag{42}$$

At a plane a distance l from the interface, the ratio of the reflected and incident waves is

$$R_1' = \frac{|e_R|e^{jk_0z}}{|e_I|e^{-jk_0z}}\bigg|_{z=-l} = R_1 e^{-2jk_0l} \tag{43}$$

since $z = -l$. We define the normalized input impedance at $z = -l$ by (42) but with R_1 replaced by R_1'; thus

$$Z_{in} = \frac{1+R_1'}{1-R_1'} = \frac{1+R_1 e^{-2jk_0l}}{1-R_1 e^{-2jk_0l}} = \frac{e^{jk_0l}+R_1 e^{-jk_0l}}{e^{jk_0l}-R_1 e^{-jk_0l}}$$

$$= \frac{(1+R_1)/(1-R_1)+j\tan k_0l}{1+[j(1+R_1)/(1-R_1)]\tan k_0l} = \frac{Z/Z_0+j\tan k_0l}{1+j(Z/Z_0)\tan k_0l}. \tag{44}$$

The input impedance transforms with distance l according to the usual transmission-line formula. The quantity $(1+|R_1|)/(1-|R_1|)$ is called the standing-wave ratio (SWR) and is equal to the ratio of the maximum to the minimum total electric field amplitudes.

The definition of wave impedance may be applied to many types of wave motion. For electromagnetic waves (both TEM waves and waves with longitudinal components), the wave impedance is defined as the ratio of the transverse electric field component to the mutually perpendicular transverse magnetic field component. In the above statement, the adjective transverse is used to mean the components of the field vectors in the plane that is perpendicular to the axis along which the wave is considered to propagate. This latter axis may not necessarily coincide with the wave normal. The concept of wave impedance is a very useful one and greatly facilitates the analysis of many wave-propagation problems; it also provides the basis for a formal analogy between the theory of wave-guiding structures and conventional low-frequency transmission-line theory. In Section 1.4 it was demonstrated that it was sufficient to match only the tangential field components across a discontinuity interface in order to ensure proper continuity of the whole solution across the interface. It is for this reason that the concept of wave impedance is so useful in practice, since this concept, together with transmission-line theory, provides the proper solution of Maxwell's equations to ensure continuity of the transverse field components. For electromagnetic waves with the z axis considered as the axis of propagation, the wave impedance is defined by

$$Z = \pm\frac{E_x}{H_y} = \mp\frac{E_y}{H_x} \tag{45}$$

with the sign chosen so that the real part is positive. An imaginary part arises in the case of a medium with finite losses and for waves that are evanescent in the direction of propagation. It should be noted that, in the application of the definition (45), a single wave propagating in either the positive z direction or the negative z direction is considered.

3.4. WAVE MATRICES

With reference to Fig. 3.1 in Section 3.3, let a TEM wave of amplitude c_1 be incident from the left and a wave of amplitude b_2 be incident from the right on the plane-discontinuity interface. In the region $z < 0$ there will be a wave, made up partly of a reflected wave and partly of a wave transmitted past the discontinuity from the right, propagating in the negative z direction. In the region $z > 0$ a similar wave propagating in the positive z direction will

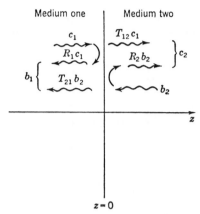

Fig. 3.2. Reflected and transmitted waves at a discontinuity interface.

exist. Let the amplitudes of the latter two waves be b_1 and c_2, respectively, as in Fig. 3.2. With reference to Fig. 3.2 we may write

$$b_1 = R_1 c_1 + T_{21} b_2 \tag{46a}$$

$$c_2 = R_2 b_2 + T_{12} c_1 \tag{46b}$$

or

$$b_1 = \left(T_{21} - \frac{R_1 R_2}{T_{12}} \right) b_2 + \frac{R_1}{T_{12}} c_2 \tag{47a}$$

$$c_1 = \frac{c_2}{T_{12}} - \frac{R_2 b_2}{T_{12}} \tag{47b}$$

where R_1 and R_2 are the interface reflection coefficients, and T_{12} and T_{21} are the transmission coefficients. These equations may be written in matrix form as follows:

$$\begin{bmatrix} c_1 \\ b_1 \end{bmatrix} = \frac{1}{T_{12}} \begin{bmatrix} 1 & -R_2 \\ R_1 & T_{12} T_{21} - R_1 R_2 \end{bmatrix} \begin{bmatrix} c_2 \\ b_2 \end{bmatrix} = \begin{bmatrix} A_{11} & A_{12} \\ A_{21} & A_{22} \end{bmatrix} \begin{bmatrix} c_2 \\ b_2 \end{bmatrix}. \tag{48}$$

We will call the matrix $[A]$ a wave-transmission chain matrix since it relates the amplitudes of the forward- and backward-propagating waves on the output side to those on the input side. The element A_{11} in the first row and column is equal to

$$\left. \frac{c_1}{c_2} \right|_{b_2=0} = \frac{1}{T_{12}}$$

which is the reciprocal of the transmission coefficient from medium 1 to medium 2. Although (46) to (48) were derived with particular reference to TEM waves and a plane discontinuity, they obviously hold for any wave-transmission system for which only a forward- and a backward-traveling wave exist on either side of the discontinuity and equivalent reflection and transmission coefficients can be defined. The equivalent reflection and transmission coefficients

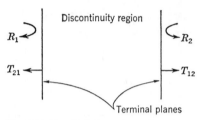

Fig. 3.3. Terminal planes for a discontinuity.

are given by the equations

$$R_1 = \frac{b_1}{c_1}\bigg|_{b_2=0} \tag{49a}$$

$$R_2 = \frac{c_2}{b_2}\bigg|_{c_1=0} \tag{49b}$$

$$T_{12} = \frac{c_2}{c_1}\bigg|_{b_2=0} \tag{49c}$$

$$T_{21} = \frac{b_1}{b_2}\bigg|_{c_1=0}. \tag{49d}$$

These equations together with (48) define the elements A_{ij} in the general case.

When there are evanescent modes or a localized diffraction field near the discontinuity, it is convenient to refer the reflection and transmission coefficients to suitably chosen terminal planes as in Fig. 3.3. The transmission coefficients then refer to transmission from one terminal plane to another. The reflection coefficients transform with distance from the terminal planes according to (43), that is, $R' = Re^{-2jkl}$, and are seen to be invariant to a shift of a multiple of $\lambda/2$ in the terminal-plane positions. The incident wave varies as e^{-jk_0z}, and the transmitted wave as e^{-jkz}. The transmission coefficients will, therefore, vary as $e^{\pm j(kl_2-k_0l_1)}$ with shifts of l_1 and l_2 in the terminal-plane positions in the input and output regions, respectively. As long as $k_0l_1 \pm kl_2$ is equal to a multiple of 2π, the transmission coefficients remain invariant.

Whenever the regions of space under consideration do not contain nonreciprocal anisotropic media, we may show that the transmission coefficients T_{12} and T_{21} are equal if we normalize the amplitudes c_1, b_1, c_2, b_2 so that the power carried by these waves is proportional to $c_1c_1^*$, $b_1b_1^*$, etc. The equality of T_{12} and T_{21} is just the condition for the reciprocity theorem to hold. We prove this result as follows. Let $\mathbf{E}_1$, $\mathbf{H}_1$ and $\mathbf{E}_2$, $\mathbf{H}_2$ be the fields which are solutions to Maxwell's equations for the given problem under two different terminal conditions. These fields are, therefore, not linearly related. According to the Lorentz reciprocity principle,

$$\oiint_S \mathbf{E}_1 \times \mathbf{H}_2 \cdot d\mathbf{S} = \oiint_S \mathbf{E}_2 \times \mathbf{H}_1 \cdot d\mathbf{S} \tag{50}$$

when there are no sources inside S. We now let $\mathbf{E}_1$, $\mathbf{H}_1$ be the solution for a wave incident from medium 1 only, and $\mathbf{E}_2$, $\mathbf{H}_2$ be the field for a wave incident from medium 2 only. Specializing to our original problem involving TEM waves, we choose for our surface S a cylindrical surface parallel to the z axis and two plane end surfaces in the xy plane at $k_0z = -2\pi$ and $kz = 2\pi$, that is, at $z = -\lambda_0$ and $z = \lambda$ as in Fig. 3.4. The contribution to the surface integral over

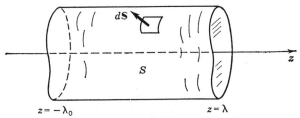

Fig. 3.4. Closed surface S for demonstration of reciprocity.

the cylindrical surface is zero since both $\mathbf{E}$ and $\mathbf{H}$ are transverse, and consequently $\mathbf{E} \times \mathbf{H}$ is perpendicular to $d\mathbf{S}$. We are left with surface integrals over the end faces only. For an incident wave,

$$\mathbf{E}_1 = c_1 \mathbf{e} e^{-jk_0 z} \tag{51a}$$

$$\mathbf{H}_1 = c_1 \bar{\mathbf{Y}}_0 \cdot \mathbf{e} e^{-jk_0 z} \tag{51b}$$

the power flow is proportional to $\mathbf{E}_1 \times \mathbf{H}_1^* \cdot \mathbf{a}_z = Y_0 c_1 c_1^* \mathbf{e} \cdot \mathbf{e}^*$. If we normalize the coefficient c_1 by choosing a new amplitude c_1' equal to $c_1 (Y_0)^{1/2}$, the power flow is proportional to $c_1' c_1'^* \mathbf{e} \cdot \mathbf{e}^*$. The coefficient c_2 may be normalized in a similar manner. Using normalized amplitudes, we may write the following expressions for the fields:

$$\begin{aligned} \mathbf{E}_1 &= Y_0^{-1/2} c_1' (1 + R_1) \mathbf{e} \\ \mathbf{H}_1 &= Y_0^{1/2} c_1' (1 - R_1)(\mathbf{a}_y \mathbf{a}_x - \mathbf{a}_x \mathbf{a}_y) \cdot \mathbf{e} \end{aligned} \qquad z = -\lambda_0 \tag{52a}$$

$$\begin{aligned} \mathbf{E}_1 &= Y_0^{-1/2} c_1' (Y_0/Y)^{1/2} T_{12}' \mathbf{e} \\ \mathbf{H}_1 &= Y^{1/2} c_1' T_{12}' (\mathbf{a}_y \mathbf{a}_x - \mathbf{a}_x \mathbf{a}_y) \cdot \mathbf{e} \end{aligned} \qquad z = \lambda \tag{52b}$$

$$\begin{aligned} \mathbf{E}_2 &= Y^{-1/2} c_2' (Y/Y_0)^{1/2} T_{21}' \mathbf{e} \\ \mathbf{H}_2 &= -Y_0^{1/2} c_2' T_{21}' (\mathbf{a}_y \mathbf{a}_x - \mathbf{a}_x \mathbf{a}_y) \cdot \mathbf{e} \end{aligned} \qquad z = -\lambda_0 \tag{52c}$$

$$\begin{aligned} \mathbf{E}_2 &= Y^{-1/2} c_2' (1 + R_2) \mathbf{e} \\ \mathbf{H}_2 &= -Y^{1/2} c_2' (1 - R_2)(\mathbf{a}_y \mathbf{a}_x - \mathbf{a}_x \mathbf{a}_y) \cdot \mathbf{e} \end{aligned} \qquad z = \lambda \tag{52d}$$

where the normalized transmission coefficients T_{12}', T_{21}' are given by

$$T_{12}' = (Y/Y_0)^{1/2} T_{12} \qquad T_{21}' = (Y_0/Y)^{1/2} T_{21}.$$

Substituting into the Lorentz reciprocity principle (50), we get

$$- \iint\limits_{z=-\lambda_0} c_1'(1 + R_1) c_2' T_{21}' [\mathbf{e} \times (\mathbf{a}_y \mathbf{a}_x - \mathbf{a}_x \mathbf{a}_y) \cdot \mathbf{e}] \cdot d\mathbf{S}$$

$$- \iint\limits_{z=\lambda} c_1' c_2' T_{12}' (1 - R_2) [\mathbf{e} \times (\mathbf{a}_y \mathbf{a}_x - \mathbf{a}_x \mathbf{a}_y) \cdot \mathbf{e}] \cdot d\mathbf{S}$$

$$= \iint\limits_{z=-\lambda_0} c_1' c_2' T_{21}' (1 - R_1) [\mathbf{e} \times (\mathbf{a}_y \mathbf{a}_x - \mathbf{a}_x \mathbf{a}_y) \cdot \mathbf{e}] \cdot d\mathbf{S}$$

$$+ \iint\limits_{z=\lambda} c_1' c_2' (1 + R_2) T_{12}' [\mathbf{e} \times (\mathbf{a}_y \mathbf{a}_x - \mathbf{a}_x \mathbf{a}_y) \cdot \mathbf{e}] \cdot d\mathbf{S}$$

which may be written as

$$T'_{21}(1 - R_1 + 1 + R_1) \iint\limits_{z=-\lambda_0} [\mathbf{e} \times (\mathbf{a}_y\mathbf{a}_x - \mathbf{a}_x\mathbf{a}_y)\cdot\mathbf{e}]\cdot d\mathbf{S}$$

$$= -T'_{12}(1 + R_2 + 1 - R_2) \iint\limits_{z=\lambda} [\mathbf{e} \times (\mathbf{a}_y\mathbf{a}_x - \mathbf{a}_x\mathbf{a}_y)\cdot\mathbf{e}]\cdot d\mathbf{S}. \qquad (53)$$

The integral at $z = \lambda$ is equal to the negative of that at $z = -\lambda_0$ because of the difference in the direction of $d\mathbf{S}$. Thus (53) gives $T'_{21} = T'_{12}$, and the stated reciprocity principle is proved. The determinant of the $[A]$ matrix is

$$A_{11}A_{22} - A_{12}A_{21} = \frac{T'_{21}}{T'_{12}} = 1 \qquad (54)$$

when the amplitudes are normalized. We shall not as a general rule use normalized amplitudes since this does not give any simplification in the mathematics for the type of problems we shall treat.

For the plane-discontinuity problem the matrix elements A_{ij} reduce to

$$A_{11} = \frac{1}{T_{12}} = \frac{Z + Z_0}{2Z} \qquad (55a)$$

$$A_{12} = -\frac{R_2}{T_{12}} = \frac{Z - Z_0}{2Z} = \frac{R_1}{T_{12}} \qquad (55b)$$

$$A_{21} = \frac{R_1}{T_{12}} \qquad (55c)$$

$$A_{22} = \frac{T_{12}T_{21} - R_1R_2}{T_{12}} = \frac{1}{T_{12}} \qquad (55d)$$

since $R_2 = -R_1$, $1 + R_1 = T_{12}$, $1 + R_2 = T_{21}$. Equation (48) may, therefore, be written as

$$\begin{bmatrix} c_1 \\ b_1 \end{bmatrix} = \frac{1}{T_{12}} \begin{bmatrix} 1 & R_1 \\ R_1 & 1 \end{bmatrix} \begin{bmatrix} c_2 \\ b_2 \end{bmatrix}. \qquad (56)$$

Before considering the cascade connection of n sections, we need to develop the wave-transmission matrix for a "length of unbounded space." Consider a forward-propagating wave $c_1 e^{-jkz}$ and a backward-propagating wave $b_1 e^{jkz}$. At $z = 0$ the amplitudes of the forward- and backward-traveling waves are c_1 and b_1, respectively. At $z = z_1$ the complex wave amplitudes are $c_1 e^{-jkz_1}$ and $b_1 e^{jkz_1}$. Let these latter two amplitudes be c_2 and b_2, respectively. Then we may write

$$c_1 = c_2 e^{jkz_1} \qquad (57a)$$

$$b_1 = b_2 e^{-jkz_1}. \qquad (57b)$$

The electrical length of the distance from $z = 0$ to $z = z_1$ is $kz_1 = \theta$. (The electrical length of a transmission line or of any medium is the angle by which the phase of a wave changes in propagating through the medium.)

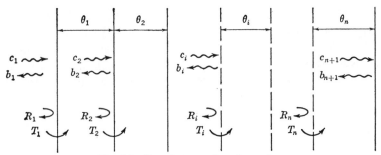

Fig. 3.5. Cascade connection of n sections.

Introducing the electrical length θ, (57) may be written in matrix form as

$$\begin{bmatrix} c_1 \\ b_1 \end{bmatrix} = \begin{bmatrix} e^{j\theta} & 0 \\ 0 & e^{-j\theta} \end{bmatrix} \begin{bmatrix} c_2 \\ b_2 \end{bmatrix}. \tag{58}$$

The diagonal matrix

$$\begin{bmatrix} e^{j\theta} & 0 \\ 0 & e^{-j\theta} \end{bmatrix}$$

is the matrix which relates the amplitudes of the forward- and backward-traveling waves at one plane to the complex amplitudes of these same waves at another terminal plane an electrical distance θ away.

Consider n sections in cascade with electrical constants ϵ_i, μ_i for the ith section. The propagation constant for the ith section is k_i. If the length of the ith section is l_i, its electrical length is $\theta_i = k_i l_i$. All the sections may differ in electrical properties and also in electrical lengths, or some may be the same without invalidating the results which we shall derive. With reference to Fig. 3.5 let R_i and T_i be the reflection and transmission coefficients looking into the ith section. That is, R_i is the reflection coefficient for a wave incident from section $i - 1$, and T_i is the amplitude transmission coefficient from section $i - 1$ to section i. The amplitudes c_2, b_2 at the input to section 2 are related to c_1 and b_1 as follows:

$$\begin{aligned}
\begin{bmatrix} c_1 \\ b_1 \end{bmatrix} &= \frac{1}{T_1} \begin{bmatrix} 1 & R_1 \\ R_1 & 1 \end{bmatrix} \begin{bmatrix} e^{j\theta_1} & 0 \\ 0 & e^{-j\theta_1} \end{bmatrix} \begin{bmatrix} c_2 \\ b_2 \end{bmatrix} \\
&= \frac{1}{T_1} \begin{bmatrix} e^{j\theta_1} & R_1 e^{-j\theta_1} \\ R_1 e^{j\theta_1} & e^{-j\theta_1} \end{bmatrix} \begin{bmatrix} c_2 \\ b_2 \end{bmatrix}.
\end{aligned} \tag{59}$$

If we express the amplitudes c_2 and b_2 in terms of c_3, b_3; c_3, b_3 in terms of c_4, b_4; and so on, we find that the output amplitudes c_{n+1}, b_{n+1} are related to c_1 and b_1 by the matrix product of the n matrices which characterize each section, i.e.,

$$\begin{aligned}
\begin{bmatrix} c_1 \\ b_1 \end{bmatrix} &= \prod_{i=1}^{n} \frac{1}{T_i} \begin{bmatrix} e^{j\theta_i} & R_i e^{-j\theta_i} \\ R_i e^{j\theta_i} & e^{-j\theta_i} \end{bmatrix} \begin{bmatrix} c_{n+1} \\ b_{n+1} \end{bmatrix} \\
&= \begin{bmatrix} A_{11} & A_{12} \\ A_{21} & A_{22} \end{bmatrix} \begin{bmatrix} c_{n+1} \\ b_{n+1} \end{bmatrix}
\end{aligned} \tag{60}$$

where $\prod\limits_{i=1}^{n}$ means the product of the n matrices, and A_{ij} are coefficients of the resultant 2×2 matrix. In carrying out this matrix product, it is important to keep the order of the matrices the same as the order of the sections since matrix multiplication is not commutative. The coefficient A_{11} is the reciprocal of the overall transmission coefficient $T_{1, n+1}$ from the input side of section 1 to the output side of section n. When the wave incident from the right on section n is zero, that is, $b_{n+1} = 0$, then

$$c_1 = A_{11} c_{n+1}$$

$$b_1 = A_{21} c_{n+1}$$

and hence the reflection coefficient at the input terminal is

$$\frac{b_1}{c_1} = \frac{A_{21}}{A_{11}}. \tag{61}$$

The above wave-transmission matrices may be applied in any type of cascade connection of transmission lines, waveguides, etc., provided there are only a forward- and a backward-traveling wave in each section, and each section is long enough so that any evanescent fields present at each discontinuity do not react with the evanescent fields of the adjacent discontinuities. The theory may also be generalized to include the case of several modes of propagation as well as the case of interacting evanescent modes.

Having considered the wave-transmission matrix in some detail, we now turn to a brief discussion of the following matrices and some of their properties:

1. scattering matrix,
2. impedance and admittance matrices, and
3. voltage- and current-transmission matrix, i.e., conventional $ABCD$ matrix.

It will be assumed that the wave-amplitude coefficients have been normalized so that the power carried by a given wave is proportional to the square of the absolute value of the amplitude coefficient. In order to have a definite example to consider, we choose the discontinuity interface discussed earlier. However, for convenience the normalized wave amplitudes will be denoted by c_1, b_1, c_2, and b_2 without the primes. Also b_2 will be chosen to represent a reflected wave and c_2 an incident wave.

The scattering matrix $[S]$ relates the amplitudes of the reflected waves b_1, b_2 to the amplitudes of the incident waves c_1, c_2 as follows:

$$\begin{bmatrix} b_1 \\ b_2 \end{bmatrix} = \begin{bmatrix} S_{11} & S_{12} \\ S_{21} & S_{22} \end{bmatrix} \begin{bmatrix} c_1 \\ c_2 \end{bmatrix}. \tag{62}$$

An examination of the above equation for the two conditions $c_1 = 0$ and $c_2 = 0$ separately shows that S_{11} is the reflection coefficient in medium 1, S_{22} is the reflection coefficient in medium 2, and S_{12}, S_{21} are transmission coefficients from medium 2 into medium 1 and vice versa, respectively. When the reciprocity principle holds, $S_{12} = S_{21}$ provided normalized amplitude coefficients are used. For general types of discontinuities the scattering-matrix elements S_{ij} are complex. If no energy loss occurs in the vicinity of the discontinuity, the energy conservation principle may be applied to derive some basic relationships among the scattering-matrix elements. The difference between the incident and the reflected power must equal the transmitted power, and hence

$$1 - S_{11}S_{11}^* = S_{21}S_{21}^* \tag{63a}$$

$$1 - S_{22}S_{22}^* = S_{21}S_{21}^* = S_{12}S_{12}^*. \tag{63b}$$

From (63) we find that $|S_{11}|$ is equal to $|S_{22}|$. The above equations expressing the conservation of energy for lossless discontinuities may be written as

$$\sum_{n=1}^{2}(c_n c_n^* - b_n b_n^*) = [c_1 \quad c_2]\begin{bmatrix} c_1 \\ c_2 \end{bmatrix}^* - [c_1 \quad c_2][S]_t[S]^*\begin{bmatrix} c_1 \\ c_2 \end{bmatrix}^* = 0 \tag{64}$$

where $[S]_t$ is the transposed matrix. The scattering matrix thus satisfies the equation

$$[U] = [S]_t[S]^*.$$

or

$$[S]_t^{-1} = [S]^* \tag{65}$$

where $[U]$ is the unit matrix. Equation (65) is the condition that $[S]$ should be a unitary matrix.[5] When the discontinuity absorbs energy, the sum of (64) is not zero, and $[S]$ is no longer a unitary matrix.

A discontinuity may be described by an impedance or admittance matrix by defining suitable equivalent terminal voltages and currents. The equivalent voltages are chosen proportional to the total transverse electric field (both incident and reflected waves), while the equivalent currents are taken proportional to the total transverse magnetic field as follows:

$$V_1 = c_1 + b_1 \tag{66a}$$

$$I_1 = c_1 - b_1 \tag{66b}$$

$$V_2 = c_2 + b_2 \tag{66c}$$

$$I_2 = c_2 - b_2. \tag{66d}$$

The voltages V_i and currents I_i are related by the impedance and admittance matrices $[Z]$ and $[Y]$ as follows:

$$\begin{bmatrix} V_1 \\ V_2 \end{bmatrix} = \begin{bmatrix} Z_{11} & Z_{12} \\ Z_{21} & Z_{22} \end{bmatrix}\begin{bmatrix} I_1 \\ I_2 \end{bmatrix}$$

or

$$[V] = [Z][I] \tag{67a}$$

$$\begin{bmatrix} I_1 \\ I_2 \end{bmatrix} = \begin{bmatrix} Y_{11} & Y_{12} \\ Y_{21} & Y_{22} \end{bmatrix}\begin{bmatrix} V_1 \\ V_2 \end{bmatrix}$$

[5] Section A.3.

or

$$[I] = [Y][V]. \tag{67b}$$

When reciprocity holds, $Z_{12} = Z_{21}$ and $Y_{12} = Y_{21}$. From (66) we get $[V] + [I] = 2[c]$ and $[V] - [I] = 2[b]$, where

$$[c] = \begin{bmatrix} c_1 \\ c_2 \end{bmatrix} \quad \text{and} \quad [b] = \begin{bmatrix} b_1 \\ b_2 \end{bmatrix}.$$

Replacing $[V]$ by $[Z][I]$ in the above relations gives

$$([Z] + [U])[I] = 2[c]$$

$$([Z] - [U])[I] = 2[b] = 2[S][c].$$

Finally, by substituting for $[c]$ from the first equation into the second, we get

$$([Z] - [U])[I] = [S]([Z] + [U])[I]$$

and hence

$$[S] = ([Z] - [U])([Z] + [U])^{-1}. \tag{68a}$$

In a similar fashion the following relation between the scattering and admittance matrices may be derived:

$$[S] = ([U] - [Y])([U] + [Y])^{-1}. \tag{68b}$$

When analyzing a cascade connection of discontinuities on a voltage and current basis, it is convenient to relate the voltage and current on the output side to those on the input side as follows:

$$\begin{bmatrix} V_1 \\ I_1 \end{bmatrix} = \begin{bmatrix} A & B \\ C & D \end{bmatrix} \begin{bmatrix} V_2 \\ I_2 \end{bmatrix}. \tag{69}$$

The matrix used here is analogous to the $ABCD$ matrix employed in conventional low-frequency circuit analysis. The $ABCD$ matrix is given in terms of the impedance-matrix elements as follows:

$$\begin{bmatrix} A & B \\ C & D \end{bmatrix} = \frac{1}{Z_{21}} \begin{bmatrix} Z_{11} & Z_{12}Z_{21} - Z_{11}Z_{22} \\ 1 & -Z_{22} \end{bmatrix}. \tag{70}$$

3.5. Transmission through Dielectric Sheets

The transmission of plane electromagnetic waves through dielectric sheets at oblique angles of incidence has been treated by several methods. However, most treatments of the subject are unnecessarily complicated. By means of wave matrices, we shall be able to give a very simple approach to the subject. We shall refer to transmission-line theory and replace the dielectric

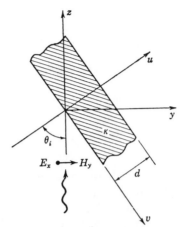

Fig. 3.6. Dielectric-sheet discontinuity.

sheet by an equivalent transmission-line circuit. We have not discussed transmission lines as such yet, but the concepts involved in this treatment were given in Section 3.3, in particular, (40) to (45). The case of lossless sheets is treated first.

Let a plane TEM wave be incident on an infinite-plane lossless dielectric sheet at an angle θ_i measured from the interface normal as in Fig. 3.6. The dielectric constant of the sheet is κ, and the index of refraction is $\eta = \kappa^{1/2}$.

In the xy plane this wave may be considered as a uniform wave propagating in the z direction, while in the uv coordinate frame it is a nonuniform wave propagating in the direction of the interface normal. It is necessary to distinguish between the two cases where the electric vector lies in the plane containing the interface normal and the direction of propagation (parallel polarization), and where the electric vector is perpendicular to this plane (perpendicular polarization). The case of perpendicular-polarized radiation will be treated first. The incident wave will be taken as

$$E_x = e^{-jk_0 z} \qquad H_y = \frac{k_0}{\omega \mu_0} e^{-jk_0 z}$$

where $k_0 = \omega(\mu_0 \epsilon_0)^{1/2} = 2\pi/\lambda_0$. This wave will be referred to a new coordinate system xvu obtained by rotating the old coordinate frame about the x axis through an angle θ_i. Simple geometry shows that the coordinate transformation is given by

$$x = x$$

$$y = v \cos \theta_i + u \sin \theta_i$$

$$z = -v \sin \theta_i + u \cos \theta_i.$$

In this new coordinate system,

$$E_x = e^{-jk_0(-v \sin \theta_i + u \cos \theta_i)}. \tag{71}$$

The components of **H** are obtained from $\nabla \times \mathbf{E} = -j\omega\mu_0\mathbf{H}$, where the curl is taken in the new coordinate frame; thus

$$j\omega\mu_0 H_u = jk_0 \sin \theta_i E_x \qquad (72a)$$

$$j\omega\mu_0 H_v = jk_0 \cos \theta_i E_x. \qquad (72b)$$

The wave impedance in the direction of increasing u (that is, normal to the surface) is

$$\frac{E_x}{H_v} = \left(\frac{\mu_0}{\epsilon_0}\right)^{1/2} \sec \theta_i = Z_1. \qquad (73)$$

This is the characteristic impedance of free space measured in a direction at an angle θ_i to the direction of propagation of the plane uniform perpendicular-polarized wave.

When this wave passes through the air-dielectric surface it is bent toward the normal and propagates at an angle θ_r to the interface normal. Referred to the xvu coordinate frame, the form of the wave will be the same with θ_i replaced by θ_r and k_0 replaced by ηk_0. In order that the tangential field components may be matched for all values of v along the air-dielectric interface, the variation with v must be the same for the wave that exists in the dielectric sheet and the wave that exists in the free-space region. This requires that

$$k_0 \sin \theta_i = \eta k_0 \sin \theta_r \qquad \text{or} \qquad \eta = \frac{\sin \theta_i}{\sin \theta_r} \qquad (74)$$

which is Snell's law of refraction. The wave impedance in the u direction in the dielectric for perpendicular polarization may be written down at once by analogy with the expression for the wave impedance in free space. It is given by

$$Z_2 = \left(\frac{\mu_0}{\kappa\epsilon_0}\right)^{1/2} \sec \theta_r. \qquad (75)$$

Usually it is convenient to normalize this impedance by dividing by the impedance for free space. Hence, we may write for the normalized characteristic impedance of the dielectric, with a perpendicularly polarized plane wave incident at an angle θ_i,

$$Z = \frac{\cos \theta_i}{(\kappa - \sin^2 \theta_i)^{1/2}} \qquad (76)$$

where $\cos \theta_r$ has been replaced by $(\kappa - \sin^2 \theta_i)^{1/2}/\eta$ as obtained from Snell's law. The equivalent circuit of a dielectric sheet of thickness d is illustrated in Fig. 3.7.

The transmission properties of such a sheet may be obtained by ordinary transmission-line theory, using the above expression for Z. It should be noted that the propagation factor in the dielectric sheet is $jk_0\eta \cos \theta_r = j(2\pi/\lambda_0)(\kappa - \sin^2 \theta_i)^{1/2}$, where λ_0 is the free-space wavelength.

The reflection coefficient for the air-dielectric interface is given by

$$R_1 = \frac{Z - 1}{Z + 1}. \qquad (77)$$

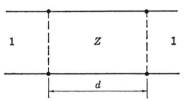

Fig. 3.7. Equivalent circuit for a dielectric sheet.

The input impedance at the air-dielectric interface is given by

$$Z_{\text{in}} = \frac{Z(1 + jZt)}{Z + jt} \tag{78}$$

where

$$t = \tan\left[\frac{2\pi}{\lambda_0}(\kappa - \sin^2 \theta_i)^{1/2} d\right].$$

For a reflectionless sheet this normalized input impedance must reduce to unity, as it does, provided $t = 0$. (Under these conditions the sheet looks like a continuation of free space for the incident wave, and no reflection takes place.) For $t = 0$, we must have

$$d = \frac{m\lambda_0}{2}(\kappa - \sin^2 \theta_i)^{-1/2}, \qquad m = 0, 1, \ldots . \tag{79}$$

For $m = 1$, the thickness is equal to a half wavelength measured in the dielectric sheet for that particular angle of oblique incidence θ_i.

The analysis for a parallel-polarized incident wave proceeds in a similar fashion and results in an equation like (78) for the input impedance, but with Z replaced by the normalized wave impedance Z' of the dielectric for a parallel-polarized wave. The wave impedance Z' may be easily determined by an application of the transformations

$$\mathbf{E}' = \left(\frac{\mu}{\epsilon}\right)^{1/2} \mathbf{H}$$

$$\mathbf{H}' = -\left(\frac{\epsilon}{\mu}\right)^{1/2} \mathbf{E}$$

as discussed in Section 1.7 in connection with Babinet's principle. The unnormalized wave impedance of free space and the dielectric medium for a parallel-polarized wave incident from free space at an angle θ_i are found as follows:

$$-\frac{E_v'}{H_x'} = \frac{(\mu_0/\epsilon_0)^{1/2}H_v}{(\epsilon_0/\mu_0)^{1/2}E_x} = \left(\frac{\mu_0}{\epsilon_0}\right)^{1/2} \cos \theta_i$$

in free space and

$$-\frac{E_v'}{H_x'} = \frac{(\mu_0/\kappa\epsilon_0)^{1/2}H_v}{(\kappa\epsilon_0/\mu_0)^{1/2}E_x} = \left(\frac{\mu_0}{\epsilon_0}\right)^{1/2} \frac{(\kappa - \sin^2 \theta_i)^{1/2}}{\kappa}$$

in the dielectric medium. The normalized wave impedance Z' is thus given by

$$Z' = \frac{(\kappa - \sin^2 \theta_i)^{1/2}}{\kappa \cos \theta_i}. \tag{80}$$

The propagation factor is the same as that for perpendicular polarization, and, hence, the dielectric sheet will be reflectionless when d has the value given by (79). The reflection coefficient at the air-dielectric interface is

$$R_1' = \frac{Z' - 1}{Z' + 1}. \tag{81}$$

When the appropriate values for the normalized impedances Z and Z' are substituted into (77) and (81), respectively, we obtain the well-known Fresnel reflection coefficients for lossless dielectric media. An examination of (80) and (81) shows that, for some angle θ_i, the reflection coefficient vanishes since Z' becomes equal to unity. Let this value of θ_i be θ_b; thus from (80)

$$\sin \theta_b = \left(\frac{\kappa}{\kappa + 1} \right)^{1/2} \tag{82}$$

which defines the Brewster angle. A similar phenomenon does not occur for perpendicular polarization unless the two media have different permeabilities.

We can now readily evaluate the transmission coefficient for a cascade section of two or more parallel but different dielectric sheets by applying the wave-matrix theory developed in Section 3.4. We must, of course, use the appropriate expressions for the reflection coefficients as given here. The transmission coefficients are given by

$$T_{12} = 1 + R_1 \tag{83a}$$

$$T_{12}' = 1 + R_1' \tag{83b}$$

for the free-space–dielectric interface.

As a typical example, consider a dielectric sheet of thickness d in front of an infinite dielectric medium as in Fig. 3.8. A parallel-polarized wave is assumed incident from free space at an angle θ_i. The dielectric constants are κ_1 and κ_2. The electrical length of the sheet of thickness d is $\phi = k_0 d(\kappa_1 - \sin^2 \theta_i)^{1/2}$. The reflection and transmission coefficients at the two interfaces are

$$R_1 = \frac{Z_1 - 1}{Z_1 + 1} \qquad T_1 = 1 + R_1$$

$$R_2 = \frac{Z_2 - Z_1}{Z_2 + Z_1} \qquad T_2 = 1 + R_2$$

where

$$Z_1 = \frac{(\kappa_1 - \sin^2 \theta_i)^{1/2}}{\kappa_1 \cos \theta_i}$$

$$Z_2 = \frac{(\kappa_2 - \sin^2 \theta_i)^{1/2}}{\kappa_2 \cos \theta_i}.$$

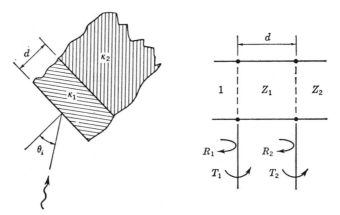

Fig. 3.8. A dielectric-sheet discontinuity and its equivalent circuit.

Let the incident wave have an amplitude c_1, the reflected wave an amplitude b_1, and the wave transmitted into the medium with dielectric constant κ_2 an amplitude c_3. Using the wave-chain-matrix theory gives

$$\begin{bmatrix} c_1 \\ b_1 \end{bmatrix} = \frac{1}{T_1 T_2} \begin{bmatrix} e^{j\phi} & R_1 e^{-j\phi} \\ R_1 e^{j\phi} & e^{-j\phi} \end{bmatrix} \begin{bmatrix} 1 & R_2 \\ R_2 & 1 \end{bmatrix} \begin{bmatrix} c_3 \\ 0 \end{bmatrix}. \tag{84}$$

Multiplying the two matrices together gives the following result for the element A_{11} in the first row and column of the resultant 2×2 matrix,

$$A_{11} = \frac{e^{j\phi} + R_1 R_2 e^{-j\phi}}{T_1 T_2}. \tag{85}$$

This element is equal to the reciprocal of the overall transmission coefficient T. The power-transmission coefficient is given by TT^*/Z_2 since T is the nonnormalized amplitude-transmission coefficient. From (85) we get

$$\frac{TT^*}{Z_2} = (A_{11}A_{11}^* Z_2)^{-1} = \frac{T_1^2 T_2^2}{Z_2(1 + R_1^2 R_2^2 + 2R_1 R_2 \cos 2\phi)}. \tag{86}$$

For a sheet which is effectively a quarter wavelength thick, $\phi = \pi/2$ and (86) reduces to

$$\frac{TT^*}{Z_2} = \frac{4Z_1^2 Z_2}{(Z_1^2 + Z_2)^2}.$$

All the incident power is transmitted into the final medium when

$$TT^* = Z_2$$

or

$$Z_1 = Z_2^{1/2}. \tag{87}$$

This is the condition for matching one dielectric sheet to free space by means of another sheet of electrical length $\pi/2$ and a normalized wave impedance equal to the square root of the normalized wave impedance of the sheet to be matched. Obviously a perfect match is obtained only at one frequency and one angle of incidence. This is the electromagnetic basis of the technique used in optics to match the surfaces of lenses to free space and is known as lens blooming. Such a matching sheet will not match both polarizations simultaneously except for normal incidence when the distinction between the two polarizations vanishes. The required value of the dielectric constant can be determined from the expressions for the impedances Z_1 and Z_2. The results for the case of perpendicular polarization may be obtained by using the appropriate expressions for the impedances.

Many dielectrics used at microwave frequencies have such small losses that the losses may be neglected. However, there are many applications where moderately lossy dielectrics must be used, e.g., in radome construction. Dielectric loss may be due to a finite conductivity of the material or to the fact that the applied frequency is too near a resonant frequency of the molecules that compose the dielectric material, so that the damping forces are such that the molecular polarization lags behind the applied field. When the material has a finite conductivity σ, the curl equation for the magnetic field vector $\mathbf{H}$ becomes

$$\nabla \times \mathbf{H} = j\omega\epsilon_0\kappa\mathbf{E} + \sigma\mathbf{E}$$

$$= j\omega\epsilon_0\kappa \left(1 - j\frac{\sigma}{\omega\kappa\epsilon_0}\right)\mathbf{E}$$

$$= j\omega\epsilon_0\kappa(1 - j\tan\delta)\mathbf{E} \tag{88}$$

where $\tan\delta = \sigma/\omega\epsilon_0\kappa$ and is called the loss tangent of the material. We shall consider only the case where $\sigma \ll \omega\epsilon_0\kappa$, so that $\tan\delta \approx \delta$ and $\delta \ll 1$. The other forms of dielectric loss may be accounted for by introducing a suitably defined loss tangent also. If we replace κ in the previously derived solutions to Maxwell's equations by the complex dielectric constant

$$\kappa(1 - j\delta)$$

we obtain at once the effect of small losses in the dielectric medium.

Using the approximation $(1 - j\delta)^{1/2} \approx 1 - j\delta/2$, the propagation constant becomes $j(2\pi/\lambda_0)\cos\theta_r(1 - j\delta/2)$. From Snell's law,

$$\cos\theta_r = \frac{[(1 - j\delta)\kappa - \sin^2\theta_i]^{1/2}}{[1 - j(\delta/2)]\eta} = \frac{\{(\kappa - \sin^2\theta_i)[1 - j\kappa\delta/(\kappa - \sin^2\theta_i)]\}^{1/2}}{[1 - j(\delta/2)]\eta}$$

$$\approx \frac{S(1 - j\kappa\delta/2S^2)}{\eta(1 - j\delta/2)} \tag{89}$$

where $S = (\kappa - \sin^2\theta_i)^{1/2}$. Hence the propagation constant becomes

$$j\frac{2\pi S}{\lambda_0} + \frac{2\pi}{\lambda_0}\frac{\kappa\delta}{2S} = j\beta + \alpha \tag{90}$$

where $j\beta$ is the phase factor, and α is the attenuation factor. For perpendicular polarization, the normalized wave impedance becomes

$$Z = \frac{\cos\theta_i}{\eta[1 + j(\delta/2)]\cos\theta_r} \approx \frac{\cos\theta_i}{S}\left(1 + j\frac{\kappa\delta}{2S^2}\right). \tag{91}$$

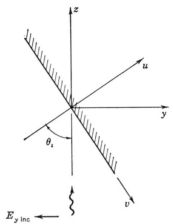

Fig. 3.9. An infinite conducting plane.

For parallel polarization, the normalized wave impedance becomes

$$Z' = \frac{\cos \theta_r}{\eta[1 - j(\delta/2)] \cos \theta_i} \approx \frac{S[1 - j(\kappa/2S^2 - 1)\delta]}{\kappa \cos \theta_i}. \tag{92}$$

The wave impedances are complex whenever the medium has finite losses. We may treat the propagation of electromagnetic waves through lossy dielectric sheets formally the same as that through nonlossy sheets. The only difference is that we must use the wave impedances and the expression for the propagation constant which are appropriate for lossy dielectric sheets.

3.6. Reflection from a Finite Conducting Plane

The theory developed for reflection from dielectric sheets, in particular, lossy dielectric sheets, may be applied to the problem of reflection from finite conducting metal planes also. For metals commonly used at microwave frequencies, the conduction current is normally much greater than the displacement current; that is, $\sigma \gg \omega\epsilon$. For copper, at 10,000 mHz, $\sigma/\omega\epsilon \approx 10^8$. Consequently, Maxwell's curl equation for $\mathbf{H}$ may be taken as $\nabla \times \mathbf{H} = j\omega\epsilon\mathbf{E} + \sigma\mathbf{E} \approx \sigma\mathbf{E}$ in metals. We may, therefore, obtain the solutions to Maxwell's equations in a metal by replacing $j\omega\epsilon_0$ in the solutions for lossless dielectric sheets by σ. The propagation constant k becomes $(-j\omega\mu_0\sigma)^{1/2} = (\omega\mu_0\sigma/2)^{1/2}(1 - j)$, where it is assumed that the permeability of the metal may be taken as that of free space.

Let a parallel-polarized wave be incident on the metallic sheet at an angle θ_i, as in Fig. 3.9. The electrical constants of the metal are assumed to be μ_0, ϵ_0, and σ. The incident magnetic field may be taken as

$$H_{xI} = H_x e^{-jk_0 z} = H_x e^{jk_0(v \sin \theta_i - u \cos \theta_i)}. \tag{93}$$

Using the curl equation for $\mathbf{H}$, we find that the incident electric field in the vux coordinate frame is

$$j\omega\epsilon_0 E_{uI} = -jk_0 \sin \theta_i H_{xI} \tag{94a}$$

$$j\omega\epsilon_0 E_{vI} = -jk_0 \cos \theta_i H_{xI}. \tag{94b}$$

The wave impedance measured in a direction normal to the surface, i.e., along u, is

$$-\frac{E_{vI}}{H_{xI}} = \left(\frac{\mu_0}{\epsilon_0}\right)^{1/2} \cos\theta_i = Z_1 = Y_1^{-1}. \tag{95}$$

By analogy, the wave transmitted into the metal is, by (36) to (39) in Section 3.3,

$$E_{vT} = -TZ_1 H_x e^{jk(v\sin\theta_r - u\cos\theta_r)} \tag{96a}$$

$$H_{xT} = -Y_m E_{vT} \tag{96b}$$

$$\sigma E_{uT} = -jk\sin\theta_r H_{xT} = jkY_m\sin\theta_r E_{vT} \tag{96c}$$

where $k = (\omega\mu_0\sigma/2)^{1/2}(1-j)$, and T is the amplitude-transmission coefficient. The wave impedance of the metal measured in the u direction is

$$Z_m = Y_m^{-1} = -\frac{E_{vT}}{H_{xT}} = \left(\frac{\omega\mu_0}{2\sigma}\right)^{1/2}(1+j)\cos\theta_r. \tag{97}$$

When σ approaches infinity, Z_m approaches zero, and the metal becomes an ideal short circuit. The attenuation factor along the u direction in the metal is $(\omega\mu_0\sigma/2)^{1/2}\cos\theta_r$ and becomes infinite with σ. Consequently, for a perfect conductor, the field cannot penetrate into the metal.

The reflected wave is given by the following expressions:

$$E_{vR} = -RZ_1 H_x e^{jk_0(v\sin\theta_i + u\cos\theta_i)} \tag{98a}$$

$$H_{xR} = Y_1 E_{vR} \tag{98b}$$

$$j\omega\epsilon_0 E_{uR} = -jk_0\sin\theta_i H_{xR} = -jk_0\sin\theta_i Y_1 E_{vR} \tag{98c}$$

where R is the reflection coefficient. The reflection and transmission coefficients are given by Eqs. (40), as follows:

$$T = 1 + R = \frac{2Z_m}{Z_m + Z_1} \tag{99a}$$

$$R = \frac{Z_m - Z_1}{Z_m + Z_1}. \tag{99b}$$

By Snell's law, $k_0\sin\theta_i = k\sin\theta_r$, and hence

$$k\cos\theta_r = (k^2 - k_0^2\sin^2\theta_i)^{1/2} \approx k$$

since $\sigma \gg \omega\epsilon_0$. This is a result of fundamental importance for it shows that, to a very high degree of approximation, the propagation constant normal to the interface may be taken as k in the conductor independent of the angle of incidence θ_i. From (97) we find that

$$Z_m = \left(\frac{\omega\mu_0}{2\sigma}\right)^{1/2}(1+j)$$

again independent of θ_i. Since $k = (\omega\mu_0\theta/2)^{1/2}(1-j)$, the attenuation coefficient in the conductor is $(\omega\mu_0\sigma/2)^{1/2}$. In a distance $\delta_s = (2/\omega\mu_0\sigma)^{1/2}$ called the skin depth, the field has decayed to e^{-1} of its value at the surface. By using the definition of δ_s it can be shown that the wave impedance of the metal may be expressed as

$$Z_m = \frac{1+j}{\delta_s\sigma}. \tag{100}$$

The resistive part is seen to be equal to the low-frequency resistance of a conducting sheet of thickness δ_s and conductivity σ. Since these latter results are independent of θ_i we may treat the boundary of an arbitrarily shaped conducting obstacle as a discontinuity surface with a surface impedance Z_m, even though the field incident on the obstacle is not a plane wave. An arbitrary field may be constructed from a superposition of plane waves. The approximation, however, does break down in the vicinity of surfaces that have a radius of curvature not much greater than the skin depth.

Returning to the problem of the conducting plane, we find that at the boundary surface the following relations must hold among the tangential components of the incident, reflected, and transmitted fields:

$$E_{vI} + E_{vR} = -(1+R)Z_1 H_x = -TZ_1 H_x = E_{vT} \tag{101a}$$

$$H_{xI} + H_{xR} = (1-R)H_x = (Z_1/Z_m)TH_x = H_{xT} \tag{101b}$$

$$j\omega\epsilon_0(E_{uI} + E_{uR}) = \sigma E_{uT}. \tag{101c}$$

Equation (101c) shows that the normal displacement flux is continuous across the surface. The total current flowing per unit width in the metal is

$$I = \int_0^\infty \sigma E_{vT}\, du = -\sigma T Z_1 H_x e^{jkv\sin\theta_r} \int_0^\infty e^{-[(1+j)/\delta_s]u\cos\theta_r}\, du$$

$$= -\frac{2Z_1 H_x}{Z_1 + Z_m} e^{jk_0v\sin\theta_i}.$$

The total current flowing is independent of σ when σ approaches infinity, since under these conditions Z_m tends to zero. Thus the current is forced to flow on the surface of the metal when $\sigma \to \infty$. The total current flowing is $2H_x e^{jk_0v\sin\theta_i}$ and is equal to the tangential magnetic field at the surface. Since the field vanishes inside the metal, the normal component of $\mathbf{D}$ must now terminate in a surface charge distribution. These two boundary conditions are easily expressed as

$$\mathbf{J}_s = \mathbf{n} \times \mathbf{H} \tag{102a}$$

$$\rho_s/\epsilon_0 = \mathbf{n}\cdot\mathbf{E} \tag{102b}$$

where $\mathbf{n}$ is the unit normal pointing away from the surface. On the surface the current density $\mathbf{J}_s$ is related to the charge density ρ_s by the equation of continuity. For the above problem,

$$\nabla_s\cdot\mathbf{J}_s = \frac{\partial}{\partial v}(-2H_x e^{jk_0v\sin\theta_i}) = -2jk_0 H_x \sin\theta_i e^{jk_0v\sin\theta_i}$$

and

$$-j\omega\rho_s = j\omega\epsilon_0(E_{uI} + E_{uR}) = -2jk_0 H_x \sin\theta_i e^{jk_0v\sin\theta_i}$$

so that $\nabla_s\cdot\mathbf{J}_s = -\partial\rho_s/\partial t$, where ∇_s is the del operator on the discontinuity surface.

The analysis for a perpendicular-polarized incident wave does not give any additional new results, so the details will not be given.

3.7. Plane Waves in Anisotropic Dielectric Media

Anisotropic media are those characterized by tensor permittivities or tensor permeabilities, or perhaps both. Initially we will consider some of the properties of plane waves in anisotropic dielectric materials. This will be followed by a brief treatment of propagation in ferrite material which exhibits a tensor permeability. By a suitable choice of axes, the tensor permeability or permittivity can be reduced to a diagonal form. The normal coordinates for the two tensors, in general, will not coincide. Also, the direction cosines of the normal coordinates relative to the original coordinates are not necessarily real, so that diagonalization of the constitutive tensors cannot always be obtained by a simple rotation of coordinates. A familiar example is an infinite ferrite medium for which the normal coordinates are clockwise- and counterclockwise-rotating coordinates (see Problem 3.7).

Let us consider a lossless dielectric material with principal dielectric constants κ_x, κ_y, κ_z along the principal axes, which are assumed to coincide with the x, y, z axes. The electric field for a uniform plane wave propagating in a direction defined by the unit normal

$$\mathbf{n} = \mathbf{a}_x n_x + \mathbf{a}_y n_y + \mathbf{a}_z n_z$$

may be represented as

$$\mathbf{E} = \mathbf{E}_0 e^{-j\beta \mathbf{n}\cdot\mathbf{r}} \tag{103}$$

where $\mathbf{E}_0$ is a constant vector, β is the propagation constant, as yet undermined, and $\mathbf{r}$ is the position vector $\mathbf{a}_x x + \mathbf{a}_y y + \mathbf{a}_z z$. Using Maxwell's equations, we find that

$$-j\omega\mu_0 \mathbf{H} = \nabla \times \mathbf{E} = \nabla \times \mathbf{E}_0 e^{-j\beta \mathbf{n}\cdot\mathbf{r}} = -\mathbf{E}_0 \times \nabla e^{-j\beta \mathbf{n}\cdot\mathbf{r}} = -j\beta \mathbf{n} \times \mathbf{E} \tag{104a}$$

$$j\omega\mathbf{D} = \nabla \times \mathbf{H} = \frac{j\beta^2}{\omega\mu_0}[\mathbf{E} - \mathbf{n}(\mathbf{n}\cdot\mathbf{E})]. \tag{104b}$$

Equation (104b) shows that $\mathbf{D}$ and $\mathbf{E}$ are not collinear unless $\mathbf{n}\cdot\mathbf{E} = 0$. In general, $\mathbf{n}\cdot\mathbf{E} \neq 0$; hence replacing the components of $\mathbf{E}$ in the first term in (104b) by

$$\frac{D_x}{\kappa_x \epsilon_0} \qquad \frac{D_y}{\kappa_y \epsilon_0} \qquad \frac{D_z}{\kappa_z \epsilon_0}$$

and solving for the components of $\mathbf{D}$ gives

$$\mathbf{D} = \frac{\beta^2}{k_0^2}\epsilon_0 \mathbf{n}\cdot\mathbf{E}\left(\frac{\mathbf{a}_x n_x k_x^2}{\beta^2 - k_x^2} + \frac{\mathbf{a}_y n_y k_y^2}{\beta^2 - k_y^2} + \frac{\mathbf{a}_z n_z k_z^2}{\beta^2 - k_z^2}\right) \tag{105}$$

where

$$k_0^2 = \omega^2 \mu\epsilon_0$$
$$k_x^2 = \kappa_x k_0^2$$

and similarly for k_y and k_z. From (104b) it is seen that $\mathbf{n} \cdot \mathbf{D}$ equals zero, so that $\mathbf{D}$ is always normal to $\mathbf{n}$. The dot product of (105) with $\mathbf{n}$ gives

$$\frac{n_x^2 k_x^2}{\beta^2 - k_x^2} + \frac{n_y^2 k_y^2}{\beta^2 - k_y^2} + \frac{n_z^2 k_z^2}{\beta^2 - k_z^2} = 0 \qquad (106)$$

which is the eigenvalue equation for determining the propagation factor β^2. This is a quadratic equation for β^2 and determines two distinct values for β^2 in general. At this time it is convenient to assume that the axes have been chosen so that $k_x > k_y > k_z$. For (106) to be zero, at least one term must be negative; at the same time not all three can be negative, and so $k_z^2 < \beta^2 < k_x^2$. If the square of the principal velocities

$$v_x^2 = \frac{v_c^2 k_0^2}{k_x^2} \qquad v_y^2 = \frac{v_c^2 k_0^2}{k_y^2} \qquad v_z^2 = \frac{v_c^2 k_0^2}{k_z^2}$$

and the wave velocity $v^2 = v_c^2 k_0^2 / \beta^2$ are introduced, (106) becomes

$$\frac{n_x^2}{v_x^2 - v^2} + \frac{n_y^2}{v_y^2 - v^2} + \frac{n_z^2}{v_z^2 - v^2} = 0 \qquad (107)$$

where v_c is the velocity of light in free space. For each direction, in the xyz coordinate frame, defined by the direction cosines n_x, n_y, n_z, two values of v^2 are determined by (107). Thus v^2 traces out a double-sheeted surface, which is called the Fresnel wave-velocity surface. The trace of the intersection of these surfaces in the coordinate planes consists of a circle and an oval. In the coordinate plane containing the largest and smallest principal velocities these curves intersect at four points. According to the way in which we choose our axes, this plane, in our case, is the xz plane. Thus, putting $n_y = 0$ in (107), we get

$$(v^2 - v_y^2)[v^2 - n_x^2 v_z^2 - (1 - n_x^2)v_x^2] = 0.$$

The two curves traced out are the circle $v = v_y$ and the oval

$$v^2 = n_x^2 v_z^2 + (1 - n_x^2)v_x^2$$

with semiaxes v_x and v_z. Since $v_x < v_y < v_z$, these curves intersect as illustrated in Fig. 3.10. The points of intersection define two directions, known as the optic axes, along which only a single value for v^2 and β^2 exists. If the positive directions of these axes are chosen as in Fig. 3.10, then the unit normals $\mathbf{N}_1$ and $\mathbf{N}_2$ along these axes are

$$\mathbf{N}_1 = \mathbf{a}_x N_x + \mathbf{a}_z N_z \qquad (108a)$$

$$\mathbf{N}_2 = \mathbf{a}_x N_x - \mathbf{a}_z N_z \qquad (108b)$$

where

$$N_x^2 = \frac{v_y^2 - v_x^2}{v_z^2 - v_x^2} \qquad (108c)$$

$$N_z^2 = \frac{v_z^2 - v_y^2}{v_z^2 - v_x^2}. \qquad (108d)$$

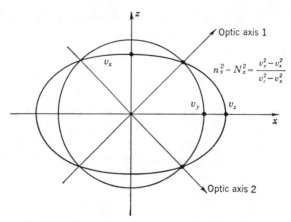

Fig. 3.10. Wave-velocity surface traces in the xz plane.

These are the only directions for which a single value for β^2 exists. The oval and circle do not intersect in the xy or yz plane since the oval's semiaxes are either both greater or both smaller than the radius of the circle, which is either v_x or v_z in these planes.

The direction cosines between the wave normal $\mathbf{n}$ and the optic axes are

$$\cos\theta_1 = \mathbf{n}\cdot\mathbf{N}_1 \tag{109a}$$

$$\cos\theta_2 = \mathbf{n}\cdot\mathbf{N}_2. \tag{109b}$$

In terms of these, we find that the two solutions to (107) are v_1^2 and v_2^2, where

$$2v_1^2 = v_z^2 + v_x^2 + (v_z^2 - v_x^2)\cos(\theta_1 - \theta_2) \tag{110a}$$

$$2v_2^2 = v_z^2 + v_x^2 + (v_z^2 - v_x^2)\cos(\theta_1 + \theta_2). \tag{110b}$$

The interested reader may readily verify that $(v^2 - v_1^2)(v^2 - v_2^2) = 0$ is identical to (107). The two values of β^2 are given by

$$\beta_i^2 = k_0^2 \frac{v_c^2}{v_i^2}, \qquad i = 1, 2.$$

Having determined the two values of the wave velocity, we return to (105) and find that the direction of $\mathbf{D}$ is fixed, i.e.,

$$D_{ix}:D_{iy}:D_{iz} = \frac{n_x}{v_i^2 - v_x^2} : \frac{n_y}{v_i^2 - v_y^2} : \frac{n_z}{v_i^2 - v_z^2} \tag{111}$$

where $\mathbf{D}_i$ is the solution for $\mathbf{D}$ corresponding to the root v_i^2. The directions of $\mathbf{D}$, in turn, determine the allowed directions of $\mathbf{E}_0$ and $\mathbf{H}$. The two solutions for $\mathbf{D}$ are mutually orthogonal since

$$\mathbf{D}_1\cdot\mathbf{D}_2 \propto \frac{n_x^2}{(v_1^2 - v_x^2)(v_2^2 - v_x^2)} + \frac{n_y^2}{(v_1^2 - v_y^2)(v_2^2 - v_y^2)} + \frac{n_z^2}{(v_1^2 - v_z^2)(v_2^2 - v_z^2)}$$

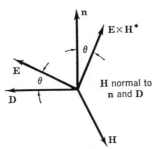

Fig. 3.11. Space relationship between the fields **E** and **D** compared with that of the wave normal and Poynting vector.

which, when the identity

$$\frac{n_x^2}{(v_1^2 - v_x^2)(v_2^2 - v_x^2)} = \frac{n_x^2}{v_2^2 - v_1^2}\left(\frac{1}{v_1^2 - v_x^2} - \frac{1}{v_2^2 - v_x^2}\right)$$

(and similarly for the other terms) is used, is seen to vanish by virtue of (107), of which v_1^2 and v_2^2 are roots.

The direction of energy flow is that of the complex Poynting vector $\mathbf{E} \times \mathbf{H}^*$ and is not in the direction of the wave normal $\mathbf{n}$ in general. In contrast to this, the cross product $\mathbf{D} \times \mathbf{H}^*$ is in the direction of $\mathbf{n}$ since both $\mathbf{D}$ and $\mathbf{H}$ are perpendicular to $\mathbf{n}$. By forming the products $\mathbf{D}\cdot\mathbf{E}/|\mathbf{D}||\mathbf{E}|$ and $\mathbf{n}\cdot\mathbf{E} \times \mathbf{H}^*/|\mathbf{E} \times \mathbf{H}^*|$ we find that the direction cosines (which are given by the above terms) between $\mathbf{D}$ and $\mathbf{E}$ and the wave normal $\mathbf{n}$ and the Poynting vector $\mathbf{E} \times \mathbf{H}^*$ are equal and, in particular, given by $(1 - |(\mathbf{n}\cdot\mathbf{E})/\mathbf{E}|^2)^{1/2}$. The situation is illustrated in Fig. 3.11. The wave propagates along the normal $\mathbf{n}$ according to $e^{-j\beta\mathbf{n}\cdot\mathbf{r}}$. If we let $\mathbf{s}$ be a unit vector along the direction of the Poynting vector, then

$$\mathbf{n} = (\mathbf{n}\cdot\mathbf{s})\mathbf{s} + [\mathbf{n} - (\mathbf{n}\cdot\mathbf{s})\mathbf{s}] = (\mathbf{n}\cdot\mathbf{s})\mathbf{s} + \tau$$

where τ is defined by this equation and is a vector perpendicular to $\mathbf{s}$. Introducing these vectors into the propagation factor it becomes $e^{-j\beta[(\mathbf{n}\cdot\mathbf{s})\mathbf{s}\cdot\mathbf{r}+\tau\cdot\mathbf{r}]}$. In the direction of $\mathbf{s}$, the velocity of propagation is

$$u = \frac{v_c k_0}{\beta \mathbf{n}\cdot\mathbf{s}} = \frac{v}{\mathbf{n}\cdot\mathbf{s}} \tag{112}$$

and is the velocity of energy propagation for the wave.

The analysis of plane-wave propagation in the interior of an anisotropic dielectric is relatively straightforward. However, the problem of determining the fields transmitted into an anisotropic medium when a plane wave is incident on an interface separating an isotropic and an anisotropic medium is much more complicated, since in this case either $\mathbf{n}$, β, nor the direction of $\mathbf{D}$ is known. A medium with two optical axes is known as a biaxial medium. When two of the principal dielectric constants are equal, say $\kappa_y = \kappa_z$, only a single optical axis exists, and the medium is said to be uniaxial. Referring to (108c) and (108d) with $v_z^2 = v_y^2$ shows that this axis coincides with the x axis in the present case. Reflection from and transmission through a uniaxial anisotropic interface are considerably easier to analyze; this interface is the next topic which we will take up.

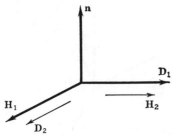

Fig. 3.12. Orientation of the magnetic field and displacement field relative to the wave normal for the two modes of propagation in a uniaxial medium.

An Anisotropic Dielectric Interface

Consider a uniaxial medium with principal dielectric constants κ_1, κ_2, and κ_2 along the x, y, and z axes, respectively. With reference to (104b), we find that, if $\mathbf{n} \cdot \mathbf{E} = 0$, $\mathbf{D}$ and $\mathbf{E}$ are collinear, and

$$\beta^2 = k_2^2 = \kappa_2 k_0^2.$$

In this case $\mathbf{E}$ must lie in the yz plane for which no anisotropic effect is present. This solution is called the ordinary wave since its behavior is the same as that of a plane wave in an isotropic medium. When

$$\mathbf{n} \cdot \mathbf{E} \neq 0$$

we find from (106) that

$$\beta^2 = \frac{k_1^2 k_2^2}{n_x^2 k_1^2 + (1 - n_x^2)k_2^2} \tag{113}$$

where $k_i^2 = \kappa_i k_0^2$, $i = 1, 2$. For this case $\mathbf{D}$ has components along all three axes in general. With reference to Fig. 3.12, let $\mathbf{D}_1$, $\mathbf{H}_1$ be the solution for the ordinary wave, and $\mathbf{D}_2$, $\mathbf{H}_2$ be the solution corresponding to the value of β^2 given by (113). This latter wave is called the extraordinary wave.

Since $\mathbf{D}_1$ and $\mathbf{H}_1$ are at right angles to each other and are both perpendicular to $\mathbf{n}$, and $\mathbf{D}_2$ is perpendicular to $\mathbf{D}_1$ and $\mathbf{n}$, then $\mathbf{D}_2$ must be along the direction of $\mathbf{H}_1$. Consequently, $\mathbf{H}_2$, which is perpendicular to $\mathbf{D}_2$ and $\mathbf{n}$, is collinear with $\mathbf{D}_1$, and hence lies in the yz plane, i.e., has no x component. These properties of the two solutions, namely, that $\mathbf{D}_1$ and $\mathbf{H}_2$ have no x components, permit us to derive the fields for the two modes of propagation from a magnetic and an electric Hertzian potential function having only an x component. The ordinary wave may be found from the magnetic Hertzian potential $\mathbf{\Pi}_h = \Pi_h \mathbf{a}_x$ as follows:

$$\mathbf{E} = -j\omega\mu_0 \nabla \times \mathbf{\Pi}_h \tag{114a}$$

$$\mathbf{H} = \nabla \times \nabla \times \mathbf{\Pi}_h \tag{114b}$$

and Π_h is a solution of

$$\nabla^2 \Pi_h + k_2^2 \Pi_h = 0. \tag{114c}$$

The extraordinary wave is found from the electric Hertzian potential $\Pi_e = \mathbf{a}_x \Pi_e$ as follows:

$$\mathbf{E} = k_0^2 \Pi_e + \kappa_2^{-1} \nabla \nabla \cdot \Pi_e \tag{115a}$$

$$\mathbf{H} = j\omega\epsilon_0 \nabla \times \Pi_e \tag{115b}$$

and Π_e is a solution of

$$\nabla^2 \Pi_e + k_1^2 \Pi_e + \frac{\kappa_1 - \kappa_2}{\kappa_2} \frac{\partial^2 \Pi_e}{\partial x^2} = 0. \tag{115c}$$

Equation (115c) may be derived as follows. Since $\nabla \times \mathbf{H} = j\omega\mathbf{D}$, we get

$$j\omega\mathbf{D} = j\omega\epsilon_0 \nabla \times \nabla \times \Pi_e = j\omega\epsilon_0 (\nabla \nabla \cdot \Pi_e - \nabla^2 \Pi_e).$$

Also, since $\nabla \times \mathbf{E} = -j\omega\mu_0\mathbf{H} = k_0^2 \nabla \times \Pi_e$, we have $\mathbf{E} = k_0^2 \Pi_e + \nabla\Phi$, where Φ is as yet arbitrary. Substituting for $\mathbf{D}$ in the above expression gives

$$j\omega\mathbf{D} = j\omega[\kappa_2\epsilon_0\mathbf{E} + (\kappa_1 - \kappa_2)\epsilon_0\mathbf{a}_x E_x]$$

$$= j\omega\kappa_2\epsilon_0 \left[k_0^2 \Pi_e + \nabla\Phi + \frac{\kappa_1 - \kappa_2}{\kappa_2} \left(k_0^2 \Pi_e + \frac{\partial\Phi}{\partial x} \right) \mathbf{a}_x \right]$$

$$= j\omega\epsilon_0 (\nabla \nabla \cdot \Pi_e - \nabla^2 \Pi_e).$$

If we now let $\nabla \cdot \Pi_e = \kappa_2\Phi = \partial\Pi_e/\partial x$, we obtain (115c) and (115a). This latter step is permissible since $\nabla \cdot \Pi_e$ and Φ have not been specified up to this point. Equation (114a) gives a solution for $\mathbf{E}$ with no x component, while (115b) gives a solution for $\mathbf{H}$ with no x component. Furthermore, (115c) yields the eigenvalue equation (113) when a form such as $e^{-j\beta\mathbf{n}\cdot\mathbf{r}}$ is assumed for Π_e. When these two modes of propagation are excited by a plane wave incident from free space onto the anisotropic medium, the two modes do not, in general, have the same wave normal.

Let a uniaxial anisotropic medium occupy the whole half space $z > 0$, and let the optic axis be along the x axis, which in turn lies in the plane infinite interface as in Fig. 3.13. A perpendicular- or parallel-polarized plane wave incident on this interface along a direction defined by the unit vector or wave normal $\mathbf{N}$ can be represented by a suitable superposition of the two modes derivable from Π_h and Π_e. Thus we are concerned with two distinct modes of propagation and will find that the interface can be represented by a four-terminal-pair network. An isotropic dielectric interface was found to be representable by a two-terminal-pair network and was consequently much simpler to analyze.

Let the Hertzian potentials for the incident and reflected plane waves in the region $z < 0$ be

$$\Pi_h = \mathbf{a}_x (A_0 e^{-jk_0\mathbf{N}\cdot\mathbf{r}} + A_1 e^{-jk_0\mathbf{N}'\cdot\mathbf{r}}) \tag{116a}$$

$$\Pi_e = \mathbf{a}_x (B_0 e^{-jk_0\mathbf{N}\cdot\mathbf{r}} + B_1 e^{-jk_0\mathbf{N}'\cdot\mathbf{r}}) \tag{116b}$$

where $k_0^2 = \omega^2\mu_0\epsilon_0$, and the unit wave normals are given by

$$\mathbf{N} = N_x\mathbf{a}_x + N_y\mathbf{a}_y + N_z\mathbf{a}_z$$

$$\mathbf{N}' = N_x\mathbf{a}_x + N_y\mathbf{a}_y - N_z\mathbf{a}_z.$$

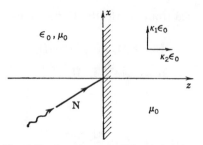

Fig. 3.13. An anisotropic dielectric interface.

In the region $z > 0$, suitable forms for the potentials are

$$\Pi_h = \mathbf{a}_x(C_1 e^{-jk_2\mathbf{n}_1\cdot\mathbf{r}} + C_0 e^{-jk_2\mathbf{n}_1'\cdot\mathbf{r}}) \tag{117a}$$

$$\Pi_e = \mathbf{a}_x(D_1 e^{-j\beta\mathbf{n}_2\cdot\mathbf{r}} + D_0 e^{-j\beta\mathbf{n}_2'\cdot\mathbf{r}}) \tag{117b}$$

where the unit wave normals $\mathbf{n}_1'$ and $\mathbf{n}_2'$ have the same relation to $\mathbf{n}_1$ and $\mathbf{n}_2$ as $\mathbf{N}'$ does to $\mathbf{N}$, that is, a change in sign for the z component.

At the interface $z = 0$, the tangential electric and magnetic fields must be continuous for all values of x and y. This is possible only if the potentials have the same variation with x and y on both sides of the interface. Hence, the following relations must hold:

$$k_0 N_x = k_2 n_{1x} = \beta n_{2x} \tag{118a}$$

$$k_0 N_y = k_2 n_{1y} = \beta n_{2y}. \tag{118b}$$

In addition, β is a solution of (113). From (113) and (118a) we get

$$\beta = k_0 \left[\kappa_1 + N_x^2 \left(1 - \frac{\kappa_1}{\kappa_2} \right) \right]^{1/2}. \tag{118c}$$

From (118b) and (118c) the component n_{2y} may be found. The components of $\mathbf{n}_1$ may be found at once from (118a) and (118b), together with the condition $n_{1z}^2 = 1 - n_{1x}^2 - n_{1y}^2$. We see that a knowledge of the wave normal $\mathbf{N}$ for the incident waves permits us to determine both the normals $\mathbf{n}_1$, $\mathbf{n}_2$ and β.

The fields are obtained from the Hertzian potentials by means of (114) and (115). When the tangential components are equated in the interface plane $z = 0$, the following four equations are obtained:

$$H_x = k_0^2(1 - N_x^2)(A_0 + A_1) = k_2^2(1 - n_{1x}^2)(C_1 + C_0) \tag{119a}$$

$$E_x = k_0^2(1 - N_x^2)(B_0 + B_1) = k_0^2 \left(1 - n_{2x}^2 \frac{\beta^2}{k_2^2} \right)(D_1 + D_0) \tag{119b}$$

$$H_y = -k_0^2 N_x N_y(A_0 + A_1) + k_0^2(1 - N_x^2)Y_{1e}(B_0 - B_1)$$

$$= -k_2^2 n_{1x} n_{1y}(C_1 + C_0) + k_0^2 \left(1 - \frac{\beta^2}{k_2^2} n_{2x}^2 \right) Y_{2e}(D_1 - D_0) \tag{119c}$$

$$E_y = k_0^2(1 - N_x^2)Z_{1m}(A_1 - A_0) - k_0^2 N_x N_y(B_0 + B_1)$$

$$= k_2^2(1 - n_{1x}^2)Z_{2m}(C_0 - C_1) - k_0^2 \frac{\beta^2}{k_2^2} n_{2x} n_{2y}(D_1 + D_0) \tag{119d}$$

TABLE 3.1

Mode	Voltage	Current
	Input Side, $z < 0$	
Π_h	$V_1 = (1 - N_x^2)Z_{1m}(A_1 - A_0)$	$I_1 = (1 - N_x^2)(-A_1 - A_0)$
Π_e	$V_2 = (1 - N_x^2)(B_0 + B_1)$	$I_2 = (1 - N_x^2)(B_0 - B_1)Y_{1e}$
	Output Side, $z > 0$	
Π_h	$V_3 = \kappa_2 Z_{2m}(1 - n_{1x}^2)(C_0 - C_1)$	$I_3 = \kappa_2(1 - n_{1x}^2)(C_1 + C_0)$
Π_e	$V_4 = (1 - n_{1x}^2)(D_1 + D_0)$	$I_4 = (1 - n_{1x}^2)(D_0 - D_1)Y_{2e}$

where

$$Y_{1m} = Z_{1m}^{-1} = \frac{1 - N_x^2}{N_z}Y_0 = \text{wave admittance for } \Pi_h \text{ mode for } z < 0$$

$$Y_{1e} = Z_{1e}^{-1} = \frac{N_z}{1 - N_x^2}Y_0 = \text{wave admittance for } \Pi_e \text{ mode for } z < 0$$

$$Y_{2m} = Z_{2m}^{-1} = \frac{k_2}{k_0}\frac{1 - n_{1x}^2}{n_{1z}}Y_0 = \text{wave admittance for } \Pi_h \text{ mode for } z > 0$$

$$Y_{2e} = Z_{2e}^{-1} = \frac{\beta n_{2z}Y_0}{k_0[1 - (\beta^2/k_2^2)n_{2x}^2]} = \text{wave admittance for } \Pi_e \text{ mode for } z > 0$$

$$Y_0 = Z_0^{-1} = \left(\frac{\epsilon_0}{\mu_0}\right)^{1/2}.$$

The wave admittances are defined as the ratio of the transverse magnetic field to the mutually orthogonal transverse electric field. These equations take on a particularly significant form when suitable equivalent transmission-line voltages and currents are introduced. The equivalent voltages are chosen proportional to the total transverse electric field, while the equivalent currents are chosen proportional to the total mutually perpendicular transverse magnetic field. The characteristic impedance of each line may be arbitrarily chosen so we will choose the characteristic impedances equal to the corresponding mode wave impedances. The equivalent voltages may be arbitrarily chosen subject to the restriction that the power carried by a given mode be equal to $K(V_i V_i^*/Z_i)$, where the constant of proportionality K is the same for all modes, V_i is the voltage on the ith line, and Z_i is the characteristic impedance of the ith line. The choice of equivalent voltages and currents in Table 3.1 leads to a very simple equivalent circuit for the interface.

Replacing βn_{2x} by $k_2 n_{1x}$ and βn_{2y} by $k_2 n_{1y}$ from (118), and also introducing the voltages and currents from Table 3.1, the set of equations (119) becomes

$$I_1 = -I_3 \tag{120a}$$

$$V_2 = V_4 \tag{120b}$$

$$\frac{N_x N_y}{1 - N_x^2}I_1 + I_2 = -\frac{n_{1x}n_{1y}}{1 - n_{1x}^2}I_3 - I_4 \tag{120c}$$

$$V_1 - \frac{N_x N_y}{1 - N_x^2}V_2 = V_3 - \frac{n_{1x}n_{1y}}{1 - n_{1x}^2}V_4. \tag{120d}$$

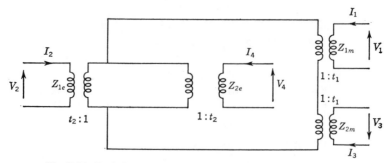

Fig. 3.14. Equivalent circuit of a uniaxial anisotropic dielectric interface.

Equation (120a) shows that lines 1 and 3 are series-coupled, while (120b) shows that lines 2 and 4 are parallel-coupled. A suitable equivalent circuit which gives rise to a set of equations of the above form is illustrated in Fig. 3.14. The transformers illustrated are considered as ideal ones with turns ratios t_1: 1 and t_2: 1. Only one of the turns ratios is independent; for example, t_2: 1 may be chosen as 1: 1.[6] The above equivalent circuit obeys the following equations:

$$I_1 = -I_3 \tag{121a}$$

$$V_2 = V_4 \tag{121b}$$

$$I_2 + I_4 = \frac{t_1}{t_2} I_3 = -\frac{t_1}{t_2} I_1 \tag{121c}$$

$$V_1 - V_3 = \frac{t_1}{t_2} V_2 = \frac{t_1}{t_2} V_4. \tag{121d}$$

Equations (120c) and (120d) may be put into the form of (121c) and (121d) by using (120a) and (120b). Thus, for (120c), we get

$$I_2 + I_4 = \left(\frac{N_x N_y}{1 - N_x^2} - \frac{n_{1x} n_{1y}}{1 - n_{1x}^2} \right) I_3$$

and, hence, comparing with (121c) shows that

$$\frac{t_1}{t_2} = \frac{N_x N_y}{1 - N_x^2} - \frac{n_{1x} n_{1y}}{1 - n_{1x}^2}. \tag{122}$$

Equation (120d) gives the same result.

If we are interested in the solution for a plane perpendicular-polarized wave incident from the region $z < 0$, the ratio of A_0 and B_0 must be chosen as follows: $A_0/B_0 = N_x N_z/N_y$, in order to make $E_z = 0$. Similarly for an incident parallel-polarized wave $H_z = 0$, and we

---

[6] An ideal transformer is one with infinite primary and secondary inductance and unity coupling. The voltages and currents on side 1 with n_1 turns are always related to those on side 2 with n_2 turns as follows: $V_1/V_2 = I_2/I_1 = n_1/n_2$. Thus, if the secondary is open-circuited, the primary is effectively an open circuit also because of its infinite self-impedance.

must choose $A_0/B_0 = N_y/N_x N_z$. When either N_x or N_y equals zero, Eqs. (119) show that there is no coupling between the Π_e and the Π_h modes. In any case, the amount of coupling is small when κ_1 and κ_2 do not differ by more than 50% or so.

When $N_y = 0$, the Π_e mode, which now corresponds to a parallel-polarized wave, is directly coupled between lines 2 and 4, with no interaction from lines 1 and 3. The normalized input impedance looking into the anisotropic medium is Z_{2e}/Z_{1e}. Using (113) and (118) to eliminate $\beta n_{2z} = \beta(1 - n_{2x}^2)^{1/2}$ gives

$$Z_{\text{in}} = \frac{Z_{2e}}{Z_{1e}} = \frac{(\kappa_2 - \sin^2 \theta_i)^{1/2}}{(\kappa_1 \kappa_2)^{1/2} \cos \theta_i} \tag{123}$$

where θ_i is the angle of incidence in the xz plane measured from the interface normal; that is, $\sin \theta_i = N_x$. A Brewster angle exists such that $Z_{\text{in}} = 1$, and no reflection takes place. From (123) this angle is given by

$$\sin \theta_b = \left[\frac{\kappa_2(\kappa_1 - 1)}{\kappa_1 \kappa_2 - 1} \right]^{1/2} \tag{124}$$

and differs from that derived in Section 3.5 for an isotropic dielectric medium. A value of θ_b less than $45°$ may exist. For isotropic dielectrics θ_b is always greater than $45°$ in the less dense medium.

3.8. TEM WAVES IN A FERRITE MEDIUM

It is not our purpose here to give a theory for the magnetic properties of ferrites. Nevertheless, considerable insight into the electromagnetic-wave-propagation properties of ferrites is obtained by considering a simple semiclassical model of the magnetization mechanism which takes place in ferromagnetic materials. Ordinary ferromagnetic materials such as iron and nickel are of little use at microwave frequencies because of their high conductivities. A class of materials known collectively as ferrites has, however, been developed which is quite suitable for use at microwave frequencies. Ferrites are a ceramiclike material with specific resistivities over a million times greater than those of metals, with relative permeabilities ranging up to several thousand and relative dielectric constants in the range 5 to 25. The magnetic properties of these materials arise principally from the magnetic moment associated with the spinning electron. By treating the spinning electron as a gyroscopic top, a classical picture of the magnetization process and, in particular, of the anisotropic magnetic properties may be obtained.[7]

The electron has an angular momentum $\mathbf{P}$ equal to $\frac{1}{2}\hbar$ or 0.527×10^{-34} joule-meters, where $\hbar$ is Planck's constant divided by 2π. The magnetic momentum associated with the spinning electron is one Bohr magneton or $m = e\hbar/2w = 9.27 \times 10^{-24}$ ampere-meter², where m is the magnetic dipole moment and w is the mass of the electron. The angular-momentum vector $\mathbf{P}$ and magnetic momentum $\mathbf{m}$ are antiparallel for the electron. The ratio of the magnetic momentum to the angular momentum is called the gyromagnetic ratio γ, that is, $\gamma = m/P$. If a spinning electron is located in a constant magnetic field $\mathbf{B}_0$, and $\mathbf{m}$ and $\mathbf{B}_0$ are not collinear, a torque $\mathbf{T} = \mathbf{m} \times \mathbf{B}_0$ is exerted on the system as in Fig. 3.15. The rate of change of angular

---

[7] The material presented here is based essentially on [3.3]–[3.6].

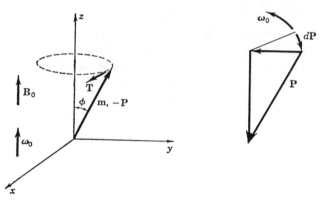

Fig. 3.15. Free precession of a spinning electron.

momentum is equal to the applied torque, and hence[8]

$$\mathbf{T} = \frac{d\mathbf{P}}{dt} = \mathbf{m} \times \mathbf{B}_0 = \boldsymbol{\omega}_0 \times \mathbf{P} = -\gamma^{-1}\boldsymbol{\omega}_0 \times \mathbf{m} \tag{125}$$

or

$$mB_0 \sin \phi = \omega_0 P \sin \phi = \gamma^{-1}\omega_0 m \sin \phi$$

where ω_0 is the vector precession angular velocity directed along $\mathbf{B}_0$, and $P \sin \phi$ is the component of $\mathbf{P}$ which is changing in direction. At first sight, Fig. 3.15 appears to give rotation in a direction opposite to that which should be produced by the acting torque, but this is only because $\mathbf{m}$ and $\mathbf{P}$ are antiparallel. From (125) the following expression for the free-precession angular velocity (Larmor frequency) is obtained:

$$\omega_0 = \gamma B_0. \tag{126}$$

For free precession, the precession angle ϕ is arbitrary.

To obtain a feeling for what happens when a spinning electron is placed in a field $\mathbf{B}_t$ which is a static field plus a small sinusoidally varying component, consider the following situation. With reference to Fig. 3.16, let $\mathbf{B}_0$ be a static field along the z axis, and let $\mathbf{B}_1$ be a constant counterclockwise-rotating vector in the xy plane. The rate of rotation is ω, and, hence, $\mathbf{B}_1$ is the same type of field which would exist in a plane negative circularly polarized wave propagating in the direction of $\mathbf{B}_0$. The total field vector $\mathbf{B}_t$ rotates about the z axis at a rate ω and with a constant angle θ. When equilibrium has been established between the field and the spinning electron, the magnetic dipole axis must precess around $\mathbf{B}_0$ in synchronism with $\mathbf{B}_t$. Thus the precession angle ϕ must be less than θ in order to obtain a torque acting in the proper direction to produce counterclockwise precession. The equation of motion (125) gives $mB_t \sin(\theta - \phi) = \omega P \sin \phi$ or $\gamma B_t(\sin \theta \cos \phi - \cos \theta \sin \phi) = \omega \sin \phi$. Replacing $B_t \sin \theta$ by B_1 and $B_t \cos \theta$ by B_0 and solving for $\sin \phi$ gives

$$\sin \phi = \frac{\gamma B_1}{[(\gamma B_0 + \omega)^2 + (\gamma B_1)^2]^{1/2}}.$$

[8]In the literature some authors include the negative sign in the definition of γ, whereas others omit the negative sign completely. This leads to an inconsistency which amounts to changing the sign in the off-diagonal terms in the susceptibility tensor or reversing the direction of the magnetic field $\mathbf{B}_0$.

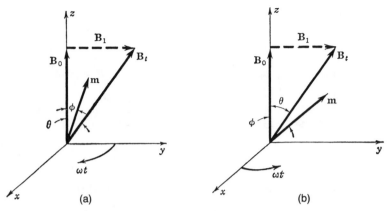

Fig. 3.16. Forced precession of a spinning electron.

The component of **m** in the xy plane which rotates in synchronism with $\mathbf{B}_1$ is $m_- = m \sin \phi = m \cos \phi \tan \phi = m_0 \tan \phi$, where m_0 is the component of **m** along $\mathbf{B}_0$. Substituting for $\tan \phi$ gives

$$m_- = \frac{\gamma \mu_0 H_1 m_0}{\gamma B_0 + \omega}. \tag{127}$$

If we have a positive circularly polarized field $\mathbf{B}_1$, the precession angle ϕ must be greater than θ, as in Fig. 3.16(b), in order to produce a torque which will make **m** precess about $\mathbf{B}_0$ in step with the total field $\mathbf{B}_t$. A significant feature is at once apparent: since the precession angle ϕ is different for the two cases, the effective permeability for negative and positive circularly polarized waves propagating in the direction along $\mathbf{B}_0$ will be different. For the latter case the equation of motion gives

$$\sin \phi = \frac{\gamma B_1}{[(\gamma B_0 - \omega)^2 + (\gamma B_1)^2]^{1/2}}$$

and hence the magnetization m_+ associated with a positive circularly polarized field $\mathbf{H}_1$ is

$$m_+ = \frac{\gamma \mu_0 m_0 H_1}{\gamma B_0 - \omega}. \tag{128}$$

A ferrite material will consist of a collection of spinning electrons. If the effective number of spinning electrons per unit volume is N, the magnetic dipole polarization per unit volume is $\mathbf{M} = N\mathbf{m}$. If a constant uniform magnetic field $\mu_0 \mathbf{H}_0$ is applied externally to an arbitrarily shaped sample of ferrite material, the interior field in the sample will be nonuniform and will differ from $\mathbf{H}_0$ because of the demagnetizing effects set up by the free magnetic poles on the surface of the sample. For ellipsoidal-shaped samples (and, of course, the degenerate forms such as spheroids, spheres, disks, and long rods) the interior field is uniform when a uniform exterior field is applied. The interior field may be evaluated quite readily and is conveniently expressed in the following form [3.9, ch. 4]:

$$\mathbf{H}_i = \mathbf{H}_o - \bar{\mathbf{L}} \cdot \mathbf{M}$$

where $\bar{\mathbf{L}}$ is a suitable demagnetization dyadic. When the coordinate axis coincides with the

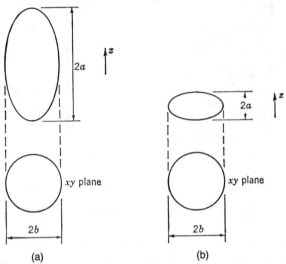

(a)   (b)

Fig. 3.17. Illustration of spheroid-shaped samples. (a) Prolate spheroid. (b) Oblate spheroid.

principal axis of the ellipsoid, $\bar{\mathbf{L}}$ is a diagonal dyadic with elements L_{xx}, L_{yy}, and L_{zz}. The sum of these elements is equal to unity. For prolate and oblate spheroids as illustrated in Fig. 3.17 we have, respectively,

$$L_{zz} = \frac{b^2}{2a^2} \frac{1}{e^3} \left(\ln \frac{1+e}{1-e} - 2e \right) \tag{129a}$$

$$L_{xx} = L_{yy} = \frac{1}{2}(1 - L_{zz}) \tag{129b}$$

where

$$e = \left(1 - \frac{b^2}{a^2} \right)^{1/2}$$

and

$$L_{zz} = \frac{ab^2}{(b^2 - a^2)^{3/2}} \left[\frac{(b^2 - a)^{1/2}}{a} - \tan^{-1} \frac{(b^2 - a^2)^{1/2}}{a} \right] \tag{130}$$

with L_{xx} and L_{yy} given by (129b) again. For a sphere,

$$L_{xx} = L_{yy} = L_{zz} = \frac{1}{3}.$$

For an ellipsoidal sample with a constant external field H_o applied along the z axis, the effective internal H field is $H_i = H_o - L_{zz}M$. The total magnetic induction field in the sample when both a static internal field $\mathbf{H}_i$ and an internal time-varying field $\mathbf{H}$ exist in the sample is $\mathbf{B} = \mu_0(\mathbf{H}_i + \mathbf{H} + \mathbf{M})$. This is not the local induction field which each spinning electron sees, however. Each spinning electron sees an induction field arising from the applied field, the demagnetization field which depends on sample size and shape, and the field due to all the neighboring spins. A reasonable approximation to make, in practice, is to assume that the partial field contributed by all the neighboring electron spins is collinear and proportional to the total magnetization $\mathbf{M}$. This partial field will not produce a torque on the magnetic moment of the electron, provided the static field is large enough to saturate the sample so that all the electron spins are aligned. It is only the fields that are perpendicular to $\mathbf{M}$ which will produce a torque. In view of these considerations, we may treat the ferrite sample as a single magnetic dipole of strength $\mathbf{M}_s$ per unit volume where the subscript s signifies the saturation dipole moment. The field producing a torque on the system is the partial field $\mathbf{B} = \mu_0(\mathbf{H}_i + \mathbf{H})$. The problem is formally the same as that of a single spinning electron with $\mathbf{m}$ replaced by $\mathbf{M}_s$ and with $\mathbf{H}_i$ and $\mathbf{H}$ as the internal applied fields. If $\mathbf{H}_i$ is directed along the z axis, and $|\mathbf{H}| \ll |\mathbf{H}_i|$, then the component of magnetization M_0 along the z axis is essentially equal to the saturation dipole moment M_s.

For a negative circularly polarized internal field $\mu_0\mathbf{H}_1$ in the xy plane, the negative circularly polarized component of magnetization will be

$$M_- = \frac{\gamma\mu_0 H_1 M_0}{\omega_0 + \omega} \tag{131a}$$

where $\omega_0 = \gamma\mu_0 H_i$. Hence, the intrinsic susceptibility for a negative circularly polarized field is

$$\chi_- = \frac{\gamma\mu_0 M_0}{\omega_0 + \omega} \tag{131b}$$

while the permeability is given by

$$\mu_- = \mu_0 \frac{\gamma\mu_0 M_0 + \omega_0 + \omega}{\omega_0 + \omega}. \tag{131c}$$

For a positive circularly polarized field, the corresponding quantities are

$$\chi_+ = \frac{\gamma\mu_0 M_0}{\omega_0 - \omega} \tag{132a}$$

$$\mu_+ = \mu_0 \frac{\gamma\mu_0 M_0 + \omega_0 - \omega}{\omega_0 - \omega}. \tag{132b}$$

As the applied angular frequency ω approaches the free-precession angular frequency ω_0, the precession angle ϕ approaches $\pi/2$, and the magnetization M_+ approaches[9] $Nm = M_s$. For circularly polarized fields the effective permeabilities are scalar constants.

The solution for the magnetization when a time-varying field $(\mathbf{a}_x H_x + \mathbf{a}_y H_y)e^{j\omega t}$, in addition to the static field $\mathbf{a}_z B_0$, is applied is readily found from the equation of motion

$$\frac{d\mathbf{M}_s}{dt} = -\gamma\mathbf{M}_s \times \mathbf{B} \tag{133}$$

where $\mathbf{M}_s = N\mathbf{m}$ and is the dipole moment per unit volume. If a solution for $\mathbf{M}_s$ of the form

[9]Damping forces are always present, and so this never happens in practice. The resonant frequency is ω_0. In terms of the external fields, the resonant frequency is given by a much more involved relation, which depends on sample shape through the demagnetization factors. For arbitrarily shaped samples the internal field is not uniform, in general, and ferromagnetic resonance does not occur simultaneously throughout the whole sample.

$\mathbf{a}_z M_0 + (\mathbf{a}_x M_x + \mathbf{a}_y M_y + \mathbf{a}_z M_z)e^{j\omega t}$ is assumed, the following three equations are obtained from (133):

$$j\omega M_x = \gamma(\mu_0 M_0 H_y - M_y \mu_0 H_i + \mu_0 M_z H_y e^{j\omega t}) \tag{134a}$$

$$j\omega M_y = \gamma(M_x \mu_0 H_i - \mu_0 M_0 H_x - \mu_0 M_z H_x e^{j\omega t}) \tag{134b}$$

$$j\omega M_z = \mu_0 \gamma(M_y H_x - M_x H_y)e^{j\omega t}. \tag{134c}$$

If H_x and H_y are very small compared with H_i, then M_z is small in comparison with M_x and M_y. Also from (134c) it is seen that M_z varies at a rate ω relative to the applied field, and, hence, we may neglect M_z for a first approximation. For the same reason the terms multiplied by $e^{j\omega t}$ in (134a) and (134b) may be neglected. With this simplification the solutions for the time-varying magnetization components obtained from (134) are

$$M_x = \frac{\mu_0 \gamma \omega_0 M_0 H_x + j\omega \gamma \mu_0 M_0 H_y}{\omega_0^2 - \omega^2} \tag{135a}$$

$$M_y = \frac{\mu_0 \gamma \omega_0 M_0 H_y - j\omega \gamma \mu_0 M_0 H_x}{\omega_0^2 - \omega^2} \tag{135b}$$

$$M_z = 0. \tag{135c}$$

Thus the effective susceptibility tensor or dyadic is

$$\bar{\mathbf{X}}_m = \mathbf{a}_x \mathbf{a}_x \frac{\gamma \mu_0 \omega_0 M_0}{\omega_0^2 - \omega^2} + \mathbf{a}_x \mathbf{a}_y \frac{j\omega \gamma \mu_0 M_0}{\omega_0^2 - \omega^2} - \mathbf{a}_y \mathbf{a}_x \frac{j\omega \gamma \mu_0 M_0}{\omega_0^2 - \omega^2} + \mathbf{a}_y \mathbf{a}_y \frac{\gamma \mu_0 \omega_0 M_0}{\omega_0^2 - \omega^2} \tag{136}$$

and is antisymmetric. This latter property accounts for the nonreciprocal properties of wave propagation in ferrites. The permeability dyadic is given by

$$\bar{\mu} = \mu_0(\bar{\mathbf{I}} + \bar{\mathbf{X}}_m) \tag{137}$$

where $\bar{\mathbf{I}}$ is the unit dyadic or idemfactor. When $H_y = 0$ and

$$H_x = \mathrm{Re}\,(H_1 e^{j\omega t}) = H_1 \cos \omega t$$

the magnetization components are given by

$$M_x = \chi_{xx} \cos \omega t \qquad \text{and} \qquad M_y = \frac{\omega}{\omega_0}\chi_{xx} \sin \omega t$$

where $\chi_{xx} = \gamma \mu_0 \omega_0 M_0/(\omega_0^2 - \omega^2)$. The two components M_x and M_y constitute a positive elliptically polarized magnetization vector, as in Fig. 3.18, and this is the projection of the total precessing magnetization vector onto the xy plane. The projection of $\mathbf{M}$ along the z axis is readily seen to result in a small z component of magnetization varying at a rate 2ω in accordance with (134c). Under the influence of large time-varying fields, harmonics are generated which have appreciable amplitudes, and ferrites may be employed as harmonic generators [3.7] (see also [3.8]).

The magnetization vector may be decomposed into a positive circularly polarized component $M_+ = (\chi_{xx}/2)(1 + \omega/\omega_0)$ and a negative circularly polarized component $M_- = (\chi_{xx}/2)(1 -$

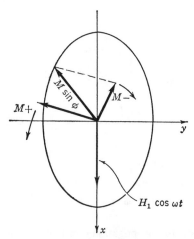

Fig. 3.18. Projection of **M** on the xy plane.

ω/ω_0) in the xy plane. Similarly, $\mathbf{H}_x$ may be decomposed into oppositely rotating vectors each of amplitude $H_1/2$. The effective susceptibilities for the two circularly polarized components of $\mathbf{H}_x$ are $\chi_+ = \chi_{xx}(1+\omega/\omega_0) = \gamma\mu_0 M_0/(\omega_0-\omega)$ and $\chi_- = \chi_{xx}(1-\omega/\omega_0) = \gamma\mu_0 M_0/(\omega_0+\omega)$, and have the same form as given earlier by (132a) and (131b). Below the resonant frequency, that is, $\omega < \omega_0$, a magnetic field $\mathbf{H}_x$ results in a magnetization along the y axis which lags the applied field by one quarter of a period.

Above the resonant frequency, i.e., for $\omega > \omega_0$, χ_{xx} is negative and, hence, χ_+ is negative but χ_- is still positive. This means that $\mathbf{M}$ and the positive circularly polarized component of the field are $180°$ out of phase as indicated in Fig. 3.19 rather than as shown in Fig. 3.16(b). This orientation is permissible since it gives a torque to cause $\mathbf{M}$ to precess in the same direction as $\mathbf{H}_t$. For the negative circularly polarized field, $\mathbf{M}$ and $\mathbf{H}_t$ are always in phase. This change in polarity of χ_+ for $\omega > \omega_0$ causes the major axis of the magnetization ellipse to be along the y axis when a linearly polarized sinusoidally varying field is applied along the x axis. Below the resonant frequency μ_+ is greater than μ_-, while the opposite is true for $\omega > \omega_0$.

For practical ferrite materials, damping forces which tend to reduce the precession angle ϕ are always present. The general form of the permeability matrix must now be represented as

$$[\mu] = \begin{bmatrix} \mu'-j\mu'' & j(\kappa'-j\kappa'') & 0 \\ -j(\kappa'-j\kappa'') & \mu'-j\mu'' & 0 \\ 0 & 0 & \mu_0 \end{bmatrix} = \begin{vmatrix} \mu & j\kappa & 0 \\ -j\kappa & \mu & 0 \\ 0 & 0 & \mu_0 \end{vmatrix} \qquad (138)$$

where $\mu = \mu'-j\mu''$, etc. The terms which are double-primed arise because of the damping forces which give rise to finite losses. The damping forces are proportional to the precession angle ϕ and, hence, are greater for a positive circularly polarized field than for a negative circularly polarized field. This leads to selective absorption of the positive circularly polarized component, particularly so if ω is close to ω_0 so that the precession angle ϕ is large.

Our earlier discussion on the magnetization resulting from circularly polarized fields leads us to expect that the only solutions to Maxwell's equations for plane waves propagating along the direction of the applied static field $\mathbf{H}_i$ will be circularly polarized waves, since for such waves the effective permeability is a scalar quantity. A linearly polarized wave may be decomposed into positive and negative circularly polarized waves, and, since the effective permeability for

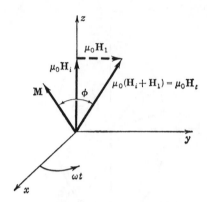

Fig. 3.19. Space relation between $\mathbf{M}$ and $\mu_0\mathbf{H}_t$ for $\omega > \omega_0$.

the two polarizations is different, the propagation constants will also be different. Thus, as the wave propagates along the direction of $\mathbf{H}_i$, a linearly varying phase difference between the two polarizations results. This causes the plane of polarization of the resultant wave to rotate, a phenomenon known as Faraday rotation. In ferrites Faraday rotation is a nonreciprocal effect, since the plane of polarization always rotates in the same direction looking in the direction of $\mathbf{H}_i$ whether the wave is propagating parallel or antiparallel to $\mathbf{H}_i$.

Consider an unbounded ferrite medium with an internal static magnetic field $\mathbf{H}_i$ applied along the z axis. For plane TEM waves propagating in the z direction, Maxwell's equations reduce to $\Gamma\mathbf{E}_t e^{-\Gamma z} = -j\omega\mathbf{a}_z \times \mathbf{B}_t e^{-\Gamma z}$ and $\mathbf{H}_t e^{-\Gamma z} = j\omega\epsilon\mathbf{a}_z \times \mathbf{E}_t e^{-\Gamma z}$, where a propagation constant Γ has been assumed and $\mathbf{E}_t$, etc., are vectors in the xy plane. Introducing the relation $\bar{\mu}\cdot\mathbf{H}_t = \mathbf{B}_t$, we get

$$\Gamma^2\bar{\mu}\cdot\mathbf{H}_t = \Gamma^2\mathbf{B}_t = j\omega\epsilon\Gamma\bar{\mu}\cdot\mathbf{a}_z \times \mathbf{E}_t = j\omega\epsilon\bar{\mu}\cdot\mathbf{a}_z \times (-j\omega\mathbf{a}_z \times \mathbf{B}_t) = -\omega^2\epsilon\bar{\mu}\cdot\mathbf{B}_t. \quad (139)$$

Equating the x and y components separately gives

$$(\Gamma^2 + k^2)B_x + j\omega^2\epsilon\kappa B_y = 0 \qquad (140a)$$

$$-j\omega^2\epsilon\kappa B_x + (\Gamma^2 + k^2)B_y = 0 \qquad (140b)$$

where $k^2 = \omega^2\epsilon\mu$. For a solution, the determinant must vanish; hence, $(\Gamma^2+k^2)^2-(\omega^2\epsilon\kappa)^2 = 0$, and thus the solutions for Γ are

$$\Gamma_\pm = j\omega\epsilon^{1/2}(\mu \pm \kappa)^{1/2}. \qquad (141)$$

For the root Γ_+, (140b) gives $B_y = -jB_x$, which represents a positive circularly polarized wave with an effective permeability $\mu + \kappa$. The root Γ_- gives $B_y = jB_x$, which is a negative circularly polarized wave with an effective permeability $\mu - \kappa$. The effective permeabilities are of the same form as those derived earlier for circularly polarized fields, provided the losses are negligible.

At the plane $z = 0$, let the time-varying magnetic field be

$$\mathbf{B} = \mathbf{a}_x 2B_1 \operatorname{Re} e^{j\omega t}$$

where Re signifies the real part. The field $\mathbf{B}$ may be decomposed into a positive circularly polarized field

$$\mathbf{B}_+ = B_1 \operatorname{Re} (\mathbf{a}_x e^{j(\omega t - \beta_+ z)} - j\mathbf{a}_y e^{j(\omega t - \beta_+ z)}) \tag{142a}$$

and a negative circularly polarized field

$$\mathbf{B}_- = B_1 \operatorname{Re} (\mathbf{a}_x e^{j(\omega t - \beta_- z)} + j\mathbf{a}_y e^{j(\omega t - \beta_- z)}) \tag{142b}$$

where $j\beta_\pm = \Gamma_\pm$. At a plane $z = l$, the resultant field is

$$\mathbf{B} = B_1 \operatorname{Re} [\mathbf{a}_x (e^{-j\beta_+ l} + e^{-j\beta_- l}) - j\mathbf{a}_y (e^{-j\beta_+ l} - e^{-j\beta_- l})] e^{j\omega t}$$

$$= 2B_1 \operatorname{Re} e^{j\omega t - j(\beta_+ + \beta_-)l/2} \left[\mathbf{a}_x \cos (\beta_+ - \beta_-)\frac{l}{2} - \mathbf{a}_y \sin (\beta_+ - \beta_-)\frac{l}{2} \right].$$

If the losses are negligible, $\beta_\pm$ is real, and the resultant field makes an angle θ given by

$$\theta = (\beta_- - \beta_+)\frac{l}{2} \tag{143}$$

with the x axis, and, hence, the plane of polarization has been rotated through an angle θ. The Faraday rotation per unit length is thus $(\frac{1}{2})(\beta_- - \beta_+)$. For $\omega > \omega_0$, the effective permeability for the positive circularly polarized wave is very small and $\beta_- > \beta_+$; hence, the Faraday rotation is in a clockwise sense about $\mathbf{H}_i$. Below the resonant frequency, $\beta_+ > \beta_-$ and the sense of rotation is reversed. For a wave propagating in the negative z direction, the signs in front of $\beta_\pm$ and $\mathbf{a}_y$ in (142a) and (142b) must be changed. A similar analysis then shows that the plane of polarization again undergoes rotation through an angle given by (143). This nonreciprocal Faraday-rotation property has been used to construct nonreciprocal microwave components called gyrators, isolators, and circulators [3.6]. It should be noted that, for $\omega > \omega_0$ and a sufficiently strong applied static field, μ_+ may be negative; hence, β_+ is imaginary, and the positive circularly polarized wave does not propagate. In practice, finite losses are always present, and so Γ_+ will always have an imaginary part, although this imaginary part may be very small under certain conditions. The region $\omega \gg \omega_0$ and weak applied static fields are favored for gyrators because of the large Faraday rotation together with small losses which may be obtained under these conditions.

3.9. Dyadic Green's Function for Layered Media

The radiation from a current element located in a homogeneous layer in a multilayered medium is important in connection with a number of physical problems such as antennas located over a stratified earth, microstrip antennas involving planar conducting patches on a substrate material, microstrip transmission lines, etc. In this section we will derive an expression for the electric field dyadic Green's function for a source located in a homogeneous layer and having one or more additional homogeneous layers on either side. In Fig. 3.20 we show a current element $\mathbf{J} = \mathbf{I}_0 \delta(x - x')\delta(y - y')\delta(z - z')$ located in a layer with electrical parameters ϵ and μ. The layer extends from $z = 0$ to $z = d$. The adjacent layers have electrical parameters ϵ_i, μ_i for the ith layer as shown in the figure.

We will first find the primary field produced by $\mathbf{J}$ as though the layer filled all of space. We can then superimpose solutions to the homogeneous vector wave equation in order to satisfy

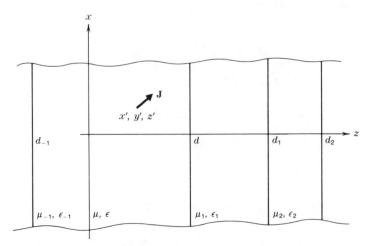

Fig. 3.20. A current source in a multilayered medium.

the boundary conditions at each interface. We will show that outside the source region the field consists of a superposition of transverse electric (TE) and transverse magnetic (TM) waves with respect to the interface normals and that the boundary conditions at the source can be split into separate conditions for the two types of waves that are excited. The ability to decouple the TE and TM waves in the source region is very convenient in practice. In some practical applications, such as to microstrip transmission-line structures, edge conditions at the edges of conducting strips will couple the TE and TM waves so that these waves do not generally exist by themselves. The microstrip problem is discussed in a later chapter. When the TE and TM waves are coupled the resulting waves are often called hybrid waves; i.e., they are no longer pure TE or TM waves.

The TE and TM waves are TEM waves that propagate at an angle to the z axis. These TEM waves were classified as parallel- and perpendicular-polarized waves in Section 3.5. The classification of these waves as TE and TM waves with respect to the z axis is a more common classification. The TE wave does not have a z component of electric field but does have a z component of magnetic field and for this reason is also called an H wave. Similarly, the TM wave is called an E wave because E_z is not zero but $H_z = 0$. A localized current source will radiate a continuous spectrum of plane waves propagating at all real angles with respect to the z axis as well as waves propagating at complex angles with respect to z. The radiated spectrum of plane waves can be synthesized by using a two-dimensional Fourier transform with respect to x and y to solve the radiation problem.

By means of a two-dimensional Fourier transform with respect to x and y we can represent the current source as sheets of current located in the $z = z'$ plane. Thus we have

$$\mathbf{J}(x, y, z) = [\mathbf{J}_{0t}(x, y) + \mathbf{J}_{0z}(x, y)]\delta(z - z')$$

$$= \mathbf{I}_0\delta(z - z')\frac{1}{4\pi^2}\int\!\!\!\int_{-\infty}^{\infty} e^{-ju(x-x')-jv(y-y')}\,du\,dv$$

where we have used

$$\delta(x - x') = \frac{1}{2\pi}\int_{-\infty}^{\infty} e^{-ju(x-x')}\,du$$

and similarly for $\delta(y - y')$. The fields have the representation

$$E(x, y, z) = \frac{1}{4\pi^2} \int\int_{-\infty}^{\infty} \hat{E}(u, v, z)e^{-jux-jvy}\,du\,dv$$

and similarly for $H(x, y, z)$. In the Fourier transform or spectral domain the current is given by

$$\hat{J}(u, v, z) = I_0 e^{jux'+jvy'}\delta(z - z') = [\hat{J}_{0t} + a_z\hat{J}_{0z}]\delta(z - z'). \tag{144}$$

The TE waves can be found from an **M**-type wave function as follows:

$$E = \nabla \times a_z\psi = -a_z \times \nabla\psi = a_x\frac{\partial\psi}{\partial y} - a_y\frac{\partial\psi}{\partial x} \tag{145a}$$

$$-j\omega\mu H = \nabla \times \nabla \times a_z\psi = \nabla\frac{\partial\psi}{\partial z} - a_z\nabla^2\psi = \nabla_t\frac{\partial\psi}{\partial z} - a_z\nabla_t^2\psi \tag{145b}$$

where ∇_t represents the x and y parts of ∇. In the Fourier transform domain

$$\hat{E}(u, v, z) = -jv\hat{\psi}(u, v, z)a_x + ju\hat{\psi}(u, v, z)a_y \tag{146a}$$

$$jkZ\hat{H} = (jua_x + jva_y)\frac{\partial\hat{\psi}}{\partial z} - a_z(u^2 + v^2)\hat{\psi}. \tag{146b}$$

The appropriate solution for $\hat{\psi}$ that will represent waves propagating away from the source is

$$\hat{\psi} = \begin{cases} C^+e^{-j\beta(z-z')}, & z > z' \\ C^-e^{j\beta(z-z')}, & z < z' \end{cases} \tag{147}$$

where $\beta = (k^2 - u^2 - v^2)^{1/2}$, $k = \omega\sqrt{\mu\epsilon}$, $Z = \sqrt{\mu/\epsilon}$.

The TM waves can be constructed from an **N**-type wave function by means of the relations

$$E = \nabla \times \nabla \times a_z\Phi = \nabla_t\frac{\partial\Phi}{\partial z} - \nabla_t^2\Phi a_z \tag{148a}$$

$$-jkZH = a_z \times \nabla^2 \nabla\Phi = -k^2a_z \times \nabla\Phi. \tag{148b}$$

In the Fourier transform domain we have

$$\hat{E} = (-jua_x - jva_y)\frac{\partial\hat{\Phi}}{\partial z} + (u^2 + v^2)a_z\hat{\Phi} \tag{149a}$$

$$-jZ\hat{H} = (-jva_x + jua_y)k\hat{\Phi}. \tag{149b}$$

The solution for $\hat{\Phi}$ is of the form

$$\hat{\Phi} = \begin{cases} D^+e^{-j\beta(z-z')}, & z > z' \\ D^-e^{j\beta(z-z')}, & z < z'. \end{cases} \tag{150}$$

For nonpropagating waves β is a negative imaginary quantity. Both ψ and Φ are solutions of the scalar Helmholtz equation.

One component of the current source is a sheet of z-directed current given by

$$\hat{J}_{0z}\delta(z - z') = I_{0z}\delta(z - z')e^{jux'+jvy'}.$$

This current sheet excites TM waves only since the TE waves do not have an axial component of electric field. A TE wave does not interact with an axial current element and by reciprocity such a current element does not radiate a TE wave. For a sheet of normally directed current the boundary condition given by (46) in Chapter 1 applies. Thus we have

$$\mathbf{a}_z \times (\mathbf{E}^+ - \mathbf{E}^-) = \frac{1}{j\omega\epsilon}\nabla \times \mathbf{a}_z J_{0z} = -\frac{1}{j\omega\epsilon}\mathbf{a}_z \times \nabla J_{0z}$$

or

$$\mathbf{a}_z \times (\hat{\mathbf{E}}^+ - \hat{\mathbf{E}}^-) = -\frac{1}{\omega\epsilon}(v\mathbf{a}_x - u\mathbf{a}_y)I_{0z}e^{jux'+jvy'}$$

in the spectral domain. When we use (149a) and (150) we obtain

$$\beta(D^+ + D^-) = -\frac{1}{\omega\epsilon}I_{0z}e^{jux'+jvy'}.$$

For the z-directed current the tangential magnetic field is continuous across the current sheet so from (149b) we must also have $D^+ = D^-$. Consequently, from the z-directed current sheet we obtain a contribution to $\hat{\Phi}$ with amplitude that we now label as D_n where

$$D_n^+ = D_n^- = D_n = -\frac{I_{0z}Z}{2k\beta}e^{jux'+jvy'}. \tag{151}$$

The subscript n signifies a contribution from the normally directed current. The current source J_{0z} acts like a series voltage source because it requires the tangential magnetic field to be continuous and hence the tangential electric field will be discontinuous across the current sheet.

The total tangential magnetic field must be discontinuous by an amount determined by the tangential component $\mathbf{J}_{0t}$ of the current sheet. For this part of the field the tangential electric field is continuous across the current sheet. In the spatial domain the first boundary condition gives

$$\mathbf{a}_z \times \left(\nabla_t\frac{\partial\psi}{\partial z} - \mathbf{a}_z\nabla_t^2\psi - k^2\mathbf{a}_z \times \nabla\Phi\right)\Bigg|_{z'_-}^{z'_+} = -jkZ\mathbf{J}_{0t}$$

or

$$\left(-\nabla_t \times \mathbf{a}_z\frac{\partial\psi}{\partial z} + k^2\nabla_t\Phi\right)\Bigg|_{z'_-}^{z'_+} = -jkZ\mathbf{J}_{0t}.$$

The transverse divergence of this equation gives

$$k\nabla_t^2\Phi\Bigg|_{z'_-}^{z'_+} = -jZ\nabla_t\cdot\mathbf{J}_{0t} \tag{152}$$

while the transverse curl gives

$$\mathbf{a}_z \nabla_t^2 \left. \frac{\partial \psi}{\partial z} \right|_{z'_-}^{z'_+} = -jkZ\nabla_t \times \mathbf{J}_{0t}. \tag{153}$$

The source term in (152) is the charge density associated with $\nabla_t \cdot \mathbf{J}_{0t}$, while the source term in (153) is a vortex source (equivalent to a z-directed magnetic current). In the spectral domain, the boundary conditions (152) and (153) become

$$k(u^2 + v^2)(D^+ - D^-) = Z(uI_{0x} + vI_{0y})e^{jux' + jvy'} \tag{154a}$$

$$\beta(u^2 + v^2)(C^+ + C^-) = jkZ(uI_{0y} - vI_{0x})e^{jux' + jvy'}. \tag{154b}$$

The continuity of the tangential electric field requires that

$$\nabla_t \frac{\partial \Phi}{\partial z} + \nabla \times \mathbf{a}_z \psi$$

be continuous across the $z = z'$ plane. Since one component is given by a gradient operation and the other by a curl operation we require each term to be separately continuous. Thus we find that

$$D^+ = -D^- \quad \text{and} \quad C^+ = C^-.$$

We will denote the solutions for the amplitude coefficients due to the transverse current sheets with a subscript t. These solutions are given by

$$D_t^+ = -D_t^- = D_t = \frac{Z(uI_{0x} + vI_{0y})}{2k(u^2 + v^2)} e^{jux' + jvy'} \tag{155a}$$

$$C_t^+ = C_t^- = C_t = \frac{jkZ(uI_{0y} - vI_{0x})}{2\beta(u^2 + v^2)} e^{jux' + jvy'}. \tag{155b}$$

The complete solutions for $\hat{\psi}$ and $\hat{\Phi}$ due to the primary excitation are

$$\hat{\psi} = C_t e^{-j\beta|z - z'|} \tag{156a}$$

$$\hat{\Phi} = [D_t \operatorname{sg}(z - z') + D_n]e^{-j\beta|z - z'|}. \tag{156b}$$

In the spectral domain the transverse electric and magnetic fields are given by (146) and (149) and are

$$\hat{\mathbf{E}}_t = (-jv\mathbf{a}_x + ju\mathbf{a}_y)C_t e^{-j\beta|z - z'|}$$
$$\quad - \beta(u\mathbf{a}_x + v\mathbf{a}_y)[D_t + D_n \operatorname{sg}(z - z')]e^{-j\beta|z - z'|} \tag{157a}$$

$$\hat{\mathbf{H}}_t = \frac{\beta}{kZ}(-ju\mathbf{a}_x - jv\mathbf{a}_y)C_t \operatorname{sg}(z - z')e^{-j\beta|z - z'|}$$
$$\quad + \frac{k}{Z}(v\mathbf{a}_x - u\mathbf{a}_y)[D_t \operatorname{sg}(z - z') + D_n]e^{-j\beta|z - z'|} \tag{157b}$$

where $\text{sg}(z - z')$ equals $+1$ for $z > z'$ and -1 for $z < z'$. The wave impedance for a TE wave is given by

$$Z_h = \frac{E_x}{H_y} = -\frac{E_y}{H_x} = \frac{kZ}{\beta} \tag{158a}$$

while for a TM wave it is given by

$$Z_e = \frac{\beta}{k} Z. \tag{158b}$$

The corresponding wave admittances are Y_h and Y_e and are the reciprocals of the wave impedances. The transverse fields for the plane wave spectrum are obtained by multiplying (157) by $(4\pi^2)^{-1} e^{-jux-jvy}$. The complete solution for the transverse fields associated with the primary excitation is obtained by integrating over u and v.

We have emphasized the solution for the transverse fields because one convenient way to obtain the solution for the fields in the layered medium is to construct an equivalent transmission-line circuit for the problem. We will present this approach before we proceed to complete the solution for the dyadic Green's function. The transmission-line approach has been extensively discussed in the text by Felsen and Marcuvitz for a wide variety of problems [3.20].

In order to develop the transmission-line model we introduce the following transverse mode functions for the E and H waves:

$$\mathbf{e}_h = \nabla_t \times \mathbf{a}_z e^{-jux-jvy} = -j(v\mathbf{a}_x - u\mathbf{a}_y)e^{-jux-jvy} \tag{159a}$$

$$\mathbf{e}_e = \nabla_t e^{-jux-jvy} = -j(u\mathbf{a}_x + v\mathbf{a}_y)e^{-jux-jvy} \tag{159b}$$

$$\mathbf{h}_h = \mathbf{a}_z \times \mathbf{e}_h \tag{159c}$$

$$\mathbf{h}_e = \mathbf{a}_z \times \mathbf{e}_e. \tag{159d}$$

The z component of current is an independent source. For this current the radiated transverse fields that are part of the spectrum of plane waves may be obtained from (157) after multiplying by $e^{-jux-jvy}$ and dividing by $4\pi^2$. We can describe this field in the following way:

$$\mathbf{E}_t = \begin{cases} V_n^+ \mathbf{e}_e e^{-j\beta(z-z')}, & z > z' \\ V_n^- \mathbf{e}_e e^{j\beta(z-z')}, & z < z' \end{cases} \tag{160a}$$

$$\mathbf{H}_t = \begin{cases} I_n^+ \mathbf{h}_e e^{-j\beta(z-z')}, & z > z' \\ -I_n^- \mathbf{h}_e e^{j\beta(z-z')}, & z < z' \end{cases} \tag{160b}$$

where $I_n^\pm = V_n^\pm Z_e$ and

$$V_n^+ = -V_n^- = \frac{-j\beta D_n}{4\pi^2} = \frac{jZI_{0z}e^{jux'+jvy'}}{8\pi^2 k} = \frac{V_g}{2}. \tag{161}$$

Apart from the mode functions, Eqs. (160) describe a transmission line with characteristic impedance equal to the wave impedance Z_e and driven by a series voltage source of strength V_g given by (161). This transmission-line circuit is shown in Fig. 3.21(a). In Fig. 3.21(b) we show the equivalent transmission-line circuit for a layered medium driven by a z-directed

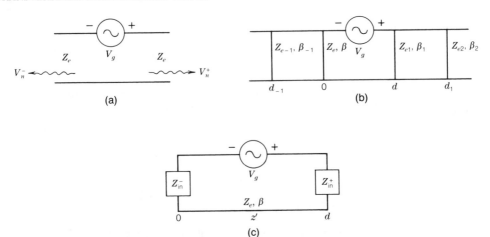

Fig. 3.21. (a) Equivalent transmission-line circuit for primary excitation by a z-directed current element. (b) Equivalent transmission-line circuit for a multilayered medium. (c) Equivalent transmission-line circuit for the layer containing the source.

current element. Each transmission-line section has a characteristic impedance equal to the wave impedance. For the ith section

$$Z_{ei} = \frac{\beta_i}{k_i} Z_i = \frac{\beta_i}{\omega \epsilon_i}$$

where the propagation constant β_i is given by

$$\beta_i = \sqrt{k_i^2 - u^2 - v^2}, \qquad k_i = \omega \sqrt{\mu_i \epsilon_i}.$$

Transmission-line theory requires the total voltage and current to be continuous across each interface or junction. This is equivalent to requiring that the tangential electric and magnetic fields be continuous across each interface. Consequently, we can use transmission-line theory and wave matrices as described in Sections 3.3–3.5 to determine the voltage and current wave amplitudes in each layer. The fields are obtained by multiplying by the appropriate mode functions for each layer. For example, if we want to find the fields in the region $0 < z < d$ we can use transmission-line theory to determine the input impedances Z_{in}^+ and Z_{in}^- at $z = d$ and $z = 0$, respectively. The equivalent circuit is thus reduced to that shown in Fig. 3.21(c). A solution for the total voltage wave in the region $0 < z < d$ is of the form

$$V(z) = \begin{cases} V_1^+ e^{-j\beta z} + V_1^- e^{j\beta z}, & z < z' \\ V_2^+ e^{-j\beta z} + V_2^- e^{j\beta z}, & z > z'. \end{cases}$$

The boundary conditions at $z = 0, d$ require that

$$\frac{V_1^+}{V_1^-} = R_1 = \frac{Z_{in}^- - Z_e}{Z_{in}^- + Z_e}$$

$$\frac{V_2^-}{V_2^+} e^{2j\beta d} = R_2 = \frac{Z_{in}^+ - Z_e}{Z_{in}^+ + Z_e}.$$

The boundary conditions at the source are

$$V_1^+ e^{-j\beta z'} + V_1^- e^{j\beta z'} + V_g = V_2^+ e^{-j\beta z'} + V_2^- e^{j\beta z'}$$
$$I_1^+ e^{-j\beta z'} - I_1^- e^{j\beta z'} = I_2^+ e^{-j\beta z'} - I_2^- e^{j\beta z'}.$$

These equations may be solved for the voltage amplitudes and give

$$V_2^+ = \frac{(e^{j\beta z'} - R_1 e^{-j\beta z'})V_g}{2(1 - R_1 R_2 e^{-2j\beta d})}, \qquad\qquad V_2^- = R_2 V_2^+ e^{-2j\beta d}$$

$$V_1^- = -\frac{(e^{-j\beta z'} - R_2 e^{-2j\beta d + j\beta z'})V_g}{2(1 - R_1 R_2 e^{-2j\beta d})}, \qquad V_1^+ = R_1 V_1^-.$$

By setting $R_1 = R_2 = 0$ we obtain the primary excitation given by (160). Note that the reflection coefficients R_1 and R_2 are functions of u and v as is the source strength V_g. The total transverse electric field in the region $0 < z < d$ may be obtained by multiplying the voltage waves by the transverse mode function $\mathbf{e}_e$ and integrating over u and v. The inverse Fourier transform involves a denominator which, in some circumstances, can equal zero for certain values of u and v. When the denominator vanishes a wave known as a surface wave is excited. We will not discuss surface waves at this point since these will be discussed in a later chapter.

The fields excited by the transverse component of the current may also be determined by using an equivalent transmission-line model. TE and TM waves can be treated as separate cases since we were able to obtain uncoupled source boundary conditions for these two classes of waves. For the TM or E waves we can deduce from (157) that

$$\mathbf{E}_t = \begin{cases} I_e Z_e \mathbf{e}_e e^{-j\beta(z-z')}, & z > z' \\ I_e Z_e \mathbf{e}_e e^{j\beta(z-z')}, & z < z' \end{cases} \tag{162a}$$

$$\mathbf{H}_t = \begin{cases} I_e \mathbf{h}_e e^{-j\beta(z-z')}, & z > z' \\ -I_e \mathbf{h}_e e^{j\beta(z-z')}, & z < z' \end{cases} \tag{162b}$$

where $I_e = I_{ge}/2$ and the shunt current source has a strength

$$I_{ge} = -j\frac{uI_{0x} + vI_{0y}}{4\pi^2(u^2 + v^2)}e^{jux' + jvy'}. \tag{163}$$

These equations describe the equivalent transmission-line circuit shown in Fig. 3.22(a). The equivalent circuit for the layered medium is shown in Fig. 3.22(b). In the region $0 \le z \le d$ the total voltage and current waves excited by the source I_{ge} are given by

$$V_e = \begin{cases} V_{1e}^+ e^{-j\beta z} + V_{1e}^- e^{j\beta z}, & z < z' \\ V_{2e}^+ e^{-j\beta z} + V_{2e}^- e^{j\beta z}, & z > z' \end{cases} \tag{164}$$

$$I_e = \begin{cases} I_{1e}^+ e^{-j\beta z} - I_{1e}^- e^{j\beta z}, & z < z' \\ I_{2e}^+ e^{-j\beta z} - I_{2e}^- e^{j\beta z}, & z > z'. \end{cases} \tag{165}$$

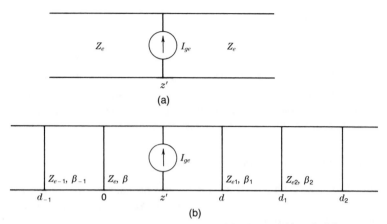

Fig. 3.22. (a) Equivalent transmission-line circuit for the primary field excited by a transverse current element. (b) Equivalent transmission-line circuit for the layered medium. For TE waves Z_{ei} and I_{ge} are replaced by Z_{hi} and I_{gh}.

The boundary conditions at $z = 0, d$ are the same as those for excitation by $\hat{J}_{0z}$. The source boundary conditions are

$$V_{1e}^{+}e^{-j\beta z'} + V_{1e}^{-}e^{j\beta z'} = V_{2e}^{+}e^{-j\beta z'} + V_{2e}^{-}e^{j\beta z'}$$

$$I_{2e}^{+}e^{-j\beta z'} - I_{2e}^{-}e^{j\beta z'} - I_{1e}^{+}e^{-j\beta z'} + I_{1e}^{-}e^{j\beta z'} = I_{ge}$$

where

$$I_{1e}^{\pm}Z_e = V_{1e}^{\pm}, \qquad I_{2e}^{\pm}Z_e = V_{2e}^{\pm}.$$

By means of transmission-line theory we readily find that

$$V_{2e}^{+} = \frac{I_{ge}Z_e(e^{j\beta z'} + R_1 e^{-j\beta z'})}{2(1 - R_1 R_2 e^{-2j\beta d})}, \qquad V_{2e}^{-} = R_2 e^{-2j\beta d} V_{2e}^{+}$$

$$V_{1e}^{-} = \frac{I_{ge}Z_e(e^{-j\beta z'} + R_2 e^{-2j\beta d + j\beta z'})}{2(1 - R_1 R_2 e^{-2j\beta d})}, \qquad V_{1e}^{+} = R_1 V_{1e}^{-}.$$

Transmission-line theory may also be used to find the voltage and current waves in any other layer. The transverse fields are obtained by multiplying by the mode functions given by (159). The complete spectrum of excited waves is obtained by integrating over u and v.

The excitation of TE or H waves by the transverse component of the current may be found in a similar way from an equivalent transmission-line circuit. The relevant equations for the transverse fields are the same as (162) but with the mode functions $\mathbf{e}_e$ and $\mathbf{h}_e$ replaced by $\mathbf{e}_h$ and $\mathbf{h}_h$ and with the equivalent current amplitude I_e relabeled as I_h. The shunt current source now has a strength I_{gh} given by

$$I_{gh} = -\frac{j(vI_{0x} - uI_{0y})}{4\pi^2(u^2 + v^2)} e^{jux' + jvy'}. \tag{166}$$

In addition, the characteristic impedance of each transmission-line section for the multilayered structure is the corresponding wave impedance for TE waves in that section. These are given by

$$Z_{hi} = \frac{k_i Z_i}{\beta_i} = \frac{\omega \mu_i}{\beta_i}. \tag{167}$$

The reflection coefficients R_1 and R_2 also change when Z_e is replaced by Z_h. The solution for the total voltage waves in the region $0 \leq z \leq d$ is the same as that for TM waves after the appropriate changes in characteristic impedances for each layer are made, i.e., replace all Z_{ei} by Z_{hi}. For this reason we do not include those equations here.

Dyadic Green's Function for Primary Excitation

The electric field dyadic Green's function for the primary excitation (current source in a homogeneous medium) is readily found in terms of **M**, **N**, and **L** functions. These vector eigenfunctions are obtained from the equations

$$\mathbf{L} = \nabla \psi \tag{168a}$$

$$\mathbf{M} = \nabla \times \mathbf{a}_z \psi \tag{168b}$$

$$\sqrt{\lambda}\,\mathbf{N} = \nabla \times \nabla \times \mathbf{a}_z \psi \tag{168c}$$

where ψ is a solution of

$$(\nabla^2 + \lambda)\psi = 0. \tag{169}$$

The appropriate solution for ψ is

$$\psi = e^{-jux - jvy - jwz}.$$

The solution for the electric field is given by

$$\mathbf{E}(\mathbf{r}) = \int\!\!\!\int\!\!\!\int_{-\infty}^{\infty} (A\mathbf{M} + B\mathbf{N} + C\mathbf{L})\, du\, dv\, dw \tag{170}$$

where A, B, and C are expansion coefficients that are functions of u, v, and w. The functions **L**, **M**, and **N** are easily shown to be mutually orthogonal when integrated over all values of x, y, and z (see Problem 3.12). The normalization integrals are

$$\int\!\!\!\int\!\!\!\int_V \mathbf{L}(u, v, w, \mathbf{r}) \cdot \mathbf{L}^*(u', v', w', \mathbf{r})\, dV$$

$$= 8\pi^3 \lambda \delta(u - u')\delta(v - v')\delta(w - w')$$

$$\int\!\!\!\int\!\!\!\int_V \mathbf{M} \cdot \mathbf{M}^*\, dV = \int\!\!\!\int\!\!\!\int_V \mathbf{N} \cdot \mathbf{N}^*\, dV$$

$$= 8\pi^3 (\lambda - w^2)\delta(u - u')\delta(v - v')\delta(w - w')$$

where $\lambda = u^2 + v^2 + w^2$.

We now substitute the expansion (170) into the equation

$$(\nabla \times \nabla \times \mathbf{E} - k^2\mathbf{E}) = -j\omega\mu\mathbf{J}$$

and use the orthogonality of the eigenfunctions to obtain

$$\mathbf{E}(\mathbf{r}) = \iiint\limits_{V} \iiint\limits_{-\infty}^{\infty} \frac{du\,dv\,dw}{8\pi^3} \left[\frac{-\mathbf{L}(\mathbf{r})\mathbf{L}^*(\mathbf{r}')}{\lambda k^2} \right.$$

$$\left. + \frac{\mathbf{M}(\mathbf{r})\mathbf{M}^*(\mathbf{r}') + \mathbf{N}(\mathbf{r})\mathbf{N}^*(\mathbf{r}')}{(\lambda - w^2)(\lambda - k^2)} \right] \cdot [-j\omega\mu\mathbf{J}(\mathbf{r}')\,dV']. \tag{171}$$

From this expression we can readily identify the dyadic Green's function operator. The integration over w is readily carried out so as to obtain a modal expansion of $\bar{\mathbf{G}}_e$. In the w plane the poles are located where $\lambda = 0$ or $w = \pm j(u^2 + v^2)^{1/2}$ and at $\lambda = k^2$ or $w = \pm(k^2 - u^2 - v^2)^{1/2} = \pm\beta$. In view of the definition (168c) the $\mathbf{NN}^*$ term also has poles at $\lambda = 0$. The residues at these poles cancel the contribution from the $\mathbf{L}$ functions outside the source region. The integral to be evaluated is

$$\bar{\mathbf{G}}_e(\mathbf{r}, \mathbf{r}') = \frac{1}{8\pi^3} \iint_{-\infty}^{\infty} e^{-ju(x-x')-jv(y-y')}\,du\,dv$$

$$\cdot \int_{-\infty}^{\infty} \left[\frac{(u\mathbf{a}_x + v\mathbf{a}_y + w\mathbf{a}_z)(u\mathbf{a}_x + v\mathbf{a}_y + w\mathbf{a}_z)}{-k^2(u^2 + v^2 + w^2)} + \frac{(v\mathbf{a}_x - u\mathbf{a}_y)(v\mathbf{a}_x - u\mathbf{a}_y)}{(u^2 + v^2)(w^2 - \beta^2)} \right.$$

$$\left. + \frac{[w(u\mathbf{a}_x + v\mathbf{a}_y) - (u^2 + v^2)\mathbf{a}_z][w(u\mathbf{a}_x + v\mathbf{a}_y) - (u^2 + v^2)\mathbf{a}_z]}{(u^2 + v^2)(u^2 + v^2 + w^2)(w^2 - \beta^2)} \right]$$

$$\cdot e^{-jw(z-z')}\,dw. \tag{172}$$

The contour runs above the pole at $w = \beta$ and below the pole at $w = -\beta$. The contour may be closed by a semicircle of infinite radius in the upper half plane for $z < z'$ and in the lower half plane for $z > z'$. The integral can be evaluated using residue theory for all values of z with one exception. When $z = z'$ one of the terms coming from the $\mathbf{L}$ functions is

$$\frac{-w^2\mathbf{a}_z\mathbf{a}_z}{k^2(w^2 + u^2 + v^2)}$$

for which residue theory cannot be used because Jordan's lemma will not apply. However, this term can be expressed in the form

$$-\frac{\mathbf{a}_z\mathbf{a}_z}{k^2} + \frac{(u^2 + v^2)\mathbf{a}_z\mathbf{a}_z}{k^2(w^2 + u^2 + v^2)}.$$

The first term will then yield a contribution $-(\mathbf{a}_z\mathbf{a}_z/k^2)\delta(\mathbf{r} - \mathbf{r}')$ to $\bar{\mathbf{G}}_e$. The second term can be evaluated using residue theory. We readily find that

$$\bar{\mathbf{G}}_e(\mathbf{r}, \mathbf{r}') = -\frac{\mathbf{a}_z\mathbf{a}_z}{k^2}\delta(\mathbf{r} - \mathbf{r}') - \frac{j}{4\pi^2} \iint_{-\infty}^{\infty} [\mathbf{M}_\sigma(\mathbf{r})\tilde{\mathbf{M}}_\sigma(\mathbf{r}') + \mathbf{N}_\sigma(\mathbf{r})\tilde{\mathbf{N}}_\sigma(\mathbf{r}')]\,du\,dv \tag{173}$$

where the normalized $\mathbf{M}_\sigma$ and $\mathbf{N}_\sigma$ functions are given by

$$\mathbf{M}_\sigma(\mathbf{r}) = \frac{\nabla \times \mathbf{a}_z e^{-jux-jvy}}{[2\beta(u^2+v^2)]^{1/2}} \begin{cases} e^{-j\beta z}, & z > z' \\ e^{j\beta z}, & z < z' \end{cases} = \begin{cases} \mathbf{M}^+(\mathbf{r}), & z > z' \\ \mathbf{M}^-(\mathbf{r}), & z < z' \end{cases}$$

$$\mathbf{N}_\sigma(\mathbf{r}) = \frac{\nabla \times \nabla \times \mathbf{a}_z e^{-jux-jvy}}{k[2\beta(u^2+v^2)]^{1/2}} \begin{cases} e^{-j\beta z}, & z > z' \\ e^{j\beta z}, & z < z' \end{cases} = \begin{cases} \mathbf{N}^+(\mathbf{r}), & z > z' \\ \mathbf{N}^-(\mathbf{r}), & z < z'. \end{cases}$$

The functions $\tilde{\mathbf{M}}_\sigma(\mathbf{r}')$ and $\tilde{\mathbf{N}}_\sigma(\mathbf{r}')$ are given by

$$\tilde{\mathbf{M}}_\sigma(\mathbf{r}') = \frac{\nabla' \times \mathbf{a}_z e^{jux'+jvy'}}{[2\beta(u^2+v^2)]^{1/2}} \begin{cases} e^{-j\beta z'}, & z' > z \\ e^{j\beta z'}, & z' < z \end{cases} = \begin{cases} \tilde{\mathbf{M}}^+(\mathbf{r}'), & z' > z \\ \tilde{\mathbf{M}}^-(\mathbf{r}'), & z' < z \end{cases}$$

$$\tilde{\mathbf{N}}_\sigma(\mathbf{r}') = \frac{\nabla' \times \nabla' \times \mathbf{a}_z e^{jux'+jvy'}}{k[2\beta(u^2+v^2)]^{1/2}} \begin{cases} e^{-j\beta z'}, & z' > z \\ e^{j\beta z'}, & z' < z \end{cases} = \begin{cases} \tilde{\mathbf{N}}^+(\mathbf{r}'), & z' > z \\ \tilde{\mathbf{N}}^-(\mathbf{r}'), & z' < z. \end{cases}$$

These functions differ from the $\mathbf{M}_\sigma(\mathbf{r})$ and $\mathbf{N}_\sigma(\mathbf{r})$ by having x and y replaced by $-x'$ and $-y'$ and z replaced by z'. The σ indicator designates the forward- or reverse-propagating modes according to the rules set forth above.

The reader can readily verify that when (173) is used in (171) we obtain the same solution as found earlier using equivalent transmission-line models.

The integrals over u and v in (173) can be reexpressed as a spectrum of cylindrical waves in the transverse direction by converting the integral to one in cylindrical coordinates in uv space. The result of such a conversion is the expression (184) given in Chapter 2 for the dyadic Green's function.

For the layered medium we must add solutions to the homogeneous equation for $\bar{\mathbf{G}}_e$ in each layer so that the boundary conditions at each interface will be satisfied. For the general case the solution becomes quite complex so it is usually not worthwhile to derive the solution in each layer for an arbitrary number of layers. Most practical problems involve only one or two layers so it is preferable to consider each specific problem by itself in order to take advantage of any simplifications that the specific problem may offer. In order to illustrate the procedure we will find $\bar{\mathbf{G}}_e$ only in the layer that lies in $0 \leq z \leq d$.

In uv space the primary excitation by the $\mathbf{M}$-type modes is proportional to $\mathbf{M}_\sigma(\mathbf{r})\tilde{\mathbf{M}}_\sigma(\mathbf{r}')\cdot\mathbf{J}(\mathbf{r}')$ to which we add the homogeneous solution $V^+\mathbf{M}^+(\mathbf{r}) + V^-\mathbf{M}^-(\mathbf{r})$. The total incident fields at $z = 0$ and d are, respectively, $\mathbf{M}_\sigma(\mathbf{r})\tilde{\mathbf{M}}_\sigma(\mathbf{r}')\cdot\mathbf{J}(\mathbf{r}') + V^-\mathbf{M}^-(\mathbf{r})$ and $\mathbf{M}_\sigma(\mathbf{r})\tilde{\mathbf{M}}_\sigma(\mathbf{r}')\cdot\mathbf{J}(\mathbf{r}') + V^+\mathbf{M}^+(\mathbf{r})$. The corresponding reflected fields are $V^+\mathbf{M}^+(\mathbf{r})$ at $z = 0$ and $V^-\mathbf{M}^-(\mathbf{r})$ at $z = d$. By introducing the reflection coefficients[10] R_1' and R_2' for TE waves we can solve for V^+ and V^-. At $z = 0$ we have

$$V^+\mathbf{M}^+(\mathbf{r}) = R_1'[\mathbf{M}^-(\mathbf{r})\tilde{\mathbf{M}}^+(\mathbf{r}')\cdot\mathbf{J}(\mathbf{r}') + V^-\mathbf{M}^-(\mathbf{r})]$$

which, upon deleting all common factors gives,

$$V^+ = R_1'[\tilde{\mathbf{M}}^+(\mathbf{r}')\cdot\mathbf{J}(\mathbf{r}') + V^-].$$

---

[10]R_1' and R_2' can be obtained from R_1 and R_2 by replacing all Z_{ei} by Z_{hi}.

The corresponding equation at $z = d$ is

$$V^- e^{j\beta d} = R'_2[\tilde{\mathbf{M}}^-(\mathbf{r}') \cdot \mathbf{J}(\mathbf{r}') + V^+]e^{-j\beta d}.$$

From these two equations we obtain

$$V^+ = \frac{R'_1[R'_2 e^{-2j\beta d} \tilde{\mathbf{M}}^-(\mathbf{r}') + \tilde{\mathbf{M}}^+(\mathbf{r}')] \cdot \mathbf{J}(\mathbf{r}')}{1 - R'_1 R'_2 e^{-2j\beta d}}$$

$$V^- = \frac{R'_2 e^{-2j\beta d}[R'_1 \tilde{\mathbf{M}}^+(\mathbf{r}') + \tilde{\mathbf{M}}^-(\mathbf{r}')] \cdot \mathbf{J}(\mathbf{r}')}{1 - R'_1 R'_2 e^{-2j\beta d}}.$$

We can find the homogeneous solution involving the $\mathbf{N}$ functions the same way. When the results are collected we get the following solution for the complete dyadic Green's function in the layer $0 \le z \le d$:

$$\bar{\mathbf{G}}_e(\mathbf{r}, \mathbf{r}')$$

$$= -\frac{\mathbf{a}_z \mathbf{a}_z}{k^2} \delta(\mathbf{r} - \mathbf{r}') - \frac{j}{4\pi^2} \int\int_{-\infty}^{\infty} \left\{ \mathbf{M}_\sigma(\mathbf{r}) \tilde{\mathbf{M}}_\sigma(\mathbf{r}') + \mathbf{N}_\sigma(\mathbf{r}) \tilde{\mathbf{N}}_\sigma(\mathbf{r}') \right.$$

$$+ \frac{R'_1 \mathbf{M}^+(\mathbf{r}) \tilde{\mathbf{M}}^+(\mathbf{r}') + R'_2 e^{-2j\beta d}[\mathbf{M}^-(\mathbf{r}) \tilde{\mathbf{M}}^-(\mathbf{r}') + R'_1 \mathbf{M}^+(\mathbf{r}) \tilde{\mathbf{M}}^-(\mathbf{r}') + R'_1 \mathbf{M}^-(\mathbf{r}) \tilde{\mathbf{M}}^+(\mathbf{r}')]}{1 - R'_1 R'_2 e^{-j2\beta d}}$$

$$\left. - \frac{R_1 \mathbf{N}^+(\mathbf{r}) \tilde{\mathbf{N}}^+(\mathbf{r}') + R_2 e^{-2j\beta d}[\mathbf{N}^-(\mathbf{r}) \tilde{\mathbf{N}}^-(\mathbf{r}') - R_1 \mathbf{N}^+(\mathbf{r}) \tilde{\mathbf{N}}^-(\mathbf{r}') - R_1 \mathbf{N}^-(\mathbf{r}) \tilde{\mathbf{N}}^+(\mathbf{r}')]}{1 - R_1 R_2 e^{-j2\beta d}} \right\} du\, dv.$$

$$(174)$$

It should be apparent from the complexity of this expression that the use of equivalent transmission-line models provides a more transparent method of analysis than the use of dyadic Green's functions does. Some simplification of (174) is, however, possible by combining the primary field with the homogeneous solutions.

A solution for the dyadic Green's function in terms of cylindrical waves in the transverse direction for a three-layered medium has been given by Cheng [3.21]. An extensive body of literature exists on the problem of radiation in and propagation along planar stratified media. The reader is referred to the books by Wait [3.22], Galejs [3.23], Baños [3.24], and Brekhovskikh [3.25]. The dyadic Green's function for a three-layered medium containing a uniaxial anisotropic middle layer has been worked out by Lee and Kong [3.26].

The general problem of radiation in layered anisotropic media is straightforward in principle but gets quite complex from an algebraic point of view because anisotropy generally results in a coupling between the TE and TM waves at each interface.

3.10. Wave Velocities [3.27]–[3.33]

A propagating electromagnetic wave has a number of different velocities associated with it. In this section we will describe the different velocities and show their interrelationships.

Group, Signal, and Phase Velocities

The group velocity is defined as the velocity with which a signal composed of a narrow band or group of frequency components propagates, the definition being based on the time delay the packet undergoes in propagating a distance r. Consider a source located at the origin. The field radiated can be described by a superposition of plane waves. Let a component plane wave at the origin be

$$\mathbf{E}(t, 0) = \mathbf{C}f(t) \cos \omega_0 t \qquad (175)$$

where $\mathbf{C}$ is a constant vector and the modulating function $f(t)$ has a narrow-band frequency spectrum $g(\omega)$ which is zero outside the range $|\omega| > |\omega_1|$ and where $\omega_1 \ll \omega_0$. The spectrum of $\mathbf{E}(t, 0)$ is

$$\mathbf{E}_0(\omega, 0) = \mathbf{C}\int_{-\infty}^{\infty} \frac{1}{2}f(t)[e^{-j(\omega-\omega_0)t} + e^{-j(\omega+\omega_0)t}]\,dt$$

$$= \frac{1}{2}\mathbf{C}[g(\omega - \omega_0) + g(\omega + \omega_0)]. \qquad (176a)$$

The inverse transform relation is

$$\mathbf{E}(t, 0) = \frac{1}{2\pi}\int_{-\infty}^{\infty} \mathbf{E}_0(\omega, 0)e^{j\omega t}\,d\omega$$

$$= \operatorname{Re} \mathbf{C}\frac{1}{2\pi}\int_0^{\infty} g(\omega - \omega_0)e^{j\omega t}\,d\omega \qquad (176b)$$

since $g(-\omega) = g^*(\omega)$ for a real signal $f(t)$.

At a point $\mathbf{r}$ the plane wave field is given by

$$\mathbf{E}(t, \mathbf{r}) = \operatorname{Re} \mathbf{C}\frac{1}{2\pi}\int_0^{\infty} g(\omega - \omega_0)e^{-j\boldsymbol{\beta}(\omega)\cdot\mathbf{r}+j\omega t}\,d\omega.$$

For a narrow-band signal we can approximate $\boldsymbol{\beta}(\omega)$ by the first two terms of a Taylor series expansion; thus

$$\boldsymbol{\beta}(\omega) \approx \boldsymbol{\beta}(\omega_0) + \boldsymbol{\beta}'(\omega_0)(\omega - \omega_0), \qquad \boldsymbol{\beta}' = \frac{\partial\boldsymbol{\beta}}{\partial\omega}.$$

We now obtain

$$\mathbf{E}(t, \mathbf{r}) = \operatorname{Re} \mathbf{C}\frac{1}{2\pi}e^{-j\boldsymbol{\beta}(\omega_0)\cdot\mathbf{r}+j\omega_0 t}\int_0^{\infty} g(\omega - \omega_0)e^{j(\omega-\omega_0)(t-\boldsymbol{\beta}'\cdot\mathbf{r})}\,d\omega$$

$$= \operatorname{Re} \mathbf{C}e^{-j\boldsymbol{\beta}\cdot\mathbf{r}+j\omega_0 t}f(t - \boldsymbol{\beta}'\cdot\mathbf{r})$$

$$= \mathbf{C}f(t - \boldsymbol{\beta}'\cdot\mathbf{r}) \cos(\omega_0 t - \boldsymbol{\beta}\cdot\mathbf{r}). \qquad (177)$$

The signal or modulating function is reproduced at $\mathbf{r}$ without distortion but with a time delay

$\tau = \boldsymbol{\beta}'(\omega_0) \cdot \mathbf{r}$. The group speed is thus given by

$$v_g = \frac{r}{\tau} = \frac{r}{\dfrac{\partial \boldsymbol{\beta}}{\partial \omega} \cdot \mathbf{r}}. \tag{178}$$

The plane wave propagates in the direction of $\boldsymbol{\beta}$ and if we place $\mathbf{r} = x\mathbf{a}_x$ we obtain the component of the group velocity in the x direction, i.e.,

$$v_{gx} = \frac{x}{x\dfrac{\partial \beta_x}{\partial \omega}} = \frac{\partial \omega}{\partial \beta_x}.$$

Similarly, we find that

$$v_{gy} = \frac{\partial \omega}{\partial \beta_y}, \qquad v_{gz} = \frac{\partial \omega}{\partial \beta_z}$$

and hence in vector form the group velocity is given by

$$\mathbf{v}_g = \nabla_\beta \omega \tag{179}$$

where ∇_β designates the gradient in $\boldsymbol{\beta}$ space. The gradient of ω in $\boldsymbol{\beta}$ space is evaluated from the dispersion equation.

When $\boldsymbol{\beta}$ can be approximated by a linear function of ω then there is no signal distortion so the *signal velocity* can be defined as equal to the group velocity. In the case when there is signal distortion it is not possible to define a unique signal velocity.

The conditions under which $\boldsymbol{\beta}$ may be approximated by a linear function of ω, for which only a time delay but no distortion occurs, depend on how rapidly $\boldsymbol{\beta}$ varies with ω, the width of the signal frequency band, and the magnitude of $\mathbf{r}$. If $\mathbf{r}$ is sufficiently large additional terms in the expansion will always be required, leading to signal distortion. In this case the concept of group and signal velocities is not applicable because the signal is dispersed in the time domain. If the signal was originally a localized wave packet in both time and space it would eventually be dispersed in both time and space when $\boldsymbol{\beta}$ is a nonlinear function of ω.

The phase velocity is the velocity associated with the propagation of the constant phase fronts of a strictly time-harmonic field. It is equal to ω divided by the propagation constant β for a lossless medium, i.e.,

$$v_p = \omega/\beta. \tag{180}$$

Wavefront Velocity

Consider a current source $I(t)\mathbf{a}_z$ which equals zero for $t < 0$. A Fourier transform of $I(t)$ gives $I(\omega)$. The vector potential $A_z(r, \omega)$ is a solution of

$$(\nabla^2 + k^2)A_z(r, \omega) = -\mu I(\omega)\delta(\mathbf{r}) \tag{181}$$

from which we can obtain the solution in the time domain

$$A_z(\mathbf{r}, t) = \frac{\mu}{4\pi r} \frac{1}{2\pi} \int_C e^{j\omega t} I(\omega) e^{-j\omega \sqrt{\mu \epsilon} r} \, d\omega. \tag{182}$$

Note that $k = \omega\sqrt{\mu\epsilon}$. For a physical medium, μ and ϵ approach the free-space values μ_0 and ϵ_0 as $\omega \to \infty$. Because $I(t) = 0$ for $t < 0$ the transform $I(\omega)$ is free of singularities in the lower half complex ω plane (lhp). Thus for

$$t - \sqrt{\mu_0\epsilon_0}r < 0$$

we can close the contour by a semicircle of infinite radius in the lhp and the integral (182) yields zero. Hence no information will arrive at the point r before

$$t = \sqrt{\mu_0\epsilon_0}r = r/c$$

and the velocity of light is therefore the wavefront velocity.

Energy Transport Velocity

The velocity of energy transport has the direction of the Poynting vector and is equal to the group velocity. The variational theorem given by (29) in Chapter 1 can be used to derive an expression for the velocity of energy transport in an electromagnetic wave. The energy flow velocity is defined by the relation

$$(U_e + U_m)\mathbf{v}_g = \tfrac{1}{2}\,\mathrm{Re}\,\mathbf{E} \times \mathbf{H}^* \tag{183}$$

where U_e and U_m are the electric and magnetic energy densities (joules/m^3). We now assume a loss-free medium and apply the variational theorem to the plane wave function

$$\mathbf{E} = \mathbf{E}_0(\omega)e^{-j\boldsymbol{\beta}(\omega)\cdot\mathbf{r}}$$

$$\mathbf{H} = \mathbf{H}_0(\omega)e^{-j\boldsymbol{\beta}(\omega)\cdot\mathbf{r}}$$

where the propagation vector $\boldsymbol{\beta}(\omega)$ is real. By using (29) and (30) from Chapter 1 we obtain

$$\nabla\cdot\left[\mathbf{E}_0 \times \frac{\partial\mathbf{H}_0^*}{\partial\omega} + j\mathbf{r}\cdot\frac{\partial\boldsymbol{\beta}}{\partial\omega}(\mathbf{E}_0 \times \mathbf{H}_0^* + \mathbf{E}_0^* \times \mathbf{H}_0) + \frac{\partial\mathbf{E}_0^*}{\partial\omega} \times \mathbf{H}_0\right]$$

$$= j\nabla\cdot\left[\mathbf{r}\cdot\frac{\partial\boldsymbol{\beta}}{\partial\omega}2\,\mathrm{Re}\,\mathbf{E}_0 \times \mathbf{H}_0^*\right] = 4j(U_e + U_m)$$

since the terms not involving $\mathbf{r}$ are constants with zero divergence. We may rewrite the above in the form

$$\frac{1}{2}\,\mathrm{Re}\,\mathbf{E}_0 \times \mathbf{H}_0^*\cdot\nabla\left(\mathbf{r}\cdot\frac{\partial\boldsymbol{\beta}}{\partial\omega}\right) = \frac{1}{2}\,\mathrm{Re}\,\mathbf{E}_0 \times \mathbf{H}_0^*\cdot\frac{\partial\boldsymbol{\beta}}{\partial\omega} = U_e + U_m$$

since

$$\nabla\left(\mathbf{r}\cdot\frac{\partial\boldsymbol{\beta}}{\partial\omega}\right) = \nabla\left(x\frac{\partial\beta_x}{\partial\omega} + \cdots\right) = \frac{\partial\boldsymbol{\beta}}{\partial\omega}.$$

By using (183) in the above we obtain

$$\mathbf{v}_g\cdot\frac{\partial\boldsymbol{\beta}}{\partial\omega} = 1. \tag{184}$$

The dispersion equation for $\boldsymbol{\beta}$ gives a functional relationship between the components β_x, β_y, β_z of $\boldsymbol{\beta}$ and ω of the form $D(\beta_x, \beta_y, \beta_z, \omega) = 0$ with β_x, β_y, β_z not explicit functions of ω. A change $\delta\omega$ does not uniquely specify $\delta\boldsymbol{\beta}$ since $D = 0$ only requires that the total variation in D with changes $\delta\omega$ in ω, $\delta\beta_x$ in β_x, etc., equal zero, i.e.,

$$\delta D = \frac{\partial D}{\partial \beta_x}\delta\beta_x + \frac{\partial D}{\partial \beta_y}\delta\beta_y + \frac{\partial D}{\partial \beta_z}\delta\beta_z + \frac{\partial D}{\partial \omega}\delta\omega = 0.$$

We may hence choose $\delta\beta_y = \delta\beta_z = 0$ in which case $\delta\beta_x/\delta\omega = \partial\beta_x/\partial\omega$ and $\mathbf{v}_g \cdot (\partial\boldsymbol{\beta}/\partial\omega) = 1$ yields

$$v_{gx}\frac{\partial\beta_x}{\partial\omega} = 1 = v_{gx}\frac{\delta\beta_x}{\delta\omega} = -v_{gx}\frac{\partial D/\partial\omega}{\partial D/\partial\beta_x}.$$

Hence

$$v_{gx} = \frac{\partial\omega}{\partial\beta_x} = -\frac{\partial D/\partial\beta_x}{\partial D/\partial\omega}$$

with $\partial\omega/\partial\beta_x$ evaluated for β_y and β_z held constant. By combining similar results for v_{gy} and v_{gz} we obtain

$$\mathbf{v}_g = \nabla_\beta\omega = -\frac{\nabla_\beta D}{\partial D/\partial\omega} \tag{185}$$

where ∇_β is the gradient in $\boldsymbol{\beta}$ space, i.e.,

$$\nabla_\beta\omega = \mathbf{a}_x\frac{\partial\omega}{\partial\beta_x} + \mathbf{a}_y\frac{\partial\omega}{\partial\beta_y} + \mathbf{a}_z\frac{\partial\omega}{\partial\beta_z}.$$

For an anisotropic dielectric medium (106) gives

$$\frac{k_x^2\beta_x^2}{\beta^2 - k_x^2} + \frac{k_y^2\beta_y^2}{\beta^2 - k_y^2} + \frac{k_z^2\beta_z^2}{\beta^2 - k_z^2} = 0 = D(\beta_x, \beta_y, \beta_z, \omega) \tag{186}$$

where we have also put $n_x = \beta_x/\beta$, etc., and multiplied through by β^2. This is a typical dispersion relation and determines $\boldsymbol{\beta}$ as a function of ω (since $k_0 = \omega\sqrt{\mu_0\epsilon_0}$ and the κ_i are functions of ω) and direction in space. For each value of ω (186) determines two surfaces for $\boldsymbol{\beta}$ vs. direction. A given value of $\boldsymbol{\beta}$ and ω determine a point on this surface. Now from (185) the energy flow is in the direction of $\nabla_\beta D$ which is along the normal to the surface. Thus unless the surface is spherical $\boldsymbol{\beta}$ and $\mathbf{v}_g$ are not in the same direction as shown in Fig. 3.23. For certain media it is possible to have a component of $\boldsymbol{\beta}$ and $\mathbf{v}_g$ oppositely directed as shown in Fig. 3.23(b). In media of this type the proper boundary condition at infinity is that the energy flow velocity must be directed toward infinity even if it means that the phase velocity is directed inward.

The so-called ray velocity is the velocity of energy flow and is $\mathbf{v}_g$. The ray direction is along $\mathbf{v}_g$, i.e., along the direction of the Poynting vector. If we let $\mathbf{s} = s_x\mathbf{a}_x + s_y\mathbf{a}_y + s_z\mathbf{a}_z$ be a unit vector along the ray then one can show that the equation for $\mathbf{v}_g$ in an anisotropic

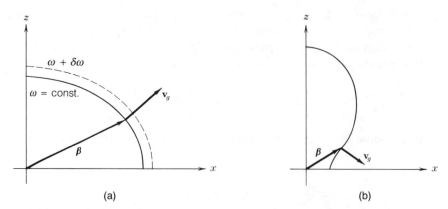

Fig. 3.23. (a) The β surface for ω = constant and for $\omega + \delta\omega$. (b) A dispersion surface for which β_z and v_{gz} are oppositely directed.

nondispersive medium (κ_i not functions of ω) is [3.15]

$$\frac{v_x^2 s_x^2}{v_g^2 - v_x^2} + \frac{v_y^2 s_y^2}{v_g^2 - v_y^2} + \frac{v_z^2 s_z^2}{v_g^2 - v_z^2}. \tag{187}$$

This is an equation for a two-sheeted surface called the ray surface. In general, one finds two directions in which the two roots for v_g are equal. These define the *ray* optic axes which normally do not coincide with the optic axes defined in terms of the wave phase velocity.

3.11. POINT SOURCE RADIATION IN ANISOTROPIC MEDIA [3.33]–[3.37]

Let the current $\mathbf{J} = \mathbf{a}\delta(\mathbf{r})$, $\mathbf{a}$ = unit vector; then Maxwell's equations require that

$$\nabla \times \mathbf{E} = -j\omega\mu_0\mathbf{H}, \qquad \nabla \times \mathbf{H} = j\omega\epsilon_0\boldsymbol{\kappa} \cdot \mathbf{E} + \mathbf{a}\delta(\mathbf{r}). \tag{188}$$

We take a three-dimensional Fourier transform to obtain

$$-j\boldsymbol{\beta} \times \mathbf{E}_0 = -j\omega\mu_0\mathbf{H}_0, \qquad -j\boldsymbol{\beta} \times \mathbf{H}_0 = j\omega\epsilon_0\boldsymbol{\kappa} \cdot \mathbf{E}_0 + \mathbf{a} \tag{189}$$

where

$$\mathbf{E}_0(\boldsymbol{\beta}) = \int\limits_{-\infty}^{\infty}\!\!\!\int\!\!\!\int \mathbf{E}(x, y, z)e^{j\boldsymbol{\beta}\cdot\mathbf{r}}\,dx\,dy\,dz$$

and similarly for $\mathbf{H}_0$. Note that we assume $\boldsymbol{\kappa} = \kappa_x\mathbf{a}_x\mathbf{a}_x + \kappa_y\mathbf{a}_y\mathbf{a}_y + \kappa_z\mathbf{a}_z\mathbf{a}_z$. When we eliminate $\mathbf{H}_0$ we obtain

$$\bar{\boldsymbol{\Gamma}} \cdot \mathbf{E}_0 = -j\omega\mu_0\mathbf{a} \tag{190}$$

where

$$\bar{\boldsymbol{\Gamma}} = \beta^2\bar{\mathbf{I}} - \boldsymbol{\beta}\boldsymbol{\beta} - k_0^2\boldsymbol{\kappa}$$

or in matrix form

$$[\Gamma] = \begin{bmatrix} \beta^2 - k_x^2 - \beta_x^2 & -\beta_x\beta_y & -\beta_x\beta_z \\ -\beta_x\beta_y & \beta^2 - k_y^2 - \beta_y^2 & -\beta_y\beta_z \\ -\beta_x\beta_z & -\beta_y\beta_z & \beta^2 - k_z^2 - \beta_z^2 \end{bmatrix} \tag{191}$$

where $k_x = \sqrt{\kappa_x}k_0$, etc.
  The solution for $\mathbf{E}_0$ is

$$\mathbf{E}_0 = -j\omega\mu_0\bar{\boldsymbol{\Gamma}}^{-1}\cdot\mathbf{a} = -j\omega\mu_0\frac{\bar{\mathbf{L}}\cdot\mathbf{a}}{\Delta} \tag{192}$$

where $\Delta = |\bar{\boldsymbol{\Gamma}}|$ is the determinant of $\bar{\boldsymbol{\Gamma}}$ and $\bar{\mathbf{L}}$ has factors made up of the cofactors of $\bar{\boldsymbol{\Gamma}}$; e.g.,

$$L_{xy} = (-1)[-\beta_x\beta_y(\beta^2 - k_z^2 - \beta_z^2) - \beta_x\beta_z\beta_y\beta_z] = \text{cofactor}$$

obtained by striking out row x and column y in the $[\Gamma]$ matrix. We note that the L_{ij} are polynomials in β_x, β_y, β_z. When we invert to get $\mathbf{E}(\mathbf{r})$ we can use the result

$$j\frac{\partial}{\partial x}\int_{-\infty}^{\infty} e^{-j\beta_x x}f(\beta_x, \beta_y, \beta_z)\,d\beta_x = \int_{-\infty}^{\infty} \beta_x f e^{-j\beta_x x}\,d\beta_x$$

and similar ones for the y and z derivatives, in repeated form, to replace $\bar{\mathbf{L}}(\beta_x, \beta_y, \beta_z)$ by a differential operator outside the integral. Thus we have

$$\mathbf{E}(\mathbf{r}) = -\frac{j\omega\mu_0}{8\pi^3}\bar{\mathbf{L}}\left(j\frac{\partial}{\partial x}, j\frac{\partial}{\partial y}, j\frac{\partial}{\partial z}\right)\cdot\mathbf{a}\int\!\!\!\int\!\!\!\int_{-\infty}^{\infty} \frac{e^{-j\boldsymbol{\beta}\cdot\mathbf{r}}}{\Delta}\,d\beta_x\,d\beta_y\,d\beta_z. \tag{193}$$

Hence the nature of the solution is determined by the integral

$$I = \int\!\!\!\int\!\!\!\int_{-\infty}^{\infty} \frac{e^{-j\boldsymbol{\beta}\cdot\mathbf{r}}}{\Delta}\,d\beta_x\,d\beta_y\,d\beta_z. \tag{194}$$

When we expand Δ we obtain

$$-\beta^4[n_x^2 k_x^2 + n_y^2 k_y^2 + n_z^2 k_z^2] + \beta^2[n_x^2 k_x^2(k_y^2 + k_z^2)$$
$$+ n_y^2 k_y^2(k_x^2 + k_z^2) + n_z^2 k_z^2(k_x^2 + k_y^2)] - k_x^2 k_y^2 k_z^2 = \Delta \tag{195}$$

where $\mathbf{n} = \boldsymbol{\beta}/\beta$.
  We can express this as an equation for $\beta_z^2 = \beta^2 n_z^2$. Thus with $\beta_t^2 = \beta_x^2 + \beta_y^2$ we obtain

$$\beta_z^4 k_z^2 + \beta_z^2[k_z^2\beta_t^2 + k_x^2\beta_x^2 + k_y^2\beta_y^2 - k_x^2 k_z^2 - k_y^2 k_z^2]$$
$$+ \beta_t^2[\beta_x^2 k_x^2 + \beta_y^2 k_y^2] - \beta_x^2 k_x^2(k_y^2 + k_z^2)$$
$$- \beta_y^2 k_y^2(k_x^2 + k_z^2) + k_x^2 k_y^2 k_z^2 = -\Delta. \tag{196}$$

We can now write

$$-\Delta = k_z^2(\beta_z^2 - \beta_{1z}^2)(\beta_z^2 - \beta_{2z}^2) \tag{197}$$

where β_{1z}^2 and β_{2z}^2 are the two roots for β_z^2 that make $\Delta = 0$. These roots are functions of β_x and β_y. We now obtain

$$\frac{1}{\Delta} = \frac{-1}{k_z^2(\beta_{1z}^2 - \beta_{2z}^2)}\left[\frac{1}{\beta_z^2 - \beta_{1z}^2} - \frac{1}{\beta_z^2 - \beta_{2z}^2}\right] \tag{198}$$

so our integral I splits into two terms

$$I = \int\int_{-\infty}^{\infty} \frac{-1}{k_z^2(\beta_{1z}^2 - \beta_{2z}^2)}(I_1 - I_2)\,d\beta_x\,d\beta_y \tag{199}$$

where

$$I_i = \int_{-\infty}^{\infty} \frac{e^{-j\boldsymbol{\beta}\cdot\mathbf{r}}}{\beta_z^2 - \beta_{iz}^2}\,d\beta_z, \qquad i = 1, 2. \tag{200}$$

For $z > 0$ the contour may be closed in the lower half plane (lph) and evaluated in terms of the residue at the enclosed pole. For an isotropic medium we would require an outward-propagating wave of the form $e^{-j\beta_{iz}z}$ and hence would choose the inversion contour to run below the pole at $-\beta_{iz}$ and above the pole at β_{iz}. In this case we would find that

$$I_1 = -2\pi j\left[e^{-j\boldsymbol{\beta}_t\cdot\mathbf{r}}\frac{e^{-j\beta_{1z}|z|}}{2\beta_{1z}}\right] \tag{201}$$

which is valid for all values of z. We will use this form but will find that for some cases it may be necessary to replace β_{1z} by $-\beta_{1z}$. This happens when β_{1z} and v_{gz} are oppositely directed. Since β_{1z} and β_{2z} are functions of β_x and β_y the appropriate sign for β_{1z} and β_{2z} can be determined from the analysis given below, which specifies the propagation vector at the stationary phase points. To evaluate the integral over β_x and β_y for large r we use the method of stationary phase.[11] This gives an asymptotic solution valid as $r \to \infty$; i.e., it gives the radiation field.

Let $\mathbf{r} = r(N_x\mathbf{a}_x + N_y\mathbf{a}_y + N_z\mathbf{a}_z)$ and let $h(\beta_x, \beta_y) = N_x\beta_x + N_y\beta_y + N_z\beta_{1z}$. Then

$$e^{-j\boldsymbol{\beta}_t\cdot\mathbf{r}-j\beta_{1z}z}$$

has a stationary phase where $\boldsymbol{\beta}_t\cdot\mathbf{r} + \beta_{1z}z = hr$ is stationary. The stationary phase point is a solution of

$$\frac{\partial h}{\partial \beta_x} = \frac{\partial h}{\partial \beta_y} = 0 \quad \text{or} \quad N_x + N_z\frac{\partial \beta_{1z}}{\partial \beta_x} = 0,\, N_y + N_z\frac{\partial \beta_{1z}}{\partial \beta_y} = 0.$$

[11] See Mathematical Appendix.

Let the solutions be β_{x0}, β_{y0}. Then to first order

$$h = N_x\beta_{x0} + N_y\beta_{y0} + N_z\beta_{z0} + \frac{1}{2}\frac{\partial^2 h}{\partial\beta_x^2}(\beta_x - \beta_{x0})^2$$

$$+ \frac{1}{2}\frac{\partial^2 h}{\partial\beta_y^2}(\beta_y - \beta_{y0})^2 + \frac{\partial^2 h}{\partial\beta_x\partial\beta_y}(\beta_y - \beta_{y0})(\beta_x - \beta_{x0}) \qquad (202)$$

where β_{z0} is the value of β_{1z} for $\beta_x = \beta_{x0}$, $\beta_y = \beta_{y0}$ and all derivatives are evaluated at β_{x0}, β_{y0}. Let us write for h

$$h = \boldsymbol{\beta}_0 \cdot \mathbf{N} + h_1(\beta_x - \beta_{x0})^2 + h_2(\beta_x - \beta_{x0})(\beta_y - \beta_{y0}) + h_3(\beta_y - \beta_{y0})^2 \qquad (203)$$

where h_1, h_2, h_3 can be identified from the earlier expression. We then have

$$\left[\frac{1}{k_z^2(\beta_{2z}^2 - \beta_{1z}^2)}\right]\left(\frac{-\pi j}{\beta_{z0}}\right) e^{-j\boldsymbol{\beta}_0 \cdot \mathbf{N}r}\int\int_{-\infty}^{\infty} e^{-jr[h_1 u^2 + h_2 uv + h_3 v^2]}\,d\beta_x\,d\beta_y$$

for the first term in (199), where $\beta_x - \beta_{x0} = u$, $\beta_y - \beta_{y0} = v$. We now complete the square to get

$$h_1 u^2 + h_2 uv + h_3 v^2 = h_1(u - u_0)^2 + h_3 v^2 - h_1 u_0^2$$

where

$$-2h_1 u_0 = h_2 v.$$

If we use the result

$$\int_{-\infty}^{\infty} e^{-jrh_1(u - u_0)^2}\,du = \int_{-\infty}^{\infty} e^{-jrh_1\lambda^2}\,d\lambda = \sqrt{\frac{\pi}{jh_1 r}}$$

we find that the first term in (199) is

$$\frac{-j\pi}{k_z^2(\beta_{2z}^2 - \beta_{1z}^2)|_0\beta_{z0}} e^{-j\boldsymbol{\beta}_0 \cdot \mathbf{N}r}\int\int_{-\infty}^{\infty} e^{-jrh_1(u - u_0)^2}\,du\,e^{-jr(h_3 - h_1(h_2^2/4h_1^2))v^2}\,dv$$

$$= \frac{-j\pi}{k_z^2(\beta_{2z}^2 - \beta_{1z}^2)|_0\beta_{z0}} e^{-j\boldsymbol{\beta}_0 \cdot \mathbf{N}r}\frac{\pi}{jr}\frac{1}{[h_1(h_3 - h_2^2/4h_1)]^{1/2}}.$$

A similar solution is obtained for the second term in (199).

We note that $\beta_{1z} - \beta_{1z}(\beta_x, \beta_y) = 0$ is the equation for the $\boldsymbol{\beta}$ surface. The normal to the surface is given by

$$\nabla_\beta[\beta_{1z} - \beta_{1z}(\beta_x, \beta_y)] = \mathbf{a}_z - \frac{\partial\beta_{1z}}{\partial\beta_x}\mathbf{a}_x - \frac{\partial\beta_{1z}}{\partial\beta_y}\mathbf{a}_y$$

since $\Delta = 0$ gives the $\boldsymbol{\beta}$ surface and this may be expressed as an equation giving β_{1z} as a

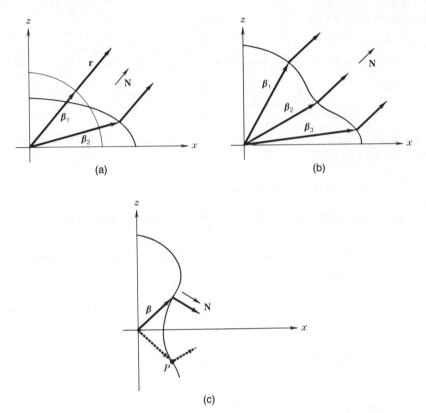

Fig. 3.24. (a) and (b) Illustration of values of β that give contributing waves in the radiation zone. (c) A dispersion surface for which β_z and v_{gz} are oppositely directed.

function of β_x, β_y, i.e., $\beta_{1z} = \beta_{1z}(\beta_x, \beta_y)$ from which $\beta_{1z} - \beta_{1z}(\beta_x, \beta_y) = 0$. The components of the normal τ to the β surface are in the ratio

$$\frac{\tau_x}{\tau_z} = -\frac{\partial \beta_{1z}}{\partial \beta_x}, \qquad \frac{\tau_y}{\tau_z} = -\frac{\partial \beta_{1z}}{\partial \beta_y}.$$

But from the solution for the stationary phase point

$$\frac{N_x}{N_z} = -\frac{\partial \beta_{1z}}{\partial \beta_x}\bigg|_{\beta_{x0}, \beta_{y0}} \qquad \frac{N_y}{N_z} = \frac{\partial \beta_{1z}}{\partial \beta_y}\bigg|_{\beta_{x0}, \beta_{y0}}.$$

Hence

$$e^{-j\boldsymbol{\beta}_0 \cdot \mathbf{N}r}$$

has a propagation vector $\boldsymbol{\beta}_0$ which corresponds to the value of $\boldsymbol{\beta}$ for which $\mathbf{N}$ is normal to the β surface. Thus the field at $\mathbf{r}$ is made up of the two or more waves for which $\boldsymbol{\beta}$ has the values that make $\mathbf{N}$ equal to the normal at that point on the β surface (see Fig. 3.24 for illustration of this feature of the solution).

There may be more than one solution for the stationary phase point. Each solution will yield a contributing field with a solution of the form given above. In the evaluation of the integral in (200) we assumed that for $z > 0$ the pole at β_{1z} would be enclosed and this gave us the result shown in (201). For a given direction $\mathbf{r}$ the unit vector $\mathbf{N}$ is specified. Hence when we solve the equations for the stationary phase point for a medium characterized by a dispersion surface like that shown in Fig. 3.24(c) we will find that β_{1z} may, in fact, be negative when N_z is positive. Thus the particular pole that we enclose when carrying out the integral over β_z is immaterial since the solution for β at the stationary phase point will result in the correct sign for β_{1z} to become specified. In Fig. 3.24(c) the stationary phase point would be the point P when $N_z > 0$. Thus the form (201) can be used for all values of z with the proviso that the sign of β_{1z} is specified by $\boldsymbol{\beta}_0$ at the stationary phase point. Similar remarks apply to the field contributed by the stationary phase points associated with the surface $\beta_{2z} = \beta_{2z}(\beta_x, \beta_y)$.

REFERENCES AND BIBLIOGRAPHY

[3.1] N. W. McLachlan, *Bessel Functions for Engineers*, 2nd ed.   New York, NY: Oxford University Press, 1946, chs. 1 and 2.

[3.2] F. Bowman, *Introduction to Bessel Functions*.   New York, NY: Dover Publications, 1958.

[3.3] G. P. Harnwell, *Principles of Electricity and Electromagnetism*, 2nd ed.   New York, NY: McGraw-Hill Book Company, Inc., 1949, chs. 10 and 11.

[3.4] C. Kittel, "On the theory of ferromagnetic resonance absorption," *Phys. Rev.*, vol. 73, pp. 155–161, Jan. 1948.

[3.5] D. Polder, "On the theory of ferromagnetic resonance," *Phil. Mag.*, vol. 40, pp. 99–114, Jan. 1949.

[3.6] C. L. Hogan, "The microwave gyrator," *Bell Syst. Tech. J.*, vol. 31, pp. 1–31, Jan. 1952.

[3.7] J. L. Melchor *et al.*, "Microwave frequency doubling from 9 to 18 kmc in ferrites," *Proc. IRE*, vol. 45, pp. 643–646, May 1957.

[3.8] J. L. Melchor, W. P. Ayres, and P. H. Vartanian, "Frequency doubling in ferrites," *J. Appl. Phys.*, vol. 27, p. 188, 1956.

TEM Waves in General

[3.9] J. A. Stratton, *Electromagnetic Theory*.   New York, NY: McGraw-Hill Book Company, Inc., 1941.

[3.10] S. Ramo and J. R. Whinnery, *Fields and Waves in Modern Radio*, 2nd ed.   New York, NY: John Wiley & Sons, Inc., 1953.

Wave Matrices and Wave-Impedance Concepts

[3.11] C. G. Montgomery, R. H. Dicke, and E. M. Purcell, *Principles of Microwave Circuits*, vol. 8 of MIT Rad. Lab. Series.   New York, NY: McGraw-Hill Book Company, Inc., 1948.

[3.12] E. F. Bolinder, "Note on the matrix representation of linear two port networks," *IRE Trans. Circuit Theory*, vol. CT-4, p. 337, Dec. 1957.

[3.13] S. A. Schelkunoff, "The impedance concept and its application to problems of reflection, refraction, shielding and power absorption," *Bell Syst. Tech. J.*, vol. 17, pp. 17–48, Jan. 1938.

[3.14] H. G. Booker, "The elements of wave propagation using the impedance concept," *J. IEE (London)*, vol. 94, part IIIA, pp. 171–198, May 1947.

TEM Waves in Anisotropic Dielectric Media

[3.15] G. Joos, *Theoretical Physics*, 2nd ed.   Glasgow: Blackie & Sons, Ltd., 1951.

[3.16] J. Valasek, *Introduction to Theoretical and Experimental Optics*.   New York, NY: John Wiley & Sons, Inc., 1949.

[3.17] R. E. Collin, "A simple artificial anisotropic dielectric medium," *IRE Trans. Microwave Theory Tech.*, vol. MTT-6, pp. 206–209, Apr. 1958.

The literature of ferrites is extensive. Two particularly good sources of review and original papers on the subject are:

[3.18] Ferrites Issue, *Proc. IRE*, vol. 44, Oct. 1956.

[3.19] *IRE Trans. Microwave Theory Tech.*, vol. MTT-6, Jan. 1958.

Radiation and Propagation in Layered Media

[3.20] L. B. Felsen and N. Marcuvitz, *Radiation and Scattering of Waves*.   Englewood Cliffs, NJ: Prentice-Hall, 1973.

[3.21]   D. H. S. Cheng, "On the formulation of the dyadic Green's function in a layered medium," *Electromagnetics J.*, vol. 6, no. 2, pp. 171–182, 1986.

[3.22]   J. R. Wait, *Electromagnetic Waves in Stratified Media*, 2nd ed.   Oxford: Pergamon Press, 1970.

[3.23]   J. Galejs, *Antennas in Inhomogeneous Media*.   New York, NY: Pergamon Press, 1969.

[3.24]   A. Baños, Jr., *Dipole Radiation in the Presence of a Conducting Half-space*.   Elmsford, NY: Pergamon Press, 1966.

[3.25]   L. M. Brekhovskikh, *Waves in Layered Media*.   New York, NY: Academic Press, Inc., 1960.

[3.26]   J. K. Lee and J. A. Kong, "Dyadic Green's functions for layered anisotropic medium," *Electromagnetics J.*, vol. 3, no. 2, pp. 111–130, 1983.

Wave Velocities and Radiation in Anisotropic Media

[3.27]   L. Brillouin, *Wave Propagation and Group Velocity*.   New York, NY: Academic Press, Inc., 1960.

[3.28]   C. O. Hines, "Wave packets, the Poynting vector, and energy flow," *J. Geophys. Res.*, vol. 56, Part I, pp. 63–72, Part II, pp. 191–220, 1951.

[3.29]   S. M. Rytov, "Some theorems concerning the group velocity of electromagnetic waves," *Zh. Eksp. Teor. Fiz.*, vol. 17, pp. 930–936, 1947.

[3.30]   V. L. Ginzburg, "On the electrodynamics of anisotropic media," *J. Exp. Theor. Phys.*, vol. 10, pp. 601–607, 1940.

[3.31]   V. L. Ginzburg, *Propagation of Electromagnetic Waves in Plasmas*.   New York, NY: Gordon and Breach Science Publishers, Inc., 1961, sects. 22 and 24. (Translated from the 1960 Russian edition.)

[3.32]   K. G. Budden, *Radio Waves in the Ionosphere*.   Cambridge: Cambridge University Press, 1961, sect. 13.18.

[3.33]   F. V. Bunkin, "On radiation in anisotropic media," *J. Exp. Theor. Phys.*, vol. 32, pp. 338–346, 1957.

[3.34]   H. Kogelnik, "On electromagnetic radiation in magneto-ionic media," *J. Res. Nat. Bur. Stand.*, vol. 64D, pp. 515–523, 1960.

[3.35]   W. C. Meecham, "Source and reflection problems in magneto-ionic media," *J. Phys. Fluids*, vol. 4, pp. 1517–1524, 1961.

[3.36]   E. C. Jordan (Ed.), *Electromagnetic Theory and Antennas*, part I.   New York, NY: Pergamon Press, 1963. See papers by Arbel and Felsen; Clemmow, Kogelnik, and Motz; and Mittra and Deschamps.

[3.37]   Electromagnetic Wave Propagation in Anisotropic Media, Special Issue, *Radio Sci.*, vol. 2, Aug. 1967.

PROBLEMS

3.1. Find the solution for a uniform cylindrical TEM wave with components E_θ, H_z. Obtain the same solution by using the transformation $\mathbf{E}' = Z_0\mathbf{H}$, $\mathbf{H}' = -Y_0\mathbf{E}$, where $\mathbf{E}$, $\mathbf{H}$ is the solution for a uniform cylindrical TEM wave with components H_θ, E_z.

Answer:

$$E_\theta = AH_1^2(k_0r)$$

$$H_z = \frac{jk}{\omega\mu_0}AH_0^2(k_0r).$$

3.2. Consider an infinite biconical transmission line as illustrated in Fig. P3.2. For TEM waves a solution for $\mathbf{E}_t$ of the form $\mathbf{E}_t = -\nabla_t\psi(\theta,\phi)g(r)$ may be found. In the spherical coordinate system $r\theta\phi$, the function ψ is a solution of

$$\sin\theta\frac{\partial}{\partial\theta}\left(\sin\theta\frac{\partial\psi}{\partial\phi}\right) + \frac{\partial^2\psi}{\partial\phi^2} = 0.$$

Make the transformation $u = \ln\tan(\theta/2)$, and show that the function ψ is now a solution of $\partial^2\psi/\partial u^2 + \partial^2\psi/\partial\phi^2 = 0$. The above transformation maps the cross sections S_1 and S_2 into the rectangular coordinate plane $u\phi$ as illustrated. A solution to Laplace's equation for the field between S_1 and S_2 in the $u\phi$ plane thus provides the solution for ψ. Show that, when the cross sections S_1 and S_2 are the same, then the mapping is symmetrical about $u = 0$. When S_1 and S_2 are circles defined by $\theta = \theta_0$ and $\theta = \pi - \theta_0$, find a solution for the capacitance between S_1 and S_2 in the $u\phi$ plane. (There is no fringing effect since the mapping is periodic in ϕ.) Using the relation $Z_c = (v_cC)^{-1}$, where $v_c = (\mu_0\epsilon_0)^{-1/2}$ is the velocity of light, and C is the capacitance per unit length, obtain the characteristic impedance Z_c of the biconical line.

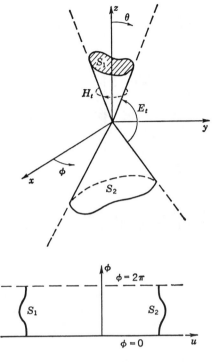

Fig. P3.2.

Answer:

$$Z_c = \frac{1}{\pi} \left(\frac{\mu_0}{\epsilon_0} \right)^{1/2} \tan \frac{\theta_0}{2}.$$

3.3. Find the solutions for the field of a TEM wave with components E_θ, H_ϕ for a biconical line with circular cross section as in Problem 3.2. Integrate H_ϕ around one conductor to find the total radial current I on the conductor. Also integrate E_θ along a path $r = $ constant from one conductor to the next to obtain the potential difference V. Show that $V/I = Z_c$ is the same as in Problem 3.2. Show that the line integral of E_θ along any closed curve on the surface of a sphere $r = r_0$ is zero.

3.4. Show that the impedance matrix is always symmetrical independently of whether normalized or unnormalized amplitudes are used, provided the scattering matrix is symmetrical when normalized amplitudes are used.

3.5. Consider an infinite anisotropic dielectric medium located in the half space $z > 0$. Let the principal dielectric constants be κ_1 along the x axis and κ_2 along the y and z axes. For a parallel-polarized wave incident from $z < 0$ along a direction defined by $N_x = N_y = \frac{1}{2}$, find the power transmitted as an ordinary wave and that transmitted as an extraordinary wave in the dielectric. Find also the reflected power in the reflected parallel- and perpendicular-polarized waves. The interface is in the xy plane at $z = 0$.

3.6. A ferrite medium occupies the half space $z > 0$ and is magnetized along the y axis by a static magnetic field. Let a TEM wave with the electric vector along the y axis be incident from $z < 0$ at an angle θ_i. Find the amplitude of the wave reflected in the direction $-\theta_i$ and also the field transmitted into the ferrite medium. Show that, for a wave incident from the direction $-\theta_i$, the reflected field in the direction θ_i is different from that obtained for the first case. This is an example of nonreciprocal behavior. Assume a permeability tensor of the form

$$[\mu] = \begin{bmatrix} \mu & 0 & -j\kappa \\ 0 & \mu_0 & 0 \\ +j\kappa & 0 & \mu \end{bmatrix} \quad \text{with } \mu \text{ and } \kappa \text{ real.}$$

(See Fig. P3.6.)

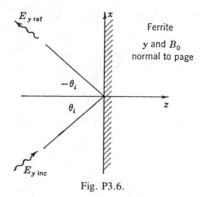

Fig. P3.6.

Answer: The propagation factor in ferrite is $e^{-jk(x\sin\theta_r + z\cos\theta_r)}$, where $k^2 = \omega^2\epsilon\mu(1 - \kappa^2/\mu^2)$. For a wave incident from θ_i, the reflection coefficient in the direction $-\theta_i$ is $R(\theta_i, -\theta_i) = (Z_{\text{in}} - 1)/(Z_{\text{in}} + 1)$, where

$$Z_{\text{in}} = \frac{(\mu^2 - \kappa^2)k_0 \cos\theta_i}{\mu_0[\mu(k^2 - k_0^2 \sin^2\theta_i)^{1/2} - jk_0 \sin\theta_i]}$$

and $k_0^2 = \omega^2\mu_0\epsilon_0$. The reflection coefficient $R(-\theta_i, \theta_i)$ is equal to the complex conjugate of $R(\theta_i, -\theta_i)$.

3.7. Find the principal axis (normal coordinates) for the permeability tensor given in Problem 3.6. Show that these may be interpreted as rectangular coordinates rotating clockwise and counterclockwise around the y axis. Transform Maxwell's equations into this new coordinate system, and obtain the solutions for TEM waves propagating along the y axis in an infinite ferrite medium.

3.8. Obtain the reflection and transmission coefficients for right and left circularly polarized TEM waves incident normally on an infinite ferrite–air interface. Assume the static magnetic field to be applied in the direction of the interface normal.

3.9. A medium is characterized by a permeability μ_1 along the x axis and μ_2 along the y and z axes. Determine suitable potential functions from which the ordinary and extraordinary plane wave solutions may be found. Obtain the eigenvalue equations for the propagation phase constants. The dielectric constant is κ along all three axes.

3.10. Consider a time-harmonic current element $I_0\mathbf{a}_z\delta(x)\delta(y)\delta(z - z')$ radiating in free space above a dielectric half space in $z \leq 0$. Show that for $z > 0$ the radiated electric field is given by

$$\mathbf{E} = \frac{-I_0 Z_0}{8\pi^2 k_0}\nabla \times \nabla \times \mathbf{a}_z \int\limits_{-\infty}^{\infty}\!\!\int [e^{-j\beta_0|z-z'|} - R_1 e^{-j\beta_0(z+z')}]\frac{e^{-jux-jvy}}{\beta_0}\,du\,dv$$

where $R_1 = (\beta_1 - \kappa\beta_0)/(\beta_1 + \kappa\beta_0)$, $\beta_i = (\kappa_i k_0^2 - u^2 - v^2)^{1/2}$, and $\kappa_i = 1$ for $z > 0$ and equals κ for $z < 0$. The dielectric constant of the medium in $z < 0$ is κ.

3.11. Let $u = \gamma\cos\alpha$, $v = \gamma\sin\alpha$, and convert the integral over u and v to one in cylindrical coordinates γ, α to show that the electric field in (Problem 3.10) can be expressed as

$$\mathbf{E} = -\frac{I_0 Z_0}{4\pi k_0}\nabla \times \nabla \times \mathbf{a}_z \int_0^{\infty}[e^{-j\beta_0|z-z'|} - R_1 e^{-j\beta_0(z+z')}]J_0(\gamma r)\frac{\gamma\,d\gamma}{\beta_0}.$$

Use $x = r\cos\theta$ and $y = r\sin\theta$. $J_0(\gamma r)$ is the 0th-order Bessel function. This result was due originally to Sommerfeld.

3.12. Introduce the vector $\boldsymbol{\beta} = u\mathbf{a}_x + v\mathbf{a}_y + w\mathbf{a}_z$. Show that the **L**, **M**, and **N** functions in Section 3.9, Eq. (168) consist of three mutually orthogonal vectors. Note that $\nabla = -j\boldsymbol{\beta}$. Use this property to show that $\mathbf{L}(\boldsymbol{\beta}, \mathbf{r})$ is orthogonal to $\mathbf{M}(\boldsymbol{\beta}, \mathbf{r})$ and $\mathbf{N}(\boldsymbol{\beta}, \mathbf{r})$ when integrated over $\mathbf{r}$. Similarly, show that $\mathbf{N}(\boldsymbol{\beta}, \mathbf{r})$ and $\mathbf{M}(\boldsymbol{\beta}, \mathbf{r})$ are orthogonal to $\mathbf{M}(\boldsymbol{\beta}', \mathbf{r})$ and $\mathbf{N}(\boldsymbol{\beta}', \mathbf{r})$, respectively.

3.13. With reference to Problem 2.31, express $k_0^2\bar{\mathbf{I}} - \mathbf{kk}$ in terms of components along the three mutually perpendicular vectors **k**, $\mathbf{k} \times \mathbf{a}_z$, and $\mathbf{k} \times \mathbf{k} \times \mathbf{a}_z$ and thus express $\bar{\mathbf{G}}$ in terms of **L**, **M**, and **N** functions. Show that the

L function contribution is

$$\frac{-1}{8\pi^3 k_0^2} \int\!\!\!\int\!\!\!\int_{-\infty}^{\infty} \frac{\mathbf{kk}}{k^2} e^{-j\mathbf{k}\cdot(\mathbf{r}-\mathbf{r}')}\, dk_x\, dk_y\, dk_z = \frac{-\nabla\nabla'}{8\pi^3 k_0^2} \int\!\!\!\int\!\!\!\int_{-\infty}^{\infty} \frac{e^{-j\mathbf{k}\cdot(\mathbf{r}-\mathbf{r}')}}{k^2}\, dk_x\, dk_y\, dk_z.$$

Let $k^2 = k_t^2 + k_z^2$ and complete the integral over k_z using residue theory to obtain

$$\frac{-\nabla\nabla'}{8\pi^2 k_0^2} \int\!\!\!\int \frac{e^{-j\mathbf{k}_t\cdot(\mathbf{r}-\mathbf{r}')-k_t|z-z'|}}{k_t}\, dk_x\, dk_y.$$

Show that the operation $\mathbf{a}_z\mathbf{a}_z\partial^2/\partial z\partial z'$ produces a contribution $-\mathbf{a}_z\mathbf{a}_z\delta(\mathbf{r}-\mathbf{r}')/k_0^2$ upon differentiating the exponential function $e^{-k_t|z-z'|}$ and then completing the integrals over k_x and k_y.

3.14. A horizontal current element $I_0\mathbf{a}_x\delta(x)\delta(y)\delta(z-z')$ radiates above a dielectric half space in $z < 0$. Show that the radiated electric field in $z > 0$ is given by

$$\mathbf{E} = \frac{-k_0 Z_0 I_0}{8\pi^3} \int\!\!\!\int_{-\infty}^{\infty} \frac{du\, dv}{\beta_0(u^2 + v^2)} \left\{ jv\nabla \times \mathbf{a}_z[e^{-j\beta_0|z-z'|} + R_1'e^{-j\beta_0(z+z')}] \right.$$

$$\left. - \frac{u\beta_0}{k_0^2}\nabla \times \nabla \times \mathbf{a}_z[e^{-j\beta_0|z-z'|}\, \mathrm{sg}\,(z-z') + R_1 e^{-j\beta_0(z+z')}]e^{-jux-jvy} \right\}$$

where R_1 is given in Problem 3.10 and $R_1' = (\beta_0 - \beta_1)/(\beta_0 + \beta_1)$. If a ground plane is inserted at $z = d$ it is only necessary to replace R_1 and R_1' by the new reflection coefficients seen at $z = 0$ looking into the grounded slab.

3.15. Consider a uniaxial medium with $k_x = k_y$ and $k_z > k_x$. Show that the two solutions for β_z are given by

$$\beta_x^2 + \beta_z^2 = k_x^2 \qquad \text{and} \qquad \frac{\beta_x^2}{k_z^2} + \frac{\beta_z^2}{k_x^2} = 1.$$

The traces in the xz plane may be rotated about the z axis to generate the two β surfaces. For radiation in the direction given by $\mathbf{N} = (\mathbf{a}_x + \mathbf{a}_z)/\sqrt{2}$ show that the two contributing values of β are

$$\beta = \frac{k_x(\mathbf{a}_x + \mathbf{a}_z)}{\sqrt{2}} \qquad \text{and} \qquad \beta = \frac{\mathbf{a}_x k_z^2 + \mathbf{a}_z k_x^2}{\sqrt{k_x^2 + k_z^2}}.$$

Find the far zone-radiated electric field from a z-directed current element located at the origin.

3.16. For a biaxial medium with the dispersion equation (186) show that the group velocity is given by

$$\mathbf{v}_g = \frac{\dfrac{k_x^2\beta_x(k_x^2 - \beta_z^2 - \beta_x^2)\mathbf{a}_x}{(\beta^2 - k_x^2)^2} + \dfrac{k_y^2\beta_y(k_y^2 - \beta_z^2 - \beta_y^2)\mathbf{a}_y}{(\beta^2 - k_y^2)^2} + \dfrac{k_z^2\beta_z(k_z^2 - \beta_x^2 - \beta_y^2)\mathbf{a}_z}{(\beta^2 - k_z^2)^2}}{\dfrac{\beta^2}{\omega}\left[\dfrac{k_x^2\beta_x^2}{(\beta^2 - k_x^2)^2} + \dfrac{k_y^2\beta_y^2}{(\beta^2 - k_y^2)^2} + \dfrac{k_z^2\beta_z^2}{(\beta^2 - k_z^2)^2}\right]}.$$

4

Transmission Lines

One of the most familiar wave-guiding structures is the conventional transmission line such as the two-wire line and the coaxial line. The fundamental mode of propagation on a transmission line is essentially a transverse electromagnetic wave. In the ideal case, when the conductors can be considered to have infinite conductivity, the basic mode is a TEM mode. Practical lines have finite conductivity, and this results in a perturbation or change from a TEM mode to a mode that has a small axial component of electric field. Many practical lines have such small losses that we may analyze their behavior by considering them as ideal lines, and then make a simple perturbation calculation to obtain the effects of the finite conductivity. Transmission-line theory has two aspects. In one case we are given the characteristic parameters of the line, and our problem is to study the propagation of electromagnetic waves along the line. In the other case we know only the conductor configuration, and our problem is to determine the line parameters such as characteristic impedance, attenuation constant, propagation constant, and shunt conductance. The object of this latter aspect is ultimately to determine the suitability of the transmission line for a given application. It is this second aspect of transmission-line theory that we shall be most concerned with in this chapter.

The conformal mapping method and the variational methods for obtaining the capacitance of a transmission line will be developed in detail in this chapter. The latter part of the chapter will discuss planar transmission lines consisting of conducting strips on a dielectric substrate. These nonhomogeneous transmission lines do not support a TEM mode of propagation except at zero frequency. Integral equation techniques for the solution of the current and charge distributions and the propagation constants for these transmission lines will be developed.

4.1. GENERAL TRANSMISSION-LINE THEORY

The Ideal Two-Conductor Line

Consider two separate perfectly conducting conductors with uniform cross sections, infinitely long and oriented parallel to the z axis as in Fig. 4.1. The cross sections of the two conductors are denoted by the two open or closed curves S_1 and S_2 in the xy plane. The medium surrounding the conductors is assumed homogeneous and isotropic with electric parameters μ and ϵ. A further assumption will be made that all other material bodies are sufficiently remote so as not to perturb the field around the conductors.

For an ideal transmission line as illustrated, a TEM mode of propagation is possible. In the preceding chapter it was found that for TEM waves Maxwell's equations reduce to

$$\mathbf{E}_t = -\nabla_t \Phi(x, y) e^{\pm jkz} \tag{1a}$$

$$\mathbf{H}_t = \mp \bar{\mathbf{Y}} \cdot \mathbf{E}_t \tag{1b}$$

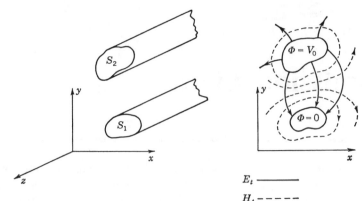

$$E_t \, \text{———}$$
$$H_t \, \text{-----}$$

Fig. 4.1. Two-conductor transmission line.

where $\mathbf{E}_t$ and $\mathbf{H}_t$ are transverse vectors, ∇_t is the transverse operator $\mathbf{a}_x(\partial/\partial x) + \mathbf{a}_y(\partial/\partial y)$, $k^2 = \omega^2\mu\epsilon = 4\pi^2/\lambda^2$, $\bar{\mathbf{Y}}$ is the dyadic wave admittance $Y(\mathbf{a}_y\mathbf{a}_x - \mathbf{a}_x\mathbf{a}_y)$, and $Y = (\epsilon/\mu)^{1/2}$ is the intrinsic admittance of the medium surrounding the conductors. The negative sign in (1a) is chosen only to conform with the convention used in electrostatics. The potential function Φ is a solution of Laplace's equation in the xy plane. A solution for Φ exists only if one conductor is at a potential which is different from that of the other one. If the system as a whole is electrically neutral, we may take one conductor at potential $V_0/2$ and the other at potential $-V_0/2$. Since the field $\mathbf{E}_t$ is invariant to a change in Φ corresponding to the addition of a constant, we may, for convenience, take the conductor S_1 at zero potential and S_2 at a potential V_0 instead. In the transverse plane $\nabla_t \times \mathbf{E}_t$ is identically zero, and, hence, the line integral of $\mathbf{E}_t$ from S_1 to S_2 is unique and does not depend on the path of integration. It is this property which permits us to introduce a unique potential between the two conductors. Physically this property arises because of the absence of axial magnetic flux, and, hence, there is no induced potential around any closed curve in the xy plane. If we let $\mathbf{e}$ be equal to $\mathbf{E}_t$ without the factor $\mathbf{e}^{\pm jkz}$, we have

$$V_0 = -\int_{S_1}^{S_2} \mathbf{e}\cdot d\mathbf{l} = \int_{S_1}^{S_2} \nabla_t\Phi\cdot d\mathbf{l} = \int_{S_1}^{S_2} d\Phi. \tag{2}$$

Associated with the transverse electric field wave $\mathbf{E}_t = -\nabla_t\Phi e^{-jkz}$ is a unique voltage wave $V = V_0 e^{-jkz}$.

Since the conductors S_1 and S_2 are assumed to be perfectly conducting, the normal component of $\mathbf{e}$ is discontinuous at these surfaces, and a surface distribution of charge given by $\rho_s = \epsilon\mathbf{n}\cdot\mathbf{e}$ exists on the conductors, where $\mathbf{n}$ is the unit outward normal from the conductors. For the same reason the tangential magnetic field is discontinuous at the conductor surfaces, and a surface current density given by $\mathbf{n} \times \mathbf{H}_t = \mathbf{J}_s$ exists on the conductors. The direction of $\mathbf{J}_s$ is along the z axis since both $\mathbf{n}$ and $\mathbf{H}_t$ are vectors in the xy plane. At the conductor surfaces, the normal component of $\mathbf{H}_t$ is zero, and the tangential component of $\mathbf{E}_t$ is zero, and, hence, $|\mathbf{J}_s| = |\mathbf{H}_t| = Y|\mathbf{n}\cdot\mathbf{E}_t| = Y\rho_s/\epsilon = \rho_s/(\mu\epsilon)^{1/2}$. The total current flowing on the conductor S_2 is given by the line integral of the current density around S_2, and, hence,

$$I_0 = \oint_{S_2} |\mathbf{J}_s|\, dl = \oint_{S_2} \frac{\rho_s\, dl}{(\mu\epsilon)^{1/2}} = Qv_c \tag{3}$$

where I_0 is the magnitude of the total axial current on S_2, Q is the charge on S_2 per unit length, and v_c is the velocity of light in a medium with parameters μ, ϵ. Since the unit normal is oppositely directed on the two conductors, the current I_0 flows down the line on one conductor and back on the other. The total current on each conductor is the same since the total charge (positive on one conductor and negative on the other) is the same on both. Thus, associated with the magnetic field wave $\mathbf{H}_t$ is a unique current wave $I = I_0 e^{-jkz}$. The ratio of the potential difference between the conductors to the total current flowing on the conductors for a wave propagating in either the positive or negative z direction is defined as the characteristic impedance Z_c of the transmission line, i.e.,

$$Z_c = \frac{V_0}{I_0} = \frac{V_0}{Q}(\mu\epsilon)^{1/2} = \frac{1}{Cv_c} = \frac{\epsilon}{C}Z \tag{4}$$

where $Z = (\mu/\epsilon)^{1/2}$ is the intrinsic impedance of the medium surrounding the conductors, and C is the electrostatic capacitance between the two conductors per unit length. The characteristic impedance differs from the intrinsic or wave impedance by a factor which is a function of the geometry of the line only. Even though the frequency of the wave may be thousands of megahertz, the capacitance function in the expression for Z_c is that for a static electric field existing between S_1 and S_2. This latter property follows again only because $\nabla_t \times \mathbf{E}_t$ is identically zero. The ideal transmission line is unique among wave-guiding structures in that its parameters are those associated with static field distributions in the transverse plane.

The time-average electric energy per unit length of line in a single propagating TEM wave is given by

$$W_e = \frac{1}{4}\epsilon \iint \mathbf{E}_t \cdot \mathbf{E}_t^* \, dx \, dy = \frac{1}{4}\epsilon \iint \nabla_t \Phi \cdot \nabla_t \Phi \, dx \, dy \tag{5}$$

where the integral is over the whole xy plane, and ϵ is assumed real. The two-dimensional form of Green's first identity is[1]

$$\iint_S (\nabla_t \Phi \cdot \nabla_t \Phi + \Phi \nabla_t^2 \Phi) \, dS = \oint_C \Phi \frac{\partial \Phi}{\partial n} \, dl$$

where C is a closed contour bounding the surface S, and $\mathbf{n}$ is the unit outward normal to the contour C. By introducing suitable cuts as in Fig. 4.2 and using Green's first identity, the expression for W_e becomes

$$W_e = \frac{1}{4}\epsilon \oint_{S_2} V_0 \frac{\partial \Phi}{\partial n} \, dl \tag{6}$$

since $\nabla_t^2 \Phi = 0$, $\Phi = 0$ on S_1, $\Phi = V_0$ on S_2 and, as r becomes large, $\Phi(\partial \Phi/\partial n)$ vanishes at least as fast as r^{-2}, when the net charge on the two conductors is zero. On the conductor S_2 we have $\partial \Phi/\partial n = \rho_s/\epsilon$, since $\mathbf{n}$ points toward the conductor and, hence, $\partial \Phi/\partial n = |\mathbf{e}|$, where $\mathbf{e}$ is the outward-directed normal electric field at the surface. Using this result in (6) shows that

$$W_e = \frac{1}{4}\epsilon \oint_{S_2} \frac{V_0}{\epsilon} \rho_s \, dl = \frac{1}{4}V_0 Q = \frac{1}{4}V_0^2 C \tag{7}$$

[1] Section A.1c.

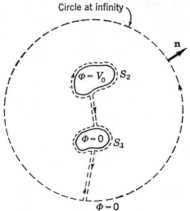

Fig. 4.2. Contour for evaluation of W_e.

where C is the electrostatic capacitance per unit length. The additional factor $\frac{1}{2}$ arises because of averaging over one period in time.

The time-average magnetic energy in the wave per unit length is

$$W_m = \frac{1}{4}\mu \iint \mathbf{H}_t \cdot \mathbf{H}_t^* \, dS = \frac{1}{4}\mu Y^2 \iint \mathbf{E}_t \cdot \mathbf{E}_t^* \, dS = W_e \qquad (8)$$

since $\mathbf{H}_t = \bar{\mathbf{Y}} \cdot \mathbf{E}_t$, and $\mu Y^2 = \epsilon$. The inductance L per unit length of line is most suitably defined by the relation

$$\frac{1}{4}LI_0^2 = W_m. \qquad (9)$$

Since $I_0 Z_c = V_0$, we get $LI_0^2 = LV_0^2/Z_c^2 = 4W_m = 4W_e = V_0^2 C$, and, hence,

$$Z_c = \left(\frac{L}{C}\right)^{1/2} \qquad (10a)$$

and, using the relation $Z_c = (Cv_c)^{-1}$, we also get

$$(LC)^{1/2} = (\mu\epsilon)^{1/2} = v_c^{-1}. \qquad (10b)$$

The definition of inductance given in (9) may be easily shown to give the same results as the low-frequency definition in terms of flux linkages per unit current.

The power flowing along the line is given by

$$P = \frac{1}{2}\,\mathrm{Re}\iint \mathbf{E}_t \times \mathbf{H}_t^* \cdot \mathbf{a}_z \, dx\, dy$$

$$= \frac{1}{2}\,\mathrm{Re}\iint Y\mathbf{E}_t \cdot \mathbf{E}_t^* \, dx\, dy$$

$$= \frac{2YW_e}{\epsilon}$$

$$= v_c(W_e + W_m) \qquad (11)$$

and is seen to be equal to the sum of the electric and magnetic energy per unit length multiplied by the velocity of propagation. This result is to be expected since the wave is propagating with a velocity v_c and power is a rate of flow of energy. Equation (11) may be put into the following form as well:

$$P = \frac{1}{2}Z_c I_0^2 = \frac{1}{2}V_0 I_0. \tag{12}$$

The result obtained so far has demonstrated that the ideal transmission line behaves in many respects as a low-frequency circuit element. This accounts for the great success of the conventional transmission-line theory based on the concept of distributed parameters and using a circuit-analysis approach. The voltage and current waves $V = V_0 e^{-jkz}$, $I = I_0 e^{-jkz}$ are readily shown to be solutions to the differential equations

$$\frac{d^2 V}{dz^2} + \omega^2 LCV = 0 \tag{13a}$$

$$\frac{d^2 I}{dz^2} + \omega^2 LCI = 0. \tag{13b}$$

These are the conventional telegraphists' or transmission-line equations for an ideal line.

If we have a transmission line with N separate conductors as in Fig. 4.3, there will be $N-1$ basic TEM modes of propagation. A suitable set of modes is obtained by solving Laplace's equation in the xy plane for the boundary conditions

$$\Phi = V_1 \text{ on } S_1 \quad \text{and} \quad \Phi = 0 \text{ on } S_i, \; i = 2, 3, \ldots, N$$
$$\Phi = V_2 \text{ on } S_2 \quad \text{and} \quad \Phi = 0 \text{ on } S_i, \; i = 1, 3, \ldots, N$$
$$\Phi = V_N \text{ on } S_N \quad \text{and} \quad \Phi = 0 \text{ on } S_i, \; i = 1, 2, \ldots, N-1.$$

Since the potential is arbitrary to within an additive constant, any one of the above solutions can be expressed in terms of the remainder, and so there are only $N-1$ independent solutions. A superposition of these basic modes will permit the potential of all the conductors to be arbitrarily specified. For a three-wire line there are just two independent TEM modes of propagation, as shown in Fig. 4.4.

The Two-Conductor Line with Small Losses

If the medium in which the conductors are located has finite losses, then ϵ must be replaced by $\epsilon(1 - j \tan \delta_l)$, where $\tan \delta_l$ is the loss tangent of the material. The effective conductivity

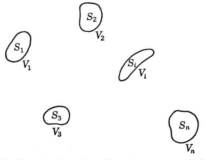

Fig. 4.3. Cross section of an N-conductor transmission line.

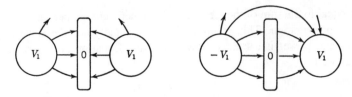

Symmetrical mode   Unsymmetrical mode

Fig. 4.4. Basic modes of propagation for a symmetrical three-wire transmission line.

of the medium is σ, where

$$\sigma = \omega\epsilon \tan \delta_l. \tag{14}$$

If the conductors are still assumed to be perfect conductors, the fundamental mode of propagation is still a pure TEM mode, since making ϵ complex does not alter the solution to
Maxwell's equations, apart from the change in ϵ throughout. The total shunt current per unit
length of line flowing from conductor S_2 across to the conductor S_1 is

$$I_s = \oint_{S_2} \sigma \mathbf{n} \cdot \mathbf{e} \, dl$$

$$= \oint_{S_2} \frac{\sigma}{\epsilon} \rho_s \, dl = \frac{\sigma Q}{\epsilon}. \tag{15}$$

The shunt conductance G is given by

$$G = \frac{I_s}{V_0} = \frac{\sigma C}{\epsilon} = \omega C \tan \delta_l. \tag{16}$$

Although G is given by an expression which is similar to that for a static field problem, it
is a function of frequency because of the factor ω arising in the expression for σ, and also
because $\tan \delta_l$ is a function of frequency. The propagation constant k now becomes

$$jk = j\omega[\mu\epsilon(1 - j \tan \delta_l)]^{1/2} = j\omega \left[\mu\epsilon \left(1 - j\frac{G}{\omega C} \right) \right]^{1/2}$$

$$\approx j\omega(\mu\epsilon)^{1/2} \left(1 - j\frac{1}{2} \tan \delta_l \right) = j\beta + \alpha \tag{17a}$$

since $\tan \delta_l \ll 1$ in any practical case and where β is the phase constant and α the attenuation
constant. Using (10b) and (16) shows that

$$j\beta + \alpha = j\omega(LC)^{1/2} + \frac{1}{2} \left(\frac{L}{C} \right)^{1/2} G. \tag{17b}$$

The characteristic impedance now becomes

$$Z_c = \left[\frac{L}{C(1 - jG/\omega C)} \right]^{1/2} = \left(\frac{j\omega L}{j\omega C + G} \right)^{1/2}.$$

When the conductivity of the conductors is finite, power is absorbed in the conductors, and, hence, there must be a component of the Poynting vector directed into the conductors. This, in turn, implies at least a longitudinal component of electric field. In general, longitudinal components of both electric and magnetic fields will exist. A longitudinal component of magnetic field will have associated with it transverse currents on the conductors. Since these transverse currents arise only because of the perturbation of the TEM mode into a mode with longitudinal field components, they are small in comparison with the longitudinal currents. Thus the losses associated with the transverse currents are also small in comparison with the power loss arising from the longitudinal current. For a first approximation, we may assume that the transverse currents are negligible. This qualitative argument is supported to some extent by the rigorous analysis of some simple conductor configurations such as the single-wire line [4.1]. The single-wire line has circular symmetry and does allow a solution for a mode with no axial magnetic field component, but such symmetry is not present in all transmission lines used in practice, and we may expect small transverse currents to exist, in general.

If we assume that the current on the conductors flows only in the z direction, and also that the medium surrounding the conductors is homogeneous and isotropic with constant parameters μ and ϵ, the field may be derived from a vector potential function $\mathbf{A}$, where $\mathbf{A}$ has only a z component. From Section 1.6, the relevant equations are found to be

$$\mathbf{B} = \nabla \times \mathbf{A} \tag{18a}$$

$$\mathbf{E} = -j\omega\mathbf{A} + \frac{\nabla\nabla \cdot \mathbf{A}}{j\omega\epsilon\mu} \tag{18b}$$

$$\nabla^2\mathbf{A} + k^2\mathbf{A} = 0. \tag{18c}$$

Equation (18c) holds only in the region external to the conductors. The potential $\mathbf{A}$ has only a z component, because the current is only in the z direction, and no other components of $\mathbf{A}$ are required to satisfy the boundary conditions. It is seen from (18a) that $\mathbf{B}$ and, hence, $\mathbf{H}$ have transverse components only.

In Section 3.6 on the discussion of reflection of plane waves from a conducting plane, it was demonstrated that the effect of finite conductivity in good conductors may be accounted for by introducing the surface impedance Z_m of the metal, where

$$Z_m = \frac{1+j}{\sigma\delta_s} = R_m + jX_m.$$

If the metal is assumed to have the same permeability as free space, the skin depth δ_s is given by $(2/\omega\mu_0\sigma)^{1/2}$. At the conductor surface, the current density $\mathbf{J}_s$ and the longitudinal electric field are related as follows:

$$Z_m\mathbf{J}_s = E_z. \tag{19}$$

The current $\mathbf{J}_s$ is given by

$$\mathbf{J}_s = \mathbf{n} \times \mathbf{H}_t = \mu^{-1}\mathbf{n} \times (\nabla \times \mathbf{A})$$

$$= \mu^{-1}[\nabla\mathbf{n} \cdot \mathbf{A} - (\mathbf{n} \cdot \nabla)\mathbf{A}] = -\mu^{-1}\frac{\partial\mathbf{A}}{\partial n} \tag{20}$$

where $\mathbf{n}$ is the unit outward normal from the conductors, and $\mathbf{n} \cdot \mathbf{A} = 0$, since $\mathbf{n}$ and $\mathbf{A}$ are perpendicular. From (18b) it is seen that

$$\mathbf{a}_z E_z = \frac{k^2 + \gamma^2}{j\omega\epsilon\mu} \mathbf{A} \tag{21}$$

if it is assumed that $\mathbf{A}$ propagates along the line according to the factor $e^{-\gamma z}$. Combining (19), (20), and (21) shows that $\mathbf{A}$ must satisfy the impedance boundary condition

$$\frac{j\omega\epsilon Z_m}{k_c^2} \frac{\partial \mathbf{A}}{\partial n} = -\mathbf{A} \tag{22}$$

where $k_c^2 = k^2 + \gamma^2$. A solution for $\mathbf{A}$ which satisfies (18c) and the boundary condition (22) is extremely difficult, if not impossible, to obtain for a general two-conductor transmission line. We are thus forced to consider techniques that will give us approximate solutions for the quantities in which we are interested. A variational principle for obtaining an approximate solution for the propagation constant γ will be developed, but first we will give the usual derivation of the parameters of a line with small losses.

If the field distribution around the line and the current flowing on the line are assumed to be the same as in the ideal transmission line, then the power loss in the conductors per unit length is

$$P_L = \frac{1}{2} R_m \oint_{S_1 + S_2} |\mathbf{J}_s|^2 \, dl \tag{23}$$

where $|\mathbf{J}_s|$ is the magnitude of the current density for the unperturbed mode. A resistance R per unit length of line may be defined by the relation $\frac{1}{2} I_0^2 R = P_L$, or

$$R = \frac{R_m \oint_{S_1 + S_2} |\mathbf{J}_s|^2 \, dl}{\left(\oint_{S_2} |\mathbf{J}_s| \, dl \right)^2}. \tag{24}$$

This latter expression can be evaluated when the field distribution, and, hence, $\mathbf{J}_s$, is known. The attenuation constant α can be found from the rate of power loss with distance along the line. If the propagation constant is $\gamma = j\beta + \alpha$, the power carried along the line relative to that at $z = 0$ is given by $P = P_0 e^{-2\alpha z}$, and the rate of decrease of power is $-dP/dz = 2\alpha P$. The rate of decrease of power is equal to the power loss per unit length, and, hence,

$$\alpha = P_L / 2P \tag{25}$$

where P_L includes both the series resistance loss $\frac{1}{2} R I_0^2$ and the shunt conductance loss $\frac{1}{2} V_0^2 G$ if present. Just as the shunt conductance can be associated with a complex value of dielectric constant, the series resistance can be associated with a complex value of permeability. The characteristic impedance of the line is thus given by

$$Z_c = \left[\frac{L(1 - jR/\omega L)}{C(1 - jG/\omega C)} \right]^{1/2} = \left(\frac{R + j\omega L}{G + j\omega C} \right)^{1/2} \approx \left(\frac{L}{C} \right)^{1/2} \tag{26}$$

since $G \ll \omega C$, $R \ll \omega L$ when the losses are small. Equation (12) for power flow along the line is still valid to a first approximation, and, hence, (25) gives

$$\alpha = \frac{1}{2}(RY_c + GZ_c).$$ (27a)

The propagation constant γ is now given by

$$\gamma = j\beta + \alpha = j\omega(\mu\epsilon)^{1/2}\left(1 - \frac{jR}{\omega L}\right)^{1/2}\left(1 - \frac{jG}{\omega C}\right)^{1/2}$$

$$\approx j\omega(LC)^{1/2} + \frac{1}{2}(RY_c + GZ_c).$$ (27b)

The approximation of taking Z_c real is inconsistent from an energy-conservation point of view but has negligible effect on the calculation of the reflection coefficient and input impedance of a terminated line and is, therefore, frequently made in practice.

The above simple analysis of the transmission line with small losses gives results which are sufficiently accurate for most engineering purposes but is somewhat lacking in rigor. The variational principle to be given next will provide a firmer foundation for the above approach.

The starting point for developing a variational expression for the eigenvalues or propagation constants for a wave function is the *wave equation*. If it is assumed that the only currents on the conductors are in the z direction, the vector potential $\mathbf{A}$ may be represented as follows:

$$\mathbf{A} = \mathbf{a}_z \psi(x, y)e^{-\gamma z}$$ (28)

where ψ is a solution of the scalar Helmholtz equation

$$\nabla_t^2 \psi + k_c^2 \psi = 0$$ (29)

and satisfies the impedance boundary condition

$$g\frac{\partial \psi}{\partial n} - \psi = 0$$ (30)

where $k_c^2 = k^2 + \gamma^2$ and $g = j\omega\epsilon Z_m/k_c^2$, and the normal $\mathbf{n}$ is assumed directed into the conductor rather than outward. Multiplying (29) by ψ and integrating over the whole xy plane gives

$$k_c^2 = -\frac{\iint \psi \nabla_t^2 \psi \, dS}{\iint \psi^2 \, dS}.$$ (31)

To determine whether this expression gives a solution for k_c^2 whose first variation is zero, when we perturb the functional form of ψ by a small amount $\delta\psi$ we conduct the variation to obtain

$$2k_c\delta k_c \iint \psi^2 \, dS + 2k_c^2 \iint \psi \, \delta\psi \, dS = -\iint (\delta\psi \, \nabla_t^2 \psi + \psi \, \nabla_t^2 \, \delta\psi) \, dS.$$

If Green's second identity[2] is used, which in the present case gives

$$\iint (\psi \nabla_t^2 \delta\psi - \delta\psi \nabla_t^2 \psi) \, dS = \oint_C \left(\psi \frac{\partial \delta\psi}{\partial n} - \delta\psi \frac{\partial \psi}{\partial n} \right) dl$$

where $\mathbf{n}$ is directed outward from the closed contour C as illustrated previously in Fig. 4.2, the variation in k_c^2 is found to be

$$2k_c \, \delta k_c \iint \psi^2 \, dS = -2 \iint \delta\psi(k_c^2\psi + \nabla_t^2\psi) \, dS - \oint_C \left(\psi \frac{\partial \delta\psi}{\partial n} - \delta\psi \frac{\partial \psi}{\partial n} \right) dl. \quad (32)$$

The surface integral vanishes since ψ satisfies the scalar Helmholtz equation. If the contour integral would also vanish, then the variation δk_c in k_c^2 would vanish as well. In this case (31) would provide a variational expression which would permit us to calculate a value for k_c^2 accurate to the second order for an assumed approximate form for ψ accurate to the first order only. The logical approximation to use for ψ would be the potential function for the same transmission line but with perfect conductors. To ensure that our variation $\delta\psi$ in ψ would make the contour integral vanish is a difficult condition to meet in practice. We therefore attempt to find a suitable term to add to (31) such that the first variation in k_c^2 will vanish independently of what boundary conditions $\delta\psi$ satisfies. The term that we add to the numerator of (31) must be such that it does not affect the value of k_c^2 when the true eigenfunction ψ is used in the resultant expression, and must also be such that its first variation will cancel the contour integral in (32). Any function multiplied by the expression $j\omega\epsilon Z_m(\partial\psi/\partial n) - k_c^2\psi$ from (30) will satisfy the first condition. After a few trial substitutions involving combinations of ψ and $\partial\psi/\partial n$, we find that a suitable expression to add to the numerator in (31) is

$$\oint_C \left(\frac{j\omega\epsilon Z_m}{k_c^2} \frac{\partial\psi}{\partial n} - \psi \right) \frac{\partial\psi}{\partial n} \, dl.$$

In place of (32), we now obtain

$$2\delta k_c \left[k_c \iint \psi^2 \, dS + \frac{g}{k_c} \oint_C \left(\frac{\partial\psi}{\partial n} \right)^2 dl \right]$$

$$= -2 \iint \delta\psi(k_c^2\psi + \nabla_t^2\psi) \, dS + 2 \oint_C \left(g \frac{\partial\psi}{\partial n} - \psi \right) \frac{\partial \delta\psi}{\partial n} \, dl = 0$$

since ψ is a solution of (29) and (30). Thus an appropriate variational integral or expression for k_c^2 is

$$k_c^2 = -\frac{\iint \psi \nabla_t^2\psi \, dS - \oint_C \left(\frac{j\omega\epsilon Z_m}{k_c^2} \frac{\partial\psi}{\partial n} - \psi \right) \frac{\partial\psi}{\partial n} \, dl}{\iint \psi^2 \, dS}. \quad (33)$$

The integral in the denominator serves as a normalization term, and, in practice, we may always choose our approximate eigenfunction, say ψ_0, such that $\iint \psi_0^2 \, dS$ is equal to unity or some other convenient factor.

---

[2] Section A.1c.

For the approximation to ψ we choose the correct potential function ψ_0 for the ideal transmission line. This function is a solution of Laplace's equation $\nabla_t^2 \psi_0 = 0$ in the xy plane. In
terms of ψ_0, the fields are given by $H_x = \mu^{-1} \partial\psi_0/\partial y$, $H_y = -\mu^{-1}\partial\psi_0/\partial x$, $E_x = ZH_y$,
and $E_y = -ZH_x$, where $Z = (\mu/\epsilon)^{1/2}$. In terms of the potential function Φ introduced for
the analysis of the ideal line, we had $H_x = Y\partial\Phi/\partial y$ and $H_y = -Y\partial\Phi/\partial x$, and, hence,
apart from any normalization factor, we may identify ψ_0 with $\mu Y \Phi$. The current density on
the conductors is given by $\mu|\mathbf{J}_s| = |\partial\psi_0/\partial n|$. On the conductor S_2 we have $\psi_0 = \mu Y V_0/2$,
while on S_1 we have

$$\psi_0 = \frac{-\mu Y V_0}{2}.$$

For later convenience it will be assumed that ψ_0 has been normalized so that $\iint \psi_0^2 \, dS = \mu\epsilon$,
and thus $\iint \Phi^2 \, dS$ equals unity. Introducing the above values of ψ_0 and $\partial\psi_0/\partial n$ at the conductor
surfaces, the expression for k_c^2 becomes

$$k_c^2 = -Z\frac{V_0}{2}\left(\oint_{S_1} |\mathbf{J}_s|\, dl + \oint_{S_2} |\mathbf{J}_s|\, dl\right) + \frac{jkZZ_m}{k_c^2}\oint_{S_1+S_2} |\mathbf{J}_s|^2 \, dl \qquad (34)$$

since $\partial\psi_0/\partial n$ is positive on S_2 and negative on S_1. The surface integral vanishes since ψ_0
is a solution of Laplace's equation in the xy plane. The first two integrals give $ZV_0 I_0$,
which is equal to $2ZP$, where P is the power flowing along the line. In the last integral
$R_m \oint_{S_1+S_2} |\mathbf{J}_s|^2 \, dl = 2P_L$, where P_L is the power loss in the conductors. Since the conductivity
is finite, the magnetic field penetrates into the conductor, and, hence, there is a net amount
of magnetic energy stored internal to the conductor surface. At the surface the magnetic field
is equal to $\mathbf{J}_s$ in magnitude and decays exponentially with the distance u into the conductor
according to the factor e^{-u/δ_s}. Thus the internal magnetic energy is

$$W_{mi} = \frac{1}{4}\mu_0 \int_0^{u_0} \oint_{S_1+S_2} |\mathbf{J}_s|^2 e^{-2u/\delta_s} \, du\, dl \qquad (35)$$

where the integral over u need be taken only to some interior point u_0 where the field is
negligible. From (35) we get

$$W_{mi} = \frac{1}{4}\mu_0\frac{\delta_s}{2}\oint_{S_1+S_2} |\mathbf{J}_s|^2 \, dl$$

$$= \frac{1}{4}(\omega\sigma\delta_s)^{-1}\oint_{S_1+S_2} |\mathbf{J}_s|^2 \, dl$$

$$= \frac{1}{2\omega}P_L$$

when $(2/\omega\mu_0\sigma)^{1/2}$ is substituted for δ_s. An internal inductance L_i may be defined by the relation
$\frac{1}{4}I_0^2 L_i = W_{mi}$, and thus $L_i = R/\omega$, where R is the series resistance of the line per unit length.
With this result we are able to replace the last integral in (34) by $-2kZ(-jP_L + 2\omega W_{mi})/k_c^2$,
and, hence, the equation for k_c^2 becomes

$$k_c^4 + 2ZPk_c^2 + 2kZ(-jP_L + 2\omega W_{mi}) = 0. \qquad (36)$$

The solution for k_c^2 is

$$k_c^2 = -ZP + [(ZP)^2 + 2kZ(jP_L - 2\omega W_{mi})]^{1/2}$$
$$\approx \frac{k(jP_L - 2\omega W_{mi})}{P} \tag{37}$$

since the second term in the brackets is small compared with the first, and the expression to the one-half power may be expanded according to the binomial expansion. Replacing k_c^2 by $k^2 + (j\beta + \alpha)^2$, we get

$$2\beta\alpha = \frac{kP_L}{P} \tag{38a}$$

$$k^2 - \beta^2 + \alpha^2 = -\frac{2W_{mi}}{P}\omega k. \tag{38b}$$

Since the losses are small, we must have $\beta \approx k$, and, hence, (38a) gives essentially

$$\alpha = \frac{P_L}{2P}. \tag{39a}$$

In (38b) we may replace $k^2 - \beta^2$ by $(k + \beta)(k - \beta) \approx 2k(k - \beta)$. Also, since the right-hand side of (38b) is of magnitude $2k\alpha$, we see that α^2 is negligible in comparison, and, hence,

$$2k(k - \beta) = -\frac{2\omega k W_{mi}}{P} - \alpha^2 \approx -\frac{2\omega k W_{mi}}{P}$$

and thus

$$\beta = k + \frac{\omega W_{mi}}{P}. \tag{39b}$$

The use of the variational principle has established that the simple approach employed to calculate the parameters of the line with losses is valid and gives results which are correct to the second order for an assumed field distribution correct to only the first order, provided the transverse currents may be neglected. It also shows that the change in β from the unperturbed value k is principally due to an increase in the magnetic energy in the region internal to the conductors. Equations (39) may also be written in the following form when the approximation $Z_c = (L/C)^{1/2}$ is made:

$$\alpha = \frac{I_0^2 R}{2I_0^2 Z_c} = \frac{R}{2Z_c} \tag{40a}$$

$$\beta = \omega(LC)^{1/2} + \frac{\omega L_i}{2Z_c} \approx \omega[C(L + L_i)]^{1/2}. \tag{40b}$$

In practical lines at high frequencies the internal inductance L_i is negligible in comparison with the external inductance L. When ϵ is made complex to account for the shunt conductance, we will obtain the complete expression for α corresponding to (27) given earlier. Since the internal inductance and series resistance arise because of the surface impedance Z_m of the conductor, we may regard the series resistance as an imaginary part added to the inductance and, hence, also consider the effect of the series resistance as equivalent to making the permeability μ complex.

4.2. THE CHARACTERISTIC IMPEDANCE OF TRANSMISSION LINES

We have seen that if we can determine the capacitance C of a transmission line, we can then obtain the characteristic impedance from the relation $Z_c = (\mu\epsilon)^{1/2}C^{-1}$. There are many methods for the solution of Laplace's equation in two dimensions. We shall discuss only the conformal transformation method and the variational methods which yield upper and lower bounds on Z_c. Many configurations that occur in practice are extremely difficult, if not impossible, to obtain exact solutions for, so that an approximate solution such as can be obtained by variational methods has to suffice. The big advantage of a variational method is that the capacitance is correct to the second order for an approximate solution for the field distribution correct to only the first order.

Characteristic Impedance by Conformal Mapping

Consider the solution of Laplace's equation $\nabla_t^2 \Phi$ for the two-dimensional electrostatic problem illustrated in Fig. 4.5, with the boundary conditions $\Phi = \Phi_1$ on S_1 and $\Phi = \Phi_2$ on S_2. The curves S_1 and S_2 can represent the two conductors of a transmission line. In general, the solution to this problem is difficult unless S_1 and S_2 coincide with constant coordinate curves. For a more suitable choice of coordinates this may actually be the case, and we would then obviously use this set of coordinates in preference to the x, y coordinate system. Consider, therefore, a transformation of coordinates, from the x, y coordinates over to the general curvilinear coordinates, u, v, where

$$u = u(x, y) \tag{41a}$$

$$v = v(x, y). \tag{41b}$$

We will take the unit vectors along the u and v coordinate curves as $\mathbf{a}_1$ and $\mathbf{a}_2$, respectively. We will restrict our coordinate transformations to those that can be effected through the complex function transformation

$$W = F(Z) = F(x + jy) = u + jv. \tag{42}$$

The function $F(Z)$ is restricted to be an analytic function of Z; that is, $dW/dZ = dF/dZ$

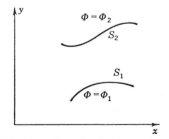

Fig. 4.5. A two-dimensional electrostatic problem.

must have a unique value at each point in the xy plane. The condition that F should be analytic is ensured if u and v satisfy the Cauchy–Riemann equations[3]

$$\frac{\partial u}{\partial x} = \frac{\partial v}{\partial y} \qquad \frac{\partial u}{\partial y} = -\frac{\partial v}{\partial x}. \tag{43}$$

Conversely, if $F(Z)$ is analytic, the Cauchy–Riemann equations are satisfied. Since $\nabla_t u = \mathbf{a}_x \, \partial u/\partial x + \mathbf{a}_y \, \partial u/\partial y$ and

$$\nabla_t v = \mathbf{a}_x \frac{\partial v}{\partial x} + \mathbf{a}_y \frac{\partial v}{\partial y} = -\mathbf{a}_x \frac{\partial u}{\partial y} + \mathbf{a}_y \frac{\partial u}{\partial x}$$

from (43), we have the result that $\nabla_t u \cdot \nabla_t v = 0$ and, hence, u and v form an orthogonal coordinate system. The scale factors h_1 and h_2 are given by

$$\frac{1}{h_1^2} = \left(\frac{\partial u}{\partial x}\right)^2 + \left(\frac{\partial u}{\partial y}\right)^2 = \left(\frac{\partial v}{\partial y}\right)^2 + \left(\frac{\partial v}{\partial x}\right)^2 = \frac{1}{h_2^2} = \frac{1}{h^2} = \left|\frac{dW}{dZ}\right|^2 \tag{44}$$

and are equal by virtue of the Cauchy–Riemann equations. The coordinate transformation effected by the complex function transformation (42) is a very special type of coordinate transformation for which $h_1 = h_2$. Laplace's equation in the uv coordinate system is

$$\frac{\partial}{\partial u} \frac{h_2}{h_1} \frac{\partial \Phi}{\partial u} + \frac{\partial}{\partial v} \frac{h_1}{h_2} \frac{\partial \Phi}{\partial v} = \frac{\partial^2 \Phi}{\partial u^2} + \frac{\partial^2 \Phi}{\partial v^2} = 0 \tag{45}$$

since $h_1 = h_2 = h$. The potential function Φ, therefore, satisfies the same equation in the u, v coordinate system that it does in the x, y coordinate system. If in the u, v coordinate system the curves S_1 and S_2 can be represented by constant coordinate curves as illustrated in Fig. 4.6, the solution to the equation $\partial^2 \Phi/\partial u^2 + \partial^2 \Phi/\partial v^2 = 0$ will be much simpler than the solution to $\partial^2 \Phi/\partial x^2 + \partial^2 \Phi/\partial y^2 = 0$. The energy stored in the electrostatic field is

$$W_e = \frac{1}{2}\epsilon \iint_{\substack{xy \\ \text{plane}}} \left[\left(\frac{\partial \Phi}{\partial x}\right)^2 + \left(\frac{\partial \Phi}{\partial y}\right)^2 \right] dx\, dy = \frac{1}{2}\epsilon \iint_{\substack{xy \\ \text{plane}}} |\nabla_t \Phi|^2 \, dx\, dy$$

$$= \frac{1}{2}\epsilon \iint_{\substack{uv \\ \text{plane}}} \left[\frac{1}{h_1^2}\left(\frac{\partial \Phi}{\partial u}\right)^2 + \frac{1}{h_2^2}\left(\frac{\partial \Phi}{\partial v}\right)^2 \right] h_1 \, du\, h_2 \, dv$$

$$= \frac{1}{2}\epsilon \iint_{\substack{uv \\ \text{plane}}} \left[\left(\frac{\partial \Phi}{\partial u}\right)^2 + \left(\frac{\partial \Phi}{\partial v}\right)^2 \right] du\, dv = \frac{1}{2}C(\Phi_2 - \Phi_1)^2 \tag{46}$$

where C is the capacitance of the line per unit length perpendicular to the xy plane. Equations (45) and (46) show that we may treat u and v as rectangular coordinates, and, hence, we may replot the curves $S_1(x, y)$ and $S_2(x, y)$ onto a rectangular uv grid by expressing the x and y

---

[3] The continuity of u and v and their first-order partial derivatives is also required.

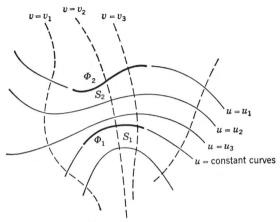

Fig. 4.6. Illustration of constant coordinate curves.

values along S_1 and S_2 as u and v values by means of the coordinate transformation equations (41). This mapping of the original curves S_1 and S_2 into new curves in the uv coordinate plane, i.e., the complex W plane, is called a conformal transformation. Equation (46) shows that the capacitance of the conductor configuration in the uv plane is the same as that in the xy plane. Thus, provided we can solve Laplace's equation for the conformal mapping of the original configuration, we can then transform back to the x, y coordinates to obtain the field distribution for our actual physical problem. If we are interested only in the capacitance, we do not even need to transform back to our original x, y coordinates.

As a simple example, consider the problem of two coaxial cylinders as illustrated in Fig. 4.7. The solution to the equation

$$\frac{\partial^2 \Phi}{\partial x^2} + \frac{\partial^2 \Phi}{\partial y^2} = 0$$

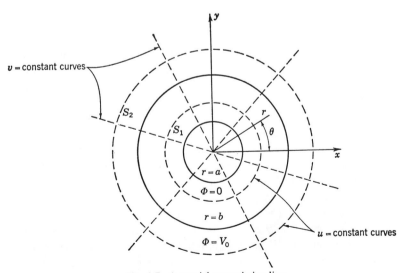

Fig. 4.7. A coaxial transmission line.

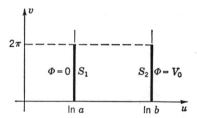

Fig. 4.8. Conformal mapping of a coaxial-line cross section.

is not simple for this configuration. Our natural choice of coordinates would be cylindrical coordinates. This coordinate transformation is obtained through the complex function transformation $W = \ln Z = \ln re^{j\theta} = \ln r + j\theta$ and, hence, $u = \ln r$, $v = \theta$. A plot of the constant uv coordinate curves is given in Fig. 4.7 also. The scale factor h is equal to r. If we treat u and v as rectangular coordinates, the curves S_1 and S_2 defined by $r = a$, $r = b$, $0 < \theta < 2\pi$, map into the curves $u = \ln a$, $u = \ln b$, $0 < v < 2\pi$, as in Fig. 4.8.

The region between the two coaxial cylinders maps into the region inside the rectangle bounded by S_1, S_2 and $v = 0$, $v = 2\pi$. As θ ranges from $-\infty$ to $+\infty$, v does also, and we obtain a periodic mapping of the region between the two coaxial cylinders into the whole infinite strip bounded by the lines $u = \ln a$ and $u = \ln b$. The field in the region $2n\pi < v < 2(n+1)\pi$ is the same as that in the region $0 < v < 2\pi$. The equivalent electrostatic problem in the complex W plane is, therefore, just the simple problem of finding the potential distribution between the two infinite planes. The solution is obviously

$$\Phi = \frac{u - \ln a}{\ln b - \ln a} V_0.$$

The energy stored in the field is

$$W_e = \frac{1}{2}\epsilon \int_0^{2\pi} \int_a^b \left(\frac{\partial \Phi}{\partial u}\right)^2 du = \frac{\pi \epsilon V_0^2}{\ln(b/a)} = \frac{1}{2}CV_0^2. \tag{47}$$

For the equivalent transmission-line problem with a simple-harmonic time variation, the time-average electric energy stored in the field is one-half the above value. The capacitance of the coaxial cylinders is $C = 2\pi\epsilon/\ln(b/a)$ per unit length. When used as a transmission line, the characteristic impedance is $Z_c = (\mu\epsilon)^{1/2}/C = (1/2\pi)(\mu/\epsilon)^{1/2} \ln(b/a)$. For an air-filled line, $(\mu_0/\epsilon_0)^{1/2} = 120\pi$ and $Z_c = 60 \ln(b/a)$ ohms.

We may regard the mapping of the curves S_1 and S_2 into the complex W plane as a distorted representation of our original S_1 and S_2 curves and the uv coordinate curves. In this mapping, the curvilinear coordinates have been straightened out into a rectangular grid. Obviously, any constant potential curve $\Phi = K$ in the xy plane is a constant potential curve $\Phi = K$ in the uv coordinate plane also, since we have not changed the physical problem. Similarly, any curve $\nabla_t \Phi = 0$ goes over into a curve $\nabla_t \Phi = 0$ in the uv coordinate frame. At any point in the xy plane, $\nabla_t \Phi = \mathbf{K}$ goes over into $(\mathbf{a}_1/h)(\partial\Phi/\partial u) + (\mathbf{a}_2/h)(\partial\Phi/\partial v) = \mathbf{K}$ in the u, v curvilinear coordinate system, and hence $\mathbf{a}_1 \partial\Phi/\partial u + \mathbf{a}_2 \partial\Phi/\partial v = h\mathbf{K}$. Therefore, the gradient of Φ, when calculated in the complex W plane according to the relation $\mathbf{a}_1 \partial\Phi/\partial u + \mathbf{a}_2 \partial\Phi/\partial v$, must be divided by the scale factor h to obtain the value of the gradient of Φ at the corresponding point in the xy plane.

If we differentiate (43) with respect to x and y, we get

$$\frac{\partial^2 u}{\partial x^2} - \frac{\partial^2 v}{\partial x\, \partial y} = \frac{\partial^2 u}{\partial x^2} + \frac{\partial^2 u}{\partial y^2} = 0$$

and similarly $\nabla_t^2 v = 0$. Hence, if in our coordinate transformation the curves S_1 and S_2 map into constant coordinate curves, we then have the result that u (or v) gives the potential distribution directly; that is, $\Phi = u$ (or $\Phi = v$) is the required solution. It is not necessary for the original S_1 and S_2 curves to map into constant-value coordinate curves, although the method is usually of only limited help if this does not happen. Several transformations in succession may be used in order to obtain the desired final configuration. In practice, the main difficulty is in obtaining the desired transformation. Problems that can be solved by conformal-mapping techniques can be solved by other methods also, although in many cases by considerably more labor and dexterity on the part of the individual.

4.3. The Schwarz–Christoffel Transformation

The Schwarz–Christoffel transformation is a conformal transformation that will map the real axis in the Z plane into a general polygon in the W plane with the upper half of the Z plane mapping into the region interior to the polygon. To derive the basic transformation, we consider first a curve S in the Z plane and its conformal mapping S_1 in the W plane as illustrated in Fig. 4.9.

Let the unit tangent to S in the Z plane be t_0 and the unit tangent to S_1 in the W plane be τ_0, where

$$t_0 = \lim_{\Delta Z \to 0} \frac{\Delta Z}{|\Delta Z|}$$

$$\tau_0 = \lim_{\Delta W \to 0} \frac{\Delta W}{|\Delta W|}.$$

If the mapping function is $W = F(Z)$, we have

$$\frac{dW}{dZ} = F'(Z) = \lim_{\Delta Z \to 0} \frac{\Delta W}{\Delta Z} = \lim_{\Delta Z \to 0} \frac{\Delta W/|\Delta W|}{\Delta Z/|\Delta Z|} \left| \frac{\Delta W}{\Delta Z} \right| = \frac{\tau_0}{t_0} |F'(Z)|$$

and, hence,

$$\tau_0 = \frac{F'(Z)}{|F'(Z)|} t_0. \tag{48}$$

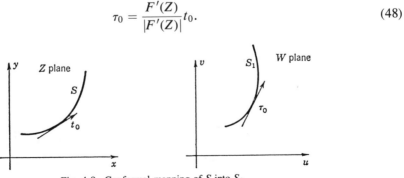

Fig. 4.9. Conformal mapping of S into S_1.

The angle that τ_0 makes with the u axis is given by

$$\angle\tau_0 = \angle t_0 + \angle F'(Z). \tag{49}$$

Consider next the function

$$F'(Z) = A(Z - x_1)^{-k_1}(Z - x_2)^{-k_2} \cdots (Z - x_N)^{-k_N} = A\prod_{i=1}^{N}(Z - x_i)^{-k_i} \tag{50}$$

where A is an arbitrary constant, k_i is a real number, and $x_1 < x_2 < \cdots < x_N$. If the x axis is chosen as the curve S in the Z plane, the angle of the unit tangent to the mapping of S, that is, to S_1, in the W plane will be

$$\angle\tau_0 = \angle F' = \angle A - \sum_{i=1}^{N} k_i \angle(Z - x_i). \tag{51}$$

For $x < x_i$, we have $\angle(x - x_i) = \pi$, and, for $x > x_i$, we have

$$\angle(x - x_i) = 0.$$

Thus, as each point x_i is passed in the Z plane, the angle of τ_0 changes discontinuously by an amount $k_i\pi$, and a polygon such as that illustrated in Fig. 4.10 is traced out. The exterior angles to the polygon are $k_i\pi$, and these angles must add up to 2π if the polygon is to be closed; hence,

$$\sum_{i=1}^{N} k_i = 2. \tag{52}$$

The constant A serves to rotate and magnify the figure in the W plane. The points x_i map into the points $W_i = F(x_i)$. The mapping function is given by

$$W = \int^{Z} F'(Z)\,dZ + B = A\int^{Z} \prod_{i=1}^{N}(Z - x_i)^{-k_i}\,dZ + B \tag{53}$$

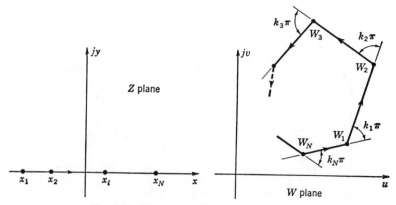

Fig. 4.10. The Schwarz–Christoffel transformation.

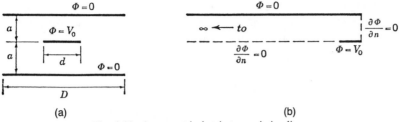

(a)

(b)

Fig. 4.11. A symmetrical-strip transmission line.

where B is an arbitrary constant which serves to translate the figure in the W plane. The integration of (53) is usually not possible unless $|k_i| = 0, \frac{1}{2}, 1, \frac{3}{2}$, or 2. It should be noted that in this transformation there is a factor for each vertex corresponding to a finite value of x, but no terms for the points $x = \pm\infty$.

In practice, we would normally wish to map a given polygon in the W plane into the x axis. This requires the inverse mapping function giving Z as a function of W. Generally, it is difficult to obtain this inverse transformation, and so we would usually proceed by trial and error to set up a transformation of the form (53) which will map the x axis into the given polygon. In this procedure we are aided by the condition that three of the points x_i may be arbitrarily chosen.

To illustrate the use of the Schwarz–Christoffel transformation, we will find the capacitance per unit length of the symmetrical-strip transmission line illustrated in Fig. 4.11(a). If the spacing a is small compared with the width D of the two outer strips, the field is essentially zero at the edges of the outer strips, and, hence, we may simplify the problem by assuming that the outer strips are infinitely wide. The center strip is assumed to have negligible thickness and the medium between the ground planes to be free space with electrical parameters ϵ_0, μ_0.

From a consideration of symmetry we may reduce the electrostatic boundary-value problem to that illustrated in Fig. 4.11(b). Along a portion of the boundary, the normal gradient of the potential Φ is zero. With reference to Fig. 4.12(a) the capacitor configuration is seen to be obtained as the points W_1 and W_4 approach $-\infty$. As a first step toward the solution of this boundary-value problem, we will map the boundary of the polygon in the W plane into the real axis in the Z plane, as in Fig. 4.12(b). The external angles of the polygon are equal to $\pi/2$, and so all the parameters k_i are equal to $+\frac{1}{2}$. The mapping function is of the form

$$W = A' \int^Z (Z - x_1)^{-1/2}(Z - x_2)^{-1/2}(Z - x_3)^{-1/2}(Z - x_4)^{-1/2} \, dZ + B$$

$$= A \int^Z \left[\left(1 - \frac{Z}{x_1}\right)\left(1 - \frac{Z}{x_4}\right)(Z - x_3)(Z - x_2) \right]^{-1/2} dZ + B$$

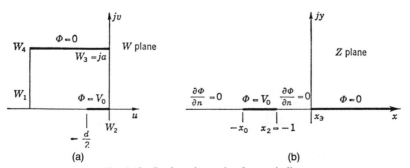

(a)

(b)

Fig. 4.12. Conformal mapping for a strip line.

where $A = A'(x_1 x_4)^{-1/2}$. In order to cover the whole x axis, we let x_1 and x_4 tend to infinity, and the corresponding terms in the integral may be replaced by unity. If we also choose $x_2 = -1$ and $x_3 = 0$, our mapping function becomes

$$W = A \int^Z \frac{dZ}{[Z(Z+1)]^{1/2}} + B$$

$$= 2A \ln[Z^{1/2} + (Z+1)^{1/2}] + B. \tag{54}$$

When $W = 0$, $Z = -1$, and, when $W = ja$, $Z = 0$; so (54) gives

$$2A \ln(-1)^{1/2} = j\pi A = -B$$

$$ja = B$$

and, hence, our final transformation becomes

$$W = -\frac{2a}{\pi} \ln[Z^{1/2} + (Z+1)^{1/2}] + ja. \tag{55}$$

When $W = -d/2$, let the corresponding value of Z be $-x_0$. From (55) we then get

$$x_0^{1/2} + (x_0 - 1)^{1/2} = e^{d\pi/4a}$$

or

$$x_0 = \cosh^2(\pi d/4a) \tag{56}$$

since $Z = -x_0$ is a negative number.

At this point in the analysis we are really no better off than we were at the beginning since the boundary-value problem in the Z plane [Fig. 4.12(b)] is just as difficult to solve as the original one. We therefore will attempt to find another mapping function which will map the x axis in the Z plane into a rectangle in the W' plane as illustrated in Fig. 4.13. If this can be done, we may determine the capacitance at once, since the final boundary-value problem in the W' plane is an elementary one. The capacitance is obviously given by $C = \epsilon_0/v_0$, where v_0 is the plate spacing in the W' plane. In the W' plane we make $W_1' = 0$ correspond to the point $Z = -x_0$, $W_2' = 1$ correspond to $Z = -1$, and $W_3' = 1 + jv_0$ correspond to $Z = 0$. Our required mapping function thus has the form

$$W' = A_1 \int^Z \frac{dZ}{[Z(Z+x_0)(Z+1)]^{1/2}} + B_0. \tag{57}$$

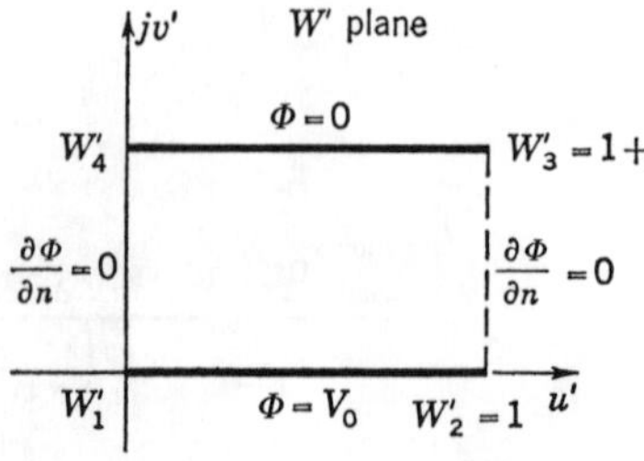

Fig. 4.13. Final mapping for a strip-line cross section.

This integral cannot be evaluated in terms of elementary functions. It does, however, represent an inverse elliptic function, and tables are available for its evaluation. To evaluate the constants A_1 and B_0, we substitute the values $Z = -x_0, -1, 0$ and the corresponding values of W' into (57). Since the upper limit of integration is a negative number or zero, we make the following change in variables, $Z = -\lambda^2$, and (57) becomes

$$W' = \frac{2jA_1}{x_0^{1/2}} \int^{(-Z)^{1/2}} \frac{d\lambda}{[(1-\lambda^2)(1-\lambda^2/x_0)]^{1/2}} + B_0. \tag{58}$$

The inverse elliptic function $\mathrm{sn}^{-1}(x, k)$ is defined by the integral

$$\mathrm{sn}^{-1}(x, k) = \int_0^x \frac{d\lambda}{[(1-\lambda^2)(1-k^2\lambda^2)]^{1/2}} \tag{59}$$

where x may be a complex variable, and k is called the modulus of the elliptic function. Since B_0 is as yet arbitrary in (58), we may put in a lower limit of integration and, by introducing a new constant

$$A_0 = 2jA_1/x_0^{1/2}$$

we get

$$W' = A_0 \, \mathrm{sn}^{-1}[(-Z)^{1/2}, x_0^{-1/2}] + B_0 \tag{60}$$

since $x_0^{-1/2}$ corresponds to k in (59). When $k^2 = 0$, that is, x_0 infinite, the integration in (59) is easily performed and gives

$$\int_0^{(-Z)^{1/2}} \frac{d\lambda}{(1-\lambda^2)^{1/2}} = \sin^{-1}(-Z)^{1/2}.$$

The particular elliptic function occurring in (60) is a generalization of the inverse sine function and the notation $\mathrm{sn}^{-1}[(-Z)^{1/2}, k]$ has been chosen for this reason.[4] The elliptic function $x = f = \mathrm{sn}\, W$, where W is given by the integral in (59), is a *doubly periodic function*, i.e., a function that is periodic along two different directions in the complex plane. In a unit cell or rectangle with sides corresponding to the two periods, the function $f = \mathrm{sn}\, W$ takes on all its possible values as W varies throughout the unit cell. Consequently, as f takes on all its possible values, the inverse function $W = \mathrm{sn}^{-1} f$ varies throughout the unit cell or rectangle. It is for these reasons that inverse elliptic functions always occur in the mapping of finite rectangular polygons. As one of the sides of the polygon or rectangle becomes infinite, the inverse elliptic functions degenerate into inverse trigonometric or hyperbolic functions.

Returning to our problem and putting in the boundary conditions $Z = -x_0$ for $W' = 0$, $Z = -1$ for $W' = 1$, and $Z = 0$ for $W' = 1 + jv_0$, we get

$$A_0 \, \mathrm{sn}^{-1}(x_0^{1/2}, x_0^{-1/2}) + B_0 = 0 \tag{61a}$$

$$A_0 \, \mathrm{sn}^{-1}(1, x_0^{-1/2}) + B_0 = 1 \tag{61b}$$

$$B_0 = 1 + jv_0. \tag{61c}$$

---

[4]For a brief treatment of the elliptic functions, see [4.2]. For a more complete treatment see [4.3].

TABLE 4.1

d/a	x_0	K	K'	v_0	Z_c	$Z_{c,\text{approx}}$
2	6.3	1.64	2.26	0.695	65.5	94.2
4	134	1.57	3.84	0.41	38.6	47.1
8	$0.25e^{4\pi}$	1.57	6.98	0.225	21.2	23.6
10	$0.25e^{5\pi}$	1.57	8.55	0.184	17.3	18.8
20	$0.25e^{10\pi}$	1.57	16.4	0.0958	9.03	9.4

Solving (61a) and (61b) for B_0 and substituting into (61c) gives

$$v_0 = \frac{-j \, \text{sn}^{-1}(1, x_0^{-1/2})}{\text{sn}^{-1}(x_0^{1/2}, x_0^{-1/2}) - \text{sn}^{-1}(1, x_0^{-1/2})} . \tag{62}$$

The function $\text{sn}(x + jy)$ is periodic in x with period $4K$, where

$$K = \int_0^1 \frac{d\lambda}{[(1 - \lambda^2)(1 - k^2\lambda^2)]^{1/2}} \tag{63}$$

and periodic in y with period $2K'$, where

$$K' = \int_0^1 \frac{d\lambda}{[(1 - \lambda^2)(1 - \lambda^2 + k^2\lambda^2)]^{1/2}} = \int_1^{1/k} \frac{d\lambda}{[(\lambda^2 - 1)(1 - k^2\lambda^2)]^{1/2}} . \tag{64}$$

Also $\text{sn } K = 1$, and $\text{sn}(K - jK') = 1/k$; so $\text{sn}^{-1}(1, k) = K$, and $\text{sn}^{-1}(1/k, k) = K - jK'$. Hence, (62) gives

$$v_0 = \frac{K}{K'} = \frac{K(k)}{K(k')} \tag{65}$$

where $k' = (1 - k^2)^{1/2}$, and K as given by (63) is called the complete elliptic integral of the first kind and has been tabulated in [4.4].[5] In Table 4.1 the values of x_0, K, K', and v_0 are given for several typical values of the parameter d/a. The value of x_0 for a given value of d/a is determined by (56). The capacitance per unit length of the strip line is four times that of the mapping of one quarter section in the W' plane, that is, $C = 4\epsilon_0/v_0$. The characteristic impedance is given by $Z_c = 30\pi v_0$ for an air-filled line. Values of Z_c are also given in the table as well as the approximate values $Z_{c,\text{approx}}$ determined by considering only the parallel-plate capacitance $2d\epsilon_0/a$. For $d/a > 4$, the error is much less than 1% if K is replaced by $\pi/2$ and K' by $\ln 2 + \pi d/4a$. It is thus seen that, for d/a sufficiently large, the $\ln 2$ term is negligible, and the parallel-plate capacitance, neglecting fringing fields, is all that needs to be taken into account.

Once the required conformal transformations have been obtained, the capacitance per unit length and the characteristic impedance are quite straightforward to evaluate. Unfortunately, this is not true for the series resistance of the line since this depends on $(\partial\Phi/\partial n)^2$. The charge density as obtained, in say the W' plane, according to the relation $\epsilon_0(\partial\Phi/\partial v')$ must be multi-

---

[5]This reference also contains several relief drawings which clearly illustrate the double-periodicity property of the Jacobi elliptic functions.

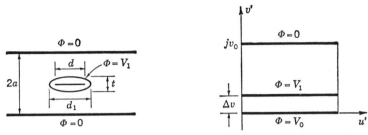

Fig. 4.14. Mapping of an equipotential contour corresponding to a conductor cross section of finite thickness.

plied by $|dW'/dZ||dZ/dW| = |dW'/dW|$ to obtain the true charge density corresponding to that on the actual conductors. The resultant expression is often very complicated to integrate in closed form. In practice, it is not realistic to deal with infinitely thin conducting strips with acute angles, since the charge density becomes infinite at such an edge and leads to values of attenuation and conductor losses which are too large and may even become infinite.[6] For the purpose of evaluating the capacitance, the resulting error introduced by replacing a thin strip by an infinitely thin strip is usually negligible. For evaluating the attenuation we may replace the infinitely thin strip by a strip which coincides with one of the original equipotential contours surrounding the thin strip. The equipotential contour has rounded corners and is often a better approximation to the conductor cross section than an ideal rectangular cross section. In addition, it has the great advantage that the same mapping function used to map the infinitely thin strip also maps the desired equipotential contour, and the final mapping of the conductor cross section is normally a simple contour in the final mapped representation of the strip-line cross section. The above ideas will be clarified by an application to the strip line originally considered.

Figure 4.14 illustrates the original strip-line cross section and an equipotential contour $\Phi = V_1$, which is now taken to be the cross section of the actual center conductor. The mapping of one quadrant of this equipotential contour onto the W' plane is also illustrated in the same figure. If the separation of the two contours $\Phi = V_0$ and $\Phi = V_1$ is designated as Δv in the W' plane, the conductor elliptical cross section in the uv plane is determined by finding the u and v values corresponding to $jv' = j \Delta v$, $0 < u' < 1$, by means of the two mapping functions (55) and (60). The spacing Δv in the W' plane has to be chosen so as to give the correct thickness t for the conductor cross section in the actual line. It should be noted that the conductor width has also been modified from its zero-thickness width d to a new width d_1 which again depends on Δv. If $\Delta v \ll v_0$, the capacitance in the W' plane is changed by a negligible amount; that is, $\Delta C = \epsilon_0 \Delta v[v_0(v_0 - \Delta v)]^{-1}$, and the thickness t may be assumed zero for the purpose of evaluating the capacitance.

From (23) the conductor loss is given by

$$P_L = \frac{1}{2}R_m \oint_W |\mathbf{J}_s|^2 |dW| = \frac{1}{2}R_m Y_0^2 \oint_W \left|\frac{\partial \Phi}{\partial n}\right|^2 |dW| \tag{66}$$

since

$$|\mathbf{J}_s| = Y_0 \left|\frac{\partial \Phi}{\partial n}\right|$$

[6]See Problem 4.6.

where $\partial\Phi/\partial n$ is the normal gradient at the conductor surface, and the integral is taken around the conductor cross sections in the W plane with $|dW| = dl$ being the element of path length. The value of $\partial\Phi/\partial n$ is given by

$$\left|\frac{\partial\Phi}{\partial n}\right| = \left|\frac{\partial\Phi}{\partial v'}\right|\left|\frac{dW'}{dW}\right| = \frac{V_0}{v_0}\left|\frac{dW'}{dW}\right| \tag{67}$$

since

$$\frac{\partial\Phi}{\partial v'} = \frac{\partial}{\partial v'}\frac{V_1 v'}{v_0 - \Delta v} = \frac{V_1}{v_0 - \Delta v} = \frac{V_0}{v_0}.$$

The conductor losses are, therefore, given by

$$P_L = \frac{1}{2}R_m Y_0^2 \frac{V_0^2}{v_0^2}I \tag{68}$$

where

$$I = \oint_W \left|\frac{dW'}{dW}\right|^2 |dW|.$$

The integral I may also be evaluated in either the Z or W' plane. Since

$$\left|\frac{dW'}{dW}\right||dW| = \left|\frac{dW'}{dW}\right|\left|\frac{dW}{dZ}\right||dZ| = \left|\frac{dW'}{dZ}\right||dZ| = |dW'|$$

we get

$$I = \oint_W \left|\frac{dW'}{dW}\right|^2 |dW| = 4\oint_Z \left|\frac{dW'}{dW}\right|\left|\frac{dW'}{dZ}\right||dZ| = 4\oint_{W'}\left|\frac{dW'}{dW}\right||dW'|. \tag{69}$$

The factor 4 arises in the integration in the Z and W' planes because only one quadrant of the strip-line cross section has been mapped. The particular plane in which to carry out the integration is dependent on the complexity of the integrand. From (55), we get

$$\left|\frac{dW}{dZ}\right| = \frac{a}{\pi}|[Z(Z+1)]^{-1/2}|$$

while, from (60),

$$\left|\frac{dW'}{dZ}\right| = \frac{|A_0|x_0^{1/2}}{2}|[Z(Z+x_0)(Z+1)]^{-1/2}|$$

where $A_0 = [\operatorname{sn}^{-1}(1, k) - \operatorname{sn}^{-1}(k^{-1}, k)]^{-1} = (jK')^{-1}$, with $k^2 = x_0^{-1}$. Substituting into (69) gives

$$I = \frac{\pi}{a(kK')^2}\oint_Z \frac{|dZ|}{|(Z+x_0)[Z(Z+1)]^{1/2}|} = 4\oint_{W'}\left|\frac{dW'}{dZ}\right|\left|\frac{dZ}{dW}\right||dW'|$$

$$= \frac{2\pi}{K'a}\oint_{W'}\frac{|dW'|}{|[1 - k^2\operatorname{sn}^2(jK'W' - jK' + K, k)]^{1/2}|} \tag{70}$$

since, from (60), $(-Z)^{1/2} = \text{sn}[(W'-B_0)/A_0] = \text{sn}(jK'W' - jK' + K, k)$. By means of the formulas and relations between the various elliptic functions tabulated in the book by Jahnke and Emde, we shall be able to perform the integration in the W' plane along the contour $v' = v_0$, $0 < u' < 1$, corresponding to the ground planes in the strip line. By considering Δv very small, we shall also be able to obtain an approximate value for the integral along $v' = \Delta v$, $0 < u' < 1$, corresponding to the center conductor.

For $W' = u' + jv_0 = u' + jK/K'$, we have

$$\text{sn}(+jK'W' - jK' + K, k) = \text{sn}(+jK'u' - jK', k)$$
$$= [k\,\text{sn}(jK'u', k)]^{-1} = [jk\,\text{tn}(K'u', k')]^{-1}$$

where k' is the complementary modulus $(1-k^2)^{1/2}$. Along this contour the integrand becomes

$$[1 - k^2\,\text{sn}(jK'W' - jK' + K, k)]^{-1/2} = \text{sn}(K'u', k')$$

upon using the relations $\text{tn}(K'u', k') = \text{sn}(K'u', k')/\text{cn}(K'u', k')$ and $\text{sn}^2(K'u', k') + \text{cn}^2(K'u', k') = 1$. If we denote the part of I corresponding to the integration along the $v' = v_0$ contour by I_1, we have

$$I_1 = \frac{2\pi}{aK'}\int_0^1 \text{sn}(K'u', k')\,du' = \frac{2\pi}{aK'^2}\int_0^{K'} \text{sn}(K'u', k')\,d(K'u')$$

$$= \frac{2\pi}{ak'K'^2}\ln\frac{\text{dn}(K', k') - k'\,\text{cn}(K', k')}{1 - k'} \tag{71}$$

where $\text{dn}^2(K', k') = 1 - k'^2\,\text{sn}^2(K', k')$. Now $\text{dn}(K', k') = k$,

$$\text{cn}(K', k') = 0,$$

(71) becomes

$$I_1 = \frac{2\pi}{ak'K'^2}\ln\frac{k}{1-k'}, \tag{72}$$

and the conductor loss arising from the ground planes is

$$P_{L,1} = \frac{\pi R_m Y_0^2 V_0^2}{aK^2 k'}\ln\frac{k}{1-k'} \tag{73}$$

where K/K' has been substituted for v_0. For $d \geq 4a$, this expression simplifies considerably since under these conditions we have, to an accuracy better than 1%,

$$k = x_0^{-1/2} = 2e^{-\pi d/4a}$$

$$k' = 1 - 2e^{-\pi d/2a} \approx 1$$

$$\frac{k}{1-k'} = e^{\pi d/4a}$$

$$K = \frac{\pi}{2}$$

and (73) becomes

$$P_{L,1} = R_m Y_0^2 \left(\frac{V_0}{a}\right)^2 d \tag{74}$$

which is just the loss which would arise for a uniform field existing between the center strip and ground planes and a zero field outside this region.

For the second part of the integral, which we will denote as I_2, the contour is along $v' = \Delta v$, $0 < u' < 1$. In this case

$$\mathrm{sn}[jK'(u' + j\,\Delta v) - jK' + K, k] = k^{-1}\,\mathrm{dn}(K'u' + jK'\,\Delta v, k')$$

and, since $1 - \mathrm{dn}^2(\ ) = k'^2\,\mathrm{sn}^2(\ )$, we get

$$\begin{aligned}
I_2 &= \frac{2\pi}{aK'k'}\int_0^1 \frac{du'}{|\mathrm{sn}(K'u' + jK'\,\Delta v, k')|}\\
&= \frac{2\pi}{ak'K'^2}\int_0^{K'} \frac{du}{|\mathrm{sn}(u + jK'\,\Delta v, k')|}.
\end{aligned} \tag{75}$$

If we choose $\Delta v \le 0.1v_0$ and also assume that $d \ge 4a$, then

$$K'\,\Delta v = K\frac{\Delta v}{v_0} \le \frac{\pi}{20}$$

and $\mathrm{sn}(u + jK'\,\Delta v, k') \approx \mathrm{sn}(u, k') + jK'\,\Delta v\,\mathrm{cn}(u, k')\,\mathrm{dn}(u, k')$. The modulus of this term is

$$[\mathrm{sn}^2(u, k') + (K'\,\Delta v)^2\,\mathrm{cn}^2(u, k')\,\mathrm{dn}^2(u, k')]^{1/2} \approx [\mathrm{sn}^2(u, k') + (K'\,\Delta v)^2]^{1/2}$$

since, for u near zero, $\mathrm{cn}(u, k')$ and $\mathrm{dn}(u, k')$ are nearly equal to unity. When $u = K'$, $\mathrm{cn}(K', k')$ equals zero, but now $\mathrm{sn}(K', k')$ equals unity, and so there is negligible error in still having the term $(K'\,\Delta v)^2$, which is less than 0.03, present. For values of u up to u_1, where u_1 is about 0.4 or less, $\mathrm{sn}(u, k') \approx u$. For values of $u > u_1$, we may neglect the term $(K'\,\Delta v)^2$ in comparison with $\mathrm{sn}^2(u, k')$, and this also removes the error at the upper end of the range, where $u = K'$. The integral I_2 is thus evaluated in two parts, corresponding to the two ranges in u, where different approximations are used. This artifice permits us to carry out the integrations readily, and we get

$$\begin{aligned}
I_2 &= \frac{2}{ak'K'^2}\left[\int_0^{u_1} \frac{du}{[u^2 + (K'\,\Delta v)^2]^{1/2}} + \int_{u_1}^{K'} \frac{du}{\mathrm{sn}(u, k')}\right]\\
&= \frac{2\pi}{ak'K'^2}\left\{\ln\frac{u_1 + [u_1^2 + (K'\,\Delta v)^2]^{1/2}}{K'\,\Delta v} + \ln\frac{\mathrm{cn}(u_1, k') + \mathrm{dn}(u_1, k')}{k\,\mathrm{sn}(u_1, k')}\right\}\\
&\approx \frac{2\pi}{ak'K'^2}\left(\ln\frac{2u_1}{K'\,\Delta v} + \ln\frac{2}{ku_1}\right)\\
&= \frac{2\pi}{ak'K'^2}\ln\frac{4}{kK'\,\Delta v} \tag{76}
\end{aligned}$$

since u_1 is small and $\ln kK' \Delta v$ is the dominant term. Before proceeding any further, it is necessary to establish the value of Δv in terms of the thickness t of the center conductor. Provided t and Δv are small, this may be done directly from the relation $\Delta W = (dW/dW') \Delta W'$, where $\Delta W = j(t/2)$ and $\Delta W' = j \Delta v$ for small displacements from the origin along the v axis. Employing the same approximations as those used to arrive at (76), we get

$$K' \Delta v = \frac{\pi t}{4a}. \tag{77}$$

The conductor loss associated with the center conductor is now readily evaluated and is given by

$$P_{L,2} = \frac{R_m Y_0^2 V_0^2}{\pi a} \left(4 \ln \frac{8a}{\pi t} + \frac{\pi d}{a} \right). \tag{78}$$

The second term in parentheses is the loss which would arise on the assumption of a uniform field between the center strip and the ground planes, while the logarithmic term gives the increase in the losses arising from the concentration of the current at the edges of the center strip.

The power flowing along the line is given by $\frac{1}{2} V_0^2 / Z_c$, and, hence, by using (25), the attenuation factor arising from conductor losses is found to be

$$\alpha = \frac{R_m}{2Z_0 a} \frac{\pi d/2a + \ln(8a/\pi t)}{\ln 2 + \pi d/4a} \quad \text{nepers/unit length}, \tag{79}$$

a result valid for $d \geq 4a$ and $t \leq a/5$.

4.4. Characteristic Impedance by Variational Methods

Variational Expression for Lower Bound on Z_{c0}

The electrostatic energy per unit length stored in the field surrounding the two conducting surfaces represented by the curves S_1 and S_2 in Fig. 4.1 is given by

$$W_e = \frac{1}{2} \epsilon \iint \left[\left(\frac{\partial \Phi}{\partial x} \right)^2 + \left(\frac{\partial \Phi}{\partial y} \right)^2 \right] dx\, dy = \frac{1}{2} C_0 (\Phi_2 - \Phi_1)^2 = \frac{1}{2} C_0 V_0^2 \tag{80}$$

where C_0 is the capacitance of the line per unit length, and $V_0 = \Phi_2 - \Phi_1$ and is the potential difference between the two conductors. Without loss in generality, we may take $\Phi_1 = 0$.

If we compute the first-order variation in W_e due to a first-order variation in the functional form of Φ, subject to the condition that $\Phi = 0$ on S_1 and $\Phi = \Phi_2$ on S_2, we find, by applying the Euler–Lagrange equations to the integrand, that this first-order variation vanishes, provided Φ satisfies the equation[7]

$$\frac{\partial^2 \Phi}{\partial x^2} + \frac{\partial^2 \Phi}{\partial y^2} = \nabla_t^2 \Phi = 0. \tag{81}$$

---

[7] Section A.4.

(The result also follows from Thompson's theorem given in Section 1.3.) Since we know that Φ is a solution of Laplace's equation, we obtain at once the following variational expression for the capacitance and, hence, for the characteristic impedance Z_{c0}:

$$\frac{C_0}{(\epsilon\mu)^{1/2}} = \frac{1}{Z_{c0}} = \frac{\displaystyle\iint \nabla_t\Phi \cdot \nabla_t\Phi \, dx\, dy}{(\mu/\epsilon)^{1/2} V_0^2}. \tag{82}$$

The integral is always positive, and, hence, the stationary value is an absolute minimum.[8] Therefore, if we substitute an approximate value for $\nabla_t\Phi$ into the integral, the calculated value C for the capacitance will always be too large; that is, $C > C_0$, $Z_c < Z_{c0}$. The calculated value of Z_c is always too small, i.e., is a lower bound to the true value Z_{c0}. In practice, we choose a functional form for our approximate solution that has several parameters $\alpha_1, \alpha_2, \ldots, \alpha_N$, by which we can change the functional form of Φ. When we compute the integral, we get a value $C(\alpha_1, \alpha_2, \ldots, \alpha_N)$ for the capacitance that is an ordinary function of the variational parameters α_i. We obtain the best possible solution for C by choosing the minimum value of C that varying the parameters α_i can produce. This required minimum is obtained by treating the α_i as independent variables and equating all $\partial C/\partial\alpha_i$ equal to zero; thus

$$\frac{\partial C}{\partial \alpha_i} = 0, \qquad i = 1, 2, \ldots, N. \tag{83}$$

Equations (83) are a set of N homogeneous equations which, together with the equation $\Phi = V_0$ on S_2, allows us to solve for the α_i and, hence, compute the minimum value of C. Equations (83) only give the ratio of the α_i to any one, say α_1, and, hence, the requirement of the additional condition $\Phi = V_0$ on S_2. The variational parameters α_i may be the coefficients in a Fourier series expansion of Φ.

We may illustrate what the above procedure does as follows. Let Φ_0 be the correct solution for the problem, and $\Phi(\alpha_1)$, a function of one variational parameter only, be the approximate solution. Let $\delta\Phi$ be a variation in the functional form of Φ_0. If we plot the value of C as a function of $\Phi_0 + \delta\Phi$ for a range of functions $\delta\Phi$, we get a curve as illustrated in Fig. 4.15. The true value C_0 is obtained when $\delta\Phi = 0$. The broken curve is a plot of the calculated value of the capacitance as a function of the parameter α_1 in our approximate solution. The best possible solution is the minimum value as obtained by equating $dC(\alpha_1)/d\alpha_1$ to zero. If our function $\Phi(\alpha_1)$ is sufficiently flexible, it will be able to approximate Φ_0 closely, and, since C varies slowly in the vicinity of Φ_0 [δC is of the order of $\iint |(\nabla_t \delta\Phi)|^2 \, dx\, dy$], the calculated value of C will be an excellent approximation to C_0. The more variational parameters α_i we include in our approximation, the more nearly will Φ be able to approximate Φ_0 and yield a closer approximation to C_0. Unfortunately, the numerical work involved increases rapidly with an increase in the number of parameters α_i.

As a simple example, we will find the capacitance of the two coaxial cylinders illustrated in Fig. 4.7. For a first approximation, we take $\Phi = \alpha_0 + \alpha_1 r$. Rather than have to choose our Φ so that $\Phi = 0$ on the inner conductor, and $\Phi = V_0$ on the outer conductor, we can modify (82) to

$$C = \epsilon \frac{\displaystyle\iint |\nabla_t\Phi|^2 \, dx\, dy}{\left(\displaystyle\int_{S_1}^{S_2} \nabla_t\Phi \cdot d\mathbf{l}\right)^2} \tag{84}$$

---

[8]See Section 1.3, where it is shown that the change in W_e is always positive.

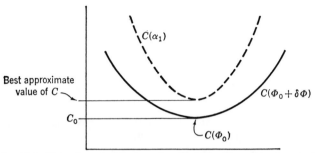

Fig. 4.15. Illustration of minimum value obtained with variational method.

since the potential difference existing between the conductors S_1 and S_2 is simply the line integral of $\nabla_t \Phi$ from S_1 to S_2. Since $\nabla_t \Phi = \alpha_1$ for the above approximation, we have

$$C = \epsilon \frac{\displaystyle\int_0^{2\pi} \int_a^b \alpha_1^2 r \, dr \, d\theta}{\left(\displaystyle\int_a^b \alpha_1 \, dr\right)^2} = \epsilon \pi \frac{b+a}{b-a}. \tag{85}$$

The true value of capacitance is

$$C_0 = \frac{2\pi\epsilon}{\ln(b/a)}.$$

For $b > a$, we have

$$\ln \frac{b}{a} = 2 \left[\frac{b-a}{b+a} + \frac{1}{3} \left(\frac{b-a}{b+a} \right)^3 + \cdots \right] \approx 2 \frac{b-a}{b+a}$$

when b is not much greater than a. Thus we have

$$\frac{\pi\epsilon(b+a)}{b-a} = C \approx C_0$$

when $b - a$ is small.

If $b = 2a$, $C = 3\pi\epsilon$, $C_0 \approx 2.9\pi\epsilon$, and so the error is around 3%. We note that α_0 does not enter into the expression for C. The reason is that the potential is unique only to within an arbitrary constant. The only value of any significance is the difference in potential.

As a second approximation, take $\Phi = \alpha_0 + \alpha_1 r + \alpha_2 r^2$; thus

$$C = \epsilon \frac{\displaystyle\int_0^{2\pi} \int_a^b (\alpha_1 + 2\alpha_2 r)^2 r \, dr \, d\theta}{\left[\displaystyle\int_a^b (\alpha_1 + 2\alpha_2 r) \, dr \right]^2}$$

$$= \frac{2\pi\epsilon \left[\dfrac{b^2 - a^2}{2} \alpha_1^2 + \dfrac{b^3 - a^3}{3} 4\alpha_1 \alpha_2 + (b^4 - a^4)\alpha_2^2 \right]}{(b-a)^2 \alpha_1^2 + (b^2 - a^2)(b-a) 2\alpha_1 \alpha_2 + (b^2 - a^2)^2 \alpha_2^2}. \tag{86a}$$

Hence,

$$\frac{C}{\pi\epsilon}[(b-a)^2\alpha_1^2 + (b^2-a^2)(b-a)2\alpha_1\alpha_2 + (b^2-a^2)^2\alpha_2^2]$$

$$-\left[(b^2-a^2)\alpha_1^2 + \frac{8}{3}(b^3-a^3)\alpha_1\alpha_2 + 2(b^4-a^4)\alpha_2^2\right] = 0. \quad (86b)$$

In order to obtain the best possible value of C with this approximation, we equate $\partial/\partial\alpha_i$ of (86b) to zero. This is equivalent to equating $\partial C/\partial\alpha_i$ given by (86a) to zero. Let D be the denominator in (86a) and N be the numerator; thus

$$\frac{\partial C}{\partial\alpha_i} = \frac{D(\partial N/\partial\alpha_i) - N(\partial D/\partial\alpha_i)}{D^2} = \frac{1}{D}\left(\frac{\partial N}{\partial\alpha_i} - \frac{N}{D}\frac{\partial D}{\partial\alpha_i}\right) = 0$$

and hence $\partial N/\partial\alpha_i - C\,\partial D/\partial\alpha_i = 0$, which is just the derivative of (86b). Carrying out the differentiation, we obtain the following two homogeneous equations in α_1 and α_2:

$$\alpha_1\left[\frac{2C}{\pi\epsilon}(b-a)^2 - 2(b^2-a^2)\right] + \alpha_2\left[\frac{2C}{\pi\epsilon}(b^2-a^2)(b-a) - \frac{8}{3}(b^3-a^3)\right] = 0$$

$$\alpha_1\left[\frac{2C}{\pi\epsilon}(b^2-a^2)(b-a) - \frac{8}{3}(b^3-a^3)\right] + \alpha_2\left[\frac{2C}{\pi\epsilon}(b^2-a^2)^2 - 4(b^4-a^4)\right] = 0.$$

For a solution for the α_i to exist, the determinant of the above equations must vanish, and this gives directly the value of C without the necessity of evaluating the α_i. By performing the following operations on this determinant:

1. divide first row by $2(b-a)^2$,
2. divide second row by $2(b^2-a^2)(b-a)$,
3. divide second column by $b+a$,
4. subtract first row from second

it reduces to

$$\begin{vmatrix} \dfrac{C}{\pi\epsilon} - \dfrac{b+a}{b-a} & \dfrac{C}{\pi\epsilon} - \dfrac{4}{3}\dfrac{b^3-a^3}{(b^2-a^2)(b-a)} \\[3mm] \dfrac{b+a}{b-a} - \dfrac{4}{3}\dfrac{b^3-a^3}{(b^2-a^2)(b-a)} & \dfrac{4}{3}\dfrac{b^3-a^3}{(b^2-a^2)(b-a)} - 2\dfrac{b^4-a^4}{(b^2-a^2)^2} \end{vmatrix} = 0.$$

If we expand this determinant in terms of the elements of the first row, we get

$$\frac{C}{\pi\epsilon}\begin{vmatrix} 1 & 1 \\[2mm] \dfrac{b+a}{b-a} - \dfrac{4}{3}\dfrac{b^3-a^3}{(b^2-a^2)(b-a)} & \dfrac{(4/3)(b^3-a^3)}{(b^2-a^2)(b-a)} - 2\dfrac{b^4-a^4}{(b^2-a^2)^2} \end{vmatrix}$$

$$= \left(\frac{b+a}{b-a}\right)^2\begin{vmatrix} 1 & \dfrac{4}{3}\dfrac{b^3-a^3}{(b^2-a^2)(b+a)} \\[2mm] 1 - \dfrac{4}{3}\dfrac{b^3-a^3}{(b^2-a^2)(b+a)} & \dfrac{(4/3)(b^3-a^3)}{(b^2-a^2)(b+a)} - \dfrac{2(b^4-a^4)}{(b^2-a^2)(b+a)^2} \end{vmatrix}. \quad (87)$$

There is only one root for C. When $b/a = 2$, this second approximation gives $C = 2.889\pi\epsilon$. The true value is $C_0 = 2.886\pi\epsilon$, so the error in C is only 0.1%. However, this accuracy has been obtained at the expense of considerably more labor.

Variational Expression for Upper Bound on Z_{c0}

The previous section presented a method whereby the upper bound on C and, hence, a lower bound on Z_{c0} could be obtained. By solving the electrostatic problem in terms of the unknown charge distribution, we shall be able to obtain a variational expression which will give a lower bound on the capacitance and, hence, an upper bound on the characteristic impedance Z_{c0}. The method utilizes the Green's function technique for solving boundary-value problems. By comparing the upper and lower bounds for Z_{c0}, we can find at once the maximum possible error in our approximate values.

Consider the two-dimensional electrostatic problem illustrated in Fig. 4.16. The two conductors S_1 and S_2 are of infinite length and parallel to the z axis. We may take S_1 at zero potential and S_2 at a potential V_0. In order that S_2 should be at a potential V_0, there must exist a distribution of positive electric charge on its surface. The only way in which the conductor S_2 can make its presence known to the field surrounding it is by this charge distribution on its surface. The magnitude of the charge is equal to $\epsilon E_n = -\epsilon(\partial\Phi/\partial n)$, where E_n is the normal electric field at the surface of S_2 and is equal to the normal derivative of the potential function Φ on the surface S_2. Provided we leave the charge distribution undisturbed, we may remove the conductor S_2 without disturbing the field. We are then left with the problem of finding the potential function Φ that satisfies Poisson's equation

$$\nabla_t^2 \Phi = -\frac{1}{\epsilon}\rho(x', y') \tag{88}$$

and the boundary condition that it vanishes on S_1. The coordinates x', y' are the values of x and y that defined the original conductor surface S_2, and $\rho(x', y')$ is the charge distribution that existed on S_2. As a first step toward the solution of (88), we will find the solution to the problem of a unit charge located at (x', y') in the presence of S_1 and such that the potential vanishes on S_1. [Since the problem is a two-dimensional one, we mean by a unit charge at (x', y') a unit line charge extending from $-\infty$ to $+\infty$ along the z axis.] The most convenient way of representing a unit charge mathematically is by means of the delta function, which is zero for all values of x and y except x', y'. At the point (x', y'), the function is such that if we integrate it over a small interval including the point (x', y'), then the value of the integral

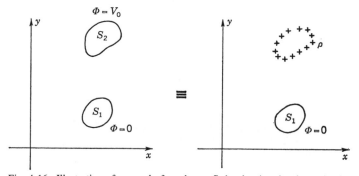

Fig. 4.16. Illustration of removal of conductor S_2 but leaving the charge in place.

is unity. We therefore require a solution of the equation

$$\nabla_t^2 \Phi(x, y) = -\frac{1}{\epsilon} \delta(x - x') \delta(y - y') \tag{89}$$

subject to the boundary condition that Φ equals zero on S_1 and at infinity. The above equation defines the two-dimensional Green's function of the first kind for the given boundary.[9] Let the solution to (89) be represented by $G(x, y|x', y')$; by superposition it follows that the solution to (88) is

$$\Phi(x, y) = \oint_{S_2} G(x, y|x', y') \rho(x', y') \, dl'. \tag{90}$$

On the surface S_1, the potential Φ must reduce to zero, while, on S_2, it reduces to V_0. Hence, from (90), we have

$$\Phi(x, y) = V_0 = \oint_{S_2} G(x, y|x', y') \rho(x', y') \, dl' \qquad x, y \text{ on } S_2. \tag{91}$$

Equation (91) is an integral equation involving the unknown charge distribution $\rho(x', y')$, whose solution would give $\rho(x', y')$ and, hence, would determine the potential function $\Phi(x, y)$ everywhere by means of (90). This is the condition that must be imposed on the charge distribution in order that Φ should reduce to V_0 on S_2. Since G vanishes on S_1 and at infinity, the potential Φ does also. Equation (91) is the basic equation from which we will now derive our variational expression for the capacitance C_0. Multiply (91) by $\rho(x, y)$, and integrate over the surface S_2; thus

$$V_0 \oint_{S_2} \rho(x, y) \, dl = V_0 Q = \oint \oint_{S_2} G(x, y|x', y') \rho(x, y) \rho(x', y') \, dl \, dl' \tag{92}$$

where Q is the total charge on S_2. The capacitance C_0 is equal to Q/V_0, and, hence, $QV_0 = Q^2/C_0$, and we get from (92) the result

$$\frac{1}{C_0} = \frac{\displaystyle\oint \oint_{S_2} G(x, y|x', y') \rho(x, y) \rho(x', y') \, dl \, dl'}{\left[\displaystyle\oint_{S_2} \rho(x, y) \, dl\right]^2} \tag{93}$$

which is our required variational expression. To show that this expression is a stationary expression for $1/C_0$ for arbitrary first-order changes in the functional form of $\rho(x, y)$, let ρ change by $\delta\rho$. Thus

$$\left[\oint_{S_2} \rho(x, y) \, dl\right]^2 \delta\left(\frac{1}{C_0}\right) + \frac{2}{C_0}\left[\oint_{S_2} \rho(x, y) \, dl \oint_{S_2} \delta\rho \, dl\right]$$

$$= \oint \oint_{S_2} G(x, y|x', y') \rho(x, y) \delta\rho(x', y') \, dl \, dl'$$

$$+ \oint \oint_{S_2} G(x, y|x', y') \rho(x', y') \, \delta\rho(x, y) \, dl \, dl'$$

$$= 2 \oint \oint_{S_2} G(x, y|x', y') \rho(x', y') \, \delta\rho(x, y) \, dl \, dl'$$

[9]The Green's function for Poisson's equation may or may not include the factor ϵ. We have included it here for later convenience, but this is not always done.

since G is symmetrical in the variables x, y and x', y', and similarly for ρ and $\delta\rho$. Introducing the total charge Q, we may write the above equations as

$$Q^2 \delta\left(\frac{1}{C_0}\right) = 2\left\{\oint_{S_2} \delta\rho(x, y)\left[-\frac{Q}{C_0} + \oint_{S_2} G(x, y|x', y')\rho(x', y')\,dl'\right]dl\right\}$$

which shows at once that the variation in $1/C_0$ vanishes by virtue of (91), and that $Q = C_0 V_0$. We have, therefore, proved that (93) is a stationary expression for C_0^{-1}, and, hence, it follows that, if we can find an approximate solution for $\rho(x', y')$, we will be able to compute a value for C^{-1} which is correct to the second order. Since G is a symmetrical function, the integral in (93) is always positive, and the stationary value must be an absolute minimum. Consequently, our approximate value of C^{-1} is too large, and, hence, $C < C_0$, $Z_c > Z_{c0}$, and we have an upper bound for our characteristic impedance Z_{c0}.

In practice, the charge distribution would be approximated by a function containing one or more variational parameters, and the best possible solution for C_0 would be obtained by minimizing (93) with respect to the adjustable variational parameters. The procedure to be followed is the same as that used to treat (84).

4.5. CHARACTERISTIC IMPEDANCE OF A STRIP LINE DETERMINED BY VARIATIONAL METHODS

Lower Bound to Z_{c0}

As an application of the variational methods given in the preceding section, the upper and lower bounds to the characteristic impedance of the shielded strip line with a rectangular center conductor will be obtained. The dimensions of the line are given in Fig. 4.17. In view of the symmetry of the line, we need only find the field distribution for one quarter of the cross section as in Fig. 4.17(b), with the potential Φ satisfying the following boundary conditions:

$$\Phi = 0 \qquad \text{for } y = 0$$

$$\Phi = V_0 \qquad \text{on inner conductor}$$

$$-E_x = \frac{\partial\Phi}{\partial x} = 0, \qquad x = 0,\ 0 \leq y \leq a$$

$$-E_y = \frac{\partial\Phi}{\partial y} = 0, \qquad y = d,\ l \leq x.$$

In each region Φ is a solution of Laplace's equation $\nabla_t^2\Phi = 0$. In region 1, a suitable

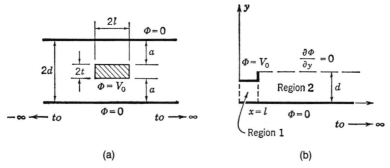

Fig. 4.17. Strip line with rectangular inner conductor.

Fourier series expansion of the potential is[10]

$$\Phi = \frac{V_0 y}{a} + \sum_{n=1}^{\infty} a_n \cosh \frac{n\pi x}{a} \sin \frac{n\pi y}{a}. \tag{94}$$

Each term $\cosh(n\pi x/a) \sin(n\pi y/a)$ satisfies Laplace's equation, and the derivative with respect to x vanishes at $x = 0$. The term $V_0 y/a$ is included to satisfy the boundary condition at $y = a$, and the a_n are unknown coefficients to be determined. In region 2, a suitable expansion of the potential is

$$\Phi = \sum_{n=1,3,\ldots}^{\infty} b_n \sin \frac{n\pi y}{2d} e^{-n\pi x/2d} \tag{95}$$

since $\cos(n\pi y/2d)$ vanishes at $y = d$ for odd values of n, and, hence, $\partial\Phi/\partial y$ does also. The potential vanishes as $x \to \infty$ because of the decaying exponential terms. The unknown coefficients a_n and b_n must be determined so that the electric field and potential are continuous at $x = l$, $0 \le y \le a$. Continuity of the potential at $x = l$ gives

$$V_0 \frac{y}{a} + \sum_{n=1}^{\infty} a_n \cosh \frac{n\pi l}{a} \sin \frac{n\pi y}{a} = \sum_{1,3,\ldots}^{\infty} b_n e^{-n\pi l/2d} \sin \frac{n\pi y}{2d}. \tag{96}$$

Since we have made Φ continuous at $x = l$, then $\partial\Phi/\partial y$, and, hence, E_y, is also continuous, as may be seen at once by differentiating (96) with respect to y. The continuity of E_x is ensured by making $\partial\Phi/\partial x$ continuous, i.e.,

$$\sum_{1}^{\infty} \frac{n\pi}{a} a_n \sinh \frac{n\pi l}{a} \sin \frac{n\pi y}{a} = -\sum_{1,3,\ldots}^{\infty} \frac{n\pi}{2d} b_n e^{-n\pi l/2d} \sin \frac{n\pi y}{2d}, \qquad 0 \le y \le a. \tag{97}$$

A rigorous solution to (96) and (97) is difficult to obtain, and, hence, we appeal to the stationary property of the electrostatic energy integral to obtain an approximate value for the capacitance of the line. The electrostatic energy stored in region 1 is

$$\frac{1}{2}\epsilon_0 \int_0^a \int_0^l \left[\left(\sum_{1}^{\infty} \frac{n\pi}{a} a_n \sinh \frac{n\pi x}{a} \sin \frac{n\pi y}{a} \right)^2 \right.$$
$$\left. + \left(\frac{V_0}{a} + \sum_{1}^{\infty} \frac{n\pi}{a} a_n \cosh \frac{n\pi x}{a} \cos \frac{n\pi y}{a} \right)^2 \right] dx\,dy = W_{e1}.$$

Using the orthogonal properties of the $\sin(n\pi y/a)$ and $\cos(n\pi y/a)$ functions, i.e.,

$$\int_0^a \sin \frac{n\pi y}{a} \sin \frac{m\pi y}{a} dy = \begin{cases} 0, & n \neq m \\ \dfrac{a}{2}, & n = m \end{cases}$$

[10]This form of solution is obtained at once by using the finite Fourier sine transform. See [4.5].

and similarly for $\cos(n\pi y/a)$, the integration is readily carried out to give

$$W_{e1} = \frac{1}{2}\epsilon_0 \left(\frac{V_0^2 l}{a} + \frac{\pi}{4} \sum_1^\infty a_n^2 n \sinh \frac{2n\pi l}{a} \right). \tag{98}$$

Similarly, the electrostatic energy stored in region 2 is found to be

$$W_{e2} = \frac{1}{2}\epsilon_0 \left(\frac{\pi}{4} \sum_{1,3,\dots}^\infty b_n^2 n e^{-n\pi l/d} \right). \tag{99}$$

Let the unknown potential distribution at $x = l$ be

$$G(y) = \begin{cases} V_0\left(\dfrac{y}{a}\right) + g(y), & 0 \le y \le a \\ V_0, & a \le y \le d. \end{cases} \tag{100}$$

We may express the unknown coefficients a_n and b_n in terms of this potential distribution by Fourier series analysis since both sides of (96) are equal to the potential distribution given by Eqs. (100); thus

$$a_n \cosh \frac{n\pi l}{a} = \frac{2}{a} \int_0^a g(y) \sin \frac{n\pi y}{a} \, dy \tag{101a}$$

$$b_n e^{-n\pi l/2d} = \frac{2}{d} \int_0^d G(y) \sin \frac{n\pi y}{d} \, dy. \tag{101b}$$

Substituting into (98) and (99), we get

$$W_e = W_{e1} + W_{e2} = \frac{1}{2}\epsilon_0 \frac{V_0^2 l}{a}$$

$$+ \frac{\pi\epsilon_0}{a^2} \sum_1^\infty n \tanh \frac{n\pi l}{a} \left[\int_0^a g(y) \sin \frac{n\pi y}{a} \, dy \right]^2$$

$$+ \frac{\pi\epsilon_0}{2d^2} \sum_{1,3,\dots}^\infty n \left[\int_0^d G(y) \sin \frac{n\pi y}{2d} \, dy \right]^2 \tag{102}$$

where the identity $\sinh 2\theta = 2 \sinh\theta \cosh\theta$ was used in (98). We may readily show that (102) is stationary with respect to arbitrary first-order variations in the functional form $G(y)$, that is, in $g(y)$. The total energy stored in the space surrounding the strip line in Fig. 4.17 is $4W_e$ per unit length. Hence the capacitance per unit length is $C = 8W_e/V_0^2$. For a first approximation to C we will take $g(y)$ as zero; thus, $a_n = 0$, and

$$b_n = \frac{2}{d} e^{n\pi l/2d} \left(\int_0^a \frac{V_0 y}{a} \sin \frac{n\pi y}{2d} \, dy + \int_a^d V_0 \sin \frac{n\pi y}{2d} \, dy \right) = \frac{8V_0 d}{n^2\pi^2 a} e^{-n\pi l/2d} \sin \frac{n\pi a}{2d}.$$

Hence

$$W_e = \frac{1}{2}\epsilon_0 \frac{V_0^2 l}{a} + \frac{8\epsilon_0 V_0^2 d^2}{\pi^3 a^2} \sum_{1,3,\dots}^\infty \frac{1}{n^3} \sin^2 \frac{n\pi a}{2d}. \tag{103}$$

The series may be summed by the methods discussed in the Mathematical Appendix to give

$$S_0 = \sum_{1,3,\ldots}^{\infty} \frac{1}{n^3} \sin^2 \frac{n\pi a}{2d} = \sum_{n=1,3,\ldots}^{\infty} \frac{1}{n^3} \cos^2 \frac{n\pi t}{2d}$$

$$= \frac{1}{2} \left(\frac{\pi a}{2d} \right)^2 \left(1.5 - \ln \frac{\pi a}{2d} \right) - \frac{1}{36} \left(\frac{\pi a}{2d} \right)^4 - \frac{7}{2700} \left(\frac{\pi a}{2d} \right)^6 - \cdots$$

$$= \sum_{1,3,\ldots}^{\infty} \frac{1}{n^3} - \frac{1}{2} \left(\frac{\pi t}{2d} \right)^2 \left(1.5 - \ln \frac{\pi t}{2d} \right) + \frac{1}{36} \left(\frac{\pi t}{2d} \right)^4 + \frac{7}{2700} \left(\frac{\pi t}{2d} \right)^6 + \cdots . \tag{104}$$

If we let $t = a = d/2$ and subtract the two expressions for the sum S_0, we get the interesting result

$$\sum_{1,3,\ldots}^{\infty} \frac{1}{n^3} \le \frac{\pi^2}{16} \left[1.5 + \ln \frac{4}{\pi} - \frac{\pi^2}{288} - \frac{14}{2700} \left(\frac{\pi}{4} \right)^4 - \cdots \right]. \tag{105}$$

If the first N terms in (105) are summed directly, a comparison with the right-hand side of (105) gives at once the maximum error involved, since the sum of $N + P$ terms is greater than the sum of N terms but yet is less than the right-hand side of (105) because all the higher order terms are preceded by minus signs. The value of the sum is 1.052 to four figures.[11] It is convenient to have the two forms for the sum S_0, depending on whether t is much smaller than a or vice versa. The capacitance C is given by $C = 4\epsilon_0 l / a + (64\epsilon_0 d^2 / \pi^3 a^2)S_0$, and, hence, the characteristic impedance is given by

$$Z_c = Z_0 \left(\frac{4l}{a} + \frac{64d^2 S_0}{\pi^3 a^2} \right)^{-1} \tag{106}$$

and is a lower bound to the true value Z_{c0}. The term $4\epsilon_0 l / a$ is just the ordinary parallel-plate capacitance, while the term involving S_0 is the increase in capacitance due to the field existing past the edges of the center strip. Higher order approximations to C may be obtained by taking

$$g(y) = \sum_{n=1}^{N} c_n \sin \frac{n\pi y}{a} \tag{107}$$

and making the energy integral stationary with respect to the N unknown variational parameters (coefficients) c_n. Needless to say, the numerical work increases rapidly, so we will not pursue the problem any further here.

In practice, the upper and lower plates are of finite width. The field decays as $e^{-n\pi x/2d}$, so, provided the plates are wide enough so that the first mode $e^{-\pi x/2d}$ has decayed to a negligible value, the perturbing effect of finite-width plates is small. A value of x equal to $l + 3d$ makes the exponential term about 1% of what its value is at $x = l$. Thus a total width of $2l + 6d$ is sufficient to make the effect of finite-width plates negligible. The upper and lower plates should extend past the center strip by approximately one and a half times the spacing of the upper and lower strips.

[11]The sum may also be evaluated in terms of the Riemann zeta function. See [4.4].

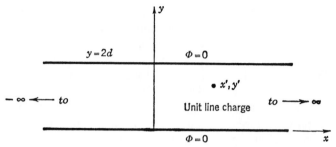

Fig. 4.18. Line charge between two infinite parallel planes.

Upper Bound for Z_{c0}

If we do not disturb the charge distribution that exists on the center conductor, we may remove the center conductor, leaving the charge behind. The electrostatic problem is thus to find a suitable solution to Poisson's equation. As a first step, we require the potential (Green's function) due to a unit line charge at x', y' and subject to the following boundary conditions (see Fig. 4.18):

$$\Phi = 0, \qquad y = 0, 2d$$

$$\Phi \to 0, \qquad |x| \to \infty.$$

The Green's function satisfies the equation

$$\frac{\partial^2 G}{\partial x^2} + \frac{\partial^2 G}{\partial y^2} = -\frac{1}{\epsilon_0} \delta(x - x') \delta(y - y'). \tag{108}$$

A suitable solution may be obtained by the methods given in Section 2.6.

The required solution for the Green's function is

$$G(x, y | x', y) = \frac{1}{\epsilon_0 \pi} \begin{cases} \displaystyle\sum_1^\infty \frac{1}{n} \sin \frac{n\pi y}{2d} \sin \frac{n\pi y'}{2d} e^{-(n\pi/2d)(x'-x)}, & x \le x' \\[2ex] \displaystyle\sum_1^\infty \frac{1}{n} \sin \frac{n\pi y}{2d} \sin \frac{n\pi y'}{2d} e^{(n\pi/2d)(x'-x)}, & x \ge x' \end{cases}$$

$$= \frac{1}{\epsilon_0 \pi} \sum_1^\infty \frac{1}{n} \sin \frac{n\pi y}{2d} \sin \frac{n\pi y'}{2d} e^{-(n\pi/2d)|x-x'|}. \tag{109}$$

The reason why the derivative of G is discontinuous at (x', y') is that the derivative is equal to the negative electric field component that terminates on the unit line charge. Whenever the electric field terminates on a charge, it is discontinuous by the amount $1/\epsilon_0$ times the charge density.

From (93), the capacitance of the strip line is given by

$$\frac{1}{C} = \frac{\displaystyle\oint_{S_2} \oint G(x, y | x', y') \rho(x, y) \rho(x', y') \, dl \, dl'}{\left[\displaystyle\oint_{S_2} \rho(x, y) \, dl \right]^2} \tag{110}$$

where $\rho(x, y)$ is the charge distribution on the center conductor, and the integration is carried out around the center conductor. The simplest approximation to ρ which we can make is to assume that it is constant. We may take the magnitude of ρ as unity since (110) does not depend on the absolute magnitude of ρ. It depends only on the functional form of ρ. Thus the denominator in (110) is simply $16(l + t)^2$. We will integrate the numerator over the variables x', y' first. The integral from $-l$ to l is the sum of two terms, one term for $y' = d - t$ and the other for $y' = d + t$. We may carry out the integration for the nth term in the Green's function and then sum this typical term over all values of n. Bearing in mind that the Green's function has two different forms, depending on whether $x' > x$ or $x' < x$ [see (109)], we obtain

$$\sin \frac{n\pi y}{2d} \int_{-l}^{x} \left(\sin n\pi \frac{d + t}{2d} + \sin n\pi \frac{d - t}{2d} \right) e^{-n\pi x/2d} e^{n\pi x'/2d} \, dx'$$

$$+ \sin n\pi \frac{y}{2d} \int_{x}^{l} \left(\sin n\pi \frac{d + t}{2d} + \sin n\pi \frac{d - t}{2d} \right) e^{n\pi x/2d} e^{-n\pi x'/2d} \, dx'$$

$$+ \int_{d-t}^{d+t} \sin n\pi \frac{y}{2d} \sin n\pi \frac{y'}{2d} e^{-n\pi l/2d} (e^{-n\pi x/2d} + e^{n\pi x/2d}) \, dy'$$

where the Green's function for $x > x'$ is used when $x' = -l$, and the Green's function for $x < x'$ is used when $x' = l$ in the integral over y'. The integration gives the following result:

$$\frac{8d}{n\pi} \sin n\pi \frac{y}{2d} \sin \frac{n\pi}{2} \left[\cos n\pi \frac{t}{2d} \left(1 - e^{-n\pi l/2d} \cosh n\pi \frac{x}{2d} \right) \right.$$

$$\left. + \sin n\pi \frac{t}{2d} e^{-n\pi l/2d} \cosh n\pi \frac{x}{2d} \right].$$

Owing to the factor $\sin(n\pi/2)$ all the even terms vanish. Integrating over the variables x and y, we obtain the following result for the typical nth term:

$$\frac{32d^2}{n^2\pi^2} \left[\frac{n\pi l}{2d} \cos^2 n\pi \frac{t}{2d} + \sin n\pi \frac{t}{d} - \cos n\pi \frac{t}{d} + e^{-n\pi l/d} \left(1 - \sin n\pi \frac{t}{d} \right) \right].$$

Multiplying by $1/\epsilon_0 n\pi$ and summing over n, we obtain the following final result for the upper bound to Z_{c0}:

$$Z_c = \frac{(\mu_0 \epsilon_0)^{1/2}}{C} = \frac{2d^2 Z_0}{\pi^3 (l + t)^2} (S_1 + S_2 - S_3 + S_4) \tag{111}$$

where

$$S_1 = \frac{\pi l}{d} \sum_{1,3,\ldots}^{\infty} \frac{\cos^2 n\pi(t/2d)}{n^2}$$

$$S_2 = \sum_{1,3,\ldots}^{\infty} \frac{\sin n\pi(t/d)}{n^3} = \sum_{1,3,\ldots}^{\infty} \frac{\sin n\pi(a/d)}{n^3}$$

$$S_3 = \sum_{1,3,\ldots}^{\infty} \frac{\cos n\pi(t/d)}{n^3} = -\sum_{1,3,\ldots}^{\infty} \frac{\cos n\pi(a/d)}{n^3}$$

$$S_4 = \sum_{1,3,\ldots}^{\infty} \frac{e^{-n\pi l/d}}{n^3} \left(1 - \sin n\pi \frac{t}{d}\right).$$

The series may be summed by the methods described in the Mathematical Appendix to yield the following results:

$$S_1 = \frac{\pi^3}{8} \frac{l}{d} \left(1 - \frac{t}{d}\right)$$

$$S_2 = \frac{\pi^3}{8} \frac{t}{d} \left(1 - \frac{t}{d}\right)$$

$$S_3 = \sum_{1,3,\ldots}^{\infty} \frac{1}{n^3} + \frac{1}{4} \left[\left(\frac{\pi t}{d}\right)^2 \ln \frac{\pi t}{2d} - \frac{3}{2} \left(\frac{\pi t}{d}\right)^2 + \frac{1}{72} \left(\frac{\pi t}{d}\right)^4 + \cdots \right]$$

$$= -\sum_{1,3,\ldots}^{\infty} \frac{1}{n^3} - \frac{1}{4} \left[\left(\frac{\pi a}{d}\right)^2 \ln \frac{\pi a}{2d} - \frac{3}{2} \left(\frac{\pi a}{d}\right)^2 + \frac{1}{72} \left(\frac{\pi a}{d}\right)^4 + \frac{7}{21,600} \left(\frac{\pi a}{d}\right)^6 + \cdots \right].$$

The series S_4 is readily summed term by term since it converges very rapidly because of the decaying exponential term.

In Fig. 4.19 the upper and lower bounds, as well as the average value, of the characteristic impedance are plotted as a function of l/d for a fixed value of t/d equal to 0.2. For all values of l/d greater than 0.5 the maximum error in the average value of the characteristic impedance is less than 3%, decreasing as l/d becomes large.

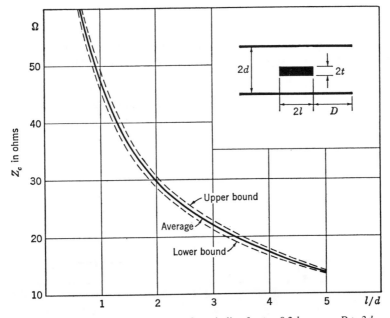

Fig. 4.19. Characteristic impedance of a strip line for $t = 0.2d$, $\epsilon = \epsilon_0$, $D > 3d$.

4.6. Integral Equations for Planar Transmission Lines

A variety of microwave transmission lines involve thin conducting strips placed on a dielectric substrate material. The dominant mode of propagation on these transmission lines is a quasi-TEM mode in that as ω tends to zero the mode becomes a TEM mode. The propagation constant is not a linear function of ω for lines with conductors surrounded by nonhomogeneous dielectric media. These transmission-line problems cannot be solved using conformal mapping methods except in an approximate way. In practice it is found that an effective way to obtain a solution is to formulate integral equations for the current and charge on the conducting strips and to solve these integral equations numerically using the method of moments. The numerical accuracy can be improved without increasing the computation time if a priori knowledge of the behavior of the charge and current densities at the conductor edges are utilized in the expansion of the unknown densities. It is also advantageous to be able to express the dominant or singular part of the Green's function as an orthogonal series over the conducting strips.

The purpose of this section is to formulate the basic integral equation for the current on a single strip and a coupled strip transmission line such as those shown in Fig. 4.20. We will use conformal mapping techniques to establish the edge behavior of the current and also to diagonalize the static Green's function kernel. These results will be used in Sections 4.9 and 4.10 to treat the problem of the microstrip line and the coupled microstrip line, both of which involve conducting strips on a dielectric substrate.

For the transmission line shown in Fig. 4.20(a) a TEM mode of propagation exists. The line consists of a conducting strip of width $2w$ at a height h above a ground plane. Two conducting sidewalls are symmetrically placed at $x = \pm a$. The current on the strip is $J(x)e^{-jk_0 z}$ and is in the z direction. The vector potential A_z is a solution of

$$\left(\frac{\partial^2}{\partial x^2} + \frac{\partial^2}{\partial y^2} \right) A_z = -\mu_0 J(x) \tag{112}$$

$$A_z = 0, \quad x = \pm a, \quad y = 0$$

$$A_z = \text{constant on strip.}$$

An integral equation for A_z can be obtained by introducing a Green's function $G(x, y, x', y')$ that represents the potential from a line source located at x', y'. The Green's function is a solution of

$$\left(\frac{\partial^2}{\partial x^2} + \frac{\partial^2}{\partial y^2} \right) G = -\delta(x - x')\delta(y - y') \tag{113}$$

$$G = 0, \quad x = \pm a, \quad y = 0.$$

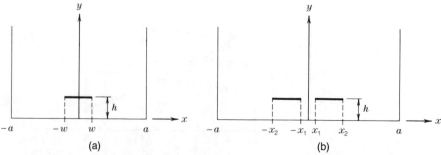

Fig. 4.20. (a) A partially shielded strip line. (b) A partially shielded coupled strip line.

The Green's function is readily found using the methods described in Chapter 2. It has a part that is an even function of x and a part that is an odd function of x. Since $J(x)$ is an even function of x we only need the even part, which we will designate as G_e. We can assume that

$$G_e = \sum_{n=1,3,\ldots}^{\infty} f_n(y) \cos \frac{n\pi x}{2a}$$

which satisfies the boundary conditions at $x = \pm a$. When we substitute this form into (113) and carry out a Fourier analysis we obtain

$$\left[\frac{\partial^2}{\partial y^2} - \left(\frac{n\pi}{2a} \right)^2 \right] a f_n(y) = -\cos \frac{n\pi x'}{2a} \delta(y - y').$$

We now assume that

$$f_n(y) = \begin{cases} C_n \sinh \dfrac{n\pi y}{2a}, & 0 \le y \le y' \\ D_n e^{-n\pi y/2a}, & y \ge y' \end{cases}$$

which satisfies the boundary conditions at $y = 0$ and infinity. We require $f_n(y)$ to be continuous at $y = y'$ and the discontinuity condition

$$a \left. \frac{\partial f_n}{\partial y} \right|_{y'_-}^{y'_+} = -\cos \frac{n\pi x'}{2a}$$

must also hold. The solution can be completed in a straightforward manner to obtain

$$G_e(x, y, x', y') = \sum_{n=1,3,\ldots}^{\infty} \frac{2}{n\pi} \cos \frac{n\pi x}{2a} \cos \frac{n\pi x'}{2a} \sinh \frac{n\pi y_<}{2a} e^{-n\pi y_> /2a}. \qquad (114)$$

By using superposition the solution for A_z is found to be

$$A_z(x, y) = \mu_0 \int_{-w}^{w} G_e(x, y, x', h) J(x') \, dx'.$$

The required integral equation for $J(x')$ is obtained by setting A_z equal to a constant, which we choose as μ_0, on the strip; thus

$$\int_{-w}^{w} G_e(x, h, x', h) J(x') \, dx' = 1. \qquad (115)$$

A straightforward way to solve this integral equation is to use the method of moments. For a numerical solution we first approximate $J(x)$ by an expansion in terms of a finite number of basis functions $\psi_m(x)$, i.e.,

$$J(x) \approx \sum_{m=1}^{M} I_m \psi_m(x). \qquad (116)$$

When we use this expansion in (115) we obtain

$$\sum_{m=1}^{M} G_m(x) I_m \approx 1 \tag{117}$$

where

$$G_m(x) = \int_{-w}^{w} G_e(x, x') \psi_m(x') \, dx'.$$

We can generate a matrix equation for I_m by multiplying both sides by a set of M weighting functions $W_n(x)$ and equating the integrated residual error to zero for each W_n; thus

$$\sum_{m=1}^{M} I_m \int_{-w}^{w} G_m(x) W_n(x) \, dx - \int_{-w}^{w} W_n(x) \, dx = 0, \qquad n = 1, 2, \ldots, M$$

which gives

$$\sum_{m=1}^{M} G_{nm} I_m = S_n, \qquad n = 1, 2, \ldots, M \tag{118}$$

where the matrix elements G_{nm} are given by

$$G_{nm} = \iint_{-w}^{w} G_e(x, x') \psi_m(x') W_n(x) \, dx' \, dx$$

and the source terms are given by

$$S_n = \int_{-w}^{w} W_n(x) \, dx.$$

If we choose $W_n = \psi_n$ then the method of solution is called Galerkin's method. If we choose $W_n = \delta(x - x_n)$ we obtain a solution corresponding to point matching (collocation) whereby the two sides of (117) are matched at M values (usually equispaced) of x. If we choose $W_m(x)$ to be equal to $G_m(x)$ we obtain a least-squares solution having the property that the integrated square of the difference of the two sides in (117) is minimized (see Problem 4.13).

The efficiency and accuracy with which a numerical solution to (115) can be obtained are critically dependent on the expansion used for the unknown current and on our ability to represent the singular part of the Green's function in closed form or as a rapidly converging series. In the present case we note that for large values of n, from (114),

$$G_e \sim \sum_{n=1,3,\ldots}^{\infty} \frac{1}{n\pi} \cos \frac{n\pi x}{2a} \cos \frac{n\pi x'}{2a} = G_{ed} \tag{119}$$

which is the dominant part of the Green's function for $y = y' = h$. This series is readily summed, as shown in the Mathematical Appendix, to give

$$G_{ed} = \frac{1}{2\pi} \operatorname{Re} \sum_{n=1,3,\ldots}^{\infty} \frac{1}{n} [e^{jn\pi(x-x')/2a} + e^{jn\pi(x+x')/2a}]$$

$$= -\frac{1}{4\pi} \ln \left[\tan \frac{\pi |x - x'|}{4a} \tan \frac{\pi |x + x'|}{4a} \right]. \tag{120}$$

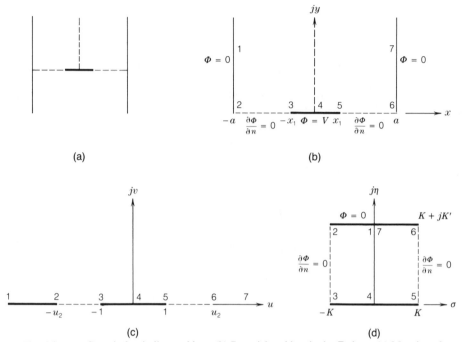

Fig. 4.21. (a) Canonical strip line problem. (b) Potential problem in the Z plane. (c) Mapping of a boundary onto the W plane. (d) Potential problem in the ζ plane.

This closed-form expression for the dominant part of the Green's function is not very useful in practice because of the difficulty of integrating the product of $J(x')$ and (120) over the strip. However, we will show that (120) can be re-expanded into a series of functions that are orthogonal over the strip. In order to accomplish this task we note that the dominant part of G_e is the Green's function for a strip between two sidewalls in the absence of the ground plane. We can map the boundary shown in Fig. 4.21(b) into the real axis in the complex $W = u + jv$ plane shown in Fig. 4.21(c) by means of the mapping function

$$\frac{dZ}{dW} = \frac{A}{\sqrt{u_2^2 - W^2}}. \tag{121}$$

We may integrate this function to obtain

$$Z = A \, \sin^{-1} \frac{W}{u_2} + B$$

where A and B are constants. We now make $Z = 0$ map into the point $W = 0$ and make $Z = x_1$ map into $W = 1$. The specification of these two points allows us to find A and B and thereby obtain

$$Z = x_1 \frac{\sin^{-1} (W/u_2)}{\sin^{-1} (1/u_2)}. \tag{122}$$

The mapping of the real axis in the W plane into a rectangle in the $\zeta = \sigma + j\eta$ plane is described by

$$\frac{d\zeta}{dW} = \frac{u_2 A}{\sqrt{(1 - W^2)(u_2^2 - W^2)}} = \frac{A}{\sqrt{(1 - W^2)(1 - k^2 W^2)}} \tag{123}$$

where $k = 1/u_2$. The integral of this expression is the inverse elliptic sine function with modulus k; thus, if we let $A = 1$

$$\zeta = \text{sn}^{-1}(W, k). \tag{124}$$

When $W = \pm 1$ we have $\sigma = \pm K$ where

$$K = \int_0^1 \frac{d\lambda}{\sqrt{(1 - \lambda^2)(1 - k^2 \lambda^2)}}. \tag{125}$$

When $W = \pm u_2 = \pm k^{-1}$ we have $\sigma = \pm K + jK'$ where

$$K' = \int_1^{1/k} \frac{d\lambda}{\sqrt{(\lambda^2 - 1)(1 - k^2 \lambda^2)}}. \tag{126}$$

In general,

$$\zeta = \int_0^W \frac{d\lambda}{\sqrt{(1 - \lambda^2)(1 - k^2 \lambda^2)}}. \tag{127}$$

The mapping from the Z plane into the ζ plane represents a coordinate transformation from the x, y coordinates into the σ, η coordinates. Consequently, we can represent the Green's function G_{ed} given by (119) or (120) as a function of σ and η by solving for the potential from a line source at σ', η' inside the rectangle shown in Fig. 4.21(d). With the boundary conditions $G = 0$ at $\eta = K'$, $\partial G/\partial \sigma = 0$ at $\sigma = \pm K$, G will have reflection symmetry about the σ axis. The Green's function for the configuration shown in Fig. 4.21(a) is equal to $2G_{ed}$. Hence we find that

$$G_{ed}(\sigma, \eta, \sigma', \eta') = \frac{K' - \eta_>}{4K} + \sum_{n=1}^{\infty} \frac{1}{2n\pi} \cos\frac{n\pi\sigma}{K} \cos\frac{n\pi\sigma'}{K} \frac{\cosh\dfrac{n\pi\eta_<}{K} \sinh\dfrac{n\pi}{K}(K' - \eta_>)}{\cosh\dfrac{n\pi K'}{K}} \tag{128}$$

after solving for the Green's function for the problem shown in Fig. 4.21(d). Note the presence of a term independent of σ which arises because of Neumann boundary conditions at $\sigma = \pm K$. This is the part of the Green's function that has even symmetry about the η axis. When we set $\eta = \eta' = 0$ we get

$$G_{ed}(\sigma, \sigma') = \frac{K'}{4K} + \sum_{n=1}^{\infty} \frac{1}{2n\pi} \cos\frac{n\pi\sigma}{K} \cos\frac{n\pi\sigma'}{K} \tanh\frac{n\pi K'}{K} \tag{129}$$

which is a representation of the function given in (120) but in terms of the variable σ. The connection between the variables x and σ can be found from the mapping functions and is

$$\mathrm{sn}\,(\sigma, k) = \frac{1}{k}\,\sin\left(\frac{x}{w}\,\sin^{-1} k\right) \tag{130}$$

where we have replaced x_1 by the half width w of the strip and $k = \sin(\pi w/2a)$. For a numerical analysis a lookup table can be readily constructed that will give the values of x in the interval $-w \le x \le w$ in terms of selected values of σ in the interval $-K \le \sigma \le K$. We can express $J(x)\,dx$ in the form

$$J(x)\,dx = 2J(\sigma)\frac{dx}{d\sigma}\,d\sigma = 2J(\sigma)\left|\frac{dZ}{dW}\frac{dW}{d\varsigma}\right|\,d\sigma$$

$$= J(\sigma)\frac{4a}{\pi}\left[\sin^2\frac{\pi W}{2a} - \sin^2\frac{\pi x}{2a}\right]^{1/2}\,d\sigma = 2I(\sigma)\,d\sigma \tag{131}$$

where $I(\sigma) = J(\sigma)\,dx/d\sigma$. The current on the two sides of the strip is $J(x)$, whereas $J(\sigma)$ is the current on one side only.

The integral equation (115) can be expressed as

$$\int_{-w}^{w} G_{ed}(x, x')J(x')\,dx' + \int_{-w}^{w} [G_e(x, x') - G_{ed}(x, x')]J(x')\,dx' = 1$$

where the second integral is a correction to the dominant part. When we introduce the new variable σ we obtain

$$\int_{-K}^{K}\left[\sum_{n=1}^{\infty}\frac{1}{n\pi}I(\sigma')\cos\frac{n\pi\sigma}{K}\cos\frac{n\pi\sigma'}{K}\tanh\frac{n\pi K'}{K} + \frac{K'}{2K}I(\sigma')\right]\,d\sigma'$$

$$- \sum_{n=1,3,\ldots}^{\infty} 2\frac{e^{-n\pi h/a}}{n\pi}\int_{-K}^{K} I(\sigma')\cos\frac{n\pi x(\sigma)}{2a}\cos\frac{n\pi x'(\sigma')}{2a}\,d\sigma' = 1. \tag{132}$$

The second series is a rapidly converging series. The integrals in this series have to be evaluated numerically using (130). If we neglect the correction series we see immediately that the solution to the integral equation is

$$I(\sigma') = \frac{1}{K'} \tag{133}$$

since the potential, and hence $I(\sigma')$, must be a constant for $|\sigma| \le K$. The total current on the strip equals I_0 where

$$I_0 = \int_{-K}^{K} 2I(\sigma)\,d\sigma = 4\frac{K}{K'} \tag{134}$$

since there are two surfaces to the strip. The inductance per unit length is μ_0/I_0 or $L = \mu_0 K'/4K$. The transmission-line capacitance per unit length is given by

$$C = \frac{\epsilon_0\mu_0}{L} = \epsilon_0\frac{4K}{K'}. \tag{135}$$

This solution can be seen to be similar to that derived for the strip line in Section 4.3.

We can include the correction series but will then need to use a series expansion for $I(\sigma)$. The constant term is the dominant one so good accuracy can be achieved using three or four terms, i.e.,

$$I(\sigma) = \sum_{i=0}^{3} C_i \cos \frac{i\pi\sigma}{K}. \tag{136}$$

The integral equation is readily solved using Galerkin's method. The dominant part can be integrated easily in view of the orthogonal properties of the functions involved. The integrals in the correction series are readily done numerically on a computer.

The beauty of the results we have obtained is that the dominant part of the Green's function has been diagonalized, i.e., represented in terms of a series of orthogonal functions on the strip. This enables us to find the exact solution to the dominant part of the integral equation. It also leads to an efficient evaluation of the exact integral equation since only a small correction is needed to the dominant part of the current distribution and the correction series converges rapidly. We also find, by this conformal mapping procedure, the dominant behavior of the current distribution. From (131) and (133) we obtain

$$J(x) = 2I(\sigma)\frac{d\sigma}{dx} = \frac{\pi}{aK'}\left[\sin^2 \frac{\pi w}{2a} - \sin^2 \frac{\pi x}{2a}\right]^{-1/2} \approx \frac{2}{K'\sqrt{w^2 - x^2}} \tag{137}$$

where the last term is the approximate result valid for large values of a/w.

If desired, the correction series in (132) can also be summed into closed form to give

$$\sum_{n=1,3,\ldots}^{\infty} \frac{e^{-n\pi h/a}}{n\pi} \cos \frac{n\pi x}{2a} \cos \frac{n\pi x'}{2a}$$

$$= \frac{1}{8\pi} \ln \left[\frac{\cosh \frac{\pi h}{a} + \cos \frac{\pi}{2a}(x - x') \cosh \frac{\pi h}{a} + \cos \frac{\pi}{2a}(x + x')}{\cosh \frac{\pi h}{a} - \cos \frac{\pi}{2a}(x - x') \cosh \frac{\pi h}{a} - \cos \frac{\pi}{2a}(x + x')} \right]. \tag{138}$$

This expression along with the variable transformation (130) can be used to numerically evaluate the correction term in the integral equation (132).

The coupled strip transmission-line problem can be analyzed in a similar way. The main difference is in the conformal mapping details. For the even mode (both strips have identical currents) we can place a magnetic wall between the two strips for the line shown in Fig. 4.20(b). For this mode we again only need the even part of the Green's function. The required Green's function is the same as that for the single strip.

We can map the boundary shown in Fig. 4.22(a) into the real axis in the W plane using

$$W = -\cos \frac{\pi Z}{a} + u_1 \tag{139}$$

where $u_1 = \cos(\pi x_1/a)$ and $u_2 = \cos(\pi x_2/a)$. For mapping into the ζ plane we choose

$$\frac{d\zeta}{dW} = \frac{A}{\sqrt{W(W - u_1 + u_2)(W - 1 - u_1)}}.$$

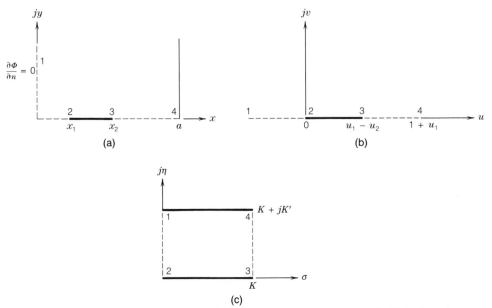

Fig. 4.22. Conformal mapping for a coupled strip line for the even mode. (a) Coupled microstrip line and boundary conditions for the even mode. (b) Mapped boundaries in the W plane. (c) Mapped boundaries in the ζ plane.

Now let $\lambda^2 = W/(u_1 - u_2)$; then

$$\frac{d\zeta}{d\lambda} = \frac{d\zeta}{dW}\frac{dW}{d\lambda} = \frac{1}{\sqrt{(1-\lambda^2)(1-k^2\lambda^2)}} \tag{140}$$

where we have set $2A = \sqrt{1 + u_1}$ and

$$k^2 = \frac{u_1 - u_2}{1 + u_1} = \frac{\sin^2\dfrac{\pi x_2}{2a} - \sin^2\dfrac{\pi x_1}{2a}}{\cos^2\dfrac{\pi x_1}{2a}} \tag{141}$$

$$\lambda = \left[\frac{\sin^2\dfrac{\pi x}{2a} - \sin^2\dfrac{\pi x_1}{2a}}{\sin^2\dfrac{\pi x_2}{2a} - \sin^2\dfrac{\pi x_1}{2a}}\right]^{1/2}. \tag{142}$$

Thus we have

$$\zeta = \mathrm{sn}^{-1}(\lambda, k). \tag{143}$$

When $x_1 = 0$ the above results reduce to those derived earlier for a single strip of width $2w = 2x_2$. The new boundaries in the ζ plane are shown in Fig. 4.23(c).

The dominant part of the Green's function G_e is readily constructed. When the source point

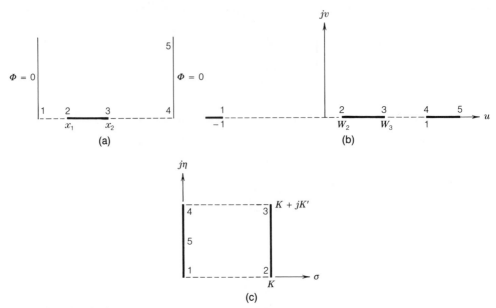

Fig. 4.23. Conformal mapping for a coupled strip line for the odd mode. (a) Coupled microstrip line and boundary conditions for the odd mode. (b) Mapped boundaries in the W plane. (c) Mapped boundaries in the ζ plane.

and the field point lie on the strip at $\eta = 0$ it is given by [the Green's function for Fig. 4.23(c) equals $4G_{ed}$]:

$$G_{ed}(\sigma, \sigma') = \frac{K'}{4K} + \sum_{n=1}^{\infty} \frac{1}{2n\pi} \tanh \frac{n\pi K'}{K} \cos \frac{n\pi\sigma}{K} \cos \frac{n\pi\sigma'}{K}. \qquad (144)$$

When x_1 is equal to zero this expression reduces to that given by (129).

The dominant current on the strip again corresponds to $I(\sigma) = J(\sigma)\,dx/d\sigma$ being equal to a constant which in this case is $1/K'$. As a function of x the dominant part of the current is given by

$$J(x) = 2I(\sigma)\frac{d\sigma}{dx} = \frac{2}{K'}\left|\frac{d\zeta}{dZ}\right| = \frac{2}{K'}\left|\frac{d\zeta}{d\lambda}\frac{d\lambda}{dW}\frac{dW}{dZ}\right|$$

$$= \frac{\pi \cos \dfrac{\pi x_1}{2a} \sin \dfrac{\pi x}{2a}}{aK'\left[\sin^2 \dfrac{\pi x_2}{2a} - \sin^2 \dfrac{\pi x}{2a}\right]^{1/2}\left[\sin^2 \dfrac{\pi x}{2a} - \sin^2 \dfrac{\pi x_1}{2a}\right]^{1/2}} \qquad (145)$$

where $J(x)$ is the current on both sides of the strip. This result is easiest to interpret for the case when a is very large. The expression for $J(x)$ then reduces to

$$J(x) = \frac{2x}{K'[(x^2 - x_1^2)(x_2^2 - x^2)]^{1/2}}. \qquad (146)$$

The current distribution is seen to have the expected "one over the square root of the distance from the edge" behavior. But, in addition, it is weighted by the factor x in the numerator. This reduces the amplitude of the singular behavior near x_1 relative to that at x_2 by a factor of $(x_1/x_2)^{1/2}$. The factor of x in the numerator also makes the current distribution merge smoothly into that for a single strip of width $2x_2$ as x_1 tends to zero.

The inductance per unit length for a single strip in the coupled strip line without a ground plane and with even excitation is given by

$$L_e = \frac{\mu_0}{I_0} = \frac{\mu_0 K'}{2K}. \tag{147a}$$

The corresponding capacitance is

$$C_e = \epsilon_0 \frac{2K}{K'}. \tag{147b}$$

For odd excitation of the coupled strip line we need a Green's function that is an odd function of x. This Green's function can be constructed the same way as G_e in (114) was. It is given by

$$G_o = \sum_{n=1}^{\infty} \frac{1}{n\pi} \sin \frac{n\pi x}{a} \sin \frac{n\pi x'}{a} \sinh \frac{n\pi y_<}{a} e^{-n\pi y_>/a}. \tag{148}$$

The dominant part G_{od} can also be summed into closed form to give, for $y = y' = h$,

$$G_{od}(x, x') = \sum_{n=1}^{\infty} \frac{1}{2n\pi} \sin \frac{n\pi x}{a} \sin \frac{n\pi x'}{a} = -\frac{1}{4\pi} \ln \left| \frac{\sin \dfrac{\pi(x - x')}{2a}}{\sin \dfrac{\pi(x + x')}{2a}} \right|. \tag{149}$$

We can use conformal mapping to re-express G_{od} in terms of a series of functions that are orthogonal over a single strip. The boundary to be mapped is shown in Fig. 4.23(a). The mapping into the real axis in the W plane is accomplished using $W = -\cos(\pi Z/a)$. The points $Z = 0, a$ are mapped into the points $W = \mp 1$ as shown in Fig. 4.23(b). Consider next the mapping into the ζ plane which is described by

$$\frac{d\zeta}{dW} = \frac{AC}{[(1 - W^2)(W - W_2)(W - W_3)]^{1/2}}.$$

We now use $W = -\cos(\pi Z/a) = 1 - 2\cos^2(\pi Z/2a)$, $W_2 = 1 - 2\cos^2(\pi x_1/2a)$, and $W_3 = 1 - 2\cos^2(\pi x_2/2a)$. As a next step we write

$$\frac{d\zeta}{dW} = \frac{d\zeta}{dZ} \frac{dZ}{dW}$$

to obtain

$$\frac{d\zeta}{dZ} = \frac{C}{[(W - W_2)(W - W_3)]^{1/2}}.$$

We now substitute for W, W_2, W_3 in terms of Z, x_1, x_2 and use the relation $\sec^2 \theta = 1 + \tan^2 \theta$

to obtain

$$\frac{d\zeta}{dZ} = \frac{C}{2\cos^2 \dfrac{\pi Z}{2a}[(1+t^2)C_1^2 - 1]^{1/2}[(1+t^2)C_2^2 - 1]^{1/2}}$$

where $t = \tan(\pi Z/2a)$, $C_1 = \cos(\pi x_1/2a)$, and $C_2 = \cos(\pi x_2/2a)$. If the following substitutions are made

$$\lambda = \frac{\tan(\pi Z/2a)}{\tan(\pi x_1/2a)} \tag{150a}$$

$$k = \frac{\tan(\pi x_1/2a)}{\tan(\pi x_2/2a)} \tag{150b}$$

$$\frac{d\lambda}{dZ} = \frac{\pi}{2at}\sec^2(\pi Z/2a) \tag{150c}$$

$$C = \frac{\pi\sin(\pi x_1/2a)\sin(\pi x_2/2a)}{a\tan(\pi x_1/2a)} \tag{150d}$$

we finally obtain

$$\zeta = \text{sn}^{-1}(\lambda, k). \tag{151}$$

With this mapping function we have

$$
\begin{aligned}
Z &= 0, & \lambda &= 0, & \zeta &= \text{sn}^{-1}(0, k) = 0 \\
Z &= x_1, & \lambda &= 1, & \zeta &= \text{sn}^{-1}(1, k) = K \\
Z &= x_2, & \lambda &= 1/k, & \zeta &= \text{sn}^{-1}(1/k, k) = K + jK' \\
Z &= a, & \lambda &= \infty, & \zeta &= \text{sn}^{-1}(\infty, k) = jK'.
\end{aligned}
$$

The boundaries in the ζ plane are shown in Fig. 4.23(c).

The dominant part G_{od} of the Green's function is now readily found in terms of an orthogonal series over the strip that lies along $\sigma = K$, $0 \le \eta \le K'$. It is given by

$$G_{od}(\eta, \eta') = \frac{K}{4K'} + \sum_{n=1}^{\infty} \frac{1}{2n\pi}\tanh\frac{n\pi K}{K'}\cos\frac{n\pi\eta}{K'}\cos\frac{n\pi\eta'}{K'}. \tag{152}$$

In the absence of the ground plane the current $I(\eta) = J(\eta)\, dx/d\eta$ is a constant equal to $1/K$. Thus the inductance per unit length for one strip and for odd excitation is

$$L_o = \frac{\mu_0 K}{2K'} \tag{153a}$$

and the corresponding capacitance is

$$C_o = \epsilon_0 \frac{2K'}{K}. \tag{153b}$$

The behavior of the dominant part of the current $J(x)$ as a function of x may be found using $J(x) = 2I(\eta)\, d\eta/dx$ and is given by

$$J(x) = \frac{2}{K}\left|\frac{d\zeta}{dZ}\right| = \frac{\pi \cos \dfrac{\pi x_1}{2a} \sin \dfrac{\pi x_2}{2a}}{Ka \left[\sin^2 \dfrac{\pi x}{2a} - \sin^2 \dfrac{\pi x_1}{2a}\right]^{1/2} \left[\sin^2 \dfrac{\pi x_2}{2a} - \sin^2 \dfrac{\pi x}{2a}\right]^{1/2}} \tag{154}$$

which for large values of a relative to x_1 and x_2 reduces to

$$J(x) = \frac{2x_2}{K[(x^2 - x_1^2)(x_2^2 - x^2)]^{1/2}}. \tag{155}$$

When x_1 approaches zero the current has a singular behavior approaching $1/x$ near the edge at x_1. This is caused by the strong electric field in a very narrow slit with the adjacent conductors at opposite potential.

The integral equation for the current on the strips in a coupled strip transmission line can be solved numerically using the same approach as that for the single strip. By extracting the dominant part of the Green's function and introducing the diagonalized form for it, a very accurate solution can be found using a three- or four-term expansion for the current in terms of the orthogonal function basis on the strip. The correction series will involve integrals that have to be evaluated numerically, but this can be done with relatively little computational effort.

4.7. INHOMOGENEOUS TRANSMISSION LINES

The technology used to produce printed circuit boards has been extended to make a variety of planar transmission-line structures. These transmission-line structures usually include a dielectric substrate material with a ground plane and one or more conducting strips on the upper side. The open microstrip line consists of a single strip as shown in Fig. 4.24(a). The microstrip with a cover plate and the shielded microstrip line shown in Figs. 4.24(b) and (c), respectively, are also widely used. In Fig. 4.24(d) we show the cross section of a coupled microstrip line. The slot line and coplanar transmission line shown in Figs. 4.24(e) and (f), respectively, are also used in a number of applications. A variety of impedance transforming structures, filter elements, directional couplers, tees, bends, etc., can be readily fabricated using the same techniques used to construct the transmission lines.

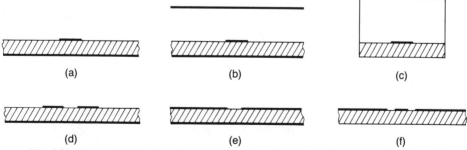

Fig. 4.24. (a) A microstrip line. (b) A microstrip line with a top shield. (c) A shielded microstrip line. (d) A coupled microstrip line. (e) A slot line. (f) A coplanar line.

TABLE 4.2

Material	κ	κ_y
PTFE/microfiberglass	2.26	2.20
PTFE/woven glass	2.84	2.45
Epsilam 10	13	10.3
Duroid 6006	6.36	6
Boron nitride	5.12	3.4
Alumina	9.7	9.7
Silicon	11.7	11.7
Germanium	16	16
Gallium arsenide	12.9	12.9
Sapphire	9.4	11.6

The dielectric substrate material needs to have a low loss tangent and a relatively large dielectric constant in order to achieve a very compact microwave circuit. Many of the substrates used are anisotropic. A popular low-cost substrate material is fiberglass-impregnated Teflon®. These substrates are usually anisotropic due to the general alignment of the glass fibers in the plane of the material. The dielectric constants for a number of common substrate materials are listed in Table 4.2 where the dielectric constant in the plane of the material is called κ, while that normal to the interface is labeled κ_y.

Since the dielectric does not completely surround the conductors, the fundamental mode of propagation on any of the planar transmission lines shown in Fig. 4.24 is not a TEM mode. At low frequencies, typically below 1 GHz for practical lines, the fundamental mode can be approximated as a TEM mode. The propagation constant and characteristic impedance can then be found in terms of the capacitance and inductance per unit length. The inductance is the same as that for an air-filled line since the dielectric has no effect on the low-frequency magnetic field. For an air-filled line we have the relationship $k_0 = \omega\sqrt{\mu_0\epsilon_0} = \omega\sqrt{L_oC_o}$ so $L_o = \mu_0\epsilon_0/C_o$ where L_o and C_o are the inductance and capacitance per meter for the line. If we let C be the capacitance per meter for the line with the dielectric present, then the propagation constant β is given by

$$\beta = \omega\sqrt{L_oC} = \sqrt{C/C_o}k_0 = \sqrt{\kappa_e}k_0 \tag{156}$$

and the characteristic impedance is given by

$$Z_c = \sqrt{L_o/C} = \frac{1}{\sqrt{\kappa_e}}\sqrt{L_o/C_o} \tag{157}$$

where κ_e is called the effective dielectric constant. As the frequency is increased the effective dielectric constant κ_e increases in a nonlinear way and approaches that of the substrate in an asymptotic way for very high frequencies. The transmission line is said to exhibit dispersion since κ_e is a function of ω. For a dispersive transmission line the group velocity, which equals the velocity of energy transport, is given by

$$v_g = \frac{1}{d\beta/d\omega} = \frac{c/\sqrt{\kappa_e}}{1 + \frac{\omega}{2}\frac{d\ln\kappa_e}{d\omega}}. \tag{158}$$

The quality factor or Q of a transmission-line resonator is often quoted in the literature as

$\beta/2\alpha$ where α is the attenuation constant. For a dispersive line the correct expression is

$$Q = \omega \frac{\text{energy per unit length}}{\text{power loss per unit length}} = \frac{\omega P / v_g}{P_L} = \frac{\omega}{2 v_g \alpha} \tag{159}$$

where P_L is the power loss per meter and the power flow P equals the energy density multiplied by the energy transport velocity v_g.

The early work on planar transmission-line structures was almost all directed toward evaluating the low-frequency capacitance and inductance, and from the latter, the effective dielectric constant and characteristic impedance. A useful set of approximate relationships was derived by Wheeler [4.17].

In more recent years an extensive amount of work has been carried out on full-wave high-frequency analysis of planar transmission lines. The most popular method of analysis used is the spectral-domain method which will be described in Section 4.8. Solutions can also be obtained in an efficient way using potential theory which will be described in Section 4.9. Numerical methods such as the finite-element method, the finite-difference method, and the method of lines have also been used. The literature on the subject is very extensive with many papers published in the *IEEE Transactions on Microwave Theory and Techniques*. Since the published papers are readily available no attempt at compiling a bibliography on this subject is made in this book. An account of some of the work that has been done can be found in the book by Gupta, Garg, and Bahl [4.37].

There are no simple analytic solutions for the modes of propagation on planar transmission lines of the type under consideration. Therefore, design data have to be generated numerically from formal analytical solutions. A number of efficient numerical computer codes have been developed for this purpose.

4.8. SPECTRAL-DOMAIN GALERKIN METHOD [4.25]–[4.30]

In the Fourier-transform or spectral-domain method the required Green's function and boundary conditions are formulated in the spectral domain. The integral equation for the current on the strips is also solved in the spectral domain. The popularity of the method is due mainly to the feature that the Green's function is a relatively simple algebraic expression in the spectral domain. The method appears to have originated in an early paper by Yamashita and Mittra [4.23] and was made more specific in a later publication by Itoh and Mittra [4.25]. In this section the salient features of the method will be developed in connection with the open microstrip line shown in Fig. 4.25.

The natural modes of propagation on a dielectric slab on a ground plane are surface wave modes that are either TE or TM modes with respect to the interface normal. These modes are also called the LSE and LSM modes as described in Section 3.9. A detailed discussion of surface wave modes is given in Chapter 11 and will not, for this reason, be presented here. For the dominant mode on the microstrip line we can assume that all the field components have a z dependence of the form $e^{-j\beta z}$. The equations describing the LSE and LSM modes are

$$\mathbf{E} = \nabla \times \mathbf{a}_y \psi_h(x, y, z) \qquad \text{LSE modes} \tag{160a}$$

$$\mathbf{H} = \nabla \times \mathbf{a}_y \psi_e(x, y, z) \qquad \text{LSM modes.} \tag{160b}$$

In the absence of the strip each longitudinal section mode can exist by itself. The presence of

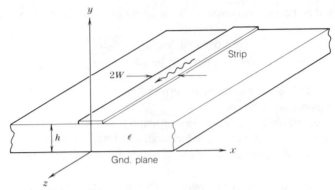

Fig. 4.25. An open microstrip line.

one or more strips couples the modes. We will define functions f and g as follows:

$$e^{-j\beta z} f(y, w) = \int_{-\infty}^{\infty} e^{jwx - j\beta z} \psi_h(x, y) \, dx \tag{161a}$$

$$e^{-j\beta z} g(y, w) = \int_{-\infty}^{\infty} e^{jwx - j\beta z} \psi_e(x, y) \, dx \tag{161b}$$

where f and g are solutions of

$$\left[\frac{d^2}{dy^2} + (\kappa k_0^2 - \gamma^2) \right] \left\{ \begin{array}{c} f \\ g \end{array} \right\} = 0, \qquad 0 \leq y < h$$

$$\left[\frac{d^2}{dy^2} + (k_0^2 - \gamma^2) \right] \left\{ \begin{array}{c} f \\ g \end{array} \right\} = 0, \qquad y > h$$

with $\gamma^2 = w^2 + \beta^2$.

In the presence of an infinitely thin strip, of width $2W$ and which is perfectly conducting, the following boundary conditions must hold:

$$E_x = E_z = 0 \text{ on strip,} \qquad -W \leq x \leq W$$
$$H_x^+ - H_x^- = -J_z \text{ at } y = h, \qquad -W \leq x \leq W \tag{162}$$
$$H_z^+ - H_z^- = J_x \text{ at } y = h, \qquad -W \leq x \leq W$$

where J_x, J_z are the components of the current density on the strip.

From ψ_h we find that

$$E_x = -\frac{\partial \psi_h}{\partial z}, \qquad E_z = \frac{\partial \psi_h}{\partial x}, \qquad j\omega\mu_0 H_x = -\frac{\partial E_z}{\partial y}$$

$$j\omega\mu_0 H_y = \frac{\partial E_z}{\partial x} - \frac{\partial E_x}{\partial z}, \qquad j\omega\mu_0 H_z = \frac{\partial E_x}{\partial y}.$$

The continuity of the tangential fields E_x, E_y at $y = h$ requires $f(y, w)$ to be continuous at

$y = h$ and to vanish at $y = 0$. Hence $f(y, w)$ is of the form

$$f(y, w) = \begin{cases} A(w) \sin \ell y, & y \le h \\ A(w) \sin \ell h \, e^{-p(y-h)}, & y \ge h \end{cases}$$

where $\ell^2 = \kappa k_0^2 - \gamma^2$, $p^2 = \gamma^2 - k_0^2$. If there were no strip we would require H_x, H_z to be continuous and then df/dy would be continuous at $y = h$. For this case we would obtain

$$\ell A \cos \ell h = -pA \sin \ell h$$

or

$$\tan \ell h = -\ell/p, \qquad \ell^2 + p^2 = (\kappa - 1)k_0^2$$

which determines the LSE surface wave propagation constants.

From ψ_e we obtain

$$H_x = -\frac{\partial \psi_e}{\partial z}, \quad H_z = \frac{\partial \psi_e}{\partial x}, \quad j\omega\epsilon_0\kappa(y)E_x = \frac{\partial H_z}{\partial y}$$

$$j\omega\epsilon_0\kappa(y)E_y = \frac{\partial H_x}{\partial z} - \frac{\partial H_z}{\partial x}, \quad j\omega\epsilon_0\kappa(y)E_z = -\frac{\partial H_x}{\partial y}.$$

The continuity of E_x, E_z at $y = h$ and the fact that $E_x = E_z = 0$ at $y = 0$ require g to have the form

$$g(y, w) = \begin{cases} B(w) \cos \ell y, & y < h \\ \dfrac{\ell B(w)}{\kappa p} \sin \ell h \, e^{-p(y-h)}, & y > h \end{cases}$$

which makes $(1/\kappa(y))(dg/dy)$ continuous at $y = h$.

In the absence of a strip we would also require g to be continuous at $y = h$, which gives

$$\tan \ell h = \frac{\kappa p}{\ell}, \qquad \ell^2 + p^2 = (\kappa - 1)k_0^2.$$

The solutions of these equations give the LSM surface wave propagation constants.

From this point on we suppress the $e^{-j\beta z}$ factor. We now define $J_x(x)$, $J_z(x)$ to be functions that are *identically zero* for $|x| > W$. This will allow us to express the boundary conditions on H_x, H_z in the Fourier (spectral) domain. Let

$$\hat{J}_x(w) = \mathfrak{F}J_x(x)$$

$$\hat{J}_z(w) = \mathfrak{F}J_z(x).$$

We require $\hat{H}_x^+ - \hat{H}_x^- = -\hat{J}_z$ and $\hat{H}_z^+ - \hat{H}_z^- = \hat{J}_x$. The symbol ($\hat{}$) denotes the Fourier transform. We now express $\hat{H}_x^\pm, \hat{H}_z^\pm$ in terms of f and g and obtain

$$\hat{H}_x = \frac{w}{k_0 Z_0} \frac{df}{dy} + j\beta g$$

$$\hat{H}_z = \frac{\beta}{k_0 Z_0} \frac{df}{dy} - jwg \tag{163}$$

and thus

$$A(w) \frac{w}{k_0 z_0} (p \sin \ell h + \ell \cos \ell h) + j\beta B(w) \left(\cos \ell h - \frac{\ell}{\kappa p} \sin \ell h \right) = \hat{J}_z$$

$$A(w) \frac{\beta}{k_0 Z_0} (p \sin \ell h + \ell \cos \ell h) - jw B(w) \left(\cos \ell h - \frac{\ell}{\kappa p} \sin \ell h \right) = -\hat{J}_x.$$

Now let

$$A' = \frac{A}{k_0 Z_0} (p \sin \ell h + \ell \cos \ell h)$$

$$B' = B \left(\cos \ell h - \frac{\ell}{\kappa p} \sin \ell h \right);$$

then

$$wA' + j\beta B' = \hat{J}_z$$

$$\beta A' - jw B' = -\hat{J}_x$$

from which we obtain

$$A' = \frac{w\hat{J}_z - \beta \hat{J}_x}{\beta^2 + w^2}$$

$$\tag{164}$$

$$B' = -\frac{j(\beta \hat{J}_z + w\hat{J}_x)}{\beta^2 + w^2}.$$

In the Fourier domain $\nabla \cdot \mathbf{J} = -j\omega\rho$ becomes $-j(w\hat{J}_x + \beta\hat{J}_z) = -j\omega\hat{\rho}$ so B' is proportional to the charge on the conducting strip. The boundary conditions $E_x = E_z = 0$ on the strip will determine J_x, J_z.

Spectral-Domain Solution (Galerkin's Method)

In the spatial domain the solution for the transverse electric field $\mathbf{E}_t(x, y) = E_x \mathbf{a}_x + E_y \mathbf{a}_y$ can be expressed in the form

$$\mathbf{E}_t(x, y) = \int_{-W}^{W} \bar{\mathbf{G}}(x - x', y, h) \cdot \mathbf{J}(x') \, dx' \tag{165}$$

where $\bar{\mathbf{G}}$ is a suitable Green's dyadic function.

Let J_x, J_z be expanded in a suitable set of basis functions (a finite sum gives an approximate representation):

$$J_x(x) = \sum_{n=1}^{N} I_{xn} \Phi_{xn}(x) \tag{166a}$$

$$J_z(x) = \sum_{n=1}^{N} I_{zn} \Phi_{zn}(x) \tag{166b}$$

where Φ_{xn}, Φ_{zn} are defined to be identically zero for $|x| > W$. The expansions are substituted into (165) and then both sides of the equation are tested by multiplying by Φ_{xn} and Φ_{zn}, $n = 1, 2, \ldots, N$, in turn, integrating over x, and setting both sides equal to zero at $y = h$ in order to enforce the boundary condition $E_t = 0$ on the strip at $y = h$. This solution method generates $2N$ equations for I_{xn}, I_{zn} and since it is a homogeneous set a solution exists only if the determinant vanishes. The vanishing of the determinant gives the values of β for the modes along the microstrip line. In view of the finite support of the Φ_{xn}, Φ_{zn} the integration can be written as extending from $x = -\infty$ to ∞. A typical integral to be evaluated is

$$\sum_{n=1}^{N} I_{xn} \int\!\!\!\int_{-\infty}^{\infty} \Phi_{xm}(x) G_{xx}(x - x') \Phi_{xn}(x') \, dx' \, dx$$

$$+ \sum_{n=1}^{N} I_{zn} \int\!\!\!\int_{-\infty}^{\infty} \Phi_{xm}(x) G_{xz}(x - x') \Phi_{zn}(x') \, dx' \, dx = 0 \tag{167}$$

where $m = 1, 2, \ldots, N$.

This equation expresses the condition $E_x = 0$ on $|x| \leq W$. Parseval's theorem can now be used to evaluate the integral in the spectral domain. The Fourier transform of

$$\int_{-\infty}^{\infty} G_{ij}(x - x') \Phi_{jn}(x') \, dx'$$

is $\hat{G}_{ij}(w)\hat{\Phi}_{jn}(w)$ where $\hat{G}_{ij}(w)$ relates $\hat{E}_i(w)$ to $\hat{J}_j(w)$; $i, j = x, z$, in the spectral domain. By using Parseval's theorem, (167) becomes

$$\sum_{n=1}^{N} I_{xn} \int_{-\infty}^{\infty} \hat{\Phi}_{xm}(-w)\hat{G}_{xx}(w)\hat{\Phi}_{xn}(w) \, dw$$

$$+ \sum_{n=1}^{N} I_{zn} \int_{-\infty}^{\infty} \hat{\Phi}_{xm}(-w)\hat{G}_{xz}(w)\hat{\Phi}_{zn}(w) \, dw = 0, \qquad m = 1, 2, \ldots, N. \tag{168}$$

The significance of the spectral-domain method is that $\hat{G}_{ij}(w)$ is easy to find and much simpler than the spatial Green's function which can only be expressed as an inverse Fourier transform. We will now obtain the expressions for $\hat{G}_{ij}(w)$. In the spectral domain, at $y = h$,

$$\hat{E}_x = \left[j\beta A(w) + \frac{w\ell Z_0}{\kappa k_0} B(w) \right] \sin \ell h \tag{169a}$$

$$\hat{E}_z = \left[-jw A(w) + \frac{\beta Z_0 \ell}{\kappa k_0} B(w) \right] \sin \ell h. \tag{169b}$$

By using the solutions for $A'(w)$, $B'(w)$ we obtain

$$
\hat{G}(w) = \frac{jZ_0 \sin \ell h}{k_0[p \sin \ell h + \ell \cos \ell h][\kappa p \cos \ell h - \ell \sin \ell h]}
$$
$$
\times \{\mathbf{a}_x\mathbf{a}_x[(w^2 - k^2)p \cos \ell h - (w^2 - k_0^2)\ell \sin \ell h]
$$
$$
+ (\mathbf{a}_x\mathbf{a}_z + \mathbf{a}_z\mathbf{a}_x)[w\beta(p \cos \ell h - \ell \sin \ell h)]
$$
$$
+ \mathbf{a}_z\mathbf{a}_z[(\beta^2 - k^2)p \cos \ell h - (\beta^2 - k_0^2)\ell \sin \ell h]\} \tag{170}
$$

where $k^2 = \kappa k_0^2$, $\ell^2 = k^2 - w^2 - \beta^2$, $p^2 = \beta^2 + w^2 - k_0^2$.

In order to keep the number of basis functions to a minimum, the currents J_z and J_x should be expressed in a form that includes the correct zero-frequency behavior of the currents at the edge of the strip (see Section 4.6).

For $J_z(x)$ a useful form for which the Fourier transform is also readily found is

$$
J_z(x) = \sum_{n=0}^{N} I_{zn} \frac{T_{2n}(x/W)}{\sqrt{1 - (x/W)^2}} \tag{171}
$$

where T_{2n} is a Chebyshev polynomial. The first few polynomials are

$$
T_0(u) = 1
$$
$$
T_1(u) = u
$$
$$
T_2(u) = 2u^2 - 1
$$
$$
T_3(u) = 4u^3 - 3u.
$$

These polynomials satisfy the recurrence relation

$$
T_n(u) = 2uT_{n-1}(u) - T_{n-2}(u).
$$

If we let $u = \cos \theta$ for $|u| \le 1$ and let $u = \cosh \theta$ for $|u| > 1$, then the following useful relations hold:

$$
T_n(\cos \theta) = \cos n\theta
$$
$$
T_n(\cosh \theta) = \cosh n\theta.
$$

By using $x = W \cos \theta$ and $dx = -W\sqrt{1 - x^2/W^2}\, d\theta$ the Fourier transform is found to be given by

$$
\int_0^\pi W \cos 2n\theta\, e^{jwW \cos \theta}\, d\theta
$$

since $T_{2n}(\cos \theta) = \cos 2n\theta$. Now use $e^{jwW \cos \theta} = J_0(wW) - 2[J_2(wW) \cos 2\theta - J_4(wW) \cos 4\theta + \cdots] + 2j[J_1(wW) \cos \theta - J_3(wW) \cos 3\theta \ldots]$. The Fourier transform is thus found to be $\pi W(-1)^n J_{2n}(wW)$.

Since $J_x(x)$ is an odd function of x and at $x = \pm W$ must vanish like $[1 - (x/W)^2]^{1/2}$ a suitable expansion is

$$J_x(x) = \sqrt{1 - (x/W)^2} \sum_{n=1}^{N} j I_{xn} U_{2n-1}(x/W) \tag{172}$$

where $U_{2n-1}(x/W)$ is a Chebyshev polynomial of the second kind. Some useful relationships are

$$U_0(u) = 1, \quad U_1 = 2u, \quad U_2 = 4u^2 - 1, \quad U_3 = 8u^3 - 4u$$

$$U_{n+1} = 2uU_n - U_{n-1}, \quad U_n(\cos\theta) = [\sin(n+1)\theta]/\sin\theta.$$

By using $x = W\cos\theta$ the Fourier transform of a typical term is found to be

$$jW \int_0^\pi \sin\theta \, \sin(2n\theta) e^{jwW\cos\theta} \, d\theta$$

$$= \frac{W}{2} j \int_0^\pi [\cos(2n-1)\theta - \cos(2n+1)\theta] e^{jwW\cos\theta} \, d\theta$$

$$= (-1)^n \frac{W\pi}{2}[J_{2n-1}(wW) + J_{2n+1}(wW)] = (-1)^n \frac{2n\pi}{w} J_{2n}(wW).$$

The Fourier transforms of the basis functions may be substituted into (168). The spectral integrals over w are then evaluated numerically. There is another equation similar to (168) that expresses the condition $E_z = 0$ on $|x| \leq W$. The determinant of the homogeneous set of equations is equated to zero in order to find the propagation constant β. Since the $\hat{G}_{ij}$ are functions of β in practice the determinant must be evaluated for a sequence of assumed values of β until that value which makes the determinant equal to zero is found. Typical results for the effective dielectric constants of microstrip and coupled microstrip lines can be found in the paper by Jansen [4.26]. Jansen has reported that the use of a two-term expansion for J_z and a single term for J_x produces results for the effective dielectric constant $\kappa_e = \beta^2/k_0^2$ accurate to 1% or better. Some authors have used pulse functions for the expansion of the current and have not taken the edge behavior of the current into account explicitly. In this case a considerably larger number of terms for the current expansions has to be used and the resultant system of equations is of much higher order. In general, it is preferable to take the edge behavior of the current into account explicitly since this results in greater accuracy with fewer terms in the expansion. The required computational effort is also reduced by this means.

The even and odd modes on the coupled microstrip line may be analyzed using the same basic formulation. The only required change is to include both the even and odd Chebyshev polynomials in the expansions for J_x and J_z since the currents are no longer odd and even functions, respectively, about the center of the strip. Of course, the origin for the current expansions must be shifted to the center of the strip. In the spectral domain the Fourier transform of the current is simply multiplied by a factor e^{jwx_0} when the origin is shifted by an amount x_0.

4.9. POTENTIAL THEORY FOR MICROSTRIP LINES [4.35]

In the spectral-domain method discussed in the previous section attention was focused on the electric field as a function of the current on the strip. An alternative approach to the analysis of a microstrip line is to formulate integral equations for the axial current J_z and

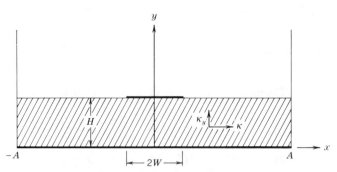

Fig. 4.26. A microstrip line with sidewalls.

the charge density ρ on the strip as determined by boundary conditions imposed on the axial magnetic potential A_z and the scalar potential Φ. This approach has several advantages which are: (a) independent scalar integral equations for J_z and ρ are obtained, (b) the equations can be solved numerically in an efficient way using an iterative technique, and (c) the propagation constant β is determined by a simple algebraic equation relating the total axial current I_z to the total charge per unit length on the strip.

The microstrip line under consideration is shown in Fig. 4.26. It consists of a conducting strip of width $2W$ located on a substrate of thickness H on a ground plane. Two conducting sidewalls are placed at $x = \pm A$ in order to facilitate the analysis through use of Fourier series. When numerical results are computed A is set equal to the larger of $15W$ or $15H$ for which case the sidewalls have a negligible influence on the characteristics of the microstrip line. The substrate is assumed to be anisotropic with a dielectric constant κ_y in the direction perpendicular to the substrate and κ in the x and z directions.

We will derive the basic equations for the vector potential $\mathbf{A}$ and scalar potential Φ by initially regarding κ and κ_y to be functions of y.

The dielectric constant tensor is represented as

$$\boldsymbol{\kappa}(y) = \kappa(y)\bar{\mathbf{I}} + [\kappa_y(y) - \kappa(y)]\mathbf{a}_y\mathbf{a}_y \tag{173}$$

where $\bar{\mathbf{I}}$ is the unit dyad. We let $\mathbf{B} = \nabla \times \mathbf{A}$ and then from Maxwell's equation $\nabla \times \mathbf{E} = -j\omega\nabla \times \mathbf{A}$ we obtain

$$\mathbf{E} = -j\omega\mathbf{A} - \nabla\Phi. \tag{174}$$

From the equation $\nabla \times \mathbf{B} = j\omega\epsilon_0\mu_0\boldsymbol{\kappa}\cdot\mathbf{E} + \mu_0\mathbf{J} = \nabla \times \nabla \times \mathbf{A}$ we find, upon using (174), that

$$\nabla\nabla\cdot\mathbf{A} - \nabla^2\mathbf{A} = j\omega\epsilon_0\mu_0[-j\omega\kappa\mathbf{A} - \kappa\nabla\Phi - j\omega(\kappa_y - \kappa)A_y\mathbf{a}_y$$

$$- (\kappa_y - \kappa)\nabla\Phi\cdot\mathbf{a}_y\mathbf{a}_y] + \mu_0\mathbf{J}$$

$$= \kappa k_0^2\mathbf{A} - j\omega\mu_0\epsilon_0\nabla(\kappa\Phi) - (\kappa_y - \kappa)\left(j\omega\mu_0\epsilon_0\frac{\partial\Phi}{\partial y} - k_0^2A_y\right)\mathbf{a}_y$$

$$+ j\omega\mu_0\epsilon_0\Phi\frac{\partial\kappa}{\partial y}\mathbf{a}_y + \mu_0\mathbf{J}.$$

The two gradient terms are now equated to yield the conventional Lorentz condition and the following equations for the components of $\mathbf{A}$ in terms of the axial current J_z, the transverse current J_x, and the scalar potential Φ:

$$\nabla \cdot \mathbf{A} = -j\omega\mu_0\epsilon_0\kappa(y)\Phi \tag{175a}$$

$$\nabla^2 A_z + \kappa k_0^2 A_z = -\mu_0 J_z \tag{175b}$$

$$\nabla^2 A_x + \kappa k_0^2 A_x = -\mu_0 J_x \tag{175c}$$

$$\nabla^2 A_y + \kappa_y k_0^2 A_y = -j\omega\mu_0\epsilon_0\Phi\frac{\partial\kappa}{\partial y} + j\omega\mu_0\epsilon_0(\kappa_y - \kappa)\frac{\partial\Phi}{\partial y}$$

$$= j\omega\mu_0\epsilon_0\left[(\kappa_y - \kappa)\frac{\partial\Phi}{\partial y} + \Phi(H)(\kappa - 1)\delta(y - H)\right] \tag{175d}$$

where we have taken the step change in $\kappa(y)$ at $y = H$ into account and put $\partial\kappa/\partial y = (1-\kappa)\delta(y-H)$ with $\delta(y-H)$ being the delta function. By using Gauss' law $\nabla \cdot (\boldsymbol{\kappa} \cdot \mathbf{E}) = \rho/\epsilon_0$, (174), and the Lorentz condition, the following equation for Φ is obtained:

$$\kappa\left[\frac{\partial^2\Phi}{\partial x^2} + \frac{\partial^2\Phi}{\partial z^2}\right] + \frac{\partial}{\partial y}\kappa_y\frac{\partial\Phi}{\partial y} + \kappa^2 k_0^2\Phi$$

$$= -\frac{\rho}{\epsilon_0} + j\omega(\kappa_y - 1)\delta(y - H)A_y - j\omega(\kappa_y - \kappa)\frac{\partial A_y}{\partial y}. \tag{176}$$

The equations for A_y and Φ are coupled and must be solved together. The coupling occurs through boundary values at the air–dielectric interface. The differential equations are all second order in y and hence all the potential functions must be continuous at $y = h$ since the equations hold for all values of y. If a potential function were not continuous at $y = h$, additional delta function terms would appear on the right-hand side. In addition, $\partial A_x/\partial y$ and $\partial A_z/\partial y$ are also continuous across the interface. From the differential equations (175d) and (176) we find, by integrating over a vanishingly small interval along y centered on the interface, that

$$\left.\frac{\partial A_y}{\partial y}\right|_-^+ = j\omega\mu_0\epsilon_0(\kappa - 1)\Phi(H) \tag{177a}$$

$$\kappa_y(y)\left.\frac{\partial\Phi}{\partial y}\right|_-^+ = \left.\frac{\partial\Phi}{\partial y}\right|_+ - \kappa_y\left.\frac{\partial\Phi}{\partial y}\right|_- = -\frac{\rho}{\epsilon_0} + j\omega(\kappa_y - 1)A_y(H) \tag{177b}$$

where the negative sign refers to y just below the interface and the positive sign refers to y just above the conducting strip. In addition to these discontinuity conditions, the boundary conditions on the perfectly conducting infinitely thin microstrip are

$$E_z = 0 \quad \text{or} \quad \omega A_z = \beta\Phi \tag{178a}$$

$$E_x = 0 \quad \text{or} \quad -j\omega A_x = \frac{\partial\Phi}{\partial x} \tag{178b}$$

where we have now assumed that all the fields have an $e^{-j\beta z}$ dependence on z so that $\partial/\partial z$

can be replaced by $-j\beta$. One additional relationship that will be used comes from integrating the continuity equation

$$\nabla_s \cdot \mathbf{J} = -j\omega\rho$$

across the strip from $x = -W$ to W, and is

$$\beta I_{\text{TOT}} = \omega Q \tag{179}$$

where I_{TOT} is the total z-directed current and Q is the total charge per unit length on the strip. The integral of (178b) on the strip gives

$$\Phi(x, H) = -j\omega \int_0^x A_x \, dx + \Phi(0, H)$$

which can be expressed in the form

$$\epsilon_0 \Phi(x, H) = 1 - j\omega\mu_0\epsilon_0 \int_0^x \int_{-W}^W G(x, x')J_x(x') \, dx' \, dx = 1 + S(x) \tag{180}$$

for $0 < x < W$. In (180) we have set $\Phi(0, H)$ equal to $1/\epsilon_0$. The second term in (180) involving the integral of J_x and the Green's function G for (175c) is a small perturbation to the boundary condition for Φ since at the lower frequencies, and in particular for narrow strips, J_x is very small. Numerical computations show that even for extreme conditions this term is not large.

The integral equations for ρ and J_z are

$$\epsilon_0 \Phi(x, H) = 2 \int_0^W G_2(x, x')\rho(x') \, dx' = 1 + S(x) \tag{181a}$$

$$\mu_0^{-1} A_z(x, H) = 2 \int_0^W G_1(x, x')J_z(x') \, dx' = \frac{\beta}{\omega\mu_0\epsilon_0}[1 + S(x)] \tag{181b}$$

since on the strip the boundary condition $E_z = 0$ gives $\mu_0^{-1} A_z = (\beta/\omega\mu_0\epsilon_0)\epsilon_0\Phi$. Instead of solving (181b) we can introduce relative values of A_z and J_z, denoted by A_{zr}, J_{zr}, and solve the equation

$$2 \int_0^W G_1(x, x')J_{zr}(x') \, dx' = 1 + S(x). \tag{182}$$

The actual current on the strip is then given by

$$J_z = \frac{\beta}{\omega\mu_0\epsilon_0} J_{zr} = K J_{zr}. \tag{183}$$

At low frequencies the function $S(x)$ is negligible. The required Green's functions G_1 and G_2 are determined below.

We will obtain solutions for J_z and ρ by expressing the fields and the required Green's functions as Fourier series in x.

The current J_z and the charge ρ can be represented by the following Fourier series:

$$J_z(x) = \sum_{n=1,3,\ldots}^{\infty} J_n \cos w_n x \qquad (184a)$$

$$\rho(x) = \sum_{n=1,3,\ldots}^{\infty} \rho_n \cos w_n x \qquad (184b)$$

where

$$J_n = \frac{2}{a} \int_0^1 J_z(x) \cos w_n x \, dx \qquad (184c)$$

$$\rho_n = \frac{2}{a} \int_0^1 \rho(x) \cos w_n x \, dx \qquad (184d)$$

and normalized dimensions have been introduced so that $W = 1$ meter, $a = A/W$, and $\alpha = H/W$. Also $w_n = n\pi/2a$ in (184). The potentials are also represented by Fourier series in the form

$$\Phi = \sum_{n=1,3,\ldots}^{\infty} \Phi_n(y) \cos w_n x \qquad (185a)$$

$$A_z = \sum_{n=1,3,\ldots}^{\infty} A_{zn}(y) \cos w_n x \qquad (185b)$$

$$A_y = \sum_{n=1,3,\ldots}^{\infty} A_{yn}(y) \cos w_n x \qquad (185c)$$

$$A_x = \sum_{n=1,3,\ldots}^{\infty} A_{xn}(y) \sin w_n x \qquad (185d)$$

where the factor $e^{-j\beta z}$ has been suppressed. In (184) we have taken advantage of the fact that J_z and ρ are even functions of x. In essence we are solving the problem with a magnetic wall in the yz plane.

The equations for Φ and A_y must be solved simultaneously. We assume that

$$\Phi_n = C \sinh \gamma y, \quad A_{yn} = D \cosh \gamma y, \qquad y < \alpha$$

and substitute into the differential equations (175d) and (176). For the nth Fourier term we obtain

$$j\omega\mu_0\epsilon_0(\kappa_y - \kappa)\gamma C + (\beta^2 + w_n^2 - \kappa_y k_0^2 - \gamma^2)D = 0$$

$$\left(\beta^2 + w_n^2 - \kappa k_0^2 - \frac{\kappa_y}{\kappa}\gamma^2\right)C - j\omega\gamma\frac{\kappa_y - \kappa}{\kappa}D = 0.$$

These homogeneous equations have a solution only if the determinant vanishes. From this requirement we find that there are two possible values for γ which are given by

$$\gamma_{1n} = (\beta^2 - \kappa k_0^2 + w_n^2)^{1/2} \tag{186a}$$

$$\gamma_{2n} = \left(\frac{\kappa}{\kappa_y}\right)^{1/2} (\beta^2 - \kappa_y k_0^2 + w_n^2)^{1/2} \tag{186b}$$

for each value of n. The corresponding solutions for the ratio D/C are $D_1 = -(\gamma_1/j\omega)C_1$ and $D_2 = (-j\kappa k_0^2/\omega\gamma_2)C_2$. Note that since E_x and $\partial E_y/\partial y$ must vanish on the ground plane, Φ and $\partial A_y/\partial y$ must be zero on the ground plane.

The general solutions for Φ_n and A_{yn} are now chosen as

$$\Phi_n = \begin{cases} C_1 \sinh \gamma_{1n} y + C_2 \sinh \gamma_{2n} y, & y \leq \alpha \\ C_3 e^{-\gamma_n y}, & y \geq \alpha \end{cases} \tag{187a}$$

$$A_{yn} = \begin{cases} D_1 \cosh \gamma_{1n} y + D_2 \cosh \gamma_{2n} y, & y \leq \alpha \\ D_3 e^{-\gamma_n y}, & y \geq \alpha \end{cases} \tag{187b}$$

where $\gamma_n = (\beta^2 - k_0^2 + w_n^2)^{1/2}$.

The potentials must be continuous at $y = \alpha$ and satisfy the discontinuity conditions given by (177). The use of these boundary conditions gives

$$C_1 \sinh \gamma_{1n}\alpha + C_2 \sinh \gamma_{2n}\alpha = C_3 e^{-\gamma_n \alpha}$$

$$D_1 \cosh \gamma_{1n}\alpha + D_2 \cosh \gamma_{2n}\alpha = D_3 e^{-\gamma_n \alpha}$$

and

$$\gamma_n D_3 e^{-\gamma_n\alpha} + \gamma_{1n}D_1 \sinh \gamma_{1n}\alpha + \gamma_{2n}D_2 \sinh \gamma_{2n}\alpha = -j\omega\mu_0\epsilon_0(\kappa - 1)C_3 e^{-\gamma_n\alpha}$$

$$\gamma_n C_3 e^{-\gamma_n\alpha} + \kappa_y\gamma_{1n}C_1 \cosh \gamma_{1n}\alpha + \kappa_y\gamma_{2n}C_2 \sinh \gamma_{2n}\alpha = \frac{\rho_n}{\epsilon_0} - j\omega(\kappa_y - 1)D_n e^{-\gamma_n\alpha}.$$

These equations involve six unknowns but two more relations are given by the ratios D_1/C_1 and D_2/C_2 above. Hence the system of equations can be solved to give

$$C_1 = \frac{\gamma_{2n}\rho_n k_0^2}{\epsilon_0 \Delta}(\gamma_{2n} \sinh \gamma_{2n}\alpha + \kappa\gamma_n \cosh \gamma_{2n}\alpha)$$

$$C_2 = \frac{\gamma_{2n}^2\gamma_n\rho_n}{\epsilon_0 \Delta}(\gamma_n \sinh \gamma_{1n}\alpha + \gamma_{1n} \cosh \gamma_{1n}\alpha)$$

where

$$\Delta = (k_0^2 + \gamma_n^2)\gamma_{2n}(\gamma_n \sinh \gamma_{1n}\alpha + \gamma_{1n} \cosh \gamma_{1n}\alpha) \times (\gamma_{2n} \sinh \gamma_{2n}\alpha + \kappa\gamma_n \cosh \gamma_{2n}\alpha).$$

The solution for $\Phi_n(\alpha)$ is now found to be given by

$$\epsilon_0 \Phi_n(\alpha) = \left[\frac{w_1(n)w_3(n)Sh_3(n)}{w_3(n)Sh_3(n) + \kappa w_1(n)Ch_3(n)} \right.$$

$$\left. + \frac{k_0^2 Sh_2(n)}{w_1(n)Sh_2(n) + w_2(n)Ch_2(n)} \right] \frac{\rho_n}{\beta^2 + w_n^2} \quad (188)$$

where the following shorthand notation has been introduced:

$$w_1(n) = \gamma_n \qquad Sh_i(n) = \sinh w_i(n)\alpha$$

$$w_2(n) = \gamma_{1n} \qquad Ch_i(n) = \cosh w_i(n)\alpha$$

$$w_3(n) = \gamma_{2n} \qquad i = 1, 2, 3.$$

The solutions for A_{yn} in $y < \alpha$ are readily found by using the relationship between C_1, C_2 and D_1, D_2 given above. In the solution for Φ_n we can replace β^2 using the relation

$$\beta^2 = \kappa_e k_0^2 \quad (189)$$

where κ_e is the effective dielectric constant. The coefficient of ρ_n in (188) is the nth coefficient of the Green's function G_2 needed to solve for Φ in terms of ρ after eliminating A_y.

When $k_0 = 0$ we have

$$\epsilon_0 \Phi_n(\alpha) = \frac{\rho_n \sinh w_n \alpha'}{w_n(\sinh w_n \alpha' + \sqrt{\kappa \kappa_y} \cosh w_n \alpha')} \quad (190)$$

where $\alpha' = (\kappa/\kappa_y)^{1/2}\alpha$.

The result shown in (190) expresses the property that the potential Φ for an anisotropic substrate is the same as that for an isotropic substrate with dielectric constant $(\kappa \kappa_y)^{1/2}$ and normalized thickness α'. However, A_z is still the same as that for a line with normalized height α for the strip so the zero-frequency inductance per unit length must be found for the unscaled line. This can be expressed in terms of the capacitance of the air-filled unscaled line.

The solution for A_{zn} is much easier to find and is, at $y = \alpha$,

$$A_{zn} = \frac{\mu_0 J_{zn} Sh_2(n)}{w_1(n)Sh_2(n) + w_2(n)Ch_2(n)} \quad (191)$$

which reduces, for $k_0 = 0$, to

$$A_{zn} = \mu_0 J_{zn} e^{-w_n \alpha} \frac{\sinh w_n \alpha}{w_n}. \quad (192)$$

The coefficient of $\mu_0 J_{zn}$ is the nth Fourier coefficient of the Green's function G_1 needed to solve for A_z in terms of J_z.

At $y = \alpha$, the Green's functions G_1 and G_2 have the representations

$$G_i(x, ') = \frac{1}{a} \sum_{n=1,3,\ldots}^{\infty} G_{in} \cos w_n x \cos w_n x', \qquad i = 1, 2 \quad (193)$$

where G_{1n} is the coefficient of $\mu_0 J_{zn}$ in (191) or (192) and G_{2n} is the coefficient of ρ_n in (188) or (190).

The Fourier series solution for A_x can be found directly from the Lorentz condition (175a) so J_x does not need to be solved for explicitly. An alternative procedure is to use the continuity equation to find J_x.

We readily find that

$$J_x = \sum_{n=1,3,\ldots}^{\infty} \frac{j\omega}{w_n} \left(\frac{\beta}{\omega} J_{zn} - \rho_n \right) \sin w_n x. \tag{194}$$

In the iteration solution to be presented we will need the static Green's functions G_1^0 and G_2^0. We can sum the dominant or singular part of the series for the G_i^0 into closed form, which can later be integrated exactly as the dominant part of the integral equations that will determine ρ and J_z.

The static Green's function G_1^0, at $y = \alpha$, is given by

$$G_1^0(x, x') = \sum_{n=1,3,\ldots}^{\infty} \frac{\cos w_n x \cos w_n x'(1 - e^{-2w_n \alpha})}{n\pi}. \tag{195}$$

The two series can be summed to give (see Section 4.6)

$$G_1^0 = -\frac{1}{4\pi} \ln \left(\tan \frac{\pi}{4a} |x - x'| \tan \frac{\pi}{4a} |x + x'| \right)$$

$$-\frac{1}{8\pi} \ln \left[\frac{\cosh \dfrac{\pi\alpha}{a} + \cos \dfrac{\pi}{2a}(x - x')}{\cosh \dfrac{\pi\alpha}{a} - \cos \dfrac{\pi}{2a}(x - x')} \right] \left[\frac{\cosh \dfrac{\pi\alpha}{a} + \cos \dfrac{\pi}{2a}(x + x')}{\cosh \dfrac{\pi\alpha}{a} - \cos \dfrac{\pi}{2a}(x + x')} \right]. \tag{196}$$

We now assume that a is set equal to the larger of 15 or 15α in which case α/a and $(x \pm x')/a$ are small. Hence, we can use the small argument approximations for the cosine and hyperbolic cosine terms and also expand the logarithm to obtain

$$G_1^0 = -\frac{1}{4\pi} \ln |x^2 - x'^2| + \frac{1}{8\pi} \ln \{[(2\alpha)^2 + (x - x')^2][(2\alpha)^2 + (x + x')^2]\} - \frac{\pi\alpha^2}{24a^2}. \tag{197}$$

The dominant term in α/a has been retained in (197) in order to determine how large a has to be before the sidewalls have a negligible effect. The numerical results verify the condition given after (196) to be sufficient in order to neglect the effects of finite a.

The static Green's function G_2^0 at $y = \alpha$ is given by

$$G_2^0(x, x') = \sum_{n=1,3,\ldots}^{\infty} \frac{2 \cos w_n x \cos w_n x' \sinh w_n \alpha'}{n\pi(\sinh w_n \alpha' + \kappa_g \cosh w_n \alpha')} \tag{198}$$

where $\kappa_g = (\kappa \kappa_y)^{1/2}$ and $\alpha' = (\kappa/\kappa_y)^{1/2}\alpha$. The above series may be expressed in the form

$$G_2^0(x, x') = \sum_{n=1,3,\ldots}^{\infty} \frac{2 \cos w_n x \cos w_n x'(1 - e^{-2w_n \alpha'})}{(\kappa_g + 1)n\pi(1 + \eta e^{-2w_n \alpha'})} \tag{199}$$

where $\eta = (\kappa_g - 1)/(\kappa_g + 1)$. We now expand the factor $(1 + \eta e^{-2w_n\alpha'})^{-1}$ into a power series and can then carry out the summation over n for each term to obtain

$$G_2^0(x, x') = \frac{2}{\kappa_g + 1} G_1^0(x, x', \alpha') - \frac{\kappa_g}{2\pi(\kappa_g + 1)^2}$$

$$\times \left\{ \frac{\kappa_g - 1}{2\kappa_g} \ln \left[\frac{\cosh \frac{\pi\alpha'}{a} + \cos \frac{\pi}{2a}(x - x')}{\cosh \frac{\pi\alpha'}{a} - \cos \frac{\pi}{2a}(x - x')} \right] \right.$$

$$\times \left[\frac{\cosh \frac{\pi\alpha'}{a} + \cos \frac{\pi}{2a}(x + x')}{\cosh \frac{\pi\alpha'}{a} - \cos \frac{\pi}{2a}(x + x')} \right]$$

$$+ \sum_{m=1}^{\infty} (-\eta)^m \ln \left[\frac{\cosh(m + 1)\frac{\pi\alpha'}{a} + \cos \frac{\pi}{2a}(x - x')}{\cosh(m + 1)\frac{\pi\alpha'}{a} - \cos \frac{\pi}{2a}(x - x')} \right]$$

$$\times \left[\frac{\cosh(m + 1)\frac{\pi\alpha'}{a} + \cos \frac{\pi}{2a}(x + x')}{\cosh(m + 1)\frac{\pi\alpha'}{a} - \cos \frac{\pi}{2a}(x + x')} \right] \right\}. \qquad (200)$$

The alternating series represents a correction to the dominant part of G_2^0 which is expressed in closed form. In the numerical evaluation of the integral involving G_2^0 it was established that truncating the series when the mth term was smaller than 10^{-3} times the sum of the first $m - 1$ terms resulted in an insignificant error. For narrow strips with κ_g less than 5 it is found that 8 to 10 terms are sufficient, while for strips as wide as $10H$ and with $\kappa_g = 12$ a total of 45 terms is required.

The dominant series is of the type considered in Section 4.6. For large values of a the variable transformation given by (130) is simplified since when a becomes infinite the modulus k of the elliptic sine function becomes zero and (130) reduces to

$$x = w \sin \sigma = w \sin \theta$$

where σ is now replaced by the variable θ. The parameter K given by (125) becomes $K = \pi/2$ and K' given by (126) becomes infinite like $\ln(4/k)$. With reference to Fig. 4.21(d) we note that when the boundary at jK' moves out to infinity the $n = 0$ term in the Green's function given by (129) approaches $(1/2\pi)[\ln(2/k) + \ln 2]$. The correction series given by (138) contains a term $-(1/2\pi) \ln(4a/\pi)$. Since $k = \pi/2a$ the $\ln(4a/\pi)$ terms cancel. These terms have also been canceled in (197). Hence, when we change to the variable $\sigma = \theta$ the first term in (197) can be replaced by (129) after the following changes are made in (129). Replace the constant term $K'/4K$ by $(\ln 2)/2\pi$, set $K = \pi/2$, and replace the hyperbolic tangent functions by unity. Thus we can replace (197) by

$$G_1^0 = \frac{1}{2\pi} \ln 2 + \sum_{n=1}^{\infty} \frac{1}{2n\pi} \cos 2n\theta \cos 2n\theta'$$

$$+ \frac{1}{8\pi} \ln \{[(2\alpha)^2 - (x - x')^2][(2\alpha)^2 + (x + x')^2]\} \qquad (201)$$

where $x = \sin\theta$, $x' = \sin\theta'$. The dominant part of G_2^0 is given by this expression multiplied by the factor $2/(\kappa_g + 1)$.

Solution by Iteration

In the iteration method the current and charge densities and the effective dielectric constant κ_e are first found at zero frequency. For this case $\epsilon_0\Phi$ is set equal to 1 on the strip and A_{zr}/μ_0 is also set equal to 1 on the strip where A_{zr} is a reference value for the vector potential. The integral equations to be solved are

$$2\int_0^1 G_1^0(x, x')J_{zr}(x')\,dx' = 1 \tag{202a}$$

$$2\int_0^1 G_2^0(x, x')\rho(x')\,dx' = 1. \tag{202b}$$

We can expand the current J_z in terms of four basis functions in the form

$$J_z = \frac{I_0 - I_1 T_2(x) + I_2 T_4(x) - I_3 T_6(x)}{\sqrt{1 - x^2}}$$

where $T_n(x)$ is a Chebyshev polynomial. In terms of the variable θ

$$J_z(x)\,dx = (I_0 + I_1 \cos 2\theta + I_3 \cos 4\theta + I_4 \cos 6\theta)\,d\theta. \tag{203}$$

A similar expansion for ρ is used with coefficients Q_n, $n = 0, 1, 2, 3$. The integrals in (202) can be evaluated analytically for the dominant parts of G_1^0 and G_2^0. The correction terms are easily evaluated by numerical integration using Simpson's rule. The resulting equations can be converted to matrix equations by using point matching at the four points $x = \sin(2i + 1)\pi/16$, $i = 0, 1, 2, 3$. Test cases using Galerkin's method and the method of least squares have been evaluated with essentially the same results. The use of point matching requires somewhat less computational effort. The numerical results show that I_2 and Q_2 are small and I_3 and Q_3 are almost negligible so four basis functions are sufficient.

After the expansion coefficients I_n and Q_n for ρ have been found the total current and charge on the strip are given by $I_T = \pi I_0$ and $Q_T = \pi Q_0$. The total current is a relative value so we now multiply I_T by a constant K and enforce the two conditions (178a) and (179) to obtain

$$\omega\epsilon_0 K A_{zr} = \beta\epsilon_0\Phi = \beta = \omega\mu_0\epsilon_0 K$$

and

$$\beta K I_T = \omega Q_T$$

which then give

$$\frac{\beta^2}{k_0^2} = \kappa_e = \frac{Q_T}{I_T} \tag{204}$$

$$Z_c = \frac{1/\epsilon_0}{KI_T} = \frac{Z_0}{I_T\sqrt{\kappa_e}} \tag{205}$$

where Z_c is the characteristic impedance of the strip line and $Z_0 = (\mu_0/\epsilon_0)^{1/2}$ and $\beta K = \omega\kappa_e$.

The next step that must be carried out is to evaluate the Fourier coefficients J_{zn} and ρ_n using the static current and charge distributions. The Green's functions at the first frequency increment (typically steps of 1 or 2 GHz) may be approximated by using the static value of κ_e and the following integral equations can then be solved:

$$2\int_0^1 G_1^1(x, x')J_{zr}(x')\,dx' = 1 + S(x) \tag{206a}$$

$$2\int_0^1 G_2^1(x, x')\rho(x')\,dx' = 1 + S(x) \tag{206b}$$

where $S(x)$ is the correction to the static boundary value for the potentials as given by (180). From the continuity equation we obtain $w_n J_{xn} = j\omega(\kappa_e J_{zn} - \rho_n)$. Also, the Green's function for A_x has the same Fourier coefficients G_{1n} but with $\sin w_n x$ replacing $\cos w_n x$. By using these relations the integral in (180) is readily evaluated to give

$$S(x) = \sum_{n=1,3,\ldots} k_0^2 G_{1n}^0 \frac{\cos w_n x - 1}{w_n^2}(\rho_n - \kappa_e J_{zn}). \tag{207}$$

Note that $\rho_n - \kappa_e J_{zn}$ is proportional to the nth Fourier coefficient of J_x and is therefore quite small. The Green's functions G_1^1 and G_2^1 at the first iteration (denoted by the superscript 1) can be expressed as

$$G_i^1 = G_i^0 + (G_i^1 - G_i^0).$$

The first part is the static Green's function or dominant part, while the second part is a correction and may be represented in a Fourier series form. By means of this technique it turns out that good numerical convergence is obtained by using only 30 terms in the Fourier series for the correction term and also for $S(x)$. The integral equations in (206) may be solved and new values of I_n and Q_n at the first frequency increment are thus obtained. A corrected value of κ_e may then be found using the same relation given earlier by (204).

The iteration is now repeated by calculating a more accurate value for $S(x)$ using the new value of κ_e and new computed values for ρ_n and J_{zn}. The new value of κ_e is also used in the Green's functions G_1^2 and G_2^2 for the second iteration. These Green's functions are expressed in the form

$$G_i^2 = G_i^1 + (G_i^2 - G_i^1).$$

The first term is known from the earlier computation and the second term is a rapidly converging Fourier series correction term. This iteration procedure is repeated until successive values of κ_e do not change by more than 0.1%. When convergence has been obtained the frequency is incremented to the next value. Linear extrapolation can be used to obtain a good initial value for κ_e at the new frequency. By this means the iteration converges very fast and typically one, two, or three iterations are all that is required at each frequency when the frequency increment is 2 GHz. A smaller frequency increment requires fewer iterations at each new frequency.

At each frequency the characteristic impedance of the microstrip line can be evaluated after a converged value of κ_e has been found. The following definition for the characteristic impedance Z_c can be used:

$$
Z_c = \frac{1}{I_{\text{TOT}}} \int_0^\alpha -E_y \, dy = \frac{1}{I_T} \left[\int_0^\alpha \left(\frac{\partial \Phi}{\partial y} + j\omega A_y \right) dy \right]
$$

$$
= \frac{1}{I_{\text{TOT}}} \left[\frac{1}{\epsilon_0} + j\omega \int_0^\alpha A_y \, dy \right]
\tag{208}
$$

where I_{TOT} is the total z-directed current on the strip and the integral of E_y is carried out at $x = 0$ to obtain an equivalent voltage across the line. The vector potential function A_y can be evaluated in Fourier series form so that (208) can be expressed as

$$
Z_c = \frac{120\pi}{I_T \sqrt{\kappa_e}} \left\{ 1 + \sum_{n=1,3,\ldots} \frac{k_0^2 \rho_n}{\kappa_e k_0^2 + w_n^2} \left[\frac{\kappa w_1(n) Sh_3(n)}{w_3(n)[w_3(n)Sh_3(n) + \kappa w_1(n)Ch_3(n)]} \right. \right.
$$

$$
\left. \left. - \frac{Sh_2(n)}{w_1(n)Sh_2(n) + w_2(n)Ch_2(n)} \right] \right\}
\tag{209}
$$

where I_T is the total relative value of the z-directed current obtained from the solution of an equation like (206a) in the last iteration at the frequency of interest. A total of 30 terms gives an accurate value for the sum in (209).

Typical numerical results are shown in Figs. 4.27 and 4.28 for the case of a gallium arsenide substrate with $\kappa = 12.9$ and in Figs. 4.29 and 4.30 for the case of a sapphire substrate with

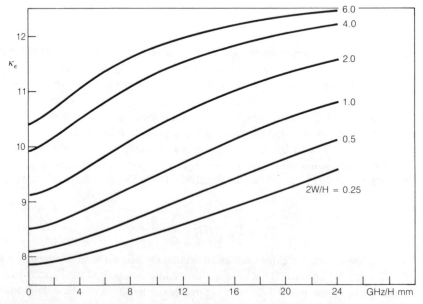

Fig. 4.27. Effective dielectric constant for a microstrip line on a GaAs substrate. $\kappa = 12.9$ and substrate thickness $H = 1$ mm.

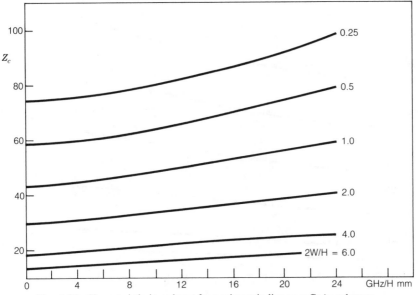

Fig. 4.28. Characteristic impedance for a microstrip line on a GaAs substrate.

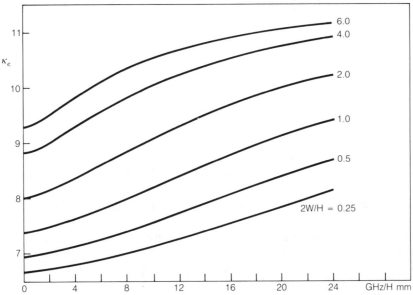

Fig. 4.29. Effective dielectric constant for a microstrip line on a sapphire substrate. $\kappa = 9.4$, $\kappa_y = 11.6$.

$\kappa = 9.4$ and $\kappa_y = 11.6$. Other computed results can be found in a paper by Kretch and Collin [4.35].

The mode of propagation on a microstrip line is not a TEM mode and hence the line integral of the electric field between the ground plane and the strip is dependent on the path. Thus a unique value of voltage does not exist and consequently the expression for characteristic impedance used above is also not unique. Alternative definitions based on power considerations

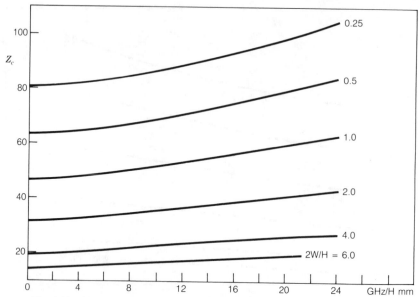

Fig. 4.30. Characteristic impedance for a microstrip line on a sapphire substrate.

have been proposed. We could define Z_c by any of the following three relationships:

$$Z_c = \frac{V}{I}$$

$$P = \frac{1}{2}Z_c|I|^2$$

$$P = \frac{1}{2Z_c^*}|V|^2$$

where I is the total z-directed current on the microstrip, P is the complex power on the line, and V is a defined voltage such as the line integral of the electric field from the ground plane to the center of the strip line. The three definitions do not give the same values for Z_c except in the limit of zero frequency.

For modes of propagation that are not TEM modes the definitions of voltage and current are arbitrary in the equivalent transmission-line model. Either the equivalent voltage or the equivalent current can be chosen arbitrarily. If the other quantity is then chosen so that the power flow is given by

$$P = \tfrac{1}{2}VI^*$$

then all three definitions given above for the characteristic impedance are equivalent [4.6]. However, the value of the characteristic impedance will depend on the definition used for the voltage or current. When we characterize the junction of two different microstrip lines by an equivalent circuit or scattering matrix, it is essential that the voltage and current be defined so that the expression for power flow will be proportional to the true value of power flow with the same proportionality constant for both the input and output lines. This is necessary in order to have power conservation for the equivalent circuit.

In many instances we can obtain a useful equivalent network description of a microstrip discontinuity using arbitrary definitions for the characteristic impedance. For example, consider a microstrip line in which there is a step change in the width. We will assume that we have been able to determine the reflection coefficient Γ and the transmission coefficient T for the total axially directed current on the input and output lines. The incident and reflected powers on the input line are given by (we now assume loss free lines)

$$P_{\text{inc}} - P_{\text{ref}} = \frac{K_1}{2} |I^+|^2 Z_1 (1 - |\Gamma|^2)$$

and must equal the transmitted power

$$P_t = \frac{K_2}{2} |TI^+|^2 Z_2$$

where Z_1 and Z_2 are arbitrarily chosen characteristic impedances. The constants K_1 and K_2 must be chosen so that power conservation holds in a relative sense. This requires that

$$\frac{K_2}{K_1} = \frac{Z_1 (1 - |\Gamma|^2)}{Z_2 |T|^2}.$$

Usually it is only the relative power that needs to be known so we can choose K_1 equal to unity. The parameter K_2 is then determined from the equation given above. A new transmission coefficient $T' = \sqrt{K_2} T$ can now be defined so that power conservation holds true in a relative sense.

4.10. POTENTIAL THEORY FOR COUPLED MICROSTRIP LINES

The method of analysis presented in the previous section is readily extended to treat the coupled microstrip line shown in Fig. 4.24(d). The case of even excitation is a relatively simple extension to the problem of a single strip. If we assume the substrate to be anisotropic and place sidewalls at $x = \pm A$ as in Fig. 4.26 then the same Green's functions as used in Section 4.9 apply. The two strips extend over the intervals $x_1 \leq x \leq x_2$ and $-x_2 \leq x \leq -x_1$. In order to conform with the normalized dimensions used in Section 4.9 we assume that each strip is of width w with $w = x_2 - x_1$ equal to one unit. When $x_1 = 0$ we then obtain the single-strip case. The center of the strip is located at $x_c = (x_2 - x_1)/2$. The reference point for the potential Φ is chosen to be at x_c. The integral of the boundary condition $E_x = -j\omega A_x - \partial \Phi / \partial x = 0$ from x to x_c gives

$$\epsilon_0 \Phi(x) = \epsilon_0 \Phi(x_c) - j\omega\mu_0\epsilon_0 \int_{x_1}^{x_2} \int_{x_c}^{x} 2G(x, x') J_x(x') \, dx' \, dx. \qquad (210)$$

For convenience we choose $\epsilon_0 \Phi(x_c)$ equal to one. The factor of 2 accounts for the contribution to A_x from the current J_x on the strip in the interval $-x_2$ to $-x_1$. This equation replaces (180) for the boundary condition for $\Phi(x)$ on the strip. The expression (210) can be evaluated the same way as in Section 4.9 and in place of (207) we obtain

$$\epsilon_0 \Phi(x) = 1 + S(x)$$

with

$$S(x) = \sum_{n=1,3,\dots}^{\infty} k_0^2 G_{1n}^0 \frac{\cos w_n x - \cos w_n x_c}{w_n^2} (\rho_n - \kappa_e J_{zn}) \tag{211}$$

where the G_{1n}^0 are the same as in (207).

The two integral equations to be solved are

$$2 \int_{x_1}^{x_2} G_1(x, x') J_{zr}(x') \, dx' = 1 + S(x) \tag{212a}$$

$$2 \int_{x_1}^{x_2} G_2(x, x') \rho(x') \, dx' = 1 + S(x). \tag{212b}$$

The static limits of G_1 and G_2 are given by G_1^0 and G_2^0 shown in (196) and (200). These are now expressed as functions of a new variable σ that will give an orthogonal series representation for the singular parts over the strip interval. The appropriate variable transformation has been described in Section 4.6 and is given by (142) and (143) in the form

$$\sigma = \mathrm{sn}^{-1}(\lambda, k) \tag{213a}$$

$$\lambda = \left[\frac{\sin^2 \dfrac{\pi x}{2a} - \sin^2 \dfrac{\pi x_1}{2a}}{\sin^2 \dfrac{\pi x_2}{2a} - \sin^2 \dfrac{\pi x_1}{2a}} \right]^{1/2} \tag{213b}$$

and k is given by (141). For numerical evaluations we need to know x as a function of σ. This is given by

$$x = \frac{2a}{\pi} \sin^{-1} \left[\left(\sin^2 \frac{\pi x_2}{2a} - \sin^2 \frac{\pi x_1}{2a} \right) \mathrm{sn}^2(\sigma, k) + \sin^2 \frac{\pi x_1}{2a} \right]^{1/2}. \tag{213c}$$

The dominant part of G_1^0 is given by (144). The current J_{zr} multiplied by dx may be expanded in the form

$$J_{zr}(x) \, dx = \sum_{n=0}^{N} I_n \cos \frac{n\pi\sigma}{K} \, d\sigma. \tag{214}$$

In practice it is found that the use of four terms in the expansion of the current gives values for the effective dielectric constant accurate to better than 1%. The dominant part of the integral equation can be evaluated analytically. For the first iteration the integral equation (212a) becomes

$$\frac{K'}{2} I_0 + \sum_{n=1}^{3} \frac{I_n K}{4n\pi} \tanh \frac{n\pi K'}{K} \cos \frac{n\pi\sigma}{K} - \frac{1}{4\pi} \sum_{n=0}^{3} I_n \int_0^K \cos \frac{n\pi\sigma'}{K}$$

$$\times \ln \left[\frac{\cosh \dfrac{\pi\alpha}{a} + \cos \dfrac{\pi}{2a}(x - x') \cosh \dfrac{\pi\alpha}{a} + \cos \dfrac{\pi}{2a}(x + x')}{\cosh \dfrac{\pi\alpha}{a} - \cos \dfrac{\pi}{2a}(x - x') \cosh \dfrac{\pi\alpha}{a} - \cos \dfrac{\pi}{2a}(x + x')} \right] d\sigma' = 1 \tag{215}$$

where the dominant part has been integrated and the remaining part has to be evaluated numerically using (213c) to generate values of x' in terms of σ'. Point matching at $\sigma = (2j - 1)K/8$, $j = 1, 2, 3, 4$ can be used to obtain four algebraic equations to determine the I_n. The integral equation for ρ can be solved the same way at the first iteration step.

The Fourier coefficients J_{zn} are given by

$$J_{zn} = \frac{2}{a} \int_{x_1}^{x_2} J_{zr}(x) \cos w_n x \, dx$$

$$= \frac{2}{a} \sum_{m=0}^{3} I_m \int_0^K \cos \frac{m\pi\sigma}{K} \cos w_n x \, d\sigma \tag{216}$$

with a similar formula for ρ_n. The basic integrals in (216) must be evaluated numerically using (213c) to express x as a function of σ. The same integrals occur in the evaluation of ρ_n. These integrals may be conveniently stored as elements in an array in a computer program since they are needed in the successive steps of the iteration process. At each step of the iteration, corrected values for κ_e are obtained using (204). These are then employed to evaluate the corrections to the static Green's functions so as to obtain the dynamic Green's functions. The iteration process is the same as that described in Section 4.9.

The solution for the case of odd excitation is very similar to that described above. A basic change that is required in the construction of Green's functions for J_z and ρ is replacement of $\cos(n\pi x/2a) \cos(n\pi x'/2a)$ in (193) by $\frac{1}{2} \sin(n\pi x/a) \sin(n\pi x'/a)$ and summation over all integer values of n. This substitution will generate the required odd Green's functions that are needed. The static Green's function for A_z can be summed into closed form as shown in Section 4.9 and Problem 4.14. The integral equations for J_{zr} and ρ are the same as in (212) with the appropriate odd Green's functions used. The boundary source term $S(x)$ is given by (211) after replacing $\cos w_n x - \cos w_n x_c$ by $\sin w_n x - \sin w_n x_c$, where now $w_n = n\pi/a$ and the sum is taken over all integer values of n.

The singular part of the static Green's functions can be transformed into a series of orthogonal functions on the strip by using the variable transformation given by (150) and (151) in Section 4.6, i.e.,

$$x = \frac{2a}{\pi} \tan^{-1} \left[\tan \frac{\pi x_1}{2a} \operatorname{sn}(K + j\eta, k) \right]. \tag{217}$$

The transformed singular part of the Green's functions is given by (152) and the correction term is given in Problem 4.14. The current J_{zr} and charge ρ may be expanded as series in terms of the $\cos(n\pi\eta/K')$ functions that appear in (152), e.g.,

$$J_{zr}(x) \, dx = \sum_{n=0}^{N} I_n \cos \frac{n\pi\eta}{K'} \, d\eta.$$

Apart from the changes indicated above the solution for the odd-mode case is carried out the same way as that for the case of even excitation and that for the single strip. A detailed description of the procedure may be found in the thesis by Morich [4.36]. When $x_1 \ll x_2$ the odd-mode solution is essentially that for a slot line above a ground plane. If also $x_1 \ll H$, then the case of an isolated slot line is obtained.

In Fig. 4.31 we show some typical computed results for a coupled microstrip line on a gallium arsenide substrate with a dielectric constant of 12.9. These results were obtained by Morich [4.36] and included the effect of a cover plate at a height b above the ground plane. The substrate thickness $H = 1$ mm, $2x_1 = 1$ mm, the sidewall spacing $2A = 6$ cm, and the strip width is 1 mm. The values of b given in the figure are also in millimeters. The results shown can be scaled to be applied at other frequencies. A notable feature is that the effective dielectric constant for even-mode excitation is significantly larger than that for the odd mode at high frequencies. It can also be seen that a top shield has relatively little effect if it is placed a distance equal to four or more times the substrate thickness above the ground plane.

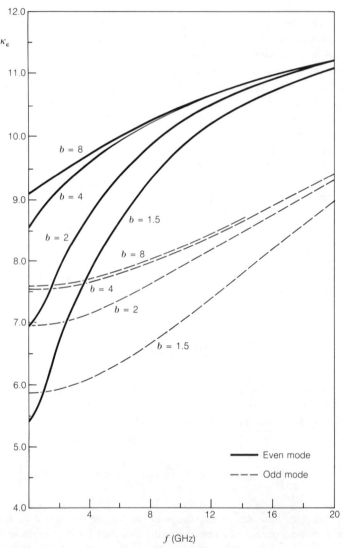

Fig. 4.31. Effective dielectric constants for the even and odd modes of a coupled microstrip line on a GaAs substrate. $H = 1$ mm, $A = 30$ mm, and b is given in millimeters.

REFERENCES AND BIBLIOGRAPHY

[4.1]   J. A. Stratton, *Electromagnetic Theory*,   New York, NY: McGraw-Hill Book Company, Inc., 1941, ch. 9.

[4.2]   P. M. Morse and H. Feshbach, *Methods of Theoretical Physics*, part I.   New York, NY: McGraw-Hill Book Company, Inc., 1953, ch. 4.

[4.3]   H. Hancock, *Lectures on the Theory of Elliptic Functions*.   New York, NY: Dover Publications, 1958.

[4.4]   E. Jahnke and F. Emde, *Tables of Functions*, 4th ed.   New York, NY: Dover Publications, 1945.

[4.5]   R. V. Churchill, *Operational Mathematics*, 2nd ed.   New York, NY: McGraw-Hill Book Company, Inc., 1958, ch. 10.

[4.6]   J. R. Brews, "Characteristic impedance of microstrip lines," *IEEE Trans. Microwave Theory Tech.*, vol. MTT-35, pp. 30–34, 1987.

General Theory of Transmission Lines

[4.7]   R. W. P. King, *Transmission Line Theory*.   New York, NY: McGraw-Hill Book Company, Inc., 1955.

[4.8]   R. K. Moore, *Traveling-Wave Engineering*.   New York, NY: McGraw-Hill Book Company, Inc., 1960.

[4.9]   L. N. Dworsky, *Modern Transmission Line Theory and Applications*.   New York, NY: John Wiley & Sons, Inc., 1979.

Theory of Conformal Mapping

[4.10]   R. V. Churchill, *Introduction to Complex Variables and Applications*.   New York, NY: McGraw-Hill Book Company, Inc., 1948.

[4.11]   H. Kober, *Dictionary of Conformal Representation*, 2nd ed.   New York, NY: Dover Publications, 1957.

Application of Conformal Mapping to Planar Transmission Lines

[4.12]   Symposium on Microstrip Circuits, *IRE Trans. Microwave Theory Tech.*, vol. MTT-3, Mar. 1955.

[4.13]   S. B. Cohn, "Shielded coupled strip transmission line," *IRE Trans. Microwave Theory Tech.*, vol. MTT-3, pp. 29–38, Oct. 1955.

[4.14]   B. A. Dahlman, "A double ground plane strip line system for microwaves," *IRE Trans. Microwave Theory Tech.*, vol. MTT-3, pp. 52–57, Oct. 1955. [This paper treats the same problem as that in Sect. 4.3 but approximates the center conductor by a semi-infinite strip and solves for the fringing field. Equation (21) is in error; when corrected, it gives the same value for the attenuation constant as that obtained here in Sect. 4.3.]

[4.15]   D. Park, "Planar transmission lines," *IRE Trans. Microwave Theory Tech.*, vol. MTT-3, part I, pp. 8–12, Apr. 1955; part II, pp. 7–11, Oct. 1955.

[4.16]   W. H. Hayt, Jr., "Potential solution of a homogeneous strip line of finite width," *IRE Trans. Microwave Theory Tech.*, vol. MTT-3, pp. 16–18, July 1955. (This paper gives a solution to the same problem as that treated here in Sect. 4.3 but does not assume that ground planes are infinite.)

[4.17]   H. A. Wheeler, "Transmission-line properties of parallel strips separated by a dielectric sheet," *IEEE Trans. Microwave Theory Tech.*, vol. MTT-13, pp. 172–185, 1965.

Finite-Difference and Finite-Element Methods

[4.18]   H. E. Green, "The numerical solution of some important transmission line problems," *IEEE Trans. Microwave Theory Tech.*, vol. MTT-13, pp. 676–692, 1965.

[4.19]   P. Silvester, "TEM properties of microstrip transmission lines," *Proc. IEE*, vol. 115, pp. 43–48, 1968.

[4.20]   Z. Pantic and R. Mittra, "Quasi-TEM analysis of microwave transmission lines by the finite element method," *IEEE Trans. Microwave Theory Tech.*, vol. MTT-34, pp. 1086–1103, 1986.

Application of Variational Methods to Transmission Lines

[4.21]   R. E. Collin, "Characteristic impedance of a slotted coaxial line," *IRE Trans. Microwave Theory Tech.*, vol. MTT-4, pp. 4–8, Jan. 1956.

[4.22]   R. M. Chrisholm, "The characteristic impedance of trough and slab lines," *IRE Trans. Microwave Theory Tech.*, vol. MTT-4, pp. 166–172, July 1956.

[4.23]   E. Yamashita and R. Mittra, "Variational method for the analysis of microstrip lines," *IEEE Trans. Microwave Theory Tech.*, vol. MTT-16, pp. 251–256, 1968.

[4.24]   E. Yamashita, "Variational method for the analysis of microstrip-like transmission lines," *IEEE Trans. Microwave Theory Tech.*, vol. MTT-16, pp. 529–535, 1968.

Spectral-Domain Methods

[4.25] T. Itoh and R. Mittra, "Spectral domain approach for calculating the dispersion characteristics of microstrip lines," *IEEE Trans. Microwave Theory Tech.*, vol. MTT-21, pp. 496–499, 1973.

[4.26] R. H. Jansen, "High speed computation of single and coupled microstrip parameters including dispersion, high order modes, loss and finite strip thickness," *IEEE Trans. Microwave Theory Tech.*, vol. MTT-26, pp. 75–82, 1978.

[4.27] T. Kitazawa and Y. Hayashi, "Propagation characteristics of striplines with multilayered anisotropic media," *IEEE Trans. Microwave Theory Tech.*, vol. MTT-31, pp. 429–433, June 1983.

[4.28] N. G. Alexopoulos and C. M. Krowne, "Characteristics of single and coupled microstrips on anisotropic substrates," *IEEE Trans. Microwave Theory Tech.*, vol. MTT-26, pp. 387–393, June 1978.

[4.29] N. G. Alexopoulos, "Integrated-circuit structures on anisotropic substrates," *IEEE Trans. Microwave Theory Tech.*, vol. MTT-33, pp. 847–881, 1985.

[4.30] R. H. Jansen, "The spectral-domain approach for microwave integrated circuits," *IEEE Trans. Microwave Theory Tech.*, vol. MTT-33, pp. 1043–1056, 1985.

Potential Theory for Microstrip Lines

[4.31] Y. Fujiki, Y. Hayashi, and M. Suzuki, "Analysis of strip transmission lines by iteration method," *J. Inst. Electron. Commun. Eng. Japan*, vol. 55-B, pp. 212–219, 1972.

[4.32] E. F. Kuester and D. C. Chang, "Theory of dispersion in microstrip of arbitrary width," *IEEE Trans. Microwave Theory Tech.*, vol. MTT-28, pp. 259–265, 1980.

[4.33] T. Itoh and R. Mittra, "Analysis of microstrip transmission lines," Sci. Rep. 72-5, Antenna Lab., University of Illinois, Urbana, June 1972.

[4.34] D. C. Chang and E. Kuester, "An analytic theory for narrow open microstrip," *Arch. Elek. Übertragung.*, vol. 33, pp. 199–206, 1979.

[4.35] B. E. Kretch and R. E. Collin, "Microstrip dispersion including anisotropic substrates," *IEEE Trans. Microwave Theory Tech.*, vol. MTT-35, pp. 710–718, 1987.

[4.36] M. A. Morich, "Broadband dispersion analysis of coupled microstrip on anisotropic substrates by perturbation-iteration theory," MS thesis, Case Western Reserve University, Cleveland, OH, May 1987.

General Reference

[4.37] K. C. Gupta, R. Garg, and I. J. Bahl, *Microstrip Lines and Slotlines*.   Dedham, MA: Artech House, Inc., 1979.

PROBLEMS

4.1. Find the characteristic impedance and attenuation constant arising from conductor loss for a coaxial line with inner radius a and outer radius b.

Answer:

$$Z_c = \frac{Z_0}{2\pi} \ln \frac{b}{a} \qquad \alpha = \frac{R_m}{4\pi Z_c} \left(\frac{1}{a} + \frac{1}{b} \right).$$

4.2. Show that the conformal transformation $W = \ln[(Z + a)/(Z - a)]$ maps the cross section of a two-wire line into the strips $u = \pm u_1$, $0 \le v \le 2\pi$, in the W plane. The conductor radius and spacing are given by $R = a(\sinh u_1)^{-1}$, $D = 2a \coth u_1$. Find the characteristic impedance of the line. By a suitable integration along the conductor strips in the W plane, determine the power loss in the conductors and, hence, the attenuation constant for the line. (See Fig. P4.2.)

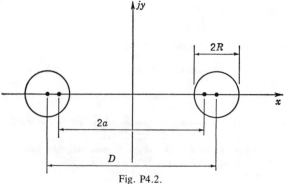

Fig. P4.2.

Answer:

$$Z_c = \frac{Z_0}{\pi} \cosh^{-1} \frac{D}{2R} \qquad \alpha = \frac{R_m}{\pi Z_c 2R} \frac{D/2R}{[(D/2R)^2 - 1]^{1/2}}.$$

4.3. For a constant value of outer radius b, show that the optimum ratio of outer to inner conductor radius, i.e., b/a, which will make the attenuation constant of a coaxial line a minimum satisfies the equation $x \ln x = 1 + x$, where $x = b/a$. Find the corresponding value of the characteristic impedance.

4.4. By a suitable conformal transformation, determine the characteristic impedance of the strip line illustrated in Fig. P4.4. Assume that the ground planes are infinitely wide and that the thickness of the center conductor is negligible.

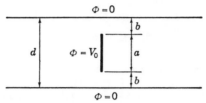

Fig. P4.4.

4.5. Consider a semi-infinite parallel-plate capacitor as illustrated in Fig. P4.5. In the W plane the polygon corresponding to the two plates is obtained as the points w_1 and w_3 tend to infinity. Find a suitable conformal transformation which will map the capacitor plates into the real axis in the Z plane. Integrate the normal gradient of the potential along the x axis over that region which corresponds to the inner and outer surfaces of one plate extending in a distance l from the edge. Find the capacitance $C(l)$ as a function of l. Use this result to determine the approximate characteristic impedance of a parallel-plate transmission line with plate spacing d and total width $2l$.

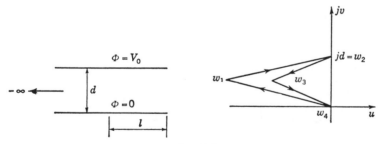

Fig. P4.5.

Answer: A suitable mapping function is $W = (d/\pi)((1 - Z^2)/2 + \ln Z)$. Charge $Q(l) = (\epsilon_0 V_0/\pi) \ln(x_2/x_1)$, where x_2 and x_1 are determined by $-l\pi/d = (1 - x_i^2)/2 + \ln x_i$. For $l \gg d$,

$$\ln x_1 \approx \frac{-l\pi}{d} - \frac{1}{2}$$

$$x_2^2 \approx \frac{2l\pi}{d} + 1.$$

$$\text{Capacitance } C(l) \approx \frac{\epsilon_0 l}{d} \left[1 + \frac{d}{2\pi l} \ln \left(1 + \frac{2\pi l}{d} \right) \right] \qquad Z_c = \frac{(\mu_0 \epsilon_0)^{1/2}}{2C(l)}.$$

4.6. Consider a conducting wedge of internal angle α and a line charge located at $Z = 1$ as illustrated in Fig. P4.6. By means of the transformation $W = Z^\beta$, where $\beta = \pi/(2\pi - \alpha)$, map this into a plane and a line charge in the W plane. This latter problem is readily solved by image theory to give $\Phi = K \ln |(W - a)/(W + a)|$, where K is

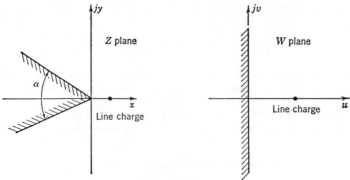

Fig. P4.6.

a suitable constant. Find the charge density on the wedge, and show that near the edge the surface charge ρ_s varies with the distance r from the edge according to $r^{\beta-1}$. Compare this singularity condition with that given in Chapter 1, Eq. (51). Note that, for $\alpha > 0$, the integral of $(\partial\Phi/\partial n)^2$ over the wedge surface is finite.

4.7. Obtain upper and lower bounds to the characteristic impedance of the strip line illustrated in Problem 4.4. In the variational expressions use the approximations that the potential varies linearly from the center conductor to the ground plane and that the charge density is constant on the center strip. Compare the value of Z_c obtained for $a = 0.5d$ with the correct value obtained in Problem 4.4.

4.8. Consider a series of three two-conductor transmission lines with cross sections S, S_1; S, S_2; and S, S_3, where the cross section S_1 completely overlaps S_2, and S_2 completely overlaps S_3 as illustrated in Fig. P4.8. Let the

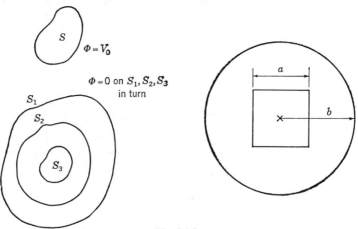

Fig. P4.8.

capacitance per unit length for the three cross sections be C_1, C_2, and C_3, respectively. By means of the variational expression for capacitance show that $C_1 > C_2 > C_3$.

HINT: For an approximation for the potential Φ_3 for the cross section S, S_3, use the correct expression Φ_2 for the potential appropriate for the cross section S, S_2 in the region external to S_2, and zero in the region between S_2 and S_3. The variational integral now gives C_2 as the upper bound on C_3, and hence $C_3 < C_2$. A similar procedure may be used to show $C_2 < C_1$.

Use this result to obtain upper and lower bounds for the characteristic impedance of a coaxial line with a square center conductor. What is the maximum error in the average value for Z_c when $b = 3a$?

4.9. Apply Taylor's expansion

$$F = F_0 + \sum_i \Delta x_i \left.\frac{\partial F}{\partial x_i}\right|_0 + \frac{1}{2}\sum_i\sum_j \Delta x_i\,\Delta x_j \left.\frac{\partial^2 F}{\partial x_i\,\partial x_j}\right|_0 + \cdots$$

to the expression (98) plus (99) to show that the first variation in $W_e = W_{e1} + W_{e2}$ is zero, the second variation is positive, and all higher order variations are zero. Let the variation of each coefficient a_n and b_n about the correct value be Δa_n and Δb_n.

4.10. A coaxial line is terminated by a thin resistive disk having a resistance R_0 ohms per unit square and a perfect short circuit a distance l away. Determine the required values of R_0 and l in order to provide a matched termination, i.e., to prevent a reflected wave from the termination. (See Fig. P4.10.)

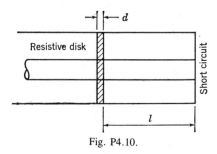

Fig. P4.10.

HINT: Consider a thin disk of thickness d with parameters μ_2, σ. When $l = \lambda_0/4$, show that the condition for a match is $Z = Z_0 \tanh kd$, where $Z = (j\omega\mu_2/\sigma)^{1/2}$ and $k = (j\omega\mu_2\sigma)^{1/2}$.

Answer:

$$R_0 = \left(\frac{\mu_0}{\epsilon_0}\right)^{1/2} \qquad l = \frac{\lambda_0}{4}.$$

4.11. From (35) it was found that $\omega L_i = R$. Equation (35) shows that the internal inductance L_i arises because the magnetic field penetrates an effective distance $0.5\delta_s$ into the conductor. Thus an alternative way to evaluate L_i is to find the change in the external inductance L when all conductor surfaces are pushed in by an amount $0.5\delta_s$, that is, $L_i = (\delta_s/2)\partial L/\partial n$, where n is the inward normal to the conductor surface. This is the incremental inductance rule proposed by Wheeler (*Proc. IRE*, vol. 30, p. 412, 1942). Use this rule to find the series resistance of the coaxial line and that of the two-wire line in Problems 4.1 and 4.2, respectively. For the coaxial line, differentiate L with respect to $-a$ and $+b$. For the two-wire line, differentiate L with respect to $-R$. Note that $v_c L = Z_c$ and that $\alpha = R/2Z_c$. Comment on the limitations of this rule.

4.12. Let $\rho(x, y)$ be an arbitrary charge distribution on a surface S_2 such as in Fig. 4.16. On S_1 the potential $\Phi = 0$. The equation for Φ is $\nabla_t^2 \Phi = -\rho/\epsilon_0$. Show that

$$\iint_S (\nabla_t \Phi)^2 \, dS = \iint_S \nabla_t \cdot (\Phi \nabla_t \Phi) \, dS + \oint_{S_2} \frac{\Phi\rho}{\epsilon_0} \, dl$$

where S is the surface between S_1, S_2 and a cylindrical contour at infinity. Use the divergence theorem to show that

$$\iint_S (\nabla_t \Phi)^2 \, dS = \oint_{S_2} \frac{\Phi\rho}{\epsilon_0} \, dl > 0.$$

Now use (90) to show that

$$\oint \oint_{S_2} G(x, y|x', y')\rho(x, y)\rho(x', y') \, dl \, dl' > 0.$$

Thus G is a positive definite operator. Show that the second-order variation in $1/C_0$ for (93) is proportional to

$$\oint \oint_{S_2} G(x, y|x', y')\delta\rho(x, y)\delta\rho(x', y') \, dl \, dl' \geq 0$$

and hence (93) is a lower bound on C_0.

4.13. Minimize the expression

$$\int_{-w}^{+w} \left[\sum_{m=1}^{M} I_m G_m(x) - 1 \right]^2 dx$$

and show that the solutions for I_m correspond to using $G_m(x)$ as weighting functions in the method of moments.

4.14. Show that the correction series for G_{od} in Section 4.6 can be summed to give

$$-\sum_{n=1}^{\infty} \frac{e^{-2n\pi h/a}}{2n\pi} \sin \frac{n\pi x}{a} \sin \frac{n\pi x'}{a} = \frac{1}{8\pi} \ln \frac{\cosh \dfrac{2\pi h}{a} - \cos \dfrac{\pi(x - x')}{a}}{\cosh \dfrac{2\pi h}{a} - \cos \dfrac{\pi(x + x')}{a}}.$$

4.15. Show that by summing the series

$$\ln \left| \sin^2 \theta - \sin^2 \theta' \right| = \ln \left| \frac{\cos 2\theta - \cos 2\theta'}{2} \right| = -\ln 4 - \sum_{n=1}^{\infty} \frac{2}{n} \cos 2n\theta \cos 2n\theta'.$$

4.16. With reference to Fig. 4.12 and using $J \propto |dW'/dW|$ show that the current on a symmetrical strip line is given by

$$J = \frac{C}{\left[\cosh^2 \dfrac{\pi d}{4a} - \cosh^2 \dfrac{\pi u}{2a} \right]^{1/2}}$$

where C is a constant.

4.17. Consider a microstrip line as shown in Fig. 4.24(a). Under static conditions the vector potential A_z satisfies the boundary condition $A_z = A_0$ on the strip. Consider a contour C with two sides along z and located on the ground plane and the strip. The two ends join these segments and are perpendicular to the ground plane. By using Stokes' theorem

$$\oint_C \mathbf{A} \cdot d\mathbf{l} = \int_S \nabla \times \mathbf{A} \cdot \mathbf{n} \, dS = \int_S \mathbf{B} \cdot \mathbf{n} \, dS$$

show that the magnetic flux per unit length linking the strip equals A_0. Hence the inductance per unit length is given by $L = A_0/I$. Use this result and the analogy with (110) to derive the following variational expression for L:

$$L \left[\int_{-W}^{W} J_z(x) \, dx \right]^2 = \int\int_{-W}^{W} G(x, x') J_z(x) J_z(x') \, dx \, dx'.$$

If the dielectric is air the characteristic impedance is given by $Z_c = L/\sqrt{\mu_0 \epsilon_0}$ so the variational expression gives an upper-bound solution.

Show that when the current is expanded in terms of a finite set of basis functions that the expansion coefficients obtained from the variational expression are the same as those obtained using Galerkin's method of moments. Thus Galerkin's method is equivalent to a variational method for this problem.

5
Waveguides and Cavities

In this chapter we shall be concerned with the fundamental properties of electromagnetic waves in the interior of uniform cylindrical waveguides. The type of waveguide to be considered is a hollow conducting cylindrical tube with a cross section that is uniform along the direction of propagation. The axis of the guide will be taken as the z axis. Initially it will be assumed that the walls have infinite conductivity. Also it will be assumed that the guide is either empty or else filled with a homogeneous isotropic medium with electrical parameters ϵ, μ.

After the basic solutions for the fields have been developed we will take into account the finite conductivity of the walls and calculate the attenuation constant for waveguides and also examine the coupling of degenerate modes by the finite conductivity of the walls. The next major topic will be the development of dyadic Green's functions for waveguides.

The latter part of the chapter will be devoted to the theory of electromagnetic resonators or cavities. We will present a theory for the modes of a cavity which can be used to expand the electric and magnetic fields inside the cavity. The excitation problem will then be discussed along with the derivation for the dyadic Green's functions for cavities. It will be found that the basic structure of the cavity dyadic Green's function is essentially the same as the eigenfunction expansion of the dyadic Green's function for free space. The chapter will conclude with a description of a variational method for calculating the eigenvalues for a cavity and a perturbation theory that is useful for calculating the change in the resonant frequency of a cavity when a small object is placed inside the cavity.

In the interior of a waveguide such as the one described above Maxwell's equations can be divided into two basic sets of solutions or modes. For one set of modes, no longitudinal or axial magnetic field component exists. These modes have an axial component of electric field, however, and are called "electric type," that is, E modes or "transverse magnetic," that is, TM modes. The field components for these modes may be derived from an electric-type Hertzian potential having a single component along the axis of the guide. The other basic set of modes has an axial magnetic field but no axial component of electric field; therefore, they are referred to as "magnetic type," that is, H or "transverse electric," that is, TE modes. These modes may be derived from a magnetic-type Hertzian potential again having only an axial component. In a hollow cylindrical waveguide, a TEM mode of propagation is not possible since the assumption of transverse fields leaves neither an axial current or axial components of electric displacement nor magnetic flux to generate the transverse field components. From another point of view, we can say that a solution to Laplace's equation in the transverse plane in the interior of a closed boundary on which the potential is constant does not exist (apart from perhaps a constant) unless a singularity is located within the boundary, and, hence, a potential function from which the TEM mode can be derived does not exist.

The division into E and H modes is not necessary, and in some cases it is more convenient to choose a different set of basic modes. This latter set can, however, be represented as a linear combination of the E and H modes because the E and H modes form a complete set in terms of which an arbitrary field can be represented.

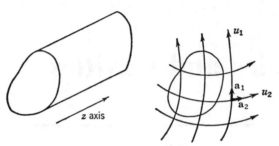

Fig. 5.1. A general cylindrical waveguide.

5.1. GENERAL PROPERTIES OF CYLINDRICAL WAVEGUIDES

All cylindrical waveguides have many properties in common. For this reason we will begin our discussion by considering a guide with an arbitrary cross section as illustrated in Fig. 5.1. The frame of reference will be an orthogonal curvilinear cylindrical coordinate system u_1, u_2, z with unit vectors $\mathbf{a}_1$, $\mathbf{a}_2$, and $\mathbf{a}_z$ directed along the coordinate curves. The interior of the guide is assumed to be free space. All fields are time harmonic with a time variation according to $e^{j\omega t}$.

Transverse Electric Modes

The TE modes may be derived from a magnetic Hertzian potential $\mathbf{\Pi}_h = \mathbf{a}_z \Pi_h$ by means of the following equations:

$$\mathbf{E} = -j\omega\mu_0 \nabla \times \mathbf{\Pi}_h \tag{1a}$$

$$\mathbf{H} = k_0^2 \mathbf{\Pi}_h + \nabla\nabla\cdot\mathbf{\Pi}_h = \nabla \times \nabla \times \mathbf{\Pi}_h \tag{1b}$$

where $\mathbf{\Pi}_h$ is a solution of the equation

$$\nabla^2\mathbf{\Pi}_h + k_0^2\mathbf{\Pi}_h = \nabla\nabla\cdot\mathbf{\Pi}_h - \nabla \times \nabla \times \mathbf{\Pi}_h + k_0^2\mathbf{\Pi}_h = 0 \tag{1c}$$

and $k_0^2 = \omega^2\mu_0\epsilon_0 = 4\pi^2/\lambda_0^2$, with λ_0 being the free-space wavelength for TEM waves. Since we are looking for propagating waves, a solution for $\mathbf{\Pi}_h$ of the form

$$\mathbf{\Pi}_h = \mathbf{a}_z\psi_h(u_1, u_2)e^{\pm\Gamma z} \tag{2}$$

will be assumed. The function ψ_h satisfies the two-dimensional scalar Helmholtz equation

$$\nabla_t^2\psi_h + k_c^2\psi_h = 0 \tag{3}$$

where $k_c^2 = k_0^2 + \Gamma^2$, and ∇_t^2 is the transverse part of the ∇^2 operator. In the u_1u_2z coordinate frame with scale factors h_1, h_2, and unity, (3) becomes[1]

$$\frac{1}{h_1h_2}\frac{\partial}{\partial u_1}\frac{h_2}{h_1}\frac{\partial\psi_h}{\partial u_1} + \frac{1}{h_1h_2}\frac{\partial}{\partial u_2}\frac{h_1}{h_2}\frac{\partial\psi_h}{\partial u_2} + k_c^2\psi_h = 0. \tag{4}$$

[1] Section A.1e.

In general, the solution to this equation is difficult to obtain unless it is separable so that a solution of the form $\psi_h = U_1(u_1)U_2(u_2)$ may be found. If such a solution for ψ_h exists, it will still be a difficult task to make ψ_h satisfy the required boundary conditions unless the waveguide boundary coincides with the constant coordinate curves. The allowed values of k_c^2 and, hence, the propagation constants Γ are determined by the boundary conditions on ψ_h. An infinite number of discrete values for k_c^2 exist and, hence, an infinite number of modes exist. Only a finite number of the propagation constants are imaginary, and thus only a finite number of the modes will propagate freely along the guide.

By using the expressions for the curl and divergence in orthogonal curvilinear coordinates, the following equations for the field components are obtained from (1):

$$H_{u_1} = \frac{\pm\Gamma}{h_1}\frac{\partial\psi_h}{\partial u_1}e^{\pm\Gamma z} \tag{5a}$$

$$H_{u_2} = \pm\frac{\Gamma}{h_2}\frac{\partial\psi_h}{\partial u_2}e^{\pm\Gamma z} \tag{5b}$$

$$H_z = -\frac{1}{h_1 h_2}\left(\frac{\partial}{\partial u_1}\frac{h_2}{h_1}\frac{\partial\psi_h}{\partial u_1} + \frac{\partial}{\partial u_2}\frac{h_1}{h_2}\frac{\partial\psi_h}{\partial u_2}\right)e^{\pm\Gamma z} \tag{5c}$$

$$E_{u_1} = -j\omega\mu_0\frac{1}{h_2}\frac{\partial\psi_h}{\partial u_2}e^{\pm\Gamma z} \tag{5d}$$

$$E_{u_2} = j\omega\mu_0\frac{1}{h_1}\frac{\partial\psi_h}{\partial u_1}e^{\pm\Gamma z}. \tag{5e}$$

An examination of these equations shows that they may be more compactly written as

$$\mathbf{H}_t = \pm\Gamma e^{\pm\Gamma z}\,\nabla_t\psi_h \tag{6a}$$

$$H_z = -\nabla_t^2\psi_h e^{\pm\Gamma z} = k_c^2\psi_h e^{\pm\Gamma z} \tag{6b}$$

$$\mathbf{E}_t = \mp\bar{\mathbf{Z}}_h\cdot\mathbf{H}_t = \pm Z_h\mathbf{a}_z\times\mathbf{H}_t \tag{6c}$$

where the subscript t means the transverse part, $\bar{\mathbf{Z}}_h$ is the dyadic wave impedance given by

$$\bar{\mathbf{Z}}_h = Z_h(\mathbf{a}_1\mathbf{a}_2 - \mathbf{a}_2\mathbf{a}_1) \tag{7}$$

and $Z_h = j\omega\mu_0/\Gamma = (jk_0/\Gamma)Z_0$ is the scalar wave impedance, with $Z_0 = (\mu_0/\epsilon_0)^{1/2}$ being the intrinsic impedance of free space. For a propagating mode, Γ is imaginary and will be taken as $j\beta$. The wave impedance Z_h is now real and equal to $Z_0 k_0/\beta$. For a nonpropagating mode, the wave impedance is imaginary and inductive in nature.

For a waveguide with perfectingly conducting walls, ψ_h must satisfy the boundary condition $\partial\psi_h/\partial n = 0$, in order to make the tangential electric field and normal magnetic field vanish on the boundary. This boundary condition determines the allowed values of k_c^2 and Γ^2. The parameter k_c is called the cutoff wavenumber, since it determines the wavelength at which free propagation ceases, i.e., for which Γ becomes equal to zero. This occurs when λ_0 is chosen so that $k_c = k_0$ for a given mode, and, hence, the cutoff wavelength is given by

$$\lambda_c = \frac{2\pi}{k_c}. \tag{8a}$$

For a guide filled with a medium with parameters ϵ, μ, cutoff occurs when $k_c = \omega(\mu\epsilon)^{1/2} = k$, and, hence, in general the free-space cutoff wavelength is given by

$$\lambda_c = \frac{2\pi}{k_c} \left(\frac{\mu\epsilon}{\mu_0\epsilon_0} \right)^{1/2}. \tag{8b}$$

When the cutoff wavelength (which is a function of the guide cross section and $\epsilon\mu$ only) has been determined for a particular propagating mode, the value of β and the guide wavelength may be found. The guide wavelength λ_g is the distance the wave must propagate to undergo a phase change of 2π radians, and hence

$$\lambda_g = \frac{2\pi}{\beta}. \tag{9a}$$

Using the relation $k_0^2 - \beta^2 = k_c^2$, we may solve for λ_g to get

$$\lambda_g = \frac{\lambda_0}{[1 - (\lambda_0/\lambda_c)^2]^{1/2}} \tag{9b}$$

or, for a guide filled with material with parameters ϵ, μ,

$$\lambda_g = \frac{\lambda_0(\mu_0\epsilon_0/\mu\epsilon)^{1/2}}{[1 - (\lambda_0/\lambda_c)^2]^{1/2}} \tag{9c}$$

where λ_c is now given by (8b), and λ_0 is still the free-space wavelength. For all values of $\lambda_0 < \lambda_c$, free propagation of the mode occurs. As λ_0 approaches λ_c, the guide wavelength approaches infinity until, at $\lambda_0 = \lambda_c$, free propagation stops. For $\lambda_0 > \lambda_c$ the mode is exponentially damped with distance along the guide and is commonly referred to as an evanescent mode.

Transverse Magnetic Modes

The TM or E modes may be derived from an electric-type Hertzian potential $\mathbf{\Pi}_e = \mathbf{a}_z \Pi_e$ by means of the equations

$$\mathbf{E} = k_0^2\mathbf{\Pi}_e + \nabla\nabla\cdot\mathbf{\Pi}_e = \nabla \times \nabla \times \mathbf{\Pi}_e \tag{10a}$$

$$\mathbf{H} = j\omega\epsilon_0\nabla \times \mathbf{\Pi}_e \tag{10b}$$

where $\mathbf{\Pi}_e$ is a solution of

$$\nabla^2\mathbf{\Pi}_e + k_0^2\mathbf{\Pi}_e = \nabla\nabla\cdot\mathbf{\Pi}_e - \nabla \times \nabla \times \mathbf{\Pi}_e + k_0^2\mathbf{\Pi}_e = 0. \tag{10c}$$

If a solution for $\mathbf{\Pi}_e$ of the form $\mathbf{a}_z\psi_e(u_1, u_2)e^{\pm\Gamma z}$ is assumed, a set of equations similar to (6) for the field components is obtained. These are

$$\mathbf{E}_t = \pm\Gamma\nabla_t\psi_e e^{\pm\Gamma z} \tag{11a}$$

$$E_z = -\nabla_t^2\psi_e e^{\pm\Gamma z} = k_c^2\psi_e e^{\pm\Gamma z} \tag{11b}$$

$$\mathbf{H}_t = \mp\bar{\mathbf{Y}}_e\cdot\mathbf{E}_t = \mp Y_e\mathbf{a}_z \times \mathbf{E}_t \tag{11c}$$

where $\bar{\mathbf{Y}}_e$ is the dyadic wave admittance given by

$$\bar{\mathbf{Y}}_e = Y_e(\mathbf{a}_2\mathbf{a}_1 - \mathbf{a}_1\mathbf{a}_2). \tag{11d}$$

In (11d), Y_e is the scalar wave admittance equal to jk_0Y_0/Γ, and Y_0 is the intrinsic admittance of free space given by $(\epsilon_0/\mu_0)^{1/2}$. For perfectly conducting walls, E_z must vanish on the boundary, and, hence, ψ_e must vanish on the boundary also. Again an infinite number of values for k_c^2 exist. Also a cutoff wavelength λ_c for each mode may be found, and the guide wavelength is given by (9). For $\Gamma = j\beta$, the wave admittance Y_e is real and equal to Y_0k_0/β, while, for Γ real, i.e., for a nonpropagating mode, it is capacitive.

5.2. Orthogonal Properties of the Modes

In a waveguide with perfectly conducting walls, the E and H modes have several interesting and useful orthogonal properties. Let ψ_i and ψ_j be the solutions for the ith and jth E modes or H modes with corresponding propagation constants Γ_i and Γ_j. Multiplying the equation satisfied by ψ_i by ψ_j gives

$$\psi_j \nabla_t^2 \psi_i + (k_0^2 + \Gamma_i^2)\psi_i\psi_j = 0. \tag{12a}$$

Similarly, multiplying the equation for ψ_j by ψ_i gives

$$\psi_i \nabla_t^2 \psi_j + (k_0^2 + \Gamma_j^2)\psi_i\psi_j = 0. \tag{12b}$$

Subtracting these two equations and substituting into the two-dimensional form of Green's second identity gives[2]

$$(\Gamma_i^2 - \Gamma_j^2) \iint_S \psi_i\psi_j \, dS = \iint_S (\psi_i \nabla_t^2 \psi_j - \psi_j \nabla_t^2 \psi_i) \, dS$$

$$= \oint_C \left(\psi_i \frac{\partial \psi_j}{\partial n} - \psi_j \frac{\partial \psi_i}{\partial n} \right) dl \tag{13}$$

where S denotes the cross-sectional surface of the guide, and C denotes the guide boundary. For E modes, ψ_i and ψ_j vanish on the boundary, while, for H modes, $\partial\psi_i/\partial n$ and $\partial\psi_j/\partial n$ vanish on the boundary. Therefore, in either case the contour integral vanishes, and, provided $\Gamma_i^2 \neq \Gamma_j^2$, we obtain the following orthogonal property for both the E and H modes:

$$\iint_S \psi_i\psi_j \, dS = 0, \qquad i \neq j. \tag{14}$$

Thus the axial components of the field for two different modes are orthogonal.

If the two modes are degenerate such that $\Gamma_i = \Gamma_j$, the proof just given breaks down. However, for degenerate modes we may choose a suitable linear combination of the degenerate modes such that this subset of modes is an orthogonal set. For example, if ψ_1 and ψ_2 are two degenerate modes with eigenvalue Γ^2, we may choose a new subset of modes as follows: $\psi_1' = \psi_1$, $\psi_2' = \psi_2 + \alpha\psi_1$, where α is a constant to be determined so that $\int\int_S \psi_1'\psi_2' \, dS = 0$.

[2] Section A.1c.

If we let $\int\int_S \psi_j \psi_i \, dS = P_{ij}$, $i, j = 1, 2$, we find that $\alpha = -P_{12}/P_{11}$. With this value of α, the two modes ψ_1' and ψ_2' are orthogonal. This procedure may be generalized so that any subset of n degenerate modes can be converted into a new subset of n mutually orthogonal modes. Thus we may, in a general discussion, assume that $\Gamma_i^2 \neq \Gamma_j^2$ without invalidating the final result.

The transverse electric fields for two different E modes or for two different H modes, as well as the transverse electric field for one E mode and one H mode, are mutually orthogonal. The same is true for the transverse magnetic field for any two (E, H, or E and H) modes. The proof for the case of the transverse electric fields will be given; that for the transverse magnetic fields may be constructed along similar lines.

For two different E modes, the transverse electric fields are orthogonal, provided the following integral vanishes:

$$I = \iint_S \nabla_t \psi_i \cdot \nabla_t \psi_j \, dS \tag{15}$$

where ψ_i and ψ_j are two different eigenfunctions from which the E modes may be derived by means of (11). Using the two-dimensional form of Green's first identity, the surface integral becomes

$$\iint_S \nabla_t \psi_i \cdot \nabla_t \psi_j \, dS = -\iint_S \psi_i \, \nabla_t^2 \psi_j \, dS + \oint_C \psi_i \frac{\partial \psi_j}{\partial n} \, dl. \tag{16}$$

The contour integral vanishes since $\psi_i = 0$ on the boundary. In the surface integral on the right $\nabla_t^2 \psi_j$ may be replaced by $-k_c^2 \psi_j$, and, using the result that ψ_i and ψ_j are orthogonal, it is seen to vanish also. Consequently, (15) vanishes, and the stated orthogonal property is proved.

For two different H modes the transverse electric fields are orthogonal, provided

$$\iint_S (\mathbf{a}_z \times \nabla_t \psi_i) \cdot (\mathbf{a}_z \times \nabla_t \psi_j) \, dS \tag{17}$$

vanishes. In expression (17), ψ_i and ψ_j are two eigenfunctions, giving rise to two H modes, and the expressions for the transverse electric fields have been obtained from (6). The integrand in (17) may be written as $\mathbf{a}_z \cdot [\nabla_t \psi_i \times (\mathbf{a}_z \times \nabla_t \psi_j)]$ and expanded to give

$$\mathbf{a}_z \cdot [(\nabla_t \psi_i \cdot \nabla_t \psi_j) \mathbf{a}_z - (\nabla_t \psi_i \cdot \mathbf{a}_z) \nabla_t \psi_j] = \nabla_t \psi_i \cdot \nabla_t \psi_j$$

since $\mathbf{a}_z \cdot \nabla_t \psi_i = 0$, because these two vectors are perpendicular. Replacing the integrand by $\nabla_t \psi_i \cdot \nabla_t \psi_j$ and using Green's first identity gives a result similar to (16). For H modes, $\partial \psi_i / \partial n$ vanishes on the boundary, and, since ψ_i and ψ_j are orthogonal, the vanishing of the integral (17) follows as a consequence.

Finally, for the transverse electric fields for one E mode and one H mode we must show that

$$\iint_S (\mathbf{a}_z \times \nabla_t \psi_{hi}) \cdot \nabla_t \psi_{ej} \, dS \tag{18}$$

vanishes. To avoid confusion between the eigenfunctions for the E and H modes in (18), the subscripts h and e have been introduced again. Consider the following expression:

$$\nabla_t \cdot (\mathbf{a}_z \psi_{hi} \times \nabla_t \psi_{ej}) = (\nabla_t \times \mathbf{a}_z \psi_{hi}) \cdot \nabla_t \psi_{ej} - \mathbf{a}_z \psi_{hi} \cdot \nabla_t \times \nabla_t \psi_{ej}$$

$$= -(\mathbf{a}_z \times \nabla_t \psi_{hi}) \cdot \nabla_t \psi_{ej}$$

since $\nabla_t \times \nabla_t \psi_{ej}$ is identically zero, and $\nabla_t \times \mathbf{a}_z \psi_{hi} = -\mathbf{a}_z \times \nabla_t \psi_{hi}$. Using this result, the integral in (18) may be replaced by

$$-\iint_S \nabla_t \cdot (\mathbf{a}_z \psi_{hi} \times \nabla_t \psi_{ej}) \, dS = \oint_C (\mathbf{a}_z \times \nabla_t \psi_{ej}) \cdot \mathbf{n} \psi_{hi} \, dl$$

where $\mathbf{n}$ is the unit normal to the waveguide wall and directed inward. In the contour integral, $\nabla_t \psi_{ej}$ is proportional to the transverse electric field and, hence, has only a normal component at the guide wall. Thus $\mathbf{a}_z \times \nabla_t \psi_{ej}$ is a factor that is tangential to the guide wall (it is a vector proportional to the transverse magnetic field), and, hence, the dot product with $\mathbf{n}$ and also the integrand vanish. This establishes the orthogonality of the transverse electric fields corresponding to one E mode and one H mode.

The above orthogonality properties are useful in practice because they enable a given arbitrary field to be expanded into a series of E and H modes. These orthogonal properties are not, however, completely general. For a guide with finite conducting walls, the proofs given above are not valid because the eigenfunctions ψ_e and ψ_h no longer satisfy the assumed boundary conditions; that is, $\psi_e = 0$ and $\partial \psi_h / \partial n = 0$ no longer hold on the guide boundary. The presence of finite conductivity results in a cross coupling between the various E and H modes. This phenomenon is considered later in the section on attenuation. In inhomogeneously filled waveguides it is found, in many instances, that the normal modes of propagation are neither E nor H modes but linear combinations of these modes. A more general orthogonal principle must be resorted to in order to expand an arbitrary field into a series of these latter normal modes. This more general orthogonal relation follows as a result of the Lorentz reciprocity principle and will be demonstrated below.

Let $\mathbf{H}_{tn}, \mathbf{E}_{tn}$ and $\mathbf{H}_{tm}, \mathbf{E}_{tm}$ be the transverse fields for two linearly independent solutions to Maxwell's equations. The curl equations for the electric field give $\nabla \times \mathbf{E}_n = -j\omega\mu\mathbf{H}_n$, $\nabla \times \mathbf{E}_m = -j\omega\mu\mathbf{H}_m$, where $\mathbf{E}_n, \mathbf{H}_n$ and $\mathbf{E}_m, \mathbf{H}_m$ are the total fields, i.e., both the transverse and axial components. Scalar multiplying the above equations by $\mathbf{H}_m$ and $\mathbf{H}_n$, respectively, and subtracting gives $\mathbf{H}_m \cdot \nabla \times \mathbf{E}_n - \mathbf{H}_n \cdot \nabla \times \mathbf{E}_m = 0$. Scalar multiplying the curl equations for $\mathbf{H}_n, \mathbf{H}_m$ by $\mathbf{E}_m, \mathbf{E}_n$ and subtracting gives a similar result, but with the roles of $\mathbf{E}$ and $\mathbf{H}$ interchanged; thus $\mathbf{E}_m \cdot \nabla \times \mathbf{H}_n - \mathbf{E}_n \cdot \nabla \times \mathbf{H}_m = 0$. Adding these equations gives $\nabla \cdot (\mathbf{E}_n \times \mathbf{H}_m - \mathbf{E}_m \times \mathbf{H}_n) = 0$, since

$$\mathbf{H}_m \cdot \nabla \times \mathbf{E}_n - \mathbf{E}_n \cdot \nabla \times \mathbf{H}_m = \nabla \cdot \mathbf{E}_n \times \mathbf{H}_m \text{ etc.}$$

If $\mathbf{E}_n, \mathbf{H}_n$ depend on the coordinate z only as $e^{-\Gamma_n z}$ and $\mathbf{E}_m, \mathbf{H}_m$ as $e^{-\Gamma_m z}$, this last relation reduces to

$$\nabla \cdot (\mathbf{E}_n \times \mathbf{H}_m - \mathbf{E}_m \times \mathbf{H}_n) = \nabla_t \cdot (\mathbf{E}_n \times \mathbf{H}_m - \mathbf{E}_m \times \mathbf{H}_n)$$

$$+ \mathbf{a}_z \frac{\partial}{\partial z} \cdot (\mathbf{E}_n \times \mathbf{H}_m - \mathbf{E}_m \times \mathbf{H}_n)$$

$$= \nabla_t \cdot (\mathbf{E}_n \times \mathbf{H}_m - \mathbf{E}_m \times \mathbf{H}_n)$$

$$- (\Gamma_n + \Gamma_m)\mathbf{a}_z \cdot (\mathbf{E}_{tn} \times \mathbf{H}_{tm} - \mathbf{E}_{tm} \times \mathbf{H}_{tn})$$

since the terms involving the axial components of the fields vanish upon taking the scalar product with $\mathbf{a}_z$. Using the two-dimensional form of the divergence theorem now gives

$$\iint\limits_S \nabla_t \cdot (\mathbf{E}_n \times \mathbf{H}_m - \mathbf{E}_m \times \mathbf{H}_n)\, dS = \oint_C \mathbf{n} \cdot (\mathbf{E}_n \times \mathbf{H}_m - \mathbf{E}_m \times \mathbf{H}_n)\, dl$$

$$= (\Gamma_n + \Gamma_m) \iint\limits_S \mathbf{a}_z \cdot (\mathbf{E}_{tn} \times \mathbf{H}_{tm} - \mathbf{E}_{tm} \times \mathbf{H}_{tn})\, dS.$$

$$(19)$$

The contour integral vanishes since $\mathbf{n} \times \mathbf{E}_n$ and $\mathbf{n} \times \mathbf{E}_m$ vanish on the perfectly conducting guide walls. Thus we obtain

$$(\Gamma_n + \Gamma_m) \iint\limits_S \mathbf{a}_z \cdot (\mathbf{E}_{tn} \times \mathbf{H}_{tm} - \mathbf{E}_{tm} \times \mathbf{H}_{tn})\, dS = 0. \qquad (20)$$

Equation (20) also holds when the walls are imperfect conductors and we use the impedance boundary condition $\mathbf{E}_t = Z_m \mathbf{n} \times \mathbf{H}$ since the integrand will be zero on the waveguide walls when the impedance boundary condition is imposed.

It is convenient to write the expression for the transverse fields in the form

$$\mathbf{H}_{tn} = \mathbf{h}_n(u_1, u_2)e^{-\Gamma_n z} \qquad (21a)$$

$$\mathbf{E}_{tn} = \mathbf{e}_n(u_1, u_2)e^{-\Gamma_n z} \qquad (21b)$$

where $\mathbf{h}_n$ and $\mathbf{e}_n$ are transverse vector functions of the transverse coordinates u_1, u_2. Introducing these, (20) becomes

$$(\Gamma_n + \Gamma_m) \iint\limits_S \mathbf{a}_z \cdot (\mathbf{e}_n \times \mathbf{h}_m - \mathbf{e}_m \times \mathbf{h}_n)\, dS = 0 \qquad (22)$$

since the common exponential terms can be canceled. To show that each term vanishes separately, we consider the two solutions $\mathbf{E}_n$, $\mathbf{H}_n$ and $\mathbf{E}'_m$, $\mathbf{H}'_m$, where $\mathbf{E}'_m$, $\mathbf{H}'_m$ is the same mode as considered previously, but depending on z according to $e^{\Gamma_m z}$ rather than $e^{-\Gamma_m z}$. This corresponds to a reversal in the direction of propagation, and, consequently, the direction of the transverse magnetic field is reversed; that is, $\mathbf{E}'_{tm} = \mathbf{e}_m e^{\Gamma_m z}$, $\mathbf{H}'_{tm} = -\mathbf{h}_m e^{\Gamma_m z}$. The equation corresponding to (22) is

$$(\Gamma_n - \Gamma_m) \iint\limits_S \mathbf{a}_z \cdot (-\mathbf{e}_n \times \mathbf{h}_m - \mathbf{e}_m \times \mathbf{h}_n)\, dS = 0. \qquad (23)$$

Addition and subtraction of (22) and (23) gives

$$\iint\limits_S \mathbf{e}_n \times \mathbf{h}_m \cdot \mathbf{a}_z\, dS = 0 \qquad (24a)$$

$$\iint\limits_S \mathbf{e}_m \times \mathbf{h}_n \cdot \mathbf{a}_z\, dS = 0 \qquad (24b)$$

which is the desired orthogonality relation. When there are no losses present, a similar derivation shows that

$$\iint\limits_{S} \mathbf{e}_n \times \mathbf{h}_m^* \cdot \mathbf{a}_z \, dS = 0 \tag{25}$$

where $\mathbf{h}_m^*$ is the complex conjugate of $\mathbf{h}_m$. This latter relation shows that the power flow in a lossless guide is the sum of the power carried by each mode individually. These results will be used in the derivation of Green's functions for general cylindrical waveguides.

The results given by (24) are also true for the case when the two degenerate modes are an E mode and an H mode. If $\mathbf{e}_n$, $\mathbf{h}_n$ is an E mode then $Z_{en}\mathbf{h}_n = \mathbf{a}_z \times \mathbf{e}_n$ and if $\mathbf{e}_m$, $\mathbf{h}_m$ is an H mode we have $Z_{hm}\mathbf{h}_m = \mathbf{a}_z \times \mathbf{e}_m$. Since $Z_{en} \neq Z_{hm}$ even when $\Gamma_n = \Gamma_m$, the integrand in (22) becomes $(Z_{en} - Z_{hm})\mathbf{h}_n \cdot \mathbf{h}_m$ and hence the integral of $\mathbf{h}_n \cdot \mathbf{h}_m$ over the guide cross section must vanish, which is equivalent to the relations (24). Even when the degenerate modes are of the same type the relations (24) are often true even though the proof is no longer valid (this is the case for the E_{nm} and E_{mn} modes in a square waveguide).

5.3. Power, Energy, and Attenuation

In a waveguide with perfectly conducting walls, each mode propagates power along the guide independently of the presence of other modes. That this should be so is apparent from the orthogonal properties of the modes, in particular the orthogonal property expressed by (25). The power flow along the guide is given by the real part of the integral of the complex Poynting vector over the guide cross section.

For a single propagating H mode, the time-average power flow is given by

$$P = \frac{1}{2}\operatorname{Re}\iint\limits_{S} \mathbf{E} \times \mathbf{H}^* \cdot \mathbf{a}_z \, dS = \frac{1}{2}\operatorname{Re}\iint\limits_{S} \mathbf{E}_t \times \mathbf{H}_t^* \cdot \mathbf{a}_z \, dS$$

$$= -\frac{1}{2}\operatorname{Re}\iint\limits_{S} Z_h\beta^2[(\mathbf{a}_z \times \nabla_t\psi_h) \times \nabla_t\psi_h] \cdot \mathbf{a}_z \, dS$$

by using (6) for the transverse fields and where Γ has been replaced by $j\beta$. The integrand may be expanded to give

$$(\mathbf{a}_z \times \nabla_t\psi_h) \times \nabla_t\psi_h = -(\nabla_t\psi_h \cdot \nabla_t\psi_h)\mathbf{a}_z + (\nabla_t\psi_h \cdot \mathbf{a}_z)\nabla_t\psi_h$$

which when scalar multiplied by $\mathbf{a}_z$ gives $-\nabla_t\psi_h \cdot \nabla_t\psi_h$. Using Green's first identity, the expression for power flow becomes

$$\frac{1}{2}Z_h\beta^2\iint\limits_{S} \nabla_t\psi_h \cdot \nabla_t\psi_h \, dS = -\frac{1}{2}Z_h\beta^2\iint\limits_{S} \psi_h \nabla_t^2\psi_h \, dS + \frac{1}{2}Z_h\beta^2\oint_C \psi_h \frac{\partial\psi_h}{\partial n} \, dl.$$

On the guide boundary, $\partial\psi_h/\partial n$ vanishes, and, replacing $\nabla_t^2\psi_h$ by $-k_c^2\psi_h$, and Z_h by $(k_0/\beta)Z_0$, we finally get

$$P = \frac{1}{2}Z_0k_0\beta k_c^2\iint\limits_{S} \psi_h^2 \, dS \tag{26}$$

for the time-average power flow in a propagating H mode.

For a single E mode propagating along the guide, the time-average power flow is given by

$$P = \frac{1}{2} \iint_S Y_e \beta^2 [\nabla_{te} \psi \times (\mathbf{a}_z \times \nabla_t \psi_e)] \cdot \mathbf{a}_z \, dS$$

$$= \frac{1}{2} Y_0 k_0 \beta k_c^2 \iint_S \psi_e^2 \, dS. \tag{27}$$

Equation (27) is obtained by expanding the integrand and using Green's first identity again and the boundary condition $\psi_e = 0$ on the guide walls.

For a propagating mode, the time-average electric and magnetic energies associated with the mode are equal. This is readily demonstrated by integrating the complex Poynting vector over a closed surface consisting of two cross-sectional planes separated by an arbitrary distance and the guide walls. Since the walls are assumed perfectly conducting, the only contribution to the surface integral is from the two cross-sectional planes. From Chapter 1, Eq. (25), we have

$$\frac{1}{2} \iint_S \mathbf{E} \times \mathbf{H}^* \cdot d\mathbf{S} = 2j\omega(W_m - W_e) + P_L = 0$$

since $\mathbf{E} \times \mathbf{H}^* \cdot \mathbf{a}_z$ is real, and just as much power enters at one cross-sectional plane as leaves at the other cross-sectional plane for a lossless guide. In the volume bounded by these cross-sectional planes, it is thus necessary that $W_m = W_e$ and $P_L = 0$.

For an H mode, the time-average electric energy per unit length of guide is given by

$$W_e = \frac{1}{4} \epsilon_0 \iint_S \mathbf{E}_t \cdot \mathbf{E}_t^* \, dS$$

$$= \frac{1}{4} \epsilon_0 \iint_S (Z_h \beta)^2 (\mathbf{a}_z \times \nabla_t \psi_h)^2 \, dS$$

$$= \frac{1}{4} \epsilon_0 (Z_0 k_0)^2 k_c^2 \iint_S \psi_h^2 \, da = W_m \tag{28}$$

this latter result being obtained by expanding the integrand and using Green's first identity. In a similar fashion, we find that the time-average magnetic energy per unit length of guide in a propagating E mode is given by

$$W_m = \frac{1}{4} \mu_0 (Y_0 k_0)^2 k_c^2 \iint_S \psi_e^2 \, dS = W_e. \tag{29}$$

The power flowing along the guide is equal to the product of the total energy per unit length and the velocity v_g of energy propagation. Using (26), (28), and the relation $P = 2W_e v_g$, we may solve for the velocity of energy propagation to get

$$v_g = \frac{\beta}{k_0} v_c = \frac{\lambda_0}{\lambda_g} v_c \tag{30}$$

where $v_c = (\mu_0\epsilon_0)^{-1/2}$ is the velocity of TEM waves in free space. The phase velocity of a given mode is that velocity with which an observer would have to move to keep $\beta z - \omega t = $ constant. Differentiating gives

$$\frac{dz}{dt} = v_p = \frac{\omega}{\beta} = \frac{\omega(\mu_0\epsilon_0)^{1/2}}{\beta}v_c = \frac{k_0}{\beta}v_c \tag{31}$$

where v_p is the phase velocity. Comparison of (30) and (31) shows that

$$v_p v_g = v_c^2. \tag{32}$$

The velocity of energy propagation is never greater than the velocity of light v_c. In Chapter 7 it will be shown that v_g is also equal to the group velocity of the signal, and the subscript g was chosen for this reason. The use of (27) and (29) shows that the velocity of energy propagation for E modes is also given by (30). Equations (31) and (32) are valid for E modes as well.

Energy in Evanescent Modes

We now turn to a consideration of the energy relations for nonpropagating E and H modes. Consider a section of waveguide bounded by two planes S_1 and S_2 a distance l apart, as in Fig. 5.2. We wish to determine the time-average net amount of energy, electric or magnetic, which is associated with a nonpropagating mode in the volume bounded by the two cross-sectional surfaces S_1 and S_2. The volume V is assumed to contain no lossy material. If the time-average complex Poynting vector is integrated over the closed surface surrounding the volume V, we get

$$\frac{1}{2}\iint_{S_1} \mathbf{E}_t \times \mathbf{H}_t^* \cdot \mathbf{a}_z \, dS - \frac{1}{2}\iint_{S_2} \mathbf{E}_t \times \mathbf{H}_t^* \cdot \mathbf{a}_z \, dS = 2j\omega(W_m - W_e) \tag{33}$$

since $\mathbf{E}_t \times \mathbf{H}_t^*$ is zero on the guide walls, and the vector element of area on S_2 is directed in the negative z direction. The integrals over S_1 and S_2 are similar to those in (26) and (27) with the exception that Γ is now real and, consequently, an exponential term $e^{-2\Gamma z}$ is present and Z_h and Y_e are imaginary. An analysis similar to that used to obtain (26) and (27) shows that, for H modes, (33) becomes

$$\frac{1}{2}jZ_0k_0\Gamma k_c^2(e^{-2\Gamma z_1} - e^{-2\Gamma z_2})\iint_{S_1} \psi_h^2 \, dS = 2j\omega(W_m - W_e)$$

and, hence,

$$W_m - W_e = \frac{Z_0(\mu_0\epsilon_0)^{1/2}\Gamma k_c^2}{4}e^{-2\Gamma z_1}(1 - e^{-2\Gamma l})\iint_{S_1} \psi_h^2 \, dS. \tag{34}$$

This result shows that, for nonpropagating H modes, the time-average magnetic energy stored in a given length of guide is greater than the time-average electric energy stored in the same length of guide. For E modes, we find that there is a net amount of electric energy stored

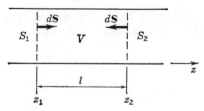

Fig. 5.2. Volume used to evaluate energy in evanescent modes.

in a given length of guide for a nonpropagating mode; that is, $W_e > W_m$. For E modes, the expression corresponding to (34) is

$$W_e - W_m = \frac{Y_0(\mu_0\epsilon_0)^{1/2}k_c^2}{4}e^{-2\Gamma z_1}(1 - e^{-2\Gamma l})\iint\limits_{S_1} \psi_e^2 \, dS. \tag{35}$$

The method of derivation is the same as that for (34). The predominance of magnetic energy in the case of nonpropagating H modes and electric energy in the case of nonpropagating E modes is, of course, correlated with the inductive and capacitive nature of the wave impedances for these nonpropagating modes. In the analysis of discontinuities in waveguides, we find that the inductive or capacitive nature of a given obstacle depends on the net amount of energy present in the nonpropagating modes in the vicinity of the obstacle. The above energy relations are of importance for this reason.

Attenuation in Waveguides

The attenuation constant for electromagnetic waves propagating in the interior of a waveguide filled with homogeneous isotropic dielectric material having a loss tangent $\tan \delta_l$ is easily evaluated. The only modification required in the previously derived formulas is the replacement of ϵ_0 by $\epsilon(1 - j \tan \delta_l)$. The equation for the propagation constant $\Gamma = j\beta + \alpha$ becomes

$$\Gamma^2 = (j\beta + \alpha)^2 = k_c^2 - \omega^2\mu_0\epsilon(1 - j \tan \delta_l) \tag{36}$$

where β is the phase constant, and α the attenuation constant. When the losses are small, $\delta_l \ll 1$, and (36) may be approximated by

$$-\beta^2 + 2j\alpha\beta = k_c^2 - \omega^2\mu_0\epsilon + j\omega^2\mu_0\epsilon \tan \delta_l$$

for frequencies sufficiently above the cutoff frequency so that $\alpha \ll \beta$. Hence,

$$\beta^2 = \omega^2\mu_0\epsilon - k_c^2 \tag{37a}$$

$$\alpha = \frac{k_0^2\kappa \tan \delta_l}{2\beta} \tag{37b}$$

where $\kappa = \epsilon/\epsilon_0$ is the relative dielectric constant.

A first approximation to the attenuation arising from the finite conductivity of the guide walls may be obtained in the same way as for transmission lines with small losses. In this method it is assumed that the currents flowing on the guide wall for a given mode of propagation in the presence of finite conductivity are essentially the same as when the conductivity is infinite.

This permits us to evaluate the losses in the walls from the known current distribution of the unperturbed mode, and use of the relation $2\alpha P = P_L$ then leads to a value for the attenuation constant. In the above relation, P is the power flow along the guide for a given mode, and P_L is the associated power loss in the walls. The surface current density on the walls is equal to the tangential magnetic field. Thus, for H modes we obtain, by integrating around the guide boundary C, the following results for the losses:

$$P_L = \frac{1}{2} R_m \oint_C [(\beta)^2 |\mathbf{n} \times \nabla_t \psi_h|^2 + k_c^4 \psi_h^2] \, dl$$

per unit length of guide, where $R_m = (\omega\mu/2\sigma)^{1/2}$ is the skin-effect surface resistance of a metal with permeability μ. The attenuation constant for H modes is hence given by

$$\alpha = \frac{R_m \oint_C [(\beta)^2 |\mathbf{n} \times \nabla_t \psi_h|^2 + k_c^4 \psi_h^2] \, dl}{2Z_0 k_0 \beta k_c^2 \iint_S \psi_h^2 \, dS} \tag{38}$$

upon using (26) for the power flow along the guide. The analogous expression for E modes is

$$\alpha = \frac{R_m k_0 Y_0}{2\beta k_c^2} \frac{\oint_C |\nabla_t \psi_e|^2 \, dl}{\iint_S \psi_e^2 \, dS}. \tag{39}$$

In (38), $\mathbf{n}$ is the unit normal to the surface of the waveguide wall. The expressions for α may be evaluated for a given guide when the eigenfunctions ψ_m and ψ_e have been found. This simple theory is adequate for computing the attenuation of practical waveguides, provided the frequency is not too close to the cutoff value, and the skin depth is considerably greater than the average surface irregularities. In practice, it is found that, at the shorter wavelengths, normal surfaces cannot be considered as smooth surfaces, and the attenuation is greater than the theoretical formulas predict. This is attributed directly to the effects of surface roughness [5.1].

   At the cutoff frequency, $\beta = 0$ and (38) and (39) predict infinite values of attenuation. On the basis of the simple theory, what has happened is that propagation of power along the guide has ceased, while, at the same time, the losses in the walls have remained finite. A more careful analysis shows that infinite values of α do not occur. The presence of finite conductivity results in a coupling between the E and H modes, and propagation along the guide does not stop at a particular frequency. Rather, there is a range of frequencies, over which transition from a propagating mode to a more and more highly attenuated mode takes place, and only as the frequency tends to zero does propagation cease. In the particular case of a lossless guide filled with lossy dielectric material, this phenomenon is readily understood. Equation (36) may be written as

$$\Gamma = j\beta + \alpha = [(k_c^2 - \omega^2 \mu_0 \epsilon)^2 + (\omega^2 \mu_0 \epsilon \tan \delta_l)^2]^{1/2} e^{j\theta/2} \tag{40}$$

where

$$\theta = \tan^{-1} \frac{\omega^2 \mu_0 \epsilon \tan \delta_l}{k_c^2 - \omega^2 \mu_0 \epsilon}.$$

For frequencies well above the no-loss cutoff frequency, i.e., for $\omega^2 \mu_0 \epsilon \gg k_c^2$, the angle θ is close to π, and (40) is very nearly a pure-imaginary quantity; hence β is large, and α is small. At the no-loss cutoff frequency, $k_c^2 = \omega^2 \mu_0 \epsilon$, and $\theta = \pi/2$, and from (40) it is seen that

$$\beta = \alpha = 0.707 \omega^2 \mu_0 \epsilon \tan \delta_l.$$

Finally, for frequencies below the no-loss cutoff frequency, θ becomes progressively smaller, and thus α becomes progressively larger than β, until, when $\theta = 0$, $\beta = 0$ also. It is apparent that a sharp cutoff frequency does not exist as long as $\tan \delta_l$ does not vanish. For most practical purposes, the attenuation becomes so large for frequencies below the no-loss cutoff frequency that this latter frequency may be regarded as the practical lower frequency bound for useful propagation of a given mode.

At high frequencies, where the skin depth δ_s is extremely small, we may replace a guide with lossy walls by a guide with perfectly conducting walls, lined with a resistance sheet of thickness a few times greater than δ_s, and having a volume conductivity σ. In the resulting inhomogeneously filled guide it is found, in most cases, that a pure E or H mode of propagation is not possible. One exception is the circular cylindrical guide where a circularly symmetric E or H mode solution may be found [5.2]. For noncircular symmetry, each allowed mode of propagation is a combination of an E and an H mode. Also for the circular guide the problem of propagation in the case of finite conducting walls may be rigorously solved. This is not true for the rectangular guide and other general shapes, so a perturbation technique must be employed. Even for the circular guide a perturbation technique is preferable, since it leads to a simpler analysis. A perturbation solution for propagation in a waveguide with wall losses has been given by Papadopoulos [5.3]. A variational method for mode coupling in lossy waveguides and cavities has been developed by Gustincic [5.4]. This method will be used here and will show that mode coupling effects for degenerate modes are quite strong and produce changes in the attenuation constants as large as 50% or more.

In a waveguide with finite conductivity the tangential electric field at the boundary is related to the current density by the relation $\mathbf{E}_t = Z_m \mathbf{J}_s$ where Z_m is the surface impedance of the metal as given by $Z_m = (1 + j)/\sigma \delta_s$. If the scalar product $\mathbf{E}_1 \cdot \mathbf{J}_2^*$ of the electric field $\mathbf{E}_1$ of mode 1 with the current $\mathbf{J}_2^*$ of mode 2, when integrated around the boundary of the waveguide, is not zero, then the current $\mathbf{J}_2$ will deliver power to mode 1. Similarly, the current $\mathbf{J}_1$ of mode 1 will deliver power to mode 2 since $\mathbf{E}_2 \cdot \mathbf{J}_1^* = Z_m \mathbf{J}_2 \cdot \mathbf{J}_1^*$ will not be zero if $\mathbf{E}_1 \cdot \mathbf{J}_2^* = Z_m \mathbf{J}_1 \cdot \mathbf{J}_2^*$ is not zero. In this instance there will be coupling between the two modes because of the finite impedance of the waveguide walls. It was also noted earlier that power orthogonality did not necessarily hold for degenerate modes. If we have N degenerate modes for which power orthogonality does not hold or which are coupled together by the finite wall impedances, we can define new modes that are a linear combination of the old modes so as to obtain a set of uncoupled modes. The details for finding this new set of modes are described below. Later on we will show that the results come from a variational principle.

Consider a set of N degenerate modes and let

$$\mathbf{E}_n = (\mathbf{e}_n + \mathbf{e}_{zn})e^{-\Gamma z} \tag{41a}$$

$$\mathbf{H}_n = (\mathbf{h}_n + \mathbf{h}_{zn})e^{-\Gamma z} \tag{41b}$$

be the fields of the nth mode. In (41) $\mathbf{e}_n$ and $\mathbf{h}_n$ represent the transverse fields, while $\mathbf{e}_{zn}$ and $\mathbf{h}_{zn}$ are the axial components. All modes have the common propagation factor or eigenvalue

Γ. The surface current for the nth mode is given by

$$\mathbf{J}_n = \mathbf{n} \times \mathbf{H}_n. \tag{42}$$

We will define the following interaction terms for these modes:

$$P_{nm} = \frac{1}{2} \iint_S \mathbf{E}_n \times \mathbf{H}_m^* \cdot \mathbf{a}_z \, dS \tag{43a}$$

$$W_{nm} = \frac{Z_m}{2} \oint_C \mathbf{J}_n \cdot \mathbf{J}_m^* \, dl \tag{43b}$$

which represent power flow coupling and surface current coupling, respectively. Our objective now is to introduce a new set of modes that are a linear combination of the old set and for which the interaction terms for different modes are zero.

Let the new modes be defined as follows:

$$\mathbf{E}_s' = \sum_{n=1}^{N} C_n^s \mathbf{E}_n \tag{44a}$$

$$\mathbf{H}_s' = \sum_{n=1}^{N} C_n^s \mathbf{H}_n. \tag{44b}$$

The interaction terms for the new modes are

$$P_{sr}' = \frac{1}{2} \iint_S \mathbf{E}_s' \times \mathbf{H}_r'^* \cdot \mathbf{a}_z \, dS \tag{45a}$$

$$W_{sr}' = \frac{Z_m}{2} \oint_C \mathbf{J}_s' \cdot \mathbf{J}_r'^* \, dl. \tag{45b}$$

The waveguide is assumed to have small losses and the fields given in (41) are the fields of the degenerate modes in the loss-free guide. For these modes, assumed to be propagating, $\Gamma = j\beta$ and $\mathbf{e}_n$ and $\mathbf{h}_n$ can be chosen to be real functions of the transverse coordinates and $\mathbf{e}_{zn}$ and $\mathbf{h}_{zn}$ are then imaginary. Since $\mathbf{E}_n \times \mathbf{H}_m^* \cdot \mathbf{a}_z = \mathbf{e}_n \times \mathbf{h}_m \cdot \mathbf{a}_z$ is real the P_{nm} are real. By using (22) we see that $P_{nm} = P_{mn}$. We note that $Z_{wn}\mathbf{h}_n = \mathbf{a}_z \times \mathbf{e}_n$ where Z_{wn} is the wave impedance of the nth mode. If the degeneracy is between an E mode and an H mode then $P_{nm} = 0$ as shown by the discussion at the end of the last section. For degenerate E modes or degenerate H modes the wave impedances are equal. We now see that $P_{nm} = P_{mn}$ for degenerate E modes or degenerate H modes, but for a degenerate E and H mode $P_{nm} = 0$ for $n \neq m$. The surface interaction terms $\mathbf{J}_n \cdot \mathbf{J}_m^*$ are equal to $\mathbf{H}_n \cdot \mathbf{H}_m^*$ and are thus also real and $W_{nm} = W_{mn}$. Thus the $W_{nm}/(1 + j)$ are real and symmetric in the indices n and m.

The condition that the new modes are uncoupled, i.e., $W_{sr}' = P_{sr}' = 0$ for $r \neq s$, may be stated in the form

$$\sum_{n=1}^{N} \sum_{m=1}^{M} C_n^s P_{nm} (C_m^r)^* = P_{sr}' \delta_{sr}$$

$$\sum_{n=1}^{N} \sum_{m=1}^{M} C_n^s W_{nm} (C_m^r)^* = W_{sr}' \delta_{sr}.$$

We now introduce the matrices

$$[P] = \begin{bmatrix} P_{11} & P_{12} & \cdots & P_{1N} \\ P_{12} & \cdots & \cdots & \cdots \\ \vdots & & & \\ P_{1N} & \cdots & \cdots & P_{NN} \end{bmatrix}$$

$$[W] = \begin{bmatrix} W_{11} & W_{12} & \cdots & W_{1N} \\ W_{12} & \cdots & \cdots & \cdots \\ \vdots & & & \\ W_{1N} & \cdots & \cdots & W_{NN} \end{bmatrix}$$

$$[C^s] = \begin{bmatrix} C_1^s \\ C_2^s \\ \vdots \\ C_N^s \end{bmatrix}$$

and the transposed vector

$$[C^s]_t = [C_1^s \ C_2^s \ldots C_N^s].$$

The conditions given may be stated in matrix form as follows:

$$[C^s]_t[P][C^r]^* = P'_{sr}\delta_{sr} \tag{46a}$$

$$[C^s]_t[W][C^r]^* = W'_{sr}\delta_{sr} \tag{46b}$$

We now multiply (46a) by $W'_{rr}/P'_{rr}Z_m = \lambda_r$ and subtract the resultant equation from (46b) divided by Z_m to obtain

$$[C^s]_t\{Y_m[W] - \lambda_r[P]\}[C^r]^* = 0$$

where $Y_m = Z_m^{-1} = \sigma\delta_s/(1+j)$. Since $[C^s]_t$ is nonzero we must have

$$\{Y_m[W] - \lambda_r[P]\}[C^r]^* = 0. \tag{47}$$

The matrix eigenvalue equation (47) involves real symmetric matrices and is of the type discussed in Section A.3 of the Mathematical Appendix. From that discussion we can state that there will be N real roots for λ_r and for each root an eigenvector $[C^r]$ exists. Also, the eigenvectors form an orthogonal set in that

$$[C^r]_t[P][C^s] = [C^r]_t[W][C^s] = 0, \qquad r \neq s.$$

The roots λ_r are determined by the vanishing of the following determinant:

$$\begin{vmatrix} Y_m W_{11} - \lambda_r P_{11} & Y_m W_{12} - \lambda_r P_{12} & \cdots & Y_m W_{1N} - \lambda_r P_{1N} \\ Y_m W_{12} - \lambda_r P_{12} & \cdots & & \\ \vdots & & & \\ Y_m W_{1N} - \lambda_r P_{1N} & \cdots & & Y_m W_{NN} - \lambda_r P_{NN} \end{vmatrix} = 0. \qquad (48)$$

We note that for the rth root

$$\lambda_r = \frac{W'_{rr}}{Z_m P'_{rr}}$$

and since the attenuation constant for the new rth mode is given by

$$\alpha_r = \mathrm{Re}\, \frac{W'_{rr}}{2P'_{rr}} = \frac{R_m \lambda_r}{2} \qquad (49)$$

where $R_m = 1/\sigma\delta_s$ we can obtain the attenuation constants for the new uncoupled modes directly from the roots of the eigenvalue equation (48). The imaginary part of $Z_m \lambda_r / 2$ is the perturbation in the phase constant β_r caused by the additional magnetic energy stored in the inductive part of the surface impedance Z_m. Hence if $j\beta$ is the propagation constant for the original set of degenerate modes the new propagation constants are given by

$$\gamma_r = j\beta + Z_m \frac{\lambda_r}{2}. \qquad (50)$$

An application of the above theory to a rectangular waveguide will be made in the next section. In particular for the E_{11} and H_{11} modes it is found that the attenuation constants are significantly modified by the coupling between the two modes.

For nondegenerate modes the effect of coupling by wall losses is either zero or very small. A variational expression for the propagation constant in a lossy waveguide can be derived and is similar to that for a lossy transmission line as given in Section 4.1. We can express the magnetic field of a mode in a lossy waveguide in the form

$$\mathbf{H} = (\mathbf{h} + \mathbf{h}_z)e^{-\gamma z}$$

where $\mathbf{h}$ is the transverse part and $\mathbf{h}_z$ is the axial part. The unknown propagation constant is γ. The magnetic field is a solution of

$$\nabla^2 \mathbf{H} + k_0^2 \mathbf{H} = 0 = (\nabla_t^2 + k_c^2)(\mathbf{h} + \mathbf{h}_z) = 0$$

where $k_c^2 = k_0^2 + \gamma^2$. We now express ∇_t^2 in the form $\nabla_t^2 = \nabla_t \nabla_t \cdot - \nabla_t \times \nabla_t \times$ and then find that

$$-\nabla_t \nabla_t \cdot \mathbf{h} + \nabla_t \times \nabla_t \times \mathbf{h} - k_c^2 \mathbf{h} = 0 \qquad (51a)$$

$$\nabla_t \times \nabla_t \times \mathbf{h}_z - k_c^2 \mathbf{h}_z = 0. \qquad (51b)$$

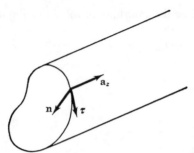

Fig. 5.3. Illustration showing unit normal to waveguide surface.

By separating the equation $\nabla \times \mathbf{H} = j\omega\epsilon_0\mathbf{E} = jk_0Y_0\mathbf{E}$ into transverse and axial components we obtain

$$\nabla_t \times \mathbf{h} = jk_0Y_0\mathbf{e}_z \tag{52a}$$

$$\nabla_t \times \mathbf{h}_z - \gamma\mathbf{a}_z \times \mathbf{h} = jk_0Y_0\mathbf{e} \tag{52b}$$

$$\gamma h_z = \nabla_t \cdot \mathbf{h} \tag{52c}$$

where the electric field of the mode has been written as

$$\mathbf{E} = (\mathbf{e} + \mathbf{e}_z)e^{-\gamma z}.$$

The boundary conditions for $\mathbf{E}$ are

$$jk_0Y_0\mathbf{e}_z = jk_0Y_0Z_m\mathbf{n} \times \mathbf{h} = \nabla_t \times \mathbf{h} \tag{53a}$$

$$jk_0Y_0\mathbf{e} = jk_0Y_0Z_m\mathbf{n} \times \mathbf{h}_z = \nabla_t \times \mathbf{h}_z - \gamma\mathbf{a}_z \times \mathbf{h} \tag{53b}$$

where $\mathbf{n}$ is the inward normal as shown in Fig. 5.3.

Consider now the variation in the integral

$$\delta \iint\limits_{S} [\nabla_t \times \mathbf{h}\cdot\nabla_t \times \mathbf{h} + \nabla_t\cdot\mathbf{h}\nabla_t\cdot\mathbf{h} - \nabla_t \times \mathbf{h}_z\cdot\nabla_t \times \mathbf{h}_z - k_c^2(\mathbf{h}\cdot\mathbf{h} - \mathbf{h}_z\cdot\mathbf{h}_z)]\, dS$$

which is given by

$$2\iint\limits_{S} [\nabla_t \times \delta\mathbf{h}\cdot\nabla_t \times \mathbf{h} - \nabla_t\cdot\delta\mathbf{h}\nabla_t\cdot\mathbf{h} - \nabla_t \times \delta\mathbf{h}_z\cdot\nabla_t \times \mathbf{h}_z$$
$$-k_c^2(\delta\mathbf{h}\cdot\mathbf{h} - \delta\mathbf{h}_z\cdot\mathbf{h}_z)]\, dS - 2k_c\,\delta k_c \iint\limits_{S} (\mathbf{h}\cdot\mathbf{h} - \mathbf{h}_z\cdot\mathbf{h}_z)\, dS.$$

We can use the vector operations

$$\nabla_t\cdot[\delta\mathbf{h} \times \nabla_t \times \mathbf{h}] = \nabla_t \times \delta\mathbf{h}\cdot\nabla_t \times \mathbf{h} - \delta\mathbf{h}\cdot\nabla_t \times \nabla_t \times \mathbf{h}$$

$$\nabla_t\cdot[\delta\mathbf{h}\nabla_t\cdot\mathbf{h}] = \nabla_t\cdot\delta\mathbf{h}\nabla_t\cdot\mathbf{h} + \delta\mathbf{h}\cdot\nabla_t\nabla_t\cdot\mathbf{h}$$

along with the divergence theorem to express the variation in the following form:

$$
2 \iint_{S} [\delta \mathbf{h} \cdot (\nabla_t \times \nabla_t \times \mathbf{h} - \nabla_t \nabla_t \cdot \mathbf{h} - k_c^2 \mathbf{h})
$$

$$
- \delta \mathbf{h}_z \cdot (\nabla_t \times \nabla_t \times \mathbf{h}_z - k_c^2 \mathbf{h}_z)] \, dS - 2k_c \, \delta k_c \iint_{S} (\mathbf{h} \cdot \mathbf{h} - \mathbf{h}_z \cdot \mathbf{h}_z) \, dS
$$

$$
- 2 \oint_{C} [\mathbf{n} \cdot \delta \mathbf{h} \times \nabla_t \times \mathbf{h} + \mathbf{n} \cdot \delta \mathbf{h} \nabla_t \cdot \mathbf{h}
$$

$$
- \mathbf{n} \cdot \delta \mathbf{h}_z \times \nabla_t \times \mathbf{h}_z] \, dl.
$$

By virtue of (51) we see that the first surface integral is zero. By using (53) the integrand in the contour integral can be expressed in the form

$$
jk_0 Y_0 Z_m [\mathbf{n} \cdot \delta \mathbf{h} \times (\mathbf{n} \times \mathbf{h}) - \mathbf{n} \cdot \delta \mathbf{h}_z \times (\mathbf{n} \times \mathbf{h}_z)]
$$

$$
- \gamma \mathbf{n} \cdot \delta \mathbf{h}_z \times (\mathbf{a}_z \times \mathbf{h}) + \gamma (\mathbf{n} \cdot \delta \mathbf{h}) h_z
$$

$$
= \tfrac{1}{2} j k_0 Y_0 Z_m \delta [(\mathbf{n} \times \mathbf{h}) \cdot (\mathbf{n} \times \mathbf{h}) - (\mathbf{n} \times \mathbf{h}_z) \cdot (\mathbf{n} \times \mathbf{h}_z)]
$$

$$
+ \gamma \delta [(\mathbf{n} \cdot \mathbf{h}) h_z] = \delta F - \delta \gamma \, \mathbf{n} \cdot \mathbf{h} h_z.
$$

The last result can be verified by evaluating the first variation of the functional in brackets. We see that the contour integral can be expressed as the variation of a functional F. If we add the contour integral of F to the original surface integral then we obtain an expression whose first variation in $k_c^2 = k_0^2 + \gamma^2$ and γ, i.e., in γ, is zero. Hence our required variational expression is

$$
I = \iint_{S} [(\nabla_t \times \mathbf{h})^2 + (\nabla_t \cdot \mathbf{h})^2 - (\nabla_t \times \mathbf{h}_z)^2 - k_c^2 (\mathbf{h}^2 - \mathbf{h}_z^2)] \, dS
$$

$$
+ j k_0 Y_0 Z_m \oint_{C} [(\mathbf{n} \times \mathbf{h})^2 - (\mathbf{n} \times \mathbf{h}_z)^2] \, dl + 2\gamma \oint_{C} \mathbf{n} \cdot \mathbf{h} h_z \, dl. \quad (54)
$$

It can be shown that for the correct field $I = 0$.

We now approximate $\mathbf{h} + \mathbf{h}_z$ by a finite sum of the propagating modal functions for a loss-free guide (these have propagation constants Γ_n):

$$
\mathbf{h} + \mathbf{h}_z = \sum_{n=1}^{N} C_n (\mathbf{h}_n + \mathbf{h}_{zn}).
$$

We note that $\mathbf{h}_n$ can be chosen as real and $\mathbf{h}_{zn}$ is then pure imaginary. Hence $(\mathbf{n} \times \mathbf{h}_n) \cdot (\mathbf{n} \times \mathbf{h}_m) - (\mathbf{n} \times \mathbf{h}_{zn}) \cdot (\mathbf{n} \times \mathbf{h}_{zm}) = \mathbf{J}_n \cdot \mathbf{J}_m^*$ where $\mathbf{J}_n$ and $\mathbf{J}_m$ are the surface currents for the nth and mth modes. We can show that the approximate field has the property that for nondegenerate modes the cross-product terms in the surface integral integrate to zero. Also $\mathbf{n} \cdot \mathbf{h}_n = 0$ so our variational expression becomes

$$
I = \sum_{n=1}^{N} C_n^2 \frac{(\Gamma_n^2 - \gamma^2)}{\Gamma_n} \iint_{S} \mathbf{e}_n \times \mathbf{h}_n \cdot \mathbf{a}_z \, dS + Z_m \sum_{n=1}^{N} \sum_{m=1}^{N} C_n C_m \oint_{C} \mathbf{J}_n \cdot \mathbf{J}_m^* \, dl \quad (55)
$$

upon using the properties that for the modes in a loss-free guide

$$\iint_S [(\nabla_t \times \mathbf{h}_n)^2 - (\nabla_t \times \mathbf{h}_{zn})^2 + (\nabla_t \cdot \mathbf{h}_n)^2 - (k_0^2 + \Gamma_n^2)(\mathbf{h}_n^2 - \mathbf{h}_{zn}^2)] \, dS = 0$$

and

$$\iint_S (\mathbf{h}_n^2 - \mathbf{h}_{zn}^2) \, dS = \frac{jk_0 Y_0}{\Gamma_n} \iint_S \mathbf{e}_n \times \mathbf{h}_n \cdot \mathbf{a}_z \, dS.$$

The first result following (55) can be proved by expressing the integral in the same form as was done for the first variation and noting that with $\delta \mathbf{h} = \mathbf{h}_n$ and $\delta \mathbf{h}_z = \mathbf{h}_{zn}$ the contour integral is zero because of the boundary conditions satisfied by the loss-free modes. The second result follows by using (52b) to form the product $\mathbf{e}_n \times \mathbf{h}_n \cdot \mathbf{a}_z$ and integrating over the cross section S. In the derivation the integral

$$\iint_S (\nabla_t \times \mathbf{h}_{zn}) \times \mathbf{h}_n \cdot \mathbf{a}_z \, dS$$

is reduced by using $\nabla_t \cdot [\mathbf{h}_{zn} \times (\mathbf{h}_n \times \mathbf{a}_z)] = \nabla_t \times \mathbf{h}_{zn} \cdot (\mathbf{h}_n \times \mathbf{a}_z) - \mathbf{h}_{zn} \cdot \nabla_t \times (\mathbf{h}_n \times \mathbf{a}_z) = \nabla_t \times \mathbf{h}_{zn} \cdot (\mathbf{h}_n \times \mathbf{a}_z) + h_{zn} \nabla_t \cdot \mathbf{h}_n$ to obtain

$$-\Gamma_n \iint_S h_{zn}^2 \, dS + \oint_C \mathbf{n} \cdot \mathbf{h}_{zn} \times (\mathbf{h}_n \times \mathbf{a}_z) \, dl = -\Gamma_n \iint_S h_{zn}^2 \, dS$$

since $\mathbf{n} \cdot \mathbf{h}_{zn} \times (\mathbf{h}_n \times \mathbf{a}_z) = \mathbf{n} \cdot (h_{zn} \mathbf{h}_n) = 0$ on the boundary C.

The variation in (55) gives

$$C_n \frac{\Gamma_n^2 - \gamma^2}{\Gamma_n} P_{nn} + \frac{1}{2} Z_m \sum_{m=1}^{N} C_m \oint_C \mathbf{J}_n \cdot \mathbf{J}_m^* \, dl = 0$$

or

$$C_n \frac{\Gamma_n^2 - \gamma^2}{\Gamma_n} P_{nn} + \sum_{m=1}^{N} C_m W_{nm} = 0 \tag{56}$$

where the P_{nn} and W_{nm} are defined as in (43).

If there is no surface current coupling, then $W_{nm} = 0$ for $n \neq m$ and (56) gives

$$\frac{\gamma^2 - \Gamma_n^2}{\Gamma_n} = (\gamma - \Gamma_n) \frac{(\gamma + \Gamma_n)}{\Gamma_n} \approx 2(\gamma - \Gamma_n) = \frac{W_{nn}}{P_{nn}}$$

or

$$\gamma = \gamma_n = j\beta + \alpha = j\beta_n + (1 + j) \frac{P_L}{2P_{nn}} \tag{57}$$

for the propagation constant of the nth perturbed mode where P_L is the power loss in the wall

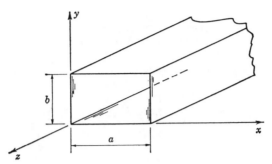

Fig. 5.4. The rectangular waveguide.

given by

$$P_L = \frac{R_m}{2} \oint_C \mathbf{J}_n \cdot \mathbf{J}_n^* \, dl$$

and P_{nn} is the power flow. This result is the standard power loss result for the attenuation.

For the case of N degenerate modes where all $\Gamma_m = \Gamma$ the system (56) becomes the same as that in (47) if we identify $\lambda_r Z_m$ with $(\gamma^2 - \Gamma^2)/\Gamma \approx 2(\gamma - \Gamma)$. Thus the variational principle provides a rigorous basis for the method presented earlier for evaluating the coupling effects associated with degenerate modes.

5.4. The Rectangular Waveguide

The most commonly used waveguide is the rectangular guide. Some of the reasons for this are a good bandwidth of operation for single-mode propagation, reasonably low attenuation, and good mode stability for the fundamental mode of propagation. The dimensions of the common rectangular guide are such that the width a is approximately equal to twice the height b, as illustrated in Fig. 5.4.

Transverse Electric or H Modes

The transverse electric or H modes may be derived from the scalar function

$$\psi_{h,nm} = \cos\frac{n\pi x}{a} \cos\frac{m\pi y}{b}, \qquad \begin{array}{l} n = 0, 1, 2, \ldots \\ m = 0, 1, 2, \ldots \\ m \neq n = 0 \end{array} \tag{58}$$

by means of (6). It is readily seen that ψ_h as given is a solution to the scalar Helmholtz equation (3) and satisfies the boundary condition $\partial\psi_h/\partial n = 0$, provided the propagation constant is chosen as

$$\Gamma_{nm}^2 = \left(\frac{n\pi}{a}\right)^2 + \left(\frac{m\pi}{b}\right)^2 - k_0^2 = k_c^2 - k_0^2. \tag{59}$$

To each set of integers n, m, a solution or mode exists, and these modes are designated as

TABLE 5.1
COMMON PROPERTIES OF PROPAGATING MODES IN EMPTY
CYLINDRICAL WAVEGUIDES

Property	H modes	E modes
Wave impedance	$Z_h = (k_0/\beta)Z_0$	$Z_e = (\beta/k_0)Z_0$
Propagation constant	$\Gamma = j\beta = (k_c^2 - k_0^2)^{1/2}$	Same
Cutoff wavelength	$\lambda_c = 2\pi/k_c$	Same
Guide wavelength	$\lambda_g = \lambda_0/(1 - \lambda_0^2/\lambda_c^2)^{1/2}$	Same
Group velocity	$v_g = v_c\lambda_0/\lambda_g$	Same
Phase velocity	$v_p = v_c\lambda_g/\lambda_0$	Same

Note: λ_0 = free-space wavelength; $Z_0 = (\mu_0/\epsilon_0)^{1/2}$ = intrinsic impedance of free space; $v_c = (\mu_0\epsilon_0)^{-1/2}$ = velocity of light in free space.

the TE_{nm} or H_{nm} modes. The cutoff wavelength $\lambda_{c,nm}$ for the nmth mode is given by

$$\lambda_{c,nm} = \frac{2\pi}{k_c} = \frac{2ab}{(n^2b^2 + m^2a^2)^{1/2}}. \tag{60}$$

The dominant mode of propagation (mode with the largest cutoff wavelength) is the H_{10} mode when $a > b$, and its cutoff wavelength is $2a$. If $a = 2b$, the next mode to propagate is the H_{01} mode, which has a cutoff wavelength equal to a.

The properties of H modes which are common to all guides are summarized in Table 5.1, while those properties of H modes which are valid for the rectangular guide only are summarized in Table 5.2. The expression given for the attenuation constant in Table 5.2 is that computed by the power-loss method, i.e., (38).

Transverse Magnetic or E Modes

The appropriate solution for ψ_e which satisfies the boundary condition $\psi_e = 0$ on the guide walls is

$$\psi_{e,nm} = \sin\frac{n\pi x}{a} \sin\frac{m\pi y}{b}, \qquad \begin{matrix} n = 1, 2, \ldots \\ m = 1, 2, \ldots \end{matrix} \tag{61}$$

and $\Gamma_{nm}^2 = (n\pi/a)^2 + (m\pi/b)^2 - k_0^2$. The fields may be derived from $\psi_{e,nm}$ by means of (11). The modes are designated as TM_{nm} or E_{nm} modes. The dominant E mode is the E_{11} mode with a cutoff wavelength $2ab/(a^2 + b^2)^{1/2}$, which is equal to $2a/5^{1/2}$ for $a = 2b$. The properties of E modes in rectangular guides are summarized in Tables 5.1 and 5.2. The attenuation constant given is based on (39), that is, that determined by the power-loss method. It should be noted that the H_{nm} and E_{nm} modes are degenerate and have the same propagation constant Γ_{nm}.

Coupling of Modes in Lossy Rectangular Guides

The normal mode functions $\mathbf{e}_n$ introduced in Section 5.3 to represent the transverse electric field distribution will be written as $\mathbf{e}_{nm}$ for the H_{nm} modes and as $\mathbf{e}'_{nm}$ for the E_{nm} modes. The corresponding transverse and longitudinal magnetic field components will be written as

TABLE 5.2
PROPERTIES OF MODES IN EMPTY RECTANGULAR WAVEGUIDES

Property	H_{nm} modes	E_{nm} modes
Generating function	$\psi_{h,nm} = \cos\dfrac{n\pi x}{a}\cos\dfrac{m\pi y}{b}$	$\psi_{e,nm} = \sin\dfrac{n\pi x}{a}\sin\dfrac{m\pi y}{b}$
Cutoff wavenumber	$k_{c,nm}^2 = \left(\dfrac{n\pi}{a}\right)^2 + \left(\dfrac{m\pi}{b}\right)^2$	Same
Propagation constant	$\Gamma_{nm}^2 = k_{c,nm}^2 - k_0^2$	Same
Dyadic wave impedance	$\bar{\mathbf{Z}}_h = \dfrac{jk_0}{\Gamma_{nm}} Z_0(\mathbf{a}_x\mathbf{a}_y - \mathbf{a}_y\mathbf{a}_x)$	$\bar{\mathbf{Z}}_e = \dfrac{\Gamma_{nm}}{jk_0} Z_0(\mathbf{a}_x\mathbf{a}_y - \mathbf{a}_y\mathbf{a}_x)$
Dyadic wave admittance	$\bar{\mathbf{Y}}_h = \dfrac{\Gamma_{nm}}{jk_0} Y_0(\mathbf{a}_y\mathbf{a}_x - \mathbf{a}_x\mathbf{a}_y)$	$\bar{\mathbf{Y}}_e = \dfrac{jk_0}{\Gamma_{nm}} Y_0(\mathbf{a}_y\mathbf{a}_x - \mathbf{a}_x\mathbf{a}_y)$
Axial magnetic field	$k_{c,nm}^2 \psi_{h,nm} e^{\pm\Gamma_{nm}z}$	Zero
Axial electric field	Zero	$k_{c,nm}^2 \psi_{e,nm} e^{\pm\Gamma_{nm}z}$
Transverse magnetic field	$\mathbf{H}_t = \pm\Gamma_{nm} e^{\pm\Gamma_{nm}z}\nabla_t\psi_{h,nm}$	$\mp\bar{\mathbf{Y}}_e \cdot \mathbf{E}_t$
Transverse electric field	$\mp\bar{\mathbf{Z}}_h \cdot \mathbf{H}_t$	$\mathbf{E}_t = \pm\Gamma_{nm} e^{\pm\Gamma_{nm}z}\nabla_t\psi_{e,nm}$
Power flow, Eqs. (26), (27); $\Gamma_{nm} = j\beta_{nm}$	$\dfrac{abZ_0 k_0\beta_{nm}k_{c,nm}^2}{2\epsilon_{0n}\epsilon_{0m}}$	$\dfrac{abY_0 k_0\beta_{nm}k_{c,nm}^2}{2\epsilon_{0n}\epsilon_{0m}}$
Attenuation, Eqs. (38), (39); $R_m = (\omega\mu/2\sigma)^{1/2}$, α in nepers per meter	$\dfrac{2R_m}{bZ_0(1 - k_{c,nm}^2/k_0^2)^{1/2}}\left[\left(1+\dfrac{b}{a}\right)\dfrac{k_{c,nm}^2}{k_0^2} + \dfrac{b}{a}\left(\dfrac{\epsilon_{0m}}{2} - \dfrac{k_{c,nm}^2}{k_0^2}\right)\left(\dfrac{n^2ab+m^2a^2}{n^2b^2+n^2a^2}\right)\right]$ $\epsilon_{0n} = \begin{cases} 1, & n=0, \\ 2, & n>0 \end{cases}$	$\dfrac{2R_m}{bZ_0(1 - k_{c,nm}^2/k_0^2)^{1/2}}\left(\dfrac{n^2b^3+m^2a^3}{n^2b^2a+m^2a^3}\right)$

$\mathbf{h}_{nm}$, $\mathbf{h}_{znm}$ for the H_{nm} modes and as $\mathbf{h}'_{nm}$ for the E_{nm} modes. The surface currents on the guide wall will be designated by $\mathbf{J}_{tnm}$, $\mathbf{J}_{znm}$, and $\mathbf{J}'_{znm}$, where the prime again corresponds to the E_{nm} modes. The surface currents are given by

$$\mathbf{J}_{tnm} = \mathbf{n} \times \mathbf{h}_{znm}$$

$$\mathbf{J}_{znm} = \mathbf{n} \times \mathbf{h}_{nm} \tag{62}$$

where $\mathbf{n}$ is the unit inward normal to the guide surface. A similar set of equations holds for the E_{nm} modes with all quantities replaced by primed quantities. Note that $\mathbf{e}_{znm}$, $\mathbf{h}'_{znm}$, $\mathbf{J}'_{tnm}$ are all zero, and that the propagation factor $e^{-\Gamma_{nm}z}$ is not included in the above.

The normal mode functions are readily obtained from Table 5.2 in terms of $\psi_{h,nm}$ and $\psi_{e,nm}$ and when normalized are given by the following equations:

for H_{nm} modes

$$\mathbf{e}_{nm} = \left[\left(\frac{m\pi}{b} \right)^2 + \left(\frac{n\pi}{a} \right)^2 \right]^{-1/2} \left(\frac{\epsilon_{0n}\epsilon_{0m}}{ab} \right)^{1/2}$$

$$\cdot \left(\mathbf{a}_x \frac{m\pi}{b} \cos \frac{n\pi x}{a} \sin \frac{m\pi y}{b} - \mathbf{a}_y \frac{n\pi}{a} \sin \frac{n\pi x}{a} \cos \frac{m\pi y}{b} \right); \tag{63a}$$

for E_{nm} modes

$$\mathbf{e}'_{nm} = \left[\left(\frac{m\pi}{b} \right)^2 + \left(\frac{n\pi}{a} \right)^2 \right]^{-1/2} \left(\frac{\epsilon_{0n}\epsilon_{0m}}{ab} \right)^{1/2}$$

$$\cdot \left(\mathbf{a}_x \frac{n\pi}{a} \cos \frac{n\pi x}{a} \sin \frac{m\pi y}{b} + \mathbf{a}_y \frac{m\pi}{b} \sin \frac{n\pi x}{a} \cos \frac{m\pi y}{b} \right). \tag{63b}$$

The magnetic field may be obtained by means of equations that are the dual of (52), i.e.,

$$\nabla_t \times \mathbf{e}_{nm} = -jk_0 Z_0 \mathbf{h}_{znm} \tag{64a}$$

$$-jk_0 Z_0 \mathbf{h}_{nm} = -\Gamma \mathbf{a}_z \times \mathbf{e}_{nm} + \nabla_t \times \mathbf{e}_{znm} \tag{64b}$$

$$\Gamma \mathbf{e}_{znm} = \nabla_t \cdot \mathbf{e}_{nm}. \tag{64c}$$

The surface currents are found by means of (62). For the H_{nm} modes the surface currents are

$$\mathbf{J}_{tnm} = \frac{\left[\left(\frac{n\pi}{a} \right)^2 + \left(\frac{m\pi}{b} \right)^2 \right]^{1/2} \left(\frac{\epsilon_{0n}\epsilon_{0m}}{ab} \right)^{1/2}}{j\omega\mu_0} \begin{cases} -\mathbf{a}_y \cos \frac{m\pi y}{b}, & x = 0 \\ \mathbf{a}_y \cos n\pi \cos \frac{m\pi y}{b}, & x = a \\ \mathbf{a}_x \cos \frac{n\pi x}{a}, & y = 0 \\ -\mathbf{a}_x \cos m\pi \cos \frac{n\pi x}{a}, & y = b \end{cases} \tag{65a}$$

$$
J_{znm} = \frac{\beta_{nm}\left(\dfrac{\epsilon_{0n}\epsilon_{0m}}{ab}\right)^{1/2}}{\omega\mu_0\left[\left(\dfrac{n\pi}{a}\right)^2+\left(\dfrac{m\pi}{b}\right)^2\right]^{1/2}}
\begin{cases}
\dfrac{m\pi}{b}\sin\dfrac{m\pi y}{b}, & x=0\\[2mm]
-\dfrac{m\pi}{b}\cos n\pi\sin\dfrac{m\pi y}{b}, & x=a\\[2mm]
-\dfrac{n\pi}{a}\sin\dfrac{n\pi x}{a}, & y=0\\[2mm]
\dfrac{n\pi}{a}\cos m\pi\sin\dfrac{n\pi x}{a}, & y=b
\end{cases}
\tag{65b}
$$

where $j\beta_{nm}=\Gamma_{nm}$, and ϵ_{0n} is the Neumann factor which equals unity for $n=0$ and equals 2 otherwise. For the E_{nm} modes, there are no transverse currents, the longitudinal current is given by

$$
J'_{znm} = \frac{\left[\left(\dfrac{n\pi}{a}\right)^2+\left(\dfrac{m\pi}{b}\right)^2\right]^{-1/2}\left(\dfrac{\epsilon_{0n}\epsilon_{0m}}{ab}\right)^{1/2}k_0^2}{\omega\mu_0\beta_{nm}}
\begin{cases}
\dfrac{n\pi}{a}\sin m\pi\dfrac{y}{b}, & x=0\\[2mm]
-\dfrac{n\pi}{a}\cos n\pi\sin\dfrac{m\pi y}{b}, & x=a\\[2mm]
\dfrac{m\pi}{b}\sin\dfrac{n\pi x}{a}, & y=0\\[2mm]
-\dfrac{m\pi}{b}\cos m\pi\sin\dfrac{n\pi x}{a}, & y=b.
\end{cases}
\tag{65c}
$$

An examination of the above expressions for the surface currents shows that, for H_{nm} and E_{rs} modes, with $r\neq n$ and $s\neq m$, the currents are orthogonal, and, hence, these modes are not coupled. The coupling between E and H modes is brought about by the axial currents only. From (56) a first-order solution for the propagation constant γ may be obtained by including only the E_{nm} and H_{nm} modes since $\gamma\approx\Gamma_{nm}$, and all the coefficients in the expansion of the transverse electric field are small, except the coefficients of the two modes $\mathbf{e}_{nm}$ and $\mathbf{e}'_{nm}$. The coefficients for these two modes will be designated as a_{nm} and a'_{nm}. Substituting (65) into (56) and performing the integrations gives the following two homogeneous equations:

$$
\left\{\frac{(\gamma^2+\beta_{nm}^2)j\omega\mu_0 ab}{Z_m\epsilon_{0n}\epsilon_{0m}}+k_{c,nm}^2\left(\frac{2b}{\epsilon_{0m}}+\frac{2a}{\epsilon_{0n}}\right)+\frac{\beta_{nm}^2}{k_{c,nm}^2}\right.
$$
$$
\left.\cdot\left[b\left(\frac{m\pi}{b}\right)^2+a\left(\frac{n\pi}{a}\right)^2\right]\right\}a_{nm}+\frac{k_0^2}{k_{c,nm}^2}\frac{mn\pi^2}{ab}(b-a)a'_{nm}=0 \tag{66a}
$$

$$
\frac{\beta_{nm}^2}{k_{c,nm}^2}\frac{mn\pi^2}{ab}(b-a)a_{nm}+\left\{\frac{(\gamma^2+\beta_{nm}^2)j\omega\mu_0 ab}{Z_m\epsilon_{0n}\epsilon_{0m}}+\frac{k_0^2}{k_{c,nm}^2}\right.
$$
$$
\left.\cdot\left[\left(\frac{n\pi}{a}\right)^2 b+\left(\frac{m\pi}{b}\right)^2 a\right]\right\}a'_{nm}=0, \qquad n=m\neq 0. \tag{66b}
$$

One interesting result immediately apparent from (66) is the absence of coupling between E and H modes in a square guide for which $a=b$ and the off-diagonal terms in (66) vanish. For $a\neq b$, (66) determines a solution for a_{nm} and a'_{nm}, provided the determinant vanishes.

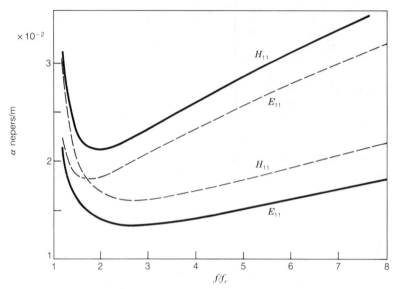

Fig. 5.5. Attenuation in a copper waveguide with $a = 2b = 1$ inch. Solid line, variational solution; broken line, power-loss-method solution.

The vanishing of the determinant leads to two roots for γ^2. For one root $a_{nm} > a'_{nm}$, and this is the propagation constant for a mode corresponding to a perturbed H_{nm} mode. The other root corresponds to a perturbed E_{nm} mode, for which $a'_{nm} > a_{nm}$. In Fig. 5.5 the attenuation constants for the perturbed E_{11} and H_{11} modes are plotted as a function of frequency for $a = 2b$ and a conductivity corresponding to that of copper (5.8×10^7 siemens per meter). The broken curves in the same figure give the attenuation constants for the same two modes as computed by the power-loss method. There is an appreciable difference between the results as obtained by the two methods. We can conclude from this example that the power-loss method does not give accurate results in those cases where degenerate E and H modes are coupled together.

If $m = 0$ and $n = 1$, (66a) reduces to a single term and gives the phase constant and attenuation constant for the H_{10} mode. A plot of α and β versus frequency is given in Fig. 5.6 for $a = 2b = 1$ inch and a copper waveguide. For this mode, the power-loss method gives essentially the same results except in the immediate vicinity of the no-loss cutoff frequency $\omega_c/2\pi$.

5.5. CIRCULAR CYLINDRICAL WAVEGUIDES

In a circular cylindrical waveguide of radius a, the generating functions ψ_h and ψ_e for the H and E modes are solutions of the following equation:

$$\frac{\partial^2 \psi_i}{\partial r^2} + \frac{1}{r}\frac{\partial \psi_i}{\partial r} + \frac{1}{r^2}\frac{\partial^2 \psi_i}{\partial \theta^2} + k_c^2 \psi_i = 0, \qquad i = h, e \tag{67}$$

where $k_c^2 = \Gamma^2 + k_0^2$ and $\psi_i = 0$ at $r = a$ for E modes, and $\partial \psi_i/\partial r = 0$ at $r = a$ for H modes. Figure 5.7 illustrates a circular guide and the coordinates r, θ, z. The solution to (67), which

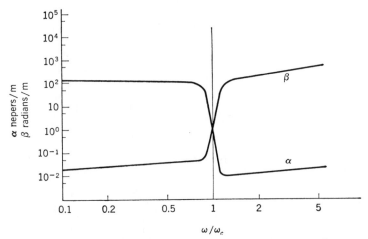

Fig. 5.6. Attenuation and phase constant for H_{10} mode in a copper rectangular waveguide. $a = 2b = 1$ in.

is finite at the origin and single-valued throughout the interior, is

$$\psi_i = J_n(k_c r) \begin{cases} \sin n\theta, \\ \cos n\theta, \end{cases} \qquad n = 0, 1, 2, \ldots. \tag{68}$$

For H modes, $dJ_n(k_c r)/dr = 0$ at $r = a$. This equation determines an infinite number of roots for $k_c a$ which will be designated by p'_{nm}. The corresponding H mode will be labeled H_{nm} or TE_{nm}, where the first subscript refers to the number of cyclic variations with θ, and the second subscript refers to the mth root of the Bessel function. A number of the roots p'_{nm} are listed in Table 5.3. The dominant, or first, mode to propagate in a circular guide is the TE_{11} mode.

For E modes, $J_n(k_c a) = 0$ determines the allowed values of k_c and, hence, the propagation constants. The roots of this equation are designated by p_{nm}, and the corresponding modes are labeled E_{nm} or TM_{nm}. Table 5.4 gives the first few roots of the above equation.

The properties of E and H modes in circular waveguides are summarized in Table 5.5. In circular cylindrical coordinates the transverse operator ∇_t is given by

$$\nabla_t = \mathbf{a}_r \frac{\partial}{\partial r} + \frac{\mathbf{a}_\theta}{r} \frac{\partial}{\partial \theta}. \tag{69}$$

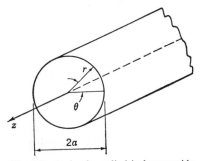

Fig. 5.7. A circular cylindrical waveguide.

TABLE 5.3
ROOTS OF $(dJ_n(k_c r)/dr)|_{r=a} = 0$

n	p'_{n1}	p'_{n2}	p'_{n3}	p'_{n4}
0	3.832	7.016	10.174	13.324
1	1.841	5.331	8.536	11.706
2	3.054	6.706	9.970	13.170

TABLE 5.4
ROOTS OF $J_n(k_c a) = 0$

n	p_{n1}	p_{n2}	p_{n3}	p_{n4}
0	2.405	5.520	8.654	11.792
1	3.832	7.016	10.174	13.324
2	5.135	8.417	11.620	14.796

The derivatives of the Bessel function $J_n(k_c r)$ are given by the relation

$$\frac{1}{k_c}\frac{d}{dr}J_n(k_c r) = \frac{-n}{k_c r}J_n(k_c r) + J_{n-1}(k_c r) = \frac{n}{k_c r}J_n(k_c r) - J_{n+1}(k_c r). \tag{70}$$

The integral of the product of two Bessel functions of the same order n and of different argument is given by

$$\int_0^a J_n(k_1 r)J_n(k_2 r)r\,dr = \frac{a}{k_1^2 - k_2^2}[k_2 J_n(k_1 a)J_{n-1}(k_2 a) - k_1 J_{n-1}(k_1 a)J_n(k_2 a)] \tag{71a}$$

and, when $k_1 = k_2$, the value of the integral is

$$\int_0^a r[J_n(k_1 r)]^2\,dr = \frac{a^2}{2}\left\{\left[\frac{1}{k_1}\frac{dJ_n(k_1 a)}{dr}\right]^2 + \left(1 - \frac{n^2}{k_1^2 a^2}\right)J_n^2(k_1 a)\right\}. \tag{71b}$$

If $k_1 a$ and $k_2 a$ correspond to two of the roots given in Table 5.3 or 5.4, the value of the integral (71a) is zero. This result follows in view of the orthogonality properties of the modes in a cylindrical waveguide and may be proved directly from Bessel's differential equation by the methods given in Section 5.2.

An examination of the formula for the attenuation constant for H_{nm} modes shows that the H_{01} mode has the unique property that its attenuation decreases like $f^{-3/2}$. It is for this reason that considerable work has been done on techniques and components for the utilization of the H_{01} mode in low-loss long-distance communication links.[3]

5.6. GREEN'S FUNCTIONS

The general dyadic Green's function for a cylindrical guide is a solution of one of the

---

[3]See, for example, [5.5], [5.6].

TABLE 5.5
PROPERTIES OF MODES IN EMPTY CIRCULAR WAVEGUIDES

Property	H_{nm} modes	E_{nm} modes
Generating function	$\psi_{h,nm} = J_n(k_{c,nm}r)\begin{Bmatrix}\sin n\theta\\\cos n\theta\end{Bmatrix}$	$\psi_{e,nm} = J_n(k_{c,nm}r)\begin{Bmatrix}\sin n\theta\\\cos n\theta\end{Bmatrix}$
Cutoff wavenumber	$k_{c,nm} = \dfrac{p'_{nm}}{a}$	$k_{c,nm} = \dfrac{p_{nm}}{a}$
Propagation constant	$\Gamma^2_{nm} = \left(\dfrac{p'_{nm}}{a}\right)^2 - k_0^2$	$\Gamma^2_{nm} = \left(\dfrac{p_{nm}}{a}\right)^2 - k_0^2$
Dyadic wave impedance	$\bar{\mathbf{Z}}_h = \dfrac{jk_0}{\Gamma_{nm}}Z_0(\mathbf{a}_\theta\mathbf{a}_\theta - \mathbf{a}_\theta\mathbf{a}_r)$	$\bar{\mathbf{Z}}_e = \dfrac{\Gamma_{nm}}{jk_0}Z_0(\mathbf{a}_r\mathbf{a}_\theta - \mathbf{a}_\theta\mathbf{a}_r)$
Dyadic wave admittance	$\bar{\mathbf{Y}}_h = \dfrac{\Gamma_{nm}}{jk_0}Y_0(\mathbf{a}_\theta\mathbf{a}_r - \mathbf{a}_r\mathbf{a}_\theta)$	$\bar{\mathbf{Y}}_e = \dfrac{jk_0}{\Gamma_{nm}}Y_0(\mathbf{a}_\theta\mathbf{a}_r - \mathbf{a}_r\mathbf{a}_\theta)$
Axial magnetic field	$k_{c,nm}^2\psi_{h,nm}e^{\pm\Gamma_{nm}z}$	Zero
Axial electric field	Zero	$k_{c,nm}^2\psi_{e,nm}e^{\pm\Gamma_{nm}z}$
Transverse magnetic field	$\mathbf{H}_t = \pm\Gamma_{nm}e^{\pm\Gamma_{nm}z}\nabla_t\psi_{h,nm}$	$\mp\bar{\mathbf{Y}}_e\cdot\mathbf{E}_t$
Transverse electric field	$\mp\bar{\mathbf{Z}}_h\cdot\mathbf{H}_t$	$\mathbf{E}_t = \pm\Gamma_{nm}e^{\pm\Gamma_{nm}z}\nabla_t\psi_{e,nm}$
Power flow, Eqs. (26), (27); $\Gamma_{nm} = j\beta_{nm}$	$\dfrac{Z_0k_0\beta_{nm}\pi}{2\epsilon_{0n}}(p'^2_{nm} - n^2)J_n'^2(p'_{nm})$	$\dfrac{Y_0k_0\beta_{nm}\pi}{2\epsilon_{0n}}p_{nm}^2\left[\dfrac{dJ_n(k_cr)}{dk_cr}\right]^2_a$
Attenuation in nepers per meter, Eqs. (38), (39); $R_m = (\omega\mu/2\sigma)^{1/2}$	$\dfrac{R_m}{aZ_0}\left(1 - \dfrac{k_{c,nm}^2}{k_0^2}\right)^{-1/2}\left(\dfrac{k_{c,nm}^2}{k_0^2} + \dfrac{n^2}{p'^2_{nm} - n^2}\right)$	$\dfrac{R_m}{aZ_0}\left(1 - \dfrac{k_{c,nm}^2}{k_0^2}\right)^{-1/2}$

following two equations:

$$\nabla \times \nabla \times \bar{\mathbf{G}} - k^2\bar{\mathbf{G}} = \bar{\mathbf{I}}\delta(x - x')\delta(y - y')\delta(z - z') \tag{72a}$$

$$\nabla^2\bar{\mathbf{G}} + k^2\bar{\mathbf{G}} = -\bar{\mathbf{I}}\delta(x - x')\delta(y - y')\delta(z - z') \tag{72b}$$

subject to certain boundary conditions on the guide walls. Very often, in practice, the field can be derived from a single scalar function, and then a scalar Green's function is all that is required. For arbitrary known current distributions it is usually easier to compute the radiated field directly than to first obtain the solution for the appropriate dyadic Green's function. In this section we will consider first the scalar Green's function for H_{n0} modes in rectangular waveguides, and then a general method for the derivation of the field radiated by arbitrary current filaments in cylindrical guides.

Green's Function for H_{n0} Modes in a Rectangular Guide

Consider a uniform unit line current extending across the rectangular guide, parallel with the y axis and located at (x', z'), as in Fig. 5.8. The current is uniform along the y coordinate, and, hence, the field will not vary with y. We will use the vector potential $\mathbf{A}$ from which to derive our field components, where $\mathbf{A} = \mathbf{a}_y A_y$, and

$$\frac{\partial^2 A_y}{\partial z^2} + \frac{\partial^2 A_y}{\partial x^2} + k_0^2 A_y = -\mu_o\delta(x - x')\delta(z - z') \tag{73}$$

since the current has only a y component, and, hence, $\mathbf{A}$ has only a y component. This Green's function problem was treated in Section 2.7. When we put $E_y = -j\omega A_y$ and use (60) from Section 2.7 we obtain

$$E_y = G(x, z|x', z') = \frac{-j\omega\mu_0}{a}\sum_{n=1}^{\infty}\frac{1}{\Gamma_n}\sin\frac{n\pi x}{a}\sin\frac{n\pi x'}{a}e^{-\Gamma_n|z-z'|}. \tag{74}$$

An alternative representation in terms of an image series is given by (91) in Section 2.9. To obtain the electric field we must multiply G in that equation by $-j\omega\mu_0$. The image series is not very useful in practice since it does not converge very rapidly. In many practical problems, when this particular Green's function is encountered it is useful to sum the dominant series, obtained from (74) by setting $k_0 = 0$, into closed form as described by (69) in Section 2.7.

Green's Functions for General Cylindrical Guides

The method to be used to find the field radiated by an arbitrary current filament in a cylindrical guide is to expand the radiated field in terms of a suitable set of normal modes, and to determine the amplitude coefficients in this expansion by an application of the Lorentz reciprocity theorem. With reference to Fig. 5.9, let C represent an arbitrary infinitely thin unit current element, with the current flowing in the direction of the unit vector τ along C. Such a current filament must be maintained by some external field or source, but in the evaluation of the Green's function we are concerned only with the radiated fields, and consequently the field or source that maintains the specified current does not enter into the picture here. In calculating the fields radiated by probes and loops in waveguides, we must take into account

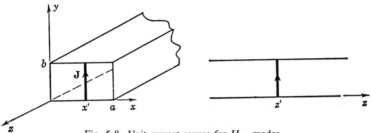

Fig. 5.8. Unit current source for H_{n0} modes.

the source that gives rise to the currents on the probe or loop antenna. The Green's functions derived here will be found useful in the analysis of such problems.

An arbitrary field in a waveguide can be represented as an infinite series of the normal modes for the guide. These modes may be chosen as the E and H modes discussed earlier or any other convenient set of modes. The field for any one mode may be expressed in terms of the transverse electric field $\mathbf{e}_n e^{\pm \Gamma_n z}$ by means of (64). For propagation in the positive z direction, let the fields for the nth mode be represented as follows:

$$\mathbf{E}_n^+ = (\mathbf{e}_n + \mathbf{e}_{zn})e^{-\Gamma_n z} \tag{75a}$$

$$\mathbf{H}_n^+ = (\mathbf{h}_n + \mathbf{h}_{zn})e^{-\Gamma_n z} \tag{75b}$$

where for E modes $\mathbf{h}_{zn}$ is zero, and for H modes $\mathbf{e}_{zn}$ is zero. For propagation along the negative z direction, the fields for the nth mode are then given by

$$\mathbf{E}_n^- = (\mathbf{e}_n - \mathbf{e}_{zn})e^{\Gamma_n z} \tag{75c}$$

$$\mathbf{H}_n^- = (-\mathbf{h}_n + \mathbf{h}_{zn})e^{\Gamma_n z}. \tag{75d}$$

In (75), $\mathbf{e}_n$ and $\mathbf{h}_n$ are transverse vector functions of the transverse coordinates, while $\mathbf{e}_{zn}$ and $\mathbf{h}_{zn}$ are axial vector functions of the transverse coordinates. It will be assumed that the normal mode functions have been normalized, so that

$$\iint_S \mathbf{e}_n \times \mathbf{h}_n \cdot \mathbf{a}_z \, dS = 1. \tag{76a}$$

The modes are orthogonal, so we also have

$$\iint_S \mathbf{e}_n \times \mathbf{h}_m \cdot \mathbf{a}_z \, dS = 0, \qquad n \neq m. \tag{76b}$$

In (76), the integration is over the guide cross section S. An examination of (22) and (23) shows that the above orthogonal property is valid for degenerate modes also.

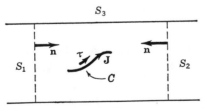

Fig. 5.9. An arbitrary current filament in a waveguide.

Let the field radiated in the positive z direction by the current filament be represented by

$$\mathbf{E}^+ = \sum_n a_n \mathbf{E}_n^+ \tag{77a}$$

$$\mathbf{H}^+ = \sum_n a_n \mathbf{H}_n^+ \tag{77b}$$

and the field radiated in the negative z direction be represented by

$$\mathbf{E}^- = \sum_n b_n \mathbf{E}_n^- \tag{78a}$$

$$\mathbf{H}^- = \sum_n b_n \mathbf{H}_n^- . \tag{78b}$$

In particular, the expansion (77) gives the radiated field on the cross-sectional plane S_2, while (78) gives the radiated field on S_1.

To determine the expansion coefficients a_n and b_n, we need the Lorentz reciprocity principle, which is valid for a region containing a current source. Let $\mathbf{E}$ and $\mathbf{H}$ be the fields radiated by the current $\mathbf{J}$. These fields are solutions of the following equations: $\nabla \times \mathbf{E} = -j\omega\mu_0\mathbf{H}$, $\nabla \times \mathbf{H} = j\omega\epsilon_0\mathbf{E} + \mathbf{J}$. The normal mode function $\mathbf{E}_n^\pm$, $\mathbf{H}_n^\pm$ is a solution of the source-free equations. Consider the expansion

$$\nabla\cdot(\mathbf{E}_n^\pm \times \mathbf{H} - \mathbf{E} \times \mathbf{H}_n^\pm) = \mathbf{H}\cdot\nabla \times \mathbf{E}_n^\pm - \mathbf{E}_n^\pm\cdot\nabla \times \mathbf{H} - \mathbf{H}_n^\pm\cdot\nabla \times \mathbf{E} + \mathbf{E}\cdot\nabla \times \mathbf{H}_n^\pm = -\mathbf{J}\cdot\mathbf{E}_n^\pm$$

when the terms involving the curl are replaced by their equivalents as obtained from Maxwell's equations. Integrating over a volume bounded by a closed surface S, and converting the volume integral of the divergence to a surface integral, we get the desired form of the Lorentz reciprocity principle:

$$\oiint_S (\mathbf{E}_n^\pm \times \mathbf{H} - \mathbf{E} \times \mathbf{H}_n^\pm)\cdot\mathbf{n}\, dS = \iiint \mathbf{J}\cdot\mathbf{E}_n^\pm\, dV$$

where $\mathbf{n}$ has been chosen as the inward-directed normal to S.

We now choose as our volume that bounded by the perfectly conducting guide walls and the two cross-sectional planes S_1 and S_2 as in Fig. 5.9. There is no contribution to the surface integral arising from the guide walls since $\mathbf{n} \times \mathbf{E}$ and $\mathbf{n} \times \mathbf{E}_n^\pm$ are zero on the guide wall. For the normal mode function $\mathbf{E}_n^+$, $\mathbf{H}_n^+$, it is readily found that

$$\iint_{S_1} (\mathbf{E}_n^+ \times \mathbf{H} - \mathbf{E} \times \mathbf{H}_n^+)\cdot\mathbf{a}_z\, dS = -2b_n \tag{79a}$$

$$-\iint_{S_2} (\mathbf{E}_n^+ \times \mathbf{H} - \mathbf{E} \times \mathbf{H}_n^+)\cdot\mathbf{a}_z\, dS = a_n e^{-2\Gamma_n z_2} - a_n e^{-2\Gamma_n z_2} = 0 \tag{79b}$$

when the expansion (77) and (78) for $\mathbf{H}, \mathbf{E}$ and the orthogonal property (76b) are used. Consequently, (79) gives

$$2b_n = -\int_C \boldsymbol{\tau}\cdot\mathbf{E}_n^+\, dl \tag{80}$$

since it is assumed that the total current is of unit magnitude and concentrated in an infinitely thin filament. The integral in (80) may be regarded as the voltage impressed along C by the normal mode field $\mathbf{E}_n^+$, and, because of the reciprocity principle which holds for the field, the current generates a mode with this same amplitude. If we begin with the normal mode function $\mathbf{E}_n^-$, $\mathbf{H}_n^-$ instead, we find in a similar manner that the coefficient a_n is given by

$$2a_n = -\int_C \boldsymbol{\tau} \cdot \mathbf{E}_n^- \, dl. \tag{81}$$

If the current is a function of the distance along C, the integrands in (80) and (81) must be replaced by $I(l)\boldsymbol{\tau} \cdot \mathbf{E}_n^{\pm}$, where $I(l)$ is the total current along C.

If $\boldsymbol{\tau}$ lies in a transverse plane (80) and (81) show that $a_n = b_n$, and, consequently, the transverse electric field (analogous to voltage) is continuous at the current filament, and the current filament behaves as a shunt source. If $\boldsymbol{\tau}$ is along the z direction, $a_n = -b_n$, the transverse magnetic field (analogous to current) is continuous, and the current element acts as a series source.

If C is a small closed loop, we may modify (80) and (81) as follows:

$$2b_n = -\oint_C \mathbf{E}_n^+ \cdot \boldsymbol{\tau} \, dl = -\iint_{S_0} \nabla \times \mathbf{E}_n^+ \cdot d\mathbf{S}$$

$$= j\omega\mu_0 \iint_{S_0} \mathbf{H}_n^+ \cdot d\mathbf{S}$$

$$2a_n = j\omega\mu_0 \iint_{S_0} \mathbf{H}_n^- \cdot d\mathbf{S}$$

where S_0 is the surface spanning the closed contour C. The integrals represent the negative induced voltage in the loop C arising from the rate of change of the magnetic flux of the nth normal mode through C. Let the induced voltages be

$$V_n^+ = -j\omega \iint_{S_0} \mu_0 \mathbf{H}_n^+ \cdot d\mathbf{S} \tag{82a}$$

and

$$V_n^- = -j\omega \iint_{S_0} \mu_0 \mathbf{H}_n^- \cdot d\mathbf{S}; \tag{82b}$$

we may write

$$2b_n = -V_n^+ \tag{83a}$$

$$2a_n = -V_n^-. \tag{83b}$$

For an infinitesimal loop having a vector area $\mathbf{A}$ and carrying a current I, the magnetic moment $\mathbf{m}$ is given by

$$\mathbf{m} = I\mathbf{A}. \tag{84}$$

Introducing this concept we find that we may also write

$$2b_n = j\omega\mu_0\mathbf{H}_n^+\cdot\mathbf{m} \tag{85a}$$

$$2a_n = j\omega\mu_0\mathbf{H}_n^-\cdot\mathbf{m}. \tag{85b}$$

If the field $\mathbf{H}_n^{\pm}$ cannot be considered constant over the area of the loop, (85) must be replaced by an integral as in (82). These latter results are useful for finding the Green's functions from magnetic dipole sources. Using (75) we get

$$2b_n = j\omega\mu_0(\mathbf{h}_n + \mathbf{h}_{zn})\cdot\mathbf{m}e^{-\Gamma_n z} \tag{86a}$$

$$2a_n = j\omega\mu_0(-\mathbf{h}_n + \mathbf{h}_{zn})\cdot\mathbf{m}e^{\Gamma_n z} \tag{86b}$$

for an infinitesimal current loop of moment $\mathbf{m}$. When $\mathbf{m}$ is in the z direction, the loop appears as a shunt source, but, if $\mathbf{m}$ lies in a transverse plane, it appears as a series source. The Green's function is the expansion (77) and (78), with the coefficients a_n and b_n given by the appropriate expression above. In an empty guide or a homogeneously filled one, a current element in the z direction excites only E modes, while a magnetic dipole in the z direction (a current loop in the transverse plane) excites only H modes.

The dyadic Green's function is the operator which gives the electric field when the scalar product $-j\omega\mu_0\bar{\mathbf{G}}_e(\mathbf{r}, \mathbf{r}')\cdot\mathbf{J}(\mathbf{r}')$ is formed. From the results obtained above it is clear that

$$\bar{\mathbf{G}}_e(\mathbf{r}, \mathbf{r}') = \frac{1}{2j\omega\mu_0}\sum_{n=1}^{\infty}\mathbf{E}_{n\sigma}(\mathbf{r})\mathbf{E}_{n\sigma}(\mathbf{r}'), \qquad r \neq r' \tag{87}$$

where

$$\mathbf{E}_{n\sigma}(\mathbf{r}) = \begin{cases} \mathbf{E}_n^+(\mathbf{r}), & z > z' \\ \mathbf{E}_n^-(\mathbf{r}), & z < z' \end{cases}$$

$$\mathbf{E}_{n\sigma}(\mathbf{r}') = \begin{cases} \mathbf{E}_n^+(\mathbf{r}'), & z' > z \\ \mathbf{E}_n^-(\mathbf{r}'), & z' < z. \end{cases}$$

This modal expansion for the electric field dyadic Green's function is valid outside the source region. If the electric field inside the source region is to be found the term

$$\frac{-\mathbf{a}_z\mathbf{a}_z}{k_0^2}\delta(\mathbf{r} - \mathbf{r}')$$

should be added to the right-hand side of (87) [see (184) in Section 2.16].

The above discussion has been restricted to the case of an empty waveguide. The extension to the case of a homogeneously or inhomogeneously filled guide is obtained by using the appropriate normal mode functions for such a guide. It should also be kept in mind that, even though a given current filament may be in a single direction, say along the z axis, it may be necessary to use a vector potential with both a z component and a transverse component, in order to obtain a field which will have the proper continuity across surfaces along which the electrical parameters of the medium change discontinuously. When the method of expansion in terms of normal mode functions is used, such discontinuity surfaces are taken care of by the normal mode functions that satisfy the appropriate continuity conditions.

Eigenfunction Expansion of the Dyadic Green's Function

For some problems it is important to have the complete eigenfunction expansion of the dyadic Green's function. Outside of the source region the eigenfunction expansion is equivalent to the mode expansion given in (87). The eigenfunction expansion can be developed in a manner analogous to that presented in Section 2.16. The $\mathbf{M}$ and $\mathbf{N}$ functions correspond to the H and E modes, respectively. In a waveguide we will need to introduce three scalar functions, all solutions of the scalar Helmholtz equation, from which the transverse $\mathbf{M}$ and $\mathbf{N}$ modes and the longitudinal $\mathbf{L}$ modes can be constructed. We can take a Fourier transform with respect to z which is equivalent to choosing a z dependence of e^{-jwz} for the scalar eigenfunctions. We will choose the following scalar functions:

$$\psi_{en}(u_1, u_2)e^{-jwz}$$

$$\psi_{hn}(u_1, u_2)e^{-jwz}$$

$$\psi_{\ell n}(u_1, u_2)e^{-jwz}$$

where u_1, u_2 represent orthogonal curvilinear coordinates for the waveguide cross section and n is a summation integer. The function ψ_{en} is a solution of

$$(\nabla^2 + \lambda_n)\psi_{en}e^{-jwz} = 0$$

or

$$(\nabla_t^2 + p_n^2)\psi_{en} = 0, \qquad p_n^2 = \lambda_n - w^2$$

with the boundary condition $\psi_{en} = 0$ on C, the waveguide boundary. The vector eigenfunction $\mathbf{N}_n$ is given by

$$\sqrt{\lambda_n}\mathbf{N}_n = \nabla \times \nabla \times \mathbf{a}_z\psi_{en}e^{-jwz}.$$

Solutions for ψ_{en} exist for a double infinite sequence of discrete values of p_n, but for convenience we will use a single index n to represent all eigenvalues. For example, in a rectangular waveguide

$$\psi_{enm} = \sin\frac{n\pi x}{a} \sin\frac{m\pi y}{b}$$

and the two sets of integers n, m can be ordered into a single set in any convenient way. The function ψ_{hn} is a solution of

$$(\nabla_t^2 + q_n^2)\psi_{hn} = 0$$

with the boundary condition $\partial\psi_{hn}/\partial n = 0$ on C. From this function we obtain the $\mathbf{M}$-type vector eigenfunction as follows[4]:

$$\mathbf{M}_n = \nabla \times \mathbf{a}_z\psi_{hn}e^{-jwz}.$$

---

[4]The solution $\psi_{h0} = $ constant, $q_0 = 0$, is not needed in the expansion of the electric field Green's function. In a multiple connected waveguide such as a coaxial transmission line, the solution ψ_{e0} with $p_0 = 0$ and with boundary conditions ψ_{e0} equal to different constant values on the two conductors is needed to generate the TEM mode. For the magnetic field Green's function expansion the subscripts e and h and the boundary conditions are interchanged. For this expansion the solution $\psi_{l0} = $ constant, $l_0 = 0$, is required as part of the $\mathbf{L}_n$ set. For a coaxial transmission line ψ_{e0}, $q_0 = 0$ is needed as part of the $\mathbf{M}_n$ set to obtain the TEM mode.

The $\mathbf{L}_n$ functions are generated from $\psi_{\ell n}$ where

$$(\nabla_t^2 + \ell_n^2)\psi_{\ell n} = 0$$

with the boundary condition $\psi_{\ell n} = 0$ on C. We note that $\psi_{\ell n}$ must be equal to ψ_{en} so the latter can be used to generate the $\mathbf{L}_n$ using

$$\mathbf{L}_n = \nabla\psi_{en}e^{-jwz}.$$

It is readily verified that the $\mathbf{M}_n$, $\mathbf{N}_n$, and $\mathbf{L}_n$ functions are all mutually orthogonal when integrated over all values of u_1, u_2 and z. The normalization integral for the $\mathbf{L}_n$ functions is

$$\iint\limits_S \int_{-\infty}^{\infty} \nabla\psi_{en}\, e^{-jwz} \cdot \nabla\psi_{en}\, e^{jw'z}\, dz\, dS = 2\pi\delta(w - w')\iint\limits_S [\nabla_t\psi_{en}\cdot\nabla_t\psi_{en} + w^2\psi_{en}^2]\, dS.$$

We now use $\nabla_t\cdot(\psi_{en}\,\nabla_t\psi_{en}) = \nabla_t\psi_{en}\cdot\nabla_t\psi_{en} + \psi_{en}\,\nabla_t^2\psi_{en}$ to get

$$2\pi\delta(w - w')\left[\oint_C \psi_{en}\frac{\partial\psi_{en}}{\partial n}\, dl + (p_n^2 + w^2)\iint\limits_S \psi_{en}^2\, dS\right].$$

The contour integral is zero and we will assume that the ψ_{en} are normalized so that

$$\iint\limits_S \psi_{en}^2\, dS = 1.$$

We now have

$$\iint\limits_S \int_{-\infty}^{\infty} \mathbf{L}_n(\mathbf{r}, w)\cdot\mathbf{L}_n(\mathbf{r}, -w')\, dz\, dS = 2\pi\delta(w - w')(p_n^2 + w^2).$$

The normalization integral for the $\mathbf{N}$ functions involves $\sqrt{\lambda_n}\mathbf{N}_n = (\nabla\nabla\cdot - \nabla^2)\mathbf{a}_z\psi_{en}e^{-jwz} = [-jw\,\nabla_t\psi_{en} + p_n^2\mathbf{a}_z\psi_{en}]e^{-jwz}$. The normalization is thus

$$\frac{1}{p_n^2 + w^2}\iint\limits_S \int_{-\infty}^{\infty} [ww'\,\nabla_t\psi_{en}\cdot\nabla_t\psi_{en} + p_n^4\psi_{en}^2]e^{-j(w-w')z}\, dz\, dS$$

$$= \iint\limits_S \int_{-\infty}^{\infty} \mathbf{N}_n(\mathbf{r}, w)\cdot\mathbf{N}_n(\mathbf{r}, -w')\, dz\, dS = 2\pi p_n^2\delta(w - w').$$

The normalization integral for the $\mathbf{M}_n$ functions is

$$\iint\limits_S \int_{-\infty}^{\infty} \mathbf{M}_n(\mathbf{r}, w)\cdot\mathbf{M}_n(\mathbf{r}, -w')\, dz\, dS = 2\pi\delta(w - w')\iint\limits_S \nabla_t\psi_{hn}\cdot\nabla_t\psi_{hn}\, dS$$

$$= 2\pi\delta(w - w')\iint\limits_S q_n^2\psi_{hn}^2\, dS = 2\pi q_n^2\delta(w - w')$$

upon using the same vector integration by parts that was used in the $\mathbf{L}_n$ normalization integral and choosing the normalization of the ψ_{hn} such that

$$\iint_S \psi_{hn}^2 \, dS = 1.$$

The solution of (72a) for the dyadic Green's function is now chosen as the following expansion:

$$\bar{\mathbf{G}}_e(\mathbf{r}, \mathbf{r}') = \sum_n \int_{-\infty}^{\infty} [\mathbf{M}_n(\mathbf{r}, w')\mathbf{A}_n + \mathbf{N}_n(\mathbf{r}, w')\mathbf{B}_n + \mathbf{L}_n(\mathbf{r}, w')\mathbf{C}_n] \, dw'$$

where $\mathbf{A}_n, \mathbf{B}_n, \mathbf{C}_n$ are vector coefficients. By following the standard procedure of substituting this expansion into (72a), scalar multiplying both sides by $\mathbf{M}_n(\mathbf{r}, -w)$, $\mathbf{N}_n(\mathbf{r}, -w)$, and $\mathbf{L}_n(\mathbf{r}, -w)$ in turn, and integrating over the total volume, we obtain

$$\bar{\mathbf{G}}_e(\mathbf{r}, \mathbf{r}') = \frac{1}{2\pi} \sum_n \int_{-\infty}^{\infty} \left[\frac{\mathbf{M}_n(\mathbf{r}, w)\mathbf{M}_n(\mathbf{r}', -w)}{q_n^2(q_n^2 + w^2 - k_0^2)} \right.$$

$$\left. + \frac{\mathbf{N}_n(\mathbf{r}, w)\mathbf{N}_n(\mathbf{r}', -w)}{p_n^2(p_n^2 + w^2 - k_0^2)} - \frac{\mathbf{L}_n(\mathbf{r}, w)\mathbf{L}_n(\mathbf{r}', -w)}{k_0^2(p_n^2 + w^2)} \right] dw. \quad (88)$$

In deriving (88) we have used the orthogonal property of the vector eigenfunctions, the normalization integrals, and the relations

$$\nabla \times \nabla \times \mathbf{N}_n = (p_n^2 + w^2)\mathbf{N}_n$$

$$\nabla \times \nabla \times \mathbf{M}_n = (q_n^2 + w^2)\mathbf{M}_n.$$

The mode expansion can be recovered by evaluating the integral over w. Residue theory can be used except for the $\mathbf{a}_z\mathbf{a}_z$ contribution from the $\mathbf{L}_n$ functions. This term involves the integral

$$\frac{1}{2\pi} \int_{-\infty}^{\infty} \frac{w^2}{k_0^2(w^2 + p_n^2)} e^{-jw(z-z')} \, dw = \frac{1}{2\pi k_0^2} \int_{-\infty}^{\infty} e^{-jw(z-z')} \, dw - \frac{p_n^2}{2\pi k_0^2} \int_{-\infty}^{\infty} \frac{e^{-jw(z-z')}}{w^2 + p_n^2} \, dw.$$

The first integral on the right-hand side gives $\delta(z - z')/k_0^2$ and since

$$\sum_n \psi_{en}(u_1, u_2)\psi_{en}(u_1', u_2') = -\frac{\delta(u_1 - u_1')\delta(u_2 - u_2')}{h_1 h_2}$$

where h_1 and h_2 are metric coefficients, we obtain the expected $-\mathbf{a}_z\mathbf{a}_z\delta(\mathbf{r}-\mathbf{r}')/k_0^2$ contribution. The residues at the poles $w = \pm p_n$ cancel the corresponding residues coming from the $\mathbf{N}_n$ contribution to $\bar{\mathbf{G}}_e$. Note that the product $\mathbf{N}_n(\mathbf{r}, w)\mathbf{N}_n(\mathbf{r}, -w)$ has a factor $p_n^2 + w^2$ in the denominator. The residues at the poles $w = \pm(k_0^2 - p_n^2)^{1/2}$ and $w = \pm(k_0^2 - q_n^2)^{1/2}$ give the expected E and H mode contributions outside the source region.

For the rectangular waveguide the normalized scalar functions are

$$\psi_{\ell nm} = \psi_{enm} = \frac{2}{\sqrt{ab}} \sin \frac{n\pi x}{a} \sin \frac{m\pi y}{b} \tag{89a}$$

$$\psi_{hnm} = \sqrt{\frac{\epsilon_{0n}\epsilon_{0m}}{ab}} \cos \frac{n\pi x}{a} \cos \frac{m\pi y}{b}. \tag{89b}$$

For the circular waveguide the corresponding functions are

$$\psi_{\ell nm} = \psi_{enm} = \sqrt{\frac{\epsilon_{0n}}{\pi}} \frac{J_n(p_{nm}r/a)}{aJ_n'(p_{nm})} \frac{\cos}{\sin} n\phi \tag{89c}$$

$$\psi_{hnm} = \sqrt{\frac{\epsilon_{0n}}{\pi}} \frac{J_n(p_{nm}'r/a)}{a(1 - n^2/p_{nm}'^2)^{1/2}J_n(p_{nm}')} \frac{\cos}{\sin} n\phi. \tag{89d}$$

The generic form for the dyadic Green's function in terms of a mode expansion is

$$\bar{G}_e(\mathbf{r}, \mathbf{r}') = -\frac{\mathbf{a}_z\mathbf{a}_z}{k_0^2}\delta(\mathbf{r} - \mathbf{r}') + \sum_n \left[\frac{\mathbf{M}_n(\mathbf{r}, \mp j\Gamma_n)\mathbf{M}_n(\mathbf{r}', \pm j\Gamma_n)}{2\Gamma_n q_n^2} \right.$$

$$\left. + \frac{\mathbf{N}_n(\mathbf{r}, \mp j\gamma_n)\mathbf{N}_n(\mathbf{r}', \pm j\gamma_n)}{2\gamma_n p_n^2} \right] \tag{90}$$

where the upper signs apply for $z > z'$ and the lower signs apply for $z < z'$. The propagation constants are given by

$$\gamma_n = (p_n^2 - k_0^2)^{1/2}, \qquad E \text{ modes} \tag{91a}$$

$$\Gamma_n = (q_n^2 - k_0^2)^{1/2}, \qquad H \text{ modes} \tag{91b}$$

For the rectangular waveguide, $p_n^2 = q_n^2 = (n\pi/a)^2 + (m\pi/b)^2$. For the circular waveguide, $p_n = p_{nm}/a$ and $q_n = p_{nm}'/a$. The mode functions are given by

$$\mathbf{M}_n(\mathbf{r}, w) = \nabla \times \mathbf{a}_z\psi_{hn}e^{-jwz} \tag{91c}$$

$$k_0\mathbf{N}_n(r, w) = \nabla \times \nabla \times \mathbf{a}_z\psi_{en}e^{-jwz} \tag{91d}$$

with w replaced by $\pm j\Gamma_n$ or $\pm j\gamma_n$ as indicated. The q_n and p_n correspond to the respective cutoff wavenumbers k_{cn} for the H and E modes.

If the integral over w in (88) is not completed, then one of the series in (88) can be summed in closed form to yield an alternative representation for the Green's dyadic function. For example, in a circular waveguide the double series involved for the $\mathbf{N}_n$ functions, before the curl operations are carried out, is

$$\sum_{n=0}^{\infty} \sum_{m=1}^{\infty} \frac{J_n(p_{nm}r/a)J_n(p_{nm}r'/a)}{p_{nm}^2(p_{nm}^2 + w^2a^2)[p_{nm}^2 + (w^2 - k_0^2)a^2][J_n'(p_{nm})]^2} \times \begin{pmatrix} \cos n\phi & \cos n\phi' \\ \sin n\phi & \sin n\phi' \end{pmatrix}.$$

By using a partial fraction decomposition the series to be summed over m becomes

$$S = \sum_{m=1}^{\infty} \frac{J_n(p_{nm}r/a)J_n(p_{nm}r'/a)}{a^4[J_n'(p_{nm})]^2} \left[\frac{1}{w^2(w^2 - k_0^2)} \frac{1}{p_{nm}^2} \right.$$

$$\left. + \frac{1}{k_0^2 w^2} \frac{1}{(p_{nm}^2 + w^2 a^2)} - \frac{1}{k_0^2(w^2 - k_0^2)} \frac{1}{[p_{nm}^2 + (w^2 - k_0^2)a^2]} \right]. \quad (92)$$

Note that for the circular guide the parameter p_n in (88), after multiplication by a, is equal to p_{nm} which is the Bessel function root. Also the first and last series cancel when $w^2 \to k_0^2$ so $w = \pm k_0$ are not poles of S and neither is $w = 0$ a pole.

Consider the one-dimensional Green's function problem

$$\frac{1}{r} \frac{d}{dr} r \frac{dg_n}{dr} - \frac{n^2}{r^2} g_n + p^2 g_n = -\frac{\delta(r - r')}{r'}$$

with the boundary condition $g_n(a) = 0$. We can solve for g_n using Method I or II as described in Section 2.4. Thus we have

$$g_n(r) = \sum_{m=1}^{\infty} 2 \frac{J_n(p_{nm}r_</a)J_n(p_{nm}r_>/a)}{(p_{nm}^2 - p^2 a^2)[J_n'(p_{nm})]^2}$$

$$= \frac{\pi}{2J_n(pa)} \{J_n(pr_<)[J_n(pr_>)Y_n(pa) - J_n(pa)Y_n(pr_>)]\}. \quad (93)$$

By using this result we are able to express the original series S in closed form by identifying p with zero, jw, or $(k_0^2 - w^2)^{1/2}$. For the first series in S we must take the limit as p approaches zero to obtain

$$\lim_{p \to 0} g_n(r) = -\sum_{m=1}^{\infty} \frac{2}{a^2} \frac{J_n(p_{nm}r_</a)J_n(p_{nm}r_>/a)}{p_{nm}^2[J_n'(p_{nm})]^2}$$

$$= -\frac{1}{2n} \left(\frac{r_<}{a}\right)^n \left[\left(\frac{r_>}{a}\right)^n - \left(\frac{a}{r_>}\right)^n\right]. \quad (94)$$

When one of the series in (88) is summed it is usually not possible to evaluate the integral over w analytically.

The technique outlined above can be applied in general to sum one of the series in the dyadic Green's function as given by (88) into closed form. The method also provides sums for some interesting series that occur in many boundary-value problems. The one-dimensional Green's function problem for a rectangular waveguide can also be solved in terms of an image series. Thus a variety of representations for the dyadic Green's function are possible.

5.7. ANALOGY WITH TRANSMISSION LINES

The similarity between the normal mode solutions in a waveguide and the TEM mode on a transmission line permits us to set up a system of equations representing the fields in a waveguide which is formally identical with the current and voltage relations on a system of transmission lines.

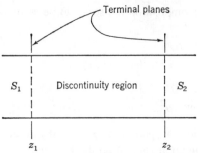

Fig. 5.10. Terminal planes in a guide containing a discontinuity.

With reference to Fig. 5.10, let the region between the two arbitrarily chosen reference planes S_1 and S_2 be a region in which a localized discontinuity (such as a post or diaphragm) exists. If it is assumed (this assumption is not necessary and is made only to simplify the development of the basic ideas involved) that the input and output guides are the same, the following expansion for the transverse fields in the guide may be written.

For $z \leq z_1$, the transverse electric and magnetic fields are

$$\sum_{n=1}^{N} a_n \mathbf{e}_n e^{-\Gamma_n(z-z_1)} + \sum_{n=1}^{\infty} b_n \mathbf{e}_n e^{\Gamma_n(z-z_1)} \tag{95a}$$

$$\sum_{n=1}^{N} a_n \mathbf{h}_n e^{-\Gamma_n(z-z_1)} - \sum_{n=1}^{\infty} b_n \mathbf{h}_n e^{\Gamma_n(z-z_1)} \tag{95b}$$

while, for $z \geq z_2$, the corresponding fields are

$$\sum_{n=1}^{M} c_n \mathbf{e}_n e^{\Gamma_n(z-z_2)} + \sum_{n=1}^{\infty} d_n \mathbf{e}_n e^{-\Gamma_n(z-z_2)} \tag{96a}$$

$$-\sum_{n=1}^{M} c_n \mathbf{h}_n e^{\Gamma_n(z-z_2)} + \sum_{n=1}^{\infty} d_n \mathbf{h}_n e^{-\Gamma_n(z-z_2)}. \tag{96b}$$

In the above expressions, the finite sums represent incident modes, while the infinite sums represent reflected modes, which are set up by the discontinuity, or transmitted past the discontinuity because of modes incident from the other side. The normal mode functions $\mathbf{e}_n$ and $\mathbf{h}_n$ are transverse vector functions of the transverse coordinates. These normal mode functions may be E and H modes or suitable combinations of E and H modes. The normal mode functions $\mathbf{e}_n$ and $\mathbf{h}_n$ are related by means of the dyadic wave impedance $\bar{Z}_n$ or dyadic wave admittance $\bar{\mathbf{Y}}_n$ for the mode as follows:

$$\mathbf{e}_n = \bar{Z}_n \cdot \mathbf{h}_n \tag{97a}$$

$$\mathbf{h}_n = \bar{\mathbf{Y}}_n \cdot \mathbf{e}_n \tag{97b}$$

where

$$\bar{Z}_n = Z_n(\mathbf{a}_x \mathbf{a}_y - \mathbf{a}_y \mathbf{a}_x)$$

$$\bar{\mathbf{Y}}_n = Y_n(\mathbf{a}_y \mathbf{a}_x - \mathbf{a}_x \mathbf{a}_y)$$

and Z_n and Y_n are the scalar wave impedance and admittance, respectively. If the following equivalent transmission-line voltages and currents are introduced:

$$\left.\begin{aligned} V_{n1}^+ &= a_n \\ V_{n1}^- &= b_n \\ I_{n1}^+ &= Y_n a_n = Y_n V_{n1}^+ \\ I_{n1}^- &= Y_n b_n = Y_n V_{n1}^- \end{aligned}\right\} \quad z \le z_1 \tag{98a}$$

$$\left.\begin{aligned} V_{n2}^+ &= c_n \\ V_{n2}^- &= d_n \\ I_{n2}^+ &= -Y_n c_n = -Y_n V_{n2}^+ \\ I_{n2}^- &= -Y_n d_n = -Y_n V_{n2}^- \end{aligned}\right\} \quad z \ge z_2 \tag{98b}$$

the following set of relations may be written in place of those given in (95) and (96):

$$\sum_{n=1}^N V_{n1}^+ e^{-\Gamma_n(z-z_1)} + \sum_{n=1}^\infty V_{n1}^- e^{\Gamma_n(z-z_1)} \tag{99a}$$

$$\sum_{n=1}^N I_{n1}^+ e^{-\Gamma_n(z-z_1)} - \sum_{n=1}^\infty I_{n1}^- e^{\Gamma_n(z-z_1)} \tag{99b}$$

$$\sum_{n=1}^M V_{n2}^+ e^{\Gamma_n(z-z_2)} + \sum_{n=1}^\infty V_{n2}^- e^{-\Gamma_n(z-z_2)} \tag{99c}$$

$$\sum_{n=1}^M I_{n2}^+ e^{\Gamma_n(z-z_2)} - \sum_{n=1}^\infty I_{n2}^- e^{-\Gamma_n(z-z_2)}. \tag{99d}$$

When the currents and voltages are known, we may obtain the expression for the fields at the terminal planes by means of (95) to (98). The above equations are formally the same as those which would be used to describe the behavior of a junction of transmission lines (infinite in number), with waves incident on N lines for $z < z_1$, and on M lines for $z > z_2$. The currents and voltages on one side may be related to those on the other side by a suitable impedance, admittance, scattering, or other matrix. Each of these matrices may be used to define an equivalent lumped-parameter network to represent the discontinuity.

The usual situation, in practice, is the one where the frequency of operation is chosen so that only the dominant mode propagates. It is then convenient to choose terminal planes sufficiently far removed from the discontinuity so that all evanescent modes have decayed to a negligible value at these planes. Even when several modes propagate, it is convenient to do this, since it is then possible to describe the effect of the discontinuity on the propagating modes by an equivalent transmission-line circuit, which couples together the various transmission lines used to represent each propagating mode. It should also be noted that other definitions of the equivalent voltages and currents may be chosen in place of those given by (98). It is not necessary that $\frac{1}{2}V_n^+I_n^{+*}$ represent the power carried by the mode. However, it is necessary that the constant of proportionality between the power carried by each mode and that carried by its equivalent transmission line be the same for all modes, if there is to be conservation of power and energy in the equivalent circuit.

In the discussion to follow, it will be assumed that only one mode propagates, and that the terminal planes are chosen far enough from the discontinuity so that all evanescent modes are of negligible amplitude at these planes. In place of (99), we now have

$$
\begin{aligned}
V_1^+ e^{-\Gamma(z-z_1)} + V_1^- e^{\Gamma(z-z_1)}, \\
I_1^+ e^{-\Gamma(z-z_1)} + I_1^- e^{\Gamma(z-z_1)},
\end{aligned}
\qquad z \le z_1
\qquad (100\text{a})
$$

$$
\begin{aligned}
V_2^+ e^{\Gamma(z-z_2)} + V_2^- e^{-\Gamma(z-z_2)}, \\
I_2^+ e^{\Gamma(z-z_2)} - I_2^- e^{-\Gamma(z-z_2)},
\end{aligned}
\qquad z \ge z_2
\qquad (100\text{b})
$$

where the subscript n has been dropped (if $n = 1$ is the propagating mode, this would be a common subscript on all quantities and may be dropped for convenience). In view of the linearity of Maxwell's equations, the relationship between the equivalent voltages and currents at the terminal planes on the input and output sides is a linear one. Thus, we may write

$$
V_1 = V_1^+ + V_1^- = I_1 Z_{11} + I_2 Z_{12}
\qquad (101\text{a})
$$

$$
V_2 = V_2^+ + V_2^- = I_1 Z_{21} + I_2 Z_{22}
\qquad (101\text{b})
$$

where V_1 is the total voltage at $z = z_1$, which is the location of the terminal plane on the input side, V_2 is the total voltage at the terminal plane located at $z = z_2$ on the output side, I_1 is the total current $I_1^+ - I_1^-$, and I_2 is the total current $I_2^+ - I_2^-$, again measured at the respective terminal planes. In (101), the parameters $Z_{11}, Z_{12}, Z_{21}, Z_{22}$ may be identified with an equivalent-T-network representation of the discontinuity. For a reciprocal structure, $Z_{12} = Z_{21}$, and the equivalent circuit is as illustrated in Fig. 5.11.

If the following reflection and transmission coefficients are introduced,

$$
R_1 = \left. \frac{V_1^-}{V_1^+} \right|_{V_2^+=0}
\qquad (102\text{a})
$$

$$
R_2 = \left. \frac{V_2^-}{V_2^+} \right|_{V_1^+=0}
\qquad (102\text{b})
$$

$$
T_{12} = \left. \frac{V_2^-}{V_1^+} \right|_{V_2^+=0}
\qquad (102\text{c})
$$

$$
T_{21} = \left. \frac{V_1^-}{V_2^+} \right|_{V_1^+=0}
\qquad (102\text{d})
$$

we may write in place of (101) the equations below, which are valid when $V_2^+ = 0$,

$$
1 + R_1 = Y_1(1 - R_1)Z_{11} - Y_1 T_{12} Z_{12}
\qquad (103\text{a})
$$

$$
T_{12} = Y_1(1 - R_1)Z_{12} - Y_1 T_{12} Z_{22}
\qquad (103\text{b})
$$

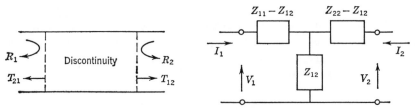

Fig. 5.11. Equivalent circuit for a waveguide discontinuity.

since $I_1^\pm = Y_1 V_1^\pm$ and $I_2^- = -Y_1 V_2^-$. Solving these equations for R_1 and T_{12} gives

$$R_1 = \frac{Y_1(Z_{11}Z_{22} - Z_{12}^2) + Z_{11} - (Z_1 + Z_{22})}{Y_1(Z_{11}Z_{22} - Z_{12}^2) + Z_{11} + (Z_1 + Z_{22})} \tag{104a}$$

$$T_{12} = \frac{2Z_{12}}{Y_1(Z_{11}Z_{22} - Z_{12}^2) + Z_{11} + (Z_1 + Z_{22})}. \tag{104b}$$

Expressions for R_2 and T_{21} may be obtained by considering the special case $V_1^+ = 0$ and $V_2^+ \neq 0$, and are

$$R_2 = \frac{Y_1(Z_{11}Z_{22} - Z_{12}^2) + Z_{22} - (Z_1 + Z_{11})}{Y_1(Z_{11}Z_{22} - Z_{12}^2) + Z_{22} + (Z_1 + Z_{11})} \tag{105a}$$

$$T_{21} = T_{12}. \tag{105b}$$

The equality of T_{12} and T_{21} follows because $Z_{12} = Z_{21}$, and the wave impedance for the mode is the same on both sides of the discontinuity. If the wave impedance on the input side were Z_1, and on the output side Z_2, then the equation we would obtain in place of (105b) is

$$T_{12}' = T_{12} \left(\frac{Z_1}{Z_2}\right)^{1/2} = T_{21}' = T_{21} \left(\frac{Z_2}{Z_1}\right)^{1/2} \tag{106}$$

where T_{12}' and T_{21}' are normalized transmission coefficients. The reflection coefficients are referred to their respective terminal planes, while the transmission coefficients represent transmission from one terminal plane to the other. The expressions for R_1, R_2, and T_{12} must be modified when the input and output transmission lines do not have the same characteristic impedance.

The input impedance at the terminal plane at $z = z_1$ is

$$Z_{\text{in}} = \frac{V_1}{I_1} = \frac{1 + R_1}{1 - R_1} Z_1 = \frac{(Z_{11}Z_{22} - Z_{12}^2) + Z_1 Z_{11}}{Z_1 + Z_{22}} \tag{107}$$

a result which may be derived from (104a) or from the equivalent circuit in Fig. 5.11 by the usual methods of circuit analysis. If the discontinuity is lossless, no power is absorbed in the circuit elements Z_{ij}, and, hence, they must be pure-imaginary elements. Also, for a lossless discontinuity, $|R_1| = |R_2|$ and $1 - |R_1|^2 = |T_{12}|^2$. From (104a) and (105b), we get $(R_1 - R_2)/T_{12} = (Z_{11} - Z_{22})/Z_{12}$, and thus, when $R_1 = R_2$, the equivalent network is symmetrical; that is, $Z_{11} = Z_{22}$. For a lossless network, the right-hand side of the above relation is real, and, hence, the left-hand side must be real also. Let $R_1 = Re^{j\theta_1}$, $R_2 = Re^{j\theta_2}$, and $T_{12} = (1 - R^2)^{1/2}e^{j\phi}$. It is necessary for $\sin(\theta_2 - \phi)$ to equal $\sin(\theta_1 - \phi)$ if $(R_1 - R_2)/T_{12}$ is to be real. Thus we get $\theta_2 - \phi = \theta_1 - \phi + 2n\pi$ or $\theta_2 - \phi = -(\theta_1 - \phi) + (2n + 1)\pi$, and, hence, $\theta_2 = \theta_1 + 2n\pi$, or $\theta_2 = -\theta_1 + 2\phi + (2n + 1)\pi$. Since R_i is invariant to a change in its phase by $2n\pi$ radians, the solution $\theta_1 = \theta_2$ applies to a symmetrical structure only. For a nonsymmetrical lossless structure,

$$R_2 = Re^{j\theta_2} = -R_1 e^{j2(\phi - \theta_1)} \tag{108a}$$

$$2\angle T_{12} = \angle R_1 + \angle R_2 \pm \pi. \tag{108b}$$

The above theory may be extended to cover the case of the junction of more than two wave-guides as well. If there are two or more propagating modes, a transmission line is introduced for each mode, and the discontinuity is now represented by a multiterminal network coupling together the various transmission lines. The equivalent-circuit representation of a discontinuity neglects all knowledge of the detailed structure of the field in the vicinity of the discontinuity, and contents itself with a knowledge of the fields at the terminal planes only. On the other hand, the parameters of the equivalent circuit can be determined only by a detailed solution of Maxwell's field equations at the discontinuity, or by measurements of the fields at the terminal planes. The analytical approach will be covered in later chapters.

A further insight into the relationship between the field and the equivalent circuit may be obtained by an application of the complex Poynting vector result given in Section 1.3. If the complex Poynting vector is integrated over the closed surface consisting of the two terminal planes S_1, S_2 and the perfectly conducting guide wall, we get

$$\frac{1}{2}a_1 a_1^*(1 + R_1)(1 - R_1^*)\iint_{S_1} \mathbf{e}_1 \times \mathbf{h}_1^* \cdot d\mathbf{S}$$

$$= 2j\omega(W_m - W_e) + P_L + \frac{1}{2}a_1 a_1^* T_{12} T_{12}^* \iint_{S_2} \mathbf{e}_1 \times \mathbf{h}_1^* \cdot d\mathbf{S} \tag{109}$$

where it is assumed that we have a guide in which only one mode propagates; that the only field incident on the discontinuity is the $n = 1$ mode from the region $z < z_1$; and that W_m, W_e, P_L are, respectively, the time-average magnetic energy, electric energy, and power loss associated with the fields in the vicinity of the discontinuity, i.e., between the terminal planes. The last integrals in (109) are equal to the power P_t transmitted past the discontinuity. The first integrals may be modified as follows:

$$\frac{1}{2}a_1 a_1^*(1 + R_1)(1 - R_1^*)\iint_{S_1} \mathbf{e}_1 \times \mathbf{h}_1^* \cdot d\mathbf{S}$$

$$= \frac{1 + R_1}{1 - R_1}\frac{1}{2}a_1 a_1^*(1 - R_1)(1 - R_1^*)\iint_{S_1} \mathbf{e}_1 \times \mathbf{h}_1^* \cdot d\mathbf{S}$$

$$= \frac{1 + R_1}{1 - R_1}Z_1\left\{\frac{1}{2}a_1 a_1^*(1 - R_1)(1 - R_1^*)\iint_{S_1} [(\mathbf{a}_x \mathbf{a}_y - \mathbf{a}_y \mathbf{a}_x) \cdot \mathbf{h}_1 \times \mathbf{h}_1^*] \cdot d\mathbf{S}\right\}$$

$$= \frac{1 + R_1}{1 - R_1}Z_1 P_i \frac{1}{2}I_1 I_1^*$$

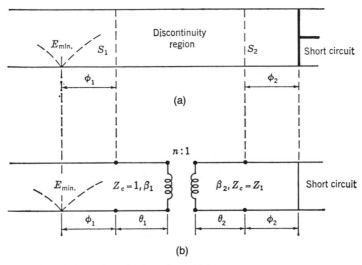

Fig. 5.12. Equivalent circuit for a lossless discontinuity. (a) Illustration of terminal planes. (b) Equivalent circuit of discontinuity.

where P_i may also be written as $P_i = \iint_{S_1} \mathbf{h}_1 \cdot \mathbf{h}_1^* \, dS$, and I_1 is the total equivalent transmission-line current at the terminal plane S_1. Substituting into (109), the following result for the input impedance at the terminal plane S_1 is obtained:

$$Z_{\text{in}} = \frac{1+R_1}{1-R_1} Z_1 = \frac{4j\omega(W_m - W_e)}{P_i I_1 I_1^*} + 2\frac{P_L + P_t}{P_i I_1 I_1^*}. \qquad (110)$$

In this equation, P_i serves as a normalization factor. It is seen that the reactive part of the input impedance is determined by the difference in the time-average magnetic and electric energy stored in the field around the discontinuity, while the resistive part arises from the power loss in the discontinuity and the power transmitted past the discontinuity. Equation (110) is formally the same as that derived in Section 1.3 for the input impedance of a simple RLC series network.

5.8. THE TANGENT METHOD FOR THE EXPERIMENTAL DETERMINATION OF THE EQUIVALENT-CIRCUIT PARAMETERS

In this section a brief treatment of the tangent method for determining the equivalent-circuit parameters of a lossless discontinuity will be given. This method is discussed here because in Chapter 8 an analytical method, which is equivalent to this experimental method, will be given. The tangent method, or nodal-shift method, as it is sometimes called, was developed independently by Weissfloch [5.7] and Feenberg [5,8].[5]

Consider a guide containing a lossless discontinuity, located between the terminal planes S_1 and S_2. This section of guide and the discontinuity may be represented by an equivalent circuit containing three independent parameters. A convenient circuit to choose is one consisting of two lengths of transmission line of electrical length θ_1 and θ_2, connected together by an ideal transformer of turns ratio $n : 1$, as in Fig. 5.12. It will be assumed for generality that the

[5] Good descriptions of the method may also be found in [5.9], [5.32, Sect. 3.4].

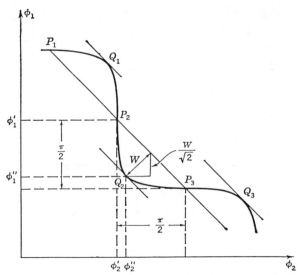

Fig. 5.13. A plot of ϕ_1 versus ϕ_2.

transmission line on the output side has a characteristic impedance of Z_1, and that the input line has a characteristic impedance of unity. The propagation constants of the input and output guides are $j\beta_1$ and $j\beta_2$, respectively. In practice, the measurements must be made at points sufficiently far away from the discontinuity so that only the dominant modes are present. In theoretical work it is convenient to shift the terminal planes used in the experimental measurements back toward the discontinuity by integral multiples of a half guide wavelength so as to coincide more closely with the actual discontinuity. The equivalent-circuit parameters are invariant under such a shift in the terminal-plane positions.

A plot of the electrical field null position ϕ_1 as a function of the short-circuit position ϕ_2 yields a curve of the form illustrated in Fig. 5.13. In this figure, ϕ_1 and ϕ_2 are electrical lengths rather than physical lengths. An analysis of the equivalent circuit in Fig. 5.12 yields expressions for the equivalent-circuit parameters θ_1, θ_2, and n: 1 in terms of the parameters of the curve of ϕ_1 vs. ϕ_2.

With reference to Fig. 5.12(b), we find from conventional transmission-line theory that the field minimum position ϕ_1 (this corresponds to the point where the input impedance is zero) is given by

$$\tan(\phi_1 + \theta_1) = -N^2 \tan(\phi_2 + \theta_2) \tag{111}$$

where $N^2 = n^2 Z_1$. A plot of ϕ_1 vs. ϕ_2 as determined by this relation results in an oscillatory curve symmetrical about a line of slope -1 and with a period π for both variables as in Fig. 5.13. From (111) it is readily found that, if $N < 1$, the slope of the curve at P_1 is $-N^2$, and the slope at P_2 is $-1/N^2$. If $N > 1$, the reverse is true. The slope of the curve is given by

$$\frac{d\phi_1}{d\phi_2} = -N^2 \frac{\sec^2(\phi_2 + \theta_2)}{\sec^2(\phi_1 + \theta_1)}. \tag{112}$$

At the points Q_1, Q_2, Q_3, the slope is equal to -1. Whenever the curve crosses the straight line, we have $\theta_1 + \phi_1 = s\pi/2$, $\theta_2 + \phi_2 = n\pi/2$, where s and n are suitable integers. When s

and n are even, (112) shows that the slope is $-N^2$, and, hence, if $N > 1$, the coordinates of the point P_2 are $\phi_1' = n_1\pi - \theta_1$, $\phi_2' = n_2\pi - \theta_2$, where n_1 and n_2 are suitable integers. If $N < 1$, the coordinates of the point P_2 are

$$\phi_1' = n_1'\pi + \frac{\pi}{2} - \theta_1 \quad \text{and} \quad \phi_2' = n_2'\pi + \frac{\pi}{2} - \theta_2$$

where n_1' and n_2' are again a suitable set of integers. Equating (112) to -1 and solving this equation and (111) simultaneously for N gives

$$N = \pm \cot(\phi_2'' + \theta_2) \tag{113}$$

where ϕ_2'' is the value of ϕ_2 where the slope equals -1, say at Q_2. If the positive root is chosen, the value of $\phi_2 + \theta_2$ must be equal to an even multiple of $\pi/2$ at P_2 since the cotangent function must be positive, and this is possible only if the angle increases from some multiple of π. In this case, the value of N obtained is greater than unity, as the preceding discussion shows. If the negative root is chosen, the corresponding value of $\phi_2 + \theta_2$ at P_2 must be equal to an odd multiple of $\pi/2$, and N is less than unity. Either root may be chosen and leads to a correct equivalent circuit. Along the straight line, and in particular at P_2, we have $\phi_1 + \theta_1 + \phi_2 + \theta_2 = n\pi$, where n is a suitable integer. At the point Q_2 we have $\phi_1'' + \theta_1 + \phi_2'' + \theta_2 = -2^{1/2}W + n\pi$, where W is the peak amplitude of the curve in radians, measured from the straight line. Expanding the tangent of this angle gives

$$\begin{aligned}
\tan 2^{1/2}W &= -\frac{\tan(\phi_1'' + \theta_1) + \tan(\phi_2'' + \theta_1)}{1 - \tan(\phi_1'' + \theta_1)\tan(\phi_2'' + \theta_2)} \\
&= \pm \frac{N^2 - 1}{2N} = \pm \frac{1 - 1/N^2}{2/N}
\end{aligned} \tag{114}$$

when (111) and (113) are used. Now

$$\tan 2^{1/2}W = \cot\left(\frac{\pi}{2} - 2^{1/2}W\right) = \frac{1 - \tan^2(\pi/4 - 2^{1/2}W/2)}{2\tan(\pi/4 - 2^{1/2}W/2)}$$

and, hence, upon comparison with (114), we get for one solution

$$N = \cot\left(\frac{\pi}{4} - 2^{-1/2}W\right). \tag{115}$$

Equation (115) determines the value of N in terms of the amplitude W of the curve. Furthermore, the value of N given by (115) is greater than unity, and, therefore, the point P_2 has coordinates $\phi_1' = n_1\pi - \theta_1$, $\phi_2' = n_2\pi - \theta_2$. From the measured values of ϕ_1' and ϕ_2', the equivalent-circuit parameters θ_1 and θ_2 are now readily found to within a multiple of π.

If the negative sign in (114) is chosen, we get in place of (115) the solution

$$N = \tan\left(\frac{\pi}{4} - 2^{1/2}W\right) \tag{116}$$

and, hence, N is less than unity, and the coordinates of the point P_2 are $\phi_1' = n_1'\pi + \pi/2 - \theta_1$,

$\phi_2' = n_2'\pi + \pi/2 - \theta_2$. Either solution may be chosen; however, it is important to note that the coordinates of the point P_2 are different, and, hence, the values of θ_1 and θ_2 are different, depending on whether N is chosen greater or smaller than unity.

In analytical work we generally obtain a bilinear relation between ϕ_1 and ϕ_2 of the following form [5.10]:

$$\tan \phi_1 = \frac{A + B \tan \phi_2}{C + D \tan \phi_2} \tag{117}$$

where A, B, C, and D are known constants. It is now convenient to be able to determine the equivalent-circuit parameters in terms of the known constants, without having to construct the curve of ϕ_1 vs. ϕ_2 and analyze this curve by the previous method. Equating the slope as obtained from (117) to -1, and using (117) to eliminate $\tan \phi_1$, we get

$$(D^2 + B^2 - AD + BC) \tan^2 \phi_2 + (2CD + 2AB) \tan \phi_2$$

$$+ (C^2 + A^2 + BC - AD) = 0. \tag{118}$$

This equation may be solved for ϕ_2, and determines the points Q_1, Q_2, Q_3, etc. The solution is of the form

$$\tan \phi_2 = x \pm y. \tag{119}$$

If the two solutions for ϕ_2 are averaged, either the point P_2 or P_3 is determined. If the two solutions for ϕ_2 given by (119) are designated ϕ_{21} and ϕ_{22}, we get

$$\tan (\phi_{21} + \phi_{22}) = \frac{\tan \phi_{21} + \tan \phi_{22}}{1 - \tan \phi_{21} \tan \phi_{22}} = \frac{2x}{1 - x^2 + y^2}.$$

When x and y are obtained from (118), we get

$$\phi_2' = \frac{\phi_{21} + \phi_{22}}{2} = \frac{1}{2} \tan^{-1} \frac{2(AB + CD)}{A^2 + C^2 - B^2 - D^2}. \tag{120}$$

The corresponding value of ϕ_1' may be found from (117) and is

$$\phi_1' = \tan^{-1} \frac{A + B \tan \phi_2'}{C + D \tan \phi_2'}. \tag{121}$$

If the derivative $d\phi_1/d\phi_2$ is equated to $-N^2$ at the point (ϕ_1', ϕ_2'), we get

$$N^2 = n^2 Z_1 = \frac{(AD - BC)(1 + \tan^2 \phi_2')}{(C + D \tan \phi_2')^2 (1 + \tan^2 \phi_1')}. \tag{122}$$

If (112) determines N as greater than unity, then we have

$$\theta_1 = \phi_1' + n_1 \pi \tag{123a}$$

$$\theta_2 = -\phi_2' + n_2 \pi \tag{123b}$$

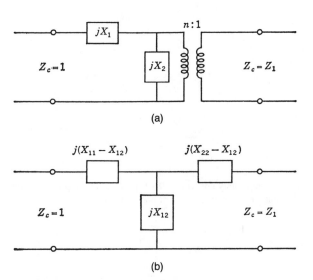

Fig. 5.14. Alternative equivalent circuits.

whereas if N is found to be less than unity, $\pm\pi/2$ must be added to the parameters θ_1 and θ_2 as given by (123).

The parameters of alternative equivalent circuits as illustrated in Fig. 5.14 may also be obtained in terms of the constants A, B, C, and D. For Fig. 5.14(a), we get

$$X_1 = \frac{-A}{C} \tag{124a}$$

$$X_2 = \frac{A}{C} - \frac{B}{D} \tag{124b}$$

$$n^2 Z_1 = \frac{DA}{C^2} - \frac{B}{C} \tag{124c}$$

while, for Fig. 5.14(b),

$$X_{11} = \frac{-B}{D} \tag{125a}$$

$$X_{22} = \frac{CZ_1}{D} \tag{125b}$$

$$X_{12} = \pm \left(\frac{AZ_1}{D} - \frac{BCZ_1}{D^2} \right)^{1/2}. \tag{125c}$$

5.9. Electromagnetic Cavities

An electromagnetic cavity or resonator is a metallic enclosure, such as shown in Figs. 5.15 and 5.16, in which modes of free oscillation can exist at an infinite number of discrete frequencies. A cavity is usually coupled to the outside source by means of a small aperture providing coupling to a waveguide or by means of a probe or loop that terminates a coaxial

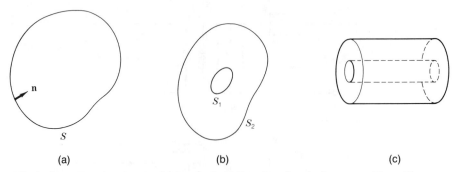

Fig. 5.15. (a) Type 1 cavity, simply connected. (b) Type 2 cavity, simply connected but with two surfaces. (c) Type 3 cavity, not simply connected, one surface.

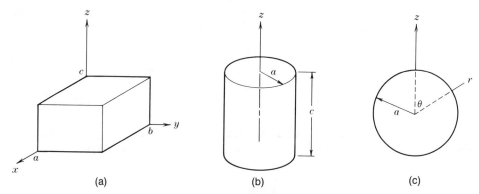

Fig. 5.16. (a) Rectangular box cavity. (b) Cylindrical cavity. (c) Spherical cavity.

transmission line. A notable feature of cavities in the microwave frequency range is the very high quality factor or Q that is exhibited. The Q may be as large as 10^4 or more. Two common forms of cavities are sections of a rectangular or circular waveguide with short-circuit plates inserted a multiple of one-half guide wavelength apart. The field in a cavity of this type consists of a waveguide standing-wave mode. In this and following sections we will develop the basic mode theory for resonators, the effect of finite conducting walls, and the general formulation for the dyadic Green's function for a cavity. In a later section we will develop a perturbation theory for cavities that is useful for determining the change in resonant frequency and Q when a small object is inserted into the cavity. This perturbation theory is the basis for a method of measuring the dielectric constant and loss tangent of a small dielectric sample placed in the cavity.

The first complete theory for the spectrum of modes in a cavity was presented by Kurokawa [5.41]. The modes consist of both irrotational and solenoidal modes. According to Helmholtz's theorem the electric field in the interior of a volume V bounded by a closed surface S can be expressed in the form (Mathematical Appendix, Section A.1):

$$\mathbf{E}(\mathbf{r}) = -\nabla \left[\iiint_V \frac{\nabla_0 \cdot \mathbf{E}(\mathbf{r}_0)}{4\pi R} \, dV_0 + \oiint_S \frac{\mathbf{n} \cdot \mathbf{E}(\mathbf{r}_0)}{4\pi R} \, dS_0 \right]$$

$$+ \nabla \times \left[\iiint_V \frac{\nabla_0 \times \mathbf{E}(\mathbf{r}_0)}{4\pi R} \, dV_0 + \oiint_S \frac{\mathbf{n} \times \mathbf{E}(\mathbf{r}_0)}{4\pi R} \, dS_0 \right]$$

where $R = |\mathbf{r} - \mathbf{r}_0|$ and $\mathbf{n}$ is the unit inward normal to the surface S. This theorem gives the conditions for which the electric field can be a pure solenoidal or a pure irrotational field. A pure solenoidal field must satisfy the conditions $\nabla \cdot \mathbf{E} = 0$ in V and $\mathbf{n} \cdot \mathbf{E} = 0$ on S, in which case there is no volume or surface charge associated with the field. An example of a pure solenoidal electric field in a cylindrical cavity such as that illustrated in Fig. 5.16(b) is

$$\mathbf{E} = \mathbf{a}_\phi J_1(p'r/a) \sin \frac{\pi z}{d}$$

where $J_1(p') = 0$. An example of an electric field that satisfies $\nabla \cdot \mathbf{E} = 0$ in V but for which $\mathbf{n} \cdot \mathbf{E}$ is not zero everywhere on S is

$$\mathbf{E} = \mathbf{a}_z J_0(pr/a)$$

where $J_0(p) = 0$. This electric field is solenoidal in the volume V but is not a pure solenoidal field. In a similar way there are two conditions that must be met in order for a field to be a pure irrotational or lamellar field, namely, $\nabla \times \mathbf{E} = 0$ in V and $\mathbf{n} \times \mathbf{E} = 0$ on S. If we have a cavity with perfectly conducting walls the boundary condition $\mathbf{n} \times \mathbf{E} = 0$ must hold on S. For a time-dependent field $\nabla \times \mathbf{E}$ is not zero. In general, $\mathbf{n} \cdot \mathbf{E}$ does not vanish, and is not required to vanish, on S. Hence the electric field is generally not a pure solenoidal or lamellar field. In many cavity problems the value of $\mathbf{n} \cdot \mathbf{E}$ on the boundary is determined by the solution and is a natural boundary condition for the electric field.

It would seem natural to expand the electric field in terms of pure solenoidal and pure irrotational vector eigenmodes. However, in practice this is not an optimum procedure because such modes are difficult to find analytically. Analytic solutions of simple form are possible only for cavities with simple geometric shapes such that the eigenfunction set can be decomposed into $\mathbf{M}$-, $\mathbf{N}$-, and $\mathbf{L}$-type functions. The $\mathbf{L}$ functions are given by $\nabla \psi_L$ and when the boundary condition $\mathbf{n} \times \mathbf{L} = 0$ on S is imposed these are pure irrotational modes. We can find two sets of solenoidal functions by means of the operations

$$\mathbf{M} = \nabla \times \mathbf{a}_z \psi_M$$

$$\mathbf{N} = \nabla \times \nabla \times \mathbf{a}_z \psi_N$$

for the cases of cylindrical and rectangular box cavities. If we impose the boundary condition $\mathbf{n} \cdot \mathbf{M} = \mathbf{n} \cdot \mathbf{N} = 0$ on S we obtain pure solenoidal modes. Unfortunately, making the normal component vanish is not compatible with the required boundary condition $\mathbf{n} \times \mathbf{E} = 0$ on S. If we make $\mathbf{n} \times \mathbf{M} = \mathbf{n} \times \mathbf{N} = 0$ on S, then in general the normal components are not zero and the eigenmodes are not pure solenoidal modes. The series expansion for the electric field will converge much better if we use eigenfunctions that satisfy the same boundary conditions as the field we are expanding. From a mathematical point of view it really does not matter whether the eigenfunctions are pure solenoidal or pure irrotational as long as they form a complete set. The set of functions we will use for the expansion of the electric field will be called the short-circuit modes because they satisfy the boundary condition $\mathbf{n} \times \mathbf{E} = 0$ on S. Those modes which also satisfy the condition $\nabla \times \mathbf{E} = 0$ will be pure irrotational modes. The modes that satisfy the condition $\nabla \cdot \mathbf{E} = 0$ in V will be referred to as solenoidal modes but in general will not be pure solenoidal modes because the boundary condition $\mathbf{n} \cdot \mathbf{E} = 0$ will not hold.

The magnetic field will satisfy the boundary condition $\mathbf{n} \cdot \mathbf{H} = 0$ and the value of $\mathbf{n} \times \mathbf{H}$ will be determined by the solution. Thus $\mathbf{M}$ and $\mathbf{N}$ modes used to expand H will be pure solenoidal modes with $\mathbf{n} \times \mathbf{M}$ and $\mathbf{n} \times \mathbf{N}$ not specified but with $\mathbf{n} \cdot \mathbf{M} = \mathbf{n} \cdot \mathbf{N} = 0$ on S being a required boundary condition. The electric field is obtained from the curl of the magnetic field so the boundary condition that will actually be used is $\mathbf{n} \times \nabla \times \mathbf{M} = \mathbf{n} \times \nabla \times \mathbf{N} = 0$ on S. These boundary conditions will automatically make $\mathbf{n} \cdot \mathbf{M} = \mathbf{n} \cdot \mathbf{N} = 0$ on S.

In an aperture-coupled cavity $\mathbf{n} \cdot \mathbf{H}$ will not always be zero in the aperture opening. As a consequence, we will find that it is necessary to include irrotational modes also for a complete expansion of the magnetic field. The irrotational modes for the magnetic field will be subjected to the boundary condition $\mathbf{n} \cdot \mathbf{L} = \mathbf{n} \cdot \nabla \psi = 0$ on S. These modes will not satisfy the condition $\mathbf{n} \times \mathbf{L} = 0$ on S and hence are not pure irrotational modes.

In view of the lack of purity in the short-circuit modes it will turn out that in many instances non-pure solenoidal modes will be sufficient to expand certain electric field distributions for which $\mathbf{n} \cdot \mathbf{E}$ is not zero on S. For similar reasons we may need irrotational modes in the expansion of the magnetic field in a restricted region even though $\mathbf{H}$ is a pure solenoidal field. The field in a cavity can also be expanded in terms of open-circuit modes in which case there is an interchange in the eigenfunctions used to expand the electric and magnetic fields.

For the expansion of the electric field we will introduce solenoidal modes $\mathbf{E}_n$ that are solutions of the equations

$$(-\nabla \times \nabla \times + k_n^2)\mathbf{E}_n = (\nabla^2 + k_n^2)\mathbf{E}_n = 0 \tag{126a}$$

$$\nabla \cdot \mathbf{E}_n = 0 \tag{126b}$$

$$\mathbf{n} \times \mathbf{E}_n = 0 \qquad \text{on } S \tag{126c}$$

and irrotational modes $\mathbf{F}_n$ that are solutions of

$$(\nabla^2 + \ell_n^2)\mathbf{F}_n = 0 \tag{127a}$$

$$\nabla \times \mathbf{F}_n = 0 \tag{127b}$$

$$\mathbf{n} \times \mathbf{F}_n = 0 \qquad \text{on } S. \tag{127c}$$

The solenoidal modes consist of the usual $\mathbf{M}$ and $\mathbf{N}$ functions for cavities for which analytical solutions can be derived. The irrotational modes are generated from scalar functions Φ_n that are solutions of

$$\nabla^2 \Phi_n + \ell_n^2 \Phi_n = 0 \tag{128a}$$

$$\Phi_n = 0 \qquad \text{on } S \tag{128b}$$

from which we obtain

$$\ell_n \mathbf{F}_n = \nabla \Phi_n. \tag{129}$$

The factor ℓ_n is included so that the $\mathbf{F}_n$ are normalized when the Φ_n are normalized. For $n = 0$, $\ell_0 = 0$ and we use $\mathbf{F}_0 = \nabla \Phi_0$.

There are three basic cavity types that must be considered as shown in Fig. 5.15. The type

1 cavity is a simply connected volume with a single enclosing surface and does not support a zero-frequency mode. The type 2 cavity is simply connected and can support a zero-frequency (electrostatic) mode if we assign two different but constant values for Φ_0 on the two surfaces ($n = 0$ designates this mode). There is also a zero-frequency solenoidal mode with $k_0 = 0$ and for which $\nabla^2 \mathbf{E}_0 = 0$. But since

$$\nabla^2 \mathbf{E}_0 = \nabla \nabla \cdot \mathbf{E}_0 - \nabla \times \nabla \times \mathbf{E}_0 = 0$$

we see that $\nabla \times \mathbf{E}_0 = 0$ and hence $\mathbf{E}_0 = \nabla \psi$ where ψ is a scalar function. Thus the $\mathbf{E}_0$ mode is not distinguishable from the $\mathbf{F}_0$ mode so it is not necessary to include both. The type 3 cavity has a single surface but is not simply connected. It will support a zero-frequency magnetic mode. The theory to be developed will be for a type 1 cavity. It will also apply to the other cavity types when the additional zero-frequency modes are added to the expansions.

We will assume that the $\mathbf{E}_n$ modes are normalized so that

$$\iiint\limits_V \mathbf{E}_n \cdot \mathbf{E}_n \, dV = 1 \tag{130}$$

and likewise we will assume that the Φ_n are normalized, i.e.,

$$\iiint\limits_V \Phi_n^2 \, dV = 1 \tag{131}$$

where the integration in both cases is over the volume V of the cavity. For the $\mathbf{F}_n$ modes we have

$$\iiint\limits_V \mathbf{F}_n \cdot \mathbf{F}_n \, dV = \iiint\limits_V \ell_n^{-2} \, \nabla \Phi_n \cdot \nabla \Phi_n \, dV.$$

We now use $\nabla \cdot \Phi_n \nabla \Phi_n = \Phi_n \nabla^2 \Phi_n + \nabla \Phi_n \cdot \nabla \Phi_n$, the Helmholtz equation for Φ_n, and the divergence theorem to obtain ($\mathbf{n}$ is the unit inward normal):

$$\iiint\limits_V \mathbf{F}_n \cdot \mathbf{F}_n \, dV = \iiint\limits_V \Phi_n^2 \, dV - \frac{1}{\ell_n^2} \oiint\limits_S \Phi_n \frac{\partial \Phi_n}{\partial n} \, dS = 1 \tag{132}$$

since $\Phi_n = 0$ on the cavity surface S. Thus the $\mathbf{F}_n$ modes as defined are also normalized.

The $\mathbf{E}_n$ and $\mathbf{F}_n$ modes are orthogonal as we show below. We begin with the relation $\nabla \cdot \mathbf{F}_m \times \nabla \times \mathbf{E}_n = \nabla \times \mathbf{F}_m \cdot \nabla \times \mathbf{E}_n - \mathbf{F}_m \cdot \nabla \times \nabla \times \mathbf{E}_n = -k_n^2 \mathbf{F}_m \cdot \mathbf{E}_n$. Thus we find, upon using the divergence theorem, that

$$k_n^2 \iiint\limits_V \mathbf{F}_m \cdot \mathbf{E}_n \, dV = \oiint\limits_S \mathbf{n} \times \mathbf{F}_m \cdot \nabla \times \mathbf{E}_n \, dS = 0$$

which gives the desired result. The modes $\mathbf{E}_n$ are also mutually orthogonal when they are not degenerate. By subtracting the equations satisfied by $\mathbf{E}_n$ and $\mathbf{E}_m$ we obtain

$$\iiint_V (\mathbf{E}_n \cdot \nabla \times \nabla \times \mathbf{E}_m - \mathbf{E}_m \cdot \nabla \times \nabla \times \mathbf{E}_n)\, dV$$

$$= (k_m^2 - k_n^2) \iiint_V \mathbf{E}_n \cdot \mathbf{E}_m \, dV$$

$$= \iiint_V \nabla \cdot (\mathbf{E}_m \times \nabla \times \mathbf{E}_n - \mathbf{E}_n \times \nabla \times \mathbf{E}_m)\, dV$$

$$= \oiint_S (\mathbf{n} \times \mathbf{E}_n \cdot \nabla \times \mathbf{E}_m - \mathbf{n} \times \mathbf{E}_m \cdot \nabla \times \mathbf{E}_n)\, dS = 0.$$

When $k_m^2 \neq k_n^2$ the modes $\mathbf{E}_n$ and $\mathbf{E}_m$ are orthogonal. For degenerate modes we can use the Gram–Schmidt orthogonalization procedure to construct a new subset of orthogonal modes (Section 2.3).

Electric Field Expansion

The electric field due to a current $\mathbf{J}(\mathbf{r})$ in the cavity is a solution of

$$\nabla \times \nabla \times \mathbf{E} - k_0^2 \mathbf{E} = -j\omega\mu_0 \mathbf{J}. \tag{133}$$

We can solve this equation by expanding $\mathbf{E}$ in terms of the $\mathbf{E}_n$ and $\mathbf{F}_n$ modes. We have

$$\mathbf{E} = \sum_n (A_n \mathbf{E}_n + B_n \mathbf{F}_n).$$

When this expansion is used the differential equation gives

$$\sum_n [(k_n^2 - k_0^2) A_n \mathbf{E}_n - k_0^2 B_n \mathbf{F}_n] = -j\omega\mu_0 \mathbf{J}.$$

We scalar multiply by $\mathbf{E}_n$ and $\mathbf{F}_n$ in turn and integrate over the volume V to obtain

$$(k_n^2 - k_0^2) A_n = -j\omega\mu_0 \iiint_V \mathbf{E}_n(\mathbf{r}') \cdot \mathbf{J}(\mathbf{r}')\, dV'$$

$$-k_0^2 B_n = -j\omega\mu_0 \iiint_V \mathbf{F}_n(\mathbf{r}') \cdot \mathbf{J}(\mathbf{r}')\, dV'.$$

The solution for E can now be completed and is

$$\mathbf{E}(\mathbf{r}) = -j\omega\mu_0 \iiint_V \left[\frac{\mathbf{E}_n(\mathbf{r})\mathbf{E}_n(\mathbf{r}')}{k_n^2 - k_0^2} - \frac{\mathbf{F}_n(\mathbf{r})\mathbf{F}_n(\mathbf{r}')}{k_0^2} \right] \cdot \mathbf{J}(\mathbf{r}')\, dV'. \tag{134}$$

The quantity in brackets is the dyadic Green's function for the electric field in the cavity; thus

$$\bar{\mathbf{G}}_e(\mathbf{r}, \mathbf{r}') = \sum \left[\frac{\mathbf{E}_n(\mathbf{r})\mathbf{E}_n(\mathbf{r}')}{k_n^2 - k_0^2} - \frac{\mathbf{F}_n(\mathbf{r})\mathbf{F}_n(\mathbf{r}')}{k_0^2} \right]. \quad (135)$$

The integer n represents an ordered arrangement of a triple set of integers and the series in (135) is in reality a triple series. In practice analytical solutions for the mode functions $\mathbf{E}_n$ and $\mathbf{F}_n$ are limited to those for cavities with simple geometrical shapes such as the rectangular box cavity, the cylindrical cavity, and the spherical cavity. For these the solenoidal modes $\mathbf{E}_n$ can be split into the $\mathbf{M}_n$- and $\mathbf{N}_n$-type modes which are similar to those which occur in waveguide problems and the dyadic Green's function expansion in spherical modes for the free-space problem. The irrotational modes $\mathbf{F}_n$ correspond to the modes $\mathbf{L}_n$ used previously.

For the rectangular box cavity shown in Fig. 5.16(a) with dimensions $a \times b \times c$ along x, y, and z the scalar functions from which the $\mathbf{M}_n$ modes are obtained are

$$\psi_{Mrst} = \left[\left(\frac{r\pi}{a}\right)^2 + \left(\frac{s\pi}{b}\right)^2 \right]^{-1/2} \left(\frac{\epsilon_{0r}\epsilon_{0s}\epsilon_{0t}}{abc}\right)^{1/2} \cos\frac{r\pi x}{a} \cos\frac{s\pi y}{b} \sin\frac{t\pi z}{c} \quad (136a)$$

and

$$\mathbf{M}_{rst} = \nabla \times \mathbf{a}_z \psi_{Mrst}. \quad (136b)$$

The corresponding equations for the $\mathbf{N}_n$ modes are

$$\psi_{Nrst} = \left(\frac{\epsilon_{0r}\epsilon_{0s}\epsilon_{0t}}{abc}\right)^{1/2} \left[\left(\frac{r\pi}{a}\right)^2 + \left(\frac{s\pi}{b}\right)^2 \right]^{-1/2} \sin\frac{r\pi x}{a} \sin\frac{s\pi y}{b} \cos\frac{t\pi z}{c} \quad (137a)$$

$$k_{rst}\mathbf{N}_{rst} = \nabla \times \nabla \times \mathbf{a}_z \psi_{Nrst} \quad (137b)$$

where $k_{rst} = [(r\pi/a)^2 + (s\pi/b)^2 + (t\pi/c)^2]^{1/2}$ and r, s, t are integers. The $\mathbf{F}_n$ or $\mathbf{L}_n$ modes are given by

$$\psi_{Lrst} = \left(\frac{\epsilon_{0r}\epsilon_{0s}\epsilon_{0t}}{abc}\right)^{1/2} \sin\frac{r\pi x}{a} \sin\frac{s\pi y}{b} \sin\frac{t\pi z}{c} \quad (138a)$$

$$k_{rst}\mathbf{L}_n = \nabla\psi_{Lrst}. \quad (138b)$$

Thus for the rectangular box cavity the Green's dyadic is given by

$$\bar{\mathbf{G}}_e(\mathbf{r}, \mathbf{r}') = \sum_{r=0}^{\infty}\sum_{s=0}^{\infty}\sum_{t=0}^{\infty} \left[\frac{\mathbf{M}_{rst}(\mathbf{r})\mathbf{M}_{rst}(\mathbf{r}') + \mathbf{N}_{rst}(\mathbf{r})\mathbf{N}_{rst}(\mathbf{r}')}{k_{rst}^2 - k_0^2} - \frac{1}{k_0^2}\mathbf{L}_{rst}(\mathbf{r})\mathbf{L}_{rst}(\mathbf{r}') \right]. \quad (139)$$

Alternative representations may be obtained by summing one of the series into closed form. If the sum is carried out over the variable t, which can be easily done before applying the vector ∇ operations, it will be found that outside the source region the $\mathbf{L}_n$ modes are canceled by contributions from the $\mathbf{N}_n$ mode series. The residual term is $-\mathbf{a}_z\mathbf{a}_z\delta(\mathbf{r} - \mathbf{r}')/k_0^2$, the same as in a waveguide. This alternative representation is an expansion in terms of the E and H modes in a rectangular waveguide with short-circuit plates at $z = 0$ and d. The $\mathbf{a}_z \mathbf{a}_z$ term in the $\mathbf{L}_n \mathbf{L}_n$ series is

$$\sum_{t=0}^{\infty} \frac{\epsilon_{0t}}{c} \frac{(\pi t/c)^2}{k_{rst}^2} \cos \frac{t\pi z}{c} \cos \frac{t\pi z'}{c} = \sum_{t=0}^{\infty} \frac{\epsilon_{0t}}{c} \left[\cos \frac{t\pi z}{c} \cos \frac{t\pi z'}{c} \right] \left[1 - \frac{(\pi r/a)^2 + (\pi s/b)^2}{k_{rst}^2} \right].$$

The first series equals $\delta(z - z')$ and when the additional sums over r and s are completed the delta function term given above is obtained. For the rectangular cavity all of the modes have the same k_{rst} eigenvalues so this cavity has a large number of degenerate modes, particularly so if a, b, and c are in the ratio of integers.

The mode spectrum for the circular cylinder cavity shown in Fig. 5.16(b) is very similar to that for the rectangular box cavity. The properly normalized scalar functions from which the $\mathbf{N}_n$, $\mathbf{M}_n$, and $\mathbf{L}_n$ modes can be found are

$$\psi_{Nnmt} = \left(\frac{\epsilon_{0n} \epsilon_{0t}}{\pi c} \right)^{1/2} \frac{J_n(p_{nm} r/a) \cos \dfrac{t\pi z}{c}}{J_n'(p_{nm}) p_{nm} k_{nmt}} \cos_{\sin} n\phi \tag{140a}$$

$$\psi_{Mnmt} = \left(\frac{\epsilon_{0n} \epsilon_{0t}}{\pi c} \right)^{1/2} \frac{J_n(p_{nm}' r/a) \sin \dfrac{t\pi z}{c}}{(1 - n^2/p_{nm}'^2)^{1/2} J_n(p_{nm}') p_{nm}'} \cos_{\sin} n\phi \tag{140b}$$

$$\psi_{Lnmt} = \left(\frac{\epsilon_{0n} \epsilon_{0t}}{\pi c} \right)^{1/2} \frac{J_n(p_{nm} r/a) \sin \dfrac{t\pi z}{c}}{a J_n'(p_{nm})} \cos_{\sin} n\phi. \tag{140c}$$

The $\mathbf{N}_{nmt}$ and $\mathbf{L}_{nmt}$ have the same eigenvalues or resonant wavenumbers k_{nmt} given by

$$k_{nmt} = [(p_{nm}/a)^2 + (t\pi/c)^2]^{1/2}. \tag{141a}$$

The eigenvalues for the $\mathbf{M}_{nmt}$ modes are given by

$$q_{nmt} = [(p_{nm}'/a)^2 + (t\pi/c)^2]^{1/2} \tag{141b}$$

where $J_n(p_{nm}) = J_n'(p_{nm}') = 0$ and the prime on J_n' denotes the derivative with respect to the argument.

The dyadic Green's function is given by

$$\bar{\mathbf{G}}_e(\mathbf{r}, \mathbf{r}') = \sum_{n=0}^{\infty} \sum_{m=1}^{\infty} \sum_{t=0}^{\infty} \left[\frac{\mathbf{M}_{nmt}(\mathbf{r}) \mathbf{M}_{nmt}(\mathbf{r}')}{q_{nmt}^2 - k_0^2} \right.$$
$$\left. + \frac{\mathbf{N}_{nmt}(\mathbf{r}) \mathbf{N}_{nmt}(\mathbf{r}')}{k_{nmt}^2 - k_0^2} - \frac{1}{k_0^2} \mathbf{L}_{nmt}(\mathbf{r}) \mathbf{L}_{nmt}(\mathbf{r}') \right]. \tag{142}$$

Alternative representations for the dyadic Green's function for the circular cylinder cavity can be obtained by summing the Bessel function series in the same manner as was shown for the circular waveguide, or by summing the series over the variable t. If the latter choice is made a delta function term $- \mathbf{a}_z \mathbf{a}_z \delta(\mathbf{r} - \mathbf{r}')/k_0^2$ will remain after the remainder of the $\mathbf{L}_n \mathbf{L}_n$ series has been canceled by terms arising from the $\mathbf{N}_n \mathbf{N}_n$ series. When the Bessel function series is summed the $\mathbf{L}_n$ functions are again canceled apart from a delta function term $-\mathbf{a}_r\mathbf{a}_r\delta(\mathbf{r}-\mathbf{r}')/k_0^2$. The basic structure of the cavity dyadic Green's function is very similar to that for free space when developed as an eigenfunction expansion.

For the spherical cavity shown in Fig. 5.16(c) the vector eigenfunctions are given by

$$\mathbf{M}_{nmt} = \nabla \times r\mathbf{a}_r\psi_{Mnmt} \tag{143a}$$

$$k_{nmt}\mathbf{N}_{nmt} = \nabla \times \nabla \times r\mathbf{a}_r\psi_{Nnmt} \tag{143b}$$

$$k_{nmt}\mathbf{L}_{nmt} = \nabla\psi_{Lnmt}. \tag{143c}$$

The normalized scalar generating functions are given by

$$\psi_{Mnmt} = \frac{\sqrt{\pi}P_n^m(\cos\theta)j_n(k_{nt}r)}{[n(n+1)Q_{nm}a^3]^{1/2}k_{nt}j_n'(k_{nt}a)}\begin{matrix}\cos\\\sin\end{matrix}m\phi \tag{144a}$$

$$\psi_{Nnmt} = \frac{\sqrt{\pi}P_n^m(\cos\theta)j_n(k_{nt}'r)}{[n(n+1)Q_{nm}a^3]^{1/2}[1-n(n+1)/k_{nt}'^2a^2]^{1/2}k_{nt}'j_n(k_{nt}'a)}\begin{matrix}\cos\\\sin\end{matrix}m\phi \tag{144b}$$

$$\psi_{Lnmt} = \frac{\sqrt{\pi}P_n^m(\cos\theta)j_n(k_{nt}r)}{[Q_{nm}a^3]^{1/2}k_{nt}j_n'(k_{nt}a)}\begin{matrix}\cos\\\sin\end{matrix}m\phi \tag{144c}$$

where j_n is the spherical Bessel function, P_n^m is the Legendre polynomial, and

$$Q_{nm} = \frac{\epsilon_{0m}\pi}{2n+1}\frac{(n+m)!}{(n-m)!}.$$

The eigenvalues are the roots of the equations

$$j_n(k_{nt}a) = 0 \qquad \text{and} \qquad d[aj_n(k_{nt}'a)]/da = 0.$$

For the spherical cavity, the $\mathbf{M}_n$ and $\mathbf{L}_n$ modes have the same resonant wavenumbers k_{nt}, while the k_{nt}' are the eigenvalues for the $\mathbf{N}_n$ modes. The dyadic Green's function is given by

$$\bar{\mathbf{G}}_e(\mathbf{r},\mathbf{r}') = \sum_{n=0}^{\infty}\sum_{m=0}^{\infty}\sum_{t=1}^{\infty}\left[\frac{\mathbf{M}_{nmt}(\mathbf{r})\mathbf{M}_{nmt}(\mathbf{r}')}{k_{nt}^2-k_0^2} + \frac{\mathbf{N}_{nmt}(\mathbf{r})\mathbf{N}_{nmt}(\mathbf{r}')}{k_{nt}'^2-k_0^2} - \frac{\mathbf{L}_{nmt}(\mathbf{r})\mathbf{L}_{nmt}(\mathbf{r}')}{k_0^2}\right]. \tag{145}$$

Alternative forms can also be constructed by summing the Bessel function series into closed form (see Problems 5.16–5.18).

For the cavity the unit dyadic source can be represented in the form

$$\bar{\mathbf{I}}\delta(\mathbf{r}-\mathbf{r}') = \sum_n[\mathbf{E}_n(\mathbf{r})\mathbf{E}_n(\mathbf{r}') + \mathbf{F}_n(\mathbf{r})\mathbf{F}_n(\mathbf{r}')] \tag{146}$$

and is the delta function operator for vector functions $\mathbf{E}$ that satisfy the boundary condition $\mathbf{n}\times\mathbf{E} = 0$ on S. By using this relation we can eliminate the irrotational modes in (135) and express the dyadic Green's function in the form

$$\bar{\mathbf{G}}_e(\mathbf{r},\mathbf{r}') = \sum_n\frac{k_n^2}{k_0^2(k_n^2-k_0^2)}\mathbf{E}_n(\mathbf{r})\mathbf{E}_n(\mathbf{r}') - \frac{1}{k_0^2}\bar{\mathbf{I}}\delta(\mathbf{r}-\mathbf{r}'). \tag{147}$$

However, the series now has poorer convergence properties.

Detailed expressions for various forms of cavity dyadic Green's functions can be found in the cited references at the end of the chapter. In addition, a number of useful results are given in the problems at the end of the chapter.

Magnetic Field Expansion

The expansion of the magnetic field requires solenoidal modes $\mathbf{H}_n$ and irrotational modes $\mathbf{G}_n$. The $\mathbf{H}_n$ modes are solutions of

$$-\nabla \times \nabla \times \mathbf{H}_n + k_n^2 \mathbf{H}_n = (\nabla^2 + k_n^2)\mathbf{H}_n = 0 \tag{148a}$$

$$\mathbf{n} \times (\nabla \times \mathbf{H}_n) = 0 \qquad \text{on } S \tag{148b}$$

$$\nabla \cdot \mathbf{H}_n = 0. \tag{148c}$$

These modes can be found from the electric field solenoidal modes $\mathbf{E}_n$ by the relation

$$k_n \mathbf{H}_n = \nabla \times \mathbf{E}_n. \tag{149a}$$

The dual relationship

$$k_n \mathbf{H}_n = \nabla \times \mathbf{E}_n \tag{149b}$$

will also hold. The curl of (126a) shows that the modes defined by (149a) satisfy the vector Helmholtz equation (148a). Since $\nabla \times \nabla \times \mathbf{E}_n = k_n^2 \mathbf{E}_n$ the boundary condition (148b) is satisfied because $\mathbf{n} \times \mathbf{E}_n = 0$ on S. When the $\mathbf{E}_n$ are normalized the $\mathbf{H}_n$ are also normalized since $k_n^2 \mathbf{H}_n \cdot \mathbf{H}_n = \nabla \times \mathbf{E}_n \cdot \nabla \times \mathbf{E}_n = \nabla \cdot (\mathbf{E}_n \times \nabla \times \mathbf{E}_n) + \mathbf{E}_n \cdot \nabla \times \nabla \times \mathbf{E}_n = \nabla \cdot (\mathbf{E}_n \times \nabla \times \mathbf{E}_n) + k_n^2 \mathbf{E}_n \cdot \mathbf{E}_n$. By integrating over the volume and using the divergence theorem and boundary condition $\mathbf{n} \times \mathbf{E}_n = 0$ we obtain

$$\iiint_V \mathbf{H}_n \cdot \mathbf{H}_n \, dV = \iiint_V \mathbf{E}_n \cdot \mathbf{E}_n \, dV. \tag{150}$$

The equations for the irrotational modes are

$$(\nabla^2 + p_n^2)\psi_n = 0 \tag{151a}$$

$$p_n \mathbf{G}_n = \nabla \psi_n \tag{151b}$$

$$\mathbf{n} \cdot \mathbf{G}_n = \frac{\partial \psi_n}{\partial n} = 0 \qquad \text{on } S. \tag{151c}$$

For $n = 0$, $p_0 = 0$ and in place of (151b) we choose $\mathbf{G}_0 = \nabla \psi_0$. The irrotational and solenoidal modes form a mutually orthogonal set. Also, for nondegenerate modes, $\mathbf{G}_n$ and $\mathbf{G}_m$, and likewise $\mathbf{H}_n$ and $\mathbf{H}_m$, are orthogonal. For electric current sources $\mathbf{J}$ in a cavity the

$\mathbf{G}_n$ modes are not required. The magnetic field is a solution of

$$\nabla \times \nabla \times \mathbf{H} - k_0^2 \mathbf{H} = \nabla \times \mathbf{J}. \tag{152}$$

The coupling to the $\mathbf{G}_n$ modes is given by

$$\iiint_V \mathbf{G}_n \cdot \nabla \times \mathbf{J} \, dV = \iiint_V [\mathbf{J} \cdot \nabla \times \mathbf{G}_n - \nabla \cdot (\mathbf{G}_n \times \mathbf{J})] \, dV$$

$$= \oiint_S \mathbf{n} \cdot \mathbf{G}_n \times \mathbf{J} \, dS = 0$$

since $\nabla \times \mathbf{G}_n = 0$ and on S the source term $\mathbf{J} = 0$. For aperture-coupled cavities where the impressed magnetic field in the aperture has a nonzero normal component the $\mathbf{G}_n$ modes will be excited. A normal magnetic field component in an aperture opening can be viewed as being equivalent to a layer of magnetic charge ρ_m. If we have a magnetic current $\mathbf{J}_m$ in the cavity then $\mathbf{H}$ is a solution of [see Chapter 1, Eq. (40)]

$$\nabla \times \nabla \times \mathbf{H} - k_0^2 \mathbf{H} = -j\omega\epsilon_0 \mathbf{J}_m. \tag{153}$$

The solution of (153) requires both the $\mathbf{H}_n$ and $\mathbf{G}_n$ modes in the expansion of $\mathbf{H}$. By analogy with (135) we can readily infer that the magnetic field dyadic Green's function for a cavity is

$$\bar{\mathbf{G}}_m(\mathbf{r}, \mathbf{r}') = \sum_n \left[\frac{\mathbf{H}_n(\mathbf{r})\mathbf{H}_n(\mathbf{r}')}{k_n^2 - k_0^2} - \frac{\mathbf{G}_n(\mathbf{r})\mathbf{G}_n(\mathbf{r}')}{k_0^2} \right]. \tag{154}$$

The construction of the modal functions $\mathbf{H}_n$ and $\mathbf{G}_n$ for specific cavities is essentially the same as that for the $\mathbf{E}_n$ and $\mathbf{F}_n$ modes. We can again introduce $\mathbf{M}_n$, $\mathbf{N}_n$, and $\mathbf{L}_n$ modes and the only change required is a change in the scalar generating functions due to a change in the boundary conditions. The scalar functions ψ_M, ψ_N, and ψ_L must be chosen so that $\mathbf{n} \times \nabla \times \mathbf{H}_n = \mathbf{n} \cdot \mathbf{G}_n = 0$ on the boundary. As noted earlier, the $\mathbf{H}_n$ can be found from the $\mathbf{E}_n$ using (149a). The scalar function ψ_L must satisfy the Neumann boundary condition $\partial \psi_L / \partial n = 0$ on S.

5.10. Cavity with Lossy Walls

For the cavity with perfectly conducting walls the eigenvalues k_n are real. Consequently, if the frequency of a time-harmonic current source is such that k_0 equals k_n the corresponding mode $\mathbf{E}_n$, $\mathbf{H}_n$ is excited with an infinite amplitude as reference to (135) and (154) shows. In practice the finite conductivity of the walls will make the eigenvalues have a small imaginary term so the resonant mode is excited with a large but finite amplitude. The theory for attenuation in a cavity can be developed by perturbation methods similar to those used for waveguides. A commonly used approach is based on an energy balance principle which we present below.

The time-average magnetic energy stored in the cavity volume for the nth mode is

$$W_m = \frac{\mu_0}{4} \iiint_V \mathbf{H}_n \cdot \mathbf{H}_n^* \, dV \tag{155}$$

and equals the time-average stored electric energy at the resonant frequency where $k_0 = k_n$.

From Maxwell's equations and assuming that $\mathbf{H} = \mathbf{H}_n$

$$\nabla \times \mathbf{E} = -j\omega\mu_0\mathbf{H}_n = -jk_0Z_0\mathbf{H}_n$$

$$\nabla \times \mathbf{H}_n = j\omega\epsilon_0\mathbf{E} = k_n\mathbf{E}_n$$

$$\nabla \times \nabla \times \mathbf{H}_n = k_0^2\mathbf{H}_n = k_n^2\mathbf{H}_n$$

and thus we must have $k_0^2 = k_n^2$. That is, $\mathbf{H}_n$ can be the magnetic field in the cavity only if $k_0 = k_n$. The stored electric energy is thus given by

$$W_e = \frac{\epsilon_0}{4} \iiint_V \mathbf{E} \cdot \mathbf{E}^* \, dV = \frac{\mu_0}{4} \iiint_V \mathbf{E}_n \cdot \mathbf{E}_n^* \, dV = W_m \tag{156}$$

since $\mathbf{E} = (k_n/jk_0Y_0)\mathbf{E}_n$ because of the manner in which the cavity modes are normalized. The electric field modes $\mathbf{E}_n$ and also the $\mathbf{H}_n$ can be chosen as real functions. Maxwell's equations then require that $\mathbf{E} = (-jk_n/k_0Y_0)\mathbf{E}_n$ so $\mathbf{E}$ is imaginary. Hence at resonance the electric and magnetic fields are $90°$ out of phase.

The currents flowing on the cavity walls for the nth mode are given by

$$\mathbf{J}_s = \mathbf{n} \times \mathbf{H} = \mathbf{n} \times \mathbf{H}_n.$$

The resultant power loss is

$$P_L = \frac{R_m}{2} \oiint_S \mathbf{J}_s \cdot \mathbf{J}_s^* \, dS = \frac{R_m}{2} \oiint_S \mathbf{H}_n \cdot \mathbf{H}_n^* \, dS. \tag{157}$$

The rate of decrease of average stored energy must be equal to the power loss; hence

$$-\frac{dW}{dt} = P_L$$

where $W = W_e + W_m$. But the power loss and stored energy are both proportional to the square of the field strength; thus the power loss at any instant of time is proportional to the stored energy at that time. Hence we must have

$$P_L = -\frac{dW}{dt} = 2\alpha W$$

which has the solution

$$W = W_0 e^{-2\alpha t}$$

where

$$\alpha = P_L/2W \tag{158}$$

and W_0 is the stored energy at $t = 0$. The corresponding decay in the field strength is according to $e^{-\alpha t}$. Consequently, a free oscillation in the cavity at the resonant frequency ω_n should have the following time dependence:

$$\mathbf{E} \propto e^{j\omega_n t - \alpha t}.$$

The quality factor or Q is defined by the relation

$$Q = \omega \frac{\text{average stored energy}}{\text{energy loss per second}} = \frac{\omega W}{P_L} = \frac{\omega}{2\alpha}. \tag{159}$$

For the nth mode we can then set

$$\alpha_n = \omega_n / 2Q_n \tag{160}$$

and the new perturbed resonant frequency is given by

$$\omega_n' = \omega_n (1 - j/2Q_n) \tag{161a}$$

or

$$k_n' = \frac{\omega_n}{c}(1 - j/2Q_n). \tag{161b}$$

When $k_0 = \omega_n/c$ the amplitude of the excited resonant mode will be proportional to

$$(k_n'^2 - k_0^2)^{-1} \approx Q_n / k_0^2.$$

Since Q_n can be very large the resonant mode is excited with a large amplitude relative to the other modes and is therefore the dominant mode in the cavity. An interesting feature of this resonance phenomenon is that it requires only a small coupling aperture or probe to excite a field with sufficient amplitude so that the power loss in the walls absorbs all of the incident power. When all of the incident power is absorbed the cavity is said to be critically coupled to the input waveguide or transmission line.

The energy balance method presented above can be substantiated by means of a variational formulation for the eigenvalues. The magnetic field mode in a cavity with lossy walls is a solution of

$$\nabla \times \nabla \times \mathbf{H} - k^2 \mathbf{H} = 0. \tag{162}$$

The corresponding electric field is given by

$$\mathbf{E} = -\frac{j}{kY_0} \nabla \times \mathbf{H}.$$

On the boundary we will require that $\mathbf{n} \times \mathbf{E} = Z_m \mathbf{n} \times \mathbf{J}_s = Z_m \mathbf{n} \times (\mathbf{n} \times \mathbf{H}) = Z_m(\mathbf{n} \cdot \mathbf{H} \mathbf{n} - \mathbf{H}) = -Z_m \mathbf{H}_t$ where the tangential component of the magnetic field is denoted by $\mathbf{H}_t$. The boundary condition can be stated in the form

$$\mathbf{n} \times \nabla \times \mathbf{H} = -jY_0 k Z_m \mathbf{H}_t. \tag{163}$$

The source-free modes in a lossy cavity exist only for complex values of ω. In (162) the eigenvalue parameter k replaces k_0 in Maxwell's equations. The surface impedance Z_m is a function of ω and should be evaluated for the complex ω. However, for practical cavities the imaginary part of ω is very small relative to the real part so we can neglect the imaginary part of ω in evaluating Z_m.

Consider the functional equation

$$I = \iiint_V (\nabla \times \mathbf{H} \cdot \nabla \times \mathbf{H}\, dV - k^2 \mathbf{H} \cdot \mathbf{H})\, dV + jY_0 k Z_m \oiint_S \mathbf{H}_t \cdot \mathbf{H}_t\, dS \qquad (164)$$

where $\mathbf{H}_t$ is the tangential component of the magnetic field on the boundary and k is the unknown eigenvalue. The variation of this integral is

$$\delta I = 2\iiint_V (\nabla \times \delta\mathbf{H} \cdot \nabla \times \mathbf{H} - k^2\, \delta\mathbf{H} \cdot \mathbf{H} - k\, \delta k \mathbf{H} \cdot \mathbf{H})\, dV$$

$$+ jY_0 k Z_m \oiint_S 2\, \delta\mathbf{H}_t \cdot \mathbf{H}_t\, dS + \delta k\, jY_0 Z_m \oiint_S \mathbf{H}_t \cdot \mathbf{H}_t\, dS.$$

We now use $\nabla \cdot (\delta\mathbf{H} \times \nabla \times \mathbf{H}) = \nabla \times \delta\mathbf{H} \cdot \nabla \times \mathbf{H} - \delta\mathbf{H} \cdot \nabla \times \nabla \times \mathbf{H} = \nabla \times \delta\mathbf{H} \cdot \nabla \times \mathbf{H} - k^2 \delta\mathbf{H} \cdot \mathbf{H}$ and the divergence theorem to obtain ($\mathbf{n}$ is the unit inward normal)

$$\delta I = -2k\, \delta k \iiint_V \mathbf{H} \cdot \mathbf{H}\, dV + \oiint_S \delta\mathbf{H} \cdot \mathbf{n} \times \nabla \times \mathbf{H}\, dS$$

$$+ jY_0 k Z_m \oiint_S \delta\mathbf{H}_t \cdot \mathbf{H}_t\, dS + jY_0\, \delta k\, Z_m \oiint_S \mathbf{H}_t \cdot \mathbf{H}_t\, dS$$

$$= \delta k \left[-2k \iiint_V \mathbf{H} \cdot \mathbf{H}\, dV + jY_0 Z_m \oiint_S \mathbf{H}_t \cdot \mathbf{H}_t\, dS \right]$$

since on the boundary $\mathbf{n} \times \nabla \times \mathbf{H} = -jY_0 k Z_m \mathbf{H}_t$. For the correct field $\mathbf{H}$ the expression $I = 0$ and hence $\delta I = 0$ and we can conclude that δk is then also zero (see Problem 5.19). Thus (164) is a variational expression for the eigenvalue k.

We now assume that $\mathbf{H}$ can be expanded in terms of the loss-free modes of the cavity; thus

$$\mathbf{H} = \sum_n^N C_n \mathbf{H}_n$$

where only a finite number of modes are included in the trial field. We will define energy storage elements W_{nm} by the integrals

$$\mu_0 W_{nm} = \frac{\mu_0}{2} \iiint_V \mathbf{H}_n \cdot \mathbf{H}_m^*\, dV = \frac{\mu_0}{2} \iiint_V \mathbf{H}_n \cdot \mathbf{H}_m\, dV \qquad (165a)$$

since we can assume that the mode functions are chosen to be real. We note that $\nabla \times \mathbf{H}_n \cdot \nabla \times \mathbf{H}_m = \nabla \cdot (\mathbf{H}_n \times \nabla \times \mathbf{H}_m) + \mathbf{H}_n \cdot \nabla \times \nabla \times \mathbf{H}_m = k_m^2 \mathbf{H}_n \cdot \mathbf{H}_m +$

$\nabla \cdot (\mathbf{H}_n \times \nabla \times \mathbf{H}_m)$ so an equivalent expression for W_{nm} is

$$W_{nm} = \frac{1}{2k_m^2} \iiint\limits_V \nabla \times \mathbf{H}_n \cdot \nabla \times \mathbf{H}_m \, dV. \qquad (165b)$$

The volume integral of the divergence was converted to a surface integral which vanished because $\mathbf{n} \times \nabla \times \mathbf{H}_m = 0$ on S. We will also define a power loss element

$$P_{nm} = \frac{R_m}{2} \oiint\limits_S \mathbf{H}_n \cdot \mathbf{H}_m^* \, dS = \frac{R_m}{2} \oiint\limits_S \mathbf{J}_n \cdot \mathbf{J}_m^* \, dS \qquad (166)$$

where $\mathbf{J}_n$ and $\mathbf{J}_m$ are the surface currents for the nth and mth modes and we have used the properties that $\mathbf{n} \cdot \mathbf{H}_n = \mathbf{n} \cdot \mathbf{H}_m = 0$ on S and the $\mathbf{H}_n$ represent an actual magnetic field. With these definitions the variational integral (164) can be expressed as

$$I = \sum_{n=1}^{N} \sum_{m=1}^{N} [(k^2 - k_m^2)C_n C_m W_{nm} + (1-j)kY_0 P_{nm} C_n C_m].$$

The vanishing of the first variation requires that

$$\sum_{n=1}^{N} C_m[(k^2 - k_m^2)W_{nm} + (1-j)kY_0 P_{nm}] = 0.$$

The elements W_{nm} and P_{nm} are real and symmetric in the indices n and m so the system of equations will determine a set of orthogonal modes or eigenvectors C_m^r with the properties that $[C^r]_t[W][C^s] = 0$ and $[C^r]_t[P][C^s] = 0$, $r \neq s$ as shown in Section A.3 of the Mathematical Appendix. The eigenvector $[C^r]$ is the vector of coefficients corresponding to the rth root of the following determinant:

$$\begin{vmatrix} (k^2 - k_1^2)W_{11} + (1-j)Y_0 kP_{11} & \cdots & (k^2 - k_N^2)W_{1N} + (1-j)Y_0 kP_{1N} \\ \vdots & & \vdots \\ (k^2 - k_1^2)W_{1N} + (1-j)Y_0 kP_{1N} & \cdots & (k^2 - k_N^2)W_{NN} + (1-j)Y_0 kP_{NN} \end{vmatrix} = 0. \quad (167)$$

If the loss-free modes are not degenerate then they are orthogonal and $W_{nm} = 0$ for $n \neq m$. In this case there is usually no mode coupling due to the wall currents and $P_{nm} = 0$ for $n \neq m$. For uncoupled modes the solution for the mth perturbed eigenvalue is

$$k \approx k_m - \frac{(1-j)Y_0 P_{mm}}{2W_{mm}}$$

upon using $k^2 - k_m^2 = (k - k_m)(k + k_m) \approx 2k_m(k - k_m)$. The term P_{mm} is the power loss P_L and $\mu_0 W_{mm}$ is the average stored energy. If we introduce the Q from (159), i.e., $P_L/W = \omega_m/Q_m$, and note that $\mu_0 Y_0 \omega_m = k_m$ we find that

$$k \approx k_m \left(1 - \frac{1-j}{2Q_m}\right). \qquad (168)$$

Thus the attenuation constant is $k_m/2Q_m$ as found earlier. The real part of the eigenvalue k is also reduced by an amount $k_m/2Q_m$ due to the additional magnetic energy stored in the inductive reactance of the surface impedance.

For the case of N degenerate modes all having the same eigenvalue and which are coupled together because of nonorthogonal surface currents or because of a lack of orthogonality, we can define new modes by a linear combination of the old modes; thus

$$\mathbf{H}'_r = \sum_{n=1}^{N} C^r_n \mathbf{H}_n.$$

In order for these new modes to be uncoupled it is necessary that

$$\iiint_V \mathbf{H}'_r \cdot \mathbf{H}'_s \, dV = 0, \qquad r \neq s$$

$$\oiint_S \mathbf{H}'_r \cdot \mathbf{H}'_s \, dS = 0, \qquad r \neq s.$$

This is equivalent to the conditions

$$\sum_{n=1}^{N}\sum_{m=1}^{N} C^s_n C^r_m W_{nm} = 0, \qquad r \neq s$$

$$\sum_{n=1}^{N}\sum_{m=1}^{N} C^s_n C^r_m P_{nm} = 0, \qquad r \neq s.$$

The solutions to the system of equations given above (167) have these properties so the roots of (167) will give the eigenvalues for the new set of uncoupled modes.

Mode Expansion of Maxwell's Equations

We will consider a cavity in which electric and magnetic current sources exist. Maxwell's equations are

$$\nabla \times \mathbf{E} = -j\omega\mu_0\mathbf{H} - \mathbf{J}_m \tag{169a}$$

$$\nabla \times \mathbf{H} = j\omega\epsilon_0\mathbf{E} + \mathbf{J}_e. \tag{169b}$$

We now assume that the fields can be expanded in terms of the cavity modes as follows:

$$\mathbf{E} = \sum_n (e_n\mathbf{E}_n + f_n\mathbf{F}_n) \tag{170a}$$

$$\mathbf{H} = \sum_n (h_n\mathbf{H}_n + g_n\mathbf{G}_n) \tag{170b}$$

where e_n, f_n, h_n, and g_n are amplitude constants. On the boundary $\mathbf{E}_t = Z_m\mathbf{n} \times \mathbf{H}$ but since $\mathbf{n} \times \mathbf{E}_n = \mathbf{n} \times \mathbf{F}_n = 0$ the expansion for the electric field will not be uniformly convergent on the boundary (analogous to the Fourier sine series expansion of a square wave having odd symmetry about the origin). The lack of nonuniform convergence means that the above series cannot be differentiated term by term. To overcome this difficulty we integrate Maxwell's equations by parts after scalar multiplication by one of the mode functions. Thus consider first

$$\iiint_V \mathbf{H}_n \cdot \nabla \times \mathbf{E}\, dV = \iiint_V -j\omega\mu_0 \mathbf{H} \cdot \mathbf{H}_n\, dV - \iiint_V \mathbf{H}_n \cdot \mathbf{J}_m\, dV.$$

We can use $\nabla \cdot (\mathbf{H}_n \times \mathbf{E}) = \nabla \times \mathbf{H}_n \cdot \mathbf{E} - \mathbf{H}_n \cdot \nabla \times \mathbf{E} = k_n \mathbf{E}_n \cdot \mathbf{E} - \mathbf{H}_n \cdot \nabla \times \mathbf{E}$ and the divergence theorem to obtain

$$\iiint_V k_n \mathbf{E}_n \cdot \mathbf{E}\, dV + \oiint_S -\mathbf{H}_n \cdot \mathbf{n} \times \mathbf{E}\, dS$$

$$= \iiint_V k_n \mathbf{E}_n \cdot \mathbf{E}\, dV + Z_m \oiint_S \mathbf{H}_n \cdot \mathbf{H}_t\, dS$$

$$= -j\omega\mu_0 \iiint_V \mathbf{H}_n \cdot \mathbf{H}\, dV - \iiint_V \mathbf{H}_n \cdot \mathbf{J}_m\, dV. \tag{171a}$$

In a similar way we can obtain the following equations from (169):

$$\iiint_V k_n \mathbf{H}_n \cdot \mathbf{H}\, dV = j\omega\epsilon_0 \iiint_V \mathbf{E}_n \cdot \mathbf{E}\, dV + \iiint_V \mathbf{E}_n \cdot \mathbf{J}_e\, dV \tag{171b}$$

$$Z_m \oiint_S \mathbf{G}_n \cdot \mathbf{H}_t\, dS = -j\omega\mu_0 \iiint_V \mathbf{G}_n \cdot \mathbf{H}\, dV - \iiint_V \mathbf{G}_n \cdot \mathbf{J}_m\, dV \tag{171c}$$

$$0 = j\omega\epsilon_0 \iiint_V \mathbf{F}_n \cdot \mathbf{E}\, dV + \iiint_V \mathbf{F}_n \cdot \mathbf{J}_e\, dV. \tag{171d}$$

We can substitute the expansions for $\mathbf{E}$ and $\mathbf{H}$ into these equations to obtain

$$k_n e_n + Z_m \oiint_S \mathbf{H}_n \cdot \mathbf{H}_t\, dS = -j\omega\mu_0 h_n - \iiint_V \mathbf{H}_n \cdot \mathbf{J}_m\, dV \tag{172a}$$

$$k_n h_n = j\omega\epsilon_0 e_n + \iiint_V \mathbf{E}_n \cdot \mathbf{J}_e\, dV \tag{172b}$$

$$j\omega\mu_0 g_n = -Z_m \oiint_S \mathbf{G}_n \cdot \mathbf{H}_t\, dS - \iiint_V \mathbf{G}_n \cdot \mathbf{J}_m\, dV \tag{172c}$$

$$j\omega\epsilon_0 f_n = -\iiint_V \mathbf{F}_n \cdot \mathbf{J}_e\, dV. \tag{172d}$$

We have shown above that in a lossy cavity we can choose our cavity modes such that they are uncoupled and hence the surface integral of $\mathbf{H}_t \cdot \mathbf{H}_n$ reduces to that of $h_n \mathbf{H}_n \cdot \mathbf{H}_n$. The stored energy in the nth mode is $\mu_0/2$ and the power loss at the frequency ω is given by

$$P_L = \frac{R_m}{2} \oiint_S \mathbf{H}_n \cdot \mathbf{H}_n \, dS = \frac{\omega W}{Q_n} = \frac{\omega \mu_0}{2Q_n}$$

and hence

$$Z_m \oiint_S \mathbf{H}_n \cdot \mathbf{H}_n \, dS = 2(1+j)P_L = (1+j)\frac{\omega \mu_0}{Q_n}.$$

The first two equations in the set (172) become

$$k_n e_n = -j\omega\mu_0 h_n - (1+j)\frac{\omega \mu_0}{Q_n} h_n - \iiint_V \mathbf{H}_n \cdot \mathbf{J}_m \, dV$$

$$k_n h_n = j\omega\epsilon_0 e_n + \iiint_V \mathbf{E}_n \cdot \mathbf{J}_e \, dV$$

and have the solutions

$$e_n = \frac{-jk_0 Z_0 \left(1 + \dfrac{1-j}{Q_n}\right) \displaystyle\iiint_V \mathbf{E}_n \cdot \mathbf{J}_e \, dV - k_n \displaystyle\iiint_V \mathbf{H}_n \cdot \mathbf{J}_m \, dV}{k_n^2 - k_0^2 \left(1 + \dfrac{1-j}{Q_n}\right)}$$

$$\approx \frac{\displaystyle\iiint_V (-jk_0 Z_0 \mathbf{E}_n \cdot \mathbf{J}_e - k_n \mathbf{H}_n \cdot \mathbf{J}_m)\, dV}{k_n^2 - k_0^2 \left(1 + \dfrac{1-j}{Q_n}\right)} \tag{173a}$$

$$h_n \approx \frac{\displaystyle\iiint_V (k_n \mathbf{E}_n \cdot \mathbf{J}_e - jk_0 Y_0 \mathbf{H}_n \cdot \mathbf{J}_m)\, dV}{k_n^2 - k_0^2 \left(1 + \dfrac{1-j}{Q_n}\right)}. \tag{173b}$$

A clearer interpretation of the integrals for g_n and f_n can be obtained by introducing the scalar functions that generate the irrotational modes $\mathbf{G}_n$ and $\mathbf{F}_n$. We can express (172d) in the form

$$j\omega\epsilon_0 \ell_n f_n = -\iiint_V \mathbf{J}_e \cdot \nabla \Phi_n \, dV = \iiint_V [\Phi_n \nabla \cdot \mathbf{J}_e - \nabla \cdot (\Phi_n \mathbf{J}_e)]\, dV$$

$$= \iiint_V \Phi_n \nabla \cdot \mathbf{J}_e \, dV + \oiint_S \Phi_n \mathbf{n} \cdot \mathbf{J}_e \, dS.$$

This result shows that if $\nabla \cdot \mathbf{J}_e = 0$ in V but $\mathbf{n} \cdot \mathbf{J}_e \neq 0$ on S and we have a type 2 cavity where Φ_n does not vanish on both surfaces then f_n will not be zero. The functions Φ_n can be viewed as scalar potential functions used to expand the scalar potential in terms of volume sources $\nabla \cdot \mathbf{J}_e$ and surface sources $\mathbf{n} \cdot \mathbf{J}_e$. For an aperture-coupled cavity the surface source can be approximated as a normal electric dipole in the aperture. In a similar way the scalar functions ψ_n that generate the $\mathbf{G}_n$ can be viewed as the functions needed to expand the magnetic scalar potential arising from equivalent magnetic charge sources. A tangential magnetic current or dipole in an aperture would couple to the $\mathbf{G}_n$ modes since $\mathbf{n} \times \mathbf{G}_n$ is not zero on the boundary S. Several examples of cavity excitation are given in the problems at the end of the chapter. We will use these formulas in Section 5.12 to develop a perturbation theory for small objects in a cavity.

5.11. Variational Formulation for Cavity Eigenvalues

For cavity shapes for which exact analytical expressions for the mode functions cannot be obtained it is useful to have available a variational formula from which approximations for the eigenvalues can be found. Such variational expressions are readily formulated. In the discussion to follow we will assume that we are dealing with an ideal cavity for which the boundary condition $\mathbf{n} \times \mathbf{E} = 0$ on S holds.

The electric field is a solution of

$$\nabla \times \nabla \times \mathbf{E} - \kappa k_0^2 \mathbf{E} = 0 \qquad (174a)$$

$$\nabla \cdot \kappa \mathbf{E} = 0 \qquad (174b)$$

where the dielectric constant κ may be a function of position. We scalar multiply this equation by $\mathbf{E}_1$ and integrate over the cavity volume to obtain

$$\iiint_V \mathbf{E}_1 \cdot (\nabla \times \nabla \times \mathbf{E} - \kappa \mathbf{E}) \, dV = 0.$$

We now use $\nabla \cdot (\mathbf{E}_1 \times \nabla \times \mathbf{E}) = \nabla \times \mathbf{E}_1 \cdot \nabla \times \mathbf{E} - \mathbf{E}_1 \cdot \nabla \times \nabla \times \mathbf{E}$ and the divergence theorem to get

$$k_0^2 \iiint_V \kappa \mathbf{E}_1 \cdot \mathbf{E} \, dV = \iiint_V \nabla \times \mathbf{E}_1 \cdot \nabla \times \mathbf{E} \, dV + \oiint_S \mathbf{n} \times \mathbf{E}_1 \cdot \nabla \times \mathbf{E} \, dS$$

where $\mathbf{n}$ is the unit inward normal. We will choose $\mathbf{n} \times \mathbf{E}_1 = 0$ on S so as to eliminate the surface integral. The variation in the above integrals for small changes $\delta\mathbf{E}$ in $\mathbf{E}$ is

$$2k_0 \delta k_0 \iiint_V \kappa \mathbf{E}_1 \cdot \mathbf{E} \, dV = \iiint_V (\nabla \times \mathbf{E}_1 \cdot \nabla \times \delta\mathbf{E} - k_0^2 \kappa \mathbf{E}_1 \cdot \delta\mathbf{E}) \, dV$$

$$= \iiint_V \delta\mathbf{E} \cdot (\nabla \times \nabla \times \mathbf{E}_1 - \kappa k_0^2 \mathbf{E}_1) \, dV - \oiint_S \mathbf{n} \times \delta\mathbf{E} \cdot \nabla \times \mathbf{E}_1 \, dS.$$

We see that provided $\mathbf{E}_1$ is chosen so as to satisfy (174a), the boundary condition $\mathbf{n} \times \mathbf{E}_1 = 0$ on S, and $\mathbf{n} \times \delta\mathbf{E} = 0$ on S, the variation in k_0 will be zero. Subject to these conditions we obtain the following variational expression for k_0^2 by choosing $\mathbf{E}_1 = \mathbf{E}$:

$$k_0^2 = \frac{\iiint_V \nabla \times \mathbf{E} \cdot \nabla \times \mathbf{E} \, dV}{\iiint_V \kappa \mathbf{E} \cdot \mathbf{E} \, dV}. \tag{175}$$

The trial function $\mathbf{E}_a$ to be used in this equation must be chosen so that $\mathbf{n} \times \mathbf{E}_a = 0$ on S. It should also satisfy the condition

$$\nabla \cdot \kappa \mathbf{E}_a = 0$$

in order that spurious solutions for k_0 do not occur. This point will be discussed in greater detail later on. If the approximating function is chosen as a finite series of trial functions such as

$$\mathbf{E}_a = \sum_{n=1}^{N} C_n \mathbf{E}_n$$

then enforcing the stationary condition will lead to a system of N homogeneous equations for the unknown coefficients C_n. In order to obtain a solution the determinant of the system of equations is equated to zero and this will determine N roots or values for k_0^2. These different values correspond to the eigenvalues for N different cavity modes.

When the proper boundary conditions are imposed unique solutions are obtained when analytic functions are used. However, in numerical solutions, such as those obtained using the method of finite differences or finite elements, the boundary conditions are enforced at a finite number of discrete points only. This approximate satisfaction of the boundary conditions is not sufficient to guarantee unique solutions. For many cavities with complex shapes it is very difficult to find trial functions $\mathbf{E}_n$ that satisfy the boundary condition $\mathbf{n} \times \mathbf{E}_n = 0$ on S. Consequently, one must evaluate the variational expression (170) numerically. The numerical evaluation of a functional that is required to be stationary for the correct field solution is called the method of finite elements. We will illustrate the basic procedure by considering the numerical solution for the cubical cavity shown in Fig. 5.17(a). The cavity is of unit length on all sides. We will look for a solution for the mode where E_z has fourfold symmetry. The analytic solution for E_z is

$$E_z = \sin \pi x \, \sin \pi y \, \cos \pi z$$

and $k_0^2 = 3\pi^2$ so $k_0 = 5.44$. For the numerical solution we will assume that we know the z dependence but not the dependence on x and y. Thus we assume that

$$E_x = F(x, y) \sin \pi z$$

$$E_y = G(x, y) \sin \pi z$$

$$E_z = H(x, y) \cos \pi z.$$

The cross section of the cavity is divided into nine cells as shown in Fig. 5.17(b). The

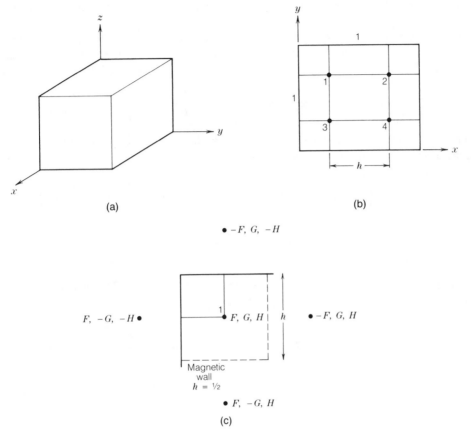

Fig. 5.17. (a) Cubical cavity. (b) Mesh for numerical solution. (c) Image nodes.

components of $\mathbf{E}$ at the ith node are represented by F_i, G_i, and H_i. The numerical evaluation of (175) then results in an equation of the form

$$\sum_i \sum_j \{[a_{ij}F_iF_j + b_{ij}G_iG_j + c_{ij}H_iH_j + d_{ij}F_iG_j$$

$$+ e_{ij}F_iH_j + f_{ij}G_iH_j] - k_0^2[g_{ii}F_i^2 + h_{ii}G_i^2$$

$$+ k_{ii}H_i^2]\delta_{ij}\} = 0 \qquad (176)$$

as we will show later on. The a_{ij}, b_{ij}, etc., are known parameters and $\delta_{ij} = 1$ if $i = j$ and equals zero otherwise. The partial derivatives $\partial/\partial F_i$, $\partial/\partial G_i$, $\partial/\partial H_i$ are now set equal to zero. This gives a set of linear equations that has a solution only if the determinant vanishes, which is possible only for certain values of k_0^2. These values give the resonant frequencies.

The condition $\nabla \cdot \mathbf{E} = 0$ would place a constraint on the relationship among the F_i, G_i, H_i at each node. No such constraint is imposed in arriving at (176) from a direct numerical implementation of (175) and consequently spurious solutions with $\nabla \cdot \mathbf{E} \neq 0$ arise. Numerical solutions with $\nabla \times \mathbf{E} \neq 0$ and $\nabla \cdot \mathbf{E} \neq 0$ produce incorrect eigenvalues (nonphysical solutions). For a system with N nodes there are $3N$ unknowns and $3N$ solutions for k_0^2. If the condition $\nabla \cdot \mathbf{E} = 0$ is imposed only $2N$ unknowns remain and at least N spurious solutions are eliminated.

For the present problem, taking the assumed symmetry into account, we will obtain the values for the field components at the image nodes that are shown in Fig. 5.17(c). We need to evaluate

$$\nabla \times \mathbf{E} = \mathbf{a}_x \left(\frac{\partial E_z}{\partial y} - \frac{\partial E_y}{\partial z} \right) + \mathbf{a}_y \left(\frac{\partial E_x}{\partial z} - \frac{\partial E_z}{\partial x} \right) + \mathbf{a}_z \left(\frac{\partial E_y}{\partial x} - \frac{\partial E_x}{\partial y} \right)$$

numerically and proceed as follows. We approximate the fields by second-degree polynomials with the origin chosen at the node point; thus

$$E_x = 16F \left(\frac{1}{4} - x \right) \left(\frac{1}{4} - y \right) S, \qquad E_y = 16G \left(x + \frac{1}{4} \right) \left(y + \frac{1}{4} \right) S$$

$$E_z = 16H \left(x + \frac{1}{4} \right) \left(\frac{1}{4} - y \right) C.$$

The $\sin \pi z$ and $\cos \pi z$ functions have been denoted by S and C. The quantity $\nabla \times \mathbf{E}$ is readily evaluated and then used in (175) to obtain

$$\frac{32}{3} H^2 + \left(\frac{4}{9} \pi^2 + \frac{16}{3} \right) (F^2 + G^2) + \frac{8\pi}{3} H(G - F) + 8FG - \frac{4}{9} k_0^2 (F^2 + G^2 + H^2).$$

The partial derivatives are now equated to zero. The vanishing of the determinant of the resultant set of linear equations gives

$$(k_0^2 - \pi^2 - 21)[k_0^4 - k_0^2(27 + \pi^2) + (6\pi^2 + 72)].$$

The first root $k_0^2 = \pi^2 + 21$ gives $k_0 = 5.556$ which is about 2% larger than the correct value of 5.44. The other two roots are spurious ones. Their values are 5.734 and 1.998 for k_0.

The condition $\nabla \cdot \mathbf{E} = 0$ cannot be imposed over the whole square because of the chosen dependence of $\mathbf{E}$ on x, y, z. If we impose this condition at the node point it gives $H = (4/\pi)(G - F)$. With this constraint we can eliminate H. The resultant pair of equations gives rise to the following eigenvalue equation:

$$(k_0^2 - \pi^2 - 21)[(32 + \pi^2)k_0^2 - (48 \times 16 + 51\pi^2 + \pi^4)] = 0.$$

One root is the same as that obtained before, $k_0 = 5.556$. The spurious root is $k_0 = 5.718$ which is quite close to one of the previous spurious roots. The low-frequency root at $k_0 = 1.998$ has been eliminated.

The problem of spurious modes also exists in the numerical solution of waveguide problems. Various techniques have been proposed to reduce the number of spurious modes that occur but a theory giving the necessary and sufficient conditions for eliminating all spurious modes is not available [5.11].

In a homogeneously filled cavity where $\nabla \cdot \mathbf{E} = 0$ we can add $-\nabla \nabla \cdot \mathbf{E}$ to (169a) to get

$$\nabla \times \nabla \times \mathbf{E} - \nabla \nabla \cdot \mathbf{E} - k_0^2 \kappa \mathbf{E} = 0 \qquad \text{or} \qquad (\nabla^2 + \kappa k_0^2)\mathbf{E} = 0.$$

For this equation the resultant variational expression for k_0^2 is

$$k_0^2 = \frac{\iiint\limits_V [|\nabla \times \mathbf{E}|^2 + |\nabla\cdot\mathbf{E}|^2]\,dV}{\iiint\limits_V \kappa|\mathbf{E}|^2\,dV}. \tag{177}$$

When the functional $|\nabla \times \mathbf{E}|^2 + |\nabla\cdot\mathbf{E}|^2 - \kappa k_0^2|\mathbf{E}|^2$ is used it is found that for the numerical solution described earlier both of the spurious roots are eliminated. However, the solution for k_0^2 is now $24 - \pi^2$ so $k_0 = 5.82$. The approximation is poorer because the trial function is more constrained. The derivation of (177) is given below.

We begin with the equation

$$\nabla \times \nabla \times \mathbf{E} - \nabla\nabla\cdot\mathbf{E} - k^2\mathbf{E} = 0 \tag{178}$$

and scalar multiply by $\mathbf{E}_1$ and integrate over the cavity volume to obtain

$$k^2 \iiint\limits_V \mathbf{E}_1\cdot\mathbf{E}\,dV = \iiint\limits_V (\mathbf{E}_1\cdot\nabla \times \nabla \times \mathbf{E} - \mathbf{E}_1\cdot\nabla\nabla\cdot\mathbf{E})\,dV.$$

Consider now the variation in k^2 due to a variation $\delta\mathbf{E}$ in $\mathbf{E}$.

$$\delta k^2 \iiint\limits_V \mathbf{E}\cdot\mathbf{E}_1\,dV = \iiint\limits_V (\mathbf{E}_1\cdot\nabla \times \nabla \times \delta\mathbf{E} - \mathbf{E}_1\cdot\nabla\nabla\cdot\delta\mathbf{E} - k^2\mathbf{E}_1\cdot\delta\mathbf{E})\,dV.$$

We now use

$$\nabla\cdot(\mathbf{E}_1 \times \nabla \times \delta\mathbf{E}) = \nabla \times \mathbf{E}_1\cdot\nabla \times \delta\mathbf{E} - \mathbf{E}_1\cdot\nabla \times \nabla \times \delta\mathbf{E}$$

$$\nabla\cdot(\mathbf{E}_1\nabla\cdot\delta\mathbf{E}) = \nabla\cdot\mathbf{E}_1\nabla\cdot\delta\mathbf{E} + \mathbf{E}_1\cdot\nabla\nabla\cdot\delta\mathbf{E}$$

$$\nabla\cdot(\mathbf{E}_1 \times \nabla \times \delta\mathbf{E}) - \nabla\cdot(\delta\mathbf{E} \times \nabla \times \mathbf{E}_1) = \delta\mathbf{E}\cdot\nabla \times \nabla \times \mathbf{E}_1 - \mathbf{E}_1\cdot\nabla \times \nabla \times \delta\mathbf{E}$$

$$\nabla\cdot(\mathbf{E}_1\nabla\cdot\delta\mathbf{E}) - \nabla\cdot(\delta\mathbf{E}\nabla\cdot\mathbf{E}_1) = \mathbf{E}_1\cdot\nabla\nabla\cdot\delta\mathbf{E} - \delta\mathbf{E}\cdot\nabla\nabla\cdot\mathbf{E}_1.$$

We then find, upon using the divergence theorem, that

$$\delta k^2 \iiint\limits_V \mathbf{E}\cdot\mathbf{E}_1\,dV = \iiint\limits_V \delta\mathbf{E}\cdot[\nabla \times \nabla \times \mathbf{E}_1 - \nabla\nabla\cdot\mathbf{E}_1 - k^2\mathbf{E}_1]\,dV$$

$$+ \oiint\limits_S [\mathbf{n} \times \delta\mathbf{E}\cdot\nabla \times \mathbf{E}_1 - \mathbf{n} \times \mathbf{E}_1\cdot\nabla \times \delta\mathbf{E}$$

$$+ \mathbf{n}\cdot\mathbf{E}_1\nabla\cdot\delta\mathbf{E} - \mathbf{n}\cdot\delta\mathbf{E}\nabla\cdot\mathbf{E}_1]\,dS.$$

If we choose $\mathbf{E}_1$ to be a solution of (178) and such that $\nabla\cdot\mathbf{E}_1 = 0$ on S and to also satisfy the boundary condition $\mathbf{n} \times \mathbf{E}_1 = 0$ on S, then the variation in k^2 will vanish for an arbitrary $\delta\mathbf{E}$ that satisfies $\mathbf{n} \times \delta\mathbf{E} = \nabla\cdot\delta\mathbf{E} = 0$ on S.

We can rewrite the equation

$$k^2 \iiint\limits_V \mathbf{E}_1 \cdot \mathbf{E} \, dV = \iiint\limits_V (\mathbf{E}_1 \cdot \nabla \times \nabla \times \mathbf{E} - \mathbf{E}_1 \cdot \nabla\nabla \cdot \mathbf{E}) \, dV$$

as

$$k^2 \iiint\limits_V \mathbf{E}_1 \cdot \mathbf{E} \, dV = \iiint\limits_V (\nabla \times \mathbf{E}_1 \cdot \nabla \times \mathbf{E} + \nabla \cdot \mathbf{E} \nabla \cdot \mathbf{E}_1) \, dV$$

$$+ \oiint\limits_S (\mathbf{n} \times \mathbf{E}_1 \cdot \nabla \times \mathbf{E} - \mathbf{n} \cdot \mathbf{E}_1 \nabla \cdot \mathbf{E}) \, dS.$$

If we identify $\mathbf{E}_1$ with $\mathbf{E}$ for which $\mathbf{n} \times \mathbf{E} = \nabla \cdot \mathbf{E} = 0$ on S, then a stationary expression for k^2 is

$$k^2 = \frac{\iiint\limits_V (|\nabla \times \mathbf{E}|^2 + |\nabla \cdot \mathbf{E}|^2) \, dV}{\iiint\limits_V |\mathbf{E}|^2 \, dV}$$

which is (177) with $\kappa k_0^2 = k^2$. For this expression the variation in k^2 due to a variation $\delta\mathbf{E}$ in $\mathbf{E}$ is

$$\frac{1}{2} \delta k^2 \int_V \kappa \mathbf{E} \cdot \mathbf{E} \, dV = \int_V \delta\mathbf{E} \cdot [\nabla \times \nabla \times \mathbf{E} - \nabla\nabla \cdot \mathbf{E} - k^2 \mathbf{E}] \, dV$$

$$+ \oiint\limits_S (\mathbf{n} \times \delta\mathbf{E} \cdot \nabla \times \mathbf{E} + \mathbf{n} \cdot \delta\mathbf{E} \nabla \cdot \mathbf{E}) \, dS.$$

This shows that $\delta k^2 = 0$ provided $\mathbf{n} \times \delta\mathbf{E} = 0$ on S and either $\nabla \cdot \mathbf{E} = 0$ on S or $\mathbf{n} \cdot \delta\mathbf{E} = 0$ on S. The condition $\mathbf{n} \cdot \delta\mathbf{E} = 0$ on S would be difficult to enforce for a trial function since it is equivalent to knowing the correct value of $\mathbf{n} \cdot \mathbf{E}$ on S. Consequently, the field $\mathbf{E}$ needs to be subjected to the two boundary conditions $\mathbf{n} \times \mathbf{E} = 0$ and $\nabla \cdot \mathbf{E} = 0$ on S. The correct field does satisfy the condition $\nabla \cdot \mathbf{E} = 0$ on S so the trial function $\mathbf{E}_a$ only needs to satisfy the condition $\mathbf{n} \times \mathbf{E}_a = 0$ on S which will then make $\mathbf{n} \times \delta\mathbf{E} = 0$ on S.

5.12. Cavity Perturbation Theory

Consider a cavity resonating in the nth mode and which contains a small object which can be a dielectric, magnetizable, or conducting object. The object is considered to be so small that it perturbs the resonant frequency of the nth mode by only a few percent. Such a small object can be characterized by its electric and magnetic dipole moments according to

$$\mathbf{M} = \bar{\alpha}_m \cdot \mathbf{H}, \qquad \mathbf{P} = \epsilon_0 \bar{\alpha}_e \cdot \mathbf{E}$$

where $\bar{\alpha}_m, \bar{\alpha}_e$ are the polarizability tensors for the object. For a dielectric sphere of radius a we have

$$\bar{\alpha}_m = 0, \qquad \bar{\alpha}_e = 4\pi a^3 \frac{\kappa - 1}{k + 2} \bar{\mathbf{I}}$$

while for a perfectly conducting sphere

$$\bar{\alpha}_m = -2\pi a^3 \mathbf{I}, \qquad \bar{\alpha}_e = 4\pi a^3 \mathbf{I}.$$

The polarizability tensors for other objects will be given in Chapter 12.

The dipole moments of small objects are equivalent to current elements according to the relationships

$$\mathbf{J}_e = j\omega \mathbf{P}$$

$$\mathbf{J}_m = j\omega\mu_0 \mathbf{M}.$$

The field scattered into the cavity by the induced dipole moments of the object can be found using (173). Thus $\mathbf{E} = e_n \mathbf{E}_n$, $\mathbf{H} = h_n \mathbf{H}_n$ where

$$h_n = (j\omega k_n \mathbf{E}_n \cdot \mathbf{P} + k_0^2 \mathbf{H}_n \cdot \mathbf{M}) / \left[k_n^2 - k_0^2 \left(1 + \frac{1-j}{Q_n} \right) \right]$$

$$e_n = (-j\omega\mu_0 \mathbf{H}_n \cdot \mathbf{M} + \omega^2 \mu_0 \mathbf{E}_n \cdot \mathbf{P}) / \left[k_n^2 - k_0^2 \left(1 + \frac{1-j}{Q_n} \right) \right].$$

Now use $\mathbf{M} = h_n \bar{\alpha}_m \cdot \mathbf{H}_n$, $\mathbf{P} = e_n \epsilon_0 \bar{\alpha}_e \cdot \mathbf{E}_n$. We then obtain a pair of homogeneous equations for e_n and h_n which will have a solution only if the determinant is zero. Setting the determinant equal to zero gives (self-consistent solution)

$$[k_n^2 - k_0^2(1+\delta) - k_0^2 \Delta_m][k_n^2 - k_0^2(1+\delta) - k_0^2 \Delta_e] - k_0^2 k_n^4 \Delta_e \Delta_m = 0$$

where $\delta = (1-j)/Q_n$, $\Delta_e = \mathbf{E}_n \cdot \bar{\alpha}_e \cdot \mathbf{E}_n$, $\Delta_m = \mathbf{H}_n \cdot \bar{\alpha}_m \cdot \mathbf{H}_n$. This is a quadratic equation in k_0^2 with a solution given by

$$\frac{k_0^2}{k_n^2} = \frac{(2 + 2\delta + \Delta_e + \Delta_m + \Delta_e \Delta_m) \pm \sqrt{(\Delta_e + \Delta_m)^2 + \Delta_e \Delta_m (\Delta_e \Delta_m + 2\Delta_e + 2\Delta_m + 4\delta)}}{2(1 + \delta + \Delta_m)(1 + \delta + \Delta_e)}.$$

$$(179)$$

If we use the binomial expansion and keep first-order terms only then

$$k_0^2 = k_n^2 (1 - \delta - \Delta_e - \Delta_m) \qquad \text{and} \qquad k_0 \approx k_n \left(1 - \frac{\delta + \Delta_e + \Delta_m}{2} \right). \qquad (180)$$

The perturbation terms Δ_e and Δ_m are of order $\Delta V / V$ where ΔV is the volume of the object and V is the cavity volume. Thus the change in the resonant frequency is also of order $\Delta V / V$. Of course the object could be placed where $\mathbf{E}_n$ or $\mathbf{H}_n$ is zero in which case Δ_e or Δ_m will vanish to first order.

When the object is a small lossy dielectric sphere

$$k_0 = k_n \left(1 - \frac{1-j}{2Q_n} - 2\pi a^3 \mathbf{E}_n \cdot \mathbf{E}_n \frac{\kappa - 1}{\kappa + 2} \right).$$

When we use $\kappa = \kappa' - j\kappa''$ we obtain

$$k_0 = k_n\left[1 - \frac{1}{2Q_n} - 2\pi a^3 \mathbf{E}_n \cdot \mathbf{E}_n \frac{(\kappa' - 1)(\kappa' + 2) + (\kappa'')^2}{(\kappa' + 2)^2 + (\kappa'')^2}\right.$$
$$\left. + \frac{j}{2}\left(\frac{1}{Q_n} + 2\pi a^3 \mathbf{E}_n \cdot \mathbf{E}_n \frac{3\kappa''}{(\kappa' + 2)^2 + (\kappa'')^2}\right)\right].$$

In this formula $\mathbf{E}_n$ is the normalized cavity electric field at the location of the center of the sphere when the sphere is absent. The loaded Q of the cavity is

$$Q = \frac{Q_n Q_d}{Q_n + Q_d}$$

where

$$Q_d = \frac{(\kappa' + 2)^2 + (\kappa'')^2}{2\pi a^3 \mathbf{E}_n \cdot \mathbf{E}_n (2\kappa' + 1)\kappa''}.$$

By measuring the change in resonant frequency and Q when the sample is inserted into the cavity the complex dielectric constant can be calculated.

REFERENCES AND BIBLIOGRAPHY

[5.1]   J. Allison and F. A. Benson, "Surface roughness and attenuation of precision drawn, chemically polished, electropolished, electroplated and electroformed waveguides," *Proc. IEE (London)*, vol. 102, part B, pp. 251–259, 1955.

[5.2]   J. A. Stratton, *Electromagnetic Theory*.   New York, NY: McGraw-Hill Book Company, Inc., 1941, sects. 9.15–9.18.

[5.3]   V. M. Papadopoulos, "Propagation of electromagnetic waves in cylindrical waveguides with imperfectly conducting walls," *Quart. J. Mech. Appl. Math.*, vol. 7, pp. 325–334, 1954.

[5.4]   J. J. Gustincic, "A general power loss method for attenuation of cavities and waveguides," *IEEE Trans. Microwave Theory Tech.*, vol. MTT-11, pp. 83–87, 1963.

[5.5]   S. E. Miller, "Waveguide as a communication medium," *Bell Syst. Tech. J.*, vol. 33, pp. 1209–1265, Nov. 1954.

[5.6]   H. E. M. Barlow and E. G. Effemey, "Propagation characteristics of low loss tubular waveguides," *Proc. IEE (London)*, vol. 104, part B, pp. 254–260, May 1957.

[5.7]   A. Weissfloch, "Ein transformationssatz verlustose vierpole und sein anwendung auf die experimentelle untersuchung van dezimeter- und zentimeterwellen-schaltungen," *Hochfreq. Elektroakust.*, vol. 60, pp. 67–73, Sept. 1942.

[5.8]   E. Feenberg, "The relation between nodal position and standing-wave ratio in a composite transmission system," *J. Appl. Phys.*, vol. 17, pp. 530–532, 1946.

[5.9]   E. L. Ginzton, *Microwave Measurements*.   New York, NY: McGraw-Hill Book Company, Inc., 1957, sect. 5.8.

[5.10] R. E. Collin, "Determination of equivalent circuit parameters," *IRE Trans. Microwave Theory Tech.*, vol. MTT-5, pp. 266–267, Oct. 1957.

[5.11] J. B. Davies, F. A. Fernandez, and G. Y. Philippou, "Finite element analysis of all modes in cavities with circular symmetry," *IEEE Trans. Microwave Theory Tech.*, vol. MTT-30, pp. 1975–1980, Nov. 1982.

General References on Waveguides

[5.12] S. Ramo, J. R. Whinnery, and T. Van Duzer, *Fields and Waves in Modern Radio*.   New York, NY: John Wiley & Sons, Inc., 1984.

[5.13] G. Southworth, *Principles and Application of Waveguide Transmission*.   Princeton, NJ: D. Van Nostrand Company, Inc., 1950.

[5.14] R. F. Harrington, *Time Harmonic Electromagnetic Fields*.   New York, NY: McGraw-Hill Book Company, Inc., 1961.

[5.15] R. E. Collin, *Foundations for Microwave Engineering*. New York, NY: McGraw-Hill Book Company, Inc., 1965.

The following papers are largely of historical interest now, but nevertheless contain a good deal of the fundamental theory and principles of waveguides:

[5.16] Lord Rayleigh, "On the passage of electric waves through tubes, or the vibration of dielectric cylinders," *Phil. Mag.*, vol. 43, pp. 125–132, Feb. 1897.

[5.17] J. R. Carson, S. P. Mead, and S. A. Schelkunoff, "Hyper-frequency waveguides: Mathematical theory," *Bell Syst. Tech. J.*, vol. 15, pp. 310–333, Apr. 1936.

[5.18] G. Southworth, "Hyper-frequency waveguides: General considerations and experimental results," *Bell Syst. Tech. J.*, vol. 15, pp. 284–309, Apr. 1936.

[5.19] W. L. Barrow, "Transmission of electromagnetic waves in hollow tubes of metal," *Proc. IRE*, vol. 24, pp. 1298–1328, Oct. 1936.

Attenuation in Waveguides

[5.20] S. Kuhn, "Calculation of attenuation in waveguides," *Proc. IEE (London)*, vol. 93, part IIIA, pp. 663–678, 1946.

[5.21] A. E. Karbowiak, "Theory of imperfect waveguides, the effect of wall impedance," *Proc. IEE (London)*, vol. 102, part B, pp. 698–707, 1955.

[5.22] P. N. Butcher, "A new treatment of lossy periodic waveguides," *Proc. IEE (London)*, vol. 103, part B, pp. 301–306, 1956.

Mode Completeness and Green's Functions

[5.23] G. M. Roe, "Normal modes in the theory of waveguides," *Phys. Rev.*, vol. 69, p. 255, Mar. 1946.

[5.24] J. Van Bladel, "Expandability of a waveguide field in terms of normal modes," *J. Appl. Phys.*, vol. 22, pp. 68–69, Jan. 1951.

[5.25] J. Van Bladel, "Field expandability in normal modes for a multilayered rectangular or circular waveguide," *J. Franklin Inst.*, vol. 253, pp. 313–321, 1952.

[5.26] J. J. Freeman, "The field generated by an arbitrary current within a waveguide," *J. Res. Nat. Bur. Stand.*, vol. 44, pp. 193–198, Feb. 1950.

[5.27] C. T. Tai, *Dyadic Green's Functions in Electromagnetic Theory*. Scranton, PA: Intext Educational Publishers, 1971.

[5.28] C. T. Tai, "On the eigenfunction expansion of dyadic Green's functions," *Proc. IEEE*, vol. 61, pp. 480–481, 1973.

[5.29] R. E. Collin, "On the incompleteness of E and H modes in waveguides," *Can. J. Phys.*, vol. 51, pp. 1135–1140, 1973.

[5.30] H. J. Butterweck, "Ueber die Anregung Elektromagnetischer Wellenleiter" ("On the excitation of electromagnetic waveguides"), *Arch. Elek. Übertragung.*, vol. 16, pp. 498–552, 1962.

[5.31] R. E. Collin, "Electromagnetic potentials and field expansions for plasma radiation in waveguides," *IEEE Trans. Microwave Theory Tech.*, vol. MTT-13, pp. 413–420, 1965.

Transmission-Line Analogy and Equivalent Circuits

[5.32] N. Marcuvitz, *Waveguide Handbook*, vol. 10 of MIT Rad. Lab. Series. New York, NY: McGraw-Hill Book Company, Inc., 1951.

[5.33] C. G. Montgomery, R. H. Dicke, and E. M. Purcell, *Principles of Microwave Circuits*, vol. 8 of MIT Rad. Lab. Series. New York, NY: McGraw-Hill Book Company, Inc., 1948.

Variational Methods

[5.34] R. M. Soria and T. J. Higgins, "A critical study of variational and finite difference methods for calculating the operating characteristics of waveguides and other electromagnetic devices," *Proc. Nat. Electronics Conf.*, vol. 3, pp. 670–679, 1947.

[5.35] D. S. Jones, *Methods in Electromagnetic Wave Propagation*. Oxford: Clarendon Press, 1979.

[5.36] K. Kurokawa, "Electromagnetic waves in waveguides with wall impedance," *IEEE Trans. Microwave Theory Tech.*, vol. MTT-10, pp. 314–320, 1962.

[5.37] K. Morishita and N. Kumagai, "Unified approach to the derivation of variational expressions for electromagnetic fields," *IEEE Trans. Microwave Theory Tech.*, vol. MTT-25, pp. 34–40, 1977.

[5.38] L. Cairo and T. Kahan, *Variational Techniques in Electromagnetism*. New York, NY: Gordon and Breach, 1965.

[5.39]   A. D. Berk, "Variational principles for electromagnetic resonator and waveguides," *IRE Trans. Antennas Propagat.*, vol. AP-4, pp. 104–111, 1956.

Electromagnetic Cavities
(In addition to the references listed below, see [5.15].)

[5.40]   C. T. Tai and P. Rozenfeld, "Different representations of dyadic Green's functions for a rectangular cavity," *IEEE Trans. Microwave Theory Tech.*, vol. MTT-24, pp. 597–601, 1976.

[5.41]   K. Kurokawa, "The expansions of electromagnetic fields in cavities," *IRE Trans. Microwave Theory Tech.*, vol. MTT-6, pp. 178–187, 1958.

[5.42]   Y. Rahmat-Samii, "On the question of computation of the dyadic Green's function at the source region in waveguides and cavities," *IEEE Trans. Microwave Theory Tech.*, vol. MTT-23, pp. 762–765, 1975.

[5.43]   M. Kisliuk, "The dyadic Green's functions for cylindrical waveguides and cavities," *IEEE Trans. Microwave Theory Tech.*, vol. MTT-28, pp. 894–898, 1980. (In this paper the complete splitting of the dyadic Green's functions into the longitudinal and transverse parts is not achieved since the eigenfunction expansion method is not used.)

[5.44]   M. Bressan and G. Conciauro, "Singularity extraction for cylindrical waveguides and cavities," *IEEE Trans. Microwave Theory Tech.*, vol. MTT-33, pp. 407–414, 1985.

[5.45]   R. E. Collin, "Dyadic Green's function expansions in spherical coordinates," *Electromagnetics*, vol. 6, pp. 183–207, 1986.

[5.46]   G. Goubau, *Electromagnetic Waveguides and Cavities.*   New York, NY: Pergamon Press, 1961.

Problems

5.1. Show that the following integral expression is a stationary expression for the cutoff wavenumber k_c; that is, show that the first variation in k_c is zero for arbitrary first-order variations in ψ, subject to the boundary constraints that the variation in ψ vanishes on the perfectly conducting guide wall for E modes, and that the variation in $\partial\psi/\partial n$ vanishes on the boundary for H modes.

$$k_c^2 = \frac{\displaystyle\iint_S \nabla_t\psi\cdot\nabla_t\psi\,dS}{\displaystyle\iint_S \psi^2\,dS}.$$

The integration is over the guide cross section.

5.2. Show that, by a conformal transformation of the guide cross section, say $W = u + jv = F(x + jy)$, the wave equation for ψ becomes

$$\frac{\partial^2\psi}{\partial u^2} + \frac{\partial^2\psi}{\partial v^2} + h^2 k_c^2\psi = 0 \qquad \text{where } h = \left|\frac{dF}{dZ}\right|^{-1}.$$

The transformation $W = \ln Z$ maps a circular guide in the Z plane into an infinitely wide rectangular guide in the uv plane.

5.3. Derive the orthogonality relation $\iint_S \mathbf{E}_{tn} \times \mathbf{H}_{tm}^*\cdot dS = 0$ given in (25).

HINT: Begin with the expression $\mathbf{E}_n\cdot\nabla\times\mathbf{H}_m^* + \mathbf{E}_m^*\cdot\nabla\times\mathbf{H}_n$ and the similar one with the roles of $\mathbf{E}$ and $\mathbf{H}$ interchanged.

5.4. Show that, in a rectangular guide lined with a thin resistance sheet at the wall $x = a$ (see Fig. P5.4), the

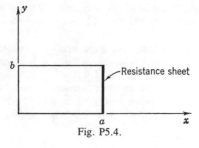

Fig. P5.4.

propagation constants for the H_{n0} modes are given by a solution of the equations

$$\tan k_c a = -k_c (j\omega\mu_0\sigma)^{1/2} \quad \text{and} \quad \Gamma^2 = k_c^2 - k_0^2$$

where σ is the conductivity of the resistance sheet.

5.5. Show that (49) gives (38) and (39) for the attenuation of nondegenerate E and H modes. Note that

$$P = \frac{1}{2} \iint_S \mathbf{e}_n \times \mathbf{h}_n^* \cdot dS = \frac{1}{2Z_n} \iint_S \mathbf{e}_n \cdot \mathbf{e}_n \, dS = \frac{1}{2Z_n} \quad \text{where } \mathbf{a}_z \times \mathbf{e}_n = Z_n \mathbf{h}_n.$$

5.6. Derive the expressions for attenuation given in Tables 5.2 and 5.5.

5.7. Use the method of expansion in normal modes to derive the Green's function for H_{n0} modes [Eq. (74)] in a rectangular guide.

5.8. Find the Green's function representing the E modes excited by an axial current element $\mathbf{a}_z \delta(x - x')\delta(y - y')\delta(z - z')$. Use the method of solving for the vector potential function A_z first. Note that the second derivative with respect to z will generate a delta function term.

5.9. Find the position and orientation of two small loops in a rectangular guide which will radiate in one direction only. Assume that only the H_{10} mode can propagate. The two loops are to be located in the same transverse plane. The loop currents are I and $Ie^{j\theta}$. Specify the required value of the phase angle θ also.

5.10. Complete the integration over w in (88) to obtain the mode expansion of the dyadic Green's function for the rectangular waveguide. Verify your solution by finding the dyadic Green's function using (87). See also Problems 5.12 and 5.14.

5.11. Find the reflection and transmission coefficients for the transmission-line circuit illustrated in Fig. P5.11. Prove the relation $T_{12}Z_1 = T_{21}Z_2$.

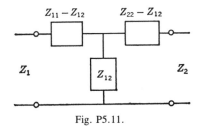

Fig. P5.11.

5.12. (a) Combine the $\mathbf{N}_n$ and $\mathbf{L}_n$ contributions to $\bar{\mathbf{G}}_e$ in (88) and show that their contribution can be expressed in the form

$$\frac{4}{2\pi abk_0^2} \int_{-\infty}^{\infty} e^{-jw(z-z')} \sum_{n=1}^{\infty}\sum_{m=1}^{\infty} \left[\frac{(k_0^2 - w^2)\mathbf{a}_z\mathbf{a}_z + jw(\mathbf{a}_z\nabla_t' - \nabla_t\mathbf{a}_z)}{(p_n^2 + w^2 - k_0^2)} \right.$$
$$\left. + \frac{(k_0^2 - p_n^2)\nabla_t\nabla_t'}{p_n^2(p_n^2 + w^2 - k_0^2)} \right] \sin\frac{n\pi x}{a} \sin\frac{n\pi x'}{a} \sin\frac{m\pi y}{b} \sin\frac{m\pi y'}{b} \, dw$$

where $p_n^2 = [n\pi/a]^2 + [m\pi/b]^2$.

(b) Combine the second group of terms involving $\nabla_t\nabla_t'$ with the $\mathbf{M}_n$ contributions and show that the combined contribution to $\bar{\mathbf{G}}_e$ is

$$\frac{1}{2\pi k_0^2} \int_{-\infty}^{\infty} e^{-jw(z-z')} \sum_{n=0}^{\infty}\sum_{m=0}^{\infty} \frac{\epsilon_{0n}\epsilon_{0m}}{ab} \left[\mathbf{a}_x\mathbf{a}_x \left(k_0^2 + \frac{\partial^2}{\partial x^2} \right) + \mathbf{a}_y\mathbf{a}_y \left(k_0^2 + \frac{\partial^2}{\partial y^2} \right) \right.$$
$$\left. + \mathbf{a}_x\mathbf{a}_y \frac{\partial^2}{\partial y\,\partial x'} + \mathbf{a}_y\mathbf{a}_x \frac{\partial^2}{\partial x\,\partial y'} \right] \cos\frac{n\pi x}{a} \cos\frac{m\pi y}{b} \cos\frac{n\pi x'}{a} \cos\frac{m\pi y'}{b}$$
$$\times \frac{dw}{p_n^2 + w^2 - k_0^2}.$$

(c) Show that

$$g = \sum_{n=0}^{\infty} \frac{\epsilon_{0n}}{a} \frac{\cos \dfrac{n\pi x}{a} \cos \dfrac{n\pi x'}{a}}{\left(\dfrac{n\pi}{a}\right)^2 + l_m^2}$$

$$= \frac{\cosh l_m x_< \cosh l_m (a - x_>)}{l_m \sinh l_m a}$$

where $l_m^2 = (m\pi/b)^2 + w^2 - k_0^2$. Note that

$$\frac{d^2 g}{dx^2} = -\delta(x - x') - l_m^2 g.$$

(d) In Part (b) show that the term involving the second derivative with respect to x will have one term which is a delta function term

$$\mathbf{a}_x \mathbf{a}_x \frac{1}{2\pi k_0^2} \int_{-\infty}^{\infty} \sum_{m=0}^{\infty} \frac{\epsilon_{0m}}{b} \cos \frac{m\pi y}{b} \cos \frac{m\pi y'}{b} e^{-jw(z-z')} [-\delta(x - x')]\, dw = -\mathbf{a}_x \mathbf{a}_x \frac{\delta(\mathbf{r} - \mathbf{r}')}{k_0^2}.$$

5.13. Find the dyadic Green's function (corresponding to the vector potential from a unit current source) which is a solution of the equation

$$\nabla^2 \bar{\mathbf{G}} + k^2 \bar{\mathbf{G}} = -\bar{\mathbf{I}} \delta(x - x')\delta(y - y')\delta(z - z')$$

and satisfies the boundary conditions $\mathbf{n} \cdot (\nabla \times \bar{\mathbf{G}}) = 0$, $\nabla \cdot \bar{\mathbf{G}} = 0$, in a rectangular guide. The solution may be constructed from the three components of the vector potential $\mathbf{A}$, where $\nabla^2 A_i + k^2 A_i = -\mu a_i \delta(x - x')\delta(y - y')\delta(z - z')$, $i = x, y, z$. The required Green's dyadic function is $\mu\bar{\mathbf{G}} = \mathbf{a}_x \mathbf{a}_x A_x + \mathbf{a}_y \mathbf{a}_y A_y + \mathbf{a}_z \mathbf{a}_z A_z$. Note that $\bar{\mathbf{G}}$ is a symmetric dyadic, so that $\bar{\mathbf{G}} \cdot \mathbf{J} = \mathbf{J} \cdot \bar{\mathbf{G}}$ and also

$$\bar{\mathbf{G}}(x, \ldots |x', \ldots) = \bar{\mathbf{G}}(x', \ldots |x, \ldots).$$

Answer:

$$\bar{\mathbf{G}} = \sum_{n=0}^{\infty} \sum_{m=0}^{\infty} \frac{\epsilon_{0n}\epsilon_{0m}}{2ab\Gamma_{nm}} e^{-\Gamma_{nm}|z-z'|} \left[\mathbf{a}_x \mathbf{a}_x \left(\cos \frac{n\pi x}{a} \cos \frac{n\pi x'}{a} \sin \frac{m\pi y}{b} \sin \frac{m\pi y'}{b} \right) + \mathbf{a}_y \mathbf{a}_y \right.$$

$$\cdot \left. \left(\sin \frac{n\pi x}{a} \sin \frac{n\pi x'}{a} \cos \frac{m\pi y}{b} \cos \frac{m\pi y'}{b} \right) + \mathbf{a}_z \mathbf{a}_z \left(\sin \frac{n\pi x}{a} \sin \frac{n\pi x'}{a} \sin \frac{m\pi y}{b} \sin \frac{m\pi y'}{b} \right) \right].$$

5.14. Since $\mu\bar{\mathbf{G}}$ in Problem 5.13 corresponds to a vector potential, the electric field $\mathbf{E}$ from an arbitrary current source is given by

$$\mathbf{E}(x, y, z) = \frac{-j\omega\mu}{k^2} \iiint_V (k^2 + \nabla\nabla \cdot) \bar{\mathbf{G}}(x, y, z|x', y', z') \cdot \mathbf{J}(x', y', z')\, dV'$$

$$= -j\omega\mu \iiint_V \bar{\mathbf{G}}_1(x, y, z|x', y', z') \cdot \mathbf{J}(x', y', z')\, dV'$$

where

$$\bar{\mathbf{G}}_1 = \left(1 + \frac{1}{k^2} \nabla\nabla \cdot \right) \bar{\mathbf{G}}(x, y, z|x', y', z')$$

and is the dyadic Green's function which satisfies the same equation as **E**. Show that $\bar{\mathbf{G}}_1$ is a solution of the equation

$$\nabla \times \nabla \times \bar{\mathbf{G}}_1 - k^2 \bar{\mathbf{G}}_1 = \bar{\mathbf{I}} \delta(x - x') \delta(y - y') \delta(z - z')$$

by using the vector identity $\nabla \times \nabla \times\ = \nabla\nabla\cdot - \nabla^2$. Also show that $\bar{\mathbf{G}}_1$ satisfies the boundary condition $\mathbf{n} \times \bar{\mathbf{G}}_1 = 0$ on the guide wall. By direct differentiation of the result given in Problem 5.13, show that

$$\nabla\nabla\cdot\bar{\mathbf{G}}(x, y, z|x', y', z') = \bar{\mathbf{g}}(x, y, z|x', y', z')$$

where $\bar{\mathbf{g}}$ is a dyadic with the property

$$\bar{\mathbf{g}}(x, y, z|x', y', z') = \bar{\mathbf{g}}_t(x', y', z'|x, y, z)$$

in which the subscript t means the transposed dyadic; i.e., replace g_{xy} by g_{yx}, etc. Thus the expression for the electric field may also be written as

$$\mathbf{E}(x, y, z) = -j\omega\mu \iiint_V \mathbf{J}(x', \ldots)\cdot\bar{\mathbf{G}}_1(x', y', z'|x, y, z)\,dV'$$

but not as

$$-j\omega\mu \iiint_V \mathbf{J}(x', \ldots)\cdot\bar{\mathbf{G}}_1(x, y, z|x', y', z')\,dV'$$

in general.

5.15. For a cavity, show that when a current source $\mathbf{J}$ has zero divergence, i.e., $\nabla\cdot\mathbf{J} = 0$, the coupling to the $\mathbf{L}_n$ modes is zero.

HINT: Consider the integral

$$\iiint_V \mathbf{J}\cdot\nabla\Phi\,dV$$

and use $\nabla\cdot(\mathbf{J}\Phi) = \Phi\nabla\cdot\mathbf{J} + \mathbf{J}\cdot\nabla\Phi$. The result is valid only if $\mathbf{n}\cdot\mathbf{J} = 0$ on the boundary S_0 of the volume V_0 that contains $\mathbf{J}$.

5.16. (a) By summing the Bessel function series show that the contribution to $\bar{\mathbf{G}}_e$ in a spherical cavity from the $\mathbf{L}_n$ modes is given by (see Problem 2.24)

$$\bar{\mathbf{G}}_{eL} = \sum_{n, m, e, o} \frac{\pi}{2k_0^2 Q_{nm}} \nabla\nabla' P_n^m(\cos\theta) P_n^m(\cos\theta') \begin{matrix}\cos\\\sin\end{matrix} m\phi \begin{matrix}\cos\\\sin\end{matrix} m\phi' \frac{r_<^n}{(2n+1)a^n}\left(\frac{a^n}{r_>^{n+1}} - \frac{r_>^n}{a^{n+1}}\right).$$

(b) Show that the $\mathbf{M}_n$ function contribution is

$$\bar{\mathbf{G}}_{eM} = -\sum_{n, m, e, o} \frac{\pi}{2n(n+1)Q_{nm}}\left[\left(\nabla \times \mathbf{a}_r r P_n^m \begin{matrix}\cos\\\sin\end{matrix} m\phi\right)\left(\nabla' \times \mathbf{a}_{r'} r' P_n^m \begin{matrix}\cos\\\sin\end{matrix} m\phi'\right)\right]$$

$$\cdot \frac{k_0 j_n(k_0 r_<)[j_n(k_0 a)y_n(k_0 r_>) - j_n(k_0 r_>)y_n(k_0 a)]}{j_n(k_0 a)}.$$

(c) Show that the $\mathbf{N}_n$ contribution is (see Problem 2.26)

$$\bar{\mathbf{G}}_{eN} = \sum_{n,m,e,o} \frac{\pi}{2n(2n+1)Q_{nm}} \left[(\nabla \times \nabla \times \mathbf{a}_r)(\nabla \times \nabla \times \mathbf{a}_{r'})rr' \right.$$

$$\left. \cdot P_n^m(\cos\theta)P_n^m(\cos\theta')\frac{\cos}{\sin}m\phi\frac{\cos}{\sin}m\phi' \right]$$

$$\cdot \left\{ \frac{k_0^2 j_n(k_0 r_<)\{y_n(k_0 r_>)[a j_n(k_0 a)]' - j_n(k_0 r_>)[a y_n(k_0 a)]'\}}{[k_0 a j_n(k_0 a)]'} \right.$$

$$\left. + \frac{r_<^n}{a^n} \frac{(n+1)\dfrac{a^n}{r_>^{n+1}} + n\dfrac{r_>^n}{a^{n+1}}}{(n+1)(2n+1)} \right\}.$$

Note that the second series consists of zero-frequency or static-like modes.

5.17. In Problem 5.16(a) show that the $\partial^2/\partial r\,\partial r'$ operator in the $\mathbf{a}_r\mathbf{a}_{r'}$ term in the $\mathbf{L}_n$ contribution produces a delta function term

$$-\sum_{n,m,e,o} \frac{\pi}{2k_0^2 Q_{nm}} \mathbf{a}_r \mathbf{a}_r P_n^m \frac{\cos}{\sin}m\phi P_n^m \frac{\cos}{\sin}m\phi' \frac{\delta(r-r')}{r^2} = -\frac{1}{k_0^2}\mathbf{a}_r\mathbf{a}_r\delta(\mathbf{r}-\mathbf{r}').$$

5.18. In Problem 5.16(c) show that the $\mathbf{N}_n$ mode contribution has two canceling delta function terms coming from the $\partial^2/\partial r\,\partial r'$ operation. For this proof you will need to use the Wronskian relation $j_n y_n' - y_n j_n' = (k_0 r)^{-2}$. Show also that the static-like terms in the $\mathbf{N}_n$ contribution cancel the $\mathbf{L}_n$ terms and that $\bar{\mathbf{G}}_e$ can be expressed in terms of contributions only from the $\mathbf{N}_n$ and $\mathbf{M}_n$ modes plus the delta function term given in Problem 5.17. Thus

$$\bar{\mathbf{G}}_e = -\sum_{n,m,e,o} \frac{\pi}{2n(n+1)Q_{nm}k_0} \left[\frac{\mathbf{M}_{j\sigma}(\mathbf{r})\mathbf{M}_{j\sigma}(\mathbf{r}')}{j_n(k_0 a)} + \frac{\mathbf{N}_{j\sigma}(\mathbf{r})\mathbf{N}_{j\sigma}(\mathbf{r}')}{[k_0 a j_n(k_0 a)]'} \right] - \frac{\mathbf{a}_r\mathbf{a}_r}{k_0^2}\delta(\mathbf{r}-\mathbf{r}')$$

where j stands for a particular combination of n, m, e, o and

$$\mathbf{M}_{j\sigma}(\mathbf{r}) = \nabla \times \mathbf{a}_r k_0 r P_n^m(\cos\theta)\frac{\cos}{\sin}m\phi \begin{cases} j_n(k_0 r), & r < r' \\ z_n(k_0 r), & r > r' \end{cases}$$

$$k_0\mathbf{N}_{j\sigma}(\mathbf{r}) = \nabla \times \nabla \times \mathbf{a}_r k_0 r P_n^m(\cos\theta)\frac{\cos}{\sin}m\phi \begin{cases} j_n(k_0 r), & r < r' \\ w_n(k_0 r), & r > r' \end{cases}$$

$$z_n(k_0 r) = j_n(k_0 a)y_n(k_0 r) - j_n(k_0 r)y_n(k_0 a)$$

$$w_n(k_0 r) = [k_0 a j_n(k_0 a)]' y_n(k_0 r) - [k_0 a y_n(k_0 a)]' j_n(k_0 r).$$

The functions of r' are the same with θ, ϕ, r replaced by θ', ϕ', r' throughout. The σ indicator specifies which radial function to use for $r > r'$ and $r < r'$ according to the rules given above.

5.19. For the field that satisfies (162) and the boundary condition (163) show that the expression (164) is zero. HINT: Integrate the $\nabla \times \mathbf{H} \cdot \nabla \times \mathbf{H}$ term by parts.

5.20. Consider a cylindrical cavity of radius b and height H that is excited with the TE_{010} mode having

$$E_z = CJ_1(p_{11}r/b).$$

A small dielectric sphere of radius a is placed on the axis inside the cavity. Derive a formula for the change in resonant frequency. C is a constant.

5.21. In a rectangular cavity like that illustrated in Fig. 5.16(a) a current distribution given by

$$\mathbf{J} = \mathbf{a}_z \delta(z - z') \sin(\pi x/a) \sin(\pi y/b)$$

exists. Find the spectrum of $\mathbf{E}_n$, $\mathbf{F}_n$, $\mathbf{H}_n$, and $\mathbf{G}_n$ modes that are excited.

HINT: Use (167) to find the expansion coefficients. Assume that $Z_m = 0$.

5.22. The cavity in Problem 5.21 is excited by a small circular aperture located in the center of the $z = c$ wall. The aperture field is equivalent to an electric dipole source $-P_0\mathbf{a}_z\delta(x - a/2)\delta(y - b/2)$ and a magnetic dipole source $M_0\mathbf{a}_x\delta(x - a/2)\delta(y - b/2)$. Find the expansion coefficients e_n, f_n, h_n, and g_n.

HINT: Note the following relations between current elements and dipole sources:

$$j\omega\mathbf{P}_e = \mathbf{J}_e$$

$$j\omega\mu_0\mathbf{M} = \mathbf{J}_m.$$

5.23. For a cavity with an aperture, but otherwise perfectly conducting walls, show that the equation for the coefficient g_n [see (172c)] is

$$j\omega\mu_0 g_n = \oiint_S \mathbf{n} \times \mathbf{E} \cdot \mathbf{G}_n \, dS$$

where $\mathbf{n} \times \mathbf{E}$ is the tangential electric field in the aperture. Assume that $\mathbf{n} \times \mathbf{E}$ is equivalent to a magnetic current $-\mathbf{J}_m$ in the aperture and that on the outside of this magnetic current sheet the aperture is closed by a perfectly conducting wall. Show that now

$$j\omega\mu_0 g_n = -\iiint_V \mathbf{G}_n \cdot \mathbf{J}_m \, dV$$

reduces in the limit of a current sheet to the previous formula.

5.24. Consider a cavity with an aperture in which $\mathbf{n} \times \mathbf{E}$ and $\mathbf{n} \times \mathbf{H}$ are not zero. The cavity walls are perfectly conducting. Show that the equations for the mode amplitude coefficients are

$$k_n e_n + j\omega\mu_0 h_n = -\iint_{S_a} \mathbf{n} \times \mathbf{E} \cdot \mathbf{H}_n \, dS$$

$$j\omega\epsilon_0 e_n - k_n h_n = 0$$

$$f_n = 0$$

$$j\omega\mu_0 g_n = -\iint_{S_a} \mathbf{n} \times \mathbf{E} \cdot \mathbf{G}_n \, dS$$

where S_a is the aperture surface. Note the absence of any contribution from $\mathbf{n} \times \mathbf{H}$.

5.25. Consider a short-circuited coaxial transmission line with inner radius a, outer radius b, short circuit at $z = d$, and an input plane at $z = 0$. On the input plane the impressed electric field is $\mathbf{E}_r = V\mathbf{a}_r/r \ln(b/a)$. Find the circularly symmetric field modes $\mathbf{H}_n$, $\mathbf{G}_n$ that satisfy the boundary conditions $\mathbf{n} \cdot \mathbf{H}_n = \mathbf{n} \cdot \mathbf{G}_n = 0$ on all surfaces including the input plane. The $\mathbf{H}_n$ can be generated from the scalar functions Ψ_{Mn} and Ψ_{Nn}. The modes needed for this problem can be found from unnormalized functions given by

$$\Psi_{Mn} = \ln r \cos(n\pi z/d), \qquad \mathbf{n} \times (\nabla \times \nabla \times \mathbf{a}_z\Psi_{Mn}) = 0 \text{ on } S.$$

Show that

$$\mathbf{H}_n = \mathbf{a}_\phi \frac{1}{r \sqrt{\pi d \, \ln(b/a)}} \cos \frac{n\pi z}{d}, \qquad n = 1, 2, 3 \ldots .$$

Show that a $\mathbf{G}_0$ mode also exists and is given by

$$\mathbf{G}_0 = \frac{1}{r \sqrt{2\pi d \, \ln(b/a)}} \mathbf{a}_\phi.$$

Note that $k_n = n\pi/d$ and that for $n = 0$, $k_n = 0$. Thus the $\mathbf{H}_0$ mode is the same as the $\mathbf{G}_0$ mode when properly normalized. Only one needs to be included. Find the expansion coefficients h_n and g_0 and then find the input current on the center conductor. Show that

$$Y_{\text{in}} = \frac{I_{\text{in}}}{V} = \frac{-2j\omega}{\mu_0(\omega^2 - \omega_n^2)d} - \frac{j}{\omega\mu_0 d}$$

$$= \frac{-j}{\omega\mu_0 d} \left[1 + \sum_{n=1}^{\infty} \frac{2}{1 - \left(\dfrac{n\pi}{k_0 d}\right)^2} \right] = -jY_0 \cot k_0 d$$

where $k_0 = \omega\sqrt{\mu_0\epsilon_0}$, $Y_0 = \sqrt{\epsilon_0/\mu_0}$. If the $\mathbf{G}_0$ function or zero-frequency eigenfunction $\mathbf{H}_0$ had not been included the correct result for Y_{in} would not have been obtained. (K. Kurokawa, "The expansions of electromagnetic fields in cavities," *IRE Trans. Microwave Theory Tech.*, vol. MTT-6, pp. 178–187, 1958.)

5.26. Consider a spherical cavity of radius a. A current element $\mathbf{a}_r I$ is located on the polar axis (z axis) at $r = a$. Find the field excited in the cavity in terms of the $\mathbf{N}_{j\sigma}$ modes in Problem 5.18. Also find the excited field using the modes $\mathbf{E}_n$ and $\mathbf{F}_n$. Show that the two results are the same.

6
Inhomogeneously Filled Waveguides and Dielectric Resonators

Waveguides partially filled with dielectric slabs find application in a variety of waveguide components such as phase changers, matching transformers, and quarter-wave plates. The first few sections of this chapter will be devoted to an examination of some of the basic properties of inhomogeneously filled waveguides. It will be assumed that the dielectric loss is very small and may be neglected. When this assumption is not valid, the solution may be obtained by replacing the permittivity ϵ by $\epsilon(1 - j \tan \delta_l)$ in the solutions for the lossless case. Some of the properties of guides loaded with ferrite material will be discussed in the latter part of the chapter.

The development of optical frequency sources such as lasers has resulted in widespread use of dielectric waveguides known as optical fibers. These generally consist of a circular core, often with a radially varying dielectric constant, surrounded by a cladding of dielectric material. An extensive body of literature dealing with optical fibers exists, including several books. Space does not permit a treatment of this type of wave-guiding medium although some of the analytical methods that we will discuss are applicable to the analysis of optical fibers. For some millimeter wave applications, as well as for optical devices, a variety of dielectric-type waveguides have been investigated. For open boundary structures, the guided modes are surface waves which will be discussed in a more complete manner in Chapter 11. The surface wave is bound to the dielectric guide and its field decays exponentially away from the surface. As a consequence, this type of waveguide can be enclosed in a metallic waveguide having a much larger cross section without significant perturbation of the surface wave modes. Some of the more important techniques suitable for the analysis of this type of waveguide will be outlined in this chapter.

Very high dielectric constant materials with low loss have been developed and are finding increasing use as resonator elements in microwave circuits. A dielectric resonator made of material with a dielectric constant of 100 or more is about 1/10th of the size of a conventional cavity resonator and for this reason is an attractive alternative. The field is confined primarily to the interior of the dielectric so the radiation loss is quite small. Quality factors of several hundred can be obtained so these resonators are useful for a variety of applications. The last part of the chapter will give an introductory treatment of dielectric resonators.

6.1. Dielectric-Slab–Loaded Rectangular Guides

Figure 6.1 illustrates the cross-sectional view of several typical slab-loaded rectangular waveguides. For the configurations in (a)–(d) it is a relatively simple task to find the expressions for the eigenfunctions and propagation constants. The general case illustrated in Fig. 6.1(e) cannot be solved in closed form even if the cross section of the dielectric slab is rec-

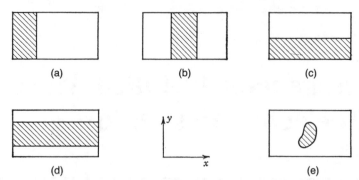

Fig. 6.1. Dielectric-slab–loaded rectangular guides. (a) Dielectric slab along side wall of a waveguide. (b) Centered dielectric slab in a waveguide. (c) Dielectric slab along bottom wall of a waveguide. (d) Dielectric slab centered between upper and lower walls of a waveguide. (e) Dielectric rod with an arbitrary cross-section in a waveguide.

tangular. When the guide is loaded with two or more slabs, the solution becomes difficult to handle because of the involved transcendental equations which occur for the determination of the eigenvalues. In such cases a perturbation or variational technique may be employed to advantage. One such technique, the Rayleigh–Ritz method, will be discussed in a later section.

The normal modes of propagation for guides loaded with dielectric slabs as in Fig. 6.1 are not, in general, either E or H modes, but combinations of an E and an H mode. An exception is the case of H_{n0} modes with the electric field parallel to the slab and no variation of the fields along the air–dielectric interface. For a rectangular guide terminated by a dielectric plug as in Fig. 6.2, it is readily found that the boundary conditions at the air–dielectric interface can be satisfied by E modes alone or H modes alone. The configurations illustrated in Fig. 6.1 are essentially the same, except that the air–dielectric interface lies in the xz or yz plane, instead of in the xy plane as in Fig. 6.2. This suggests that the basic modes of propagation for slab-loaded rectangular guides may be derived from magnetic and electric types of Hertzian potential functions, having single components directed normal to the air–dielectric interface. This is indeed the case, and the resultant modes may be classified as E or H modes with respect to the interface normal (x or y axis). From the magnetic Hertzian potential, we obtain the solution for a mode that has no component of electric field normal to the interface. The electric field thus lies in the longitudinal interface plane, and the mode is referred to as a longitudinal-section electric (LSE) mode. From the electric Hertzian potential, a mode with no magnetic field component normal to the interface is obtained. This mode is called a longitudinal-section magnetic (LSM) mode.

As a first example we will consider the solution for an asymmetrically loaded guide as illustrated in Fig. 6.3. The thickness of the slab is t, and its relative dielectric constant is κ.

LSE Modes

For a magnetic-type Hertzian potential $\Pi_h = \mathbf{a}_x \psi_h(x, y)e^{-\gamma z}$, the electric and magnetic fields are given by

$$\mathbf{E} = -j\omega\mu_0 \nabla \times \Pi_h \tag{1a}$$

$$\mathbf{H} = \nabla \times \nabla \times \Pi_h = \kappa(x)k_0^2\Pi_h + \nabla\nabla\cdot\Pi_h \tag{1b}$$

Fig. 6.2. A dielectric plug in a rectangular waveguide.

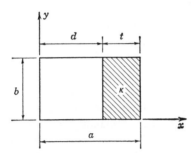

Fig. 6.3. Asymmetrical dielectric-slab-loaded guide.

and

$$\nabla_t^2 \psi_h + [\gamma^2 + \kappa(x)k_0^2]\psi_h = 0 \tag{1c}$$

where

$$\kappa(x) = \begin{cases} 1, & 0 \le x \le d \\ \kappa, & d \le x \le a. \end{cases}$$

Propagation along the guide may be assumed to be according to the exponential factor $e^{-\gamma z}$ (or $e^{\gamma z}$) and must be the same in the empty portion and dielectric-filled portion of the guide, if the boundary conditions at the interface are to be satisfied for all arbitrary values of z. A solution for ψ_h which satisfies (1c) in the two regions of interest, and also gives an electric field whose tangential components vanish on the perfectly conducting guide walls, is readily found to be

$$\psi_h = \begin{cases} A \sin hx \cos \dfrac{m\pi y}{b}, & 0 \le x \le d \\ B \sin l(a - x) \cos \dfrac{m\pi y}{b}, & d \le x \le a \end{cases} \tag{2}$$

where

$$\gamma^2 = l^2 + \left(\frac{m\pi}{b}\right)^2 - \kappa k_0^2 = h^2 + \left(\frac{m\pi}{b}\right)^2 - k_0^2 \tag{3}$$

and A and B are amplitude constants. The variation with y must be the same in both regions, in order to satisfy the boundary conditions for all y along the interface. The solution given in (22) is valid for all integer values of m. The ratio of the coefficients, the wavenumbers l, h, and propagation constant γ will be determined by the boundary conditions which must hold at $x = d$. From (1a), it is found that $E_z = j\omega\mu_0 e^{-\gamma z} \partial\psi_h/\partial y$, while $H_y = (\partial^2\psi_h/\partial y\, \partial x)e^{-\gamma z}$.

In order for E_z and H_y to be continuous at the interface, the following conditions are imposed on the solution given by (2):

$$A \sin hd = B \sin lt \tag{4a}$$

$$Ah \cos hd = -Bl \cos lt. \tag{4b}$$

By dividing (4a) by (4b), the transcendental equation

$$h \tan lt = -l \tan hd \tag{5a}$$

is obtained, which together with the relation

$$l^2 = h^2 + (\kappa - 1)k_0^2 \tag{5b}$$

derivable from (3) determines an infinite number of solutions for the wavenumbers l and h. An examination of the expressions for the remaining field components shows that they satisfy the proper boundary conditions at the interface when E_z and H_y do. When l and h become large, they approach equality, and the nth solution to (5a) approaches $n\pi/a$; hence l_n and h_n approach $n\pi/a$ for large n. By eliminating the coefficient B by means of (4a), the solution for the nmth LSE mode may be written as

$$\psi_{h,nm}e^{-\gamma_{nm}z} = A_{nm}e^{-\gamma_{nm}z} \begin{cases} \sin h_n x \cos \dfrac{m\pi y}{b}, & 0 \le x \le d \\[2ex] \dfrac{\sin h_n d}{\sin l_n t} \sin l_n(a-x) \cos \dfrac{m\pi y}{b}, & d \le x \le a \end{cases} \tag{6}$$

where A_{nm} may be chosen arbitrarily. When $m = 0$, both E_z and H_y are zero, and the only field components present are H_x, E_y, and H_z. For this special case, the LSE mode degenerates into an H_{n0} mode.

The wave impedance measured in the x direction is given by

$$-\frac{E_z}{H_y} = \frac{E_y}{H_z} = -\frac{j\omega\mu_0\psi_h}{\partial\psi_h/\partial x}. \tag{7}$$

The eigenvalue equation (5a) may be obtained directly by considering the equivalent circuit of the loaded guide in the transverse plane as illustrated in Fig. 6.4. By considering propagation to take place along the x direction, the equivalent transmission-line circuit is a junction of two lines with characteristic impedance proportional to the wave impedance (inversely proportional to the wavenumbers) in the two sections and short-circuited at the ends. The wavenumbers l and h are determined by the condition that the zero impedance load (short circuit) at $x = a$ must transform into a zero impedance at $x = 0$. By conventional transmission-line theory we find that the input impedance at $x = d$ is $jl^{-1} \tan lt$, and that at $x = 0$ is

$$Z_{\text{in}} = \frac{1}{h} \frac{jl^{-1} \tan lt + jh^{-1} \tan hd}{h^{-1} - l^{-1} \tan lt \tan hd}.$$

The condition that Z_{in} should equal zero is found at once to be that given by (5a). This transverse-resonance procedure is useful for finding the eigenvalue equation for guides that are loaded with several dielectric slabs.

The eigenvalue equation for LSE modes for an inhomogeneously filled guide as illustrated

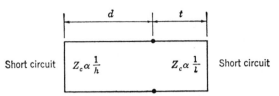

Fig. 6.4. Transverse equivalent transmission-line circuit.

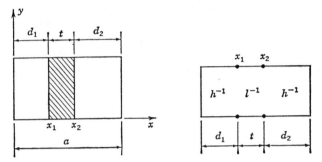

Fig. 6.5. Illustration for transverse resonance method.

in Fig. 6.5 will be determined by this transverse-resonance method. At $x = x_2$, the input impedance is $jh^{-1} \tan hd_2$. At $x = x_1$, the input impedance looking to the right must be equal to the negative value of the input impedance seen at this point looking to the left. Hence, we get

$$- jh^{-1} \tan hd_1 = \frac{1}{l} \frac{jh^{-1} \tan hd_2 + jl^{-1} \tan lt}{l^{-1} - h^{-1} \tan hd_2 \tan lt}$$

which may be simplified to

$$h^2 \tan lt + hl \tan hd_2 + lh \tan hd_1 - l^2 \tan lt \tan hd_1 \tan hd_2 = 0. \qquad (8)$$

This equation together with (3) determines the allowed solutions for l, h, and γ.

For the special case of a symmetrically placed slab, $d_1 = d_2 = d$, and our transverse-resonance condition is equivalent to the condition that a short circuit should appear at the center of the slab for asymmetrical modes, and that an open circuit should appear at $x = a/2$ for symmetrical modes. The corresponding eigenvalue equations are, respectively,

$$l \tan hd = -h \tan \frac{lt}{2} \qquad (9a)$$

$$h \cot hd = l \tan \frac{lt}{2}. \qquad (9b)$$

These equations may be obtained from (8), but are more readily determined directly by using the appropriate transverse-resonance conditions given above. The symmetrical modes are those which have a symmetrical variation of E_y with respect to x about the point $x = a/2$.

LSM Modes

The LSM modes may be derived from an electric type of Hertzian potential of the form $\Pi_e = \mathbf{a}_x \psi_e(x, y)e^{-\gamma z}$, for the case of a guide inhomogeneously filled as in Fig. 6.3. The equations determining the fields are

$$\mathbf{H} = j\omega\epsilon_0\kappa(x)\nabla \times \boldsymbol{\Pi}_e \tag{10a}$$

$$\mathbf{E} = \nabla \times \nabla \times \boldsymbol{\Pi}_e = \kappa(x)k_0^2\boldsymbol{\Pi}_e + \nabla\nabla\cdot\boldsymbol{\Pi}_e \tag{10b}$$

$$\nabla_t^2\psi_e + [\gamma^2 + \kappa(x)k_0^2]\psi_e = 0. \tag{10c}$$

These equations are valid in each region over which $\kappa(x)$ is constant, but do not hold at the interface across which $\kappa(x)$ changes discontinuously. However, we may solve for the fields in each section and match the tangential components at the interface to obtain the solution valid for all x.

Appropriate solutions for ψ_e in each region, such that the tangential electric field will vanish on the guide boundary, are

$$\psi_e = \begin{cases} A \cos hx \sin \dfrac{m\pi y}{b}, & 0 \leq x \leq d \\[2ex] B \cos l(a-x)\sin \dfrac{m\pi y}{b}, & d \leq x \leq a. \end{cases} \tag{11}$$

The field component E_z is proportional to $\partial\psi_e/\partial x$, while H_y is proportional to $\kappa(x)\psi_e$ in both regions. Continuity of these tangential field components at $x = d$ imposes the following restriction on the solution given by (11):

$$Ah \sin hd = -Bl \sin lt \tag{12a}$$

$$A \cos hd = \kappa B \cos lt. \tag{12b}$$

Dividing (12a) by (12b) yields the eigenvalue equation

$$\kappa h \tan hd = -l \tan lt \tag{13}$$

which, together with the relation $\gamma^2 - (m\pi/b)^2 = h^2 - k_0^2 = l^2 - \kappa k_0^2$, determines the allowed values of h, l, and γ. Again an infinite number of discrete solutions exist, and both h_n and l_n approach $n\pi/a$ for large n. From (12), the ratio of the coefficients A and B may be found, and, when substituted into (11), completes the solution.

For more general slab arrangements, the eigenvalue equation may be determined by the transverse-resonance method. It should be noted that the wave impedances in the x direction are proportional to h and $l\kappa^{-1}$ in the empty and dielectric-loaded sections, respectively, for LSM modes. For rectangular guides loaded with dielectric slabs which fill the guide uniformly along the x direction, but discontinuously along the y direction, the field may be found from Hertzian potentials directed along the y axis.

Orthogonality of Modes

The various orthogonality relations derived for the E and H modes in an empty guide in the preceding chapter are not all valid for the LSM and LSE modes in an inhomogeneously filled guide. However, the following orthogonality relation does hold:

$$\iint_S \mathbf{E}_{tn} \times \mathbf{H}_{tm}\cdot\mathbf{a}_z\, dS = 0, \qquad n \neq m$$

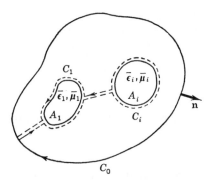

Fig. 6.6. Cross section of a general inhomogeneously filled guide.

where $\mathbf{E}_{tn}$ is the transverse electric field for the nth mode and $\mathbf{H}_{tm}$ is the transverse magnetic field for the mth mode. If the two fields involved are for one LSE and one LSM mode, the above orthogonality property holds for $n = m$ also, since these latter fields are independent nondegenerate solutions of Maxwell's equations. The above relation will be proved as a special case of a more general orthogonality property.

Consider a general cylindrical guide, with perfectly conducting walls, and inhomogeneously filled, as in Fig. 6.6. Over each cross section A_i, the medium filling that section is characterized by a general permittivity tensor $\bar{\epsilon}_i$ and a permeability tensor $\bar{\mu}_i$. In general, the modes cannot be classified as E, H, LSM, or LSE modes, but nevertheless an infinite number of discrete modes may exist.

Let $\mathbf{E}_n$, $\mathbf{H}_n$ be the solution for the nth mode, and $\mathbf{E}'_m$, $\mathbf{H}'_m$ be the solution for the mth mode, with $\bar{\mu}_i$ and $\bar{\epsilon}_i$ replaced by their transpose dyadics $(\bar{\mu}_i)_t$ and $(\bar{\epsilon}_i)_t$. Over each cross section, the curl equations for the fields are

$$\nabla \times \mathbf{E}_n = -j\omega\bar{\mu}_i \cdot \mathbf{H}_n \qquad \nabla \times \mathbf{H}_n = j\omega\bar{\epsilon}_i \cdot \mathbf{E}_n$$

$$\nabla \times \mathbf{E}'_m = -j\omega(\bar{\mu}_i)_t \cdot \mathbf{H}'_m \qquad \nabla \times \mathbf{H}'_m = j\omega(\bar{\epsilon}_i)_t \cdot \mathbf{E}'_m.$$

By forming the dot products indicated below, we get

$$j\omega[\mathbf{E}'_m \cdot \bar{\epsilon}_i \cdot \mathbf{E}_n - \mathbf{E}_n \cdot (\bar{\epsilon}_i)_t \cdot \mathbf{E}'_m - \mathbf{H}'_m \cdot \bar{\mu}_i \cdot \mathbf{H}_n + \mathbf{H}_n \cdot (\bar{\mu}_i)_t \cdot \mathbf{H}'_m]$$
$$= \nabla \cdot (\mathbf{E}_n \times \mathbf{H}'_m - \mathbf{E}'_m \times \mathbf{H}_m) = 0$$

since $\mathbf{E}'_m \cdot \bar{\epsilon}_i \cdot \mathbf{E}_n = \mathbf{E}_n \cdot (\bar{\epsilon}_i)_t \cdot \mathbf{E}'_m$, etc. Let the z dependence of the fields $\mathbf{E}_n$, $\mathbf{H}_n$ and $\mathbf{E}'_m$, $\mathbf{H}'_m$ be according to $e^{-\gamma_n z}$ and $e^{-\gamma'_m z}$, respectively. The divergence relation now gives

$$\nabla_t \cdot (\mathbf{E}_n \times \mathbf{H}'_m - \mathbf{E}'_m \times \mathbf{H}_n) = (\gamma_n + \gamma'_m)\mathbf{a}_z \cdot (\mathbf{E}_n \times \mathbf{H}'_m - \mathbf{E}'_m \times \mathbf{H}_n).$$

Integrating over the cross section A_i and converting the two-dimensional divergence integral to a contour integral gives

$$\oint_{C_i} (\mathbf{n} \times \mathbf{E}_n \cdot \mathbf{H}'_m - \mathbf{n} \times \mathbf{E}'_m \cdot \mathbf{H}_n)\, dl = (\gamma_n + \gamma'_m)\iint_{A_i} \mathbf{a}_z \cdot (\mathbf{E}_n \times \mathbf{H}'_m - \mathbf{E}'_m \times \mathbf{H}_n)\, dS.$$

The cross product $\mathbf{n} \times \mathbf{E}_n$ selects the component of $\mathbf{E}_n$ tangential to the bounding surface of section A_i. The dot product with $\mathbf{H}'_m$ subsequently selects only the tangential component of $\mathbf{H}'_m$. The tangential components are continuous, and hence we may sum over all cross-sectional areas, and the contribution from all the interior contours C_i will vanish since they are traversed twice, but in opposite senses. On the exterior contour, $\mathbf{n} \times \mathbf{E}_n$ and $\mathbf{n} \times \mathbf{E}'_m$ vanish, and hence the total contour integral is zero. Thus we obtain the final result

$$(\gamma_n + \gamma'_m) \iint_S \mathbf{a}_z \cdot (\mathbf{E}_n \times \mathbf{H}'_m - \mathbf{E}'_m \times \mathbf{H}_n)\, dS = 0. \tag{14}$$

This result is the general statement of the reciprocity principle for the electromagnetic field in a source-free guide bounded by perfectly conducting walls. For general media such as ferrites, which are characterized by nonsymmetrical tensors, the reciprocity principle involves fields which are solutions to the related problem, with media characterized by the transposed permittivity and permeability tensors.

If a guide exhibits reflection symmetry with respect to the z coordinate, then corresponding to the solutions $\mathbf{E}_n$, $\mathbf{H}_n$; $\mathbf{E}'_m$, $\mathbf{H}'_m$ with transverse and longitudinal components $\mathbf{E}_{tn}$, $\mathbf{E}_{zn}$; $\mathbf{H}_{tn}$, $\mathbf{H}_{zn}$ and $\mathbf{E}'_{tm}$, $\mathbf{E}'_{zm}$; $\mathbf{H}'_{tm}$, $\mathbf{H}'_{zm}$, respectively, there exist solutions $\mathbf{E}_{tn}$, $-\mathbf{E}_{zn}$; $-\mathbf{H}_{tn}$, $\mathbf{H}_{zn}$ and $\mathbf{E}'_{tm}$, $-\mathbf{E}'_{zm}$; $-\mathbf{H}'_{tm}$, $\mathbf{H}'_{zm}$ with eigenvalues $-\gamma_n$ and $-\gamma'_m$, respectively. This is just the definition of reflection symmetry.

For guides of the above type, we obtain from (14) the result

$$(\gamma_n - \gamma'_m) \iint_S \mathbf{a}_z \cdot (\mathbf{E}_n \times \mathbf{H}'_m + \mathbf{E}'_m \times \mathbf{H}_n)\, dS = 0 \tag{15}$$

upon replacing the solution $(\mathbf{H}'_{tm}, \mathbf{H}'_{zm}, \gamma'_m)$ by $(-\mathbf{H}'_{tm}, \mathbf{H}'_{zm}, -\gamma'_m)$. The combination of (14) and (15) gives

$$\iint_S \mathbf{a}_z \cdot \mathbf{E}_{tn} \times \mathbf{H}'_{tm}\, dS = 0 \tag{16}$$

after canceling the factors $\gamma_n + \gamma'_m$, which is permissible when $\gamma_n \neq \pm \gamma'_m$. Only the transverse field components enter into the triple scalar products in (14) and (15), the terms involving the longitudinal fields being orthogonal to $\mathbf{a}_z$.

All uniform guides inhomogeneously filled with isotropic media have reflection symmetry. The same is true of guides filled with media that have transverse anisotropy only; for example, $\mu_{xz} = \mu_{zx} = \mu_{zy} = \mu_{yz} = 0$. When the medium filling the guide is lossless, the following orthogonality relation is also valid:

$$\iint_S \mathbf{a}_z \cdot \mathbf{E}_{tn} \times \mathbf{H}^*_{tm}\, dS = 0 \tag{17}$$

where $*$ denotes the complex conjugate value.[1] In the case of the inhomogeneously filled rectangular guide, the transverse electric fields for two different LSE modes are orthogonal. The same is true for the transverse magnetic fields and the longitudinal field components.

---

[1] Much more complete discussions of general orthogonality properties of waveguides can be found in the book by Felsen and Marcuvitz [6.21].

For the LSM modes, none of these orthogonal properties hold. In place of these we have orthogonality with respect to a suitable weighting function, a topic to be considered in the next section.

The orthogonality relations (14) and (16) for guides with reflection symmetry may be used to construct the dyadic Green's function outside the source region in terms of the waveguide modes as was done in Section 5.6. The solution has the same form as that given by (87) in Chapter 5 for waveguides with reflection symmetry. At the source region a delta function term $-\mathbf{a}_z\mathbf{a}_z\delta(\mathbf{r}-\mathbf{r}')/k^2$ should be added where k is the wavenumber in the region in which the source is located. This singularity occurs because in the mode expansion each term represents the radiation from a current sheet located in the $z = z'$ plane. According to the source boundary condition given after (46) in Chapter 1 the normal electric field has a delta function singularity in the source region.

The dyadic Green's function can also be constructed using the method of scattering superposition in the manner employed in Section 3.9. Another approach is to derive solutions for the vector and scalar potential functions and then express the fields in terms of these. In general, it is not possible to find a set of $\mathbf{L}$, $\mathbf{M}$, and $\mathbf{N}$ functions in which to expand the electric field since the electric field is not usually a pure solenoidal field outside the source region in an inhomogeneously filled waveguide.

6.2. The Rayleigh–Ritz Method

The Rayleigh–Ritz method is a systematic scheme for determining a finite set of approximate eigenfunctions and eigenvalues to a given differential equation and its boundary conditions. These are obtained by means of a variational integral, whose stationary values correspond to the true eigenvalues, when the true eigenfunctions are employed in the integrand. The details of this method will be developed in this section.

As noted in the previous section, the equations for the LSM modes do not hold at the interface where $\kappa(x)$ changes discontinuously. For the LSE modes, both ψ_h and $\partial\psi_h/\partial x$ are continuous, and so the equations given for these modes are valid at the interface. To avoid the complications associated with a piecewise-continuous function $\kappa(x)$, we will assume that $\kappa(x)$ is continuous and specialize to the case of a discontinuous dielectric constant after the general equations have been derived. The theory will be developed for LSM modes, and only the final equations will be given for LSE modes.

As a first step, we need to derive the differential equation for $\mathbf{\Pi}_e = \mathbf{a}_x\psi_e e^{-\gamma z}$ for a rectangular guide filled with a dielectric whose dielectric constant $\kappa = \kappa(x)$ is a continuous function of x. Maxwell's equations in a source-free region are

$$\nabla \times \mathbf{H} = j\omega\epsilon_0\kappa\mathbf{E} \tag{18a}$$

$$\nabla \times \mathbf{E} = -j\omega\mu_0\mathbf{H} \tag{18b}$$

$$\nabla\cdot\mathbf{B} = 0 \tag{18c}$$

$$\nabla\cdot\mathbf{D} = 0 = \nabla\cdot\kappa\mathbf{E}. \tag{18d}$$

Since μ_0 is a constant $\nabla\cdot\mathbf{H} = 0$, so we may take

$$\mathbf{H} = j\omega\epsilon_0\nabla \times \mathbf{\Pi}_e. \tag{19}$$

Note that this equation differs from (10a) by the absence of the factor $\kappa(x)$. From (18b) we get $\nabla \times \mathbf{E} = k_0^2 \nabla \times \boldsymbol{\Pi}_e$, which integrates to

$$\mathbf{E} = k_0^2 \boldsymbol{\Pi}_e + \nabla \Phi$$

where Φ is as yet an arbitrary scalar function. Taking the curl of (19) [the factor $\kappa(x)$ was omitted from this equation so that the curl could be readily taken], we get

$$\nabla \times \mathbf{H} = j\omega\epsilon_0 \nabla \times \nabla \times \boldsymbol{\Pi}_e = j\omega\epsilon_0 \kappa(x)\mathbf{E}.$$

Expanding and substituting for $\mathbf{E}$ gives

$$\nabla\nabla\cdot\boldsymbol{\Pi}_e - \nabla^2\boldsymbol{\Pi}_e = \kappa(x)k_0^2\boldsymbol{\Pi}_e + \kappa(x)\,\nabla\Phi$$
$$= \kappa k_0^2\boldsymbol{\Pi}_e + \nabla\kappa\Phi - \Phi\,\nabla\kappa.$$

Since $\nabla\cdot\boldsymbol{\Pi}_e$ and Φ are arbitrary up to this point, we choose the following relation between them:

$$\nabla\nabla\cdot\boldsymbol{\Pi}_e = \nabla\kappa\Phi$$

which apart from an irrelevant constant integrates to $\kappa\Phi = \nabla\cdot\boldsymbol{\Pi}_e$. Thus, the following equation for $\boldsymbol{\Pi}_e$ is obtained:

$$\nabla^2\boldsymbol{\Pi}_e - \kappa^{-1}(\nabla\kappa)\nabla\cdot\boldsymbol{\Pi}_e + \kappa k_0^2\boldsymbol{\Pi}_e = 0. \tag{20}$$

The equation for $\mathbf{E}$ becomes

$$\mathbf{E} = k_0^2\boldsymbol{\Pi}_e + \nabla(\kappa^{-1}\nabla\cdot\boldsymbol{\Pi}_e)$$
$$= k_0^2\boldsymbol{\Pi}_e + \nabla\cdot\boldsymbol{\Pi}_e\nabla\kappa^{-1} + \kappa^{-1}\nabla\nabla\cdot\boldsymbol{\Pi}_e$$
$$= \kappa^{-1}\nabla \times \nabla \times \boldsymbol{\Pi}_e \tag{21}$$

when substitution from (20) is made. From (21) it is readily seen that $\nabla\cdot\kappa\mathbf{E} = 0$, since the divergence of the curl of any vector is identically zero. If now it is assumed that $\boldsymbol{\Pi}_e$ is of the form

$$\boldsymbol{\Pi}_e = \mathbf{a}_x \Phi_e(x) \sin \frac{m\pi y}{b} e^{-\gamma z}$$

we find that Φ_e must satisfy the equation

$$\frac{d^2\Phi_e}{dx^2} - \kappa^{-1}\frac{d\kappa}{dx}\frac{d\Phi_e}{dx} + (\kappa k_0^2 + \gamma^2 - h^2)\Phi_e = 0$$

or, equivalently,

$$\frac{d}{dx}\frac{1}{\kappa}\frac{d\Phi_e}{dx} + \frac{1}{\kappa}(\kappa k_0^2 + \gamma^2 - h^2)\Phi_e = 0 \tag{22}$$

where h^2 has been written for $(m\pi/b)^2$. In order that the tangential electric field derivable from Π_e shall vanish on the guide boundary at $x = 0, a$, the function Φ_e must satisfy the boundary conditions

$$\frac{d\Phi_e}{dx} = 0, \qquad x = 0, a. \tag{23}$$

The differential equation (22), together with the boundary conditions (23), constitutes what is known as a Sturm–Liouville system. Such a system has, among others, the following important properties:

1. An infinite number of eigenfunctions Φ_{en} with discrete eigenvalues γ_n^2 exists.
2. The eigenfunctions Φ_{en} form an orthogonal set over the closed interval $0 \leq x \leq a$, with respect to the weighting function κ^{-1}.
3. The eigenfunctions form a complete set, in terms of which a piecewise-continuous function of x defined over the interval $0 \leq x \leq a$ may be expanded.
4. The solution to the system may be formulated equivalently as the problem of finding the function which will render a certain integral stationary.

The orthogonality of the eigenfunctions is readily proved as follows. Multiply the equation for Φ_{en} by Φ_{em}, and the equation for Φ_{em} by Φ_{en}; subtracting the two resultant equations and integrating over the interval $0 \leq x \leq a$, we get

$$\int_0^a \left(\Phi_{em} \frac{d}{dx} \frac{1}{\kappa} \frac{d\Phi_{en}}{dx} - \Phi_{en} \frac{d}{dx} \frac{1}{\kappa} \frac{d\Phi_{em}}{dx} \right) dx$$

$$= (\gamma_m^2 - \gamma_n^2) \int_0^a \kappa^{-1} \Phi_{em} \Phi_{en} \, dx$$

$$= \left(\frac{\Phi_{em}}{\kappa} \frac{d\Phi_{en}}{dx} - \frac{\Phi_{en}}{\kappa} \frac{d\Phi_{em}}{dx} \right) \Big|_0^a$$

$$- \int_0^a \kappa^{-1} \left(\frac{d\Phi_{em}}{dx} \frac{d\Phi_{en}}{dx} - \frac{d\Phi_{en}}{dx} \frac{d\Phi_{em}}{dx} \right) dx = 0 \tag{24}$$

upon integrating by parts once and using the boundary conditions (23). The eigenfunctions are thus seen to be orthogonal with respect to the weighting function κ^{-1}, since $\gamma_n^2 \neq \gamma_m^2$. The eigenfunctions may be normalized so that they form an orthonormal set, i.e.,

$$\int_0^a \kappa^{-1} \Phi_{en} \Phi_{em} \, dx = \delta_{nm} \tag{25}$$

where δ_{nm} is the Kronecker delta which equals unity for $n = m$, and zero otherwise. An arbitrary piecewise-continuous function of x, say $g(x)$, may be expanded into an infinite series of these eigenfunctions as follows:

$$g(x) = \sum_{n=0}^{\infty} a_n \Phi_{en} \tag{26a}$$

where a_n is determined by multiplying both sides by $\kappa^{-1} \Phi_{en}$, integrating over $0 \leq x \leq a$, and using (25) to get

$$a_n = \int_0^a \kappa^{-1} g \Phi_{en} \, dx. \tag{26b}$$

Minimum Characterization of the Eigenvalues

If (22) is multiplied by Φ_e and integrated over the interval $0 \leq x \leq a$, we get

$$\gamma^2 \int_0^a \kappa^{-1} \Phi_e^2 \, dx = \int_0^a \kappa^{-1} \left[\left(\frac{d\Phi_e}{dx} \right)^2 - (\kappa k_0^2 - h^2) \, \Phi_e^2 \right] dx \qquad (27)$$

by integrating by parts once. The integrated term vanishes by virtue of the boundary conditions on $d\Phi_e/dx$ at $x = 0, a$. Equation (27) is a variational expression for the eigenvalue γ^2. The condition that (27) should be stationary may be obtained by imposing the condition that the first-order change or variation in γ^2 should vanish for arbitrary first-order variations $\delta\Phi_e$ in Φ_e. This, of course, is equivalent to making the integrand satisfy the Euler–Lagrange equation from the calculus of variations, and imposes on Φ_e the condition that it satisfy the differential equation (22).[2]

An extremization of (27) by that class of functions $\Phi(x)$ which are continuous, with at least a piecewise-continuous first derivative which vanishes at $x = 0, a$, and are orthogonal to the first K true eigenfunctions Φ_{en}, yields an upper bound on the true eigenvalue γ_K^2. By means of the expansion theorem for a complete set of functions, i.e., (26), we may write

$$\Phi = \sum_{n=0}^{\infty} a_n \Phi_{en}. \qquad (28a)$$

The orthogonality conditions imposed on Φ give

$$a_n = 0, \qquad n = 0, 1, 2, \ldots, K - 1. \qquad (28b)$$

Note that the Kth eigenfunction is the one labeled $K - 1$. Substituting into (27) gives

$$\gamma^2 \sum_{n=K}^{\infty} a_n^2 = \int_0^a \kappa^{-1} \sum_{s=K}^{\infty} \sum_{n=K}^{\infty} \left[\frac{d\Phi_{en}}{dx} \frac{d\Phi_{es}}{dx} - (\kappa k_0^2 - h^2) \Phi_{en} \Phi_{es} \right] a_n a_s \, dx$$

$$= \sum_{s=K}^{\infty} \sum_{n=K}^{\infty} a_n a_s \left\{ \kappa^{-1} \Phi_{es} \frac{d\Phi_{en}}{dx} \bigg|_0^a \right.$$

$$\left. - \int_0^a \left[\Phi_{es} \frac{d}{dx} \frac{1}{\kappa} \frac{d\Phi_{en}}{dx} + \kappa^{-1} (\kappa k_0^2 - h^2) \Phi_{es} \Phi_{en} \right] dx \right\}. \qquad (29)$$

The integrated term vanishes, and, substituting into (29) from the differential equation (22), we get

$$\gamma^2 \sum_{n=K}^{\infty} a_n^2 = \sum_{n=K}^{\infty} a_n^2 \gamma_n^2. \qquad (30)$$

If it is assumed that the eigenvalues have been arranged into a monotonically increasing sequence, so that $\gamma_0^2 < \gamma_1^2 < \gamma_2^2 < \gamma_3^2 < \cdots < \gamma_K^2$, the result (30) may be written as

$$\gamma^2 = \frac{\displaystyle\sum_{n=K}^{\infty} a_n^2 \gamma_n^2}{\displaystyle\sum_{n=K}^{\infty} a_n^2} = \gamma_K^2 + \frac{\displaystyle\sum_{n=K+1}^{\infty} a_n^2 (\gamma_n^2 - \gamma_K^2)}{\displaystyle\sum_{n=K}^{\infty} a_n^2} \geq \gamma_K^2 \qquad (31)$$

[2] Section A.1d.

since $\gamma_n^2 > \gamma_K^2$ for all $n > K$. Only when all $a_n = 0$ for $n > K$, that is, $\Phi \equiv \Phi_{eK}$, will $\gamma^2 = \gamma_K^2$. In general, the approximate eigenvalue is too large, that is, γ^2 is an upper bound on γ_K^2. Suitable functions to use for the extremization of (27) are the corresponding eigenfunctions for the empty guide. For the LSM modes these are

$$f_n(x) = \left(\frac{\epsilon_{0n}}{a}\right)^{1/2} \cos \frac{n\pi x}{a}, \qquad n = 0, 1, 2, \ldots \tag{32}$$

where $\epsilon_{0n} = 1$ for $n = 0$, and 2 otherwise.

The Approximate Eigenfunctions

In many practical cases the true eigenfunctions are very complex, and there is considerable advantage in working with a finite set of approximate eigenfunctions instead. In constructing this set of eigenfunctions, we do not have available any of the true eigenfunctions to make the nth approximate eigenfunction orthogonal too. However, it turns out that the $(K + 1)$st extremization of the integral by that class of functions which are orthogonal to the first K approximate eigenfunctions again yields an upper bound on the true eigenvalue γ_K^2. Also, if the eigenfunctions given by (32) for the empty guide are used for the approximation, the approximate eigenfunctions are automatically orthogonal, and, hence, the orthogonality conditions originally imposed on the function used for the $(K + 1)$st extremization may be dispensed with. Both these statements will be proved, but first we accept the results and construct a set of $N + 1$ approximate eigenfunctions. The approximate eigenfunctions and eigenvalues will be written as $\bar\phi_{en}$ and $\bar\gamma_n^2$.

For the $(K + 1)$st approximate eigenfunction, we choose the following series:

$$\bar\phi_{eK} = \sum_{n=0}^{N} a_{nK} f_n(x) \tag{33}$$

where the a_{nK} are a set of coefficients to be determined, subject to the normalization condition

$$\int_0^a \kappa^{-1} \bar\phi_{eK}^2 \, dx = \sum_{n=0}^{N} \sum_{s=0}^{N} a_{nK} a_{sK} P_{sn} = 1 \tag{34}$$

where

$$P_{sn} = \int_0^a \kappa^{-1} f_n f_s \, dx = P_{ns}. \tag{35a}$$

Substituting into (27) gives

$$\sum_{n=0}^{N} \sum_{s=0}^{N} \int_0^a \kappa^{-1} \left[\frac{df_n}{dx} \frac{df_s}{dx} - (\kappa k_0^2 - h^2 + \bar\gamma_K^2) f_n f_s \right] a_{nK} a_{sK} \, dx$$

$$= \text{a stationary quantity.} \tag{35b}$$

Let

$$T_{sn} = T_{ns} = \int_0^a \kappa^{-1} \left[\frac{df_n}{dx} \frac{df_s}{dx} - (\kappa k_0^2 - h^2) f_n f_s \right] dx \tag{36}$$

and (35b) may be written as

$$\sum_{n=0}^{N}\sum_{s=0}^{N} a_{nK} a_{sK} (T_{sn} - \bar{\gamma}_K^2 P_{sn}) = \text{a stationary quantity.} \tag{37}$$

To render (37) stationary, all the partial derivatives $\partial/\partial a_{nK}$ for $n = 0, 1, \ldots, N$ are equated to zero, and the following set of homogeneous equations is obtained:

$$\sum_{n=0}^{N} a_{nK} (T_{sn} - \bar{\gamma}_K^2 P_{sn}) = 0, \qquad s = 0, 1, 2, \ldots, N. \tag{38}$$

This set of equations constitutes a matrix eigenvalue problem of the type discussed in Section A.3. Because of the symmetry of the T_{sn} and P_{sn} matrices, all the eigenvalues $\bar{\gamma}_K^2$ are real, and the eigenvectors, whose components are the coefficients a_{nK}, form an orthogonal set with respect to the weighting factors P_{sn}. The proofs of these properties are given in the Mathematical Appendix. For (38) to hold, the determinant of the coefficients must vanish. The vanishing of this determinant determines $N + 1$ roots for $\bar{\gamma}_K^2$, which are the first $N + 1$ eigenvalues. For each root, say $\bar{\gamma}_K^2$, a set of coefficients (eigenvector) a_{nK} may be determined. This set of coefficients is unique when subjected to the normalization condition (34). Since the totality of eigenvectors so determined forms an orthonormal set, it is clear that there is no need to impose any orthogonality restriction on the functions used to extremize (27). The orthogonality relations that hold are

$$\sum_{n=0}^{N}\sum_{s=0}^{N} a_{nK} a_{sL} P_{sn} = \delta_{KL}. \tag{39}$$

Having found the coefficients a_{nK}, the approximate eigenfunction is given by (33). If N were chosen as infinite, the eigenfunctions determined by the above method would be identical with the set of true eigenfunctions, since the set of functions f_n is complete, and hence able to represent the set Φ_{en}.

To prove that $\bar{\gamma}_K^2 \geq \gamma_K^2$ requires the introduction of the concept of a class of functions. The complete set of true eigenfunctions $\Phi_{e0}, \Phi_{e1}, \ldots, \Phi_{en}, \ldots$, with eigenvalues $\gamma_0^2, \gamma_1^2, \ldots, \gamma_n^2, \ldots$, where n runs from zero to infinity, will be called the class C. The partial set C_K consists of the functions Φ_{en}, with n running from 0 to K. The set of approximate eigenfunctions $\bar{\phi}_{e0}, \ldots, \bar{\phi}_{eN}$, with eigenvalues $\bar{\gamma}_0^2, \ldots, \bar{\gamma}_N^2$, will be called the class $\bar{C}_N$. The partial set consisting of the first $K + 1$ approximate eigenfunctions will be called the class $\bar{C}_K$. We also construct one further class of functions by removing from the class $\bar{C}_N$ those functions which are orthogonal to the first K true eigenfunctions $\Phi_{e0}, \ldots, \Phi_{e(K-1)}$, and call this the class $\bar{C}_N - C_{K-1}$.

The variational integral for γ^2 may be written as follows:

$$\gamma^2(g) = \frac{\displaystyle\int_0^a \kappa^{-1}\left[\left(\frac{dg}{dx}\right)^2 - \left(\kappa k_0^2 - h^2\right) g^2\right] dx}{\displaystyle\int_0^a \kappa^{-1} g^2 \, dx}. \tag{40}$$

Consider a function $g(x)$ in the class $\bar{C}_K - C_{K-1}$. This function may be written as

$$g = \sum_{n=0}^{K} b_n \bar{\phi}_{en} \tag{41a}$$

and must satisfy the K orthogonality conditions

$$\int_0^a \kappa^{-1} g \Phi_{es} \, dx = \sum_{n=0}^{K} b_n \int_0^a \kappa^{-1} \bar{\phi}_{en} \Phi_{es} \, dx$$

$$= \sum_{n=0}^{K} b_n Q_{sn} = 0, \qquad s = 0, 1, \ldots, K-1 \tag{41b}$$

where

$$Q_{sn} = \int_0^a \kappa^{-1} \bar{\phi}_{en} \Phi_{es} \, dx.$$

These orthogonality conditions ensure that g is in the class $\bar{C}_K - C_{K-1}$. For convenience, it will also be assumed that g has been normalized so that

$$\int_0^a \kappa^{-1} g^2 \, dx = \sum_{n=0}^{K} b_n^2 = 1. \tag{42}$$

Equations (41) and (42) are a set of $K+1$ equations with $K+1$ unknowns and may be solved for the coefficients b_n. Substituting into (40), integrating, and making use of (36), (38), and (39), we readily find that

$$\gamma^2(\bar{C}_K - C_{K-1}) = \gamma^2(g) = \sum_{n=0}^{K} b_n^2 \bar{\gamma}_n^2$$

$$= \sum_{n=0}^{K} b_n^2 \bar{\gamma}_K^2 + \sum_{n=0}^{K} b_n^2 (\bar{\gamma}_n^2 - \bar{\gamma}_K^2)$$

$$= \bar{\gamma}_K^2 - \sum_{n=0}^{K} b_n^2 (\bar{\gamma}_K^2 - \bar{\gamma}_n^2) \le \bar{\gamma}_K^2. \tag{43}$$

We know that any function in the class $C - C_{K-1}$, that is, any function $\sum_{n=K}^{\infty} b_n \Phi_{en}$ with b_n arbitrary, yields $\gamma^2 \ge \gamma_K^2$. Any function in the class $\bar{C}_K - C_{K-1}$ does not have as many degrees of freedom, i.e., is not so flexible, as a function in the class $C - C_{K-1}$, and hence cannot yield a value for γ^2 less than γ_K^2. Thus we are able to state that

$$\gamma^2(\bar{C}_K - C_{K-1}) \ge \gamma^2(C - C_{K-1})$$

and, hence, by using (43), we have

$$\gamma^2(C - C_{K-1}) \le \gamma^2(\bar{C}_K - C_{K-1}) \le \bar{\gamma}_K^2. \tag{44}$$

If the function in the class $C - C_{K-1}$ is chosen as the true eigenfunction Φ_{eK}, (44) gives at once $\gamma_K^2 \leq \bar{\gamma}_K^2$, and, therefore, the approximate eigenvalues are approximations from above, i.e., upper bounds on the true eigenvalues.

Proof of Completeness

Since the machinery for proving that the set of true eigenfunctions is a complete set is now available, it seems worthwhile to give this proof here. For an arbitrary function $G(x)$ such that $dG/dx = 0$, $x = 0, a$, the set of eigenfunctions Φ_{en} is complete provided

$$\lim_{N \to \infty} \int_0^a \kappa^{-1} \left[G(x) - \sum_{n=0}^N a_n \Phi_{en} \right]^2 dx = 0$$

where

$$a_n = \int_0^a \kappa^{-1} G \Phi_{en} \, dx.$$

Since some of the first few eigenvalues γ_n^2 may be negative, we remove from $G(x)$ the part which has a component along the eigenfunctions with negative eigenvalues. (The reason for doing this will be pointed out later.) A value of M may be chosen so that all $\gamma_n^2 < 0$ for $n < M$ and all $\gamma_n^2 > 0$ for $n \geq M$. Thus, consider the modified function

$$g(x) = G(x) - \sum_{n=0}^{M-1} a_n \Phi_{en}.$$

Also let

$$g_N(x) = \frac{1}{\Delta_N} \left(g - \sum_{n=M}^N a_n \Phi_{en} \right)$$

where Δ_N is a normalization factor such that

$$\Delta_N^2 = \int_0^a \left(g - \sum_{n=M}^N a_n \Phi_{en} \right)^2 \kappa^{-1} \, dx.$$

From the definition of Δ_N it is seen that, provided we can prove Δ_N approaches zero as N goes to infinity, the completeness property will have been proved. Substitution of the function g_N into (40) gives

$$\gamma^2(g_N) = \frac{1}{\Delta_N^2} \left[\gamma^2(g) \int_0^a \kappa^{-1} g^2 \, dx - \sum_{n=M}^N a_n^2 \gamma_n^2 \right] \tag{45}$$

after integration by parts and use of the series expansion of g to arrive at the second term. The function g_N is eligible for the $(N+2)$nd minimization of the integral (40), and hence

$\gamma^2(g_N) \geq \gamma^2_{N+1}$. Since all $\gamma^2_n > 0$ for $n \geq M$, we obtain from (45) the result

$$\gamma^2(g) \int_0^a \kappa^{-1} g^2 \, dx \geq \Delta^2_N \gamma^2_{N+1} + \sum_{n=M}^N a^2_n \gamma^2_n.$$

The sum is a positive term, and hence

$$\Delta^2_N \leq \frac{\gamma^2(g) \displaystyle\int_0^a \kappa^{-1} g^2 \, dx}{\gamma^2_{N+1}}. \tag{46}$$

The terms with negative eigenvalues were removed in order to leave a positive term to permit the step leading to expression (46). The numerator is independent of N and γ^2_{N+1} approaches infinity, and consequently Δ^2_N vanishes, as N becomes infinite.

Since the functions $\sin(m\pi y/b)$ form a complete set also, we are able to state that, for a rectangular guide inhomogeneously filled in the manner considered here, the LSM modes form a complete set. Any arbitrary physical field having no x component of magnetic field may be expanded as a series of LSM modes.

Equations for LSE Modes

The equations given in Section 6.1 for the LSE modes are valid when κ is any arbitrary function of x, since these modes do not have an x component of electric field. The generating function ψ_h may be taken as

$$\psi_h(x, y) = \Phi_h(x) \cos \frac{m\pi y}{b}. \tag{47}$$

The eigenfunction Φ_h is a solution of

$$\frac{d^2 \Phi_h}{dx^2} + [\gamma^2 - h^2 + \kappa(x)k_0^2] \Phi_h = 0 \tag{48}$$

and satisfies the boundary conditions $\Phi_h = 0$, $x = 0, a$. The variational integral for γ^2 is

$$\gamma^2 \int_0^a \Phi_h^2 \, dx = \int_0^a \left[\left(\frac{d\Phi_h}{dx}\right)^2 - \left(\kappa k_0^2 - h^2\right) \Phi_h^2 \right] dx. \tag{49}$$

The approximate eigenfunctions may be found by substituting into (49) a series of the corresponding eigenfunctions for the empty guide, i.e., a series of the functions

$$f_n(x) = \left(\frac{2}{a}\right)^{1/2} \sin \frac{n\pi x}{a}, \qquad n = 1, 2, \ldots. \tag{50}$$

Thus, if we take

$$\bar{\phi}_{hK} = \sum_{n=1}^N a_{nK} f_n \tag{51a}$$

subject to the normalization condition

$$\int_0^a \bar{\phi}_{hK}^2 \, dx = \sum_{n=1}^N a_{nK}^2 = 1 \tag{51b}$$

the following set of homogeneous equations is obtained:

$$\sum_{n=1}^N a_{nK}(T_{sn} - \bar{\gamma}_K^2 \delta_{sn}) = 0, \qquad s = 1, 2, \ldots, N \tag{52}$$

where

$$T_{sn} = T_{ns} = \int_0^a \left[\frac{df_n}{dx} \frac{df_s}{dx} - (\kappa k_0^2 - h^2) f_n f_s \right] dx.$$

The approximate eigenfunctions form an orthonormal set, i.e.,

$$\sum_{n=1}^N a_{nK} a_{nL} = \delta_{KL}. \tag{53}$$

In the preceding equations, h is used to represent the quantity $m\pi/b$.

Application to Dielectric-Slab–Loaded Guides

In deriving the variational expression (27) for the propagation constant γ^2, the term $\Phi_e(d/dx)(1/\kappa)(d\Phi_e/dx)$ was integrated by parts, and the integrated term $\Phi_e(1/\kappa)(d\Phi_e/dx)$ vanished at $x = 0, a$. When κ is a discontinuous function of x, the boundary conditions on Φ_e require that both Φ_e and $(1/\kappa)(d\Phi_e/dx)$ be continuous for all values of x, and that $d\Phi_e/dx = 0$ at $x = 0, a$. Consequently, the integrated term will also vanish when κ is a discontinuous function of x, and, therefore, (27) is a valid variational expression for γ^2 for this case as well. All the previously derived formulas for both LSM and LSE modes may be applied to find the approximate eigenvalues and eigenfunctions for dielectric-slab–loaded rectangular guides.

Very often, in practice, all we want to know is the propagation constant for the dominant mode. Expressions for the propagation constant for the dominant mode for rectangular guides asymmetrically filled as in Fig. 6.7 will be given below. By using a two-mode approximation, results for γ accurate to within a few percent are obtained in most cases.

For Fig. 6.7(a), with variation along the x direction according to $\sin(\pi x/a)$, the propagation constant γ_0^2 is given by (38) and is

$$\gamma_0^2 = \frac{Q}{2|P|} - \left(\frac{Q^2}{4|P|^2} - \frac{|T|}{|P|} \right)^{1/2} \tag{54}$$

where

$$Q = T_{11}P_{00} + T_{00}P_{11} - 2T_{01}P_{01}$$
$$|P| = P_{11}P_{00} - P_{01}^2$$
$$|T| = T_{11}T_{00} - T_{01}^2.$$

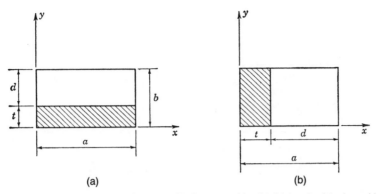

Fig. 6.7. (a) Dielectric slab along bottom wall of a waveguide. (b) Dielectric slab along side wall of a waveguide.

The matrix elements P_{ij} are given by (34) and are

$$P_{00} = \frac{1}{b} \int_0^b \kappa^{-1} \, dy = \frac{1}{b} \left(\int_0^b dy - \frac{\kappa - 1}{\kappa} \int_0^t dy \right) = 1 - \frac{\kappa - 1}{\kappa} \frac{t}{b} \qquad (55a)$$

$$P_{01} = \frac{2}{b} \left(\int_0^b \cos \frac{\pi y}{b} \, dy - \frac{\kappa - 1}{\kappa} \int_0^t \cos \frac{\pi y}{b} \, dy \right)$$

$$= -\frac{\kappa - 1}{\kappa} \frac{2}{\pi} \sin \frac{\pi t}{b} \qquad (55b)$$

$$P_{11} = \frac{2}{b} \left(\int_0^b \cos^2 \frac{\pi y}{b} \, dy - \frac{\kappa - 1}{\kappa} \int_0^t \cos^2 \frac{\pi y}{b} \, dy \right)$$

$$= P_{00} - \frac{\kappa - 1}{2\pi\kappa} \sin \frac{2\pi t}{b}. \qquad (55c)$$

From (36) the elements T_{ij} are found to be

$$T_{00} = \left(\frac{\pi}{a} \right)^2 P_{00} - k_0^2 \qquad (56a)$$

$$T_{01} = \left(\frac{\pi}{a} \right)^2 P_{01} \qquad (56b)$$

$$T_{11} = \left(\frac{\pi}{a} \right)^2 P_{11} - k_0^2 + \left(\frac{\pi}{b} \right)^2 (2P_{00} - P_{11}). \qquad (56c)$$

Because of the symmetry of the dominant mode, it follows by image theory that the above expression for γ_0^2 also gives the propagation constant for the dominant mode in a guide inhomogeneously filled as in Figs. 6.8(a) and (b).

For Fig. 6.7(b) the propagation constant for the dominant mode (H_{10} mode) is given by (52) as follows:

$$\gamma_1^2 = \frac{T_{11} + T_{22}}{2} - \left[\frac{(T_{11} + T_{22})^2}{4} - |T| \right]^{1/2} \qquad (57)$$

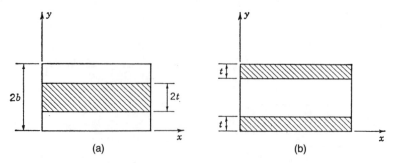

Fig. 6.8. Symmetrically loaded guides obtained by image theory.

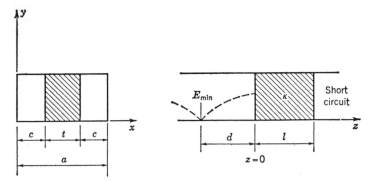

Fig. 6.9. A dielectric step discontinuity.

where $|T| = T_{11}T_{22} - T_{12}^2$. The matrix elements are

$$T_{11} = \left(\frac{\pi}{a}\right)^2 - k_0^2 - (\kappa - 1)k_0^2 \left(\frac{t}{a} - \frac{1}{2\pi} \sin \frac{2\pi t}{a}\right) \tag{58a}$$

$$T_{12} = (\kappa - 1)k_0^2 \left(\frac{1}{3\pi} \sin \frac{3\pi t}{a} - \frac{1}{\pi} \sin \frac{\pi t}{a}\right) \tag{58b}$$

$$T_{22} = 4\left(\frac{\pi}{a}\right)^2 - k_0^2 - (\kappa - 1)k_0^2 \left(\frac{t}{a} - \frac{1}{4\pi} \sin \frac{4\pi t}{a}\right). \tag{58c}$$

6.3. A Dielectric Step Discontinuity

In this section an example of how the Rayleigh–Ritz method may be employed to obtain an approximate solution for the equivalent circuit of the junction of an empty guide and inhomogeneously filled rectangular guide will be given. The problem to be treated is illustrated in Fig. 6.9. For $z > 0$, the guide is symmetrically loaded with a dielectric slab of thickness t and dielectric constant κ. An H_{10} mode is assumed incident on the step from the region $z < 0$. From symmetry considerations it is apparent that only those modes which are symmetrical about $x = a/2$ will be excited by the discontinuity. It will be assumed that only the dominant mode propagates. The only modes present are the H_{n0} modes (degenerate LSE modes) with n odd. The field components present are E_y, H_x, and H_z. The electric field E_y is proportional to the eigenfunctions $\Phi_h(x)e^{-\gamma z}$, while H_x is proportional to $\partial E_y/\partial z$. In order to make E_y and H_x continuous at $z = 0$, it is thus necessary to make E_y and $\partial E_y/\partial z$ continuous at $z = 0$.

As a first step, we have to determine a set of approximate eigenfunctions for the region $z > 0$. Only two modes will be taken into account, in order to keep the analysis relatively simple. The approximate eigenfunctions are taken as

$$\Phi_1 e^{\pm\gamma_1 z} = e^{\pm\gamma_1 z}(a_{11}f_1 + a_{31}f_3) \tag{59a}$$

$$\Phi_3 e^{\pm\gamma_3 z} = e^{\pm\gamma_3 z}(a_{13}f_1 + a_{33}f_3) \tag{59b}$$

where $f_n = (2/a)^{1/2} \sin(n\pi x/a)$. The eigenvector components a_{ij} and propagation constants γ_j are determined by a solution of the matrix eigenvalue system

$$\sum_{n=1,3} a_{nj}(T_{nj} - \gamma^2 \delta_{nj}) = 0, \qquad j = 1, 3 \tag{60}$$

subject to the normalization conditions

$$\sum_{n=1,3} a_{nj}^2 = a_{1j}^2 + a_{3j}^2 = 1, \qquad j = 1, 3.$$

The matrix elements T_{nj} are given by

$$T_{11} = \left(\frac{\pi}{a}\right)^2 - k_0^2 - (\kappa - 1)k_0^2 \left(\frac{t}{a} + \frac{1}{\pi} \sin \frac{\pi t}{a}\right) \tag{61a}$$

$$T_{13} = (\kappa - 1)k_0^2 \left(\frac{1}{\pi} \sin \frac{\pi t}{a} + \frac{1}{2\pi} \sin \frac{2\pi t}{a}\right) \tag{61b}$$

$$T_{33} = 9\left(\frac{\pi}{a}\right)^2 - k_0^2 - (\kappa - 1)k_0^2 \left(\frac{t}{a} + \frac{1}{3\pi} \sin \frac{3\pi t}{a}\right). \tag{61c}$$

The propagation constants γ_j^2 are the roots of the determinant of the coefficients a_{nj} in (60). The two modes for the region $z < 0$ corresponding to (59) are

$$f_1 e^{\pm\Gamma_1 z} \tag{62a}$$

$$f_3 e^{\pm\Gamma_3 z} \tag{62b}$$

where $\Gamma_n^2 = (n\pi/a)^2 - k_0^2$. The electric field in the two regions $z > 0$ and $z < 0$ may be written as a series expansion in terms of the functions given in (59) and (62). If the output guide is assumed to extend to infinity, the amplitude coefficients and reflection and transmission coefficients are complex. To avoid this complication in the subsequent numerical work, it is preferable to terminate the output guide at $z = l$ by a short circuit. All amplitude coefficients now turn out to be real quantities.

A suitable expansion of the electric field in the two regions $z < 0$ and $z > 0$ is

$$E_y = \begin{cases} A_1 \sin |\Gamma_1|(z + d)f_1(x) + A_3 f_3(x)e^{\Gamma_3 z}, & z < 0 \tag{63a} \\[2mm] B_1 \sin |\gamma_1|(z - l)\Phi_1(x) + B_3 \Phi_3(x)e^{-\gamma_3 z}, & z > 0 \tag{63b} \end{cases}$$

where d is seen to correspond to a position where the electric field is zero in the region $z < 0$. It is assumed that l is chosen large enough so that the mode $\Phi_3 e^{-\gamma_3 z}$ is negligible at $z = l$. If this is not the case, $e^{-\gamma_3 z}$ must be replaced by $\sinh \gamma_3(z - l)$. Equations (63) are, of course, only an approximation, since, in general, an infinite number of modes are present. However, this simple two-term approximation does yield useful results in practical cases. To determine the amplitude coefficients A_i and B_i, E_y and $\partial E_y/\partial z$ are made continuous at $z = 0$. These boundary conditions yield the two equations

$$A_1 f_1 \sin |\Gamma_1| d + A_3 f_3 = -B_1 \Phi_1 \sin |\gamma_1| l + B_3 \Phi_3$$

$$|\Gamma_1| A_1 f_1 \cos |\Gamma_1| d + A_3 \Gamma_3 f_3 = |\gamma_1| B_1 \Phi_1 \cos |\gamma_1| l - \gamma_3 B_3 \Phi_3.$$

By substituting for Φ_i from (59), and letting $A_1 \cos |\Gamma_1| d = A_1'$, and $B_1 \cos |\gamma_1| l = B_1'$, the coefficients of f_1 and f_3 may be equated, and we get

$$A_1' \tan |\Gamma_1| d + B_1' a_{11} \tan |\gamma_1| l - a_{13} B_3 = 0 \tag{64a}$$

$$A_3 + B_1' a_{31} \tan |\gamma_1| l - a_{33} B_3 = 0 \tag{64b}$$

$$|\Gamma_1| A_1' - |\gamma_1| a_{11} B_1' + \gamma_3 a_{13} B_3 = 0 \tag{64c}$$

$$\Gamma_3 A_3 - |\gamma_1| a_{31} B_1' + \gamma_3 a_{33} B_3 = 0. \tag{64d}$$

For a solution, the determinant of the coefficients must vanish, and this gives

$$\tan |\Gamma_1| d = \frac{A + B \tan |\gamma_1| l}{C + D \tan |\gamma_1| l} \tag{65}$$

where

$$A = -|\Gamma_1 \gamma_1| a_{13} a_{31}$$

$$B = |\Gamma_1| [\gamma_3 a_{11} a_{33} + \Gamma_3 (a_{11} a_{33} - a_{13} a_{31})]$$

$$C = |\gamma_1| [\gamma_3 (a_{31} a_{13} - a_{11} a_{33}) - \Gamma_3 a_{11} a_{33}]$$

$$D = \Gamma_3 \gamma_3 a_{13} a_{31}.$$

An examination of these latter equations shows that (65) is independent of any normalization condition on the a_{ij}, and thus we may take $a_{11} = a_{33} = 1$. Since the approximate eigenfunctions are orthogonal, $a_{13} = -a_{31}$, and hence the following simplified equations are obtained:

$$A = |\Gamma_1 \gamma_1| a_{13}^2 \tag{66a}$$

$$B = |\Gamma_1| [\gamma_3 + \Gamma_3 (1 + a_{13}^2)] \tag{66b}$$

$$C = -|\gamma_1| [\gamma_3 (1 + a_{13}^2) + \Gamma_3] \tag{66c}$$

$$D = -\Gamma_3 \gamma_3 a_{13}^2. \tag{66d}$$

From the coefficients A, B, C, D, a variety of equivalent circuits may be determined. Some of these were given at the end of Chapter 5. For the problem considered here, an equivalent circuit of the form given in Fig. 6.10 is chosen. The normalized characteristic impedance of the output line is chosen equal to Γ_1/γ_1, which is equal to the ratio of the wave impedances of the dominant modes on the input and output sides. From Chapter 5, the equivalent-circuit

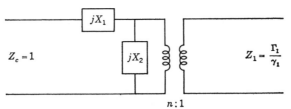

Fig. 6.10. Equivalent circuit for a dielectric step.

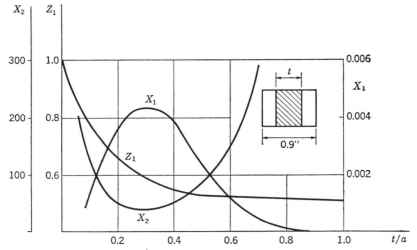

Fig. 6.11. Equivalent-circuit parameters for a dielectric step; $\kappa = 2.56$, $k_0 = 2$ rads/cm.

parameters are given by

$$X_1 = -\frac{A}{C}$$

$$X_2 = \frac{A}{C} - \frac{B}{D}$$

$$n^2 Z_1 = \frac{AD}{C^2} - \frac{B}{C}.$$

The equivalent-circuit parameters have been computed for $k_0^2 = 4$, $\kappa = 2.56$, $a = 0.90$ inch, and t ranging from 0 to a, and are plotted as a function of t/a in Fig. 6.11. The turns ratio $n:1$ is found to be equal to unity for all values of t. The series reactance X_1 is very small, while the shunt reactance X_2 is large. For all practical purposes, the equivalent circuit is just a junction of two transmission lines, with characteristic impedances proportional to the wave impedances of the H_{10} mode in the two regions.

A number of different junction discontinuity problems have been analyzed, using the method described above but with many more modes, by Villeneuve. It has been found that for most cases the ideal transformer turns ratio is very close to unity [6.1].

6.4. Ferrite Slabs in Rectangular Guides

The general theory of wave propagation in waveguides inhomogeneously filled with anisotropic ferrite material is very complex, and no attempt will be made to present such

---

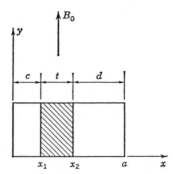

Fig. 6.12. Ferrite slab in a rectangular guide.

a theory here. There are special cases, however, which have considerable practical importance, and at the same time are quite readily analyzed. One particular case which has received considerable attention is illustrated in Fig. 6.12. The static magnetizing field B_0 is applied along the y axis, and hence the permeability dyadic is of the form

$$\bar{\mu} = \mathbf{a}_x\mathbf{a}_x\mu - \mathbf{a}_x\mathbf{a}_z j\kappa + \mathbf{a}_y\mathbf{a}_y\mu_0 + \mathbf{a}_z\mathbf{a}_x j\kappa + \mathbf{a}_z\mathbf{a}_z\mu. \tag{67}$$

The dielectric permittivity of the ferrite is ϵ, and will be written as ϵ throughout this section. The Greek letter κ is reserved for representing the off-diagonal components in the permeability dyadic. Any losses which are present can be taken into account by making μ, κ, and ϵ complex.

The off-diagonal terms in $\bar{\mu}$ result in a coupling between the field components H_x and H_z. This coupling is independent of y, and the structure is uniform along the y direction, and, consequently, a solution for modes of the H_{n0} type may be found. These modes may be derived from the single electric field component E_y which is present. Since solutions corresponding to waves propagating along the z direction are possible, we may take

$$E_y = \psi(x)e^{-\gamma z}.$$

From the curl equation for the electric field, it is then found that, in the ferrite slab,

$$\mathbf{a}_y \times \nabla[\psi(x)e^{-\gamma z}] = j\omega\bar{\mu}\cdot\mathbf{H}. \tag{68}$$

From this equation we obtain

$$j\omega\mu H_x + \kappa\omega H_z = -\gamma\psi e^{-\gamma z} \tag{69a}$$

$$-\kappa\omega H_x + j\omega\mu H_z = -\frac{d\psi}{dx}e^{-\gamma z}. \tag{69b}$$

Solving for H_x and H_z gives

$$H_x = \frac{j\omega\mu\gamma\psi - \kappa\omega(d\psi/dx)}{\omega^2(\mu^2 - \kappa^2)}e^{-\gamma z} \tag{70a}$$

$$H_z = \frac{\gamma\kappa\omega\psi + j\omega\mu(d\psi/dx)}{\omega^2(\mu^2 - \kappa^2)}e^{-\gamma z}. \tag{70b}$$

To obtain the equation satisfied by ψ in the ferrite medium, scalar multiply (68) by the transposed dyadic $\bar{\mu}_t$ to get

$$\bar{\mu}_t \cdot \mathbf{a}_y \times \nabla(\psi e^{-\gamma z}) = j\omega\bar{\mu}_t \cdot \bar{\mu} \cdot \mathbf{H} = j\omega(\mu^2 - \kappa^2)\mathbf{H}$$

since $\mathbf{H}$ has no y component. Taking the curl of both sides and substituting $\mathbf{a}_y j\omega\epsilon\psi e^{-\gamma z}$ for $\nabla \times \mathbf{H}$ gives

$$\nabla \times [\bar{\mu}_t \cdot \mathbf{a}_y \times \nabla(\psi e^{-\gamma z})] = -\mathbf{a}_y\omega^2\epsilon(\mu^2 - \kappa^2)\psi e^{-\gamma z}.$$

Expansion of the left-hand side of this equation yields the final result, which is

$$\frac{d^2\psi}{dx^2} + \left(\gamma^2 + \frac{\mu^2 - \kappa^2}{\mu}\omega^2\epsilon\right)\psi = 0. \tag{71}$$

The parameter $(\mu^2 - \kappa^2)/\mu$ will be called the effective permeability, and will be written as μ_e. In the empty part of the guide, the solution is obtained by replacing μ, ϵ by μ_0, ϵ_0 and putting $\kappa = 0$ in (70) and (71).

A general solution for ψ, which vanishes at $x = 0, a$, is readily found to be

$$\psi = \begin{cases} D \sin hx, & 0 \leq x \leq x_1 \\ A \sin l(x - x_1) + B \sin l(x - x_2), & x_1 \leq x \leq x_2 \\ C \sin h(a - x), & x_2 \leq x \leq a. \end{cases} \tag{72}$$

The particular form for ψ in the region $x_1 \leq x \leq x_2$ was chosen to facilitate matching the tangential fields at $x = x_1, x_2$. In order for (72) to be a solution of (71) and the corresponding equation for the empty guide, the following restrictions on the transverse wavenumbers h and l must be imposed:

$$\gamma^2 = h^2 - k_0^2 = l^2 - k^2 \tag{73}$$

where $k^2 = \omega^2\mu_e\epsilon$.

At $x = x_1, x_2$, E_y and hence ψ must be continuous; therefore

$$D \sin hc = -B \sin lt \tag{74a}$$

$$C \sin hd = A \sin lt. \tag{74b}$$

Also, at $x = x_1, x_2$, the tangential magnetic field component H_z must be continuous. Using (70b) and (72) the following two equations are obtained:

$$-Dh \cos hc = -\frac{\mu_0}{\mu^2 - \kappa^2}(j\gamma\kappa B \sin lt + \mu lB \cos lt + \mu lA) \tag{75a}$$

$$-hC \cos hd = \frac{\mu_0}{\mu^2 - \kappa^2}(-j\gamma\kappa A \sin lt + \mu lA \cos lt + \mu lB). \tag{75b}$$

A solution for the coefficients A, B, C, D exists only if the determinant of the set of equations (74) and (75) vanishes. The resulting equation is the eigenvalue equation which, together with (73), determines the wavenumbers l, h and the propagation factor γ. The eigenvalue equation is easily set up by substituting for B and C from (74) and (75), and is

$$h \cot hc \left(\frac{\mu_0}{\mu_e} l \cot lt - \frac{j\gamma\kappa\mu_0}{\mu\mu_e} \right) + h \cot hd \left(\frac{\mu_0}{\mu_e} l \cot lt + \frac{\mu_0}{\mu\mu_e} j\gamma\kappa \right)$$

$$+ h^2 \cot hc \cot hd + \left(\frac{\mu_0}{\mu_e} \right)^2 \left(\frac{\gamma^2\kappa^2}{\mu^2} - l^2 \right) = 0. \quad (76)$$

For the special case when $c = 0$, (76) reduces to

$$\frac{\mu_0}{\mu_e} l \cot lt - \frac{j\gamma\kappa\mu_0}{\mu\mu_e} + h \cot hd = 0 \qquad (77a)$$

and, when $d = 0$, it becomes

$$\frac{\mu_0}{\mu_e} l \cot lt + \frac{j\gamma\kappa\mu_0}{\mu\mu_e} + h \cot hc = 0. \qquad (77b)$$

These two eigenvalue equations are not the same and show that the propagation factor in the positive z direction will be different for a ferrite slab placed along the guide wall at $x = 0$, as compared with the propagation factor obtained with the slab against the guide wall at $x = a$. For propagation in the negative z direction, γ is replaced by $-\gamma$ in (76) and (77). This has the effect of interchanging the two equations (77a) and (77b). This nonreciprocal propagation behavior may be understood in a qualitative way from a consideration of the properties of the magnetic field for an H_{10} mode in an empty guide. In the empty guide the magnetic field for the H_{10} mode is given by

$$j\omega\mu_0 H_x = \pm \Gamma \sin \frac{\pi x}{a} e^{\pm \Gamma z}$$

$$j\omega\mu_0 H_z = -\frac{\pi}{a} \cos \frac{\pi x}{a} e^{\pm \Gamma z}$$

where

$$\Gamma^2 = \left(\frac{\pi}{a} \right)^2 - k_0^2.$$

Since Γ is imaginary, the magnetic field is circularly polarized, but of opposite sense, at the two positions where $|H_x| = |H_z|$. These positions are determined by a solution of the equation

$$\sin \frac{\pi x}{a} = \frac{\lambda_0}{2a}.$$

Since the transverse field H_x changes sign when the direction of propagation is reversed, the positions where the total magnetic field is positive and negative circularly polarized are interchanged. If a ferrite slab is placed in a circularly polarized magnetic field, its effective permeability is different for the two circular polarizations, and hence the perturbation of the field by the ferrite slab depends on the direction of propagation. These effects exist when the magnetic field is elliptically polarized as well. The nonreciprocal propagation is brought about because of the reversal in sign of H_x, but not H_z, when the direction of propagation is changed. A reversal in direction of the applied static field B_0 has the same effect as changing

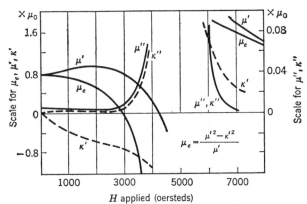

Fig. 6.13. Permeability components for Ferramic R_1 at 9 gHz, $\epsilon = (11.9 - j4 \times 10^{-4})\epsilon_0$; the double-primed parameters arise from the losses in the material. (Adapted by permission from C. E. Fay, "Ferrite-tuned resonant cavities," *Proc. IRE*, vol. 44, p. 1446, Fig. 1, Oct. 1956.)

the direction of propagation or changing the position of the ferrite slab to the other side of the center position $x = a/2$. A guide symmetrically loaded with ferrite slabs, and with the applied static field in the same direction for all slabs, will not exhibit a nonreciprocal propagation property.

The particular case of a ferrite slab placed along one wall, i.e., $c = 0$, will now be examined in greater detail. In order to obtain numerical values for the various parameters of interest, a knowledge of the intrinsic parameters μ, κ, and ϵ for the ferrite is required. Typical measured values are illustrated in Fig. 6.13. For a frequency well above the resonant frequency of the magnetization (weak applied static field), the losses are negligible. We may choose a value of applied static field such that the effective permeability μ_e is positive, zero, or even negative. Thus the product $\mu_e\epsilon$ may be chosen greater or smaller than the free-space value $\mu_0\epsilon_0$.

The wavenumbers l and h are solutions of the two equations (73) and (77a). For numerical evaluation of these parameters it is convenient to introduce the two functions

$$f_1(\beta) = l \cot lt + \frac{\mu_e}{\mu_0} h \cot hd \tag{78a}$$

$$f_2(\beta) = \frac{-\beta\kappa}{\mu} \tag{78b}$$

where γ has been placed equal to $j\beta$. When $f_1 = f_2$, (77a) is satisfied. By choosing a value of β, both l and h are determined, and f_1 and f_2 may be evaluated. A plot of f_1 and f_2 as a function of β for two values of t equal to $a/4$ and $a/2$ is given in Fig. 6.14. The assumed parameters for this example are $\mu' = \mu = 0.8\mu_0$, $\kappa' = \kappa = 0.6\mu_0$, $\epsilon = 10\epsilon_0$, $k_0 = 2$, $a = 2.286$ centimeters, and the losses are neglected so that $\mu'' = \kappa'' = 0$.

The intersection of the curves $f_1 = \pm f_2$ determines the propagation constants for the forward- and backward-propagating dominant modes. These are designated as β_+ and β_- in Fig. 6.14. For large values of t, only a single forward-and-backward–propagating mode exists. For small values of t, we find that two forward-and-backward–propagating modes may exist. The modes for $\beta > 3.72$ have both l and h imaginary, and hence have transverse field components which are described by the hyperbolic functions in both the empty and the filled portions of the guide. These modes bear a close resemblance to the surface waves, which

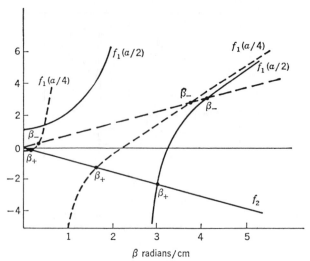

Fig. 6.14. Plot of f_1 and f_2 for finding the propagation constants for a ferrite-loaded rectangular guide.

may be propagated along the surface of a thin ferrite sheet backed by a conducting plane and located in free space. The modes for $\beta < 3.72$ are the normal perturbed waveguide modes.

Equation (77a) does not have a solution for real values of γ. Consequently, all the higher order modes have complex propagation constants. When γ is complex, the transverse wavenumbers l and h are also complex, and this makes the numerical evaluation of the eigenvalues very laborious to carry out.

A Variational Formulation

A general study of a rectangular guide partially filled with a transversely magnetized slab is facilitated if it is assumed that the parameters μ, κ, and ϵ are all continuous functions of x. The resulting differential equation for the transverse electric field is considerably more complicated in this case, since it must take into account the behavior of the field at the air–ferrite interface, if such an interface exists. The derivation of the required equation may proceed directly from the field equations provided we take μ, κ, and ϵ as functions of x. The following alternative procedure leads to the desired end result in a somewhat simpler manner.

For any propagating mode in a lossless guide, the time-average electric and magnetic energy densities are equal. By minimizing the integral of the energy function $W_m - W_e$ over the guide cross section, we are led to a variational integral formulation for the problem. The condition imposed on the minimizing function, in order that it shall render the integral of $W_m - W_e$ stationary, is that it be a solution of the required differential equation that we are seeking.

Let the transverse electric field in the guide be represented by $E_y = \psi(x)e^{-\gamma z}$. The magnetic field and magnetic flux density components are given by

$$j\omega B_x = -\gamma\psi e^{-\gamma z} \qquad j\omega B_z = -\psi' e^{-\gamma z}$$

$$\omega^2(\mu^2 - \kappa^2)H_x = (j\omega\mu\gamma\psi - \kappa\omega\psi')e^{-\gamma z}$$

$$\omega^2(\mu^2 - \kappa^2)H_z = (\gamma\kappa\omega\psi + j\omega\mu\psi')e^{-\gamma z}$$

where $\psi' = d\psi/dx$. For a propagating mode ψ is a real function. The time-average magnetic energy density is

$$W_m = \frac{1}{4}\mathbf{B}\cdot\mathbf{H}^* = \frac{\beta^2\mu\psi^2 + 2\kappa\beta\psi\psi' + \mu(\psi')^2}{4\omega^2(\mu^2 - \kappa^2)}$$

where γ has been replaced by $j\beta$. The time-average electric energy density is $W_e = (\epsilon/4)\psi^2$. Forming the difference function $4(W_m - W_e)$, introducing the effective permeability $\mu_e = (\mu^2 - \kappa^2)/\mu$, and integrating over x, we get

$$I = \int_0^a (\omega^2\mu_e)^{-1}\left[\beta^2\psi^2 + \frac{2\kappa}{\mu}\beta\psi\psi' + (\psi')^2 - \omega^2\mu_e\epsilon\psi^2\right]dx. \tag{79}$$

We will now impose on the function ψ the condition that it shall render the integral I stationary. Let ψ vary by a small amount $\delta\psi$. The variation δI in I becomes

$$\delta I = 2\int_0^a (\omega^2\mu_e)^{-1}\left[(\beta^2 - \omega^2\mu_e\epsilon)\psi\,\delta\psi + \psi'\,\delta\psi' + \frac{\kappa}{\mu}\beta\psi\,\delta\psi' + \frac{\kappa}{\mu}\beta\psi'\,\delta\psi\right]dx.$$

Consider the term

$$2\int_0^a (\omega^2\mu_e)^{-1}\left(\psi' + \frac{\kappa}{\mu}\beta\psi\right)\delta\psi'\,dx = 2(\omega^2\mu_e)^{-1}\left(\psi' + \frac{\kappa}{\mu}\beta\psi\right)\delta\psi\Big|_0^a$$

$$-2\int_0^a \left[\frac{d}{dx}\frac{\psi' + (\kappa/\mu)\beta\psi}{\omega^2\mu_e}\right]\delta\psi\,dx$$

upon integrating by parts once. The integrated term is proportional to H_z times $\delta\psi$. Since H_z is continuous, this term will vanish, provided $\delta\psi$ is continuous and vanishes at $x = 0, a$. Using this latter result in the expression for δI gives

$$\delta I = 2\int_0^a \delta\psi\left[\frac{\kappa}{\omega^2\mu\mu_e}\beta\psi' - \frac{d}{dx}\frac{\psi' + \beta\psi\kappa/\mu}{\omega^2\mu_e} + \frac{(\beta^2 - \omega^2\mu_e\epsilon)\psi}{\omega^2\mu_e}\right]dx.$$

Since $\delta\psi$ is arbitrary, the variation δI will vanish provided

$$\frac{d}{dx}\left(\frac{\psi'}{\mu_e} + \frac{\kappa}{\mu\mu_e}\beta\psi\right) - \frac{\kappa}{\mu\mu_e}\beta\psi' + \frac{k^2 - \beta^2}{\mu_e}\psi = 0 \tag{80}$$

where $k^2 = \omega^2\mu_e\epsilon$. Equation (80) is the required equation for the function ψ. In the derivation we considered only propagating modes for which $\gamma\gamma^* = +\beta^2$. However, (80) is perfectly general and holds for all the modes, and β in this equation may be replaced by $-j\gamma$. For nonpropagating modes both γ and ψ are complex. Replacing β by $-j\gamma$, we obtain

$$\frac{d}{dx}\left(\frac{\psi'}{\mu_e} - j\gamma\frac{\kappa}{\mu\mu_e}\psi\right) + \frac{j\gamma\kappa}{\mu\mu_e}\psi' + \frac{\gamma^2 + k^2}{\mu_e}\psi = 0. \tag{81}$$

The above equation clearly shows that, if $\psi(x)$ with an eigenvalue γ is a solution, then $\psi(x)$ with an eigenvalue $-\gamma$ is a solution to the problem with a medium characterized by the transposed permeability dyadic. A transposition of the permeability dyadic is equivalent to changing the sign of κ (reversing the d-c magnetizing field changes the sign of κ), and, if the sign of γ is changed at the same time, the differential equation remains invariant, since γ and κ occur as a product in the differential equation. We may also readily see from the form of the equation that a solution for γ pure-real cannot exist.

Let $\mathcal{L}$ be the differential operator occurring in (80), such that $\mathcal{L}\psi = 0$ represents the differential equation (80). The adjoint differential equation corresponding to the adjoint differential operator $\mathcal{L}_a$ operating on a function Φ may be defined according to the relation

$$\int_0^a \Phi \mathcal{L}\psi \, dx = \int_0^a \psi \mathcal{L}_a \Phi \, dx. \tag{82}$$

Multiplying (80) by Φ and integrating by parts once gives

$$\Phi\left(\frac{\psi'}{\mu_e} + \sigma\beta\psi\right)\Big|_0^a - \int_0^a \left[\Phi'\frac{\psi'}{\mu_e} + \sigma\beta(\psi\Phi' + \Phi\psi') - \frac{k^2 - \beta^2}{\mu_e}\psi\Phi\right] dx = 0$$

where $\kappa/\mu\mu_e$ has been denoted by σ. A further integration by parts to eliminate ψ' yields

$$\frac{\Phi\psi' - \psi\Phi'}{\mu_e}\Big|_0^a + \int_0^a \psi\left[\frac{d}{dx}\left(\frac{\Phi'}{\mu_e} + \sigma\beta\Phi\right) - \sigma\beta\Phi' - \frac{k^2 - \beta^2}{\mu_e}\Phi\right] dx = 0. \tag{83}$$

Adding and subtracting $\beta\sigma\Phi\psi$ from the integrated part gives

$$\left[\Phi\left(\frac{\psi'}{\mu_e} + \sigma\beta\psi\right) - \psi\left(\frac{\Phi'}{\mu_e} + \sigma\beta\Phi\right)\right]\Big|_0^a.$$

These terms will vanish provided Φ and $\Phi'/\mu_e + \beta\sigma\Phi$ are continuous functions of x and $\Phi(0) = \Phi(a) = 0$. This is true for the function ψ, since ψ is proportional to the electric field E_y, and $\psi'/\mu_e + \beta\sigma\psi$ is proportional to the magnetic field H_z. Thus we find by comparing (83) and (82) that the adjoint operator $\mathcal{L}_a$ is equal to the operator $\mathcal{L}$. Since Φ and ψ also satisfy identical boundary conditions, the differential equation (80) is self-adjoint.

Because of the self-adjoint property of (80) we may obtain an orthogonality relation for the eigenfunctions ψ_m of this equation. If the equation were not self-adjoint, the orthogonality relations would involve the eigenfunctions of the adjoint operator $\mathcal{L}_a$ as well. Let ψ_n, ψ_m be two eigenfunctions of (80) with nonidentical eigenvalues β_n, β_m. Multiplying the equation for ψ_n by ψ_m, the equation for ψ_m by ψ_n, subtracting, and performing an integration by parts, we get

$$\left[\psi_m\left(\frac{\psi_n'}{\mu_e} + \sigma\beta_n\psi_n\right) - \psi_n\left(\frac{\psi_m'}{\mu_e} + \sigma\beta_m\psi_m\right)\right]\Big|_0^a$$

$$- \int_0^a \left[\psi_m'\left(\frac{\psi_n'}{\mu_e} + \sigma\beta_n\psi_n\right) - \psi_n'\left(\frac{\psi_m'}{\mu_e} + \sigma\beta_m\psi_m\right) + \sigma\beta_n\psi_m\psi_n'\right.$$

$$\left. - \sigma\beta_m\psi_n\psi_m' + \frac{\beta_n^2 - \beta_m^2}{\mu_e}\psi_n\psi_m\right] dx = 0.$$

The integrated part vanishes and the remaining terms may be combined to give

$$(\beta_n - \beta_m)\int_0^a \left[\sigma(\psi_n\psi_m' + \psi_m\psi_n') + (\beta_n + \beta_m)\frac{\psi_n\psi_m}{\mu_e}\right] dx = 0.$$

Rearranging the terms, we may write this latter result as

$$\int_0^a \left[\psi_n \left(\sigma\psi_m' + \frac{\beta_m\psi_m}{\mu_e} \right) + \psi_m \left(\sigma\psi_n' + \frac{\beta_n\psi_n}{\mu_e} \right) \right] dx = 0. \tag{84}$$

Referring to (70a) shows that $\sigma\psi_n' + \beta_n\psi_n/\mu_e$ is proportional to the transverse magnetic field component H_x for the nth mode, and similarly for the other term in parentheses. If we note that the solution for the nth mode when the permeability dyadic is transposed is ψ_n, with an eigenvalue $-\beta_n$, and a transverse magnetic field proportional to $-\sigma\psi_n' - \beta_n\psi_n/\mu_e$, then (84) will be seen to be the same orthogonality property which was deduced directly from Maxwell's equations, i.e., equivalent to (14). The rather involved orthogonality relation given by (84) is a consequence of the fact that the differential equation (80) is a more general form than the Sturm–Liouville equation discussed in Section 6.2. It is indicative of the complex nature of electromagnetic-wave propagation in guides filled with anisotropic media. Note that the rectangular waveguide with the ferrite slab does not have reflection symmetry unless the ferrite slab is centered in the guide.

6.5. DIELECTRIC WAVEGUIDES

Dielectric waveguides are very important in integrated optical systems, in optical communications (optical fibers), and also at the shorter millimeter wavelengths. The cross sections of some typical dielectric waveguides are shown in Fig. 6.15. Analytical solutions are available for the dielectric rod of circular cross section (also with cladding) and the uniform elliptic rod. These solutions involve complicated transcendental dispersion equations that require lengthy numerical analysis to determine the propagation constants. Consequently, most of the recent work on dielectric waveguides has been based on numerical methods. The most commonly used methods are the finite-element method, the finite-difference method, and the boundary-element method (an integral equation technique).

The finite-difference method is based on the use of a finite-difference approximation to the governing partial differential equation. For the vector problem this method is not used as often as the finite-element method, which appears to be superior. The finite-element method is based on the numerical evaluation of a variational expression for the parameter k_0^2. The cross section of the waveguide is subdivided into a large number of finite rectangular or triangular elements as shown in Fig. 6.16. Each field component is approximated by a suitable polynomial function of relatively low degree and the functional to be evaluated is integrated over each element followed by a summation over all elements. The stationary property is then enforced which leads to a system of linear equations. The determinant of this system is equated to zero and gives the dispersion equation from which k_0 can be found for assumed values of the propagation constant β. An important feature associated with the finite-element method is that a variable dielectric constant is easily incorporated by assigning constant but different values for it over each element. By using a sufficiently fine mesh, the result is a good approximation to the actual values of the dielectric constant over the guide cross section.

The boundary-element method is an integral equation technique that is well suited to problems involving guides in which the dielectric constant does not vary in each section of the waveguide. It is an exact method of providing matching of the tangential field components across arbitrary surfaces joining regions with different dielectric media. The boundary-element method has not been widely used since its use for eigenvalue problems is of more recent origin. In many respects it would appear to be a more efficient method than the finite-element method in terms of the computational complexity and no doubt will find increasing use.

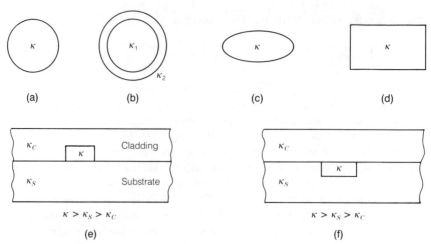

Fig. 6.15. Dielectric waveguides. (a) Circular rod. (b) Circular rod with cladding. (c) Elliptic rod. (d) Rectangular rod. (e) Embossed guide. (f) Channel guide.

A number of approximate mode-matching and field-matching methods have been employed in special circumstances. Since these methods are not general and the accuracy of the approximations is not always easy to assess we will not discuss them (some examples can be found in [6.25]–[6.28]). We will only discuss the finite-element and boundary-element methods. However, even these methods will not be developed in great detail because of space limitations.

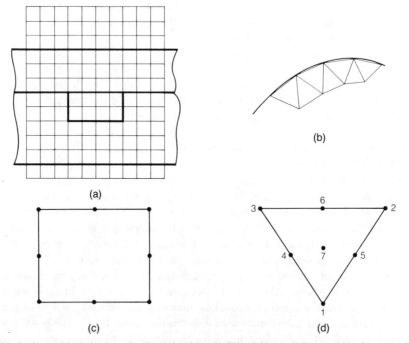

Fig. 6.16. (a) Discretization of a channel guide. (b) Triangular mesh for a guide with a curved boundary. (c) and (d) Basic elements illustrating nodes at which the unknown field amplitudes are assigned.

The Method of Finite Elements

The method of finite elements employs a variational formulation of the boundary-value problem and the numerical solution of it. Below we describe the functionals that have been found useful in treating dielectric waveguides.

Electric Field Functional: Maxwell's equations are

$$\nabla \times \mathbf{H} = j\omega\epsilon_0\bar{\kappa}\cdot\mathbf{E}, \quad \nabla \times \mathbf{E} = -j\omega\mu_0\mathbf{H}, \quad \nabla\cdot\mathbf{H} = \nabla\cdot\bar{\kappa}\cdot\mathbf{E} = 0.$$

We will assume that the dielectric tensor $\bar{\kappa}$ is real and symmetric, but may be a function of the transverse coordinates. The inverse of $\bar{\kappa}$ will be denoted by $\bar{\tau}$, i.e., $\bar{\kappa}^{-1} = \bar{\tau}$. We readily find that

$$\nabla \times (\bar{\tau}\cdot\nabla \times \mathbf{H}) = j\omega\epsilon_0\nabla \times \mathbf{E} = k_0^2\mathbf{H} \tag{85}$$

$$\nabla \times \nabla \times \mathbf{E} = k_0^2\bar{\kappa}\cdot\mathbf{E}. \tag{86}$$

We now examine the conditions under which (86) represents a self-adjoint operator using the scalar product

$$\iiint_V \mathbf{E}\cdot\mathbf{E}^* \, dV.$$

Consider

$$\iiint_V (\mathbf{E}^*\cdot\nabla \times \nabla \times \mathbf{E} - k_0^2\mathbf{E}^*\cdot\bar{\kappa}\cdot\mathbf{E}) \, dV$$

and use

$$\nabla\cdot[\mathbf{E}^* \times \nabla \times \mathbf{E} - \mathbf{E} \times \nabla \times \mathbf{E}^*] = \mathbf{E}\cdot\nabla \times \nabla \times \mathbf{E}^* - \mathbf{E}^*\cdot\nabla \times \nabla \times \mathbf{E}$$

to transfer the operator onto $\mathbf{E}^*$. Thus we get

$$0 = \iiint_V \mathbf{E}\cdot[\nabla \times \nabla \times \mathbf{E}^* - k_0^2\bar{\kappa}\cdot\mathbf{E}^*] \, dV$$

$$-\oiint_S (\mathbf{n} \times \mathbf{E}^*\cdot\nabla \times \mathbf{E} - \mathbf{n} \times \mathbf{E}\cdot\nabla \times \mathbf{E}^*) \, dS. \tag{87}$$

Now $\mathbf{n} \times \mathbf{E}$ vanishes on a perfect conductor and is continuous across dielectric interfaces. Thus, since $\mathbf{E}^*$ satisfies the same differential equation, the operator is self-adjoint if $\mathbf{E}^*$ satisfies the same boundary conditions. (A radiation condition is replaced by the adjoint condition.) In (87) we can use $\nabla\cdot(\mathbf{E}^* \times \nabla \times \mathbf{E}) = \nabla \times \mathbf{E}^*\cdot\nabla \times \mathbf{E} - \mathbf{E}^*\cdot\nabla \times \nabla \times \mathbf{E}$ to get

$$\iiint_V (\nabla \times \mathbf{E}^*\cdot\nabla \times \mathbf{E} - k_0^2\mathbf{E}^*\cdot\bar{\kappa}\cdot\mathbf{E}) \, dV - \oiint_S \mathbf{n} \times \mathbf{E}^*\cdot\nabla \times \mathbf{E} \, dS = 0.$$

On the planes $z = \pm L$ the integrals cancel for a waveguide when the fields vary like $e^{-j\beta z}$

since $\mathbf{n} \times \mathbf{E}^* \cdot \nabla \times \mathbf{E}$ is independent of z. The integral over the transverse plane will vanish as long as $\mathbf{n} \times \mathbf{E}^*$ and $\mathbf{n} \times \nabla \times \mathbf{E} \propto \mathbf{n} \times \mathbf{H}$ are continuous across all interfaces and one or the other vanishes on the outer boundary. The electric field functional whose minimization will give the solution for $\mathbf{E}$ is

$$F_1 = \nabla \times \mathbf{E}^* \cdot \nabla \times \mathbf{E} - k_0^2 \mathbf{E}^* \cdot \bar{\kappa} \cdot \mathbf{E}. \tag{88}$$

This functional is equivalent to an energy functional $W_m - W_e$ such as that used to obtain (79). Consider a variation in k_0^2 due to a variation in $\mathbf{E}^*$ of

$$I = \iiint_V (\nabla \times \mathbf{E}^* \cdot \nabla \times \mathbf{E} - k_0^2 \mathbf{E}^* \cdot \bar{\kappa} \cdot \mathbf{E}) \, dV. \tag{89}$$

We have

$$\iiint_V (\nabla \times \delta\mathbf{E}^* \cdot \nabla \times \mathbf{E} - k_0^2 \, \delta\mathbf{E}^* \cdot \bar{\kappa} \cdot \mathbf{E}) \, dV$$

$$= \delta k_0^2 \iiint_V \mathbf{E}^* \cdot \bar{\kappa} \cdot \mathbf{E} \, dV$$

$$= \iiint_V \delta\mathbf{E}^* \cdot [\nabla \times \nabla \times \mathbf{E} - k_0^2 \bar{\kappa} \cdot \mathbf{E}] \, dV + \oiint_S \mathbf{n} \times \delta\mathbf{E}^* \cdot \nabla \times \mathbf{E} \, dV.$$

The variation will vanish provided $\mathbf{n} \times \delta\mathbf{E}^*$ is continuous across all dielectric interfaces and vanishes on the outer boundary. Equivalent results are obtained if the symmetric scalar product involving $\mathbf{E} \cdot \mathbf{E}$ is used.

Equation (89) is solved numerically by first dividing the guide cross section into small squares or triangles (triangles are more appropriate to approximate curved boundaries) as shown previously in Fig. 6.16. For a shielded waveguide the mesh terminates at the boundary of the metallic shield. For an open waveguide structure such as the elliptic or rectangular dielectric rod the field could be assumed to have decayed to a negligible value at a distance from the boundary equal to several times the linear transverse dimensions of the guide. This termination results in a small error except for values of β close to the cutoff point where the field is loosely bound to the guide and extends a considerable distance from the surface. If the medium surrounding the dielectric rod is air then the guided modes have propagation constants between that of free space and that of the dielectric rod, namely,

$$k_0 < \beta < \sqrt{\kappa} k_0 = k.$$

The radial decay constant is $\alpha = (\beta^2 - k_0^2)^{1/2}$ and becomes small when β approaches the cutoff value k_0.

After the mesh has been defined it is necessary to construct a trial function to represent each component of the electric field. The trial function must satisfy the conditions that the tangential components of the electric field and the normal component of electric field multiplied by the dielectric constant be continuous across each dielectric interface. The latter condition is not an explicit requirement when the functional F_1 is used but it is good practice to incorporate it. The trial functions for each field component can be expressed in terms of interpolating polynomials. If a field component is represented by a second-degree polynomial such as

$$A_1 + A_2 x + A_3 y + A_4 xy + A_5 x^2 + A_6 y^2$$

the coefficients A_1 to A_6 can be expressed in terms of the assigned values at six nodes, usually at the corners and the midpoints of each side of the triangular element, i.e., nodes 1 through 6 in Fig. 6.16(d). The functional can be evaluated and integrated over each subdomain. The stationary property is invoked next and the partial derivatives with respect to each unknown field amplitude are equated to zero. The determinant of the resultant homogeneous set of linear equations is equated to zero and the eigenvalues k_0^2 are solved for. For a waveguide an assumed value is used for β.

The difficulty with using the functional F_1 given by (88) is that the assignment of unconstrained values to all three components of $\mathbf{E}$ will not guarantee that the solution will satisfy $\nabla \cdot \bar{\kappa} \cdot \mathbf{E} = 0$. As a result, many spurious nonphysical numerical solutions occur for which $\nabla \cdot \bar{\kappa} \cdot \mathbf{E} \neq 0$.

The constraint $\nabla \cdot \bar{\kappa} \cdot \mathbf{E} = 0$ can be introduced into the variational problem using a Lagrange multiplier λ. Thus, instead of the functional F_1 we can consider using

$$F_2 = \nabla \times \mathbf{E}^* \cdot \nabla \times \mathbf{E} - k_0^2 \mathbf{E}^* \cdot \bar{\kappa} \cdot \mathbf{E} + \lambda (\nabla \cdot \bar{\kappa} \cdot \mathbf{E}^*)(\nabla \cdot \bar{\kappa} \cdot \mathbf{E}).$$

Rahman and Davies have referred to this technique as the use of a penalty function [6.2]. However, their formulation is in terms of the magnetic field $\mathbf{H}$, which we will discuss next. It is a simpler formulation for dielectric guides.

Magnetic Field Functional: Consider the scalar product of (85) with $\mathbf{H}^*$:

$$\iiint_V (\mathbf{H}^* \cdot \nabla \times \bar{\tau} \cdot \nabla \times \mathbf{H} - k_0^2 \mathbf{H}^* \cdot \mathbf{H}) \, dV = 0.$$

Now use $\nabla \cdot (\mathbf{H}^* \times \bar{\tau} \cdot \nabla \times \mathbf{H} - \mathbf{H} \times \bar{\tau} \cdot \nabla \times \mathbf{H}^*) = \nabla \times \mathbf{H}^* \cdot \bar{\tau} \cdot \nabla \times \mathbf{H} - \nabla \times \mathbf{H} \cdot \bar{\tau} \cdot \nabla \times \mathbf{H}^* - \mathbf{H}^* \cdot \nabla \times \bar{\tau} \cdot \nabla \times \mathbf{H} + \mathbf{H} \cdot \nabla \times \bar{\tau} \cdot \nabla \times \mathbf{H}^*$. Since $\bar{\tau}$ is symmetric the first two terms cancel. Thus we obtain

$$\iiint_V \mathbf{H} \cdot (\nabla \times \bar{\tau} \cdot \nabla \times \mathbf{H}^* - k_0^2 \mathbf{H}^*) \, dV - \oiint_S (\mathbf{n} \times \mathbf{H}^* \cdot \bar{\tau} \cdot \nabla \times \mathbf{H} - \mathbf{n} \times \mathbf{H} \cdot \bar{\tau} \cdot \nabla \times \mathbf{H}^*) \, dS.$$

The surface integral vanishes provided $\mathbf{n} \times \mathbf{H}$, $\mathbf{n} \times \mathbf{H}^*$, $\mathbf{n} \times \bar{\tau} \cdot \nabla \times \mathbf{H} \propto \mathbf{n} \times \mathbf{E}$, and $\mathbf{n} \times \bar{\tau} \cdot \nabla \times \mathbf{H}^*$ are continuous across all dielectric interfaces. With these boundary conditions the operator is self-adjoint.

The same application of the divergence theorem allows us to obtain

$$\iiint_V (\nabla \times \mathbf{H}^* \cdot \bar{\tau} \cdot \nabla \times \mathbf{H} - k_0^2 \mathbf{H}^* \cdot \mathbf{H}) \, dV = 0. \tag{90}$$

The magnetic field functional is seen to be

$$F_3 = \nabla \times \mathbf{H}^* \cdot \bar{\tau} \cdot \nabla \times \mathbf{H} - k_0^2 \mathbf{H}^* \cdot \mathbf{H}. \tag{91}$$

The use of this functional in the finite-element method again produces spurious nonphys-

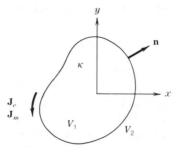

Fig. 6.17. A homogeneous dielectric guide.

ical solutions for which $\nabla\cdot\mathbf{H}\neq0$. Rahman and Davies have added the penalty function $(\alpha/\epsilon_0)\nabla\cdot\mathbf{H}^*\nabla\cdot\mathbf{H}$ to this functional and have found that for $\alpha\approx1$ many of the spurious solutions are eliminated. Koshiba *et al.* have added $\nabla\cdot\mathbf{H}^*\nabla\cdot\mathbf{H}$ to the functional F_3 and have found that the spurious modes that remain have $\beta<k_0$ and are easily identified [6.3]. The preferred functional to use is therefore

$$F_4 = \nabla\times\mathbf{H}^*\cdot\bar{\tau}\cdot\nabla\times\mathbf{H} + \nabla\cdot\mathbf{H}^*\nabla\cdot\mathbf{H} - k_0^2\mathbf{H}^*\cdot\mathbf{H}. \tag{92}$$

Chew and Nasir have derived functionals that involve only the transverse electric field components or the transverse magnetic field components and which give variational solutions for the propagation constant. Their formulation applies to waveguides that contain inhomogeneous anisotropic media and that have reflection symmetry. In a specific application of their formulation they found that spurious solutions did not occur [6.34]. However, it is not known if this is true in general.

The Boundary-Element Method

The boundary-element method is best suited to waveguides that have a constant value of dielectric constant in each region. We will illustrate the method for a waveguide such as that shown in Fig. 6.17.

Let $\mathbf{E}$, $\mathbf{H}$ be the electric and magnetic fields of a mode having the z dependence $e^{-j\beta z}$. The field equivalence principle states that the field in V_2 can be found in terms of electric currents $\mathbf{J}_e = \mathbf{n}\times\mathbf{H}$ and magnetic currents $\mathbf{J}_m = -\mathbf{n}\times\mathbf{E}$ on the boundary S. Likewise, the field in V_1 can be found from currents $-\mathbf{J}_e$, $-\mathbf{J}_m$ on S.

In V_1, $\mathbf{E}_1$, $\mathbf{H}_1$ are solutions of

$$\nabla\times\nabla\times\mathbf{E}_1 - k^2\mathbf{E}_1 = 0 \tag{93a}$$

$$\nabla\times\nabla\times\mathbf{H}_1 - k^2\mathbf{H}_1 = 0 \tag{93b}$$

where $k^2 = \kappa k_0^2$.

Consider also a Green's dyadic function $\bar{\mathbf{G}}_1$ that is a solution of (all of space is considered to have $k^2 = \kappa k_0^2$)

$$\nabla\times\nabla\times\bar{\mathbf{G}}_1 - k^2\bar{\mathbf{G}}_1 = \bar{\mathbf{I}}\delta(\mathbf{r}-\mathbf{r}'). \tag{94}$$

We now form the expression

$$\iiint_V [\mathbf{E}_1(\mathbf{r})\cdot\nabla\times\nabla\times\bar{\mathbf{G}}_1(\mathbf{r},\mathbf{r}') - \nabla\times\nabla\times\mathbf{E}_1(\mathbf{r})\cdot\bar{\mathbf{G}}_1(\mathbf{r},\mathbf{r}')]\,d\mathbf{r} = \iiint_V \mathbf{E}_1(\mathbf{r})\delta(\mathbf{r}-\mathbf{r}')\,d\mathbf{r}.$$

We can use $\nabla \cdot [\mathbf{E}_1 \times \nabla \times \bar{\mathbf{G}}_1 + (\nabla \times \mathbf{E}_1) \times \bar{\mathbf{G}}_1] = \nabla \times \mathbf{E}_1 \cdot \nabla \times \bar{\mathbf{G}}_1 - \mathbf{E}_1 \cdot \nabla \times \nabla \times \bar{\mathbf{G}}_1 + (\nabla \times \nabla \times \mathbf{E}_1) \cdot \bar{\mathbf{G}}_1 - \nabla \times \mathbf{E}_1 \cdot \nabla \times \mathbf{G}_1$ and the divergence theorem to obtain

$$\mathbf{E}_1(\mathbf{r}') = -\oiint_S (\mathbf{n} \times \mathbf{E}_1 \cdot \nabla \times \bar{\mathbf{G}}_1 + \mathbf{n} \times \nabla \times \mathbf{E}_1 \cdot \bar{\mathbf{G}}_1) \, dS, \qquad \mathbf{r} \text{ in } V_1.$$

But $\nabla \times \mathbf{E}_1 = -j\omega\mu_0 \mathbf{H}_1$ so

$$\mathbf{E}_1(\mathbf{r}') = -\oiint_S (\mathbf{n} \times \mathbf{E}_1 \cdot \nabla \times \bar{\mathbf{G}}_1 - j\omega\mu_0 \mathbf{n} \times \mathbf{H}_1 \cdot \bar{\mathbf{G}}_1) \, dS \tag{95}$$

which expresses $\mathbf{E}_1(\mathbf{r}')$ in terms of the boundary values $-\mathbf{n} \times \mathbf{E}_1 = \mathbf{J}_m$ and $\mathbf{n} \times \mathbf{H}_1 = \mathbf{J}_e$ on S.

For field points outside S we use a Green's dyadic that satisfies (94) with k^2 replaced by k_0^2. A similar derivation then shows that for $\mathbf{r}'$ in V_2

$$\mathbf{E}_2(\mathbf{r}') = \oiint_S (\mathbf{n} \times \mathbf{E}_2 \cdot \nabla \times \bar{\mathbf{G}}_2 - j\omega\mu_0 \mathbf{n} \times \mathbf{H}_2 \cdot \bar{\mathbf{G}}_2) \, dS. \tag{96}$$

The corresponding solutions for the magnetic field are

$$\mathbf{H}_1(\mathbf{r}') = -\oiint_S (\mathbf{n} \times \mathbf{H}_1 \cdot \nabla \times \mathbf{G}_1 + j\omega\epsilon_0\kappa \mathbf{n} \times \mathbf{E}_1 \cdot \bar{\mathbf{G}}_1) \, dS \tag{97}$$

$$\mathbf{H}_2(\mathbf{r}') = \iint_S (\mathbf{n} \times \mathbf{H}_2 \cdot \nabla \times \mathbf{G}_2 + j\omega\epsilon_0 \mathbf{n} \times \mathbf{E}_2 \cdot \bar{\mathbf{G}}_2) \, dS. \tag{98}$$

The boundary conditions require that on S $\mathbf{n} \times \mathbf{E}_1 = \mathbf{n} \times \mathbf{E}_2 = \mathbf{n} \times \mathbf{E}$, $\mathbf{n} \times \mathbf{H}_1 = \mathbf{n} \times \mathbf{H}_2 = \mathbf{n} \times \mathbf{H}$. Hence we get the following pair of integral equations that must hold on the boundary S (use $\mathbf{n} \times \mathbf{E}_1 - \mathbf{n} \times \mathbf{E}_2 = 0$, etc.):

$$0 = \oiint_S \{\mathbf{n} \times \mathbf{E}(\mathbf{r}) \cdot [\nabla \times \bar{\mathbf{G}}_1(\mathbf{r}, \mathbf{r}') + \nabla \times \bar{\mathbf{G}}_2(\mathbf{r}, \mathbf{r}')] \times \mathbf{n}$$
$$- j\omega\mu_0 \mathbf{n} \times \mathbf{H}(\mathbf{r}) \cdot [\bar{\mathbf{G}}_1(\mathbf{r}, \mathbf{r}') + \bar{\mathbf{G}}_2(\mathbf{r}, \mathbf{r}')] \times \mathbf{n}\} \, dS \tag{99a}$$

$$0 = \oiint_S \{\mathbf{n} \times \mathbf{H}(\mathbf{r}) \cdot [\nabla \times \bar{\mathbf{G}}_1(\mathbf{r}, \mathbf{r}') + \nabla \times \bar{\mathbf{G}}_2(\mathbf{r}, \mathbf{r}')] \times \mathbf{n}$$
$$+ j\omega\epsilon_0 \mathbf{n} \times \mathbf{E}(\mathbf{r}) \cdot [\kappa \bar{\mathbf{G}}_1(\mathbf{r}, \mathbf{r}') + \bar{\mathbf{G}}_2(\mathbf{r}, \mathbf{r}')] \times \mathbf{n}\} \, dS \tag{99b}$$

for $\mathbf{r}'$ approaching S from the interior or exterior side, respectively, for the interior or exterior fields. The second unit normal $\mathbf{n}$ in these integrals is evaluated at the point $\mathbf{r}'$.

Since the fields vary like $e^{-j\beta z}$ the z integration is readily done and gives the Fourier transform of the Green's functions at the spectral wavenumber β. Equations (99) can be converted to a homogeneous set of algebraic equations whose solutions require the determinant to vanish, which results in a dispersion equation for β.

One way to treat (99) is to expand $\mathbf{n} \times \mathbf{E}$ and $\mathbf{n} \times \mathbf{H}$ in a set of vector basis functions on S,

$$\mathbf{n} \times \mathbf{E} = \sum_1^N V_n \mathbf{e}_n \qquad \mathbf{n} \times \mathbf{H} = \sum_1^N I_n \mathbf{h}_n.$$

These may be used in (99) and then point matching applied at N points or the equations may be tested with $\mathbf{e}_m$, $\mathbf{h}_m$ in turn (Galerkin's method) so as to generate a total of $2N$ algebraic equations.

Circular Symmetric Modes on a Dielectric Rod

As a simple example of the use of (99) we will consider circular symmetric modes on a circular dielectric rod of radius a. The dyadic Green's function given by (182) in Section 2.16 can be used. Only the $n = 0$ term is required. The integration over z replaces w by $-\beta$ and multiplies the expressions given by 2π. In order to avoid introducing additional symbols we will continue to use $\bar{\mathbf{G}}_1$ and $\bar{\mathbf{G}}_2$ which will now stand for the Fourier transforms of the Green's functions. Also $\mathbf{E}$ and $\mathbf{H}$ will now be understood to represent the fields without the factor $e^{-j\beta z}$ which is the assumed z dependence. The surface integrals in (99) will then be integrals taken around the contour C of the dielectric waveguide. The required Green's dyadic $\bar{\mathbf{G}}_1$ is readily obtained and is

$$\bar{\mathbf{G}}_1(\mathbf{r}, \mathbf{r}') = -\frac{j}{4}\left[H_0'(\gamma r)J_0'(\gamma r')\mathbf{a}_\phi\mathbf{a}_{\phi'} + \left(\frac{j\beta H_0'(\gamma r)}{k}\mathbf{a}_r + \frac{\gamma H_0(\gamma r)}{k}\mathbf{a}_z\right)\right.$$
$$\left.\times \left(\frac{-j\beta J_0'(\gamma r')}{k}\mathbf{a}_{r'} + \frac{\gamma J_0(\gamma r')}{k}\mathbf{a}_{z'}\right)\right] \tag{100a}$$

$$\bar{\mathbf{G}}_1 \times \mathbf{a}_{r'} = -\frac{j}{4}\left[-H_0'J_0'\mathbf{a}_\phi\mathbf{a}_{z'} + \frac{\gamma^2}{k^2}H_0J_0\mathbf{a}_z\mathbf{a}_{\phi'}\right] \tag{100b}$$

where only the tangential components have been retained in (100b). The primes denote derivatives of the Bessel functions with respect to the arguments. We also need the quantities

$$\nabla \times \bar{\mathbf{G}}_1(\mathbf{r}, \mathbf{r}') = -\frac{j}{4}\left[-(j\beta H_0'\mathbf{a}_r + \gamma H_0\mathbf{a}_z)(J_0'\mathbf{a}_{\phi'})\right.$$
$$\left.+ (kH_0'\mathbf{a}_\phi)\left(\frac{j\beta J_0'}{k}\mathbf{a}_{r'} - \frac{\gamma J_0}{k}\mathbf{a}_{z'}\right)\right] \tag{101a}$$

$$\nabla \times \bar{\mathbf{G}}_1 \times \mathbf{a}_{r'} = -\frac{j}{4}[\gamma H_0J_0'\mathbf{a}_z\mathbf{a}_{z'} - \gamma H_0'J_0\mathbf{a}_\phi\mathbf{a}_{\phi'}] \tag{101b}$$

where again only the tangential components have been retained in (101b). The parameter $\gamma = \sqrt{k^2 - \beta^2}$. For $\bar{\mathbf{G}}_2$ we replace γ by γ_0 where $\gamma_0 = \sqrt{k_0^2 - \beta^2}$ and interchange the roles of H_0 and J_0 since now $r' > r$.

For this problem complete expansions for $\mathbf{n} \times \mathbf{E}$ and $\mathbf{n} \times \mathbf{H}$ are

$$\mathbf{n} \times \mathbf{E} = V_1\mathbf{a}_z + V_2\mathbf{a}_\phi \qquad \mathbf{n} \times \mathbf{H} = I_1\mathbf{a}_z + I_2\mathbf{a}_\phi.$$

When we use these in (99) and denote $H_0(\gamma_0 a)$ by $\bar{H}_0$ and similarly for $J_0(\gamma_0 a)$, we obtain

$$0 = V_1[(\gamma H_0J_0' + \gamma_0\bar{J}_0\bar{H}_0')\mathbf{a}_z] + V_2[(-\gamma H_0'J_0 - \gamma_0\bar{J}_0'\bar{H}_0)\mathbf{a}_{\phi'}]$$
$$- j\omega\mu_0 I_1\left[\left(\frac{\gamma^2}{k^2}H_0J_0 + \frac{\gamma_0^2}{k_0^2}\bar{H}_0\bar{J}_0\right)\mathbf{a}_{\phi'}\right]$$
$$- j\omega\mu_0 I_2[(-H_0'J_0' - \bar{H}_0'\bar{J}_0')\mathbf{a}_{z'}] \tag{102a}$$

$$0 = I_1[(\gamma H_0 J_0' + \gamma_0 \bar{J}_0 \bar{H}_0')\mathbf{a}_{z'}] + I_2[(-\gamma H_0' J_0 - \gamma_0 \bar{J}_0' \bar{H}_0)\mathbf{a}_{\phi'}]$$

$$+ j\omega\epsilon_0 V_1 \left[\left(\frac{\kappa\gamma^2}{k^2} H_0 J_0 + \frac{\gamma_0^2}{k_0^2} \bar{H}_0 \bar{J}_0 \right) \mathbf{a}_{\phi'} \right]$$

$$+ j\omega\epsilon_0 V_2 [(-\kappa H_0' J_0' - \bar{H}_0' \bar{J}_0')\mathbf{a}_{z'}]. \tag{102b}$$

The vector components are equated to zero to obtain a pair of equations for V_1, I_2 and another pair for V_2, I_1. For the first set the vanishing of the determinant gives

$$\left[\frac{J_0'(\gamma_0 a)}{\gamma_0 J_0(\gamma_0 a)} - \frac{H_0'(\gamma a)}{\gamma H_0(\gamma a)} \right] \left[\frac{J_0'(\gamma a)}{\gamma J_0(\gamma a)} - \frac{H_0'(\gamma_0 a)}{\gamma_0 H_0(\gamma_0 a)} \right] = 0 \tag{103a}$$

while for the second set

$$\left[\frac{J_0'(\gamma_0 a)}{\gamma_0 J_0(\gamma_0 a)} - \frac{\kappa H_0'(\gamma a)}{\gamma H_0(\gamma a)} \right] \left[\frac{\kappa J_0'(\gamma a)}{\gamma J_0(\gamma a)} - \frac{H_0'(\gamma_0 a)}{\gamma_0 H_0(\gamma_0 a)} \right] = 0. \tag{103b}$$

For the surface wave modes $k_0 < \beta < k$ and $\gamma_0 = -jp$. For this case the first factor in (103a) and (103b) is not zero and we get

$$\frac{J_0'(\gamma a)}{\gamma J_0(\gamma a)} = -\frac{K_0'(pa)}{p K_0(pa)}, \qquad \text{TE modes} \tag{104a}$$

$$\frac{\kappa J_0'(\gamma a)}{\gamma J_0(\gamma a)} = -\frac{K_0'(pa)}{p K_0(pa)}, \qquad \text{TM modes} \tag{104b}$$

where K_0 is the modified Bessel function of the second kind. These are the well-known dispersion equations for the circular symmetric modes on a dielectric rod. The numerical evaluation of (104) will be given in Chapter 11 for a typical dielectric rod. The solutions of (104) represent the propagation constants for circular symmetric surface wave modes.

There are several important features of this problem which we will now discuss. If in place of a dielectric rod we had a circular tunnel of radius a in an infinite dielectric medium we would interchange the roles of γ and γ_0, k and k_0, and multiply $\bar{G}_2$ in (99b) by κ instead of multiplying $\bar{G}_1$ by this factor. If these changes are made in (103) the effect is to interchange the first and second factors. Consequently, the dispersion relations that have been obtained apply to both the original problem and the complementary problem in which the air and dielectric regions are interchanged. There is, therefore, a lack of uniqueness in the solution. Surface waves do not exist along an air tunnel in an infinite dielectric medium, but if a surrounding metal shield of radius b were introduced there would be a second set of modes corresponding to those in a dielectric-filled circular waveguide with the dielectric having a circular hole in the middle. Consequently, when the boundary-element method is used some care must be exercised in identifying the modes that correspond to different values of β that are found [6.40].

A variety of choices for the internal dyadic Green's function $\bar{G}_1$ are available. We could choose $\bar{G}_1$ so that the boundary condition $\mathbf{n} \times \bar{G}_1 = 0$ at $r = a$ is satisfied. If this were done we would have a sum over all of the radial eigenfunctions involving $J_0(p_{0m}r/a)$ in (99). In addition, this choice for the dyadic Green's function would have the property that the two terms $\mathbf{n} \times \mathbf{H} \cdot \bar{G}_1$ and $\mathbf{n} \times \mathbf{E} \cdot \bar{G}_1$ would be zero on the boundary at $r = a$ and hence would not be present in (99). Also, $\bar{\mathbf{G}}_1$ would have the property that

$$\lim_{r'\to a} \oint_C \mathbf{n}(\mathbf{r}) \times \mathbf{E}(\mathbf{r}) \cdot \nabla \times \bar{\mathbf{G}}_1 \times \mathbf{n}(\mathbf{r}')\, dl = \mathbf{n} \times \mathbf{E}(a) \tag{105}$$

where C is the boundary at $r = a$. The two terms involving $\nabla \times \bar{\mathbf{G}}_1$ in (99) would reduce simply to $\mathbf{n} \times \mathbf{E}$ and $\mathbf{n} \times \mathbf{H}$ evaluated on the boundary. The resultant system of equations is simpler, to the point of not reflecting at all the properties of the internal medium in $r < a$ explicitly since $\bar{\mathbf{G}}_1$ no longer enters into the equations. Thus we cannot use the boundary condition $\mathbf{n} \times \bar{\mathbf{G}}_1 = 0$ at $r = a$ in solving for both the internal electric and magnetic fields. However, we can use a Green's dyadic function $\bar{\mathbf{G}}_{1e}$ with boundary condition $\mathbf{n} \times \bar{\mathbf{G}}_{1e} = 0$ at $r = a$ to solve for the electric field, and a Green's dyadic function $\bar{\mathbf{G}}_{1m}$ with boundary condition $\mathbf{n} \times \nabla \times \bar{\mathbf{G}}_{1m} = 0$ at $r = a$ to solve for the internal magnetic field. In this case the two terms $\mathbf{n} \times \mathbf{H} \cdot \bar{\mathbf{G}}_1$ and $\mathbf{n} \times \mathbf{E} \cdot \nabla \times \bar{\mathbf{G}}_1$ in (99) are eliminated and the integral of $\mathbf{n} \times \mathbf{E} \cdot \nabla \times \bar{\mathbf{G}}_1 \times \mathbf{n}$ becomes $\mathbf{n} \times \mathbf{E}$ as noted above. In place of (99) we now have (the second unit normal is evaluated at the point $\mathbf{r}'$ in each integral)

$$\mathbf{n} \times \mathbf{E} + \oint_C \mathbf{n} \times \mathbf{E} \cdot \nabla \times \bar{\mathbf{G}}_2 \times \mathbf{n}\, dl - j\omega\mu_0 \oint_C \mathbf{n} \times \mathbf{H} \cdot \bar{\mathbf{G}}_2 \times \mathbf{n}\, dl = 0 \tag{106a}$$

$$\oint_C \mathbf{n} \times \mathbf{H} \cdot \nabla \times \bar{\mathbf{G}}_2 \times \mathbf{n}\, dl + j\omega\epsilon_0 \oint_C \mathbf{n} \times \mathbf{E} \cdot (\kappa\bar{\mathbf{G}}_{1m} + \bar{\mathbf{G}}_2) \times \mathbf{n}\, dl = 0. \tag{106b}$$

An alternative choice would be to use the boundary conditions $\mathbf{n} \times \nabla \times \bar{\mathbf{G}}_{1e} = 0$ and $\mathbf{n} \times \bar{\mathbf{G}}_{1m} = 0$ at $r = a$.

It will be instructive to carry out the analysis prescribed by (106) for two reasons. First, it provides an opportunity to show by a concrete example that the property displayed by (105) is indeed true. Second, it will bring out the requirement that the longitudinal modes must be kept in the Fourier transform of the eigenfunction expansion of the dyadic Green's function and furthermore it will show that by summing over the radial eigenfunctions the longitudinal modes can be eliminated. Since we will be concerned with modes having circular symmetry only the parts of the Green's dyadic functions that do not depend on the angle ϕ will be retained in the discussion that follows.

The dyadic Green's function and the required scalar generating functions are given by (88) and (89), respectively, in Chapter 5. We are interested in the quantity $\mathbf{a}_r \times \nabla \times \bar{\mathbf{G}}_{1e} \times \mathbf{a}_{r'}$ which is readily found from (88) and (89) in Chapter 5. It is given by

$$\mathbf{a}_r \times \nabla \times \bar{\mathbf{G}}_{1e} \times \mathbf{a}_{r'} = \frac{1}{\pi a^3} \sum_{m=1}^{\infty} \left\{ \frac{p'_{0m} J_0(p'_{0m}r/a) J_1(p'_{0m}r'/a)}{J_0^2(p'_{0m})[(p'_{0m}/a)^2 - \gamma^2]} \mathbf{a}_\phi \mathbf{a}_{z'} \right.$$

$$\left. + \frac{p_{0m} J_1(p_{0m}r/a) J_0(p_{0m}r'/a)}{J_1^2(p_{0m})[(p_{0m}/a)^2 - \gamma^2]} \mathbf{a}_z \mathbf{a}_{\phi'} \right\} \tag{107}$$

where $\gamma^2 = k^2 - \beta^2$ and $J_0(p_{0m}) = J_1(p'_{0m}) = 0$. We note that the above series has coefficients that will be proportional to $1/m$ for large m. Therefore, the series will not converge uniformly for $r = a$ and r' approaching a. If we set $r = r' = a$ we obtain zero. In (99) we have to evaluate the terms involving $\bar{\mathbf{G}}_1$ as r' approaches a from the interior and evaluate the terms involving $\bar{\mathbf{G}}_2$ as r' approaches a from the exterior.

If we solve the one-dimensional Green's function problems

$$\frac{1}{r}\frac{d}{dr}r\frac{dg}{dr} + \gamma^2 g = \frac{\delta(r - r')}{r}, \qquad \frac{dg}{dr} = 0, \ r = a$$

$$\frac{1}{r}\frac{d}{dr}r\frac{dg}{dr} + \gamma^2 g = \frac{\delta(r - r')}{r}, \qquad g = 0, \ r = a$$

by Methods I and II described in Section 2.4, we can establish the summation formulas

$$\sum_{m=1} \frac{J_0(p'_{0m}r/a)J_0(p'_{0m}r'/a)}{J_0^2(p'_{0m})[(p'_{0m}/a)^2 - \gamma^2]} = \frac{\pi a^2}{4}\frac{J_0(\gamma r_<)}{J_1(\gamma a)}[J_0(\gamma r_>)Y_1(\gamma a) - J_1(\gamma a)Y_0(\gamma r_>)] \quad \text{(108a)}$$

$$\sum_{m=1}^{\infty} \frac{J_0(p_{0m}r/a)J_0(p_{0m}r'/a)}{J_1^2(p_{0m})[(p_{0m}/a)^2 - \gamma^2]} = \frac{\pi a^2}{4}\frac{J_0(\gamma r_<)}{J_0(\gamma a)}[J_0(\gamma r_>)Y_0(\gamma a) - J_0(\gamma a)Y_0(\gamma r_>)]. \quad \text{(108b)}$$

We now take the derivative of (108a) with respect to r' and that of (108b) with respect to r using $r' < r$. This will generate the series that occurs in (107). After the derivatives are carried out we can set $r' = r = a$. Then upon using the Wronskian relation

$$J_1(\gamma a)Y_0(\gamma a) - J_0(\gamma a)Y_1(\gamma a) = \frac{2}{\pi\gamma a}$$

we find that (107) simplifies to

$$\mathbf{a}_r \times \nabla \times \bar{\mathbf{G}}_{1e} \times \mathbf{a}_{r'} = \frac{1}{2\pi a}[-\mathbf{a}_\phi \mathbf{a}_{z'} + \mathbf{a}_z \mathbf{a}_{\phi'}]. \tag{109}$$

We now note that

$$\int_0^{2\pi} \mathbf{E}\cdot\mathbf{a}_r \times \nabla \times \bar{\mathbf{G}}_{1e} \times \mathbf{a}_{r'}a\,d\phi = -E_\phi \mathbf{a}_{z'} + E_z \mathbf{a}_{\phi'} = -\mathbf{a}_{r'} \times \mathbf{E}$$

in accordance with the property set forth in (105). Consequently, we have shown that the integral

$$\lim_{r' \to a} \oint_C \mathbf{a}_r \times \mathbf{E}\cdot\nabla \times \bar{\mathbf{G}}_{1e} \times \mathbf{a}_{r'}\,dl$$

reproduces the value $\mathbf{a}_{r'} \times \mathbf{E}$ on the boundary when $\mathbf{a}_r \times \bar{\mathbf{G}}_{1e} = 0$ on the boundary. The function of r' given by this integral is discontinuous at $r' = a$. It equals $\mathbf{a}_r \times \mathbf{E}(a)$ as r' approaches a from the interior and equals zero when $r' = a$. If we differentiate (108) as specified but use $r' > r$ then we can substantiate that upon putting $r' = r = a$ the right-hand sides of (108a) and (108b) vanish.

In order to evaluate the other terms in (106) we need $\mathbf{a}_r \times \bar{\mathbf{G}}_{1m} \times \mathbf{a}_{r'}$ where $\bar{\mathbf{G}}_{1m}$ satisfies the boundary condition $\mathbf{a}_r \times \nabla \times \bar{\mathbf{G}}_{1m} = 0$ at $r = a$. In order to construct this Green's function we use the scalar functions ψ_{e0m} given by (89d) in Chapter 5 to generate the $\mathbf{N}_n$ and $\mathbf{L}_n$ functions. The required $\mathbf{M}_n$ functions are obtained using (89c) in Chapter 5. It is readily found that

$$\mathbf{a}_r \times \bar{\mathbf{G}}_{1m} \times \mathbf{a}_{r'} = -\frac{1}{\pi a^2} \sum_{m=1}^{\infty} \left\{ \frac{J_1(p_{0m}r/a)J_1(p_{0m}r'/a)}{J_1^2(p_{0m})[(p_{0m}/a)^2 - \gamma^2]} \mathbf{a}_z \mathbf{a}_{z'} \right.$$

$$+ \frac{J_0(p'_{0m}r/a)J_0(p'_{0m}r'/a)}{J_0^2(p'_{0m})[(p'_{0m}/a)^2 + \beta^2]} \left[\frac{(p'_{0m}/a)^2 \mathbf{a}_\phi \mathbf{a}_{\phi'}}{(p'_{0m}/a)^2 - \gamma^2} \right.$$

$$\left. \left. - \frac{\beta^2 \mathbf{a}_\phi \mathbf{a}_{\phi'}}{k^2} \right] \right\}. \tag{110}$$

We can combine the $\mathbf{N}_n$ and $\mathbf{L}_n$ contributions to obtain

$$-\frac{1}{\pi a^2} \sum_{m=1}^{\infty} \frac{\gamma^2}{k^2} \frac{J_0(p'_{0m}r/a)J_0(p'_{0m}r'/a)}{J_0^2(p'_{0m})[(p'_{0m}/a)^2 - \gamma^2]} \mathbf{a}_\phi \mathbf{a}_{\phi'}.$$

The sum of this series can be found using (108a). If the $\mathbf{N}_n$ and $\mathbf{L}_n$ mode series were summed separately, terms involving Bessel functions with the argument $j\beta a$ would arise. Such terms are absent in the combined series because of cancellation of the $\mathbf{L}_n$ mode series by a part of the $\mathbf{N}_n$ mode series.

In order to sum the series multiplied by $\mathbf{a}_z \mathbf{a}_{z'}$ in (110) we proceed as follows. We differentiate both sides of (108b) with respect to r and r' and note that the derivative of the right-hand side with respect to r leaves a function that is discontinuous at $r' = r$. Consequently, the derivative with respect to r' will produce a delta function term $\delta(r - r')$ which is evaluated using the Wronskian relation given earlier. By this means we obtain

$$\sum_{m=1}^{\infty} \frac{(p_{0m}/a)^2 J_1(p_{0m}r/a)J_1(p_{0m}r'/a)}{J_1^2(p_{0m})[(p_{0m}/a)^2 - \gamma^2]}$$

$$= \frac{\pi a^2 \gamma^2}{4} \frac{J_1(\gamma r_<)}{J_0(\gamma a)} [J_1(\gamma r_>)Y_0(\gamma a) - J_0(\gamma a)Y_1(\gamma r_>)] + \frac{a^2}{2r'} \delta(r - r').$$

The series has a factor

$$\frac{(p_{0m}/a)^2}{(p_{0m}/a)^2 - \gamma^2} = 1 + \frac{\gamma^2}{(p_{0m}/a)^2 - \gamma^2}.$$

The expansion of the delta function is given by

$$\frac{\delta(r - r')}{r'} = \sum_{m=1}^{\infty} \frac{2}{a^2} \frac{J_1(p_{0m}r/a)J_1(p_{0m}r'/a)}{J_1^2(p_{0m})}$$

where we have used the fact that the functions $J_1(p_{0m}r/a)$ form a complete orthogonal set. By using the above results we find that our required summation formula is

$$\sum_{m=1}^{\infty} \frac{J_1(p_{0m}r/a)J_1(p_{0m}r'/a)}{J_1^2(p_{0m})[(p_{0m}/a)^2 - \gamma^2]} = \frac{\pi a^2}{4} \frac{J_1(\gamma r_<)}{J_0(\gamma a)} [J_1(\gamma r_>)Y_0(\gamma a) - J_1(\gamma a)Y_0(\gamma r_>)]. \tag{111}$$

By summing the series in (110), setting $r = r' = a$, and using the Wronskian relation, we obtain

$$\mathbf{a}_r \times \bar{\mathbf{G}}_{1m} \times \mathbf{a}_r = -\frac{J_1(\gamma a)}{2\pi a \gamma J_0(\gamma a)} \mathbf{a}_z \mathbf{a}_{z'} + \frac{\gamma J_0(\gamma a)}{2\pi a k^2 J_1(\gamma a)} \mathbf{a}_\phi \mathbf{a}_{\phi'}. \tag{112}$$

We now return to (106) and introduce the expansions $\mathbf{a}_r \times \mathbf{E} = V_1 \mathbf{a}_z + V_2 \mathbf{a}_\phi$ and $\mathbf{a}_r \times \mathbf{H} = I_1 \mathbf{a}_z + I_2 \mathbf{a}_\phi$ and then obtain the equations

$$\left[1 - \frac{j\pi a}{2}\gamma_0 J_0(\gamma_0 a) H_0'(\gamma_0 a)\right] V_1 + \left[\frac{\pi a}{2} k_0 J_0'(\gamma_0 a) H_0'(\gamma_0 a)\right] Z_0 I_2 = 0 \tag{113a}$$

$$\left[\frac{\gamma_0^2}{4k_0} H_0(\gamma_0 a) J_0(\gamma_0 a) + j\frac{\gamma}{2\pi a k_0}\frac{J_0(\gamma a)}{J_0'(\gamma a)}\right] V_1 + \left[\frac{j\gamma_0}{4} J_0'(\gamma_0 a) H_0(\gamma_0 a)\right] Z_0 I_2 = 0 \tag{113b}$$

$$\left[1 + \frac{j\pi a}{2}\gamma_0 J_0'(\gamma_0 a) H_0(\gamma_0 a)\right] V_2 - \left[\frac{\pi a}{2}\frac{\gamma_0^2}{k_0} J_0(\gamma_0 a) H_0(\gamma_0 a)\right] Z_0 I_1 = 0 \tag{114a}$$

$$\left[-\frac{k_0}{4} J_0'(\gamma_0 a) H_0'(\gamma_0 a) + \frac{j\kappa k_0}{2\pi a \gamma}\frac{J_0'(\gamma a)}{J_0(\gamma a)}\right] V_2 - \left[\frac{j\gamma_0}{4} J_0(\gamma_0 a) H_0'(\gamma_0 a)\right] Z_0 I_1 = 0. \tag{114b}$$

By equating the determinant of the pair of equations (113) to zero we obtain the dispersion equation for TE waves

$$J_0'(\gamma_0 a)\left[\gamma_0 H_0(\gamma_0 a) - \gamma H_0'(\gamma_0 a)\frac{J_0(\gamma a)}{J_0'(\gamma a)}\right]. \tag{115}$$

The factor in brackets can be expressed in the same form as (104a) and thus gives the correct dispersion equation for TE waves. The multiplying factor $J_0'(\gamma_0 a)$ has zeros for $\gamma_0 a = p_{0m}'$ which gives $\beta = [k_0^2 - (p_{0m}'/a)^2]^{1/2}$. These are the propagation constants for TE waves in a hollow waveguide of radius a and having perfectly conducting walls. These roots are spurious ones and do not represent modes that can exist on a circular dielectric cylinder. By examining (113) we readily see that the solutions corresponding to the spurious roots require $V_1 = 0$ but do not specify the value of I_2.

For the second set of equations (114) the dispersion equation is

$$\gamma_0 J_0(\gamma_0 a)\left[H_0'(\gamma_0 a) - \kappa\frac{\gamma_0}{\gamma} H_0(\gamma_0 a)\frac{J_0'(\gamma a)}{J_0(\gamma a)}\right] = 0. \tag{116}$$

The factor in brackets, when equated to zero, gives the same equation as (104b) and is the correct dispersion equation for TM modes. The multiplying factor $J_0(\gamma_0 a)$ has zeros corresponding to $\beta = [k_0^2 - (p_{0m}/a)^2]^{1/2}$ which are the propagation constants for TM waves in a circular waveguide with perfectly conducting walls. These roots are also spurious ones.

A conclusion that can be drawn from the above example is that the boundary-element method will also suffer from the problem of giving spurious solutions. Furthermore, the spurious solutions that arise will be dependent on the choice of dyadic Green's functions that are used to solve the problem. It also should be apparent that if the longitudinal modes had been inadvertently left out of the Green's function the correct dispersion equations would not have been obtained. If the four equations, obtained from the tangential components of (95)–(98) are solved instead of the reduced pair of equations (99a) and (99b) a unique solution is obtained [6.4].

Before leaving the subject of waves on circular rods we need to point out an interesting phenomenon that occurs for hybrid modes (non-circularly symmetric modes consisting of E and H modes coupled by the boundary conditions at the air–dielectric interface) in a circular waveguide containing a coaxial dielectric rod. This phenomenon is the occurrence of a backward wave having oppositely directed group and phase velocities. The backward wave occurs when conditions exist such that the E_{11} and H_{11} modes have the same cutoff frequency. The hybrid wave consisting of coupled E_{11} and H_{11} modes splits into pure E_{11} and H_{11} modes at cutoff. The degeneracy occurs for a dielectric constant somewhat larger than 9. Away from cutoff the H_{11} mode goes over into a backward wave provided the dielectric constant of the rod exceeds the critical value. Several papers dealing with the backward-wave phenomenon have been published by Clarricoats and co-workers [6.41]–[6.43] as well as others [6.44], [6.45]. The same phenomenon has been found to occur for rectangular and square waveguides having a high-dielectric-constant central core [6.46]. The reader is referred to these papers for a detailed treatment.

In addition to backward-wave modes, waveguides partially filled with dielectric material can have complex mode solutions. These modes occur in pairs with the propagation constants being complex conjugates of each other. These complex modes do not carry any real power and behave more like evanescent modes. A review of the literature on complex mode solutions and a general analysis of such waveguides may be found in a paper by Omar and Schünemann [6.5]. These authors point out that in a partially filled circular waveguide complex modes occur only for fields that have a dependence on the azimuthal angle. Complex modes also do not exist in rectangular waveguides that are inhomogeneously filled with dielectric slabs and support LSE and LSM modes.

Shielded Rectangular Dielectric Waveguide

In Fig. 6.18 we show a rectangular dielectric waveguide of dimensions $2a \times 2b$ placed inside a larger metallic rectangular waveguide of dimensions $2A \times 2B$. The propagation constants for the surface wave modes on the dielectric guide can be found by using a modification of the theory presented above for the dielectric rod. To find the electric field in the dielectric we can use a Green's dyadic function for the large waveguide filled with dielectric that satisfies the boundary conditions $\mathbf{n} \times \overline{\mathbf{G}}_{1e} = 0$ on the metal walls. As alternatives we can use dyadic Green's functions that satisfy electric wall boundary conditions $\mathbf{n} \times \overline{\mathbf{G}}_{1e} = 0$ on the surfaces

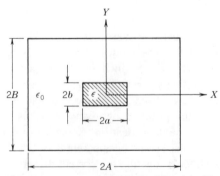

Fig. 6.18. A shielded rectangular dielectric guide.

of the dielectric rod, or that satisfy magnetic wall boundary conditions $\mathbf{n} \times \nabla \times \bar{G}_{1e} = 0$ on the surface of the dielectric rod. Similar choices are available for determining the magnetic field inside the dielectric rod. A convenient choice is to use the dyadic Green's function for the larger waveguide, using $k = \sqrt{\kappa} k_0$ for the interior problem and k_0 for the exterior one, since this requires the construction of fewer dyadic Green's functions. We will then find that spurious roots will occur and correspond to modes propagating in a dielectric-filled rectangular waveguide with the dielectric having a rectangular hole in the middle. The interior and exterior fields are given by

$$\mathbf{E}_1(\mathbf{r}') = -\oint_C [\mathbf{n} \times \mathbf{E}_1(\mathbf{r}) \cdot \nabla \times \bar{G}_{1e} - j\omega\mu_0 \mathbf{n} \times \mathbf{H}_1(\mathbf{r}) \cdot \bar{G}_{1e}] \, dl \qquad (117a)$$

$$\mathbf{H}_1(\mathbf{r}') = -\oint_C [\mathbf{n} \times \mathbf{H}_1(\mathbf{r}) \cdot \nabla \times \bar{G}_{1m} + j\omega\epsilon_0\kappa \mathbf{n} \times \mathbf{E}_1(\mathbf{r}) \cdot \bar{G}_{1m}] \, dl \qquad (117b)$$

$$\mathbf{E}_2(\mathbf{r}') = \oint_C [\mathbf{n} \times \mathbf{E}_2(\mathbf{r}) \cdot \nabla \times \bar{G}_{2e} - j\omega\mu_0 \mathbf{n} \times \mathbf{H}_2(\mathbf{r}) \cdot \bar{G}_{2e}] \, dl \qquad (117c)$$

$$\mathbf{H}_2(\mathbf{r}') = \oint_C [\mathbf{n} \times \mathbf{H}_2(\mathbf{r}) \cdot \nabla \times \bar{G}_{2m} + j\omega\epsilon_0 \mathbf{n} \times \mathbf{E}_2(\mathbf{r}) \cdot \bar{G}_{2m}] \, dl \qquad (117d)$$

where $\bar{G}_{1m}$ and $\bar{G}_{2m}$ are dyadic Green's functions for the magnetic field and satisfy the boundary conditions $\mathbf{n} \times \nabla \times \bar{G}_{1m} = \mathbf{n} \times \nabla \times \bar{G}_{2m} = 0$ on the metal walls. The functions labeled with one and two as subscripts differ only in that the former has k and the latter has k_0 as a parameter. Also, the above dyadic Green's functions are the Fourier transforms with respect to the z and z' variables. The integration is around the contour C of the dielectric rod. It should be noted that both the electric- and magnetic-type dyadic Green's functions must include the longitudinal modes as well as the transverse modes. The integrals involving the curls of the Green's dyadic functions are discontinuous across the boundary C so the tangential fields must be evaluated as the observation point $\mathbf{r}'$ approaches the boundary C from the interior for the interior fields and from the exterior for the exterior fields. The discontinuous behavior of these integrals may be described mathematically as follows:

$$\lim_{\mathbf{r}' \to S^-} \oint_C \mathbf{n}(\mathbf{r}) \times \mathbf{E}(\mathbf{r}) \cdot \nabla \times \bar{G}_{1e}(\mathbf{r}, \mathbf{r}') \times \mathbf{n}(\mathbf{r}') \, dl$$

$$= \lim_{\mathbf{r}' \to S^+} \oint_C \mathbf{n}(\mathbf{r}) \times \mathbf{E}(\mathbf{r}) \cdot \nabla \times \bar{G}_{1e}(\mathbf{r}, \mathbf{r}') \times \mathbf{n}(\mathbf{r}') \, dl$$

$$+ \mathbf{n} \times (\mathbf{E}^- - \mathbf{E}^+) \qquad (118)$$

where S^- denotes the interior side of S, S^+ denotes the exterior side, and $\mathbf{n} \times \mathbf{E}^-$ and $\mathbf{n} \times \mathbf{E}^+$ are the limiting values of the tangential electric fields as the surface S is approached from the interior and exterior sides, respectively. The equivalent magnetic current $\mathbf{J}_m$ on the surface is $\mathbf{n} \times (\mathbf{E}^- - \mathbf{E}^+)$. When we equate the tangential field components given by (117) we can interchange the sides of S that $\mathbf{r}'$ approaches for the interior and exterior fields since the integrals involving $\nabla \times \bar{G}_{1e}$ and $\nabla \times \bar{G}_{2e}$ have equal and opposite discontinuities across the boundary and therefore cancel. As pointed out by Collin and Ksienski [6.40] we can also place $\mathbf{r}'$ directly on the contour C and thus eliminate the need to take limiting values of the non-uniformly converging series that arise. We will demonstrate these various options later on.

The analysis of the various modes of propagation for a rectangular dielectric waveguide can

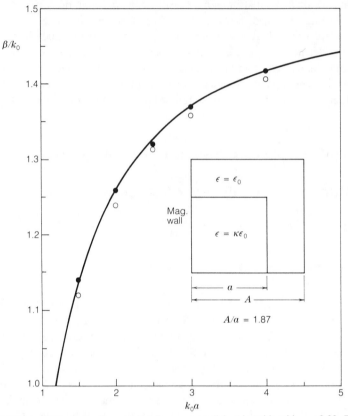

Fig. 6.19. Normalized propagation constant for a square dielectric guide with $\kappa = 2.22$. The solid circles are values obtained by mode matching and the open circles are values obtained from a finite-difference solution [6.29].

be simplified by considering separately modes having specific symmetry properties. For the guide shown in Fig. 6.18 there are modes for which the symmetry planes containing the x and y axes correspond to electric walls, magnetic walls, or a combination of the two. A mode of practical interest is the one where the yz plane is a magnetic wall and the xz plane is an electric wall. For this mode the problem reduces to a consideration of one-quarter of the total cross section as shown in Fig. 6.19. For this mode the scalar functions from which the $\mathbf{M}_n$, $\mathbf{N}_n$, and $\mathbf{L}_n$ modes in the Green's function $\overline{\mathbf{G}}_{1e}$ are generated are

$$\psi_{M,nm} = \sqrt{\frac{\epsilon_{0m}}{AB}}\ \sin\frac{n\pi x}{2A}\ \cos\frac{m\pi y}{B}e^{-jwz}$$

$$n = 1, 3, 5, \ldots; \quad m = 0, 1, 2, \ldots$$

$$\psi_{N,nm} = \sqrt{\frac{2}{AB}}\ \cos\frac{n\pi x}{2A}\ \sin\frac{m\pi y}{B}e^{-jwz} = \psi_{L,nm}.$$

The generic form for the Fourier transform of the dyadic Green's function is given by the integrand in (88) in Chapter 5. It is a straightforward matter to construct $\nabla \times \overline{\mathbf{G}}_{1e}$. Upon replacing w by $-\beta$ we have

$$\nabla \times \bar{\mathbf{G}}_{1e} = \sum_{n=1,3,\ldots}^{\infty} \sum_{m=0,1,\ldots}^{\infty} \frac{\nabla \times \mathbf{M}_{nm}(\mathbf{r}, -\beta)\mathbf{M}_{nm}(\mathbf{r}', \beta) + \nabla \times \mathbf{N}_{nm}(\mathbf{r}, -\beta)\mathbf{N}_{nm}(\mathbf{r}', \beta)}{k_{nm}^2(k_{nm}^2 - \gamma^2)}$$

where $k_{nm}^2 = (n\pi/2A)^2 + (m\pi/B)^2$. We will examine the tangential components of this function on the upper surface $y = b$ of the dielectric waveguide. These are given by

$$\frac{2}{AB} \sum_{n=1,3,\ldots}^{\infty} \sum_{m=1,2,\ldots}^{\infty}$$

$$\left[\frac{\left(k_{nm}^2\mathbf{a}_z + j\beta\mathbf{a}_x\dfrac{\partial}{\partial x}\right)\left(\sin\dfrac{n\pi x}{2A}\cos\dfrac{m\pi y}{B}\right)\left(\mathbf{a}_x\dfrac{\partial}{\partial y'}\sin\dfrac{n\pi x'}{2A}\cos\dfrac{m\pi y'}{B}\right)}{k_{nm}^2\left(k_{nm}^2 - \gamma^2\right)}\right.$$

$$\left.+\frac{\left(\mathbf{a}_x\dfrac{\partial}{\partial y}\sin\dfrac{n\pi x}{a}\cos\dfrac{m\pi y}{B}\right)\left(k_{nm}^2\mathbf{a}_z - j\beta\mathbf{a}_x\dfrac{\partial}{\partial x'}\right)\left(\sin\dfrac{n\pi x'}{2A}\cos\dfrac{m\pi y'}{B}\right)}{k_{nm}^2\left(k_{nm}^2 - \gamma^2\right)}\right].$$

$$(119)$$

The series to be summed over m for the $\mathbf{a}_z\mathbf{a}_x$ term is

$$\frac{\partial}{\partial y'}\sum_{m=1}^{\infty}\frac{\cos\dfrac{m\pi y}{B}\cos\dfrac{m\pi y'}{B}}{\left(\dfrac{m\pi}{B}\right)^2 - \Gamma_n^2} = \frac{1}{2}\frac{\partial}{\partial y'}\sum_{m=1}^{\infty}\frac{\cos\dfrac{m\pi}{B}(y+y') + \cos\dfrac{m\pi}{B}(y-y')}{\left(\dfrac{m\pi}{B}\right)^2 - \Gamma_n^2}$$

where $\Gamma_n^2 = \gamma^2 - (n\pi/2A)^2$. The sum of this series is given in the Mathematical Appendix so we find that we get

$$\left(\frac{B}{\pi}\right)^2\frac{\partial}{\partial y'}\left[\frac{\pi^2}{2\Gamma_n^2 B^2} - \frac{\pi^2}{4\Gamma_n B}\frac{\cos\Gamma_n(y+y'-B) + \cos\Gamma_n(|y-y'|-B)}{\sin\Gamma_n B}\right]$$

where the magnitude $|y - y'|$ has been introduced because the series involving the terms $\cos m\pi(y - y')/B$ is an even function of $y - y'$ with a derivative with respect to y' that is zero at $y' = y$ but is nonzero and of opposite sign as y' approaches y from the left and right sides. Our final result is

$$\frac{B}{4\sin\Gamma_n B}[\sin\Gamma_n(y+y'-B) - \mathrm{sg}\,(y-y')\sin\Gamma_n(|y-y'|-B)]$$

with $\mathrm{sg}\,(y - y')$ being the sign of $y - y'$. For $y = b$ and y' approaching b from the left we get

$$\frac{B}{4} + \frac{B\sin\Gamma_n(2b-B)}{4\sin\Gamma_n B}$$

while for y' approaching b from the right we get

$$-\frac{B}{4} + \frac{B \sin \Gamma_n(2b - B)}{4 \sin \Gamma_n B}.$$

If we set $y = y' = b$ after differentiating the original series term by term (this is called for in the construction of the $\mathbf{M}_{nm}$ functions) we get the average value of the two limiting values. The discontinuity across the line $y = b$ is $-B/2$. When we multiply this discontinuity by the functions of x and sum over n we obtain

$$-\mathbf{a}_z\mathbf{a}_x \sum_{n=1,3,\ldots}^{\infty} \frac{1}{A} \sin \frac{n\pi x}{2A} \sin \frac{n\pi x'}{2A} = -\mathbf{a}_z\mathbf{a}_x\delta(x - x').$$

If we now scalar multiply by $\mathbf{a}_y \times \mathbf{E}$ and integrate over x we find that

$$\int_0^a \mathbf{E}\cdot\mathbf{a}_y \times \mathbf{a}_z\mathbf{a}_x\delta(x - x')\,dx = E_x(x')\mathbf{a}_x$$

which shows that for one term in the integral

$$\oint_C \mathbf{n} \times \mathbf{E}\cdot\nabla \times \bar{\mathbf{G}}_{1e}\,dl$$

there is a discontinuous change across the boundary equal to the tangential value of E_x. The series associated with the $\mathbf{a}_x\mathbf{a}_z$ term has a discontinuity of opposite sign because the derivative is with respect to y. It will give a term $\mathbf{E}_z(x')\mathbf{a}_z$ when the discontinuity in the integral of $\mathbf{n} \times \mathbf{E}\cdot\nabla \times \bar{\mathbf{G}}_{1e}$ along the $y = b$ surface is evaluated. The series associated with $\mathbf{a}_x\mathbf{a}_x$ has a coefficient proportional to m^{-3} for large values of m. It may, therefore, be inferred that this is a uniformly convergent series for all values of y and y'. A direct proof follows readily by using the identity

$$\frac{1}{k_{nm}^2(k_{nm}^2 - \gamma^2)} = \frac{1}{\gamma^2}\left[\frac{1}{k_{nm}^2 - \gamma^2} - \frac{1}{k_{nm}^2}\right]$$

and summing the two series. It should be apparent that the discontinuities will cancel since the two series are of opposite sign. The same kind of discontinuous behavior occurs on the $x = a$ surface and may be established by summing the series that occurs on that part of the boundary over the integer n. The functions $\nabla \times \bar{\mathbf{G}}_{2e}$, $\nabla \times \bar{\mathbf{G}}_{1m}$, and $\nabla \times \bar{\mathbf{G}}_{2m}$ have the same type of behavior. In view of these results it is quite clear that our earlier statement, that in the integral equations for the boundary values of $\mathbf{n} \times \mathbf{E}$ and $\mathbf{n} \times \mathbf{H}$ we can interchange the sides of S that the observation point $\mathbf{r}'$ approaches or set both $\mathbf{r}$ and $\mathbf{r}'$ on the boundary, is valid. In the first instance the discontinuities cancel and the last choice is equivalent to an average of the results obtained when $\mathbf{r}'$ approaches S^- and S^+.

The detailed solution to the rectangular dielectric rod waveguide will be left incomplete since it is mostly a matter of carrying out the formal steps along lines already presented for the circular rod. An example of results that are obtained is shown in Fig. 6.19 based on computations made by Collin and Ksienski [6.40]. For the mode under consideration the following basis functions (also used as testing functions) were used to describe the tangential fields on the guide surface:

$$\mathbf{n} \times \mathbf{E} = \begin{cases} V_1\mathbf{a}_x \cos \dfrac{\pi x}{2d} + V_2\mathbf{a}_z \sin \dfrac{\pi x}{2d}, & y = b \\[2ex] -V_1\mathbf{a}_y \sin \dfrac{\pi y}{2d} + V_2\mathbf{a}_z \cos \dfrac{\pi y}{2d}, & x = a \end{cases} \tag{120a}$$

$$\mathbf{n} \times \mathbf{H} = \begin{cases} I_1\mathbf{a}_x \sin \dfrac{\pi x}{2d} + I_2\mathbf{a}_z \cos \dfrac{\pi x}{2d}, & y = b \\[2ex] -I_1\mathbf{a}_y \cos \dfrac{\pi y}{2d} + I_2\mathbf{a}_z \sin \dfrac{\pi y}{2d}, & x = a \end{cases} \tag{120b}$$

where $d = a + b$ is the total length of the contour C. The equivalent magnetic and electric currents on the surface were chosen to be continuous at the corner $x = a$, $y = b$. As can be seen from the results shown in Fig. 6.19 the use of four global basis functions gives values of β for the mode under consideration that agree closely with results obtained by using the method of mode matching or the method of finite differences. The evaluation of β from the 4×4 determinant required only modest computational effort.

If the cross section of the dielectric guide is not rectangular or circular it will generally be necessary to evaluate the contour integrals numerically. This will add some additional computational burden which could be offset by expending additional time on the analysis and summing the various series over one of the integers n or m. The boundary-element method is an exact procedure and can be adapted to treat a variety of dielectric waveguides. In common with many other methods it has the disadvantage of giving spurious roots or values for the propagation constants. However, in many cases it would not be difficult to identify the unwanted solutions. The boundary-element method is formulated in terms of the tangential values of the fields on the surface. The tangential magnetic field components are not singular at the edge of a dielectric structure. However, the tangential electric field component normal to the edge can be singular (Section 1.5). It would be preferable to incorporate the edge singularity into the basis functions in order to achieve a better approximation to the exact surface field.

6.6. DIELECTRIC RESONATORS

The common shapes used for high-dielectric-constant dielectric resonators are the sphere, the cylinder, and the rectangular box-like structure as shown in Fig. 6.20. The resonator is often placed on a ground plane or on a dielectric substrate material. The methods used to analyze dielectric waveguides can also be used to treat dielectric resonators. In particular, the

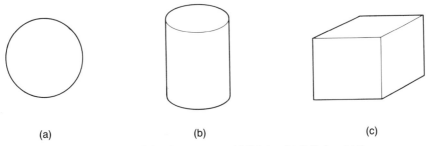

(a)   (b)   (c)

Fig. 6.20. Common dielectric resonators. (a) Sphere. (b) Cylinder. (c) Box.

boundary-element method is an exact method in principle and can be used for many practical resonator structures. The method does allow considerable latitude in the choice of dyadic Green's functions to be used.

The dielectric sphere, as a resonator problem, can be solved exactly using the spherical vector wave functions which have been described in Chapter 2. The dielectric sphere resonator was analyzed as early as 1939 by Richtmyer [6.47]. In the interior of the sphere we can choose the electric field to be described by a single mode such as the following TE mode:

$$\mathbf{E} = A_{nm}\mathbf{M}_{nm}(\mathbf{r}, k), \qquad r < a. \tag{121a}$$

The magnetic field is then given by

$$\mathbf{H} = \frac{j}{k_0 Z_0}\nabla \times \mathbf{E} = \frac{j}{k_0 Z_0}A_{nm}\nabla \times \mathbf{M}_{nm} = \frac{jk}{k_0 Z_0}A_{nm}\mathbf{N}_{nm}(\mathbf{r}, k), \qquad r < a. \tag{121b}$$

For the field outside the sphere we choose the outward-propagating modes as solutions; thus

$$\mathbf{E} = B_{nm}\mathbf{M}_{nm}^+(\mathbf{r}, k_0), \qquad r > a \tag{121c}$$

$$\mathbf{H} = j\frac{B_{nm}}{Z_0}\mathbf{N}_{nm}^+(\mathbf{r}, k_0), \qquad r > a. \tag{121d}$$

The mode functions are given in (161) and (165) in Chapter 2 [see also the relations (135) in Chapter 2].

For free oscillations of the sphere the boundary conditions at $r = a$ require that the tangential field components match at the air–dielectric interface. By using the expressions for the mode functions it is readily found that the following equations must hold:

$$kj_n(ka)A_{nm} - k_0 h_n^2(k_0 a)B_{nm} = 0$$

$$\frac{k}{k_0}\frac{d[kaj_n(ka)]}{dka}A_{nm} - \frac{d[k_0 a h_n^2(k_0 a)]}{dk_0 a}B_{nm} = 0.$$

For a solution the determinant must vanish and this yields the eigenvalue equation

$$j_n(ka)\frac{d[k_0 a h_n^2(k_0 a)]}{dk_0 a} - h_n^2(k_0 a)\frac{d[kaj_n(ka)]}{dka} = 0. \tag{122}$$

This equation is independent of m so consequently there are many degenerate solutions corresponding to even and odd modes with different azimuthal dependence on the angle. The spherical Bessel functions consist of a finite number of terms. For $n = 0, 1$ we have

$$j_0(x) = \frac{\sin x}{x}$$

$$h_0^2(x) = \frac{\sin x + j\cos x}{x}$$

$$j_1(x) = \frac{\sin x}{x^2} - \frac{\cos x}{x}$$

$$h_1^2(x) = \frac{(1 + jx)\sin x + (j - x)\cos x}{x^2}.$$

For $n = 0$ the $\mathbf{M}_{nm}$ and $\mathbf{N}_{nm}$ modes vanish. For $n = 1$ the eigenvalue equation becomes

$$(1 + jk_0a)\kappa \sin ka - \sin ka + ka \cos ka = 0. \tag{123}$$

When κ is large an approximate solution may be obtained by using $ka = \pi$ as the first approximation and expanding the eigenvalue equation about this point. This procedure gives

$$ka = \pi \left(1 - \frac{1}{N^2 + jN\pi} \right) \tag{124}$$

where $N^2 = \kappa$. For $\kappa = 86$ we obtain $ka = 3.109 + j0.0111$. The Q of the mode is given by (we assume that κ is real)

$$Q = \frac{\text{Re}\,ka}{2\,\text{Im}\,ka} = \frac{3.109}{0.0222} = 140. \tag{125}$$

The $n = 1$ TE mode is denoted by TE_{1m1} for the smallest root of (123).

Dual modes with the electric and magnetic fields interchanged also exist. For these TM modes the eigenvalue equation is

$$\kappa j_n(ka)\frac{d[k_0ah_n^2(k_0a)]}{dk_0a} - h_n^2(k_0a)\frac{d[kaj_n(ka)]}{dka} = 0. \tag{126}$$

For $n = 1$ the eigenvalue equation has the explicit form

$$[N^3 - N(ka)^2 - N + j(N^2 - 1)ka][\sin ka - ka \cos ka]$$
$$+[N(ka)^2 + j(ka)^3]\sin ka = 0. \tag{127}$$

The approximate solution is given by $\sin ka - ka \cos ka = 0$ and gives $ka = 4.4934$. By means of a Taylor series expansion about this point we get

$$ka = x_0 - \frac{(Nx_0 + jx_0^2)}{(N^3 + 2N - Nx_0^2) + j(N^2x_0 + 3x_0)} \tag{128}$$

where $x_0 = 4.4934$ is the first approximation. For $\kappa = 86$ we find that the second approximation gives $ka = 4.432 + j0.00714$ and a Q equal to 310. These results agree with those obtained by Gastine et al. [6.52] for the real part of ka, but the value of Q is smaller than the value of 330 which is obtained by solving (126) more exactly. Thus (128) is of adequate accuracy when the dielectric constant is of order 100 or more and has the distinct advantage that it is not a transcendental equation.

The Q that comes from radiation loss is an upper bound since dielectric dissipation will result in a lower quality factor. We can define the total Q, which we will call Q_0, by relation (159) in Chapter 5; thus

$$Q_0 = \frac{\omega W}{P_L} = \frac{\omega(W_1 + W_2)}{P_1 + P_2} \tag{129}$$

where W_1 is the average stored energy in the interior of the sphere, W_2 is the average stored energy in the field external to the sphere, and P_L is the total power dissipation P_1 plus radiated power P_2. The imaginary part of ka gives the Q taking radiation loss only into account. Hence this Q is given by

$$Q = \frac{\omega W}{P_2}. \tag{130}$$

For TM modes the external stored electric energy $W_{e,e}$ is greater than the stored magnetic energy $W_{m,e}$. The external Q, denoted Q_e, for the $n = 1$ TE and TM modes is given by [6.60]

$$Q_e = \frac{N^3}{(ka)^3} + \frac{N}{ka}. \tag{131}$$

For our example this gives $Q_e = 2\omega W_{e,e}/P_2 = 11.26$ which shows that the internal stored electric energy is much greater. The stored internal electric energy is $W_{e,i}$ and the dissipated power $P_1 = (2\omega\epsilon''/\epsilon')W_{e,i}$ where ϵ' and ϵ'' are the real and imaginary parts of the permittivity. The internal Q, which we will call Q_i, is given by

$$Q_i = \frac{2\omega W_{e,i}}{P_1} = \frac{\epsilon'}{\epsilon''}. \tag{132}$$

We can now put down the following relations:

$$\omega W = 2\omega W_{e,e} + 2\omega W_{e,i} = Q_e P_2 + Q_i P_1 = Q_0(P_1 + P_2) = QP_2.$$

From these we find that $P_1 Q_0 = (Q - Q_0)P_2$ and we are able to eliminate P_1 and P_2 to obtain

$$Q_0 = \frac{QQ_i}{Q - Q_e + Q_i} \approx \frac{QQ_i}{Q + Q_i}, \qquad \text{TM modes} \tag{133}$$

since Q_e is typically small relative to Q and Q_i. For our example, assuming that $Q = Q_i = 330$, we find that $Q_0 = 168$.

For TE waves the external stored magnetic energy is greater than the stored electric energy. Hence for the internal field the opposite is true since at resonance $W_{e,i} + W_{e,e} = W_{m,i} + W_{m,e}$. Consequently, the internal power dissipation will be somewhat smaller than the quantity $(2\omega\epsilon''/\epsilon')W_{m,i}$. Since Q_e is based on $W_{m,e}$ and Q_0 uses $W_{m,i} + W_{m,e}$ we have to work with the magnetic field energy and therefore cannot specify P_1 in an exact manner. If we define Q_i as

$$Q_i = \frac{2\omega W_{m,i}}{P_1}$$

this is larger than ϵ'/ϵ''. However, since $W_{m,e} \ll W_{m,i}$ we can use ϵ'/ϵ'' for Q_i with a relatively small error with the consequence that (133) is a good approximation for Q_0 for TE modes, for high-dielectric-constant resonators.

The total Q_0 is, of course, available from the analytic solution for ka. When the dielectric constant is complex the resonant complex frequency is given by $\omega' + j\omega'' =$

$(ka)/[\sqrt{\epsilon_0\mu_0(\kappa' - j\kappa'')}a]$ where ka is the complex eigenvalue. If we let $ka = x + jy$ then

$$\omega' + j\omega'' = \frac{x + jy}{a\sqrt{\epsilon_0\mu_0(\kappa' - j\kappa'')^{1/2}}} \approx \frac{x + jy + j\kappa''/2\kappa'}{a\sqrt{\kappa'\epsilon_0\mu_0}}$$

since in most cases $y \ll x$ and $\kappa'' \ll \kappa'$. The total Q_0 is given by

$$Q_0 = \frac{\omega'}{2\omega''} = \frac{x}{2y + \kappa''/\kappa'} = \frac{QQ_i}{Q + Q_i} \tag{134}$$

which agrees with (133). The definition of quality factor used in (134) in terms of the real and imaginary parts of the complex frequency is not exactly the same as that based on stored energy and dissipated power. For high-Q systems both definitions give nearly the same value for the quality factor.

The equation that describes the continuity of the tangential magnetic field for TM modes is

$$Nj_n(ka)A_{nm} - h_n^2(k_0a)B_{nm} = 0.$$

For these modes $j_n(ka)$ is not small at the resonant frequencies so we see that in the limit as the index of refraction N tends to infinity the coefficient B_{nm} must vanish. Thus the field is confined to the interior of the sphere. This confinement of the modes for large values of N does not occur for the TE modes since $j_n(ka)$ for TE modes approaches zero at resonance for large values of N. From another point of view, the radial magnetic flux B_r must be continuous across the boundary so if it has a finite value in the interior it will also have a finite value on the exterior of the surface. The radial electric field, on the other hand, can terminate on the polarization charge on the air–dielectric surface. Van Bladel has shown [6.48] that for dielectric bodies of revolution that are more general than the spherical shape the only confined modes are those that have only an azimuthal magnetic field and do not depend on the azimuthal angle ϕ. Furthermore, Van Bladel has shown, through the use of a power series expansion of the field in inverse power of N, that for high-dielectric-constant resonators the field outside the resonator may be approximated by quasi-static fields. This property can be inferred from the fact that ka is a finite eigenvalue and hence when N is very large k_0a must be small and the external field is then a quasi-static field.

Cylindrical Dielectric Resonator

For a practical dielectric resonator the finite-length cylinder or the box-like structure is preferred over the sphere because, unlike the spherical resonator, these structures do not have the large number of degenerate modes and are easier to fabricate and mount in a microwave circuit. One method that has been used to analyze the cylindrical resonator is to expand the internal fields in a series of $\mathbf{M}_{nm}$ and $\mathbf{N}_{nm}$ modes and to expand the external fields in terms of $\mathbf{M}_{nm}^+$ and $\mathbf{N}_{nm}^+$ modes. A finite number of modes is used and the boundary conditions requiring matching of the tangential field components are enforced in a mean square sense. Tsuji et al. obtained very good results using 10 modes for a cylindrical resonator with a length-to-diameter ratio of one [6.57], [6.58]. The expansion in terms of spherical vector modes is valid inside the largest inscribed sphere inside the resonator and outside the smallest circumscribed sphere outside the resonator. Between the two spherical surfaces so described, and this region includes the resonator boundary, the completeness of the expansions is questionable [6.27].

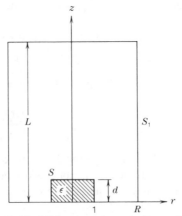

Fig. 6.21. A cylindrical dielectric resonator inside a cylindrical cavity.

The real part of the resonant frequency for a cylindrical resonator can be found with good accuracy by placing the dielectric resonator in the interior of a much larger cylindrical cavity having metal walls on which $\mathbf{n} \times \mathbf{E} = 0$. In this case the dispersion equation is real and the eigenvalues are real. This method of analysis does not give any information on the Q of the resonant modes. It also introduces all the cavity resonances as additional solutions. Many of the cavity resonant frequencies can be close to those of the dielectric resonator so significant mode coupling with its associated shifts in the resonant frequencies will occur. Many times in practice the resonator will be located inside a closed shielded box and coupling with the resonant modes of the enclosure will then be important to analyze.

In Fig. 6.21 we show a dielectric resonator inside a cylindrical cavity. This resonator has been analyzed using the boundary-element method along with basis functions appropriate for the TE_{012} mode [6.40]. This mode corresponds to the fundamental mode in a cylindrical resonator of length equal to 2 units and having only an azimuthal electric field component that has odd symmetry about the midplane. The cavity–resonator problem is equivalent to using a resonator of unit length placed on the bottom plate of the enclosing cylindrical cavity.

One suitable choice for the dyadic Green's function for calculating the field in the internal dielectric region is an expansion in terms of the cavity modes of the small cylindrical region using magnetic wall boundary conditions at the air–dielectric surface and using the electric wall boundary conditions on the lower $z = 0$ surface. With this choice the expressions for the fields in the dielectric become

$$\mathbf{E}(\mathbf{r}') = j\omega\mu_0 \oiint_S \mathbf{n} \times \mathbf{H} \cdot \bar{\mathbf{G}}_{1e}\, dS$$

$$\mathbf{H}(\mathbf{r}') = -\oiint_S \mathbf{n} \times \mathbf{H} \cdot \nabla \times \bar{\mathbf{G}}_{1m}\, dS$$

where $\mathbf{n} \times \nabla \times \bar{\mathbf{G}}_{1e} = \mathbf{n} \times \bar{\mathbf{G}}_{1m} = 0$ on the air–dielectric surfaces and $\mathbf{n} \times \bar{\mathbf{G}}_{1e} = \mathbf{n} \times \nabla \times \bar{\mathbf{G}}_{1m} = 0$ on the $z = 0$ surface. When $\mathbf{r}'$ approaches the surface S the second equation above gives $\mathbf{n} \times \mathbf{H}$ equal to the boundary value so only the first equation needs to be evaluated. For the TE_{012} mode where $\mathbf{n} \times \mathbf{H}$ has only an azimuthally independent ϕ component only the $\mathbf{M}_{nm}$ functions are needed for $\bar{\mathbf{G}}_{1e}$. These functions can be generated from the scalar functions

$$\psi_{M,nm} = \frac{2}{\sqrt{\pi d}} \frac{J_0(p_{0m}r)}{p_{0m}J_0'(p_{0m})} \sin\frac{n\pi z}{2d}, \qquad n = 1, 3, \ldots$$

where $J_0(p_{0m}) = 0$ and the resonator radius is taken to be of unit length and with a normalized axial length of d.

In order to evaluate the fields outside the dielectric resonator the dyadic Green's functions for the large cylindrical cavity are needed. These functions satisfy the boundary conditions $\mathbf{n} \times \bar{\mathbf{G}}_{2e} = \mathbf{n} \times \nabla \times \bar{\mathbf{G}}_{2m} = 0$ on the metal walls. The required generating functions are given by (140) in Chapter 5 with $c = L$, $a = R$, and keeping only the $n = 0$ terms. For $\bar{\mathbf{G}}_{2e}$ only the circularly symmetric $\mathbf{M}_{nm}$ functions are needed. For $\bar{\mathbf{G}}_{2m}$ only the contributions from the $\mathbf{N}_{nm}$ and $\mathbf{L}_{nm}$ functions are required. The generating functions for these are

$$\psi_{N,nm} = \frac{2}{\sqrt{\pi L}} \frac{J_0(p'_{0m}r)}{J_0(p'_{0m})} \sin \frac{n\pi z}{d}$$

$$\psi_{L,nm} = \sqrt{\frac{2\epsilon_{0n}}{\pi L}} \frac{J_0(p'_{0m}r)}{J_0(p'_{0m})} \cos \frac{n\pi z}{d}.$$

The generic form for the dyadic Green's functions is given by (142) in Chapter 5.

By using dyadic Green's functions constructed as outlined above the boundary-element method gives the following two coupled integral equations:

$$\iint\limits_S \mathbf{n} \times \mathbf{E} \cdot \nabla \times \bar{\mathbf{G}}_{2e} \times \mathbf{n} \, dS - j\omega\mu_0 \iint\limits_S \mathbf{n} \times \mathbf{H} \cdot (\bar{\mathbf{G}}_{1e} + \bar{\mathbf{G}}_{2e}) \times \mathbf{n} \, dS = 0 \quad (135a)$$

$$\mathbf{n} \times \mathbf{H} + \iint\limits_S \mathbf{n} \times \mathbf{H} \cdot \nabla \times \bar{\mathbf{G}}_{2m} \times \mathbf{n} \, dS + j\omega\epsilon_0 \iint\limits_S \mathbf{n} \times \mathbf{E} \cdot (\bar{\mathbf{G}}_{2m} \times \mathbf{n}) \, dS = 0 \quad (135b)$$

where S is the air–dielectric surface. The integrals involving $\nabla \times \bar{\mathbf{G}}_{2e}$ and $\nabla \times \bar{\mathbf{G}}_{2m}$ must be evaluated in the limit as $\mathbf{r}'$ approaches the surface S from the exterior region since the series that occur in these terms are not uniformly convergent when $\mathbf{r}'$ approaches $\mathbf{r}$.

In [6.40] the integral equations are solved using the basis functions

$$\mathbf{E} \times \mathbf{n} = \begin{cases} V\mathbf{a}_r J_1(p_{01}r), & z = d \\ -V\mathbf{a}_z J_1(p_{01}) \sin \dfrac{\pi z}{2d}, & r = 1 \end{cases} \quad (136a)$$

$$\mathbf{H} \times \mathbf{n} = \begin{cases} I\mathbf{a}_\phi J_1(p_{01}r), & z = d \\ AI_1\mathbf{a}_\phi J_1(p_{01}) \sin \dfrac{\pi z}{2d}, & r = 1. \end{cases} \quad (136b)$$

The equations are also tested with these functions (Galerkin's method). The field E_ϕ is continuous along the boundary so at the edge $r = 1$, $z = d$, the equivalent current $\mathbf{J}_m$ is made continuous along the boundary. The projection of $\mathbf{H}$ onto the surface $r = 1$ and that onto the surface $z = d$ do not have to be the same and hence the equivalent current $J_{e\phi}$ does not have to be continuous along the contour at the edge. The parameter A was introduced so that the relative values of $J_{e\phi}$ on the two surfaces could be adjusted. By choosing $A = 0.1$ the eigenvalue k exhibited a stationary value. The numerical evaluation was simplified by summing the series over n when $\mathbf{r}$ and $\mathbf{r}'$ were on the surface $z = d$ and summing the series over m when $\mathbf{r}$ and $\mathbf{r}'$ were on the $r = 1$ surface. Figure 6.22 shows the TE_{012} eigenvalue as a

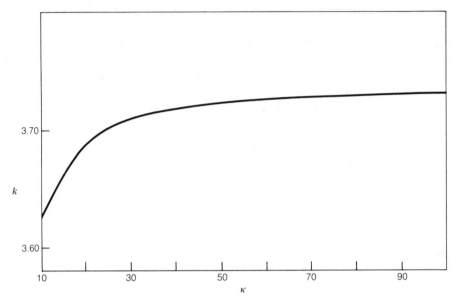

Fig. 6.22. The eigenvalue k as a function of dielectric constant.

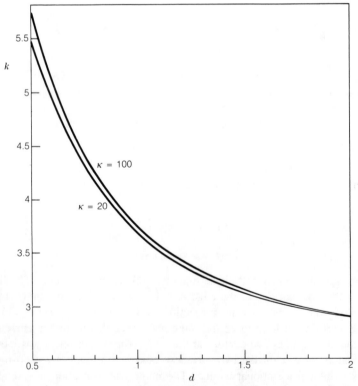

Fig. 6.23. The eigenvalue k as a function of normalized axial length of a cylindrical dielectric resonator.

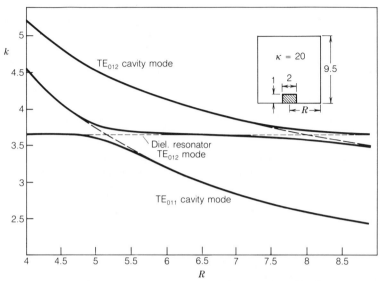

Fig. 6.24. Dispersion curves as a function of cavity radius R.

function of the dielectric constant. It can be seen that for κ greater than 50 the eigenvalue becomes essentially independent of the dielectric constant. This phenomenon is a reflection of the property that for large dielectric constants the fields outside the resonator are quasi-static and hence not dependent on frequency. In Fig. 6.23 we show the dependence of k on the normalized resonator length d. Figure 6.24 shows k as a function of the radius R of the enclosing cylindrical cavity. As R varies, the resonant frequencies of the TE_{011} and TE_{012} cavity modes become coincident with that of the TE_{012} mode for the dielectric resonator. Since the resonator and cavity modes interact the plot of k versus R shows the classical splitting apart of the degeneracy. The dielectric resonator curve of k vs. R goes over into the cavity mode curve of k vs. R, and vice versa, as each coincident resonant frequency is encountered. If the dielectric resonator has a radius a, then the plots of k should be interpreted as plots of ka.

It should be noted that the boundary-element method when applied to dielectric resonators may also produce spurious solutions.

REFERENCES AND BIBLIOGRAPHY

[6.1]   A. T. Villeneuve, "Equivalent circuits of junctions of slab-loaded rectangular waveguides," *IEEE Trans. Microwave Theory Tech.*, vol. MTT-33, pp. 1196–1203, 1985.

[6.2]   B. M. Azizur Rahman and J. B. Davies, "Penalty function improvement of waveguide solution by finite elements," *IEEE Trans. Microwave Theory Tech.*, vol. MTT-32, pp. 922–928, Aug. 1984.

[6.3]   M. Koshiba, K. Hayata, and M. Suzuki, "Improved finite element formulation in terms of the magnetic-field vector for dielectric waveguides," *IEEE Trans. Microwave Theory Tech.*, vol. MTT-33, pp. 227–233, Mar. 1985.

[6.4]   D. Ksienski, private communication.

[6.5]   A. S. Omar and K. F. Schünemann, "Complex and backward-wave modes in inhomogeneously and anisotropically filled waveguides," *IEEE Trans. Microwave Theory Tech.*, vol. MTT-35, pp. 268–275, 1987.

Guides Partially Filled with Dielectric Media

(In addition to the references listed below, see [5.25].)

[6.6]   L. Pincherle, "Electromagnetic waves in metal tubes filled longitudinally with two dielectrics," *Phys. Rev.*, vol. 66, pp. 118–130, Sept. 1944.

[6.7]   L. G. Chambers, "Propagation in waveguides filled longitudinally with two or more dielectrics," *Brit. J. Appl. Phys.*, vol. 4, pp. 39–45, Feb. 1953. (This is a review article containing 16 references.)

[6.8]   P. H. Vartanian, W. P. Ayers, and A. L. Helgesson, "Propagation in dielectric slab loaded rectangular waveguide," *IRE Trans. Microwave Theory Tech.*, vol. MTT-6, pp. 215–222, 1958.

[6.9]   F. E. Gardiol, "Higher-order modes in dielectrically loaded rectangular waveguides," *IEEE Trans. Microwave Theory Tech.*, vol. MTT-16, pp. 919–924, 1968.

[6.10] G. N. Tsandoulas, "Bandwidth enhancement in dielectric-lined circular waveguides," *IEEE Trans. Microwave Theory Tech.*, vol. MTT-21, pp. 651–654, 1973.

Variational Techniques and the Rayleigh–Ritz Method

[6.11] L. G. Chambers, "An approximate method for the calculation of propagation constants for inhomogeneously filled waveguides," *Quart. J. Mech. Appl. Math.*, vol. 7, pp. 299–316, Sept. 1954.

[6.12] R. E. Collin and R. M. Vaillancourt, "Application of the Rayleigh-Ritz method to dielectric steps in waveguides," *IRE Trans. Microwave Theory Tech.*, vol. MTT-5, pp. 177–184, July 1957.

[6.13] A. D. Berk, "Variational principles for electromagnetic resonators and waveguides," *IRE Trans. Antennas Propagat.*, vol. AP-4, pp. 104–111, Apr. 1956.

[6.14] C. H. Chen and C.-D. Lien, "The variational principle for non-self-adjoint electromagnetic problems," *IEEE Trans. Microwave Theory Tech.*, vol. MTT-28, pp. 878–887, 1980.

[6.15] W. J. English, "Vector variational solutions of inhomogeneously loaded cylindrical waveguide structures," *IEEE Trans. Microwave Theory Tech.*, vol. MTT-19, pp. 9–18, 1971.

[6.16] W. J. English and F. J. Young, "An E vector variational formulation of the Maxwell equations for cylindrical waveguide problems," *IEEE Trans. Microwave Theory Tech.*, vol. MTT-19, pp. 40–46, 1971.

Guides Partially Filled with Ferrite Media

The literature on propagation in ferrite-filled guides is extensive. Representative papers are listed below:

[6.17] A. A. Th. M. Van Trier, "Guided electromagnetic waves in anisotropic media," *Appl. Sci. Res.*, vol. 3B, pp. 305–371, 1953.

[6.18] H. Suhl and L. R. Walker, "Topics in guided wave propagation through gyromagnetic media," *Bell Syst. Tech. J.*, vol. 33. Part I, "The completely filled cylindrical guide," pp. 579–659 (May); Part II, "Transverse magnetization and the non-reciprocal helix," pp. 939–986 (July); Part III, "Perturbation theory and miscellaneous results," pp. 1133–1194 (Sept.), 1954.

[6.19] M. L. Kales, "Modes in waveguides containing ferrites," *J. Appl. Phys.*, vol. 24, pp. 604–608, May 1953.

[6.20] Ferrites Issue, *Proc. IRE*, vol. 44, Oct. 1956.

[6.21] L. B. Felsen and N. Marcuvitz, *Radiation and Scattering of Waves*. Englewood Cliffs, NJ: Prentice-Hall, Inc., 1973.

[6.22] B. Lax and K. J. Button, *Microwave Ferrites and Ferrimagnetics*. New York, NY: McGraw-Hill Book Company, Inc., 1962.

[6.23] R. F. Soohoo, *Theory and Application of Ferrites*. Englewood Cliffs, NJ: Prentice-Hall, Inc., 1960.

[6.24] P. Clarricoats, *Microwave Ferrites*. New York, NY: John Wiley & Sons, Inc., 1961.

Dielectric Waveguides

  Mode-Matching Methods:

[6.25] J. E. Goell, "A circular-harmonic computer analysis of rectangular dielectric waveguides," *Bell Syst. Tech. J.*, vol. 48, pp. 2133–2160, 1969.

[6.26] A. L. Cullen, O. Ozkan, and L. A. Jackson, "Point matching technique for rectangular cross-section dielectric rod," *Electron. Lett.*, vol. 7, pp. 497–499, 1971.

[6.27] R. H. T. Bates, J. R. James, I. N. L. Gallet, and R. F. Millar, "An overview of point matching," *Radio Electron. Eng.*, vol. 43, pp. 193–200, 1973.

[6.28] K. Solbach and I. Wolff, "The electromagnetic fields and the phase constants of dielectric image lines," *IEEE Trans. Microwave Theory Tech.*, vol. MTT-26, pp. 266–274, 1978.

  Finite-Difference and Finite-Element Methods:

[6.29] E. Schweig and W. B. Bridges, "Computer analysis of dielectric waveguides: A finite difference method," *IEEE Trans. Microwave Theory Tech.*, vol. MTT-32, pp. 531–541, 1984.

[6.30] S. M. Saad, "Review of numerical methods for the analysis of arbitrarily-shaped microwave and optical dielectric waveguides," *IEEE Trans. Microwave Theory Tech.*, vol. MTT-33, pp. 894–899, 1985.

[6.31] T. Angkaew, M. Matsuhara, and N. Kumagai, "Finite-element analysis of waveguide modes: A novel approach that eliminates spurious modes," *IEEE Trans. Microwave Theory Tech.*, vol. MTT-35, pp. 117–123, 1987.

[6.32] M. Hoshiba, K. Hayata, and M. Suzuki, "Improved finite element formulation in terms of the magnetic field vector for dielectric waveguides," *IEEE Trans. Microwave Theory Tech.*, vol. MTT-33, pp. 227–233, 1985.

[6.33] B. M. A. Rahman and J. B. Davies, "Finite-element analysis of optical and microwave waveguide problems," *IEEE Trans. Microwave Theory Tech.*, vol. MTT-32, pp. 20–28, 1984.

[6.34] W. C. Chew and M. A. Nasir, "A variational analysis of anisotropic, inhomogeneous dielectric waveguides," *IEEE Trans. Microwave Theory Tech.*, vol. 37, pp. 661–668, 1989.

Boundary-Element Methods:

[6.35] C. A. Brebbia and S. Walker, *Boundary Element Techniques in Engineering*. London: Butterworths, 1980.

[6.36] K. Yashiro, M. Miyazaki, and S. Ohkawa, "Boundary element method approach to magnetostatic wave problems," *IEEE Trans. Microwave Theory Tech.*, vol. MTT-33, pp. 248–253, 1985.

[6.37] C. C. Su, "A surface integral equations method for homogeneous optical fibers and coupled image lines of arbitrary cross sections," *IEEE Trans. Microwave Theory Tech.*, vol. MTT-33, pp. 1114–1119, 1985.

[6.38] C. C. Su, "A combined method for dielectric waveguides using the finite-element technique and the surface integral equations method," *IEEE Trans. Microwave Theory Tech.*, vol. MTT-34, pp. 1140–1146, 1986.

[6.39] N. Morita, "A method extending the boundary condition for analyzing guided modes of dielectric waveguides of arbitrary cross-sectional shape," *IEEE Trans. Microwave Theory Tech.*, vol. MTT-30, pp. 6–12, 1982.

[6.40] R. E. Collin and D. A. Ksienski, "Boundary element method for dielectric resonators and waveguides," *Radio Sci.*, vol. 22, pp. 1155–1167, 1987.

[6.41] P. J. B. Clarricoats and R. A. Waldron, "Non-periodic slow wave and backward wave structures," *J. Electron. Contr.*, vol. 8, pp. 455–458, 1960.

[6.42] P. J. B. Clarricoats, "Backward waveguides containing dielectrics," *Proc. IEE*, vol. 108C, pp. 496–501, 1961.

[6.43] P. J. B. Clarricoats, "Propagation along unbounded and bounded dielectric rods," *Proc. IEE*, vol. 107C, pp. 170–186, 1960.

[6.44] V. Ya. Smorgonskiy, "Calculation of the double-value part of the dispersion curve of a circular waveguide with dielectric rod," *Radio Eng. Electron. Phys. (USSR)*, vol. 13, pp. 1809–1810, 1968.

[6.45] Yu. A. Ilarionov, "Ambiguity in the dispersion curves of waves of a circular, partially filled waveguide," *Isv. Vuz Radioelektron.*, vol. 16, pp. 30–37, 1973.

[6.46] G. N. Tsandoulas, "Propagation in dielectric lined square waveguides," *IEEE Trans. Microwave Theory Tech.*, vol. MTT-23, pp. 406–410, 1975.

Dielectric Resonators

[6.47] R. D. Richtmyer, "Dielectric resonators," *J. Appl. Phys.*, vol. 10, pp. 391–398, 1939.

[6.48] J. Van Bladel, "On the resonances of a dielectric resonator of very high permittivity," *IEEE Trans. Microwave Theory Tech.*, vol. MTT-23, pp. 199–208, 1975.

[6.49] J. Van Bladel, "The excitation of dielectric resonators of very high permittivity," *IEEE Trans. Microwave Theory Tech.*, vol. MTT-23, pp. 208–217, 1975.

[6.50] M. Verplanken and J. Van Bladel, "The electric-dipole resonances of ring resonators of very high permittivity," *IEEE Trans. Microwave Theory Tech.*, vol. MTT-24, pp. 108–112, 1976.

[6.51] M. Verplanken and J. Van Bladel, "The magnetic-dipole resonances of ring resonators of very high permittivity," *IEEE Trans. Microwave Theory Tech.*, vol. MTT-27, pp. 328–333, 1979.

[6.52] M. Gastine, L. Courtois, and J. L. Dormann, "Electromagnetic resonances of free dielectric spheres," *IEEE Trans. Microwave Theory Tech.*, vol. MTT-15, pp. 694–700, 1967.

[6.53] H. Y. Yee, "Natural resonant frequencies of microwave dielectric resonators," *IEEE Trans. Microwave Theory Tech.*, vol. MTT-13, p. 256, 1965.

[6.54] K. K. Chow, "On the solution and field pattern of cylindrical dielectric resonators," *IEEE Trans. Microwave Theory Tech.*, vol. MTT-14, p. 439, 1966.

[6.55] T. Itoh and R. Rudokas, "New method for computing the resonant frequencies of dielectric resonators," *IEEE Trans. Microwave Theory Tech.*, vol. MTT-25, pp. 52–54, 1977. (The method described is essentially that presented earlier by K. K. Chow; see [6.54].)

[6.56] Y. Konishi, N. Hoshino, and Y. Utsumi, "Resonant frequency of a $TE_{01\delta}{}^{\circ}$ dielectric resonator," *IEEE Trans. Microwave Theory Tech.*, vol. MTT-24, pp. 112–114, 1976.

[6.57] M. Tsuji, H. Shigesawa, and K. Takiyama, "On the complex resonant frequency of open dielectric resonators," *IEEE Trans. Microwave Theory Tech.*, vol. MTT-31, pp. 392–396, 1983.

[6.58] M. Tsuji, H. Shigesawa, and K. Takiyama, "Analytical and experimental investigations on several resonant modes in open dielectric resonators," *IEEE Trans. Microwave Theory Tech.*, vol. MTT-32, pp. 628–633, 1984.

[6.59] A. W. Glisson, D. Kajfez, and J. James, "Evaluation of modes in dielectric resonators using a surface integral equation formulation," *IEEE Trans. Microwave Theory Tech.*, vol. MTT-31, pp. 1023–1029, 1983.

[6.60] R. E. Collin and S. Rothschild, "Evaluation of antenna Q," *IEEE Trans. Antennas Propagat.*, vol. AP-12, pp. 23–27, 1964.

PROBLEMS

6.1. Show that the LSM and LSE modes are a linear combination of the E and H modes.

6.2. Derive the eigenvalue equations for the symmetrical and unsymmetrical modes (both LSM and LSE modes) for rectangular guides inhomogeneously filled with lossless dielectric material as illustrated in Fig. P6.2.

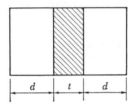 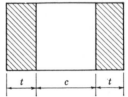

Fig. P6.2.

6.3. Use a two-term approximation in the Rayleigh–Ritz method to compute an approximate value for the propagation constant γ_1 for the dominant mode in a guide loaded with a centered dielectric slab as in Problem 6.2. Compare the approximate eigenvalue with the true eigenvalue. Note that, since γ_1 is imaginary, the absolute value of the approximate eigenvalue will be too small. Evaluate the approximate normalized eigenfunction, and compare it with the true eigenfunction by plotting them as functions of x. Carry out the computation only for the LSE mode and take $a = 0.9$ inch, $t = 0.3a$, $\kappa = 2.56$, and $\lambda_0 = 3.14$ centimeters.

6.4. Show that (27) is a stationary expression for γ^2 by evaluating the variation $\delta\gamma$ and also by applying the Euler–Lagrange equations from the calculus of variations.

6.5. Give a detailed derivation of (43) and (45).

6.6. Show that the variational expressions for the eigenvalues γ^2 for the LSE and LSM modes are equivalent to minimizing the integral over the guide cross section of the energy function $4(W_m - W_e) = \mu_0 \mathbf{H} \cdot \mathbf{H}^* - \kappa(x)\epsilon_0 \mathbf{E} \cdot \mathbf{E}^*$.
HINT: Express $W_m - W_e$ in terms of the Hertzian potential function.

6.7. Use a two-mode approximation in the Rayleigh–Ritz method to compute the equivalent-circuit parameters of the asymmetrical dielectric step in a rectangular guide. Assume that an H_{10} mode is incident and that $a = 0.9$ inch, $t = 0.3a$, $\kappa = 2.56$, and $\lambda_0 = 3.14$ centimeters, and find X_1, X_2, $n^2 Z_1$.

6.8. A circular guide of radius b is inhomogeneously filled with a dielectric cylinder of radius t, where $t < b$. The relative dielectric constant is κ. Obtain the solutions for circularly symmetric E and H modes.

6.9. For the structure of Problem 6.8, show that all modes which have angular variation are not pure E or H modes but a linear combination of the two. Obtain the eigenvalue equation for these hybrid modes.

6.10. Derive suitable variational expressions for the propagation constants of the circularly symmetric E and H modes for the inhomogeneously filled circular guide of Problem 6.8.

6.11. A rectangular guide is completely filled with a ferrite medium. The d-c magnetizing field is applied along the longitudinal axis. Determine the possible modes of propagation in the guide.

6.12. Use a three-mode approximation to find the equivalent-circuit parameters of the dielectric step discontinuity shown in Fig. 6.9. Compare your results with those given in the text. Assume that $t/a = 0.4$.

6.13. Use the boundary-element method to derive the solutions for the propagation constants of the LSE and LSM modes for the slab-loaded rectangular waveguide shown in Fig. 6.1(a). The dyadic Green's functions that are needed do not have any y dependence. Use Green's functions that are appropriate for the rectangular waveguide and that satisfy the boundary conditions

$$\mathbf{n} \times \bar{\mathbf{G}}_{1e} = \mathbf{n} \times \bar{\mathbf{G}}_{2e} = \mathbf{n} \times \nabla \times \bar{\mathbf{G}}_{1m} = \mathbf{n} \times \nabla \times \bar{\mathbf{G}}_{2m} = 0$$

on the waveguide boundary.

6.14. Use the boundary-element method to derive the eigenvalue equations for circularly symmetric modes of a dielectric resonator. Use the mode expansion for the dyadic Green's functions for unbounded space for both the interior and exterior problems. Compare your results with (122) and (126).

7
Excitation of Waveguides and Cavities

In this chapter we will examine the basic problems of radiation from a probe antenna and a small loop antenna in a rectangular waveguide. The impedance of an antenna in a waveguide is significantly different from that of an antenna located in free space. A very common type of waveguide antenna is a probe connected to a coaxial transmission line as shown in Fig. 7.1. By choosing the antenna length d and the short-circuit position ℓ correctly the input impedance can be made equal to the characteristic impedance Z_c of the input coaxial transmission line over a fairly broad range of impedances.

Two waveguides may be coupled by means of probe or loop antennas or a combination of the two. However, it is usually simpler to couple two waveguides by one or more apertures in the common wall. An approximate theory of coupling by small apertures was developed by Bethe [7.5]. We will give an improved version of this small-aperture theory that will provide for conservation of power and hence give more physically meaningful equivalent circuits.

In the latter part of the chapter we will develop the small-aperture theory for waveguide-to-cavity coupling. A cavity can also be excited by means of a small probe or loop antenna coupled to an input coaxial transmission line. The theory is similar to that for antennas in waveguides and for this reason will not be covered.

7.1. The Probe Antenna

The type of coaxial-line probe antenna to be analyzed is shown in Fig. 7.1. It consists of a small coaxial line, terminated in the center of the broad face of a rectangular waveguide, with its inner conductor extending a distance d into the waveguide. In order to have the antenna radiate in one direction only a short-circuiting plunger is placed a distance ℓ to the left of the probe. The probe has a radius r which we will assume to be small relative to the guide height b. However, for typical probes r is large enough that it is necessary to account for the probe thickness if accurate results for the antenna impedance are to be obtained. The field in the waveguide is excited by the unknown aperture fields in the coaxial-line opening. For an exact analysis we should express the field in the waveguide and coaxial line in terms of the unknown aperture electric field and impose the boundary conditions that require the tangential magnetic field to be continuous across the aperture opening. This procedure would lead to equations for determining the electric field in the aperture opening and the current on the probe.

In a typical arrangement the coaxial line is relatively small in diameter and the amplitudes of the higher order coaxial-line modes that are excited are small. A reasonable approximation is to assume that only an incident and a reflected TEM mode exist in the coaxial line. Thus

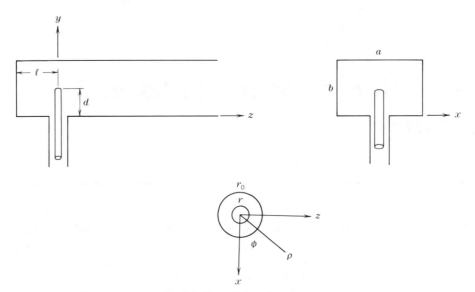

Fig. 7.1. Waveguide probe antenna.

we will assume that in the coaxial line, at the aperture opening,

$$E_r = (V^+ + V^-)\frac{1}{\rho \ln r_0/r} \tag{1a}$$

$$H_\phi = \frac{V^+ - V^-}{Z_c}\frac{1}{2\pi\rho} \tag{1b}$$

where r_0 is the outer-conductor radius, ρ is the radial coordinate, Z_c is the characteristic impedance, and V^+ and V^- are the amplitudes of the incident and reflected TEM waves. The total input current at the probe input equals $2\pi r H_\phi(r) = I_0$ and the antenna input impedance equals

$$Z_{\text{in}} = (V^+ + V^-)/I_0 = V/I_0.$$

If we introduce a dyadic Green's function $\bar{\mathbf{G}}_e$ that satisfies the boundary condition $\mathbf{n} \times \bar{\mathbf{G}}_e = 0$ on the waveguide walls and the radiation condition as z approaches infinity, we can express the electric field in the form

$$\mathbf{E}(\mathbf{r}') = \iint\limits_{S_a} \mathbf{n} \times \mathbf{E}(\mathbf{r}) \cdot \nabla \times \bar{\mathbf{G}}_e(\mathbf{r}, \mathbf{r}')\, dS - j\omega\mu_0 \iint\limits_{S_0} \mathbf{J}(\mathbf{r}) \cdot \bar{\mathbf{G}}_e(\mathbf{r}, \mathbf{r}')\, dS \tag{2}$$

where S_a is the coaxial-line aperture and S_0 is the surface of the probe. This result follows from applying Green's theorem to the surface shown in Fig. 7.2 and making use of the boundary conditions satisfied by $\mathbf{E}$ and $\bar{\mathbf{G}}_e$ on the waveguide walls, the probe surface, and at infinity. The first integral over the surface S_a gives the applied field acting on the probe which will be called $\mathbf{E}_a$. The boundary condition on the probe is $\mathbf{n} \times \mathbf{E} = 0$ and leads to the

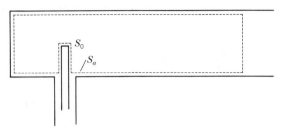

Fig. 7.2. Surface of integration for Green's theorem.

following integral equation for the probe current $\mathbf{J}$:

$$\iint\limits_{S_0} \mathbf{J}(\mathbf{r})\cdot\bar{\mathbf{G}}(\mathbf{r}, \mathbf{r}')\,dS \times \mathbf{n} = -\mathbf{E}_a \times \mathbf{n}, \qquad \mathbf{r}' \text{ on } S_0 \tag{3}$$

where $\bar{\mathbf{G}} = -j\omega\mu_0\bar{\mathbf{G}}_e$.

We will develop a variational expression for an impedance functional Z from which we can find the antenna input impedance Z_{in}. We will then show that this variational method gives the same solution as that obtained by solving the integral equation (3) using the method of moments and Galerkin's procedure. The variational method for antenna impedance was developed by Storer[1] and was formulated for the case when the applied electric field acted over a vanishingly small length of the antenna, essentially a delta function source. We will generalize the method so as to take account of an applied electric field that acts along the full length of the antenna.

The integral equation (3) involves the tangential components along the probe. Instead of taking a cross product with the unit normal $\mathbf{n}$ we will take a scalar product with $\mathbf{J}(\mathbf{r}')$ and integrate over the probe surface, which gives

$$\iint\limits_{S_0} \iint\limits_{S_0} \mathbf{J}(\mathbf{r})\cdot\bar{\mathbf{G}}(\mathbf{r}, \mathbf{r}')\cdot\mathbf{J}(\mathbf{r})\,dS\,dS' = -\iint\limits_{S_0} \mathbf{E}_a(\mathbf{r}')\cdot\mathbf{J}(\mathbf{r}')\,dS'.$$

We now define the impedance functional

$$Z = \frac{V^2}{\displaystyle\iint\limits_{S_0} \mathbf{J}\cdot\mathbf{E}_a\,dS} \tag{4}$$

and consider its variation δZ using

$$\iint\limits_{S_0} \iint\limits_{S_0} \mathbf{J}(\mathbf{r})\cdot\bar{\mathbf{G}}(\mathbf{r}, \mathbf{r}')\cdot\mathbf{J}(\mathbf{r}')\,dS\,dS' + \frac{Z}{V^2}\left[\iint\limits_{S_0} \mathbf{J}(\mathbf{r}')\cdot\mathbf{E}_a(\mathbf{r}')\,dS'\right]^2 = 0. \tag{5}$$

[1]An account of Storer's method may be found in [7.1].

The variation δZ in Z due to a variation $\delta \mathbf{J}$ in the current distribution is given by

$$\frac{\delta Z}{V^2} \left[\iint_{S_0} \mathbf{J} \cdot \mathbf{E}_a \, dS' \right]^2 = - \iint_{S_0} \iint_{S_0} \left[\delta \mathbf{J} \cdot \bar{\mathbf{G}} \cdot \mathbf{J} + \mathbf{J} \cdot \bar{\mathbf{G}} \cdot \delta \mathbf{J} \right.$$

$$\left. + \frac{Z}{V^2} 2 \delta \mathbf{J} \cdot \mathbf{E}_a \mathbf{J} \cdot \mathbf{E}_a \right] dS \, dS'. \quad (6)$$

By using the symmetry property of $\bar{\mathbf{G}}$, namely that $\mathbf{J}(\mathbf{r}) \cdot \bar{\mathbf{G}}(\mathbf{r}, \mathbf{r}') \cdot \delta \mathbf{J}(\mathbf{r}') = \delta \mathbf{J}(\mathbf{r}') \cdot \bar{\mathbf{G}}(\mathbf{r}', \mathbf{r}) \cdot \mathbf{J}(\mathbf{r})$, and relabeling the variables $\mathbf{r}$ and $\mathbf{r}'$ as appropriate we obtain

$$\frac{\delta Z}{V^2} \left[\iint_{S_0} \mathbf{J} \cdot \mathbf{E}_a \, dS' \right]^2 = -2 \iint_{S_0} \left[\iint_{S_0} \mathbf{J}(\mathbf{r}) \cdot \bar{\mathbf{G}}(\mathbf{r}, \mathbf{r}') \, dS + \mathbf{E}_a(\mathbf{r}') \right] \cdot \delta \mathbf{J}(\mathbf{r}') \, dS' \quad (7)$$

where we have also made use of (4). Since the current $\mathbf{J}$ is a solution of the integral equation (3) we see that $\delta Z = 0$. Hence (5) is a variational expression for the impedance functional Z.

At this point we need to establish the relationship between Z and the antenna input impedance Z_{in}. If the applied field was a delta function source $V\delta(y)$ then (4) gives

$$Z = \frac{V^2}{\iint_{S_0} V\delta(y) J_y(y) \, dy \, r \, d\phi} = \frac{V}{\int_0^{2\pi} J_y(0) r \, d\phi} = \frac{V}{I_0} = Z_{\text{in}} \quad (8)$$

so Z equals Z_{in} in this special situation. The electric field $\mathbf{E}_a(y) \cdot \mathbf{a}_y$ acting along the probe will be linearly proportional to V, the voltage associated with the aperture field in the coaxial-line opening. Hence we can write

$$\mathbf{E}_a(y) \cdot \mathbf{a}_y = V e_a(y)$$

where $e_a(y)$ is a normalized electric field along the probe. From the definitions of Z and Z_{in} we have

$$Z_{\text{in}} = \frac{V}{I_0} = \frac{V^2}{\iint_{S_0} \mathbf{E}_a \cdot \mathbf{J} \, dS} \frac{\iint_{S_0} e_a J_y \, dS}{I_0} = \frac{Z}{I_0} \int_0^d e_a(y) I_y(y) \, dy \quad (9)$$

where $I_y(y) = \int_0^{2\pi} J_y(y) r \, d\phi = 2\pi r J_y(y)$ is the total current on the probe. We have assumed that the current $J_y(y)$ is uniform around the circumference of the probe.

We can expand the current I_y in terms of a suitable set of basis functions $\psi_n(y)$ so that

$$I_y(y) = \sum_{n=1}^{N} I_n \psi_n(y). \quad (10)$$

We can always choose the $\psi_n(y)$ so that $\psi_n(0) = 1$. Hence we will have

$$I_0 = \sum_{n=1}^{N} I_n. \tag{11}$$

When we use this expansion in (5) and also make use of (9) we obtain the following expression for Z_{in}:

$$Z_{in} = \frac{\displaystyle\sum_{n=1}^{N}\sum_{m=1}^{N} G_{nm} I_n I_m}{\left(\displaystyle\sum_{n=1}^{N} f_n I_n\right)^2} \frac{\displaystyle\sum_{n=1}^{N} f_n I_n}{\displaystyle\sum_{n=1}^{N} I_n} = \frac{\displaystyle\sum_{n=1}^{N}\sum_{m=1}^{N} G_{nm} I_n I_m}{\displaystyle\sum_{n=1}^{N} I_n \sum_{n=1}^{N} f_n I_n} \tag{12}$$

where

$$G_{nm} = \frac{-1}{(2\pi r)^2} \iint_{S_0} \iint_{S_0} \psi_n(y) G_{yy}(r, r') \psi_m(y')\, dS\, dS'$$

$$f_n = \int_0^d \psi_n(y) e_a(y)\, dy.$$

The equations for determining the I_n are obtained by using the current expansion in (5) and setting $\delta Z = 0$; thus,

$$\sum_{n=1}^{N} G_{mn} I_n = Z f_m \sum_{n=1}^{N} f_n I_n = C f_m, \qquad m = 1, 2, \ldots, N \tag{13}$$

where $C = Z \sum_{n=1}^{N} f_n I_n$ is a constant for each equation. We will show next that Galerkin's method leads to the same solution for Z_{in}.

Galerkin's Method of Solution

In the method-of-moments solution we substitute the expansion for $J_y(y) = I_y(y)/2\pi r$ into the integral equation and then test the resultant equation with each of the basis functions $\psi_m(y)$ in turn. This procedure results in the following system of equations:

$$\sum_{n=1}^{N} G_{mn} I_n = \frac{1}{2\pi r} \iint_{S_0} V e_a(y) \psi_m(y)\, dy\, r\, d\phi = f_m V, \qquad m = 1, 2, \ldots, N. \tag{14}$$

This system of equations is the same as that given by (13) with the possible exception of the normalization of the I_n. We can multiply (14) by I_m, sum over m, and solve for V. We then find that

$$Z_{in} = \frac{V}{I_0} = \frac{\displaystyle\sum_{n=1}^{N}\sum_{m=1}^{N} G_{mn} I_m I_n}{\displaystyle\sum_{n=1}^{N} I_n \sum_{m=1}^{N} f_m I_m} = \frac{V}{\displaystyle\sum_{n=1}^{N} I_n} \tag{15}$$

which is the same solution as given by (12). This solution for Z_{in} is not dependent on how the I_n are normalized. The solution for Z_{in} is, in general, not a stationary value. The impedance functional Z given by (4) is stationary for the correct current distribution, but since Z_{in} is given by the product of Z and a functional of $J_y(y)$ that is not stationary, Z_{in} is not stationary. Variational expressions for complex quantities such as Z do not have any unusual properties as regards the stationary values. These are not absolute maxima or minima. In actuality, if the current distribution contains several variational parameters we could adjust these so that the variational expression for Z would give the exact value for Z, if we knew that value a priori. This current distribution will generally not be the same as the one obtained by setting the first variation equal to zero. What one can infer from the variational expression for Z is that if the trial function chosen for $J(y)$ is a good approximation to the true current distribution then the error in Z is of second order. If the trial function is a poor approximation very little can be said about the accuracy of the stationary value of Z relative to the true value. On the other hand, it has been found in practice that in the case of a delta function source term the variational solution for Z_{in} agrees closely with that obtained by other methods of analysis. Consequently, we can anticipate that Galerkin's method will also give good estimates for Z_{in} using relatively low order expansions for $I_y(y)$.

Solution for Z_{in}

We only need the G_{yy} component of the Green's dyadic function when we assume that $\mathbf{J}$ has only a y component. This is tantamount to neglecting the radial current on the end of the probe, which is justifiable for a thin probe. We can readily construct the required Green's function from the vector potential A_y due to a y-directed current element. The required Green's function that will give the electric field E_y from a y-directed current element is

$$G = \frac{2jZ_0}{abk_0} \sum_{n=1}^{\infty} \sum_{m=0}^{\infty} \frac{\epsilon_{0m} k_m^2}{\Gamma_{nm}} \sin \frac{n\pi x}{a} \sin \frac{n\pi x'}{a} \cos \frac{m\pi y}{b}$$

$$\cdot \cos \frac{m\pi y'}{b} e^{-\Gamma_{nm}(z_> + \ell)} \sinh \Gamma_{nm}(z_< + \ell) \quad (16)$$

where $k_m^2 = (m\pi/b)^2 - k_0^2$ and $\Gamma_{nm}^2 = (n\pi/a)^2 + k_m^2$. For a centered probe we only need the terms for $n = 1, 3, 5, \ldots$ in (16). Also, with the exception of the $n = 1$, $m = 0$, and possibly the $n = 2$, $m = 0$, and $n = 1$, $m = 1$ modes, we can replace the hyperbolic sine function by $\frac{1}{2} e^{\Gamma_{nm}(z_< + \ell)}$. In order to obtain a more rapidly converging series than the one over n in (16) and to also facilitate the integration around the circumference of the probe, we introduce the image series given by (91) in Chapter 2. For the $m > 0$ modes where $k_m^2 > 0$ we have

$$\sum_{n=1}^{\infty} \frac{\sin \dfrac{n\pi x}{a} \sin \dfrac{n\pi x'}{a}}{\Gamma_{nm} a} e^{-\Gamma_{nm}|z-z'|}$$

$$= -\frac{j}{4} H_0^2(-jk_m \sqrt{(x - x')^2 + (z - z')^2})$$

$$+ \frac{j}{4} H_0^2(-jk_m \sqrt{(x + x')^2 + (z - z')^2})$$

$$- \frac{j}{4} \sum_{n=-\infty}^{\infty}{}' [H_0^2(-jk_m \sqrt{(x - x' - 2na)^2 + (z - z')^2})$$

$$- H_0^2(-jk_m \sqrt{(x + x' + 2na)^2 + (z - z')^2})] \quad (17)$$

where the prime means that the $n = 0$ term is excluded from the sum. In all terms but the first we can put $x = x' = a/2$ and $z = z' = 0$ with little error. In the first term we use $x = (a/2) + r \cos \phi$, $z = r \sin \phi$ and similar expressions for x' and z'. We then find that

$$\sum_{n=1,3,\ldots}^{\infty} \frac{\sin \dfrac{n\pi x}{a} \sin \dfrac{n\pi x'}{a}}{\Gamma_{nm} a} e^{-\Gamma_{nm}|z-z'|} = \frac{1}{2\pi} \left\{ K_0 \left(2k_m r \left| \sin \frac{\phi - \phi'}{2} \right| \right) \right.$$

$$\left. + \sum_{n=1}^{\infty} [4K_0(2k_m na) - 2K_0(k_m na)] \right\} \quad (18)$$

where K_0 is the modified Bessel function of the second kind. The average value of the integral of the first term over the angles ϕ and ϕ' is given by

$$\frac{1}{(2\pi)^2} \int\!\!\!\int_0^{2\pi} K_0 \left(2k_m r \left| \sin \frac{\phi - \phi'}{2} \right| \right) d\phi' \, d\phi = I_0(k_m r) K_0(k_m r) \quad (19)$$

as obtained by using $x = \sin(\phi - \phi')/2$ and

$$\int_0^1 \frac{K_0(2k_m rx)}{\sqrt{1 - x^2}} \, dx = \frac{\pi}{2} I_0(k_m r) K_0(k_m r).$$

The Bessel function series converges rapidly because of the exponential decay of the K_0 functions. For the $m = 0$ terms $-jk_m$ becomes equal to k_0 and the image series involves a slowly converging series of Hankel functions. We can use the same approximations for x and x' in all terms but the first. We also use $z = r \sin \phi$, $z' = r \sin \phi'$ in the first term but use $z - z' = r$ in all the higher order terms so as to generate the following series, after averaging over ϕ and ϕ':

$$-\frac{j}{4}[J_0(k_0 r) - 1]H_0^2(k_0 r) - \frac{j}{4}H_0^2(k_0 r) + \frac{j}{4}H_0^2(k_0 \sqrt{a^2 + r^2})$$

$$-\frac{j}{4} \sum_{n=-\infty}^{\infty}{}' [H_0^2(k_0 \sqrt{(2na)^2 + r^2}) - H_0^2(k_0 \sqrt{(2n+1)^2 a^2 + r^2})]$$

$$= -\frac{j}{4}[J_0(k_0 r) - 1]H_0^2(k_0 r) + \sum_{n=1,3,\ldots}^{\infty} \frac{e^{-\Gamma_{n0} r}}{\Gamma_{n0} a}. \quad (20)$$

We next use $J_0(k_0 r) - 1 \approx -(k_0 r/2)^2$, $H_0^2(k_0 r) \approx 1 - j(2/\pi)(\gamma + \ln k_0 r/2)$ where $\gamma = 0.5772$, and express the series in the form

$$\sum_{n=1,3,\ldots}^{\infty} \left[\frac{e^{-n\pi r/a}}{n\pi} + \left(\frac{e^{-\Gamma_{n0} r}}{\Gamma_{n0} a} - \frac{e^{-n\pi r/a}}{n\pi} \right) \right]$$

$$= -\frac{1}{2\pi} \ln \tanh \frac{\pi r}{2a} + \sum_{n=1,3,\ldots}^{\infty} \left(\frac{e^{-\Gamma_{n0} r}}{\Gamma_{n0} a} - \frac{e^{-n\pi r/a}}{n\pi} \right). \quad (21)$$

We can also approximate $\tanh \pi r/2a$ by its argument. By this means the $m = 0$ terms, averaged over the angles ϕ, ϕ', become

$$\frac{jk_0^2 r^2}{16} + \frac{k_0^2 r^2}{8\pi}\left(\gamma + \ln\frac{k_0 r}{2}\right) - \frac{1}{2\pi}\ln\frac{\pi r}{2a} + \sum_{n=1,3,\ldots}^{\infty}\left(\frac{e^{-\Gamma_{n0} r}}{\Gamma_{n0} a} - \frac{e^{-n\pi r/a}}{n\pi}\right).$$

The terms corresponding to the modes reflected from the short circuit have the factor

$$\sin\frac{n\pi x}{a}\sin\frac{n\pi x'}{a}e^{-2\Gamma_{nm}\ell - \Gamma_{nm}(z+z')}.$$

For these terms we use $x = (a/2) + r\cos\phi$, $z = r\sin\phi$ and similarly for x' and z'. By means of a Taylor series expansion about $x = x' = a/2$ and $z = z' = 0$ we get

$$e^{-2\Gamma_{nm}\ell}\left[1 - \left(\frac{n\pi}{a}\right)^2\frac{r^2}{2}(\cos^2\phi + \cos^2\phi')\right.$$

$$\left. + \Gamma_{nm}^2\frac{r^2}{2}(\sin\phi + \sin\phi')^2 - \Gamma_{nm}r(\sin\phi + \sin\phi')\right]$$

to order r^2. The average over ϕ and ϕ' gives

$$e^{-2\Gamma_{nm}\ell}\left[1 + \left(\Gamma_{nm}^2 - \frac{n^2\pi^2}{a^2}\right)\frac{r^2}{2}\right] = e^{-2\Gamma_{nm}\ell}\left(1 + k_m^2\frac{r^2}{2}\right).$$

Since these terms are significant only for the lowest values of n and m, the approximation is adequate for the terms we need.

We now collect all of these results so as to rewrite the fundamental integral equation for the probe current $I(y)$ in the form

$$\frac{jZ_0 k_0}{b}\left[\frac{jk_0^2 r^2}{16} + \frac{k_0^2 r^2}{8\pi}\left(\gamma + \ln\frac{k_0 r}{2}\right) - \frac{1}{2\pi}\ln\frac{\pi r}{2a} + \sum_{n=1,3,\ldots}^{\infty}\left(\frac{e^{-\Gamma_{n0} r}}{\Gamma_{n0} a} - \frac{e^{-n\pi r/a}}{n\pi}\right)\right.$$

$$\left. + \left(1 - \frac{k_0^2 r^2}{2}\right)\sum_{n=1,3,\ldots}^{\infty}\frac{e^{-2\Gamma_{n0}\ell}}{\Gamma_{n0} a}\right]\int_0^d I(y)\,dy - \frac{2jZ_0}{k_0 b}$$

$$\cdot\sum_{m=1}^{\infty}\left[\frac{1}{2\pi}I_0(k_m r)K_0(k_m r) + \sum_{n=1}^{\infty}\frac{4K_0(k_m 2na) - 2K_0(k_m na)}{2\pi}\right.$$

$$\left. - \left(1 + \frac{k_m^2 r^2}{2}\right)\sum_{n=1,3,\ldots}^{\infty}\frac{e^{-2\Gamma_{nm}\ell}}{\Gamma_{nm} a}\right]k_m^2\cos\frac{m\pi y'}{b}\int_0^d I(y)\cos\frac{m\pi y}{b}\,dy = V e_a(y').$$

$$(22)$$

All the series in this expression converge very rapidly so only a few terms need to be retained.

The basis functions that will be used to represent $I(y)$ are

$$\psi_1(y) = \frac{\sin k_0(d - y)}{\sin k_0 d}\tag{23a}$$

$$\psi_2(y) = \frac{1 - \cos k_0(d - y)}{1 - \cos k_0 d}.\tag{23b}$$

For these functions

$$\int_0^d \psi_1(y) \cos \frac{m\pi y}{b} \, dy = P_m = \frac{k_0 \left(\cos k_0 d - \cos \dfrac{m\pi d}{b} \right)}{k_m^2 \sin k_0 d} \tag{24a}$$

$$\int_0^d \psi_2(y) \cos \frac{m\pi y}{b} \, dy = Q_m = \frac{k_0 \sin k_0 d - (k_0^2 b/m\pi) \sin m\pi d/b}{k_m^2 (1 - \cos k_0 d)}. \tag{24b}$$

By using these expressions the matrix elements G_{nm} are readily evaluated. In order to simplify the expressions we will express (22) in the form

$$g_0 \int_0^d I(y) \, dy + \sum_{m=1}^{\infty} g_m \cos \frac{m\pi y'}{b} \int_0^d I(y) \cos \frac{m\pi y}{b} \, dy = V e_a(y')$$

and then

$$G_{11} = g_0 P_0^2 + \sum_{m=1}^{\infty} g_m P_m^2 \tag{25a}$$

$$G_{12} = G_{21} = g_0 P_0 Q_0 + \sum_{m=1}^{\infty} g_m P_m Q_m \tag{25b}$$

$$G_{22} = g_0 Q_0^2 + \sum_{m=1}^{\infty} g_m Q_m^2. \tag{25c}$$

The solution for Z_{in} is given by (15). Upon solving (14) for I_1 and I_2 we obtain

$$Z_{\text{in}} = \frac{G_{11} G_{22} - G_{12}^2}{f_1 G_{22} + f_2 G_{11} - (f_1 + f_2) G_{12}}. \tag{26}$$

In order to evaluate this expression we need to find the coefficients f_1 and f_2. The field $e_a(y')$ is given by

$$\int_0^{2\pi} \int_r^{r_0} \mathbf{a}_y \times \mathbf{a}_\rho \frac{1}{\ln r_0/r} \cdot \nabla \times \bar{\mathbf{G}}_e \cdot \mathbf{a}_y \, d\rho \, d\phi.$$

The field $e_a(y)$ is localized near the probe input at $y = 0$ so we can approximate it by using a quasi-static solution. The curl of the free-space dyadic Green's function and its image in the $y = 0$ plane in the quasi-static approximation is simply

$$\nabla \times \frac{\bar{\mathbf{I}}}{2\pi R} = \frac{1}{2\pi} \nabla \frac{1}{R} \times \bar{\mathbf{I}}.$$

Hence

$$
e_a(y) \approx \frac{-1}{2\pi \ln r_0/r} \int_r^{r_0} \int_0^{2\pi} \frac{\partial}{\partial \rho} \frac{1}{R} \, d\phi \, d\rho
$$

$$
= \frac{-2}{\pi \ln r_0/r} \int_0^{\pi/2} \left[\frac{1}{\sqrt{y^2 + (r_0 + r)^2 - 4r_0 r \cos^2 u}} - \frac{1}{\sqrt{y^2 + 4r^2 - 4r^2 \cos^2 u}} \right] du
$$

where we have put $\phi - \phi' = 2u$ and used $R = |\mathbf{a}_y y + \mathbf{a}_x (\rho \cos \phi - r \cos \phi') + \mathbf{a}_z (\rho \sin \phi - r \sin \phi')|$.

The parameters f_1 and f_2 are given by

$$
f_1 = \int_0^d \psi_1(y) e_a(y) \, dy \tag{27a}
$$

$$
f_2 = \int_0^d \psi_2(y) e_a(y) \, dy \tag{27b}
$$

and are evaluated numerically using the approximate solution for $e_a(y)$. The solution for $e_a(y)$ can be expressed in terms of elliptic integrals if desired; thus

$$
e_a(y) = \frac{-2}{\pi \ln r_0/r} \left[\frac{k_1 K(k_1)}{2\sqrt{rr_0}} - \frac{k_2 K(k_2)}{2r} \right]
$$

where $k_1^2 = 4rr_0/[y^2 + (r + r_0)^2]$, $k_2^2 = 4r^2/[y^2 + 4r^2]$ and K is the complete elliptic integral of the first kind. When y approaches zero k_2 approaches one and the field $e_a(y)$ takes on the limiting value

$$
e_a(y) \sim \frac{-1}{r\pi \ln r_0/r} \ln \frac{y}{8r}.
$$

In view of this logarithmic singularity the numerical evaluation of (27) must be done carefully. A convenient procedure is to use

$$
\int_0^d \psi_i(y) e_a(y) \, dy = \int_0^d [\psi_i(y) - 1] e_a(y) \, dy + \int_0^d e_a(y) \, dy
$$

where we have used $\psi_i(0) = 1$, $i = 1, 2$. The last integral is given by

$$
\frac{2}{\pi \ln r_0/r} \left[\int_0^{\pi/2} \ln \frac{d + \sqrt{d^2 + (r_0 + r)^2 - 4rr_0 \cos^2 u}}{d + \sqrt{d^2 + 4r^2 - 4r^2 \cos^2 u}} \, du \right.
$$

$$
\left. + \int_0^{\pi/2} \ln 2r \sin u \, du - \frac{1}{2} \int_0^{\pi/2} \ln[(r_0 + r)^2 - 4rr_0 \cos^2 u] \, du \right]
$$

$$
= -1 + \frac{2}{\pi \ln r_0/r} \int_0^{\pi/2} \ln \frac{d + \sqrt{d^2 + (r_0 + r)^2 - 4rr_0 \cos^2 u}}{d + \sqrt{d^2 + 4r^2 - 4r^2 \cos^2 u}} \, du.
$$

This procedure eliminates the troublesome singularity that occurs at $y = 0$. The integral involving $\psi_i(y) - 1$ has a vanishing integrand at $y = 0$ and can be evaluated numerically in a straightforward manner.

A worthwhile improvement in the accuracy of the solution for the input impedance is obtained by including a frequency-dependent contribution to the applied electric field. The following approximation can be used:

$$\frac{e^{-jk_0R}}{2\pi R} = \frac{1}{2\pi R} - \frac{jk_0}{2\pi} - \frac{k_0^2 R}{4\pi} + \frac{jk_0^3 R^2}{12\pi}.$$

Since $e_a(y)$ is generated by taking a derivative with respect to ρ followed by an integral over ρ the corrections to the constants f_1 and f_2 are readily evaluated and are

$$\Delta f_i = -\frac{k_0^2}{\pi \ln(r_0/r)} \int_0^d \int_0^{\pi/2} \psi_i(y) [\sqrt{y^2 + (r_0 + r)^2 - 4rr_0 \cos^2 u}$$

$$- \sqrt{y^2 + 4r^2 - 4r^2 \cos^2 u}] \, du \, dy$$

$$+ \frac{jk_0^3(r_0^2 - r^2)}{6 \ln(r_0/r)} \begin{cases} P(0), & i = 1 \\ Q(0), & i = 2. \end{cases}$$

These corrections produce a 5 to 10% change in the computed values of Z_{in} and thus demonstrate that Z_{in} is quite sensitive to the values of the applied electric field that acts on the probe.

In Fig. 7.3 we show some computed values for the probe resistance R and reactance X as a function of frequency. The results shown are based on the formulas given above and include the corrections Δf_i. The data apply for a standard X-band waveguide with $a = 2.286$ cm, $b = 1.016$ cm, and a probe length $d = 0.62$ cm. For the two probe radii of 1 and 1.5 mm the short-circuit position $\ell = 0.495$ cm. For the thin probe having $r = 0.5$ mm the short-circuit position is at $\ell = 0.505$ cm. In Fig. 7.4 the return loss, given by

$$\text{Loss} = 20 \log |\Gamma|$$

where Γ is the reflection coefficient, is shown. For both figures the coaxial line has a characteristic impedance of 50 ohms. From Fig. 7.4 it is quite clear that the thick probe provides a broader bandwidth of operation. At the -30 dB return loss level the thick probe has a bandwidth of 9%, while the corresponding bandwidth of the thin probe is about 4.5%.

In Fig. 7.5 we show the probe impedance for various probe lengths. The probe radius is 1 mm and the short-circuit position ℓ is chosen so that $2\beta\ell = \pi/2$ at 10 GHz, i.e., $\ell = 0.609$ cm. This choice of ℓ maximizes the inductive loading from the H_{10} standing wave at 10 GHz. Of particular interest are the decreasing values of probe reactance with increasing frequency when the probe length is greater than about 0.65 cm. This feature can be used to design a broadband waveguide to waveguide probe–probe coupling system of the type shown in Fig. 7.6. At the center of the frequency band of interest the transmission-line length is chosen so as to transform the impedance of one probe into the complex conjugate of the impedance of the second probe. When the frequency changes, the changing electrical length of the transmission line is compensated for by the manner in which the reactance of the probes changes with frequency. This compensation results in a relatively well matched system over a broad band of frequencies.

The input impedance was also computed using three basis functions to expand the current.

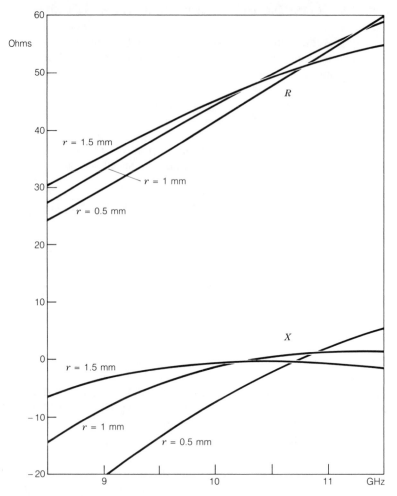

Fig. 7.3. Probe input resistance and reactance as a function of frequency for $d = 0.62$ cm, $\ell = 0.495$ cm, $a = 2.286$ cm, $b = 1.016$ cm. For the thin probe $r = 0.5$ mm, $\ell = 0.505$ cm.

The first was $\psi_1(y)$ as already specified and the other two were

$$\psi_2 = \cos 3\pi y/2d$$

$$\psi_3 = \cos 5\pi y/2d.$$

Some representative results are shown in Fig. 7.5 as broken curves for $d = 0.5$ cm and 0.8 cm. For short probes the use of two basis functions appears to be adequate, but for the longer probes the use of three basis functions makes a significant difference in the computed values of the probe impedance.

There is very little experimental data on waveguide probe antenna impedance that can be used to check the accuracy of the approximate theory given above. Jarem has made calculations similar to the above using the same basis functions for the expansion of the current and found the agreement with experimental results to be quite good for a moderately thick probe [7.17].

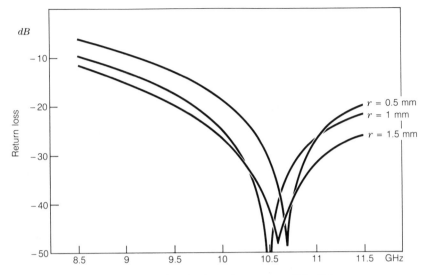

Fig. 7.4. Return loss for the probe antennas of Fig. 7.3.

Ragan has described a probe antenna in [7.20] for which $a = 7.2136$ cm, $b = 3.4036$ cm, $r_0 = 1.9393$ cm, $r = 0.7937$ cm, $\ell = 2.55$ cm, $d = 1.91$ cm and that is impedance matched at 2.747 GHz. For this case our theory gives $Z_{in} = 57.3 - j18$. The line impedance is 53.4 ohms so the return loss is only -15.72 dB. By increasing the probe length to 2.12 cm and reducing ℓ to 1.95 cm our theory predicts an impedance match (return loss of -51 dB). If we use $\cos(3\pi y/2d)$ for the second basis function $\psi_2(y)$ we find that $d = 1.994$ cm and $\ell = 2.17$ cm for an impedance match. These values are in closer agreement with those given by Ragan and thus indicate that the second choice is a better one for the basis function ψ_2.

In another example involving an X-band waveguide the optimum values of d and ℓ are given as $d = 0.635$ cm and $\ell = 0.787$ cm at 9.09 GHz. Our theory requires $d = 0.67$ cm and $\ell = 0.55$ cm. If we use the alternative basis function ψ_2 given above we find that we require $d = 0.632$ cm and $\ell = 0.62$ cm. These values are in better agreement with the experimental ones. When we use three basis functions we find that the optimum value of ℓ is 0.601 cm for $d = 0.635$ cm. The evidence suggests that the theory predicts an optimum value of ℓ that is too small. Furthermore, the limitation does not appear to be the number of basis functions used to expand the current. The probes for which the theory is being compared against experimental data are thick probes. It can be expected that for these cases it will be necessary to include the effects of higher order modes in the coaxial transmission line and to use a better approximation for the applied field $e_a(y)$.

The theory as given does not appear to be of high accuracy for probes that are as thick as those used in typical coaxial-line–waveguide transitions. The results given by Jarem, for which good agreement with experimental data was obtained, involved a probe of radius approximately equal to $0.03a$. For the examples discussed above the probe radii are $0.11a$ and $0.07a$, respectively, which are significantly larger.

7.2. THE LOOP ANTENNA

As an example of a loop antenna, we consider a coaxial line terminated in the narrow wall of an infinitely long rectangular guide and with its center conductor bent into a semicircular

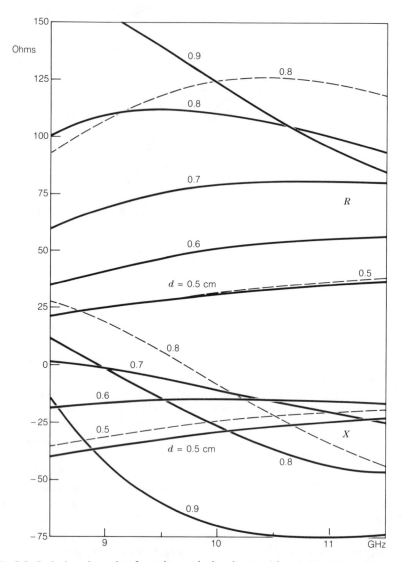

Fig. 7.5. Probe impedance data for various probe lengths. a and b as in Fig. 7.3, $\ell = 0.609$ cm, $r = 1$ mm. The broken curves are based on the use of three basis functions.

loop of radius d, as in Fig. 7.7. The center of the loop is located midway between the top and bottom walls of the guide and coincides with the origin of the coordinate system to be used. The plane of the loop is parallel to the xy plane, and hence only those modes (H_{nm} modes) with an axial magnetic field component are excited. Provided the size of the loop is small compared with a wavelength, say d less than $0.1\lambda_0$, the current in the loop may be assumed constant.

For the purpose of computing the power radiated into the guide, the current may be considered concentrated in a thin filament along the center of the conductor. The evaluation of the

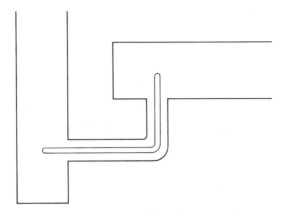

Fig. 7.6. Probe–probe coupling of two waveguides.

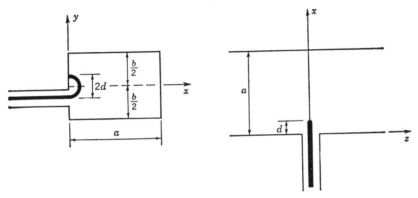

Fig. 7.7. Coaxial-line loop antenna.

radiated power leads directly to an expression for the input resistance (radiation resistance) of the antenna. The radiated power is readily evaluated by finding the coupling coefficient between the propagating H_{10} mode (all other modes are assumed to be evanescent) and the loop antenna, according to the methods given in Section 5.6. To evaluate the input reactance to the antenna, the finite radius r of the conductor must be taken into account, or else we run into the embarrassing situation of an infinite reactance. To evaluate the reactance term, the antenna and guide will be replaced by a double infinite array of image antennas. Our calculation then reduces to finding the self-inductance of a single loop in free space and the mutual inductance between the loop at the origin and all the image loops. In this formulation it is quite clear what approximations can be made with negligible error. An analysis based entirely on an expansion in terms of the normal waveguide modes could also be carried out, but the approximations that can be made are not so apparent in this case.

The normalized H_{nm} modes may be derived from the scalar function

$$\psi_{nm}(x, y) = \left(\frac{\epsilon_{0n}\epsilon_{0m}}{abjk_0Z_0\Gamma_{nm}k_{c,nm}^2} \right)^{1/2} \cos \frac{n\pi x}{a} \cos m\pi \frac{2y - b}{b} \qquad (28)$$

by means of the following equations:

$$\mathbf{h}_{znm} = \mathbf{a}_z k_{c,nm}^2 \psi_{nm} \tag{29a}$$

$$\mathbf{h}_{nm} = -\Gamma_{nm} \nabla_t \psi_{nm} \tag{29b}$$

$$\mathbf{e}_{nm} = \frac{jk_0 Z_0}{\Gamma_{nm}} (\mathbf{a}_x \mathbf{a}_y - \mathbf{a}_y \mathbf{a}_x) \cdot \mathbf{h}_{nm} \tag{29c}$$

$$\mathbf{E}_{nm}^{\pm} = \mathbf{e}_{nm} e^{\mp \Gamma_{nm} z} \tag{29d}$$

$$\mathbf{H}_{nm}^{\pm} = (\pm \mathbf{h}_{nm} + \mathbf{h}_{znm}) e^{\mp \Gamma_{nm} z} \tag{29e}$$

where $k_{c,nm}^2 = (m\pi/b)^2 + (n\pi/a)^2$, $\mathbf{h}_{nm}$ and $\mathbf{e}_{nm}$ are the transverse magnetic and electric normal mode fields, and $\Gamma_{nm}^2 = k_{c,nm}^2 - k_0^2$. The normalization in (28) has been chosen so that

$$\int_0^a \int_0^b \mathbf{e}_{nm} \times \mathbf{h}_{nm} \cdot \mathbf{a}_z \, dx \, dy = 1. \tag{30}$$

The axial magnetic field in the two regions $z < 0$ and $z > 0$ may be represented as an expansion in terms of the above normal mode functions as follows:

$$\mathbf{H}_z = \begin{cases} \displaystyle\sum_{n=0}^{\infty} \sum_{m=0}^{\infty} C_{nm} \mathbf{h}_{znm} e^{-\Gamma_{nm} z}, & z > 0 \\[2ex] \displaystyle\sum_{n=0}^{\infty} \sum_{m=0}^{\infty} D_{nm} \mathbf{h}_{znm} e^{\Gamma_{nm} z}, & z < 0. \end{cases} \tag{31}$$

From Chapter 5, Eqs. (82) and (83), the coefficients are given by

$$C_{nm} = D_{nm} = \frac{j\omega\mu_0 I_0}{2} \iint_{S_0} \mathbf{h}_{znm} \cdot \mathbf{a}_z \, dx \, dy \tag{32}$$

where S_0 is the surface spanned by the loop, and I_0 is the total loop current assumed concentrated in a thin filament. For the H_{10} mode, the coupling coefficient C_{10} is given by

$$C_{10} = \frac{j\omega\mu_0 I_0}{2} \left(\frac{\pi}{a}\right)^2 \left(\frac{2a}{jk_0 Z_0 \Gamma_{10} \pi^2 b}\right)^{1/2} \iint_{S_0} \cos\frac{\pi x}{a} \, dx. \tag{33}$$

Introducing a local cylindrical coordinate system $t\theta$ as in Fig. 7.8 and expanding $\cos(\pi x/a)$ since $\pi d/a$ is small, the integral in (33) becomes

$$\int_0^d \int_0^\pi \left(1 - \frac{1}{2}\frac{\pi^2}{a^2} t^2 \cos^2\theta\right) t \, dt \, d\theta = \frac{\pi d^2}{2}\left(1 - \frac{\pi^2 d^2}{8a^2}\right) \approx \frac{\pi d^2}{2}.$$

The power radiated into the guide is twice that radiated in one direction and, hence, is given

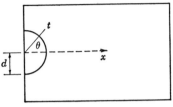

Fig. 7.8.

by

$$P = \int_0^a \int_0^b C_{10}C_{10}^* \mathbf{e}_{10} \times \mathbf{h}_{10}^* \cdot \mathbf{a}_z \, dx \, dy = C_{10}C_{10}^*$$

since, from the definition of the normal mode functions and the normalization condition (30), we have

$$\int_0^a \int_0^b \mathbf{e}_{nm} \times \mathbf{h}_{nm}^* \cdot \mathbf{a}_z \, dx \, dy = \begin{cases} 1 & \Gamma_{nm} \text{ imaginary} \\ j & \Gamma_{nm} \text{ real.} \end{cases}$$

Substituting for C_{10}, we get

$$P = \frac{k_0 Z_0}{2\beta_{10}ab} \left(\frac{\pi}{a}\right)^2 \left(\frac{\pi d^2}{2}\right)^2 I_0^2 \tag{34}$$

where $\beta_{10}^2 = k_0^2 - (\pi/a)^2$. If the higher order modes in the coaxial line are neglected, the integral of the complex Poynting vector over the coaxial-line opening gives $\frac{1}{2}Z_{in}I_0^2$, since I_0 may be assumed real. The input impedance is given by

$$Z_{in} = \frac{P + 2j\omega(W_m - W_e)}{\frac{1}{2}I_0^2}$$

and, hence, the input resistance, i.e., radiation resistance, is

$$R = \frac{k_0 Z_0}{ab\beta_{10}} \left(\frac{\pi}{a}\right)^2 \left(\frac{\pi d^2}{2}\right)^2. \tag{35}$$

The radiation resistance is seen to be proportional to the square of the loop area and for a small loop is quite small; typical values range from 10 to 30 ohms.

To evaluate the input reactance, we replace the semicircular loop and the guide by a lattice of loop antennas as illustrated in Fig. 7.9. After we have found the field radiated by a single circular loop, it will be seen that the system of loop antennas in Fig. 7.9 radiates a total field which has a zero tangential electric field on the guide boundary. The semicircular loop and its image in the $x = 0$ guide wall are illustrated in Fig. 7.10 and are equivalent to a circular loop driven by a voltage generator having twice the voltage required to maintain the current in the semicircular loop.

To evaluate the input reactance to the antenna, we must compute the total flux linking the loop at the origin due to all of the image antennas, plus the self-flux linkage arising from the magnetic field set up by its own current. The time rate of change of the flux linking the antenna

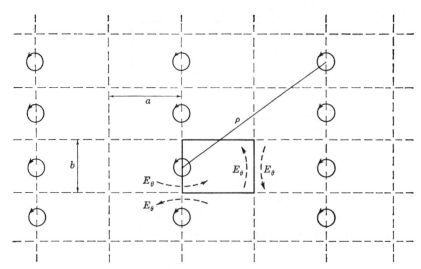

Fig. 7.9. The loop antenna and its images.

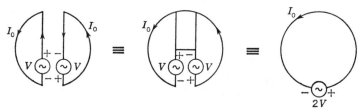

Fig. 7.10. Semicircular loop, its image, and their equivalent loop configuration.

results in a back emf being induced. The applied voltage must be equal and opposite to this induced emf, in order to maintain the current I_0 in the loop. The flux linking the antenna and in phase with the current I_0 leads to an induced emf in phase quadrature with the current. It is this flux which gives rise to the inductive reactance term. The flux linking the antenna and in phase quadrature with the current gives rise to an emf in phase with the current, and hence contributes to the radiation resistance. Since we have already obtained an expression for the radiation resistance, we are interested only in that part of the flux linkage which is in phase with the current I_0.

The field from the image antennas at the position of the loop located at the origin may be found with negligible error by assuming the current I_0 to be concentrated in a thin filament. Each image loop may be replaced by a z-directed magnetic dipole of moment $M = I_0 \pi d^2$. To find the field radiated by this array of magnetic dipoles it is convenient to first find a magnetic-type Hertzian potential Π_z having only a z component. The relevant equations are

$$\nabla^2 \Pi_z + k_0^2 \Pi_z = -M$$

$$H_z = k_0^2 \Pi_z + \frac{\partial^2 \Pi_z}{\partial z^2}.$$

For a single loop antenna located at the origin as in Fig. 7.11 the solution is

$$\Pi = \frac{M}{4\pi\rho} e^{-jk_0\rho}$$

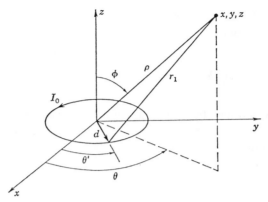

Fig. 7.11. Loop antenna in the xy plane.

where $\rho = (x^2 + y^2 + z^2)^{1/2}$. For a row of dipoles located at $y = mb$, $m = 0, \pm 1, \pm 2, \ldots,$ the potential is given by

$$\Pi_1 = \frac{M}{4\pi} \sum_{m=-\infty}^{\infty} \frac{\exp\{-jk_0[x^2 + z^2 + (y - mb)^2]^{1/2}\}}{[x^2 + z^2 + (y - mb)^2]^{1/2}}. \quad (36)$$

This series is readily converted to a more rapidly converging one by using the Poisson summation formula. The required Fourier transform is

$$\int_{-\infty}^{\infty} e^{jwu} \frac{\exp\{-jk_0[x^2 + z^2 + (y - u)^2]^{1/2}\}}{[x^2 + z^2 + (y - u)^2]^{1/2}} \, du = 2e^{jwy} K_0[(x^2 + z^2)(w^2 - k_0^2)]^{1/2} \quad (37)$$

where K_0 is the modified Bessel function of the second kind. Using the Poisson summation formula now converts (36) into the following:

$$\frac{M}{4\pi} \sum_{m=-\infty}^{\infty} \frac{\exp\{-jk_0[x^2 + z^2 + (y - mb)^2]^{1/2}\}}{[x^2 + z^2 + (y - mb)^2]^{1/2}}$$

$$= \frac{M}{2\pi b} \sum_{m=-\infty}^{\infty} e^{j2m\pi y/b} K_0[\Gamma_{0m}(x^2 + z^2)^{1/2}]$$

$$= \frac{M}{\pi b} \left\{ \frac{1}{2} K_0[jk_0(x^2 + z^2)^{1/2}] + \sum_{m=1}^{\infty} \cos \frac{2m\pi y}{b} K_0[\Gamma_{0m}(x^2 + z^2)^{1/2}] \right\} \quad (38)$$

where $\Gamma_{0m} = [(2m\pi/b)^2 - k_0^2]^{1/2}$.

The potential arising from all the image dipoles is obtained by replacing x by $x - 2na$ in (38) and summing over all n but excluding the contribution from the dipole at the origin (the driven loop). The flux linkage will be computed by evaluating the magnetic field at the center of the loop and multiplying by the loop area. The evaluation of H_z involves derivatives with respect to z only, and so we may place x and y equal to zero at this point. For the rows of dipoles located at $x = 2na$, $n = \pm 1, \pm 2, \ldots,$ we get, from (38),

$$\Pi_2 = \frac{M}{2\pi b} \sum_{n=-\infty}^{\infty}{}' K_0[jk_0(z^2 + 4n^2a^2)^{1/2}] + \frac{2M}{\pi b} \sum_{n=1}^{\infty} \sum_{m=1}^{\infty} K_0[\Gamma_{0m}(z^2 + 4n^2a^2)^{1/2}]. \quad (39a)$$

To the above we must add the contribution from dipoles located at $x = z = 0$, $y = mb$, $m = \pm 1, \pm 2, \ldots$. This latter contribution is, from (36),

$$\frac{M}{2\pi} \sum_{m=1}^{\infty} \frac{\exp[-jk_0(z^2 + m^2 b^2)^{1/2}]}{(z^2 + m^2 b^2)^{1/2}}. \tag{39b}$$

In (39a) the double series may be neglected, since, for the range of parameters involved here, $K_0[\Gamma_{0m}(z^2 + 4n^2 a^2)^{1/2}]$ is extremely small. The first series, however, converges very slowly because the argument is imaginary. By using the Poisson summation formula again, this series may be transformed as follows [the required transform may be obtained by inverting (37)]:

$$\frac{M}{2\pi b} \sum_{n=-\infty}^{\infty} K_0[jk_0(z^2 + 4n^2 a^2)^{1/2}] - \frac{M}{2\pi b} K_0(jk_0 z) = \frac{M}{4ab} \sum_{n=-\infty}^{\infty} \frac{e^{-\Gamma_{n0} z}}{\Gamma_{n0}} - \frac{M}{2\pi b} K_0(jk_0 z)$$

where $\Gamma_{n0} = [(n\pi/a)^2 - k_0^2]^{1/2}$. Our final expression for the total Hertzian potential is

$$\Pi_z = \frac{M}{4ab} \sum_{n=-\infty}^{\infty} \frac{e^{-\Gamma_{n0} z}}{\Gamma_{n0}} - \frac{M}{2\pi b} K_0(jk_0 z) + \frac{M}{2\pi} \sum_{m=1}^{\infty} \frac{\exp[-jk_0(z^2 + m^2 b^2)^{1/2}]}{(z^2 + m^2 b^2)^{1/2}}. \tag{40}$$

The field H_z, at the origin, is given by $k_0^2 \Pi_z + \partial^2 \Pi_z / \partial z^2$ evaluated for $z = 0$. The first term gives a contribution

$$\frac{k_0^2 M}{4ab} \sum_{n=-\infty}^{\infty} \frac{e^{-\Gamma_{n0} z}}{\Gamma_{n0}} - \frac{k_0^2 M}{2\pi b} K_0(jk_0 z) + \frac{k_0^2 M}{2\pi} \sum_{m=1}^{\infty} \frac{e^{-jk_0 mb}}{mb}$$

where the limit as z tends to zero is to be taken. We have $\Gamma_{n0}^{-1} \approx a/n\pi$, and hence the first series is a dominant series

$$\frac{k_0^2 M}{2ab} \sum_{m=1}^{\infty} \frac{a e^{-n\pi z/a}}{n\pi}$$

together with the correction series

$$\frac{k_0^2 M}{2ab} \left[\sum_{n=1}^{\infty} \left(\frac{1}{\Gamma_{n0}} - \frac{a}{n\pi} \right) - \frac{j}{2k_0} \right].$$

The dominant series is readily summed by the methods given in the Mathematical Appendix to give

$$-\frac{k_0^2 a M}{2\pi ab} \left(\ln 2 \sinh \frac{\pi z}{2a} - \frac{\pi z}{2a} \right).$$

The logarithmic singularity occurring here is canceled by that arising from the term $K_0(jk_0 z)$ since for z small we have $K_0(jk_0 z) \rightarrow -[\gamma + \ln(jk_0 z/2)]$, where $\gamma = 0.577$ and is Euler's constant. If the series in m is also summed and the real part of all terms taken, we finally obtain for the part of $k_0^2 \Pi_z$ which is in phase with the current I_0 the result

$$\mathrm{Re}\,(k_0^2 \Pi_z) = \frac{k_0^2 M}{2\pi b}\left[\gamma + \ln \frac{k_0 a}{2\pi} - \ln 2 \sin \frac{k_0 b}{2} + \frac{\pi}{a}\sum_{n=2}^{\infty}\left(\frac{1}{\Gamma_{n0}} - \frac{a}{n\pi}\right) - 1\right]. \quad (41)$$

To find the second part of H_z, we first evaluate $\partial^2 \Pi_z / \partial z^2$ to get

$$\frac{\partial^2 \Pi_z}{\partial z^2} = \frac{M}{4ab}\sum_{n=-\infty}^{\infty}\Gamma_{n0}e^{-\Gamma_{n0}z} + \frac{Mk_0^2}{2\pi b}\left[K_0(jk_0 z) + \frac{K_1(jk_0 z)}{jk_0 z}\right]$$

$$+ \frac{M}{2\pi}\sum_{m=1}^{\infty}\frac{e^{-jk_0 r_m}}{r_m}\left[\frac{jk_0 z}{r_m^2} + \frac{2z}{r_m^3} - \frac{k_0^2 z}{r_m} + \frac{jk_0 z}{r_m^2} - \left(jk_0 + \frac{1}{r_m}\right)\left(\frac{1}{r_m} - \frac{z^2}{r_m^3}\right)\right]$$

where $r_m = (z^2 + m^2 b^2)^{1/2}$.

The first series consists of a dominant part

$$\frac{M}{2ab}\left[\sum_{n=1}^{\infty}e^{-n\pi z/a}\left(\frac{n\pi}{a} - \frac{k_0^2 a}{2\pi n}\right)\right]$$

plus a correction series

$$\frac{M}{2ab}\left[\sum_{n=1}^{\infty}\left(\Gamma_{n0} - \frac{n\pi}{a} + \frac{k_0^2 a}{2\pi n}\right) + \frac{jk_0}{2}\right]$$

since $\Gamma_{n0} \approx n\pi/a - k_0^2 a/2\pi n$. The dominant part of the series sums to

$$\frac{M}{2ab}\left[\frac{\pi}{4a \sinh^2(\pi z/2a)} + \frac{k_0^2 a}{2\pi}\left(\ln 2 \sinh \frac{\pi z}{2a} - \frac{\pi z}{2a}\right)\right].$$

As z approaches zero, the singularities occurring in the above expression are canceled by those arising from the Bessel functions K_0 and K_1. As x approaches zero, we have $K_0(x) \to -[\gamma + \ln(x/2)]$, and $K_1(x) \to x^{-1} + [\ln(x/2) + \gamma - \frac{1}{2}]x/2$, and these expansions may be used to evaluate the contribution from the terms $K_0(jk_0 z)$ and $K_1(jk_0 z)$.

In the series over m, we may place z equal to zero, and the resultant series are then readily summed. The required summation formulas are tabulated in the Mathematical Appendix, so we will write down only the final results that we require. After performing the indicated steps, we obtain for the real part of $\partial^2 \Pi_z / \partial z^2$ the following result:

$$\mathrm{Re}\,\frac{\partial^2 \Pi_z}{\partial z^2} = \frac{M}{2ab}\left[\sum_{n=2}^{\infty}\left(\Gamma_{n0} - \frac{n\pi}{a} + \frac{k_0^2 a}{2\pi n}\right) - \frac{13}{12}\frac{\pi}{a} + \frac{k_0^2 a}{4\pi} - \frac{\gamma k_0^2 a}{2\pi}\right.$$

$$\left. + \frac{k_0^2 a}{2\pi}\ln\frac{2\pi}{k_0 a}\right] - \frac{M}{2\pi b^3}\left(1.2 + \frac{k_0^2 b^2}{4} + \frac{3k_0^4 b^4}{288} - \frac{k_0^2 b^2}{2}\ln k_0 b\right). \quad (42)$$

The total in phase flux linking the loop antenna at the origin is obtained by multiplying the real part of the field H_z from the image antennas by $\mu_0 \pi d^2$, that is, multiplying the sum of (41) and (42) by $\mu_0 \pi d^2$. Thus, the in phase flux linkage arising from the mutual coupling with

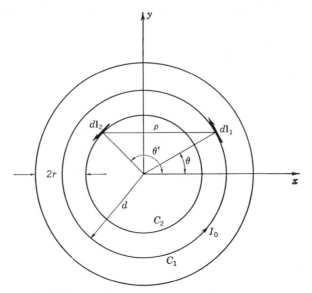

Fig. 7.12. Coordinate system for a single loop antenna.

the image antennas is given by

$$\sum_{n,m}' F_{nm} = \mu_0 \pi d^2 I_0 \pi d^2 \left[\frac{k_0^2}{4\pi b} \ln \frac{k_0^2 ab}{4\pi(1 - \cos k_0 b)} - \frac{k_0^2}{4\pi b}(2 - \gamma) - \frac{13\pi}{24a^2 b} \right.$$

$$\left. - \frac{0.6}{\pi b^3} - \frac{3k_0^4 b}{576\pi} + \frac{1}{2ab} \sum_{n=2}^{\infty} \left(\frac{k_0^2}{\Gamma_{n0}} + \Gamma_{n0} - \frac{n\pi}{a} - \frac{k_0^2 a}{2n\pi} \right) \right] \quad (43)$$

where M has been replaced by $I_0 \pi d^2$.

The only flux linkage left to be evaluated is that due to the current in the loop at the origin, i.e., the flux giving rise to the self-inductance of the loop. The vector potential is given by

$$\mathbf{A} = \frac{\mu_0}{4\pi} \iint_{S_0} \mathbf{J}_0 \frac{e^{-jk_0\rho}}{\rho} \, dS$$

where $\mathbf{J}_0$ is the uniform circumferential current density on the loop antenna surface S_0, and ρ is the radial distance from a current element to the point in space where $\mathbf{A}$ is computed. For a small loop with $\rho \leq 2d$, the exponential term does not vary significantly across the domain of the loop. To be consistent with our assumption of a uniform loop current, we place $e^{-jk_0\rho}$ equal to unity, and we are then left with the static field problem of evaluating the self-inductance of a small circular loop of radius d with a conductor radius r. The real part of the flux linkage will be correct to order $(k_0 d)^2$, since the second term in the expansion of $e^{-jk_0\rho}$ is $-jk_0\rho$, and contributes to the radiation resistance only. With reference to Fig. 7.12, the field at the surface of the conductor is $H = J_0 = I_0/2\pi r$, and is the same as that produced by a current filament I_0 at the center, i.e., located along the contour C_1. The magnetic field does not penetrate into the conductor, and, hence, we only require the flux linking the contour

C_2 which coincides with the inner surface of the conductor. This flux linkage is given by

$$F_{00} = \iint \mathbf{B} \cdot d\mathbf{S} = \iint \nabla \times \mathbf{A} \cdot d\mathbf{S} = \oint_{C_2} \mathbf{A} \cdot d\mathbf{l}_2.$$

Substituting for $\mathbf{A}$ gives

$$F_{00} = \frac{\mu_0 I_0}{4\pi} \oint_{C_2} \oint_{C_1} \frac{d\mathbf{l}_1 \cdot d\mathbf{l}_2}{\rho}. \tag{44}$$

From the figure it is seen that

$$\rho = [d^2 + (d - r)^2 - 2d(d - r) \cos(\theta' - \theta)]^{1/2}$$

and $d\mathbf{l}_1 \cdot d\mathbf{l}_2 = d(d - r) \cos(\theta' - \theta) \, d\theta \, d\theta'$. If we integrate over θ first, we must get a result independent of θ' in view of the symmetry involved. We may, therefore, place θ' equal to zero, and replace the integration over θ' by a factor 2π. Hence we get

$$F_{00} = \frac{\mu_0 I_0}{2} \int_0^{2\pi} \frac{d(d - r) \cos \theta \, d\theta}{[d^2 + (d - r)^2 - 2d(d - r) \cos \theta]^{1/2}}$$

$$= -\mu_0 I_0 d(d - r) \int_0^{\pi} \frac{\cos \theta \, d\theta}{[d^2 + (d - r)^2 + 2d(d - r) \cos \theta]^{1/2}}.$$

This integral may be reduced to the standard form for elliptic integrals with the following substitutions: $\theta = 2\phi$, $\cos \theta = 1 - 2 \sin^2 \phi$, and

$$k^2 = \frac{4d(d - r)}{(2d - r)^2}.$$

We now get

$$-\frac{2d - r}{2} \mu_0 I_0 k^2 \int_0^{\pi/2} \frac{1 - 2 \sin^2 \phi}{(1 - k^2 \sin^2 \phi)^{1/2}} d\phi$$

$$= (2d - r)\mu_0 I_0 \left[\left(1 - \frac{k^2}{2}\right) \int_0^{\pi/2} \frac{d\phi}{(1 - k^2 \sin^2 \phi)^{1/2}} - \int_0^{\pi/2} \frac{1 - k^2 \sin^2 \phi}{(1 - k^2 \sin^2 \phi)^{1/2}} d\phi \right]$$

$$= (2d - r)\mu_0 I_0 \left[\left(1 - \frac{k^2}{2}\right) K(k) - E(k) \right] \tag{45}$$

where K and E are complete elliptic integrals of modulus k [7.2]. Referring back to the definition of k, it is seen that $k \approx 1$ and, hence, $K(k) \approx \ln[4/(1 - k^2)^{1/2}]$, $E(k) \approx 1$. Replacing $(1 - k^2)^{1/2}$ by $r/2d$, we get

$$F_{00} = \mu_0 I_0 d \left(\ln \frac{8d}{r} - 2 \right). \tag{46}$$

If F is the total flux linkage, the total induced back emf will be $j\omega F$ and must equal the applied voltage $2V$. Hence we have

$$2V = j\omega F$$

$$2VI_0 = 2I_0^2 Z_{\text{in}} = j\omega F I_0$$

$$Z_{\text{in}} = R + jX = \frac{j\omega F}{2I_0}$$

and

$$X = \operatorname{Re} \frac{\omega F}{2I_0}.$$

Adding together all the contributions to F gives the following expression for the input reactance to the small-loop antenna:

$$
X = \frac{Z_0 k_0 d}{2} \left(\ln \frac{8d}{r} - 2 \right) - \frac{\pi^2 d^4 k_0 Z_0}{2} \left[\frac{k_0^2}{4\pi b}(2 - \gamma) + \frac{13\pi}{24a^2 b} + \frac{0.6}{\pi b^3} + \frac{3k_0^4 b}{576\pi} \right.
$$
$$
\left. - \frac{k_0^2}{4\pi b} \ln \frac{k_0^2 ab}{4\pi(1 - \cos k_0 b)} - \frac{1}{2ab} \sum_{n=2}^{\infty} \left(\frac{k_0^2}{\Gamma_{n0}} + \Gamma_{n0} - \frac{n\pi}{a} - \frac{k_0^2 a}{2n\pi} \right) \right]. \quad (47)
$$

For a small loop and a thin conductor (r small), the self-inductance term predominates.

For the purpose of obtaining an antenna radiating in one direction only, a short-circuiting plunger may be placed a distance l away from the loop. This has negligible effect on the evanescent modes excited by the antenna, since these will have decayed to a negligible value at the short-circuit position for $l > \lambda_0/4$. The effect on the dominant mode is the same as superimposing the H_{10} mode radiated by an image antenna carrying a current $-I_0$ and located at $z = -2l$. The total H_{10}-mode axial magnetic field in the guide for $z > 0$ becomes

$$C_{10}\mathbf{h}_{z10}(e^{-\Gamma_{10}z} - e^{-\Gamma_{10}z - 2\Gamma_{10}l}) = 2jC_{10}\mathbf{h}_{z10}e^{-\Gamma_{10}(z+l)} \sin \beta_{10}l$$

where $\beta_{10} = |\Gamma_{10}|$. The amplitude of the wave propagating in the positive z direction is increased by a factor $2 \sin \beta_{10}l$, and the power flow is increased by a factor $4 \sin^2 \beta_{10}l$ in the positive z direction and reduced to zero in the negative z direction. The total radiated power, and hence the radiation resistance, becomes

$$R_0 = \frac{k_0 Z_0}{ab\beta_{10}} \left(\frac{\pi}{a} \right)^2 \frac{(\pi d^2)^2}{2} \sin^2 \beta_{10}l. \quad (48)$$

The reactive energy associated with the H_{10}-mode standing wave between the antenna and short-circuiting plunger is given by

$$2j\omega(W_m - W_e)_{10} = \frac{1}{2} \int_0^a \int_0^b E_y H_x^* \, dx \, dy$$

where, for $z = 0$, we have

$$\mathbf{a}_y E_y = C_{10}(1 - e^{-j2\beta_{10}l})\mathbf{e}_{10}$$

$$\mathbf{a}_x H_x = -C_{10}(1 + e^{-j2\beta_{10}l})\mathbf{h}_{10}.$$

Since

$$\int_0^a \int_0^b \mathbf{e}_{10} \times \mathbf{h}_{10}^* \cdot \mathbf{a}_z \, dx \, dy = 1$$

we get, after substituting for C_{10},

$$2j\omega(W_m - W_e)_{10} = jC_{10}C_{10}^* \sin 2\beta_{10}l = jP \sin 2\beta_{10}l$$

where P is given by (34). The additional contribution to the input reactance is obtained by dividing by $\frac{1}{2}I_0^2$, and the total antenna reactance now becomes

$$X_0 = \frac{k_0 Z_0}{ab\beta_{10}} \left(\frac{\pi}{a}\right)^2 \left(\frac{\pi d^2}{2}\right)^2 \sin 2\beta_{10}l + X \qquad (49)$$

where X is given by (47). To partially cancel the inductive term X, the value of l should be between $\lambda_g/4$ and $\lambda_g/2$, where λ_g is the guide wavelength.

In Fig. 7.13 the input reactance X and radiation resistance R of a small loop in an infinite guide are plotted as a function of loop radius d for the following parameters: $k_0 = 2$, $a = 0.9$ inch, $b = 0.4$ inch, $r = 0.5$ mm. The self-inductance part is very large (for $d = 0.4$ cm, it is equal to 324 ohms), and, consequently, placing a short-circuiting plunger in the guide will

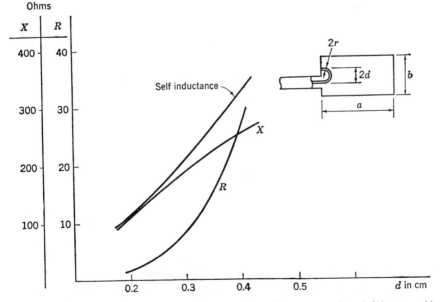

Fig. 7.13. Radiation resistance and input reactance of a small-loop antenna in an infinite waveguide plotted as a function of loop radius d for $a = 0.9$ inch, $b = 0.4$ inch, $r = 0.5$ mm, $\lambda_0 = 3.14$ cm.

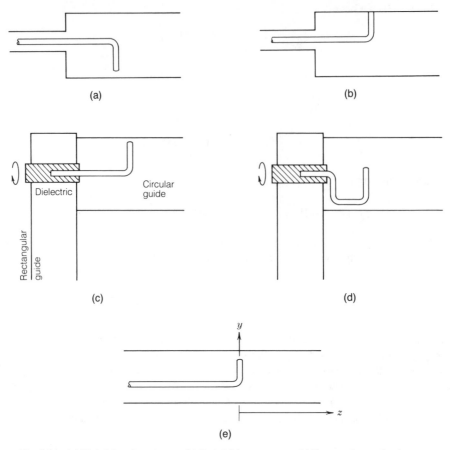

Fig. 7.14. (a) End-fed probe antenna. (b) End-fed loop antenna. (c) Rectangular to circular waveguide coupling with a rotatable probe system. (d) Broadband rectangular to circular waveguide
probe coupling system. (e) Transmission-line–fed waveguide probe antenna.

not produce enough capacitive loading on the loop to tune out the inductive reactance. The
maximum capacitive reactance which can be obtained is equal to the radiation resistance R as
plotted, although, with one end of the guide short-circuited, the radiation resistance R must
be multiplied by $2 \sin^2 \beta_{10} l$ to obtain the radiation resistance R_0 which is applicable now.

Other Waveguide Probe and Loop Antennas

There is an almost endless variety of probe and loop antenna configurations that can be used
in a waveguide. In Figs. 7.14(a) and (b) we show basic end-fed probe and loop antennas. This
particular loop antenna configuration in a circular waveguide has been analyzed in an approximate way by Deshpande and Das [7.19]. These authors have assumed that the current on the
loop is that associated with the dominant transmission-line standing-wave current distribution
and have neglected mutual coupling between the two sections of the loop antenna. In Figs.
7.14(c) and (d) we show probe antenna systems that are used to couple circular and rectangular
waveguides. The probes can be rotated by means of a small motor. These systems are used as
part of the prime focus feed in satellite receive-only earth stations and make it easy to orient

the probe for the reception of a signal having a particular polarization. The probe system shown in Fig. 7.14(c) is a narrow-band system unless additional matching elements are added [7.3]. The system shown in Fig. 7.14(d), on the other hand, has an unusually broad band response due to the particular transmission-line arrangement that is employed [7.4]. The basic antenna configuration that is representative of the above systems is shown in Fig. 7.14(e). We will outline the formulation of the theory for this latter antenna system in order to illustrate the main features involved in the analytical solution.

If we consider a rectangular waveguide then the vector potentials from y- and z-directed current elements are given by

$$A_y = \frac{\mu_0}{ab} \sum_{n=1}^{\infty} \sum_{m=0}^{\infty} \epsilon_{0m} \sin \frac{n\pi x}{a} \sin \frac{n\pi x'}{a} \cos \frac{m\pi y}{b} \cos \frac{m\pi y'}{b} \frac{e^{-\Gamma_{nm}|z-z'|}}{\Gamma_{nm}} \tag{50a}$$

$$A_z = \frac{2\mu_0}{ab} \sum_{n=1}^{\infty} \sum_{m=1}^{\infty} \sin \frac{n\pi x}{a} \sin \frac{n\pi x'}{a} \sin \frac{m\pi y}{b} \sin \frac{m\pi y'}{b} \frac{e^{-\Gamma_{nm}|z-z'|}}{\Gamma_{nm}}. \tag{50b}$$

The components of the Green's dyadic function that will give the y and z components of the electric field are given by

$$G_{yy} = \frac{1}{j\omega\mu_0\epsilon_0} \left(k_0^2 + \frac{\partial^2}{\partial y^2} \right) A_y \tag{51a}$$

$$G_{yz} = \frac{1}{j\omega\mu_0\epsilon_0} \frac{\partial^2}{\partial y \partial z} A_z \tag{51b}$$

$$G_{zy} = \frac{1}{j\omega\mu_0\epsilon_0} \frac{\partial^2}{\partial y \partial z} A_y \tag{51c}$$

$$G_{zz} = \frac{1}{j\omega\mu_0\epsilon_0} \left(k_0^2 + \frac{\partial^2}{\partial z^2} \right) A_z. \tag{51d}$$

The electric field radiated by the current on the transmission line and probe sections is given by

$$\mathbf{E}(\mathbf{r}) = \iint_S \bar{\mathbf{G}}(\mathbf{r}, \mathbf{r}') \cdot \mathbf{J}(\mathbf{r}') \, dS'.$$

The tangential component of the electric field must vanish along the transmission-line conductor and on the probe. This boundary condition gives the integral equation

$$\mathbf{n} \times \iint_S \bar{\mathbf{G}} \cdot \mathbf{J} \, dS' = 0, \qquad \mathbf{r} \text{ on } S$$

where S is the conductor surface.

The current on the transmission line consists of the TEM-mode current plus some additional current that is localized near the probe:

$$\frac{I^+ e^{-jk_0 z}}{2\pi r} - \frac{I^- e^{jk_0 z}}{2\pi r} + J_z$$

where we have assumed that the current is uniform around the conductor and I^+, I^- are the incident and reflected TEM-mode current amplitudes.

We now consider the electric field acting along the probe due to the TEM-mode currents. The latter is given by

$$\frac{1}{2\pi} \int_{-\infty}^{0} \int_{0}^{2\pi} G_{yz}(I^+ e^{-jk_0 z'} - I^- e^{jk_0 z'}) d\phi' dz'$$

where the integration over ϕ' is around the periphery of the transmission-line conductor. A typical term to be evaluated is

$$\int_{-\infty}^{0} e^{-\Gamma_{nm}|z-z'| \mp jk_0 z'} dz'.$$

After performing the z' integration we find the following contribution to A_z:

$$\frac{\mu_0}{\pi ab} \sum_{n=1}^{\infty} \sum_{m=1}^{\infty} \sin \frac{n\pi x}{a} \sin \frac{m\pi y}{b} \int_{0}^{2\pi} \sin \frac{n\pi x'}{a} \sin \frac{m\pi y'}{b} d\phi' \frac{1}{\Gamma_{nm}^2 + k_0^2}$$
$$\cdot \left[2I^+ e^{-jk_0 z} - 2I^- e^{jk_0 z} + \frac{jk_0}{\Gamma_{nm}}(I^+ + I^-)e^{\Gamma_{nm} z} - (I^+ - I^-)e^{\Gamma_{nm} z} \right].$$

At $z = 0$ the electric field E_y is given by

$$E_y = \frac{jZ_0}{2\pi k_0 ab} \sum_{n=1}^{\infty} \sum_{m=1}^{\infty} \frac{m\pi}{b} \sin \frac{n\pi x}{a} \cos \frac{m\pi y}{b}$$
$$\cdot \int_{0}^{2\pi} \sin \frac{n\pi x'}{a} \sin \frac{m\pi y'}{b} d\phi' \frac{jk_0(I^+ + I^-) + \Gamma_{nm}(I^+ - I^-)}{\Gamma_{nm}^2 + k_0^2}. \quad (52)$$

The term proportional to $I^+ + I^- = V/Z_c$ is the TEM-mode electric field acting on the probe. We can view this as the applied electric field and consider V to be a given voltage. By using this as the forcing function the integral equation for $\mathbf{J}$ can be formulated as a pair of coupled integral equations by setting $E_y = 0$ on the probe and $E_z = 0$ on the transmission-line conductor. The solution for the current can be obtained using Galerkin's method. The probe impedance is given by

$$Z_{\text{in}} = \frac{V}{I^+ - I^-} = \frac{V}{I}.$$

At $z = 0$ the total z-directed current must be made equal to the total y-directed current. This junction boundary condition requires that

$$2\pi r J_z(0) + I^+ - I^- = 2\pi r J_y(0).$$

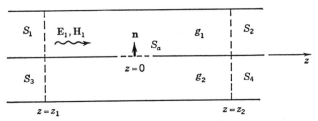

Fig. 7.15. Two guides coupled by a small aperture.

7.3. Coupling by Small Apertures

Electromagnetic energy may be coupled from one waveguide into another guide or into a cavity resonator by a small aperture located at a suitable position in the common wall. For apertures whose linear dimensions are small compared with the wavelength, an approximate theory is available which states that the aperture is equivalent to a combination of radiating electric and magnetic dipoles, whose dipole moments are respectively proportional to the normal electric field and the tangential magnetic field of the incident wave. This theory, originally developed by Bethe [7.5], will be presented here. The approach to be used here differs from that of Bethe, and has been chosen because it leads to the final result in a somewhat more direct manner. We will also modify the Bethe theory so that power conservation will hold.

Consider the problem of coupling between two waveguides by means of a small aperture in a common sidewall as shown in Fig. 7.15. The solution procedure to be followed, which is based on one of Schelkunoff's field equivalence principles, consists of the following sequence of steps. We first close the aperture by a perfect magnetic wall. The incident field, which is chosen as the field in the absence of the aperture, will induce a magnetic current $\mathbf{J}_m$ and a magnetic charge ρ_m on the magnetic wall surface S_a. These sources produce a scattered field that can be expressed as a field radiated by the dipole moments of the source distribution. After the field scattered into guide g_1 has been found the aperture is opened and a magnetic current $-\mathbf{J}_m$ is placed in the aperture. The field radiated by this source into the two guides, together with the specified incident field and the scattered field in g_1 found earlier, represents the total unique solution to the coupling problem. The total field as given is readily shown to have tangential electric and magnetic field components that are continuous across the aperture opening. Since all of the required boundary conditions are satisfied the solution is the correct and unique one.

With the aperture closed by a magnetic wall, let a normal mode $\mathbf{E}_1$, $\mathbf{H}_1$ be incident from the left. Because of the magnetic wall discontinuity, a scattered field $\mathbf{E}_s$, $\mathbf{H}_s$ will be excited such that the total field satisfies the following boundary conditions on the magnetic wall in the aperture S_a:

$$\mathbf{n} \times \mathbf{H}_s = -\mathbf{n} \times \mathbf{H}_1 \quad \text{or} \quad \mathbf{n} \times (\mathbf{H}_s + \mathbf{H}_1) = 0 \tag{53a}$$

$$\mathbf{n} \cdot \mathbf{E}_s = -\mathbf{n} \cdot \mathbf{E}_1 \quad \text{or} \quad \mathbf{n} \cdot (\mathbf{E}_s + \mathbf{E}_1) = 0. \tag{53b}$$

The normal magnetic field and the tangential electric field are not equal to zero on S_a, and, hence, a magnetic charge and a magnetic current distribution given by

$$\rho_m = \mu_0 \mathbf{n} \cdot \mathbf{H}_s \tag{54a}$$

$$\mathbf{J}_m = -\mathbf{n} \times \mathbf{E}_s \tag{54b}$$

will exist on S_a. For the incident mode, we have $\mathbf{n} \cdot \mathbf{H}_1 = \mathbf{n} \times \mathbf{E}_1 = 0$ on S_a. The scattered field may be expanded in terms of the normal waveguide modes as follows (see Section 5.6):

$$\mathbf{E}_s = \sum a_n \mathbf{E}_n^+, \qquad z > 0 \tag{55a}$$

$$\mathbf{H}_s = \sum a_n \mathbf{H}_n^+, \qquad z > 0 \tag{55b}$$

$$\mathbf{E}_s = \sum b_n \mathbf{E}_n^-, \qquad z < 0 \tag{55c}$$

$$\mathbf{H}_s = \sum b_n \mathbf{H}_n^-, \qquad z < 0. \tag{55d}$$

The scattered field is a solution of the equations

$$\nabla \times \mathbf{E}_s = -j\omega\mu_0 \mathbf{H}_s - \mathbf{J}_m$$

$$\nabla \times \mathbf{H}_s = j\omega\epsilon_0 \mathbf{E}_s$$

while each normal mode function $\mathbf{E}_n$, $\mathbf{H}_n$ is a solution of the source-free equations. If we apply the Lorentz reciprocity theorem along the same lines as was done in Section 5.6, we find that the expansion coefficients a_n, b_n are given by

$$2b_n = \iint\limits_{S_a} \mathbf{H}_n^+ \cdot \mathbf{J}_m \, dS \tag{56a}$$

$$2a_n = \iint\limits_{S_a} \mathbf{H}_n^- \cdot \mathbf{J}_m \, dS. \tag{56b}$$

For a small aperture we will show that Eqs. (56) represent coupling to an electric and magnetic dipole plus a magnetic quadrupole.

With reference to Fig. 7.16 the following parameters are introduced:

1. $\boldsymbol{\tau}$ a unit vector tangent to the aperture contour C,
2. $\mathbf{n}_1$ a unit vector normal to $\boldsymbol{\tau}$ and in the plane of the aperture,
3. $\mathbf{n}$ a unit vector perpendicular to S_a and directed into guide g_1,
4. u, v, w a localized rectangular coordinate system with the origin at the center of the aperture and w directed along $\mathbf{n}$, and
5. $\mathbf{n}_0$ a unit vector normal to S_a and directed from guide g_1 into guide g_2.

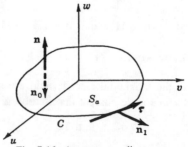

Fig. 7.16. Aperture coordinates.

The scattered field satisfies the boundary condition $\boldsymbol{\tau}\cdot\mathbf{E}_s = 0$, and, hence, $\mathbf{n}_1\cdot\mathbf{J}_m$ equals zero on C; that is, there is no normal component of magnetic current at the boundary of the aperture. Since $\rho_m = \mu_0\mathbf{n}\cdot\mathbf{H}_s$, we have

$$\mu_0^{-1}\iint\limits_{S_a}\rho_m\,dS = \iint\limits_{S_a}\mathbf{H}_s\cdot\mathbf{n}\,dS = \iint\limits_{S_a}\frac{\nabla\times\mathbf{E}_s\cdot\mathbf{n}\,dS}{-j\omega\mu_0} = -\frac{1}{j\omega\mu_0}\oint_C\mathbf{E}_s\cdot\boldsymbol{\tau}\,dl = 0$$

which shows that there is no net magnetic charge on S_a. For a small aperture we may expand $\mathbf{H}_n$ in a Taylor series about the origin to get

$$\mathbf{H}_n(u,\,v) = \mathbf{H}_n(0) + \frac{u\,\partial\mathbf{H}_n}{\partial u}\bigg|_0 + \frac{v\,\partial\mathbf{H}_n}{\partial v}\bigg|_0 + \cdots \approx \mathbf{H}_n(0) + \mathbf{r}\cdot\nabla\mathbf{H}_n$$

where $\mathbf{r} = \mathbf{a}_u u + \mathbf{a}_v v$, and $\nabla\mathbf{H}_n$ is to be evaluated at the origin. The coupling coefficient b_n is given by

$$2b_n = \iint\limits_{S_a}\mathbf{H}_n^+\cdot\mathbf{J}_m\,dS = \mathbf{H}_n^+(0)\cdot\iint\limits_{S_a}\mathbf{J}_m\,dS + \iint\limits_{S_a}(\mathbf{r}\cdot\nabla\mathbf{H}_n^+)\cdot\mathbf{J}_m\,dS. \tag{57}$$

Consider the following integral:

$$\iint\limits_{S_a}\nabla\cdot\phi\mathbf{J}_m\,dS = \iint\limits_{S_a}(\phi\nabla\cdot\mathbf{J}_m + \mathbf{J}_m\cdot\nabla\phi)\,dS = \oint_C\phi\mathbf{J}_m\cdot\mathbf{n}_1\,dl = 0$$

since $\mathbf{J}_m\cdot\mathbf{n}_1 = 0$, and ϕ is an arbitrary scalar function. Let $\phi = u$, and the above result gives

$$\mathbf{a}_u\iint\limits_{S_a}\mathbf{J}_m\cdot\nabla u\,dS = \mathbf{a}_u\iint\limits_{S_a}\mathbf{J}_m\cdot\mathbf{a}_u\,dS = -\mathbf{a}_u\iint\limits_{S_a}u\nabla\cdot\mathbf{J}_m\,dS. \tag{58}$$

A similar result is obtained if we let $\phi = v$. Combining the two results for $\phi = u$ and v gives

$$\iint\limits_{S_a}\mathbf{J}_m\,dS = -\iint\limits_{S_a}\mathbf{r}\nabla\cdot\mathbf{J}_m\,dS = j\omega\iint\limits_{S_a}\mathbf{r}\rho_m\,dS = j\omega\mu_0\mathbf{M} \tag{59}$$

since $\nabla\cdot\mathbf{J}_m = -j\omega\rho_m$ from the continuity equation relating current and charge, and the integral $(1/\mu_0)\iint_{S_a}\mathbf{r}\rho_m\,dS$ defines an equivalent magnetic dipole moment $\mathbf{M}$ in direct analogy with the electric dipole moment arising from electric charge ρ. The first term on the right-hand side in (57) is thus seen to represent coupling with a magnetic dipole $\mathbf{M}$.

The second integral on the right-hand side in (57) is readily evaluated by writing the integrand in component form. We have

$$\mathbf{r}\cdot\nabla\mathbf{H}_n^+\cdot\mathbf{J}_m = (\mathbf{a}_u u + \mathbf{a}_v v)\cdot\left(\mathbf{a}_u\mathbf{a}_u\frac{\partial H_{nu}^+}{\partial u} + \mathbf{a}_u\mathbf{a}_v\frac{\partial H_{nv}^+}{\partial u} + \mathbf{a}_v\mathbf{a}_u\frac{\partial H_{nu}^+}{\partial v}\right.$$

$$\left. + \mathbf{a}_v\mathbf{a}_v\frac{\partial H_{nv}^+}{\partial v}\right)\cdot(\mathbf{a}_u J_{mu} + \mathbf{a}_v J_{mv})$$

$$= u\frac{\partial H_{nu}^+}{\partial u}J_{mu} + u\frac{\partial H_{nv}^+}{\partial u}J_{mv} + v\frac{\partial H_{nu}^+}{\partial v}J_{mu} + v\frac{\partial H_{nv}^+}{\partial v}J_{mv}.$$

Subtracting and adding similar terms, this may be rewritten as

$$
\frac{u}{2}J_{mv}\left(\frac{\partial H_{nv}^{+}}{\partial u}-\frac{\partial H_{nu}^{+}}{\partial v}\right)-\frac{vJ_{mu}}{2}\left(\frac{\partial H_{nv}^{+}}{\partial u}-\frac{\partial H_{nu}^{+}}{\partial v}\right)+\frac{uJ_{mv}}{2}\frac{\partial H_{nu}^{+}}{\partial v}
$$

$$
+\frac{vJ_{mu}}{2}\frac{\partial H_{nv}^{+}}{\partial u}+uJ_{mu}\frac{\partial H_{nu}^{+}}{\partial u}+vJ_{mv}\frac{\partial H_{nv}^{+}}{\partial v}+\frac{uJ_{mv}}{2}\frac{\partial H_{nv}^{+}}{\partial u}+\frac{vJ_{mu}}{2}\frac{\partial H_{nu}^{+}}{\partial v}. \quad (60)
$$

The first two terms are readily recognized as being equal to $j\omega\epsilon_0\mathbf{E}_n^{+}(0)\cdot(\mathbf{r}\times\mathbf{J}_m)/2$. The magnetic dipole moment of an electric current distribution is defined by the integral

$$
\iiint\frac{\mathbf{r}\times\mathbf{J}}{2}\,dV.
$$

By analogy, we now see that

$$
-j\omega\epsilon_0\mathbf{E}_n^{+}(0)\cdot\iint\limits_{S_a}\frac{-\mathbf{r}\times\mathbf{J}_m}{2}\,dS=-j\omega\mathbf{E}_n^{+}(0)\cdot\mathbf{P}
$$

where $\mathbf{P}$ is the equivalent electric dipole moment of the circulating magnetic current. This dipole is directed normal to the aperture.

If we let $\phi=u^2/2$, $v^2/2$, and uv in turn, then in (58) we get the results

$$
\iint\limits_{S_a}uJ_{mu}\,dS=-\iint\limits_{S_a}\frac{u^2}{2}\nabla\cdot\mathbf{J}_m\,dS=\frac{j\omega}{2}\iint\limits_{S_a}u^2\rho_m\,dS
$$

$$
\iint\limits_{S_a}vJ_{mv}\,dS=\frac{j\omega}{2}\iint\limits_{S_a}v^2\rho_m\,dS
$$

$$
\iint\limits_{S_a}(vJ_{mu}+uJ_{mv})\,dS=j\omega\iint\limits_{S_a}uv\rho_m\,dS.
$$

The components of a dyadic magnetic quadrupole $\bar{\mathbf{Q}}$ are given by

$$
Q_{uu}=\frac{1}{\mu_0}\iint\limits_{S_a}u^2\rho_m\,dS
$$

$$
Q_{vu}=Q_{uv}=\frac{1}{\mu_0}\iint\limits_{S_a}uv\rho_m\,dS \qquad \text{etc.}
$$

and, hence, the last six terms in (60) may be combined by using the above relations to give finally

$$
\frac{j\omega\mu_0}{2}\nabla\mathbf{H}_n^{+}:\bar{\mathbf{Q}}
$$

where the double-dot product is taken between the two dyadics $\nabla\mathbf{H}_n^{+}$ and $\bar{\mathbf{Q}}$.

The expansion coefficients a_n and b_n are thus seen to be given by

$$2a_n = j\omega \left(\mu_0 \mathbf{H}_n^- \cdot \mathbf{M} - \mathbf{E}_n^- \cdot \mathbf{P} + \frac{\mu_0}{2} \nabla \mathbf{H}_n^- : \bar{\mathbf{Q}} \right) \qquad (61a)$$

$$2b_n = j\omega \left(\mu_0 \mathbf{H}_n^+ \cdot \mathbf{M} - \mathbf{E}_n^+ \cdot \mathbf{P} + \frac{\mu_0}{2} \nabla \mathbf{H}_n^+ : \bar{\mathbf{Q}} \right). \qquad (61b)$$

In the definition of the dipole moments, the origin for the radius vector $\mathbf{r}$ is arbitrary, since there is no net charge in the aperture. The magnetic dipole moments have been defined to be dimensionally the same as $\mathbf{H}$ times length cubed. In most applications the quadrupole term can be neglected, since it represents a small quantity depending on the fourth power of the aperture dimension.

Although we have written the expansion coefficients in terms of equivalent dipole moments, we do not know these before we find a solution for the current $\mathbf{J}_m$ and the charge ρ_m in the aperture. This solution is difficult to obtain in general. However, we may find a static field solution for the dipole moments of small elliptic- and circular-shaped apertures quite readily. In the practical application of the theory, it is found that the static field solution gives results of acceptable accuracy for small apertures. To further enhance the applicability of the theory, Cohn has described an electrolytic-tank method for measuring polarizabilities of arbitrarily shaped apertures [7.6].

Before presenting the static field solution for the dipole moments, the rest of the theory required to obtain the field coupled through the aperture into guide g_2 will be given. When the aperture is open, the field in g_2 is that radiated by an equivalent current $-\mathbf{J}_m$ on S_a. The field in g_1 is the sum of the field radiated by $-\mathbf{J}_m$ in the aperture and the field $\mathbf{E}_1 + \mathbf{E}_s$, $\mathbf{H}_1 + \mathbf{H}_s$ existing in g_1 with the aperture closed by a magnetic wall. Let the current $-\mathbf{J}_m$ radiate a field $\mathbf{E}_{s1}$, $\mathbf{H}_{s1}$ into g_1, and $\mathbf{E}_{s2}$, $\mathbf{H}_{s2}$ into g_2. At the aperture, the total tangential magnetic field must be continuous, and, hence, $\mathbf{n}_0 \times \mathbf{H}_{s2} = \mathbf{n}_0 \times (\mathbf{H}_{s1} + \mathbf{H}_s + \mathbf{H}_1) = \mathbf{n}_0 \times \mathbf{H}_{s1}$ from (53a). The total tangential electric field must also be continuous, and so we have $\mathbf{n}_0 \times \mathbf{E}_{s2} = \mathbf{n}_0 \times (\mathbf{E}_{s1} + \mathbf{E}_s + \mathbf{E}_1)$, or, by using (54b), $\mathbf{n}_0 \times (\mathbf{E}_{s1} - \mathbf{E}_{s2}) = -\mathbf{J}_m$. The field in g_2 may be expanded in terms of the normal mode functions for this guide as follows:

$$\mathbf{E}_{s2} = \sum c_n \mathbf{E}_{n2}^+, \qquad z > 0 \qquad (62a)$$

$$\mathbf{H}_{s2} = \sum c_n \mathbf{H}_{n2}^+, \qquad z > 0 \qquad (62b)$$

$$\mathbf{E}_{s2} = \sum d_n \mathbf{E}_{n2}^-, \qquad z < 0 \qquad (62c)$$

$$\mathbf{H}_{s2} = \sum d_n \mathbf{H}_{n2}^-, \qquad z < 0 \qquad (62d)$$

where the additional subscript 2 is used to distinguish between the normal modes for the two guides. If the two guides are identical, then the current $-\mathbf{J}_m$ will radiate identical fields into the two guides. The expansion coefficients may be found in the same way as for the field $\mathbf{E}_s$, $\mathbf{H}_s$, but it must be borne in mind that the tangential electric field and normal magnetic field in one guide will undergo only one-half the discontinuous change across the source region that the field $\mathbf{E}_s$, $\mathbf{H}_s$ did. The effective dipole moments associated with the aperture current $-\mathbf{J}_m$ and charge $-\rho_m$ for radiation into guide g_2 will be $-\mathbf{M}/2$ and $-\mathbf{P}/2$. One-half the

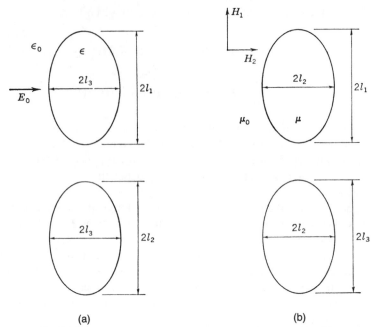

(a)   (b)

Fig. 7.17. (a) Dielectric ellipsoid. (b) Permeable magnetic ellipsoid.

required discontinuity across the source is provided by $\mathbf{E}_{s2}$, $\mathbf{H}_{s2}$; the other half by $\mathbf{E}_{s1}$, $\mathbf{H}_{s1}$. The total field in g_1 is now seen to be equivalent to the sum of the incident fields $\mathbf{E}_1$, $\mathbf{H}_1$ and the field radiated by dipoles of strength $\mathbf{M}/2$ and $\mathbf{P}/2$ since the field radiated by $-\mathbf{M}/2$, $-\mathbf{P}/2$ cancels one-half of the field $\mathbf{E}_s$, $\mathbf{H}_s$ radiated by $\mathbf{M}$ and $\mathbf{P}$.

In the more general case, when the two guides are not identical, this symmetry argument no longer applies. We will present a method later on that will enable the problem of coupling between dissimilar regions to also be solved as two uncoupled excitation problems. Before we proceed with that task we will derive expressions for the polarizabilities of small elliptical-shaped apertures.

Dipole Moments of Elliptic Apertures

The dipole moments of an elliptic aperture may be obtained from the static dipole moments of general ellipsoidal dielectric and permeable magnetic bodies, placed in uniform static electric and magnetic fields, by suitable limiting processes. Consider a dielectric ellipsoid with semiaxes l_1, l_2, l_3 placed in a uniform electric field E_0 directed along the l_3 axis as in Fig. 7.17(a).

The problem of the ellipsoid in a uniform field is readily solved in ellipsoidal coordinates [7.7]. We find that the total field in the interior is uniform and parallel to the applied field if the latter is directed along one of the axes of the ellipsoid. The dipole polarization per unit volume, therefore, has zero divergence, and the equivalent polarization volume charge in the interior is zero. On the surface, an equivalent surface-polarization charge density equal to the discontinuity in the normal component of the polarization vector exists. The total dipole moment of the ellipsoid may be calculated by a volume integral either in terms of the polarization density or in terms of the equivalent polarization surface charge. The field outside the ellipsoid is the sum of the applied field plus a dipole field arising from an elementary dipole at the origin and having a moment equal to the dipole moment of the ellipsoid. According to Stratton, the dipole moment P_3, when a uniform field E_0 is applied along the l_3 axis, is

$$P_3 = \frac{\epsilon_0 E_0 V}{L_3 + \epsilon_0/(\epsilon - \epsilon_0)} \qquad (63)$$

where

$$L_3 = \frac{l_1 l_2 l_3}{2} \int_0^\infty \frac{ds}{(s + l_3^2)[(s + l_1^2)(s + l_2^2)(s + l_3^2)]^{1/2}}$$

and $V = \frac{4}{3}\pi l_1 l_2 l_3$ is the volume of the ellipsoid. If we let ϵ be equal to zero, we obtain the dual of a perfect diamagnetic material. For $\epsilon = 0$, the internal polarization sets up a field equal and opposite to the applied field, so that the net internal displacement flux D is zero. The surface polarization charge now gives rise to an external field, which cancels the normal component of the applied field E_0 at the surface of the ellipsoid. This is precisely the boundary condition that the electric field must satisfy at a magnetic wall. Hence, a dielectric body with $\epsilon = 0$ is equivalent to a body with a magnetic wall surface. If we now let l_3 approach zero, (63) gives the dipole moment of an elliptic disk which is made from a perfect "magnetic conductor." The dipole moment of the aperture is only one-half that of a disk, since only one side of the aperture is under consideration while the disk has two sides with polarization charge on both. Finally, the effective radiating dipole moment in the aperture is one-half of the moment of an aperture closed by a magnetic wall, and hence equals one-quarter of the moment of a complete disk. Before evaluating the integral for L_3, we will give the expressions for the magnetic dipole moments of an ellipsoid.

Figure 7.17(b) illustrates a permeable ellipsoid placed in a uniform magnetic field. There are two dipole moments to consider: M_1 due to a field H_1 along the l_1 axis, and M_2 due to a field H_2 along the l_2 axis. Each case may be treated separately, and each solution is essentially the same as for the dielectric case. We find that

$$M_1 = \frac{V H_1}{L_1 + \mu_0/(\mu - \mu_0)} \qquad (64a)$$

where

$$L_1 = \frac{l_1 l_2 l_3}{2} \int_0^\infty \frac{ds}{(s + l_1^2)[(s + l_1^2)(s + l_2^2)(s + l_3^2)]^{1/2}}$$

and

$$M_2 = \frac{V H_2}{L_2 + \mu_0/(\mu - \mu_0)} \qquad (64b)$$

where

$$L_2 = \frac{l_1 l_2 l_3}{2} \int_0^\infty \frac{ds}{(s + l_2^2)[(s + l_2^2)(s + l_3^2)(s + l_1^2)^2]^{1/2}}.$$

If we let μ approach infinity, the internal magnetic field vanishes, since $\mathbf{n} \cdot \mu_0 \mathbf{H}_e$ equals $\mathbf{n} \cdot \mu \mathbf{H}_i$, where $\mathbf{H}_e$ is the external field and $\mathbf{H}_i$ is the internal field, and, hence, the product $\mu \mathbf{H}_i$ is finite and $\mathbf{H}_i$ tends to zero as μ tends to infinity. For an external field applied along a principal axis, the internal field is uniform and parallel with the applied field, and, since it vanishes at the surface, it follows that the induced external field just cancels the tangential component of the applied field at the surface. Again we have boundary conditions corresponding to those at a magnetic wall, so that, by letting l_3 approach zero and μ tend to infinity, we obtain the magnetic dipole moments of an elliptical disk having perfect magnetic conductivity.

The integrals for L_1, L_2, and L_3 may be evaluated in terms of elliptic integrals. First we show that $L_1 + L_2 + L_3 = 1$, and so we need evaluate only two of the integrals. If a new variable

$$u^2 = (s + l_1^2)(s + l_2^2)(s + l_3^2)$$

is introduced, we get

$$L_1 + L_2 + L_3 = \frac{l_1 l_2 l_3}{2} \int_0^\infty \frac{(s + l_2^2)(s + l_3^2) + (s + l_3^2)(s + l_1^2) + (s + l_1^2)(s + l_2^2)}{u^3}\, ds.$$

Using the familiar rule for the differential of a product, it is seen at once that the numerator is equal to $2u\, du$. Putting in the proper limits of integration for the variable u, the integral becomes

$$\frac{l_1 l_2 l_3}{2} \int_{l_1 l_2 l_3}^\infty \frac{2\, du}{u^2} = 1.$$

The integrals for L_1 and L_2 are somewhat easier to evaluate than that for L_3. The procedure used will be given only for the case of L_2, the evaluation of L_1 being similar. We may take $l_1 > l_2 > l_3$ without any loss in generality. In the integral for L_2, we introduce a new variable t according to the relation $s + l_2^2 = t^2$ and get

$$L_2 = l_1 l_2 l_3 \int_{l_2}^\infty \frac{dt}{t^2 (t^2 + l_1^2 - l_2^2)^{1/2}(t^2 - l_2^2 + l_3^2)^{1/2}}.$$

Multiplying the numerator by $[(t^2 + l_1^2 - l_2^2) - t^2]/(l_1^2 - l_2^2)$ the integral may be split into two parts as follows:

$$\frac{l_1 l_2 l_3}{l_1^2 - l_2^2} \left[\int_{l_2}^\infty \frac{(t^2 + l_1^2 - l_2^2)^{1/2}}{t^2 (t^2 - l_2^2 + l_3^2)^{1/2}}\, dt - \int_{l_2}^\infty \frac{dt}{(t^2 + l_1^2 - l_2^2)^{1/2}(t^2 - l_2^2 + l_3^2)^{1/2}} \right].$$

Apart from the limits of integration, the integrals occurring here are of the form that Jahnke and Emde tabulated the solutions for.[2] In order to use the solutions presented, the integrals here are rewritten as the sum of two integrals with limits chosen as indicated below:

$$\int_{l_2}^x = \int_{(l_2^2 - l_3^2)^{1/2}}^x - \int_{(l_2^2 - l_3^2)^{1/2}}^{l_2}.$$

For our application we will set l_3 equal to zero eventually, and so the second integral will vanish and, therefore, will not be evaluated. The value of the required integral is

$$\frac{l_1 l_2 l_3}{l_1^2 - l_2^2} \left[\frac{(l_1^2 - l_3^2)^{1/2}}{l_2^2 - l_3^2} \mathcal{E}(k, \phi) - \frac{1}{(l_1^2 - l_3^2)^{1/2}} F(k, \phi) \right]$$

---

[2] See [7.2, p. 58]. Fifth formula pair for evaluation of L_2, and last pair for evaluation of L_1.

where

$$k = \left(\frac{l_1^2 - l_2^2}{l_1^2 - l_3^2}\right)^{1/2} \qquad \cos\phi = \frac{(l_2^2 - l_3^2)^{1/2}}{x}$$

and F and $\mathcal{E}$ are the incomplete elliptic integrals of the first and second kind. As x tends to infinity, ϕ becomes equal to $\pi/2$, and, when l_3 is placed equal to zero, we get

$$\frac{L_2}{V} = \frac{3}{4\pi l_1^3 e^2}[(1 - e^2)^{-1}E(e) - K(e)] \tag{65}$$

where K and E are complete elliptic integrals of the first and second kind with modulus e equal to the eccentricity $(1 - l_2^2/l_1^2)^{1/2}$ of the ellipse.

In a similar fashion we find that

$$\frac{L_1}{V} = \frac{3}{4\pi l_1^3 e^2}[K(e) - E(e)] \tag{66}$$

$$\frac{L_3 - 1}{V} = -\frac{L_1 + L_2}{V} = \frac{-3E(e)}{4\pi l_1^3(1 - e^2)}. \tag{67}$$

Substituting into (63) and (64), the dipole moments are obtained:

$$P_3 = -\frac{4\pi l_1^3(1 - e^2)}{3E(e)}\epsilon_0 E_0 \tag{68a}$$

$$M_1 = \frac{4\pi l_1^3 e^2}{3[K(e) - E(e)]}H_1 \tag{68b}$$

$$M_2 = \frac{4\pi l_1^3 e^2(1 - e^2)}{3[E(e) - (1 - e^2)K(e)]}H_2. \tag{68c}$$

Returning to our waveguide-coupling problem, let $\mathbf{E}_1$, $\mathbf{H}_1$ be the incident mode in guide g_1. The scattered field in this guide is that radiated by an electric dipole $\mathbf{P}_0 = \frac{1}{2}\mathbf{P}$ and a magnetic dipole $\mathbf{M}_0 = \frac{1}{2}\mathbf{M}$. In the Bethe theory these dipole strengths are chosen equal to the static values induced by the incident fields; thus

$$\mathbf{P}_0 = \epsilon_0 \alpha_e \mathbf{nn} \cdot \mathbf{E}_1 \tag{69a}$$

$$\mathbf{M}_0 = \bar{\alpha}_m \cdot \mathbf{H}_1 \tag{69b}$$

where from (68) we have, after dividing by 4,

$$\alpha_e = -\frac{\pi l_1^3(1 - e^2)}{3E(e)} \tag{70a}$$

$$\bar{\alpha}_m = \mathbf{a}_u\mathbf{a}_u\frac{\pi l_1^3 e^2}{3[K(e) - E(e)]} + \mathbf{a}_v\mathbf{a}_v\frac{\pi l_1^3 e^2(1 - e^2)}{3[E(e) - (1 - e^2)K(e)]}. \tag{70b}$$

In (70a), α_e is the electric polarizability of the aperture, while, in (70b), $\bar{\alpha}_m$ is the dyadic magnetic polarizability of the aperture. The aperture coordinates u and v are to be oriented with u along the major axis of the ellipse. For small values of e, the following formulas may be used to evaluate the elliptic integrals:

$$E = \frac{\pi}{2}\left(1 - \frac{e^2}{4}\right)$$

$$K = \frac{\pi}{2}\left(1 + \frac{e^2}{4}\right)$$

while, for values of e approaching unity,

$$E = 1$$

$$K = \ln 4\frac{l_1}{l_2}.$$

For e equal to zero, we obtain a circular aperture with polarizabilities

$$\alpha_e = -\tfrac{2}{3}l^3 \tag{71}$$

$$\bar{\alpha}_m = \tfrac{4}{3}l^3(\mathbf{a}_u\mathbf{a}_u + \mathbf{a}_v\mathbf{a}_v) \tag{72}$$

where l is the radius of the aperture. The field radiated into guide g_2 is that due to radiating dipoles $-\mathbf{P}_0$ and $-\mathbf{M}_0$ in the aperture.

Radiation Reaction Fields

When we try to determine an equivalent network to represent the small aperture using the static polarizability tensors to determine the dipole strengths we find that we do not obtain a physically meaningful circuit and power is not conserved. One procedure to overcome these difficulties is to derive expressions for dynamic polarizability tensors. This can be done using a power series expansion in k_0 of the incident and scattered fields. Such a procedure requires an expansion up to terms including k_0^3 and the resultant expressions for $\bar{\alpha}_e$ and $\bar{\alpha}_m$ will depend on the geometrical shapes of the input and output waveguides. Thus the polarizabilities of a given aperture are no longer characteristic parameters of the aperture alone. Fortunately, there is an alternative procedure that can be followed which still allows one to use the static polarizabilities but corrects for the lack of power conservation by introducing the radiation reaction fields as part of the polarizing field. The basic concept will be described by considering the problem of the scattering of a plane wave by a small conducting sphere of radius a [7.8].

Consider a plane wave

$$\mathbf{E} = E_0e^{jk_0x}\mathbf{a}_z$$

$$\mathbf{H} = H_0e^{jk_0x}\mathbf{a}_y$$

incident on a conducting sphere located at the origin. When $k_0a \ll 1$ we can assume that a constant electric field $E_0\mathbf{a}_z$ acts on the sphere. This static field problem can be readily solved and we find that a charge distribution is induced on the sphere such that the dipole moment has a value $\mathbf{P} = 4\pi\epsilon_0 a^3 E_0\mathbf{a}_z$.

Under dynamic conditions the normal component of the total magnetic field must vanish at the conducting surface. The equivalent static field problem is that of a perfect diamagnetic sphere immersed in a field $H_0\mathbf{a}_y$. The induced magnetic dipole moment is $\mathbf{M} = -2\pi a^3 H_0\mathbf{a}_y$ as will be shown in Chapter 12. If we determine static fields in the region $r > a$ from these dipoles located at the origin then the boundary conditions $\mathbf{n} \times (\mathbf{E}_0 + \mathbf{E}_i) = 0$ and $\mathbf{n}\cdot(\mathbf{H}_0 + \mathbf{H}_i) = 0$ at $r = a$ will be satisfied where $\mathbf{E}_i$ and $\mathbf{H}_i$ are the induced or scattered fields. However, if we determine dynamic fields from these dipoles we will find that the boundary conditions are violated.

The dynamic electric field produced by a time-harmonic electric dipole P is given by [Eq. (79), Chapter 1]

$$\mathbf{E}_s = \left(k_0^2\mathbf{a}_z + \nabla\frac{\partial}{\partial z}\right)\frac{Pe^{-jk_0 r}}{4\pi\epsilon_0 r}.$$

When we expand this field in a power series in k_0 we obtain

$$\mathbf{E}_s = \frac{P}{4\pi\epsilon_0}\left[\left(\frac{2\mathbf{a}_r\cos\theta + \mathbf{a}_\theta\sin\theta}{r^3}\right) + \frac{k_0^2}{2r}(2\mathbf{a}_r\cos\theta - \mathbf{a}_\theta\sin\theta) - jk_0^3\frac{2}{3}\mathbf{a}_z + \cdots\right].$$

The first term is the static-like field whose tangential component cancels that of the applied field. The second term is a small correction of order $k_0^2 r^2$. The third term is also a small correction but it is the leading imaginary term. This term represents a uniform field that is not canceled by the applied field at the surface of the sphere. In order to cancel this term we can include it as part of the applied field in a self-consistent manner. Thus we determine the dipole moment using the expression

$$\mathbf{P} = \epsilon_0\bar{\alpha}_e\cdot\left[E_0\mathbf{a}_z - jk_0^3\frac{2}{3}\mathbf{a}_z\frac{P}{4\pi\epsilon_0}\right] = \epsilon_0\bar{\alpha}_e\cdot\mathbf{E}_0 - \frac{jk_0^3}{6\pi}\bar{\alpha}_e\cdot\mathbf{P}$$

where the polarizability tensor for the sphere is $\bar{\alpha}_e = 4\pi a^3\bar{\mathbf{I}}$. We call $-jk_0^3\mathbf{P}/6\pi\epsilon_0$ the radiation reaction field $\mathbf{E}_r$. When we solve the above for the dipole moment $\mathbf{P}$ we obtain a scattered field $\mathbf{E}_s$ with the property that the tangential component of the first term, i.e., $P\sin\theta/4\pi\epsilon_0 r^3$, plus that of the third-order term, namely, $jPk_0^3\sin\theta/6\pi\epsilon_0$, cancels that of the uniform applied field at $r = a$.

A time-harmonic dipole $\mathbf{P}$ is equivalent to a polarization current $j\omega\mathbf{P}$. The rate at which this current radiates energy is given by

$$-\frac{1}{2}\operatorname{Re}(j\omega\mathbf{P})^*\cdot\mathbf{E} = \frac{1}{2}\operatorname{Re}(j\omega\mathbf{P}^*\cdot\mathbf{E}_0 + j\omega\mathbf{P}^*\cdot\mathbf{E}_s).$$

The first term represents power extracted from the incident field and the second term represents the scattered power. For a lossless scatterer these two powers must balance. If we use the static solution for $\mathbf{P}$ then $\mathbf{P}$ and $\mathbf{E}_0$ are in phase and no power is extracted from the incident wave. The scattered power equals $\omega k_0^3\mathbf{P}\cdot\mathbf{P}^*/12\pi\epsilon_0$ so clearly we do not have power conservation.

However, with the corrected expression for $\mathbf{P}$ the power extracted from the incident wave is

$$\frac{1}{2} \operatorname{Re} \frac{j\omega(\epsilon_0 \alpha_e \mathbf{E}_0 \cdot \mathbf{E}_0^*)}{(1 - jk_0^3 2a^3/3)} = \frac{-\omega k_0^3}{12\pi\epsilon_0} \mathbf{P} \cdot \mathbf{P}^*$$

which clearly now balances the scattered power.

The induced magnetic dipole moment is also corrected in the same way; thus we use

$$\mathbf{M} = \bar{\alpha}_m \cdot (\mathbf{H}_0 + \mathbf{H}_r)$$

where $\mathbf{H}_r$ is the magnetic radiation reaction field. For the sphere problem this field is given by

$$H_r = -jk_0^3 \frac{M}{6\pi} = -jk_0^3 \frac{\alpha_m \cdot \mathbf{H}_0}{6\pi}.$$

In waveguide-coupling problems the leading imaginary terms in the scattered fields are the propagating modes that are excited. These are the only imaginary terms and are included as part of the polarizing fields in order to obtain corrected expressions for the aperture dipole moments. If we define the generator fields $\mathbf{E}_{g1}$ and $\mathbf{H}_{g1}$ to be the incident field for the input waveguide in the absence of the aperture, and similarly for $\mathbf{E}_{g2}$ and $\mathbf{H}_{g2}$, then the coupling problem is solved by specifying that the dipole strengths for radiation into guides 1 (input) and 2 (output) are

$$\mathbf{M}_0 = \mp \alpha_m \cdot (\mathbf{H}_{g1} - \mathbf{H}_{g2} + \mathbf{H}_{1r} - \mathbf{H}_{2r}) \tag{73a}$$

$$\mathbf{P}_0 = \mp \epsilon_0 \alpha_e \cdot (\mathbf{E}_{g1} - \mathbf{E}_{g2} + \mathbf{E}_{1r} - \mathbf{E}_{2r}) \tag{73b}$$

where the minus sign applies for radiation into guide 2 and the plus sign is used for guide 1. The radiation reaction fields $\mathbf{E}_{1r}$, $\mathbf{H}_{1r}$ are the dominant mode fields produced by dipoles $\mathbf{P}_0$ and $\mathbf{M}_0$ in guide 1 and evaluated at the center of the aperture. The radiation reaction fields $\mathbf{E}_{2r}$, $\mathbf{H}_{2r}$ are those produced by dipoles $-\mathbf{P}_0$, $-\mathbf{M}_0$ radiating in guide 2. The examples that follow will illustrate how this modified Bethe small-aperture theory is implemented in practice. The use of (73) also accounts correctly for the two waveguides being dissimilar in shape. The basis for the formulas given by (73) is as follows: The scattered fields on the two sides of the aperture can be expressed in terms of the tangential value of the unknown electric field $\mathbf{E}_a$ in the aperture by means of the formulas [Eq. (202a), Chapter 2]

$$\mathbf{E}_{s1} = \iint\limits_{S_a} \mathbf{n} \times \mathbf{E}_a \cdot \nabla \times \bar{\mathbf{G}}_{e1} \, dS$$

$$\mathbf{E}_{s2} = \iint\limits_{S_a} \mathbf{n}_0 \times \mathbf{E}_a \cdot \nabla \times \bar{\mathbf{G}}_{e2} \, dS$$

where the dyadic Green's functions satisfy the radiation conditions and the boundary conditions

$$\mathbf{n} \times \bar{\mathbf{G}}_{e1} = \mathbf{n} \times \bar{\mathbf{G}}_{e2} = 0$$

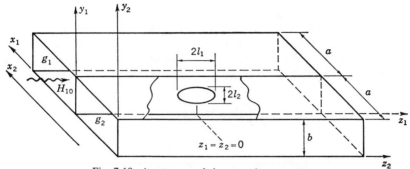

Fig. 7.18. Aperture-coupled rectangular waveguides.

on the waveguide walls and the aperture surface S_a. The boundary-value source $\mathbf{n} \times \mathbf{E}_a$ can be viewed as an equivalent magnetic current $-\mathbf{J}_m$ in the aperture. We can carry out a multipole expansion of the expressions for the scattered field, in the manner done for (56), and we then readily see that scattering into either guide is due to dipoles of equal strength but opposite sign. When we include the radiation reaction fields we are led to the symmetrical forms for the dipole strengths as given by (73). The polarizability tensors in (73) are equal to one-quarter of those for the complete two-sided aperture disk.

Aperture Coupling in Rectangular Guides

Consider an elliptic aperture in the common sidewall of two rectangular guides as in Fig. 7.18. The coordinates x_1, y_1, z_1 pertain to guide g_1, and x_2, y_2, z_2 to guide g_2. The center of the aperture is at $y_1 = b/2$, $z_1 = 0$. The major and minor semiaxes are l_1 and l_2. An H_{10} mode is incident from the left in guide g_1. We will assume that only the H_{10} mode propagates, and hence we will evaluate only the amplitudes of the scattered H_{10} modes. The normalized mode functions for the H_{10} mode are

$$\mathbf{E}_{10}^+ = \mathbf{e}_{10}e^{-\Gamma_{10}z}$$

$$\mathbf{E}_{10}^- = \mathbf{e}_{10}e^{\Gamma_{10}z}$$

$$\mathbf{H}_{10}^+ = (\mathbf{h}_{10} + \mathbf{h}_{z10})e^{-\Gamma_{10}z}$$

$$\mathbf{H}_{10}^- = (-\mathbf{h}_{10} + \mathbf{h}_{z10})e^{\Gamma_{10}z}$$

$$\mathbf{e}_{10} = -jk_0Z_0\left(\frac{2}{jabk_0Z_0\Gamma_{10}}\right)^{1/2}\sin\frac{\pi x}{a}\mathbf{a}_y$$

$$\mathbf{h}_{10} = \Gamma_{10}\left(\frac{2}{jabk_0Z_0\Gamma_{10}}\right)^{1/2}\sin\frac{\pi x}{a}\mathbf{a}_x$$

$$\mathbf{h}_{z10} = \left(\frac{2}{jabk_0Z_0\Gamma_{10}}\right)^{1/2}\frac{\pi}{a}\cos\frac{\pi x}{a}\mathbf{a}_z.$$

Let the incident mode be $A_1\mathbf{E}_{10}^+$, $A_1\mathbf{H}_{10}^+$, where A_1 is the amplitude constant. The incident mode does not have a component of electric field normal to the aperture, and so there will not be an induced electric dipole $\mathbf{P}_0$. The incident magnetic field has a tangential component along the major axis of the ellipse only, and so the induced magnetic dipole moment is in the z direction. In guide g_1 a dipole $M_0\mathbf{a}_z$ will produce a scattered field with amplitudes determined by (61). Thus in g_1

$$\mathbf{H}_{s1} = \frac{j\omega\mu_0}{2} \begin{cases} \mathbf{H}_{10}^- \cdot M_0\mathbf{a}_z\mathbf{H}_{10}^+, & z > 0 \\ \mathbf{H}_{10}^+ \cdot M_0\mathbf{a}_z\mathbf{H}_{10}^-, & z < 0. \end{cases}$$

This field has the following z component at the center of the aperture:

$$H_{s1,z} = \frac{\pi^2 M_0}{\Gamma_{10}a^3 b} = H_{r1}.$$

A dipole $-M_0\mathbf{a}_z$ radiating into guide g_2 produces a scattered field $\mathbf{H}_{s2}$ given by a similar expression. At the center of the aperture it has the value

$$H_{s2,z} = -\frac{\pi^2 M_0}{\Gamma_{10}a^3 b} = H_{r2}.$$

By using (73a) the dipole strength M_0 is found to be given by

$$M_0 = \frac{\pi l_1^3 e^2}{3[K(e) - E(e)]} \left[A_1 \frac{\pi}{a} \left(\frac{2}{jabk_0 Z_0 \Gamma_{10}} \right)^{1/2} + \frac{2\pi^2 M_0}{\Gamma_{10}a^3 b} \right]$$

or

$$M_0 = \frac{\alpha_m A_1 \dfrac{\pi}{a} \left(\dfrac{2}{jabk_0 Z_0 \Gamma_{10}} \right)^{1/2}}{1 - \alpha_m \dfrac{2\pi^2}{\Gamma_{10}a^3 b}}. \tag{74}$$

Let the scattered field in g_1 be

$$a_1\mathbf{E}_{10}^+, a_1\mathbf{H}_{10}^+, \qquad z_1 > 0$$

$$b_1\mathbf{E}_{10}^-, b_1\mathbf{H}_{10}^-, \qquad z_1 < 0.$$

Equations (61) are applicable in the present case, and hence

$$2b_1 = j\omega\mu_0\mathbf{H}_{10}^+ \cdot \mathbf{M}_0 = j\omega\mu_0\mathbf{h}_{z10} \cdot \mathbf{M}_0 \tag{75a}$$

$$2a_1 = j\omega\mu_0\mathbf{H}_{10}^- \cdot \mathbf{M}_0 = j\omega\mu_0\mathbf{h}_{z10} \cdot \mathbf{M}_0 = 2b_1. \tag{75b}$$

The total field in g_1 is the sum of the incident field plus the scattered field. If a reflection coefficient Γ is defined by the relation $b_1 = \Gamma A_1$, we find that

$$\Gamma = \frac{-j\alpha_m\pi^2/(\beta_{10}a^3 b)}{1 + 2j\alpha_m\pi^2/(\beta_{10}a^3 b)} \tag{76}$$

where $\Gamma_{10} = j\beta_{10}$ has been used and α_m is given by the first term in (70b).

In guide g_2 let the scattered field be

$$c_1 \mathbf{E}_{10}^+, c_1 \mathbf{H}_{10}^+, \qquad z_2 > 0$$

$$d_1 \mathbf{E}_{10}^-, d_1 \mathbf{H}_{10}^-, \qquad z_2 < 0.$$

If we had a current element $\mathbf{J}$ in g_2, an application of the Lorentz reciprocity theorem would give

$$2c_1 = -\iiint \mathbf{E}_{10}^- \cdot \mathbf{J} \, dV.$$

If $\mathbf{J}$ consists of a small linear element $\mathbf{J}_1$ and a circulating element $J_2 \boldsymbol{\tau}$, where $\boldsymbol{\tau}$ is a unit vector tangent to the contour around which J_2 flows, we have

$$2c_1 = -\int \mathbf{E}_{10}^- \cdot \mathbf{J}_1 \, dl - \oint J_2 \mathbf{E}_{10}^- \cdot \boldsymbol{\tau} \, dl.$$

The second integral is equal to

$$-J_2 \iint \nabla \times \mathbf{E}_{10}^- \cdot d\mathbf{S} = j\omega\mu_0 J_2 \iint \mathbf{H}_{10}^- \cdot d\mathbf{S}$$

or, for a very small loop,

$$j\omega\mu_0 \mathbf{H}_{10}^- \cdot \iint J_2 \, d\mathbf{S}.$$

This latter integral defines a magnetic dipole $\mathbf{M}$, and, since $\int \mathbf{J}_1 \, dl$ is equivalent to an electric dipole $j\omega\mathbf{P}$, we get

$$2c_1 = -j\omega \mathbf{E}_{10}^- \cdot \mathbf{P} + j\omega\mu_0 \mathbf{H}_{10}^- \cdot \mathbf{M}.$$

For radiation into g_2 the effective aperture dipoles are $-\mathbf{P}_0$, $-\mathbf{M}_0$ in general. In the present case $\mathbf{P}_0 = 0$, and so we have

$$2c_1 = -j\omega\mu_0 \mathbf{H}_{10}^- \cdot \mathbf{M}_0 = -j\omega\mu_0 \mathbf{h}_{z10} \cdot \mathbf{M}_0$$

$$2d_1 = -j\omega\mu_0 \mathbf{H}_{10}^+ \cdot \mathbf{M}_0 = 2c_1.$$

In the coordinate system used for g_2, we have, at $z_2 = 0$, $x_2 = a$,

$$\mathbf{h}_{z10} = -\left(\frac{2}{jabk_0 Z_0 \Gamma_{10}}\right)^{1/2} \frac{\pi}{a} \mathbf{a}_z$$

and hence $c_1 = d_1 = a_1 = b_1$. Note that the tangential magnetic fields in the scattered waves are of opposite sign at the aperture. This must be the case since their difference must give the nonzero value equal to the tangential component of the incident magnetic field.

The transmission coefficient into guide 2 for the dominant modes equals Γ. The transmission

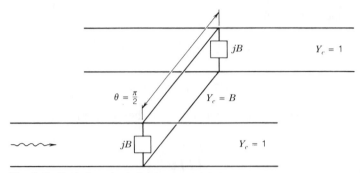

Fig. 7.19. Equivalent circuit for the two-coupled waveguides shown in Fig. 7.18.

coefficient into the output end of guide 1 equals $1 + \Gamma$. An equivalent circuit having these properties and the required fourfold symmetry is shown in Fig. 7.19. For this circuit the input admittance presented to guide 1 is

$$ Y_{\text{in}} = 1 + jB + \frac{B^2}{2 + jB} = \frac{2 + 3jB}{2 + jB}. $$

The reflection coefficient is given by

$$ \Gamma = \frac{1 - Y_{\text{in}}}{1 + Y_{\text{in}}} = \frac{-2jB}{4 + 4jB} = \frac{-j(B/2)}{1 + jB}. $$

When we compare this with (76) we find that the normalized susceptance B is given by

$$ B = \frac{2\alpha_m \pi^2}{\beta_{10} a^3 b} = \frac{2\pi^3 l_1^3 e^2}{3\beta_{10} a^3 b [K(e) - E(e)]} \tag{77} $$

where $e = (1 - l_2^2 / l_1^2)^{1/2}$. Since the equivalent circuit is a physically meaningful one it follows that power is conserved in the circuit. If the reaction fields had not been included then the reflection coefficient would have been given by the numerator in (76). A physical circuit having this form for the reflection coefficient and the transmission coefficients into guide g_2 does not exist. The equivalent circuit shows that there is a 90° phase angle associated with coupling from guide 1 to guide 2 when the aperture susceptance B is very small (strictly speaking, in the limit as B vanishes) because the coupling line has an electrical length θ equal to $\pi/2$.

A second straightforward example is the coupling of two rectangular waveguides by means of a small circular aperture in a common transverse wall as shown in Fig. 7.20. The input waveguide has dimensions $a \times b$ and the output waveguide $a' \times b'$. The incident generator fields in the input guide are

$$ \mathbf{E}_{1g} = A(\mathbf{E}_{10}^{+} - \mathbf{E}_{10}^{-}) $$

$$ \mathbf{H}_{1g} = A(\mathbf{H}_{10}^{+} - \mathbf{H}_{10}^{-}) $$

where the waveguide mode functions are those given earlier. The incident field has a nonzero x component of magnetic field so a magnetic dipole in the x direction is induced in the aperture. We can find the field radiated in the input guide by a dipole $M_0 \mathbf{a}_x$ by using image theory.

Fig. 7.20. Coupling of two rectangular waveguides by a circular aperture in a common transverse wall.

This gives a field which is the same as that radiated by a dipole $2M_0\mathbf{a}_x$ in the unbounded waveguide. The radiation reaction field for the input waveguide is

$$H_{1r} = j\omega\mu_0\mathbf{H}_{10}^+\cdot\mathbf{M}_0\mathbf{H}_{10}^-\cdot\mathbf{a}_x = -\frac{2j\beta_{10}}{ab}M_0.$$

For the output waveguide the radiation reaction field produced by a dipole $-M_0\mathbf{a}_x$ is

$$H_{2r} = j\omega\mu_0\mathbf{H}_{10}^-\cdot(-M_0\mathbf{a}_x)\mathbf{H}_{10}^+\cdot\mathbf{a}_x = \frac{2j\beta_{10}'}{a'b'}M_0$$

where β_{10}' is the dominant mode propagation constant for the output waveguide. Upon using (73a) we obtain

$$M_0 = \alpha_m\left[2A\Gamma_{10}\left(\frac{2}{jabk_0Z_0\Gamma_{10}}\right)^{1/2} - \frac{2j\beta_{10}}{ab}M_0 - \frac{2j\beta_{10}'}{a'b'}M_0\right]$$

which gives

$$M_0 = \frac{2A\alpha_m\Gamma_{10}\left(\dfrac{2}{jabk_0Z_0\Gamma_{10}}\right)^{1/2}}{1 + j\alpha_m\left(\dfrac{2\beta_{10}}{ab} + \dfrac{2\beta_{10}'}{a'b'}\right)}$$

where $\alpha_m = 4l^3/3$ and l is the radius of the aperture. The reflected wave in the input guide has an amplitude given by

$$-A + b_1 = j\omega\mu_0\mathbf{M}_0\cdot\mathbf{H}_{10}^+ - A.$$

The reflection coefficient Γ equals $(b_1 - A)/A$ and is

$$\Gamma = \frac{4\dfrac{j\alpha_m\beta_{10}}{ab}}{1 + 2j\alpha_m\left(\dfrac{\beta_{10}}{ab} + \dfrac{\beta_{10}'}{a'b'}\right)} - 1.$$

The aperture admittance presented to the input waveguide is

$$Y_{\text{in}} = \frac{1-\Gamma}{1+\Gamma} = \frac{\beta_{10}'ab}{\beta_{10}a'b'} - j\frac{ab}{2\beta_{10}\alpha_m}. \tag{78}$$

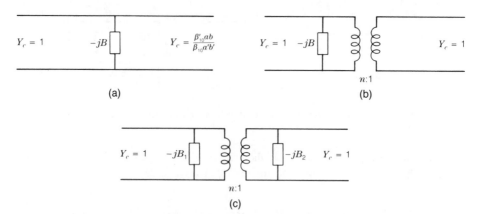

Fig. 7.21. (a) Equivalent circuit for the coupled waveguides shown in Fig. 7.20. (b) Susceptance B located on input side. (c) Susceptance B split into two components.

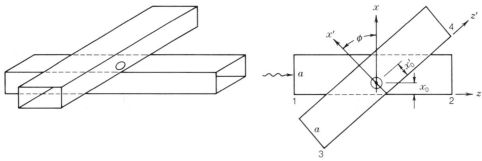

Fig. 7.22. Two waveguides coupled by an offset aperture in the broadwall.

The equivalent circuit consists of an output transmission line with normalized characteristic admittance equal to $(\beta'_{10}ab)/(\beta_{10}a'b')$ with an inductive susceptance of normalized value $ab/2\beta_{10}\alpha_m = 3ab/8\beta_{10}l^3$ in shunt with the junction. The slot susceptance can be placed on one side of the ideal transformer as shown in Fig. 7.21(b) or split into two parts, one on each side, as shown in Fig. 7.21(c). In the latter circuit $B_1 = B/2$ and $B_2 = B/2n^2$ where $n^2 = \beta_{10}a'b'/\beta'_{10}ab$.

A coupling problem that is considerably more complex is that between two crossed rectangular waveguides coupled by means of an offset circular aperture in the common broadwall as shown in Fig. 7.22. In the input waveguide the center of the aperture is at $x = x_0$, $z = 0$, while in the output waveguide the center is at $x' = x'_0$, $z' = 0$. The axes of the two waveguides are inclined at an angle ϕ. The incident generator fields will be chosen as

$$\mathbf{E}_{g1} = A_1 \mathbf{E}^+_{10} + A_2 \mathbf{E}^-_{10}, \quad \mathbf{H}_{g1} = A_1 \mathbf{H}^+_{10} + A_2 \mathbf{H}^-_{10}$$

$$\mathbf{E}_{g2} = A_3 \mathbf{E}^+_{10} + A_4 \mathbf{E}^-_{10}, \quad \mathbf{H}_{g2} = A_3 \mathbf{H}^+_{10} + A_4 \mathbf{H}^-_{10}.$$

A full complement of dipoles P_{0y}, M_{0x}, M_{0y} is induced in the aperture. In order to evaluate the radiation reaction fields at the center of the aperture we note that M_{0x} produces a field for which E_y and H_z have odd symmetry about $z = 0$ and hence are zero at the center of the aperture. The dipoles P_{0y}, M_{0z} produce a field H_x that is odd about $z = 0$ and hence is zero at the center of the aperture. The nonvanishing contributions to the radiation reaction field in the input waveguide are

$$\mathbf{E}_{1r} = \frac{j\omega\mu_0}{2} M_{0z}\mathbf{a}_z \cdot \mathbf{H}_{10}^- \mathbf{E}_{10}^+ - \frac{j\omega}{2}\mathbf{P}_0 \cdot \mathbf{E}_{10}^- \mathbf{E}_{10}^+$$

$$\mathbf{H}_{1r} = \frac{j\omega\mu_0}{2}[M_{0x}\mathbf{a}_x \cdot \mathbf{H}_{10}^- H_{10x}^+\mathbf{a}_x + M_{0z}\mathbf{a}_z \cdot \mathbf{H}_{10}^- H_{10z}^+\mathbf{a}_z] - \frac{j\omega}{2}\mathbf{P}_0 \cdot \mathbf{E}_{10}^- H_{10z}^+\mathbf{a}_z$$

or in component form

$$E_{1ry} = -\frac{P_{0y}}{\epsilon_0}\left[\frac{jk_0^2}{\beta_{10}ab}\sin^2\frac{\pi x_0}{a}\right] - M_{0z}\left[\frac{\pi k_0 Z_0}{\beta_{10}a^2 b}\sin\frac{\pi x_0}{a}\cos\frac{\pi x_0}{a}\right] \tag{79a}$$

$$H_{1rx} = -M_{0x}\frac{j\beta_{10}}{ab}\sin^2\frac{\pi x_0}{a} \tag{79b}$$

$$H_{1rz} = \frac{P_{0y}}{\epsilon_0}\left[\frac{\pi k_0 Y_0}{\beta_{10}a^2 b}\sin\frac{\pi x_0}{a}\cos\frac{\pi x_0}{a}\right] - M_{0z}\left[\frac{j\pi^2}{\beta_{10}a^3 b}\cos^2\frac{\pi x_0}{a}\right]. \tag{79c}$$

In the output waveguide the radiation reaction fields are given by similar expressions apart from a change in sign and replacement of x_0 by x_0'; thus

$$E_{2ry} = \frac{P_{0y}}{\epsilon_0}\left[\frac{jk_0^2}{\beta_{10}ab}\sin^2\frac{\pi x_0'}{a}\right] + M_{0z'}\left[\frac{\pi k_0 Z_0}{\beta_{10}a^2 b}\sin\frac{\pi x_0'}{a}\cos\frac{\pi x_0'}{a}\right] \tag{80a}$$

$$H_{2rx'} = M_{0x'}\frac{j\beta_{10}}{ab}\sin^2\frac{\pi x_0'}{a} \tag{80b}$$

$$H_{2rz'} = -\frac{P_{0y}}{\epsilon_0}\left[\frac{\pi k_0 Y_0}{\beta_{10}a^2 b}\sin\frac{\pi x_0'}{a}\cos\frac{\pi x_0'}{a}\right] + M_{0z'}\left[\frac{j\pi^2}{\beta_{10}a^3 b}\cos^2\frac{\pi x_0'}{a}\right]. \tag{80c}$$

In order to evaluate the dipole strengths using (73) we make use of the following relationships:

$$\mathbf{H}_{2r} = (H_{2rx'}\cos\phi + H_{2rz'}\sin\phi)\mathbf{a}_x + (H_{2rz'}\cos\phi - H_{2rx'}\sin\phi)\mathbf{a}_z$$

$$\mathbf{M}_0 = (M_{0x}\cos\phi - M_{0z}\sin\phi)\mathbf{a}_{x'} + (M_{0z}\cos\phi + M_{0x}\sin\phi)\mathbf{a}_{z'} = M_{0x'}\mathbf{a}_{x'} + M_{0z'}\mathbf{a}_{z'}.$$

The equations for the dipole strengths are

$$P_{0y} = \epsilon_0\alpha_e\left\{-jk_0 Z_0 N(A_1 + A_2)\sin\frac{\pi x_0}{a} + jk_0 Z_0 N(A_3 + A_4)\sin\frac{\pi x_0'}{a}\right.$$

$$- \frac{P_{0y}}{\epsilon_0}\frac{jk_0^2}{\beta_{10}ab}\left(\sin^2\frac{\pi x_0}{a} + \sin^2\frac{\pi x_0'}{a}\right) - \frac{\pi k_0 Z_0}{\beta_{10}a^2 b}$$

$$\left. \cdot \left[M_{0z}\sin\frac{\pi x_0}{a}\cos\frac{\pi x_0}{a} + (M_{0z}\cos\phi + M_{0x}\sin\phi)\sin\frac{\pi x_0'}{a}\cos\frac{\pi x_0'}{a}\right]\right\} \tag{81a}$$

$$M_{0x} = \alpha_m \left\{ j\beta_{10}N(A_1 - A_2)\sin\frac{\pi x_0}{a} - j\beta_{10}N(A_3 - A_4)\sin\frac{\pi x_0'}{a}\cos\phi \right.$$

$$- \frac{\pi}{a}N(A_3 + A_4)\cos\frac{\pi x_0'}{a}\sin\phi - \frac{j\beta_{10}}{ab}M_{0x}\sin^2\frac{\pi x_0}{a}$$

$$+ \frac{P_{0y}}{\epsilon_0}\frac{\pi k_0 Y_0}{\beta_{10}a^2 b}\sin\phi\sin\frac{\pi x_0'}{a}\cos\frac{\pi x_0'}{a} - M_{0x}$$

$$\cdot \left(\frac{j\beta_{10}}{ab}\cos^2\phi\sin^2\frac{\pi x_0'}{a} + \frac{j\pi^2}{\beta_{10}a^3 b}\sin^2\phi\cos^2\frac{\pi x_0'}{a} \right)$$

$$\left. + M_{0z}\sin\phi\cos\phi\left(\frac{j\beta_{10}}{ab}\sin^2\frac{\pi x_0'}{a} - \frac{j\pi^2}{\beta_{10}a^3 b}\cos^2\frac{\pi x_0'}{a} \right) \right\} \tag{81b}$$

$$M_{0z} = \alpha_m \left\{ \frac{\pi N}{a}(A_1 + A_2)\cos\frac{\pi x_0}{a} - \frac{\pi N}{a}(A_3 + A_4)\cos\phi\cos\frac{\pi x_0'}{a} \right.$$

$$+ j\beta_{10}N(A_3 - A_4)\sin\phi\sin\frac{\pi x_0'}{a} + \frac{P_{0y}}{\epsilon_0}\frac{\pi k_0 Y_0}{\beta_{10}a^2 b}$$

$$\cdot \left(\sin\frac{\pi x_0}{a}\cos\frac{\pi x_0}{a} + \cos\phi\sin\frac{\pi x_0'}{a}\cos\frac{\pi x_0'}{a} \right)$$

$$- M_{0x}\cos\phi\sin\phi\left(\frac{j\pi^2}{\beta_{10}a^3 b}\cos^2\frac{\pi x_0'}{a} - \frac{j\beta_{10}}{ab}\sin^2\frac{\pi x_0'}{a} \right)$$

$$- \frac{j\pi^2}{\beta_{10}a^3 b}M_{0z}\left(\cos^2\frac{\pi x_0}{a} + \cos^2\phi\cos^2\frac{\pi x_0'}{a} \right)$$

$$\left. - \frac{j\beta_{10}}{ab}M_{0z}\sin^2\phi\sin^2\frac{\pi x_0'}{a} \right\} \tag{81c}$$

where the normalization constant $N = (2/jabk_0 Z_0\Gamma_{10})^{1/2}$. Note that there is significant interaction between the dipoles.

If we call the amplitudes of the dominant mode scattered fields a_1, a_2, a_3, and a_4, then these are given by

$$a_1 = \frac{j\omega\mu_0}{2}\mathbf{M}_0\cdot\mathbf{H}_{10}^+ - \frac{j\omega}{2}\mathbf{P}_0\cdot\mathbf{E}_{10}^+ \tag{82a}$$

$$a_2 = \frac{j\omega\mu_0}{2}\mathbf{M}_0\cdot\mathbf{H}_{10}^- - \frac{j\omega}{2}\mathbf{P}_0\cdot\mathbf{E}_{10}^- \tag{82b}$$

$$a_3 = -\frac{j\omega\mu_0}{2}\mathbf{M}_0'\cdot\mathbf{H}_{10}^+ + \frac{j\omega}{2}\mathbf{P}_0\cdot\mathbf{E}_{10}^+ \tag{82c}$$

$$a_4 = -\frac{j\omega\mu_0}{2}\mathbf{M}_0'\cdot\mathbf{H}_{10}^- + \frac{j\omega}{2}\mathbf{P}_0\cdot\mathbf{E}_{10}^-. \tag{82d}$$

In the output waveguide $\mathbf{M}_0' = M_{0x'}\mathbf{a}_{x'} + M_{0z'}\mathbf{a}_{z'} = (M_{0x}\cos\phi - M_{0z}\sin\phi)\mathbf{a}_{x'} +$

$(M_{0z} \cos \phi + M_{0x} \sin \phi)\mathbf{a}_{z'}$. Also, the mode functions $\mathbf{E}_{10}^{\pm}$, $\mathbf{H}_{10}^{\pm}$ for the output waveguide are different but of the same form as those for the input waveguide.

We can identify the following equivalent incident and total scattered wave voltages at each of the four ports of the junction being analyzed:

$$V_1^+ = A_1, \qquad V_1^- = A_2 + a_1$$

$$V_2^+ = A_2, \qquad V_2^- = A_1 + a_2$$

$$V_3^+ = A_3, \qquad V_3^- = A_4 + a_3$$

$$V_4^+ = A_4, \qquad V_4^- = A_3 + a_4.$$

The system of equations (81) can be solved for the dipole strengths. We can solve (82) for the amplitudes a_i, $i = 1, \ldots, 4$ of the scattered waves. With this information the scattering-matrix elements S_{ij} for the four-port junction can be evaluated using the defined equivalent voltages and setting up the linear system

$$\begin{bmatrix} V_1^- \\ \vdots \\ V_4^- \end{bmatrix} = [S] \begin{bmatrix} V_1^+ \\ \vdots \\ V_4^+ \end{bmatrix}.$$

There is a considerable amount of algebra involved so we will limit the specific solutions to the special case $\phi = \pi/2$ and with the aperture centered for both waveguides so that $x_0 = x_0' = a/2$ and the case $\phi = 0$. In this instance the junction has fourfold symmetry and can be described by a relatively simple network.

We will assume that $A_3 = A_4 = 0$. The equations for the dipole strengths, with the assumption of a centered aperture, give

$$P_{0y} = \frac{-jk_0 Z_0 N \epsilon_0 \alpha_e (A_1 + A_2)}{1 + 2 \dfrac{jk_0^2 \alpha_e}{\beta_{10} ab}} \tag{83a}$$

$$M_{0x} = \frac{j\beta_{10} N \alpha_m (A_1 - A_2)}{1 + \dfrac{j\beta_{10}}{ab} \alpha_m}. \tag{83b}$$

The scattered wave in the input guide has the amplitude a_1 where

$$a_1 = \frac{j\omega\mu_0}{2} \Gamma_{10} N M_{0x} - \frac{\omega k_0 Z_0}{2} N P_{0y} = \frac{\dfrac{j\beta_{10}\alpha_m}{ab}(A_1 - A_2)}{1 + \dfrac{j\beta_{10}\alpha_m}{ab}} - \frac{\dfrac{jk_0^2\alpha_e}{\beta_{10}ab}(A_1 + A_2)}{1 + \dfrac{2jk_0^2\alpha_e}{\beta_{10}ab}}.$$

For even excitation $A_2 = A_1 = A$ and the reflection coefficient is

$$\Gamma_e = 1 + \frac{a_1}{A} = \frac{1}{1 + \dfrac{2jk_0^2\alpha_e}{\beta_{10}ab}}. \tag{84}$$

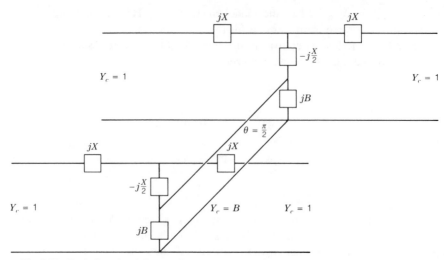

Fig. 7.23. Equivalent circuit for coupling between two orthogonal waveguides by a centered circular aperture in the broadwall.

For odd excitation $A_2 = -A_1 = -A$ and the reflection coefficient is

$$\Gamma_o = -1 + \frac{a_1}{A} = \frac{\dfrac{j\beta_{10}\alpha_m}{ab} - 1}{\dfrac{j\beta_{10}\alpha_m}{ab} + 1}. \tag{85}$$

The equivalent circuit that has the **required properties** is shown in Fig. 7.23. In this circuit the series reactance $X = \beta_{10}\alpha_m/ab$ and the shunt susceptance $B = 2k_0^2\alpha_e/\beta_{10}ab$. The reader can readily verify that with an open circuit at the midplane the circuit shown in Fig. 7.23 has an input admittance given by

$$Y_{in,e} = \frac{1 - \Gamma_e}{1 + \Gamma_e} = \frac{\dfrac{jk_0^2\alpha_e}{\beta_{10}ab}}{1 + \dfrac{jk_0^2\alpha_e}{\beta_{10}ab}} \tag{86a}$$

and with a short circuit at the **midplane**

$$Y_{in,o} = \frac{1 - \Gamma_o}{1 + \Gamma_o} = \frac{1}{j\beta_{10}\alpha_m/ab}. \tag{86b}$$

The amplitudes of the dominant **modes** coupled into the output waveguide are given by

$$a_3 = a_4 = \frac{\omega k_0 Z_0}{2} N P_{0y} = \frac{\dfrac{jk_0^2\alpha_e}{\beta_{10}ab}(A_1 + A_2)}{1 + 2j\dfrac{k_0^2\alpha_e}{\beta_{10}ab}}.$$

When the two waveguides are **aligned** ($\phi = 0$) but the aperture is not centered there is interaction between P_{0y} and M_{0z}. The reader can verify that for this case the solutions for the dipole strengths are ($A_3 = A_4 = 0$ is assumed)

$$\frac{P_{0y}}{\epsilon_0} = \frac{-jk_0Z_0NS\alpha_e}{\Delta}(A_1 + A_2)$$
(87a)

$$M_{0z} = \frac{\pi NC\alpha_m}{a\Delta}(A_1 + A_2)$$
(87b)

$$M_{0x} = \frac{j\alpha_m\beta_{10}NS(A_1 - A_2)}{1 + 2j\dfrac{\alpha_m\beta_{10}}{ab}S^2}$$
(87c)

where

$$S = \sin\frac{\pi x_0}{a}, \qquad C = \cos\frac{\pi x_0}{a}$$

and

$$\Delta = 1 + 2j\frac{\alpha_e k_0^2 S^2}{\beta_{10}ab} + 2j\frac{\alpha_m\pi^2 C^2}{\beta_{10}a^3 b}.$$

By considering the two special cases of even excitation with $A_1 = A_2 = A$ and odd excitation with $A_2 = -A_1 = -A$ the even and odd reflection coefficients are found to be

$$\Gamma_e = \frac{1}{1 + jB}$$
(88a)

$$\Gamma_o = \frac{-1}{1 + jX}$$
(88b)

where

$$B = \frac{2k_0^2\alpha_e S^2}{\beta_{10}ab} + \frac{2\pi^2\alpha_m C^2}{\beta_{10}a^3 b}$$
(89a)

$$X = \frac{2\beta_{10}\alpha_m S^2}{ab}.$$
(89b)

The equivalent circuits for even and odd excitation are shown in Fig. 7.24. For a wave incident at port 1 only, i.e., $A_2 = A_3 = A_4 = 0$, the amplitudes of the waves coupled into ports 3 and 4 are

$$a_3 = \left(\frac{-jX/2}{1 + jX} + \frac{jB/2}{1 + jB}\right)A_1$$
(90a)

$$a_4 = \left(\frac{jX/2}{1 + jX} + \frac{jB/2}{1 + jB}\right)A_1.$$
(90b)

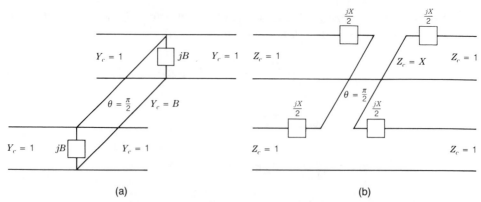

(a)   (b)

Fig. 7.24. (a) Equivalent circuit for two parallel waveguides coupled by an offset circular aperture in the common broadwall for even excitation. (b) Equivalent circuit for odd excitation.

The dipoles P_{0y} and M_{0z} radiate symmetrically into the upper waveguide, while the dipole M_{0x} radiates an antisymmetrical field. If we choose the aperture position such that $X = B$ then there will be zero transmission into port 3. The junction now has the properties of a directional coupler (Bethe hole coupler). The required position of the circular aperture is given by

$$\sin \frac{\pi x_0}{a} = \frac{\lambda_0}{\sqrt{6a}}. \tag{91}$$

The field at port 4 could be canceled by choosing

$$\sin \frac{\pi x_0}{a} = \frac{\lambda_0}{\sqrt{2(\lambda_0^2 - a^2)}} \tag{92}$$

if the radiation reaction field could be neglected. This equation has a solution only when $\lambda_0 \geq \sqrt{2}a$.

When the radiation reaction field is included and we make $X = -B$, i.e., impose the condition given by (92), then

$$a_3 = \frac{jB}{1 + B^2} A_1$$

$$a_4 = \frac{B^2}{1 + B^2} A_1.$$

The coupling coefficient C is thus given by

$$C = 20 \log \left| \frac{A_1}{a_3} \right| = 20 \log \frac{1 + B^2}{B} \approx 20 \log \frac{1}{B} \, dB.$$

The directivity is given by

$$D = 20 \log \left| \frac{a_3}{a_4} \right| = 20 \log \frac{1}{B} \approx C \, dB.$$

It has generally been thought that a_4 could be made equal to zero, but the more general aperture coupling theory that includes the reaction fields shows that this is not possible. The more complete theory shows that only a finite directivity can be obtained.

A directional coupler can also be obtained by using a centered circular aperture and non-aligned waveguides (see Problem 7.10). For this type of coupler the theoretical directivity is also somewhat less than the coupling. Experimentally it is found that the attenuation caused by the finite thickness of the waveguide wall in which the aperture is located reduces the directivity below the theoretical value [7.36].

7.4. Cavity Coupling by Small Apertures

It is common practice to couple cavities to waveguides by means of small apertures. The small-aperture coupling theory presented in the previous section applies equally well to cavity coupling when the reaction fields coming from the cavity are identified to be the resonant mode fields in the cavity. We will illustrate the theory by considering two examples.

End Excited Cavity

A rectangular cavity is formed by placing a transverse wall with a small centered circular aperture into a rectangular waveguide, a distance d from the short-circuited end as shown in Fig. 7.25. The cavity is assumed to be resonant for the TE_{101} mode. The normalized TE_{101} cavity mode fields are

$$\mathbf{E}_{101} = \left(\frac{4}{abd}\right)^{1/2} \sin\frac{\pi x}{a} \sin\frac{\pi z}{d}\mathbf{a}_y$$

$$\mathbf{H}_{101} = \frac{1}{k_{101}}\nabla \times \mathbf{E}_{101} = \left(\frac{4}{abd}\right)^{1/2} k_{101}^{-1}\left[-\frac{\pi}{d}\sin\frac{\pi x}{a}\cos\frac{\pi z}{d}\mathbf{a}_x + \frac{\pi}{a}\cos\frac{\pi x}{a}\sin\frac{\pi z}{d}\mathbf{a}_z\right].$$

A magnetic dipole $-M_0\mathbf{a}_x$ will excite this mode with an amplitude given by (173) in Chapter

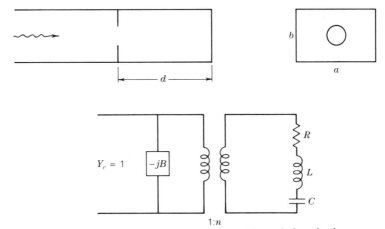

Fig. 7.25. End excited rectangular cavity and its equivalent circuit.

5; thus

$$h_{101} = -\frac{k_0^2 M_0 \mathbf{a}_x \cdot \mathbf{H}_{101}}{k_{101}^2 - k_0^2 \left(1 + \dfrac{1-j}{Q}\right)} = \frac{\left(\dfrac{4}{abd}\right)^{1/2} k_0^2 M_0 (\pi/k_{101} d)}{k_{101}^2 - k_0^2 \left(1 + \dfrac{1-j}{Q}\right)}$$

where $k_{101}^2 = (\pi/a)^2 + (\pi/d)^2$ and Q is the quality factor for the mode. The x component of the magnetic field at the center of the aperture is

$$-h_{101} \left(\frac{4}{abd}\right)^{1/2} \frac{\pi}{k_{101} d}.$$

This field is taken as the reaction field because it is large when the cavity is tuned to resonate for the TE_{101} mode. In the rectangular waveguide the radiation reaction field is the same as it is for the aperture in a transverse wall, which was analyzed in the previous section. Thus the equation for M_0 is

$$M_0 = \alpha_m \left\{ 2Aj\beta_{10}N - \frac{2j\beta_{10}}{ab} M_0 + \frac{4k_0^2 \pi^2 M_0}{k_{101}^2 abd^3 \left[k_{101}^2 - k_0^2 \left(1 + \dfrac{1-j}{Q}\right)\right]} \right\}$$

which gives

$$M_0 = \frac{2\alpha_m j\beta_{10}NA}{1 + jX + W}$$

where $X = 2\alpha_m \beta_{10}/ab$,

$$W = -\frac{4\alpha_m k_0^2 \pi^2}{k_{101}^2 abd^3 \left[k_{101}^2 - k_0^2 \left(1 + \dfrac{1-j}{Q}\right)\right]}$$

and A is the amplitude of the incident TE_{10} mode.

The total reflected TE_{10} mode amplitude is

$$-A + j\omega\mu_0 M_0 \mathbf{a}_x \cdot \mathbf{H}_{10}^+ = -A + \frac{4j\alpha_m \beta_{10}A/ab}{1 + jX + W}.$$

The reflection coefficient is obtained by dividing by A. The input admittance is

$$Y_{\text{in}} = \frac{1-\Gamma}{1+\Gamma} = \frac{1+W}{jX} = -jB + Y \tag{93a}$$

where

$$B = \frac{ab}{2\alpha_m \beta_{10}} \tag{93b}$$

$$Y = \frac{j2k_0^2\pi^2}{\beta_{10}k_{101}^2d^3\left[k_{101}^2 - k_0^2\left(1 + \frac{1-j}{Q}\right)\right]}. \tag{93c}$$

In the equivalent circuit shown in Fig. 7.25 the inductance L is the sum of that for the cavity, L_c, plus L_s arising from the surface impedance of the cavity walls. The input admittance is

$$Y_{in} = \frac{n^2}{j\omega L_c + j\omega L_s + R - j/\omega C} - jB.$$

We can express the denominator in the form

$$\frac{j}{\omega\omega_0^2 C}(\omega^2 + \omega^2\omega_0^2 L_s C - \omega_0^2 - j\omega\omega_0^2 RC)$$

where $\omega_0 = (L_c C)^{-1/2}$ is the loss-free cavity resonant frequency. We now take note of the relations

$$\frac{R}{\omega_0 L_c} = \omega_0 RC = \frac{1}{Q} = \frac{\omega_0 L_s}{\omega_0 L_c} = \omega_0^2 L_s C$$

and use the approximation $\omega\omega_0^2 RC \approx \omega^2/Q$ to obtain

$$Y_{in} = -jB + \frac{j\omega\omega_0^2 C n^2}{\omega_0^2 - \omega^2\left(1 + \frac{1-j}{Q}\right)}. \tag{94}$$

In order to determine the parameter n^2 we need to specify the impedance level of the cavity. We will do this by considering the line integral of the electric field across the center of the cavity as the equivalent voltage V_c across the capacitor C in the equivalent circuit. The electric field is given by

$$E_y = \frac{V_c}{b}\sin\frac{\pi x}{a}\sin\frac{\pi z}{d}.$$

The stored electric energy is

$$W_e = \frac{\epsilon_0}{4}\int_0^a\int_0^b\int_0^d E_y^2\,dz\,dy\,dx = \frac{\epsilon_0 ad}{16b}V_c^2 = \frac{1}{4}CV_c^2$$

and hence $C = \epsilon_0 ad/4b$. We now obtain

$$Y_{in} = -jB + \frac{jk_0 Y_0\omega_0^2 ad n^2}{4b\left[\omega_0^2 - \omega^2\left(1 + \frac{1-j}{Q}\right)\right]} \tag{95}$$

upon using $\omega\epsilon_0 = k_0 Y_0$. When we compare this expression with Y given by (93c) we find that

$$n^2 = \frac{8\pi^2 b Z_0}{k_0\beta_{10}k_{101}^2 ad^4}. \tag{96}$$

We will define the resonant frequency of the coupled cavity as the frequency at which Y_{in} is real. From (93) we find that the imaginary part of Y_{in} vanishes when

$$\left(\frac{1+Q}{Q}k_0^2 - k_{101}^2\right)\left[B\beta_{10}k_{101}^4 d^3 - k_0^2\left(2\pi^2 + B\beta_{10}k_{101}^2 d^3 \frac{1+Q}{Q}\right)\right] = \frac{\beta_{101}k_{101}^4 d^3 B}{Q^2}. \quad (97)$$

For a high-Q cavity we can set the term on the right equal to zero; thus

$$k_0^2 = \frac{k_{101}^2}{1 + \frac{2\pi^2}{B\beta_{10}k_{101}^2 d^3}} \approx k_{101}^2 - \frac{2\pi^2}{B\beta_{10}d^3} = k_{101}^2 - \frac{4\pi^2\alpha_m}{abd^3}.$$

We now use $k_0^2 - k_{101}^2 \approx 2k_{101}(k_0 - k_{101})$ to obtain

$$k_0 \approx k_{101} - \frac{2\pi^2\alpha_m}{k_{101}abd^3}. \quad (98)$$

In order to obtain critical coupling we require $Y_{in} = 1$ at resonance. From (93a) we see that this requires Y to have a unit conductance. This condition gives

$$\frac{2k_0^4\pi^2}{\beta_{10}k_{101}^4 d^3 Q} = \left(k_{101}^2 - k_0^2\frac{1+Q}{Q}\right)^2 + \frac{k_0^4}{Q^2} \approx \left(\frac{4\pi^2\alpha_m}{abd^3}\right)^2 + \frac{k_0^4}{Q^2} \approx \left(\frac{4\pi^2\alpha_m}{abd^3}\right)^2$$

which can be solved for α_m to yield

$$\alpha_m = \frac{k_0^2 abd}{\pi k_{101}\sqrt{8\beta_{10}Q/d}} = \frac{4}{3}l^3. \quad (99)$$

We can use $k_0 = k_{101}$ to get the first approximation to α_m and then use (98) to calculate a corrected value for k_0 to use in (99).

As a typical example consider a cavity with $d = a = 2.2$ cm, $b = 1$ cm, and having a Q of 6000. For this cavity $k_{101} = 202$. We now use $k_0 \approx k_{101}$, $\beta_{10} = [k_0^2 - (\pi/a)^2]^{1/2} \approx [k_{101}^2 - (\pi/a)^2]^{1/2} = \pi/d$ and from (99) obtain $\alpha_m \approx 17.63 \times 10^{-9}$. The corrected value for k_0 is $202 - 0.735$ which is not a large enough change to warrant calculating a corrected value for α_m. The required aperture radius is $l = 0.236$ cm. The reader can readily verify that the approximations made to obtain (98) and (99) are fully justified since the detuning effect of the aperture susceptance $-jB$ is very small.

Two-Port Cavity

The two-port cavity shown in Fig. 7.26 has the property that maximum power is transmitted through the cavity at resonance. We will assume that the cavity resonates in the TE$_{101}$ mode. In the apertures x-directed magnetic dipoles are induced. At the center of the input aperture the incident TE$_{10}$ mode produces the following x-directed generator field H_{g1}:

$$H_{g1} = 2j\beta_{10}NA$$

where A is the amplitude of the incident mode. The resonant frequency of the cavity is given

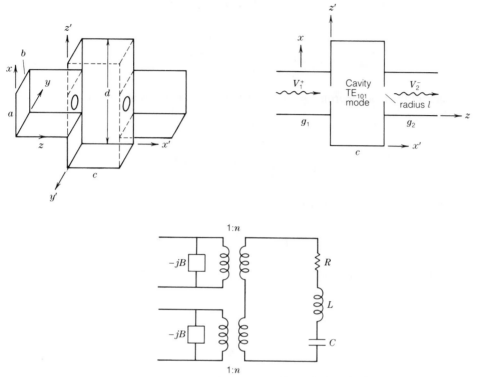

Fig. 7.26. A two-port cavity system and its equivalent circuit.

by

$$\omega_0 = \omega_{101} = k_{101}(\mu_0\epsilon_0)^{-1/2}$$

where

$$k_{101} = [(\pi/c)^2 + (\pi/d)^2]^{1/2}.$$

The cavity field will excite a magnetic dipole in the output aperture which will cause some power to be radiated into the output guide g_2. In g_1 and g_2 a dipole $M_0\mathbf{a}_x$ radiates a TE_{10} mode with H_x at the center of the aperture given by $H_{1rx} = H_{2rx} = -2j\beta_{10}M_0/ab$. In order to find the reaction field from the cavity consider dipoles $M_{z'1}$ and $M_{z'2}$ in the input and output apertures. These produce a TE_{101} mode with amplitude h_{101} for the $\mathbf{H}_{101}$ mode field where

$$\mathbf{H}_{101} = \sqrt{\frac{4}{bcd}}k_{101}^{-1}\left[-\frac{\pi}{d}\sin\frac{\pi x'}{c}\cos\frac{\pi z'}{d}\mathbf{a}_{x'} + \frac{\pi}{c}\cos\frac{\pi x'}{c}\sin\frac{\pi z'}{d}\mathbf{a}_{z'}\right]$$

$$h_{101} = \frac{k_0^2}{k_{101}^2 - k_0^2\left(1 + \frac{1-j}{Q}\right)}\sqrt{\frac{4}{bcd}}\frac{\pi}{k_{101}c}(M_{z'1} - M_{z'2}).$$

Note that $k_{101}^2 = \omega_0^2 k_0^2/\omega^2$, $\omega_0 = \omega_{101}$. The field $H_{z'}$ is given by

$$H_{z'} = H_{z'r} = \frac{\omega^2}{\omega_0^2 - \omega^2 \left(1 + \dfrac{1-j}{Q}\right)} \frac{4}{bcd} \left(\frac{\pi}{k_{101}c}\right)^2 (M_{z'1} - M_{z'2})$$

at the center of the input aperture and the negative of this at the center of the output aperture. Let the factor W be

$$W = \frac{\omega^2}{\omega_0^2 - \omega^2 \left(1 + \dfrac{1-j}{Q}\right)} \frac{4\pi^2}{bcd(k_{101}c)^2}. \tag{100}$$

We can also write $M_{z'1} = M_{x1}$, $M_{z'2} = M_{x2}$. For the input aperture we now have

$$M_0 = M_{x1} = \alpha_m \left(H_{g1} - \frac{2j\beta_{10}}{ab} M_{x1} + W M_{x1} - W M_{x2}\right) \tag{101a}$$

while for the output aperture we have

$$M_{x2} = \alpha_m \left(-\frac{2j\beta_{10}}{ab} M_{x2} + W M_{x2} - W M_{x1}\right) \tag{101b}$$

since H_{g2} is zero. We can solve the second equation for M_{x2} and the first equation then gives

$$M_{x1} \left[1 + \alpha_m \left(\frac{2j\beta_{10}}{ab} - W\right) - \frac{\alpha_m^2 W^2}{1 + \alpha_m \left(\dfrac{2j\beta_{10}}{ab} - W\right)}\right] = \alpha_m H_{g1}. \tag{102}$$

The total reflected field in g_1 is $-A + a_1 = -A - k_0 Z_0 \beta_{10} N M_{x1}$, and $\Gamma = -1 + a_1/A$ is given by

$$\Gamma = -1 + \frac{2jX(1 + \alpha_m U)}{(1 + \alpha_m U)^2 - \alpha_m^2 W^2}$$

where

$$X = \frac{2\alpha_m \beta_{10}}{ab}, \qquad U = \frac{2j\beta_{10}}{ab} W.$$

We now find that

$$Y_{in} = \frac{1 - \Gamma}{1 + \Gamma} = \frac{1}{jX} - \frac{\alpha_m W}{jX \left(1 - \dfrac{\alpha_m W}{1 + jX}\right)}. \tag{103}$$

Let $K = (4\alpha_m/bcd)(\pi/k_{101}c)^2$; then we get

$$Y_{in} = \frac{1}{jX} - \frac{K\omega^2}{jX \left[\omega_0^2 - \omega^2 \left(1 + \dfrac{1-j}{Q}\right)\right] - \dfrac{\omega^2 K}{1 + (1/jX)}}. \tag{104}$$

The equivalent circuit for the two-port cavity is shown in Fig. 7.26. For this circuit

$$Y_{in} = -jB + \cfrac{n^2}{j\omega L + \cfrac{1}{j\omega C} + R + \cfrac{n^2}{1 - jB}}$$

$$= -jB + \cfrac{n^2\omega^2}{j\omega L\left(\omega^2 - \omega_0^2 - j\cfrac{R}{\omega L}\omega^2\right) + \cfrac{n^2\omega^2}{1 - jB}} \qquad (105)$$

where $\omega_0^2 = 1/LC$. Now $\omega L/R = Q$ so we have

$$Y_{in} = -jB - \cfrac{n^2\omega^2}{j\omega L\left(\omega_0^2 - \omega^2 + \omega^2\cfrac{j}{Q}\right) - \cfrac{n^2\omega^2}{1 - jB}}. \qquad (106)$$

Hence we must choose

$$\frac{1}{jX} = -jB$$

so $B = ab/2\beta_{10}\alpha_m$. In addition,

$$\frac{n^2}{\omega L} = \frac{K}{X} = \frac{2ab}{\beta_{10}bcd}\left(\frac{\pi}{k_{101}c}\right)^2.$$

The input admittance Y_{in} for the cavity has the factor $\omega_0^2 - \omega^2((1 - j)/Q) - \omega^2/Q$. The last term comes from the additional inductance due to the surface impedance $(1 + j)/\sigma\delta_s$. This can be included in the equivalent circuit as a series inductance $j\omega L_s = R$. Then $(R + j\omega L_s)/j\omega L$ becomes $(1 + j)/Q$ and Y_{in} becomes

$$Y_{in} = -jB - \cfrac{n^2\omega^2}{j\omega L\left(\omega_0^2 - \omega^2 + \cfrac{\omega^2 j}{Q} - \cfrac{\omega^2}{Q}\right) - \cfrac{n^2\omega^2}{1 - jB}} \qquad (107)$$

which is the same function of ω as occurs in (104). We need to add L_s in series with L and then the new resonant frequency is

$$\omega_0' = \frac{1}{\sqrt{C(L + L_s)}} = \sqrt{\frac{Q}{1 + Q}}\omega_0.$$

For convenience, we assume that L_s is included in L.

The field radiated into g_2 is given by

$$a_2 = \beta_{10}k_0Z_0NM_{x2}.$$

Now $M_{x2} = -\alpha_m W M_{x1}/[1 + \alpha_m(2j\beta_{10}/ab - W)]$ and since M_{x1} radiates a wave with amplitude $-\beta_{10}k_0Z_0NM_{x1} = (1 + \Gamma)A$ we see that

$$a_2 = \frac{\alpha_m W}{1 + \alpha_m \left(\dfrac{2j\beta_{10}}{ab} - W\right)}(1 + \Gamma)A.$$

The factor $A(1 + \Gamma)$ is the input voltage in the equivalent circuit. This appears as a series voltage $A(1 + \Gamma)n$ in the resonator circuit. The total loop impedance is

$$j\omega(L + L_s) + \frac{1}{j\omega C} + R + \frac{n^2}{1 - jB} = Z_L.$$

A fraction

$$\left(\frac{-n^2}{1 - jB}\right)\frac{1}{Z_L}$$

appears across the output transformer and produces an output voltage $1/n$ smaller. Thus

$$a_2 = \frac{-n(1 + \Gamma)An}{(1 - jB)Z_L} = \frac{-n^2 A(1 + \Gamma)}{(1 - jB)Z_L}.$$

This is the same result obtained by the small-aperture coupling theory.

The parameter $n^2/\omega L$ occurs as a single parameter. n^2 is arbitrary which corresponds to the arbitrary choice of the impedance level $\sqrt{L/C}$ of the equivalent resonant circuit representing the cavity. A useful choice in practice is to make the voltage V_c across the capacitor C correspond to the line integral of the electric field across the cavity at the center. We then find from energy considerations that

$$C = \frac{\epsilon_0 cd}{4b}.$$

The equivalent inductance is given by the resonance condition

$$L = \frac{1}{\omega_0^2 C}$$

when the surface impedance term L_s is excluded. The parameter n^2 is given by

$$n^2 = \frac{2a}{\beta_{10}cd}\left(\frac{\pi}{k_{101}c}\right)^2 \omega_0 L = \frac{8ab}{\beta_{10}k_0c^2d^2}\left(\frac{\pi}{k_{101}c}\right)^2 Z_0 \qquad (108)$$

which is of the same form as (96).

The two-port cavity represents a lossy network that couples the input and output waveguides together. Consequently, it is not possible to obtain complete power transmission through the cavity. The output waveguide presents a resistive loading of the cavity. This loading should be large relative to the cavity losses so that most of the power will be transmitted to the output waveguide. We note that (105) can be expressed as

$$Y_{\text{in}} = -jB + \frac{n^2}{j\omega L + \dfrac{1}{j\omega C} + \dfrac{jn^2 B}{1 + B^2} + R + \dfrac{n^2}{1 + B^2}} \cdot$$

Since normally B is very large the effective series loading resistance R_e equals n^2/B^2. If we make $n^2/B^2 \gg R$ then only a small fraction of the input power is dissipated in the cavity. The loading resistance is given by

$$R_e = \frac{n^2}{B^2} = \frac{8\alpha_m^2 \beta_{10} Z_0}{k_0 ab(cd)^2} \left(\frac{\pi}{k_{101} c}\right)^2. \tag{109}$$

The external Q_e associated with this resistance is

$$Q_e = \frac{B^2}{n^2 \omega C} = \frac{acdb^2}{8\alpha_m^2 \beta_{10}} \left(\frac{k_{101} c}{\pi}\right)^2. \tag{110}$$

In order to illustrate typical values that can be obtained consider a waveguide with $a = 2.3$ cm, $b = 1$ cm, $c = 2.5$ cm, $d = 3$ cm, and an aperture radius $l = 0.25$ cm. For this case $Q_e = 9353$. It is quite clear that it will be difficult to achieve a high degree of loading unless a large aperture is used. If we increase l to 0.35 cm then $Q_e = 1242$ which would normally be significantly smaller than the unloaded Q of the cavity. However, an aperture radius of this size is large enough to raise questions as to the validity of the small-aperture theory.

7.5. General Remarks on Aperture Coupling

If the medium in g_1 has electrical constitutive parameters ϵ_1 and μ_1 and that in guide g_2 has ϵ_2, μ_2 then the dipole strengths for radiation into g_1 are $\mathbf{P}_1$, $\mathbf{M}_1$, where [7.30]

$$(\epsilon_1 + \epsilon_2)\mathbf{P}_1 = 2\epsilon_1 \bar{\alpha}_e \cdot [\epsilon_1 \mathbf{E}_{g1} - \epsilon_2 \mathbf{E}_{g2} + \epsilon_1(\bar{\mathbf{A}}_1 \cdot \mathbf{P}_1 + \bar{\mathbf{B}}_1 \cdot \mathbf{M}_1)$$

$$- \epsilon_2(\bar{\mathbf{A}}_2 \cdot \mathbf{P}_2 + \bar{\mathbf{B}}_2 \cdot \mathbf{M}_2)] \tag{111a}$$

$$(\mu_1 + \mu_2)\mathbf{M}_1 = 2\mu_2 \bar{\alpha}_m \cdot \left[\mathbf{H}_{g1} - \mathbf{H}_{g2} + \left(\bar{\mathbf{C}}_1 + \frac{\epsilon_2}{\epsilon_1} \bar{\mathbf{C}}_2\right) \cdot \mathbf{P}_1\right.$$

$$\left. - \left(\bar{\mathbf{D}}_1 + \frac{\mu_1}{\mu_2} \bar{\mathbf{D}}_2\right) \cdot \mathbf{M}_1\right]. \tag{111b}$$

For radiation into guide g_2 the dipole strengths are $-\mathbf{P}_2$ and $-\mathbf{M}_2$ where

$$\epsilon_1 \mathbf{P}_2 = \epsilon_2 \mathbf{P}_1 \tag{111c}$$

$$\mu_2 \mathbf{M}_2 = \mu_1 \mathbf{M}_1. \tag{111d}$$

The tensors $\bar{\mathbf{A}}_1$, $\bar{\mathbf{B}}_1$, $\bar{\mathbf{C}}_1$, and $\bar{\mathbf{D}}_1$ are defined by expressing the radiation reaction fields in g_1 due to $\mathbf{P}_1$ and $\mathbf{M}_1$ in the form

$$\mathbf{E}_{1r} = \bar{\mathbf{A}}_1 \cdot \mathbf{P}_1 + \bar{\mathbf{B}}_1 \cdot \mathbf{M}_1 \tag{112a}$$

$$\mathbf{H}_{1r} = \bar{\mathbf{C}}_1 \cdot \mathbf{P}_1 + \bar{\mathbf{D}}_1 \cdot \mathbf{M}_1. \tag{112b}$$

The tensors $\bar{\mathbf{A}}_2$, $\bar{\mathbf{B}}_2$, $\bar{\mathbf{C}}_2$, and $\bar{\mathbf{D}}_2$ are defined so that the radiation reaction fields in g_2 due to $-\mathbf{P}_2$ and $-\mathbf{M}_2$ are given by

$$\mathbf{E}_{2r} = -\bar{\mathbf{A}}_2 \cdot \mathbf{P}_2 - \bar{\mathbf{B}}_2 \cdot \mathbf{M}_2 \tag{112c}$$

$$\mathbf{H}_{2r} = -\bar{\mathbf{C}}_2 \cdot \mathbf{P}_2 - \bar{\mathbf{D}}_2 \cdot \mathbf{M}_2. \tag{112d}$$

The small-aperture theory does not take into account the reduction of the coupling due to the finite thickness of the wall that the aperture is located in. This reduction in coupling can be as large as 1 or 2 dB in typical waveguide walls. Some theoretical results for the effects of finite wall thickness may be found in the papers by McDonald [7.33] and Leviatan *et al.* [7.32].

The problem of coupling through an aperture in a thick wall may be formulated in an exact way. With reference to Fig. 7.27 consider an aperture in a wall of thickness t. Let $\bar{\mathbf{G}}_{e1}$ and $\bar{\mathbf{G}}_{e2}$ be electric-type dyadic Green's functions for the input and output guides. These satisfy the radiation conditions and the boundary conditions $\mathbf{n} \times \bar{\mathbf{G}}_{e1} = \mathbf{n} \times \bar{\mathbf{G}}_{e2} = 0$ on the waveguide walls and the aperture surfaces S_1 and S_2. Let the unknown tangential electric fields on S_1 and S_2 be $\mathbf{E}_1$ and $\mathbf{E}_2$. The equivalent magnetic currents are

$$\mathbf{J}_{m1} = -\mathbf{n}_1 \times \mathbf{E}_1, \qquad \mathbf{J}_{m2} = -\mathbf{n}_2 \times \mathbf{E}_2.$$

The scattered fields in g_1 and g_2 are given by [Eq. (202a), Chapter 2]

$$\mathbf{E}_{s1} = -\iint_{S_1} \mathbf{J}_{m1} \cdot \nabla \times \bar{\mathbf{G}}_{e1} \, dS \tag{113a}$$

$$\mathbf{E}_{s2} = -\iint_{S_2} \mathbf{J}_{m2} \cdot \nabla \times \bar{\mathbf{G}}_{e2} \, dS. \tag{113b}$$

We also need an equation that connects $\mathbf{J}_{m1}$ and $\mathbf{J}_{m2}$. For this purpose we introduce a dyadic Green's function $\bar{\mathbf{G}}_e$ for the cavity formed by the two surfaces S_1, S_2 and the sidewalls of the aperture opening. Inside the cavity the scattered field is given by

$$\mathbf{E}_s = \iint_{S_1} \mathbf{J}_{m1} \cdot \nabla \times \bar{\mathbf{G}}_e \, dS + \iint_{S_2} \mathbf{J}_{m2} \cdot \nabla \times \bar{\mathbf{G}}_e \, dS. \tag{114}$$

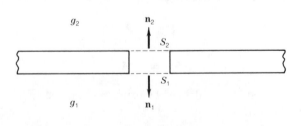

Fig. 7.27. Aperture in a thick wall.

Equations (113) and (114) will ensure the continuity of the tangential electric field through the aperture opening. The equivalent magnetic currents must be determined so that the tangential magnetic field is also continuous through the aperture. This requirement can be stated in the form

$$\mathbf{n} \times (\mathbf{H}_{g1} + \mathbf{H}_{s1}) = \mathbf{n} \times \mathbf{H}_s \qquad \text{on } S_1 \tag{115a}$$

$$\mathbf{n} \times (\mathbf{H}_{g2} + \mathbf{H}_{s2}) = \mathbf{n} \times \mathbf{H}_s \qquad \text{on } S_2 \tag{115b}$$

where $\mathbf{H}_{g1}$ and $\mathbf{H}_{g2}$ are the incident generator fields and the scattered magnetic fields are obtained from the corresponding electric fields by using Maxwell's equations. For the aperture cavity the dyadic Green's function $\bar{\mathbf{G}}_e$ can be approximated by its quasi-static value. For a circular aperture it can be constructed from the waveguide modes in a circular waveguide below cutoff. The two most important and least attenuated modes are the TE_{11} and TM_{01} modes which are the ones needed to describe coupling from the magnetic and electric dipoles, respectively. For small apertures a multipole expansion of the integrals given by (113) and (114) can be carried out in the same way that (56) was expanded. In general, the available theoretical results for coupling by apertures in thick walls is very limited. The edge conditions are different for an aperture in a thick wall as compared to an aperture in an infinitely thin wall. Consequently, some difference in the polarizability of the aperture would be expected. A detailed analysis of this problem is not available at this time.

7.6. Transients in Waveguides

In any waveguide, the phase velocity v_p is a function of frequency. As a consequence, all signals having a finite frequency spectrum will undergo dispersion when transmitted through a length of guide. The phase relationship between the frequency components of the original signal at the feeding point continually changes as the signal progresses along the guide. The analysis of this phenomenon belongs in the domain of transient analysis, and may be handled by the conventional techniques utilizing Laplace and Fourier transforms. Any realistic analysis should take into account the frequency characteristics of the antenna or aperture that couples the signal into the guide as well as the characteristics of the circuit elements used to extract the signal at the receiving end. In this section we shall consider only the properties of the guide itself. The effect of losses and their variation will also be neglected. In practice, this does not lead to significant errors, because in most cases the frequency bandwidth of the signal is relatively narrow, and the mid-band frequency of operation is usually chosen far enough above the cutoff frequency so that the attenuation curve is approximately constant throughout the band.

Before some time, which we choose as the time origin $t = 0$, the disturbance in the guide is zero. When a current element is introduced into the guide, a disturbance or signal is generated. This signal is a solution of the time-dependent field equations. Let $\mathcal{E}(\mathbf{r}, t)$, $\mathcal{H}(\mathbf{r}, t)$ be the time-dependent field vectors. For a unit impulse current element $\delta(t - t')$ applied at time $t = t'$ and located at the point (x', y', z'), the field vectors are a solution of

$$\nabla \times \mathcal{E} = -\mu_0 \frac{\partial \mathcal{H}}{\partial t} \tag{116a}$$

$$\nabla \times \mathcal{H} = \epsilon_0 \frac{\partial \mathcal{E}}{\partial t} + \tau \delta(t - t') \delta(\mathbf{r} - \mathbf{r}') \tag{116b}$$

where τ is a unit vector giving the direction of the current element, and $\mathbf{r}$ designates the field point (x, y, z), while $\mathbf{r}'$ designates the source point or location. If the Laplace transform of these equations is taken, we get

$$\nabla \times \mathbf{E}(\mathbf{r}, p) = -\mu_0 p \mathbf{H}(\mathbf{r}, p) \tag{117a}$$

$$\nabla \times \mathbf{H}(\mathbf{r}, p) = \epsilon_0 p \mathbf{E}(\mathbf{r}, p) + \tau e^{-pt'}\delta(\mathbf{r} - \mathbf{r}') \tag{117b}$$

where

$$\mathbf{E}(\mathbf{r}, p) = \int_0^\infty \mathcal{E}(\mathbf{r}, t)e^{-pt}\, dt$$

with a similar definition for $\mathbf{H}(\mathbf{r}, p)$. These equations are formally the same as those obtained by assuming a time dependence $e^{j\omega t}$, and the solution may be obtained by the methods we have previously discussed. All our previous solutions may be converted into solutions of (117) by replacing $j\omega$ by p and $e^{j\omega t}$ by $e^{-pt'}$. The Laplace transform has the effect of suppressing the time variable. The solution to (117) constitutes the Laplace transform of the time-dependent Green's function. Inverting the transform yields the time-dependent Green's function. For an arbitrary spatial and time variation, the solution may be obtained by a superposition integral.

If we restrict ourselves to a line current extending across the narrow dimension of a rectangular guide, the appropriate Green's function is obtained from Chapter 5, Eq. (74). We have

$$E_y(\mathbf{r}, p) = G(\mathbf{r}|\mathbf{r}', p) = \frac{-\mu_0}{a}\sum_{n=1}^\infty \Phi_n(x)\Phi_n(x')$$

$$\cdot \frac{p\,\exp\{-pt' - [(n\pi/a)^2 + \mu_0\epsilon_0 p^2]^{1/2}|z - z'|\}}{[(n\pi/a)^2 + \mu_0\epsilon_0 p^2]^{1/2}} \tag{118}$$

where $\Phi_n(x) = \sin(n\pi x/a)$. The inversion of this expression gives us the time-dependent Green's function corresponding to the disturbance set up in the waveguide by an impulse line current located at x', z'. The Laplace transform of the time derivative of a function $f(t)$ is

$$\mathcal{L}\frac{df}{dt} = \int_0^\infty e^{-pt}\frac{df}{dt}\,dt = fe^{-pt}\Big|_0^\infty + p\int_0^\infty e^{-pt}f\,dt.$$

If f vanishes at $t = 0$, we have $\mathcal{L}\,df/dt = p\mathcal{L}f$, where $\mathcal{L}$ indicates the operation of taking the Laplace transform of the function. Using this result, we may invert each term in (118) disregarding the factor p in the numerator, and then differentiate the result with respect to t. A typical term from (118) gives

$$\frac{1}{2\pi j}\int_C \frac{\exp\{p(t - t') - [(n\pi/a)^2 + \mu_0\epsilon_0 p^2]^{1/2}|z - z'|\}}{[(n\pi/a)^2 + \mu_0\epsilon_0 p^2]^{1/2}}\,dp$$

$$= \begin{cases} v_c J_0\left\{\dfrac{n\pi}{a}[v_c^2(t - t')^2 - (z - z')^2]^{1/2}\right\}, & 0 < \dfrac{|z - z'|}{v_c} < t - t' \\ 0, & \text{otherwise} \end{cases}$$

where J_0 is the Bessel function of the first kind, and $v_c = (\mu_0\epsilon_0)^{-1/2}$ is the velocity of light in a vacuum.[3] At any given distance $|z - z'|$ from the source, the disturbance is zero until a time $t = t' + |z - z'|/v_c$ is reached when the presence of the signal first becomes known to the observer at this position. No information reaches the observer in a time interval less than the time required to propagate a disturbance with the velocity of light. The velocity of light is, therefore, the wavefront velocity.

The time derivative of the above function is

$$\frac{-v_c^3(n\pi/a)(t-t')}{[v_c^2(t-t')^2 - (z-z')^2]^{1/2}} J_1 \left\{ \frac{n\pi}{a} [v_c^2(t-t')^2 - (z-z')^2]^{1/2} \right\} + v_c\delta\left(t - t' - \frac{|z-z'|}{v_c}\right)$$

where J_1 is the Bessel function of the first kind and order 1. The solution for the time-dependent Green's function which is equal to $\mathcal{E}_y(\mathbf{r}, t)$ becomes

$$\mathcal{G}(\mathbf{r}|\mathbf{r}', t - t') = \frac{Z_0}{a} \sum_{n=1}^{\infty} \Phi_n(x)\Phi_n(x')$$

$$\cdot \left[v_c^2 \frac{n\pi}{a}(t - t') \frac{J_1\{(n\pi/a)[v_c^2(t-t')^2 - (z-z')^2]^{1/2}\}}{[v_c^2(t-t')^2 - (z-z')^2]^{1/2}} - \delta\left(t - t' - \frac{|z-z'|}{v_c}\right) \right] \quad (119)$$

for $t' < t' + |z - z'|/v_c \leq t$. Outside this range for t the Green's function is zero. In the interpretation of (119) it should be noted that

$$\lim_{u\to 0} \frac{J_1(ku)}{u} = \lim_{u\to 0} \frac{dJ_1}{du} = \lim_{u\to 0} \frac{k}{2}[J_0(ku) - J_2(ku)] = \frac{k}{2}.$$

Alternatively, the recurrence relation $J_1(ku) = (ku/2)[J_2(ku) + J_0(ku)]$ may be used to eliminate the factor $[v_c^2(t - t')^2 - (z - z')^2]^{1/2}$ in the denominator. The field contributed by each mode first makes its appearance as an impulse plus a step change, and thereafter oscillates similar to a damped sinusoidal wave. Thus what was originally a pulse localized in both time and space has become a disturbance distributed throughout both space and time.

For an impressed current $\mathbf{J}(\mathbf{r}, t)$ which varies arbitrarily with time for $t > 0$, the response is obtained by a superposition integral, i.e.,

$$\iiint_0^t \mathcal{G}(\mathbf{r}|\mathbf{r}', t - t')J(\mathbf{r}', t') \, dt' \, dS' \quad (120)$$

when $J(t) = 0$ for $t < 0$. For a step input, the solution is readily found since the Laplace transform of a unit step applied at $t = t'$ is $p^{-1}e^{-pt'}$. This corresponds to (118) with the factor p deleted from the numerator. The terms that arise are those obtained in arriving at (119) before the derivative with respect to time was taken.

Impulse and step inputs are not typical of the signal we wish to transmit along a waveguide. More realistic signals are those consisting of a narrow spectrum of frequencies, centered around some high frequency ω_0, such as is encountered in amplitude-modulated sinusoids. For an analysis of these we turn to an application of Fourier transforms.

---

[3] This transform is listed by Churchill in [7.9].

Let the input current at $x = a/2$, $z = 0$ be of the form of an amplitude-modulated signal

$$I = I_0[1 + f(t)] \cos \omega_0 t \tag{121}$$

where $f(t)$ is an arbitrary function of time subject to the restriction that it contains frequencies in the range $0 \leq \omega \leq \omega_1$ with $\omega_1 \ll \omega_0$. Thus the resultant signal has frequency components in the range $\omega_0 - \omega_1 \leq \omega \leq \omega_0 + \omega_1$, and ω_0 corresponds to the carrier frequency. Furthermore, let ω_1 be chosen so that $(\pi/a)v_c < \omega < (2\pi/a)v_c$. This latter restriction ensures us that only the H_{10} mode in a rectangular guide of width a will propagate.

The current input is the real part of $(1+f)e^{j\omega_0 t}$. The function $f(t)$ has a frequency spectrum given by the Fourier transform of $f(t)$:

$$g(\omega) = \int_{-\infty}^{\infty} e^{-j\omega t} f(t)\, dt. \tag{122}$$

The function $f(t)$ may be recovered by the inverse transform relation

$$f(t) = \frac{1}{2\pi} \int_{-\infty}^{\infty} e^{j\omega t} g(\omega)\, d\omega = \frac{1}{2\pi} \int_{-\omega_1}^{\omega_1} e^{j\omega t} g(\omega)\, d\omega. \tag{123}$$

The transform of $e^{j\omega_0 t}$ is $2\pi\delta(\omega - \omega_0)$, and, hence, for the function $(1+f)e^{j\omega_0 t}$ the frequency spectrum is given by $2\pi\delta(\omega - \omega_0) + g(\omega - \omega_0)$. Each frequency component propagates along the guide in a manner described by the steady-state solution corresponding to an impressed current varying with time according to $e^{j\omega t}$. For the H_{10} mode, the y component of electric field, at a distance z from the origin, arising from a single frequency component is

$$-\frac{j\omega\mu_0}{a} \Phi_1(x) \frac{\exp\{j\omega t - j|z|[(\omega/v_c)^2 - (\pi/a)^2]^{1/2}\}}{[(\omega/v_c)^2 - (\pi/a)^2]^{1/2}} = -\frac{j\omega\mu_0}{a} \Phi_1(x) \frac{e^{j\omega t - j\beta(\omega)|z|}}{\beta(\omega)}$$

where $\beta(\omega) = [(\omega/v_c)^2 - (\pi/a)^2]^{1/2}$. This expression without the factor $e^{j\omega t}$ may be considered as the transfer function, in the frequency domain, relating the response at $|z|$ to the input at $z = 0$. With the specified input current, the total response is

$$\mathcal{E}_y = -\mathrm{Re}\frac{\mu_0 I_0}{a2\pi} \Phi_1 \int_{-\infty}^{\infty} \frac{j\omega e^{-j\beta(\omega)|z|}}{\beta(\omega)} [2\pi\delta(\omega - \omega_0) + g(\omega - \omega_0)]e^{j\omega t}\, d\omega$$

$$= \frac{-\mu_0 I_0}{2\pi a} \Phi_1 \,\mathrm{Re} \left[\frac{j2\pi\omega_0 e^{j\omega_0 t - j\beta_0|z|}}{\beta_0} + \int_{-\infty}^{\infty} \frac{j\omega e^{j\omega t - j\beta(\omega)|z|}}{\beta(\omega)} g(\omega - \omega_0)\, d\omega \right] \tag{124}$$

where $\beta_0 = \beta(\omega_0)$. The first term in (124) corresponds to the unmodulated carrier and does not contain any information. The second term is the signal containing the original modulation-frequency components. In this term, the integration need be extended only over the range $\omega_0 - \omega_1$ to $\omega_0 + \omega_1$, since $g(\omega - \omega_0)$ is zero outside this range.

To evaluate the integral, we expand $\beta(\omega)$ in a Taylor series about the point ω_0 to get

$$\beta(\omega) = \beta_0 + \beta_0'(\omega - \omega_0) + \frac{1}{2!}\beta_0''(\omega - \omega_0)^2 + \cdots \tag{125}$$

where $\beta_0' = d\beta/d\omega|_{\omega=\omega_0}$, etc. For a signal containing only a narrow band of frequencies, we retain only the first two terms in (125) and approximate the factor ω/β by ω_0/β_0. For this case we then get

$$\frac{1}{2\pi}\int_{-\infty}^{\infty}\frac{j\omega e^{j\omega t - j\beta|z|}}{\beta}g(\omega - \omega_0)\,d\omega \approx \frac{j\omega_0}{2\pi\beta_0}e^{j\omega_0 t - j\beta_0|z|}\int_{\omega_0 - \omega_1}^{\omega_0 + \omega_1}e^{j(\omega - \omega_0)(t - \beta_0'|z|)}g(\omega - \omega_0)\,d\omega$$

$$= \frac{j\omega_0}{\beta_0}e^{j\omega_0 t - j\beta_0|z|}f(t - \beta_0'|z|). \quad (126)$$

The latter result follows by making the change of variables $\omega - \omega_0 = \omega'$ and comparing with (123). Combining the two terms and taking the real part, we obtain the total response at $|z|$:

$$\mathcal{E}_y = \frac{\mu_0 I_0}{a}\Phi_1(x)\frac{\omega_0}{\beta_1}[1 + f(t - \beta_0'|z|)]\sin\omega_0\left(t - \frac{\beta_0}{\omega_0}|z|\right). \quad (127)$$

To this order of approximation the original modulation is reproduced without distortion but delayed in time by an amount $\beta_0'|z|$. The velocity with which the signal propagates is equal to the distance $|z|$ divided by the time delay, and hence is given by

$$v_g = \frac{1}{\beta_0'} = \left[\frac{d\beta(\omega)}{d\omega}\bigg|_{\omega_0}\right]^{-1} = \frac{v_c^2\beta_0}{\omega_0} = \frac{v_c^2}{v_p} \quad (128)$$

where the carrier phase velocity v_p is equal to $\omega_0/\beta_0 = (k_0/\beta_0)v_c$. This particular definition of signal velocity is called the group velocity, since it corresponds to the velocity with which a narrow band or group of frequency components is propagated. It is also equal to the velocity of energy propagation introduced in Chapter 3 and is always less than the velocity of light for a uniform hollow waveguide.

If the band of frequencies involved is too large for only two terms of the Taylor series expansion of $\beta(\omega)$ to give a good representation of $\beta(\omega)$ throughout the band, then additional terms must be included. These higher order terms always lead to distortion of the signal. In general, these terms are difficult to evaluate unless $f(t)$ happens to be of such a form that the inverse transforms can be found. Forrer has presented an analysis for the propagation of a pulse of the form $e^{-(t+t_0)^2}$. In this case terms up to and including $\frac{1}{2}\beta_0''(\omega - \omega_0)^2$ in the expansion of β may be evaluated, since the transforms involved are of the type that may be readily inverted [7.10]. Several other papers that treat various types of transients in waveguides have also been published. Several of these papers are listed in the general references below.

References and Bibliography

[7.1]   R. W. P. King, *The Theory of Linear Antennas.*   Cambridge, MA: Harvard University Press, 1956, sect. II.39.

[7.2]   E. Jahnke and F. Emde, *Tables of Functions*, 4th ed.   New York, NY: Dover Publications, 1945.

[7.3]   E. P. Augustin, U.S. Patent 4,528,528, July 9, 1985.

[7.4]   H. T. Howard, U.S. Patent 4,414,516, Nov. 8, 1983.

[7.5]   H. A. Bethe, "Theory of diffraction by small holes," *Phys. Rev.*, vol. 66, pp. 163–182, 1944.

[7.6]   S. B. Cohn, "Determination of aperture parameters by electrolytic tank measurements," *Proc. IRE*, vol. 39, pp. 1416–1421, Nov. 1951.

[7.7]   J. A. Stratton, *Electromagnetic Theory.*   New York, NY: McGraw-Hill Book Company, Inc., 1941, sect. 3.27.

[7.8]   R. E. Collin, "Rayleigh scattering and power conservation," *IEEE Trans. Antennas Propagat.*, vol. AP-29, pp. 795–798, 1981.

[7.9]   R. V. Churchill, *Operational Mathematics*, 2nd ed.   New York, NY: McGraw-Hill Book Company, Inc., 1958, p. 329.
[7.10] M. P. Forrer, "Analysis of millimicrosecond RF pulse transmission," *Proc. IRE*, vol. 46, pp. 1830–1835, Nov. 1958.

Waveguide Antennas

[7.11] L. Infeld, A. F. Stevenson, and J. L. Synge, "Contributions to the theory of waveguides," *Can. J. Res.*, vol. 27, pp. 68–129, July 1949.
[7.12] H. Motz, *Electromagnetic Problems of Microwave Theory*, Methuen Monograph.   London: Methuen & Co., Ltd., 1951.
[7.13] L. Lewin, "A contribution to the theory of probes in waveguides," *Proc. IEE (London)*, vol. 105, part C, pp. 109–116, Mar. 1958.
[7.14] M. J. Al-Hakkak, "Experimental investigation of the input impedance characteristics of an antenna in a rectangular waveguide," *Electron. Lett.*, vol. 5, pp. 513–514, 1969.
[7.15] A. G. Williamson and D. V. Otto, "Coaxially fed hollow cylindrical monopole in a rectangular waveguide," *Electron. Lett.*, vol. 9, pp. 218–220, 1973.
[7.16] A. G. Williamson, "Coaxially fed hollow probe in a rectangular waveguide," *Proc. IEE*, vol. 132, part H, pp. 273–285, 1985.
[7.17] J. M. Jarem, "A multifilament method-of-moments solution for the input impedance of a probe-excited semi-infinite waveguide," *IEEE Trans. Microwave Theory Tech.*, vol. MTT-35, pp. 14–19, 1987.
[7.18] M. D. Deshpande and B. N. Das, "Input impedance of coaxial line to circular waveguide feed," *IEEE Trans. Microwave Theory Tech.*, vol. MTT-25, pp. 954–957, 1977.
[7.19] M. D. Deshpande and B. N. Das, "Analysis of an end launcher for a circular cylindrical waveguide," *IEEE Trans. Microwave Theory Tech.*, vol. MTT-26, pp. 672–675, 1978.
[7.20] G. L. Ragan, *Microwave Transmission Circuits*, vol. 9 of Rad. Lab. Series.   New York, NY: McGraw-Hill Book Company, Inc., 1948.

Aperture Coupling

[7.21] L. B. Felson and N. Marcuvitz, "Slot coupling of rectangular and spherical waveguides," *J. Appl. Phys.*, vol. 24, pp. 755–770, June 1953.
[7.22] S. B. Cohn, "The electric polarizability of apertures of arbitrary shape," *Proc. IRE*, vol. 40, pp. 1069–1071, Sept. 1952.
[7.23] S. B. Cohn, "Microwave coupling by large apertures," *Proc. IRE*, vol. 40, pp. 696–699, 1952.
[7.24] E. Arvas and R. F. Harrington, "Computation of the magnetic polarizability of conducting disks and the electric polarizabilities of apertures," *IEEE Trans. Antennas Propagat.*, vol. AP-31, pp. 719–725, 1983.
[7.25] N. A. McDonald, "Polynomial approximations for the electric polarizabilities of some small apertures," *IEEE Trans. Microwave Theory Tech.*, vol. MTT-33, pp. 1146–1149, 1985.
[7.26] J. Van Bladel, "Small hole coupling of resonant cavities and waveguides," *Proc. IEE*, vol. 117, pp. 1098–1104, 1970.
[7.27] J. Van Bladel, "Small holes in a waveguide wall," *Proc. IEE*, vol. 118, pp. 43–50, 1971.
[7.28] R. F. Harrington and J. R. Mautz, "A generalized network formulation for aperture problems," *IEEE Trans. Antennas Propagat.*, vol. AP-24, pp. 870–873, 1976.
[7.29] Y. Rahmat-Samii and R. Mittra, "Electromagnetic coupling through small apertures in a conducting screen," *IEEE Trans. Antennas Propagat.*, vol. AP-25, pp. 180–187, 1977.
[7.30] R. E. Collin, "Small aperture coupling between dissimilar regions," *Electromagnetics*, vol. 2, pp. 1–24, 1982. (In this reference the polarizability tensors are twice as large and the radiation reaction fields in the output waveguide are defined with the opposite sign.)
[7.31] C. H. Liang and D. K. Cheng, "Generalized network representations for small-aperture coupling between dissimilar regions," *IEEE Trans. Antennas Propagat.*, vol. AP-31, pp. 177–182, 1983.
[7.32] Y. Leviatan, R. F. Harrington, and J. R. Mautz, "Electromagnetic transmission through apertures in a cavity in a thick conductor," *IEEE Trans. Antennas Propagat.*, vol. AP-30, pp. 1153–1165, 1982.
[7.33] N. A. McDonald, "Electric and magnetic coupling through small apertures in shield walls of any thickness," *IEEE Trans. Microwave Theory Tech.*, vol. MTT-20, pp. 689–695, 1972.
[7.34] R. Levy, "Analysis and synthesis of waveguide multi-aperture directional couplers," *IEEE Trans. Microwave Theory Tech.*, vol. MTT-16, pp. 995–1006, 1968.

[7.35]   R. Levy, "Improved single and multiaperture waveguide coupling theory, including explanation of mutual interactions," *IEEE Trans. Microwave Theory Tech.*, vol. MTT-28, pp. 331–338, 1980.

[7.36]   C. G. Montgomery, *Technique of Microwave Measurements*, vol. 11 of Rad. Lab. Series. New York, NY: McGraw-Hill Book Company, Inc., 1947. Sec. 14.3.

Waveguide Transients

[7.37]   M. Cerillo, "Transient phenomena in waveguides," Electronics Tech. Rep. 33, MIT Research Lab., 1948.

[7.38]   G. I. Cohn, "Electromagnetic transients in waveguides," in *Proc. Nat. Electronics Conf.*, vol. 8, pp. 284–295, 1952.

[7.39]   M. Cotte, "Propagation of a pulse in a waveguide," *Onde Elec.*, vol. 34, pp. 143–146, 1954.

[7.40]   R. S. Elliott, "Pulse waveform degradation due to dispersion in waveguides," *IRE Trans. Microwave Theory Tech.*, vol. MTT-5, pp. 245–257, 1957.

[7.41]   A. E. Karbowiak, "Propagation of transients in waveguides," *Proc. IEE (London)*, vol. 104, part C, pp. 339–349, 1957.

PROBLEMS

7.1. A small probe antenna is located in the center of the end wall of a rectangular guide (see Fig. P7.1). The

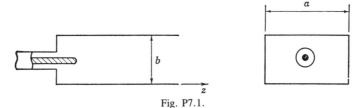

Fig. P7.1.

guide dimensions are such that the only E mode that will propagate is the E_{11} mode. Obtain an expression for the radiation resistance of the antenna.

7.2. Consider a small circular loop antenna of radius d, carrying a uniform current I_0 and located in free space. Expand $e^{-jk_0\rho}$ to get $1 - jk_0\rho - \frac{1}{2}k_0^2\rho^2 + (jk_0^3/6)\rho^3 + \cdots$. Evaluate the self-flux linkage through the loop contributed by the terms in phase quadrature with the current, and show that the term with the factor $jk_0\rho$ does not contribute while the term with the factor $(jk_0^3/6)\rho^3$ gives $(-j\mu_0\pi I_0/6k_0)(k_0d)^4$. Evaluate the induced emf around the loop due to this portion of the flux linkage, and show that it leads to an input resistance $Z_0(k_0d)^4\pi/6$. Compute the power radiated by the loop by integrating the complex Poynting vector over the surface of a large sphere. If P is the radiated power, use the relation $\frac{1}{2}RI_0^2 = P$ to define a radiation resistance, and show that this is equal to the input resistance obtained from a consideration of the flux linkages.

7.3. Perform the summations leading to (38).

7.4. Sum the following series directly, and compare with the sum obtained by using the Poisson summation formula:

$$\sum_{n=1}^{\infty} \frac{e^{-(n^2+1)^{1/2}}}{(n^2+1)^{1/2}}.$$

7.5. Use the Poisson summation formula to sum the following series:

$$\sum_{n=1}^{\infty} e^{-\alpha(n^2+\beta^2)^{1/2}} \qquad \sum_{n=1}^{\infty} e^{-\pi n^2/1000}.$$

Note the large difference in the rate of convergence of the second series and its transformed equivalent. Consult a table of Fourier transforms for the required transforms.

7.6. Modify the small-aperture theory for waveguides to obtain an approximate theory for the reflected and transmitted fields from a small aperture in an infinite-plane conducting screen. Assume a parallel-polarized TEM wave incident at an angle θ_i with respect to the normal (see Fig. P7.6).

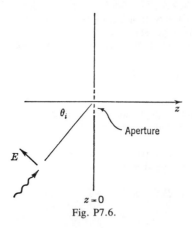

Fig. P7.6.

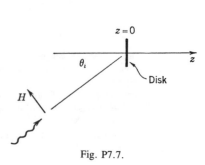

Fig. P7.7.

7.7. Develop a theory, similar to that in Problem 7.6, for the scattered field from a small conducting disk when a perpendicular-polarized TEM wave is incident at an angle θ_i (see Fig. P7.7). (For an expansion of the vector potential which gives the radiation from a small-volume distribution of current as radiation from an electric and magnetic dipole and other higher order multipoles see J. A. Stratton, *Electromagnetic Theory*. New York, NY: McGraw-Hill Book Company, Inc., 1941, sect. 8.4.) Use Babinet's principle to obtain the solution for a small aperture in a conducting screen from the solution of the disk problem. Show that this leads to the solution obtained in Problem 7.6.

7.8. Show that an elliptic aperture in a transverse wall in a rectangular guide as illustrated in Fig. P7.8 is

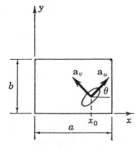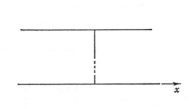

Fig. P7.8.

equivalent to a shunt inductive susceptance

$$B = \left(\frac{-2\beta_{10}}{ab} \sin^2 \frac{\pi x_0}{a} \bar{\alpha}_m : \mathbf{a}_x \mathbf{a}_x \right)^{-1}$$

where $\bar{\alpha}_m : \mathbf{a}_x \mathbf{a}_x$ is the double-dot product of $\bar{\alpha}_m$ with the unit vectors $\mathbf{a}_x$ and x_0 is the coordinate of the center of the aperture.

7.9. A small circular aperture of radius l is located in the center of the transverse wall in a circular guide of radius a (see Fig. P7.9). Show that, for a TE_{11} mode, the aperture is equivalent to an inductive susceptance

$$B = \frac{-0.955a^2\lambda_g}{4\alpha_m}$$

Fig. P7.9.

where λ_g is the guide wavelength, and $\alpha_m = \frac{4}{3}l^3$. For a TM_{01} mode, show that the aperture is equivalent to a shunt capacitive susceptance

$$B = \frac{0.92a^4}{|\alpha_e|\lambda_g}, \qquad |\alpha_e| = \frac{2}{3}l^3.$$

7.10. The Bethe hole directional coupler consists of two rectangular guides with their axes oriented with an angle θ between them and coupled through a small circular aperture in the center of the broadwall (see Fig. P7.10). Find

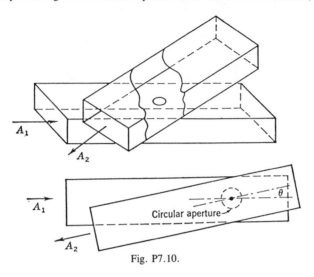

Fig. P7.10.

the required angle θ such that the electric and magnetic dipole coupling results in a wave propagating in the direction of arrow A_2 when a wave is incident along the direction of arrow A_1. Find approximate expressions for the coupling coefficient and the directivity of this coupler. Note that when the radiation reaction fields are included the wave propagating in the direction opposite to that of arrow A_2 cannot be made to vanish, although its amplitude can be made small.

Answer:

$$a_{3,4} = \left[\frac{\mp 2jX\cos\theta}{4 + 4jX - X^2\sin^2\theta} + \frac{jB}{2 + 2jB} \right] A_1$$

where $X = 2\beta_{10}\alpha_m/ab$ and $B = 2k_0^2\alpha_e/\beta_{10}ab$. Let $B = -X\cos\theta$, then $C \approx 20 \log (1/X\cos\theta)$ and $D \approx C + 20 \log [2\cos\theta/(1+\cos\theta)]$.

7.11. If it is desirable to maintain the angle θ between the two guide axes equal to zero in Problem 7.10, a similar type of coupler may be built, using a properly oriented elliptical aperture or an offset circular aperture as illustrated in Fig. P7.11. Determine the orientation angle θ of the elliptical aperture and the offset d of the circular aperture in order to obtain a directional coupler.

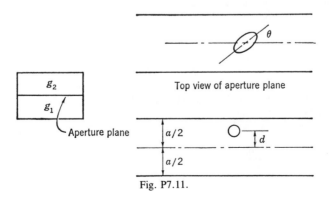

Fig. P7.11.

Answer:

$$\alpha_{mu} \sin^2 \theta + \alpha_{mv} \cos^2 \theta = \frac{k_0^2}{\beta_0^2} |\alpha_e|.$$

7.12. Consider a rectangular guide closed at $z = 0$ by an electric wall in which a small magnetic disk carrying a linear magnetic current of density $\mathbf{J}_m$ is located (see Fig. P7.12). Let the field radiated by the magnetic current in

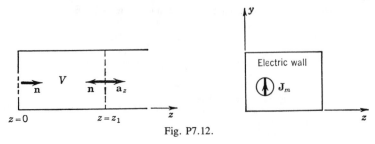

Fig. P7.12.

the region $z > 0$ be

$$\mathbf{E}_s = \sum c_n \mathbf{E}_n e^{-\Gamma_n z}$$

$$\mathbf{H}_s = \sum c_n \mathbf{H}_n e^{-\Gamma_n z}.$$

In the volume bounded by the guide walls and transverse planes at $z = 0$ and $z = z_1$, the field $\mathbf{E}_s$, $\mathbf{H}_s$ is source-free. On the electric wall at $z = 0$ we have $\mathbf{n} \times \mathbf{E}_s = 0$, while on the surface of the magnetic disk $\mathbf{n} \times \mathbf{E}_s = -\mathbf{J}_m$. Let $\mathbf{E}$, $\mathbf{H}$ be a normal-mode standing wave:

$$\mathbf{E} = \mathbf{e}_m(e^{-\Gamma_m z} - e^{\Gamma_m z}) + \mathbf{e}_{zm}(e^{-\Gamma_m z} + e^{\Gamma_m z})$$

$$\mathbf{H} = \mathbf{h}_m(e^{-\Gamma_m z} + e^{\Gamma_m z}) + \mathbf{h}_{zm}(e^{-\Gamma_m z} - e^{\Gamma_m z}).$$

This field is also source-free in V. The Lorentz reciprocity theorem gives

$$\oiint_S \mathbf{n} \cdot (\mathbf{E}_s \times \mathbf{H} - \mathbf{E} \times \mathbf{H}_s) \, dS = 0.$$

Show that, over the plane $z = z_1$, the integral gives $-2c_m \iint \mathbf{e}_m \times \mathbf{h}_m \cdot \mathbf{a}_z \, dS$, while, over the $z = 0$ plane, the result is $-\iint \mathbf{J}_m \cdot \mathbf{H} \, dS = -2 \iint \mathbf{J}_m \cdot \mathbf{h}_m \, dS$. A linear magnetic current element is equivalent to a magnetic dipole, and, if $\mathbf{h}_m$ is assumed constant over the magnetic disk, the coefficient c_m is found to be given by

$$2c_m \iint \mathbf{e}_m \times \mathbf{h}_m \cdot \mathbf{a}_z \, dS = -j\omega\mu_0 2\mathbf{M}_0 \cdot \mathbf{h}_m$$

where $\mathbf{M}_0$ is the magnetic dipole moment associated with the current $\mathbf{J}_m$.

7.13. Consider a z-directed magnetic dipole $M\mathbf{a}_z$ located at the origin between two infinite conducting planes placed parallel with the xz plane at $y = \pm b/2$. Obtain the solution for the magnetic Hertzian potential Π_z by solving the scalar Helmholtz equation

$$\nabla^2 \Pi_z + k_0^2 \Pi_z = -M\delta(x)\delta(y)\delta(z)$$

in cylindrical coordinates. The solution is the same as that given by (38).

HINT: Expand Π_z in a Fourier series of the form

$$\Pi_z = \sum_{m=0}^{\infty} a_m \cos \frac{2m\pi y}{b} K_0(\Gamma_m r) \qquad \Gamma_m^2 = \left(\frac{2m\pi}{b}\right)^2 - k_0^2.$$

Next use Fourier analysis to obtain

$$\left(\frac{1}{r}\frac{\partial}{\partial r}r\frac{\partial}{\partial r} - \Gamma_m^2\right)a_mK_0(\Gamma_mr) = -\frac{M\epsilon_{0m}}{b}\delta(r).$$

Each term must have a logarithmic singularity $-(M\epsilon_{0m}/2b\pi)\ln r$, and hence $a_m = M\epsilon_{0m}/2\pi b$, $\epsilon_{0m} = 1$, $m = 0$; $\epsilon_{0m} = 2$, $m > 0$.

7.14. Figure P7.14 shows a directional coupler consisting of two rectangular waveguides coupled by small circular

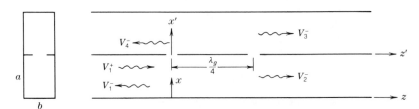

Fig. P7.14.

apertures in the common sidewall. Find the amplitudes V_1^-, V_2^-, V_3^-, V_4^- at the four ports when a TE_{10} mode with amplitude V_1^+ is incident at port 1.

HINT: Assume dipoles of strength M_{z1} and M_{z2} in the two apertures and find the total H_z reaction fields from these.

Answer: Generator fields at the two apertures are $-Nk_cV_1^+$ and $jNk_cV_1^+$. Equations for the dipole strengths are

$$M_{z1} = \alpha_m[-Nk_cV_1^+ + jk_0Z_0k_c^2N^2(M_{z1} - jM_{z2})]$$

$$M_{z2} = \alpha_m[jNk_cV_1^+ + jk_0Z_0k_c^2N^2(M_{z2} - jM_{z1})]$$

$$M_{z1} = -\alpha_mk_cNV_1^+(1 + jB)/(1 + 2jB - 2B^2)$$

$$M_{z2} = j\alpha_mk_cNV_1^+/(1 + 2jB - 2B^2) = -jM_{z1}/(1 + jB)$$

$$V_1^- = V_4^- = k_0Z_0Nk_cBM_{z1}/(1 + 2jB)$$

$$V_3^- = -k_0Z_0Nk_cM_{z1}(1 + jB)/(1 + 2jB) = V_2^- + jV_1^+$$

where $k_c = \pi/a$, $B = 2\alpha_mk_c^2/\beta_{10}ab$.

7.15. The equivalent circuit for the directional coupler in Problem 7.14 is shown in Fig. P7.15. Show that this circuit gives the same scattered field amplitudes when a wave of amplitude V_1^+ is incident at port 1.

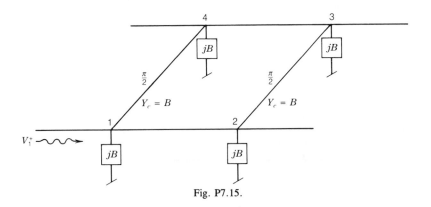

Fig. P7.15.

7.16. Repeat the analysis of the directional coupler in Problem 7.14 but let the apertures be spaced a distance d apart so that $\beta_{10}d = \theta$. Find the value of θ that will make $V_4^+ = 0$.

Answer: $M_{z1} = -M_{z2}e^{-j\theta}$, $\theta = -\tan^{-1} B + \pi/2$.

7.17. A rectangular cavity is coupled to a rectangular waveguide by a small circular aperture in the common sidewall as shown in Fig. P7.17. Find the equivalent circuit for this system. The cavity resonates in the TE_{101} mode.

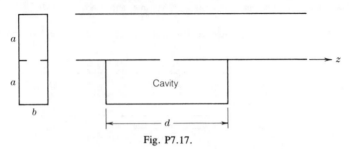

Fig. P7.17.

7.18. Figure P7.18(a) shows a cavity coupled to a rectangular guide by a circular aperture with radius r_0 and which is centered in the broad wall of the guide and cavity. The cavity is excited in the TE_{101} mode. The incident

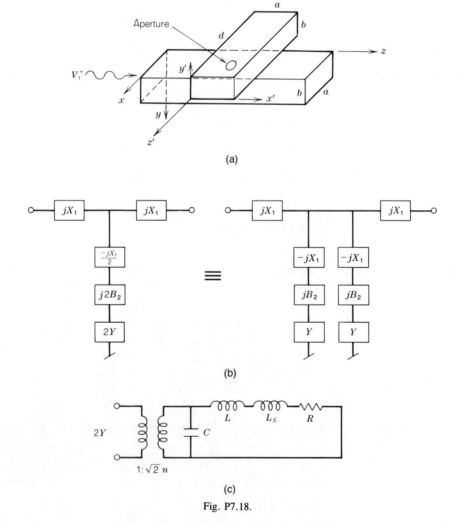

(a)

(b)

(c)

Fig. P7.18.

field has

$$E_y = V_1^+ N \sin \frac{\pi x}{a} e^{-j\beta z}, \qquad H_x = -Y_w E_y$$

where $N = (2Z_w/ab)^{1/2}$.

Find the amplitudes of the TE_{101} cavity mode and the reflected and transmitted TE_{10} modes in the waveguide. Note that dipoles P_y and M_x are excited but only P_y couples to the TE_{101} mode (why?).

Verify that

$$M_x = -\frac{\alpha_m N Y_w V_1^+}{2 + j\alpha_m \beta/ab}, \quad P_y = \frac{\epsilon_0 \alpha_e N V_1^+}{2 + \alpha_e W + j\alpha_e k_0^2/ab}, \quad W = \frac{4\omega^2}{abd \left[\omega^2 \left(1 + \frac{1-j}{Q} \right) - \omega_0^2 \right]}$$

where ω_0 is the resonant frequency of the cavity mode.

The equivalent circuit for the coupled cavity is also shown in Fig. P7.18(b). Show that for even and odd excitation the input reflection coefficients are

$$\Gamma_e = 1 - \frac{2jB_2}{1 + jB_2 + jB_2/Y}, \qquad \Gamma_0 = -1 + \frac{2jX_1}{1 + jX_1}.$$

By superimposing the two solutions show that the input reflection coefficient is given by

$$\Gamma = \frac{1}{2}(\Gamma_e + \Gamma_o) = \frac{jX_1}{1 + jX_1} - \frac{jB_2}{1 + jB_2 + jB_2/Y}.$$

By comparing this expression with your analytical solution show that

$$X_1 = \frac{\alpha_m \beta}{2ab}, \quad B_2 = \frac{k_0^2 \alpha_e}{2\beta ab}, \quad Y = \frac{jk_0^2 d}{2\omega^2 \beta} \left[\omega^2 \left(1 + \frac{1-j}{Q} \right) - \omega_0^2 \right].$$

Verify that for the resonant circuit shown in Fig. P7.18(c)

$$Y \approx \frac{j\omega C n^2}{\omega^2} \left[\omega^2 \left(1 + \frac{1-j}{Q} \right) - \omega_0^2 \right]$$

where $\omega_0^2 LC = 1$, $C = \epsilon_0 ad/4b$, $n^2 = k_0 b Z_0/\beta a$.

8
Variational Methods for Waveguide Discontinuities

When waveguides are used in practice, it is often necessary to introduce discontinuities such as diaphragms or irises, metallic posts, dielectric slabs, and so forth, for the purpose of matching the waveguide to a given load or termination, to obtain a phase shift in the transmitted wave, or for a variety of other reasons. The presence of a discontinuity gives rise to a reflected wave and a storage of reactive energy in the vicinity of the discontinuity because of excitation of higher order waveguide modes which are evanescent, i.e., decay exponentially with distance from the discontinuity. Under most conditions, in practice, only the dominant mode can propagate in the guide. Under these conditions three is the maximum number of parameters required to describe the effect of the discontinuity on the propagating mode if the discontinuity is lossless. These parameters are the modulus ρ and the phase angle θ of the reflection coefficient R and the phase angle α of the transmission coefficient T. The modulus of T is given by $TT^* = 1 - \rho^2$, a result which follows at once from the conservation-of-energy principle. Alternatively, we may describe the discontinuity by an equivalent transmission-line circuit which would give rise to a reflected and a transmitted wave of magnitudes proportional to R and T, respectively. This latter description is most commonly used.

As a general rule, the exact solution of Maxwell's equations for the field distribution existing near a discontinuity is very difficult, if not impossible, to obtain. However, since we are interested only in the effect produced on the dominant mode, the detailed nature of the diffraction field around the discontinuity is not required. The higher order modes excited act as a perturbation, giving rise to a phase shift in the reflected and transmitted waves, and an impedance transformation at the discontinuity. When the magnitude of these higher order modes is small, approximate methods may be used to determine the parameters of the equivalent circuit for the discontinuity. A very useful method capable of handling a large variety of problems was introduced by Schwinger during the period 1940 to 1945 [8.1]. This method, referred to as the variational method, has a variety of forms, some of which are preferable to others in certain cases. Some of the various formulations possible will be illustrated by application to a specific problem.

8.1. OUTLINE OF VARIATIONAL METHODS

The problem of evaluating the equivalent-circuit parameters for the junction between an empty rectangular waveguide and one inhomogeneously filled with an asymmetrical dielectric slab will serve to illustrate the various variational techniques that may be used. The output guide for $z > 0$ is inhomogeneously filled with a dielectric slab of thickness t and relative dielectric constant κ, as illustrated in Fig. 8.1. The junction is assumed lossless, which at microwave frequencies is usually a very good approximation.

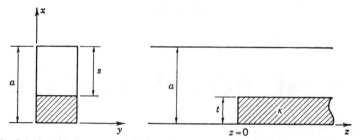

Fig. 8.1. Junction between an empty and an inhomogeneously filled rectangular guide.

The incident wave from the left is an H_{10} mode, and, since the discontinuity is uniform along the y axis, the only types of higher order modes that are excited at the junction are H_{n0} modes. It is therefore possible to derive the various field components from the one scalar component E_y by means of the following equations:

$$j\omega\mu_0 H_z = -\frac{\partial E_y}{\partial x} \qquad j\omega\mu_0 H_x = \frac{\partial E_y}{\partial z} \tag{1}$$

where E_y satisfies the scalar Helmholtz equation

$$\frac{\partial^2 E_y}{\partial x^2} + \frac{\partial^2 E_y}{\partial z^2} + k_0^2 \kappa E_y = 0 \tag{2}$$

and

$$k_0^2 = \omega^2 \mu_0 \epsilon_0 = \left(\frac{2\pi}{\lambda_0}\right)^2.$$

The solutions to these equations in the empty guide are

$$E_y = \sin\frac{n\pi x}{a} e^{\mp\Gamma_n z} \tag{3a}$$

$$H_x = \mp\frac{\Gamma_n}{jZ_0 k_0} \sin\frac{n\pi x}{a} e^{\mp\Gamma_n z} \tag{3b}$$

$$H_z = \frac{jn\pi}{aZ_0 k_0} \cos\frac{n\pi x}{a} e^{\mp\Gamma_n z} \tag{3c}$$

where $\Gamma_n^2 = (n\pi/a)^2 - k_0^2$ and $Z_0 = (\mu_0/\epsilon_0)^{1/2}$ is the wave impedance of free space. The term $\Gamma_n/jZ_0 k_0$ has the dimensions of an admittance, and is the wave admittance of the nth mode in the empty guide. It is equal to $\pm H_x/E_y$; we see that the transverse magnetic and electric field components play the roles of a current and a voltage, respectively, on a fictitious transmission line. When the mode propagates, Γ_n is imaginary, and the wave admittance Y_n is real. When the mode is evanescent, Y_n is imaginary. The following convention will be adopted in regard to the sign of Y_n. For modes that propagate, Y_n will be taken as a positive-real quantity. For evanescent H_{n0} modes, Y_n will be taken as negative-imaginary, since evanescent H_{n0} modes store more magnetic energy than electric energy, corresponding to a net inductive impedance.

Let the amplitude of the wave incident from the left on the junction be a_1. Owing to the discontinuity, a reflected wave of amplitude $R_1 a_1$ and an infinite number of higher order modes of amplitudes a_n are excited. Thus, in the region $z \leq 0$, we may write the following expressions for the total transverse fields:

$$E_y = a_1(e^{-\Gamma_1 z} + R_1 e^{\Gamma_1 z})\Phi_1 + \sum_2^\infty a_n \Phi_n e^{\Gamma_n z} \tag{4a}$$

$$H_x = -Y_1 a_1(e^{-\Gamma_1 z} - R_1 e^{\Gamma_1 z})\Phi_1 + \sum_2^\infty a_n Y_n \Phi_n e^{\Gamma_n z} \tag{4b}$$

where $\Phi_n = (2/a)^{1/2} \sin(n\pi x/a)$. The factor $(2/a)^{1/2}$ is introduced for normalization purposes only. The analogy to waves propagating on a transmission line is readily seen when the following definitions for the equivalent transmission-line currents and voltages are made:

$$V_{11}^+ = a_1 \qquad V_{11}^- = a_1 R_1 \qquad V_{n1}^- = a_n \tag{5a}$$

$$I_{11}^+ = a_1 Y_1 \quad I_{11}^- = -a_1 R_1 Y_1 \quad I_{n1}^- = -a_n Y_n \tag{5b}$$

$$V_1 = V_{11}^+ e^{-\Gamma_1 z} + V_{11}^- e^{\Gamma_1 z} \tag{5c}$$

$$I_1 = I_{11}^+ e^{-\Gamma_1 z} + I_{11}^- e^{\Gamma_1 z}. \tag{5d}$$

The power flow along the guide for the nth mode is

$$\text{Re} \, \frac{1}{2} \int_0^a \int_0^b a_n a_n^* Y_n^* \Phi_n^2 \, dx \, dy = \frac{b}{2} \, \text{Re}(a_n a_n^* Y_n^*) = \text{Re}\left(\frac{b}{2} V_n I_n^*\right). \tag{6}$$

For an evanescent mode the real power flow is zero.

At a distance d in front of the junction, an input impedance may be defined by the relation

$$Z_{\text{in}}(d) = \frac{V_1(d)}{I_1(d)} = \frac{e^{\Gamma_1 d} + R_1 e^{-\Gamma_1 d}}{e^{\Gamma_1 d} - R_1 e^{-\Gamma_1 d}} Z_1 = Z_1 \frac{Z_L + j Z_1 \tan \beta_1 d}{Z_1 + j Z_L \tan \beta_1 d} \tag{7}$$

where $\beta_1 = |\Gamma_1|$, and Z_L is the equivalent input impedance referred to the junction plane $z = 0$, that is,

$$Z_L = \frac{1 + R_1}{1 - R_1} Z_1. \tag{8}$$

A description in terms of an equivalent transmission-line circuit is not necessary, but does provide very convenient terminology for the description of a junction. Usually it is the practice to describe a junction in terms of normalized impedance and reactive elements. In the present case, the normalized input impedance of the junction is

$$\frac{Z_L}{Z_1} = \frac{1 + R_1}{1 - R_1}.$$

Knowledge of the normalized input impedance permits the physically meaningful reflection coefficient to be computed at once.

So far we have written down expansions for the transverse fields in the input guide only. In the output guide an infinite number of H_{n0} modes are excited. The method of solution for these has been covered in Chapter 6. For the purpose of discussion here, all we need to know is that an infinite number of H_{n0} modes exist; that these modes form a complete set, in terms of which an arbitrary electromagnetic field having only a y component of electric field, and no variation with y, may be expanded; and, finally, that these functions may be normalized so that the eigenfunction set forms an orthonormal system. We will let $\psi_m(x)$ be the mth eigenfunction for the inhomogeneously filled guide. These functions satisfy the orthogonal relations

$$\int_0^a \psi_m(x)\psi_n(x)\,dx = \delta_{nm} \tag{9}$$

where the Kronecker delta δ_{nm} equals unity for $n = m$, and zero otherwise.

The transverse fields for $z \geq 0$ are given by

$$E_y = \sum_{m=1}^\infty b_m \psi_m e^{-\gamma_m z} \tag{10a}$$

$$H_x = -\sum_{m=1}^\infty b_m Y_{0m}\psi_m e^{-\gamma_m z} \tag{10b}$$

where b_m are amplitude coefficients, Y_{0m} is the mode admittance γ_m/jk_0Z_0, and γ_m is the propagation constant. The propagation constants are determined by the equations

$$\frac{\tan l_m t}{l_m} = \frac{-\tan k_m s}{k_m} \tag{11a}$$

$$\gamma_m^2 = l_m^2 - \kappa k_0^2 = k_m^2 - k_0^2. \tag{11b}$$

It will be assumed that the guide dimensions and frequency have been chosen so that only the $m = 1$ mode propagates. Thus all Γ_n and γ_n are real except Γ_1 and γ_1, which are pure-imaginary. For the output guide, the equivalent transmission-line voltages and currents will be defined by

$$V_{n2}^+ = b_n$$

$$I_{n2}^+ = Y_{0n}b_n.$$

At the junction $z = 0$, the transverse fields must be continuous. This condition leads to the following two equations:

$$a_1(1+R_1)\Phi_1 + \sum_{n=2}^\infty a_n \Phi_n = \sum_{m=1}^\infty b_m \psi_m \tag{12a}$$

$$a_1(1-R_1)Y_1\Phi_1 - \sum_{n=2}^\infty a_n Y_n \Phi_n = \sum_{m=1}^\infty b_m Y_{0m}\psi_m. \tag{12b}$$

A simultaneous solution of these two equations would give us a complete description of the field. However, it is quite apparent that such a solution would be extremely difficult to obtain, and recourse to an approximate method must be made. The variational method is capable of giving an accurate approximate solution with a minimum of computational labor, so we will now proceed to develop this method.

Method 1

The two fundamental continuity equations, i.e., Eqs. (12), will be repeated here for convenience with the amplitude of the incident wave assumed to be unity. Thus

$$(1 + R_1)\Phi_1 + \sum_2^\infty a_n \Phi_n = \sum_1^\infty b_m \psi_m \tag{13a}$$

$$(1 - R_1)Y_1\Phi_1 = \sum_2^\infty a_n Y_n \Phi_n + \sum_1^\infty b_m Y_{0m} \psi_m. \tag{13b}$$

In the aperture plane $z = 0$, let the electric field distribution be $\mathcal{E}(x)$. Since Φ_n and ψ_m form complete sets of orthonormal functions, we may develop $\mathcal{E}(x)$ into a Fourier series in terms of them; i.e.,

$$\mathcal{E}(x) = (1 + R_1)\Phi_1 + \sum_2^\infty a_n \Phi_n = \sum_1^\infty b_m \psi_m$$

so that

$$1 + R_1 = \int_0^a \mathcal{E}\Phi_1 \, dx$$

$$a_n = \int_0^a \mathcal{E}\Phi_n \, dx$$

$$b_m = \int_0^a \mathcal{E}\psi_m \, dx.$$

The coefficients a_n and b_m are thus expressible in terms of the aperture electric field distribution. The continuity of the electric field in the aperture is automatically ensured since both sides of the first equation in (13) have been equated to $\mathcal{E}(x)$. If these values of the coefficients are substituted into the second equation, we obtain

$$(1 - R_1)Y_1\Phi_1 \frac{\displaystyle\int_0^a \mathcal{E}\Phi_1 \, dx}{1 + R_1} = \sum_2^\infty Y_n \Phi_n \int_0^a \mathcal{E}\Phi_n \, dx + \sum_1^\infty Y_{0m} \psi_m \int_0^a \mathcal{E}\psi_m \, dx. \tag{14a}$$

Interchanging the order of integration and summation gives

$$\frac{1 - R_1}{1 + R_1} Y_1 \Phi_1 \int_0^a \mathcal{E}\Phi_1 \, dx = \int_0^a \mathcal{E}(x')G_1(x|x') \, dx' \tag{14b}$$

where

$$G_1(x|x') = \sum_2^\infty Y_n \Phi_n(x)\Phi_n(x') + \sum_1^\infty Y_{0m}\psi_m(x)\psi_m(x').$$

The function $G_1(x|x')$ is symmetrical in the variables x and x', and for convenience it will be called a Green's function, although, strictly speaking, it is not a true Green's function in the normal sense. Equation (14b) is an identity, and we may, therefore, multiply both sides by an arbitrary function of x and integrate from $0 \leq x \leq a$. Let this arbitrary function of x be $\mathcal{E}_1(x)$; thus

$$Y_{in}Y_1 \int\!\!\int_0^a \mathcal{E}_1(x)\mathcal{E}(x')\Phi_1(x)\Phi_1(x')\,dx\,dx' = \int\!\!\int_0^a \mathcal{E}_1(x)\mathcal{E}(x')G_1(x|x')\,dx\,dx'$$

where Y_{in} has been written for $(1-R_1)/(1+R_1)$, since it is the normalized input admittance at the plane $z = 0$. Let us calculate the variation in Y_{in} due to a small variation in the functional form of the function $\mathcal{E}_1(x)$. We have

$$Y_{in}Y_1 \int_0^a \delta\mathcal{E}_1 \Phi_1\,dx \int_0^a \mathcal{E}\Phi_1\,dx + \delta Y_{in}Y_1 \int_0^a \mathcal{E}_1\Phi_1\,dx \int_0^a \mathcal{E}\Phi_1\,dx$$

$$= \int_0^a \delta\mathcal{E}_1(x)\int_0^a G_1(x|x')\mathcal{E}(x')\,dx'\,dx$$

which may be written as

$$\int_0^a \delta\mathcal{E}_1 \left[Y_{in}Y_1\Phi_1 \int_0^a \mathcal{E}\Phi_1\,dx - \int_0^a G_1(x|x')\mathcal{E}(x')\,dx' \right] dx$$

$$+ \delta Y_{in}Y_1 \int_0^a \mathcal{E}_1\Phi_1\,dx \int_0^a \mathcal{E}\Phi_1\,dx = 0.$$

By virtue of (14b), the integrand vanishes, and, hence, δY_{in} is zero for a variation in $\mathcal{E}_1$. Since we want to make δY_{in} zero for an arbitrary small variation in $\mathcal{E}$ also, we conduct the variation with respect to $\mathcal{E}$ to obtain

$$\int_0^a \delta\mathcal{E} \left[Y_{in}Y_1\Phi_1 \int_0^a \mathcal{E}_1\Phi_1\,dx - \int_0^a G_1(x|x')\mathcal{E}_1(x)\,dx \right] dx'$$

$$+ \delta Y_{in}Y_1 \int_0^a \mathcal{E}_1\Phi_1\,dx \int_0^a \mathcal{E}\Phi_1\,dx = 0$$

as the condition to be imposed on $\mathcal{E}_1$. Examination of this equation shows that δY_{in} will equal zero if $\mathcal{E}_1 = \mathcal{E}$.

We, therefore, have the following variational expression for Y_{in}:

$$Y_{in} = \frac{\displaystyle\int\!\!\int_0^a G_1(x|x')\mathcal{E}(x)\mathcal{E}(x')\,dx\,dx'}{Y_1 \left[\displaystyle\int_0^a \mathcal{E}(x)\Phi_1\,dx \right]^2} \tag{15}$$

which has the property that Y_{in} is stationary for small arbitrary variations in the electric field distribution about its correct value. It, therefore, follows that a first-order approximation to the electric field distribution will yield a second-order approximation to the input admittance. Also, the expression for Y_{in} is homogeneous in the sense that it does not depend on the amplitude of the electric field distribution $\mathcal{E}(x)$, but only on its functional form.

As an alternative to the above formulation, we may derive a similar expression for the normalized input impedance $(1 + R_1)/(1 - R_1)$. If we let the transverse magnetic field in the aperture plane be $\mathcal{H}(x)$, then

$$1 - R_1 = Z_1 \int_0^a \mathcal{H}\Phi_1 \, dx$$

$$a_n = -Z_n \int_0^a \mathcal{H}\Phi_n \, dx$$

$$b_m = Z_{0m} \int_0^a \mathcal{H}\psi_m \, dx.$$

Using the same procedure as before, we find that

$$Z_{\text{in}} = \frac{\displaystyle\int\!\!\int_0^a G_2(x|x')\mathcal{H}(x)\mathcal{H}(x') \, dx \, dx'}{Z_1 \left[\displaystyle\int_0^a \mathcal{H}(x)\Phi_1 \, dx\right]^2} \tag{16}$$

where

$$G_2(x|x') = \sum_2^\infty Z_n \Phi_n(x)\Phi_n(x') + \sum_1^\infty Z_{0m}\psi_m(x)\psi_m(x').$$

We will now show how, in practice, we may obtain a suitable approximation to the aperture fields and, hence, to Y_{in} and Z_{in}. First it will be necessary to consider some of the properties of the expansion of one set of complete orthonormal functions in terms of another set. We may expand each function ψ_m in a Fourier series consisting of the functions Φ_n as follows:

$$\psi_m = \sum_{n=1}^\infty P_{nm}\Phi_n \tag{17}$$

where the coefficients P_{nm} are given by $\int_0^a \Phi_n \psi_m \, dx = P_{nm}$. Similarly, we may expand Φ_n in terms of ψ_m as follows:

$$\Phi_n = \sum_{m=1}^\infty P_{nm}\psi_m \tag{18}$$

where

$$P_{nm} = \int_0^a \Phi_n \psi_m \, dx.$$

Substituting the expansion for ψ_m into (18) gives

$$\Phi_n = \sum_{m=1}^{\infty} P_{nm} \sum_{s=1}^{\infty} P_{sm} \Phi_s = \sum_{s=1}^{\infty} \Phi_s \sum_{m=1}^{\infty} P_{nm} P_{sm} \qquad (19)$$

from which we conclude that

$$\sum_{m=1}^{\infty} P_{nm} P_{sm} = \delta_{sn}$$

and hence

$$\sum_{m=1}^{\infty} P_{nm}^2 = 1 \quad \text{and} \quad \sum_{m=1}^{\infty} P_{nm} P_{sm} = 0, s \neq n.$$

Similarly, we find that

$$\sum_{n=1}^{\infty} P_{nm} P_{nr} = \delta_{mr}.$$

The transformation from one set of orthonormal functions to another set is analogous to the transformation from one orthogonal coordinate system to another one. The matrix of the transformation is composed of the elements P_{nm}. The product of the matrix $[P_{nm}]$ with its transpose $[P_{mn}]$ is equal to the unit matrix so that the matrix $[P_{nm}]$ is unitary. Since the elements are all real, the matrix of the transformation is also orthogonal.

For a suitable approximation to the aperture electric field $\mathcal{E}(x)$, we may take a finite series of the functions $\Phi_n(x)$, say

$$\sum_{1}^{N} a_n \Phi_n.$$

Since Y_{in} depends only on the functional form of $\mathcal{E}$, we may take a_1 equal to unity without any loss in generality. The variational expression for Y_{in} involves the integral of $G_1(x|x')$ times the functions $\Phi_n(x)\Phi_n(x')$ and is computed as follows:

$$\sum_{s=1}^{N} \sum_{r=1}^{N} \int\!\!\int_0^a G_1(x|x')\Phi_s(x)\Phi_r(x')a_r a_s \, dx \, dx'$$

$$= \sum_{s=1}^{N} \sum_{r=1}^{N} a_s a_r \int\!\!\int_0^a \Phi_r(x')\Phi_s(x) \left[\sum_{n=2}^{\infty} Y_n \Phi_n(x)\Phi_n(x') + \sum_{m=1}^{\infty} Y_{0m}\psi_m(x)\psi_m(x') \right] dx \, dx'$$

$$= \sum_{n=2}^{N} Y_n a_n^2 + \sum_{s=1}^{N} \sum_{r=1}^{N} a_r a_s \sum_{m=1}^{\infty} Y_{0m} P_{rm} P_{sm}$$

$$= \sum_{s=1}^{N} \sum_{r=1}^{N} a_s a_r \left(\sum_{n=2}^{\infty} Y_n \delta_{ns}\delta_{nr} + \sum_{m=1}^{\infty} Y_{0m} P_{sm} P_{rm} \right). \qquad (20)$$

For brevity, we will denote the quantity

$$\sum_{n=2}^{\infty} Y_n \delta_{ns} \delta_{nr} + \sum_{m=1}^{\infty} Y_{0m} P_{sm} P_{rm}$$

by g_{sr}. It should be noted that $g_{sr} = g_{rs}$. Thus, (15) for Y_{in} becomes

$$Y_{\text{in}} = \frac{1}{Y_1} \sum_{s=1}^{N} \sum_{r=1}^{N} a_s a_r g_{sr}. \tag{21}$$

Since P_{rm} and P_{sm} approach zero as m becomes large when r and s are different from m, the summation over m in the evaluation of g_{sr} may be terminated at a given finite value of m consistent with the required accuracy. We must now impose on (21) the condition that Y_{in} should be stationary for arbitrary variations in $\mathcal{E}$, that is, in a_s. This condition is obtained by equating the partial derivatives of the unknown coefficients a_s to zero and gives the best possible solution for Y_{in} that can be obtained with the assumed approximation for $\mathcal{E}$. From (21) we get

$$\sum_{r=1}^{N} a_r g_{sr} = 0, \qquad s = 2, 3, 4, \ldots, N \tag{22}$$

since g_{sr} is symmetrical in the indices s and r. The coefficient a_1 has been taken equal to unity, and hence $\partial Y_{\text{in}}/\partial a_1$ is automatically zero. Equation (22) is a set of $N - 1$ equations which may be solved for the $N - 1$ unknowns a_r in terms of $g_{11}a_1$ or g_{11}. These values of a_r are now substituted back into (21), and Y_{in} is evaluated. We may handle (16) for Z_{in} in an entirely analogous manner. The following points are of interest:

1. Since Y_{01} is real while all Y_n and Y_{0m} for n and m greater than one are imaginary, the factors g_{sr} are complex. Consequently, the coefficients a_r obtained from (22) are complex also.
2. Since the a_r are complex, Y_{in} is complex, as, of course, it must be, since it represents a transmission of power past the discontinuity plus a storage of reactive energy at the junction.
3. Because of the complex nature of Y_{in}, the stationary value of Y_{in} is neither a maximum nor a minimum. The same is true of the stationary value of Z_{in}. Therefore, a comparison of the calculated values of Y_{in} and Z_{in}^{-1} will not give upper and lower bounds to the true value of Y_{in}.
4. The calculation of Y_{in} gives only two of the three parameters required to specify the junction completely. To compute the third parameter, the above analysis has to be repeated for a wave incident on the junction from the inhomogeneously filled region.
5. The necessity of carrying out the analysis twice and the fact that the coefficients a_r are complex make the numerical computation laborious and lengthy.

We may organize the numerical work in a more compact fashion by treating (15) for Y_{in} [similarly with (16) for Z_{in}] in a somewhat different manner. As before, we approximate $\mathcal{E}(x)$ by a finite series of the Φ_n, but we do not equate a_1 to unity; rather we leave it as arbitrary for the present. In place of (21), we then get

$$Y_{\text{in}}Y_1 a_1^2 - \sum_{s=1}^{N}\sum_{r=1}^{N} a_s a_r g_{sr} = 0. \tag{23}$$

When we equate the partial derivatives with respect to a_s equal to zero, we get

$$2Y_{\text{in}}Y_1 a_1 - 2\sum_{r=1}^{N} a_r g_{1r} = 0 \tag{24a}$$

$$-2\sum_{r=1}^{N} a_r g_{sr} = 0, \qquad s = 2, 3, \ldots, N. \tag{24b}$$

This is a set of N homogeneous equations for N unknowns, which has a solution only if the determinant vanishes. The vanishing of the determinant gives the solution for Y_{in} without the necessity of computing the values of the coefficients. We have

$$\begin{vmatrix} Y_{\text{in}}Y_1 - g_{11} & -g_{12} & \cdots & -g_{1N} \\ -g_{21} & \cdots & \cdots & \cdots \\ \cdots\cdots\cdots\cdots\cdots\cdots\cdots\cdots \\ -g_{N1} & -g_{N2} & \cdots & -g_{NN} \end{vmatrix} = 0.$$

This determinant may be factored into the sum of two determinants as follows:

$$\begin{vmatrix} Y_{\text{in}}Y_1 & -g_{12} & \cdots & -g_{1N} \\ 0 & -g_{22} & \cdots & \cdots \\ \cdots\cdots\cdots\cdots\cdots\cdots\cdots \\ 0 & -g_{N2} & \cdots & -g_{NN} \end{vmatrix} + (-1)^N \begin{vmatrix} g_{11} & \cdots & g_{1N} \\ g_{21} & \cdots & \cdots \\ \cdots\cdots\cdots\cdots \\ g_{N1} & \cdots & g_{NN} \end{vmatrix}$$

and, hence,

$$Y_{\text{in}} = \frac{1}{Y_1} \frac{\begin{vmatrix} g_{11} & \cdots & g_{1N} \\ \cdots\cdots\cdots\cdots\cdots \\ g_{N1} & \cdots & g_{NN} \end{vmatrix}}{\begin{vmatrix} g_{22} & \cdots & g_{2N} \\ \cdots\cdots\cdots\cdots\cdots \\ g_{N2} & \cdots & g_{NN} \end{vmatrix}}. \tag{25}$$

Thus we have reduced the computation of Y_{in} to the evaluation of two determinants of order N and $N-1$.

Method 2

The main objections to the previously derived variational solution are those listed in the previous section as points 3, 4, and 5. All these objections can be removed by a modification of the method. If we terminate the output guide in a short circuit located at $z = l$, we have a pure reactive termination since all of the incident power is reflected. Under these conditions Y_{in} is pure-imaginary, and the electric field is everywhere in time quadrature with the magnetic field. We also find that the field distributions in the aperture plane, apart from a constant multiplier which may be complex, may be represented by a real function. Since the variational expressions depend only on the functional form of the aperture fields, this constant multiplier may be dropped. We will assume that only the dominant mode propagates in the output guide, and the short circuit will be placed at $z = l + n\lambda_{0g}/2$, where n is chosen large enough so that all evanescent modes have decayed to a negligible value at the short circuit. The guide wavelength in the inhomogeneously filled guide is denoted by λ_{0g}.

In order for the electric field to vanish at the short circuit, the field of the dominant mode in the output guide must be

$$b_1 \sin \beta_{01}(z - l)\psi_1(x) \tag{26}$$

where $\beta_{01} = |\gamma_1|$ is the phase constant for the output guide. From (1), the transverse magnetic field is

$$-jY_{01}b_1\psi_1 \cos \beta_{01}(z - l). \tag{27}$$

If the aperture electric field is $\mathcal{E}(x)$, the coefficient b_1 is given by

$$b_1 = -\frac{1}{\sin \beta_{01}l} \int_0^a \mathcal{E}\psi_1 \, dx.$$

Since all the incident energy is reflected, the modulus of the reflection coefficient R_1 is unity. We may therefore put R_1 equal to $e^{j\theta}$, and, hence,

$$Y_{in} = \frac{1 - e^{j\theta}}{1 + e^{j\theta}} = -j \tan \frac{\theta}{2}.$$

In the input guide, the dominant-mode electric field will have a series of maxima and minima as a function of z. The first minimum occurs when the incident and reflected fields are out of phase, i.e., when

$$e^{-j\beta_1 z} + e^{j(\theta + \beta_1 z)} = 0.$$

Let the corresponding value of $|z|$ be d, and we get $\theta = \pi + 2\beta_1 d$, and, hence,

$$Y_{in} = -j \tan \left(\frac{\pi}{2} + \beta_1 d \right) = j \cot \beta_1 d. \tag{28}$$

The input admittance at the junction is given by (28) in terms of the location or position d of the electric field minimum from the junction. We may now carry through the same analysis as before, and, in place of (15), we get

$$\cot \beta_1 d = -\frac{\displaystyle\iint_0^a G_1'(x|x')\mathcal{E}(x)\mathcal{E}(x') \, dx \, dx'}{Y_1 \left(\displaystyle\int_0^a \mathcal{E}\Phi_1 \, dx \right)^2} \tag{29}$$

where

$$G'_1(x|x') = \sum_{n=2}^{\infty} j[Y_n\Phi_n(x)\Phi_n(x')] + \sum_{m=2}^{\infty} j[Y_{0m}\psi_m(x)\psi_m(x')] + Y_{01}\psi_1(x)\psi_1(x')\cot\beta_{01}l.$$

We may handle (29) in the same manner as (15). If we define g'_{sr} as

$$g'_{sr} = \int\int_0^a G'_1(x|x')\Phi_s(x)\Phi_r(x')\,dx\,dx' = j\left(\sum_{n=2}^{\infty} Y_n\delta_{ns}\delta_{nr}\right.$$

$$\left. + \sum_{m=2}^{\infty} Y_{0m}P_{sm}P_{rm}\right) + Y_{01}P_{s1}P_{r1}\cot\beta_{01}l$$

we get, in place of (21),

$$\cot\beta_1 d = -\frac{1}{Y_1 a_1^2}\sum_{s=1}^{N}\sum_{r=1}^{N} a_s a_r g'_{sr}. \tag{30}$$

As before, we may take a_1 equal to unity, and then determine the other $N-1$ coefficients so that $\cot\beta_1 d$ is stationary. Let us compute the second-order variation in $\cot\beta_1 d$ due to arbitrary, but small, variations in the coefficients a_s.

Taylor's expansion of a function of N variables may be written symbolically as

$$F(x_1, x_2, \ldots, x_N) = \left(\exp\sum_{s=1}^{N}\Delta x_s\frac{\partial}{\partial x_s}\right)F(x_1,\ldots,x_N)\Bigg|_{x_i=x_{i0}}$$

$$= \left(1 + \sum_{s=1}^{N}\Delta x_s\frac{\partial}{\partial x_s} + \frac{1}{2!}\sum_{s=1}^{N}\sum_{r=1}^{N}\Delta x_s\Delta x_r\frac{\partial}{\partial x_s}\frac{\partial}{\partial x_r} + \cdots +\right)$$

$$\times F(x_1,\ldots,x_N)|_{x_i=x_{i0}}$$

where all the partial derivatives are to be evaluated at the point $x_i = x_{i0}$, $i = 1, 2, \ldots, N$, with x_{i0} being the initial values of the variables. This expansion gives the variation or change in the function F correct to the second order if we retain the first three terms in the expansion of the exponential. Applying this result to $\cot\beta_1 d$, we get

$$Y_1[\cot\beta_1 d - \cot\beta_1 d|_{a_s=a_{s0}}] = Y_1\delta^2\cot\beta_1 d$$

$$= -\left(\sum_{i=2}^{N}\Delta a_i\frac{\partial}{\partial a_i} + \frac{1}{2}\sum_{i=2}^{N}\sum_{k=2}^{N}\Delta a_i\Delta a_k\frac{\partial}{\partial a_i}\frac{\partial}{\partial a_k}\right)$$

$$\cdot\sum_{s=1}^{N}\sum_{r=1}^{N} a_s a_r g'_{sr}|_{a_i=a_{i0}}.$$

The operation $\sum_{i=2}^{N}\Delta a_i(\partial/\partial a_i)$ gives the first variation of $\cot\beta_1 d$, and, since it is assumed that we have chosen the a_s so that $\cot\beta_1 d$ is stationary, the first variation vanishes, and we are left with

$$Y_1\delta^2 \cot \beta_1 d = -\frac{1}{2}\left(\sum_{i=2}^{N}\sum_{k=2}^{N} \Delta a_i \,\Delta a_k \,\frac{\partial}{\partial a_i}\frac{\partial}{\partial a_k}\right)\sum_{s=1}^{N}\sum_{r=1}^{N} a_s a_r g'_{sr}\big|_{a_i=a_{i0}}$$

$$= -\sum_{i=2}^{N}\Delta a_i \,\frac{\partial}{\partial a_i}\sum_{k=2}^{N}\Delta a_k \sum_{s=1}^{N} a_s g'_{ks}\big|_{a_i=a_{i0}} = -\sum_{k=2}^{N}\sum_{i=2}^{N}\Delta a_i \,\Delta a_k \,g'_{ki}. \quad (31)$$

This is, in fact, the total variation in $\cot \beta_1 d$, since there are now no coefficients a_s left on which higher order operators in the Taylor-series expansion can operate.

In (31) the second variation in $\cot \beta_1 d$ may be written as

$$-\sum_{k=2}^{N}\sum_{i=2}^{N} \Delta a_k \,\Delta a_i \left[\sum_{n=2}^{\infty}(jY_n\delta_{nk}\delta_{ni}+jY_{0n}P_{kn}P_{in})+Y_{01}P_{k1}P_{i1}\cot \beta_{01}l\right]$$

$$= -\sum_{n=2}^{\infty}\left[jY_n\left(\sum_{k=2}^{N}\Delta a_k \,\delta_{nk}\right)^2+jY_{0n}\left(\sum_{k=2}^{N}\Delta a_k \,P_{kn}\right)^2\right]$$

$$- Y_{01}\cot \beta_{01}l\left(\sum_{k=2}^{N}\Delta a_k \,P_{k1}\right)^2.$$

When $\cot \beta_{01}l$ is positive, the second variation is always a negative real quantity since jY_n, jY_{0n}, and Y_{01} are real and positive. Apart from the minus sign, the second variation is a positive definite quadratic form. The same is true for (30). Therefore, any variation in the coefficients a_s from their true value will decrease the value of $\cot \beta_1 d$. Thus our approximation for $\mathcal{E}(x)$ will give a value for $\cot \beta_1 d$ which is always algebraically too small, except for that particular choice of trial functions that gives a correct functional form for $\mathcal{E}(x)$.

We may solve our problem in a similar fashion for the input impedance involving the transverse magnetic aperture field $\mathcal{H}(x)$. We would then get

$$\tan \beta_1 d = -\frac{1}{Z_1 a_1^2}\sum_{s=1}^{N}\sum_{r=1}^{N} a_s a_r h'_{sr} \quad (32)$$

where

$$h'_{sr} = -j\left(\sum_{n=2}^{\infty}Z_n\delta_{ns}\delta_{nr}+\sum_{m=2}^{\infty}Z_{0m}P_{rm}P_{sm}\right)+Z_{01}P_{s1}P_{r1}\tan \beta_{01}l.$$

This expression gives a lower algebraic bound on $\tan \beta_1 d$ and, consequently, an upper bound on $(\tan \beta_1 d)^{-1} = \cot \beta_1 d$. We thus have available two expressions giving upper and lower bounds on $\cot \beta_1 d$ and, therefore, are able to find the maximum error in our approximate solution.

Equation (30) or (32) is the analytic equivalent of the experimental tangent method for finding the equivalent-circuit parameters of a lossless junction. The solutions given by (30) or

(32) are always of the form

$$\tan \beta_1 d = \frac{A + B \tan \beta_{01} l}{C + D \tan \beta_{01} l}$$

and the equivalent-circuit parameters may be found in terms of the coefficients A, B, C, and D by the methods given in Section 5.8.

As in the first method, it is not necessary to solve for the unknown coefficients. We may obtain the values of $\cot \beta_1 d$ and $\tan \beta_1 d$ from the ratio of two determinants. Carrying out the analysis gives

$$\cot \beta_1 d \begin{vmatrix} g'_{22} & \cdots & g'_{2N} \\ \cdots\cdots\cdots\cdots\cdots \\ g'_{N2} & \cdots & g'_{NN} \end{vmatrix} = -Z_1 \begin{vmatrix} g'_{11} & \cdots & g'_{1N} \\ \cdots\cdots\cdots\cdots\cdots \\ g'_{N1} & \cdots & g'_{NN} \end{vmatrix} \tag{33a}$$

$$\tan \beta_1 d \begin{vmatrix} h'_{22} & \cdots & h'_{2N} \\ \cdots\cdots\cdots\cdots\cdots \\ h'_{N2} & \cdots & h'_{NN} \end{vmatrix} = -Y_1 \begin{vmatrix} h'_{11} & \cdots & h'_{1N} \\ \cdots\cdots\cdots\cdots\cdots \\ h'_{N1} & \cdots & h'_{NN} \end{vmatrix}. \tag{33b}$$

The advantages of this second method over the first method are:

1. Upper and lower bounds for the equivalent-circuit parameters are obtained and, hence, knowledge of the maximum error involved is also obtained.
2. All the functions needed in the computation are real, and this simplifies the numerical work.
3. The method gives a convenient solution for all three parameters required for a complete description of the junction.

Method 3

The third method which we shall illustrate gives a direct solution for the impedance elements in the equivalent T-network representation of the junction or, alternatively, the admittance elements in the π-network representation.

If we assume an incident dominant mode of amplitude b_1 from the region $z > 0$, as well as an incident mode from the region $z < 0$, the two continuity equations at the aperture plane $z = 0$ become

$$[(1 + R_1)a_1 + T_{21}b_1]\Phi_1 + \sum_2^\infty a_n \Phi_n = [(1 + R_2)b_1 + T_{12}a_1]\psi_1 + \sum_2^\infty b_m \psi_m \tag{34a}$$

$$[(1 - R_1)a_1 - T_{21}b_1]Y_1\Phi_1 - \sum_2^\infty a_n Y_n \Phi_n = -[(1 - R_2)b_1 - T_{12}a_1]Y_{01}\psi_1 + \sum_2^\infty b_m Y_{0m}\psi_m$$

$$\tag{34b}$$

where R_2 is the reflection coefficient in the inhomogeneously filled guide, T_{12} is the transmission coefficient from the empty guide into the partially filled guide, and T_{21} is the transmission coefficient in the reverse direction. Equations (34) have been written so as to show explicitly that the dominant mode propagating away from the junction on each side depends linearly on the amplitudes of the incident modes on both sides of the junction. The following equivalent transmission-line voltages and currents are now introduced:

$$V_1 = (1 + R_1)a_1 + T_{21}b_1 \qquad\qquad V_2 = (1 + R_2)b_1 + T_{12}a_1$$

$$I_1 = (1 - R_1)Y_1a_1 - T_{21}Y_1b_1 \qquad I_2 = -(1 - R_2)Y_{01}b_1 + T_{21}Y_{01}a_1.$$

The amplitudes a_1, b_1 may be chosen so that I_2 equals zero. The aperture electric field, which is proportional to a_1 and b_1 and, hence, proportional to I_1 and I_2 in general, in this case is proportional to I_1 only. If incident amplitudes are chosen so that I_1 equals zero, the aperture electric field will be proportional to I_2 only. For arbitrary amplitudes for the incident modes the aperture electric field is just a linear superposition of the fields which are proportional to I_1 and I_2, and may be written as

$$\mathcal{E}(x) = I_1\mathcal{E}_1(x) - I_2\mathcal{E}_2(x). \tag{35}$$

By the usual method of Fourier analysis, the equivalent voltages V_1, V_2 and the amplitude coefficients a_n, b_m are found to be given by

$$V_1 = \int_0^a (I_1\mathcal{E}_1 - I_2\mathcal{E}_2)\Phi_1 \, dx \tag{36a}$$

$$V_2 = \int_0^a (I_1\mathcal{E}_1 - I_2\mathcal{E}_2)\psi_1 \, dx \tag{36b}$$

$$a_n = \int_0^a (I_1\mathcal{E}_1 - I_2\mathcal{E}_2)\Phi_n \, dx \tag{36c}$$

$$b_m = \int_0^a (I_1\mathcal{E}_1 - I_2\mathcal{E}_2)\psi_m \, dx. \tag{36d}$$

The first two equations show that the voltages and currents are related by linear relations of the form

$$V_1 = Z_{11}I_1 - Z_{12}I_2 \tag{37a}$$

$$V_2 = Z_{21}I_1 - Z_{22}I_2. \tag{37b}$$

It will be shown that $Z_{12} = Z_{21}$, and hence (37) are a system of equations which may be interpreted as those describing an equivalent T network for the junction as illustrated in Fig. 8.2. The impedance elements are given by

$$Z_{11} = \int_0^a \mathcal{E}_1\Phi_1 \, dx \tag{38a}$$

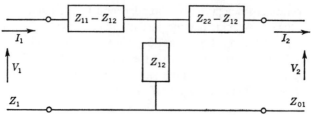

Fig. 8.2. Equivalent T network for a junction.

$$Z_{12} = \int_0^a \mathcal{E}_2 \Phi_1 \, dx \tag{38b}$$

$$Z_{21} = \int_0^a \mathcal{E}_1 \psi_1 \, dx \tag{38c}$$

$$Z_{22} = \int_0^a \mathcal{E}_2 \psi_1 \, dx. \tag{38d}$$

Substituting for a_n, b_m from (36) into (34b) gives

$$I_1 \Phi_1 - I_2 \psi_1 = -j \int_0^a [I_1 \mathcal{E}_1(x') - I_2 \mathcal{E}_2(x')] G_1(x|x') \, dx' \tag{39}$$

where, for convenience, we have defined

$$-jG_1 = \sum_2^\infty [Y_n \Phi_n(x)\Phi_n(x') + Y_{0n}\psi_n(x)\psi_n(x')]$$

$$= -j \sum_2^\infty [|Y_n|\Phi_n(x)\Phi_n(x') + |Y_{0n}|\psi_n(x)\psi_n(x')].$$

It should be noted that this definition of G_1 differs from that used in Method 1. For $n \geq 2$, the mode admittances Y_n and Y_{0n} are negative-imaginary, and hence G_1 is a real function. Since I_1 and I_2 have been chosen as the linearly independent variables, (39) is equivalent to the two separate equations

$$\Phi_1 = -j \int_0^a \mathcal{E}_1(x') G_1(x|x') \, dx' \tag{40a}$$

$$\psi_1 = -j \int_0^a \mathcal{E}_2(x') G_1(x|x') \, dx'. \tag{40b}$$

The required variational expressions for the impedance elements are obtained as follows. Multiplying (40a) by $\mathcal{E}_1(x)$ and integrating over x, dividing both sides by $(\int_0^a \mathcal{E}_1 \Phi_1 \, dx)^2$, and comparing the result with (38a), we get

$$\frac{1}{\int_0^a \mathcal{E}_1 \Phi_1 \, dx} = \frac{1}{Z_{11}} = -j \frac{\int_0^a \int_0^a G_1(x|x')\mathcal{E}_1(x)\mathcal{E}_1(x') \, dx \, dx'}{\left(\int_0^a \mathcal{E}_1 \Phi_1 \, dx\right)^2}. \tag{41a}$$

In a similar fashion we get from (40b)

$$\frac{1}{Z_{22}} = -j \frac{\displaystyle\int\!\!\!\int_0^a G_1(x|x')\mathcal{E}_2(x)\mathcal{E}_2(x')\,dx\,dx'}{\left(\displaystyle\int_0^a \mathcal{E}_2\psi_1\,dx\right)^2}. \tag{41b}$$

For Z_{21} we multiply (40a) by $\mathcal{E}_2(x)$, integrate over x, and divide by $\int_0^a \mathcal{E}_2\Phi_1\,dx\,\int_0^a \mathcal{E}_1\psi_1\,dx$ to get

$$\frac{1}{Z_{21}} = -j \frac{\displaystyle\int\!\!\!\int_0^a G_1(x|x')\mathcal{E}_2(x)\mathcal{E}_1(x')\,dx\,dx'}{\displaystyle\int\!\!\!\int_0^a \mathcal{E}_1(x)\mathcal{E}_2(x')\Phi_1(x')\psi_1(x)\,dx\,dx'}. \tag{41c}$$

If we begin with (40b), we obtain an expression for Z_{12} which is

$$\frac{1}{Z_{12}} = -j \frac{\displaystyle\int\!\!\!\int_0^a G_1(x|x')\mathcal{E}_2(x')\mathcal{E}_1(x)\,dx\,dx'}{\displaystyle\int\!\!\!\int_0^a \mathcal{E}_1(x)\mathcal{E}_2(x')\Phi_1(x')\psi_1(x)\,dx\,dx'} = \frac{1}{Z_{21}} \tag{41d}$$

and is identical with (41c), since $G_1(x|x') = G_1(x'|x)$; that is, it is symmetrical in the variables x, x'. For later convenience a new set of real impedance elements will be defined by the relations $Z_{ij}' = -jZ_{ij}; i, j = 1, 2.$

We can readily prove that the integral expressions for the impedance elements are stationary with respect to arbitrary first-order variations in the aperture field functions $\mathcal{E}_1$ and $\mathcal{E}_2$. For example, consider (41a), which gives the value of Z_{11}. If we vary the function $\mathcal{E}_1$ from its true value, we get

$$\left(\int_0^a \mathcal{E}_1\Phi_1\,dx\right)^2 \delta\frac{1}{Z_{11}} + \frac{2}{Z_{11}}\int\!\!\!\int_0^a \mathcal{E}_1(x')\delta\mathcal{E}_1(x)\Phi_1(x)\Phi_1(x')\,dx\,dx'$$

$$= -2j\int\!\!\!\int_0^a G_1(x|x')\mathcal{E}_1(x')\delta\mathcal{E}_1(x)\,dx\,dx'$$

or

$$\left(\int_0^a \mathcal{E}_1\Phi_1\,dx\right)^2 \delta\frac{1}{Z_{11}} = 2\int_0^a \delta\mathcal{E}_1(x')\left[-j\int_0^a G_1(x|x')\mathcal{E}_1(x)\,dx - \Phi_1(x')\right]dx'$$

after canceling $\int_0^a \mathcal{E}_1\Phi_1\,dx$ with Z_{11}. An examination of the integrand on the right-hand side shows that it is identically zero, by virtue of (40a). Therefore, we again have the property that a first-order approximation to the aperture fields will yield a second-order approximation to the impedance elements Z_{ij}.

The expressions (41) are homogeneous in the sense that they are independent of the magnitudes of the aperture fields $\mathcal{E}_1$ and $\mathcal{E}_2$. The impedance elements depend only on the functional forms of $\mathcal{E}_1$ and $\mathcal{E}_2$. Since Z_{11}', Z_{22}', and Z_{12}' are real quantities, and G_1 is real, the functions $\mathcal{E}_1$ and $\mathcal{E}_2$ are also real, with the possible exception of a constant complex multiplier, which cancels out and does not affect the magnitude of Z_{ij}'. We may hence show that the expressions for $(Z_{11}')^{-1}$ and $(Z_{22}')^{-1}$ are absolute minima for the correct field distributions. Thus approximate expressions for $\mathcal{E}_1$ and $\mathcal{E}_2$ will yield values for $(Z_{11}')^{-1}$ and $(Z_{22}')^{-1}$ which are numerically too large, i.e., values for Z_{11}' and Z_{22}' which are too small. The proof is as follows. Since Z_{11}' depends only on the functional form of $\mathcal{E}_1$, we may choose $\mathcal{E}_1$ so that $\int_0^a \mathcal{E}_1 \Phi_1 \, dx = 1$. We have therefore placed a constraint on $\mathcal{E}_1$ so that, when we vary $\mathcal{E}_1$, the variation will be orthogonal to Φ_1, in order to satisfy the condition $\int_0^a \mathcal{E}_1 \Phi_1 \, dx = 1$. Let $(Z_{11}')_0$ be the value of Z_{11}' for the correct field distribution. Thus we have

$$\delta \frac{1}{Z_{11}'} = \frac{1}{Z_{11}'} - \frac{1}{(Z_{11}')_0} = \frac{(Z_{11}')_0 - Z_{11}'}{Z_{11}'(Z_{11}')_0}$$

$$= \int\!\!\!\int_0^a G_1(x|x')[\mathcal{E}_1(x) + \delta\mathcal{E}_1(x)][\mathcal{E}_1(x') + \delta\mathcal{E}_1(x')] \, dx \, dx'$$

$$- \int\!\!\!\int_0^a G_1(x|x')\mathcal{E}_1(x)\mathcal{E}_1(x') \, dx \, dx'$$

$$= \int\!\!\!\int_0^a G_1(x|x')[\mathcal{E}_1(x)\delta\mathcal{E}_1(x') + \mathcal{E}_1(x')\delta\mathcal{E}_1(x)] \, dx \, dx'$$

$$+ \int\!\!\!\int_0^a G_1(x|x')\delta\mathcal{E}_1(x)\delta\mathcal{E}_1(x') \, dx \, dx'.$$

The first integral on the right is equal to $j \int_0^a \delta\mathcal{E}_1 \Phi_1 \, dx + j \int_0^a \delta\mathcal{E}_1 \Phi_1 \, dx$ in view of (40a) and vanishes since we have constrained $\delta\mathcal{E}_1$ to be orthogonal to Φ_1. Thus our result reduces to

$$\frac{(Z_{11}')_0 - Z_{11}'}{Z_{11}'(Z_{11}')_0} = \int\!\!\!\int_0^a G_1(x|x')\delta\mathcal{E}_1(x')\delta\mathcal{E}_1(x) \, dx \, dx'$$

which is a positive-definite quadratic form. Hence $(Z_{11}')_0 - Z_{11}'$ is positive, and $Z_{11}' \leq (Z_{11}')_0$, which proves the stated minimum property. The proof for Z_{12}' does not hold since the quantity

$$\int\!\!\!\int_0^a G_1(x|x')\delta\mathcal{E}_1(x)\delta\mathcal{E}_2(x') \, dx \, dx'$$

is not a positive-definite quadratic form; i.e., it may be either positive or negative (for example, if $\delta\mathcal{E}_1 = -\delta\mathcal{E}_2$, it will be negative).

The evaluation of the impedance elements may proceed along lines similar to those used in the previous two variational methods. Suitable approximations for the aperture fields $\mathcal{E}_1$ and $\mathcal{E}_2$ which lead to a straightforward computational procedure are

$$\mathcal{E}_1 = \sum_1^N a_n \Phi_n \qquad \mathcal{E}_2 = \sum_1^N b_m \psi_m.$$

For the impedance elements Z'_{11} and Z'_{22} we get

$$Z'_{11} \begin{vmatrix} g_{11} & \cdots & g_{1N} \\ \cdots\cdots\cdots\cdots\cdots \\ g_{N1} & \cdots & g_{NN} \end{vmatrix} = \begin{vmatrix} g_{22} & \cdots & g_{2N} \\ \cdots\cdots\cdots\cdots\cdots \\ g_{N2} & \cdots & g_{NN} \end{vmatrix} \tag{42a}$$

$$Z'_{22} \begin{vmatrix} t_{11} & \cdots & t_{1N} \\ \cdots\cdots\cdots\cdots\cdots \\ t_{N1} & \cdots & t_{NN} \end{vmatrix} = \begin{vmatrix} t_{22} & \cdots & t_{2N} \\ \cdots\cdots\cdots\cdots\cdots \\ t_{N2} & \cdots & t_{NN} \end{vmatrix} \tag{42b}$$

where

$$g_{sr} = \sum_{n=2}^{\infty} (|Y_n|\delta_{sn}\delta_{rn} + |Y_{0n}|P_{rn}P_{sn})$$

$$t_{sr} = \sum_{n=2}^{\infty} (|Y_n|P_{ns}P_{nr} + |Y_{0n}|\delta_{sn}\delta_{rn})$$

and P_{sn} is given by $\int_0^a \Phi_s \psi_n \, dx$.

The computation of Z'_{12} is not so straightforward, so we will work it out in more detail. From (41d), we have

$$(Z'_{12})^{-1} \int_0^a \sum_{s=1}^N a_s \Phi_s \psi_1 \, dx \int_0^a \sum_{r=1}^N b_r \psi_r \Phi_1 \, dx$$

$$= \frac{1}{Z'_{12}} \sum_{s=1}^N \sum_{r=1}^N a_s b_r P_{s1} P_{1r}$$

$$= \int\!\!\int_0^a \sum_{n=2}^{\infty} [|Y_n|\Phi_n(x)\Phi_n(x') + |Y_{0n}|\psi_n(x)\psi_n(x')]$$

$$\cdot \sum_{s=1}^N a_s \Phi_s(x') \sum_{r=1}^N b_r \psi_r(x) \, dx \, dx'$$

$$= \sum_{s=1}^N \sum_{r=1}^N a_s b_r \left[\sum_{n=2}^{\infty} (|Y_n|\delta_{ns}P_{nr} + |Y_{0n}|\delta_{nr}P_{sn}) \right]$$

$$= \sum_{s=1}^N \sum_{r=1}^N [a_s b_r P_{sr}(|Y_s| + |Y_{0r}|)]$$

and, hence,

$$\sum_{s=1}^{N}\sum_{r=1}^{N}a_s b_r P_{s1}P_{1r}\left[\frac{1}{Z'_{12}}-\frac{P_{sr}}{P_{s1}P_{1r}}(|Y_s|+|Y_{0r}|)\right]=\sum_{s=1}^{N}\sum_{r=1}^{N}a_s b_r P_{s1}P_{1r}\left(\frac{1}{Z'_{12}}-Q_{sr}\right)=0$$

where $Q_{sr}=(P_{sr}/P_{s1}P_{1r})(|Y_s|+|Y_{0r}|)$. It should be noted that Q_{sr} is not symmetrical in the indices s and r since $Q_{sr}\neq Q_{rs}$. When we equate the partial derivatives with respect to a_s and b_r to zero, we obtain a homogeneous set of $2N$ equations with the determinant

$$\begin{vmatrix}
\dfrac{1}{Z'_{12}}-Q_{11} & \cdots & \dfrac{1}{Z'_{12}}-Q_{N1} & \vline & 0 & \cdots & 0 \\
\cdots\cdots\cdots\cdots\cdots & & & \vline & \cdots\cdots\cdots\cdots\cdots & & \\
\dfrac{1}{Z'_{12}}-Q_{1N} & \cdots & \dfrac{1}{Z'_{12}}-Q_{NN} & \vline & 0 & \cdots & 0 \\
\hline
0 & \cdots & 0 & \vline & \dfrac{1}{Z'_{12}}-Q_{11} & \cdots & \dfrac{1}{Z'_{12}}-Q_{1N} \\
\cdots\cdots\cdots\cdots\cdots & & & \vline & \cdots\cdots\cdots\cdots\cdots & & \\
0 & \cdots & 0 & \vline & \dfrac{1}{Z'_{12}}-Q_{N1} & \cdots & \dfrac{1}{Z'_{12}}-Q_{NN}
\end{vmatrix}.$$

The vanishing of this determinant gives the solution for Z'_{12}. The above $2N\times 2N$ determinant is equal to the product of the determinant in the upper left-hand corner and that in the lower right-hand corner. We may interchange the rows and columns of the determinant in the lower right-hand corner without changing its value. This transposition of the rows and columns makes the two determinants equal, and thus Z'_{12} is given by the condition that the following determinant shall vanish:

$$\begin{vmatrix}
\dfrac{1}{Z'_{12}}-Q_{11} & \cdots & \dfrac{1}{Z'_{12}}-Q_{N1} \\
\dfrac{1}{Z'_{12}}-Q_{1N} & \cdots & \dfrac{1}{Z'_{12}}-Q_{NN}
\end{vmatrix}=0.$$

If the first row is subtracted from all of the other rows, the resultant determinant may be factored into the sum of two determinants which gives the solution for Z'_{12} in explicit form as follows:

$$Z'_{12}\begin{vmatrix}
Q_{11} & Q_{21} & \cdots & Q_{N1} \\
Q_{11}-Q_{12} & Q_{21}-Q_{22} & \cdots & Q_{N1}-Q_{N2} \\
\cdots\cdots\cdots\cdots\cdots\cdots\cdots\cdots\cdots\cdots\cdots \\
Q_{11}-Q_{1N} & \cdots & \cdots & Q_{N1}-Q_{NN}
\end{vmatrix}$$

$$=\begin{vmatrix}
1 & 1 & \cdots & 1 \\
Q_{11}-Q_{12} & Q_{21}-Q_{22} & \cdots & Q_{N1}-Q_{N2} \\
\cdots\cdots\cdots\cdots\cdots\cdots\cdots\cdots\cdots\cdots\cdots \\
Q_{11}-Q_{1N} & \cdots & \cdots & Q_{N1}-Q_{NN}
\end{vmatrix}. \quad (43)$$

A similar analysis may be carried out to obtain variational expressions for the admittance elements in the equivalent π-network representation of the junction. Our point of departure is again the set of equations (34) expressing the continuity of the tangential fields in the aperture plane. As before, a set of equivalent transmission-line voltages and currents is introduced. The incident mode amplitudes may be chosen so that the voltages V_1 and V_2 are the independent variables. The tangential magnetic field in the aperture plane may be chosen as the linear superposition of two partial fields proportional to V_1 and V_2, and thus

$$\mathfrak{K}(x) = V_1 \mathfrak{K}_1(x) - V_2 \mathfrak{K}_2(x).$$

The unknown coefficients are then given by Fourier analysis as follows:

$$I_1 = \int_0^a (V_1 \mathfrak{K}_1 - V_2 \mathfrak{K}_2) \Phi_1 \, dx \tag{44a}$$

$$I_2 = \int_0^a (V_1 \mathfrak{K}_1 - V_2 \mathfrak{K}_2) \psi_1 \, dx \tag{44b}$$

$$a_n = -Z_n \int_0^a (V_1 \mathfrak{K}_1 - V_2 \mathfrak{K}_2) \Phi_n \, dx \tag{44c}$$

$$b_n = Z_{0n} \int_0^a (V_1 \mathfrak{K}_1 - V_2 \mathfrak{K}_2) \psi_n \, dx. \tag{44d}$$

The first two equations give a linear relationship between the currents and voltages of the form

$$I_1 = Y_{11} V_1 - Y_{12} V_2 \tag{45a}$$

$$I_2 = Y_{21} V_1 - Y_{22} V_2 \tag{45b}$$

where the admittance elements are given by

$$Y_{11} = \int_0^a \mathfrak{K}_1 \Phi_1 \, dx \tag{46a}$$

$$Y_{12} = \int_0^a \mathfrak{K}_2 \Phi_1 \, dx \tag{46b}$$

$$Y_{21} = \int_0^a \mathfrak{K}_1 \Phi_1 \, dx \tag{46c}$$

$$Y_{22} = \int_0^a \mathfrak{K}_2 \Phi_1 \, dx. \tag{46d}$$

Further along in the analysis we will find that $Y_{12} = Y_{21}$, and hence (45) are equations which describe the behavior of the equivalent π network illustrated in Fig. 8.3. If we invert the set

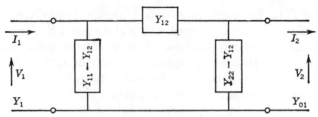

Fig. 8.3. Equivalent π network for a junction.

of Eqs. (37), we get

$$I_1 = \frac{Z_{22}}{\Delta} V_1 - \frac{Z_{12}}{\Delta} V_2$$

$$I_2 = \frac{Z_{12}}{\Delta} V_1 - \frac{Z_{11}}{\Delta} V_2$$

where $\Delta = Z_{11}Z_{22} - Z_{12}^2$. Thus we see that $Y_{11} = Z_{22}/\Delta$, $Y_{12} = Z_{12}/\Delta$, $Y_{22} = Z_{11}/\Delta$. The admittance elements Y_{ij} are pure-imaginary for a lossless function, and, for convenience, we will define the admittance elements Y'_{ij} as jY_{ij}. The admittance elements Y'_{ij} as defined are all real. Substituting for the unknown coefficients from (44) into (34a) gives

$$V_1 \Phi_1 - V_2 \psi_1 = j \int_0^a (V_1 \mathfrak{K}_1 - V_2 \mathfrak{K}_2) G_2(x|x') \, dx'$$

where

$$G_2(x|x') = \sum_{n=2}^{\infty} [|Z_n| \Phi_n(x)\Phi_n(x') + |Z_{0n}|\psi_n(x)\psi_n(x')].$$

Since V_1 and V_2 are linearly independent, we must have

$$\Phi_1 = j \int_0^a \mathfrak{K}_1 G_2 \, dx' \qquad \psi_1 = j \int_0^a \mathfrak{K}_2 G_2 \, dx'.$$

These latter two equations may be used to construct the stationary expressions for the admittance elements in the same manner as was done for the impedance elements. The results are

$$\frac{1}{Y'_{11}} = -\frac{j}{Y_{11}} = \frac{\displaystyle\iint_0^a G_2(x|x')\mathfrak{K}_1(x)\mathfrak{K}_1(x') \, dx \, dx'}{\left(\displaystyle\int_0^a \mathfrak{K}_1 \Phi_1 \, dx\right)^2} \tag{47a}$$

$$\frac{1}{Y'_{22}} = -\frac{j}{Y_{22}} = \frac{\displaystyle\iint_0^a G_2(x|x')\mathfrak{K}_2(x)\mathfrak{K}_2(x') \, dx \, dx'}{\left(\displaystyle\int_0^a \mathfrak{K}_2 \psi_1 \, dx\right)^2} \tag{47b}$$

$$\frac{1}{Y'_{12}} = -\frac{j}{Y_{12}} = -\frac{j}{Y_{21}} = \frac{\displaystyle\iint_0^a G_2(x|x')\mathfrak{K}_1(x)\mathfrak{K}_2(x')\,dx\,dx'}{\displaystyle\int_0^a \mathfrak{K}_1\psi_1\,dx \int_0^a \mathfrak{K}_2\Phi_1\,dx}. \tag{47c}$$

The evaluation of Y'_{ij} proceeds in a manner identical to that of Z'_{ij}. As before, we may show that (47a) and (47b) are absolute minima for the correct field distributions, and hence give lower bounds on Y'_{11} and Y'_{22}, respectively, when the approximate field distributions are used in the integrands. The functions $\mathfrak{K}_1$ and $\mathfrak{K}_2$ may be taken as real.

Although we can obtain lower bounds on Y'_{11} and Y'_{22}, and thus upper bounds on $-Z'_{22}/\Delta$ and $-Z'_{11}/\Delta$, we do not obtain upper and lower bounds on the impedance elements Z'_{11} and Z'_{22} by comparing the lower bounds obtained for Z'_{11} and Z'_{22} with the upper bounds on $-Z'_{22}/\Delta$ and $-Z'_{11}/\Delta$, since we do not know if Δ is too large or too small. The bounds on Δ are not known, since the error in Z_{12} may be either positive or negative.

If we have a plane discontinuity, i.e., a discontinuity localized in a plane such as $z = 0$, and the input and output guides are identical, then $\Phi_n = \psi_n$, $Z_n = Z_{0n}$, and the functions $\mathcal{E}_1$, $\mathfrak{K}_1$ are identical with $\mathcal{E}_2$ and $\mathfrak{K}_2$, respectively. Thus $Z_{11} = Z_{12} = Z_{22} = Z_s$, and $Y_{11} = Y_{12} = Y_{22} = Y_s$, and the equivalent circuit reduces to a pure shunt reactive element. Furthermore $Y_s = 1/Z_s$, and we thus can obtain upper and lower bounds on the shunt impedance element, and, consequently, we know the maximum error involved in our calculations.

The main objection to the variational method for the T- or π-network parameters is that it does not give an estimate of the error involved in the calculations. Also, three separate calculations must be made to determine all three elements. When the equivalent circuit reduces to a shunt element, only two calculations are required to obtain upper and lower bounds on the shunt impedance element, and this variational technique is then a convenient one to use.

The above three methods are by no means exhaustive, but they are perhaps the most useful ones. Under some conditions of symmetry, when a discontinuity can be represented by two parameters only, it is convenient to modify Method 3 slightly. We will illustrate this modified technique in connection with the analysis of the thick inductive window. All the above methods may be extended in a straightforward manner to the vector problem, i.e., to the case where the functions $\mathcal{E}$ and $\mathfrak{K}$ are vector functions rather than scalar functions. Under these conditions, the Green's function becomes a dyadic quantity. As a general rule, for the vector problem positive-definite quadratic forms for the equivalent-circuit parameters are not obtained because a mixture of E and H modes is excited. For such cases we obtain stationary expressions whose stationary values are neither absolute minima nor absolute maxima. Positive-definite quadratic forms are obtained only when the modes excited all store a net amount of either electric energy or magnetic energy.

8.2. Capacitive Diaphragm

As a first example of a discontinuity problem for which a variational solution will be obtained, we consider an asymmetrical capacitive diaphragm in an infinitely wide parallel-plate transmission line as in Fig. 8.4. From the solution of this problem we are able to obtain the solution to the problem of a symmetrical diaphragm in a parallel-plate transmission line, as well as the solution for similar diaphragms placed in a rectangular guide.

The diaphragm is assumed to be infinitely thin and lossless. The incident mode is a TEM mode, incident from the region $z < 0$, and polarized with the electric vector along the y

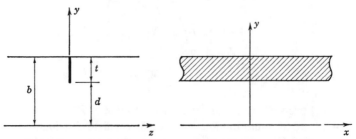

Fig. 8.4. Capacitive diaphragm in a parallel-plate transmission line.

direction. The incident field is independent of the x coordinate, and, because of the uniformity of the discontinuity, all the higher order modes that are excited are also independent of x. The only higher order modes that are independent of x and can be coupled with the incident mode in the aperture plane are the TM or E modes. If the frequency of operation is chosen so that $\lambda_0 > 2b$, none of the higher order modes will propagate. The excited evanescent modes store a net amount of reactive electric energy in the vicinity of the diaphragm, and, since the total electric field of the dominant mode is continuous across the diaphragm, the discontinuity is equivalent to a shunt capacitive reactance. Since the discontinuity requires only a single parameter to describe it, the general variational formulation presented in the preceding section is not required. It will be just as expedient to proceed from basic principles in setting up the solution for the shunt susceptance.

For a capacitive diaphragm in a rectangular guide, the higher order modes that are excited are the longitudinal-section electric (LSE) modes. When there is no variation with x, these modes reduce to the E modes. For generality we will, therefore, derive the required modes from a magnetic-type Hertzian potential having a single component in the x direction rather than from a z-directed electric-type Hertzian potential which is required for E modes. The relevant equations are

$$\mathbf{E} = -j\omega\mu_0 \nabla \times \mathbf{\Pi}_h \tag{48a}$$

$$\mathbf{H} = k_0^2 \mathbf{\Pi}_h + \nabla\nabla\cdot\mathbf{\Pi}_h \tag{48b}$$

$$\nabla^2\mathbf{\Pi}_h + k_0^2\mathbf{\Pi}_h = 0. \tag{48c}$$

For $\mathbf{\Pi}_h$ we may choose a form such as $\mathbf{a}_x\psi(y)e^{\pm\Gamma z}$. A solution for ψ that results in an electric field with a vanishing tangential component on the parallel plates is

$$\psi_n = \cos\frac{n\pi y}{b}e^{\pm\Gamma_n z}, \qquad n = 0, 1, 2, \ldots$$

where $\Gamma_n^2 = (n\pi/b)^2 - k_0^2$. From these solutions the following expansions for the transverse electric and magnetic fields are readily established:

$$E_y = \begin{cases} a_0(e^{-jk_0z} + Re^{jk_0z}) + \displaystyle\sum_1^\infty a_n \cos\frac{n\pi y}{b}e^{\Gamma_n z}, & z < 0 \\[4mm] Ta_0e^{-jk_0z} + \displaystyle\sum_1^\infty b_n \cos\frac{n\pi y}{b}e^{-\Gamma_n z}, & z > 0 \end{cases} \tag{49a}$$
$$\tag{49b}$$

$$
H_x = \begin{cases} -a_0 Y_0 (e^{-jk_0 z} - R e^{jk_0 z}) + \sum_1^\infty a_n Y_n \cos \dfrac{n\pi y}{b} e^{\Gamma_n z}, & z < 0 \qquad \text{(49c)} \\[4mm] -T a_0 Y_0 e^{-jk_0 z} - \sum_1^\infty b_n Y_n \cos \dfrac{n\pi y}{b} e^{-\Gamma_n z}, & z > 0 \qquad \text{(49d)} \end{cases}
$$

where R is the reflection coefficient and T is the transmission coefficient. In (49) the factor $-j\omega\mu_0\Gamma_n$, which arises when the electric field is derived from the Hertzian potential, has been absorbed in the amplitude coefficients a_n, b_n. The mode admittances Y_n are given by

$$
Y_n = \frac{jk_0}{\Gamma_n} Y_0 = \frac{jk_0}{\Gamma_n} \left(\frac{\epsilon_0}{\mu_0} \right)^{1/2} = j|Y_n|, \qquad n > 0 \tag{50}
$$

and are positive-imaginary for $n > 0$. For $n = 0$, $\Gamma_n = jk_0$, and $Y_n = Y_0$, the characteristic admittance of free space.

The equations expressing the continuity of the transverse fields in the aperture plane $z = 0$ are

$$
(1 + R)a_0 + \sum_1^\infty a_n \cos \frac{n\pi y}{b} = T a_0 + \sum_1^\infty b_n \cos \frac{n\pi y}{b}, \qquad 0 \le y \le b \tag{51a}
$$

$$
(1 - R)a_0 Y_0 - \sum_1^\infty a_n Y_n \cos \frac{n\pi y}{b} = T a_0 Y_0 + \sum_1^\infty b_n Y_n \cos \frac{n\pi y}{b}, \qquad 0 \le y \le d. \tag{51b}
$$

The total transverse electric field vanishes on the diaphragm, and hence (51a) is valid over the total range of y. The tangential magnetic field is discontinuous across the diaphragm by an amount equal to the total current on the diaphragm, and hence (51b) holds only in the aperture opening. Let the aperture electric field be $\mathcal{E}(y)$, where $\mathcal{E}(y)$ vanishes on the diaphragm. A Fourier analysis of (51a), holding for $0 \le y \le b$, now gives

$$
a_0(1 + R) = T a_0 = \frac{1}{b} \int_0^b \mathcal{E}(y') \, dy' \tag{52a}
$$

$$
a_n = b_n = \frac{2}{b} \int_0^b \mathcal{E}(y') \cos \frac{n\pi y'}{b} \, dy'. \tag{52b}
$$

Replacing T by $1 + R$ and substituting the integral expressions for the unknown coefficients into (51b) gives

$$
-\frac{2R}{1+R} Y_0 \int_0^b \mathcal{E}(y') \, dy' = 4 \sum_1^\infty Y_n \cos \frac{n\pi y}{b} \int_0^b \mathcal{E}(y') \cos \frac{n\pi y'}{b} \, dy', \qquad 0 \le y \le d. \tag{53}
$$

A shunt susceptance jB across a transmission line of unit characteristic impedance gives rise to a reflection coefficient (see Fig. 8.5):

$$
R = \frac{1 - Y_{\text{in}}}{1 + Y_{\text{in}}} = -\frac{jB}{2 + jB}.
$$

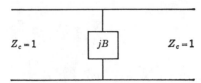

Fig. 8.5. Equivalent circuit for a capacitive diaphragm.

Solving for jB gives $jB = -2R/(1 + R)$. Thus (53) represents an integral equation whose solution would give the normalized susceptance B.

A variational integral for B may be obtained by multiplying (53) by $\mathcal{E}(y)$ and integrating over the aperture region $0 \le y \le d$. Since $\mathcal{E}(y)$ vanishes on the diaphragm, the integrals for the unknown amplitude coefficients need be extended over the aperture opening only. Thus we get

$$B = 4k_0 \frac{\displaystyle\sum_{n=1}^{\infty} \frac{1}{\Gamma_n} \int\!\!\!\int_0^d \mathcal{E}(y)\mathcal{E}(y') \cos\frac{n\pi y}{b} \cos\frac{n\pi y'}{b}\, dy\, dy'}{\left[\displaystyle\int_0^d \mathcal{E}(y)\, dy\right]^2} \tag{54}$$

after replacing $-2R/(1 + R)$ by jB and Y_n/Y_0 by jk_0/Γ_n. It is easily demonstrated by the usual method that (54) has the required stationary property.

As it stands, (54) is not directly amenable to a simple solution for B. By a rather ingenious transformation of variables, Schwinger was able to transform (54) into a dominant term consisting of a series of functions orthogonal over the range of integration plus a small correction series [8.1]. As a first step, the order of integration and summation is interchanged. The series may be written as a dominant series plus a rapidly converging series, as follows:

$$\frac{b}{\pi}\sum_1^{\infty} \frac{1}{n} \cos\frac{n\pi y}{b} \cos\frac{n\pi y'}{b} + \sum_1^{\infty}\left(\frac{1}{\Gamma_n} - \frac{b}{n\pi}\right) \cos\frac{n\pi y}{b} \cos\frac{n\pi y'}{b}.$$

The product of the two cosine terms in the first series is equal to

$$\frac{1}{2}\left[\cos\frac{n\pi}{b}(y - y') + \cos\frac{n\pi}{b}(y + y')\right].$$

With this substitution this series may be summed according to the methods given in the Mathematical Appendix to get

$$\frac{b}{\pi}\sum_1^{\infty} \frac{1}{n} \cos\frac{n\pi y}{b} \cos\frac{n\pi y'}{b} = -\frac{b}{2\pi} \ln 4\left[\sin\frac{\pi}{2b}(y - y') \sin\frac{\pi}{2b}(y + y')\right]$$

$$= -\frac{b}{2\pi} \ln 2\left(\cos\frac{\pi y}{b} - \cos\frac{\pi y'}{b}\right). \tag{55}$$

As a next step, the following changes of variables are made:

$$\cos\frac{\pi y}{b} = \alpha_1 + \alpha_2 \cos\theta \tag{56a}$$

$$\cos \frac{\pi y'}{b} = \alpha_1 + \alpha_2 \cos \theta' \tag{56b}$$

with the constants α_1 and α_2 chosen so that $\theta = 0, \pi$, when $y = 0, d$, respectively. Substitution of these particular values of θ and y into (56) gives the following solutions for α_1 and α_2:

$$\alpha_1 = \frac{1}{2}\left(1 + \cos \frac{\pi d}{b}\right) = \cos^2 \frac{\pi d}{2b} \tag{57a}$$

$$\alpha_2 = 1 - \alpha_1 = \sin^2 \frac{\pi d}{2b}. \tag{57b}$$

Substituting into (55) now gives

$$\frac{b}{\pi} \sum_1^\infty \frac{1}{n} \cos \frac{n\pi y'}{b} \cos \frac{n\pi y}{b} = -\frac{b}{2\pi}[\ln \alpha_2 + \ln 2(\cos \theta - \cos \theta')]$$

$$= -\frac{b}{\pi} \ln \sin \frac{\pi d}{2b} + \frac{b}{\pi} \sum_1^\infty \frac{1}{n} \cos n\theta \cos n\theta' \tag{58}$$

where the logarithmic term involving the factor $2(\cos \theta - \cos \theta')$ has been expanded into an infinite series by direct analogy with the original series which was summed. The beauty of this result is that it gives a series which is orthogonal over the range of integration, i.e., orthogonal in the aperture domain.

The correction series cannot be transformed in this manner; however, this is not of great consequence, since it converges rapidly, and each term may be transformed individually. For example, we have

$$\cos \frac{2\pi y}{b} = 2 \cos^2 \frac{\pi y}{b} - 1 = 2(\alpha_1 + \alpha_2 \cos \theta)^2 - 1$$

$$= \alpha_2^2 \cos 2\theta + 4\alpha_1 \alpha_2 \cos \theta + 2\alpha_1^2 + \alpha_2^2 - 1.$$

To complete the change in variables it should be noted that

$$-\frac{\pi}{b} \sin \frac{\pi y}{b} \, dy = -\alpha_2 \sin \theta \, d\theta$$

and hence

$$dy = \frac{b}{\pi} \csc \frac{\pi y}{b} \sin \theta \, d\theta = \frac{dy}{d\theta} \, d\theta.$$

Thus $\mathcal{E}(y) \, dy$ becomes $\mathcal{E}(y)(dy/d\theta) \, d\theta = \mathcal{F}(\theta) \, d\theta$, where $\mathcal{F}(\theta)$ is equal to $\mathcal{E}(y) \, dy/d\theta$ expressed as a function of θ. Each term $\cos(n\pi y/b)$ may be expressed as a series of the terms $\cos m\theta$, with m running from 0 up to n. For convenience we will write

$$\cos \frac{n\pi y}{b} = \sum_{m=0}^n P_{nm} \cos m\theta \tag{59}$$

where the P_{nm} are suitably determined coefficients.

When all these substitutions are made in (54), we get

$$B = \frac{4k_0b}{\pi} \frac{\displaystyle\int\int_0^\pi \mathcal{F}(\theta)\mathcal{F}(\theta') \left[\ln \csc \frac{\pi d}{2b} + \sum_1^\infty \frac{1}{n} \cos n\theta \cos n\theta' \right. }{\left[\displaystyle\int_0^\pi \mathcal{F}(\theta)\, d\theta \right]^2}$$

$$\left. + \sum_{n=1}^\infty \left(\frac{\pi}{b\Gamma_n} - \frac{1}{n} \right) \sum_{m=0}^n \sum_{s=0}^n P_{nm}P_{ns} \cos m\theta \cos s\theta' \right] d\theta\, d\theta' . \tag{60}$$

Since a completely rigorous solution of (60) is difficult to obtain, we appeal to its stationary property and approximate $\mathcal{F}(\theta)$ by a constant which may be chosen equal to unity. It turns out that this yields a surprisingly accurate result for B for the usual range of parameters occurring in practice. The only terms in (60) which do not integrate to zero are the constant terms, and hence we get

$$B = \frac{4k_0b}{\pi} \left[\ln \csc \frac{\pi d}{2b} + \sum_{n=1}^\infty \left(\frac{\pi}{b\Gamma_n} - \frac{1}{n} \right) P_{n0}^2 \right]. \tag{61}$$

It may be shown that all P_{n0}^2 are less than unity, and, hence, terms beyond n equal to 3 or 4 are usually entirely negligible since the correction series converges as n^{-3}. The coefficients P_{n0} may be obtained with reasonable facility by using available tabulated expansions of the Tchebysheff[1] polynomials [8.2]. We have

$$\cos \frac{n\pi y}{b} = T_n \left(\cos \frac{\pi y}{b} \right) = T_n(\alpha_1 + \alpha_2 \cos \theta).$$

The expansion of the Tchebysheff polynomial gives a power series in $\cos \theta$. Each term $(\cos \theta)^m$ may, in turn, be expressed as a series of Tchebysheff polynomials from which the coefficients P_{nm} may be readily deduced. The required expansions for $n \leq 6$ are listed in Table 8.1. With the aid of these relations the first four coefficients P_{n0} are found to be

$$P_{10} = \alpha_1$$

$$P_{20} = 2\alpha_1^2 + \alpha_2^2 - 1$$

$$P_{30} = 4\alpha_1^3 + 6\alpha_1\alpha_2^2 - 3\alpha_1$$

$$P_{40} = 8\alpha_1^4 + 3\alpha_2^4 + 24\alpha_1^2\alpha_2^2 - 8\alpha_1^2 - 4\alpha_2^2 + 1.$$

Equation (61) gives an upper bound on the true value of the susceptance. A lower bound may be obtained by solving the problem in terms of the magnetic field in the aperture, i.e., in terms of the current on the diaphragm. For this particular example, we may also obtain a lower bound by omitting the higher order terms in the correction series, since the correction series always results in a positive contribution.

---

[1] Several spellings for the name Tchebysheff are in current use.

TABLE 8.1

TCHEBYSHEFF POLYNOMIALS

$u = \cos x$	$T_n(u) = T_n(\cos x) = \cos nx$
$T_1(u)$	u
$T_2(u)$	$2u^2 - 1$
$T_3(u)$	$4u^3 - 3u$
$T_4(u)$	$8u^4 - 8u^2 + 1$
$T_5(u)$	$16u^5 - 20u^3 + 5u$
$T_6(u)$	$32u^6 - 48u^4 + 18u^2 - 1$
$T_n(u)$	$2uT_{n-1} - T_{n-2}$
u	$T_1(u)$
u^2	$\frac{1}{2}T_2 + \frac{1}{2}$
u^3	$\frac{1}{4}T_3 + \frac{3}{4}T_1$
u^4	$\frac{1}{8}T_4 + \frac{1}{2}T_2 + \frac{3}{8}$
u^5	$\frac{1}{16}T_5 + \frac{5}{16}T_3 + \frac{5}{8}T_1$
u^6	$\frac{1}{32}T_6 + \frac{3}{16}T_4 + \frac{15}{32}T_2 + \frac{5}{16}$

The function $\mathcal{F}(\theta)$ may be completely represented by the infinite expansion $\sum_{m=0}^{\infty} C_m \cos m\theta$. If in (60) we neglect all the terms in the correction series, we get

$$B = \frac{4k_0 b}{\pi C_0^2} \left(C_0^2 \ln \csc \frac{\pi d}{2b} + \frac{1}{4} \sum_{m=1}^{\infty} \frac{C_m^2}{m} \right).$$

Equating all the partial derivatives $\partial/\partial C_m$ to zero shows that $C_m = 0$, $m > 0$, and hence

$$B = \frac{4k_0 b}{\pi} \ln \csc \frac{\pi d}{2b}. \tag{62}$$

This result is, therefore, a rigorous solution for the low-frequency limit since, as the frequency approaches zero, the correction series vanishes. For the range of parameters encountered in practice (62) is not very accurate, although it is a lower bound. A significant improvement is obtained by retaining the first term in the correction series. We now get

$$B = \frac{4k_0 b}{\pi C_0^2} \left[C_0^2 \ln \csc \frac{\pi d}{2b} + \frac{1}{4} \sum_{m=1}^{\infty} \frac{C_m^2}{m} + \left(\frac{\pi}{b\Gamma_1} - 1 \right) \left(\alpha_1 C_0 + \frac{\alpha_2 C_1}{2} \right)^2 \right].$$

Again, equating the partial derivatives with respect to C_m equal to zero, we find that $C_m = 0$, $m > 1$. The first two partial derivatives, when equated to zero, yield two homogeneous equations for the coefficients C_0, C_1. The solution for these exists only if the determinant vanishes. Equating the determinant to zero gives the following solution for B, which is a lower bound on the true value:

$$B = \frac{4k_0 b}{\pi} \left[\ln \csc \frac{\pi d}{2b} + \frac{(\pi/b\Gamma_1 - 1)\alpha_1^2}{1 + (\pi/b\Gamma_1 - 1)\alpha_2^2} \right]. \tag{63}$$

That (63) does indeed present a lower bound on the susceptance will now be proved. Let B_0 be the true value of susceptance, and

$$\sum_{m=0}^{\infty} C_{0m} \cos m\theta$$

be the correct aperture field distribution. The susceptance B_0 is given by

$$B_0 = \frac{4k_0 b}{\pi C_{00}^2} \left[C_{00}^2 \ln \csc \frac{\pi d}{2b} + \frac{1}{4} \sum_{m=1}^{\infty} \frac{C_{0m}^2}{m} + \frac{1}{4} \sum_{n=1}^{\infty} \left(\frac{\pi}{b\Gamma_n} - \frac{1}{n} \right) \left(\sum_{m=0}^{n} \frac{2}{\epsilon_{0m}} P_{nm} C_{0m} \right)^2 \right].$$

This expression for B_0 is an absolute minimum, so that any variation in the coefficients C_{0m} will increase B_0. If we drop all the terms in the correction series for $n > N$, then clearly

$$B_1 = \frac{4k_0 b}{\pi C_{00}^2} \left[C_{00}^2 \ln \csc \frac{\pi d}{2b} + \frac{1}{4} \sum_{m=1}^{\infty} \frac{C_{0m}^2}{m} + \frac{1}{4} \sum_{n=1}^{N} \left(\frac{\pi}{b\Gamma_n} - \frac{1}{n} \right) \right.$$

$$\left. \cdot \left(\sum_{m=0}^{n} \frac{2}{\epsilon_{0m}} P_{nm} C_{0m} \right)^2 \right] < B_0.$$

Now $\partial B_1 / \partial C_{0m}$ will not, in general, equal zero, since the higher order correction terms have been dropped, and therefore B_1 is not a minimum. However, if we compute a new set of values for C_{0m}, say C_m, so that $\partial B_1 / \partial C_m = 0$, we will minimize the expression for B_1 to get a new susceptance $B < B_1 < B_0$. This latter procedure is, in fact, that used to obtain (63) and, therefore, demonstrates that (63) yields a lower bound on the true susceptance.

As an indication of the accuracy of (61) and (63), the upper and lower bounds on B have been computed as a function of d/b for $b = 0.4\lambda_0$ and are plotted in Fig. 8.6. For $d/b < 0.5$, the two bounds on B differ by less than 4%, and hence the average value is in error by less than 2%. For $d/b = 0.8$, the difference amounts to about 10%, but this occurs in a range for d/b which is not commonly encountered in practice.

Symmetrical Capacitive Diaphragms

Figures 8.7(a) and (b) illustrate symmetrical capacitive diaphragms in parallel-plate transmission lines. An application of image theory reduces the solutions of both these problems to that for the asymmetrical diaphragm when a TEM mode is incident from the left. The symmetry of the structure results only in the excitation of those higher order modes having a zero component of longitudinal electric field on the symmetry plane. A conducting plane may, therefore, be placed along the symmetry plane, and the problem is thereby reduced to that of the asymmetrical diaphragm. The normalized shunt capacitive susceptance is given by (61) and (63), with d and b replaced by $d/2$ and $b/2$, respectively.

Capacitive Diaphragms in a Rectangular Waveguide

Figures 8.8(a)–(c) illustrate capacitive diaphragms in a rectangular guide of height b and width a. The solution to these diaphragm problems may be obtained from the corresponding

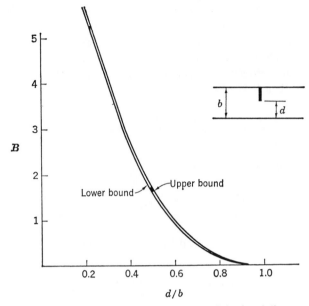

Fig. 8.6. Normalized capacitive susceptance B for $b = 0.4\lambda_0$.

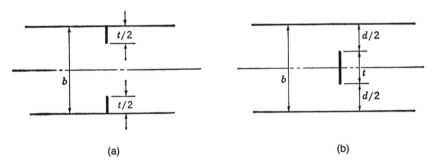

(a)   (b)

Fig. 8.7. Symmetrical capacitive diaphragms in parallel-plate transmission lines.

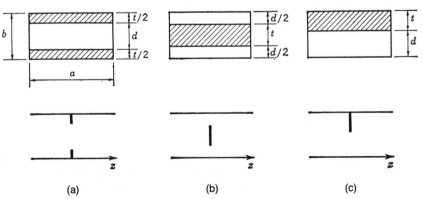

(a)   (b)   (c)

Fig. 8.8. Capacitive diaphragms in a rectangular waveguide.

solutions for similar diaphragms in a parallel-plate transmission line of height b and with a infinite.

In the parallel-plate transmission-line problem (a infinite), let the total Hertzian potential function, from which all the field components may be derived, be $\Pi_h(y, z; k_0^2)$. For the equivalent-waveguide problem, the solution is given by the following Hertzian potential function when the incident field is an H_{10} mode:

$$\Pi_h' = \sin\frac{\pi x}{a}\Pi_h\left(y, z; k_0^2 - \frac{\pi^2}{a^2}\right).$$

Clearly $\Pi_h(y, z; k_0^2 - \pi^2/a^2)$ is a solution of

$$\left(\frac{\partial^2}{\partial y^2} + \frac{\partial^2}{\partial z^2} + k_0^2 - \frac{\pi^2}{a^2}\right)\Pi_h\left(y, z; k_0^2 - \frac{\pi^2}{a^2}\right) = 0$$

and hence Π_h' is a solution of the required Helmholtz equation

$$\left(\frac{\partial^2}{\partial x^2} + \frac{\partial^2}{\partial y^2} + \frac{\partial^2}{\partial z^2} + k_0^2\right)\Pi_h' = 0$$

since the second partial derivative with respect to x introduces a term $-(\pi/a)^2$. The function $\Pi_h(y, z; k_0^2)$ has been determined so that $-j\omega\mu\,\partial\Pi_h(y, z; k_0^2)/\partial z$, corresponding to the y component of electric field, is continuous in the aperture and vanishes on the diaphragm. Also, $k_0^2\Pi_h(y, z; k_0^2)$, corresponding to the x component of magnetic field, is continuous in the aperture and is discontinuous with respect to z across the diaphragm. These boundary conditions hold for all values of k_0^2, and in particular when k_0^2 is replaced by $k_0^2 - (\pi/a)^2$. Multiplying the function by $\sin(\pi x/a)$ does not change the boundary conditions which are satisfied in the aperture plane. In the waveguide problem, the only new field components entering into the problem are y and z components of magnetic field, and these are easily shown to satisfy the proper boundary conditions since they are derivable from the curl of the electric field. Hence, the potential function Π_h', as defined, gives a solution which satisfies all the required boundary conditions for the waveguide problem. Therefore the normalized capacitive susceptance for a diaphragm in a rectangular guide may be obtained by replacing k_0^2 in the formulas (61) and (63) by $\beta_{10}^2 = k_0^2 - (\pi/a)^2$. For the symmetrical diaphragms, d and b must also be replaced by $d/2$ and $b/2$ for the same reason that these changes were made to obtain formulas applicable to symmetrical diaphragms in a parallel-plate transmission line. The propagation constants Γ_n^2 in the correction series become $(n\pi/b)^2 + (\pi/a)^2 - k_0^2$ in the waveguide case. This modification results in a more rapidly converging series.

8.3. Thin Inductive Diaphragm in a Rectangular Guide

Initially we will consider a general infinitely thin inductive diaphragm as illustrated in Fig 8.9. The solutions for a symmetrical or asymmetrical diaphragm are just special cases of the general solution.

With an H_{10} mode incident from $z < 0$, only higher order H_{n0} modes are excited, since the discontinuity is uniform along the y direction. The problem may be formulated in the same manner as that for the dielectric step considered in Section 8.1 in connection with the general discussion on variational methods. The particular problem under consideration here is simpler

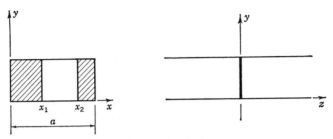

Fig. 8.9. Inductive diaphragm.

since the waveguides on both sides of the diaphragm are the same. Equations (41a) to (41d) are all applicable and, in actual fact, reduce to one single equation because of the identical input and output guides. The equivalent T network reduces to a single shunt element since $Z_{11} = Z_{22} = Z_{12}$. The normalized shunt inductive susceptance of the diaphragm is given by

$$
B = \frac{jZ_1}{Z_{12}} = \frac{2 \displaystyle\int\!\!\!\int_{x_1}^{x_2} \sum_{n=2}^{\infty} \frac{|Y_n|}{Y_1} \sin\frac{n\pi x}{a} \sin\frac{n\pi x'}{a} \mathcal{E}(x)\mathcal{E}(x')\,dx\,dx'}{\left[\displaystyle\int_{x_1}^{x_2} \mathcal{E}(x)\sin\frac{\pi x}{a}\,dx\right]^2}
\tag{64}
$$

where $|Y_n|/Y_1 = \Gamma_n/\beta_1$, $\Gamma_n^2 = (n\pi/a)^2 - k_0^2$, and $\beta_1 = |\Gamma_1|$. Equation (64) may also be readily set up from fundamental principles in much the same way as the capacitive-diaphragm problem was handled. The series in (64) may be written as a dominant series plus a correction series. The dominant series $\sum_2^\infty n \sin(n\pi x/a)\sin(n\pi x'/a)$ may be summed, and then a change of variables may be introduced in order to obtain a series of functions that are orthogonal over the range of integration. It is somewhat more convenient to first integrate each term by parts once and thereby obtain a series similar to that occurring in the capacitive-diaphragm problem. We have

$$
\int_{x_1}^{x_2} \mathcal{E}(x)\sin\frac{n\pi x}{a}\,dx = -\frac{a}{n\pi}\cos\frac{n\pi x}{a}\mathcal{E}(x)\Big|_{x_1}^{x_2} + \frac{a}{n\pi}\int_{x_1}^{x_2}\frac{d\mathcal{E}(x)}{dx}\cos\frac{n\pi x}{a}\,dx.
$$

Since the tangential electric field is zero at the edge of a thin conducting strip, the integrated term vanishes. Thus (64) is transformed to

$$
B = \frac{2}{\beta_1}\frac{\displaystyle\int\!\!\!\int_{x_1}^{x_2} \sum_{n=2}^{\infty} \frac{\Gamma_n}{n^2}\cos\frac{n\pi x}{a}\cos\frac{n\pi x'}{a}\mathcal{E}'(x)\mathcal{E}'(x')\,dx'\,dx}{\left[\displaystyle\int_{x_1}^{x_2}\mathcal{E}'(x)\cos\frac{\pi x}{a}\,dx\right]^2}
\tag{65}
$$

where $\mathcal{E}'(x) = d\mathcal{E}(x)/dx$. The series may be written as

$$
\frac{\pi}{a}\sum_{n=1}^{\infty}\frac{1}{n}\cos\frac{n\pi x}{a}\cos\frac{n\pi x'}{a} - \frac{\pi}{a}\cos\frac{\pi x}{a}\cos\frac{\pi x'}{a} + \sum_{n=2}^{\infty}\frac{\Gamma_n - n(\pi/a)}{n^2}\cos\frac{n\pi x}{a}\cos\frac{n\pi x'}{a}
$$

where the last series converges rapidly since Γ_n approaches $n\pi/a$ for large n. The first series sums to $(-\pi/2a) \ln 2[\cos(\pi x/a) - \cos(\pi x'/a)]$. The following changes in variables are now introduced:

$$\cos \frac{\pi x}{a} = \alpha_1 + \alpha_2 \cos \theta$$

$$\cos \frac{\pi x'}{a} = \alpha_1 + \alpha_2 \cos \theta'$$

with α_1 and α_2 chosen so that $\theta = 0, \pi$, when $x = x_1, x_2$, respectively. Substituting these particular values into the transformation and solving for α_1 and α_2 gives

$$\alpha_1 = \frac{1}{2} \left(\cos \frac{\pi x_1}{a} + \cos \frac{\pi x_2}{a} \right)$$

$$= \cos \pi \frac{x_2 - x_1}{2a} \cos \pi \frac{x_2 + x_1}{2a}$$

$$= \cos \frac{\pi d}{2a} \cos \pi \frac{2x_1 + d}{2a} \tag{66a}$$

$$\alpha_2 = \frac{1}{2} \left(\cos \frac{\pi x_1}{a} - \cos \frac{\pi x_2}{a} \right)$$

$$= \sin \frac{\pi d}{2a} \sin \pi \frac{2x_1 + d}{2a}. \tag{66b}$$

The summed series transforms to

$$-\frac{\pi}{2a} \ln 2\alpha_2(\cos \theta - \cos \theta') = -\frac{\pi}{2a} \ln \alpha_2 + \frac{\pi}{a} \sum_{n=1}^{\infty} \frac{1}{n} \cos n\theta \cos n\theta'$$

after the logarithmic term is expanded. With the introduction of the new variables, $\mathcal{E}'(x) dx$ is replaced by $\mathcal{E}'(x)(dx/d\theta) d\theta = \mathcal{F}(\theta) d\theta$. Also each term $\cos(n\pi x/a)$ in the correction series is replaced by $\sum_{m=0}^{n} P_{nm} \cos m\theta$, where the P_{nm} are suitably determined coefficients. With all these modifications substituted into (65), the variational expression for B becomes

$$B = \frac{2\pi}{\beta_1 a} \frac{\displaystyle\int\!\!\!\int_0^\pi \mathcal{F}(\theta)\mathcal{F}(\theta') \left[-\frac{1}{2} \ln \alpha_2 + \sum_{n=1}^{\infty} \frac{1}{n} \cos n\theta \cos n\theta' \right.}{\left[\displaystyle\int_0^\pi (\alpha_1 + \alpha_2 \cos \theta)\mathcal{F}(\theta) d\theta \right]^2}$$

$$-(\alpha_1 + \alpha_2 \cos \theta)(\alpha_1 + \alpha_2 \cos \theta')$$

$$\left. + \sum_{n=2}^{\infty} \frac{(a/\pi)\Gamma_n - n}{n^2} \sum_{m=0}^{n}\sum_{s=0}^{n} P_{nm}P_{ns} \cos m\theta \cos s\theta' \right] d\theta\, d\theta'$$

$$. \tag{67}$$

Since (67) is a stationary expression for B, we may obtain a good approximation for B (upper bound) by using a finite series in $\cos m\theta$ for $\mathcal{F}(\theta)$. In order to simplify the details, we will restrict the analysis at this point to that applicable for a symmetrical diaphragm. For this case $2x_1 + d = a$ and $\alpha_1 = 0$, $\alpha_2 = \sin(\pi d/2a)$. If we substitute a general series $\sum_{m=0}^{N} C_m \cos m\theta$ for $\mathcal{F}(\theta)$ into (67), we find that, when the minimization is carried out, all the even coefficients are zero. This we could anticipate in view of the symmetry. The electric field $\mathcal{E}(x)$ is an even function about $x = a/2$; hence $d\mathcal{E}/dx$ is an odd function about $x = a/2$, and therefore $\mathcal{E}'(x)\,dx/d\theta = \mathcal{F}(\theta)$ will contain only the terms $\cos m\theta$ with m odd. In the correction series, $\cos(n\pi x/a)$ becomes a series in $\cos m\theta$ with m only odd or even, depending on whether n is odd or even, respectively. This would not be true, in general, if α_1 were not equal to zero.

If we approximate $\mathcal{F}(\theta)$ by a single term, that is, $\cos\theta$, then (67) gives the following result for the symmetrical diaphragm with an aperture opening d:

$$B = \frac{2\pi}{\beta_1 a}\left(\csc\frac{\pi d}{2a}\right)^2\left[\cos^2\frac{\pi d}{2a} + \sum_{n=3,5,\dots}^{\infty}\frac{(a/\pi)\Gamma_n - n}{n^2}P_{n1}^2\right]. \tag{68}$$

The coefficients P_{n1} may be found by using the properties of the Tchebysheff polynomials listed in Table 8.1. For example,

$$\cos\frac{3\pi x}{a} = T_3\left(\cos\frac{\pi x}{a}\right) = T_3(\alpha_2\cos\theta) = 4(\alpha_2\cos\theta)^3 - 3\alpha_2\cos\theta.$$

Now $(\cos\theta)^3 = \frac{1}{4}T_3(\cos\theta) + \frac{3}{4}T_1(\cos\theta) = \frac{1}{4}\cos 3\theta + \frac{3}{4}\cos\theta$, and hence $\cos(3\pi x/a)$ is transformed to $\alpha_2^3\cos 3\theta + (3\alpha_2^3 - 3\alpha_2)\cos\theta$. The coefficient P_{31} is therefore given by

$$P_{31} = 3\alpha_2(\alpha_2^2 - 1). \tag{69a}$$

Similarly we find

$$P_{51} = 5\alpha_2(2\alpha_2^4 - 3\alpha_2^2 - 1). \tag{69b}$$

Terms beyond $n = 5$ are usually negligible, and the correction series may be terminated at this point.

The correction series for the inductive diaphragm gives a negative contribution since $\Gamma_n < n\pi/a$, and, hence, neglecting this series does not result in a lower bound to the susceptance B as was the case for the capacitive diaphragm. A lower bound may, however, be obtained by solving the problem in terms of the current on the diaphragm. This solution is left for a problem at the end of this chapter.

8.4. THICK INDUCTIVE WINDOW

The thick inductive window illustrated in Fig. 8.10 is an example of a symmetrical discontinuity which requires two parameters to completely describe it when the losses are negligible. The total thickness or length of the window is $2l$, and the aperture opening is d. The dimension a is chosen so that only the H_{10} mode propagates. However, d is left arbitrary so that an H_{10} mode may or may not propagate in the restricted section. The discontinuity is symmetrical about the $z = 0$ and $x = a/2$ planes. If we choose terminal reference planes at $z = \pm l$, the equivalent circuit for the window is a symmetrical T network as illustrated in Fig. 8.11.

Let H_{10} modes of amplitude a_1 be incident on the diaphragm from the left and right simultaneously. For this particular choice of incident modes we have $V_1 = (1+R_1)a_1 + T_{21}a_1 =$

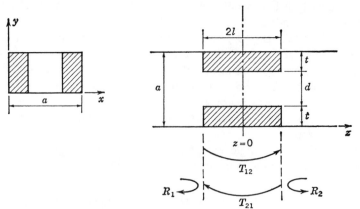

Fig. 8.10. Thick inductive window.

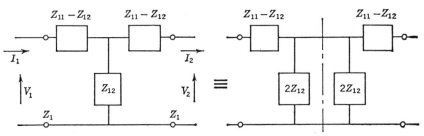

Fig. 8.11. Equivalent circuit for a thick inductive window.

$(1+R_2)a_1 + T_{12}a_1 = V_2$, since, from the symmetry of the structure, $R_1 = R_2$ and $T_{12} = T_{21}$. In view of the symmetrical excitation, the electric field E_y must be symmetrical about $z = 0$. Since H_x is proportional to $\partial E_y / \partial z$, it is antisymmetrical about $z = 0$ and, in particular, equal to zero on the symmetry plane. A magnetic wall may, therefore, be placed at $z = 0$. The circuit equations for the T network are

$$V_1 = Z_{11}I_1 - Z_{12}I_2$$

$$V_2 = Z_{12}I_1 - Z_{11}I_2$$

and, when $V_1 = V_2$, we have $I_1 = -I_2$. The input impedance to the network with an open circuit (magnetic wall) in the symmetry plane is

$$(Z_{in})_{oc} = \frac{V_1}{I_1} = Z_{11} + Z_{12}. \tag{70}$$

If we choose incident waves with amplitudes a_1 and $-a_1$, we will have E_y equal to zero on the symmetry plane because of the antisymmetrical method of excitation. In this case the input impedance is given by

$$(Z_{in})_{sc} = Z_{11} - Z_{12}. \tag{71}$$

Hence, in order to find $Z_{11} + Z_{12}$, we solve the problem with a magnetic wall in the symmetry

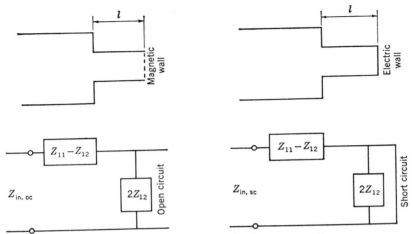

Fig. 8.12. Illustration of open-circuit and short-circuit problems.

plane, while, in order to find $Z_{11} - Z_{12}$, we solve the problem with a short circuit in the symmetry plane. From the two solutions we may then obtain Z_{11} and Z_{12} by addition and subtraction. The reduced problems are illustrated in Fig. 8.12. The field in the region $z \leq -l$ will be given by an expansion in terms of an infinite set of H_{n0} modes as follows:

$$E_y = a_1 \Phi_1 e^{-\Gamma_1(z+l)} + R_1 a_1 \Phi_1 e^{\Gamma_1(z+l)} + \sum_{n=3,5,\dots}^{\infty} a_n \Phi_n e^{\Gamma_n(z+l)} \tag{72a}$$

$$H_x = -Y_1 a_1 \Phi_1 e^{-\Gamma_1(z+l)} + R_1 a_1 Y_1 \Phi_1 e^{\Gamma_1(z+l)} + \sum_{n=3,5,\dots}^{\infty} a_n Y_n \Phi_n e^{\Gamma_n(z+l)} \tag{72b}$$

where $\Phi_n = \sin(n\pi x/a)$, $\Gamma_n^2 = n^2\pi^2/a^2 - k_0^2$, $Y_n = (-j\Gamma_n/k_0)Y_0$, and the origin for the waves is taken at $z = -l$. In the region $-l \leq z \leq 0$, we have a combination of forward- and backward-traveling waves chosen so that, at $z = 0$, the transverse magnetic field vanishes. For the nth mode, we take E_y equal to

$$b_n \psi_n \frac{e^{-\gamma_n z} + e^{\gamma_n z}}{2} = b_n \psi_n \cosh \gamma_n z.$$

The transverse magnetic field is given by $(-j/\omega\mu_0)(\partial E_y/\partial z)$. Thus we may write the following expansion for the fields in the region $-l \leq z \leq 0$:

$$E_y = \sum_{n=1,3,\dots}^{\infty} b_n \psi_n(x) \cosh \gamma_n z \tag{73a}$$

$$H_x = \sum_{n=1,3,\dots}^{\infty} b_n Y_{0n} \psi_n(x) \sinh \gamma_n z \tag{73b}$$

where $\psi_n(x) = \sin(n\pi(x - t)/d)$, $\gamma_n^2 = n^2\pi^2/d^2 - k_0^2$, and

$$Y_{0n} = \frac{-j\gamma_n}{k_0} Y_0.$$

At $z = -l$, the transverse fields must be continuous, and so we get

$$(1 + R_1)a_1\Phi_1 + \sum_{n=3,5,\dots}^{\infty} a_n\Phi_n = \sum_{n=1,3,\dots}^{\infty} b_n\psi_n \cosh\gamma_n l \tag{74a}$$

$$(1 - R_1)Y_1 a_1\Phi_1 - \sum_{n=3,5,\dots}^{\infty} a_n Y_n\Phi_n = \sum_{n=1,3,\dots}^{\infty} b_n Y_{0n}\psi_n \sinh\gamma_n l \tag{74b}$$

where $(1 + R_1)a_1 = V_1$ and $(1 - R_1)Y_1 a_1 = I_1 = (Z_{11} + Z_{12})^{-1}V_1$.

Let the transverse electric field in the aperture at $z = -l$ be $\mathcal{E}(x)$. By Fourier analysis, we get

$$V_1 = \frac{2}{a}\int_t^{t+d} \mathcal{E}(x')\Phi_1(x')\,dx' \tag{75a}$$

$$a_n = \frac{2}{a}\int_t^{t+d} \mathcal{E}(x')\Phi_n(x')\,dx' \tag{75b}$$

$$b_n = \frac{2}{d\cosh\gamma_n l}\int_t^{t+d} \mathcal{E}(x')\psi_n(x')\,dx'. \tag{75c}$$

A variational expression for $(Z_{11} + Z_{12})^{-1}$ may be obtained by substituting the integral expressions for the amplitude coefficients into (74b), multiplying (74b) by $\mathcal{E}(x)$, integrating over the aperture, and, finally, dividing both sides by $(\int_t^{t+d}\mathcal{E}\Phi_1\,dx)^2$. By this procedure, we obtain

$$(Z_{11} + Z_{12})^{-1} = -\frac{j\displaystyle\int_t^{t+d}\!\!\int \mathcal{E}(x)\mathcal{E}(x')G_2(x|x')\,dx\,dx'}{\left[\displaystyle\int_t^{t+d}\mathcal{E}(x)\Phi_1(x)\,dx\right]^2} \tag{76}$$

where

$$-jG_2(x|x') = \sum_{n=3,5,\dots}^{\infty} Y_n\Phi_n(x)\Phi_n(x') + \frac{a}{d}\sum_{n=1,3,\dots}^{\infty} Y_{0n}\psi_n(x)\psi_n(x')\tanh\gamma_n l.$$

Equation (76) leads to a positive-definite quadratic form if all γ_n and hence Y_{0n} are real. If γ_1 is imaginary, the term $Y_{01}\tanh\gamma_1 l$ is real; and as long as l is chosen so that $\tan|\gamma_1 l|$ is positive, then (76) is still a positive-definite quadratic form. When this is the case, an approximate solution for $\mathcal{E}(x)$ used in (76) gives a value for $Z_{11} + Z_{12}$ which is too small.

In the case of an electric wall at $z = 0$, the transverse electric field for the nth mode in the region $-l \leq z \leq 0$ is taken proportional to $\psi_n \sinh\gamma_n z$ in order that it should vanish at $z = 0$. By an analysis similar to the above, we obtain the following variational expression for $(Z_{11} - Z_{12})^{-1}$:

$$(Z_{11} - Z_{12})^{-1} = \frac{-j\displaystyle\int_t^{t+d}\!\!\int G_1(x|x')\mathcal{E}(x)\mathcal{E}(x')\,dx\,dx'}{\left(\displaystyle\int_t^{t+d}\mathcal{E}\Phi_1\,dx\right)^2} \tag{77}$$

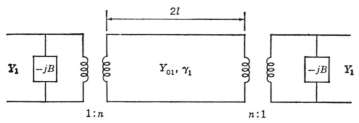

Fig. 8.13. Alternative equivalent circuit for a thick inductive window.

where

$$-jG_1(x|x') = \sum_{3,5,\ldots}^{\infty} Y_n \Phi_n(x)\Phi_n(x') + \frac{a}{d}\sum_{1,3,\ldots}^{\infty} Y_{0n}\psi_n(x)\psi_n(x') \coth\gamma_n l.$$

The aperture electric field in (77), although written as $\mathcal{E}(x)$, is not the same as that in (76).

We can obtain considerable information about the thick inductive iris from (76) and (77) without actually solving the equations. We note that $\tanh\gamma_n l$ and $\coth\gamma_n l$ approach unity rapidly as n increases. If l is not too small so that $\gamma_n l > 2$ for $n > 1$, we may replace $\tanh\gamma_n l$ by unity for $n > 1$. We may then obtain another simple equivalent circuit for the thick diaphragm. Consider the equivalent circuit illustrated in Fig. 8.13. With an open circuit at the symmetry plane, the input admittance is $Y_{\text{in}} = -jB + n^2 Y_{01}\tanh\gamma_1 l$. Since Y_{in} is also equal to $(Z_{11} + Z_{12})^{-1}$, we find upon comparison with (76) that

$$B = \frac{\displaystyle\int\!\!\!\int_t^{t+d} \mathcal{E}(x)\mathcal{E}(x') \sum_{n=3,5,\ldots}^{\infty} \left[|Y_n|\Phi_n(x)\Phi_n(x') + \frac{a}{d}|Y_{0n}|\psi_n(x)\psi_n(x')\right] dx\, dx'}{\left[\displaystyle\int_t^{t+d} \mathcal{E}(x)\Phi_1(x)\,dx\right]^2} \tag{78a}$$

$$n^2 = \frac{a}{d}\left[\frac{\displaystyle\int_t^{t+d} \mathcal{E}(x)\psi_1(x)\,dx}{\displaystyle\int_t^{t+d} \mathcal{E}(x)\Phi_1(x)\,dx}\right]^2. \tag{78b}$$

If the thickness $2l$ becomes infinite, we obtain an H-plane symmetrical step discontinuity. The inductive susceptance at the aperture plane is given by (78a) for the step, except when γ_1 is real. For γ_1 real, Y_{01} is imaginary, and we have a pure susceptive termination $-jB - jn^2|Y_{01}|$ for l infinite. For the thick diaphragm under open-circuit or short-circuit mid-plane terminations, the input admittance is always a pure susceptance whether γ_1 is imaginary or real. The two transformers in Fig. 8.13 may be removed if the characteristic admittance of the center transmission line is taken as $n^2 Y_{01}$.

When a short circuit is placed in the symmetry plane,

$$Y_{\text{in}} = -jB + n^2 Y_{01}\coth\gamma_1 l.$$

Equation (77) now gives the solutions for B and n^2 and results in the same set of equations

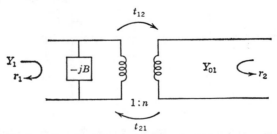

Fig. 8.14. Equivalent circuit for one junction of a thick inductive window.

as (78). If $\tanh \gamma_n l$ and $\coth \gamma_n l$ for $n > 1$ cannot be replaced by unity, the shunt susceptance B becomes a function of l. The reason is because of the interaction between the evanescent modes excited at each interface.

By means of the wave matrices, we may readily find the overall transmission coefficient through the thick diaphragm. With reference to Fig. 8.14, the following reflection and transmission coefficients, applicable to one junction in Fig. 8.13, are introduced:

$$r_1 = \frac{Y_1 - n^2 Y_{01} + jB}{Y_1 + n^2 Y_{01} - jB} \tag{79a}$$

$$r_2 = \frac{n^2 Y_{01} - Y_1 + jB}{n^2 Y_{01} + Y_1 - jB} \tag{79b}$$

$$t_{12} = (1 + r_1)n \tag{79c}$$

$$t_{21} = \frac{1 + r_2}{n}. \tag{79d}$$

The overall wave-transmission matrix which relates the amplitudes of the incident and reflected waves on the input and output sides may be found according to the methods of Section 3.4 and is

$$\begin{bmatrix} A_{11} & A_{12} \\ A_{21} & A_{22} \end{bmatrix} = \frac{1}{t_{12}t_{21}} \begin{bmatrix} 1 & -r_2 \\ r_1 & 1 + r_1 + r_2 \end{bmatrix} \begin{bmatrix} e^{2\gamma_1 l} & 0 \\ 0 & e^{-2\gamma_1 l} \end{bmatrix} \begin{bmatrix} 1 & -r_1 \\ r_2 & 1 + r_1 + r_2 \end{bmatrix}.$$

The overall voltage-transmission coefficient is

$$A_{11}^{-1} = \frac{t_{12}t_{21}e^{-2\gamma_1 l}}{1 - r_2^2 e^{-4\gamma_1 l}} = T. \tag{80}$$

When γ_1 is real, and $r_2^2 e^{-4\gamma_1 l}$ is negligible compared with unity, (80) shows that T is proportional to $e^{-2\gamma_1 l}$. The attenuation produced by the thick diaphragm is then directly proportional to l when expressed in nepers or decibels. This is the principle of the cutoff attenuator. It should be noted that the attenuation is produced by reflection at the input and not by absorption of power. A single evanescent mode cannot propagate real power, but the combination of nonpropagating modes which decay in opposite directions does lead to a transfer of power. Such interacting evanescent modes will exist between the two end faces of the thick inductive diaphragm.

Having discussed some of the general properties of the thick inductive diaphragm, we must now return to the details of evaluating the equivalent-circuit parameters. A transformation of variables, such as was used for the thin diaphragms, will not work in the present case, since two different series occur in (76) and (77). However, one of these series consists of functions that are orthogonal over the aperture domain, and it is therefore convenient to choose

$$\mathcal{E}(x) = \sum_{m=1, 3, \ldots}^{M} b_m \psi_m(x).$$

The parameter M must be chosen according to the accuracy required. If we let

$$P_{mn} = \int_t^{t+d} \psi_m \Phi_n \, dx = \frac{2m\pi}{d} \frac{\sin n\pi(t/a)}{m^2\pi^2/d^2 - n^2\pi^2/a^2}$$

for n and m both odd, then (76) becomes

$$(Z_{11} + Z_{12})^{-1} \sum_{s=1, 3, \ldots}^{M} \sum_{r=1, 3, \ldots}^{M} P_{s1} P_{r1} b_s b_r = \sum_{s=1, 3, \ldots}^{M} \sum_{r=1, 3, \ldots}^{M} b_s b_r$$

$$\cdot \left(\sum_{n=3, 5, \ldots}^{\infty} Y_n P_{sn} P_{rn} + \frac{ad}{4} \sum_{n=1, 3, \ldots}^{\infty} Y_{0m} \delta_{sm} \delta_{rm} \tanh \gamma_m l \right). \quad (81)$$

Equating the partial derivatives with respect to b_r equal to zero gives

$$(Z_{11} + Z_{12})^{-1} P_{r1} \sum_{s=1, 3, \ldots}^{M} P_{s1} b_s = -j \sum_{s=1, 3, \ldots}^{M} b_s g_{rs}, \quad r = 1, 3, \ldots, M$$

where

$$-j g_{rs} = \sum_{n=3, 5, \ldots}^{\infty} Y_n P_{sn} P_{rn} + \frac{ad}{4} \sum_{m=1, 3, \ldots}^{\infty} Y_{0m} \delta_{sm} \delta_{rm} \tanh \gamma_m l.$$

For a solution, the determinant of this set of equations must vanish. Thus denoting $(Z_{11}+Z_{12})^{-1}$ by Y_{oc}, we get

$$\begin{vmatrix} Y_{oc}P_{11}^2 + jg_{11} & Y_{oc}P_{11}P_{31} + jg_{13} & \cdots & Y_{oc}P_{11}P_{M1} + jg_{1M} \\ Y_{oc}P_{31}P_{11} + jg_{31} & \cdots & \cdots & Y_{oc}P_{31}P_{M1} + jg_{3M} \\ \cdots\cdots\cdots\cdots\cdots\cdots\cdots\cdots\cdots\cdots\cdots\cdots\cdots\cdots\cdots\cdots \\ Y_{oc}P_{M1}P_{11} + jg_{M1} & \cdots & \cdots & Y_{oc}P_{M1}^2 + jg_{MM} \end{vmatrix} = 0.$$

Performing the following operations on this determinant:

1. dividing the rth row by P_{r1}, $r = 1, 3, \ldots, M$;
2. dividing the sth column by P_{s1}, $s = 1, 3, \ldots, M$;
3. subtracting the first row from all the other rows; and
4. factoring the resultant determinant into the sum of two determinants;

we obtain the following final result for Y_{oc}:

$$
Y_{oc} \begin{vmatrix}
1 & 1 & \cdots & 1 \\
\dfrac{g_{31}}{P_{11}P_{31}} - \dfrac{g_{11}}{P_{11}^2} & \dfrac{g_{33}}{P_{31}^2} - \dfrac{g_{13}}{P_{11}P_{31}} & \cdots & \dfrac{g_{3M}}{P_{31}P_{M1}} + \dfrac{g_{1M}}{P_{11}P_{M1}} \\
\multicolumn{4}{c}{\dotfill} \\
\dfrac{g_{M1}}{P_{11}P_{M1}} + \dfrac{g_{11}}{P_{11}^2} & \cdots & \cdots & \dfrac{g_{MM}}{P_{M1}^2} - \dfrac{g_{1M}}{P_{11}P_{M1}}
\end{vmatrix}
$$

$$
= -j \begin{vmatrix}
\dfrac{g_{11}}{P_{11}^2} & \cdots & \dfrac{g_{1M}}{P_{11}P_{M1}} \\
\dfrac{g_{31}}{P_{11}P_{31}} - \dfrac{g_{11}}{P_{11}^2} & \cdots & \dfrac{g_{3M}}{P_{31}P_{M1}} - \dfrac{g_{1M}}{P_{11}P_{M1}} \\
\multicolumn{3}{c}{\dotfill} \\
\dfrac{g_{M1}}{P_{11}P_{M1}} - \dfrac{g_{11}}{P_{11}^2} & \cdots & \dfrac{g_{MM}}{P_{M1}^2} - \dfrac{g_{1M}}{P_{11}P_{M1}}
\end{vmatrix}. \quad (82)
$$

When a two-term approximation is used for $\mathcal{E}(x)$, the evaluation of $Y_{oc} = (Z_{11} + Z_{12})^{-1}$ reduces to the evaluation of two 2×2 determinants. A similar analysis may be used for determining $(Z_{11} - Z_{12})^{-1}$ and results in an equation similar to (82), with Y_{oc} replaced by $Y_{sc} = (Z_{11} - Z_{12})^{-1}$, and g_{rs} replaced by h_{rs}, where

$$
-jh_{rs} = \sum_{n=3,5,\dots}^{\infty} Y_n P_{sn} P_{rn} + \frac{ad}{4} \sum_{m=1,3,\dots}^{\infty} Y_{0m} \delta_{sm} \delta_{rm} \coth \gamma_m l.
$$

If we let $S_{rs} = \sum_{n=3,5,\dots}^{\infty} |Y_n| P_{rn} P_{sn}$, then $g_{rs} = h_{rs}$ for $s \neq r$, and

$$
g_{rr} = S_{rr} + \frac{ad}{4} jY_{0r} \tanh \gamma_r l
$$

$$
h_{rr} = S_{rr} + (ad/4) jY_{0r} \coth \gamma_r l.
$$

In view of these relations, the evaluation of Z_{11} and Z_{12} is not unduly laborious. For results accurate to within a few percent, a two-term approximation of $\mathcal{E}(x)$ is sufficient. The series S_{rs} is suitable for direct evaluation since the terms decrease as n^3 and the summation is over odd values of n only. The series is not directly summable since it is not an even rational function of the summation variable n.

8.5. A NARROW INDUCTIVE STRIP

Figure 8.15 illustrates an inductive strip of width $2t$ and centered at $x = x_1$. The strip is assumed to be perfectly conducting and of negligible thickness in the z direction. The solution to this problem will be formulated in terms of a variational expression involving the current on the obstacle.

Let an H_{10} mode be incident from the region $z < 0$. The incident electric field has only a y component given by $E_i = \sin(\pi x/a)e^{-\Gamma_1 z}$. Since the strip is uniform along the y direction, the only higher order modes excited are the H_{n0} modes. The incident field excites a current distribution $J(x)$ on the strip. This current is directed along the y axis and does not vary with

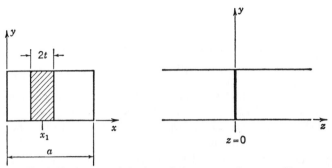

Fig. 8.15. A narrow inductive strip in a rectangular waveguide.

y since the incident field has no variation with y. The scattered field in the guide may be evaluated in terms of the currents on the strip by using the Green's function for H_{n0} modes given in Section 5.6. The scattered field is thus given by

$$E_s(x, z) = \int_S G(x, z|x')J(x')\,dx' \tag{83}$$

where S denotes the surface of the strip,

$$G(x, z|x') = -\frac{j\omega\mu_0}{a} \sum_{n=1}^{\infty} \frac{1}{\Gamma_n} \sin \frac{n\pi x}{a} \sin \frac{n\pi x'}{a} e^{-\Gamma_n|z|}$$

and $\Gamma_n^2 = (n\pi/a)^2 - k_0^2$. The total field E_t in the guide is the sum of the incident and scattered fields and must vanish on the perfectly conducting strip. The resulting integral equation for the current distribution J is

$$\sin \frac{\pi x}{a} e^{-\Gamma_1 z} + \int_S G(x, z|x')J(x')\,dx' = 0 \qquad \text{on } S. \tag{84}$$

From (83) the reflected dominant mode is seen to be given by

$$-\frac{j\omega\mu_0}{a\Gamma_1} \sin \frac{\pi x}{a} e^{\Gamma_1 z} \int_S \sin \frac{\pi x'}{a} J(x')\,dx' = R \sin \frac{\pi x}{a} e^{\Gamma_1 z}, \qquad z < 0 \tag{85}$$

where the reflection coefficient R is defined by this equation. For $z > 0$, the total dominant mode field transmitted past the obstacle is

$$\sin \frac{\pi x}{a} e^{-\Gamma_1 z} - \frac{j\omega\mu_0}{a\Gamma_1} \sin \frac{\pi x}{a} e^{-\Gamma_1 z} \int_S \sin \frac{\pi x'}{a} J(x')\,dx' = T \sin \frac{\pi x}{a} e^{-\Gamma_1 z}, \qquad z > 0 \tag{86}$$

where the transmission coefficient T is defined by this equation. From (85) and (86) we find that $1 + R = T$, and hence the obstacle will appear as an equivalent shunt element across a transmission line.

The integral equation (84) may be rewritten as

$$\sin \frac{\pi x}{a} - \frac{j\omega\mu_0}{a\Gamma_1} \sin \frac{\pi x}{a} \int_S \sin \frac{\pi x'}{a} J(x')\,dx' = \frac{j\omega\mu_0}{a} \sum_{n=2}^{\infty} \frac{1}{\Gamma_n} \sin \frac{n\pi x}{a} \int_S \sin \frac{n\pi x'}{a} J(x')\,dx'$$

since $z = 0$ on S. Using (85) now gives

$$(1 + R) \sin \frac{\pi x}{a} = \frac{j\omega\mu_0}{a} \sum_{n=2}^{\infty} \frac{1}{\Gamma_n} \sin \frac{n\pi x}{a} \int_S \sin \frac{n\pi x'}{a} J(x')\, dx'. \tag{87}$$

A shunt reactance jX across a transmission line with unit characteristic impedance produces a reflection coefficient given by $-jX = (1 + R)/2R$. Comparing with (87), it is seen that an expression for jX may be obtained by multiplying both sides of (87) by $J(x)$, integrating over the obstacle, and then dividing both sides by $[\int_S J(x) \sin(\pi x/a)\, dx]^2$ and using (85) to obtain a factor R in the denominator of the left-hand side. The indicated operations give

$$jX = -\frac{1+R}{2R} = \frac{\Gamma_1}{2} \frac{\displaystyle\sum_{n=2}^{\infty} \frac{1}{\Gamma_n} \iint_S J(x)J(x') \sin \frac{n\pi x}{a} \sin \frac{n\pi x'}{a}\, dx\, dx'}{\left[\displaystyle\int_S J(x) \sin \frac{\pi x}{a}\, dx \right]^2}. \tag{88}$$

This expression is readily shown to be a variational expression for the normalized shunt reactance jX. Furthermore, it is a positive-definite quadratic form, and so an approximate solution for J will give an upper bound on jX. The variational expression (88) is also applicable to the thin-inductive-diaphragm problems considered in Section 8.3.

At the edge of an infinitely thin perfectly conducting strip, the normal magnetic field, and hence the tangential current density, becomes infinite as $r^{-1/2}$, where r is the radial distance from the edge. However, for a sufficiently narrow strip, the integrated effect of the current should be much the same as though the current were constant over the strip. For a first approximation to jX we will, therefore, assume a constant current distribution. The integrations in (88) are readily performed and give

$$jX = \frac{\Gamma_1}{2} \frac{\displaystyle\sum_{n=2}^{\infty} \frac{1}{n^2 \Gamma_n} \left[\cos \frac{n\pi}{a}(x_1 + t) - \cos \frac{n\pi}{a}(x_1 - t) \right]^2}{\left[\cos \frac{\pi}{a}(x_1 + t) - \cos \frac{\pi}{2}(x_1 - t) \right]^2}$$

$$= \frac{\Gamma_1}{2} \frac{\displaystyle\sum_{n=2}^{\infty} \frac{1}{n^2 \Gamma_n} \left(\sin \frac{n\pi x_1}{a} \sin \frac{n\pi t}{a} \right)^2}{\left(\sin \frac{\pi x_1}{a} \sin \frac{\pi t}{a} \right)^2}. \tag{89}$$

For a centered strip $x_1 = a/2$, and (89) reduces to

$$jX = \frac{\Gamma_1}{2} \csc^2 \frac{\pi t}{a} \sum_{n=3,5,\ldots}^{\infty} \frac{1}{n^2 \Gamma_n} \sin^2 \frac{n\pi t}{a}. \tag{90}$$

If desired, the series could be written as a dominant series,

$$\sum_{n=3,5,\ldots}^{\infty} \frac{a}{\pi n^3} \sin^2 \frac{n\pi t}{a}$$

plus a correction series

$$\sum_{n=3,5,\ldots}^{\infty} \frac{1}{n^2} \left(\frac{1}{\Gamma_n} - \frac{a}{n\pi} \right) \sin^2 \frac{n\pi t}{a}.$$

The first series may be readily summed. However, there is little point in treating (90) in this manner since the original series converges very fast, and only four or five terms are required. The alternative procedure suggested above leads to just as many, if not more, terms to be evaluated.

If t is very small it is worthwhile summing the series in (90). We have

$$\sum_{n=3,4,\ldots}^{\infty} \frac{1}{n^3} \sin^2 \frac{n\pi t}{a} = \frac{1}{2} \sum_{n=1,3,\ldots}^{\infty} \frac{1}{n^3} \left(1 - \cos \frac{2n\pi t}{a} \right) - \sin^2 \frac{\pi t}{a}$$

$$= \frac{1}{2} \left(\frac{\pi t}{a} \right)^2 \left(\ln \frac{a}{\pi t} + \frac{3}{2} \right) - \left(\frac{\pi t}{a} \right)^2$$

which is correct to order $(\pi t/a)^2$. Replacing $\csc^2(\pi t/a)$ by $(a/\pi t)^2$ also, (90) gives

$$jX = \frac{a\Gamma_1}{4\pi} \left[\ln \frac{a}{\pi t} - 1 + 2 \left(\frac{a}{\pi t} \right)^2 \sum_{n=3,5,\ldots}^{\infty} \left(\frac{\pi}{a\Gamma_n} - \frac{1}{n} \right) \frac{1}{n^2} \sin^2 \frac{n\pi t}{a} \right] \tag{91}$$

for the inductive reactance.

8.6. THIN INDUCTIVE POST

An inductive post of radius t is shown in Fig. 8.16. In order to find the field scattered by this post we will use the image series for the Green's function. We can integrate the result given in Problem 2.22 over z' from minus to plus infinity to get the following Green's function for a line source, infinite along y, and located at x', z' where $x' = a/2 + t \sin \phi'$ and $z' = t \cos \phi'$:

$$G_0 = -\frac{j}{4} \sum_{n=0}^{\infty} \epsilon_{0n} J_n(k_0 r_<) H_n^2(k_0 r_>) \cos n(\phi - \phi') \tag{92}$$

where r and r' are the radial distances from the center of the post. To this dominant part we add the contributions from the images in the sidewalls of the waveguide. We will treat the images of the post as points located at $x = a/2 + na$, $n = \pm 1, \pm 2, \ldots$ and evaluate their contributions at $x = a/2$, $z = 0$. From (91) in Chapter 2 their contributions are given by

$$G_1 = \frac{j}{4} H_0^2(k_0 a) - \frac{j}{4} \sum_{-\infty}^{\infty}{}' [H_0^2(k_0 2na) - H_0^2(k_0(2n+1)a)]. \tag{93}$$

For a current J on the post the scattered electric field is given by

$$E_s = -jk_0 Z_0 \int_0^{2\pi} (G_0 + G_1) J(\phi') t \, d\phi'. \tag{94}$$

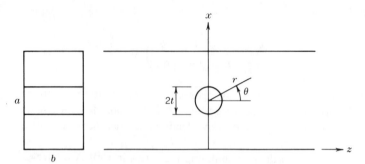

Fig. 8.16. A circular inductive post.

Note that we do not integrate over y from 0 to b because the Green's function we are using is for a line source in the waveguide.

We will choose

$$E_i = V_1^+ \sin \frac{\pi x}{a} e^{-j\beta_1 z}$$

for the incident field on the post. At the surface of the post the boundary condition $E_s = -E_i$ must hold. We now let $x = a/2 + t \sin \phi$, $z = t \cos \phi$ and expand E_i in a Taylor series about the point $x = a/2$, $z = 0$. When we retain terms up to t^2 only we obtain

$$E_i = \left(1 - \frac{\pi^2 t^2}{4a^2} - \frac{\beta_1^2 t^2}{4}\right) - j\beta_1 t \cos \phi + \left(\frac{\pi^2 t^2}{4a^2} - \frac{\beta_1^2 t^2}{4}\right) \cos 2\phi. \quad (95)$$

With this order of approximation the current density on the post can be assumed to be given by

$$J = I_0 + I_1 \cos \phi + I_2 \cos 2\phi. \quad (96)$$

By using this expansion in (94) we obtain

$$-jk_0 Z_0 2\pi t I_0 \left\{ -\frac{j}{4} J_0(k_0 t) H_0^2(k_0 t) + \frac{j}{4} H_0^2(k_0 a) - \frac{j}{4} \sum_{-\infty}^{\infty}{}' [H_0^2(k_0 2na) - H_0^2(k_0(2n+1)a)] \right\}$$

$$-jk_0 Z_0 \pi t \left[-\frac{j}{2} I_1 J_1(k_0 t) H_1^2(k_0 t) \cos \phi - \frac{j}{2} I_2 J_2(k_0 t) H_2^2(k_0 t) \cos 2\phi \right]. \quad (97)$$

The Hankel function series converges slowly so we will sum this series. We make use of the sum

$$\sum_{n=1,3,\ldots}^{\infty} \frac{e^{-\Gamma_n \alpha}}{\Gamma_n a} \sin \frac{n\pi x}{a} \sin \frac{n\pi x'}{a} = -\frac{j}{4} H_0^2(k_0 \alpha) + \frac{j}{4} H_0^2(k_0 \sqrt{a^2 + \alpha^2})$$

$$-\frac{j}{4} \sum_{-\infty}^{\infty}{}' [H_0^2(k_0 \sqrt{(2na)^2 + \alpha^2}) - H_0^2(k_0 \sqrt{(2n+1)^2 a^2 + \alpha^2})]$$

which is valid for $x = x' = a/2$. The first series is

$$\sum_{n=1,3,\ldots}^{\infty} \frac{e^{-\Gamma_n\alpha}}{\Gamma_n a} = \sum_{n=1,3,\ldots}^{\infty} \frac{e^{-n\pi\alpha/a}}{n\pi} + \sum_{n=1,3,\ldots}^{\infty} \left(\frac{e^{-\Gamma_n\alpha}}{\Gamma_n a} - \frac{e^{-n\pi\alpha/a}}{n\pi} \right)$$

$$= -\frac{1}{2\pi} \ln \tanh \frac{\pi\alpha}{2a} + \sum_{n=1,3,\ldots}^{\infty} \left(\frac{e^{-\Gamma_n\alpha}}{\Gamma_n a} - \frac{e^{-n\pi\alpha/a}}{n\pi} \right).$$

We use this result and let α tend to zero and replace $H_0^2(k_0\alpha)$ by $1 - (2j/\pi)(\gamma t \ln k_0\alpha/2)$. After some additional algebra we obtain the desired result, to be used in (97),

$$G_1 = \frac{j}{4} + \frac{\gamma}{2\pi} + \frac{1}{2\pi} \ln \frac{k_0 a}{\pi} + \sum_{n=1,3,\ldots}^{\infty} \frac{n\pi - \Gamma_n a}{\Gamma_n n\pi a}. \tag{98}$$

The series that is present in this expression converges rapidly.

We now equate the corresponding $\cos n\phi$ terms in (97) to the negative of those in (95) so as to satisfy the boundary conditions on the post. This gives the following solutions for the current amplitudes:

$$I_0 = \cfrac{-j\left[1 - \left(\dfrac{\pi t}{2a}\right)^2 - \left(\dfrac{\beta_1 t}{2}\right)^2\right] V_1^+}{k_0 Z_0 t \left[\ln \dfrac{2a}{\pi t} + \left(\dfrac{k_0 t}{2}\right)^2 \left(\gamma + \ln \dfrac{k_0 t}{2}\right) - 2 + 2\displaystyle\sum_{n=3,5\ldots}^{\infty} \dfrac{n\pi - \Gamma_n a}{\Gamma_n na} + j\dfrac{\pi}{8}(k_0 t)^2 - j\dfrac{2\pi}{\beta a}\right]} \tag{99a}$$

$$I_1 = -\frac{2\beta_1}{k_0 Z_0} V_1^+ \tag{99b}$$

$$I_2 = j\frac{\beta_1^2 t - \pi^2 t/a^2}{k_0 Z_0} V_1^+. \tag{99c}$$

In solving for I_0, I_1, and I_2 we have used the small-argument approximations for J_n and H_n^2 as given by (181) in Chapter 2. For the thicker posts the Bessel functions should be evaluated. The dominant mode scattered field in $z < 0$ is obtained from (94), thus

$$V_1^- \sin \frac{\pi x}{a} e^{\Gamma_1 z} = \frac{-jk_0 Z_0}{\Gamma_1 a} e^{\Gamma_1 z} \sin \frac{\pi x}{a} \int_0^{2\pi} e^{-\Gamma_1 z'} \sin \frac{\pi x'}{a} J(\phi') t \, d\phi'$$

which gives, upon using the expansion given in (95),

$$V_1^- = S_{11} V_1^+ = -\frac{k_0 Z_0}{\beta_1 a} 2\pi t \left[I_0 \left(1 - \frac{\pi^2 t^2}{4a^2} - \frac{\beta_1^2 t^2}{4} \right) - j\frac{\beta_1 t}{2} I_1 + \left(\frac{\pi^2 t^2}{4a^2} - \frac{\beta_1^2 t^2}{4} \right) \frac{I_2}{2} \right] \tag{100}$$

For $z > 0$ the dominant mode scattered field is also given by this equation but with a change in sign for the I_1 term. The total dominant mode field in $z > 0$ is the sum of the scattered field plus the incident field and has an amplitude $V_2^- = S_{21}V_1^+$. The scattering matrix parameter S_{21} is given by one plus the factor multiplying V_1^+ in (100) but with β_1 replaced by $-\beta_1$ for the I_1 term.

In the limit as t approaches zero we see that tI_1 and tI_2 vanish. The limiting solution for S_{11} is

$$S_{11} = \frac{-1}{1 + j\dfrac{\beta_1 a}{2\pi}\left[\ln\dfrac{2a}{\pi t} - 2 + 2\displaystyle\sum_{n=3,5,\dots}^{\infty}\dfrac{n\pi - \Gamma_n a}{\Gamma_n na}\right]}. \qquad (101)$$

A shunt reactance jX produces a reflection coefficient S_{11} given by

$$S_{11} = \frac{-1}{1 + 2jX}.$$

Hence we see that

$$jX = \frac{j\beta_1 a}{4\pi}\left(\ln\frac{a}{\pi t} - 1.3 + 2\sum_{n=3,5,\dots}^{\infty}\frac{n\pi - \Gamma_n a}{\Gamma_n na}\right). \qquad (102)$$

This result is essentially the same as that given by (91). The results would agree very closely if the equivalent radius of the post was chosen equal to 0.233 times the full width of the inductive strip.

Numerical results for the inductive post are readily computed from the formulas given above. In general, the equivalent circuit is a T network since S_{21} does not equal $1 + S_{11}$. These formulas give accurate results for t/a up to about 0.04. For thicker posts it is necessary to take into account the variation of the field from the images around the post. An analytical theory can be developed by using the Bessel function addition theorem to refer the fields produced by each image post to a new origin coinciding with the center of the primary post. The algebra is tedious and a large number of terms occur so it is more expedient to resort to a numerical solution. Such a solution has been presented by Leviatan *et al.* [8.19] in a recent paper. These authors have approximated the current on the post by N equispaced filaments and used 20 such filaments per wavelength circumference. For a post with a radius $t = 0.2a$ a total of approximately 15 to 20 filaments is required.

8.7. General Formulas for Waveguide Scattering

In this section we will develop general formulas for scattering from conducting and dielectric obstacles in waveguides. The case of scattering of incident waves by a conducting obstacle will be treated first.

Consider an obstacle that is perfectly conducting that is located in an infinitely long waveguide. The surface of the obstacle is S and the unit normal to S is $\mathbf{n}$ as shown in Fig. 8.17. The dominant mode in the waveguide will be expressed in the following form:

$$\mathbf{E}_1^+ = [\mathbf{e}_1(u_1, u_2) + \mathbf{e}_{z1}(u_1, u_2)]e^{-j\beta_1 z}$$

$$\mathbf{E}_1^- = [\mathbf{e}_1(u_1, u_2) - \mathbf{e}_{z1}(u_1, u_2)]e^{j\beta_1 z}$$

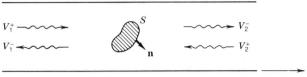

Fig. 8.17. Scattering by an obstacle in a waveguide.

where u_1, u_2 are orthogonal coordinates in the transverse plane of the waveguide. The nomenclature used here is the same as that introduced in Section 5.6.

We will let $\bar{\mathbf{G}}$ denote the dyadic Green's function operator that gives the electric field in the waveguide from a current element $\mathbf{J}$ by means of the formula $\bar{\mathbf{G}}(\mathbf{r}, \mathbf{r}') \cdot \mathbf{J}(\mathbf{r}')$. One representation for $\bar{\mathbf{G}}$ is given by (87) in Section 5.6 and is

$$\bar{\mathbf{G}}(\mathbf{r}, \mathbf{r}') = -\frac{1}{2} \sum_{n=1}^{\infty} \mathbf{E}_{n\sigma}(\mathbf{r}) \mathbf{E}_{n\sigma}(\mathbf{r}') \tag{103}$$

where $\mathbf{E}_{n\sigma}$ is given by

$$\mathbf{E}_{n\sigma}(\mathbf{r}) = \begin{cases} \mathbf{E}_n^+(\mathbf{r}), & z > z' \\ \mathbf{E}_n^-(\mathbf{r}), & z < z' \end{cases}$$

$$\mathbf{E}_{n\sigma}(\mathbf{r}') = \begin{cases} \mathbf{E}_n^+(\mathbf{r}'), & z' > z \\ \mathbf{E}_n^-(\mathbf{r}'), & z' < z. \end{cases}$$

We will let the incident field on the obstacle be

$$\mathbf{E}_i = V_1^+ \mathbf{E}_1^+ + V_2^+ \mathbf{E}_1^- .$$

The scattered field is given by

$$\mathbf{E}_s = \oiint_S \bar{\mathbf{G}}(\mathbf{r}, \mathbf{r}') \cdot \mathbf{J}(\mathbf{r}') \, dS'$$

where $\mathbf{J}(\mathbf{r}')$ is the unknown current induced on the obstacle. On the surface of the obstacle the boundary condition $\mathbf{n} \times (\mathbf{E}_i + \mathbf{E}_s) = 0$ must hold. This condition gives the integral equation

$$\mathbf{n} \times \mathbf{E}_i + \oiint_S \mathbf{n} \times \bar{\mathbf{G}} \cdot \mathbf{J} \, dS' = 0, \qquad \mathbf{r} \text{ on } S \tag{104}$$

for the current $\mathbf{J}$ on S. A practical method for obtaining an approximate solution of the integral equation is to use Galerkin's method. In this method we expand the current $\mathbf{J}$ in terms of a suitable finite set of vector basis functions defined on S; thus

$$\mathbf{J}(\mathbf{r}') = \sum_{n=1}^{N} I_n \mathbf{J}_n(\mathbf{r}')$$

where I_n are unknown amplitudes. This expansion is used in (104) but instead of taking the cross product with $\mathbf{n}$ the integral equation is scalar multiplied by $\mathbf{J}_m(\mathbf{r})$ and integrated over S for $m = 1, 2, \ldots, N$. The result is the following system of linear equations:

$$\sum_{n=1}^{N} I_n \oint\!\!\!\oint_S \oint\!\!\!\oint_S \mathbf{J}_m(\mathbf{r}) \cdot \bar{\mathbf{G}}(\mathbf{r}, \mathbf{r}') \cdot \mathbf{J}_n(\mathbf{r}') \, dS' \, dS$$

$$= \sum_{n=1}^{N} G_{mn} I_n = -\oint\!\!\!\oint_S \mathbf{E}_i(\mathbf{r}) \cdot \mathbf{J}_m(\mathbf{r}) \, dS = v_m, \qquad m = 1, 2, \ldots, N \quad (105)$$

where G_{mn} and v_m are defined by this equation.

By solving these equations for I_n, one usually obtains a good approximate solution for $\mathbf{J}$ provided a sufficiently large number of basis functions have been used. We can express v_m in the form

$$v_m = V_1^+ f_m + V_2^+ g_m \tag{106a}$$

where

$$f_m = -\oint\!\!\!\oint_S \mathbf{E}_1^+(\mathbf{r}) \cdot \mathbf{J}_m(\mathbf{r}) \, dS \tag{106b}$$

$$g_m = -\oint\!\!\!\oint_S \mathbf{E}_1^-(\mathbf{r}) \cdot \mathbf{J}_m(\mathbf{r}) \, dS. \tag{106c}$$

The current amplitudes I_n are linear functions of the amplitudes V_1^+ and V_2^+ of the incident waves.

The dominant mode fields scattered into the waveguide are given by

$$V_1^- \mathbf{E}_1^-(\mathbf{r}) = -\frac{1}{2} \mathbf{E}_1^-(\mathbf{r}) \oint\!\!\!\oint_S \sum_{n=1}^{N} I_n \mathbf{E}_1^+(\mathbf{r}') \cdot \mathbf{J}_n(\mathbf{r}') \, dS' \tag{107a}$$

$$V_2^- \mathbf{E}_1^+(\mathbf{r}) = -\frac{1}{2} \mathbf{E}_1^+(\mathbf{r}) \oint\!\!\!\oint_S \sum_{n=1}^{N} I_n \mathbf{E}_1^-(\mathbf{r}') \cdot \mathbf{J}_n(\mathbf{r}') \, dS' \tag{107b}$$

in the respective regions $z \to -\infty$ and $z \to \infty$.

From these equations and the solution for the I_n we will be able to express V_1^- and V_2^- as linear functions of V_1^+ and V_2^+ in the form

$$V_1^- = S_{11} V_1^+ + S_{12} V_2^+ \tag{108a}$$

$$V_2^- = S_{21} V_1^+ + S_{22} V_2^+ \tag{108b}$$

where the S_{ij} are the scattering matrix parameters of the obstacle. It is assumed that the equivalent voltages are defined in such a manner that the power flow in a given mode is proportional to the voltage amplitude squared. If the normalization specified by (76a) in

Section 5.6 is used this will be the case. Equivalent T and π networks for the obstacle can be found from the scattering matrix parameters.

In the implementation of the above method care must be exercised in the evaluation of the matrix elements G_{mn} since the Green's function becomes singular when $r = r'$. A variety of techniques can be used to isolate the singular part of the Green's function. Very often the series involved can be summed into closed form in the quasi-static limit where k_0 is set equal to zero. The efficient numerical evaluation of the G_{mn} is dependent on the selection of a good representation of the Green's function. A variety of techniques have been discussed in the earlier part of this chapter and in the preceding chapters as well.

The problem of scattering by a dielectric obstacle in a waveguide can be formulated using the boundary-element method described in Chapter 6 in connection with dielectric waveguides and resonators. In the region outside the obstacle the total electric and magnetic fields are given by

$$\mathbf{E}_1 = \mathbf{E}_i + \oiint_S \mathbf{n} \times \mathbf{E} \cdot \bar{\mathbf{G}}_{e1} \, dS' - j\omega\mu_0 \oiint_S \mathbf{n} \times \mathbf{H} \cdot \nabla \times \bar{\mathbf{G}}_{e1} \, dS' \qquad (109a)$$

$$\mathbf{H}_1 = \mathbf{H}_i + \oiint_S \mathbf{n} \times \mathbf{H} \cdot \bar{\mathbf{G}}_{m1} \, dS' + j\omega\epsilon_0 \oiint_S \mathbf{n} \times \mathbf{E} \cdot \nabla \times \bar{\mathbf{G}}_{m1} \, dS'. \qquad (109b)$$

The dyadic Green's functions $\bar{\mathbf{G}}_{e1}$ and $\bar{\mathbf{G}}_{m1}$ are waveguide dyadic functions that are solutions of

$$\nabla \times \nabla \times \bar{\mathbf{G}}_{e1} - k_0^2 \bar{\mathbf{G}}_{e1} = \bar{\mathbf{I}}\delta(\mathbf{r} - \mathbf{r}')$$

$$\nabla \times \nabla \times \bar{\mathbf{G}}_{m1} - k_0^2 \bar{\mathbf{G}}_{m1} = \bar{\mathbf{I}}\delta(\mathbf{r} - \mathbf{r}')$$

and satisfy the radiation conditions and the boundary conditions

$$\mathbf{n} \times \bar{\mathbf{G}}_{e1} = \mathbf{n} \times \nabla \times \bar{\mathbf{G}}_{m1} = 0$$

on the metal walls.

The fields $\mathbf{E}_2$, $\mathbf{H}_2$ in the interior of the dielectric obstacle are given by

$$\mathbf{E}_2 = -\oiint_S \mathbf{n} \times \mathbf{E} \cdot \bar{\mathbf{G}}_{e2} \, dS' + j\omega\mu_0 \oiint_S \mathbf{n} \times \mathbf{H} \cdot \nabla \times \bar{\mathbf{G}}_{e2} \, dS' \qquad (110a)$$

$$\mathbf{H}_2 = -\oiint_S \mathbf{n} \times \mathbf{H} \cdot \bar{\mathbf{G}}_{m2} \, dS' - j\omega\epsilon \oiint_S \mathbf{n} \times \mathbf{E} \cdot \nabla \times \bar{\mathbf{G}}_{m2} \, dS' \qquad (110b)$$

where the Green's functions $\bar{\mathbf{G}}_{e2}$ and $\bar{\mathbf{G}}_{m2}$ can be conveniently chosen to be the same as $\bar{\mathbf{G}}_{e1}$ and $\bar{\mathbf{G}}_{m1}$ but with k_0^2 replaced by $k^2 = \kappa k_0^2$ where $\kappa = \epsilon/\epsilon_0$ is the dielectric constant of the obstacle.

A pair of equations to determine the boundary values of $\mathbf{n} \times \mathbf{E}$ and $\mathbf{n} \times \mathbf{H}$ on S are obtained by enforcing the continuity of the tangential field components across the surface S. These boundary conditions give

$$\mathbf{n} \times \mathbf{E}_1 = \mathbf{n} \times \mathbf{E}_2 \qquad \text{on } S \qquad (111a)$$

$$\mathbf{n} \times \mathbf{H}_1 = \mathbf{n} \times \mathbf{H}_2 \qquad \text{on } S. \qquad (111b)$$

The numerical solution can be obtained by introducing a finite set of basis functions to expand the boundary values and then using Galerkin's method to obtain a system of linear equations for the unknown coefficients.

An alternative procedure is to characterize the dielectric obstacle in terms of the unknown polarization current $\mathbf{J}_p = j\omega(\epsilon - \epsilon_0)\mathbf{E}$ and to express the scattered field in terms of this current. In other words, the total electric field is a solution of

$$\nabla \times \nabla \times \mathbf{E} - k^2\mathbf{E} = -j\omega\mu_0\mathbf{J}$$

or

$$\nabla \times \nabla \times \mathbf{E} - k_0^2\mathbf{E} = (\kappa - 1)k_0^2\mathbf{E} - j\omega\mu_0\mathbf{J}$$
$$= -j\omega\mu_0 j\omega(\epsilon - \epsilon_0)\mathbf{E} - j\omega\mu_0\mathbf{J} \tag{112}$$

where $\mathbf{J}$ is the source for the incident field.

The scattered field is a solution of

$$\nabla \times \nabla \times \mathbf{E}_s - k_0^2\mathbf{E}_s = -j\omega\mu_0 j\omega(\epsilon - \epsilon_0)\mathbf{E}. \tag{113}$$

The total field $\mathbf{E}$ in the obstacle can be expanded in terms of a suitable set of basis functions. The unknown amplitude constants are determined by requiring that inside the obstacle

$$\mathbf{E}_i + \mathbf{E}_s = \mathbf{E}. \tag{114}$$

This equation can be tested with the basis functions to obtain a system of algebraic equations for the unknown amplitudes. This method is effective for small obstacles but is less efficient than the boundary-element method for large obstacles since many more basis functions are needed. However, for inhomogeneous dielectric obstacles it is usually necessary to divide the volume of the obstacle into many small regions, in each of which the dielectric permittivity can be regarded as a constant. The method outlined above is very useful for this class of problems.

REFERENCES AND BIBLIOGRAPHY

[8.1] J. Schwinger and D. S. Saxon, *Discontinuities in Waveguides: Notes on Lectures by Julian Schwinger*. New York, NY: Gordon and Breach Science Publishers, 1968.

[8.2] *Tables of Chebyshëv Polynomials*. Washington, D.C.: National Bureau of Standards, 1952.

[8.3] L. Lewin, *Advanced Theory of Waveguides*. London: Iliffe and Sons, Ltd., 1951. (This reference contains many examples of waveguide discontinuities, as well as many instructive mathematical techniques for summing a variety of infinite series which arise in practice.) See also: L. Lewin, *Theory of Waveguides*. London: Neurnes Butterworths, 1975.

[8.4] H. Motz, *Electromagnetic Problems of Microwave Theory*, Methuen Monograph. London: Methuen & Co., Ltd., 1951.

[8.5] N. Marcuvitz and J. Schwinger, "On the representation of the electric and magnetic fields produced by currents and discontinuities in waveguides," *J. Appl. Phys.*, vol. 22, pp. 806–819, June 1951.

[8.6] J. W. Miles, "The equivalent circuit for a plane discontinuity in a cylindrical guide," *Proc. IRE*, vol. 34, pp. 728–742, Oct. 1946.

[8.7] J. W. Miles, "The equivalent circuit of a corner bend in a rectangular waveguide," *Proc. IRE*, vol. 35, pp. 1313–1317, Nov. 1947.

[8.8] R. E. Collin and J. Brown, "Calculation of the equivalent circuit of an axially unsymmetrical waveguide junction," *J. IEE (London)*, vol. 103, part C, pp. 121–128, Mar. 1956.

[8.9]   N. Marcuvitz, *Waveguide Handbook*, vol. 10 of MIT Rad. Lab. Series. New York, NY: McGraw-Hill Book Company, Inc., 1951. (This is an extensive collection of theoretical and numerical results for a large variety of waveguide discontinuities.)

[8.10] F. Borgnis and C. Papas, *Randwertprobleme der Mikrowellenphysik*. Berlin: Springer-Verlag, 1955.

Solution of Waveguide and Transmission-Line Discontinuity Problems by Conformal-Mapping Techniques and Quasi-stationary Field Methods

[8.11] G. G. MacFarlane, "Quasi-stationary field theory and its application to diaphragms and junctions in transmission lines and waveguides," *J. IEE (London)*, vol. 93, part IIIA, pp. 703–719, 1946.

[8.12] J. R. Whinnery and H. W. Jamieson, "Equivalent circuits for discontinuities in transmission lines," *Proc. IRE*, vol. 32, pp. 98–114, Feb. 1944.

Additional Examples of Waveguide Discontinuity Problems

[8.13] E. D. Nielson, "Scattering by a cylindrical post of complex permittivity in a waveguide," *IEEE Trans. Microwave Theory Tech.*, vol. MTT-17, pp. 148–153, 1969.

[8.14] W. J. English, "The circular waveguide step-discontinuity mode transducer," *IEEE Trans. Microwave Theory Tech.*, vol. MTT-21, pp. 633–636, 1973.

[8.15] A. A. Oliner, "Equivalent circuits for small symmetrical longitudinal apertures and obstacles," *IRE Trans. Microwave Theory Tech.*, vol. MTT-8, pp. 72–80, 1960.

[8.16] A. Wexler, "Solution of waveguide discontinuities by modal analysis," *IEEE Trans. Microwave Theory Tech.*, vol. MTT-15, pp. 508–517, 1967.

[8.17] S. C. Wu and Y. L. Chow, "An application of the moment method to waveguide scattering problems," *IEEE Trans. Microwave Theory Tech.*, vol. MTT-20, pp. 744–749, 1972.

[8.18] J. J. H. Wang, "Analysis of a three-dimensional arbitrarily shaped dielectric biological body inside a rectangular waveguide," *IEEE Trans. Microwave Theory Tech.*, vol. MTT-26, pp. 457–462, 1978.

[8.19] Y. Leviatan, P. G. Li, A. T. Adams, and J. Perini, "Single-post inductive obstacles in rectangular waveguides," *IEEE Trans. Microwave Theory Tech.*, vol. MTT-31, pp. 806–812, 1983.

[8.20] R. Safavi-Naini and R. H. MacPhie, "On solving waveguide junction scattering problems by the conservation of complex power technique," *IEEE Trans. Microwave Theory Tech.*, vol. MTT-29, pp. 337–343, 1981.

Singular Integral Equation Method

(Lewin has used the technique of summing the dominant series of the Green's function and then with a change of variables recasting a number of waveguide discontinuity problems into the form of a singular integral equation. This technique is an alternative to the use of the Schwinger transformation and the conformal-mapping methods described in Chapter 4 for diagonalizing the static Green's function kernel.)

[8.21] L. Lewin, "On the resolution of a class of waveguide discontinuity problems by the use of singular integral equations," *IEEE Trans. Microwave Theory Tech.*, vol. MTT-9, pp. 321–332, 1961.

[8.22] L. Lewin and J. P. Montgomery, "A quasi-dynamic method of solution of a class of waveguide discontinuity problems," *IEEE Trans. Microwave Theory Tech.*, vol. MTT-20, pp. 849–852, 1972.

[8.23] J. P. Montgomery and L. Lewin, "Note on an *E*-plane waveguide step with simultaneous change of media," *IEEE Trans. Microwave Theory Tech.*, vol. MTT-20, pp. 763–764, 1972.

PROBLEMS

8.1. By using a suitable conformal transformation, obtain a solution for the fringing capacitance of an asymmetrical diaphragm in a parallel-plate transmission line (see Fig. P8.1). Assume that the lower plate is at zero potential and the upper plate and diaphragm are at a potential V. When used as a transmission line, find the normalized shunt susceptance across the line due to the fringing capacitance. Note that the characteristic impedance of the line per unit width is bZ_0.

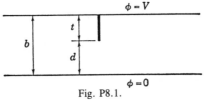

Fig. P8.1.

Answer: Fringing capacitance $C = (4\epsilon_0/\pi) \ln \csc(\pi d/2b)$. See additional comments on the method of solution in Problem 8.2.

8.2. Consider the E plane step in a parallel-plate transmission line (see Fig. P8.2). For an incident TEM mode, set up the continuity equations for the transverse fields, and show that the equivalent circuit consists of the junction of two transmission lines with characteristic impedances unity and d/b, together with a shunt capacitive susceptance at the junction. By means of a suitable conformal transformation, obtain the low-frequency value of the fringing capacitance introduced across the line by the step. From the value of this capacitance, obtain the approximate value of the shunt susceptance jB.

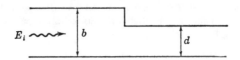

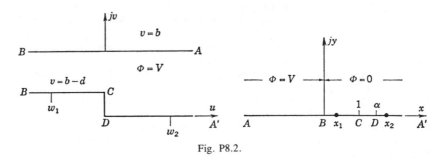

Fig. P8.2.

Answer:

$$B = \frac{k_0 b}{\pi}\left[2\ln\frac{b^2-d^2}{4bd} + \left(\frac{b}{d}+\frac{d}{b}\right)\ln\frac{d+b}{b-d}\right].$$

A suitable mapping function is

$$w = \frac{b}{\pi}\cosh^{-1}\frac{2Z-\alpha-1}{\alpha-1} - \frac{d}{\pi}\cosh^{-1}\frac{(\alpha+1)Z-2\alpha}{(\alpha-1)Z}, \qquad \alpha = \frac{b^2}{d^2}.$$

In the xy plane, $\Phi = V\theta/\pi$. Total capacitance from w_1 to w_2 is

$$C(w_2 - w_1) = C(x_2 - x_1) = \frac{\epsilon_0}{\pi}\ln\frac{x_2}{x_1}.$$

Fringing capacitance between w_1 and w_2 is $C(x_2 - x_1)$ minus parallel-plate capacitance

$$C_f(x_2 - x_1) = C(x_2 - x_1) - \frac{\epsilon_0 w_2}{b} + \frac{\epsilon_0}{d}[w_1 - j(b-d)].$$

Total fringing capacitance is

$$\lim_{\substack{x_2\to\infty\\x_1\to 0}} C_f(x_2-x_1) = \frac{\epsilon_0}{\pi}\left[\left(\frac{d}{b}+\frac{b}{d}\right)\ln\frac{b+d}{b-d} + 2\ln\frac{b^2-d^2}{4bd}\right].$$

Note that, for x_2 large,

$$w_2 \to \frac{b}{\pi}\cosh^{-1}\frac{2x_2}{\alpha-1} - \frac{d}{\pi}\cosh^{-1}\frac{\alpha+1}{\alpha-1} \to \frac{b}{\pi}\ln\frac{4x_2}{\alpha-1} - \frac{d}{\pi}\ln\frac{\alpha+1+2\alpha^{1/2}}{\alpha-1}.$$

Also, for x_1 small,

$$w_1 \to \frac{b}{\pi} \cosh^{-1} \frac{\alpha+1}{\alpha-1} - \frac{d}{\pi} \cosh^{-1} \frac{-2\alpha}{(\alpha-1)x_1} \to j(b-d) + \frac{b}{\pi} \ln \frac{\alpha+1+2\alpha}{\alpha-1} - \frac{d}{\pi} \ln \frac{4\alpha}{(\alpha-1)x_1}.$$

Also note that $(\alpha+1+2\alpha^{1/2})/(\alpha-1) = (\sqrt{\alpha}+1)/(\sqrt{\alpha}-1)$ and that $\cosh^{-1}\theta = \ln[\theta + (\theta^2-1)^{1/2}]$.

8.3. For Problem 8.2, set up a variational expression for the junction susceptance. Write the series occurring in the integrand as a dominant series plus a correction series which depends on the higher mode propagation constants. Assume a constant-aperture field, and thus obtain a value for the susceptance contributed by the correction series. The total approximate susceptance at the junction is the sum of this correction term plus the quasi-static susceptance from Problem 8.2.

Answer: The correction term gives a contribution, to the total susceptance, of the amount

$$\frac{2k_0 b^3}{d^2 \pi^3} \sum_{n=1}^{\infty} \sin^2 n\pi \frac{d}{b} \left[\left(1 - \frac{k_0^2 b^2}{n^2 \pi^2} \right)^{-1/2} - 1 \right] \frac{1}{n^3}.$$

8.4. Modify the theory of Problems 8.2 and 8.3 to obtain the equivalent-circuit parameters for a symmetrical E plane step in a parallel-plate transmission line as well as the equivalent-circuit parameters for symmetrical and asymmetrical E plane steps in a rectangular guide.

8.5. Show that the higher order E modes alone cannot satisfy the boundary conditions for a capacitive diaphragm in a rectangular guide when an H_{10} mode is incident. In particular, note that the normal magnetic field on the diaphragm will not vanish. Set up the boundary conditions for all of the field components for the LSE modes, and thus verify that these modes alone are sufficient for solving the capacitive-diaphragm and E plane-step problems in a rectangular guide.

8.6. Modify the theory for the capacitive diaphragm so as to obtain the solution for the problem of a capacitive diaphragm in a rectangular guide which is filled with a lossless dielectric medium for $z \geq 0$. The diaphragm is located at $z = 0$, and the dielectric constant is such that only the H_{10} mode propagates in the filled guide. Show that the equivalent circuit consists of a junction of two transmission lines of characteristic impedances unity and Γ_1/γ_1, where Γ_1, γ_1 are the propagation constants for the H_{10} modes, together with a shunt capacitive susceptance at the junction. Note that $jB = (1-R)/(1+R) - \gamma_1/\Gamma_1$, where R is the reflection coefficient. The variational expression obtained will give a solution for jB.

8.7. Repeat Problem 8.6 for a symmetrical inductive diaphragm in a rectangular guide filled with a lossless dielectric medium for $z > 0$.

8.8. Apply the Poisson summation formula to obtain an alternative representation for the Green's function for H_{n0} modes in a rectangular guide. Show that this alternative representation is the field radiated from the infinite number of images, of this line source, in the waveguide walls. In cylindrical coordinates, the electric field E_y radiated by an infinite unit line current located at the origin is $(-\omega\mu_0/4)H_0^2(k_0 r)$.

Answer:

$$\frac{-\omega\mu_0}{4} \sum_{-\infty}^{\infty} \{ H_0^2[k_0\sqrt{(z-z')^2 + (x-x'-2na)^2}] - H_0^2[k_0\sqrt{(z-z')^2 + (x+x'-2na)^2}] \}.$$

8.9. Show that the solution for two inductive strips in a waveguide of width $2a$ gives the solution to the problem of the inductive grating in free space with a TEM wave incident at an angle θ_i given by $\tan \theta_i = \pi/2a\beta_1$ (see Fig.

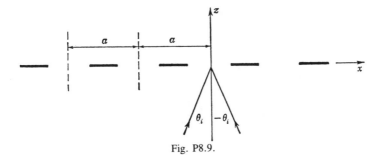

Fig. P8.9.

P8.9). Note that the H_{10} mode may be decomposed into two TEM waves propagating at angles $\pm \theta_i$. From symmetry considerations the reflection and transmission coefficients for these two composite waves will be the same.

8.10. Apply Babinet's principle to obtain the solution for a capacitive strip grating in free space from the corresponding solution of the inductive-grating problem. Show that, if B_L is the shunt susceptance of the inductive grating, and B_C is the shunt susceptance of the capacitive grating, $B_L B_C = 4Y_0^2$, where $Y_0 = (\epsilon_0/\mu_0)^{1/2}$. If R_1, T_1 and R_2, T_2 are the reflection and transmission coefficients for the two cases, show also that $T_1 + T_2 = -(R_1 + R_2) = 1$, and $T_1 T_2 = R_1 R_2$. Note that in the complementary grating the aperture and screen (strips) areas are interchanged.

8.11. For an asymmetrical E plane step in a parallel-plate transmission line, obtain a variational expression, in terms of the aperture electric field, for the position of an electric field null in the input line as a function of the short-circuit position in the output line. Assume a constant-aperture field, and find the parameters of an equivalent circuit of the form illustrated in Fig. P8.11. Repeat the analysis in terms of the aperture magnetic field (assume a constant magnetic field in the aperture for a first approximation). Compare the two solutions for the equivalent-circuit parameters for $k_0 = 2$ radians per centimeter, $2d = b = 1$ centimeter. Show that the variational expression obtained for the electric field null is equivalent to minimizing the net time-average magnetic and electric energy difference $W_m - W_e$ in the volume bounded by the short circuit and field null plane.

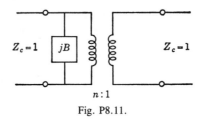

$$n:1$$

Fig. P8.11.

8.12. Find the reactance of a thin dielectric post in a rectangular waveguide. The post has a radius t, dielectric constant κ, and is located at $x = a/2$, $z = 0$ (see Fig. 8.16). Use the boundary-element method and techniques similar to those used in Section 8.6 to treat the inductive post.

8.13. For the inductive strip in a rectangular waveguide introduce the reflection coefficient R given by

$$R = -\frac{k_0 Z_0}{\beta_1 a} \int_{x_1}^{x_2} J(x') \sin \frac{\pi x'}{a} \, dx'$$

and show that the integral equation (87) for the unknown current can be expressed as

$$jX \sin \frac{\pi x}{a} \int_{x_1}^{x_2} J(x') \sin \frac{\pi x'}{a} \, dx'$$

$$= \frac{j\beta_1 a}{2} \sum_{n=2}^{\infty} \int_{x_1}^{x_2} \frac{1}{\Gamma_n} \sin \frac{n\pi x}{a} \sin \frac{n\pi x'}{a} J(x') \, dx'$$

$$= \frac{j\beta_1 a}{2} \int_{x_1}^{x_2} J(x') \left[\sum_{n=1}^{\infty} \frac{1}{n\pi} \sin \frac{n\pi x}{a} \sin \frac{n\pi x'}{a} \right.$$

$$+ \sum_{n=2}^{\infty} \left(\frac{1}{\Gamma_n a} - \frac{1}{n\pi} \right) \sin \frac{n\pi x}{a} \sin \frac{n\pi x'}{a}$$

$$\left. - \frac{1}{\pi} \sin \frac{\pi x}{a} \sin \frac{\pi x'}{a} \right] \, dx'.$$

Now introduce the transformation given by (152) in Chapter 4 and show that the integral equation becomes

$$\left(jX + \frac{j\beta_1 a}{2\pi}\right) \sin \frac{\pi x}{a} \int_0^{K'} \sin \frac{\pi x'}{a} J(\eta') \, d\eta'$$

$$= \frac{j\beta_1 a}{2} \int_0^{K'} J(\eta') \left[\frac{K}{2K'} + \sum_{n=1}^{\infty} \frac{1}{n\pi} \tanh \frac{n\pi K}{K'} \cos \frac{n\pi\eta}{K'} \cos \frac{n\pi\eta'}{K'} \right.$$

$$\left. + \sum_{n=2}^{\infty} \left(\frac{1}{\Gamma_n a} - \frac{1}{n\pi} \right) \sin \frac{n\pi x}{a} \sin \frac{n\pi x'}{a} \right] d\eta'.$$

This equation is suitable for numerical analysis but requires that the Fourier series expansions of the $\sin(n\pi x/a)$ functions be evaluated numerically, i.e.,

$$\sin \frac{n\pi x}{a} = \sum_{n=0}^{\infty} C_{nm} \cos \frac{m\pi\eta}{K'}.$$

K and K' are given by (63) and (64) in Chapter 4. The modulus of the elliptic function is given by (150b). Equation (150a) together with (151) in Chapter 4 provides the relationship between x and η.

9
Periodic Structures

Periodic structures may be classified into two basic types:

1. structures with continuous but periodically varying electrical properties, e.g., a cylindrical guide filled with a dielectric material whose dielectric constant varies in a periodic manner along the axial direction according to some periodic function such as $\kappa = \kappa_0 + \kappa_1 \cos hz$, and
2. structures with periodic boundary conditions, e.g., a waveguide loaded at regular intervals by identical diaphragms.

By far the most commonly occurring type is the class of structures with periodic boundary conditions. Periodic structures find application in a variety of devices such as linear particle accelerators, traveling-wave tubes, and microwave filter networks. Artificial dielectric media and diffraction gratings are examples of periodic structures. Periodic structures such as corrugated planes have also been used as surface wave-guiding devices for antenna applications. Periodic structures have one important characteristic in common. This is the existence of discrete passbands separated by stopbands, i.e., frequency bands for which a wave propagates freely along the structure separated by frequency bands for which the wave is highly attenuated and does not propagate along the structure.

9.1. FLOQUET'S THEOREM

Floquet's theorem is a basic theorem underlying the theory of wave propagation in periodic structures. We will illustrate this theorem by considering the problem of the propagation of an H_{10} mode in an infinitely long rectangular guide, filled with a dielectric with a relative dielectric constant given by $\kappa_0 + \kappa_1 \cos(2\pi z/p)$, as shown in Fig. 9.1. For an H_{10} mode, the field components H_x and H_z may be derived from the one scalar component of electric field E_y by the equations

$$j\omega\mu_0 H_x = \frac{\partial E_y}{\partial z} \qquad j\omega\mu_0 H_z = -\frac{\partial E_y}{\partial x}$$

and E_y satisfies the wave equation

$$\left[\frac{\partial^2}{\partial x^2} + \frac{\partial^2}{\partial z^2} + \kappa(z)k_0^2\right] E_y = 0.$$

Let $E_y = \psi(z) \sin(\pi x/a)$; hence $\psi(z)$ satisfies the equation

$$\left[\frac{d^2}{dz^2} + \left(\kappa_0 + \kappa_1 \cos\frac{2\pi z}{p}\right)k_0^2 - \frac{\pi^2}{a^2}\right]\psi(z) = 0. \tag{1}$$

605

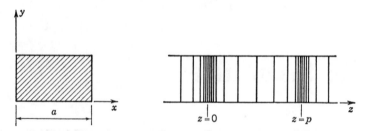

Fig. 9.1. Rectangular guide filled with a dielectric having a periodically varying dielectric constant.

If $\psi(z)$ is a solution to this equation, then $\psi(z + p)$ is also a solution, for, if we change z to $z + p$ in (1), we get

$$\left[\frac{d^2}{dz^2} + \left(\kappa_0 + \kappa_1 \cos \frac{2\pi z}{p}\right) k_0^2 - \frac{\pi^2}{a^2}\right] \psi(z + p) = 0$$

since

$$\cos 2\pi \frac{z + p}{p} = \cos \frac{2\pi z}{p}$$

and hence $\psi(z + p)$ satisfies the same equation as $\psi(z)$ and is also a solution. Equation (1) is of the second order and will have two independent solutions, which we will denote by $\psi_1(z)$ and $\psi_2(z)$. We have the result that $\psi_1(z + p)$ and $\psi_2(z + p)$ are also solutions, but, in general, $\psi_i(z + p)$ does not equal $\psi_i(z)$, since $\psi_i(z)$ is not necessarily periodic in z with a period p.

Since a second-order equation has only two linearly independent solutions, any other solution must be a linear combination of the two fundamental solutions, and, hence, we may write

$$\psi_1(z + p) = \alpha_{11}\psi_1(z) + \alpha_{12}\psi_2(z) \tag{2a}$$

$$\psi_2(z + p) = \alpha_{21}\psi_1(z) + \alpha_{22}\psi_2(z) \tag{2b}$$

where α_{ij} are suitably determined coefficients. For the particular case when ψ_i is periodic so that $\psi_i(z + p) = \psi_i(z)$, we have $\alpha_{11} = \alpha_{22} = 1$ and $\alpha_{12} = \alpha_{21} = 0$. In a similar fashion the general solution to (1) may be written as

$$F(z) = A\psi_1(z) + B\psi_2(z) \tag{3}$$

where A and B are suitably chosen coefficients. From (2) and (3), we get

$$F(z + p) = A\psi_1(z + p) + B\psi_2(z + p)$$
$$= (A\alpha_{11} + B\alpha_{21})\psi_1(z) + (A\alpha_{12} + B\alpha_{22})\psi_2(z). \tag{4}$$

If $F(z)$ is to represent a wave propagating down the structure, it is necessary that

$$F(z + p) = e^{-\gamma p}F(z) \tag{5}$$

where γ is the propagation constant. Thus (4) above must equal (3) multiplied by $e^{-\gamma p}$, and, when we equate the coefficients of A and B, we get

$$Ae^{-\gamma p} = A\alpha_{11} + B\alpha_{21}$$

$$Be^{-\gamma p} = A\alpha_{12} + B\alpha_{22}$$

or

$$A(\alpha_{11} - e^{-\gamma p}) + B\alpha_{21} = 0 \tag{6a}$$

$$A\alpha_{12} + B(\alpha_{22} - e^{-\gamma p}) = 0. \tag{6b}$$

A nontrivial solution for A and B exists only if the determinant of the above equations vanishes; thus,

$$e^{-2\gamma p} - e^{-\gamma p}(\alpha_{11} + \alpha_{22}) + \alpha_{11}\alpha_{22} - \alpha_{12}\alpha_{21} = 0. \tag{7}$$

Equation (7) is a quadratic equation whose solution will determine the allowed values of the propagation constant γ. The solution for $e^{-\gamma p}$ is

$$e^{-\gamma p} = \frac{\alpha_{11} + \alpha_{22}}{2} \pm \left[\left(\frac{\alpha_{11} + \alpha_{22}}{2}\right)^2 - (\alpha_{11}\alpha_{22} - \alpha_{12}\alpha_{21})\right]^{1/2}. \tag{8}$$

Let $\alpha_{11}\alpha_{22} - \alpha_{12}\alpha_{21} = \Delta$; then, since

$$\frac{(\alpha_{11} + \alpha_{22})^2}{4} - \left[\left(\frac{\alpha_{11} + \alpha_{22}}{2}\right)^2 - \Delta\right] = \Delta$$

we may write

$$\cosh\theta = \frac{\alpha_{11} + \alpha_{22}}{2\Delta^{1/2}} \qquad \sinh\theta = \left[\left(\frac{\alpha_{11} + \alpha_{22}}{2}\right)^2 \frac{1}{\Delta} - 1\right]^{1/2}$$

since $\cosh^2\theta - \sinh^2\theta = 1$. Hence the solution $e^{-\gamma p}$ is also given by

$$e^{-\gamma p} = \Delta^{1/2}e^{\pm\theta} = e^{\pm\theta + (1/2)\ln\Delta} \tag{9}$$

where $\cosh\theta = (\alpha_{11} + \alpha_{22})/2\Delta^{1/2}$. In general, the solution for γ may be real, imaginary, or complex. We may show that, if γ is a solution, then $-\gamma$ is also a solution, as is $\gamma \pm j2n\pi/p$, where n is any integer. This result follows since $\Delta = 1$ (see Problem 9.7).

Let $\Phi(z) = e^{\gamma z}F(z)$; we get

$$\Phi(z + p) = e^{\gamma p}e^{\gamma z}F(z + p) = e^{\gamma z}F(z) = \Phi(z)$$

by using relation (5). Hence Φ is a periodic function of z with a period p. Therefore the general solution of (1) is of the form $F(z) = e^{\pm\gamma z}\Phi(z)$, where Φ is a periodic function of z, and γ plays the role of a propagation constant. This particular result is Floquet's theorem.

We will not consider this problem any further here, since it would take us too far afield into the properties of Mathieu functions, which are the appropriate solutions to (1). The theory for periodic boundary conditions will be our main concern. The theory for problems of this type is most readily developed by matrix methods. We shall find, however, that solutions of the above form will be obtained. The more important aspects of the theory will be developed in the succeeding sections, and will be illustrated by application to a periodically loaded waveguide and the helix structure. Further examples on periodic structures will be covered in later chapters as well.

9.2. Some Properties of Lossless Microwave Quadrupoles

The basic element in the type of periodic structure we shall be concerned with will be a lossless quadrupole. Such a quadrupole may be described by the reflection and transmission coefficients, by a T or π network, by a wave matrix, or by any other matrix that is used in low-frequency circuit theory. We shall derive a few important relations between these various methods of describing a lossless quadrupole. Consider the general discontinuity illustrated in Fig. 9.2. The electrical length of the transmission line or waveguide section on either end is $\theta/2$. It is assumed that these sections are sufficiently long so all higher order modes which are excited by the discontinuity have decayed to a negligible value at the terminal planes. The transmission lines or waveguide sections on either side of the discontinuity are also assumed to be uniform and lossless, and to have identical electrical properties, in particular, a propagation phase constant β_0 and wave impedance Z_w.

Let the wave matrix describing the discontinuity be

$$[A'] = \begin{bmatrix} A'_{11} & A'_{12} \\ A'_{21} & A'_{22} \end{bmatrix} \tag{10}$$

where

$$A'_{11} = \frac{1}{t_{12}} \quad A'_{12} = -\frac{r_2}{t_{12}} \quad A'_{21} = \frac{r_1}{t_{12}} \quad A'_{22} = t_{21} - \frac{r_1 r_2}{t_{12}}.$$

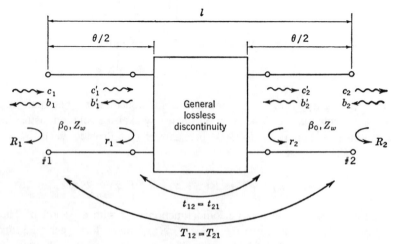

Fig. 9.2. A lossless microwave quadrupole.

The amplitudes c_1', b_1' of the forward- and backward-propagating waves on the input side are related to the amplitudes c_2', b_2' of the waves on the output side as follows:

$$\begin{bmatrix} c_1' \\ b_1' \end{bmatrix} = \begin{bmatrix} A_{11}' & A_{12}' \\ A_{21}' & A_{22}' \end{bmatrix} \begin{bmatrix} c_2' \\ b_2' \end{bmatrix}. \tag{11}$$

The wave matrix describing the structure between terminal planes 1 and 2 is

$$[A] = \begin{bmatrix} e^{j\theta/2} & 0 \\ 0 & e^{-j\theta/2} \end{bmatrix} \begin{bmatrix} A_{11}' & A_{12}' \\ A_{21}' & A_{22}' \end{bmatrix} \begin{bmatrix} e^{j\theta/2} & 0 \\ 0 & e^{-j\theta/2} \end{bmatrix} = \begin{bmatrix} A_{11} & A_{12} \\ A_{21} & A_{22} \end{bmatrix} \tag{12}$$

where

$$A_{11} = \frac{1}{T_{12}} = \frac{e^{j\theta}}{t_{12}} \qquad A_{12} = \frac{-R_2}{T_{12}} = \frac{-r_2}{t_{12}},$$

$$A_{21} = \frac{R_1}{T_{12}} = \frac{r_1}{t_{12}} \qquad A_{22} = \frac{T_{12}T_{21} - R_1 R_2}{T_{12}} = \frac{t_{12}t_{21} - r_1 r_2}{t_{12}} e^{-j\theta}.$$

From these latter relations we may readily deduce that

$$T_{12} = t_{12}e^{-j\theta} \qquad T_{21} = t_{21}e^{-j\theta}$$

$$R_1 = r_1 e^{-j\theta} \qquad R_2 = r_2 e^{-j\theta}.$$

Also we have the following relations for the reflection and transmission coefficients:

$$R_1 = \frac{A_{21}}{A_{11}} \qquad R_2 = -\frac{A_{12}}{A_{11}} \qquad T_{12} = \frac{1}{A_{11}} = T_{21}.$$

The equality of t_{12} and t_{21} and also of T_{12} and T_{21} follows from the condition of reciprocity. The quadrupole is reciprocal when the determinant of the $[A]$ matrix is equal to unity; i.e.,

$$A_{11}A_{22} - A_{12}A_{21} = A_{11}'A_{22}' - A_{12}'A_{21}' = 1.$$

For a lossless network,

$$|R_1| = |R_2| = (1 - |T_{12}|^2)^{1/2}.$$

Let

$$R_1 = \rho e^{j\phi_1} \qquad R_2 = \rho e^{j\phi_2} \qquad T_{12} = (1 - \rho^2)^{1/2} e^{j\alpha}$$

where $\phi_1 + \phi_2 = 2\alpha \pm \pi$ from Section 5.7. Hence, we have

$$A_{22} = \frac{T_{12}^2 - R_1 R_2}{T_{12}} = \frac{(1 - \rho^2)e^{2j\alpha} + \rho^2 e^{2j\alpha}}{T_{12}} = |T_{12}|^{-1}e^{j\alpha} = A_{11}^* \tag{13a}$$

$$A_{12}A_{21} = A_{11}A_{11}^* - 1 = (1 - \rho^2)^{-1} - 1 = \left|\frac{R_1}{T_{12}}\right|^2. \tag{13b}$$

With the output side matched, the input impedances at terminal planes 1 and 2 are, respectively,

$$Z_{in,1} = \frac{1 + R_1}{1 - R_1} Z_w = \frac{A_{11} + A_{21}}{A_{11} - A_{21}} Z_w \tag{14a}$$

$$Z_{in,2} = \frac{1 + R_2}{1 - R_2} Z_w = \frac{A_{11} - A_{12}}{A_{11} + A_{12}} Z_w. \tag{14b}$$

The structure between the two terminal planes may also be described by an equivalent T network as in Fig. 9.3. The equivalent voltages and currents V_i and I_j will be chosen as follows:

$$V_1 = c_1 + b_1 \quad V_2 = c_2 + b_2 \quad I_1 = (c_1 - b_1)Y_w \quad I_2 = (c_2 - b_2)Y_w$$

where c_i and b_i are the amplitudes of the transverse electric fields of the forward- and backward-traveling waves. At the terminal planes we have

$$V_1 = I_1 Z_{11} - I_2 Z_{12} = c_1 + b_1 = (c_1 - b_1)Y_w Z_{11} - (c_2 - b_2)Y_w Z_{12}$$

$$V_2 = I_1 Z_{12} - I_2 Z_{22} = c_2 + b_2 = (c_1 - b_1)Y_w Z_{12} - (c_2 - b_2)Y_w Z_{22}.$$

We may rearrange these latter equations and write them in matrix form as follows:

$$\begin{bmatrix} 1 - Y_w Z_{11} & 1 + Y_w Z_{11} \\ Y_w Z_{12} & -Y_w Z_{12} \end{bmatrix} \begin{bmatrix} c_1 \\ b_1 \end{bmatrix} = \begin{bmatrix} -Y_w Z_{12} & Y_w Z_{12} \\ 1 + Y_w Z_{22} & 1 + Y_w Z_{22} \end{bmatrix} \begin{bmatrix} c_2 \\ b_2 \end{bmatrix}. \tag{15}$$

We have

$$(1 - Y_w Z_{11})(-Y_w Z_{12}) - Y_w Z_{12}(1 + Y_w Z_{11}) = -2Y_w Z_{12}$$

and, hence, the inverse of the matrix on the left-hand side is

$$\frac{1}{2Y_w Z_{12}} \begin{bmatrix} Y_w Z_{12} & 1 + Y_w Z_{11} \\ Y_w Z_{12} & Y_w Z_{11} - 1 \end{bmatrix}$$

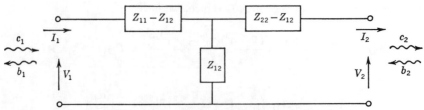

Fig. 9.3. Equivalent network for one section of a periodic structure.

so that multiplying both sides of (15) by this matrix gives

$$
\begin{bmatrix} c_1 \\ b_1 \end{bmatrix} = \frac{1}{2Y_wZ_{12}} \begin{bmatrix} Y_wZ_{12} & Y_wZ_{11}+1 \\ Y_wZ_{12} & Y_wZ_{11}-1 \end{bmatrix} \begin{bmatrix} -Y_wZ_{12} & Y_wZ_{12} \\ 1+Y_wZ_{22} & 1-Y_wZ_{22} \end{bmatrix} \begin{bmatrix} c_2 \\ b_2 \end{bmatrix}
$$

$$
= \frac{1}{2Y_wZ_{12}}
$$

$$
\times \begin{bmatrix} Y_w^2\Delta+1+Y_w(Z_{11}+Z_{22}) & -Y_w^2\Delta+1+Y_w(Z_{11}-Z_{22}) \\ Y_w^2\Delta-1+Y_w(Z_{11}-Z_{22}) & -Y_w^2\Delta-1+Y_w(Z_{11}+Z_{22}) \end{bmatrix} \begin{bmatrix} c_2 \\ b_2 \end{bmatrix}
$$

$$
= \begin{bmatrix} A_{11} & A_{12} \\ A_{21} & A_{22} \end{bmatrix} \begin{bmatrix} c_2 \\ b_2 \end{bmatrix} \tag{16}
$$

where $\Delta = Z_{11}Z_{22}-Z_{12}^2$. From (16) the following explicit expressions for the matrix elements A_{ij} may be found:

$$
A_{11} = \frac{Y_w^2\Delta+1+Y_w(Z_{11}+Z_{22})}{2Y_wZ_{12}} \tag{17a}
$$

$$
A_{12} = \frac{-Y_w^2\Delta+1+Y_w(Z_{11}-Z_{22})}{2Y_wZ_{12}} \tag{17b}
$$

$$
A_{21} = \frac{Y_w^2\Delta-1+Y_w(Z_{11}-Z_{22})}{2Y_wZ_{12}} \tag{17c}
$$

$$
A_{22} = \frac{-Y_w^2\Delta-1+Y_w(Z_{11}+Z_{22})}{2Y_wZ_{12}}. \tag{17d}
$$

For a lossless structure, the impedance elements Z_{ij} are pure-imaginary, and so $Z_{ij}^* = -Z_{ij}$. If all the impedance elements Z_{ij} are normalized by dividing them by Z_w, the expressions for A_{ij} are given by (17), provided Y_w is put equal to unity and it is understood that the Z_{ij} in these equations are normalized. The impedances of the input and output lines in Fig. 9.3 must now be taken as unity, and the currents I_i as $c_i - b_i$.

For a load Z_L connected at terminal plane 2, the input impedance at terminal plane 1 is

$$
Z_{\text{in},1} = Z_{11} - \frac{Z_{12}^2}{Z_L+Z_{22}} = \frac{Z_{11}Z_L+\Delta}{Z_L+Z_{22}}. \tag{18}
$$

This is a bilinear transformation and will map circles into circles with straight lines as limiting cases. Such a transformation has two invariant points; i.e., there are two values of Z_L in the complex Z_L plane that map into two identical points in the complex $Z_{\text{in},1}$ plane. Let these invariant values be denoted by Z_c. Thus we must have

$$
Z_c = \frac{Z_{11}Z_c+\Delta}{Z_c+Z_{22}}
$$

or

$$Z_c = \frac{Z_{11} - Z_{12}}{2} \pm \left[\left(\frac{Z_{11} + Z_{22}}{2} \right)^2 - Z_{12}^2 \right]^{1/2}. \tag{19}$$

For a load Z_L connected at terminal plane 1, the input impedance at terminal plane 2 is

$$Z_{\text{in},2} = -\frac{V_2}{I_2} = Z_{22} - \frac{Z_{12}^2}{Z_L + Z_{11}} = \frac{Z_{22}Z_L + \Delta}{Z_{11} + Z_L}. \tag{20}$$

The reversal in sign in (20) is due to the assumed positive direction of current flow for I_2. Equation (20) leads to the following expression for the characteristic values:

$$Z_c = - \left\{ \frac{Z_{11} - Z_{22}}{2} \pm \left[\left(\frac{Z_{11} + Z_{22}}{2} \right)^2 - Z_{12}^2 \right]^{1/2} \right\}. \tag{21}$$

These characteristic values Z_c are of considerable importance since they may be taken as the characteristic impedance of the infinite periodic structure composed of an infinite number of basic sections or quadrupoles in a cascade connection.

9.3. Propagation in an Infinite Periodic Structure

Consider the periodic structure illustrated in Fig. 9.4 consisting of a cascade connection of an infinite number of lossless quadrupoles separated by sections of waveguide or transmission lines of sufficient length to ensure that there is no interaction between sections due to higher order evanescent modes. This latter restriction is not necessary, and the modifications in the theory when there are higher order mode interactions will be considered later. The amplitudes at terminal plane 2 are related to those at terminal plane 1 by the matrix equation

$$\begin{bmatrix} c_1 \\ b_1 \end{bmatrix} = \begin{bmatrix} A_{11} & A_{12} \\ A_{21} & A_{22} \end{bmatrix} \begin{bmatrix} c_2 \\ b_2 \end{bmatrix}. \tag{22}$$

If a wave is to propagate down the structure, we must have

$$c_2 = e^{-\gamma l} c_1 \qquad b_2 = e^{-\gamma l} b_1$$

where γ is the propagation constant, and l is the physical length of each individual section.

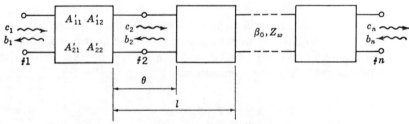

Fig. 9.4. Periodic structure consisting of a cascade connection of identical quadrupoles.

When we substitute into (22), we get two homogeneous equations for c_1 and b_1 which have a solution only if the following determinant vanishes:

$$\begin{vmatrix} A_{11} - e^{-\gamma l} & A_{12} \\ A_{21} & A_{22} - e^{-\gamma l} \end{vmatrix} = 0$$

or

$$e^{2\gamma l} - (A_{11} + A_{22})e^{\gamma l} + (A_{11}A_{22} - A_{12}A_{21}) = 0.$$

Since $A_{11}A_{22} - A_{12}A_{21} = 1$, we obtain, by multiplying through by $e^{-\gamma l}$, the following solution:

$$\cosh \gamma l = \frac{A_{11} + A_{22}}{2} = \frac{1}{2}\left(\frac{1}{T_{12}} + \frac{1}{T_{12}^*}\right) = \frac{\cos \alpha}{|T_{12}|} \tag{23a}$$

where α is the phase angle of T_{12}. When $\cos \alpha < |T_{12}|$, γ is imaginary, and the wave propagates freely, undergoing only a change in phase. When $\cos \alpha > |T_{12}|$, γ is real, and the wave is attenuated. We also see that $-\gamma$ is a solution, and so we can have a wave propagating along the structure in the negative z direction also. When $\cos \alpha = |T_{12}|$, $\gamma = 0$ or $j2n\pi/l$, and this corresponds to a pure standing wave between sections. If we substitute for A_{11} and A_{22} from (17), then the solution for γ is given by

$$\cosh \gamma l = \frac{Z_{11} + Z_{22}}{2Z_{12}}. \tag{23b}$$

We will now examine the wave solution in each section in more detail. We note that between each two discontinuities the wave solution is not a pure traveling wave but a combination of a traveling wave and a standing wave. The phase constant of the connecting lines is β_0; hence, the electric fields of the forward- and backward-traveling waves in section 1 is proportional to $c_1 e^{-j\beta_0 z}$ and $b_1 e^{j\beta_0 z}$, respectively. The ratio of the amplitudes b_1/c_1 will be denoted by R_c, where R_c is a reflection coefficient at terminal plane 1 and is characteristic of the particular periodic structure considered. The ratio of b_n to c_n at the nth terminal plane is also R_c. Hence the transverse electric field in section 1 is proportional to

$$c_1(e^{-j\beta_0 z} + R_c e^{j\beta_0 z}) = c_1[(1 + R_c)\cos \beta_0 z - j(1 - R_c)\sin \beta_0 z] \tag{24a}$$

and the transverse magnetic field is proportional to

$$c_1 Y_w(e^{-j\beta_0 z} - R_c e^{j\beta_0 z}) = c_1 Y_w[(1 - R_c)\cos \beta_0 z - j(1 + R_c)\sin \beta_0 z]. \tag{24b}$$

The impedance of the periodic wave at the nth terminal plane is

$$\frac{V_n}{I_n} = \frac{c_n + b_n}{c_n - b_n}Z_w = \frac{c_1 + b_1}{c_1 - b_1}Z_w = \frac{1 + R_c}{1 - R_c}Z_w = Z_c \tag{25}$$

and will be called the characteristic impedance of the periodic structure. It is important to note that this impedance is not unique, since it depends on the choice of location of the terminal plane. For terminals located at z, we find from (24) that

$$Z_c(z) = Z_w \frac{Z_c(0) - jZ_w \tan \beta_0 z}{Z_w - jZ_c(0) \tan \beta_0 z} \tag{26}$$

which is just the transmission-line equation for the transformation of impedance along a line. If we make the restriction that the only points at which we may sample the fields are at a terminal plane, then Z_c is truly a unique characteristic impedance for the periodic structure. One other important fact to note is that Z_c is not necessarily pure-real, but, in general, is a complex quantity. The amplitudes of the waves at terminal plane 2 are $c_1 e^{-\gamma l} = c_2$ and $b_1 e^{-\gamma l} = b_2$, and hence the transverse electric field is proportional to $c_1 e^{-\gamma l}(e^{-j\beta_0(z-l)} + R_c e^{j\beta_0(z-l)})$ in section 2, and to $c_1 e^{-n\gamma l}(e^{-j\beta_0(z-nl)} + R_c e^{j\beta_0(z-nl)})$ in section n, since, at the nth terminal plane, where $z = nl$, the electric field is proportional to

$$c_n + b_n = e^{-\gamma nl}(c_1 + b_1) = e^{-\gamma nl}c_1(1 + R_c).$$

Expressions for Z_c and R_c are readily derived from the circuit equations for the equivalent T network in Fig. 9.3. From the circuit equations we obtain, when we substitute for V_i and I_i, the relations

$$V_1 = c_1 + b_1 = Z_{11}(c_1 - b_1)Y_w - Z_{12}(c_2 - b_2)Y_w$$
$$= (c_1 - b_1)Y_w(Z_{11} - Z_{12}e^{-\gamma l})$$

$$V_2 = c_2 + b_2 = e^{-\gamma l}(c_1 + b_1)$$
$$= (c_1 - b_1)Y_w(Z_{12} - Z_{22}e^{-\gamma l})$$

and, hence,

$$Z_c = \frac{c_1 + b_1}{c_1 - b_1}Z_w = Z_{11} - Z_{12}e^{-\gamma l} = Z_{12}e^{\gamma l} - Z_{22}.$$

Adding these two expressions for Z_c and dividing by 2 gives

$$Z_c = \frac{Z_{11} - Z_{22}}{2} + Z_{12} \sinh \gamma l. \tag{27}$$

From (23) we see that

$$\sinh \gamma l = (\cosh^2 \gamma l - 1)^{1/2} = \frac{1}{Z_{12}}\left[\left(\frac{Z_{11} + Z_{22}}{2}\right)^2 - Z_{12}^2\right]^{1/2}$$

so that the characteristic impedance given by (27) is the same as that given by (19). This is to be expected since, if the impedance is to be the same at each terminal plane, it must be invariant when transformed by the T network. For a lossless network, Z_{ij} is pure-imaginary, and thus, when γ is real, Z_c is pure-imaginary, and, when γ is imaginary, Z_c is complex.

It is very important to adopt a definite convention for positive current flow if ambiguity in sign for expressions for impedances and so forth is to be avoided. We will call the complete solution for the wave that may propagate along the periodic structure a periodic wave as distinguished from the forward- and backward-traveling waves which exist in each section and make up the periodic wave. For the two traveling waves that make up the periodic wave, the coefficients c_i will always be used to denote the amplitude of the wave that propagates in the positive z direction between sections, and b_i will be used to denote the amplitude of the partial wave that propagates in the negative z direction between sections. The currents associated with c_i and b_i will be taken as $Y_w c_i$ and $-Y_w b_i$. When γ is imaginary, we will denote it by $j\beta_c$, where β_c is the characteristic phase constant for the periodic structure. Using this convention, the transverse electric field for the forward-propagating periodic wave is proportional to

$$(1 + R_c^+) c_1^+ e^{-j\beta_c nl}$$

at the nth terminal plane, while the transverse magnetic field is proportional to

$$(1 - R_c^+) Y_w c_1^+ e^{-j\beta_c nl} = Y_c^+ (1 + R_c^+) c_1^+ e^{-j\beta_c nl}$$

at the nth terminal plane, where

$$R_c^+ = \frac{Z_c^+ - Z_w}{Z_c^+ + Z_w} \qquad Z_c^+ = (Y_c^+)^{-1} = \frac{Z_{11} - Z_{22}}{2} + jZ_{12} \sin \beta_c l$$

and c_1^+ is an arbitrary amplitude coefficient. For the periodic wave propagating in the negative z direction, the transverse electric field at the nth terminal plane is proportional to

$$c_1^- (1 + R_c^-) e^{j\beta_c nl}$$

while the transverse magnetic field is proportional to

$$(1 - R_c^-) c_1^- Y_w e^{j\beta_c nl} = Y_c^- (1 + R_c^-) c_1^- e^{j\beta_c nl}$$

where

$$R_c^- = \frac{Z_c^- - Z_w}{Z_c^- + Z_w} \qquad Z_c^- = (Y_c^-)^{-1} = \frac{Z_{11} - Z_{22}}{2} - jZ_{12} \sin \beta_c l.$$

With our adopted convention we note that we must distinguish between the two cases, as Z_c is not the same for the periodic waves propagating in the positive z direction and those in the negative z direction. Figure 9.5 illustrates the adopted convention as regards the positive directions of current flow.

9.4. TERMINATED PERIODIC STRUCTURE

Let a semi-infinite periodic structure be terminated at the nth terminal plane by a load impedance Z_L. Let the incident periodic wave at the sth terminal plane be

$$E_{\text{inc}} = c_1^+ (1 + R_c^+) e^{-j\beta_c sl}$$

with a transverse magnetic field proportional to $Y_c^+ E_{\text{inc}}$. The presence of the load gives rise

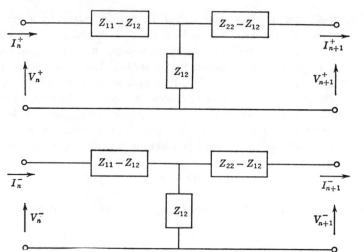

Fig. 9.5. Illustration for assumed positive directions of equivalent currents and voltages.

to a reflected periodic wave of amplitude

$$E_{\text{ref}} = c_n^- (1 + R_c^-) e^{-j\beta_c (n-s)l}$$

at the sth terminal plane and with a transverse magnetic field of amplitude proportional to $Y_c^- E_{\text{ref}}$. To determine the amplitude of the reflected periodic wave, we consider the effect of the load on the partial waves of which the periodic waves are composed. At the position of the load where $s = n$, the total forward-traveling partial wave is

$$c_1^+ e^{-j\beta_c nl} + c_n^-$$

while the total backward-traveling partial wave is

$$R_c^+ c_1^+ e^{-j\beta_c nl} + R_c^- c_n^-$$

with transverse magnetic fields proportional to

$$Y_w c_1^+ e^{-j\beta_c nl} + Y_w c_n^-$$

and

$$-Y_w R_c^+ c_1^+ e^{-j\beta_c nl} - Y_w R_c^- c_n^-$$

respectively. The load Z_L gives rise to a reflection coefficient

$$R_L = \frac{Z_L - Z_w}{Z_L + Z_w}$$

for these partial waves, and, hence, we have

$$R_L (c_1^+ e^{-j\beta_c nl} + c_n^-) = R_c^+ c_1^+ e^{-j\beta_c nl} + R_c^- c_n^-.$$

Solving this equation for c_n^- gives

$$c_n^- = -\frac{R_L - R_c^+}{R_L - R_c^-} c_1^+ e^{-j\beta_c nl}. \tag{28}$$

The incident periodic wave at the load is $c_1^+(1 + R_c^+)e^{-j\beta_c nl}$, and the reflected periodic wave at the load is, therefore,

$$c_n^-(1 + R_c^-) = (1 + R_c^-)\frac{R_c^+ - R_L}{R_L - R_c^-} c_1^+ e^{-j\beta_c nl}$$

$$= \frac{1 + R_c^-}{1 + R_c^+}\frac{R_c^+ - R_L}{R_L - R_c^-} c_1^+(1 + R_c^+)e^{-j\beta_c nl}. \tag{29}$$

The equivalent reflection coefficient Γ_n for the periodic wave is thus

$$\Gamma_n = \frac{1 + R_c^-}{1 + R_c^+}\frac{R_c^+ - R_L}{R_L - R_c^-}. \tag{30}$$

If we express Z_L, Z_c^+, and Z_c^- by their values given in terms of R_L, R_c^+, and R_c^-, we may readily show that

$$\frac{Z_L - Z_c^+}{Z_L - Z_c^-} = \frac{1 - R_c^-}{1 - R_c^+}\frac{R_L - R_c^+}{R_L - R_c^-}$$

and, hence, we get

$$\Gamma_n = -\frac{Z_c^-}{Z_c^+}\frac{Z_L - Z_c^+}{Z_L - Z_c^-}. \tag{31}$$

Thus, at the sth terminal plane, the sum of the incident plus reflected periodic waves has a transverse electric field proportional to

$$c_n^+(1 + R_c^+)(e^{-j\beta_c l(s-n)} + \Gamma_n e^{j\beta_c l(s-n)})$$

and a transverse magnetic field proportional to

$$c_n^+(1 + R_c^+)(Y_c^+ e^{-j\beta_c l(s-n)} + \Gamma_n Y_c^- e^{j\beta_c l(s-n)})$$

where

$$c_n^+ = c_1^+ e^{-j\beta_c ln}.$$

The ratio of the transverse electric field to the transverse magnetic field at the nth terminal plane is

$$\frac{1 + \Gamma_n}{Y_c^+ + Y_c^- \Gamma_n} = Z_L \tag{32}$$

a result readily proved from (31). If we let

$$\frac{Z_{11} - Z_{22}}{2} = \eta \tag{33}$$

$$Z_{12} \sinh \gamma l = j Z_{12} \sin \beta_c l = Z \tag{34}$$

we get

$$Z_c^+ = \eta + Z \tag{35a}$$

$$Z_c^- = \eta - Z \tag{35b}$$

$$\Gamma_n = -\frac{\eta - Z}{\eta + Z} \frac{(Z_L - \eta) - Z}{(Z_L - \eta) + Z}. \tag{35c}$$

For a periodic structure composed of symmetrical sections, $Z_{11} = Z_{22}$ and $\eta = 0$; so in this case

$$Z_c^+ = -Z_c^- = Z \tag{36a}$$

$$\Gamma_n = \frac{Z_L - Z}{Z_L + Z}. \tag{36b}$$

The reflection coefficient at the sth terminal plane is

$$\Gamma_s = \Gamma_n e^{2j\beta_c l(s-n)}. \tag{37}$$

The ratio of the transverse electric field to the transverse magnetic field at the sth terminal plane is

$$\frac{1 + \Gamma_s}{Y_c^+ + Y_c^- \Gamma_s} = \frac{(1 + \Gamma_s)(\eta - Z)(\eta + Z)}{(\eta - Z) + \Gamma_s(\eta + Z)} = Z_s. \tag{38}$$

With some algebraic manipulation, we can obtain the following results from (38):

$$Z_s - \eta = Z \frac{Z_c^- - Z_c^+ \Gamma_s}{Z_c^- + Z_c^+ \Gamma_s} \tag{39a}$$

and

$$\Gamma_s = -\frac{Z_c^-}{Z_c^+} \frac{(Z_s - \eta) - Z}{(Z_s - \eta) + Z}. \tag{39b}$$

Substituting for Γ_s from (37) into (39a) gives

$$\begin{aligned}
Z_s - \eta &= Z \frac{Z_c^- - Z_c^+ \Gamma_n e^{2j\beta_c l(s-n)}}{Z_c^- + Z_c^+ \Gamma_n e^{2j\beta_c l(s-n)}} \\
&= Z \frac{(Z_c^- - Z_c^+ \Gamma_n)/(Z_c^- + Z_c^+ \Gamma_n) + jZ \tan \beta_c l(n - s)}{Z + j(Z_c^- - Z_c^+ \Gamma_n)/(Z_c^- + Z_c^+ \Gamma_n) \tan \beta_c l(n - s)} \\
&= Z \frac{Z_L - \eta + jZ \tan \beta_c l(n - s)}{Z + j(Z_L - \eta) \tan \beta_c l(n - s)}.
\end{aligned} \tag{40}$$

Equation (40) gives the input impedance Z_s at the sth terminal plane with a load Z_L connected across the nth terminal plane. We note that the expressions for the reflection coefficient and the transformation of impedance become formally the same as those for a conventional line whenever the basic sections are symmetrical, so that $Z_{11} = Z_{22}$ and $\eta = 0$.

The use of equivalent transmission-line voltages and currents simplifies the derivations of a lot of the properties of periodic structures. The equivalent transmission-line voltages and currents are taken proportional to the total transverse electric and magnetic fields, respectively. Thus, for a periodic wave propagating in the positive z direction, we have

$$V_n^+ = V_1^+ e^{-j\beta_c nl} = c_1^+ (1 + R_c^+) e^{-j\beta_c nl} \tag{41a}$$

$$I_n^+ = I_1^+ e^{-j\beta_c nl} = Y_c^+ V_n^+ \tag{41b}$$

at the nth terminal plane, while, for a negatively propagating periodic wave, we have

$$V_n^- = V_1^- e^{j\beta_c nl} = c_1^- (1 + R_c^-) e^{j\beta_c nl} \tag{42a}$$

$$I_n^- = I_1^- e^{j\beta_c nl} = Y_c^- V_n^-. \tag{42b}$$

In (42a) and (42b) the positive direction of current flow is in the positive z direction. If we connect a load Z_L across the nth terminal plane, the total voltage across the load is the sum of the voltages in the incident and reflected periodic waves:

$$V_L = V_n^+ + V_n^- = V_n^+ (1 + \Gamma_n) = V_1^+ e^{-j\beta_c nl} (1 + \Gamma_n)$$

and the current flowing into the load is the sum of the currents due to the incident and reflected periodic waves:

$$I_L = I_n^+ + I_n^- = V_n^+ Y_c^+ + V_n^- Y_c^- = V_n^+ Y_c^+ \left(1 + \frac{Y_c^-}{Y_c^+} \Gamma_n \right).$$

The voltage across the load divided by the current flowing through it is equal to the load impedance, and, hence,

$$\frac{V_L}{I_L} = Z_L = \frac{1 + \Gamma_n}{Y_c^+ + Y_c^- \Gamma_n}$$

which is identical with the result obtained in (32).

From (35c) or (39b) it is seen that $Z_L = Z + \eta$ is the proper load impedance in which to terminate a periodic structure at one of the terminal planes, in order to prevent a reflected periodic wave when a periodic wave is incident from the left. For a periodic wave incident from the right along the periodic structure, the reflectionless load termination must have an impedance $Z - \eta$. These values of load impedances are equal to the characteristic impedances Z_c^+ and $-Z_c^-$.

Although the above load terminations result in a matched termination for the incident periodic waves, this does not mean that there is not a reflected component wave in the waveguide section in front of the terminating load impedance. The incident periodic wave is composed of forward- and backward-propagating component waves, and the load impedance Z_L must differ from the impedance Z_w of the waveguide or transmission line by an amount just sufficient to produce the backward-propagating component wave when the forward-propagating component wave is incident.

Often, in practice, it is desirable to excite the propagating periodic wave in a periodic structure by means of a wave incident along a uniform waveguide to which the periodic structure, together with its terminating load impedance, is connected. To prevent a reflected wave in the uniform input guide, a transition or matching section must be introduced such that the impedance seen at the input terminals to the periodic structure is transformed into the impedance Z_w of the input guide. The usual techniques employed for matching an arbitrary load impedance to a guide or transmission line may be employed. An alternative procedure is to taper the periodic structure so as to provide a gradual change from the uniform waveguide to the final periodically loaded waveguide section. This is analogous to matching by means of a tapered transmission-line section.

The required parameters of a matching T section are readily obtained by a transmission-line analysis. Consider a periodic structure that is terminated in a given load impedance, such that Z_L is the effective load impedance presented at the output terminals of the matching section. The parameters of the matching section will be distinguished from those of the basic sections of the periodic structure by primes, as in Fig. 9.6. If the matching T section formed a basic section in a periodic structure, it would be characterized by a propagation constant γ' and characteristic impedances $Z_c^{+'}, Z_c^{-'}$, where

$$\sinh \gamma' l' = \frac{1}{Z'_{12}} \left[\left(\frac{Z'_{11} + Z'_{22}}{2} \right)^2 - Z_{12}'^2 \right]^{1/2}$$

$$Z_c^{+'} = \eta' + Z'$$

$$Z_c^{-'} = \eta' - Z'$$

$$\eta' = \frac{1}{2}(Z'_{11} - Z'_{22})$$

$$Z' = Z'_{12} \sinh \gamma' l'.$$

Conventional analysis shows that the impedance seen at the input to the matching section is

$$Z_{\text{in}} = Z'_{11} - \frac{Z_{12}'^2}{Z'_{22} + Z_L}. \tag{43}$$

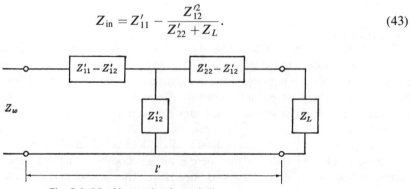

Fig. 9.6. Matching section for periodic structures.

By introducing the characteristic parameters of the matching section, (43) may be rewritten as follows [compare with (40)]:

$$Z_{\text{in}} = \eta' + Z' \frac{Z_L - \eta' + Z' \tanh \gamma' l'}{Z' + (Z_L - \eta') \tanh \gamma' l'}. \tag{44}$$

To obtain a match, we must choose the parameters Z', η', and $\gamma' l'$ so that Z_{in} becomes equal to Z_w. The particular case when Z_L is real and the matching section is symmetrical is readily handled. From (44) with $\eta' = 0$ and $\gamma' l'$ replaced by $j\theta$, we get

$$Z_{\text{in}} = Z' \frac{Z_L + jZ' \tan \theta}{Z' + jZ_L \tan \theta}.$$

If we make $\theta = \pi/2$, we obtain the equivalent of a quarter-wave transformer, and the required value of Z' for a match is given by

$$Z' = (Z_w Z_L)^{1/2}. \tag{45}$$

The required values of Z'_{11} and Z'_{12} are

$$Z'_{11} = Z'_{22} = 0 \tag{46a}$$

$$jZ'_{12} = (Z_w Z_L)^{1/2}. \tag{46b}$$

In the previous analysis we have considered each section to have a physical length l. It is not necessary to introduce a length for each section, since in all the derived formulas we always find the product γl occurring as a parameter. It is the phase shift per section which is of prime importance and not the phase shift per unit length. In some cases it is not possible to define a meaningful length for each section.

9.5. Capacitively Loaded Rectangular Waveguide

As an example, to illustrate some of the previously derived relations, we will consider a rectangular waveguide loaded with asymmetrical capacitive diaphragms at regular intervals l as illustrated in Fig. 9.7. The analysis is simplified if we choose terminal planes midway

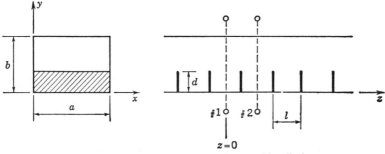

Fig. 9.7. A rectangular guide loaded with capacitive diaphragms.

between each two diaphragms, so that each section may be characterized by a symmetrical T network. The frequency will be chosen so that only the dominant H_{10} mode propagates. Initially we will also assume that the spacing l is chosen so large that all the evanescent modes excited at each diaphragm have decayed to a negligible value at the positions of the adjacent diaphragms. The first higher order mode has a propagation constant given by

$$\Gamma_{11}^2 = \left(\frac{\pi}{a}\right)^2 + \left(\frac{\pi}{b}\right)^2 - k_0^2$$

and, provided $e^{-\Gamma_{11}l} < 0.1$, interaction with this mode may usually be neglected. For our purpose here we will also assume that the normalized shunt susceptance of the diaphragm is given to sufficient accuracy by the low-frequency formula

$$B = \frac{4b\beta_0}{\pi} \ln \sec \frac{\pi d}{2b} \tag{47}$$

where β_0 is the phase constant for the H_{10} mode, and is given by

$$\beta_0^2 = k_0^2 - \left(\frac{\pi}{a}\right)^2.$$

The wave impedance Z_w of the guide for the H_{10} mode is $Z_0 k_0 / \beta_0$, where $Z_0 = (\mu_0/\epsilon_0)^{1/2}$. The susceptance B is normalized with respect to Z_w.

Using conventional transmission-line analysis, we find that the normalized input impedance at terminal plane 1 with normalized load impedance Z_L connected at terminal plane 2 as in Fig. 9.7 is

$$Z_{\text{in}} = \frac{(2t - Bt^2)/(B + 2t) + jZ_L(Bt + t^2 - 1)/(B + 2t)}{j(Bt + t^2 - 1)/(B + 2t) + Z_L}$$

where $t = \tan(\beta_0 l/2)$. Comparing with the similar equation for a symmetrical T network

$$Z_{\text{in}} = \frac{Z_{11}^2 - Z_{12}^2 + Z_{11} Z_L}{Z_{11} + Z_L}$$

shows that

$$Z_{11} = \frac{j(Bt + t^2 - 1)}{B + 2t} \tag{48a}$$

$$Z_{11}^2 - Z_{12}^2 = \frac{2t - Bt^2}{B + 2t}. \tag{48b}$$

From (48), the solutions for the characteristic impedance Z_c and propagation phase constant β_c for the infinite periodic structure are readily found. After suitable algebraic reductions, we get

$$Z_c = (Z_{11}^2 - Z_{12}^2)^{1/2} = \left(1 - \frac{2B}{2 \sin \beta_0 l + B \cos \beta_0 l + B}\right)^{1/2} \tag{49a}$$

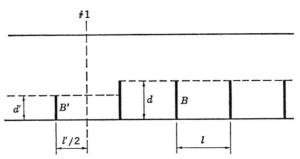

Fig. 9.8. Matching of uniform guide to periodic-loaded guide.

$$\cos \beta_c l = \frac{Z_{11}}{Z_{12}} = \cos \beta_0 l - \frac{B}{2} \sin \beta_0 l. \qquad (49b)$$

If β_c is a solution of (49b), then clearly $\beta_c \pm 2n\pi/l$ is also a solution for all integer values of n. Before considering this eigenvalue equation in detail, we will examine the matching problem.

The periodic-loaded guide may be matched to the uniform guide by means of a shunt capacitive susceptance B' placed at a distance $l'/2$ in front of the first terminal plane of the periodic structure, as shown in Fig. 9.8. We will assume that the periodic-loaded guide is terminated in its characteristic impedance at the output end. The input impedance at the first terminal plane to the loaded guide will then be Z_c. From (46), the required parameters of the matching section are $Z'_{11} = 0$, $jZ'_{12} = Z_c^{1/2}$, since all impedances are normalized relative to the uniform guide impedance Z_w. Also, we require that $\beta'_c l' = \pi/2$ for the matching section. Referring to (49b), this is seen to require that $B' = 2 \cot \beta_0 l'$. From (49a), we obtain

$$(jZ'_{12})^2 = Z_c = 1 - \frac{2B'}{2 \sin \beta_0 l' + B' + B' \cos \beta_0 l'}.$$

These two equations may be solved for B' and l' when Z_c has been specified. At the position of the matching diaphragm, the input admittance is $1 - jB'$. The addition of a shunt susceptance jB' at this point makes the input admittance real and equal to unity. The solutions for l' and B' are

$$\cos \beta_0 l' = \frac{1 - Z_c}{1 + Z_c} \qquad (50a)$$

$$B' = \frac{1 - Z_c}{Z_c^{1/2}}. \qquad (50b)$$

The k_0–β_c Diagram

The usual way of presenting the information contained in the eigenvalue equation (49b) is to plot β_c as a function of k_0. The resultant plot is called the k_0–β_c or ω–β_c diagram [note that $k_0 = \omega(\mu_0\epsilon_0)^{1/2} = \omega/v_c$]. Such a plot is given in Fig. 9.9 for the following parameters: $a = 0.9$ inch, $b = 0.2$ inch, $d = 0.15$ inch, $l = 1.0$ centimeter.

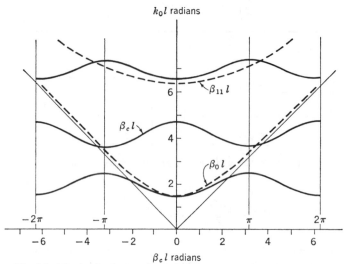

Fig. 9.9. The k_0–β_c diagram for a periodic-loaded rectangular waveguide.

The passbands and stopbands are clearly evident in Fig. 9.9. The first passband occurs in the range $1.37 < k_0 < 2.44$, the first attenuation band occurs for $2.44 < k_0 < 3.43$, the second passband occurs for $3.43 < k_0 < 4.62$, and so forth. For $k_0 > 6.19$, the first higher order mode begins to propagate, and, consequently, the curves for $\beta_c l$ have little significance for k_0 greater than about 5. From the eigenvalue equation (49b), we find that the values of β_0 for $\beta_c l = 0$ are given by

$$\tan \frac{\beta_0 l}{2} = -\frac{B}{2} \qquad \text{or} \qquad \sin \frac{\beta_0 l}{2} = 0 \tag{51a}$$

while, for $\beta_c l = \pi$, the corresponding values of β_0 are given by

$$\cot \frac{\beta_0 l}{2} = \frac{B}{2} \qquad \text{or} \qquad \cos \frac{\beta_0 l}{2} = 0. \tag{51b}$$

From (51a), it is seen that, whenever the diaphragms are spaced by a multiple of a guide wavelength, $\beta_c l$ will equal zero. The field distribution in this case is a pure standing wave between sections, with the electric field being zero at the positions of the diaphragms. It is for this reason that the diaphragms do not produce any perturbation of the standing-wave field distribution and hence permit a solution for $\beta_c l = 2n\pi = \beta_0 l$. The other solution for $\beta_c l = 0$ or $2n\pi$, as given by (51a), corresponds to a field distribution for which the transverse electric field is symmetrical about, but not equal to, zero, at each diaphragm. A similar conclusion may be reached from (51b), but with the diaphragms spaced by an odd multiple of a half guide wavelength. When the standing-wave field distribution does not have a zero transverse electric field at the diaphragms, the spacing differs from a multiple of a half guide wavelength under the conditions $\beta_c l = n\pi$ because of the perturbing effect of the diaphragms. This corresponds to the other solutions given by (51).

9.6. Energy and Power Flow

According to Floquet's theorem, the field in the loaded guide can be represented in the form

$$\mathbf{E}(x, y, z) = \sum_{n=-\infty}^{\infty} a_n \mathbf{E}_n(x, y) e^{-j(\beta_c + 2n\pi/l)z} \tag{52}$$

where a_n is a suitable amplitude coefficient and $\mathbf{E}_n(x, y)e^{-j(\beta_c + 2n\pi/l)z}$ is the field of the nth mode or space harmonic. Since the field consists, in general, of an infinite series, there is no unique phase velocity. Each space harmonic has its own phase velocity v_{pn} given by

$$v_{pn} = \frac{k_0 v_c}{\beta_c + 2n\pi/l} \tag{53}$$

where v_c is the velocity of light in free space. Some of the space harmonics will have negative phase velocities. Replacing β_c by $\beta_c + 2n\pi/l$ is equivalent to focusing attention on the nth- rather than the zeroth-order space-harmonic wave. As β_c increases from zero to $2\pi/l$, the corresponding mode changes from the $n = 0$ to the $n = 1$ mode. At the same time the $n = -1$ mode goes over into the $n = 0$ mode, etc. A single space harmonic cannot exist by itself except in special situations such as when $\beta_c l = 2m\pi$ and the diaphragms are spaced by an integral multiple of a guide wavelength. In general, all the space harmonics are coupled together through the boundary conditions which the field must satisfy. Together they make up the periodic wave.

In a passband we may readily show that, with β_c real, the time-average electric and magnetic energy stored in each section are equal. The fields at two adjacent terminal planes 1-1 and 2-2 differ only by a factor $e^{-j\beta_c l}$, and, hence, the complex Poynting vector has the same value at each terminal plane. Thus we have

$$\iint_{1\text{-}1} \mathbf{E} \times \mathbf{H}^* \cdot \mathbf{a}_z \, dS - \iint_{2\text{-}2} \mathbf{E} \times \mathbf{H}^* \cdot \mathbf{a}_z \, dS = 0$$

$$= 4j\omega(W_m - W_e) \qquad \text{and hence } W_m = W_e.$$

Since the frequency dependence of the propagation phase constant $\beta_n = \beta_c + 2n\pi/l$ for the nth space harmonic is the same as that for β_c, each space harmonic has the same group velocity. According to Chapter 7, Eq. (128), the group velocity is given by

$$v_g = \left(\frac{d\beta_c}{d\omega}\right)^{-1} = v_c \left(\frac{d\beta_c}{dk_0}\right)^{-1}. \tag{54}$$

Referring to Fig. 9.9, it is seen that, for $0 < \beta_c l < \pi$, the group velocity is positive, while, for $\pi < \beta_c l < 2\pi$, it is negative. We may also show that the power flow along the periodic structure is equal to the group velocity, divided by the period l, times the time-average electric plus magnetic energy stored per section.

The proof of the above property is readily established by an application of Green's theorem for two vector functions.[1] For two vector functions $\mathbf{A}$ and $\mathbf{B}$ which have zero divergence this

---

[1] A similar proof is given by Sensiper in [9.1].

theorem is

$$\iiint_V (\mathbf{A} \cdot \nabla^2 \mathbf{B} - \mathbf{B} \cdot \nabla^2 \mathbf{A}) \, dV = \oiint_S [\mathbf{A} \times (\nabla \times \mathbf{B}) - \mathbf{B} \times (\nabla \times \mathbf{A})] \cdot \mathbf{n} \, dS$$

where $\mathbf{n}$ is the unit outward-directed normal. Let $\mathbf{A} = \partial \mathbf{E}/\partial \omega$, $\mathbf{B} = \mathbf{E}^*$, where $\mathbf{E}$ is the total electric field of the periodic wave and is a function of ω. Since $\nabla^2 \mathbf{E} + k_0^2 \mathbf{E} = 0$, we have

$$\nabla^2 \frac{\partial \mathbf{E}}{\partial \omega} = -2\omega \mu_0 \epsilon_0 \mathbf{E} - k_0^2 \frac{\partial \mathbf{E}}{\partial \omega}$$

$$\nabla^2 \mathbf{E}^* = -k_0^2 \mathbf{E}^*.$$

Substituting in Green's theorem gives

$$\iiint_V \left(-k_0^2 \mathbf{E}^* \cdot \frac{\partial \mathbf{E}}{\partial \omega} + k_0^2 \mathbf{E}^* \cdot \frac{\partial \mathbf{E}}{\partial \omega} + 2\omega \mu_0 \epsilon_0 \mathbf{E} \cdot \mathbf{E}^* \right) dV$$

$$= 2\omega \mu_0 \epsilon_0 \iiint_V \mathbf{E} \cdot \mathbf{E}^* \, dV$$

$$= \oiint_S \left[\frac{\partial \mathbf{E}}{\partial \omega} \times (\nabla \times \mathbf{E}^*) - \mathbf{E}^* \times \left(\nabla \times \frac{\partial \mathbf{E}}{\partial \omega} \right) \right] \cdot \mathbf{n} \, dS. \qquad (55)$$

Since $\mathbf{n} \times \mathbf{E}$ vanishes on the guide walls, the surface integral reduces to an integral over the terminal planes 1 and 2 which bound one section of the periodic structure. All losses are assumed absent, and hence

$$\oiint_S \mathbf{E} \times \mathbf{H}^* \cdot d\mathbf{S} = 0 \qquad \text{or equivalently} \qquad \oiint_S \mathbf{E}^* \times (\nabla \times \mathbf{E}) \cdot \mathbf{n} \, dS = 0.$$

Differentiating with respect to ω gives

$$\oiint_S \frac{\partial \mathbf{E}^*}{\partial \omega} \times (\nabla \times \mathbf{E}) \cdot \mathbf{n} \, dS = -\oiint_S \mathbf{E}^* \times \left(\nabla \times \frac{\partial \mathbf{E}}{\partial \omega} \right) \cdot \mathbf{n} \, dS.$$

Using this result, the right-hand side of (55) becomes

$$2 \operatorname{Re} \oiint_S \frac{\partial \mathbf{E}}{\partial \omega} \times (\nabla \times \mathbf{E}^*) \cdot \mathbf{n} \, dS.$$

On terminal plane 2, the field $\mathbf{E} = \mathbf{E}_2$ is equal to $e^{-j\beta_c l} \mathbf{E}_1$, where $\mathbf{E}_1$ is the field at terminal plane 1. Thus we have

$$\frac{\partial \mathbf{E}_2}{\partial \omega} = e^{-j\beta_c l} \frac{\partial \mathbf{E}_1}{\partial \omega} - jl \frac{d\beta_c}{d\omega} \mathbf{E}_1 e^{-j\beta_c l}.$$

Substituting into the right-hand side of (55) gives

$$2 \operatorname{Re} \left[-\iint_{1\text{-}1} \frac{\partial \mathbf{E}_1}{\partial \omega} \times (\nabla \times \mathbf{E}_1^*) \cdot \mathbf{a}_z \, dS \right.$$

$$\left. + \iint_{2\text{-}2} \left(\frac{\partial \mathbf{E}_1}{\partial \omega} - jl \frac{d\beta_c}{d\omega} \mathbf{E}_1 \right) \times (\nabla \times \mathbf{E}_1^*) \cdot \mathbf{a}_z \, dS \right]$$

$$= 2 \operatorname{Re} \iint_{2\text{-}2} -jl \frac{d\beta_c}{d\omega} \mathbf{E}_1 \times \nabla \times \mathbf{E}_1^* \cdot \mathbf{a}_z \, dS$$

$$= 2 \operatorname{Re} \omega\mu_0 l \frac{d\beta_c}{d\omega} \iint_{2\text{-}2} \mathbf{E}_1 \times \mathbf{H}_1^* \cdot \mathbf{a}_z \, dS$$

$$= 2\omega\mu_0\epsilon_0 \iiint_V \mathbf{E} \cdot \mathbf{E}^* \, dV = 8\omega\mu_0 W_e$$

$$= 4\omega\mu_0(W_e + W_m)$$

since $W_e = W_m$. From this result the real power flow is seen to be given by

$$\operatorname{Re} \frac{1}{2} \iint_{2\text{-}2} \mathbf{E}_1 \times \mathbf{H}_1^* \cdot \mathbf{a}_z \, dS = \frac{W_e + W_m}{l \, d\beta_c/d\omega}$$

$$= (W_e + W_m) \frac{v_g}{l} \qquad (56)$$

and is equal to the group velocity times the spatial-average electric plus magnetic energy density per section. The method of derivation is clearly applicable to any lossless periodic structure, and hence (56) is a general result for lossless structures.

9.7. HIGHER ORDER MODE INTERACTION

For close spacing of the diaphragms, the first few least attenuated higher order modes excited by one diaphragm will not have attenuated to a negligible value at the positions of the two adjacent diaphragms. The incident field on each diaphragm is therefore a combination of a dominant mode and one or more higher order modes. The result of higher order mode interaction is to modify the characteristic propagation phase constant β_c from that computed on the basis of only dominant-mode interaction. The analysis in the general case will, however, be similar to that for single-mode interaction. For each interacting mode, a separate equivalent transmission line is introduced. The diaphragm must now be represented by a more general network which couples the various transmission lines together. The overall structure is still a periodic one, and solutions for periodic waves propagating along the structure may be found by the usual matrix methods. The analysis of systems of this type has been presented by Brown [9.2], [9.3]. A similar theory will be presented here but specialized to the particular problem of a waveguide loaded with capacitive diaphragms.

Consider a rectangular guide loaded with identical capacitive diaphragms at regular intervals l as in Fig. 9.10. The mode amplitudes at the input and output terminal planes of one basic

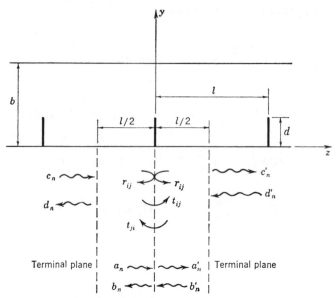

Fig. 9.10. Capacitive-loaded rectangular guide.

section are denoted by c_n, d_n, c'_n, and d'_n, where the coefficients c_n, c'_n pertain to forward-propagating modes, and d_n, d'_n to backward-propagating modes (or attenuated modes). The transverse electric and magnetic fields to the left and right of one diaphragm may be expanded into a series of the normal waveguide modes as follows:

$$\mathbf{E}_t = \sum_n a_n \mathbf{e}_n e^{-\Gamma_n z} + \sum_n b_n \mathbf{e}_n e^{\Gamma_n z}, \qquad -l \le z \le 0 \qquad (57a)$$

$$\mathbf{H}_t = \sum_n a_n \mathbf{h}_n e^{-\Gamma_n z} - \sum_n b_n \mathbf{h}_n e^{\Gamma_n z}, \qquad -l \le z \le 0 \qquad (57b)$$

$$\mathbf{E}_t = \sum_n a'_n \mathbf{e}_n e^{-\Gamma_n z} + \sum_n b'_n \mathbf{e}_n e^{\Gamma_n z}, \qquad 0 \le z \le l \qquad (57c)$$

$$\mathbf{H}_t = \sum_n a'_n \mathbf{h}_n e^{-\Gamma_n z} - \sum_n b'_n \mathbf{h}_n e^{\Gamma_n z}, \qquad 0 \le z \le l \qquad (57d)$$

where a_n, b_n, a'_n, and b'_n are the mode amplitudes at $z = 0$. It will be assumed that only the dominant mode (H_{10} mode) propagates. The higher order modes excited by each diaphragm are the longitudinal-section electric modes with all field components present except E_x. Since the mode functions $\mathbf{e}_n$ are orthogonal, and the electric field must be continuous at the diaphragm, we must have $a_n + b_n = a'_n + b'_n$. Only a finite number, say N, of the modes excited at one discontinuity will have an appreciable amplitude at the position of the next discontinuity. For any practical case, a sufficiently accurate analysis of the loaded guide will be obtained by taking into account only the first N least attenuated modes.

Each discontinuity may be characterized by a set of reflection and transmission coefficients.

The following nomenclature will be adopted:

r_{ij} = reflection coefficient for ith mode with jth mode incident

t_{ij} = transmission coefficient through diaphragm for ith mode with jth mode incident.

The indices i, j take on all values from 1 to N. Since each discontinuity is symmetrical, the above reflection and transmission coefficients hold for modes incident from either the left or the right side of the obstacle. Also, since the transverse electric field of each mode is continuous across the diaphragm, $t_{ij} = \delta_{ij} + r_{ij}$ for $i, j = 1, 2, \ldots, N$.

At $z = 0$, (57a) and (57c) give

$$\mathbf{E}_t = \sum_{n=1}^{N} a_n \mathbf{e}_n + \sum_{n=1}^{N} \mathbf{e}_n \sum_{i=1}^{N} (r_{ni} a_i + t_{ni} b_i') \tag{58a}$$

$$\mathbf{E}_t = \sum_{n=1}^{N} b_n' \mathbf{e}_n + \sum_{n=1}^{N} \mathbf{e}_n \sum_{i=1}^{N} (r_{ni} b_i' + t_{ni} a_i). \tag{58b}$$

Let A and B denote column matrices with elements a_n, b_n; $n = 1, 2, \ldots, N$, respectively, while R and T denote the following square $N \times N$ reflection and transmission coefficient matrices:

$$R = \begin{bmatrix} r_{11} & r_{12} & \cdots & r_{1N} \\ r_{21} & \cdots & \cdots & \cdots \\ & & & \\ & & & \\ r_{N1} & \cdots & \cdots & r_{NN} \end{bmatrix}$$

$$T = \begin{bmatrix} t_{11} & t_{12} & \cdots & t_{1N} \\ t_{21} & \cdots & \cdots & \cdots \\ & & & \\ & & & \\ t_{N1} & \cdots & \cdots & t_{NN} \end{bmatrix}.$$

We will also denote the column matrices with elements a_n', b_n' by A' and B', respectively. The relation between the mode amplitudes on adjacent sides of the diaphragm may be written in matrix form as

$$B = RA + TB' \tag{59a}$$

$$A' = RB' + TA \tag{59b}$$

since the amplitude of a reflected mode on the left is given by

$$b_n = \sum_{i=1}^{N} (r_{ni} a_i + t_{ni} b_i')$$

and similarly for a reflected mode on the right. Solving for A and B in terms of A' and B' gives

$$A = T^{-1}A' - T^{-1}RB' \tag{60a}$$

$$B = RT^{-1}A' + (T - RT^{-1}R)B' \tag{60b}$$

or, in partitioned matrix form,

$$\begin{bmatrix} A \\ B \end{bmatrix} = \begin{bmatrix} T^{-1} & -T^{-1}R \\ RT^{-1} & T - RT^{-1}R \end{bmatrix} \begin{bmatrix} A' \\ B' \end{bmatrix} \tag{61}$$

where T^{-1} is the inverse of the matrix T. The square matrix in (61) is the wave-amplitude transfer matrix for a single diaphragm.

A convenient set of equivalent transmission-line voltages and currents may be defined as follows:

$$V = A + B \qquad V' = A' + B'$$

$$I = A - B \qquad I' = A' - B'$$

where V, V', I, and I' are column matrices with elements v_n, v'_n, i_n, and i'_n; $n = 1, 2, \ldots, N$. From (60), the transfer matrix for the voltage and current matrices is readily found:

$$\begin{bmatrix} V \\ I \end{bmatrix} = \begin{bmatrix} U & 0 \\ 2T^{-1} - 2U & U \end{bmatrix} \begin{bmatrix} V' \\ I' \end{bmatrix} \tag{62}$$

where U is the unit matrix and the relation $U + R = T$ has been used to simplify the end result. The upper right-hand-side element in (62) written as 0 is an $N \times N$ matrix whose every element is zero. It will be convenient to deal with the voltage and current transfer matrix, instead of the wave-amplitude transfer matrix, because of its simpler form.

The mode amplitudes c_n, d_n, c'_n, d'_n at the terminal planes are related to the mode amplitudes at the diaphragm as follows: $c_n = a_n e^{\Gamma_n l/2}$, $d_n = b_n e^{-\Gamma_n l/2}$, $c'_n = a'_n e^{-\Gamma_n l/2}$, $d'_n = b'_n e^{\Gamma_n l/2}$. The terminal voltages and currents will be denoted by V_0, I_0, V'_0, I'_0 and are related to V, I, V', and I' as follows:

$$V_0 = \begin{bmatrix} c_1 + d_1 \\ c_2 + d_2 \\ \\ c_N + d_N \end{bmatrix} = \begin{bmatrix} \cosh \dfrac{\Gamma_1 l}{2} & 0 & \cdots & 0 \\ 0 & \cosh \dfrac{\Gamma_2 l}{2} & \cdots & 0 \\ & & & \\ 0 & 0 & \cdots & \cosh \dfrac{\Gamma_N l}{2} \end{bmatrix} V$$

$$+ \begin{bmatrix} \sinh \dfrac{\Gamma_1 l}{2} & 0 & \cdots & 0 \\ 0 & \sinh \dfrac{\Gamma_2 l}{2} & \cdots & 0 \\ & & & \\ 0 & 0 & \cdots & \sinh \dfrac{\Gamma_N l}{2} \end{bmatrix} I \tag{63a}$$

$$I_0 = U \sinh \frac{\Gamma l}{2} V + U \cosh \frac{\Gamma l}{2} I \tag{63b}$$

$$V' = U \cosh \frac{\Gamma l}{2} V'_0 + U \sinh \frac{\Gamma l}{2} I'_0 \tag{63c}$$

$$I' = U \sinh \frac{\Gamma l}{2} V'_0 + U \cosh \frac{\Gamma l}{2} I'_0 \tag{63d}$$

where $U \cosh(\Gamma l/2)$ and $U \sinh(\Gamma l/2)$ are used to represent diagonal matrices with elements $\cosh(\Gamma_n l/2)$, $\sinh(\Gamma_n l/2)$, $n = 1, 2, \ldots, N$, as in (63a).

In place of the system of equations (62), we now have

$$\begin{bmatrix} V_0 \\ I_0 \end{bmatrix} = \begin{bmatrix} U \cosh \frac{\Gamma l}{2} & U \sinh \frac{\Gamma l}{2} \\ U \sinh \frac{\Gamma l}{2} & U \cosh \frac{\Gamma l}{2} \end{bmatrix} \begin{bmatrix} U & 0 \\ 2T^{-1} - 2U & U \end{bmatrix}$$
$$\times \begin{bmatrix} U \cosh \frac{\Gamma l}{2} & U \sinh \frac{\Gamma l}{2} \\ U \sinh \frac{\Gamma l}{2} & U \cosh \frac{\Gamma l}{2} \end{bmatrix} \begin{bmatrix} V'_0 \\ I'_0 \end{bmatrix}. \tag{64}$$

For a propagating periodic wave we must have

$$V'_0 = U e^{-\gamma l} V_0 \qquad I'_0 = U e^{-\gamma l} I_0$$

where γ is the characteristic propagation constant. Substituting into (64) yields the following matrix eigenvalue system:

$$\left| \begin{bmatrix} U \cosh \frac{\Gamma l}{2} & U \sinh \frac{\Gamma l}{2} \\ U \sinh \frac{\Gamma l}{2} & U \cosh \frac{\Gamma l}{2} \end{bmatrix} \begin{bmatrix} U & 0 \\ 2T^{-1} - 2U & U \end{bmatrix} \right.$$
$$\left. \times \begin{bmatrix} U \cosh \frac{\Gamma l}{2} & U \sinh \frac{\Gamma l}{2} \\ U \sinh \frac{\Gamma l}{2} & U \cosh \frac{\Gamma l}{2} \end{bmatrix} - \begin{bmatrix} U e^{\gamma l} & 0 \\ 0 & U e^{\gamma l} \end{bmatrix} \right| = 0 \tag{65}$$

where the bars signify the determinant of the matrix. The solution for V and I in terms of V_0 and I_0 in (63a) and (63b) is obtained by changing l to $-l$ and interchanging V, V_0 and I, I_0. The end result shows at once that

$$\begin{bmatrix} U \cosh \frac{\Gamma l}{2} & U \sinh \frac{\Gamma l}{2} \\ U \sinh \frac{\Gamma l}{2} & U \cosh \frac{\Gamma l}{2} \end{bmatrix}^{-1} = \begin{bmatrix} U \cosh \frac{\Gamma l}{2} & -U \sinh \frac{\Gamma l}{2} \\ -U \sinh \frac{\Gamma l}{2} & U \cosh \frac{\Gamma l}{2} \end{bmatrix}.$$

The voltage–current transfer matrix for a system of transmission lines of length l is of the same form as those occurring in (63), with $l/2$ replaced by l. This transfer matrix is also given by the square of the transfer matrix for lines of length $l/2$, and, hence,

$$\begin{bmatrix} U \cosh \frac{\Gamma l}{2} & U \sinh \frac{\Gamma l}{2} \\ U \sinh \frac{\Gamma l}{2} & U \cosh \frac{\Gamma l}{2} \end{bmatrix}^2 = \begin{bmatrix} U \cosh \Gamma l & U \sinh \Gamma l \\ U \sinh \Gamma l & U \cosh \Gamma l \end{bmatrix}.$$

By premultiplying and postmultiplying (65) by the inverse of the hyperbolic sine and cosine matrix and using the above results, (65) may be converted to

$$
\left\| \begin{bmatrix} Ue^{-\gamma l} & 0 \\ (2T^{-1} - 2U)e^{-\gamma l} & Ue^{-\gamma l} \end{bmatrix} - \begin{bmatrix} U\cosh \Gamma l & -U\sinh \Gamma l \\ -U\sinh \Gamma l & U\cosh \Gamma l \end{bmatrix} \right\| = 0. \qquad (66)
$$

This result could have been derived directly by evaluating the transfer matrix from the position of one diaphragm to the same position of the next diaphragm.

If γ is an eigenvalue of (66), then $-\gamma$ is also an eigenvalue. This result may be formally proved as follows. Let V, I be an eigenvector of the matrix occurring in (66). Thus V, I satisfy the equations

$$
(Ue^{-\gamma l} + U\cosh \Gamma l)V + U\sinh \Gamma l \, I = 0 \qquad (67a)
$$

$$
[(2T^{-1} - 2U)e^{-\gamma l} + U\sinh \Gamma l]V + (Ue^{-\gamma l} - U\cosh \Gamma l)I = 0. \qquad (67b)
$$

Premultiply (67b) by $e^{\gamma l} U \sinh \Gamma l$ to obtain

$$
(U\sinh \Gamma l)(2T^{-1} - 2U + e^{\gamma l} U\sinh \Gamma l)V + (U - e^{\gamma l}U\cosh \Gamma l)U\sinh \Gamma l \, I = 0
$$

since diagonal matrices commute. Substituting for I from (67a) now gives

$$
[2U\sinh \Gamma l \, T^{-1} - 2U\sinh \Gamma l + e^{\gamma l}(U\sinh \Gamma l)^2
$$

$$
+ 2U\cosh \Gamma l - e^{-\gamma l}U - e^{\gamma l}(U\cosh \Gamma l)^2]V = 0
$$

or since $(U\cosh \Gamma l)^2 - (U\sinh \Gamma l)^2 = U$, we get

$$
(U\sinh \Gamma l \, T^{-1} - U\sinh \Gamma l + U\cosh \Gamma l - U\cosh \gamma l)V = 0. \qquad (68)
$$

Hence $\cosh \gamma l$ is an eigenvalue of the $N \times N$ square matrix

$$
[U\sinh \Gamma l \, T^{-1} - U\sinh \Gamma l + U\cosh \Gamma l].
$$

Since $\cosh \gamma l$ is an even function, both $\pm \gamma$ are eigenvalues.

The form of solution establishes the following two important properties of the periodic-loaded guide:

1. There exist $2N$ characteristic propagation constants $\pm \gamma_m$, $m = 1, 2, \ldots, N$, in the periodic-loaded guide with N interacting modes.
2. For each eigenvalue γ_m, a separate periodic wave composed of N normal waveguide modes exists.

In general, only a few of the propagation constants γ_m are imaginary, and these are associated with propagating modes. The modes with real propagation constants are evanescent modes. These modes are excited at the input region between a uniform guide and a periodic-loaded guide and are analogous to the evanescent modes which occur in a normal uniform guide at a discontinuity. For each root γ_m a voltage–current eigenvector V_m, I_m may be found to within an arbitrary amplitude constant. The elements or components v_{nm} and i_{nm}, $n = 1, 2, \ldots, N$, of V_m, I_m completely specify the field of the mth periodic wave in the loaded guide.

Let c_{nm}, d_{nm} be the mode amplitudes corresponding to the eigenvector V_m, I_m with eigenvalue γ_m. For a propagating normal waveguide mode $\mathbf{e}_n \times \mathbf{h}_n^* \cdot \mathbf{a}_z$ is real, while for an evanescent mode it is imaginary. The normal modes may be normalized so that

$$\frac{1}{2} \iint\limits_S \mathbf{e}_n \times \mathbf{h}_n^* \cdot \mathbf{a}_z \, dS = \begin{cases} 1 & \text{propagating mode} \\ j & \text{evanescent mode} \end{cases}$$

where the integration is over the guide cross section. Since the normal modes are orthogonal, the power flow associated with the mth periodic wave becomes

$$\operatorname{Re} \frac{1}{2} \sum_{n=1}^{N} (c_{nm} + d_{nm})(c_{nm}^* - d_{nm}^*) \iint\limits_S \mathbf{e}_n \times \mathbf{h}_n^* \cdot \mathbf{a}_z \, dS$$

$$= \sum_i (c_{im} c_{im}^* - d_{im} d_{im}^*) + 2 \operatorname{Im} \sum_j c_{jm} d_{jm}^* \quad (69)$$

where the first summation on the right-hand side extends over the propagating waveguide modes, and the second summation extends over the nonpropagating modes. Real power flow occurs for evanescent waveguide modes through the interaction of the fields of the forward- and backward-attenuated modes. For a nonpropagating periodic wave the real power flow is zero.

The above theory will be applied to the capacitive-loaded rectangular guide of width a and height b as in Fig. 9.10. The solution for the reflection and transmission coefficient matrices may be obtained by the methods presented in Chapter 8. For simplicity only two interacting modes will be taken into account.

With a dominant H_{10} incident mode, the higher order modes excited are longitudinal-section electric modes. These modes may be derived from an x-directed magnetic-type Hertzian potential by means of the following equations:

$$\mathbf{E} = -j\omega\mu_0 \nabla \times \mathbf{\Pi}_h$$

$$\mathbf{H} = k_0^2 \mathbf{\Pi}_h + \nabla\nabla\cdot\mathbf{\Pi}_h.$$

For the nth mode, a suitable form for Π_h is $\psi_n(y)\Phi(x)e^{\pm\Gamma_n z}$, where

$$\psi_n = \cos\frac{n\pi y}{b} \quad \Phi = \sin\frac{\pi x}{a} \quad \Gamma_n^2 = \left(\frac{n\pi}{b}\right)^2 - k^2$$

and $k^2 = k_0^2 - (\pi/a)^2$. Suitable expansions for the transverse fields E_y and H_x are readily found to be

$$E_y = \begin{cases} a_0\Phi e^{-\Gamma_0 z} + a_1\psi_1\Phi e^{-\Gamma_1 z} + \displaystyle\sum_{n=0}^{\infty} b_n\psi_n\Phi e^{\Gamma_n z}, & z \leq 0 \\ \displaystyle\sum_{n=0}^{\infty} c_n\psi_n\Phi e^{-\Gamma_n z}, & z \geq 0 \end{cases}$$

$$H_x = \begin{cases} -jY_0 \dfrac{k^2}{k_0} \left(\dfrac{a_0 \Phi}{\Gamma_0} e^{-\Gamma z} + \dfrac{a_1 \psi_1 \Phi}{\Gamma_1} e^{-\Gamma_1 z} - \displaystyle\sum_{n=0}^{\infty} \dfrac{b_n \psi_n \Phi}{\Gamma_n} e^{\Gamma_n z} \right), & z < 0 \\[3ex] -jY_0 \dfrac{k^2}{k_0} \displaystyle\sum_{n=0}^{\infty} \dfrac{c_n \psi_n \Phi}{\Gamma_n} e^{-\Gamma_n z}, & z > 0 \end{cases}$$

where a_0 and a_1 are the amplitudes of the incident modes. Let the aperture electric field be $\mathcal{E}(y)\Phi(x)$ for $d < y \leq b$, and zero for $0 \leq y < d$. By Fourier analysis, we obtain

$$a_0 + b_0 = c_0 = \frac{1}{b} \int_d^b \mathcal{E}(y') \, dy' \tag{70a}$$

$$a_1 + b_1 = c_1 = \frac{2}{b} \int_d^b \mathcal{E}(y') \psi_1(y') \, dy' \tag{70b}$$

$$b_n = c_n = \frac{2}{b} \int_d^b \mathcal{E}(y') \psi_n(y') \, dy'. \tag{70c}$$

For b sufficiently small, Γ_n may be approximated by $n\pi/b$ for $n > 0$. This approximation will be made throughout in order to simplify the analysis. The continuity condition on H_x in the aperture leads to

$$\frac{a_0}{\Gamma_0} + \frac{a_1 b \psi_1}{\pi} = \frac{1}{\Gamma_0 b} \int_d^b \mathcal{E}(y') \, dy' + \frac{2}{b} \int_d^b \mathcal{E}(y') \sum_{n=1}^{\infty} \frac{1}{n} \psi_n(y) \psi_n(y') \, dy', \qquad d \leq y \leq b. \tag{71}$$

The series is summable and yields

$$\sum_{n=1}^{\infty} \frac{1}{n} \cos \frac{n\pi y}{b} \cos \frac{n\pi y'}{b} = -\frac{1}{2} \ln \left| \cos \frac{\pi y}{b} - \cos \frac{\pi y'}{b} \right|.$$

We now introduce a new variable θ by means of the relation

$$\alpha_1 + \alpha_2 \cos \theta = \cos \frac{\pi y}{b}$$

with α_1 and α_2 chosen so that $y = d, b$ corresponds to $\theta = 0, \pi$. The solutions for α_1 and α_2 are

$$\alpha_1 = -\sin^2 \frac{\pi d}{2b} = \alpha_2 - 1$$

$$\alpha_2 = \cos^2 \frac{\pi d}{2b}.$$

A similar change in the variable θ' is also made, and thus $\mathcal{E}(y') \, dy'$ is to be replaced by $\mathcal{F}(\theta') \, d\theta'$, where $\mathcal{F}(\theta')$ is the aperture field as a function of θ'. The integral equation (71) becomes

$$\frac{a_0}{\Gamma_0} + \frac{a_1 b}{1}(\alpha_1 + \alpha_2 \cos \theta) = \left(\frac{1}{\Gamma_0 b} - \frac{\ln \alpha_2}{\pi}\right) \int_0^\pi \mathfrak{F} \, d\theta' + \frac{2}{\pi} \sum_{n=1}^\infty \frac{\cos n\theta}{n} \int_0^\pi \mathfrak{F} \cos n\theta' \, d\theta'$$

(72)

after substituting for $\cos(\pi y/b)$ and $\cos(\pi y'/b)$ in the sum of the original series and expanding the result into a series in the new variables θ, θ'.

A solution to the integral equation is obtained if we choose

$$\mathfrak{F}(\theta') = A_0 + A_1 \cos \theta'$$

where A_0 and A_1 are constants to be determined. Substituting into (72),

$$\frac{a_0}{\Gamma_0} + \frac{a_1 b}{\pi}(\alpha_1 + \alpha_2 \cos \theta) = \frac{\pi A_0}{\Gamma_0 b}\left(1 - \frac{\Gamma_0 b}{\pi} \ln \alpha_2\right) + A_1 \cos \theta.$$

Equating the coefficients of $\cos \theta$ and the constant terms gives

$$a_0 + a_1 \frac{b\alpha_1 \Gamma_0}{\pi} = \frac{\pi A_0}{b}\left(1 - \frac{\Gamma_0 b}{\pi} \ln \alpha_2\right)$$

(73a)

$$a_1 \alpha_2 = \frac{\pi}{b} A_1.$$

(73b)

Replacing y' by the new variable θ' in (70) shows that

$$a_0 + b_0 = c_0 = t_{00} a_0 + t_{01} a_1 = \frac{\pi}{b} A_0$$

$$a_1 + b_1 = c_1 = t_{10} a_0 + t_{11} a_1 = \frac{\pi}{b} A_1 \alpha_2 + \frac{2\pi}{b} A_0 \alpha_1.$$

In matrix form these relations may be written as

$$\begin{bmatrix} t_{00} & t_{01} \\ t_{10} & t_{11} \end{bmatrix} \begin{bmatrix} a_0 \\ a_1 \end{bmatrix} = \begin{bmatrix} \dfrac{\pi}{b} A_0 \\ \dfrac{\pi}{b} A_1 \alpha_2 + \dfrac{2\pi}{b} A_0 \alpha_1 \end{bmatrix}$$

$$= \frac{1}{1 - \dfrac{\Gamma_0 b}{\pi} \ln \alpha_2}$$

$$\cdot \begin{bmatrix} 1 & b\alpha_1 \dfrac{\Gamma_0}{\pi} \\ 2\alpha_1 & \alpha_2^2\left(1 - \dfrac{\Gamma_0 b}{\pi} \ln \alpha_2\right) + 2\alpha_1^2 \dfrac{b\Gamma_0}{\pi} \end{bmatrix} \begin{bmatrix} a_0 \\ a_1 \end{bmatrix}$$

(74)

after substituting for A_0 and A_1 from (73). From (74) the transmission coefficients t_{ij} are readily identified. We get

$$t_{00} = \left(1 - \frac{\Gamma_0 b}{\pi} \ln \alpha_2\right)^{-1} = \left(1 + \frac{jB}{2}\right)^{-1}$$

where B is the normalized shunt susceptance of the diaphragm as given by (47).

The inverse of the matrix T is readily found:

$$T^{-1} = (t_{00}t_{11} - t_{01}t_{10})^{-1} \begin{bmatrix} t_{11} & -t_{01} \\ -t_{10} & t_{00} \end{bmatrix}$$

$$= \begin{bmatrix} 1 - \dfrac{\Gamma_0 b}{\pi} \ln \alpha_2 + \dfrac{2\alpha_1^2}{\pi\alpha_2^2}\Gamma_0 b & -b\alpha_1 \dfrac{\Gamma_0}{\pi\alpha_2^2} \\[2ex] -2\dfrac{\alpha_1}{\alpha_2^2} & \alpha_2^{-2} \end{bmatrix}. \tag{75}$$

The eigenvalue equation for the characteristic propagation constants may now be found by substituting into (68). After simplification, we obtain[2]

$$\left[\cos \beta_0 l + \sin \beta_0 l \left(\frac{\beta_0 b}{\pi} \ln \alpha_2 - \frac{2\alpha_1^2 \beta_0 b}{\pi\alpha_2^2}\right) - \cosh \gamma l\right]$$

$$\cdot \left(\alpha_2^{-2} \sinh \frac{\pi l}{b} + e^{-\pi l/b} - \cosh \gamma l\right) + \frac{2\alpha_1^2 \beta_0 b}{\pi\alpha_2^4} \sin \beta_0 l \sinh \frac{\pi l}{b} = 0 \tag{76}$$

where $\beta_0 = |\Gamma_0|$. For determining the eigenvalue equation, it was not necessary to normalize the waveguide modes. For (69) to hold for the power-flow, the normalization of the waveguide modes, in the manner indicated in the discussion preceding (69), is necessary. As an example, the roots of (76) have been computed for $a = 0.9$ inch, $b = 0.2$ inch, $d = 0.15$ inch, $k_0 = 2$ radians per centimeter, and $l = 0.2$ centimeter. The results are $\gamma_0 = j\beta_c = j2.25$, $\gamma_1 = 24.4$. From (49b), which is based on single-mode interaction, we obtain $\gamma_0 = j2.95$. With such close spacing, it is quite apparent that the result based on single-mode interaction is considerably in error. Also, it should be noted that the first evanescent periodic wave is very highly attenuated.

For small values of l, an alternative method of solution is available. This solution is based on a transverse resonance approach and is discussed in the next chapter. The eigenvalue equation obtained is

$$(\beta_c^2 - k^2)^{1/2} b \tanh(\beta_c^2 - k^2)^{1/2} \left(b - d + \frac{1}{\pi} \ln 2\right) = k_0 b \tan k_0 \left(d - \frac{l}{\pi} \ln 2\right) \tag{77}$$

where $k^2 = k_0^2 - (\pi/a)^2 = \beta_0^2$. This equation gives a value of 2.18 for β_c for the above example. This result compares favorably with that obtained from (76). For really close spacing, (77) is more accurate than (76). In practice, the combination of (49b) for large spacing and (77) for small spacing provides a sufficiently accurate solution for all values of l, and, hence, the more complicated equation (76) is not required in order to obtain good results for γ_0. However, it is required in order to obtain a solution for γ_1 and the fields for the first evanescent mode.

---

[2] A similar result has been derived by Brown for a capacitive-loaded parallel-plate transmission line. See [9.2].

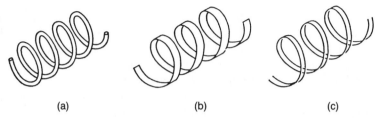

Fig. 9.11. Three typical types of helices: (a) round wire helix, (b) tape helix, and (c) bifilar (two-wire) helix.

9.8. THE SHEATH HELIX

The helix is an important structure with periodic properties which is widely used in such diverse applications as traveling-wave tubes, antennas, and delay lines. The basic property of the helix which makes it an essential component in the above devices is the relatively large reduction in the phase velocity of an electromagnetic wave propagating along it. To a first approximation, the reduction in phase velocity as compared with the velocity of light is by a factor equal to the pitch of the winding divided by the circumference of one turn. Figure 9.11 illustrates several typical forms of helices.

The Helmholtz equation is not separable in helical coordinates, and, consequently, a rigorous solution for electromagnetic-wave propagation along a helix has so far not been attained. Two approximate models which are amenable to solution have received considerable study. The simplest model replaces the actual helix by a sheath helix as in Fig. 9.12. The sheath helix is a cylindrical tube which is assumed to have infinite conductivity in the direction of the original winding and zero conductivity in a direction normal to the turns of the winding. It is essentially a cylindrical tube with anisotropic conductivity properties. If the actual helix consists of many turns in a length equal to a wavelength, the sheath helix will represent a good approximation. An analysis of the sheath helix has been given by Pierce [9.4].

The second model of a helix is an infinitely thin tape helix. An analysis of this structure has been presented by Sensiper [9.1]. The essential steps are to write a general expansion of the field in cylindrical coordinates in the regions inside and outside the helix, and then to determine the amplitude coefficients in terms of an assumed electric field in the gap, or in terms of an assumed current on the tape. The analysis becomes rather involved, so we will consider only the sheath-helix model in detail. This does not, of course, give a complete picture of the properties of a helix in general. In particular, no information on the passband–stopband characteristics is obtained.

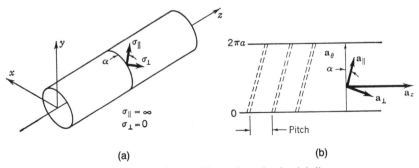

Fig. 9.12. Coordinates of the surface of a sheath helix.

The developed surface of the sheath helix is illustrated in Fig. 9.12. The pitch angle is α, and the windings of the actual helix are parallel to the unit vector $\mathbf{a}_{||}$. The radius of the sheath-helix tube is a. Let $\mathbf{E}^i_{||}$, $\mathbf{E}^e_{||}$ and $\mathbf{E}^i_{\perp}$, $\mathbf{E}^e_{\perp}$ be the tangential components of the electric field which, at the surface of the cylinder, are parallel to the unit vectors $\mathbf{a}_{||}$ and $\mathbf{a}_{\perp}$, respectively. The superscripts i and e are used to distinguish between the field interior to the boundary $r = a$ and that exterior to this boundary. The boundary conditions require continuity of the tangential electric field at $r = a$ and, in particular, the vanishing of the components $\mathbf{E}^i_{||}$, $\mathbf{E}^e_{||}$, which are in the direction of infinite conductivity. Thus we have, at $r = a$,

$$\mathbf{E}^i_{||} = \mathbf{E}^e_{||} = 0 \qquad \mathbf{E}^i_{\perp} = \mathbf{E}^e_{\perp}$$

or, equivalently,

$$E^i_z = E^e_z \tag{78a}$$

$$E^i_\theta = E^e_\theta \tag{78b}$$

$$E^{i,e}_z = -E^{i,e}_\theta \cot \alpha. \tag{78c}$$

The latter equation is merely the relation imposed on E_z and E_θ in order that $E_{||} = 0$. The boundary condition on the tangential magnetic field components is

$$H^i_{||} = H^e_{||}$$

or, equivalently,

$$H^i_z + H^i_\theta \cot \alpha = H^e_z + H^e_\theta \cot \alpha. \tag{78d}$$

The component $H_\perp$ is discontinuous across the sheath by an amount equal to the current flowing on the sheath in the direction $\mathbf{a}_{||}$. The nature of the above boundary conditions indicates that the solution for any particular mode will not be a TM or TE mode, but a combination of both. Hence, all six field components E_r, E_θ, E_z, H_r, H_θ, and H_z will be present.

The general expansions for the TE and TM modes are similar to those for the circular guide. The θ and z dependence may be assumed to be according to $e^{-jn\theta - \gamma z}$, where n is an integer, and γ is the propagation constant. If slow waves exist, they will have a propagation constant $\gamma = j\beta$, whose modulus β is greater than k_0. In anticipation of this type of solution, the appropriate Bessel functions to use in the region $r < a$ are the modified functions of the first kind $I_n(hr)$ rather than the functions $J_n(hr)$, which occur in the solution for the circular guide. Similarly, for the external region $r > a$, the solutions which decay exponentially with r are the modified Bessel functions of the second kind $K_n(hr)$. The radial wavenumber h is given by $h = (\beta^2 - k_0^2)^{1/2}$, where $j\beta = \gamma$. If the TM field is expressed in terms of E_z, and the TE field in terms of H_z, the appropriate expansion of the field in the two regions is found to be as follows:

for $r \leq a$

$$E^i_z = A_n I_n(hr) e^{-jn\theta - \gamma z}$$

$$E^i_r = \left(\frac{\gamma}{h} A_n I'_n + \frac{n\omega\mu_0}{h^2 r} B_n I_n \right) e^{-jn\theta - \gamma z}$$

$$E_\theta^i = \left(\frac{-jn\gamma}{h^2 r} A_n I_n - \frac{j\omega\mu_0}{h} B_n I_n' \right) e^{-jn\theta - \gamma z}$$

$$H_z^i = B_n I_n(hr) e^{-jn\theta - \gamma z}$$

$$H_r^i = \left(\frac{-n\omega\epsilon_0}{h^2 r} A_n I_n + \frac{\gamma}{h} B_n I_n' \right) e^{-jn\theta - \gamma z}$$

$$H_\theta^i = \left(\frac{j\omega\epsilon_0}{h} A_n I_n' - \frac{jn\gamma}{h^2 r} B_n I_n \right) e^{-jn\theta - \gamma z}$$

for $r \geq a$

$$E_z^e = C_n K_n(hr) e^{-jn\theta - \gamma z}$$

$$E_r^e = \left(\frac{\gamma}{h} C_n K_n' + \frac{n\omega\mu_0}{h^2 r} D_n K_n \right) e^{-jn\theta - \gamma z}$$

$$E_\theta^e = \left(\frac{-jn\gamma}{h^2 r} C_n K_n - \frac{j\omega\mu_0}{h} D_n K_n' \right) e^{-jn\theta - \gamma z}$$

$$H_z^e = D_n K_n(hr) e^{-jn\theta - \gamma z}$$

$$H_r^e = \left(\frac{-n\omega\epsilon_0}{h^2 r} C_n K_n + \frac{\gamma}{h} D_n K_n' \right) e^{-jn\theta - \gamma z}$$

$$H_\theta^e = \left(\frac{j\omega\epsilon_0}{h} C_n K_n' - \frac{jn\gamma}{h^2 r} D_n K_n \right) e^{-jn\theta - \gamma z}$$

where A_n, B_n, C_n, and D_n are unknown amplitude constants, and the prime means differentiation of the Bessel functions with respect to the argument hr.

From the condition $E_z = -E_\theta \cot \alpha$, the solutions for B_n and D_n in terms of A_n and C_n are found to be

$$B_n(j\omega\mu_0 ha \cot \alpha) I_n' = A_n(h^2 a - jn\gamma \cot \alpha) I_n$$

$$D_n(j\omega\mu_0 ha \cot \alpha) K_n' = C_n(h^2 a - jn\gamma \cot \alpha) K_n$$

where the Bessel functions are evaluated at $r = a$. To facilitate matching the boundary conditions at $r = a$, the following two impedance functions are formed:

$$Z_i = \frac{E_\perp^i}{H_\parallel^i} = \frac{E_z^i \cos \alpha - E_\theta^i \sin \alpha}{H_z^i \sin \alpha + H_\theta^i \cos \alpha}$$

$$= \frac{E_z^i \sec \alpha}{H_z^i \sin \alpha + H_\theta^i \cos \alpha}$$

$$Z_e = \frac{E_\perp^e}{H_\parallel^e} = \frac{E_z^e \sec \alpha}{H_z^e \sin \alpha + H_\theta^e \cos \alpha}.$$

The proper continuity of the fields is ensured if $Z_i = Z_e$ at $r = a$. Substituting for the field components, and simplifying the resultant expression, yields the following eigenvalue equation for β and h:

$$\frac{K_n'(ha)I_n'(ha)}{K_n(ha)I_n(ha)} = -\frac{(h^2a^2 + n\beta a \cot \alpha)^2}{k_0^2a^2h^2a^2 \cot^2 \alpha} \tag{79}$$

where $\beta^2 = h^2 + k_0^2$.

For the dominant mode, $n = 0$, and the eigenvalue equation reduces to

$$\frac{K_0'(ha)I_0'(ha)}{K_0(ha)I_0(ha)} = -\frac{h^2a^2}{k_0^2a^2 \cot^2 \alpha}. \tag{80}$$

A plot of the ratio $k_0a/\beta a$ as a function of k_0a is given in Fig. 9.13 for $n = 0$ and the following values of α: 10, 15, and 20°. For k_0a greater than unity, the propagation phase constant is seen to be approximately equal to $k_0 \sin \alpha$, and, hence, the reduction in phase velocity is by a factor $\sin \alpha$. The group velocity is given by $v_c(d\beta/dk_0)^{-1}$. In the region $k_0a > 1$, where $\beta = k_0 \sin \alpha$, the group velocity v_g is given by $v_c \sin \alpha$ and is equal to the phase velocity v_p.

For $n \neq 0$, the solution for h and β is somewhat more difficult to obtain. In general, imaginary values of h are not allowed, since this would lead to solutions which do not have exponential decay in the radial direction for $r > a$. Consequently, β is restricted to be less than k_0 for all modes. For any given value of k_0a, a solution for βa can be found only for certain values of n. Hence, only a finite number of modes of propagation exist at any one frequency [9.5]. This property is true, in general, for all open-boundary waveguides, and will be examined in greater detail in Chapter 11.

REFERENCES AND BIBLIOGRAPHY

[9.1]   S. Sensiper, "Electromagnetic wave propagation on helical conductors," MIT Research Lab. Electronics Tech. Rep. 194, May 16, 1951.

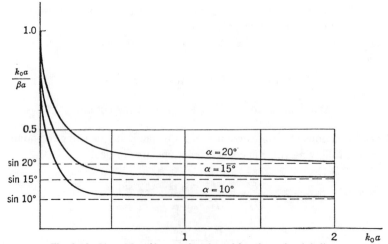

Fig. 9.13. Plot of $k_0a/\beta a$ as a function of k_0a for a sheath helix.

[9.2]   J. Brown, "A theoretical study of some artificial dielectrics," Ph.D. thesis, University of London, 1954.

[9.3]   J. Brown, "Propagation on coupled transmission line systems," *Quart. J. Mech. Appl. Math.*, vol. 11, pp. 235–243, May 1958.

[9.4]   J. R. Pierce, *Travelling Wave Tubes.* Princeton, NJ: D. Van Nostrand Company, Inc., 1950.

[9.5]   D. A. Watkins, *Topics in Electromagnetic Theory.* New York, NY: John Wiley & Sons, Inc., 1958.

[9.6]   L. Brillouin, *Wave Propagation in Periodic Structures*, 2nd ed. New York, NY: Dover Publications, 1953.

[9.7]   J. C. Slater, *Microwave Electronics.* Princeton, NJ: D. Van Nostrand Company, Inc., 1950.

[9.8]   E. L. Chu and W. W. Hansen, "Disk loaded waveguides," *J. Appl. Phys.*, vol. 18, pp. 996–1008, 1947; vol. 20, pp. 280–285, 1949.

[9.9]   J. C. Slater, "The design of linear accelerators," *Rev. Mod. Phys.*, vol. 20, pp. 473–518, July 1948.

[9.10] L. Brillouin, "Wave guides for slow waves," *J. Appl. Phys.*, vol. 19, pp. 1023–1041, Nov. 1948.

[9.11] A. W. Lines, G. R. Nicoll, and A. M. Woodward, "Some properties of waveguides with periodic structure," *Proc. IEE (London)*, vol. 97, part III, pp. 263–276, July 1950.

[9.12] A. H. W. Beck, *Space Charge Waves.* New York, NY: Pergamon Press, Inc., 1958. (This book has a good discussion of slow-wave structures for use in traveling-wave tubes.)

[9.13] R. M. Bevensee, *Electromagnetic Slow Wave Systems.* New York, NY: John Wiley & Sons, Inc., 1964.

PROBLEMS

9.1. A rectangular guide is loaded at regular intervals by symmetrical inductive diaphragms (see Fig. P9.1).

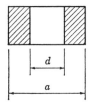
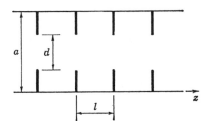

Fig. P9.1.

Derive the eigenvalue equation for the characteristic propagation constant for a periodic wave propagating along the loaded guide. Plot the k_0–β_c diagram for $a = 0.9$ inch, $d = 0.2$ inch, $l = 1.5$ centimeters. Use the approximate formula for B given in Chapter 8.

9.2. A rectangular guide is loaded by dielectric blocks of thickness t and spacing l and having a relative dielectric constant κ (see Fig. P9.2). Derive the eigenvalue equation for the propagation constant of a periodic wave. Also find

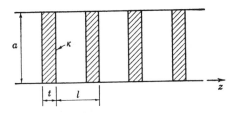

Fig. P9.2.

the required parameters of a matching section which will match the empty guide to the periodic-loaded guide. Show that, for small spacings, the loaded guide behaves the same as a guide filled with a homogeneous dielectric material having a relative dielectric constant $\kappa_e = 1 + (\kappa - 1)t/l$. Assume H_{10} mode propagation.

9.3. A rectangular guide is loaded by diaphragms at regular intervals l, each diaphragm containing a small centered circular aperture of radius r_0 (see Fig. P9.3). Derive the eigenvalue equation for the propagation constant

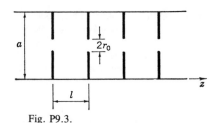

Fig. P9.3.

of a periodic wave. Use the small-aperture theory of Chapter 7 to obtain an expression for the shunt susceptance of each diaphragm. For $a = 0.9$ inch, $\lambda_0 = 3.14$ centimeters, $r_0 = 0.2$ centimeter, find the spacing l corresponding to the center of the passband. Plot the value of the propagation constant as a function of k_0 for values of k_0 extending through several stopbands and passbands. Note that the passbands are very narrow and separated by stopbands with relatively high attenuation. The periodic-loaded guide has the properties of a narrow band filter.

Answer: The center of the passband is given by $\beta_0 l = \pi/2 + \tan^{-1}(B/2)$, where β_0 is the propagation constant of the H_{10} mode.

9.4. A structure with properties similar to those of the tape helix is the parallel-plate transmission line, which has infinite conductivity in the direction of the unit vector $\mathbf{a}_{\parallel}$ and zero conductivity in the direction perpendicular to this, as illustrated in Fig. P9.4. Obtain solutions to Maxwell's equations for a structure of this type. Note that

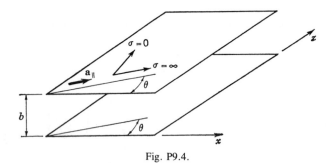

Fig. P9.4.

each mode is a combination of both an E and an H mode. Determine the eigenvalue equation for the propagation constants. Carry through a similar analysis when $\mathbf{a}_{\parallel}$ makes an angle θ with the x axis for the upper plate and an angle $-\theta$ for the lower plate.

9.5. Consider two rectangular guides propagating H_{10} modes and coupled together periodically by small circular apertures of radius r_0, spacing l, and located in the center of the common sidewall (see Fig. P9.5). Obtain the

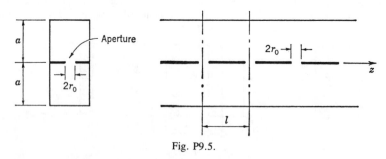

Fig. P9.5.

eigenvalue equation for the dominant modes of propagation which may exist in the composite guide. Show that the two normal modes for the composite guide are modes with an electric field which is symmetrical for one mode and antisymmetrical for the other about the common sidewall. Superimpose the two normal modes such that at one transverse plane in one guide the field is zero, while in the other guide it is maximum. Note that, as this field propagates along the composite guide, the power oscillates back and forth between the two separate guides. The phenomenon is similar to that of the mechanical one of two coupled pendulums.

9.6. A circular guide is loaded by dielectric washers of thickness t, inner radius a, and spacing l (see Fig. P9.6).

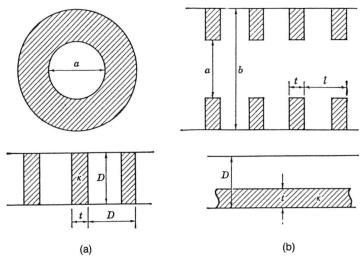

(a)   (b)

Fig. P9.6.

The dielectric constant of the material is κ. Assume that the spacing l is much smaller than a wavelength, so that the loaded portion of the guide $a \le r \le b$ may be assumed to be filled with an anisotropic dielectric material with a dielectric constant $\kappa_1 = 1 + (\kappa - 1)(t/l)$ along a transverse direction, and $\kappa_2 = (1 - ((\kappa - 1)/\kappa)(t/l))^{-1}$ along the z direction. Obtain the solutions for circularly symmetric modes (no variation with θ) in the composite structure.

The equivalent dielectric constants are obtained from the static solution for the parallel-plate capacitor loaded as illustrated. Show that the capacitance per unit area is $C = \kappa_e \epsilon_0 / D$, where the equivalent dielectric constant κ_e is equal to $1 + (\kappa - 1)t/D$ for arrangement (a) and to $(1 - ((\kappa - 1)/\kappa)(t/D))^{-1}$ for arrangement (b).

9.7. In (9) show that $\Delta = 1$. Begin by proving that the Wronskian determinant $W = \psi_1 \psi_2' - \psi_2 \psi_1'$ is equal to a constant. From Eqs. (2) the result

$$\psi_1(z + p)\psi_2'(z + p) - \psi_1'(z + p)\psi_2(z + p) = (\alpha_{11}\alpha_{22} - \alpha_{12}\alpha_{21})[\psi_1(z)\psi_2'(z) - \psi_1'(z)\psi_2(z)]$$

may be obtained. Since the Wronskian is constant, it follows that $\Delta = 1$. To prove that W is a constant, use the differential equation (1) to obtain

$$\psi_2\psi_1'' - \psi_1\psi_2'' = 0$$

where the double prime means differentiation with respect to z. By integrating once by parts the desired result is obtained.

10

Integral Transform and Function-Theoretic Techniques

There exists a class of two-part boundary-value problems which may be rigorously solved by means of integral transforms such as the Fourier transform or its equivalent, the two-sided Laplace transform. Among the problems that fall into this class are:

1. radiation of acoustic or electromagnetic waves from an infinitely thin circular guide which occupies the half space $z \leq 0$;
2. radiation from two parallel plates of infinite width and extending over $-\infty \leq z \leq 0$;
3. diffraction by a perfectly conducting half plane;
4. reflection and transmission of waves at the interface between free space and an infinite array of uniformly spaced, infinitely thin parallel plates;
5. reflection at the junction of a uniform guide and a similar guide having a resistive wall for $z \geq 0$;
6. bifurcated waveguides, e.g., a rectangular guide bifurcated or divided into two uniform regions for $z \geq 0$ by an infinitely thin conducting wall; and
7. a semidiaphragm of the capacitive or inductive type in a rectangular guide.

All the above problems are characterized by the existence of two uniform regions, one for $z \leq 0$ and the other for $z \geq 0$. At the boundary plane $z = 0$, the only discontinuity in the medium is along an infinitely thin contour such as the circular edge of a circular guide or the straight-line edge of a half plane. The solutions to all the above problems may be formulated as an integral equation of the Wiener–Hopf type [10.1]. The solution to this type of integral equation may be obtained by means of an application of Fourier or bilateral Laplace transforms, together with certain function-theoretic considerations. For those problems which have a solution consisting of an infinite series of discrete eigenfunctions with discrete eigenvalues, two other closely related methods of solution are also available. For one the fields on either side of the interface $z = 0$ are expanded into an infinite series of the proper eigenfunctions for each region. Imposition of the continuity conditions for the field components at the interface leads to an infinite set of linear equations for the unknown amplitude coefficients. A solution to the system of equations which arises may be obtained by a contour-integration technique. The other method of solution is more closely related to the formal method of Wiener and Hopf, in that the solution is formulated in terms of a suitable transform. From known properties of the solution in each region it is possible to construct directly the transform function that will solve the problem [10.2].

The concepts and techniques involved will be developed by considering a number of specific examples. The first example to be considered is a relatively simple electrostatic problem, which will serve to illustrate most of the important basic ideas involved, without the obscuring effects of too many complicated mathematical manipulations. This will be followed by the solution

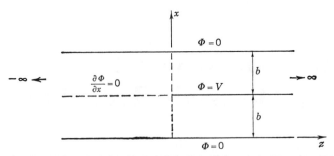

Fig. 10.1. Bifurcation of a parallel-plate region.

of a series of more elaborate problems. In addition to the problems solved here, references are given at the end of the chapter, where solutions to several other typical problems may be found as well as general discussions on the theory.

10.1. AN ELECTROSTATIC PROBLEM

Consider the two-dimensional electrostatic problem illustrated in Fig. 10.1. The two infinite outer plates are kept at zero potential. The region $z \geq 0$ is divided or bifurcated into two identical regions by an infinitely thin conducting plate held at a potential V. We require a solution for the electrostatic potential function $\Phi(x, z)$ which vanishes on the outer plates, reduces to V on the center plate for $z \geq 0$, and is regular at infinity. The solution for Φ will be obtained by the following three methods:

1. solution of an infinite set of linear algebraic equations,
2. direct construction of a transform function which will provide the solution for Φ upon inversion, and
3. solution of a "Wiener–Hopf" integral equation.

A rigorous solution may be readily found by means of a conformal transformation also, but this method is not under consideration here.

The potential Φ is a solution of Laplace's equation

$$\frac{\partial^2 \Phi}{\partial x^2} + \frac{\partial^2 \Phi}{\partial z^2} = 0.$$

In view of the symmetry involved, $\partial \Phi / \partial n = \partial \Phi / \partial x = 0$ for $x = b$, $-\infty \leq z \leq 0$. It is only necessary to solve the reduced problem in the region $x \leq b$, since, from symmetry considerations, $\Phi(2b - x)$ is the solution for $b \leq x \leq 2b$, if $\Phi(x)$ is the solution for $0 \leq x \leq b$. The potential Φ satisfies the boundary conditions

$$
\begin{aligned}
\Phi &= 0 & x &= 0 & &\text{all } z \\
\Phi &= V & x &= b & &z \geq 0 \\
\frac{\partial \Phi}{\partial x} &= 0 & x &= b & &z \leq 0 \\
\Phi &\to 0 & & & &z \to -\infty \\
\Phi &\to \frac{Vx}{b} & & & &z \to +\infty.
\end{aligned}
$$

In addition, at the edge of the bifurcation, Φ is finite, but $\partial\Phi/\partial r$ is asymptotic to $r^{-1/2}$, where r is the radial distance from the edge; that is, $r = [z^2 + (x - b)^2]^{1/2}$.

Method 1 (Residue-Calculus Method)

A suitable eigenfunction expansion of Φ in the region $z \leq 0$ is

$$\Phi(x, z) = \sum_{n=1}^{\infty} a_n \sin \frac{(2n - 1)\pi x}{2b} e^{(2n-1)\pi z/2b} \tag{1a}$$

while, for $z \geq 0$, a suitable expansion of Φ is

$$\Phi(x, z) = \frac{Vx}{b} + \sum_{m=1}^{\infty} b_m \sin \frac{m\pi x}{b} e^{-m\pi z/b}. \tag{1b}$$

In the above equations, a_n and b_m are unknown amplitude coefficients. At $z = 0$, both Φ and $\partial\Phi/\partial z$ must be continuous; hence,

$$\sum_{n=1}^{\infty} a_n \sin \frac{2n - 1}{2b}\pi x = \frac{Vx}{b} + \sum_{m=1}^{\infty} b_m \sin \frac{m\pi x}{b} \tag{2a}$$

$$\sum_{n=1}^{\infty} \frac{(2n - 1)\pi}{2b} a_n \sin \frac{2n - 1}{2b}\pi x = -\sum_{m=1}^{\infty} \frac{m\pi}{b} b_m \sin \frac{m\pi x}{b}. \tag{2b}$$

The linear term Vx/b may be developed into a Fourier series in terms of either the $\sin[(2n - 1)\pi x/2b]$ functions or the $\sin(m\pi x/b)$ functions. An expansion in terms of the first set is preferable since the resultant series will be uniformly convergent at $x = b$. By conventional Fourier analysis, it is readily found that

$$\frac{Vx}{b} = \sum_{n=1}^{\infty} \frac{2V}{b^2} \frac{\sin[(2n - 1)\pi/2]}{[(2n - 1)\pi/2b]^2} \sin \frac{2n - 1}{2b}\pi x. \tag{3}$$

A system of equations to determine the unknown coefficients b_m may be obtained as follows. Multiplying (2a) by $\sin[(2n - 1)\pi x/2b]$ and integrating over $0 \leq x \leq b$ gives

$$a_n \frac{b}{2} - \frac{V \sin[(2n - 1)\pi/2]}{b[(2n - 1)\pi/2b]^2} = \sum_{m=1}^{\infty} b_m \frac{m\pi}{b} \frac{\cos m\pi \sin[(2n - 1)\pi/2]}{[(2n - 1)\pi/2b]^2 - (m\pi/b)^2} \tag{4a}$$

and similarly, from (2b),

$$\frac{(2n - 1)\pi}{2b} a_n \frac{b}{2} = -\sum_{m=1}^{\infty} b_m \left(\frac{m\pi}{b}\right)^2 \frac{\cos m\pi \sin[(2n - 1)\pi/2]}{[(2n - 1)\pi/2b]^2 - (m\pi/b)^2}. \tag{4b}$$

For convenience, new coefficients B_m given by $B_m = b_m(m\pi/b) \cos m\pi$ will be introduced. Multiplying (4a) by $-(2n - 1)\pi/2b$ and adding the resultant equation to (4b), the coefficients

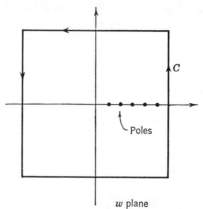

Fig. 10.2. Illustration of integration contour.

a_n are eliminated, and we get

$$\sum_{m=1}^{\infty} \frac{B_m}{m\pi/b - (2n-1)\pi/2b} - \frac{V}{b} \frac{2b}{(2n-1)\pi} = 0, \qquad n = 1, 2, 3, \ldots . \qquad (5)$$

A system of equations of the above form may be solved by constructing a function $f(w)$ of the complex variable w which will generate a set of equations formally identical with (5) when integrated around a suitable contour in the complex w plane.

In the Mathematical Appendix, series of the type $\sum_{-\infty}^{\infty}(e^{-jn\theta}/(n^2+a^2))$ are summed. The method used is to replace the sum by a contour integral

$$\oint_C \frac{e^{-jw\theta} \, dw}{(w^2+a^2)(e^{j2w\pi}-1)}$$

where C is a rectangular contour, as in Fig. 10.2, which recedes to infinity. The factor $e^{j2w\pi}-1$ is introduced to produce poles at $w = \pm n$; $n = 0, 1, 2, \ldots .$ Since the contour integral vanishes, the sum of the residues of the integrand equals zero. The residues at the simple poles $\pm n$ give rise to the series that is to be summed. The sum of the series is then given in terms of the residues at $w = \pm ja$.

In the set of equations (5) we know the sum of the series, but we do not know the form of each term, i.e., the coefficients B_m. We may, however, construct a suitable contour integral which will generate the required system of equations. Consider a contour integral of the form

$$\oint_C \frac{f(w) \, dw}{w - (2n-1)\pi/2b}, \qquad n = 1, 2, \ldots . \qquad (6)$$

To generate a series of the form (5), $f(w)$ must have simple poles at $w = m\pi/b$; $m = 1, 2, \ldots .$ In addition, $f(w)$ must have zeros at $w = (2n-1)\pi/2b$ in order to cancel the corresponding simple poles in the integrand. To produce a term of the form $(V/b)2b/(2n-1)\pi$, we must also have a simple pole at $w = 0$. The contour C is chosen as a rectangular contour which passes between the poles of the integrand. If, as C recedes to infinity, the contour integral vanishes, then the sum of the residues must also be equal to zero. With the assumed simple poles the residue expansion of the integral gives

$$\sum_{m=1}^{\infty} \frac{r(m\pi/b)}{m\pi/b - (2n-1)\pi/2b} - \frac{r(0)2b}{(2n-1)\pi} = 0, \qquad n = 1, 2, \ldots \tag{7}$$

where $r(m\pi/b)$ is the residue of $f(w)$ at the simple pole $w = m\pi/b$, that is,

$$r\left(\frac{m\pi}{b}\right) = \lim_{w \to m\pi/b} \left(w - \frac{m\pi}{b}\right) f(w).$$

The system of equations in (7) is formally the same as that in (5) if we identify B_m with $r(m\pi/b)$ and normalize $f(w)$ such that $r(0) = V/b$. Provided the function $f(w)$ can be constructed, the solution to (5) may be found. In view of the required poles and zeros, it is seen that a suitable form for $f(w)$ is

$$f(w) = \frac{p(w)\prod_{n=1}^{\infty}\left[1 - \frac{w2b}{(2n-1)\pi}\right] e^{2wb/(2n-1)\pi}}{w\prod_{m=1}^{\infty}\left(1 - \frac{wb}{m\pi}\right) e^{wb/m\pi}}. \tag{8}$$

In (8), $p(w)$ is an integral function, i.e., a function with no poles in the finite w plane, which is still to be determined. The exponential factors occurring in the infinite products are required in order to ensure the uniform convergence of these products. The necessity of these factors may be seen from a consideration of the logarithmic derivative of the infinite product. Let

$$F(w) = \prod_{m=1}^{\infty}\left(1 - \frac{w}{m}\right) e^{w/m}$$

and form the logarithmic derivative $F'(w)/F(w)$. We have

$$\frac{F'}{F} = \frac{d(\ln F)}{dw} = \sum_{m=1}^{\infty}\left(\frac{1}{w-m} + \frac{1}{m}\right).$$

If the exponential factors $e^{w/m}$ were not included in the infinite product, the terms $1/m$ would not arise, and the infinite series for F'/F would not be uniformly convergent. By including these factors, the series becomes uniformly convergent, and may be integrated term by term to yield the infinite-product function $F(w)$.

To determine $p(w)$, we must consider the asymptotic form of $f(w)$ as w approaches infinity. The function $p(w)$ must be chosen so that $f(w)$ is of algebraic growth at infinity, and, in particular, so that $f(w) \sim w^{-\alpha}$ with $\alpha > 1$, in order for the contour integral to vanish. In order to examine the behavior of $f(w)$ at infinity, it will be convenient to rewrite $f(w)$ in terms of gamma functions. To obtain the results we want, the following infinite-product expansions of the gamma function $\Gamma(u)$ and the $\sin u\pi$ function are required:

$$\Gamma(u) = \frac{e^{-\gamma u}}{u}\left[\prod_{n=1}^{\infty}\left(1 + \frac{u}{n}\right) e^{-u/n}\right]^{-1} \tag{9a}$$

where γ is Euler's constant and is equal to 0.5772, and

$$\sin u\pi = u\pi \prod_{n=1}^{\infty} \left(1 + \frac{u}{n}\right) e^{-u/n} \left(1 - \frac{u}{n}\right) e^{u/n} = u\pi \prod_{n=1}^{\infty} \left(1 - \frac{u^2}{n^2}\right). \tag{9b}$$

From the above results we may obtain the relation

$$\Gamma(u)\Gamma(-u) = -\frac{\pi}{u \sin \pi u}. \tag{10}$$

The infinite product in the numerator of $f(w)$ in (8) may be rewritten as follows:

$$\prod_{n=1}^{\infty} \left[1 - \frac{2wb}{(2n-1)\pi}\right] e^{2wb/(2n-1)\pi} = \prod_{n=1,3,\ldots}^{\infty} \left(1 - \frac{2wb}{n\pi}\right) e^{2wb/n\pi}$$

$$= \frac{\displaystyle\prod_{n=1}^{\infty} \left(1 - \frac{2wb}{n\pi}\right) e^{2wb/n\pi}}{\displaystyle\prod_{n=1}^{\infty} \left(1 - \frac{wb}{n\pi}\right) e^{wb/n\pi}}.$$

Using the result $\Gamma(u)\Gamma(-u) = -\pi/(u \sin \pi u)$, together with the infinite-product expansion of the gamma function, the above infinite product may be easily shown to be equal to

$$e^{\gamma wb/\pi} \frac{\sin 2wb}{\sin wb} \frac{\Gamma(2wb/\pi)}{\Gamma(wb/\pi)}.$$

Treating the infinite product in the denominator of $f(w)$ in (8) in a similar way, we finally arrive at the following equivalent expression for $f(w)$:

$$f(w) = \frac{2\pi p(w)}{w} \cot wb \frac{\Gamma(2wb/\pi)}{[\Gamma(wb/\pi)]^2}. \tag{11}$$

As w tends to infinity, the gamma function has the asymptotic value

$$\Gamma(w) \sim \sqrt{2\pi} e^{w \ln w} w^{-1/2} e^{-w}$$

which is valid for all w except in the vicinity of the negative-real axis, where the gamma function has poles at the negative integer values of w. Equation (10) may be used to obtain the asymptotic value of $\Gamma(w)$ on the negative-real axis. Utilizing the above results shows that the asymptotic form of $f(w)$ is

$$f(w) \sim \left(\frac{b}{w}\right)^{1/2} p(w) \cot wb \, e^{(2wb/\pi) \ln 2}. \tag{12}$$

To ensure algebraic growth at infinity, we must choose $p(w)$ equal to $e^{-(2wb/\pi) \ln 2}$ times a polynomial in w. However, in order for the contour integral to vanish, the polynomial must be a constant chosen so that $r(0) = V/b$. For the present we will let V be determined by $r(0)$

and choose $f(w)$ as follows:

$$f(w) = \frac{2\pi}{w} e^{-(2wb/\pi)\ln 2} \cot wb \frac{\Gamma(2wb/\pi)}{[\Gamma(wb/\pi)]^2}. \qquad (13)$$

The only poles that $f(w)$ has are the poles of the cotangent function at $w = m\pi/b$; $m = 0, 1, 2, \ldots$. All the other poles are canceled by zeros arising from the gamma functions. For w approaching $m\pi/b$, we have

$$\lim_{w \to m\pi/b} \left(w - \frac{m\pi}{b} \right) \cot wb = \frac{1}{b}.$$

The residue expansion (7) yields the following solution for the coefficients B_m and b_m:

$$B_m = b_m \frac{m\pi}{b} \cos m\pi = \frac{2e^{-2m\ln 2}}{m} \frac{\Gamma(2m)}{[\Gamma(m)]^2}, \qquad m = 1, 2, \ldots. \qquad (14)$$

For m very large, the coefficients B_m are approximately given by

$$B_m \sim (m\pi)^{-1/2}. \qquad (15)$$

This result follows directly from the asymptotic form of $f(w)$ given by (12).

The potential V of the center plate is given by

$$V = br(0) = b \qquad (16)$$

since $r(0)$ is equal to unity. For any other value of potential, it is necessary only to multiply the whole solution by a constant.

The solution for the coefficients a_n may be obtained from the residue expansion of the following contour integral:

$$\oint_C \frac{f(w)\,dw}{w + (2n-1)\pi/2b}, \qquad n = 1, 2, \ldots. \qquad (17)$$

We first multiply (4a) by $(2n-1)\pi/2b$, and add the resulting equation to (4b) to obtain

$$\frac{-(2n-1)\pi}{2\sin[(2n-1)\pi/2]} a_n + \frac{V}{b} \frac{2b}{(2n-1)\pi} + \sum_{m=1}^{\infty} \frac{B_m}{m\pi/b + (2n-1)\pi/2b} = 0,$$

$$n = 1, 2, \ldots. \qquad (18)$$

The contour integral (17) vanishes as C recedes to infinity, and, hence, the residue expansion of the integral, in terms of the residues at the poles $w = m\pi/b$, $m = 0, 1, 2, \ldots$, and $w = -(2n-1)\pi/2b$, gives

$$f\left(-\frac{2n-1}{2b}\pi \right) + \frac{r(0)2b}{(2n-1)\pi} + \sum_{m=1}^{\infty} \frac{r(m\pi/b)}{m\pi/b + (2n-1)\pi/2b} = 0, \qquad n = 1, 2, \ldots. \qquad (19)$$

Comparing this result with (18) shows that the solution for a_n is

$$a_n = -\frac{2 \sin[(2n-1)\pi/2]}{(2n-1)\pi} f\left(-\frac{2n-1}{2b}\pi\right). \tag{20}$$

From (8), $f(-(2n-1)\pi/2b)$ is found to be given by

$$f\left(-\frac{2n-1}{2b}\pi\right) = -\frac{b}{2\pi}e^{(2n-1)\ln 2}\frac{[\Gamma(n-1/2)]^2}{\Gamma(2n-1)} \tag{21}$$

upon using the infinite-product expansion for the gamma functions. For n large, we obtain the approximate result

$$f\left(-\frac{2n-1}{2b}\pi\right) \sim -\frac{b}{\pi}\left(\frac{2\pi}{2n-1}\right)^{1/2}$$

by using the asymptotic expansion of the gamma function. Hence, the coefficients a_n have the asymptotic value

$$a_n \sim \frac{2b}{\pi}\sqrt{\frac{2}{\pi}}\frac{\sin[(2n-1)\pi/2]}{(2n-1)^{3/2}}. \tag{22}$$

From the asymptotic value of the coefficients a_n and b_n as n tends to infinity, we may readily verify that $\partial\Phi/\partial z$ and $\partial\Phi/\partial x$ have the proper singularity at the edge of the bifurcation. For example, the series expansion for $\partial\Phi/\partial z$ for $z \geq 0$, $x = b$, has the same asymptotic behavior for z approaching zero as the series

$$\sum_{n=1}^{\infty}\frac{e^{-n\pi|z|/b}}{n^{1/2}}.$$

Using the following result,

$$\int_1^{\infty}e^{-\pi z u/b}u^{-1/2}\,du = \left(\frac{b}{z}\right)^{1/2} - 2 + \frac{2\pi z}{3b} + \cdots < \sum_{n=1}^{\infty}e^{-n\pi z/b}n^{-1/2}$$

$$< \int_0^{\infty}e^{-\pi z u/b}u^{-1/2}\,du = \left(\frac{b}{z}\right)^{1/2}$$

shows that the series, and hence $\partial\Phi/\partial z$, is asymptotic to $|z|^{-1/2}$ as z tends to zero, i.e., as the edge of the bifurcation is approached. This edge condition would not hold if $f(w)$, and hence na_n and nb_n, were not asymptotic to $w^{-1/2}$ as w approached infinity. The behavior of $f(w)$ at infinity governs the behavior of Φ at $z = 0$. When the solution to our problem has been obtained by means of an integral transform, the relation between the behavior of Φ at the origin and that of $f(w)$ at infinity will be seen to be a natural consequence arising from the basic properties of Fourier and Laplace transforms.

The above method of solution appears to have been applied only to equations of the form given in (4). For a system of equations of a more elaborate form, a similar technique could be used, provided a function $f(w)$ could be constructed so as to yield a residue expansion which would be formally identical with the equations to be solved. If each term in the series, apart from a term such as $m\pi/b \pm (2n-1)\pi/2b$ in the denominator, is not representable as a product of two factors depending on n alone and m alone, respectively, it is generally not feasible to construct the function $f(w)$ which will provide the required solution.

Method 2 (Function-Theoretic Method)

The second method of solution to be illustrated is an integral transform solution. As a first step, the variable z is suppressed by taking the bilateral Laplace transform of Φ. A solution for the transform function which satisfies the required boundary conditions and has poles which will yield the known eigenfunction expansion of Φ is then constructed. Finally, the solution for Φ is obtained by inverting the transform and evaluating the inversion integral in terms of its residues. The residue expansion of the integral is the eigenfunction expansion of the potential Φ.

Let $F(x, w)$ be the bilateral Laplace transform of $\Phi(x, z)$, that is,

$$F(x, w) = \int_{-\infty}^{\infty} \Phi(x, z)e^{-wz}\, dz. \tag{23}$$

The transform $F(x, w)$ may be split into two separate parts as follows:

$$F_+(x, w) = \int_0^{\infty} \Phi(x, z)e^{-wz}\, dz \tag{24a}$$

$$F_-(x, w) = \int_{-\infty}^0 \Phi(x, z)e^{-wz}\, dz. \tag{24b}$$

For $z \geq 0$, the potential $\Phi(x, z)$ is asymptotic to Vx/b for z large, and, hence, (24a) converges uniformly for all w in the half plane $\operatorname{Re} w > 0$. Thus (24a) determines F_+ as an analytic function of w in the half plane $\operatorname{Re} w > 0$. All the poles of F_+ must, therefore, lie in the left half plane $\operatorname{Re} w < 0$. We may invert (24a) to get

$$\Phi(x, z) = \frac{1}{2\pi j} \int_{C_+} e^{wz} F_+(x, w)\, dw, \qquad z \geq 0. \tag{25}$$

For z less than zero, (25) is equal to zero, as may be readily seen since, for $z < 0$, the contour may be closed in the right half plane; and, since no poles are enclosed, the contour integral vanishes. In (25), C_+ is a contour running parallel to the imaginary w axis in the right half plane $\operatorname{Re} w > 0$. For $z > 0$, the contour may be closed in the left half plane, and the integral evaluated in terms of its residues at the poles. This residue expansion will give the eigenfunction expansion of Φ for $z > 0$.

For $z < 0$, the potential Φ is asymptotic to $a_1 \sin(\pi x/2b)e^{-\pi|z|/2b}$ for z large and negative. Hence (24b) converges uniformly for all w in the half plane $\operatorname{Re} w < \pi/2b$, and determines F_- as an analytic function of w in that half plane. Inverting (24b) gives

$$\Phi(x, z) = \frac{1}{2\pi j} \int_{C_-} e^{wz} F_-(x, w)\, dw, \qquad z \leq 0 \tag{26}$$

where C_- is a contour parallel to the imaginary w axis and in the half plane $\operatorname{Re} w < \pi/2b$.

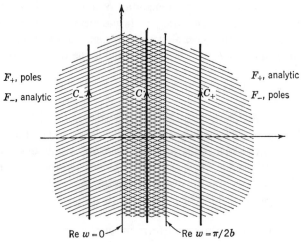

Fig. 10.3. Regions of analyticity for F_+ and F_- and the common inversion contour C.

Since $z < 0$, we may close the contour in the right half plane, and evaluate the integral in terms of its residues at the poles of $F_-(x, w)$. This series gives the eigenfunction expansion of Φ in the region $z \leq 0$. If $z > 0$ in (26), the integral is equal to zero because of the analytic properties of $F_-(x, w)$ in the left half plane.

The two functions F_+ and F_- have a common region of analyticity $0 < \operatorname{Re} w < \pi/2b$, as illustrated in Fig. 10.3. Therefore, we may choose $C_+ = C_- = C$, where C is a common contour within the common strip of analyticity and running parallel to the imaginary w axis. We may therefore invert (23) directly without separating F into two parts, and, hence,

$$\Phi(x, z) = \frac{1}{2\pi j} \int_C e^{wz} F(x, w)\, dw \qquad \text{all } z. \tag{27}$$

The bilateral Laplace transform of the equation

$$\frac{\partial^2 \Phi}{\partial z^2} + \frac{\partial^2 \Phi}{\partial x^2} = 0$$

with respect to the variable z gives

$$\frac{d^2 F(x, w)}{dx^2} + w^2 F(x, w) = 0 \tag{28}$$

valid for $0 < \operatorname{Re} w < \pi/2b$. This result follows by integrating the second partial derivative with respect to z twice by parts. The integrated terms $e^{-wz}\, \partial\Phi/\partial z$, $w\Phi e^{-wz}$ vanish at $z = \pm\infty$ for w in the specified region because of the asymptotic behavior of Φ for z large.

A general solution to (28) which vanishes for $x = 0$ is

$$F = g(w) \sin wx \tag{29}$$

where $g(w)$ is a function of w to be determined. Substituting into the inversion integral (27), we get

$$\Phi(x, z) = \frac{1}{2\pi j} \int_C e^{wz} g(w) \sin wx\, dw. \tag{30}$$

The function g may be determined uniquely from the known properties of Φ.

When z is negative, we may close the contour C in the right-half w plane. In order to obtain the eigenfunction expansion (1a) for Φ, we see that g must have simple poles at $w = (2n - 1)\pi/2b$, $n = 1, 2, \ldots$. For $z \geq 0$, we may close the contour in the left-half w plane. In order for the residue expansion of the integral to yield a solution for Φ of the form given by (1b), the function g must have simple poles at $w = -m\pi/b$, $m = 1, 2, \ldots$, in the left-half w plane. In addition, we require g to have a double pole at the origin, so that $g \sin wx$ will have a simple pole at $w = 0$ and give rise to a term of the form Vx/b. Finally, the asymptotic form of g for w approaching infinity must be such that the function $\Phi'(b, z) = \partial \Phi(b, z)/\partial z$ will be asymptotic to $z^{-1/2}$ as z tends to zero. It is only because our knowledge of Φ is as complete as it is that we are able to construct directly the required transform function $g(w)$. When this is not the case, we have to proceed by the more formal method of Wiener and Hopf, as given in the next section as Method 3.

For $x = b$, we have

$$wg(w) \sin wb = \int_{-\infty}^{\infty} e^{-wz} \Phi'(b, z) \, dz.$$

If we replace wz by u, and dz by du/w, and take the limit as w tends to infinity, we get

$$\lim_{w \to \infty} \frac{1}{w} \int_{-\infty}^{\infty} e^{-u} \Phi'\left(b, \frac{u}{w}\right) du.$$

Interchanging the order of integration and the limiting operation, we get

$$\lim_{w \to \infty} g(w)w \sin wb = \int_{-\infty}^{\infty} e^{-u} \left[\lim_{w \to \infty} \frac{1}{w} \Phi'\left(b, \frac{u}{w}\right)\right] du$$

$$= w^{-(\alpha+1)} \int_{-\infty}^{\infty} a_0 u^\alpha e^{-u} \, du$$

provided the limiting form of $\Phi'(b, z)$ as z tends to zero is $a_0 z^\alpha$. Thus, if $\Phi'(b, z)$ is asymptotic to $z^{-1/2}$ as z tends to zero, the transform of Φ will be asymptotic to $w^{-3/2}$ as w tends to infinity.

In view of the required properties of the function g, it is seen that a suitable transform function is given by

$$g(w) = -\frac{f(-w)}{w \cos wb} \tag{31}$$

where $f(w)$ is the function given by (8) or its equivalent (11). The function $f(-w)$ has poles at $w = -m\pi/b$, $m = 0, 1, 2, \ldots$, and zeros at $w = -(2n - 1)\pi/2b$, $n = 1, 2, \ldots$. Hence, $g(w)$ will have simple poles at $w = -m\pi/b$, $m = 1, 2, \ldots$, and $w = (2n - 1)\pi/2b$, $n = 1, 2, \ldots$, and a double pole at $w = 0$. Furthermore, $f(-w)$ is asymptotic to $w^{-1/2}$, and, hence, the function $g(w)$ given by (31) satisfies all the required conditions. It is interesting to note the close relationship between the function $f(w)$ used to solve the system of algebraic equations (4) and the transform function $g(w)$.

The solution for Φ may now be completed by substituting for g from (31) into (30), and evaluating the inversion integral in terms of its residues in the left-half w plane for $z \geq 0$, and in the right-half w plane for $z \leq 0$. The residue at the origin gives simply x, and this makes the potential of the center plate equal to b. To obtain a potential V, we must multiply g by

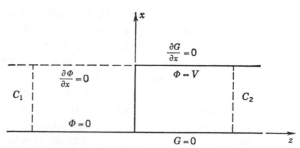

Fig. 10.4. Boundary-value problem for a Green's function.

the constant V/b. The rest of the residue expansion is the same as the series (1a) and (1b), with b_m and a_n given by (14) and (20), respectively. In view of the specified poles for the function $g(w)$, the boundary conditions on Φ for $x = b$ are satisfied for all values of z.

Method 3 (Wiener–Hopf Integral Equation)

The third method of solution to be illustrated is the method of Wiener and Hopf. In this method we find a suitable Green's function, which represents the potential due to a unit line charge at $x = b$, $z = z'$, on the center plate first. The next step is to use Green's theorem to obtain an integral equation for the charge distribution. This integral equation is of the Wiener–Hopf type and may be solved by means of a bilateral Laplace transform (or a Fourier transform). Since the method is somewhat long and involves several basic concepts, it will be advantageous to present the essential steps in the solution first. Having done this, we will proceed to work out the remaining mathematical details.

Referring to Fig. 10.4, the first step is to construct a Green's function $G(x, x', z - z')$ which is a solution of the equation

$$\frac{\partial^2 G}{\partial x^2} + \frac{\partial^2 G}{\partial z^2} = \delta(x - x')\delta(z - z') \tag{32}$$

and satisfies the boundary conditions

$$\begin{aligned} G &= 0 & x &= 0 & \text{all } z \\ \frac{\partial G}{\partial x} &= 0 & x &= b & \text{all } z \\ G &\to 0 & & & \text{as } |z| \to \infty. \end{aligned}$$

This Green's function is a function of the variable $z - z'$. Having determined the Green's function, we next use Green's second identity to obtain

$$\iint\limits_{S} (\Phi \, \nabla^2 G - G \, \nabla^2 \Phi) \, dx \, dz = \oint_C \left(\Phi \frac{\partial G}{\partial n} - G \frac{\partial \Phi}{\partial n} \right) dl$$

where C is a contour consisting of the z axis, the line $x = b$, and the two segments C_1 and C_2 at $z = -\infty, +\infty$, as in Fig. 10.4. The potential Φ is a solution of Laplace's equation, while G is a solution of (32), and, hence, the surface integral becomes simply $\Phi(x', z')$. The contour integral reduces to an integral over z from zero to infinity for $x = b$, by virtue of the boundary conditions satisfied by Φ and G. The integral over the segments C_1 and C_2 vanishes, since both Φ and G tend to zero for $z \to -\infty$, while, for $z \to +\infty$, the potential Φ approaches Vx/b, and G vanishes. Thus, we get

$$\Phi(x', z') = -\int_0^\infty G(b, x', z - z') \frac{\partial \Phi(b, z)}{\partial x} \, dz$$

or, interchanging primed and unprimed coordinates,

$$\Phi(x, z) = -\int_0^\infty G(b, x, z - z') \rho(z') \, dz' \tag{33}$$

where $\rho(z') = \partial \Phi(b, z')/\partial x'$ and is proportional to the charge density on the plate at $x = b$, $z \geq 0$. Since ρ is identically zero for $z' \leq 0$, the integration in (33) may be extended over the whole range $z' = -\infty$ to $z' = +\infty$. Putting $x = b$, we obtain the fundamental integral equation for ρ:

$$\int_{-\infty}^\infty G(z - z') \rho(z') \, dz' = -\Phi(b, z) = \begin{cases} -\Phi_-(z), & z \leq 0 \\ -\Phi_+(z) = -V, & z \geq 0 \end{cases} \tag{34}$$

where $\rho(z') = 0$, $z' < 0$.

By being able to extend the integration from minus to plus infinity, the faltung theorem may be used to obtain a simple expression for the bilateral Laplace transform of (34). The following transform functions are introduced:

$$\mathcal{G}(w) = \int_{-\infty}^\infty e^{-wz} G(z) \, dz \tag{35a}$$

$$Q(w) = Q_+(w) = \int_0^\infty e^{-wz} \rho(z) \, dz \tag{35b}$$

$$\psi_+(w) = \int_0^\infty e^{-wz} \Phi_+(z) \, dz = \int_0^\infty V e^{-wz} \, dz = \frac{V}{w} \tag{35c}$$

$$\psi_-(w) = \int_{-\infty}^0 \Phi_-(z) e^{-wz} \, dz \tag{35d}$$

$$\psi = \psi_+ + \psi_-. \tag{35e}$$

For $|z|$ large, $G(z)$ is asymptotic to $e^{-\pi|z|/2b}$, and, hence, the transform $\mathcal{G}(w)$ is an analytic function for all w in the strip $-(\pi/2b) < \operatorname{Re} w < \pi/2b$, since in this strip the integral (35a) is uniformly convergent. Since Φ_+ is asymptotic to V for $z \to +\infty$, and $\partial \Phi/\partial x'$ is asymptotic to $\partial(Vx'/b)/\partial x' = V/b$, the transforms Q_+ and ψ_+ are analytic for all w in the half plane $\operatorname{Re} w > 0$. As $z \to -\infty$, the potential Φ_- is asymptotic to $e^{-\pi|z|/2b}$, and hence ψ_- is analytic for all w in the half plane $\operatorname{Re} w < \pi/2b$. It is seen that all the transform functions in (35) are analytic in the common strip $0 < \mathrm{Re}\, w < \pi/2b$. We may therefore take the bilateral Laplace transform of (34) to get

$$\mathcal{G}(w)Q_+(w) = -\psi_+(w) - \psi_-(w). \tag{36}$$

A common strip of analyticity is necessary in order for the transform equation (36) to be valid. The faltung theorem states that the transform of an integral of the form

$$\int_{-\infty}^{\infty} G(z - z')\rho(z')\,dz'$$

is the product of the transforms of $G(z)$ and $\rho(z)$, and this result was used to obtain the left-hand side of (36).

The next essential step in the method is to factor and rewrite the transform equation (36) in the form

$$H_+(w) = H_-(w)$$

where H_+ is an analytic function with no poles in the right-half w plane and H_- is an analytic function with no poles in the left-half w plane. In addition, H_+ and H_- have a common strip of analyticity in which they are both equal. An analytic function $h(w)$ may now be defined as equal to H_+ in the right half plane and equal to H_- in the left half plane. Thus, H_- provides the analytic continuation of H_+ into the left half plane. The function $h(w)$, as defined, is free of poles in the whole finite w plane and must therefore be equal to an integral function of w. If also H_+ and H_- are bounded at infinity, then Liouville's theorem states that H_+, H_-, and h are just equal to a constant. At any rate, sufficient knowledge of the properties of the solution is generally available to determine the integral function $h(w)$, which may be a finite polynomial in w in general.

Returning to (36), we factor $\mathcal{G}$ into the form $\mathcal{G}_+/\mathcal{G}_-$, where $\mathcal{G}_+$ has no poles in the half plane $\mathrm{Re}\, w > -\pi/2b$, and $\mathcal{G}_-$ has no poles in the half plane $\mathrm{Re}\, w < \pi/2b$, and also such that both $\mathcal{G}_+$ and $\mathcal{G}_-$ have algebraic growth rather than exponential growth at infinity. We may now rewrite (36) as

$$\begin{aligned}
\mathcal{G}_+ Q_+ &= -\mathcal{G}_-\psi_- - \mathcal{G}_-\psi_+ \\
&= -\mathcal{G}_-\psi_- - \frac{\mathcal{G}_- V}{w}
\end{aligned} \tag{37}$$

upon substituting for ψ_+ from (35c). The left-hand side of (37) is analytic (free of poles) for w in the half plane $\mathrm{Re}\, w > 0$. The function $\mathcal{G}_-\psi_-$ is free of poles in the left half plane $\mathrm{Re}\, w < \pi/2b$. However, $\mathcal{G}_- V/w$ has a pole at the origin, so that the two sides of (37) do not have a common strip of analyticity. To remove this deficiency, we add $\mathcal{G}_-(0)V/w$ to both sides to get

$$\mathcal{G}_+ Q_+ + \mathcal{G}_-(0)\frac{V}{w} = -\mathcal{G}_-\psi_- - [\mathcal{G}_-(w) - \mathcal{G}_-(0)]\frac{V}{w}. \tag{38}$$

In (38), the left-hand side is free of poles for all w in the half plane $\mathrm{Re}\, w > 0$, while the right-hand side is now free of poles for all w in the half plane $\mathrm{Re}\, w < \pi/2b$. Both sides are analytic in the common strip $0 < \mathrm{Re}\, w < \pi/2b$. Together they define an analytic function

$h(w)$ with no poles in the finite w plane. In particular, both $\mathcal{G}_+$ and Q_+ are found to be bounded at infinity, and, hence, both sides of (38) must be equal to a constant A. In fact, we find that $\mathcal{G}_+$ is asymptotic to $w^{-1/2}$, and, since ρ is asymptotic to $z^{-1/2}$ as z tends to zero, Q_+ is asymptotic to $w^{-1/2}$ also as w tends to infinity. From (38) we now get

$$Q_+ = \frac{-\mathcal{G}_-(0)V/w + A}{\mathcal{G}_+(w)}$$

and it is apparent that A must equal zero if Q_+ is to be asymptotic to $w^{-1/2}$. The solution for Q_+ is then

$$Q_+ = -\frac{V}{w}\frac{\mathcal{G}_-(0)}{\mathcal{G}_+(w)}. \tag{39}$$

Having determined the transform Q_+, we return to (36) to obtain the solution for ψ, which is

$$\psi = -\mathcal{G}(w)Q_+(w) = \frac{\mathcal{G}_+}{\mathcal{G}_-}\frac{V}{w}\frac{\mathcal{G}_-(0)}{\mathcal{G}_+} = \frac{V}{w}\frac{\mathcal{G}_-(0)}{\mathcal{G}_-(w)}. \tag{40}$$

The solution for $\Phi(b, z)$ may be obtained by inverting the transform ψ. Again a residue expansion of the inversion integral leads to the series expansion for Φ.

Having outlined the basic concepts and essential steps in the solution, we must now go back and carry through the detailed analysis to obtain the Green's function, its transform, and the factors $\mathcal{G}_+$ and $\mathcal{G}_-$. The Green's function is a solution of (32). Since we require the transform of G, we will take the bilateral Laplace transform of (32) with respect to the variable $z - z'$, since G is a function of $z - z'$, and solve for $\mathcal{G}(w)$ directly. We obtain from (32)

$$\frac{d^2\mathcal{G}(w)}{dx^2} + w^2\mathcal{G}(w) = \delta(x - x') \tag{41}$$

where $\mathcal{G}$ is given by (35a). Two independent solutions to the homogeneous equation $(d^2/dx^2 + w^2)\mathcal{G} = 0$ are

$$\mathcal{G}_1 = \sin wx \qquad \mathcal{G}_2 = \cos w(b - x).$$

These solutions satisfy the required boundary conditions at $x = 0, b$, respectively. The Wronskian determinant is

$$\mathcal{G}_1\frac{d\mathcal{G}_2}{dx} - \mathcal{G}_2\frac{d\mathcal{G}_1}{dx} = -w \cos wb.$$

Using the method of Section 2.4, the solution for $\mathcal{G}$ is found to be given by

$$\mathcal{G} = (-w \cos wb)^{-1}\begin{cases} \sin wx' \cos w(b - x), & x \geq x' \\ \sin wx \cos w(b - x'), & x \leq x' \end{cases} \tag{42}$$

The above solution satisfies the required boundary conditions at $x = 0, b$. At $x' = b$, we get

$$\mathcal{G}(x, b, w) = -(w \cos wb)^{-1} \sin wx. \tag{43}$$

If the transform of (32) was taken with respect to z alone, we would obtain in place of (42) the result $e^{-wz'}\mathcal{G}(x, x', w)$. The inverse transform yields the solution for G in the form

$$G(x, x', z - z') = \frac{1}{2\pi j}\int_C e^{w(z-z')}\mathcal{G}(x, x', w)\,dw \tag{44}$$

which is clearly a function of $z - z'$. Replacing $z - z'$ by a new variable λ and taking the transform of (44) results in the function $\mathcal{G}$ as given by (42).

The transform equation (36) was obtained from the integral equation (34) by an application of the faltung theorem, which states that the Fourier or bilateral Laplace transform of an integral of the faltung (convolution) type is equal to the product of the transforms of the two functions occurring in the integrand. This theorem is proved as follows. Consider the integral

$$I = \int_{-\infty}^{\infty} G(z - z')\rho(z')\,dz'.$$

Replacing $G(z - z')$ by (44) and $\rho(z')$ by

$$\frac{1}{2\pi j}\int_C e^{w_0 z'}Q(w_0)\,dw_0$$

and interchanging the order of integration gives

$$I = \frac{1}{(2\pi j)^2}\int_C\int_C \mathcal{G}(w)Q(w_0)\int_{-\infty}^{\infty} e^{wz}e^{-z'(w-w_0)}\,dz'\,dw\,dw_0.$$

Since

$$\int_{-\infty}^{\infty} e^{-z'(w-w_0)}\,dz' = 2\pi j\delta(w - w_0)$$

we get

$$I = \frac{1}{2\pi j}\int_C\int_C \mathcal{G}(w)Q(w_0)e^{wz}\delta(w - w_0)\,dw\,dw_0$$

$$= \frac{1}{2\pi j}\int_C \mathcal{G}(w)Q(w)e^{wz}\,dw.$$

The bilateral Laplace transform of I is obtained by multiplying by $e^{-\lambda z}$ and integrating over z to get

$$\frac{1}{2\pi j}\int_C\int_{-\infty}^{\infty} \mathcal{G}(w)Q(w)e^{-z(\lambda-w)}\,dz\,dw = \mathcal{G}(\lambda)Q(\lambda).$$

This is the final result which was to be proved.

To obtain (37) and (38) we have to factor

$$\mathcal{G}(w) = \mathcal{G}(b, b, w) = -w^{-1}\tan wb$$

into the form $\mathcal{G}_+/\mathcal{G}_-$, with $\mathcal{G}_+$ analytic for all w in the half plane $\operatorname{Re} w > -\pi/2b$, and $\mathcal{G}_-$

analytic for all w in the half plane $\operatorname{Re} w < \pi/2b$. The required factorization may be carried out by inspection when $\sin wb$ and $\cos wb$ are expressed as infinite products. We have

$$-w^{-1} \tan wb = -b(wb)^{-1} \sin wb(\cos wb)^{-1}$$

$$= -b \frac{\displaystyle\prod_{n=1}^{\infty}\left(1 - \frac{wb}{n\pi}\right) e^{wb/n\pi} \prod_{n=1}^{\infty}\left(1 + \frac{wb}{n\pi}\right) e^{-wb/n\pi}}{\displaystyle\prod_{n=1}^{\infty}\left[1 - \frac{2wb}{(2n-1)\pi}\right] e^{2wb/(2n-1)\pi} \prod_{n=1}^{\infty}\left[1 + \frac{2wb}{(2n-1)\pi}\right] e^{-2wb/(2n-1)\pi}}.$$

$$(45)$$

From this result it is readily seen that suitable choices for $\mathcal{G}_+$ and $\mathcal{G}_-$ are

$$\mathcal{G}_+(w) = -bp(w) \frac{\displaystyle\prod_{n=1}^{\infty}\left(1 + \frac{wb}{n\pi}\right) e^{-wb/n\pi}}{\displaystyle\prod_{n=1}^{\infty}\left[1 + \frac{2wb}{(2n-1)\pi}\right] e^{-2wb/(2n-1)\pi}} \qquad (46a)$$

$$\mathcal{G}_-(w) = p(w) \frac{\displaystyle\prod_{n=1}^{\infty}\left[1 - \frac{2wb}{(2n-1)\pi}\right] e^{2wb/(2n-1)\pi}}{\displaystyle\prod_{n=1}^{\infty}\left(1 - \frac{wb}{n\pi}\right) e^{wb/n\pi}} \qquad (46b)$$

where $p(w)$ is chosen so that $\mathcal{G}_+$ and $\mathcal{G}_-$ have algebraic growth at infinity. By expressing $\mathcal{G}_+$ in terms of gamma functions, we readily find that $\mathcal{G}_+$ is asymptotic to $w^{-1/2} p(w) \exp[(2wb/\pi) \ln 2]$, and, hence, $p(w)$ must be chosen as follows:

$$p(w) = \exp\left(-\frac{2wb}{\pi} \ln 2\right).$$

From (46) we find that $\mathcal{G}_-(0)$ is equal to unity.

From (40) the solution for the transform $\psi(w)$ of $\Phi(b, z)$ may be found. The transform of $\Phi(x, z)$ is given by

$$\psi(w) \frac{\sin wx}{\sin wb}$$

as may be seen from (33), (34), and (43). Comparing (46a) with (8), we see that $\mathcal{G}_+(w) = b/wf(-w)$. Hence the transform of $\Phi(x, z)$ becomes

$$\frac{\psi(w) \sin wx}{\sin wb} = \frac{V}{w} \frac{\mathcal{G}_+}{\mathcal{G}_-} \frac{1}{\mathcal{G}_+} \frac{\sin wx}{\sin wb} = -V \frac{\tan wb}{wb} \frac{\sin wx}{\sin wb} f(-w)$$

$$= -\frac{V}{b} \frac{\sin wx}{w \cos wb} f(-w). \qquad (47)$$

This is the same result as given by (29) and (31), provided V is chosen equal to b in (47), as was done in (31). The eigenfunction expansion of Φ may be readily obtained from the residue expansion of the inversion integral for $\Phi(x, z)$, that is, from

$$\Phi(x, z) = \frac{1}{2\pi j} \int_C e^{wz} \frac{\psi(w) \sin wx}{\sin wb} \, dw. \tag{48}$$

It is apparent that although the three methods of solution which have been presented are distinct, there is a rather close relationship among them. For problems of the above type, i.e., those whose solution consists of an expansion in terms of a discrete set of eigenfunctions, the second method based on a direct construction of the required transform function is the simplest and most direct. However, this method does demand a knowledge of the infinite set of eigenfunctions and their eigenvalues required for the complete representation of the solution. This knowledge is usually available, and, hence, the second method of solution turns out to be a very useful one for many problems of practical interest in connection with waveguides.

Further Comments on the Wiener–Hopf Method

The inhomogeneous Wiener–Hopf equation is an equation of the form

$$\Phi(z) = v(z) + \int_0^\infty G(z - z')\Phi(z') \, dz' \tag{49}$$

where $v(z)$ is a known function and the kernel or Green's function $G(z - z')$ is a function of $z - z'$ and is also known. By introducing the functions

$$\Phi_+(z) = \begin{cases} \Phi(z), & z > 0 \\ 0, & z < 0 \end{cases}$$

$$\Phi_-(z) = \begin{cases} 0, & z > 0 \\ \Phi(z), & z < 0 \end{cases}$$

we may extend the integration over all z' to get

$$\Phi = \Phi_+ + \Phi_- = v + \int_{-\infty}^\infty G(z - z')\Phi_+(z') \, dz'. \tag{50}$$

Provided we can demonstrate that the following transforms have a common strip of analyticity, we may take the bilateral Laplace transform of (50):

$$\psi_+ = \int_0^\infty e^{-wz}\Phi_+(z) \, dz$$

$$\psi_- = \int_{-\infty}^0 e^{-wz}\Phi_-(z) \, dz$$

$$\mathcal{G} = \int_{-\infty}^\infty e^{-wz}G(z) \, dz$$

$$V_+ = \int_0^\infty v_+(z)e^{-wz}\,dz$$

$$V_- = \int_{-\infty}^0 v(z)e^{-wz}\,dz.$$

Using the faltung theorem gives

$$\psi_+ + \psi_- = V_+ + V_- + \mathcal{G}\psi_+$$

or

$$(1 - \mathcal{G})\psi_+ - V_+ = V_- - \psi_-. \tag{51}$$

The next step requires a factorization of $1 - \mathcal{G}$ to give

$$1 - \mathcal{G} = \frac{H_+(W)}{H_-(W)}$$

where H_+ is analytic in a right half plane and H_- is analytic in a left half plane. We now obtain from (51)

$$H_+\psi_+ - V_+H_- = H_-(V_- - \psi_-). \tag{52}$$

As a next step, we must factor the known function V_+H_- into the sum of two factors S_+ and S_-, where S_+ is analytic in a right half plane and S_- is analytic in a left half plane. In place of (52), we now have

$$H_+\psi_+ - S_+ = H_-(V_- - \psi_-) + S_-. \tag{53}$$

The left-hand side is analytic in a right half plane, while the right-hand side is analytic in a left half plane. In addition, if the two sides have a common strip of analyticity, then $H_+\psi_+ - S_+$ provides the analytic continuation of $H_-(V_- - \psi_-) + S_-$ into the right half plane. Together they define an integral function $h(w)$, whose form can generally be determined from the known asymptotic properties of the transforms. Having determined $h(w)$, we may solve for ψ_+ to obtain

$$\psi_+ = \frac{h + S_+}{H_+}. \tag{54}$$

The solution for ψ may now be found from (51) and is

$$\psi = V + \mathcal{G}\frac{h + S_+}{H_+}. \tag{55}$$

The solution for Φ is obtained from the inversion of the transform ψ.

The solution for the homogeneous Wiener–Hopf equation ($v = 0$) may be obtained in a similar way. Other equations of somewhat different form may be solved in an analogous manner also. In fact, the integral equation (34) differs from the type (49), but could be solved since the function $\Phi_+(z)$ is known.

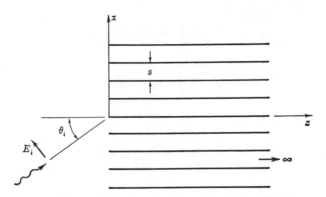

Fig. 10.5. An infinite array of equispaced parallel metallic plates.

The most difficult task in the solution is carrying out the factorization of the terms $1 - \mathcal{G}$ and $V_+ H_-$. This can only be done by inspection for certain cases. Fortunately, however, a general method of carrying out the required factorization does exist [10.21, sect. 8.5]. This method is based on the Cauchy integral representation leading to the Laurent expansion of a function which is analytic within a circular annulus or strip. The technique is described in the Mathematical Appendix.

10.2. An Infinite Array of Parallel Metallic Plates

A knowledge of the reflection and transmission coefficients at the interface between free space and an infinite array of parallel equispaced metallic plates as illustrated in Fig. 10.5 is of importance in at least three applications. These are:

1. parallel-plate microwave lens medium,
2. corrugated-plane surface waveguides for antenna applications, and
3. strip artificial dielectric medium.

In the first application, knowledge of the reflection and transmission coefficients for an incident TEM wave is all that is required. The solution for the infinite array of parallel equispaced metallic plates has been given by Carlson and Heins for both polarizations, and is based on the Wiener–Hopf method. The solution to be presented here is that based on a direct construction of a suitable transform function.

If a short-circuiting plane is located at $z = l$, the resulting structure is a corrugated plane, along which surface waves may be guided in the x direction. The surface wave is obtained by letting the angle of incidence θ_i become complex. The propagation constant in the x direction may be determined by a simple transmission-line analysis, once the equivalent-circuit parameters of the interface are known. If the spacing s is small compared with a wavelength, and the short-circuit position l is large compared with s, then only the parameters of the dominant mode are required in order to analyze the properties of the structure. When these conditions are not satisfied, one or more higher order modes must be taken into account as well, since the least attenuated of these will have a field which interacts appreciably with the short circuit.

The two-dimensional strip-type artificial dielectric medium will be treated in Chapter 12, so we will postpone all comments on this structure until later.

For the parallel-plate medium illustrated in Fig. 10.5, it is assumed that each plate is infinitely thin and perfectly conducting. Only the case of a parallel-polarized incident TEM wave, with the plane of incidence restricted to the xz plane, will be considered. The spacing s is limited to less than $\lambda_0/2$, so only the dominant modes on the two sides of the interface propagate. As a consequence of these assumptions, the field may be derived from a single scalar function $\Phi(x, z)$, which is proportional to the y component of the magnetic field. The only field components present are H_y, E_x, E_z, and these are given by

$$H_y = \Phi \quad E_x = j\frac{Z_0}{k_0}\frac{\partial \Phi}{\partial z} \quad E_z = -j\frac{Z_0}{k_0}\frac{\partial \Phi}{\partial x}.$$

Let the incident magnetic field for $z < 0$ be

$$\Phi_i = A_0 e^{-jhx-\Gamma_0 z} \tag{56a}$$

where $h = k_0 \sin \theta_i$, $\Gamma_0 = jk_0 \cos \theta_i$. The dominant plus higher order reflected modes for $z < 0$ are given by

$$\Phi_r = -R_1 A_0 e^{-jhx+\Gamma_0 z} + \sum_{n=-\infty}^{\infty}{}' B_n e^{-j(h+2n\pi/s)x+\Gamma_n z} \tag{56b}$$

where $\Gamma_n^2 = (h+2n\pi/s)^2 - k_0^2$, and R_1 is the reflection coefficient. A negative sign is introduced in front of R_1 in (56b) so that R_1 will be the usual reflection coefficient for the transverse electric field E_x. The form of solution given by (56) follows from a consideration of the periodic nature of the structure and the condition that Φ be a solution of the two-dimensional scalar Helmholtz equation. The incident field has a progressive phase delay along the x axis according to the factor e^{-jhx}. The scattered field must have this same progressive phase delay, and hence can be represented as the product of the factor e^{-jhx} and a periodic function of x with period s, in accordance with Floquet's theorem. The prime on the summation in (56b) means omission of the term $n = 0$, which has been separated out.

In the region $z > 0$, the solutions for Φ are the usual waveguide modes (E modes), and hence

$$\Phi_t = \sum_{n=0}^{\infty} C_n \cos \frac{n\pi x}{s} e^{-\gamma_n z}, \qquad 0 \le x \le s. \tag{57}$$

For any other range of x, the solution is given by

$$\Phi_t(x + ms, z) = e^{-jhms} \Phi_t(x, z)$$

where m is any integer. In (57), γ_n is given by $\gamma_n^2 = (n\pi/s)^2 - k_0^2$. The coefficient C_0 is equal to $T_{12} e^{-jhs/2} A_0 \Gamma_0/\gamma_0$, where T_{12} is the transverse electric field transmission coefficient. Since the phase of the incident wave varies with x, a phase reference point $x = s/2$ has been chosen in the definition of T_{12}.

The total field is a solution of the equation

$$\frac{\partial^2 \Phi}{\partial z^2} + \frac{\partial^2 \Phi}{\partial x^2} + k_0^2 \Phi = 0 \tag{58}$$

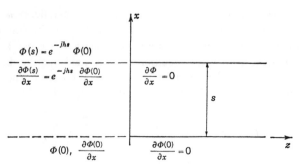

Fig. 10.6. Reduced boundary-value problem.

and satisfies the periodicity condition

$$e^{-jhms}\Phi(x, z) = \Phi(x + ms, z) \tag{59}$$

for all z and m an arbitrary integer. In addition, $\partial\Phi/\partial x$ satisfies this periodicity condition as well as the condition $\partial\Phi/\partial x = 0$ on the perfectly conducting plates in the half space $z > 0$. By virtue of these boundary conditions, the boundary-value problem reduces to that illustrated in Fig. 10.6. The range of x involved is $0_+ \leq x \leq s_-$, and, since Φ is discontinuous across each conducting plate by an amount equal to the current on each plate, the periodicity condition (59) cannot be applied to Φ for $z \geq 0$, since this condition really means

$$e^{-jhs}\Phi(0_+, z) = \Phi(s_+, z).$$

The region $x = s_+$ is outside the domain of x in the reduced problem of Fig. 10.6. However, $\partial\Phi/\partial x = 0$ for $x = 0, s; z > 0$, so the periodicity condition does hold for $\partial\Phi/\partial x$ for all values of z. At the edges of the plates, Φ is finite, but $\partial\Phi/\partial z$ is asymptotic to $z^{-1/2}$. This edge condition governs the growth of the transform of Φ at infinity.

The solution for Φ which satisfies the preceding conditions and which has the eigenfunction expansions (56) and (57) will be determined by using the known properties of Φ to construct a bilateral Laplace transform, from which Φ may be found by suitable contour integration.

Transform Solution

Let the bilateral Laplace transform of the total field Φ be

$$\psi(x, w) = \int_{-\infty}^{\infty} e^{-wz}\Phi(x, z)\, dz. \tag{60}$$

For convenience, a transform function $g(w)$, defined as follows, will also be introduced:

$$g(w) = e^{jhs}\int_{-\infty}^{\infty} \Phi(s_-, z)e^{-wz}\, dz. \tag{61}$$

For $z < 0$, Φ is asymptotic to $A_0 e^{-\Gamma_0 z} - R_1 A_0 e^{\Gamma_0 z}$. Hence, the transform ψ_- given below is analytic for all w in the half plane $\text{Re}\, w < 0$:

$$\psi_-(x, w) = \int_{-\infty}^{0} e^{-wz}\Phi(z)\, dz. \tag{62}$$

In the region $z > 0$, it will be assumed that small losses are present, so that $\gamma_0 = j|\gamma_0'| + \gamma_0''$, where γ_0'' is a small real and positive attenuation constant. The effect of small losses on the higher order modes is neglected, since these decay according to $e^{-\gamma_n z}$, with γ_n real anyway. At the end of the analysis, γ_0'' will be made zero. Thus, as z approaches plus infinity, Φ is asymptotic to $C_0 e^{-\gamma_0'' z}$, and, hence,

$$\psi_+(x, w) = \int_0^\infty e^{-wz} \Phi(x, z)\, dz \tag{63}$$

is analytic for all w in the half plane $\text{Re}\, w > -\gamma_0''$.

The assumption of small losses for $z > 0$ was made in order to obtain a finite-width strip in the w plane in which both ψ_- and ψ_+ are analytic. We may therefore invert (60) to get

$$\Phi(x, z) = \frac{1}{2\pi j} \int_C e^{wz} \psi(x, w)\, dw \tag{64}$$

where C is a contour running parallel to the imaginary w axis and in the common strip of analyticity, i.e., in the strip $-\gamma_0'' < \text{Re}\, w < 0$. The transform of (58) gives

$$\frac{d^2\psi}{dx^2} + u^2\psi = 0 \tag{65}$$

where $u^2 = w^2 + k_0^2$. The transform ψ must satisfy the boundary conditions

$$\left.\frac{d\psi}{dx}\right|_{x=s_-} = e^{-jhs}\left.\frac{d\psi}{dx}\right|_{x=0_-} \tag{66a}$$

$$\frac{d\psi}{dx} = 0 \qquad\qquad x = 0, s \qquad z > 0 \tag{66b}$$

$$\psi(s_-) = e^{-jhs}\psi(0) \qquad\qquad z < 0. \tag{66c}$$

A general solution of (65) is $\psi(x, w) = A(w) \sin ux + B(w) \cos ux$. Condition (66a) gives

$$Au \cos us - Bu \sin us = e^{-jhs} uA$$

and, hence,

$$B = -\frac{A(e^{-jhs} - \cos us)}{\sin us}.$$

From (61), we get

$$g(w) = e^{jhs}(A \sin us + B \cos us) = \frac{Ae^{jhs}}{\sin us}(\sin^2 us - e^{-jhs} \cos us + \cos^2 us).$$

Solving for A gives

$$A = \frac{g \sin us}{e^{jhs} - \cos us}.$$

The transform ψ is given in terms of g as follows:

$$\psi(x, w) = \frac{g(w)}{e^{jhs} - \cos us}[\cos u(s - x) - e^{-jhs} \cos ux]$$

after substituting for A and B. Inverting this transform, we get

$$\Phi(x, z) = \frac{1}{2\pi j}\int_C \frac{e^{wz}g(w)}{e^{jhz} - \cos us}[\cos u(s - x) - e^{-jhs} \cos ux]\,dw. \qquad (67)$$

From the condition $\Phi(s_-, z) = e^{-jhs}\Phi(0_+, z)$ for $z < 0$, we obtain the following condition on $g(w)$ [equivalent to (66c)]:

$$\int_C \frac{e^{wz}g(w)}{e^{jhs} - \cos us}(\cos hs - \cos us)\,dw = 0, \qquad z < 0. \qquad (68)$$

The condition $\partial\psi/\partial x = 0$, $x = 0, s$; $z > 0$, imposes the further condition

$$\int_C \frac{e^{wz}g(w)}{e^{jhs} - \cos us}(u \sin us)\,dw = 0, \qquad z > 0. \qquad (69)$$

Besides satisfying the conditions (68) and (69), the function $g(w)$ must be such that it will yield the solution for Φ in the form of the eigenfunction expansions (56) and (57).

In (67), the contour C may be closed in the right-half w plane for $z < 0$. Hence, g must have simple poles at $w = \Gamma_n$, $n = \pm 1, \pm 2, \ldots$, in the right half plane in order to give rise to the reflected evanescent modes for $z < 0$. Simple poles at $w = \pm \Gamma_0$ on the imaginary axis are required in order to obtain the incident and reflected dominant modes for $z < 0$. For $z > 0$, we may close the contour in the left half plane, and g must have simple poles at $w = -\gamma_n$, $n = 0, 1, 2, \ldots$, in order to produce the required eigenfunctions for $z > 0$.

For (68) to hold, g must cancel the poles of the factor $(e^{jhs} - \cos us)^{-1}$ in the right half plane, in order that the integrand shall be analytic in the right half plane, and cause the integral to vanish identically for $z < 0$. The poles of g in the right half plane do not cause any difficulty, since (68) contains the factor $\cos hs - \cos us$, which vanishes for $w = \pm \Gamma_n$; $n = \pm 1, \pm 2, \ldots$. Similarly, (69) will hold, provided g cancels the poles of $(e^{jhs} - \cos us)^{-1}$ in the left half plane as well. Again the poles of g cause no difficulty, since $\sin us = 0$ for $w = \gamma_n$ in the left-half w plane.

Taking all the above considerations into account, it is readily seen that a suitable form for the transform function g is

$$g(w) = \frac{p(w)(e^{jhs} - \cos us)}{(w^2 - \Gamma_0^2)(w + \gamma_0)\displaystyle\prod_{n=1}^{\infty}\frac{\gamma_n + w}{n\pi/s}e^{-ws/n\pi}\prod_{n=1}^{\infty}\frac{(\Gamma_n - w)(\Gamma_{-n} - w)}{(2n\pi/s)^2}e^{ws/n\pi}} \qquad (70)$$

where $p(w)$ is an integral function yet to be determined. An examination of the logarithmic derivative of the infinite products shows that the exponential factors and the terms $n\pi/s$ and $(2n\pi/s)^2$ in the denominators ensure the uniform convergence of these products, since $\gamma_n \to n\pi/s$, and $\Gamma_n, \Gamma_{-n} \to 2n\pi/s$, for n large.

The integral function $p(w)$ is to be chosen so that g has algebraic growth at infinity and is asymptotic to $w^{-3/2}$. This will ensure that Φ has the proper behavior at the edge of the plates. For n large, we have $\Gamma_n \to 2n\pi/s + h$, $\Gamma_{-n} \to 2n\pi/s - h$, $\gamma_n \to n\pi/s$. Hence, the term $e^{-ws/n\pi}(\gamma_n + w)/(n\pi/s)$ approaches $e^{-ws/n\pi}(1 + ws/n\pi)$. The other term $e^{ws/n\pi}(\Gamma_n - w)(\Gamma_{-n} - w)/(2n\pi/s)^2$ approaches

$$e^{(w+h)s/2n\pi}\left(1 - \frac{w+h}{2n\pi}s\right)e^{(w-h)s/2n\pi}\left(1 - \frac{w-h}{2n\pi}s\right).$$

Thus, the transform function g differs only by a bounded function of w from $g_1(w)$, where $g_1(w)$ is given by

$$g_1(w) = p(w)(e^{jhs} - \cos us)\left[w^3 \prod_{n=1}^{\infty}\left(1 + \frac{ws}{n\pi}\right)e^{-ws/n\pi}\right.$$

$$\times \left.\prod_{n=1}^{\infty}\left(1 - \frac{w+h}{2n\pi}s\right)e^{(w+h)s/2n\pi}\prod_{n=1}^{\infty}\left(1 - \frac{w-h}{2n\pi}s\right)e^{(w-h)s/2n\pi}\right]^{-1}$$

$$= p(w)(e^{jhs} - \cos us)\pi s\Gamma\left(\frac{ws}{\pi}\right)\left[w^2 \sin\frac{(w+h)s}{2}\sin\frac{(w-h)s}{2}\right.$$

$$\times \left.\Gamma\left(\frac{w+h}{2\pi}s\right)\Gamma\left(\frac{w-h}{2\pi}s\right)\right]^{-1}$$

upon using the infinite-product expansion for the gamma function. The asymptotic form of g_1 is readily found from the known asymptotic form of the gamma function, and is

$$g_1 \sim \frac{Kp(w)(e^{jhs} - \cos us)e^{-(ws/\pi)\ln 2}}{w^{3/2}\sin[(w+h)s/2]\sin[(w-h)s/2]}$$

where K is a suitable constant. In order for g to have algebraic growth at infinity and to be asymptotic to $w^{-3/2}$, we must choose $p(w)$ equal to a constant times $\exp[(ws/\pi)\ln 2]$. The reflection and transmission coefficients depend only on the ratio of amplitudes, and, hence, the exact value of the constant K is not required.

We may now return to (67) and evaluate Φ after substituting for g from (70) and replacing $p(w)$ by $\exp[(ws/\pi)\ln 2]$. For $z < 0$, the contour C may be closed in the right half plane. The incident dominant mode is given by the residue at $w = -\Gamma_0$, and is

$$\Phi_i = A_0 e^{-jhx-\Gamma_0 z} = \frac{j\sin hs\, e^{-(\Gamma_0 s/\pi)\ln 2}e^{-jhx-\Gamma_0 z}}{\left[\displaystyle\prod_{n=1}^{\infty}\frac{(\gamma_n - \Gamma_0)(\Gamma_n + \Gamma_0)(\Gamma_{-n} + \Gamma_0)}{4(n\pi/s)^3}\right]2\Gamma_0(\gamma_0 - \Gamma_0)}. \tag{71}$$

An additional minus sign arises because the integration is carried out in a clockwise sense when the contour C is closed in the right half plane. The reflected dominant mode is given by the residue at $w = \Gamma_0$, and is

$$\Phi_{r0} = -R_1 A_0 e^{-jhx+\Gamma_0 z} = \frac{-j\sin hs\, e^{(\Gamma_0 s/\pi)\ln 2}e^{-jhx+\Gamma_0 z}}{\left[\displaystyle\prod_{n=1}^{\infty}\frac{(\gamma_n + \Gamma_0)(\Gamma_n - \Gamma_0)(\Gamma_{-n} - \Gamma_0)}{4(n\pi/s)^3}\right]2\Gamma_0(\gamma_0 + \Gamma_0)}. \tag{72}$$

Comparing (71) and (72), we obtain the solution for the reflection coefficient R_1:

$$R_1 = \rho e^{j\theta_1} = e^{-2\Gamma_0(s/\pi)\ln 2} \frac{\gamma_0 - \Gamma_0}{\gamma_0 + \Gamma_0} \prod_{n=1}^{\infty} \frac{(\gamma_n - \Gamma_0)(\Gamma_n + \Gamma_0)(\Gamma_{-n} + \Gamma_0)}{(\gamma_n + \Gamma_0)(\Gamma_n - \Gamma_0)(\Gamma_{-n} - \Gamma_0)}. \tag{73}$$

If we now assume that γ_0'' is zero, so that γ_0 as well as Γ_0 are pure-imaginary, we see that

$$\rho = |R_1| = \frac{\gamma_0 - \Gamma_0}{\gamma_0 + \Gamma_0} \tag{74a}$$

$$\theta_1 = -2\left[|\Gamma_0|\frac{s}{\pi}\ln 2 + \sum_{n=1}^{\infty}\left(\tan^{-1}\frac{|\Gamma_0|}{\gamma_n} - \tan^{-1}\frac{|\Gamma_0|}{\Gamma_n} - \tan^{-1}\frac{|\Gamma_0|}{\Gamma_{-n}}\right)\right]. \tag{74b}$$

For $z > 0$, the contour is closed in the left half plane. The dominant-mode transmitted field is given by the residue at $w = -\gamma_0$. The transmission coefficient T_{12} is found from the relation $\gamma_0 C_0 = T_{12}A_0\Gamma_0 e^{-jhs/2}$, where C_0 is the amplitude of the transverse magnetic field of the transmitted wave. Thus,

$$T_{12} = e^{(\gamma_0 - \Gamma_0)(s/\pi)\ln 2} \sec\frac{hs}{2}\frac{2\gamma_0}{\gamma_0 + \Gamma_0}\prod_{n=1}^{\infty}\frac{(\gamma_n - \Gamma_0)(\Gamma_n + \Gamma_0)(\Gamma_{-n} + \Gamma_0)}{(\gamma_n - \gamma_0)(\Gamma_n + \gamma_0)(\Gamma_{-n} + \gamma_0)}. \tag{75}$$

If, instead of having a dominant mode incident from $z < 0$, we have a dominant mode incident from the parallel-plate region, a similar analysis may be carried out. The only modification in the transform function g that is required is replacing the factor $w + \Gamma_0$ by $w - \gamma_0$. This eliminates the incident wave for $z < 0$, and introduces an incident wave in the region $z > 0$. It is found that the reflection coefficient R_2 and the transmission coefficient T_{21} are given by

$$R_2 = -\rho e^{j\theta_2} \tag{76a}$$

where ρ is given by (74a) and

$$\theta_2 = 2\left[|\gamma_0|\frac{s}{\pi}\ln 2 + \sum_{n=1}^{\infty}\left(\tan^{-1}\frac{|\gamma_0|}{\gamma_n} - \tan^{-1}\frac{|\gamma_0|}{\Gamma_n} - \tan^{-1}\frac{|\gamma_0|}{\Gamma_{-n}}\right)\right] \tag{76b}$$

$$T_{21} = \frac{\Gamma_0}{\gamma_0}T_{12}. \tag{76c}$$

Comparing (75) with (74b) and (76b) shows that

$$\angle T_{12} = \angle T_{21} = \frac{\theta_1 + \theta_2}{2} \tag{77}$$

a result that may be deduced by more elementary means from the properties of a lossless two-terminal-pair network. The transmitted power must be equal to the difference between the incident power and the reflected power, and this relation leads directly to a simple expression for the modulus of T_{12} and T_{21}, that is,

$$|T_{12}T_{21}| = 1 - \rho^2. \tag{78}$$

When the spacing s is less than about $0.2\lambda_0$, then the series in (74b) and (76b) contribute only a small amount to the phase angles θ_1 and θ_2.

10.3. APPLICATION TO CAPACITIVE-LOADED PARALLEL-PLATE TRANSMISSION LINE

If conducting planes are placed at $z = d$ and $z = -(b - d)$ as in Fig. 10.7, the resultant structure is that of a parallel-plate transmission line of height b and loaded at regular intervals s by asymmetrical capacitive diaphragms of height d. Provided s is small compared with a wavelength, and $b - d$ and d are large compared with the spacing s, the solution for the propagation constant of the periodic wave in the loaded transmission line may be readily found from the solution of the parallel-plate medium given in Section 10.2.

Referring to Fig. 10.5, let the incident parallel-polarized TEM wave be $\Phi_i = A_0 e^{-jhx-\Gamma_0 z}$. A dominant mode is transmitted into the parallel-plate region. There is also a reflected TEM mode in the free-space region. For a small spacing s, all the higher order modes are evanescent in the z direction and, hence, localized to the region near the interface. A conducting plane placed at $z = d$ interacts only with the dominant mode, provided d is sufficiently large. With the conducting plane in place, a perfect standing wave along the z direction exists in the free-space region. For certain values of z, the x component of electric field will therefore be zero. At these positions, a second conducting plane may be placed to obtain the parallel-plate transmission line of Fig. 10.7.

The interface reflection and transmission coefficients are those given by (73) to (78). For $s \ll \lambda_0$, approximate values of these are given by

$$R_1 = \frac{\gamma_0 - \Gamma_0}{\gamma_0 + \Gamma_0} e^{-2\Gamma_0(s/\pi)\ln 2} \tag{79a}$$

$$R_2 = -\frac{\gamma_0 - \Gamma_0}{\gamma_0 + \Gamma_0} e^{-2\gamma_0(s/\pi)\ln 2} \tag{79b}$$

$$T_{12} = \frac{2\gamma_0}{\gamma_0 + \Gamma_0} e^{(\gamma_0-\Gamma_0)(s/\pi)\ln 2} \tag{79c}$$

$$T_{21} = \frac{\Gamma_0}{\gamma_0} T_{12}. \tag{79d}$$

The form of these reflection and transmission coefficients shows that the equivalent transmission-line circuit of a transverse section consists of two sections of short-circuited transmission lines of lengths $b - d + (s/\pi)\ln 2$ and $d - (s/\pi)\ln 2$, with characteristic impedances

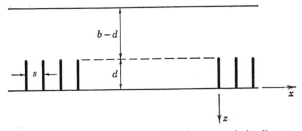

Fig. 10.7. Capacitive-loaded parallel-plate transmission line.

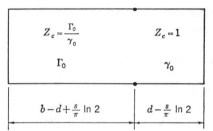

Fig. 10.8. Equivalent circuit of a transverse section.

proportional to Γ_0/γ_0 and unity. This equivalent circuit is illustrated in Fig. 10.8. For specified values of b, d, and s, the value of Γ_0 is determined by the condition that the above circuit of a transverse section should be of resonant length. By conventional transmission-line theory, it is found that the resonance condition is determined by

$$\Gamma_0 \tanh \Gamma_0 \left(b - d + \frac{s}{\pi} \ln 2\right) = k_0 \tan k_0 \left(d - \frac{s}{\pi} \ln 2\right) \tag{80}$$

since $\gamma_0 = jk_0$. The propagation phase constant along the x direction is $h = k_0 \sin \theta_i$, and h is related to Γ_0 as follows:

$$h^2 = \Gamma_0^2 + k_0^2. \tag{81}$$

The loading of the transmission line by capacitive diaphragms has an effect similar to increasing the equivalent dielectric constant of the medium. This results in h^2 being greater than k_0^2, and, hence, Γ_0 is real and the angle θ_i complex. The parameter h may be considered as the characteristic propagation phase constant for the periodically loaded line, and (80) as the eigenvalue equation for h.

For larger spacings of the diaphragms, the above analysis is no longer valid. We may now, however, apply the analysis of Section 9.5. The resultant eigenvalue equation for h is

$$\cos hs = \cos k_0 s - \frac{B}{2} \sin k_0 s \tag{82}$$

where B is the normalized shunt susceptance of a single diaphragm. An approximate value of B is given by the low-frequency formula

$$B = \frac{8bk_0}{2\pi} \ln \sec \frac{\pi d}{2b}. \tag{83}$$

The combination of (80) and (82) leads to a reasonably good approximation to h for all values of spacing s. All the above formulas become applicable to the capacitively loaded rectangular guide if k_0^2 is replaced by $k_0^2 - (\pi/a)^2$ throughout. This result follows from the same considerations used to obtain the solution for a capacitive diaphragm in a rectangular guide in Chapter 8. Image theory may be used to obtain the solution when the diaphragms are symmetrically placed along both plates or centered midway between the two plates of a transmission line.

As an example, h has been computed by the two methods for $0 < s < 1$ centimeter, $k_0 = 2$, $b = 0.5$ centimeter, $d = 0.3$ centimeter, and the results are plotted in Fig. 10.9. It is seen that,

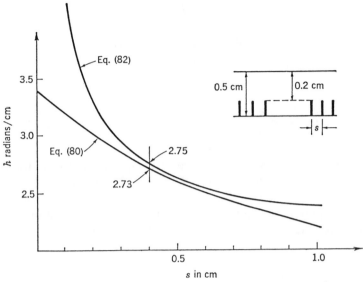

Fig. 10.9. A plot of characteristic propagation constant h versus spacing s.

in the range $0.4 < s < 0.6$, the two methods give essentially the same results (within 1%). For $s > 0.5$ we use (82), while for $s < 0.5$ we use (80), and thus obtain accurate results for the complete range of s considered. Since h is larger than k_0, the phase velocity $v_p = v_c k_0 / h$ has been reduced to a value less than that of light in free space. The maximum reduction is in the ratio 2:3.39 for a value of s approaching zero. An increase in the diaphragm height from 0.3 to 0.45 centimeter results in a maximum reduction of phase velocity by a factor 3.76 for s approaching zero.

10.4. INDUCTIVE SEMIDIAPHRAGM IN A RECTANGULAR GUIDE

The special case of an inductive diaphragm of the asymmetrical type and extending halfway across the guide as in Fig. 10.10 may be solved by any one of the three methods presented earlier. The similar problem of the capacitive semidiaphragm, as well as the grating problems which reduce to the above waveguide problems, may be solved in the same way. To date, the extension to the general diaphragm or strip-grating problems has not been achieved.

The problem under consideration may be simplified by considering the two special cases of even and odd excitation. For the odd case, let the electric field of the incident H_{10} modes be

$$E_y = -A_1 \sin \frac{\pi x}{a} e^{-\Gamma_1 z}, \qquad z < 0 \tag{84a}$$

$$E_y = A_1 \sin \frac{\pi x}{a} e^{\Gamma_1 z}, \qquad z > 0. \tag{84b}$$

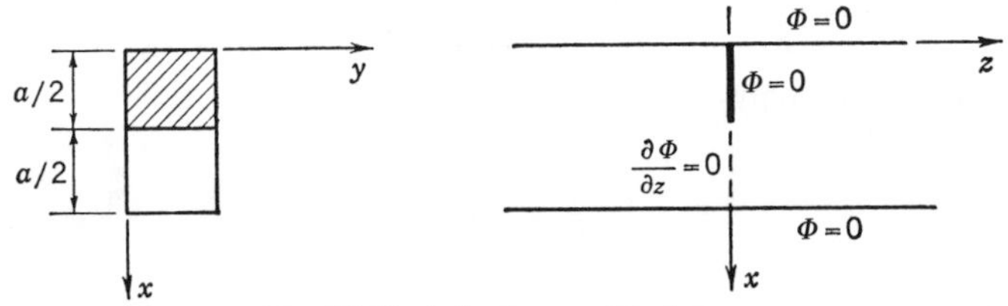

Fig. 10.10. Inductive semidiaphragm.

In view of the symmetry involved, the scattered field will be of the form

$$E_y = -A_1 R_0 \sin \frac{\pi x}{a} e^{\Gamma_1 z} - \sum_{n=2}^{\infty} A_n \sin \frac{n\pi x}{a} e^{\Gamma_n z}, \qquad z < 0 \tag{85a}$$

$$E_y = A_1 R_0 \sin \frac{\pi x}{a} e^{-\Gamma_1 z} + \sum_{n=2}^{\infty} A_n \sin \frac{n\pi x}{a} e^{-\Gamma_n z}, \qquad z > 0 \tag{85b}$$

where $\Gamma_n^2 = (n\pi/a)^2 - k_0^2$. At the position of the diaphragm $z = 0$, the electric field must be continuous and must vanish on the diaphragm $0 \le x \le a/2$. Since the $\sin(n\pi x/a)$ functions are orthogonal, this is possible only if $R_0 = -1$ and $A_n = 0$, $n > 1$. Hence, for odd excitation, we may close the aperture by an electric wall since the total electric field vanishes at $z = 0$.

For the case of even excitation, let the incident electric field be

$$E_y = A_1 \sin \frac{\pi x}{a} e^{-\Gamma_1 z}, \qquad z < 0 \tag{86a}$$

$$E_y = A_1 \sin \frac{\pi x}{a} e^{\Gamma_1 z}, \qquad z > 0. \tag{86b}$$

Symmetry considerations show that the scattered field will be of the form

$$E_y = R_e A_1 \sin \frac{\pi x}{a} e^{\Gamma_1 z} + \sum_{n=2}^{\infty} B_n \sin \frac{n\pi x}{a} e^{\Gamma_n z}, \qquad z < 0 \tag{87a}$$

$$E_y = R_e A_1 \sin \frac{\pi x}{a} e^{-\Gamma_1 z} + \sum_{n=2}^{\infty} B_n \sin \frac{n\pi x}{a} e^{-\Gamma_n z}, \qquad z > 0. \tag{87b}$$

The transverse electric field is an even function of z, and, hence, the transverse magnetic field H_x, which is proportional to $\partial E_y/\partial z$, will be an odd function of z. At the aperture $z = 0$, $a/2 \le x \le a$, the transverse magnetic field vanishes, and, hence, the aperture may be closed by a magnetic wall. The reflection coefficient R_e will not be equal to -1 in the present case, since the boundary at $z = 0$ is not a homogeneous one. The magnitude of R_e will, however, equal unity, since no power can be transmitted through either the diaphragm or the magnetic wall. For the purpose of evaluating R_e, we need to solve the reduced problem of a semi-infinite guide closed by a combination of an electric wall for $0 \le x \le a/2$ and a magnetic wall for $a/2 \le x \le a$. The general solution is obtained by superimposing the solutions for the odd and even cases, i.e., by addition of (84) and (85) to (86) and (87). The incident field for $z < 0$ will have an amplitude of $2A_1$, while for $z > 0$ there is no incident field. The reflected field has an amplitude of $(R_e - 1)A_1$, and, hence, the equivalent reflection coefficient for an H_{10} mode incident from one side only is

$$R = \frac{R_e - 1}{2}. \tag{88}$$

The normalized shunt susceptance B of the diaphragm is given by $-2jR/(1 + R)$. If we let

$R_e = e^{j\theta}$, we find that an equivalent expression for B is

$$B = 2 \tan \frac{\theta}{2}. \tag{89}$$

The electric field E_y for $z \geq 0$ may be represented by contour integrals in terms of suitable transform functions. For $0 \leq x \leq a/2$, let

$$E_y = \int_C (e^{-wz} - e^{wz}) A(w) \sin ux \, dw \tag{90a}$$

while, for $a/2 \leq x \leq a$,

$$E_y = \int_C (e^{-wz} + e^{wz}) B(w) \sin u(a - x) \, dw \tag{90b}$$

where $u^2 = w^2 + k_0^2$. This form of solution is seen to satisfy the required boundary conditions at $z = 0$,

$$E_y = 0 \qquad 0 \leq x \leq \frac{a}{2}$$

$$\frac{\partial E_y}{\partial z} = 0 \qquad \frac{a}{2} \leq x \leq a$$

as well as the boundary conditions at $x = 0, a; \; z > 0$. It is also a solution of the two-dimensional scalar Helmholtz equation. Continuity of E_y at $x = a/2$ for all z gives

$$\int_C A(e^{-wz} - e^{wz}) \sin \frac{ua}{2} \, dw = \int_C B(e^{-wz} + e^{wz}) \sin \frac{ua}{2} \, dw. \tag{91}$$

We may also represent E_y at $x = a/2$ by means of the contour integral

$$E_y \left(\frac{a}{2}, z \right) = \int_C e^{-wz} F(w) \, dw, \qquad z > 0 \tag{92}$$

where $F(w)$ is a suitable transform function. To be consistent with (91), we require that $F(w)$ be such that

$$\int_C e^{wz} F(w) \, dw = 0, \qquad z > 0$$

which is equivalent to

$$\int_C e^{-wz} F(w) \, dw = 0, \qquad z < 0.$$

Thus, $F(w)$ must be free of poles in the half plane Re $w < 0$, since the contour may be closed in this half plane, and there must be no residue of the integrand inside the contour. A form for $F(w)$ which satisfies this criterion may be found, since E_y is equal to zero for $z < 0$.

Comparing (92) with (91) shows that

$$A(w) \sin \frac{ua}{2} = F(w) \tag{93a}$$

$$B(w) \sin \frac{ua}{2} = F(w). \tag{93b}$$

The solution for E_y may be written as follows:

$$E_y = \int_C F(w) \frac{\sin ux}{\sin(ua/2)} (e^{-wz} - e^{wz}) \, dw, \qquad 0 \le x \le \frac{a}{2} \tag{94a}$$

$$E_y = \int_C F(w) \frac{\sin u(a-x)}{\sin(ua/2)} (e^{-wz} + e^{wz}) \, dw, \qquad \frac{a}{2} \le x \le a. \tag{94b}$$

Continuity of the derivative $\partial E_y/\partial x$ at $x = a/2$ imposes the following further condition on the function $F(w)$:

$$\int_C F(w) u \cot \frac{ua}{2} e^{-wz} \, dw = 0, \qquad z > 0. \tag{95}$$

If the diaphragm were not equal to $a/2$ in width, the continuity of the derivative would lead to an equation involving both e^{-wz} and e^{wz}. This appears to be the reason why the method of solution being carried out works only for the special case of the semidiaphragm.

The contour in (95) may be closed in the right-half w plane, and the integral will vanish, provided $F(w)$ has zeros at $u = 2n\pi/a$ or

$$w = \left[\left(\frac{2n\pi}{a} \right)^2 - k_0^2 \right]^{1/2}$$

$n = 1, 2, 3, \ldots$, in the right half plane, so as to cancel the poles of the cotangent function. In addition, $F(w)$ must be free of poles in the right half plane, except possibly for poles at the zeros of the cotangent function, i.e., at $w = [(n\pi/a)^2 - k_0^2]^{1/2}$, $n = 1, 2, \ldots$. For convenience, let

$$\Gamma_n^2 = \left(\frac{n\pi}{a} \right)^2 - k_0^2$$

$$\Gamma_{2n}^2 = \left(\frac{2n\pi}{a} \right)^2 - k_0^2.$$

At $x = a/2$, the eigenfunction expansion of E_y is of the form

$$(A_1 e^{\Gamma_1 z} + R_e A_1 e^{-\Gamma_1 z}) \sin \frac{\pi}{2} + \sum_{n=3,5,\ldots}^{\infty} B_n \sin \frac{n\pi}{2} e^{-\Gamma_n z}.$$

Therefore, $F(w)$ must have poles at $w = \Gamma_n$, $n = 1, 3, 5, , \ldots$, in order that (92) shall give a solution for E_y of the above form. A pole at $w = -\Gamma_1$ is also required, in order to give rise to the incident wave. Since we are assuming that only the dominant mode propagates, all Γ_n except Γ_1 are real. The poles at $\pm \Gamma_1$ lie on the imaginary w axis, and the contour C passes around these poles on the left-hand side. The function $F(w)$ must have an asymptotic form for large w that will result in a finite value for E_y at the edge of the diaphragm. The derivative $\partial E_y/\partial z$ is asymptotic to $z^{-1/2}$ at the edge of the diaphragm, and this condition governs the growth of $F(w)$ at infinity.

Taking into account all the stipulated conditions on the function $F(w)$ leads to the following suitable form for $F(w)$:

$$F(w) = \frac{p(w)}{w + \Gamma_1} \frac{\displaystyle\prod_{n=1}^{\infty}(w - \Gamma_{2n})\frac{a}{2n\pi}e^{wa/2n\pi}}{\displaystyle\prod_{n=1,3,\ldots}^{\infty}(w - \Gamma_n)\frac{a}{n\pi}e^{wa/n\pi}} \tag{96}$$

where $p(w)$ is an integral function to be determined so that $F(w)$ will have algebraic growth at infinity, in particular so that $F(w) \to w^{-3/2}$ as w approaches infinity. The asymptotic form of $F(w)$ may be found in the usual manner and is

$$F(w) \sim \frac{p(w)}{w^{3/2}}e^{-(wa/\pi)\ln 2}.$$

Hence $p(w)$ is chosen as $e^{(wa/\pi)\ln 2}$.

The solution for E_y may now be written down, and is

$$E_y = \int_C (e^{-wz} - e^{wz})\frac{\sin ux}{\sin(ua/2)}e^{(wa/\pi)\ln 2}$$

$$\times \frac{\left[\displaystyle\prod_{n=1}^{\infty}(w - \Gamma_{2n})\frac{a}{2n\pi}e^{wa/2n\pi}\right]^2}{(w + \Gamma_1)\displaystyle\prod_{n=1}^{\infty}(w - \Gamma_n)\frac{a}{n\pi}e^{wa/n\pi}}\, dw, \qquad 0 \le x \le \frac{a}{2} \tag{97}$$

after multiplying the numerator and denominator of (96) by

$$\prod_{n=2,4,\ldots}^{\infty}(w - \Gamma_n)\frac{a}{n\pi}e^{wa/n\pi}.$$

For $a/2 \le x \le a$, the solution is given by (97) with e^{wz} and $\sin ux$ replaced by $-e^{wz}$ and $\sin u(a-x)$, respectively. For $z > 0$, the contour C for the part of the solution involving the exponential function e^{wz} may be closed in the left half plane. In this half plane, the integrand has poles at the zeros of $\sin(ua/2)$, that is, at $w = -\Gamma_{2n}$, and this leads to the remainder of the eigenfunction expansion of E_y, the first part of the expansion having been obtained from the residues at the poles occurring at $w = \Gamma_n$ in the right half plane. The solution for E_y is seen to consist of two parts, one part which is an even function of x about the plane $x = a/2$

and arising from the poles of $F(w)$ in the right half plane, the other part which is an odd function of x about the plane $x = a/2$ and arising from the poles of the integrand in the left half plane. The even part consists of terms $n = 1, 3, 5, \ldots$, while the odd part consists of the remaining terms $n = 2, 4, 6, \ldots$, in the expansion (87).

The residue of the integrand at $w = -\Gamma_1$ gives rise to the incident wave, while the residue at $w = \Gamma_1$ corresponds to the reflected wave. These residues are readily evaluated from (97). It is important to note, however, that the exponential convergence factors occurring in the infinite products in the numerator and denominator in (97) cannot be canceled since the resultant infinite product is not a uniformly convergent product; that is,

$$\prod_{n=1}^{\infty} \frac{(w - \Gamma_{2n})^2}{w - \Gamma_n} \frac{a}{4n\pi}$$

is not uniformly convergent. It is preferable to rewrite (97), using the alternative form for $F(w)$ given in (96). The integrand in (97) may then be written as

$$(e^{-wz} - e^{wz}) \frac{\sin ux}{\sin(ua/2)} \frac{e^{(wa/\pi)\ln 2} \displaystyle\prod_{n=1}^{\infty} (w - \Gamma_{2n}) \frac{a}{2n\pi} e^{wa/2n\pi}}{(w + \Gamma_1) \displaystyle\prod_{n=1,3,\ldots}^{\infty} (w - \Gamma_n) \frac{a}{n\pi} e^{wa/n\pi}}.$$

Using this form for the integrand, the ratio of the residues at $w = \Gamma_1$ and $w = -\Gamma_1$ is found to be

$$R_e = -e^{(2\Gamma_1 a/\pi)(\ln 2 - 1)} \prod_{n=1}^{\infty} \frac{\Gamma_{2n} - \Gamma_1}{\Gamma_{2n} + \Gamma_1} e^{\Gamma_1 a/n\pi} \prod_{n=3,5,\ldots}^{\infty} \left(\frac{\Gamma_n + \Gamma_1}{\Gamma_n - \Gamma_1} e^{-\Gamma_1 a/n\pi} \right)$$

$$= -e^{(2\Gamma_1 a/\pi)(\ln 2 - 1)}$$

$$\times \prod_{n=1}^{\infty} \left[\frac{(\Gamma_{2n} - \Gamma_1)(\Gamma_{2n+1} + \Gamma_1)}{(\Gamma_{2n} + \Gamma_1)(\Gamma_{2n+1} - \Gamma_1)} e^{(\Gamma_1 a/\pi)[1/n - 1/(2n+1)]} \right] \qquad (98)$$

after replacing n by $2n + 1$ in the second infinite product. Since all Γ_n except Γ_1 are real, it is seen that the modulus of R_e is unity. The phase angle of R_e is given by

$$\angle R_e = \pi + \frac{2\beta_1 a}{\pi}(\ln 2 - 1) + \sum_{n=1}^{\infty} \frac{\beta_1 a}{\pi} \left(\frac{1}{n} - \frac{1}{2n + 1} \right)$$

$$- 2 \sum_{n=1}^{\infty} \left(\tan^{-1} \frac{\beta_1}{\Gamma_{2n}} - \tan^{-1} \frac{\beta_1}{\Gamma_{2n+1}} \right) \qquad (99)$$

where $\beta_1 = |\Gamma_1|$. Using the result [10.3]

$$\ln 2 = 1 - \sum_{n=1}^{\infty} \frac{1}{2n(2n + 1)}$$

and also the relation $\tan^{-1} a = \sin^{-1}[a/(1+a^2)^{1/2}]$, (99) may be rewritten as

$$\angle R_e = \theta = \pi - 2\sum_{n=1}^{\infty}\left[\sin^{-1}\frac{\beta_1 a/\pi}{(4n^2-1)^{1/2}} - \sin^{-1}\frac{\beta_1 a/\pi}{(4n^2+4n)^{1/2}}\right]. \tag{100}$$

From this result and (89), the solution for the shunt susceptance B of the semidiaphragm may be obtained.

10.5. APPLICATION TO H-PLANE BIFURCATION

Figure 10.11 illustrates a rectangular guide bifurcated by an infinitely thin perfectly conducting plane at $x = a/2$, $z \leq 0$. Since the bifurcating plane is located in a plane perpendicular to the xz plane which contains the H vector, it is referred to as an H-plane bifurcation. The solution to this problem may be found in terms of the solution for the inductive semidiaphragm by an application of the Schwarz reflection principle. It will be convenient to choose a new set of coordinates x', y, z, where $x' = x - a/2$.

As noted in Section 10.4, the solution to the problem of a rectangular guide closed by an electric wall for $0 \leq x \leq a/2$ and a magnetic wall for $a/2 \leq x \leq a$ is the sum of two parts. One part is an even function about the symmetry plane $x = a/2$ or $x' = 0$, while the other part is an odd function about $x' = 0$. Let the even part be $\Phi_e(x', z)$ and the odd part be $\Phi_o(x', z)$ in the region $-a/2 \leq x' \leq 0$, $z \geq 0$. The sum $\Phi_e + \Phi_o$ is the solution given by (97). Along the symmetry plane $x' = 0$, $z \geq 0$, the component functions satisfy the boundary conditions

$$\frac{\partial \Phi_e}{\partial x'} = \Phi_o = 0, \qquad x' = 0.$$

The Schwarz reflection principle states that the analytical continuation of a function across a plane is given by an odd reflection if the function vanishes on the plane or by an even reflection if the normal derivative of the function vanishes on the plane. With reference to Fig. 10.12(a), let $\Phi_e(x', z) + \Phi_o(x', z)$ be the solution for the lower right-hand rectangular region $CBEF$. The derivative $\partial \Phi_e/\partial x'$ is zero at $x' = 0$, and so the analytic continuation of Φ_e into the upper region $IHBC$ is given by an even reflection. The analytic continuation of Φ_o is given by an odd reflection since $\Phi_o = 0$ at $x' = 0$. This solution for $z \geq 0$ is the solution given by (97). Therefore, along the segment BE, $\Phi_e + \Phi_o$ is zero. The analytic continuation of this solution into the lower left-hand region $ABED$ is therefore given by an odd reflection, and is $-\Phi_e(x', -z) - \Phi_o(x', -z)$. Along the segment BH, the solution $\Phi_e(-x', z) - \Phi_o(-x', z)$ has a vanishing normal derivative, and its analytic continuation into the region $ABHG$ is given by an even reflection about the line BH. The function as constructed above is continuous with

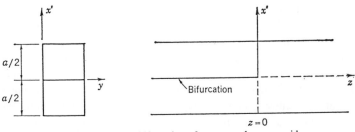

Fig. 10.11. H-plane bifurcation of a rectangular waveguide.

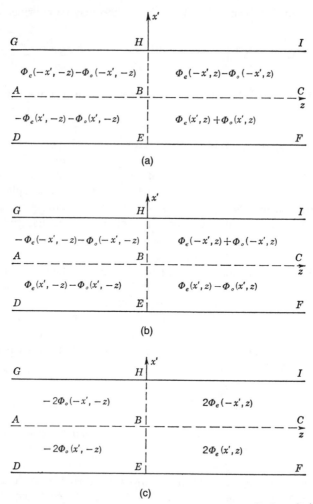

Fig. 10.12. Illustration of an application of the Schwarz reflection principle.

continuous derivatives at all points except along the line AB. Along this line both the function
and its normal derivative are discontinuous. The solution illustrated in Fig. 10.12(a) solves
the rather uninteresting problem of mixed boundary values along the plane AB.

In order to solve the bifurcation problem, we need to construct another function which is
analytic everywhere except along AB, where it must cancel the discontinuity in the function
of Fig. 10.12(a). We begin with $\Phi_e - \Phi_o$ in the region $CBEF$, as illustrated in Fig. 10.12(b).
This function is continued into the upper region $IHBC$ as $\Phi_e + \Phi_o$. Taking into account that
$\Phi_e + \Phi_o$ vanishes along BH, the analytic continuation into the region $ABHG$ is seen to be
$-\Phi_e - \Phi_o$. Similarly, the analytic continuation of $\Phi_e - \Phi_o$ across the segment BE is seen
to be given by an even reflection, and is $\Phi_e - \Phi_o$. The final function is analytic at all points
except along AB. If we now superimpose the two solutions (a) and (b), we obtain the solution
illustrated in Fig. 10.12(c). This solution is analytic everywhere except along AB. However,
along AB the solution vanishes, and we may therefore place a conducting plane along AB.
This provides the solution of the bifurcation problem. As would be expected from symmetry
considerations, the solution for $z \geq 0$ consists only of the functions $\sin(n\pi x/a)$ for n odd.

The reflection coefficient for an H_{10} mode incident on the bifurcation from the region $z > 0$ is R_e and is given by (98). The modulus of R_e is unity unless the guide width a is such that the H_{20} mode can propagate, i.e., unless $a > \lambda$.

10.6. PARALLEL-PLATE WAVEGUIDE BIFURCATION

A parallel-plate waveguide of height $c = a + b$ is shown in Fig. 10.13(a). A dielectric layer of thickness a and dielectric constant κ is placed along the bottom plate. A semi-infinite conducting plate extending from $x = -\infty$ to $x = 0$ is placed on top of the dielectric. The coupling and reflection of the modes between regions 1, 2, and 3 characterize the junction at $x = 0$. When these coupling and reflection coefficients are known the results can be used to analyze the microwave strip line having a strip of total width $2W$ shown in Fig. 10.13(b). When $2W \gg a$ only a few modes interact between the edges at $x = \pm W$ so this analysis is very appropriate for wide strips. El-Sherbiny has used the modified Wiener–Hopf method to analyze microstrip lines for both isotropic and anisotropic substrates [10.26], [10.27]. The analysis of the single junction given in this section is similar.

In all three regions the modes are uncoupled LSM and LSE modes (E modes and H modes with respect to y). In order to satisfy the edge conditions on the fields, current, and charge at $x = 0$, $y = a$, the modes are coupled together. We will assume that all fields have a z dependence of the form $e^{-j\beta z}$ where the propagation constant β satisfies the condition

$$k_0 < \beta < \sqrt{\kappa}k_0 = k. \tag{101}$$

This condition holds for the guided modes on a microstrip line. In region 1 the fields have an x dependence of the form

$$e^{\pm\sqrt{\beta^2+(n\pi/a)^2-k^2}\,x}.$$

The dominant LSM mode corresponding to $n = 0$ varies according to

$$e^{\pm j\sqrt{k^2-\beta^2}\,x}$$

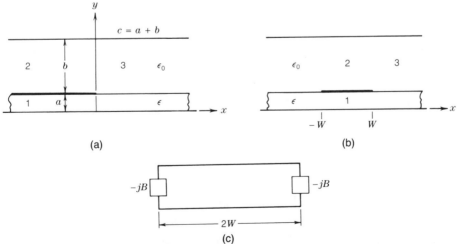

Fig. 10.13. (a) A bifurcated parallel-plate waveguide. (b) A related microstrip line. (c) Equivalent circuit of microstrip line transverse section.

and is a propagating mode. In a microstrip line a is small enough that all higher order modes are evanescent along x. Thus the dominant mode, which is an obliquely propagating TEM mode, is the major field component in the microstrip line. If a is small enough the modes in region 3 are all evanescent so the junctions at $x = \pm W$ can be characterized by reactive terminations. When the reactive elements are known we can formulate an eigenvalue equation to determine the propagation constant β. The terminations are inductive and the equivalent circuit for a wide microstrip line is that shown in Fig. 10.13(c). For modes that have E_y a maximum at $x = 0$ the impedance at $x = 0$ is infinite. For modes that have $E_y = 0$ at $x = 0$ the impedance is zero. Thus the eigenvalue equations for the even and odd modes are

$$\tan \sqrt{k^2 - \beta^2}W = B \tag{102a}$$

$$\cot \sqrt{k^2 - \beta^2}W = -B. \tag{102b}$$

For lines with narrower strips, higher order mode interaction between the two edges must be taken into account. In region 2 above the strip the dominant mode is the LSM mode with a propagation factor

$$e^{-j\beta z \pm \sqrt{\beta^2 - k_0^2}x}.$$

This mode is evanescent along x as long as $\beta > k_0$.

In order to describe the LSM and LSE modes we will follow the nomenclature in Section 4.8. We will also use the Fourier transform with respect to x; thus, for example,

$$\hat{E}_y(w, y) = \int_{-\infty}^{\infty} E_y(x, y)e^{jwx}\, dx.$$

When we take into account the shield at $y = c$ and follow the approach used in Section 4.8 then the functions f and g which give ψ_h and ψ_e, respectively, are (in the Fourier spectral domain)

$$f = \begin{cases} A(w)\dfrac{\sin \ell y}{\sin \ell a}, & 0 \leq y \leq a \\[3mm] A(w)\dfrac{\sin p(c - y)}{\sin pb}, & a \leq y \leq c \end{cases} \tag{103a}$$

$$g = \begin{cases} \kappa B(w)\dfrac{\cos \ell y}{-\ell \sin \ell a}, & 0 \leq y < a \\[3mm] B(w)\dfrac{\cos p(c - y)}{p \sin pb}, & a < y \leq c. \end{cases} \tag{103b}$$

These functions have been constructed so that the boundary conditions for the fields will be satisfied at the air–dielectric interface. The fields H_x, H_z and E_x, E_z are given by

$$H_x = \frac{-1}{j\omega\mu_0}\frac{\partial^2 \psi_h}{\partial x\, \partial y} - \frac{\partial \psi_e}{\partial z}$$

$$H_z = \frac{-1}{j\omega\mu_0}\frac{\partial^2 \psi_h}{\partial z\, \partial y} + \frac{\partial \psi_e}{\partial x}$$

$$E_x = \frac{1}{j\omega\epsilon_0\kappa(y)}\frac{\partial^2 \psi_e}{\partial x\, \partial y} - \frac{\partial \psi_h}{\partial z}$$

$$E_z = \frac{1}{j\omega\epsilon_0\kappa(y)}\frac{\partial^2 \psi_e}{\partial z\, \partial y} + \frac{\partial \psi_h}{\partial x}.$$

Introducing f, g, we have in the Fourier transform domain

$$\hat{H}_x = \frac{w}{\omega\mu_0}\frac{df}{dy} + j\beta g \tag{104a}$$

$$\hat{H}_z = \frac{\beta}{\omega\mu_0}\frac{df}{dy} - jwg \tag{104b}$$

$$\hat{E}_x = \frac{-w}{\omega\epsilon_0\kappa(y)}\frac{dg}{dy} + j\beta f \tag{104c}$$

$$\hat{E}_z = \frac{-\beta}{\omega\epsilon_0\kappa(y)}\frac{dg}{dy} - jwf. \tag{104d}$$

The structure of these equations is such that new fields can be defined that will uncouple the f and g functions. Thus it is easily found that

$$\beta\hat{H}_x - w\hat{H}_z = j(\beta^2 + w^2)g \tag{105a}$$

$$w\hat{H}_x + \beta\hat{H}_z = \frac{w^2 + \beta^2}{\omega\mu_0}\frac{df}{dy} \tag{105b}$$

$$\beta\hat{E}_x - w\hat{E}_z = j(w^2 + \beta^2)f \tag{105c}$$

$$w\hat{E}_x + \beta\hat{E}_z = -\frac{w^2 + \beta^2}{\omega\epsilon_0\kappa(y)}\frac{dg}{dy}. \tag{105d}$$

If we use $-\hat{J}_z = \hat{H}_x^+ - \hat{H}_x^-$, $\hat{J}_x = \hat{H}_z^+ - \hat{H}_z^-$ to relate the current density on the half plane to the discontinuity in the tangential magnetic field across the strip it is readily found that

$$\beta\hat{J}_x - w\hat{J}_z = \frac{w^2 + \beta^2}{\omega\mu_0}\frac{df}{dy}\bigg|_{-}^{+} = \frac{w^2 + \beta^2}{\omega\mu_0}(-\ell\cot\ell a - p\cot pb)A$$

$$\beta\hat{J}_z + w\hat{J}_x = -j(w^2 + \beta^2)g\bigg|_{-}^{+} = -j(w^2 + \beta^2)\left(\frac{\cot pb}{p} + \frac{\kappa\cot\ell a}{\ell}\right)B.$$

At $y = a$, $\beta\hat{E}_x - w\hat{E}_z = j(w^2 + \beta^2)A$ and $w\hat{E}_x + \beta\hat{E}_z = -((w^2 + \beta^2)/\omega\epsilon_0)B$. Hence we obtain

$$\left(\frac{\kappa\cot\ell a}{\ell} + \frac{\cot pb}{p}\right)j\omega\epsilon_0(w\hat{E}_x + \beta\hat{E}_z) = w\hat{J}_x + \beta\hat{J}_z = \hat{J}_1(w) \tag{106a}$$

$$(\ell\cot\ell a + p\cot pb)\frac{j}{\omega\mu_0}(\beta\hat{E}_x - w\hat{E}_z) = \beta\hat{J}_x - w\hat{J}_z = \hat{J}_2(w). \tag{106b}$$

These uncoupled functional equations can be solved by the Wiener–Hopf method or by using the function-theoretic technique. The latter method is very straightforward for this problem and is the one that we adopt.

We note the following features. The source $-j(w\hat{J}_x + \beta\hat{J}_z)$ is $\partial J_x/\partial x + \partial J_z/\partial z$ in the spatial domain and by the continuity equation equals $-j\omega\rho$ where ρ is the charge density on the strip. Consider $\mathbf{a}_y \cdot \nabla \times \mathbf{J} = \partial J_x/\partial z - \partial J_z/\partial x$ which transforms to $-j\beta\hat{J}_x + jw\hat{J}_z$. Hence it can be seen that the second source term is a vortex source equivalent to a y-directed magnetic dipole density on the strip. Thus the LSM modes are excited by the charge and the LSE modes by the vortex source on the strip.

It would appear that we could solve for f and g separately but this is not the case because the individual solutions turn out to not satisfy the required singularity (edge) conditions at $x = 0$. Thus a linear combination of f and g must be chosen so as to satisfy the edge conditions. Thus, in a microstrip line the LSE and LSM modes are coupled only by the edge conditions. This feature has been described in the spatial domain by Omar and Schünemann [10.4].

The coupling condition takes on a particularly simple form in the transform domain. If we have found solutions for $\hat{J}_1 = w\hat{J}_x + \beta\hat{J}_z$ and $\hat{J}_2 = \beta\hat{J}_x - w\hat{J}_z$ then

$$\hat{J}_x = \frac{w\hat{J}_1 + \beta\hat{J}_2}{w^2 + \beta^2}, \qquad \hat{J}_z = \frac{\beta\hat{J}_1 - w\hat{J}_2}{w^2 + \beta^2}.$$

Since J_x, J_z, J_1, and J_2 are all zero for $x > 0$ their Fourier transforms must be analytic in the lower half of the complex w plane. For example, for $x > 0$

$$J_x(x) = \frac{1}{2\pi}\int_{-\infty}^{\infty} e^{-jwx}\hat{J}_x(w)\,dw.$$

The contour can be closed by a semicircle in the lower half plane for $x > 0$. In order that we obtain $J_x(x) = 0$ for $x > 0$ the function $\hat{J}_x(w)$ must be analytic in the lower half plane. We now see that in order to not introduce poles for $\hat{J}_x$ and $\hat{J}_z$ in the lower half plane we must have $\hat{J}_x(-j\beta) = \hat{J}_z(-j\beta) = 0$. This requirement can be stated as

$$-j\beta\hat{J}_1(-j\beta) + \beta\hat{J}_2(-j\beta) = 0 \tag{107a}$$

$$\beta\hat{J}_1(-j\beta) + j\beta\hat{J}_2(-j\beta) = 0. \tag{107b}$$

The two conditions are equivalent.

The electric field components $\hat{E}_x$ and $\hat{E}_z$ vanish on the conducting half plane $x < 0$ and are analytic functions of w in the upper half plane. Since these field components are linear functions of $\hat{J}_x$ and $\hat{J}_z$ we cannot have a pole at $w = j\beta$ in the upper half plane. Hence we must also satisfy the condition

$$j\beta\hat{J}_1(j\beta) + \beta\hat{J}_2(j\beta) = 0. \tag{108}$$

Before proceeding with the solution we will discuss the eigenvalue equations and review the edge conditions.

LSM Modes

The eigenvalue equation is $(\kappa\cot\ell a)/\ell + (\cot pb)/p = 0$ or $\kappa p\tan pb = -\ell\tan\ell a$, $\ell^2 - p^2 = (\kappa - 1)k_0^2$. Let $u = pb$, $v = \ell a$; then

$$v\tan v = -\kappa\frac{a}{b}u\tan u, \qquad \left(\frac{v}{a}\right)^2 - \left(\frac{u}{b}\right)^2 = (\kappa - 1)k_0^2.$$

Case 1: ℓ is real, p is imaginary.

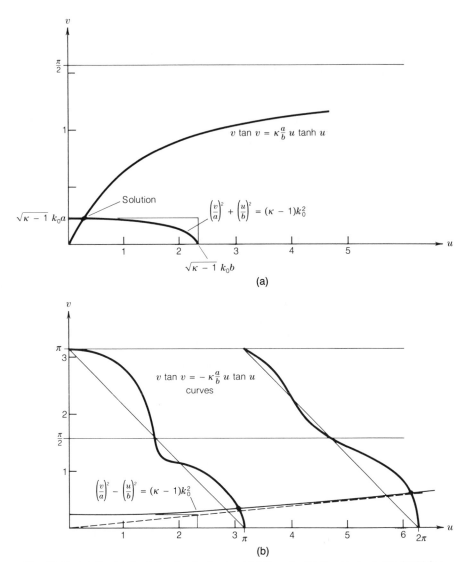

Fig. 10.14. (a) Graphical solution for the dominant LSM mode. (b) Graphical solution for higher order LSM modes.

Replace u by ju to obtain

$$v \tan v = \kappa \frac{a}{b} u \tanh u, \qquad \left(\frac{v}{a}\right)^2 + \left(\frac{u}{b}\right)^2 = (\kappa - 1)k_0^2.$$

A typical graphical solution is shown in Fig. 10.14(a). There will always be one solution with p imaginary. The dominant LSM mode will attenuate along x as long as $\ell^2 > \kappa k_0^2 - \beta^2$.

Case 2: ℓ and p are both real.

A typical graphical solution for this case is shown in Fig. 10.14(b). There will be one solution for every increment of π in u. For large v, $v = au/b$. This line intersects the line

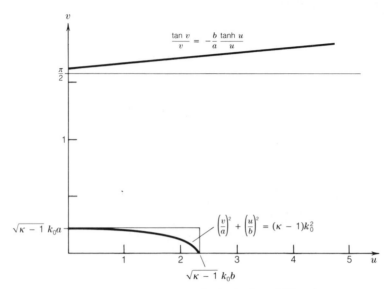

Fig. 10.15. Graphical solution for the dominant LSE mode.

$v = N\pi - u$ at $N\pi - u = au/b$ or $u = N\pi b/c$ where $c = a + b$. In this interval there are N roots so the average spacing Δu between roots is $\pi b/c$. Since $pb = u$ the average spacing Δp is π/c. Hence p_n and ℓ_n approach $n\pi/c$ for large n.

LSE Modes

For these modes the eigenvalue equation can be written as

$$\frac{\tan v}{v} = -\frac{b}{a}\frac{\tan u}{u}, \qquad \ell a = v, \ pb = u,$$

and

$$\left(\frac{v}{a}\right)^2 - \left(\frac{u}{b}\right)^2 = (\kappa - 1)k_0^2.$$

Case 1: ℓ is real, p is imaginary.
If we replace p by jp then

$$\frac{\tan v}{v} = -\frac{b}{a}\frac{\tanh u}{u}, \qquad \left(\frac{v}{a}\right)^2 + \left(\frac{u}{b}\right)^2 = (\kappa - 1)k_0^2.$$

There is no LSE surface wave mode for $\sqrt{\kappa - 1}k_0a < \pi/2$ as can be seen from Fig. 10.15.
Case 2: ℓ and p are both real.
For this case the graphical solution is similar to that for the case 2 LSM modes. When ℓ and p are large $\ell \approx p$ and then $\tan \ell a = -\tan pb$ or $\sin \ell a \cos pb + \cos \ell a \sin pb = \sin \ell c = 0$ so $\ell = n\pi/c$ for ℓ large.

A microstrip line must be designed so that the dominant LSM mode attenuates in the x direction. At cutoff $\ell^2 = \kappa k_0^2 - \beta^2$. For $b = \infty$ it is then found that the substrate thickness a

must satisfy the condition

$$a < \frac{1}{\sqrt{\kappa k_0^2 - \beta^2}} \tan^{-1} \left(\sqrt{\frac{\beta^2 - k_0^2}{\kappa k_0^2 - \beta^2}} \kappa \right).$$

This equation is obtained by solving

$$\ell \tan \ell a = \kappa p = \kappa \sqrt{(\kappa - 1)k_0^2 - \ell^2}$$

for a at cutoff.

Edge Conditions

If a function $h(x)$ behaves like x^{ν} as $x \to 0$ then its Fourier transform $H(w)$ behaves like $w^{-(1+\nu)}$ as $w \to \infty$.

For the junction the edge conditions in the spatial and spectral domains are

$$
\begin{array}{lll}
E_x \sim x^{-1/2} & \hat{E}_x \sim w^{-1/2} & \hat{J}_1 \sim w^{-1/2} \\
E_z \sim x^{1/2} & \hat{E}_z \sim w^{-3/2} & \hat{J}_2 \sim w^{1/2} \\
J_x \sim x^{1/2} & \hat{J}_x \sim w^{-3/2} & j\omega\epsilon_0(w\hat{E}_x + \beta\hat{E}_z) \sim w^{1/2} \\
J_z \sim x^{-1/2} & \hat{J}_z \sim w^{-1/2} & j\omega\epsilon_0(\beta\hat{E}_x - w\hat{E}_z) \sim w^{-1/2}.
\end{array}
$$

Analytic Solution of Functional Equations

On the half plane E_x and E_z are zero. Hence $\hat{E}_x$ and $\hat{E}_z$ are analytic functions in a suitable defined upper half of the complex w plane. We define functions $F_1^+(w)$ and $F_2^+(w)$, analytic in the upper half plane, as follows:

$$F_1^+(w) = j\omega\epsilon_0(w\hat{E}_x + \beta\hat{E}_z) \tag{109a}$$

$$F_2^+(w) = j\omega\epsilon_0(\beta\hat{E}_x - w\hat{E}_z). \tag{109b}$$

We also define functions $S_1(w)$ and $S_2(w)$ as follows:

$$S_1(w) = \frac{\kappa \cot \ell a}{\ell} + \frac{\cot pb}{p} \tag{110a}$$

$$S_2(w) = \ell \cot \ell a + p \cot pb \tag{110b}$$

where $\ell^2 = k^2 - \beta^2 - w^2$ and $p^2 = k_0^2 - \beta^2 - w^2$.

Equations (106) can be written as

$$S_1(w)F_1^+(w) = \hat{J}_1^-(w) \tag{111a}$$

$$S_2(w)F_2^+(w) = \hat{J}_2^-(w) \tag{111b}$$

where we have explicitly identified $\hat{J}_1$ and $\hat{J}_2$ as functions that are analytic in a lower half plane by the use of the minus sign as a superscript. The function $S_1(w)$ has poles when $\ell a = n\pi$, $pb = n\pi$ or at

$$w = \pm j\sqrt{\left(\frac{n\pi}{a}\right)^2 + \beta^2 - k^2}$$

$$w = \pm j\sqrt{\left(\frac{n\pi}{b}\right)^2 + \beta^2 - k_0^2}.$$

We will let $\pm j\nu_n$ and $\pm j\sigma_n$ denote these poles. There are poles for ν_0 and σ_0 also corresponding to

$$w = \pm \sqrt{k^2 - \beta^2} \quad \text{and} \quad w = \pm j\sqrt{\beta^2 - k_0^2}.$$

The parameter ν_0 is imaginary but σ_0 is real. The zeros of $S_1(w)$ correspond to the eigenvalues ℓ_n, p_n for LSM modes. These zeros will be denoted by $\pm j\gamma_n$, $n = 0, 1, 2, \ldots$. The zero for $n = 0$ gives the eigennumber $\pm j\gamma_0$ for the dominant LSM surface wave mode.

The function $S_2(w)$ has poles at $\pm j\nu_n$ and $\pm j\sigma_n$ also but none at ν_0 or σ_0. The zeros of S_2 occur at the eigenvalues ℓ_n, p_n for the LSE modes. These will be designated as $\pm j\alpha_n$.

The function $S_1(w)$ can be expressed in infinite-product form; thus

$$S_1(w) = \frac{\kappa p \sin pb \cos \ell a + \ell \sin \ell a \cos pb}{ab\left(\ell^2 \dfrac{\sin \ell a}{\ell a} p^2 \dfrac{\sin pb}{pb}\right)}$$

$$= \frac{K\displaystyle\prod_{n=1}^{\infty}\left(1 + \frac{w^2}{\gamma_n^2}\right)\left(1 + \frac{w^2}{\gamma_0^2}\right)}{(w^2 + \beta^2 - k^2)(w^2 + \beta^2 - k_0^2)\displaystyle\prod_{n=1}^{\infty}\left(1 - \frac{\ell^2 a^2}{n^2\pi^2}\right)\left(1 - \frac{p^2 b^2}{n^2\pi^2}\right)} \qquad (112)$$

where K is a constant and ℓ^2, p^2 are given by the relations following (110).

Before proceeding further we must establish what division of the complex w plane we need to use to define our upper and lower half planes. We know that F_1^+ should have poles at $w = -j\gamma_n$ in order to give rise to the spectrum of LSM modes in the region $x > 0$. All of these modes decay with increasing values of x. Similarly, F_2^+ must have poles at $w = -j\alpha_n$ to produce the spectrum of LSE modes in the region $x > 0$. These poles lie in the lower half plane. The function $\hat{J}_1^-$ must have poles at $w = j\nu_n$ and $w = j\sigma_n$, $n = 0, 1, 2, \ldots$, in order to produce the full spectrum of reflected modes in regions 1 and 2. We will also assume that the LSM mode r with a pole at $w = -j\nu_r$ is incident in region 1 and that mode t with a pole at $w = -j\sigma_t$ is incident in the region labeled 2 in Fig. 10.13(a). All of these poles lie in what we call the upper half plane. The poles associated with $\hat{J}_1^-$ and F_1^+ give rise to LSM modes. If we assume that mode i of the LSE spectrum is incident in region 1 and that mode j is incident in region 2 then $\hat{J}_2^-$ has poles at $j\nu_n$, $j\sigma_n$, $n = 1, 2, \ldots$, and at $-j\nu_i$, $-j\sigma_j$. All of these poles are also considered to lie in the upper half plane. The inversion contour that will divide the complex w plane into upper and lower half planes must thus be chosen in the

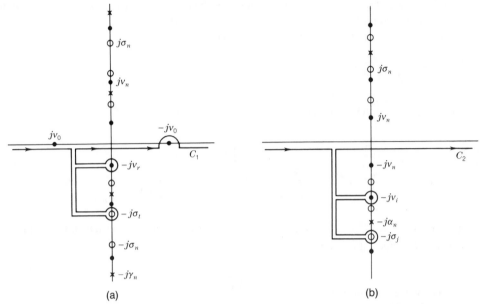

Fig. 10.16. (a) Inversion contour for LSM modes. (b) Inversion contour for LSE modes.

manner shown in Fig. 10.16 in order to conform to the above prescription for the disposition of the poles. Note that the division of the w plane is different for the LSM and LSE modes since they have different poles.

We can factor S_1 and S_2 into the forms $S_1^+ S_1^-$ and $S_2^+ S_2^-$ where $S_1^\pm$ and $S_2^\pm$ are free of zeros and poles in the upper and lower half planes, respectively, and have algebraic behavior at infinity. Equations (111) can thus be expressed in the form

$$S_1^+ F_1^+ = \frac{\hat{J}_1^-}{S_1^-} = Q_e(w) \tag{113a}$$

$$S_2^+ F_2^+ = \frac{\hat{J}_2^-}{S_2^-} = Q_h(w). \tag{113b}$$

These factored functions are analytic continuations of each other and together define functions $Q_e(w)$ and $Q_h(w)$ that are analytic in the whole complex w plane. Thus Q_e and Q_h are polynomials of finite order. The solutions for $\hat{J}_1^-$ and $\hat{J}_2^-$ are

$$\hat{J}_1^- = Q_e S_1^- \tag{114a}$$

$$\hat{J}_2^- = Q_h S_2^-. \tag{114b}$$

The coupling conditions (107) and (108) give

$$Q_e(\pm j\beta)\, S_1^-(\pm j\beta) \mp j Q_h(\pm j\beta)\, S_2^-(\pm j\beta) = 0. \tag{115}$$

The residues, upon evaluating J_1 and J_2 by contour integration, at $-jv_r$, $-j\sigma_t$, $-jv_i$, and

$-j\sigma_j$ must produce the known incident modes. We thus have six equations that will serve to determine the polynomials Q_e and Q_h as we will show next.

The factoring of S_1 and S_2 is straightforward. For S_1^- we obtain

$$S_1^- = \frac{e^{j(w/\pi)(c \ln c - a \ln a - b \ln b)}\left(1 - \dfrac{w}{j\gamma_0}\right)}{(w + \sqrt{k^2 - \beta^2})(w - j\sqrt{\beta^2 - k_0^2})(w + j\nu_r)(w + j\sigma_t)}$$

$$\times \frac{\displaystyle\prod_{n=1}^{\infty}\left(1 - \frac{w}{j\gamma_n}\right)e^{-jwc/n\pi}}{\displaystyle\prod_{n=1}^{\infty}\left(1 - \frac{w}{j\nu_n}\right)e^{-jwa/n\pi}\prod_{n=1}^{\infty}\left(1 - \frac{w}{j\sigma_n}\right)e^{-jwb/n\pi}}. \tag{116}$$

Note that the poles $-j\nu_r$, $-j\sigma_t$ are included in S_1^- because of the way in which we partitioned the w plane. The corresponding expression for S_2^- is

$$S_2^- = \frac{e^{j(w/\pi)(c \ln c - a \ln a - b \ln b)}}{(w + j\nu_i)(w + j\sigma_j)} \times \frac{\displaystyle\prod_{n=1}^{\infty}\left(1 - \frac{w}{j\alpha_n}\right)e^{-jwc/n\pi}}{\displaystyle\prod_{n=1}^{\infty}\left(1 - \frac{w}{j\nu_n}\right)e^{-jwa/n\pi}\prod_{n=1}^{\infty}\left(1 - \frac{w}{j\sigma_n}\right)e^{-jwb/n\pi}}. \tag{117}$$

The first exponential function in the numerators for S_1^- and S_2^- is introduced so that these functions will have algebraic behavior at infinity. The asymptotic behavior for large w is established by comparison with the gamma functions $\Gamma(jwc/\pi)$, $\Gamma(jwa/\pi)$, and $\Gamma(jwb/\pi)$. The function S_1^- is asymptotic to $w^{-5/2}$ and S_2^- is asymptotic to $w^{-3/2}$. Since $\hat{J}_1^-$ is asymptotic to $w^{-1/2}$ while $\hat{J}_2^-$ is asymptotic to $w^{1/2}$ we see that Q_e and Q_h are, at most, polynomials of degree 2. We now let

$$Q_e = P_0 + P_1 w + P_2 w^2 \tag{118a}$$

$$Q_h = Q_0 + Q_1 w + Q_2 w^2. \tag{118b}$$

Upon using (114), (115), and (118) we find that

$$Q_0 \pm j\beta Q_1 - \beta^2 Q_2 = (P_0 \pm j\beta P_1 - \beta^2 P_2)$$

$$\times \frac{(1 \mp \beta/\gamma_0)(\pm\beta + \nu_i)(\pm\beta + \sigma_j)}{(\pm j\beta - \sqrt{k^2 - \beta^2})(\pm j\beta - j\sqrt{\beta^2 - k_0^2})}\prod_{n=1}^{\infty}\frac{1 \mp \beta/\gamma_n}{1 \mp \beta/\alpha_n}\frac{1}{(\pm\beta + \nu_r)(\pm\beta + \sigma_t)}. \tag{119}$$

The solutions for J_1 and J_2 are given by

$$J_1(x) = \frac{1}{2\pi}\int_{C_1} Q_e(w)S_1^-(w)e^{-jwx}\,dw \tag{120a}$$

$$J_2(x) = \frac{1}{2\pi}\int_{C_2} Q_h(w)S_2^-(w)e^{-jwx}\,dw. \tag{120b}$$

These integrals can be evaluated in terms of the residues by closing the contour by a semicircle in the upper half plane. The inversion contours C_1 and C_2 are those shown in Fig. 10.16. If the dominant LSM mode is incident in region 1 then C_1 must be indented below the pole at $w = -j\nu_0 = \sqrt{k^2 - \beta^2}$. In practice it is preferable to assume only one incident mode at a time and then evaluate the reflection and coupling coefficients for this case. With only one incident mode one of the polynomials Q_e or Q_h is a constant and the other one is of degree one. In applying the theory to a microstrip line it is easier to assume a value for β and solve for the strip width $2W$. If we specify $2W$ then we have a rather complex transcendental equation to solve since the terminating reactance B in (102) is a function of β. When there are several interacting modes all the reflection and coupling coefficients must be determined as a function of β.

As a simple example we will consider a microstrip line with a very wide strip so that only the dominant LSM mode has significant interaction between the two edges at $x = \pm W$. In the expression for S_1^- we choose $\nu_r = \nu_0$ and delete the factor $w + j\sigma_t$. In the expression for S_2^- we delete the factors $(w + j\nu_i)$ and $(w + j\sigma_j)$. Then $P_2 = Q_1 = Q_2 = 0$. The reflection coefficient R for the dominant LSM mode current J_1 in region 1 is the ratio of the residues at $w = j\nu_0$ and $w = -j\nu_0$ in (120a). This ratio is (note that $j\nu_0 = -\sqrt{k^2 - \beta^2}$)

$$R = \frac{P_0 - P_1\sqrt{k^2 - \beta^2}}{P_0 + P_1\sqrt{k^2 - \beta^2}} e^{j\theta} \tag{121a}$$

where θ is given by

$$\theta = -(2/\pi)\sqrt{k^2 - \beta^2}(c \ln c - a \ln a - b \ln b)$$

$$- 2 \tan^{-1} \frac{\sqrt{k^2 - \beta^2}}{\gamma_0} - 2 \tan^{-1} \frac{\sqrt{\beta^2 - k_0^2}}{\sqrt{k^2 - \beta^2}}$$

$$- 2 \sum_{n=1}^{\infty} \left(\tan^{-1} \frac{\sqrt{k^2 - \beta^2}}{\gamma_n} - \tan^{-1} \frac{\sqrt{k^2 - \beta^2}}{\nu_n} \right.$$

$$\left. - \tan^{-1} \frac{\sqrt{k^2 - \beta^2}}{\sigma_n} \right). \tag{121b}$$

The coupling condition (119) gives

$$\frac{P_0 - j\beta P_1}{P_0 + j\beta P_1} = \frac{\gamma_0 - \beta}{\gamma_0 + \beta} \frac{\beta + \sqrt{\beta^2 - k_0^2}}{\beta - \sqrt{\beta^2 - k_0^2}} \prod_{n=1}^{\infty} \frac{\gamma_n - \beta}{\gamma_n + \beta} \frac{\alpha_n + \beta}{\alpha_n - \beta}. \tag{122}$$

The right-hand side of (122) is real. We can choose $P_0 = 1$; then in order for the left-hand side to be real P_1 must be imaginary. We can therefore infer that the magnitude $|R| = 1$ so no power is transmitted past the edges at $x = \pm W$ in the microstrip line. Since J_1 is proportional to the charge density the electric field E_y for the dominant LSM mode has the same reflection coefficient R. Thus the terminating normalized susceptance $-jB$ is given by

$$-jB = \frac{1 - R}{1 + R}. \tag{123}$$

The inductive nature of the termination is due to the inductive impedance of the dominant evanescent LSM mode in region 3. For the LSM modes the wave impedance is given by

$$Z_w = \frac{E_y}{H_z} = j\frac{\beta^2 - \gamma_n^2}{k_0\gamma_n}Z_0.$$

Since $\beta > \gamma_0$ the wave impedance of the dominant LSM mode is inductive. A sample calculation carried out for $b = 1$ cm, $a = 0.1$ cm, $f = 5$ GHz, $\kappa = 6$, gave a value of 0.34 for the normalized value of B. For this value the effective dielectric constant κ_e has the value $\beta^2/k_0^2 = \kappa_e = 5.34$ and $2W/a = 8$.

REFERENCES AND BIBLIOGRAPHY

[10.1]   R. E. A. C. Paley and N. Wiener, *Fourier Transforms in the Complex Domain*. New York, NY: Amer. Math. Soc. Colloq. Publs., vol. 19, 1934.

[10.2]   S. N. Karp, "An application of Sturm-Liouville theory to a class of two part boundary value problems," *Proc. Cambridge Phil. Soc.*, vol. 53, part 2, pp. 368–381, Apr. 1957.

[10.3]   E. T. Copson, *Theory of Functions of a Complex Variable*. New York, NY: Oxford University Press, 1935, p. 228, probs. 10 and 11.

[10.4]   A. S. Omar and K. Schünemann, "Space-domain decoupling of LSE and LSM fields in generalized planar guiding structures," *IEEE Trans. Microwave Theory Tech.*, vol. MTT-32, pp. 1626–1632, 1984.

Problems Involving the Solution of an Infinite Set of Algebraic Equations

[10.5]   L. Brillouin, "Waveguides for slow waves," *J. Appl. Phys.*, vol. 19, pp. 1023–1041, Nov. 1948.

[10.6]   F. Berz, "Reflection and refraction of microwaves at a set of parallel metallic plates," *Proc. IEE (London)*, vol. 98, part III, pp. 47–55, Jan. 1951.

[10.7]   E. A. N. Whitehead, "The theory of parallel plate media for microwave lenses," *Proc. IEE (London)*, vol. 98, part III, pp. 133–140, Jan. 1951.

[10.8]   R. A. Hurd and H. Gruenberg, "*H*-plane bifurcation of rectangular waveguides," *Can. J. Phys.*, vol. 32, pp. 694–701, Nov. 1954.

[10.9]   R. A. Hurd, "The propagation of an electromagnetic wave along an infinite corrugated surface," *Can. J. Phys.*, vol. 32, pp. 727–734, Dec. 1954.

Problems Solved by the Wiener–Hopf Method

[10.10] A. E. Heins, "The radiation and transmission properties of a pair of semi-infinite parallel plates," *Quart. Appl. Math.*, vol. 6, part I, pp. 157–166, July 1948; part II, pp. 215–220, Oct. 1948.

[10.11] J. F. Carlson and A. E. Heins, "The reflection of an electromagnetic wave by an infinite set of plates," *Quart. Appl. Math.*, part I, vol. 4, pp. 313–329, Jan. 1947; part II, vol. 5, pp. 82–88, Apr. 1947.

[10.12] L. A. Vajnshtejn, "Propagation in semi-infinite waveguides," in New York Univ. Inst. Math. Sci. Rep. EM-63 (six papers translated by J. Shmoys), 1954.

[10.13] J. D. Pearson, "Diffraction of electromagnetic waves by a semi-infinite circular waveguide," *Proc. Cambridge Phil. Soc.*, vol. 49, pp. 659–667, 1953.

[10.14] V. M. Papadopoulos, "Scattering by a semi-infinite resistive strip of dominant mode propagation in an infinite rectangular waveguide," *Proc. Cambridge Phil. Soc.*, vol. 52, pp. 553–563, July 1956.

[10.15] H. Levine and J. Schwinger, "On the radiation of sound from an unflanged circular pipe," *Phys. Rev.*, vol. 73, pp. 383–406, 1948.

Diaphragms in Waveguides

[10.16] L. A. Vajnshtejn, "Irises in waveguides," *J. Tech. Phys.*, vol. 25, no. 5, pp. 841–846, 1955.

[10.17] S. N. Karp and W. E. Williams, "Equivalence relations in diffraction theory," *Proc. Cambridge Phil. Soc.*, vol. 53, pp. 683–690, 1957.

[10.18] G. L. Baldwin and A. E. Heins, "On the diffraction of a plane wave by an infinite plane grating," *Math. Scand.*, vol. 2, pp. 103–118, 1954.

On the Theory of the Wiener–Hopf Method

[10.19] S. N. Karp, "Wiener-Hopf techniques and mixed boundary value problems," *Commun. Pure Appl. Math.*, vol. 3, pp. 411–426, Dec. 1950.

[10.20] A. E. Heins, "The scope and limitations of the method of Wiener and Hopf," *Commun. Pure Appl. Math.*, vol. 9, pp. 447–466, Aug. 1956.

[10.21] P. M. Morse and H. Feshbach, *Methods of Theoretical Physics*. New York, NY: McGraw-Hill Book Company, Inc., 1953.

[10.22] B. Noble, *Methods Based on the Wiener-Hopf Technique*. New York, NY: Pergamon Press, Inc., 1958. (This book gives an excellent treatment of the Wiener–Hopf technique, and includes a good selection of illustrative problems that can be solved.)

[10.23] R. Mittra and S. W. Lee, *Analytical Techniques in the Theory of Guided Waves*. New York, NY: The Macmillan Company, 1971. (This book contains many examples of finite range problems solved by the modified Wiener–Hopf method.)

[10.24] L. A. Weinstein, *The Theory of Diffraction and the Factorization Method*. Boulder, Colorado: The Golem Press, 1969.

[10.25] H. W. Schilling and R. E. Collin, "Circular waveguide bifurcation for asymmetric modes," *IEE Proc.*, vol. 131, part H, pp. 397–404, 1984. (This paper treats the bifurcation of a circular waveguide for asymmetric modes and is an example of coupling between E modes and H modes due to the edge conditions.)

[10.26] A. M. A. El-Sherbiny, "Exact analysis of shielded microstrip lines and bilateral fin lines," *IEEE Trans. Microwave Theory Tech.*, vol. MTT-29, pp. 669–675, 1981.

[10.27] A. M. A. El-Sherbiny, "Hybrid mode analysis of microstrip lines on anisotropic substrates," *IEEE Trans. Microwave Theory Tech.*, vol. MTT-29, pp. 1261–1265, 1981.

Problems

10.1. An infinitely wide parallel-plate transmission line is bifurcated by an infinitely thin perfect magnetic conductor in the region $z \geq 0$ (see Fig. P10.1). Find the reflection coefficient for a TEM wave incident from $z < 0$. Assume that the dimension b is such that the first mode propagates in the bifurcated region $z > 0$. Use the function-theoretic method for the solution of the infinite set of algebraic equations that determine the mode amplitudes.

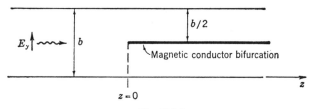

Fig. P10.1.

10.2. Repeat Problem 10.1, using the method of direct construction of a suitable transform function, and also using the Wiener–Hopf method.

10.3. Solve the H-plane bifurcation problem of Section 10.5, using the method of direct construction of a suitable transform function.

10.4. Find the reflection and transmission coefficients at the interface of an infinite set of parallel perfectly conducting plates as illustrated in Fig. 10.5. Assume an incident perpendicular-polarized TEM wave. Also assume that only the H_{10} mode propagates in the parallel-plate region. Note that, for an angle of incidence greater than some minimum value, there will be more than one propagating reflected wave in the region $z < 0$.

10.5. A rectangular guide is loaded by a thin dielectric slab of thickness t for $z \geq 0$. The relative dielectric constant is κ. Maxwell's curl equation for **H** may be written as $\nabla \times \mathbf{H} = j\omega\epsilon_0 \mathbf{E} + j\omega\epsilon_0\chi_e\mathbf{E}$, where $\chi_e = \kappa - 1$. The term $j\omega\epsilon_0\chi_e\mathbf{E}$ may be considered as the polarization current density in the slab. For a sufficiently thin slab, the electric field of an H_{n0} mode may be considered to be constant across the slab. The approximate effect of the slab is thus the same as that of an infinitely thin current sheet $j\omega\epsilon_0\chi_e t\mathbf{E}_0$, where $\mathbf{E}_0$ is the electric field of an H_{n0} mode at the center of the slab, $x = d$. Using this approximation, show that the eigenvalue equation for H_{n0} modes in the

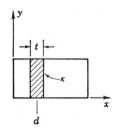
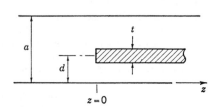

Fig. P10.5.

loaded guide is

$$\left.\frac{\partial E_y}{\partial x}\right|_{d_-}^{d_+} = \chi_e k_0^2 t E_y(d)$$

or

$$\cot hc + \cot hd = -\frac{k_0^2 \chi_e t}{h}$$

and

$$\gamma^2 = h^2 - k_0^2$$

where γ is the propagation constant, h is the transverse wavenumber, and $k_0^2 = \omega^2 \mu_0 \epsilon_0$. Obtain a solution for the reflection and transmission coefficients for an H_{n0} mode incident from $z < 0$. Use the method of direct construction of a transform function. Note that both E_y and $\partial E_y / \partial z$ are finite at $z = 0$, $x = d$, and that this condition governs the growth of the transform function at infinity.

10.6. Solve the problem of a capacitive semidiaphragm in a parallel-plate transmission line of height b (see Fig. P10.6). Assume that only the TEM mode propagates. For an incident TEM mode, show that the capacitive susceptance of the diaphragm is given by $B = 2 \tan(\theta/2)$, where

$$\theta = 2 \sum_{n=1}^{\infty} \left(\sin^{-1} \frac{b/\lambda_0}{n - \frac{1}{2}} - \sin^{-1} \frac{b/\lambda_0}{n} \right).$$

Fig. P10.6.

10.7. Use the Schwarz reflection principle to obtain the solution to Problem 10.1 from the solution to Problem 10.6.

HINT: Construct the solution in terms of the solution for the transverse magnetic field H_x of the preceding problem with the aperture closed by a magnetic wall. Assume for the present case that all modes in the bifurcated region are evanescent.

Answer: The reflection coefficient is $R = -e^{-j\theta}$, where θ is given in Problem 10.6.

10.8. A rectangular guide has a resistive wall at $x = a$, $z > 0$, as in Problem 5.4. An H_{10} mode is incident from $z < 0$. Obtain the solution for the reflection and transmission coefficients. Compare this problem with Problem 10.5 when the substitution $j\omega\epsilon_0(\kappa - 1) = \sigma$ is made.

10.9. Consider the bifurcated parallel-plate guide shown in Fig. 10.13(a). Assume that a TEM mode is incident from $x < 0$ in region 1 and that $\beta = 0$. Find the dominant mode fields transmitted into regions 2 and 3 and the reflection coefficient in region 1. For this problem only LSM modes are needed since $\beta = 0$.

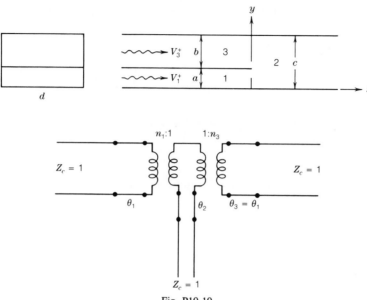

Fig. P10.10.

10.10. Consider the E-plane bifurcation of the rectangular guide shown in Fig. P10.10. Show that the equivalent circuit of the junction is that shown in the figure. This problem may be solved using LSE modes only provided the potential function is chosen as $\Psi_h(x, y, z)\mathbf{a}_x$. The residue-calculus method can be applied by expressing the continuity of E_y and H_x in the aperture. Note that if a TE_{10} mode is incident from region 2, the bifurcation has no effect so that there will be no reflected wave in region 2 and transmission into regions 1 and 3 takes place without any phase shift.

Answer: $n_1 = \sqrt{c/a}$, $n_3 = \sqrt{c/b}$, $\theta_1 + \theta_2 = 0$ or 2π, and $\theta_1 = (\beta_1/\pi)(c \ln c - a \ln a - b \ln b) - \sum_1^\infty (\tan^{-1} \beta_1/\gamma_{1n} + \tan^{-1} \beta_1/\gamma_{3n} - \tan^{-1} \beta_1/\gamma_{2n})$ where $\beta_1^2 = k_0^2 - (\pi/d)^2$, $\gamma_{1n}^2 = (n\pi/a)^2 - \beta_1^2$, $\gamma_{2n}^2 = (n\pi/c)^2 - \beta_1^2$, and $\gamma_{3n}^2 = (n\pi/b)^2 - \beta_1^2$.

10.11. For the E-plane bifurcated guide in Problem 10.10 assume that a dielectric slab of thickness a lies along the bottom wall. A dominant TE_{10} mode is incident in $x < 0$, $0 < y < a$. Find the reflection coefficient for this mode. This problem is essentially the same as the one solved in Section 10.6 with $\beta = \pi/d$ where d is the width of the guide.

10.12. Use the residue-calculus method to find the equivalent circuit of the bifurcated coaxial transmission line junction illustrated in Fig. P10.12. Assume that TEM modes are incident in which case only higher order TM_{0n} modes are excited.

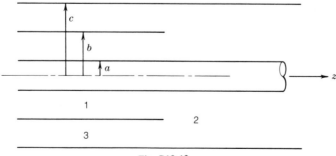

Fig. P10.12.

Answer: The equivalent circuit is the same as that shown in Fig. P10.10. The ideal transformers can be eliminated provided the characteristic impedance of region i is set equal to Z_i where $Z_1 = (Z_0/2\pi) \ln b/a$,

$Z_2 = (Z_0/2\pi) \ln c/a$, $Z_3 = (Z_0/2\pi) \ln c/b$. Note that $Z_1 + Z_3 = Z_2$. The angle θ_1 is given by

$$\theta_1 = \frac{k_0}{\pi}[(c-a)\ln(c-a) - (c-b)\ln(c-b) - (b-a)\ln(b-a)]$$

$$-\sum_{n=1}^{\infty}\left(\tan^{-1}\frac{k_0}{\gamma_{1n}} + \tan^{-1}\frac{k_0}{\gamma_{3n}} - \tan^{-1}\frac{k_0}{\gamma_{2n}}\right).$$

Let $W_i(k_i, x, r) = J_0(k_i x)Y_0(k_i r) - J_0(k_i r)Y_0(k_i x)$. Then k_{1n} are the roots of $W_1(k_1, a, b) = 0$, k_{2n} are the roots of $W_2(k_2, a, c) = 0$, and k_{3n} are the roots of $W_3(k_3, b, c) = 0$. The following Bessel function integrals are useful:

$$\int_1^2 (W_i')^2 r\, dr = \frac{r}{k_i} W_i W_i' \bigg|_1^2 + \frac{r^2}{2}(W_i'^2 + W_i^2)\bigg|_1^2$$

$$\int_1^2 W_i'(k_i r)W_j'(k_j r)r\, dr = \frac{r}{k_i} W_i W_j' \bigg|_1^2 + \frac{k_j}{k_i}\frac{r}{k_i^2 - k_j^2}(k_j W_j' W_i - k_i W_i' W_j)\bigg|_1^2 .$$

where 1 and 2 denote arbitrary lower and upper limits.

11
Surface Waveguides

In addition to the closed cylindrical conducting tube and the conventional TEM wave transmission line, there exists a class of open-boundary structures which are capable of guiding an electromagnetic wave. These structures are capable of supporting a mode that is intimately bound to the surface of the structure. The field is characterized by an exponential decay away from the surface and having the usual propagation function $e^{-j\beta z}$ along the axis of the structure. For this reason this discrete eigenvalue solution to the wave equation is called a surface wave, and the structure that guides this wave may be appropriately referred to as a surface waveguide. Typical structures that are capable of supporting a surface wave are illustrated in Fig. 11.1. These consist of dielectric-coated planes and wires and corrugated planes and wires. In addition, a surface-wave solution exists for the interface between two different media, as well as for a variety of other structures of the periodic type with open boundaries, and also for stratified plane and cylindrical structures.

Although surface waveguides have several features which are similar to those of the cylindrical conducting-tube guide, they have many characteristics which are quite different. Some of the outstanding differences are:

1. the possibility of a surface-wave mode of propagation with no low-frequency cutoff;
2. the nonexistence of an infinite number of discrete modes of propagation at a given frequency;
3. the existence of a finite number of discrete modes, together with an eigenfunction solution with a continuous eigenvalue spectrum; and
4. the possibility of mode solutions with a phase velocity less than that of light.

In this chapter some of the more important characteristics of surface waveguides will be studied by examining the solutions for typical plane and cylindrical structures. In addition, the problem of excitation of surface waves by a line source above a grounded dielectric sheet will be examined. This latter problem will provide an introduction to the role played by the continuous eigenfunction spectrum in determining the total field in the region around a surface waveguide. To evaluate the integral for the radiated field, the saddle-point method of approximate integration will be used. This is a technique that finds widespread application in radiation problems.

11.1. SURFACE WAVES ALONG A PLANE INTERFACE

The guiding of plane waves by a plane interface between two different dielectric media was investigated initially by Uller [11.1] in 1903, and later by Zenneck [11.2] in 1907. This surface wave is associated with the Brewster-angle phenomenon of total transmission across a dielectric interface. It is often referred to as a Zenneck wave.

Consider a plane interface $x = 0$, separating the air region $x > 0$ from the region $x < 0$,

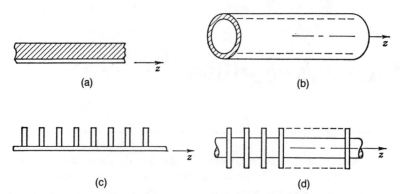

Fig. 11.1. Typical surface waveguides. (a) Dielectric-coated plane. (b) Dielectric-coated wire. (c) Corrugated plane. (d) Corrugated cylinder.

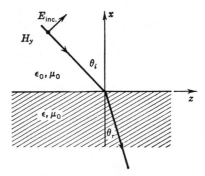

Fig. 11.2. A plane interface between free space and a dielectric-filled region.

occupied by a lossless isotropic dielectric with permittivity $\epsilon = \kappa\epsilon_0$, where κ is the relative dielectric constant. Let a parallel-polarized wave be incident on the interface from the air region at an angle θ_i to the interface normal, as in Fig. 11.2. The incident plane wave has the field components H_y, E_x, E_z, when the plane of incidence is the xz plane.

The above problem is readily solved by introducing the wave impedances and using a transmission-line analogy.[1] The incident magnetic field may be taken as

$$H_y = A \, \exp[-jk_0(z \sin \theta_i - x \cos \theta_i)], \qquad x > 0 \tag{1a}$$

where $k_0^2 = \omega^2 \mu_0 \epsilon_0$, and A is an amplitude constant. In the region $x > 0$ a reflected wave with a magnetic field

$$- RA \, \exp[-jk_0(z \sin \theta_i + x \cos \theta_i)] \tag{1b}$$

will also exist, in general. In the dielectric region, a transmitted wave

$$TA \, \exp[-jk(z \sin \theta_r - x \cos \theta_r)] \tag{1c}$$

will exist, where $k^2 = \omega^2 \mu_0 \epsilon$, θ_r is the angle of refraction, and R, T are the reflection and

[1] See Section 3.5.

transmission coefficients, respectively. The components of the electric field are given by

$$j\omega\epsilon E_x = -\frac{\partial H_y}{\partial z} \tag{2a}$$

$$j\omega\epsilon E_z = \frac{\partial H_y}{\partial x}. \tag{2b}$$

The wave impedances in the two regions are

$$Z_1 = \frac{E_z}{H_y} = Z_0 \cos \theta_i, \qquad x > 0 \tag{3a}$$

$$Z_2 = Z \cos \theta_r, \qquad x < 0 \tag{3b}$$

where $Z_0 = (\mu_0/\epsilon_0)^{1/2}$, and $Z = (\mu_0/\epsilon)^{1/2}$. In terms of these, the reflection coefficient R is given by

$$R = \frac{Z_2 - Z_1}{Z_2 + Z_1} = \frac{\cos \theta_r - \kappa^{1/2} \cos \theta_i}{\cos \theta_r + \kappa^{1/2} \cos \theta_i}. \tag{4}$$

The propagation constant along the z direction must be the same in both media (Snell's law), so that $k_0 \sin \theta_i = k \sin \theta_r$. The Brewster angle for which R vanishes is thus given by $\cos \theta_r = \kappa^{1/2} \cos \theta_i$, or, eliminating θ_r,

$$\sin \theta_i = \left(\frac{\kappa}{\kappa + 1}\right)^{1/2}. \tag{5a}$$

The corresponding angle in the dielectric medium is given by

$$\sin \theta_r = (\kappa + 1)^{-1/2}. \tag{5b}$$

For this particular angle of incidence, the solution for the magnetic field reduces to

$$H_y = \begin{cases} A \exp[-jk_0(\kappa + 1)^{-1/2}(\kappa^{1/2}z - x)], & x > 0 \\ A \exp[-jk_0(\kappa + 1)^{-1/2}(\kappa^{1/2}z - \kappa x)], & x < 0. \end{cases} \tag{6}$$

This solution is the prototype solution for a surface wave guided by a plane interface.

The above solution is for a parallel-polarized plane wave, and may be classified as a transverse magnetic (TM) wave with respect to the z axis. In the case when the dielectric has finite losses, no real angle of incidence can be found such that there is no reflection for the TM wave. However, we do not need to restrict the angle of incidence to real values only. As soon as complex angles of incidence are considered, it is found that a solution for an inhomogeneous plane wave of the TM type exists such that there is no reflection at the interface.

We will now consider the extension of the previous TM wave solution to the case of a dielectric with finite, but small, losses. Let the permittivity of the dielectric medium be $\epsilon = \epsilon' - j\epsilon''$, where $\epsilon'' \ll \epsilon'$ and represents the loss in the material. Instead of introducing the angles of incidence and refraction explicitly, the incident and transmitted waves will be represented as

$$H_y = A \, \exp(jh_1 x - j\beta z), \qquad x > 0 \tag{7a}$$

$$H_y = A \, \exp(jh_2 x - j\beta z), \qquad x < 0 \tag{7b}$$

where $h_1^2 + \beta^2 = k_0^2$, and $h_2^2 + \beta^2 = k^2$, in order that the field shall satisfy the wave equation. The wave impedances are now given by

$$Z_1 = \frac{h_1}{k_0} Z_0 \tag{8a}$$

$$Z_2 = \frac{h_2}{k} Z. \tag{8b}$$

For the surface wave, the reflection coefficient must vanish, and hence $Z_1 = Z_2$. The solutions for the transverse wavenumbers h_1 and h_2 and the propagation constant β are hence determined by the following two equations:

$$h_1^2 - k_0^2 = h_2^2 - k^2 \tag{9a}$$

$$h_1 k Z_0 = h_2 k_0 Z$$

or

$$h_1(\epsilon' - j\epsilon'') = \epsilon_0 h_2. \tag{9b}$$

For convenience, let $\epsilon' - j\epsilon'' = (\kappa' - j\kappa'')\epsilon_0 = \kappa\epsilon_0$. From (9a), we obtain $h_1 = [h_2^2 - (\kappa' - j\kappa'' - 1)k_0^2]^{1/2}$, which may be combined with (9b) to yield

$$h_2 = (\kappa' - j\kappa'')k_0(\kappa' - j\kappa'' + 1)^{-1/2}. \tag{10}$$

Since κ'' is small compared with κ', an approximate solution for h_2 is readily obtained by using the binomial expansion to get

$$h_2 \approx \frac{k_0}{(\kappa' + 1)^{1/2}} \left(\kappa' - j\frac{1}{2}\kappa'' \frac{\kappa' + 2}{\kappa' + 1} \right). \tag{11a}$$

The corresponding solution for h_1 is obtained from (9b) and (10):

$$h_1 \approx \frac{k_0}{(\kappa' + 1)^{1/2}} \left[1 + \frac{j\kappa''}{2(\kappa' + 1)} \right]. \tag{11b}$$

For small losses, it is seen that the real parts of h_1 and h_2 are the same as those in the lossless case, i.e., in the solution given by (6).

It is interesting to note from (11) that the imaginary parts of h_1 and h_2 are of opposite sign, and, hence, the surface wave has the characteristic exponential decay in the direction perpendicular and away from the interface on both sides. The solution for the magnetic field in the present case is

$$H_y = \begin{cases} A \exp[-j\beta z + (jh_1' - h_1'')x], & x > 0 \\ A \exp[-j\beta z + (jh_2' + h_2'')x], & x < 0 \end{cases} \tag{12}$$

where h_1', h_1'' and h_2', h_2'' are the real and imaginary parts of h_1 and h_2 as given by (11). The propagation constant β is given by

$$\beta^2 = k_0^2 - h_1^2 = k_0^2 - (h_1' + jh_1'')^2$$

or, since $h_1'' \ll h_1'$,

$$\beta' - j\beta'' = \beta \approx (k_0^2 - h_1'^2)^{1/2} \left(1 - j\frac{h_1'h_1''}{k_0^2 - h_1'^2}\right). \tag{13}$$

For small losses, β'' is small, and the attenuation of the wave is also small. At the same time h_1'' is small, so that the wave is rather loosely bound to the interface.

The constant-phase planes for the wave on the air side are determined by the equation

$$\beta'z - h_1'x = \text{constant} \tag{14}$$

where β' is the real part of β, as may be seen from (12). These planes make an angle θ_1 with respect to the x axis, where

$$\theta_1 = \tan^{-1}\frac{h_1'}{\beta'} = \cot \kappa' \tag{15}$$

a result which is identical to that for the lossless case. The planes of constant amplitude are determined by

$$\beta''z + h_1''x = \text{constant} \tag{16}$$

where β'' is the imaginary part of β. The constant-amplitude and constant-phase planes may be shown to be orthogonal. Further details of this problem have been given by Stratton, and hence will not be pursued any further here [11.3, Section 9.13].

A solution similar to the above for a perpendicular-polarized (transverse electric, TE) wave does not exist unless the two media have different permeabilities. When the permeabilities are the same, the wave admittances are proportional to h_1 and h_2 in the two media. In order for the reflection coefficient to vanish, h_1 must equal h_2. However, this is impossible since $h_1^2 - k_0^2$ must also be equal to $h_2^2 - k^2$. The solution for the case of unequal permeabilities may be constructed from the above solution by an application of the duality principle and is left as a problem at the end of the chapter.

11.2. Surface Waves along an Impedance Plane

By examining the possibility of surface-wave propagation along a plane that exhibits a complex surface impedance, further insight into the type of surface required to support a surface wave may be obtained. The analysis in this section will show that a TM wave requires a surface with a surface impedance having an inductive reactive term, while, in order to

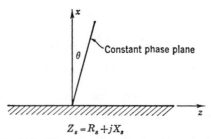

$$Z_s = R_s + jX_s$$

Fig. 11.3. Illustration of a plane impedance surface.

support a TE surface wave, the reactive part of the surface impedance must be capacitive. Since naturally occurring surfaces have an inductive reactance term in the surface impedance, the TM surface wave is of more importance than the TE wave.

Consider a plane surface in the $x = 0$ plane which has a normalized surface impedance $Z_s = R_s + jX_s$ that is independent of the angle of incidence for incident inhomogeneous plane waves. The surface impedance is normalized with respect to the intrinsic impedance Z_0 of free space. With reference to Fig. 11.3, let the magnetic field H_y of the assumed TM surface wave have the following form in the region $x > 0$:

$$H_y = A \exp(jhx - j\beta z) \tag{17}$$

where $h^2 + \beta^2 = k_0^2$. The electric field components are given by

$$j\omega\epsilon_0 E_z = \frac{\partial H_y}{\partial x} \qquad j\omega\epsilon_0 E_x = -\frac{\partial H_y}{\partial z}.$$

The wave impedance of the wave is

$$Z_1 = \frac{E_z}{H_y} = \frac{h}{\omega\epsilon_0} = \frac{h}{k_0}Z_0$$

and must be equal to the surface impedance of the plane at $x = 0$ if the field is to satisfy the required boundary conditions on this plane without a reflected field being present. Equating the normalized wave impedance h/k_0 to the surface impedance gives the required solution for h:

$$h = h' + jh'' = k_0 Z_s = k_0 R_s + jk_0 X_s. \tag{18}$$

The solution for the propagation constant β is

$$\beta = \beta' - j\beta'' = (k_0^2 - h^2)^{1/2} = k_0(1 + X_s^2 - R_s^2 - 2jR_sX_s)^{1/2}. \tag{19}$$

Equation (18) for the transverse wavenumber h clearly shows that the field will have an exponential decay normal to the surface only if the reactive part of the surface impedance is inductive. Also, the larger the inductive reactance of the surface, the more closely will the wave be bound to the surface. In order that the attenuation of the wave in the z direction will be small, the product R_sX_s should be small, so that, from (19), β'' will be small. A closely bound surface wave may be propagated along the surface with a small attenuation only if X_s is made large and R_s is made small.

The solution for the magnetic field is

$$H_y = A \, \exp[(jh' - h'')x - (j\beta' + \beta'')z] \tag{20}$$

where h', h'', β', and β'' are given by (18) and (19). The constant-phase planes are determined by the equation

$$h'x - \beta'z = \text{constant}$$

and are inclined toward the surface at an angle θ relative to the x axis, where $\tan \theta = h'/\beta'$. The wave front is tilted by an amount just sufficient to give a component of the Poynting vector in the direction of the impedance surface, to account for the power loss in the resistive part R_s of the surface impedance. The power flow per unit area into the surface at any point z is

$$
\begin{aligned}
P_s &= \frac{1}{2} \, \text{Re} \int_z^{z+1} E_z H_y^* \, dz = \frac{1}{2} R_s Z_0 \int_z^{z+1} H_y H_y^* \, dz \\
&= \frac{1}{4} R_s Z_0 A A^* e^{-2\beta'' z} \frac{1 - e^{-2\beta''}}{\beta''} \\
&\approx \frac{1}{2} R_s Z_0 A A^* e^{-2\beta'' z}
\end{aligned}
\tag{21}
$$

where β'' is small. The power flow, per unit width of the surface, in the z direction is

$$P = \frac{1}{2} \, \text{Re} \int_0^\infty E_x H_y^* \, dx = \frac{1}{4} \frac{\beta' Z_0}{h'' k_0} A A^* e^{-2\beta'' z}. \tag{22}$$

The phase velocity of the wave in the z direction is determined by the real part of β. From (19), we obtain

$$\beta' \approx k_0 (1 + X_s^2)^{1/2}$$

when $R_s \ll X_s$, and hence the phase velocity v_p is given approximately by

$$v_p = \frac{\omega}{\beta'} \approx v_c (1 + X_s^2)^{-1/2} \tag{23}$$

where v_c is the velocity of plane waves in free space. The phase velocity is seen to be less than the free-space velocity for a surface wave supported by a highly inductive surface with a small resistance. When the surface resistance is equal to or greater than the reactance, the phase velocity is equal to or greater than the free-space velocity.

The above analysis is directly applicable to the problem of a surface wave guided along a conducting plane. In Section 3.6 it was shown that, for metals at radio frequencies, the surface impedance is independent of the angle of incidence, and has an inductive part equal to the resistive part. The normalized surface impedance of a metal interface is given by

$$Z_s = (1 + j) \left(\frac{k_0}{2\sigma Z_0} \right)^{1/2} \tag{24}$$

where σ is the conductivity of the metal. The transverse wavenumber h is thus given by

$$h = k_0(1+j)\left(\frac{k_0}{2\sigma Z_0}\right)^{1/2} \tag{25a}$$

while the propagation constant β is given by

$$\beta = k_0\left(1 - \frac{jk_0}{\sigma Z_0}\right)^{1/2} \approx k_0 - \frac{jk_0^2}{2\sigma Z_0} \tag{25b}$$

since $k_0/\sigma Z_0$ is very small for any good conductor. For metals, σ is of the order of 10^7 siemens per meter, $Z_0 = 377$ ohms, and, for $\lambda_0 = 3.14$ centimeters, $k_0 = 200$ radians per meter. Hence β'' is approximately 5×10^{-6} neper per meter. This amount of attenuation is generally entirely negligible. The imaginary part h'' of h is approximately equal to 3×10^{-2} neper per meter. Thus the wave amplitude does not decrease to a negligible value before a distance of several thousand wavelengths from the surface has been reached. The wave is too loosely bound to the surface to be of much use in a transmission system of reasonable dimensions.

It has been seen that a conducting plane can support a surface wave of the TM type but that this wave is very loosely bound to the surface. To obtain a practical surface waveguide, it is necessary to increase the inductive reactance of the surface without increasing the surface resistivity significantly. Two methods for achieving this end result are (1) to coat the conductor with a thin layer of high-dielectric-constant material and (2) to corrugate the surface. These structures will be considered in the next two sections, but first we will examine the solution for a TE surface wave guided by an impedance plane.

The TE surface wave is similar in structure to the TM wave, but with the roles of electric field and magnetic field interchanged. The field is described by the following equations:

$$E_y = A \exp(jhx - j\beta z)$$

$$j\omega\mu_0 H_z = -\frac{\partial E_y}{\partial x} \qquad j\omega\mu_0 H_x = \frac{\partial E_y}{\partial z}$$

where $h^2 + \beta^2 = k_0^2$. The wave admittance is equal to

$$-\frac{H_z}{E_y} = \frac{Y_0 h}{k_0}$$

where $Y_0 = Z_0^{-1}$. The normalized wave admittance must be equal to the normalized surface admittance of the guiding plane, and, hence,

$$h = k_0 Y_s = k_0(G_s + jB_s).$$

Clearly, it is necessary that the surface susceptance B_s be positive if the imaginary part h'' of h should be positive. Thus, to obtain a field that decays exponentially with distance from the surface, the surface must have a capacitive susceptance. Dielectric-coated surfaces can be designed so as to exhibit a capacitive susceptance term in the surface admittance and will, therefore, be capable of guiding a TE surface wave.

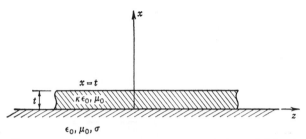

Fig. 11.4. A conducting plane coated with a thin layer of dielectric.

11.3. CONDUCTING PLANE WITH A THIN DIELECTRIC COATING

The propagation of a TM surface wave along a conducting plane coated with a thin layer of dielectric has been investigated by Attwood [11.7]. The results of this analysis have shown that a thin film of dielectric leads to a large increase in the concentration of the field near the surface. The attenuation constant of such a surface waveguide can be considerably less than that of a conventional rectangular waveguide when the parameters of the dielectric film are properly chosen.

Figure 11.4 illustrates a plane conducting surface, coated with a layer of dielectric of thickness t and having a relative complex dielectric constant $\kappa = \kappa' - j\kappa''$. Attwood solved this problem by neglecting the losses initially, in order to obtain expressions for the field components. The attenuation constant is then evaluated by finding the power dissipated in the dielectric and the conducting plane. Since the losses are small, this procedure is valid. However, we will follow an alternative approach and find the transverse wavenumbers h_1 in free space and h_2 in the dielectric, and the propagation constant β directly, by including the effect of the losses from the start. This approach is chosen in order to show more clearly how the surface reactance is increased relative to the surface resistance, by coating the conductor with a thin dielectric layer.

In the free-space region, the magnetic field H_y has the form

$$H_y = A \exp(jh_1 x - j\beta z), \qquad x > t$$

where $h_1^2 + \beta^2 = k_0^2$. In the dielectric, the field consists of two waves propagating in the positive and negative x directions and hence has the form

$$H_y = [B \exp jh_2 x + C \exp(-jh_2 x)]e^{-j\beta z}, \qquad 0 \leq x \leq t$$

where $h_2^2 + \beta^2 = k^2 = \kappa k_0^2$. The wavenumbers h_1, h_2 and the propagation constant β may be found by matching the tangential fields at the dielectric–free-space interface, and by imposing the boundary conditions at the conducting plane. The required eigenvalue equation is more readily derived from the equivalent transmission-line circuit for propagation along the x axis. The wave impedances of the wave in free space and that in the dielectric region are, respectively,

$$Z_1 = \frac{Z_0 h_1}{k_0}$$

$$Z_2 = \frac{Z_0 h_2}{\kappa k_0}.$$

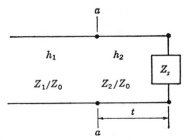

Fig. 11.5. Equivalent transmission-line circuit for propagation in the x direction for a coated conducting plane.

The normalized impedance of the conductor is

$$Z_s = R_s + jX_s = (1+j)\left(\frac{k_0}{2\sigma Z_0}\right)^{1/2}.$$

Figure 11.5 illustrates the equivalent circuit for propagation in the x direction. The surface-wave solution requires that the normalized input impedance at the dielectric interface aa be equal to Z_1/Z_0, so that there will be no reflection at this interface. Conventional transmission-line theory thus gives

$$\frac{Z_1}{Z_0} = Z_{\text{in}} = \frac{Z_2}{Z_0}\frac{Z_s + j(Z_2/Z_0)\tan h_2 t}{Z_2/Z_0 + jZ_s \tan h_2 t}. \tag{26}$$

In this equation a number of approximations may be made, since Z_s is very small at radio frequencies, and it is assumed that the thickness t is chosen so that $h_2 t$ is also very small. Substituting for the wave impedances we obtain from (26)

$$h_1 \approx k_0\left(Z_s + j\frac{h_2^2 t}{\kappa k_0} - jZ_s^2 \kappa k_0 t\right).$$

Replacing h_2^2 by $h_1^2 + (\kappa - 1)k_0^2$, and Z_s^2 by $2jR_s^2$, we get

$$h_1 = k_0\left(R_s + 2R_s^2\kappa k_0 t + j\frac{h_1^2 t}{\kappa k_0} + jk_0 t\frac{\kappa - 1}{\kappa} + jX_s\right).$$

The wavenumbers h_1 and h_2 also satisfy the relations $h_1^2 + \beta^2 = k_0^2$, $h_2^2 + \beta^2 = \kappa k_0^2$. For small losses, β is nearly equal to k_0, and hence h_1^2 is small. The wavenumber h_2, however, is of the same order of magnitude as k_0, and, hence, since $h_2 t$ is assumed to be small, $k_0 t$ is also small. In view of these considerations, we may drop the terms involving R_s^2 and h_1^2 to get the relatively simple equation

$$h_1 = k_0\left(R_s + jX_s + jk_0 t\frac{\kappa - 1}{\kappa}\right)$$

$$\approx k_0\left[\left(R_s + \frac{\kappa'' k_0 t}{\kappa'^2}\right) + j\left(X_s + \frac{\kappa' - 1}{\kappa'}k_0 t\right)\right]. \tag{27}$$

The surface resistance is thus effectively increased by the small term $\kappa'' k_0 t/\kappa'^2$, while the effective surface reactance is increased by the relatively large additional term $(\kappa' - 1)k_0 t/\kappa'$.

As a typical example, consider a copper plane ($\sigma = 5.8 \times 10^7$ siemens per meter), coated with a film of polystyrene ($\kappa = 2.56 - j0.002$) of thickness 0.001 meter at a wavelength of 0.0314 meter. The terms occurring in (27) have the following values for this example:

$$R_s = 6.7 \times 10^{-5}$$

$$\frac{\kappa'' k_0 t}{\kappa'^2} = 6 \times 10^{-5}$$

$$X_s = 6.7 \times 10^{-5}$$

$$\frac{\kappa' - 1}{\kappa'} k_0 t = 0.122 = 1.82 \times 10^3 X_s.$$

The last two terms that were dropped in arriving at (27) have the values

$$2R_s^2 \kappa k_0 t = (47 - j0.037) \times 10^{-10}$$

$$\frac{h_1^2 t}{\kappa k_0} \approx 1.2 \times 10^{-3} + j9 \times 10^{-7}.$$

These terms are seen to be negligible, and justify the approximations made. The imaginary part h_1'' of h_1 is equal to 24.4 nepers per meter, and hence the field has decayed to 10% of its value at the surface in a distance of three wavelengths from the surface. The addition of a thin film of dielectric leads to a surprisingly large increase in the decay constant away from the surface.

From (27) it is seen that a conductor coated with a thin layer of dielectric may be conveniently viewed as an impedance surface with a normalized surface impedance Z_d given by

$$Z_d = R_s \left(1 + \frac{\kappa'' k_0 t}{R_s \kappa'^2} \right) + j \left(X_s + \frac{\kappa' - 1}{\kappa'} k_0 t \right). \tag{28}$$

The propagation factor β is given by

$$\beta = \beta' - j\beta'' = (k_0^2 - h_1^2)^{1/2} = (k_0^2 - h'^2_1 + h''^2_1 - 2jh_1'h_1'')^{1/2}$$
$$\approx (k_0^2 + h''^2_1)^{1/2} - j\frac{h_1'h_1''}{k_0}$$

since $h_1' \ll h_1'' \ll k_0$. Substituting from (27), we get

$$\beta' \approx k_0 \left[1 + \left(X_s + k_0 t \frac{\kappa' - 1}{\kappa'} \right)^2 \right]^{1/2} \tag{29a}$$

$$\beta'' = k_0 \left(R_s + \frac{\kappa'' k_0 t}{k'^2} \right) \left(X_s + k_0 t \frac{\kappa' - 1}{\kappa'} \right). \tag{29b}$$

For the example considered earlier, the attenuation in the z direction is 3.1×10^{-3} neper

708 FIELD THEORY OF GUIDED WAVES

per meter or 2.7×10^{-2} decibel per meter. This compares favorably with the attenuation of conventional rectangular guides.

The preceding analysis is valid only for low-loss thin dielectric sheets over a low-loss conducting plane. For sheets with a thickness that is an appreciable fraction of a wavelength, the eigenvalue equation for the transverse wavenumber must be derived without making the approximation $h_2 t \ll 1$. The propagation of surface waves along thick dielectric slabs is considered in a later section, and so the case of thick films will not be considered here.

11.4. Surface Waves along a Corrugated Plane

The propagation of surface waves along a corrugated plane has been studied by several investigators [11.5]–[11.8]. Corrugated planes supporting surface waves have found application in end-fire antennas and are, therefore, of some practical interest [11.9]. Only the simplest case, that of infinitely thin teeth, as illustrated in Fig. 11.6, will be considered here. The corrugated plane is infinite in extent, and assumed to be perfectly conducting. The corrugations are spaced by a distance S and have a depth d.

Consider first the solution for a TM surface wave propagating along the z direction. The magnetic field has a single component H_y, from which the electric field components E_x and E_z may be found. Since the structure is a periodic one, the field has the form governed by Floquet's theorem. Above the corrugations, a suitable expansion for H_y is

$$H_y = \sum_{n=-\infty}^{\infty} A_n \exp\left[-p_n x - j\left(\beta + \frac{2n\pi}{S}\right)z\right], \qquad x > d \qquad (30)$$

where $(\beta + 2n\pi/S)^2 = k_0^2 + p_n^2$. In the corrugated region, H_y must have a form such that the electric field derivable from H_y should vanish on the sides of the corrugations, as well as on the conducting plane at $x = 0$. A suitable expansion for H_y is readily found to be

$$H_y = \sum_{n=0}^{\infty} B_n \cos \frac{n\pi z}{S} \cosh h_n x, \qquad 0 \le z \le S, \, 0 \le x \le d \qquad (31)$$

where $h_n^2 = (n\pi/S)^2 - k_0^2$. In the Nth corrugation, the field is related to that in the region $0 \le z \le S$ as follows:

$$H_y(x + NS) = e^{-j\beta NS} H_y(x) \qquad \text{all } N. \qquad (32)$$

It will be assumed that the spacing S is chosen so small that all h_n except $h_0 = jk_0$ are real. Thus, only the TEM mode in the mode expansion (31) propagates in the x direction between

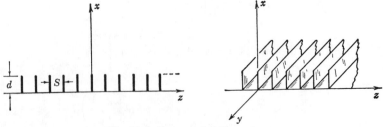

Fig. 11.6. Corrugated-plane conductor.

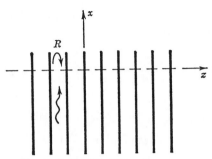

Fig. 11.7. Parallel-plate system.

corrugations. The solution to the above boundary-value problem may be found by methods similar to those discussed in Chapter 10. However, we may obtain the eigenvalue equation for β by a simple transmission-line analysis. Consider a system of parallel plates of spacing S and semi-infinite in length, as in Fig. 11.7. Let a TEM wave with a magnetic field H_y equal to

$$H_y = Ce^{-jk_0x}e^{-j\beta NS}, \qquad (N-1)S < z < NS$$

in the Nth corrugation be incident on the interface from the parallel-plate region. At the interface there is a reflected TEM wave,

$$H_y = -RCe^{jk_0x}e^{-j\beta NS}$$

in the Nth section, a dominant-mode transmitted field,

$$H_y = Ae^{-p_0(x-d)}e^{-j\beta z}$$

plus higher order evanescent modes that decay exponentially away from the interface on both sides. For the surface wave, p_0 is a positive-real wavenumber, and, hence, there is no power flow away from the interface in the positive x direction. There will, therefore, be complete reflection, and the modulus of the reflection coefficient $|R|$ must be equal to unity. In the parallel-plate region there will be a perfect standing wave, and the electric field E_z will vanish at regular intervals spaced by a half wavelength. Along any one of these planes a conducting plane may be placed, and the resulting structure is a corrugated plane. When the spacing S is sufficiently small, the conducting plane may be placed quite close to the interface without interacting appreciably with the evanescent modes. If $R = e^{-j\phi}$ is the reflection coefficient for E_z at the interface, the reflection coefficient R' a distance d from the interface is given by $e^{-j\phi-2jk_0d}$. A conducting plane may be placed at the plane where $R' = -1$, and, hence,

$$2k_0d + \phi = n\pi, \qquad n = 1, 3, \ldots .$$

The reflection coefficient at the interface was found in Section 10.2, and is given by

$$R = e^{-j\phi}$$

where

$$\phi = 2\left[\tan^{-1}\frac{k_0}{p_0} - k_0\frac{S}{\pi}\ln 2 + \sum_{n=1}^{\infty}\left(\tan^{-1}\frac{k_0}{p_n} + \tan^{-1}\frac{k_0}{p_{-n}} - \tan^{-1}\frac{k_0}{h_n}\right)\right] \qquad (33)$$

in the present notation. For spacings d less than about $0.2\lambda_0$, the sum over n in (33) contributes a negligible amount to ϕ. In this case,

$$\phi = 2\left(\tan^{-1}\frac{k_0}{p_0} - k_0\frac{S}{\pi}\ln 2\right) = n\pi - 2k_0 d$$

and, solving for p_0, we get

$$p_0 = k_0\tan k_0\left(d - \frac{S}{\pi}\ln 2\right). \tag{34}$$

A solution for positive p_0 exists only for certain ranges of k_0, and, hence, the corrugated plane exhibits the usual passband–stopband characteristics common to all periodic structures. Since the evanescent modes are TM or E modes, they give rise to a net storage of electric energy at the interface, and the parallel-plate system is effectively terminated in a capacitive susceptance. Thus the position of the first minimum may occur less than $\lambda_0/4$ from the interface. For the wave on the free-space side of the corrugations, the corrugated plane appears as a short-circuited transmission line of electrical length $k_0[d - (S/\pi)\ln 2]$ for small spacing S. As (34) shows, the solution for p_0 requires a value of d that will make the effective surface impedance an inductive reactance.

The propagation constant β is given by

$$\beta = (k_0^2 + p_0^2)^{1/2}$$

and, hence, the phase velocity $v_p = \omega/\beta$ is less than the free-space velocity v_c. Thus the corrugated plane supports a TM surface wave of the "slow wave" type.

The corrugated plane is uniform in the y direction, and the eigenvalue equation for a surface wave propagating obliquely across the corrugations may be found by a procedure similar to that used for obtaining the solution for a capacitive diaphragm in a rectangular guide from the solution for a capacitive diaphragm in a parallel-plate transmission line. As shown in Section 8.2, the mode of propagation will have all field components except E_y present, and may be classified as a longitudinal-section electric mode with respect to the z axis. If the propagation constant along the y axis is jl (propagation factor e^{-jly}), the eigenvalue equation may be obtained from (33) or (34) by replacing k_0^2 by $k_0^2 - l^2$ throughout, including the terms p_n and h_n.

For small spacings (34) is valid, but for larger spacings (33) must be used. However, (33) gives a transcendental solution for p_0, since the terms p_n, p_{-n} are not known before β has been found. The series in (33) represents only a small contribution to ϕ, and so a solution by the method of successive approximations may be applied. Equation (34) is used to find an approximate value of β, in order to find approximate values for p_n and p_{-n}. Equation (33) may then be used to solve for a second approximation to β and p_0 by using the approximate values for p_n and p_{-n} in the sum over n.

If the TM surface wave is considered to be propagating (attenuated) in the positive x direction, its wave impedance in this direction is capacitive, as in the usual waveguide situation. In the negative direction the wave impedance is inductive, and a surface exhibiting an inductive reactance is required in order to support this surface wave. For a TE surface wave, the wave impedance in the direction toward the surface is capacitive, and the surface reactance must be capacitive to support the wave. For TE modes, there would be a net storage of magnetic energy at the interface between a system of parallel plates and free space. This gives rise to an inductive reactance. In order to obtain a capacitive surface reactance, a short circuit must be placed at a suitable distance d from the interface, and the spacing S must be chosen so that the dominant mode in the parallel-plate region can propagate. The dominant mode is the TE_{10} mode, and hence S must be greater than $\lambda_0/2$. For this reason the interface will not support a TE surface wave. The edges of the plates or corrugations act as a diffraction grating, and, whenever the plates or corrugations act as a diffraction grating, and, whenever the period is greater than $\lambda_0/2$, a diffraction-grating mode that propagates away from the interface always exists. If, however, the region between the plates was filled with a high-dielectric-constant material, then the spacing could be reduced to less than $\lambda_0/2$, and it should then be possible to support a TE surface wave.

Attenuation due to Conductor Loss

A first approximation to the attenuation constant of the surface wave may be obtained by considering the field of the dominant modes only and evaluating the power loss in the conductor. The fields, in the case of zero loss, are approximately

$$H_y = Ae^{-p_0(x-d)}e^{-j\beta z}, \qquad x > d$$

$$E_x = \frac{\beta}{k_0}Z_0 H_y, \qquad x > d$$

$$H_y = A \sec k_0 d \cos k_0 x, \qquad x < d,\ 0 \le z \le S.$$

The power flow along the z axis per unit width along the y axis is

$$P = \frac{\beta Z_0}{2k_0} \int_d^\infty H_y H_y^* \, dx = \frac{AA^* \beta Z_0}{4k_0 p_0}. \tag{35}$$

The power loss per unit width in one corrugation is

$$P_L = \frac{1}{2} R_s Z_0 \left(2 \int_0^d H_y H_y^* \, dx + \int_0^S H_y H_y^* \, dz \right)$$

$$= \frac{1}{2} R_s Z_0 AA^* \sec^2 k_0 d \left(S + d + \frac{\sin 2k_0 d}{2k_0} \right) \tag{36}$$

since the current density on the conductors is equal to $|H_y|$. There are $1/S$ sections per meter, and so the power loss per meter is P_L/S. The attenuation constant α is given by $\alpha = P_L/2PS$, where P_L/S is the power loss per meter in the z direction, and P is the power flow in the z direction. From (35) and (36), we obtain

$$\alpha = \frac{R_s k_0 p_0 [S + d + (\sin 2k_0 d)/2k_0]}{\beta S \cos^2 k_0 d} \quad \text{nepers per meter} \tag{37}$$

where R_s is the normalized surface resistivity, p_0 is a solution of (33) or (34), and $\beta^2 = k_0^2 + p_0^2$. To obtain a low attenuation, the surface wave should be loosely bound to the surface, in which case p_0 is small.

As an example, consider a corrugated copper plane with $\lambda_0 = 0.0314$ meter, $S = 0.2\lambda_0$, $d = 0.159\lambda_0$, $\sigma = 5.8 \times 10^7$ siemens per meter. From (34), the first-order solution for p_0 is found to be 176 nepers per meter. Using this value of p_0 to compute β, p_n, p_{-n}, and h_n, a second approximation to p_0 may be computed from (33) by equating ϕ to $\pi - 2k_0d$. This yields a value of 182 nepers per meter for p_0. The first approximation, as given by (34), is seen to be quite good in this example. Using (37), the attenuation constant α is found to be 0.066 neper per meter. This is about 20 times greater than that for the dielectric-coated plane considered earlier. However, the wave in the present example is much more tightly bound to the surface, since p_0 is about 7 times larger than that for the dielectric-coated-plane example. A reduction of p_0 in the present case would reduce the attenuation proportionately.

11.5. SURFACE WAVES ALONG DIELECTRIC SLABS

Surface waves of both the TM and TE types may be guided along a dielectric slab or sheet. Let the thickness of a sheet be $2t$, and let the relative dielectric constant be κ, as in Fig. 11.8. The dielectric will also be assumed to have negligible loss, so that κ is real. The permeability is taken equal to μ_0. The TM modes have a single component of magnetic field H_y and electric field components E_x and E_z. The TE modes have the roles of electric field and magnetic field interchanged, and hence have components E_y, H_x, H_z. Propagation of the modes is in the z direction with a propagation factor $e^{-j\beta z}$. For both types of modes, the field decays away from the surface according to a factor $e^{-p(|x|-t)}$. Since the fields satisfy the wave equation, $\beta^2 = k_0^2 + p^2$, and hence the guide wavelength and phase velocity are both less than those for plane waves in free space.

TM Modes

The solutions for H_y in the dielectric may be of the symmetrical type (even) or antisymmetrical type (odd), where "even" and "odd" refer to the way that H_y varies with x about the symmetry plane $x = 0$. For the even solutions

$$\frac{\partial H_y}{\partial x}\bigg|_{x=0} = 0$$

and, since E_z is proportional to this derivative, the tangential electric field vanishes on the plane $x = 0$ for the even solutions. A conducting plane may be placed along $x = 0$, and the resultant solution is for a TM mode along a dielectric slab of thickness t and placed on a conducting plane. The odd solutions correspond to a magnetic wall along the symmetry plane $x = 0$.

For the even solutions, H_y will be of the following form in the dielectric region:

$$H_y = B \cos hx\, e^{-j\beta z}, \qquad |x| \le t$$

where $\beta^2 = \kappa k_0^2 - h^2$. Outside the dielectric, H_y may be represented by

$$H_y = A \exp[-p(|x| - t) - j\beta z], \qquad |x| \ge t$$

where $\beta^2 = k_0^2 + p^2$. At $x = \pm t$, H_y and E_z must be continuous. Since $j\omega\epsilon E_z = \partial H_y/\partial x$, we obtain the two equations

$$A = B \cos ht \tag{38a}$$

$$\kappa p A = hB \sin ht. \tag{38b}$$

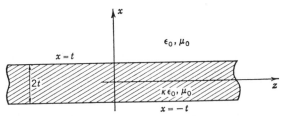

Fig. 11.8. A plane-dielectric-sheet surface waveguide.

Dividing one equation by the other yields the eigenvalue equation

$$\kappa p = h \tan ht \tag{39}$$

which, together with the relation obtained by equating the two expressions for β^2

$$(\kappa - 1)k_0^2 = p^2 + h^2 \tag{40}$$

permits a solution for h and p to be found. A solution may readily be obtained by graphical means. For this purpose it is convenient to rewrite (39) and (40) as follows:

$$\kappa pt = ht \tan ht \tag{41a}$$

$$(pt)^2 + (ht)^2 = (\kappa - 1)(k_0 t)^2. \tag{41b}$$

In this latter form the equations are independent of frequency, provided the ratio of thickness $2t$ to free-space wavelength λ_0 is kept constant.

Equation (41a) determines one relation between ht and pt, which may be plotted on an ht–pt plane. The other equation, (41b), is a circle of radius $(\kappa - 1)^{1/2}k_0 t$ in the same plane. The points of intersection between the two curves determine the eigenvalues h and p. A typical plot is given in Fig. 11.9 for a polystyrene sheet with $\kappa = 2.56$.

Several interesting properties of the modes may be deduced from the graphical solution given in Fig. 11.9. The two sets of curves will always have at least one point of intersection, even for t/λ_0 approaching zero. Hence, the first even mode, which we will label as the TM_0 mode, has no low-frequency cutoff. Since p must be positive for a surface mode, only those points of intersection that lie in the intervals $0 \leq ht \leq \pi/2$, $\pi \leq ht \leq 3\pi/2$, or, in general, $n\pi \leq ht \leq n\pi + \pi/2$, where n is any integer, correspond to surface-wave solutions. For sufficiently large values of t, there are several modes of propagation. All the higher order modes will, however, have a low-frequency cutoff. Points of intersection that occur in the range intervals $n\pi + \pi/2 \leq ht \leq (n + 1)\pi$, where n is any integer, lead to solutions with negative values of p. These modes increase exponentially away from the surface, and hence must be excluded on physical grounds since the field amplitude becomes infinite. The higher order modes are cut off for that value of t/λ_0 which causes the circle to intersect the tangent curves (41a) along the ht axis. The cutoff values are readily seen to be given by $ht = n\pi$, and hence, for the nth mode, the value of t/λ_0 at cutoff is, from (41b),

$$\left(\frac{t}{\lambda_0}\right)_{co} = \frac{n}{2(\kappa - 1)^{1/2}}, \qquad n = 0, 1, 2, \ldots . \tag{42}$$

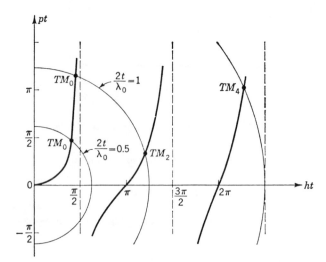

Fig. 11.9. Graphical solution for eigenvalues for even TM modes.

The surface-wave modes along a dielectric sheet are often called "trapped" modes. The reason for this terminology is that these modes may be considered as a plane wave incident from the dielectric region onto the air–dielectric interface at an angle greater than the critical angle θ_c, where $\sin \theta_c = \kappa^{-1/2}$. For incident angles greater than θ_c, there is complete reflection at the interface, and the field outside the dielectric is evanescent. Thus the mode consists of plane waves propagating along a zigzag path along the z axis and undergoing complete reflection at each interface. In other words, the field is trapped within the dielectric sheet.

For the odd modes, the magnetic field is of the following form:

$$
H_y = \begin{cases}
A \, \exp[-p(x - t) - j\beta z], & x \geq t \\
A \csc ht \, \sin hx \, e^{-j\beta z}, & -t \leq x \leq t \\
-A \, \exp[p(x + t) - j\beta z], & x \leq -t.
\end{cases}
$$

Matching the tangential fields at $x = \pm t$ leads to the eigenvalue equation for the odd modes:

$$\kappa pt = -ht \cot ht \tag{43a}$$

where

$$(pt)^2 + (ht)^2 = (\kappa - 1)(k_0 t)^2. \tag{43b}$$

The solutions for p and h are readily determined by a graphical procedure similar to that used for the even modes. A typical graphical solution is illustrated in Fig. 11.10 for a polystyrene sheet with $\kappa = 2.56$. From this plot it is clear that all the odd modes have a lower frequency cutoff. Only those points of intersection that occur in the range intervals $n\pi - \pi/2 \leq ht \leq n\pi$, where n is any integer, give rise to surface waves with p positive. The values of t/λ_0 at cutoff are given by

$$\left(\frac{t}{\lambda_0}\right)_{co} = \frac{2n + 1}{4(\kappa - 1)^{1/2}}, \qquad n = 0, 1, 2, \dots . \tag{44}$$

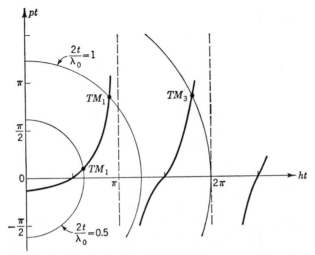

Fig. 11.10. Graphical solution for eigenvalues for odd TM modes.

The equations for the cutoff parameter for the even and odd modes may be combined to give

$$\left(\frac{2t}{\lambda_0}\right)_{co} = \frac{n}{2(\kappa - 1)^{1/2}}, \qquad n = 0, 1, 2, \ldots \tag{45}$$

where the even integers $n = 0, 2, 4, \ldots$ now refer to the even modes, while the odd integers $n = 1, 3, 5, \ldots$ refer to the odd modes.

TE Modes

The solution for the TE modes is formally the same as that for the TM modes. The single electric field component E_y will have the following forms:

$$E_y = \begin{cases} A \exp[-p(|x| - t) - j\beta z], & |x| \geq t \\ A \sec ht \cos hx \, e^{-j\beta z}, & |x| \leq t \end{cases}$$

for the even modes, and

$$E_y = \begin{cases} A \exp[-p(x - t) - j\beta z], & x \geq t \\ A \csc ht \sin hx \, e^{-j\beta z}, & |x| \leq t \\ -A \exp[p(x + t) - j\beta z], & x \leq -t \end{cases}$$

for the odd modes. The magnetic field components are given by

$$j\omega\mu_0 H_x = \frac{\partial E_y}{\partial z}$$

$$j\omega\mu_0 H_z = -\frac{\partial E_y}{\partial x}.$$

By matching the tangential fields at $x = \pm t$, the eigenvalue equations for p and h are obtained:

$$pt = ht \tan ht \qquad \text{even modes} \tag{46a}$$

$$pt = -ht \cot ht \qquad \text{odd modes} \tag{46b}$$

where

$$(pt)^2 + (ht)^2 = (\kappa - 1)(k_0 t)^2 \tag{46c}$$

$$\beta^2 = k_0^2 + p^2 = \kappa k_0^2 - h^2. \tag{46d}$$

Apart from the absence of the factor κ in (46a) and (46b), these equations are the same as those for the TM modes. Consequently, a similar graphical solution may be used. Again it is only the TE_0 mode, that is, the first even mode, that has no low-frequency cutoff. The value of $2t/\lambda_0$ at cutoff is given by the same equation as for the TM modes, i.e., by (45). For a dielectric slab of thickness t above a conducting plane, only the odd solutions are possible, since E_z does not vanish at $x = 0$ for the even modes. Thus, for a grounded dielectric slab, a TE surface mode with no low-frequency cutoff does not exist.

The Continuous Eigenvalue Spectrum

For any plane-dielectric sheet of given thickness, only a finite number of surface-wave modes exist. These modes obviously do not form a complete eigenfunction set in which the field from an arbitrary source can be expanded. It is, therefore, clear that the surface-wave modes do not represent the complete solution for the fields which may exist along a dielectric sheet. In order to simplify the problem, we will confine our attention to the TM waves along a grounded dielectric sheet of thickness t. Similar results will, however, hold for the TE surface waves along a grounded dielectric sheet, as well as for single sheets located in free space. The eigenvalue equations, (41), which apply for TM waves along a grounded dielectric sheet, have been studied in detail by Brown [11.10]. Similar results will be derived here by using a transmission-line analogy.

Consider a plane TM wave incident at an angle θ_i on a grounded dielectric sheet, as in Fig. 11.11(a). In the dielectric, a standing wave exists because of reflection at the conducting plane. Above the dielectric, a reflected wave also exists, so that the field is a standing wave along the x axis in this region also. The magnetic field in the two regions may be represented as follows:

$$H_y = A \exp[-jk_0(-x \cos \theta_i + z \sin \theta_i)]$$

$$\qquad - RA \exp -jk_0[(x - 2t) \cos \theta_i + z \sin \theta_i], \qquad x \geq t \tag{47a}$$

$$H_y = B \cos(kx \cos \theta_r) \exp(-jk_0 z \sin \theta_i), \qquad 0 \leq x \leq t \tag{47b}$$

where $k = \kappa^{1/2} k_0$, $k \sin \theta_r = k_0 \sin \theta_i$, and θ_r is the angle of refraction. The normalized wave impedances in the negative x direction in the two regions are

$$Z_1 = \frac{E_z}{Z_0 H_y} = \cos \theta_i, \qquad x > t \tag{48a}$$

$$Z_2 = \frac{\cos \theta_r}{\kappa^{1/2}} = \frac{(\kappa - \sin^2 \theta_i)^{1/2}}{\kappa}, \qquad 0 \leq x \leq t. \tag{48b}$$

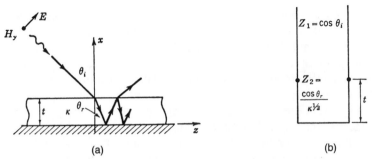

Fig. 11.11. TM wave incident on a grounded dielectric sheet.

For propagation in the x direction, the equivalent transmission-line circuit is illustrated in Fig. 11.11(b). The propagation factors in the two lines are $k_0 \cos \theta_i$ and $k \cos \theta_r$. At the dielectric interface, the normalized input impedance is

$$Z_{\text{in}} = j\kappa^{-1/2} \cos \theta_r \tan(kt \cos \theta_r) \tag{49}$$

and hence the reflection coefficient R is given by

$$R = \frac{j \cos \theta_r \tan(kt \cos \theta_r) - \kappa^{1/2} \cos \theta_i}{j \cos \theta_r \tan(kt \cos \theta_r) + \kappa^{1/2} \cos \theta_i} \tag{50}$$

and has a modulus of unity.

For convenience, let $k_0 \sin \theta_i = \beta$, $jk_0 \cos \theta_i = p$, and $k \cos \theta_r = h$. In place of (50), we now have

$$R = \frac{h \tan ht + \kappa p}{h \tan ht - \kappa p}. \tag{51}$$

From the wave equation, the following relations are obtained:

$$\beta^2 = k_0^2 + p^2 = \kappa k_0^2 - h^2. \tag{52}$$

A solution for standing waves along the x axis exists for all real angles of incidence $0 \leq \theta_i \leq \pi/2$. The corresponding ranges of p and β are

$$jk_0 \geq p \geq j0$$

$$0 \leq \beta \leq k_0.$$

This partial solution is seen to have a continuous spectrum for β in the range zero to k_0.

The standing-wave solution along the x direction will also exist for imaginary values of p greater than jk_0. This range of p corresponds to imaginary angles of incidence. Hence, for $p > jk_0$, the corresponding range for β is, from (52),

$$-j\infty < \beta < j0$$

and gives rise to a continuous spectrum of waves that is attenuated, since β is imaginary, along the z direction.

718   FIELD THEORY OF GUIDED WAVES

The next case to examine is that for real values of p. In place of a sinusoidal variation with x, we now obtain a field described by a hyperbolic sine function of x above the dielectric. Such a solution must be discarded on physical grounds since it becomes infinite for x approaching infinity. It is only for certain real values of p that a solution can exist. These solutions are the surface-wave modes discussed earlier. From (51) it is seen that the reflection coefficient becomes infinite whenever $h \tan ht = \kappa p$, which is the eigenvalue equation (39). Thus the surface wave may be considered to be produced by the pole in the expression for the reflection coefficient. Since $|R|$ tends to infinity at the pole, the amplitude of the incident wave may tend to zero, and still produce a reflected wave with finite amplitude. The answer to the question of whether a surface-wave mode should be associated with a zero or a pole of the reflection coefficient depends on whether the field above the dielectric is considered to be a reflected wave or an incident wave, since the sign of the wave impedance changes accordingly. If the field is considered as an incident wave, the solution corresponds to a zero of the reflection coefficient and not to a pole, as is the case if the field is considered as a reflected wave. Zucker has resolved this difficulty by considering the excitation problem, from which it is concluded that the pole association is the correct one, since a pole of the reflection coefficient occurs on that sheet of the Riemann surface for which the integral for the radiated field converges properly [11.11].

The complete solution for the physical field consists of one or more surface waves with $\beta > k_0$, a continuous spectrum of waves with $0 < \beta < k_0$, and a continuous spectrum of evanescent waves with $\beta = -j\alpha$ and $0 < \alpha < \infty$. The field radiated by an arbitrary source will, in general, contribute to all three of the above types of fields.

In addition to the roots leading to surface waves, the eigenvalue equations (41), (43), and (46) have an infinity of solutions with complex roots. The corresponding modes are referred to as "leaky modes," since they correspond to a flow of power away from the surface. These modes, however, do not satisfy the radiation conditions at infinity, and hence do not belong to the proper eigenvalue spectrum. Nevertheless, it is possible to utilize these modes to partially represent the radiated field from an elementary source [11.12]. In deforming the contour of integration for the radiated field into a path of "steepest descent," some of these complex poles are crossed, and their residues then contribute to the expansion of the field. Meaningful discussion is difficult without a consideration of the total field radiated by a given source; therefore, these leaky modes will be further discussed in the section on excitation of surface waves by a line source.

11.6. Surface Waves on Cylindrical Structures

The possibility of a surface-wave mode of propagation along a round wire of conducting material was demonstrated theoretically by Sommerfeld in 1899. The mode is a TM wave with components H_θ, E_r, E_z, and is axially symmetrical, i.e., independent of the angular coordinate θ. Sommerfeld's solution has been given by Stratton, and, consequently, will not be repeated here [11.3]. The mode has been found to exist only for a finite conductivity and is loosely bound to the surface. This surface-wave mode was studied by Goubau as to its suitability for a practical transmission-line system [11.13]. The results of Goubau's studies were that a single small-diameter conducting wire will propagate the axially symmetric surface-wave mode with a low attenuation, but that the field extends a considerable distance outside the wire before decaying to a negligible value. The practical use of this surface waveguide is, therefore, limited to the high frequencies $f > 10^{10}$ cycles per second. The situation here is much the same as that for the Zenneck wave above a conducting plane. In order to confine

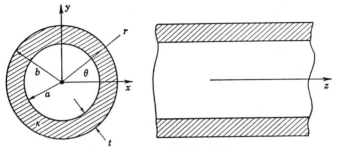

Fig. 11.12. Dielectric-coated-wire surface waveguide.

the field to the surface of the guide, the wire may be coated with a thin layer of dielectric or corrugated. These modifications were also studied by Goubau and were discussed in the same reference.

The axially symmetric surface wave is frequently called the Sommerfeld–Goubau wave, in honor of the principal investigators. In addition, higher order modes, with angular dependence, may exist on a conducting wire. These modes, however, attenuate so rapidly with distance that they are of no importance except as part of the near-zone field at the region of the exciting source.

The first analysis of a dielectric-coated wire was made by Harms in 1907 [11.14]. The solution for the axially symmetric mode will be given here for a lossless dielectric coating on a perfectly conducting cylinder, as illustrated in Fig. 11.12. The radius of the wire is a, the outside radius of the dielectric sleeve is b, and the thickness of the sleeve is $b - a = t$. For an axially symmetric mode or field, a TM solution is possible. For fields that vary with the angle θ, each mode is a combination of a TM and a TE mode with all six field components present. For the TM mode, with no variation with the angle coordinate, and with the propagation along the z axis according to $e^{-j\beta z}$, Maxwell's equations reduce to

$$E_z = \psi(r)e^{-j\beta z} \tag{53a}$$

$$E_r = -\frac{j\beta}{k_c^2}\frac{d\psi}{dr}e^{-j\beta z} \tag{53b}$$

$$H_\theta = -\frac{j\omega\epsilon}{k_c^2}\frac{d\psi}{dr}e^{-j\beta z} \tag{53c}$$

$$\frac{d^2\psi}{dr^2} + \frac{1}{r}\frac{d\psi}{dr} + k_c^2\psi = 0 \tag{53d}$$

where $k_c^2 = k_0^2 - \beta^2$ in the air region $r > b$, and is equal to $\kappa k_0^2 - \beta^2$ in the dielectric region $a < r < b$. The radial function $\psi(r)$ is a solution of Bessel's equation of order 0. In the region $a < r < b$, a suitable solution is a linear combination of the Bessel functions of the first and second kinds. In the region $r > b$, the solution is a modified Bessel function of the second kind, that is, $K_0[r(\beta^2 - k_0^2)^{1/2}]$, since this is the only solution which decays exponentially for large r. Thus we have

$$\psi = \begin{cases} AK_0[r(\beta^2 - k_0^2)^{1/2}], & r > b \\ BJ_0[r(k^2 - \beta^2)^{1/2}] + CY_0[r(k^2 - \beta^2)^{1/2}], & a < r < b \end{cases}$$

where $k^2 = \kappa k_0^2$.

At $r = a$, $E_z = 0$, and hence $\psi(a) = 0$ and the constant C is given by

$$C = -\frac{BJ_0[a(k^2 - \beta^2)^{1/2}]}{Y_0[a(k^2 - \beta^2)^{1/2}]}.$$

The eigenvalue equation for β is obtained by matching the fields tangential at the surface $r = b$, and is

$$\frac{K_1(pb)}{pK_0(pb)} = \frac{\kappa}{h}\frac{J_0(ha)Y_1(hb) - J_1(hb)Y_0(ha)}{J_0(hb)Y_0(ha) - J_0(ha)Y_0(hb)} \tag{54}$$

where $p^2 = \beta^2 - k_0^2$ and $h^2 = k^2 - \beta^2$.

The eigenvalue equation is transcendental, and graphical methods could be used to solve it, in general. However, for surface waveguides of practical interest, the thickness t is very small, the radius a and, hence, pa and ha are small, and p^2 is small compared with k_0^2, so that β is only moderately larger than k_0. In (54), the small-argument approximations

$$K_0(x) \rightarrow -\left(0.577 + \ln\frac{x}{2}\right) = -\ln 0.89x$$

$$K_1(x) \rightarrow \frac{1}{x}$$

may be used. Following Goubau, the right-hand side of (54) is simplified by expanding $J_0(ha)$, etc., in a Taylor series about $r = b$, to get

$$J_0(ha) = J_0(hb) - \frac{dJ_0(hb)}{dr}(b - a) = J_0(hb) + htJ_1(hb)$$

and similarly for $Y_0(ha)$. With these approximations, the eigenvalue equation reduces to

$$\kappa p^2 b \ln 0.89pb = -(\kappa - 1)k_0^2 t + p^2 t \approx -(\kappa - 1)k_0^2 t \tag{55}$$

since $p^2 \ll k_0^2$.

As an example, consider a copper wire with $b = 0.001$ meter, $t = 10^{-4}$ meter, $\kappa = 2.56$, and $\lambda_0 = 3.14 \times 10^{-2}$ meter. From (55), we get $p = 25.8$ nepers per meter, and, hence, $\beta = 202$ radians per meter. For large r, we have

$$K_0(pr) = \left(\frac{\pi}{2pr}\right)^{1/2} e^{-pr}$$

and thus, in a distance corresponding to a few wavelengths, the field has decayed to a negligible value. The approximations made in arriving at (55) are clearly valid for this example.

The attenuation may be computed by evaluating the losses in the conductor and the dielectric sleeve separately, adding the two, and dividing by twice the power flow along the guide in the z direction. Numerical values have been given by Goubau [11.15].

Surface Waves along a Dielectric Rod

As a last example of a structure capable of supporting a surface wave, we consider a dielectric rod of radius a, as in Fig. 11.13. Mode propagation along the dielectric rod was studied by Hondros and Debye as long ago as 1910, and later by Elsasser and Chandler as well as others [11.16]–[11.18]. Dielectric rods are similar in behavior to dielectric sheets in that a number of surface-wave modes exist. Pure TM or TE modes are possible only if the field is independent of the angular coordinate θ. As the radius of the rod increases, the number of TM and TE modes also increases. These modes do, however, have a cutoff point such that, below some minimum value of a/λ_0, the mode cannot exist any longer. When the field depends on the angular coordinate, pure TM or TE modes no longer exist. All modes with angular dependence are a combination of a TM and a TE mode, and are classified as hybrid EH or HE modes, depending on whether the TM or the TE mode predominates, respectively. All these modes, with the exception of the HE_{11} mode, exhibit cutoff phenomena similar to those of the axially symmetric modes. Since the HE_{11} mode has no low-frequency cutoff, it is the dominant mode. This mode is widely used in the dielectric-rod antenna [11.19]. For small-diameter rods, the field extends for a considerable distance beyond the surface, and the axial propagation constant β is only slightly larger than k_0. As the radius increases, the field is confined closer and closer to the rod, and β approaches $\kappa^{1/2}k_0$ in the limit of infinite radius. Since $\beta > k_0$, the phase velocity is less than that of plane waves in free space. In all of the above respects, the dielectric rod does not differ from the plane-dielectric sheet.

The field expansion around the dielectric rod is similar to that for the sheath helix, with the exception that, in the dielectric region $r < a$, the field varies according to the Bessel functions of the first kind. By analogy with the set of equations in Section 9.8 for the sheath helix, we have

$$
\left.
\begin{aligned}
E_z &= A_n J_n(hr) e^{-jn\theta - j\beta z} \\
E_r &= \left(-\frac{j\beta}{h} A_n J_n' - \frac{n\omega\mu_0}{h^2 r} B_n J_n \right) e^{-jn\theta - j\beta z} \\
E_\theta &= \left(-\frac{n\beta}{h^2 r} A_n J_n + \frac{j\omega\mu_0}{h} B_n J_n' \right) e^{-jn\theta - j\beta z} \\
H_z &= B_n J_n(hr) e^{-jn\theta - j\beta z} \\
H_r &= \left(\frac{n\omega\epsilon}{h^2 r} A_n J_n - \frac{j\beta}{h} B_n J_n' \right) e^{-jn\theta - j\beta z} \\
H_\theta &= \left(-\frac{j\omega\epsilon}{h} A_n J_n' - \frac{n\beta}{h^2 r} B_n J_n \right) e^{-jn\theta - j\beta z}
\end{aligned}
\right\} \quad r < a
$$

$$
\left.
\begin{aligned}
E_z &= C_n K_n(pr) e^{-jn\theta - j\beta z} \\
E_r &= \left(\frac{j\beta}{p} C_n K_n' + \frac{n\omega\mu_0}{p^2 r} D_n K_n \right) e^{-jn\theta - j\beta z} \\
E_\theta &= \left(\frac{n\beta}{p^2 r} C_n K_n - \frac{j\omega\mu_0}{p} D_n K_n' \right) e^{-jn\theta - j\beta z} \\
H_z &= D_n K_n(pr) e^{-jn\theta - j\beta z} \\
H_r &= \left(-\frac{n\omega\epsilon_0}{p^2 r} C_n K_n + \frac{j\beta}{p} D_n K_n' \right) e^{-jn\theta - j\beta z} \\
H_\theta &= \left(\frac{j\omega\epsilon_0}{p} C_n K_n' + \frac{n\beta}{p^2 r} D_n K_n \right) e^{-jn\theta - j\beta z}
\end{aligned}
\right\} \quad r > a
$$

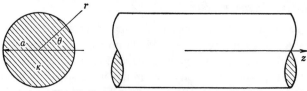

Fig. 11.13. Dielectric-rod surface waveguide.

where $\beta^2 = k_0^2 + p^2 = \kappa k_0^2 - h^2$; A_n, B_n, C_n, D_n are amplitude constants; and the prime indicates differentiation with respect to the arguments hr and pr.

Imposition of the boundary conditions at $r = a$ leads to equations for determining the relative amplitudes of the coefficients, and also the eigenvalue equation. The eigenvalue equation is

$$\left[\frac{\kappa J_n'(u_1)}{u_1 J_n(u_1)} + \frac{K_n'(u_2)}{u_2 K_n(u_2)}\right]\left[\frac{J_n'(u_1)}{u_1 J_n(u_1)} + \frac{K_n'(u_2)}{u_2 K_n(u_2)}\right] = \left[\frac{n\beta}{k_0}\frac{(u_2^2 + u_1^2)}{u_1^2 u_2^2}\right]^2 \qquad (56)$$

where $u_1 = ha$, $u_2 = pa$. When $n = 0$, the right-hand side vanishes, and each factor on the left-hand side must equal zero. These two factors give the eigenvalue equations for the axially symmetric TM and TE modes:

$$\frac{\kappa J_0'(u_1)}{u_1 J_0(u_1)} = -\frac{K_0'(u_2)}{u_2 K_0(u_2)} \qquad \text{TM modes} \qquad (57a)$$

$$\frac{J_0'(u_1)}{u_1 J_0(u_1)} = -\frac{K_0'(u_2)}{u_2 K_0(u_2)} \qquad \text{TE modes.} \qquad (57b)$$

In addition, u_1 and u_2 are related by the equation

$$u_1^2 + u_2^2 = (\kappa - 1)(k_0 a)^2. \qquad (58)$$

The solutions to the above equations can be obtained by using a root-finding algorithm on a computer. The ratio of β to k_0 as a function of $2a/\lambda_0$ for a polystyrene rod is given in Fig. 11.14 for the axially symmetric TM$_1$ and TE$_1$ modes and the HE$_{11}$ dipole mode. These

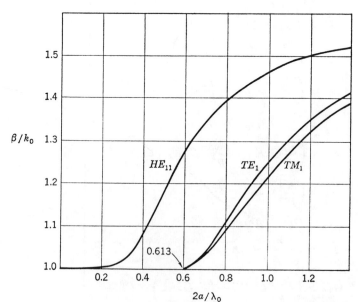

Fig. 11.14. Ratio of β to k_0 for the first three surface-wave modes on a polystyrene rod ($\kappa = 2.56$).

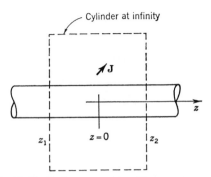

Fig. 11.15. Surface waveguide with a current source.

curves were constructed from the numerical data given by Elsasser [11.17]. Both the TM_1 and TE_1 modes are cut off for $2a < 0.613\lambda_0$, while the HE_{11} mode has no low-frequency cutoff.

11.7. FIELD ORTHOGONALITY PROPERTIES

On any surface waveguide of the type considered in this chapter the various surface-wave modes are orthogonal to one another, as well as orthogonal to the continuous eigenvalue field, i.e., radiation field. The orthogonality properties may be proved by means of the Lorentz reciprocity theorem in a manner similar to that for conventional waveguides [11.20].

Consider a surface waveguide as in Fig. 11.15. Let a current element $\mathbf{J}$ be located in the vicinity of $z = 0$. Choose a closed surface S, consisting of two transverse planes at z_1 and z_2 and the surface of a cylinder surrounding the guide at infinity. The current element radiates one or more surface-wave modes and a radiation field. The total field may be represented as follows:

$$\mathbf{E} = \sum_n a_n \mathbf{E}_n^+ + \mathbf{E}_R^+$$

$$\mathbf{H} = \sum_n a_n \mathbf{H}_n^+ + \mathbf{H}_R^+ \qquad\qquad z > 0 \qquad\qquad (59a)$$

$$\mathbf{E} = \sum_n b_n \mathbf{E}_n^- + \mathbf{E}_R^-$$

$$\mathbf{H} = \sum_n b_n \mathbf{H}_n^- + \mathbf{H}_R^- \qquad\qquad z < 0 \qquad\qquad (59b)$$

where $\mathbf{E}_R$, $\mathbf{H}_R$ is the radiation field and $\mathbf{E}_n$, $\mathbf{H}_n$ is the nth surface-wave mode field. The transverse fields of the surface-wave modes will be represented as follows:

$$\mathbf{E}_{nt}^+ = \mathbf{e}_n e^{-\Gamma_n z} \qquad\qquad (60a)$$

$$\mathbf{E}_{nt}^- = \mathbf{e}_n e^{\Gamma_n z} \qquad\qquad (60b)$$

$$\mathbf{H}_{nt}^+ = \mathbf{h}_n e^{-\Gamma_n z} \qquad\qquad (60c)$$

$$\mathbf{H}_{nt}^- = -\mathbf{h}_n e^{-\Gamma_n z} \qquad\qquad (60d)$$

where $\mathbf{e}_n$ and $\mathbf{h}_n$ are transverse vector functions of x and y.

The total field is a solution of the equations

$$\nabla \times \mathbf{E} = -j\omega\mu\mathbf{H} \qquad \nabla \times \mathbf{H} = j\omega\epsilon\mathbf{E} + \mathbf{J}$$

while each surface-wave mode is a solution of the source-free field equations. Let $\mathbf{E}_1$, $\mathbf{H}_1$ be a solution to the source-free equations. The Lorentz reciprocity theorem gives

$$\oiint_S [\mathbf{E} \times \mathbf{H}_1 - \mathbf{E}_1 \times \mathbf{H}] \cdot \mathbf{n} \, dS = \iiint \mathbf{J} \cdot \mathbf{E}_1 \, dV. \tag{61}$$

If $\mathbf{E}$, $\mathbf{H}$ is the nth surface-wave mode and $\mathbf{E}_1$, $\mathbf{H}_1$ the mth surface-wave mode, and we take $\mathbf{J} = 0$, then (61) gives

$$\iint \mathbf{e}_n \times \mathbf{h}_m \cdot \mathbf{a}_z \, dS = \iint \mathbf{e}_m \times \mathbf{h}_n \cdot \mathbf{a}_z \, dS = 0 \tag{62}$$

where the integration is over a transverse plane. The derivation of (62) is the same as that for conventional waveguides. Equation (62) provides an orthogonality relation for the surface-wave modes.

Next let $\mathbf{E}$, $\mathbf{H}$ be the total radiated field and let $\mathbf{E}_1$, $\mathbf{H}_1$ be the nth surface-wave mode field having transverse components $\mathbf{e}_n e^{-\Gamma_n z}$, $\mathbf{h}_n e^{-\Gamma_n z}$. In (61), the integral over the cylinder at infinity is zero, since the radiation field is bounded, and the surface-wave modes decay exponentially as infinity is approached. The integral over the transverse planes reduces to

$$-e^{-\Gamma_n z_1} \iint_{z_1} (\mathbf{E}_R^- \times \mathbf{h}_n - \mathbf{e}_n \times \mathbf{H}_R^-) \cdot \mathbf{a}_z \, dS$$

$$+ e^{-\Gamma_n z_2} \iint_{z_2} (\mathbf{E}_R^+ \times \mathbf{h}_n - \mathbf{e}_n \times \mathbf{H}_R^+) \cdot \mathbf{a}_z \, dS$$

$$- 2b_n \iint \mathbf{e}_n \times \mathbf{h}_n \cdot \mathbf{a}_z \, dS = \iiint \mathbf{E}_n^+ \cdot \mathbf{J} \, dV \tag{63}$$

because of the orthogonality property (62). Since z_1 and z_2 are arbitrary, the first two terms on the left-hand side must each be equal to a constant. Furthermore, the radiation field has a continuous eigenvalue spectrum, and so its dependence on z cannot annul the factors $e^{-\Gamma_n z_1}$ and $e^{-\Gamma_n z_2}$. Therefore, the constants have to be equal to zero, and (63) gives

$$- 2b_n \iint \mathbf{e}_n \times \mathbf{h}_n \cdot \mathbf{a}_z \, dS = \iiint_V \mathbf{E}_n^+ \cdot \mathbf{J} \, dV \tag{64}$$

$$\iint_{z_1} \mathbf{E}_R^- \times \mathbf{h}_n \cdot \mathbf{a}_z \, dS = \iint_{z_2} \mathbf{E}_R^+ \times \mathbf{h}_n \cdot \mathbf{a}_z \, dS = \iint_{z_1} \mathbf{e}_n \times \mathbf{H}_R^- \cdot \mathbf{a}_z \, dS$$

$$= \iint_{z_2} \mathbf{e}_n \times \mathbf{H}_R^+ \cdot \mathbf{a}_z \, dS = 0. \tag{65}$$

The results given by (65) are obtained by noting that the integrals over the transverse planes at z_1 and z_2 must also vanish when $\mathbf{h}_n$ is replaced by $-\mathbf{h}_n$, which corresponds to using $\mathbf{E}_1 = \mathbf{E}_n^-$ and $\mathbf{H}_1 = \mathbf{H}_n^-$. Thus each integral is equal to zero. If we begin with $\mathbf{E}_1$, $\mathbf{H}_1$, corresponding to the surface-wave mode $\mathbf{E}_n^-$, $\mathbf{H}_n^-$, we obtain in a similar way

$$- 2a_n \iint \mathbf{e}_n \times \mathbf{h}_n \cdot \mathbf{a}_z \, dS = \iiint \mathbf{E}_n^- \cdot \mathbf{J} \, dV. \tag{66}$$

Equations (64) and (66) provide the required relations for determining the surface-wave mode amplitudes a_n and b_n.

A short-current element of length dl is equivalent to an oscillating dipole $\mathbf{P}$, where $j\omega P = J\,dl$. If we have a small current loop $J\tau$, where τ is the unit tangent to the contour around which J flows, we have

$$\oint_C I\mathbf{E}_n\cdot\boldsymbol{\tau}\,dl = I\iint_{S_0} \nabla\times\mathbf{E}_n\cdot\mathbf{n}\,dS = -j\omega\mu_0 I\iint_{S_0}\mathbf{H}_n\cdot\mathbf{n}\,dS = -j\omega\mu_0\mathbf{H}_n\cdot\mathbf{M}$$

for a small loop of area S_0, where the magnetic dipole moment $\mathbf{M}$ is given by

$$\mathbf{M} = I\iint_{S_0}\mathbf{n}\,dS$$

and I is the total loop current. In the presence of small electric and magnetic dipoles, (64) and (66) thus become

$$2b_n\iint\mathbf{e}_n\times\mathbf{h}_n\cdot\mathbf{a}_z\,dS = -j\omega\mathbf{E}_n^+\cdot\mathbf{P} + j\omega\mu_0\mathbf{H}_n^+\cdot\mathbf{M} \tag{67a}$$

$$2a_n\iint\mathbf{e}_n\times\mathbf{h}_n\cdot\mathbf{a}_z\,dS = -j\omega\mathbf{E}_n^-\cdot\mathbf{P} + j\omega\mu_0\mathbf{H}_n^-\cdot\mathbf{M}. \tag{67b}$$

These results are of importance in determining the fields radiated by a given system of sources in the presence of a surface waveguide. In particular, they show at once whether a given source will or will not excite a particular surface-wave mode.

11.8. EXCITATION OF SURFACE WAVES

The problem of exciting surface waves has received considerable attention. Many papers and reports have been published on the subject, and a representative list of these is given at the end of the chapter. The problem differs somewhat from that of exciting a propagating mode in an ordinary closed-boundary guide, in that some of the power radiated by the source goes into the radiation field. The efficiency of an antenna in exciting a surface-wave mode is defined as the ratio of the power radiated as a surface wave to the total power radiated. Several investigators have shown that launching efficiencies of 80% and greater can be obtained.

A typical launching device is a flared horn with an aperture field chosen as nearly like the transverse field of the surface wave as possible. Other devices such as slots in conducting planes (these are equivalent to magnetic current sources), dipoles, and line sources have also been studied, and will give a good efficiency when properly oriented with respect to the surface waveguide.

In this section, the problem of a line source above a grounded dielectric sheet will be studied, in order to show the significance of the continuous eigenvalue spectrum. In addition, a discussion of the leaky modes is included. Various authors have examined this problem, and the material presented here is similar to theirs [11.21]–[11.23]. Our analysis parallels that given by Barone in several respects.

Figure 11.16 illustrates a lossless dielectric sheet, of thickness t and with a dielectric constant κ, located on a perfectly conducting plane at $x = -t$. An electric current line source is located

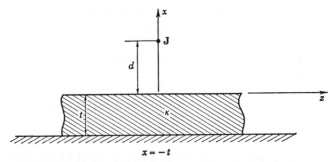

Fig. 11.16. Current line source above a grounded dielectric sheet.

at $x = d$, $z = 0$ and is parallel to the y axis. The current source may be represented as follows:

$$\mathbf{J} = \mathbf{a}_y \delta(z) \delta(x - d). \tag{68}$$

Since the current has no variation with y, the radiated field is also independent of y. The vector potential has only a y component $\psi(x, z)$, and the electric and magnetic fields are given by

$$E_y = -j\omega\psi$$

$$\mu_0 H_x = -\frac{\partial\psi}{\partial z}$$

$$\mu_0 H_z = \frac{\partial\psi}{\partial x}.$$

The function ψ is a solution of

$$\frac{\partial^2\psi}{\partial x^2} + \frac{\partial^2\psi}{\partial z^2} + k_0^2\psi = -\mu_0\delta(z)\delta(x - d), \qquad x > 0 \tag{69a}$$

$$\frac{\partial^2\psi}{\partial x^2} + \frac{\partial^2\psi}{\partial z^2} + \kappa k_0^2\psi = 0, \qquad -t < x < 0 \tag{69b}$$

and must satisfy the boundary condition that ψ and $\partial\psi/\partial x$ be continuous at $x = 0$. In addition, ψ is bounded at infinity and must represent outward-propagating waves. The solution for ψ will be obtained by means of a bilateral Laplace transform.

Let

$$g(x, \gamma) = \int_{-\infty}^{\infty} \psi(x, z) e^{\gamma z} \, dz. \tag{70}$$

We now multiply (69) by $e^{\gamma z}$, and integrate from $-\infty$ to ∞ to get

$$\frac{d^2 g}{dx^2} + (\gamma^2 + k_0^2)g = -\mu_0\delta(x - d), \qquad x > 0 \tag{71a}$$

$$\frac{d^2 g}{dx^2} + (\gamma^2 + \kappa k_0^2)g = 0, \qquad -t < x < 0 \tag{71b}$$

with g and dg/dx continuous at $x = 0$. In carrying out the integration with respect to z, the second derivative of ψ with respect to z is integrated by parts twice. The integrated terms vanish at infinity, provided we assume that k_0 has a small imaginary part, that is, $k_0 = k_0' - jk_0''$, corresponding to a small loss in the medium. Under these conditions, both $e^{\gamma z}\psi$ and $e^{\gamma z}\,\partial\psi/\partial z$ vanish at infinity for $\operatorname{Re}\gamma = 0$.

A suitable form for the function g is

$$g_1 = C_1 e^{-jl(x-d)}, \qquad\qquad\qquad\qquad\qquad x \geq d$$

$$g_2 = C_2 e^{jl(x-d)} + RC_2 e^{-jl(x+d)}, \qquad\qquad 0 \leq x \leq d$$

$$g_3 = C_3 e^{-jld}(1+R)\csc ht\,\sin h(x+t), \qquad -t \leq x \leq 0$$

where $h^2 = \gamma^2 + \kappa k_0^2$, $l^2 = \gamma^2 + k_0^2$, and C_1, C_2, C_3, R are constants to be determined. The above form of solution for g has been chosen by a consideration of the equivalent transmission-line problem for propagation along the x axis. Continuity of g at $x = 0$ gives $C_2 = C_3$, while continuity of dg/dx at $x = 0$ gives

$$R = \frac{jl - h\cot ht}{jl + h\cot ht}. \tag{72}$$

Thus R is seen to be the usual reflection coefficient at the air–dielectric interface. At $x = d$, g is continuous, but dg/dx is discontinuous. Integrating (71a) from $x = d_-$ to $x = d_+$ gives

$$\left.\frac{dg}{dx}\right|_{d_-}^{d_+} = -\mu_0$$

and this specifies the discontinuity in the derivative at the source. The boundary conditions at $x = d$ are now readily found to give

$$C_1 = \frac{\mu_0}{2jl}(1 + Re^{-2jld}) \qquad C_2 = \frac{\mu_0}{2jl}.$$

Hence, the solution for g becomes

$$g_1 = \frac{\mu_0}{2jl}(e^{-jl(x-d)} + Re^{-jl(x+d)}), \qquad\qquad x \geq d \tag{73a}$$

$$g_2 = \frac{\mu_0}{2jl}(e^{jl(x-d)} + Re^{-jl(x+d)}), \qquad\qquad 0 \leq x \leq d \tag{73b}$$

$$g_3 = \frac{\mu_0}{2jl}[e^{-jld}(1+R)\csc ht\,\sin h(x+t)], \qquad -t \leq x \leq 0. \tag{73c}$$

Equations (73a) and (73b) may be combined to give

$$g = \frac{\mu_0}{2jl}(e^{-jl|x-d|} + Re^{-jl(x+d)}), \qquad x \geq 0. \tag{74}$$

Inverting this transform gives the solution for ψ as follows:

$$\psi(x, z) = -\frac{\mu_0}{4\pi}\int_C \left(\frac{e^{-jl|x-d|}}{l} + \frac{Re^{-jl(x+d)}}{l}\right)e^{-\gamma z}\,d\gamma \tag{75}$$

where C is a contour to be determined so that ψ has the proper behavior at infinity, R is given by (72), and $l = (\gamma^2 + k_0^2)^{1/2}$.

For the reflection coefficient R, the transverse wavenumber is given by $h = (\gamma^2 + \kappa k_0^2)^{1/2}$. It does not matter which sign of the square root is chosen since $h \cot ht$ is an even function of h. For the transverse wavenumber $l = (\gamma^2 + k_0^2)^{1/2}$, the branch that leads to outward-propagating or attenuated waves must be chosen. This requires that

$$\mathrm{Re}\, l > 0$$

$$\mathrm{Im}\, l < 0.$$

The integrand in (75) is a two-valued function of γ because of the two branches of the function l. The branch points occur at

$$\pm jk_0 = \pm jk_0' \pm k_0''.$$

The complex γ plane must be cut by two branch lines running from the branch points to infinity (or joining the two branch points directly) if the integrand is to be single-valued. As far as the contour C is concerned, the branch cuts may be chosen quite arbitrarily, as long as they do not intersect the contour C. One possible construction is shown in Fig. 11.17. For any point $j\beta$ along the imaginary axis of the complex $\gamma = j\beta + \alpha$ plane, the phase angles of the two factors $(\gamma + jk_0)^{1/2}$ and $(\gamma - jk_0)^{1/2}$ lie in the ranges

$$\frac{3\pi}{4} > \mathrm{phase}(\gamma - jk_0)^{1/2} > \frac{\pi}{4}$$

$$-\frac{\pi}{4} < \mathrm{phase}(\gamma + jk_0)^{1/2} < \frac{\pi}{4}$$

since

$$\frac{3\pi}{2} > \theta_1 > \frac{\pi}{2}$$

$$-\frac{\pi}{2} < \theta_2 < \frac{\pi}{2}$$

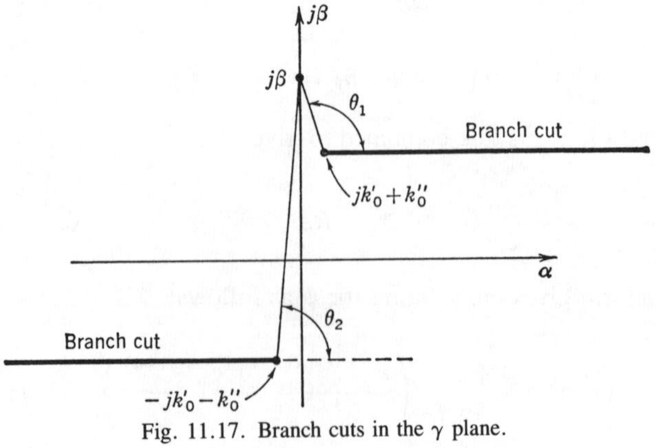

Fig. 11.17. Branch cuts in the γ plane.

as β ranges from $-\infty$ to ∞. The sum of the two angles $\theta_1 + \theta_2$ is always greater than π, but less than 2π, for all values of γ along the imaginary axis. Hence, the phase angle of $+(\gamma^2 + k_0^2)^{1/2}$ is always greater than $\pi/2$ but less than π. In order to satisfy the specified requirements on l, the negative root or branch must be chosen so that the phase angle of $l = -(\gamma - jk_0)^{1/2}(\gamma + jk_0)^{1/2}$ is in the range $3\pi/2$ to 2π for all values of γ on the imaginary axis. With this branch chosen as the proper branch, the convergence of the integral (75) at infinity is ensured. Replacing l by $-l$, we have, in place of (75), the more convenient expression

$$\psi(x, z) = \frac{\mu_0}{4\pi} \int_C \left[\frac{\exp j(\gamma^2 + k_0^2)^{1/2}|x - d|}{(\gamma^2 + k_0^2)^{1/2}} + \frac{R \exp j(\gamma^2 + k_0^2)^{1/2}(x + d)}{(\gamma^2 + k_0^2)^{1/2}} \right] e^{-\gamma z} \, d\gamma \quad (76)$$

where

$$R = \frac{(\gamma^2 + k^2)^{1/2} \cot(\gamma^2 + k^2)^{1/2} t + j(\gamma^2 + k_0^2)^{1/2}}{-(\gamma^2 + k^2)^{1/2} \cot(\gamma^2 + k^2)^{1/2} t + j(\gamma^2 + k_0^2)^{1/2}} \quad (77)$$

and $k^2 = \kappa k_0^2$. In (76) and (77), the proper branch is now the positive root of $(\gamma^2 + k_0^2)^{1/2}$. The integrand has poles whenever R becomes infinite, i.e., when

$$(\gamma^2 + k^2)^{1/2} \cot(\gamma^2 + k^2)^{1/2} t = j(\gamma^2 + k_0^2)^{1/2}. \quad (78)$$

For $\gamma = j\beta$ and $|\beta| > k_0$, we get

$$-(k^2 - \beta^2)^{1/2} \cot(k^2 - \beta^2)^{1/2} t = (\beta^2 - k_0^2)^{1/2}. \quad (79)$$

The roots of this equation lead to the surface-wave modes. These roots lie on the $j\beta$ axis between jk_0 and jk when k_0 and k are taken real, which will be done for determining the roots of (79). For $z > 0$, the contour C may be closed in the right-half γ plane, and the roots along the positive $j\beta$ axis must be enclosed, in order to give rise to the surface-wave modes. The roots along the negative $j\beta$ axis are excluded, since these give rise to a propagation factor of the form $e^{j\beta z}$, which is not appropriate for $z > 0$. These roots, however, are included when the contour C is closed in the left-half γ plane for $z < 0$. The correct contour for evaluating (76) is thus the one illustrated in Fig. 11.18. For $z > 0$, this contour may be closed in the right-half γ plane. Since the branch cut cannot be crossed, the contour must come back in from infinity on one side of the branch cut, encircle the branch point at jk_0, and recede out to infinity again along the opposite side of the branch cut. This contour is illustrated by the broken line in Fig. 11.18 and denoted by C_1 for the semicircle at infinity, and by C_b for the branch-cut integral. Since a branch cut was introduced, the integrand is single-valued, and Cauchy's residue theorem applies. The integral along $C + C_1 + C_b$ is equal to $-2\pi j$ times the enclosed residues. The negative sign arises because the contour is traversed in the clockwise sense. The original integral along C is thus equal to

$$\int_C = -\int_{C_1} - \int_{C_b} - 2\pi j \sum \text{residues}.$$

If the integral along the semicircle C_1 vanishes, then the original integral is equal to the branch-cut integral plus the residues contributed by the enclosed poles. The branch-cut integral

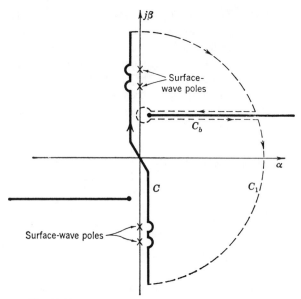

Fig. 11.18. Inversion contour C (broken line) for (76).

represents the radiation field with a continuous eigenvalue spectrum, while the residues at the enclosed poles are the surface-wave modes.

   Although the branch cut is arbitrary, as far as the original integral (76) along C is concerned, it must be chosen more carefully if the integral along C_1 is to vanish. Let us assume that the branch cut has been chosen as in Fig. 11.19. On the semicircle above the branch cut, the phase angle of $(\gamma^2 + k_0^2)^{1/2}$ is more than $\pi/2$ and less than π. So $\exp j(\gamma^2 + k_0^2)^{1/2}x$ decays exponentially with increasing positive values of x, and (76) will vanish on this portion of the semicircle. As the point P_1 moves down C_1, and in along the branch cut and back out to P_2, the angle θ_1 increases to a value greater than 2π, while θ_2 remains less than $\pi/2$. Over a portion of the semicircle below the branch cut, the phase angle of $(\gamma^2 + k_0^2)^{1/2}$ is greater

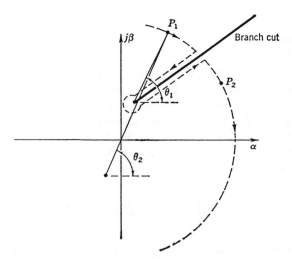

Fig. 11.19. Alternative branch cuts.

than π, and hence $\exp j(\gamma^2 + k_0^2)^{1/2}x$ is exponentially increasing for positive values of x. The exponential term in (76) becomes, for large values of γ and x,

$$\exp[(-j\alpha + \beta)x - (\alpha + j\beta)z]$$

and, unless $\beta x < \alpha z$, this term increases without limit as γ tends to infinity. The difficulty with the above choice of branch cut is that this choice does not correspond to a cut which separates the proper branch, for which

$$\mathrm{Im}(\gamma^2 + k_0^2)^{1/2} > 0 \tag{80}$$

from the improper branch, for which

$$\mathrm{Im}(\gamma^2 + k_0^2)^{1/2} < 0. \tag{81}$$

It is only on the proper branch that the integral (76) will converge at infinity. The branch cut separating the proper and improper branches occurs along the curve $\mathrm{Im}(\gamma^2 + k_0^2)^{1/2} = 0$. If $k_0 = k_0' - jk_0''$, this curve is given by

$$\alpha\beta = k_0'k_0''. \tag{82}$$

This curve is a portion of a hyperbola running from the branch point down toward the real axis in the γ plane, as illustrated in Fig. 11.20(a). As the losses in the medium decrease, k_0'' decreases, and, in the limit, the hyperbola becomes the portion of the $j\beta$ axis between 0 and jk_0 and the real α axis in the right half plane. Along the continuation of the hyperbola from the branch points along the $j\beta$ axis, the real part of $(\gamma^2 + k_0^2)^{1/2}$ is zero. In the region between the two hyperbolas [hatched region in Fig. 11.20(b)], the real part of $(\gamma^2 + k_0^2)^{1/2}$ is negative, while the imaginary part is positive. The original contour C must lie entirely in this region in order to represent outgoing waves.

With the branch cut chosen as in Fig. 11.20, the integral (76) will vanish on the semicircle in the right half plane for $z > 0$. The branch-cut integral for k_0 real gives an eigenvalue

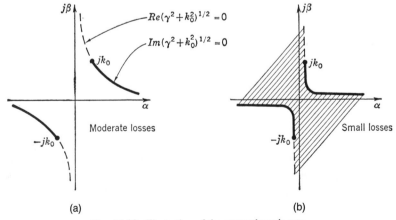

Fig. 11.20. Illustration of the proper branch cuts.

spectrum with γ in the range

$$0 \leq \gamma = \alpha < \infty$$
$$0 \leq \gamma = j\beta \leq jk_0$$

in accordance with the results of Section 11.5.

   The integrand in (76) may be conveniently viewed as defined on a two-sheeted Riemann surface. Only on the upper sheet or surface (the proper sheet), for which condition (80) holds, does the integral (76) converge. On the improper sheet, condition (81) applies, and the integral does not converge properly at infinity. The branch cuts of Fig. 11.20 provide the only cuts by which passage from one sheet to the other can be executed.

   In addition to the surface-wave poles of (78), which occur for $(\gamma^2 + k_0^2)^{1/2}$ equal to a positive-imaginary quantity, there are a number of roots for $(\gamma^2 + k_0^2)^{1/2}$ equal to a negative-imaginary quantity plus an infinite number of complex roots with negative-imaginary parts. All these latter roots or poles will lie on the improper sheet of the Riemann surface. It is only the surface-wave poles that lie on the proper sheet, since only on the proper sheet does condition (80) hold. The complex roots give rise to a class of modes called leaky modes. In evaluating (76) approximately by the saddle-point method of integration, the contour C is deformed into a contour of steepest descent. Part of this contour may lie in the improper sheet of the Riemann surface, and the leaky-wave modes then contribute to the expansion of the field. Since the integral is too complicated to be evaluated rigorously, approximate methods must be used. The saddle-point method of integration is one approximation technique which is capable of giving good results for the far-zone radiation field. Before carrying out the saddle-point integration, the roots of (78) will be determined.

Roots of (78)

   For $\gamma = j\beta$ and $k_0 < |\beta|$, let $k^2 - \beta^2 = h^2$, and $j(k_0^2 - \beta^2)^{1/2} = -p$. Thus we have

$$pt = -ht \cot ht \tag{83a}$$

and

$$(pt)^2 + (ht)^2 = (\kappa - 1)(k_0 t)^2 = a^2. \tag{83b}$$

Only the roots corresponding to positive values of p lie on the proper sheet of the Riemann surface. The solution of (83) is readily obtained graphically. For $\beta > k$, h is imaginary, and we have

$$pt = |h|t \coth|h|t$$
$$(pt)^2 - (|h|t)^2 = (\kappa - 1)(k_0 t)^2.$$

These curves are plotted as a function of $|h|$ to the left of the origin in Fig. 11.21. There will be no root for imaginary values of h unless $(\kappa - 1)^{1/2} k_0 t$ is less than unity. When this condition holds, an improper pole occurs for negative p, but there will not be a solution for a surface wave at the same time. A surface-wave mode exists only if

$$a = (\kappa - 1)^{1/2} k_0 t$$

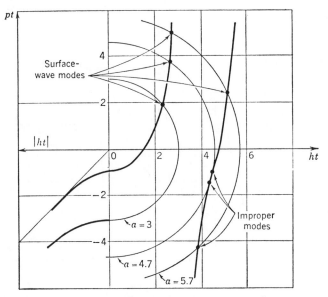

Fig. 11.21. Graphical solution for the real roots of p.

is greater than $\pi/2$. As $k_0 t$ increases, the number of surface-wave modes and improper modes continually increases.

To determine the complex roots, let pt be the complex variable

$$w = u + jv$$

and let ht be the complex variable $Z = x + jy$. The variables x, y, Z used here should not be confused with the rectangular coordinates. Equation (83a) becomes $w = -Z \cot Z$, or, in component form,

$$u = -\frac{x \tan x \,(1 - \tanh^2 y) + y \tanh y \,(1 + \tan^2 x)}{\tan^2 x + \tanh^2 y} \tag{84a}$$

$$v = \frac{-y \tan x \,(1 - \tanh^2 y) + x \tanh y \,(1 + \tan^2 x)}{\tan^2 x + \tanh^2 y}. \tag{84b}$$

Equation (83b) must also be satisfied, thus

$$u^2 - v^2 + x^2 - y^2 = a^2 \tag{84c}$$

$$uv + xy = 0. \tag{84d}$$

From (84), it is found that u is negative for all values of x and y. Consequently, the real part of p is always negative for the complex roots, and, hence, the imaginary part of $(\gamma^2 + k_0^2)^{1/2}$ is negative also since

$$p = -j(\gamma^2 + k_0^2)^{1/2}.$$

Therefore, the complex roots are located on the improper sheet of the Riemann surface.

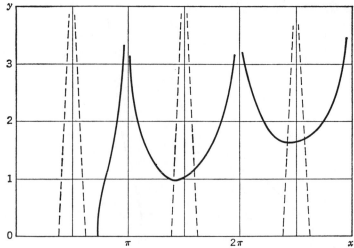

Fig. 11.22. Graphical solution for complex roots, $a = 3$. Solid lines, Eq. (85a); broken lines, Eq. (85b).

By eliminating the variable $w = pt$ in (83a) and (83b), we obtain the equation $Z = \pm a \sin Z$, or, in component form,

$$\frac{x}{\sin x} = \pm a \cosh y \tag{85a}$$

$$\cos x = \pm \frac{1}{a} \frac{y}{\sinh y}. \tag{85b}$$

A simultaneous solution of these two equations determines the complex roots for Z. From (84a) and (84b) the corresponding value of w may be found. The values of γ are found from the relation

$$w = -jt(\gamma^2 + k_0^2)^{1/2}. \tag{86}$$

From (85), it is seen that, if $x_1 + jy_1$ is a solution, then $x_1 - jy_1$, $-x_1 + jy_1$, and $-x_1 - jy_1$ are also solutions. Hence we need consider only positive values of x and y.

By plotting the two equations (85a) and (85b) on the xy plane, the roots are readily determined. Since a is greater than $\pi/2$, in order for at least one surface-wave mode to exist, a typical plot is as illustrated in Fig. 11.22. The roots are seen to be located in the vicinity of $x = n\pi - \pi/2$, where n is any integer. For large values of x, the corresponding value of y is given by $y = \ln[(2n - 1)\pi/a]$, since $\cosh y = x/a$ for $x = n\pi - \pi/2$. Simultaneous solution of (85a) and (85b) requires use of the same signs on the right-hand sides so only the roots for $n\pi < x < (2n + 1)\pi/2$, $n = 1, 2, \ldots$, are valid and satisfy the condition $xy = -uv$ from (84d). For $a = 3$, the first root is $x = \pm 1.4\pi$, $y = \pm 1$, giving $u = -1.31$, $v = \pm 3.37$. The transverse wavenumber in the x direction above the dielectric sheet is $j(\gamma^2 + k_0^2)^{1/2} = -(u + jv)/t = (1.31 \mp j3.37)/t$ and is seen to give rise to exponentially increasing waves in the x direction.

The Saddle-Point Method of Integration

The saddle-point method (also known as the method of steepest descent) for the asymptotic evaluation of an integral was introduced by Debye for obtaining asymptotic expansions of the

Hankel functions. The basic principles involved will be described in connection with the first term in (76). This term is, apart from the factor $\mu_0/4\pi$,

$$\int_C \frac{\exp[j(\gamma^2 + k_0^2)^{1/2}|x - d| - \gamma z]}{(\gamma^2 + k_0^2)^{1/2}} \, d\gamma \tag{87}$$

and represents the direct field radiated by the line source. The integral is well known and proportional to the Hankel function of the second kind and zero order, i.e., equal to $-j\pi H_0^2(k_0 r)$, where $r = [(x - d)^2 + z^2]^{1/2}$. In order to evaluate (87) asymptotically for large r by the saddle-point method, it is convenient to change from rectangular coordinates x, z to cylindrical coordinates r, θ and also to change the variable of integration γ as follows:

$$\gamma = jk_0 \sin \phi \tag{88a}$$

$$(\gamma^2 + k_0^2)^{1/2} = -k_0 \cos \phi \tag{88b}$$

$$d\gamma = jk_0 \cos \phi \, d\phi \tag{88c}$$

$$\phi = \sigma + j\eta \tag{88d}$$

$$z = r \sin \theta \tag{88e}$$

$$x - d = r \cos \theta. \tag{88f}$$

The integral (87) now becomes

$$-j \int_C e^{-jk_0 r \cos(\phi - \theta)} \, d\phi. \tag{89}$$

The details of the saddle-point integration are greatly simplified by the above transformations.

Equation (88a) represents a mapping of the complex γ plane into a strip of the complex ϕ plane. The two sheets of the Riemann surface map into a connected strip of width 2π along the σ axis. From (88a), we have

$$\gamma = \alpha + j\beta = jk_0 \sin(\sigma + j\eta)$$

or

$$\alpha = -k_0 \cos \sigma \sinh \eta \tag{90a}$$

$$\beta = k_0 \sin \sigma \cosh \eta. \tag{90b}$$

This mapping is illustrated in Fig. 11.23. The quadrants in which β and α are positive or negative are readily found from (90), and are specified in the figure. On the proper sheet of the Riemann surface (top sheet), we have $\text{Im}(\gamma^2 + k_0^2)^{1/2} > 0$, and hence, from (88b), we find that the corresponding part of the ϕ plane is determined by the condition that $\sin \sigma \sinh \eta > 0$. The four quadrants of the top and bottom sheets of the γ plane map into the regions designated $T_i, B_i, i = 1, 2, 3, 4$, respectively, in Fig. 11.23. The negative sign in (88b) was chosen so that the quadrants T_1, T_2 and T_3, T_4 would map into adjacent strips. The hyperbolas giving the branch cuts in the γ plane as in Fig. 11.20(b) map into the broken lines in Fig. 11.23. The hatched regions are corresponding regions in the two figures. It should be noted that there are no branch cuts in the ϕ plane since (88a) maps both sheets of the Riemann surface into

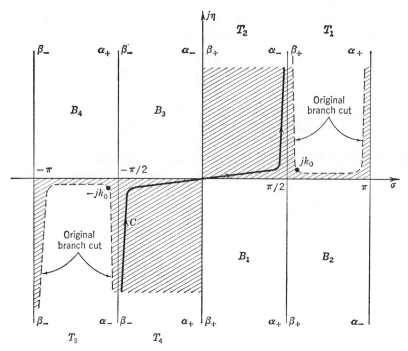

Fig. 11.23. Mapping of two sheets of the complex γ plane onto a strip of the complex ϕ plane.

a connected strip. The original contour of integration is denoted by C in Fig. 11.23. This contour may be deformed into any other convenient contour in the hatched region without changing the value of the integral (89), since no poles will be swept across.

In the integrand for (89), let

$$f(\phi) = -jk_0r \, \cos(\phi - \theta) \tag{91}$$

and the derivative with respect to ϕ,

$$\frac{df}{d\phi} = jk_0r \, \sin(\phi - \theta)$$

is equal to zero at $\phi = \theta$ or $\sigma = \theta$, $\eta = 0$. Since a complex function cannot have a maximum or a minimum, the stationary points are saddle points [11.24]. In the vicinity of the saddle point, a Taylor expansion gives

$$f(\phi) \approx f(\theta) + \frac{df}{d\phi}\bigg|_{\theta} (\phi - \theta) + \frac{1}{2} \frac{d^2f}{d\phi^2}\bigg|_{\theta} (\phi - \theta)^2$$

$$= -jk_0r + \frac{jk_0r}{2}(\sigma - \theta + j\eta)^2 \tag{92}$$

since the first derivative vanishes and the second derivative equals jk_0r at θ. Let ρ, ω be the radial and angular coordinates with origin at θ in the ϕ plane. Thus $\phi - \theta = \sigma - \theta + j\eta = \rho e^{j\omega}$,

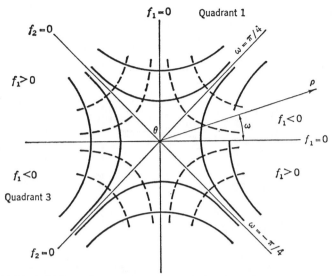

Fig. 11.24. Solid lines, f_2 = constant contours; broken lines, f_1 = constant contours.

and, if we let $f_1 + jf_2 = f(\phi) - f(\theta)$, we have $f_1 + jf_2 = (jk_0 r/2)\rho^2 e^{2j\omega}$, and, hence,

$$f_1 = \frac{-k_0 r}{2}\rho^2 \sin 2\omega \tag{93a}$$

$$f_2 = \frac{k_0 r}{2}\rho^2 \cos 2\omega. \tag{93b}$$

The contours f_1 = constant are orthogonal to the contours f_2 = constant. These contours are sketched in Fig. 11.24. The maximum rate of change of f_1 with ρ occurs for $\omega = \pi/4$ and $\pi/4 + \pi$, and also for $\omega = 3\pi/4$ and $-\pi/4$. Along these paths of steepest ascent or descent, $f_2 = 0$. In the first and third quadrants, f_1 is negative and increases rapidly in magnitude as we move away from the saddle point along the directions $\omega = \pi/4$ and $\omega = \pi/4 + \pi$. In the second and fourth quadrants, f_1 increases in the positive direction at the maximum rate along the contours $\omega = 3\pi/4$ and $-\pi/4$. If f_1 is considered to represent the elevation of the ground with respect to a reference datum at $\phi = \theta$, it is seen that the ground has the characteristics normally associated with a saddle, since it rises rapidly in two directions and falls rapidly in the two perpendicular directions. It is from this analogy that the terms saddle point, lines of steepest descent, etc., are obtained.

In view of the properties of the function f, it is clear that it is desirable to deform the contour C in (89) into a steepest-descent contour (SDC) passing through the saddle point at $\phi = \theta$, since, along this path, f_1 is negative and hence $e^{f(\phi)}$ will decrease rapidly as we move away from the saddle point. Thus, the major contribution to the integral comes from a small region or range $0 < \rho < \rho_1$ along the steepest-descent contour.

The steepest-descent contour is the contour along which $f_2 = 0$. For general values of ϕ, $f_1 + jf_2 = f(\phi) - f(\theta)$, and, hence,

$$f_2 = k_0 r[1 - \cos(\sigma - \theta)\cosh\eta]$$

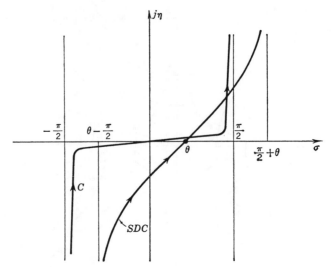

Fig. 11.25. Deformation of contour C into steepest-descent contour SDC.

and the steepest-descent contour is determined by

$$\cos(\sigma - \theta)\cosh\eta = 1. \tag{94}$$

This contour is illustrated as the contour SDC in Fig. 11.25. It passes through the saddle point at an angle $\pi/4$ in accordance with (93b), which is valid near the saddle point.

Near the saddle point,

$$f(\phi) \approx -jk_0 r + \frac{jk_0 r}{2}\rho^2 e^{2j\omega} = -jk_0 r - \frac{k_0 r}{2}\rho^2$$

along the steepest-descent contour $\omega = \pi/4$ and $\pi/4 + \pi$. Along this contour $\phi = \theta + \rho e^{j\pi/4}$ in the first quadrant, and $\phi = \theta + \rho e^{j(\pi + \pi/4)}$ in the third quadrant. Hence, we have

$$d\phi = e^{j\pi/4}\,d\rho \qquad \text{first quadrant}$$
$$d\phi = -e^{j\pi/4}\,d\rho \qquad \text{third quadrant.}$$

The integral becomes

$$-j\int_C e^f\,d\phi = -j\int_{\text{SDC}} e^f\,d\phi = -je^{-jk_0 r + j\pi/4}\left(\int_{\rho_1}^{0} -e^{-k_0 r\rho^2/2}\,d\rho + \int_0^{\rho_1} e^{-k_0 r\rho^2/2}\,d\rho\right) \tag{95}$$

since $e^{-k_0 r\rho^2/2}$ is negligible for $\rho > \rho_1$, provided $k_0 r$ is sufficiently large. Combining terms in (95) gives

$$-j\int_C e^f\,d\phi = -2je^{-j(k_0 r - \pi/4)}\int_0^{\rho_1} e^{-k_0 r\rho^2/2}\,d\rho. \tag{96}$$

Because of the rapid decay of the exponential, the integral from 0 to ρ_1 does not differ much from

$$\int_0^\infty e^{-k_0 r \rho^2/2}\, d\rho = \left(\frac{\pi}{2k_0 r}\right)^{1/2}.$$ (97)

Therefore, the original integral (87) is given by

$$-j\left(\frac{2\pi}{k_0 r}\right)^{1/2} e^{-j(k_0 r - \pi/4)}$$

for large $k_0 r$. The asymptotic expansion of the Hankel function $H_0^2(k_0 r)$ is $(2/\pi k_0 r)^{1/2} e^{-j(k_0 r - \pi/4)}$, and thus (87) is seen to be equal to $-j\pi H_0^2(k_0 r)$. The validity of the saddle-point integration depends on $k_0 r \rho^2/2$ being large for small values of ρ, and, hence, can only be used when $k_0 r$ is large, i.e., to evaluate the field far from the source. Additional terms in the asymptotic expansion may be obtained by following the method suggested in Problem 11.9. Multiplying the above result by $\mu_0/4\pi$, we obtain

$$\psi_{\text{direct}} = \frac{-j\mu_0}{4\pi}\left(\frac{2\pi}{k_0 r}\right)^{1/2} e^{-j(k_0 r - \pi/4)}, \qquad k_0 r \gg 1$$ (98)

for the far-zone direct-radiated field. In (98) the radial coordinate r is equal to $[(x - d)^2 + z^2]^{1/2}$.

Evaluation of a Reflected Field

The second term in (76), which is the field reflected from the dielectric sheet above the ground plane, may be evaluated by the saddle-point method also. With the transformation (88), and $-d$ replaced by d, this term becomes

$$\frac{-j\mu_0}{4\pi}\int_C R(\phi)\, \exp[-jk_0 r\, \cos(\phi - \theta)]\, d\phi$$ (99)

where

$$R(\phi) = \frac{jk_0 \cos\phi - h\cot ht}{jk_0 \cos\phi + h\cot ht}$$

and $h = k_0(\kappa - \sin^2\phi)^{1/2}$. The radial coordinate r now has its origin at $z = 0$, $x = -d$. As before, the contour C is deformed into a steepest-descent contour SDC passing through the saddle point $\phi = \theta$. In deforming the contour C, some of the poles of $R(\phi)$ may be encountered. When this happens, the contour should be warped as in Fig. 11.26, so as not to pass over the poles. In addition to the integral along the steepest-descent contour, the usual residue terms contribute to the field from the integration around the poles. The surface-wave poles lie on the lines $\sigma = \pi/2$, $\eta > 0$ and $\sigma = -\pi/2$, $\eta < 0$. The leaky-wave poles lie in the bottom sheet of the Riemann surface, and a finite number of these may be encountered as the contour C is deformed into the section labeled B_1. When the leaky-wave poles contribute, they do not give rise to exponentially growing waves, since, in expression (99), the real part f_1 of $-jk_0 r\, \cos(\phi - \theta)$ is negative at the location of these poles; that is, $f_1 = -k_0 r\, \sin(\sigma - \theta)\sinh\eta$

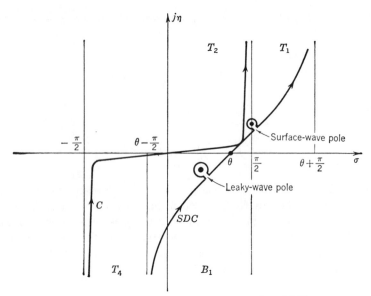

Fig. 11.26. Steepest-descent contour for a reflected field.

is negative. From Fig. 11.26, it is also seen that the surface-wave and leaky-wave residues only contribute for a range of θ greater than some minimum value θ_c, since no poles are crossed for small values of θ. Physically this means that the surface-wave and leaky-wave modes are not significant in determining the far-zone field except in the vicinity of the surface. This behavior can certainly be appreciated for the surface wave since it decays exponentially in a direction perpendicular and away from the guiding surface.

Let ϕ_s be the value of ϕ at a surface-wave pole, and let ϕ_L denote a leaky-wave pole. Also let the residue of $R(\phi)$ at a pole be denoted by $F(\phi_s)$. Thus (99) may be written as

$$\frac{-j\mu_0}{4\pi}\int_{\text{SDC}} R(\phi)e^{f(\phi)}\,d\phi - \frac{\mu_0}{2}\sum_s F(\phi_s)e^{f(\phi_s)} - \frac{\mu_0}{2}\sum_L F(\phi_L)e^{f(\phi_L)}.$$

The series arises from integration about the circles surrounding the poles. The integral along the steepest-descent contour is readily evaluated when $R(\phi)$ does not have a pole in the vicinity of the saddle point by expanding R in a Taylor series about θ. Thus

$$R(\phi) = R(\theta) + \sum_{n=1}^{\infty} R^n(\theta)\frac{(\phi - \theta)^n}{n!}$$

where $R^n(\theta) = (d^n R/d\phi^n)|_\theta$. Along the steepest-descent contour near θ,

$$\phi - \theta = \rho e^{j\pi/4} \qquad d\phi = e^{j\pi/4}\,d\rho \qquad \text{first quadrant}$$

$$\phi - \theta = -\rho e^{j\pi/4} \qquad d\phi = -e^{j\pi/4}\,d\rho \qquad \text{third quadrant.}$$

If we retain more than one term in the expansion of $R(\phi)$, then to be consistent we must use a more accurate expression for $e^{f(\phi)}$. We note that

$$e^{f(\phi)} = e^{f(\phi)+jk_0r-jk_0r(\phi-\theta)^2/2}e^{-jk_0r+jk_0r(\phi-\theta)^2/2} = e^{H(\phi)}e^{-jk_0r+jk_0r(\phi-\theta)^2/2}.$$

The function $e^{H(\phi)}$ is analytic along the SDC so we can include it with $R(\phi)$ and expand $R(\phi)e^{H(\phi)}$ in a Taylor series about $\theta = 0$. To order 4 this gives

$$R(\theta) + \sum_{n=1}^{4} R^n(\theta)\frac{(\phi - \theta)^n}{n!} - jk_0 r R(\theta)\frac{(\phi - \theta)^4}{4!}$$

since $H^1 = H^2 = H^3 = 0$ at $\phi = \theta$. In general, we can let the expansion of $R(\phi)e^{H(\phi)}$ be represented in the form

$$Re^H = \sum_{n=0}^{\infty} C_n \frac{(\phi - \theta)^n}{n!}$$

where the C_n are known constants; in particular, $C_0 = R(\theta)$, $C_1 = R^1(\theta)$, $C_2 = R^2(\theta)$, $C_3 = R^3(\theta)$, and $C_4 = R^4(\theta) - jk_0 r R(\theta)$.

Hence the integral becomes

$$-\frac{j\mu_0 e^{-j(k_0 r - \pi/4)}}{4\pi} \sum_{n=0}^{\infty} \frac{C_n(\theta)}{n!} e^{jn\pi/4}[1 + (-1)^n] \int_0^{\rho_1} \rho^n e^{-k_0 r \rho^2/2}\, d\rho.$$

The integral over ρ does not change much if ρ_1 is extended to infinity. The result

$$\int_0^{\infty} \rho^n e^{-k_0 r \rho^2/2}\, d\rho = \frac{\Gamma[(n+1)/2]}{2(k_0 r/2)^{(n+1)/2}}, \qquad n > -1 \tag{100}$$

where Γ is the gamma function, may then be used to evaluate the integral. Thus, for $k_0 r$ large and for no poles of $R(\phi)$ near θ, the reflected field is

$$\psi_{\text{ref}} = -\frac{\mu_0}{2}\left\{ \sum_s F(\phi_s) \exp[-jk_0 r \cos(\phi_s - \theta)] + \sum_L F(\phi_L) \exp[-jk_0 r \cos(\phi_L - \theta)] \right\}$$
$$-\frac{j\mu_0}{4\pi} e^{-j(k_0 r - \pi/4)} \sum_{n=0,2,\ldots} \frac{C_n(\theta)}{n!} e^{jn\pi/4} \left(\frac{2}{k_0 r}\right)^{(n+1)/2} \Gamma\left(\frac{n+1}{2}\right) \tag{101}$$

where $r^2 = (x + d)^2 + z^2$. For $k_0 r$ large, the series converges rapidly.

When $R(\phi)$ has a pole near the saddle point, the Taylor-series expansion of R is no longer valid in a sufficiently large region around the point θ. A Laurent-series expansion must be used, and this requires a further modification of the preceding method. Let $R(\phi)$ have a simple pole at $\phi = \phi_0$ with residue $F(\phi_0)$ and with ϕ_0 located near the saddle point $\phi = \theta$. In place of the Taylor-series expansion, we have

$$R(\phi) = \frac{F(\phi_0)}{\phi - \phi_0} + R_1(\phi) \tag{102}$$

where $R_1(\phi)$ is analytic in the region around θ and may be developed in a Taylor series. For R_1, we have

$$R_1(\phi) = \sum_{n=0}^{\infty} a_n(\phi - \theta)^n$$

where

$$a_n = \frac{1}{n!} \frac{d^n}{d\phi^n} \left[R(\phi) - \frac{F(\phi_0)}{\phi - \phi_0} \right] \Bigg|_{\phi=\theta}.$$

The portion of the integral for the radiated field involving R_1 leads to a contribution of the form (101).

The remaining integral to be evaluated is

$$-\frac{j\mu_0 F(\phi_0)}{4\pi} \int_C \frac{\exp[-jk_0 r \, \cos(\phi - \theta)]}{\phi - \phi_0} \, d\phi. \tag{103}$$

The asymptotic evaluation of integrals of this type was treated by Clemmow [11.43] and van der Waerden [11.42] and later by Oberhettinger. The principal results obtained by Oberhettinger will be presented here [11.25].

Consider a Laplace-type integral of the form

$$f(u) = \int_0^\infty t^\alpha g(t) e^{-ut} \, dt \tag{104}$$

where $g(t)$ is analytic at the origin and has its nearest pole at $t = t_0$, where t_0 is complex. In (104), α is a constant with $\mathrm{Re}\,\alpha > -1$. If t_0 is a pole of order m, a Laurent-series development gives

$$g(t) = b_{-m}(t - t_0)^{-m} + \cdots + b_{-1}(t - t_0)^{-1} + g_1(t)$$

where $g_1(t)$ is analytic for $|t| < t_1$, and t_1 is the next singular point with $|t_1| > |t_0|$. The function g_1 will be an analytic function in some sector $\theta_1 < \mathrm{phase}\,t < \theta_2$ and may be expanded about the origin in a Taylor series:

$$g_1(t) = \sum_{n=0}^\infty c_n t^n$$

where

$$c_n = \frac{1}{n!} \frac{d^n}{dt^n} \left[g(t) - \sum_{s=1}^m b_{-s}(t - t_0)^{-s} \right] \Bigg|_{t=0}.$$

The asymptotic behavior of $f(u)$ as $u \to \infty$ is governed by the behavior of $t^\alpha g(t)$ at the origin.

When the series expansion for $g(t)$ is substituted into (104) and integrated, term by term, we obtain

$$f(u) = \sum_{n=0}^\infty c_n \Gamma(\alpha + n + 1) u^{-\alpha-n-1}$$

$$+ \sum_{s=1}^m b_{-s} \Gamma(1 + \alpha)(-t_0)^{(\alpha-s)/2} u^{(s-\alpha-2)/2} e^{-ut_0/2} W_{-s/2-\alpha/2, \, \alpha/2-s/2+1/2}(-ut_0) \tag{105}$$

where Γ is the gamma function, and $W_{a,b}(x)$ is Whittaker's confluent hypergeometric function [11.26]. This asymptotic series is valid for $u \to \infty$ in the sector $-\pi/2 - \theta_2 <$ phase $u < \pi/2 - \theta_1$.

Some special cases of interest in diffraction work are:

(1)
$$W_{-1/4,\,-1/4}(x) = \sqrt{\pi} x^{1/4} e^{x/2} \operatorname{erfc} \sqrt{x} \qquad (106a)$$

where erfc y is the complement of the error function erf y and is given by

$$\operatorname{erfc} y = \frac{2}{\sqrt{\pi}} \int_y^\infty e^{-t^2}\, dt$$

$$\operatorname{erf} y = 1 - \operatorname{erfc} y = \frac{2}{\sqrt{\pi}} \int_0^y e^{-t^2}\, dt.$$

(2)
$$W_{-1/2,\,0}(x) = -x^{-1/2} e^{x/2} \operatorname{Ei}(-x) \qquad (106b)$$

where $\operatorname{Ei}(-x)$ is the exponential integral

$$\operatorname{Ei}(-x) = -\int_x^\infty t^{-1} e^{-t}\, dt.$$

(3)
$$W_{0,\,\nu}(x) = \left(\frac{x}{\pi} \right)^{1/2} K_\nu \left(\frac{x}{2} \right) \qquad (106c)$$

where K_ν is the modified Bessel function of the second kind.

In order to use the above results, we must convert our integral (103) into the form (104). As before, the major contribution comes from the portion of the contour in the vicinity of the saddle point θ. We may approximate $\cos(\phi - \theta)$ by $1 - (\phi - \theta)^2/2$. Replacing $d\phi$ by $e^{j\pi/4}\, d\rho$ in the first quadrant and by $-e^{j\pi/4}\, d\rho$ in the third quadrant, as in (95), and noting that

$$\phi - \phi_0 = \phi - \theta - (\phi_0 - \theta) = \rho e^{j\pi/4} - (\phi_0 - \theta)$$

in the first quadrant and equals

$$-\rho e^{j\pi/4} - (\phi_0 - \theta)$$

in the third quadrant, we get, in place of expression (103),

$$-\frac{j\mu_0 F(\phi_0)}{4\pi} e^{-jk_0 r} \left[\int_0^{\rho_1} \frac{e^{-k_0 r \rho^2/2}}{\rho - e^{-j\pi/4}(\phi_0 - \theta)}\, d\rho + \int_{\rho_1}^0 \frac{e^{-k_0 r \rho^2/2}}{\rho + e^{-j\pi/4}(\phi_0 - \theta)}\, d\rho \right]$$

$$= -\frac{j\mu_0 F(\phi_0) e^{-jk_0 r - j\pi/4}(\phi_0 - \theta)}{2\pi} \int_0^{\rho_1} \frac{e^{-k_0 r \rho^2/2}}{\rho^2 + j(\phi_0 - \theta)^2}\, d\rho \qquad (107)$$

after combining terms. Since the exponential decays rapidly, we may extend the integral over

ρ to infinity, i.e., extend ρ_1 to infinity. We now make one further change in variables and put $\rho^2 = t$. Hence, we obtain for the integral I

$$I = \int_0^\infty \frac{e^{-k_0 r \rho^2/2}}{\rho^2 + j(\phi_0 - \theta)^2}\, d\rho = \frac{1}{2}\int_0^\infty \frac{e^{-k_0 r t/2}}{t^{1/2}[t + j(\phi_0 - \theta)^2]}\, dt. \tag{108}$$

This latter integral is of the form (104). The value of the integral is given by (105) with $\alpha = -\frac{1}{2}$, $t_0 = -j(\phi_0 - \theta)^2$, and the pole is of order 1, that is, $m = 1$. In this case, the Whittaker function reduces to the result given by (106a). Hence, we have

$$I = \frac{\pi}{2}e^{jk_0 r(\phi_0 - \theta)^2/2}[j(\phi_0 - \theta)^2]^{-1/2}\, \mathrm{erfc}\left[\frac{jk_0 r}{2}(\phi_0 - \theta)^2\right]^{1/2}. \tag{109}$$

This expression must be used to evaluate the reflected field whenever the surface-wave pole at $\phi = \phi_0$ lies close to the saddle point at θ so that $(k_0 r/2)^{1/2}|\phi_0 - \theta|$ is small. When the latter quantity is large we can recover the original result given by (101).

The error function may be expanded asymptotically to give [11.26]

$$\mathrm{erfc}\, x = \frac{e^{-x^2}}{x\sqrt{\pi}}\sum_{n=0}^\infty (-1)^n \frac{\Gamma(n + \frac{1}{2})}{\Gamma(\frac{1}{2})}x^{-2n} \tag{110}$$

valid for $x \to \infty$ in the sector $-3\pi/4 < \text{phase } x < 3\pi/4$. If the phase angle of x is outside this range, it is noted that $\mathrm{erf}\, x$ is an odd function of x, and that

$$\mathrm{erfc}\, x = 1 - \mathrm{erf}\, x = 1 + \mathrm{erf}(-x) = 2 - \mathrm{erfc}(-x). \tag{111}$$

This result may be used to obtain an asymptotic expansion for x in the vicinity of the negative-real axis.

In the Mathematical Appendix a more precise asymptotic evaluation of the integral is carried out in order to show more clearly the considerations that must be taken into account whenever the pole at ϕ_0 approaches the saddle point θ and possibly also migrates across the SDC as the physical parameters of the problem change.

In our case, ϕ_0 lies either on the $\sigma = \pi/2$, $\eta > 0$ axis or else in the range $\eta < 0$. In both cases, the phase angle of $(jk_0 r/2)^{1/2}(\phi_0 - \theta)$ is in the range $-3\pi/4$ to $3\pi/4$, and, hence, by using (110), we obtain

$$I = -\frac{j\sqrt{\pi}}{(2k_0 r)^{1/2}(\phi_0 - \theta)^2}\sum_{n=0}^\infty (-1)^n \frac{\Gamma(n + \frac{1}{2})}{\Gamma(\frac{1}{2})}\left[\frac{jk_0 r(\phi_0 - \theta)^2}{2}\right]^{-n}. \tag{112}$$

Substituting (112) into (107), we obtain

$$\frac{-\mu_0 F(\phi_0)}{2\pi}e^{-jk_0 r - j\pi/4}\frac{\sqrt{\pi}}{(2k_0 r)^{1/2}(\phi_0 - \theta)}\left[1 + \frac{j}{k_0 r(\phi_0 - \theta)^2}\cdots\right]. \tag{113}$$

But this result, together with the contribution from R_1, is identical with (101) for $R_0 r(\phi_0 - \theta)$ large as expected. However, when $\phi_0 \to \theta$ only the result (109) will correctly evaluate ψ_{ref}.

REFERENCES AND BIBLIOGRAPHY

[11.1] K. Uller, dissertation, University Rostock, 1903.

[11.2] J. Zenneck, "Uber die footpflanzung ebener electromagnetischer wellen laugs einer ebener leiter flache und ihve beziehung zur drahtlosen telegraphic," *Ann. Phys.*, vol. 23, pp. 846–866, 1907.

[11.3] J. A. Stratton, *Electromagnetic Theory*. New York, NY: McGraw-Hill Book Company, Inc., 1941.

[11.4] S. S. Attwood, "Surface wave propagation over a coated plane conductor," *J. Appl. Phys.*, vol. 22, pp. 504–509, Apr. 1954.

[11.5] W. Rotman, "A study of single surface corrugated guides," *Proc. IRE*, vol. 39, pp. 952–959, Aug. 1951.

[11.6] R. S. Elliott, "On the theory of corrugated plane surfaces," *IRE Trans. Antennas Propagat.*, vol. AP-2, pp. 71–81, Apr. 1954.

[11.7] R. A. Hurd, "The propagation of an electromagnetic wave along an infinite corrugated surface," *Can. J. Phys.*, vol. 32, pp. 727–734, Dec. 1954.

[11.8] R. W. Hougardy and R. C. Hansen, "Oblique surface waves over a corrugated conductor," *IRE Trans. Antennas Propagat.*, vol. AP-6, pp. 370–376, Oct. 1958.

[11.9] D. K. Reynolds and W. S. Lucke, "Corrugated end fire antennas," *Proc. Nat. Electronics Conf.*, vol. 6, pp. 16–28, Sept. 1950.

[11.10] J. Brown, "The type of wave which may exist near a guiding surface," *Proc. IEE (London)*, vol. 100, part III, pp. 363–364, Nov. 1953.

[11.11] F. J. Zucker, "Two notes on surface wave nomenclature and classification," presented at the URSI Symposium, Toronto, Canada, June 1958.

[11.12] N. Marcuvitz, "On field representations in terms of leaky modes," *IRE Trans. Antennas Propagat.*, vol. AP-4, pp. 192–194, July 1956.

[11.13] G. Goubau, "Surface waves and their applications to transmission lines," *J. Appl. Phys.*, vol. 21, pp. 1119–1128, Nov. 1950.

[11.14] F. Harms, "Electromagnetishe wellen an einem draht mit isolierender zylindrischer hulle," *Ann. Phys.*, vol. 23, pp. 44–60, 1907.

[11.15] G. Goubau, "Single conductor surface wave transmission lines," *Proc. IRE*, vol. 39, pp. 619–624, June 1951.

[11.16] D. Hondros and P. Debye, "Electromagnetishe wellen an dielektrischen drahten," *Ann. Phys.*, vol. 32, pp. 465–476, 1910.

[11.17] W. M. Elsasser, "Attenuation in a dielectric circular rod," *J. Appl. Phys.*, vol. 20, pp. 1193–1196, Dec. 1949.

[11.18] C. H. Chandler, "Investigation of dielectric rod as waveguide," *J. Appl. Phys.*, vol. 20, pp. 1188–1192, Dec. 1949.

[11.19] D. G. Kiely, *Dielectric Aerials*. London: Methuen & Co., Ltd., 1953.

[11.20] G. Goubau, "On the excitation of surface waves," *Proc. IRE*, vol. 40, pp. 865–868, July 1952.

[11.21] R. M. Whitmer, "Fields in non-metallic waveguides," *Proc. IRE*, vol. 36, pp. 1105–1109, Sept. 1948.

[11.22] C. T. Tai, "Effect of a grounded slab on radiation from a line source," *J. Appl. Phys.*, vol. 22, pp. 405–414, Apr. 1951.

[11.23] S. Barone, "Leaky wave contribution to the field of a line source above a dielectric slab," Polytech. Inst. Brooklyn MRI Research Rep. R-532-56, Nov. 1956.

[11.24] E. A. Guillemin, *The Mathematics of Circuit Analysis*. New York, NY: John Wiley & Sons, Inc., 1949, ch. VI, art. 14; ch. VII, art. 26.

[11.25] F. Oberhettinger, "On a modification of Watson's lemma," *J. Res. Nat. Bur. Stand.*, vol. 63B, pp. 15–17, July–Sept. 1959.

[11.26] A. Erdelyi, *Higher Transcendental Functions*, vol. I. New York, NY: McGraw-Hill Book Company, Inc., 1953, p. 274.

Surface Waves in General

[11.27] H. M. Barlow and A. E. Karbowiak, "An investigation of the characteristics of cylindrical surface waves," *Proc. IEE (London)*, vol. 100, part III, pp. 321–328, Nov. 1953.

[11.28] H. M. Barlow and A. L. Cullen, "Surface waves," *Proc. IEE (London)*, vol. 100, part III, pp. 329–347, Nov. 1953.

[11.29] H. M. Barlow and A. E. Karbowiak, "An experimental investigation of the properties of corrugated cylindrical surface wave guides," *Proc. IEE (London)*, vol. 101, part III, pp. 182–188, May 1954.

[11.30] H. M. Barlow and A. E. Karbowiak, "An experimental investigation of axial cylindrical surface waves supported by capacitive surfaces," *Proc. IEE (London)*, vol. 102, part B, pp. 313–322, May 1955.

[11.31] L. A. Vajnshtejn, "Electromagnetic surface waves on a comblike structure," *Zh. Tekh. Fiz.*, vol. 26, pp. 385–397, 1956.

[11.32] S. P. Schlesinger and D. D. King, "Dielectric image lines," *IRE Trans. Microwave Theory Tech.*, vol. MTT-6, pp. 291–299, July 1958.

Excitation of Surface Waves

[11.33]   A. L. Cullen, "The excitation of plane surface waves," *Proc. IEE (London)*, vol. 101, part IV, pp. 225-234, Aug. 1954.
[11.34]   G. J. Rich, "The launching of a plane surface wave," *Proc. IEE (London)*, vol. 102, part B, pp. 237-246, Mar. 1955.
[11.35]   D. B. Brick, "The radiation of a hertzian dipole over a coated conductor," *Proc. IEE (London)*, vol. 102, part C, pp. 104-121, 1955.
[11.36]   W. M. G. Fernando and H. M. Barlow, "An investigation of the properties of radial cylindrical surface waves launched over flat reactive surfaces," *Proc. IEE (London)*, vol. 103, part B, pp. 307-318, May 1956.
[11.37]   J. R. Wait, "Excitation of surface waves on conducting, stratified, dielectric coated and corrugated surfaces," *J. Res. Nat. Bur. Stand.*, vol. 59, pp. 365-377, Dec. 1957.
[11.38]   R. H. DuHamel and J. W. Duncan, "Launching efficiency of wires and slots for a dielectric rod waveguide," *IRE Trans. Microwave Theory Tech.*, vol. MTT-6, pp. 277-284, July 1958.
[11.39]   J. W. Duncan, "The efficiency of excitation of a surface wave on a dielectric cylinder," *IRE Trans. Microwave Theory Tech.*, vol. MTT-7, pp. 257-268, Apr. 1959.
[11.40]   B. Friedman and W. E. Williams, "Excitation of surface waves," *Proc. IEE (London)*, vol. 105, part C, pp. 252-258, Mar. 1958.
[11.41]   J. Brown, "Some theoretical results for surface wave launchers," *IRE Trans. Antennas Propagat.*, vol. AP-7, Special Supplement, pp. S169-S174, 1959.

Saddle-Point Integration

[11.42]   B. L. van der Waerden, "On the method of saddle points," *Appl. Sci. Res.*, vol. 2B, pp. 33-43, 1950.
[11.43]   P. C. Clemmow, "Some extensions to the method of integration by steepest descents," *Quart. J. Mech. Appl. Math.*, vol. 3, pp. 241-260, 1950.
[11.44]   L. B. Felsen and N. Marcuvitz, *Radiation and Scattering of Waves*.   Englewood Cliffs, NJ: Prentice-Hall, 1973. (This book contains a very thorough treatment of asymptotic methods.)

Problems

11.1. Obtain the solution for a TE surface wave guided by a plane interface between free space and a medium with parameters ϵ, μ, where $\mu > \mu_0$, and $\mu/\mu_0 > \epsilon/\epsilon_0$. Assume that ϵ has a small negative-imaginary part, but that μ is real.

11.2. A vertical line source above an impedance plane or a grounded dielectric sheet radiates cylindrical waves having no angular variation. Obtain the solution for the Zenneck wave above an impedance plane, and also for a grounded dielectric sheet of thickness t. The wave is described by the function $H_0^2(\beta r)e^{-px}$, where $r^2 = y^2 + z^2$, and H_0^2 is the Hankel function of the second kind.

11.3. For a corrugated plane, find the minimum value of the short-circuit position d such that the first evanescent mode excited at the interface has decayed to less than 0.1 of its value at the interface, at the position of the short circuit. Assume a spacing S between corrugations equal to $0.2\lambda_0$. Repeat the calculation for $S = 0.4\lambda_0$.

11.4. Find the solution for a TM surface wave on a lossless dielectric tube of inner radius a and outer radius b as illustrated in Fig. P11.4.

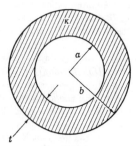

Fig. P11.4.

11.5. Evaluate the following integral by the saddle-point method:

$$\int_{-j\infty}^{j\infty} \frac{\exp[jx(\gamma^2 + k^2)^{1/2} - \gamma z]}{(\gamma^2 + k^2)^{1/2}(\gamma - 4jk)} \, d\gamma.$$

11.6. Find the field radiated by a line source located a distance d above a plane having a surface impedance jX_s. Evaluate the integral by the saddle-point method.

11.7. Repeat Problem 11.6 for a surface with reactance $-jX_s$. Note that a surface-wave mode is now excited.

11.8. For Problem 11.7, evaluate the total power radiated as a surface wave and also that radiated as a continuous spectrum. The power radiated as a continuous spectrum may be obtained by integrating the far-zone field, with respect to θ, from zero to $\pi/2$, and multiplying by 2. Find the best position of the line source in order to obtain the maximum power in the surface wave.

11.9. Show that the change in variable $\cos(\phi - \theta) = 1 - jS^2$, with $-\infty < S < \infty$, corresponds to integration along the steepest-descent contour in the ϕ plane. Show that the integral (89) becomes

$$-je^{-j(k_0 r - \pi/4)} \int_{-\infty}^{\infty} \frac{e^{-k_0 r S^2}}{(1 - jS^2/2)^{1/2}} \, dS.$$

Expand $(1 - jS^2/2)^{-1/2}$ in a power series valid for $|S| < \sqrt{2}$, and integrate the first two terms by using (100) to obtain the first two terms in the asymptotic evaluation of the integral.

Answer:

$$-j \left(\frac{2\pi}{k_0 r} \right)^{1/2} e^{-j(k_0 r - \pi/4)} \left[1 + \frac{j}{4} \frac{\Gamma(\frac{3}{2})}{(\pi k_0 r)^{1/2}} \right].$$

To obtain this result, the integration is taken from $-\infty$ to ∞, and thus an error is made, since the power series in S^2 is not valid for $|S|^2 > 2$. However, the error may be shown to be negligible for large $k_0 r$ (see B. L. van der Waerden, "On the method of saddle points," *Appl. Sci. Res.*, vol. 2B, pp. 33–43, 1950).

11.10. A vertical current element is located at a height h above a flat lossy earth as shown in Fig. P11.10. Show that the solution for the vector potential above the earth is given by

$$\Psi = \frac{\mu_0}{8\pi^2} \int_0^{2\pi} \int_0^{\infty} \frac{1}{\gamma_0} \left[1 - e^{-2\gamma_0 h} \frac{\gamma - \kappa\gamma_0}{\gamma + \kappa\gamma_0} \right] e^{-jw\rho \cos(\phi - \phi') - \gamma_0(z - h)} w \, dw \, d\phi'$$

where $\gamma_0 = j\sqrt{k_0^2 - w^2}$, $\gamma = j\sqrt{k^2 - w^2}$. Use the result

$$\int_0^{2\pi} e^{-jw\rho \cos(\phi - \phi')} \, d\phi = 2\pi J_0(w\rho)$$

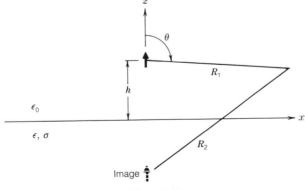

Fig. P11.10.

and show that

$$\Psi = \frac{\mu_0}{4\pi} \left[\frac{e^{-jk_0 R_1}}{R_1} - \frac{e^{-jk_0 R_2}}{R_2} + 2\kappa I \right]$$

where

$$I = \int_{-\infty}^{\infty} \frac{wH_0^2(w\rho)e^{-\gamma_0(z+h)}}{2(\gamma + \kappa\gamma_0)} \, dw.$$

Show that a pole exists at $w = w_0 = k_0 \sqrt{\kappa/(\kappa + 1)}$ where κ is the complex dielectric constant $(\epsilon - j\sigma/\omega)/\epsilon_0$ and σ is the conductivity of the earth. This pole gives rise to the Zenneck surface wave. This famous Sommerfeld problem was the subject of considerable controversy as regards the physical existence of the surface wave. The pole exists and since it lies very close to the saddle point at $\theta = \pi/2$ for typical earth parameters $(\epsilon/\epsilon_0 = 12, \sigma = 5 \times 10^{-3}$ siemens/meter) the integral must be evaluated in terms of the complement of the error function. The field at the surface is not a distinct surface-wave field for this reason. It is often called the Norton surface wave. Clearly, for $\theta = \pi/2$ the integral I is the only contribution to Ψ.

12
Artificial Dielectrics

At microwave frequencies, the wavelength is sufficiently short so that many of the techniques and principles used in optics may be carried over directly. One of the more obvious optical devices to find wide application is the lens. Lenses are used in a variety of antenna systems at microwave frequencies. The dielectric material for a microwave lens must have a small loss and an index of refraction preferably no smaller than about 1.5. One suitable material that is frequently used is polystyrene, which has an index of refraction equal to 1.6. Major disadvantages of solid dielectric materials are the excessive bulk and weight of the lens, since in many cases lens diameters greater than a meter are required. In 1948, Kock suggested the use of artificial dielectrics to replace the actual dielectric material, in order to overcome these disadvantages [12.1].

An artificial dielectric is a large-scale model of an actual dielectric, obtained by arranging a large number of identical conducting obstacles in a regular three-dimensional pattern. The obstacles are supported by a lightweight binder or filler material such as expanded polystyrene (Polyfoam). Under the action of an external applied electric field, the charges on each conducting obstacle are displaced so as to set up an induced field that will cancel the applied field at the obstacle surface. The obstacle is electrically neutral so that the dominant part of the induced field is a dipole field. Each obstacle thus simulates the behavior of a molecule (or group of molecules) in an ordinary dielectric in that it exhibits a dipole moment. The combined effect of all the obstacles in the lattice is to produce a net average dipole polarization $\mathbf{P}$ per unit volume. If $\mathbf{E}$ is the average net field in the medium, the displacement $\mathbf{D}$ is given by

$$\mathbf{D} = \epsilon_0 \mathbf{E} + \mathbf{P} = \epsilon \mathbf{E}$$

and it is apparent that the effective permittivity ϵ will be greater than ϵ_0.

When a high-frequency magnetic field is applied to a conducting obstacle, an induced circulating current is set up, such that the induced magnetic field produced will cancel the normal component of the applied field at the obstacle surface. These induced currents are equivalent to a magnetic dipole, and, therefore, in general, an artificial dielectric exhibits magnetic dipole polarization as well as electric dipole polarization. The induced magnetic dipoles always oppose the inducing field, so that the artificial dielectric is always diamagnetic when obstacles having magnetic dipole moments are used. It should be noted that certain types of obstacles, properly oriented with respect to the applied field, will not have an induced magnetic dipole. If ϵ and μ are the permittivity and permeability of the artificial dielectric, the index of refraction η is given by

$$\eta = \left(\frac{\epsilon \mu}{\epsilon_0 \mu_0} \right)^{1/2}.$$

Since μ is less than μ_0, the presence of magnetic polarization reduces the value of η, and is for this reason usually an undesirable property.

749

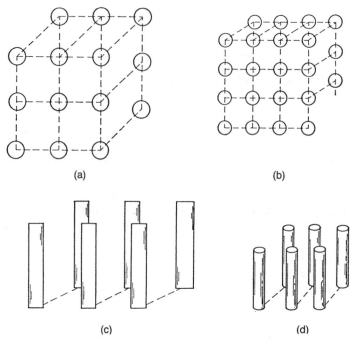

Fig. 12.1. Typical artificial dielectric structures. (a) Three-dimensional sphere medium. (b) Three-dimensional disk medium. (c) Two-dimensional strip medium. (d) Two-dimensional rod medium.

Unless the obstacles have spherical symmetry and a cubical or random lattice pattern is used, the dipole polarization for a unit applied field will be different in different directions. The artificial dielectric will then have anisotropic properties. To date, little use for such anisotropic properties has been found, except in wave polarizers, i.e., devices for converting linear polarization to circular polarization and vice versa.

Typical structures used for artificial dielectrics are illustrated in Fig. 12.1. Figure 12.1(a) illustrates a cubical array of conducting spheres, while in Fig. 12.1(b) the spheres are replaced by thin flat conducting disks. For the disk medium, no magnetic polarization is produced if the incident field is a TEM wave with the H vector in the plane of the disks. Figures 12.1(c) and (d) illustrate two-dimensional structures consisting of thin flat conducting strips and conducting cylinders or rods, respectively. With the electric field perpendicular to the strips or rods, capacitive loading of the medium is produced, and ϵ is increased. Artificial dielectrics for which the index of refraction η is greater than unity are known as phase-delay media, since the phase velocity in the medium is less than that in free space. For the two-dimensional structures of Fig. 12.1, inductive loading is produced when the applied electric field is parallel to the strips or rods. For this polarization, μ is less than μ_0, and η is less than unity. The structure is therefore referred to as a phase-advance dielectric.

The term artificial dielectric is commonly used to incorporate structures such as the parallel-plate array and stacks of square waveguides placed side by side, in addition to structures made up of arrays of discrete obstacles. Only the latter class of structures, in particular the sphere and disk media and the two-dimensional strip medium, will be examined in detail in this chapter.

There are three basic approaches used in the analysis of artificial dielectrics. The simplest is the Lorentz theory, which was originally developed for the classical theory of real dielectrics.

This theory considers only dipole interaction between obstacles, and produces accurate results for obstacle spacings limited to less than about $0.1\lambda_0$ and for obstacle sizes small compared with the spacing. When the obstacle size is not small compared with the spacing, the dipole term is not a sufficiently good approximation of the induced field. The second approach is one step more accurate than the Lorentz theory, and consists of a rigorous static field solution. Such a solution will take into account all the multipoles in the expansion of the induced field and will give good results provided the obstacle spacing is small compared with the wavelength, say no greater than $0.1\lambda_0$. Finally, the third method is to attempt a solution of Maxwell's equations. In this latter method, it is often possible to reduce the problem to that of a periodically loaded transmission line, for which the methods of analysis presented in Chapter 9 are applicable.

Many people have contributed to the development of artificial dielectrics since the pioneer work of Kock. It would require too much space to review this work here on an individual basis. However, a number of the more important papers that have been published are listed at the end of the chapter.

12.1. LORENTZ THEORY

The Lorentz theory is a static field theory which provides a solution that takes into account only the dipole term in the induced field. Consider a three-dimensional regular array of identical conducting obstacles, as in Fig. 12.2. The lattice spacings are a, b, and c along the x, y, and z directions, respectively. Let an external electrostatic field $\mathbf{E}_0$ be applied along the y direction. The obstacles will be assumed to have their principal axes of polarization coincident with the coordinate axes, so that the induced dipole moment p of one obstacle is in the y direction for an applied field in the y direction.

In the course of the analysis, several different fields will be introduced. For convenience, these are summarized below:

$$\mathbf{E}_0 = \mathbf{a}_y E_0 = \text{external applied field}$$

$$\mathbf{E}_1 = \text{dipole field produced by the obstacle located at the origin}$$

$$\mathbf{E}_p = \text{dipole field produced by all the obstacles in the}$$
$$\text{infinite array}$$

$$\mathbf{E}_e = \mathbf{E}_0 + \mathbf{E}_p - \mathbf{E}_1$$
$$= \text{effective field acting to polarize the obstacle at the origin}$$

$$\mathbf{E}_i = \mathbf{E}_p - \mathbf{E}_1 = \text{interaction field}$$

$$\mathbf{E}_{pa}, \mathbf{E}_{1a}, \mathbf{E}_{ea}, \mathbf{E}_{ia} = \text{fields defined as above but averaged over the}$$
$$\text{volume of a unit cell with dimensions } a, b, c$$

The y component of the above fields is indicated by an additional subscript y.

Since the Lorentz theory considers only dipole interaction between obstacles, the results will be valid only for obstacles with dimensions small compared with their spacing. The field acting to polarize any one obstacle may therefore be assumed to be a uniform field equal to the field at the center of the obstacle. From considerations of symmetry, it is found that the field acting to polarize the obstacle at the origin has only a y component, E_{ey}. The induced

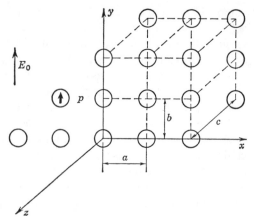

Fig. 12.2. A three-dimensional artificial dielectric medium.

dipole moment is proportional to the field, and, hence,

$$p = \alpha_e \epsilon_0 E_{ey}. \tag{1}$$

In this equation the proportionality constant α_e is called the polarizability, and is a characteristic parameter of the obstacle. For nonsymmetrical obstacles, α_e is different along different directions and, hence, is a tensor quantity in general. For an arbitrary obstacle, three principal axes will exist, such that the induced dipole moment is parallel to the applied field. In the present case, these principal axes are assumed to coincide with the coordinate axes as stated earlier. The number N of obstacles per unit volume is equal to $(abc)^{-1}$, and, hence, the dipole polarization per unit volume is

$$P = Np = \frac{p}{abc} = N\alpha_e \epsilon_0 E_{ey}. \tag{2}$$

The effective polarizing field E_{ey} is equal to $E_0 + E_{iy}$, where E_{iy} is the interaction field, i.e., the field due to all the neighboring obstacles. The interaction field is proportional to p and may be expressed by

$$E_{iy} = \frac{Cp}{\epsilon_0} \tag{3}$$

where C is called the interaction constant. In (3) it is to be understood that E_{iy} is the field at the origin or center of the obstacle under consideration. Substituting into (1) and (2), we get

$$p = \alpha_e \epsilon_0 \left(E_0 + \frac{Cp}{\epsilon_0} \right)$$

or

$$p = \frac{\alpha_e \epsilon_0 E_0}{1 - \alpha_e C} \tag{4a}$$

$$P = \frac{N\alpha_e \epsilon_0 E_0}{1 - \alpha_e C}. \tag{4b}$$

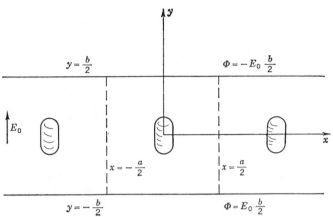

Fig. 12.3. Conducting planes inserted into the three-dimensional lattice of Fig. 12.2.

The y component of the average displacement flux is defined by

$$D_{ay} = \epsilon_0 E_{ay} + P = \kappa\epsilon_0 E_{ay} \tag{5}$$

where E_{ay} is the average value of the y component of the total field in the medium, and κ is the relative dielectric constant in the y direction. From (5), we get

$$\kappa = 1 + \frac{P}{\epsilon_0 E_{ay}} = 1 + \frac{P}{\epsilon_0(E_0 + E_{pay})}. \tag{6}$$

For an obstacle that is symmetrical about the coordinate planes passing through the center of the obstacle, the average field produced by all the induced dipoles is zero. With the applied field in the y direction, conducting planes may be inserted into the lattice at $y = \pm b/2 + mb$, $m = 0, \pm 1, \pm 2, \ldots$, without disturbing the field distribution. If the applied field is $\mathbf{a}_y E_0$, the potential of the conducting planes at $y = \pm b/2$ may be chosen as $-E_0 b/2$ and $E_0 b/2$, as in Fig. 12.3. From symmetry considerations, it follows that the total electric field has no component normal to planes at $x = \pm a/2$ and $z = \pm c/2$. Also, the field in every unit cell throughout the lattice is identical.

The potential in each unit cell may be developed into a three-dimensional Fourier series. The applied potential for the cell at the origin is

$$\Phi_0 = -E_0 y.$$

The induced potential Φ_i will be an even function of x and z and an odd function of y because of the symmetry involved. Thus we may write

$$\Phi_i = \sum_{n=0}^{\infty}\sum_{m=1}^{\infty}\sum_{s=0}^{\infty} A_{nms} \cos\frac{2n\pi x}{a} \cos\frac{2s\pi z}{c} \sin\frac{2m\pi y}{b}$$

where A_{nms} is an amplitude coefficient. The y component of induced field is thus given by

$$E_{py} = -\frac{\partial\Phi_i}{\partial y} = -\sum_{n=0}^{\infty}\sum_{m=1}^{\infty}\sum_{s=0}^{\infty} A_{nms} \frac{2m\pi}{b} \cos\frac{2n\pi x}{a} \cos\frac{2m\pi y}{b} \cos\frac{2s\pi z}{c}$$

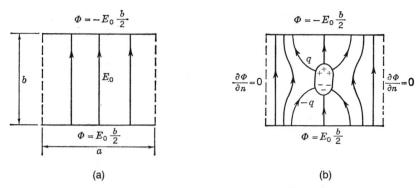

Fig. 12.4. Unit cell of an artificial dielectric medium. (a) Unloaded cell. (b) Loaded cell.

and corresponds to the total field produced by all the obstacles in the three-dimensional array. It is clear from this expression that the average induced field in a unit cell is zero since the cosine terms that depend on y have a zero average.

In view of the above discussion we may place E_{pay} equal to zero in (6). Substituting for P from (4b) now yields the classic expression known as the Clausius–Mossotti relation:

$$\kappa = 1 + \frac{N\alpha_e}{1 - \alpha_e C}. \tag{7}$$

This relation permits κ to be evaluated when the obstacle polarizability α_e and the interaction constant C are known. Practical artificial dielectrics consist of obstacles supported in a low-dielectric-constant binder, and (7) must then be interpreted as the dielectric constant relative to that of the binder or supporting medium. For simplicity, we will, however, in the remainder of this chapter assume that the obstacles are supported in a medium with a dielectric constant of unity.

In the above analysis we have assumed a y-directed applied field. A similar analysis will hold for x- and z-directed fields, but the dielectric constant along the x and z directions will not necessarily be the same, since the polarizability and interaction constant are different in different directions for general structures. In other words, the medium will have anisotropic properties in general.

12.2. ELECTROSTATIC SOLUTION

As noted in the previous section, the problem of a regular three-dimensional lattice of conducting obstacles can be reduced to that of a single obstacle located at the center of a unit cell as in Fig. 12.4(b). On the top and bottom faces, the potential is kept at $-E_0 b/2$ and $E_0 b/2$. This potential difference sets up the applied field $\mathbf{a}_y E_0$. On the sidewalls, $\partial \Phi / \partial n$ is equal to zero, where Φ is the sum of the applied potential Φ_0 and the induced potential Φ_i.

Figure 12.4(a) illustrates a unit cell with the obstacle removed. The capacitance C_0 of this cell is given by

$$C_0 = \epsilon_0 \frac{ac}{b}. \tag{8}$$

For the unloaded cell, the charge on each plate of area ac is equal to the potential difference between the plates times C_0, or $C_0 E_0 b$. For the loaded cell of Fig. 12.4(b), the charge on each plate is increased by an amount proportional to the charge that is displaced on the conducting obstacle. Let this charge be $-q$ on the upper plate, and q on the lower plate. The capacitance of the unit cell is thus increased to a value

$$C_0' = \frac{Q+q}{E_0 b} = C_0 + \frac{q}{E_0 b}.$$

The effective dielectric constant of the artificial dielectric is equal to the dielectric constant of a homogeneous dielectric that would give this same increase in capacitance. Hence, we have

$$\kappa = \frac{C_0'}{C_0} = 1 + \frac{q}{C_0 E_0 b} = 1 + \frac{q}{Q}. \tag{9}$$

The electrostatic approach thus aims at finding the change in the capacitance of a unit cell due to the presence of the obstacle. Although the problem is readily formulated, the solution is not always easy to obtain. Experimental methods may, of course, be used. One such method makes use of an electrolytic tank with insulators corresponding to the walls on which $\partial \Phi / \partial n$ equals zero. In the electrolytic tank, the current density is the analog of the electric field [12.2].

In addition to the change in capacitance of a unit cell, the change in the inductance caused by magnetic dipole polarization must be determined. At low frequencies, there is no magnetic polarization for nonferrous obstacles. However, since artificial dielectrics are normally used at high frequencies, the field does not penetrate into the conductor, and the normal magnetic field at the surface must be zero. At low frequencies, the equivalent magnetic body is a perfectly diamagnetic body with $\mu = 0$, since, for such a body, the normal magnetic field at the surface vanishes. If in Fig. 12.4(b) the applied magnetic field is H_{0x} in the x direction, this is equivalent to passing a current of density H_{0x} in the positive z direction on the upper plate at $y = b/2$, and in the negative z direction on the lower plate at $y = -b/2$. When the obstacle is absent, the inductance L_0 of the unit cell is $L_0 = \mu_0 bc/a$. With the equivalent diamagnetic body present, the inductance is reduced to a new value L_0', and the permeability of the medium may be defined by

$$\mu = \frac{aL_0'}{bc}. \tag{10}$$

The induced magnetic field may be derived from a scalar magnetic potential Φ_m, where $\Phi_m = 0$ on the sides at $x = \pm a/2$, and the normal derivative vanishes on the other four sides. From the known value of the resultant magnetic field, the inductance L_0' of the unit cell may be found by well-known standard methods.

The change in inductance may also be determined experimentally by means of an electrolytic tank. Conducting plates are used for the two sides at $x = \pm a/2$, and insulating plates (dielectric sheets) for the other sides of the unit cell. The analog of the perfectly diamagnetic body is a body with zero conductivity, and, hence, a dielectric body having the same shape as the obstacle in the artificial dielectric may be used. The ratio of the resistance of the unit cell with the dielectric body absent to that with the body present is equal to the ratio L_0'/L_0. The index of refraction of the medium is given by

$$\eta = \left(\frac{L_0' C_0'}{L_0 C_0} \right)^{1/2}.$$

12.3. Evaluation of Interaction Constants

In this section, the interaction constants for various types of lattices will be evaluated. Some of the results in this section are not of interest in connection with artificial dielectrics, but are included because of their application to aperture coupling in rectangular waveguides.

The first case to be considered is a three-dimensional array of y-directed unit dipoles located at $x = na$, $y = mb$, $z = sc$, where n, m, s are integers extending from plus infinity to minus infinity. The dipole at $x = y = z = 0$ is excluded. The y component of the electric field at the origin is the interaction field and is equal to C/ϵ_0, according to (3) for unit dipoles. The potential due to a unit dipole at x_0, y_0, z_0 is given by

$$\Phi = \frac{\mathbf{a}_y \cdot \mathbf{r}}{4\pi\epsilon_0 r^3}$$

where $r = [(x - x_0)^2 + (y - y_0)^2 + (z - z_0)^2]^{1/2}$. For the array of dipoles,

$$\Phi = \frac{1}{4\pi\epsilon_0} \sum_{n=-\infty}^{\infty} \sum_{s=-\infty}^{\infty} \sum_{m=-\infty}^{\infty}{}' \frac{y - mb}{[(x - na)^2 + (y - mb)^2 + (z - sc)^2]^{3/2}}$$

where the prime indicates omission of the term $n = m = s = 0$. The interaction constant is thus given by

$$
\begin{aligned}
C = \epsilon_0 E_i &= -\epsilon_0 \left. \frac{\partial \Phi}{\partial y} \right|_{x=y=z=0} \\
&= \frac{1}{4\pi} \sum_{n=-\infty}^{\infty} \sum_{s=-\infty}^{\infty} \sum_{m=-\infty}^{\infty}{}' \frac{2(mb)^2 - (na)^2 - (sc)^2}{[(mb)^2 + (na)^2 + (sc)^2]^{5/2}}.
\end{aligned}
\tag{11}
$$

The above series has been summed by Brown, using the Poisson summation formula [12.18]. However, the following alternative approach is instructive. By image theory, conducting planes may be inserted at $y = \pm b/2$. We then need to find the potential distribution between the two conducting plates for unit dipoles located at $x = na$, $z = sc$, $y = 0$, for $n, s = 0, \pm 1, \pm 2, \ldots$, and $n = s \neq 0$. First consider a y-directed dipole at the origin, as in Fig. 12.5. This dipole may be considered as a positive charge $q\delta(x)\delta(z)\delta(y - \tau)$ at $x = z = 0$, $y = \tau$ and a negative charge $-q\delta(x)\delta(z)\delta(y + \tau)$ at $x = z = 0$, $y = -\tau$. Using cylindrical coordinates y, r, θ, and noting that Φ is independent of θ, Poisson's equation becomes

$$\frac{\partial^2 \Phi}{\partial y^2} + \frac{1}{r}\frac{\partial}{\partial r} r \frac{\partial \Phi}{\partial r} = -\frac{\rho}{\epsilon_0} = -\frac{q}{\epsilon_0}\delta(x)\delta(z)[\delta(y - \tau) - \delta(y + \tau)]. \tag{12}$$

The potential Φ is an odd function of y, and vanishes on the conducting planes at $y = \pm b/2$. Let

$$\Phi = \sum_{m=1}^{\infty} \Phi_m = \sum_{m=1}^{\infty} A_m(r) \sin \frac{2m\pi y}{b}$$

and substitute into (12) to get

$$\sum_{m=1}^{\infty} \left[\frac{1}{r}\frac{d}{dr} r \frac{d}{dr} A_m - \left(\frac{2m\pi}{b}\right)^2 A_m \right] \sin \frac{2m\pi y}{b} = -\frac{q}{\epsilon_0}\delta(x)\delta(z)[\delta(y - \tau) - \delta(y + \tau)].$$

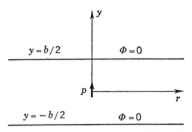

Fig. 12.5. A dipole at the origin between parallel conducting planes.

Multiply both sides by $\sin(2m\pi y/b)$, and integrate over y from $-b/2$ to $b/2$, to obtain

$$\frac{b}{2}\left[\frac{1}{r}\frac{d}{dr}r\frac{dA_m}{dr}-\left(\frac{2m\pi}{b}\right)^2 A_m\right]=-\frac{2q}{\epsilon_0}\delta(x)\delta(z)\sin\frac{2m\pi\tau}{b}.$$

The dipole moment p is equal to $2q\tau$, and, if we let τ tend to zero such that

$$\lim_{\tau\to 0}2q\tau=p$$

we obtain

$$\frac{1}{r}\frac{d}{dr}r\frac{dA_m}{dr}-\left(\frac{2m\pi}{b}\right)^2 A_m=-\frac{p}{\epsilon_0}\frac{2m\pi}{b}\frac{2}{b}\delta(x)\delta(z). \tag{13}$$

For all values of x and z not equal to zero, the solutions to (13) are the modified Bessel functions of the first and second kinds. Only the second kind of modified Bessel function vanishes as $r=(x^2+z^2)^{1/2}$ becomes large, and, hence, this is the only solution allowed. We now have

$$A_m(r)=a_m K_0\left(\frac{2m\pi r}{b}\right) \tag{14}$$

where a_m is an amplitude constant to be determined. For small r, we have

$$K_0\left(\frac{2m\pi r}{b}\right)\approx -\left(\gamma+\ln\frac{m\pi r}{b}\right)$$

where γ is Euler's constant. For large r, K_0 decays exponentially according to the relation

$$K_0\left(\frac{2m\pi r}{b}\right)\approx \left(\frac{b}{4mr}\right)^{1/2}e^{-2m\pi r/b}.$$

Equation (13) is the solution to the problem of a nonuniform line source of density $4m\pi p/b^2$ and having a variation $\sin(2m\pi y/b)$ with y. Very near this line source, i.e., for r very small, the potential Φ_m, which is a solution of (13), must have the usual line-source logarithmic singularity. Hence,

$$\lim_{r\to 0}\Phi_m=-\frac{4m\pi p}{b^2}\frac{\ln r}{2\pi\epsilon_0}$$

and since $\lim_{r\to 0} a_m K_0(2/m\pi r/b) = -a_m \ln r$, the solution for a_m is given by

$$a_m = \frac{2mp}{\epsilon_0 b^2}.$$

The solution for the potential from the dipole is thus given by

$$\Phi = \sum_{m=1}^{\infty} \frac{2mp}{\epsilon_0 b^2} \sin \frac{2m\pi y}{b} K_0 \left(\frac{2m\pi r}{b} \right). \tag{15}$$

For a dipole of unit strength $p = 1$, the y component of electric field is given by

$$E_y = -\frac{\partial \Phi}{\partial y} = -\frac{1}{\pi b \epsilon_0} \sum_{m=1}^{\infty} \left(\frac{2m\pi}{b} \right)^2 \cos \frac{2m\pi y}{b} K_0 \left(\frac{2m\pi r}{b} \right). \tag{16}$$

For the two-dimensional array of unit dipoles, excluding the dipole at the origin, the y component of electric field at the origin is

$$\begin{aligned} \epsilon_0 E_y = -\frac{1}{\pi b} &\sum_{n=-\infty}^{\infty} \sum_{s=-\infty}^{\infty} {\sum_{m=1}^{\infty}}' \left(\frac{2m\pi}{b} \right)^2 K_0 \left\{ \frac{2m\pi}{b} [(na)^2 + (sc)^2]^{1/2} \right\} \\ = -\frac{2}{\pi b} &\sum_{s=1}^{\infty} \sum_{m=1}^{\infty} \left(\frac{2m\pi}{b} \right)^2 K_0 \left(\frac{2m\pi sc}{b} \right) - \frac{2}{\pi b} \sum_{n=1}^{\infty} \sum_{m=1}^{\infty} \left(\frac{2m\pi}{b} \right)^2 \\ \times K_0 &\left(\frac{2m\pi na}{b} \right) - \frac{4}{\pi b} \sum_{n=1}^{\infty} \sum_{s=1}^{\infty} \sum_{m=1}^{\infty} \left(\frac{2m\pi}{b} \right)^2 \\ \times K_0 &\left\{ \frac{2m\pi}{b} [(na)^2 + (sc)^2]^{1/2} \right\}. \end{aligned} \tag{17}$$

Since K_0 decays exponentially at a rapid rate, only the first terms in the first two sums are required for sufficient accuracy in most cases. Thus,

$$E_y \approx \frac{-8\pi}{\epsilon_0 b^3} \left[K_0 \left(\frac{2\pi c}{b} \right) + K_0 \left(\frac{2\pi a}{b} \right) \right] \tag{18}$$

when a and c do not differ greatly. However, this is not the interaction field at the origin, since excluding the term corresponding to the dipole at the origin is equivalent to deleting all the dipoles along the line $x = z = 0$ and located at $y = mb$, $m = \pm 1, \pm 2, \ldots$. We must add to (18) the contribution from (11) with $n = s = 0$, that is,

$$\frac{1}{\pi \epsilon_0} \sum_{m=1}^{\infty} \frac{1}{(mb)^3} = \frac{1.202}{\pi \epsilon_0 b^3}$$

to obtain the total interaction field. Hence, the interaction constant C is given by

$$C = \frac{1.202}{\pi b^3} - \frac{8\pi}{b^3} \left[K_0 \left(\frac{2\pi c}{b} \right) + K_0 \left(\frac{2\pi a}{b} \right) \right]. \tag{19}$$

For a cubical lattice $a = b = c$, Lorentz obtained the value $C = (3b^3)^{-1}$ by replacing the sums in (11) by integrals. This result does not differ much from (19) for a cubical lattice. For x-directed dipoles, a and b should be interchanged in (19), while, for z-directed dipoles, b and c should be interchanged.

Summation by Means of Poisson's Formula

The Poisson summation formula gives

$$\sum_{m=-\infty}^{\infty} f(\alpha m) = \frac{1}{\alpha} \sum_{m=-\infty}^{\infty} F\left(\frac{2m\pi}{\alpha}\right) \tag{20}$$

where $F(w)$ is the Fourier transform of $f(z)$; that is,

$$F(w) = \int_{-\infty}^{\infty} e^{jwz} f(z)\, dz. \tag{21}$$

The following relationships will be required:

$$\int_{-\infty}^{\infty} e^{jwz}(z^2 + r^2)^{-\nu-1/2}\, dz = 2\left(\frac{|w|}{2r}\right)^{\nu} \pi^{1/2} \frac{K_{\nu}(|w|r)}{\Gamma(\nu + \frac{1}{2})}. \tag{22}$$

The transform of $z^2(z^2+r^2)^{-5/2}$ may be obtained from (22) by putting $\nu = 2$ and differentiating both sides twice with respect to w. Combining this with (22) and using the Bessel-function recurrence relations permits us to obtain

$$\int_{-\infty}^{\infty} e^{jwz} \frac{2z^2 - r^2}{(z^2 + r^2)^{5/2}}\, dz = -2w^2 K_0(|w|r). \tag{23}$$

Equation (11) for the interaction constant C may be written as

$$C = \frac{1}{4\pi}\left\{ 4\sum_{m=1}^{\infty} \frac{1}{(mb)^3} + 4\sum_{n=1}^{\infty}\sum_{s=1}^{\infty}\sum_{m=-\infty}^{\infty} \frac{2(mb)^2 - (na)^2 - (sc)^2}{[(mb)^2 + (na)^2 + (sc)^2]^{5/2}} \right.$$

$$+ 2\sum_{n=1}^{\infty}\sum_{m=-\infty}^{\infty} \frac{2(mb)^2 - (na)^2}{[(mb)^2 + (na)^2]^{5/2}}$$

$$\left. + 2\sum_{s=1}^{\infty}\sum_{m=-\infty}^{\infty} \frac{2(mb)^2 - (sc)^2}{[(mb)^2 + (sc)^2]^{5/2}} \right\}. \tag{24}$$

The sums over m may be totaled by using (20) and (23). For example,

$$\sum_{m=-\infty}^{\infty} \frac{2(mb)^2 - (na)^2}{[(mb)^2 + (na)^2]^{5/2}} = -\frac{2}{b}\sum_{m=-\infty}^{\infty}\left(\frac{2m\pi}{b}\right)^2 K_0\left(\frac{2|m|\pi na}{b}\right). \tag{25}$$

In (25), the $m = 0$ term on the right-hand side is zero. When the remaining sums in (24) are handled in the same way, we obtain the final result given by (19), upon making the same approximations. The results should obviously be the same, since summing over m in (24) is

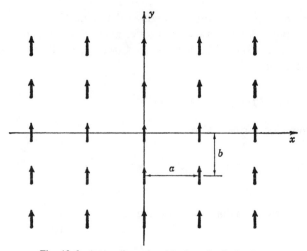

Fig. 12.6. A two-dimensional lattice of unit dipoles.

equivalent to solving the problem of a single layer of dipoles between parallel plates by the method of images.

Interaction Constant for Two-Dimensional Lattices

Consider a two-dimensional lattice of unit dipoles located at $x = na$, $y = mb$; $n, m = 0, \pm 1, \pm 2, \ldots$, as in Fig. 12.6. For y-directed dipoles, the interaction constant C_y is given by (11) with $s = 0$. Hence,

$$C_y(a, b) = \frac{1}{4\pi} \sum_{n=-\infty}^{\infty} \sum_{m=-\infty}^{\infty}{}' \frac{2(mb)^2 - (na)^2}{[(mb)^2 + (na)^2]^{5/2}}$$

$$= \frac{1}{\pi} \sum_{m=1}^{\infty} \frac{1}{(mb)^3} + \frac{1}{2\pi} \sum_{n=1}^{\infty} \sum_{m=-\infty}^{\infty} \frac{2(mb)^2 - (na)^2}{[(mb)^2 + (na)^2]^{5/2}}. \tag{26}$$

Applying the Poisson summation formula gives

$$C_y(a, b) = \frac{1.2}{\pi b^3} - \frac{2}{\pi b} \sum_{n=1}^{\infty} \sum_{m=1}^{\infty} \left(\frac{2m\pi}{b}\right)^2 K_0\left(\frac{2m\pi na}{b}\right)$$

$$\approx \frac{1.2}{\pi b^3} - \frac{8\pi}{b^3} K_0\left(\frac{2\pi a}{b}\right). \tag{27}$$

For x-directed dipoles, the interaction constant $C_x(a, b)$ is given by (27) with a and b interchanged, i.e.,

$$C_x(a, b) = C_y(b, a). \tag{28}$$

For z-directed dipoles the interaction constant $C_z(a, b)$ is given by

$$C_z(a, b) = -\frac{1}{4\pi} \sum_{n=-\infty}^{\infty} \sum_{m=-\infty}^{\infty}{}' \frac{(mb)^2 + (na)^2}{[(mb)^2 + (na)^2]^{5/2}}. \tag{29}$$

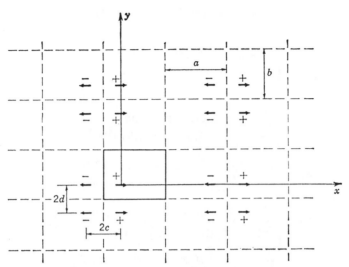

Fig. 12.7. Images of magnetic dipoles in waveguide walls.

This series may be obtained by analogy with (11) by omitting the sum over m in (11) and replacing sc by mb. In place of (29), we may write

$$C_z(a, b) = -\frac{1}{4\pi} \sum_{n=-\infty}^{\infty} \left[\sum_{m=-\infty}^{\infty}{}' \frac{2(mb)^2 - (na)^2}{[(mb)^2 + (na)^2]^{5/2}} + \sum_{m=-\infty}^{\infty}{}' \frac{2(na)^2 - (mb)^2}{[(na)^2 + (mb)^2]^{5/2}} \right].$$

Comparing with (26) gives

$$C_z(a, b) = -C_y(b, a) - C_y(a, b) = -C_x(a, b) - C_y(a, b)$$

$$= -\frac{1.2}{\pi b^3} - \frac{1.2}{\pi a^3} + \frac{8\pi}{b^3} K_0\left(\frac{2\pi a}{b}\right) + \frac{8\pi}{a^3} K_0\left(\frac{2\pi b}{a}\right). \tag{30}$$

Two-Dimensional Lattice for Waveguide Aperture Coupling

A small circular aperture in a transverse wall in a rectangular waveguide is equivalent to an x-directed magnetic dipole for H_{10} mode excitation. The image of this dipole in the guide walls must be chosen so that the field produced has a vanishing normal component of magnetic field on the guide walls. The resulting lattice of dipoles is illustrated in Fig. 12.7. The positive x-directed dipoles are located at $x = 2na$, $y = 2mb$, and $x = 2na$, $y = 2mb - 2d$, where $n, m = 0, \pm 1, \pm 2, \ldots$. The negative-directed dipoles are located at $x = 2na - 2c$, $y = 2mb$, and $x = 2na - 2c$, $y = 2mb - 2d$. The x component of magnetic field produced by a unit x-directed magnetic dipole at (x_0, y_0, z_0) is given by

$$H_x = \frac{1}{4\pi} \frac{2(x - x_0)^2 - (y - y_0)^2 - (z - z_0)^2}{[(x - x_0)^2 + (y - y_0)^2 + (z - z_0)^2]^{5/2}}.$$

The field at the origin due to all the image dipoles is the interaction field. For unit dipoles, this field is equal to the interaction constant C_m. Thus, we have

$$C_m = \frac{1}{4\pi}(S_1 + S_2 - S_3 - S_4) \tag{31}$$

where

$$S_1 = \sum_{n=-\infty}^{\infty} \sideset{}{'}\sum_{m=-\infty}^{\infty} \frac{2(2na)^2 - (2mb)^2}{[(2na)^2 + (2mb)^2]^{5/2}}$$

$$S_2 = \sum_{n=-\infty}^{\infty} \sum_{m=-\infty}^{\infty} \frac{2(2na)^2 - (2mb - 2d)^2}{[(2na)^2 + (2mb - 2d)^2]^{5/2}}$$

$$S_3 = \sum_{n=-\infty}^{\infty} \sum_{m=-\infty}^{\infty} \frac{2(2na - 2c)^2 - (2mb - 2d)^2}{[(2na - 2c)^2 + (2mb - 2d)^2]^{5/2}}$$

$$S_4 = \sum_{n=-\infty}^{\infty} \sum_{m=-\infty}^{\infty} \frac{2(2na - 2c)^2 - (2mb)^2}{[(2na - 2c)^2 + (2mb)^2]^{5/2}}.$$

The series for S_1 and S_2 are of the type already summed and are equal to

$$S_1 = \frac{1.202}{2a^3} - \frac{4}{a}\sum_{m=1}^{\infty}\sum_{n=1}^{\infty}\left(\frac{n\pi}{a}\right)^2 K_0\left(\frac{2n\pi mb}{a}\right) \tag{32a}$$

$$S_2 = -\frac{2}{a}\sum_{n=1}^{\infty}\sum_{m=-\infty}^{\infty}\left(\frac{n\pi}{a}\right)^2 K_0\left(\frac{n\pi}{a}|2mb - 2d|\right). \tag{32b}$$

To sum S_3 and S_4, it is noted that, if $F(w)$ is the Fourier transform of $f(z)$, then the Fourier transform of $f(z + r)$ is $e^{-jw\tau}F(w)$. Hence, we obtain

$$S_3 = -\frac{1}{a}\sum_{m=-\infty}^{\infty}\sum_{n=-\infty}^{\infty} e^{j2n\pi c/a}\left(\frac{n\pi}{a}\right)^2 K_0\left(\frac{|n|\pi}{a}|2mb - 2d|\right)$$

$$= -\frac{2}{a}\sum_{m=-\infty}^{\infty}\sum_{n=1}^{\infty}\left(\frac{n\pi}{a}\right)^2 \cos\frac{2n\pi c}{a} K_0\left(\frac{n\pi}{a}|2mb - 2d|\right). \tag{33a}$$

The series S_4 is equal to S_3 with d placed equal to zero. However, this makes the $m = 0$ term infinite, so this term must be summed directly. The final result is

$$S_4 = 2\sum_{n=-\infty}^{\infty}|(2na - 2c)|^{-3} - \frac{4}{a}\sum_{m=1}^{\infty}\sum_{n=1}^{\infty}\left(\frac{n\pi}{a}\right)^2 \cos\frac{2n\pi c}{a} K_0\left(\frac{2n\pi mb}{a}\right). \tag{33b}$$

In practice, only a few terms of the series are required for good accuracy.

When using the above theory for a circular aperture in a transverse wall of a rectangular guide, it must be kept in mind that it is essentially a static field theory and will yield accurate results only when the image apertures are spaced at a small distance compared with the wavelength. This is usually not the case except for an aperture close to the guide wall (c or d small). On the other hand, the effect of the image dipoles is small for small apertures not located too close to the guide wall. Thus, the theory provides a useful correction for those cases when the aperture is close to a guide wall, which is just the situation for which a correction is needed most.

12.4. SPHERE- AND DISK-TYPE ARTIFICIAL DIELECTRICS

In order to evaluate the index of refraction of an artificial dielectric by the Lorentz theory, the polarizability of the obstacle is required. For ellipsoids and the degenerate forms such as prolate and oblate spheroids, spheres, and circular and elliptical disks, the method presented in Section 7.3 may be used. A dielectric obstacle becomes equivalent to a conducting body when the permittivity ϵ is made to approach infinity. To evaluate the magnetic dipole moment, it is noted that, when μ is made to approach zero for a permeable body, it becomes a perfect diamagnetic body. The normal component of magnetic field at the surface must vanish for a perfect diamagnetic body, and this is the same boundary condition as required for the high-frequency magnetic field at the surface of a perfect conductor. The electric polarizability α_e and the magnetic polarizability α_m for typical obstacles are given in Table 12.1.

Consider a cubic array of conducting spheres with radius r. Let the spacing be denoted by a. The number N of spheres per unit volume is equal to a^{-3}. From Table 12.1, $\alpha_e = 4\pi r^3$, and $\alpha_m = -2\pi r^3$. The interaction constant C is given by (19) as

$$C = \frac{1}{\pi a^3}[1.202 - 16\pi^2 K_0(2\pi)] = \frac{1.06}{\pi a^3}.$$

From (7), the permittivity $\epsilon = \kappa\epsilon_0$ is found to be

$$\epsilon = \epsilon_0 \frac{1 + 8.32r^3/a^3}{1 - 4.24r^3/a^3}.$$

For a TEM wave applied to the lattice, induced magnetic dipoles are also produced. The effective magnetic field acting to polarize the obstacle is the sum of the applied field H_0 and the interaction field H_i. By analogy with the electric case, we have

$$m = \alpha_m(H_0 + H_i)$$

$$H_i = Cm$$

and, hence,

$$m = \frac{\alpha_m H_0}{1 - \alpha_m C}. \tag{34}$$

The magnetic polarization per unit volume is $M = Nm$, and the permeability μ is given by

$$B = \mu H_0 = \mu_0(H_0 + M)$$

or

$$\mu = \mu_0 + \frac{\mu_0 N \alpha_m}{1 - \alpha_m C}. \tag{35}$$

TABLE 12.1
POLARIZABILITY OF ELLIPSOIDS AND THEIR DEGENERATE FORMS[*]

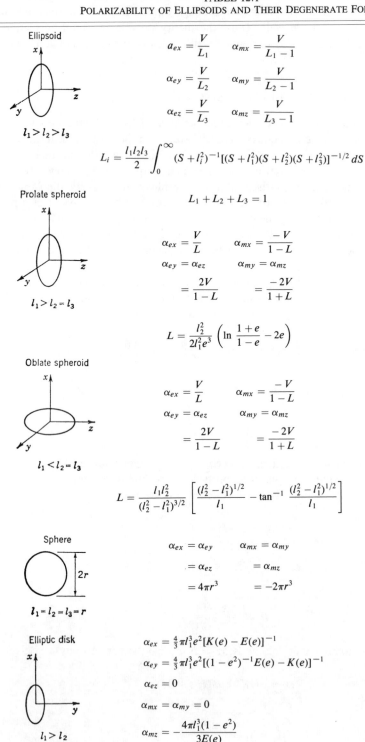

Ellipsoid

$l_1 > l_2 > l_3$

$$\alpha_{ex} = \frac{V}{L_1} \qquad \alpha_{mx} = \frac{V}{L_1 - 1}$$

$$\alpha_{ey} = \frac{V}{L_2} \qquad \alpha_{my} = \frac{V}{L_2 - 1}$$

$$\alpha_{ez} = \frac{V}{L_3} \qquad \alpha_{mz} = \frac{V}{L_3 - 1}$$

$$L_i = \frac{l_1 l_2 l_3}{2} \int_0^\infty (S + l_i^2)^{-1} [(S + l_1^2)(S + l_2^2)(S + l_3^2)]^{-1/2} \, dS$$

Prolate spheroid

$l_1 > l_2 = l_3$

$$L_1 + L_2 + L_3 = 1$$

$$\alpha_{ex} = \frac{V}{L} \qquad \alpha_{mx} = \frac{-V}{1 - L}$$

$$\alpha_{ey} = \alpha_{ez} \qquad \alpha_{my} = \alpha_{mz}$$

$$= \frac{2V}{1 - L} \qquad = \frac{-2V}{1 + L}$$

$$L = \frac{l_2^2}{2l_1^2 e^3} \left(\ln \frac{1 + e}{1 - e} - 2e \right)$$

Oblate spheroid

$l_1 < l_2 = l_3$

$$\alpha_{ex} = \frac{V}{L} \qquad \alpha_{mx} = \frac{-V}{1 - L}$$

$$\alpha_{ey} = \alpha_{ez} \qquad \alpha_{my} = \alpha_{mz}$$

$$= \frac{2V}{1 - L} \qquad = \frac{-2V}{1 + L}$$

$$L = \frac{l_1 l_2^2}{(l_2^2 - l_1^2)^{3/2}} \left[\frac{(l_2^2 - l_1^2)^{1/2}}{l_1} - \tan^{-1} \frac{(l_2^2 - l_1^2)^{1/2}}{l_1} \right]$$

Sphere

$l_1 = l_2 = l_3 = r$

$$\alpha_{ex} = \alpha_{ey} \qquad \alpha_{mx} = \alpha_{my}$$

$$= \alpha_{ez} \qquad = \alpha_{mz}$$

$$= 4\pi r^3 \qquad = -2\pi r^3$$

Elliptic disk

$l_1 > l_2$
$l_3 = 0$

$$\alpha_{ex} = \tfrac{4}{3} \pi l_1^3 e^2 [K(e) - E(e)]^{-1}$$

$$\alpha_{ey} = \tfrac{4}{3} \pi l_1^3 e^2 [(1 - e^2)^{-1} E(e) - K(e)]^{-1}$$

$$\alpha_{ez} = 0$$

$$\alpha_{mx} = \alpha_{my} = 0$$

$$\alpha_{mz} = -\frac{4\pi l_1^3 (1 - e^2)}{3E(e)}$$

$K(e)$ and $E(e)$ are complete elliptic integrals of the first and second kind, respectively.

TABLE 12.1 (continued).

Circular disk

$$\alpha_{ex} = \alpha_{ey} \qquad \alpha_{mx} = \alpha_{my} = 0$$

$$= \tfrac{16}{3} r^3 \qquad \alpha_{mz} = -\tfrac{8}{3} r^3$$

$$\alpha_{ez} = 0$$

$l_1 = l_2 = r$
$l_3 = 0$

*Semiaxes are l_1, l_2, l_3 along the x, y, z axes, respectively; eccentricity is $e = (1 - l_2^2/l_1^2)^{1/2}$; volume $V = \tfrac{4}{3}\pi l_1 l_2 l_3$; α_{ex}, α_{mx} = polarizability for fields applied along the x axis; similarly for α_{ey}, α_{my}, α_{ez}, and α_{mz}.

For the cubic array of spheres, we have

$$\mu = \mu_0 \frac{1 - 4.16 r^3/a^3}{1 + 2.12 r^3/a^3}.$$

The effective index of refraction for the medium is

$$\eta = \left(\frac{\epsilon\mu}{\epsilon_0\mu_0}\right)^{1/2} = \left[\frac{(1 + 8.32 r^3/a^3)(1 - 4.16 r^3/a^3)}{(1 - 4.24 r^3/a^3)(1 + 2.12 r^3/a^3)}\right]^{1/2} \qquad (36)$$

and is only moderately greater than unity because of the diamagnetic effect present. As an example, for $r = a/4$, we obtain $\eta = 1.046$.

The tetragonal array of circular disks illustrated in Fig. 12.8 represents a more useful medium because of the much greater index of refraction that can be obtained. The spacing of the disks in the xy plane is a, and successive planes of disks are spaced a distance c part along the z axis. A TEM wave is assumed to propagate through the lattice, along the z direction. This field does not have a component of magnetic field along the z axis, i.e., normal to the disk face, and, hence, there are no induced magnetic dipoles. The electric polarizability is $\alpha_e = 16 r^3/3$, and, from (19), the interaction constant is found to be

$$C = \frac{1.2}{\pi a^3} - \frac{8\pi}{a^3}\left[K_0(2\pi) + K_0\left(\frac{2\pi c}{a}\right)\right].$$

The relative dielectric constant is given by (7) with $N = (a^2 c)^{-1}$. The relative dielectric constant κ is plotted as a function of c/a for several values of r/a in Fig. 12.9. The results of

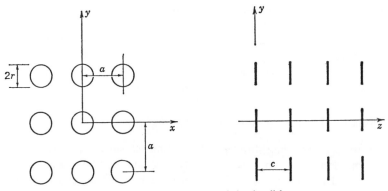

Fig. 12.8. Tetragonal array of circular disks.

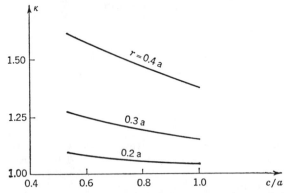

Fig. 12.9. Dielectric constant as a function of c/a for a tetragonal array of circular disks.

a low-frequency investigation carried out by Kharadly have shown that the three-dimensional Lorentz theory gives accurate results for c/a greater than about 0.8 only [12.18]. Brown has considered only the interaction between disks in a single plane. The interaction constant to be used is now that given by (27) and is equal to $0.36/a^3$. The resulting formula for κ is

$$\kappa = 1 + \frac{8a}{c[1.5(a/r)^3 - 2.88]} \tag{37}$$

and has been found to be accurate for c/a greater than 0.6.

For small values of c/a, Brown and Jackson have computed the dipole moment of a row of disks along the z axis and used the two-dimensional form of the Clausius–Mossotti relation to obtain [12.16]

$$\kappa = 1 + \frac{N\alpha_0}{1 - 0.5N\alpha_0} \tag{38}$$

where $N = a^{-2}$ and is the number of rows of disks per unit area, and α_0 is the polarizability per unit length for a single row of disks and is given by

$$\alpha_0 = 2\pi r^2 \left[\left(1 - \frac{0.22c}{r} \right)^2 + \frac{0.007c^3}{r^3} \left(1 + \frac{0.22c}{r} \right) \right].$$

As c approaches zero, α_0 approaches $2\pi r^2$, which is the polarizability per unit length of a conducting cylinder. The combination of (37) for large values of c/a and (38) for small values of c/a provides good results for the whole range of c/a values that are likely to be encountered in practice. It must be kept in mind, however, that the theory is a static field theory valid only for plane waves incident in a direction perpendicular to the plane of the disks, and with the spacing between disks less than about $0.1\lambda_0$.

12.5. Transmission-Line Approach for a Disk Medium

For a plane wave propagating through a three-dimensional array of disks, each plane of disks may be considered as a single discontinuity. For infinitely thin disks, this discontinuity is equivalent to a shunt susceptance jB connected across a transmission line. The three-

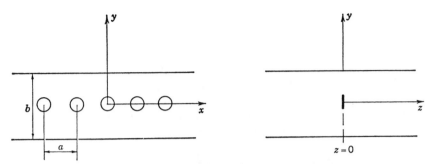

Fig. 12.10. Single row of disks in a parallel-plate transmission line.

dimensional array of disks is, in turn, equivalent to a transmission line loaded at regular intervals by identical shunt susceptances. An evaluation of the propagation constant for this equivalent periodic-loaded transmission line leads to an expression for the equivalent index of refraction of the medium.

The first step in the analysis is the evaluation of the equivalent shunt susceptance of a single plane of disks. When the disk diameters and spacings are small compared with the wavelength, the shunt susceptance may be evaluated by a method similar to the Bethe small-aperture theory discussed in Chapter 7. Let the plane of the disks be the $z = 0$ plane, let the disk spacing be a along the x direction and b along the y direction, and let the disk radius be r. The configuration is the same as that in Fig. 12.8 with the exception of unequal spacings along the x and y directions. Let a plane wave with components E_y, H_x, H_z be incident at an angle θ_i with respect to the z axis. This incident field may be represented as follows:

$$\mathbf{E}_{\text{inc}} = a_0 \mathbf{a}_y e^{-jhx - \Gamma_0 z} \tag{39a}$$

$$\mathbf{H}_{\text{inc}} = a_0 (j\omega\mu_0)^{-1} \mathbf{a}_y \times \nabla e^{-jhx - \Gamma_0 z} \tag{39b}$$

where $h = k_0 \sin \theta_i$, $\Gamma_0 = (h^2 - k_0^2)^{1/2} = jk_0 \cos \theta_i$, and a_0 is an amplitude constant. A consideration of the nature of the incident field and the symmetry of the structure shows that conducting planes parallel with the xz plane can be placed at $y = \pm b/2$. Hence we need consider only a single row of disks in a parallel-plate transmission line, as in Fig. 12.10.

For $z < 0$, there will be a dominant reflected mode given by

$$\mathbf{E}_{\text{ref}} = \mathbf{a}_y R a_0 e^{-jhx + \Gamma_0 z} \tag{40a}$$

$$\mathbf{H}_{\text{ref}} = R a_0 (j\omega\mu_0)^{-1} \mathbf{a}_y \times \nabla e^{-jhx + \Gamma_0 z}. \tag{40b}$$

In addition, there will be an infinite number of higher order evanescent modes excited. These modes have the following x and z dependence:

$$e^{-j(h + 2n\pi/a)x + \Gamma_{nm} z}, \quad n = 0, \pm 1, \cdots; \quad m = 0, 1, 2, \cdots.$$

where $\Gamma_{nm}^2 = (h + 2n\pi/a)^2 + (m\pi/b)^2 - k_0^2$. The term $m\pi/b$ arises because of variation of the fields with y. All Γ_{nm} are real for $n \neq 0$ when the spacing a is less than one-half wavelength. Since we will not require expressions for these modes in order to evaluate the reflection coefficient R, we will simply denote them as follows:

$$\mathbf{E}_{nm}^{\mp} = \mathbf{e}_{nm}(x, y)e^{\pm\Gamma_{nm}z} \mp \mathbf{e}_{znm}(x, y)e^{\pm\Gamma_{nm}z} \tag{41a}$$

$$\mathbf{H}_{nm}^{\mp} = [\mp\mathbf{h}_{nm}(x, y) + \mathbf{h}_{znm}(x, y)]e^{\pm\Gamma_{nm}z} \tag{41b}$$

where $\mathbf{e}_{nm}$, $\mathbf{h}_{nm}$ are transverse vectors and $\mathbf{e}_{znm}$, $\mathbf{h}_{znm}$ are z-directed vectors. These vector functions all have the periodicity property

$$\mathbf{e}_{nm}(x + sa, y) = e^{-jhsa}\mathbf{e}_{nm}(x, y) \tag{42}$$

as required by Floquet's theorem. Because of the periodicity property, these modes form an orthogonal set over the range $sa < x \leq (s + 1)a$, where s is any integer, and such that

$$\int_{-b/2}^{b/2}\int_{-a/2}^{a/2} \mathbf{e}_{nm} \times \mathbf{h}_{rs}^* \cdot \mathbf{a}_z \, dx \, dy = 0, \qquad n \neq r, \ m \neq s \tag{43}$$

where $\mathbf{h}_{rs}^*$ is the complex conjugate of $\mathbf{h}_{rs}$.

If we consider, first of all, odd excitation with incident fields for $z < 0$, as in (39), and with incident fields

$$\mathbf{E}_{\text{inc}} = -a_0\mathbf{a}_ye^{-jhx+\Gamma_0z}$$

$$\mathbf{H}_{\text{inc}} = -a_0(j\omega\mu_0)^{-1}\mathbf{a}_y \times \nabla e^{-jhx+\Gamma_0z}$$

for $z > 0$, the total transverse electric field in the aperture plane will vanish. The reflection coefficient in this case is simply $R_o = -1$ (see the discussion of Babinet's principle in Chapter 1).

For even excitation with incident fields chosen as

$$\mathbf{E}_{\text{inc}} = a_0\mathbf{a}_ye^{-jhx+\Gamma_0z}$$

$$\mathbf{H}_{\text{inc}} = a_0(j\omega\mu_0)^{-1}\mathbf{a}_y \times \nabla e^{-jhx+\Gamma_0z}$$

for $z > 0$, and as in (39) for $z < 0$, the total transverse magnetic field vanishes in the aperture plane $z = 0$ in the region between the disks. Let the reflection coefficient in this case be R_e. If we now superimpose the two solutions, the total incident field for $z > 0$ is zero, while that for $z < 0$ has an amplitude $2a_0$. The total reflected field for $z < 0$ has an amplitude $(R_e + R_o)a_0 = (R_e - 1)a_0 = R2a_0$, and, hence, the reflection coefficient R is given by

$$R = \frac{R_e - 1}{2}. \tag{44}$$

To find R_e, we must solve the problem of a row of conducting disks placed on a magnetic wall in the $z = 0$ plane as in Fig. 12.11. The field in the guide may be represented as a dominant-mode standing-wave field

$$\mathbf{E}_1 = a_0\mathbf{a}_ye^{-jhx} \cos \beta_0z \tag{45a}$$

$$\mathbf{H}_1 = (j\omega\mu_0)^{-1}a_0\mathbf{a}_y \times \nabla e^{-jhx} \cos \beta_0z \tag{45b}$$

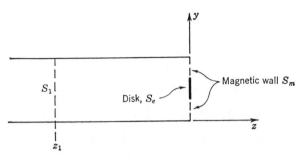

Fig. 12.11. Reduced boundary-value problem for even excitation. S_m is a magnetic wall; S_e is an electric wall (disk).

plus an additional dominant-mode reflected field

$$\mathbf{E}_2 = b_0 \mathbf{a}_y e^{-jhx+\Gamma_0 z} \tag{46a}$$

$$\mathbf{H}_2 = b_0 (j\omega\mu_0)^{-1} \mathbf{a}_y \times \nabla e^{-jhx+\Gamma_0 z} \tag{46b}$$

together with an infinite number of evanescent modes propagating in the negative z direction:

$$\mathbf{E}_3 = \sum_{n,m} b_{nm} \mathbf{E}_{nm}^- \tag{47a}$$

$$\mathbf{H}_3 = \sum_{n,m} b_{nm} \mathbf{H}_{nm}^-. \tag{47b}$$

In the above, $\beta_0 = |\Gamma_0|$, and the standing-wave field has a vanishing transverse magnetic field at $z = 0$ and, hence, satisfies the boundary conditions on the magnetic wall. The field $\mathbf{E}_1$, $\mathbf{H}_1$ is considered as the incident field, and the field $\mathbf{E}_2 + \mathbf{E}_3 = \mathbf{E}_s$, $\mathbf{H}_2 + \mathbf{H}_3 = \mathbf{H}_s$ as the total scattered field.

Let $\mathbf{E}'_{nm}$, $\mathbf{H}'_{nm}$ be a normal-mode standing-wave field:

$$\mathbf{E}'_{nm} = \mathbf{e}^*_{nm}(e^{-\Gamma_{nm}z} + e^{\Gamma_{nm}z}) + \mathbf{e}^*_{znm}(e^{-\Gamma_{nm}z} - e^{\Gamma_{nm}z}) \tag{48a}$$

$$\mathbf{H}'_{nm} = \mathbf{h}^*_{nm}(e^{-\Gamma_{nm}z} - e^{\Gamma_{nm}z}) + \mathbf{h}^*_{znm}(e^{-\Gamma_{nm}z} + e^{\Gamma_{nm}z}). \tag{48b}$$

The Lorentz reciprocity theorem gives

$$\oint_S (\mathbf{E}'_{nm} \times \mathbf{H}_s - \mathbf{E}_s \times \mathbf{H}'_{nm}) \cdot \mathbf{n} \, dS = 0$$

where S is the closed surface consisting of cross-sectional planes at $z = 0$ and $z = z_1$, the conducting planes at $y = \pm b/2$, and planes parallel to the yz plane at $x = \pm a/2$. The integral clearly vanishes on the conducting planes at $y = \pm b/2$. On the planes at $x = \pm a/2$ the integral will also vanish, since the field $\mathbf{E}_s$, $\mathbf{H}_s$ satisfies the periodicity condition (42), while $\mathbf{E}'_{nm}(x + sa, y, z) = \mathbf{E}'_{nm}(x, y, z)e^{jhsa}$, and similarly for $\mathbf{H}'_{nm}$; and the normal $\mathbf{n}$ is directed in the opposite sense on the two planes. On the cross-sectional plane at $z = z_1$, the integral reduces to

$$-2b_{nm}\int_{-b/2}^{b/2}\int_{-a/2}^{a/2}\mathbf{e}_{nm}^{*}\times\mathbf{h}_{nm}\cdot\mathbf{a}_{z}\,dx\,dy$$

provided $\mathbf{n}$ is chosen as the inward normal. On the cross-sectional plane at $z=0$, we have

$$\mathbf{n}\times\mathbf{H}_{nm}'=0\qquad\mathbf{n}\times\mathbf{H}_{s}=0\qquad\text{on }S_{m}$$

since $\mathbf{n}\times\mathbf{H}_{1}=0$ on both S_{m} and S_{e}. On S_{e}, we have $\mathbf{n}\times\mathbf{H}_{s}=\mathbf{J}$, where $\mathbf{n}$ is the normal pointing away from the disk and $\mathbf{J}$ is the current on the disk or electric wall S_{e}. Using these results, the Lorentz reciprocity theorem is found to give

$$b_{nm}\int_{-b/2}^{b/2}\int_{-a/2}^{a/2}\mathbf{e}_{nm}^{*}\times\mathbf{h}_{nm}\cdot\mathbf{a}_{z}\,dx\,dy=-\iint_{S_{e}}\mathbf{e}_{nm}^{*}\cdot\mathbf{J}\,dx\,dy \qquad (49)$$

an equation which permits the unknown amplitude coefficients b_{nm} for the scattered field to be evaluated. In particular, if we choose for the field $\mathbf{E}_{nm}'$, $\mathbf{H}_{nm}'$ a dominant-mode standing wave

$$\mathbf{E}_{00}'=\mathbf{a}_{y}e^{jhx}(e^{-\Gamma_{0}z}+e^{\Gamma_{0}z})$$

$$\mathbf{H}_{00}'=(j\omega\mu_{0})^{-1}\mathbf{a}_{y}\times\nabla e^{jhx}(e^{-\Gamma_{0}z}+e^{\Gamma_{0}z})$$

we obtain

$$b_{0}\int_{-b/2}^{b/2}\int_{-a/2}^{a/2}(j\omega\mu_{0})^{-1}\Gamma_{0}\,dx\,dy=-\iint_{S_{e}}\mathbf{a}_{y}e^{jhx}\cdot\mathbf{J}\,dx\,dy. \qquad (50)$$

The integrals on the right-hand sides of (49) and (50) may be converted into a form representing coupling with an electric dipole $\mathbf{P}$ and a magnetic dipole $\mathbf{M}$, in a manner similar to that used for the aperture problem. We obtain

$$\iint_{S_{e}}\mathbf{E}_{nm}'\cdot\mathbf{J}\,dx\,dy=j\omega\mathbf{E}_{nm}'(0)\cdot\mathbf{P}-j\omega\mu_{0}\mathbf{H}_{nm}'(0)\cdot\mathbf{M} \qquad (51)$$

where the fields are evaluated at the center of the disk, and the dipole moments are given by

$$j\omega\mathbf{P}=\iint_{S_{e}}\mathbf{J}\,dx\,dy \qquad (52a)$$

$$\mathbf{M}=\frac{1}{2}\iint_{S_{e}}\mathbf{r}\times\mathbf{J}\,dx\,dy. \qquad (52b)$$

The magnetic dipole is in the z direction, while the electric dipole lies in the xy plane. Using these results in (50), we obtain

$$b_{0}=-\frac{\omega\mu_{0}}{\beta_{0}ab}(j\omega\mathbf{a}_{y}\cdot\mathbf{P}+jh\mathbf{a}_{z}\cdot\mathbf{M}). \qquad (53)$$

The dipole moments $\mathbf{P}$ and $\mathbf{M}$ in (52) and (53) are those of one-half of the disk, since the currents flowing on one side only are involved. For small disks with large spacings, the interaction field is negligible, and $\mathbf{P}$ and $\mathbf{M}$ may be evaluated in terms of the polarizabilities of the disk and the strength of the incident field at the center of the disk. The electric and magnetic polarizabilities of one-half of a circular disk of radius r are, respectively,

$$\alpha'_e = \frac{8}{3}r^3$$

$$\alpha'_m = -\frac{4}{3}r^3.$$

The field given by (45) may be considered as the field acting to polarize the disk. For this field, we obtain

$$\mathbf{P} = \frac{8}{3}r^3\epsilon_0 a_0 \mathbf{a}_y$$

$$\mathbf{M} = -\frac{4}{3}r^3\frac{h}{\omega\mu_0}a_0 \mathbf{a}_z$$

and, hence,

$$b_0 = \frac{jh^2}{ab\beta_0}\frac{4}{3}r^3 a_0 - \frac{jk_0^2}{ab\beta_0}\frac{8}{3}r^3 a_0. \tag{54}$$

The total reflected field has an amplitude $b_0 + a_0/2$, while the incident field has an amplitude $a_0/2$, as may be seen by decomposing the standing-wave field (45) into an incident and a reflected field. The reflection coefficient R_e is thus given by

$$R_e = \frac{a_0 + 2b_0}{a_0} = 1 + \frac{2b_0}{a_0}$$

and, hence, the reflection coefficient R is, from (44),

$$R = \frac{b_0}{a_0}.$$

A shunt susceptance jB on a transmission line produces a reflection coefficient

$$R = \frac{1 - jB - 1}{2 + jB} = \frac{-jB}{2 + jB} \approx \frac{-jB}{2}$$

when B is very small.[1] From (54) we now get

$$B = \frac{8r^3}{3ab\beta_0}(2k_0^2 - h^2) = B_c - B_L \tag{55}$$

---

[1] If we had included the radiation reaction field as part of the polarizing field, as was done for the aperture problem in Chapter 7, we would have found that the expression for R was of the correct form with the solution for B given by (55).

where the capacitive part of B is given by

$$B_c = \frac{16k_0^2 r^3}{3ab\beta_0} \tag{56a}$$

while the inductive part is given by

$$B_L = \frac{8h^2 r^3}{3ab\beta_0}. \tag{56b}$$

The capacitive part arises from the electric polarization of the disk, while the inductive part arises from the magnetic dipole polarization. When the angle of incidence $\theta_i = 0$, then $h = 0$, and the inductive part B_L vanishes. For this special case,

$$B_c = \frac{16Z_0}{3ab}\omega\epsilon_0 r^3$$

where $Z_0 = (\mu_0/\epsilon_0)^{1/2}$, and is seen to have the usual frequency dependence associated with a capacitive susceptance.

More accurate results are obtained if the interaction field is also taken into account as part of the field acting to polarize the disk. This interaction field arises mainly from the nearest neighbors. If the disk spacings a and b are small compared with a wavelength, the phase of the incident wave is essentially constant over the region of several disks. Under these conditions, the interaction field may be approximated by a static interaction field. The electric and magnetic dipole moments are thus increased by factors $(1 - \alpha_e C_e)^{-1}$ and $(1 - \alpha_m C_m)^{-1}$, respectively, where C_e and C_m are the interaction constants for the electric and magnetic cases. From (27) and (30), we have

$$C_e = \frac{1.2}{\pi b^3} - \frac{8\pi}{b^3}K_0\left(\frac{2\pi a}{b}\right) \tag{57a}$$

$$C_m = -\frac{1.2}{\pi b^3} - \frac{1.2}{\pi a^3} + \frac{8\pi}{b^3}K_0\left(\frac{2\pi a}{b}\right) + \frac{8\pi}{a^3}K_0\left(\frac{2\pi b}{a}\right). \tag{57b}$$

The capacitive and inductive parts of the shunt susceptance B must now be replaced by

$$B_c = \frac{16k_0^2 r^3}{3ab\beta_0(1 - \alpha_e C_e)} \tag{58a}$$

$$B_L = \frac{8h^2 r^3}{3ab\beta_0(1 - \alpha_m C_m)} \tag{58b}$$

where $\alpha_e = 16r^3/3$, $\alpha_m = -8r^3/3$.

Having determined the shunt susceptance of a single plane of disks, we may now replace the artificial dielectric medium by a transmission line having a propagation phase constant β_0 and loaded at regular intervals by shunt susceptances B, as in Fig. 12.12. Periodic structures of this type were analyzed in Chapter 9, where it was found that the propagation phase constant β_c of the periodic wave is given by

$$\cos \beta_c c = \cos \beta_0 c - \frac{B}{2}\sin \beta_0 c. \tag{59}$$

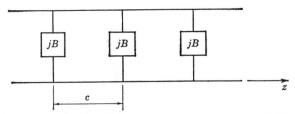

Fig. 12.12. Equivalent transmission-line circuit for a three-dimensional disk medium.

Equation (59) neglects higher order mode interaction between successive planes of disks. The least attenuated evanescent mode has a propagation constant

$$\Gamma_{-10} = \left[\left(\frac{\pi}{a} - h \right)^2 - k_0^2 \right]^{1/2}.$$

Provided $e^{-\Gamma_{-10}c} < 0.1$, interaction with higher order modes may be neglected. For $\beta_0 c$ small, (59) gives

$$\beta_c^2 = \beta_0^2 + \frac{\beta_0 B}{c} \tag{60}$$

upon using the small-argument approximations for the trigonometric functions.

According to the Clausius–Mossotti theory, the disk-type artificial dielectric medium has a dielectric constant different from unity in the x and y directions, and a permeability μ different from μ_0 in the z direction only. Let the permittivity in the y direction be $\kappa\epsilon_0$. In an anisotropic medium of this type, the solution for a plane wave with components E_y, H_x, H_z is readily found to be

$$E_y = e^{-jhx - j\beta z}$$

$$j\omega\mu_0 H_x = -j\beta E_y$$

$$j\omega\mu H_z = jh E_y$$

and $\beta^2 = \kappa k_0^2 - (\mu_0/\mu)h^2$, since E_y is a solution of

$$\frac{\partial^2 E_y}{\partial z^2} + \frac{\mu_0}{\mu} \frac{\partial^2 E_y}{\partial x^2} + \kappa k_0^2 E_y = 0.$$

If we now identify β with β_c, we obtain from (60)

$$\beta_c^2 - \beta_0^2 = \beta^2 - \beta_0^2 = (\kappa - 1)k_0^2 - \left(\frac{\mu_0}{\mu} - 1 \right) h^2 = \frac{\beta_0 B}{c}$$

where $\beta_0^2 = k_0^2 - h^2$. Substituting for B and equating the capacitive and inductive terms separately gives the following solutions for κ and μ:

$$\kappa = 1 + \frac{16r^3}{3abc(1 - \alpha_e C_e)} \tag{61a}$$

$$\frac{\mu}{\mu_0} = \left[1 + \frac{8r^3}{3abc(1 - \alpha_m C_m)} \right]^{-1} \approx 1 - \frac{8r^3}{3abc(1 - \alpha_m C_m)}. \tag{61b}$$

These results are precisely those given by the Clausius–Mossotti theory when interaction between disks in a single plane only is taken into account [see (37)]. The validity of this theory has thus been demonstrated under the condition that the spacing between disks is small. However, (59) is more general since it is valid for large values of c as well. As noted by Brown, the theory breaks down for c/a less than about 0.6 for plane waves incident normally. The reason is clear from the above transmission-line analysis, since this corresponds to a situation where higher order mode interaction between successive rows of disks can no longer be neglected.

At the shorter wavelengths, it is desirable to use disks with diameters and spacings that are an appreciable fraction of a wavelength because of the easier construction afforded and greater economy in the number of elements required. The preceding theory is not applicable in this case for several reasons, such as:

1. The incident field varies over the region of a disk sufficiently so that the polarizing field cannot be considered constant.
2. The interaction field is not well approximated by the static field. In actuality, the radiation field from each disk will be more important than the local induction field varying as the inverse third power of the distance.

A more satisfactory theory for this general case has been developed. It consists of a higher order approximation to the dipole moments of a circular disk, and hence a more accurate value for B, together with the use of dynamic-field-interaction constants [12.3]. The determination of the latter is outlined in Problems 12.7 and 12.8. The more accurate expressions for the polarizabilities of a circular disk (due to Eggimann) are ($C = 1.5$ for parallel, 2.5 for perpendicular polarization)

$$\alpha_e = \frac{16}{3}r^3 \left[1 + \frac{(k_0 r)^2}{15}(8 - C \sin^2 \theta_i) \right]$$

$$\alpha_m = -\frac{8}{3}r^3 \left[1 - \frac{(k_0 r)^2}{10}(2 + \sin^2 \theta_i) \right].$$

12.6. Two-Dimensional Strip Medium

Figure 12.13 illustrates a two-dimensional conducting-strip artificial dielectric medium. The strips are infinitely long in the y direction, and have negligible thickness. The width of the strips is a, and the spacings along the x and z directions are s and d, respectively.

For a parallel-polarized wave propagating obliquely through the medium, with the plane of propagation in the xz plane, the medium exhibits anisotropic dielectric properties. This

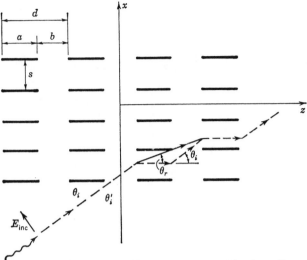

Fig. 12.13. Two-dimensional strip-type artificial dielectric medium.

is readily seen to be so, since, for an electric field applied entirely along the x direction, the strips produce no perturbation, and, hence, the dielectric constant in the x direction is unity. For a field applied along the z direction, a maximum perturbation is produced, and the effective dielectric constant in this direction will be quite large. No magnetic polarization is produced, since there is no component of magnetic field normal to the face of the strips. The medium is a phase-delay medium. For the opposite polarization, the medium exhibits magnetic anisotropy. The index of refraction will be less than unity because of the inductive loading, and, hence, the medium functions as a phase-advance dielectric for this other polarization.

A low-frequency solution may be obtained by using the Clausius–Mossotti theory and the dipole moment of the strips per unit length. A more accurate result is, however, obtained by solving the electrostatic problem with a static field applied along the z direction. The problem may be reduced to that illustrated in Fig. 12.14(a) by a consideration of the symmetry involved. By a suitable conformal mapping, the configuration in Fig. 12.14(b) may be obtained. When the strip width a is zero, the capacitance of a unit quarter cell is

$$C_0 = \frac{\epsilon_0 s}{d}.$$

When $a \neq 0$, the capacitance of a unit quarter cell is the same as that of the cell illustrated in Fig. 12.14(b), and is

$$C_0' = \frac{\epsilon_0 s'}{d'}.$$

The effective dielectric constant κ in the z direction is, thus,

$$\kappa = \frac{C_0'}{C_0} = \frac{s'd}{sd'}.$$

The conformal-mapping solution has been given by Howes and Whitehead [12.4] and has also been derived independently by Kolettis [12.5]. The solution is transcendental in nature, and

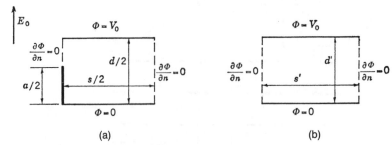

Fig. 12.14. (a) Reduced electrostatic problem for a strip medium. (b) Conformal mapping of the boundary in (a).

is given by

$$\kappa = \frac{K(k_1)K(k')}{K(k_1')K(k)} = \frac{2d}{s}\frac{K(k_1)}{K(k_1')} \tag{62}$$

where K is the complete elliptic integral of the first kind, and k', k_1' are the complementary moduli $(1 - k^2)^{1/2}$, $(1 - k_1^2)^{1/2}$. The modulus k_1 is given by

$$k_1 = \frac{1 + k^2 x}{k(1 + x)} - \left\{ \left[\frac{1 + k^2 x}{k(1 + x)} \right]^2 - 1 \right\}^{1/2}$$

and x is a solution of

$$\frac{a}{d}K(k') = \int_0^{(x^2 - 1)^{1/2}/k'x} \frac{d\lambda}{[(1 - \lambda^2)(1 - k'^2\lambda^2)]^{1/2}}$$

with k and k' determined so that

$$\frac{K(k')}{K(k)} = \frac{2d}{s}.$$

When s is small compared with d, the solution simplifies to

$$\kappa = \frac{d}{b + (s/\pi)\ln 4}. \tag{63}$$

When the spacings s and d are not small compared with a wavelength, a solution based on Maxwell's equations must be used. The starting point for a solution is the determination of the reflection and transmission coefficients of a single interface. When these are known, the artificial dielectric medium may be described in terms of a periodic cascade connection of equivalent circuits. An analysis of this periodic network yields the characteristic propagation phase constant from which the equivalent dielectric constant of the medium may be deduced. Two approaches are possible. For one approach, the x axis is considered as the direction of propagation. The reflection and transmission coefficients required are those for a plane strip grating in free space. When the spacing s is large compared with d, this approach is useful, since higher order mode interaction between adjacent gratings is negligible, and only the reflection and transmission coefficients for the dominant TEM mode are required. These may be found from the equivalent capacitive shunt susceptance of the grating.

By considering propagation to take place along the z direction, the reflection and transmission coefficients required are those for the interface between free space and an infinite array of equispaced parallel plates. This problem has been rigorously solved in Chapter 10. This second approach is well suited to the case when s is small compared with d, since, under these conditions, the evanescent modes decay rapidly, and there is negligible higher order mode interaction between adjacent interfaces. A combination of the two approaches yields accurate results for the equivalent dielectric constant for all values of s and d of interest in practice.

The case of propagation through the medium in the x direction, $\theta_i' = 0$, has been examined by Cohn [12.12] and Brown [12.6]. For this special case, no anisotropic effects exist. The more general case of oblique propagation through the medium has also been analyzed [12.7]. When $\theta_i' = 0$, the shunt susceptance of the strip grating is given by the formulas presented in Chapter 8 for a capacitive diaphragm in a parallel-plate transmission line. For oblique angles of incidence, the normalized shunt susceptance is given with reasonable accuracy by [12.8]

$$B = \frac{4d \cos \theta_i'}{\lambda_0} \left\{ \ln \csc \frac{\pi b}{2d} \right.$$
$$\left. + \frac{1}{2} \frac{(1 - \alpha^2)^2[(1 - \alpha^2/4)(A_+ + A_-) + 4\alpha^2 A_+ A_-]}{(1 - \alpha^2/4) + \alpha^2(1 + \alpha^2/2 - \alpha^4/8)(A_+ + A_-) + 2\alpha^6 A_+ A_-} \right\} \quad (64)$$

where $\alpha = \sin \pi b/2d$, and

$$A_\pm = \left[1 \pm \frac{2d}{\lambda_0} \sin \theta_t' - \left(\frac{d \cos \theta_i'}{\lambda_0} \right)^2 \right]^{-1/2} - 1.$$

The equivalent circuit for propagation in the x direction is illustrated in Fig. 12.15. The propagation phase constant for the transmission line is $k_0 \cos \theta_i'$, and the characteristic phase constant β_c' for the periodic wave propagating in the x direction is determined by a solution of

$$\cos \beta_c's = \cos(k_0 s \cos \theta_i') - \frac{B}{2} \sin(k_0 s \cos \theta_i'). \quad (65)$$

From the solution for β_c' the dielectric constant of the strip medium can be obtained and is given by (72). The required derivation will be presented later in this section.

The reflection and transmission coefficients at the interface between free space and a set of parallel plates are given in Chapter 10, Eqs. (74), (75), and (76). These results show that the interface may be represented by the equivalent circuit illustrated in Fig. 12.16 when s is

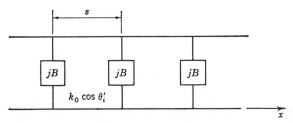

Fig. 12.15. Equivalent circuit for a strip-type artificial dielectric for modes propagating in the x direction.

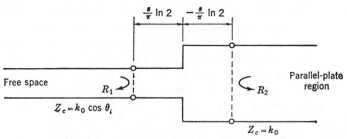

Fig. 12.16. Equivalent circuit of a parallel-plate interface.

small. From Fig. 12.16, it is seen that

$$R_1 = \frac{1 - \cos \theta_i}{1 + \cos \theta_i} \exp\left[-jk_0\left(\frac{s}{\pi} \ln 4\right) \cos \theta_i\right]$$

$$R_2 = \frac{\cos \theta_i - 1}{\cos \theta_i + 1} \exp\left(jk_0\frac{s}{\pi} \ln 4\right)$$

which are the same results as given in Chapter 10 for s small. The equivalent circuit for the artificial dielectric for modes propagating in the z direction is readily constructed, and is illustrated in Fig. 12.17. An analysis of this periodic structure shows that the characteristic phase constant β_c for a periodic wave propagating through the medium in the z direction is a solution of the following equation:

$$(1 - \rho^2) \cos \beta_c d = \cos\left[k_0(a + b \cos \theta_i) - k_0\frac{s}{\pi}(\ln 4)(1 - \cos \theta_i)\right]$$

$$-\rho^2 \cos\left[k_0(a - b \cos \theta_i) - k_0\frac{s}{\pi}(\ln 4)(1 + \cos \theta_i)\right] \quad (66)$$

where $\rho = (1 - \cos \theta_i)/(1 + \cos \theta_i)$, and θ_i is the angle of propagation for the TEM wave in the region between the strips, as in Fig. 12.13.

In order to obtain an expression for the equivalent dielectric constant of the strip medium, propagation through the artificial dielectric medium must be correlated with propagation through a uniform medium having a dielectric constant κ in the z direction and unity in the x and y directions. The latter medium is known as a uniaxial medium, and has a single optic axis in the z direction. Propagation of a TEM wave in such a medium was discussed in Section 3.7. For the extraordinary wave with a propagation factor $e^{-j\beta\mathbf{n}\cdot\mathbf{r}}$, it was found that

$$\beta^2 = \frac{\kappa k_0^2}{n_x^2 + \kappa n_z^2} \quad (67)$$

where n_x and n_z are the components of the wave normal. The angle of propagation through the medium is given by

$$\tan \theta_r = \frac{n_x}{n_z} \quad (68)$$

where θ_r is measured relative to the z axis. For the solution given by (66), we have

$$\beta_c = \beta n_z \quad \text{and} \quad k_0 \sin \theta_i = \beta n_x.$$

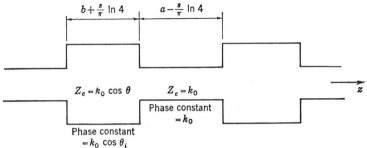

Fig. 12.17. Equivalent circuit for a strip-type artificial dielectric for modes propagating in the z direction.

Substituting into (67) and (68), we get

$$\kappa = \frac{k_0^2 \sin^2 \theta_i}{k_0^2 - \beta_c^2} \tag{69}$$

$$\tan \theta_r = \frac{k_0 \sin \theta_i}{\beta_c}. \tag{70}$$

Equation (70) gives the effective angle of propagation through the medium.

Certain limiting solutions for κ may be obtained from (66) and (69). When $d < 0.1\lambda_0$, the cosine terms may be expanded according to the relation $\cos u = 1 - u^2/2$, and (66) together with (69) gives

$$\kappa = \frac{d}{b + (s/\pi) \ln 4}$$

which is the same as the static solution given by (63) for $s \ll d$. For this case, κ is independent of the effective angle of propagation θ_r through the medium. For larger values of d, it is found that κ varies with the angle of propagation through the medium. This effect may be considered as caused by the graininess of the lattice structure. As θ_i approaches zero, ρ tends to zero, and the limiting value of κ is again the above static value, independent of the condition $d < 0.1\lambda_0$. This limiting solution is obtained by placing ρ equal to zero in (66) and solving for $\beta_c d$ to get

$$(\beta_c d)^2 = (k_0 d)^2 \left[1 - (1 - \cos \theta_i) \left(\frac{b}{d} + \frac{s}{\pi d} \ln 4 \right) \right]^2.$$

Expanding by the binomial expression and substituting into (69) now gives the end result. When $\theta_i = \pi/2$, the eigenvalue equation (66) reduces to

$$\cos \beta_c d = \cos k_0 \left(a - \frac{s}{\pi} \ln 4 \right) - \frac{k_0}{2} \left(b + \frac{s}{\pi} \ln 4 \right) \sin k_0 \left(a - \frac{s}{\pi} \ln 4 \right). \tag{71}$$

Although $\theta_i = \pi/2$, the effective angle of propagation θ_r through the medium is less than $\pi/2$. When $\theta_r = \pi/2$, $\beta_c d = 0$, and θ_i is complex with $\sin \theta_i$ greater than unity.

Returning to the first solution (65) based on modes propagating in the x direction, we have

$$\beta n_x = \beta_c' \qquad \beta n_z = k_0 \sin \theta_i' = k_0 \cos \theta_i.$$

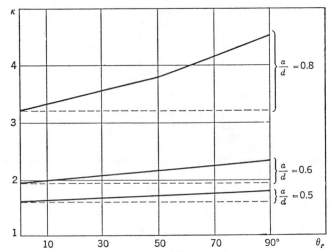

Fig. 12.18. Effective dielectric constant κ as a function of effective angle of propagation θ_r. Broken curves are static values obtained from (63), $s = 0.1\lambda_0$, $d = 0.4\lambda_0$.

Corresponding to (69) and (70), we now get

$$\kappa = \frac{\beta'^2_c}{k_0^2 \cos^2 \theta'_i} = \frac{\beta'^2_c}{k_0^2 \sin^2 \theta_i} \tag{72}$$

$$\tan \theta_r = \frac{\beta'_c}{k_0 \cos \theta_i} = \kappa^{1/2} \tan \theta_i. \tag{73}$$

As noted earlier, (72) gives accurate results for large values of s, while (69) gives accurate results for small values of s. A combination of the two methods of solution yields good results for all values of s.

Some Numerical Results

In Fig. 12.18, the effective dielectric constant κ is plotted as a function of the effective angle of propagation θ_r through the medium. These values of κ are based on the solutions given by (66) and (69). In the same figure, the static solution (63) is also plotted. For a/d greater than 0.5, it is seen that κ varies considerably with the angle of propagation through the medium.

In order to compare the results of the two methods of solution, (69) and (72), κ has been evaluated for $d = 0.4\lambda_0$ as a function of the spacing s in wavelengths, for fixed values of θ_r, and with $a = 0.75d$. Since the derived formulas do not give the end results directly, κ was evaluated first as a function of θ_i for fixed values of s/λ_0, then these curves were replotted to give κ as a function of θ_r for fixed values of s/λ_0, and finally these curves were converted into the desired family of curves. The results are plotted for $\theta_r = 30$, 50, 70, and 90° in Figs. 12.19(a)–(d). For $s > 0.2\lambda_0$ the first method gives good results, while for $s < 0.2\lambda_0$ the second method gives good results. By joining the two curves smoothly in the transition region around $s = 0.2\lambda_0$, accurate results for all values of s are obtained.

Since the strip medium is a periodic one, it will exhibit the usual passband–stopband behavior. For θ_r equal to 70 and 90°, it is noted that κ increases rapidly for s/λ_0 greater than some minimum value. This occurs because the edge of the first passband is being approached. The behavior may be understood qualitatively from the $\beta'_c s$–$k_0 s \cos \theta'_i$ diagram for the capacitive-loaded transmission line which has the equivalent circuit shown in Fig. 12.15. This diagram

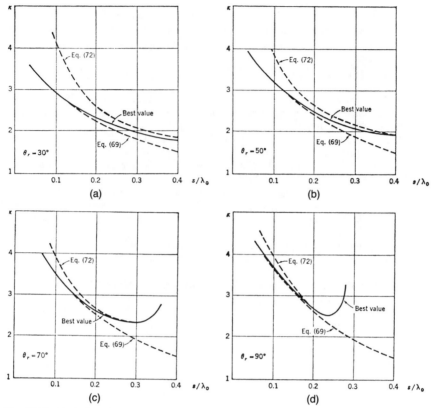

Fig. 12.19. Equivalent dielectric constant from (69) and (72) together with the best value, $d = 0.4\lambda_0$, $a = 0.75d$.

is sketched in Fig. 12.20. The effective dielectric constant κ is given by

$$\kappa = \frac{(\beta'_c s)^2}{k_0^2 s^2 \cos^2 \theta'_i}$$

and is seen to be equal to the square of the cotangent of the angle of the line joining the origin

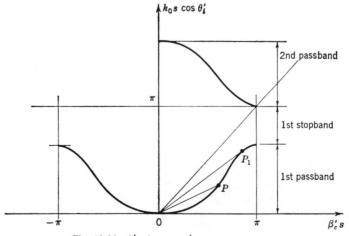

Fig. 12.20. $\beta'_c s$–$k_0 s \cos \theta'_i$ diagram for (65).

O to the point P on the curve in Fig. 12.20. As P moves up to P_1, the angle increases and κ decreases. Beyond P_1 and up to the edge of the passband the angle decreases, and hence κ has a rapid increase in value just before the edge of the passband is reached.

REFERENCES AND BIBLIOGRAPHY

[12.1]   W. E. Kock, "Metallic delay lenses," *Bell Syst. Tech. J.*, vol. 27, pp. 58–82, 1948.

[12.2]   S. B. Cohn, "Electrolytic tank measurements for microwave metallic delay lens media," *J. Appl. Phys.*, vol. 21, pp. 674–680, July 1950.

[12.3]   R. E. Collin and W. Eggimann, "Evaluation of dynamic interaction fields in a two dimensional lattice," *IRE Trans. Microwave Theory Tech.*, vol. MTT-9, pp. 110–115, 1961.

[12.4]   J. Howes and E. A. N. Whitehead, "The refractive index of a dielectric loaded with thin metal strips," Elliot Bros. Research Rep. 119, July 1949.

[12.5]   N. Kolettis, "Conformal mapping solution for equivalent relative permittivity of a strip artificial dielectric medium," Case Inst. Technol. Sci. Rep. 2, Jan. 1959. Work performed under Contract AF-19(604) 3887.

[12.6]   J. Brown, "The design of metallic delay dielectrics," *Proc. IEE (London)*, vol. 97, part III, pp. 45–48, Jan. 1950.

[12.7]   N. Kolettis, "Electric anisotropic properties of the metallic-strip-type periodic medium," Case Inst. Technol. Sci. Rep. 17, Oct. 1960. Work supported by Air Force Cambridge Research Center Contract AF-19(604) 3887.

[12.8]   N. Marcuvitz, *Waveguide Handbook*, vol. 10 of MIT Rad. Lab. Series.   New York, NY: McGraw-Hill Book Company, Inc., 1951, sect. 5.18, eq. (1*a*).

[12.9]   O. M. Stuetzer, "Development of artificial microwave optics in Germany," *Proc. IRE*, vol. 38, pp. 1053–1056, Sept. 1950.

[12.10] S. B. Cohn, "The electric and magnetic constants of metallic delay media containing obstacles of arbitrary shape and thickness," *J. Appl. Phys.*, vol. 22, pp. 628–634, May 1951.

[12.11] S. B. Cohn, "Microwave measurements on metallic delay media," *Proc. IRE*, vol. 41, pp. 1177–1183, Sept. 1953.

[12.12] S. B. Cohn, "Analysis of the metal strip delay structure for microwave lenses," *J. Appl. Phys.*, vol. 20, pp. 257–262, Mar. 1949.

[12.13] S. B. Cohn, "Experimental verification of the metal strip delay lens theory," *J. Appl. Phys.*, vol. 24, pp. 839–841, July 1953.

[12.14] S. B. Cohn, "Artificial dielectrics for microwaves," in *Modern Advances in Microwave Techniques*. J. Fox, ed., MRI Symp. Proc., vol. 4, pp. 465–480, Nov. 1954.   New York: Polytechnic Press of Polytechnic Inst. of Brooklyn. (This is a good source of additional references.)

[12.15] J. Brown, "Artificial dielectrics having refractive indices less than unity," *Proc. IEE (London)*, vol. 100, part 4, pp. 51–62, 1953.

[12.16] J. Brown and W. Jackson, "The relative permittivity of tetragonal arrays of perfectly conducting thin discs," *Proc. IEE (London)*, vol. 102, part B, pp. 37–42, Jan. 1955.

[12.17] J. Brown and W. Jackson, "The properties of artificial dielectrics at cm. wavelengths," *Proc. IEE (London)*, vol. 102, part B, pp. 11–16, Jan. 1955.

[12.18] M. M. Z. Kharadly and W. Jackson, "The properties of artificial dielectrics comprising arrays of conducting elements," *Proc. IEE (London)*, vol. 100, part III, pp. 199–212, July 1953.

[12.19] M. M. Z. Kharadly, "Some experiments on artificial dielectrics at centimeter wavelengths," *Proc. IEE (London)*, vol. 102, part B, pp. 17–25, Jan. 1955.

[12.20] R. W. Corkum, "Isotropic artificial dielectrics," *Proc. IRE*, vol. 40, pp. 574–587, May 1952.

[12.21] E. R. Wicher, "The influence of magnetic fields upon the propagation of electromagnetic waves in artificial dielectrics," *J. Appl. Phys.*, vol. 22, pp. 1327–1329, Nov. 1951.

[12.22] G. Estrin, "The effective permeability of an array of thin conducting discs," *J. Appl. Phys.*, vol. 21, pp. 667–670, July 1950.

[12.23] G. Estrin, "The effects of anisotropy in a three dimensional array of conducting discs," *Proc. IRE*, vol. 39, pp. 821–826, July 1951.

[12.24] A. L. Mikaelyan, "Methods of calculating the permittivity and permeability of artificial media," *Radiotekhnika*, vol. 10, pp. 23–36, 1955.

[12.25] W. B. Swift and T. J. Higgins, "Determination of the design constants of artificial dielectric UHF lenses by use of physical analogy," in *Proc. Nat. Electronics Conf.*, vol. 9, pp. 825–832, 1953. (This article is a good source of references on artificial dielectrics.)

[12.26] H. S. Bennett, "The electromagnetic transmission characteristics of the two dimensional lattice medium," *J. Appl. Phys.*, vol. 24, pp. 785–810, June 1953.

[12.27] L. Lewin, "Electrical constants of spherical conducting particles in a dielectric," *J. IEE (London)*, vol. 94, part III, pp. 65–68, Jan. 1947.

[12.28] Z. A. Kaprielian, "Anisotropic effects in geometrically isotropic lattices," *J. Appl. Phys.*, vol. 29, pp. 1052–1063, July 1958.

[12.29] R. E. Collin, "A simple artificial anisotropic dielectric medium," *IRE Trans. Microwave Theory Tech.*, vol. MTT-6, pp. 206–209, Apr. 1958.

[12.30] H. E. J. Neugebauer, "Clausius-Mossotti equation for certain types of anisotropic crystals," *Can. J. Phys.*, vol. 32, pp. 1–8, Jan. 1954.

[12.31] N. J. Kolettis and R. E. Collin, "Anisotropic properties of strip-type artificial dielectric, "*IRE Trans. Microwave Theory Tech.*, vol. MTT-9, pp. 436–441, 1961.

PROBLEMS

12.1. Replace the sums in (11) by integrals, and integrate from $b/2 \to \infty$, $a/2 \to \infty$, $c/2 \to \infty$; multiply by 8, and show that the interaction constant C is equal to $(3b^3)^{-1}$ for $a = b = c$.

HINT: Put $mb = y$, $dm = (1/b)\,dy$, etc., and integrate with respect to y first since the integrand is equal to $\partial\Phi/\partial y$.

12.2. Find the interaction constant for a single row of dipoles located midway between two infinite parallel conducting plates as illustrated in Fig. P12.2.

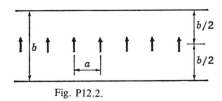

Fig. P12.2.

12.3. Repeat the analysis of Problem 12.2, but with the dipole axis parallel with the conducting planes.

12.4. An artificial dielectric medium is made up of parallel conducting plates spaced a distance c apart along the z axis (see Fig. P12.4). In each plate a rectangular array of circular holes of radius r is cut. Analyze the properties of this medium, and compare them with the properties of the circular-disk medium. Consider both normal and oblique propagation through the medium.

HINT: The shunt susceptance of one perforated plate may be obtained by duality from the susceptance of a plane of disks.

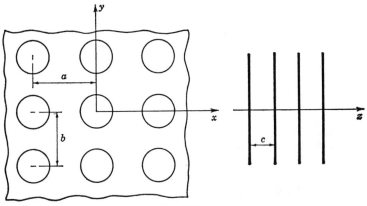

Fig. P12.4.

12.5. Consider a two-dimensional conducting-strip artificial dielectric medium with a TEM wave propagating obliquely through the lattice and having the electric field parallel to the strips. Analyze the anisotropic properties of the medium. Each inductive grating has a normalized susceptance B_L given by $B_L = 4/B$, where B is given by (64). The medium may be represented by an equivalent transmission line loaded at regular intervals by shunt inductive susceptances. Note that there is no propagation through the medium below some lower cutoff frequency f_c.

12.6. Find the relative dielectric constant of an artificial dielectric medium comprised of a cubical array of high-dielectric-constant spheres embedded in a low-dielectric-constant binder. What value of dielectric constant do the spheres have to have in order to give a medium with an effective dielectric constant 2.25 for sphere diameters equal to 0.6 of the spacing between spheres? Assume that the dielectric constant of the supporting medium is 1.02.

12.7. Consider a two-dimensional array of y-directed electric dipoles $Pe^{j\omega t}$, located at $x = na$, $n = 0, \pm 1, \pm 2, \ldots$; $y = mb$, $m = 0, \pm 1, \pm 2, \ldots$; $z = 0$. Find the y component of the dynamic interaction field, acting on the dipole at the origin, due to all the neighboring dipoles. Assume that the dipoles are excited by a plane wave with $E_y = Ae^{-jhx - \Gamma_0 z}$, $\Gamma_0 = (h^2 - k_0^2)^{1/2}$, and, hence, the row of dipoles located at $x = na$ will have a phase e^{-jhna} relative to the row located at $x = 0$.

HINT: Find a vector potential with a single y component A_y first, where A_y is a solution of

$$\nabla^2 A_y + k_0^2 A_y = -j\omega\mu_0 P\delta(x)\delta(y)\delta(z).$$

Solve this equation in cylindrical coordinates. Note that conducting planes can be placed at $y = \pm b/2$. The steps to be followed in obtaining the complete solution are similar to those used in Section 7.2 to evaluate the mutual flux linkage in a loop antenna.

Answer: The required vector potential is

$$A_y = \frac{j\omega\mu_0 P}{2\pi b}\left[\frac{\pi}{a}\sum_{m=-\infty}^{\infty}\frac{e^{-\Gamma_m|z|}}{\Gamma_m}e^{-j(h+2m\pi/a)x} - K_0(jk_0\rho)\right.$$

$$\left. + 2\sum_{m=-\infty}^{\infty}{}'\sum_{n=1}^{\infty}e^{-jhma}\cos\frac{2n\pi y}{b}K_0\{\gamma_n[z^2 + (ma - x)^2]^{1/2}\}\right]$$

$$+ \frac{j\omega\mu_0 P}{4\pi}\sum_{m=-\infty}^{\infty}{}'\frac{e^{-jk_0 r_m}}{r_m}$$

$$r_m^2 = \rho^2 + (mb - y)^2 \qquad \Gamma_m^2 = \left(h + \frac{2m\pi}{a}\right)^2 - k_0^2 \qquad \gamma_n^2 = \left(\frac{2n\pi}{b}\right)^2 - k_0^2 \qquad \rho^2 = x^2 + z^2.$$

The interaction field is

$$E_{yi} = \frac{k_0^2 P}{2ab\epsilon_0}\left[\frac{a}{\pi}\left(\gamma - \ln\frac{4\pi}{k_0 a}\right) + j\left(\frac{a}{2} - \frac{1}{|\Gamma_0|}\right) + \sum_{m=1}^{\infty}\left(\frac{1}{\Gamma_m} + \frac{1}{\Gamma_{-m}} - \frac{a}{m\pi}\right)\right]$$

$$- \frac{2P}{\pi b\epsilon_0}\sum_{n=1}^{\infty}\sum_{m=1}^{\infty}\gamma_n^2 K_0(\gamma_n ma)$$

$$+ \frac{P}{\pi\epsilon_0 b^3}\left[\left(1.2 - \frac{k_0^2 b^2}{2}\ln k_0 b + \frac{k_0^2 b^2}{4} + \frac{3k_0^4 b^4}{288}\right) - j\left(\frac{\pi k_0^2 b^2}{4} - \frac{k_0^3 b^3}{6}\right)\right].$$

12.8. For the two-dimensional array of electric dipoles considered in Problem 12.7, find the z component of the magnetic interaction field H_{zi} acting on the dipole at the origin.

Answer:

$$H_{zi} = \frac{\omega P}{2\pi b}\left[-j\frac{\pi h}{|\Gamma_0|a} + \frac{\pi}{a}\sum_{m=1}^{\infty}\left(\frac{h + 2m\pi/a}{\Gamma_m} + \frac{h - 2m\pi/a}{\Gamma_{-m}}\right) + 4\sum_{n=1}^{\infty}\sum_{m=1}^{\infty}\gamma_n K_1(\gamma_n ma)\sin hma\right].$$

This result shows that, in general, there will be coupling between the electric and magnetic dipoles in an artificial dielectric medium. The evaluation of the dynamic magnetic interaction field from z-directed magnetic dipoles is similar to the evaluation of mutual flux linkage in a loop antenna carried out in Section 7.2.

12.9. Consider a row of thin circular disks between two conducting planes at $y = \pm b/2$, as in Fig. 12.10. Let the incident field be

$$E_y = A_1 e^{-jhx - \Gamma_0 z}$$

$$H_z = A_1 Y_0 \sin \theta_i e^{-jhx - \Gamma_0 z}$$

where $h = k_0 \sin \theta_i$, $\Gamma_0 = jk_0 \cos \theta_i$ plus an x component of magnetic field. This incident field induces an electric dipole

$$P_n = \tfrac{16}{3} r^3 \epsilon_0 A_1 e^{-jhna} = P e^{-jhna}$$

and a magnetic dipole

$$M_n = -\tfrac{8}{3} r^3 A_1 Y_0 \sin \theta_i \, e^{-jhna} = M e^{-jhna}$$

in the disk located at $x = na$, where r is the disk radius. The field scattered by the y-directed electric dipoles may be found from a y-directed vector potential A_y, where

$$\nabla^2 A_y + k_0^2 A_y = -j\omega\mu_0 P \sum_{n=-\infty}^{\infty} e^{-jhna} \delta(x - na)\delta(y)\delta(z).$$

Show that a suitable solution for A_y is

$$A_y = \sum_{n=-\infty}^{\infty} \sum_{m=0,2,\ldots}^{\infty} a_{nm} \cos \frac{m\pi y}{b} e^{-j(h+2n\pi/a)x} e^{-\Gamma_{nm} z}$$

where $\Gamma_{nm}^2 = (h + 2n\pi/a)^2 + (m\pi/b)^2 - k_0^2$.

Multiply both sides of the differential equation for A_y by $e^{j(h+2n\pi/a)x}$ and integrate over $-a/2 \le x \le a/2$, $-b/2 \le y \le b/2$ to get

$$a_{00}ab \left(\frac{\partial^2}{\partial z^2} - \Gamma_{00}^2 \right) e^{-\Gamma_{00}|z|} = -j\omega\mu_0 P \delta(z).$$

Next integrate over $-\tau < z < \tau$ and let $\tau \to 0$ to obtain

$$a_{00} = \frac{j\omega\mu_0 P}{2ab\Gamma_{00}}.$$

The field scattered by the z-directed magnetic dipoles may be derived from a z-directed magnetic Hertzian potential Π_{mz}, where

$$\nabla^2 \Pi_{mz} + k_0^2 \Pi_{mz} = -M \sum_{n=-\infty}^{\infty} e^{-jhna} \delta(x - na)\delta(y)\delta(z).$$

Show that a suitable solution for Π_{mz} is

$$\sum_{n=-\infty}^{\infty} \sum_{m=0,2,\ldots}^{\infty} b_{nm} \cos \frac{m\pi y}{b} e^{-j(h+2n\pi/a)x} e^{-\Gamma_{nm} z}.$$

By analogy with the solution for a_{00} show that $b_{00} = M/2ab\Gamma_{00}$. Hence show that the reflection coefficient for the incident wave is

$$R = -j\omega \frac{a_{00}}{A_1} + \omega\mu_0 h \frac{b_{00}}{A_1}$$

by evaluating the dominant-mode electric field associated with A_y and Π_{mz}. Show that R reduces to

$$R = \frac{-j4r^3}{3ab\beta_0}(2k_0^2 - h^2), \qquad \beta_0 = |\Gamma_{00}| = |\Gamma_0|.$$

Compare this method with that used in Section 12.5.

Mathematical Appendix

In the first three parts of this section, our frame of reference will be a rectangular xyz coordinate frame. In subsection d, we will consider orthogonal curvilinear coordinate systems.

A vector quantity is an entity which requires both a magnitude and a direction to be specified at each point in space. The usual procedure is to give the components of the vector along three mutually orthogonal axes. A coordinate frame xyz is said to be right-handed or dextral if rotation of the x axis into the y axis defines a sense of rotation that would advance a right-hand screw in the positive z direction. If the sense of rotation would advance a right-hand screw along the negative z direction the coordinate system is said to be left-handed or sinistral. For our rectangular coordinate frame, the unit vectors along the coordinate axis will be represented by $\mathbf{a}_x$, $\mathbf{a}_y$, $\mathbf{a}_z$.

a. Vector Algebra

Addition and Subtraction: The addition or subtraction of two or more vectors is defined as the vector whose components are the sum or difference of the corresponding components of each individual vector; i.e.,

$$\mathbf{A} \pm \mathbf{B} = (A_x \pm B_x)\mathbf{a}_x + (A_y \pm B_y)\mathbf{a}_y + (A_z \pm B_z)\mathbf{a}_z.$$

Scalar and Vector Products: The scalar product of a vector $\mathbf{A}$ with a vector $\mathbf{B}$ is written as $\mathbf{A} \cdot \mathbf{B}$ and is defined by

$$\mathbf{A} \cdot \mathbf{B} = |A||B| \cos \theta$$

where θ is the angle between the two vectors. In rectangular coordinates, we have

$$\mathbf{A} \cdot \mathbf{B} = A_x B_x + A_y B_y + A_z B_z$$

which is the projection of $\mathbf{A}$ onto $\mathbf{B}$ multiplied by the modulus of $\mathbf{B}$.

The vector, or cross, product of a vector $\mathbf{A}$ with a vector $\mathbf{B}$ is written as $\mathbf{A} \times \mathbf{B}$ and is a vector of magnitude $|A||B| \sin \theta$, and perpendicular to both $\mathbf{A}$ and $\mathbf{B}$ with a positive direction, defined such that rotation of $\mathbf{A}$ into $\mathbf{B}$ would cause a right-hand screw to advance along it. In rectangular coordinates, we have

$$\mathbf{A} \times \mathbf{B} = \begin{vmatrix} \mathbf{a}_x & \mathbf{a}_y & \mathbf{a}_z \\ A_x & A_y & A_z \\ B_x & B_y & B_z \end{vmatrix} = -\mathbf{B} \times \mathbf{A}.$$

Scalar and Vector Higher Order Products: The scalar triple or box product is given by

$$\mathbf{A} \cdot \mathbf{B} \times \mathbf{C} = \mathbf{A} \times \mathbf{B} \cdot \mathbf{C} = \begin{vmatrix} A_x & A_y & A_z \\ B_x & B_y & B_z \\ C_x & C_y & C_z \end{vmatrix}.$$

The vector triple product is written as $\mathbf{A} \times (\mathbf{B} \times \mathbf{C})$ and is equal to $(\mathbf{A} \cdot \mathbf{C})\mathbf{B} - (\mathbf{A} \cdot \mathbf{B})\mathbf{C}$. It is important to place the parentheses correctly since, in general, $\mathbf{A} \times (\mathbf{B} \times \mathbf{C}) \neq (\mathbf{A} \times \mathbf{B}) \times \mathbf{C}$.

By an application of the above fundamental rules, we may readily obtain the following results:

$$(\mathbf{A} \times \mathbf{B})\cdot(\mathbf{C} \times \mathbf{D}) = \begin{vmatrix} \mathbf{A}\cdot\mathbf{C} & \mathbf{A}\cdot\mathbf{D} \\ \mathbf{B}\cdot\mathbf{C} & \mathbf{B}\cdot\mathbf{D} \end{vmatrix}$$

$$(\mathbf{A} \times \mathbf{B}) \times (\mathbf{C} \times \mathbf{D}) = (\mathbf{A}\cdot\mathbf{B} \times \mathbf{D})\mathbf{C} - (\mathbf{A}\cdot\mathbf{B} \times \mathbf{C})\mathbf{D}.$$

b. Differential Invariants

An invariant quantity or simply an invariant is a quantity which has a unique value at a given point in space, independent of the coordinate system to which it is referred. The fundamental vector differential operator is "del," or "nabla," and is written as

$$\nabla = \mathbf{a}_x \frac{\partial}{\partial x} + \mathbf{a}_y \frac{\partial}{\partial y} + \mathbf{a}_z \frac{\partial}{\partial z}.$$

This operator behaves like a vector in the sense that it combines with vectors according to the rules of vector algebra. In addition, it performs certain differentiations with respect to the coordinates. In common with all differential operators it is a noncommutative operator; i.e., it must always occur on the left-hand side of the quantity it operates on. The manipulation of various expressions involving this vector differential operator is greatly facilitated if the differentiation operation is suppressed until the expression has been expanded into a simpler form by treating ∇ as a vector.

Gradient and Directional Derivative: The derivative of a scalar function $\Phi(x, y, z)$ with respect to distance l along a continuous curve $x(l)$, $y(l)$, $z(l)$ is

$$\frac{d\Phi}{dl} = \frac{\partial\Phi}{\partial x}\frac{dx}{dl} + \frac{\partial\Phi}{\partial y}\frac{dy}{dl} + \frac{\partial\Phi}{\partial z}\frac{dz}{dl}$$

and is the directional derivative of Φ along the curve. The unit tangent to the curve is τ, where

$$\tau = \mathbf{a}_x \frac{dx}{dl} + \mathbf{a}_y \frac{dy}{dl} + \mathbf{a}_z \frac{dz}{dl}.$$

The gradient of Φ is given by

$$\nabla\Phi = \mathbf{a}_x \frac{\partial\Phi}{\partial x} + \mathbf{a}_y \frac{\partial\Phi}{\partial y} + \mathbf{a}_z \frac{\partial\Phi}{\partial z}$$

and, hence, the directional derivative is the projection of $\nabla\Phi$ on τ; that is,

$$\frac{d\Phi}{dl} = \nabla\Phi\cdot\tau.$$

The directional derivative has its maximum value when τ is parallel to $\nabla\Phi$. Hence, $\nabla\Phi$ is in the direction of the maximum rate of change of Φ.

Divergence and Curl: The divergence of a vector may be defined as

$$\nabla\cdot\mathbf{A} = \lim_{\Delta V \to 0} \frac{\oiint_S \mathbf{A}\cdot d\mathbf{S}}{\Delta V}.$$

The divergence of a vector field is a measure of the net outflow of flux through a closed surface S surrounding a volume ΔV, per unit volume. In rectangular coordinates,

$$\nabla\cdot\mathbf{A} = \frac{\partial A_x}{\partial x} + \frac{\partial A_y}{\partial y} + \frac{\partial A_z}{\partial z}.$$

The curl of a vector is a measure of the net circulation of the vector field around the boundary of an infinitesimal element of area. Mathematically,

$$(\text{curl } \mathbf{A})_n = (\nabla \times \mathbf{A})_n = \lim_{\Delta S \to 0} \frac{\oint_C \mathbf{A} \cdot d\mathbf{l}}{\Delta S}$$

where $(\nabla \times \mathbf{A})_n$ is the component of the curl of $\mathbf{A}$ in the direction of the normal to the surface ΔS, and C is the boundary of ΔS. In rectangular coordinates, the curl of $\mathbf{A}$ is given by the expansion of the following determinant:

$$\nabla \times \mathbf{A} = \begin{vmatrix} \mathbf{a}_x & \mathbf{a}_y & \mathbf{a}_z \\ \dfrac{\partial}{\partial x} & \dfrac{\partial}{\partial y} & \dfrac{\partial}{\partial z} \\ A_x & A_y & A_z \end{vmatrix}.$$

Some Differential Identities: The following identity is readily established:

$$\nabla \times \nabla \times \mathbf{A} = \nabla \nabla \cdot \mathbf{A} - \nabla^2 \mathbf{A}.$$

Treating ∇ as a vector and expanding the triple vector product, but always keeping ∇ in front of $\mathbf{A}$, gives the result at once. Other useful identities are

$$\nabla \times \nabla \Phi = 0$$

$$\nabla \cdot \nabla \times \mathbf{A} = 0$$

$$\nabla \cdot (\mathbf{A} \times \mathbf{B}) = \mathbf{B} \cdot \nabla \times \mathbf{A} - \mathbf{A} \cdot \nabla \times \mathbf{B}$$

$$\nabla \times (\mathbf{A} \times \mathbf{B}) = \mathbf{A} \nabla \cdot \mathbf{B} - \mathbf{B} \nabla \cdot \mathbf{A} + (\mathbf{B} \cdot \nabla) \mathbf{A} - (\mathbf{A} \cdot \nabla) \mathbf{B}.$$

The above expressions are readily obtained, as, for example,

$$\nabla \cdot (\mathbf{A} \times \mathbf{B}) = \nabla_A \cdot (\mathbf{A} \times \mathbf{B}) + \nabla_B \cdot (\mathbf{A} \times \mathbf{B})$$

$$= (\nabla_A \times \mathbf{A}) \cdot \mathbf{B} - \nabla_B \cdot (\mathbf{B} \times \mathbf{A})$$

$$= \mathbf{B} \cdot (\nabla \times \mathbf{A}) - \mathbf{A} \cdot \nabla \times \mathbf{B}$$

where ∇_A means operation on $\mathbf{A}$ only, and similarly for ∇_B. Each term may be treated as a scalar triple product, and the dot and cross may be interchanged.

c. Integral Invariants and Transformations

Gauss' Law or Divergence Theorem: This theorem states that the volume integral of the divergence of a vector is equal to the surface integral of the normal outward component of the vector over a closed surface S surrounding the volume V; that is,

$$\iiint_V \nabla \cdot \mathbf{A} \, dV = \oiint_S \mathbf{A} \cdot d\mathbf{S}.$$

Green's Identities: Green's first identity is

$$\iiint_V (\nabla v \cdot \nabla u + u \, \nabla^2 v) \, dV = \oiint_S u \, \nabla v \cdot d\mathbf{S}$$

where u and v are two scalar functions. The proof follows by applying the divergence theorem to the vector function $u \nabla v$. Interchanging u and v and subtracting gives Green's second identity or, as it is sometimes called, Green's theorem,

$$\iiint\limits_{V} (u \nabla^2 v - v \nabla^2 u)\, dV = \oiint\limits_{S} (u \nabla v - v \nabla u) \cdot d\mathbf{S}.$$

This result is of considerable use in proving the orthogonality of the eigenfunctions for the scalar wave equations. It is also of fundamental importance in the solution of certain boundary-value problems by the use of Green's functions.

Stokes' Theorem: This theorem gives the value of the integral of the curl of a vector over a surface S in terms of a line integral around a closed contour C, which bounds the surface S:

$$\iint\limits_{S} \nabla \times \mathbf{A} \cdot d\mathbf{S} = \oint\limits_{C} \mathbf{A} \cdot d\mathbf{l}.$$

The proof follows readily from the definition of the curl of a vector.

Vector Forms of Green's Identities: The vector forms of Green's first and second identities are

$$\iiint\limits_{V} (\nabla \times \mathbf{A} \cdot \nabla \times \mathbf{B} - \mathbf{A} \cdot \nabla \times \nabla \times \mathbf{B})\, dV = \iiint\limits_{V} \nabla \cdot (\mathbf{A} \times \nabla \times \mathbf{B})\, dV$$

$$= \oiint\limits_{S} \mathbf{A} \times \nabla \times \mathbf{B} \cdot d\mathbf{S}$$

$$\iiint\limits_{V} (\mathbf{B} \cdot \nabla \times \nabla \times \mathbf{A} - \mathbf{A} \cdot \nabla \times \nabla \times \mathbf{B})\, dV = \oiint\limits_{S} (\mathbf{A} \times \nabla \times \mathbf{B} - \mathbf{B} \times \nabla \times \mathbf{A}) \cdot d\mathbf{S}.$$

The vector form of the second identity is used to prove the orthogonality of the eigenfunctions of the vector wave equations and is also used in the solution of certain vector boundary-value problems by means of Green's functions.

The following modification of the vector form of Green's first identity may also be established:

$$\iiint\limits_{V} (\nabla \times \mathbf{A} \cdot \nabla \times \mathbf{B} + \nabla \cdot \mathbf{A} \nabla \cdot \mathbf{B} + \mathbf{A} \cdot \nabla^2 \mathbf{B})\, dV = \oiint\limits_{S} (\mathbf{A} \times \nabla \times \mathbf{B} + \mathbf{A} \nabla \cdot \mathbf{B}) \cdot d\mathbf{S}.$$

d. Orthogonal Curvilinear Coordinates

In the present subsection we will derive the general relationships between the components of a vector in rectangular coordinates and those of a vector referred to an orthogonal curvilinear coordinate frame, and also the transformation laws for various vector differential operations. For this purpose we will relabel our x, y, z axes as x_1, x_2, x_3 with unit vectors $\mathbf{a}_1, \mathbf{a}_2, \mathbf{a}_3$ directed along them.

Transformation of Coordinates: Consider three continuously differentiable functions of the three variables x_i,

$$u_i = u_i(x_1, x_2, x_3), \qquad i = 1, 2, 3$$

which define a transformation from the coordinates x_i to the coordinates u_i. The differentials dx_i transform as follows:

$$du_i = \sum_{s=1}^{3} \frac{\partial u_i}{\partial x_s}\, dx_s$$

which we may write in matrix form as

$$\begin{bmatrix} du_1 \\ du_2 \\ du_3 \end{bmatrix} = \begin{bmatrix} \dfrac{\partial u_1}{\partial x_1} & \dfrac{\partial u_1}{\partial x_2} & \dfrac{\partial u_1}{\partial x_3} \\ \dfrac{\partial u_2}{\partial x_1} & \dfrac{\partial u_2}{\partial x_2} & \dfrac{\partial u_2}{\partial x_3} \\ \dfrac{\partial u_3}{\partial x_1} & \dfrac{\partial u_3}{\partial x_2} & \dfrac{\partial u_3}{dx_3} \end{bmatrix} \begin{bmatrix} dx_1 \\ dx_2 \\ dx_3 \end{bmatrix} = [T] \begin{bmatrix} dx_1 \\ dx_2 \\ dx_3 \end{bmatrix}$$

where $[T]$ is the transformation matrix. The determinant of the matrix $[T]$ is called the Jacobian and is written as

$$J = |T| = \frac{\partial(u_1, u_2, u_3)}{\partial(x_1, x_2, x_3)}.$$

An expansion of the triple scalar product $\nabla u_1 \cdot \nabla u_2 \times \nabla u_3$ shows that it is equal to J. When $J \neq 0$, we can invert the above equations and solve for the dx_i as linear functions of the du_i, and also solve for the x_i as functions of the u_i; thus

$$\begin{bmatrix} dx_1 \\ dx_2 \\ dx_3 \end{bmatrix} = \begin{bmatrix} \dfrac{\partial x_1}{\partial u_1} & \dfrac{\partial x_1}{\partial u_2} & \dfrac{\partial x_1}{\partial u_3} \\ \dfrac{\partial x_2}{\partial u_1} & \dfrac{\partial x_2}{\partial u_2} & \dfrac{\partial x_2}{\partial u_3} \\ \dfrac{\partial x_3}{\partial u_1} & \dfrac{\partial x_3}{\partial u_2} & \dfrac{\partial x_3}{\partial u_3} \end{bmatrix} \begin{bmatrix} du_1 \\ du_2 \\ du_3 \end{bmatrix} = [T_i] \begin{bmatrix} du_1 \\ du_2 \\ du_3 \end{bmatrix}$$

where $[T_i]$ is the inverse transformation matrix. If we premultiply this equation by $[T]$, we get

$$[T][dx_i] = [du_i] = [T][T_i][du_i]$$

from which we conclude that $[T][T_i] = [I]$, or the unit matrix. From this latter result we get

$$\sum_{s=1}^{3} \frac{\partial u_i}{\partial x_s} \frac{\partial x_s}{\partial u_r} = \delta_{ir}$$

and

$$\sum_{s=1}^{3} \frac{\partial x_i}{\partial u_s} \frac{\partial u_s}{\partial x_r} = \delta_{ir}$$

where $\delta_{ir} = 1$ for $i = r$, and zero otherwise.

Instead of specifying a point by the coordinates x_{i0}, we may equally well specify its location by the trio of numbers $u_{i0} = u_i(x_{10}, x_{20}, x_{30})$. When the u_{i0} are given, the point in question is located at the intersection of the coordinate surfaces $u_i = u_{i0}$. These surfaces intersect along three curves called the coordinate curves. Along any one curve, say $u_2 = u_{20}, u_3 = u_{30}$, only one of the u_i will vary, in this case u_1.

At the point of intersection of the three coordinate curves, we construct a miniature rectangular coordinate frame by means of the unit vectors e_i, directed tangent to the three coordinate curves u_i increasing. We may now specify a vector by giving its components in this new reference frame. For this reason, the u_i are called curvilinear coordinates. When the three coordinate curves intersect at right angles everywhere, we have an orthogonal curvilinear coordinate frame. From this point on, we will assume that our curvilinear coordinates are orthogonal, and also that the u_i and e_i have been labeled so that e_1, e_2, e_3 form a right-handed system. Since ∇u_i is a vector normal to the surface $u_i = u_{i0}$, the condition that the curvilinear coordinates should form an orthogonal system is given by $\nabla u_s \cdot \nabla u_r = 0$ for $r \neq s$.

Line Element, or Fundamental Metric: The differentials du_i are not, in general, measures of arc length dl_i along the coordinate curves. We convert them to arc length by multiplying by suitable scale factors h_i; thus $dl_i = h_i \, du_i$. The line element, or fundamental metric, as it is called, is

$$d\mathbf{l} = \sum_{i=1}^{3} \mathbf{a}_i \, dx_i = \sum_{i=1}^{3} \mathbf{e}_i h_i \, du_i$$

and

$$dl^2 = \sum_{i=1}^{3} dx_i^2 = \sum_{i=1}^{3} h_i^2 \, du_i^2.$$

For an orthogonal coordinate system, all coordinate surfaces intersect normally, and thus ∇u_i, which is normal to the coordinate surface $u_i = u_{i0}$, must be tangent to the u_i coordinate curve, i.e., parallel to $\mathbf{e}_i$. The element of arc length along the u_i coordinate curve is dl_i. The directional derivative of u_i along u_i is, therefore,

$$\frac{du_i}{dl_i} = \nabla u_i \cdot \mathbf{e}_i$$

and hence

$$dl_i = \frac{du_i}{|\nabla u_i|} = h_i \, du_i.$$

The scale factors h_i are thus given by

$$h_i = |\nabla u_i|^{-1} = \left[\sum_{s=1}^{3} \left(\frac{\partial u_i}{\partial x_s} \right)^2 \right]^{-1/2}$$

The element of volume is

$$dV = h_1 h_2 h_3 \, du_1 \, du_2 \, du_3$$
$$= \frac{du_1 \, du_2 \, du_3}{|\nabla u_1||\nabla u_2||\nabla u_3|} = \frac{du_1 \, du_2 \, du_3}{J}$$

because of the mutual orthogonality of the gradients.

When the u_i are given as functions of the x_i, we can readily evaluate the scale factors h_i. It is convenient to have an alternative expression for h_i, which can be easily evaluated when the x_i are given as functions of the u_i, since this will eliminate the necessity of solving for the u_i as functions of the x_i first. The differentials dx_i are given by

$$dx_i = \sum_{s=1}^{3} \frac{\partial x_i}{\partial u_s} \, du_s$$

and

$$dx_i^2 = \sum_{s=1}^{3} \sum_{r=1}^{3} \frac{\partial x_i}{\partial u_s} \frac{\partial x_i}{\partial u_r} \, du_s \, du_r$$

and, hence,

$$dl^2 = \sum_{i=1}^{3} dx_i^2 = \sum_{i=1}^{3} \sum_{s=1}^{3} \sum_{r=1}^{3} \frac{\partial x_i}{\partial u_s} \frac{\partial x_i}{\partial u_r} \, du_s \, du_r$$
$$= \sum_{i=1}^{3} \sum_{s=1}^{3} \left(\frac{\partial x_i}{\partial u_s} \right)^2 du_s^2 = \sum_{s=1}^{3} du_s^2 \sum_{i=1}^{3} \left(\frac{\partial x_i}{\partial u_s} \right)^2$$

since

$$\sum_{i=1}^{3} \sum_{s=1}^{3} \sum_{r=1}^{3} \frac{\partial x_i}{\partial u_s} \frac{\partial x_i}{\partial u_r} \, du_s \, du_r = 0 \qquad \text{for } r \neq s.$$

Therefore

$$h_s^2 = \sum_{i=1}^{3} \left(\frac{\partial x_i}{\partial u_s} \right)^2$$

an expression readily evaluated when the x_i are known functions of the u_i. The proof that the triple summation above equals zero unless $r = s$ is as follows. We know that

$$h_s^{-2} = \sum_{i=1}^{3} \left(\frac{\partial u_s}{\partial x_i} \right)^2$$

and that

$$\sum_{p=1}^{3} \frac{\partial x_k}{\partial u_p} \frac{\partial u_p}{\partial x_j} = \delta_{kj}$$

from the properties of the transformation matrices $[T]$ and $[T_i]$. Multiply this latter result by $\partial u_s / \partial x_j$, and sum over j to get

$$\sum_{p=1}^{3} \frac{\partial x_k}{\partial u_p} \sum_{j=1}^{3} \frac{\partial u_p}{\partial x_j} \frac{\partial u_s}{\partial x_j} = \sum_{p=1}^{3} \frac{\partial x_k}{\partial u_p} \nabla u_p \cdot \nabla u_s$$

$$= \frac{1}{h_s^2} \frac{\partial x_k}{\partial u_s} = \sum_{j=1}^{3} \delta_{kj} \frac{\partial u_s}{\partial x_j} = \frac{\partial u_s}{\partial x_k}.$$

We now use the result $h_s^2 \, \partial u_s / \partial x_k = \partial x_k / \partial u_s$ in the triple summation to get

$$\sum_{s=1}^{3} \sum_{r=1}^{3} du_s \, du_r \sum_{i=1}^{3} h_s^2 h_r^2 \frac{\partial u_s}{\partial x_i} \frac{\partial u_r}{\partial x_i} = \sum_{s=1}^{3} h_s^4 \, du_s^2 |\nabla u_s|^2$$

since $\nabla u_s \cdot \nabla u_r = 0$ unless $s = r$. When $s \neq r$, the triple summation vanishes, and this completes the proof.

In a cylindrical coordinate frame $r\theta z$, the line elements are given by

$$h_1 = 1 \quad h_2 = r \quad h_3 = 1.$$

In a spherical coordinate system $r\theta\phi$, where θ is the polar angle, the line elements are

$$h_1 = 1 \quad h_2 = r \quad h_3 = r \sin \theta.$$

Gradient in Curvilinear Coordinates: The total differential of a scalar function $\Phi(u_1, u_2, u_3)$ is

$$d\Phi = \sum_{i=1}^{3} \frac{\partial \Phi}{\partial u_i} du_i.$$

Since du_i is a function of the x_s, we have

$$du_i = \sum_{s=1}^{3} \frac{\partial u_i}{\partial x_s} dx_s = \nabla u_i \cdot d\mathbf{l}$$

where $d\mathbf{l} = \mathbf{a}_1\, dx_1 + \mathbf{a}_2\, dx_2 + \mathbf{a}_3\, dx_3$. Hence, we get

$$d\Phi = \sum_{i=1}^{3} \frac{\partial \Phi}{\partial u_i}\, \nabla u_i \cdot d\mathbf{l}$$

and

$$\frac{d\Phi}{dl} = \left(\sum_{i=1}^{3} \frac{\partial \Phi}{\partial u_i}\, \nabla u_i \right) \cdot \frac{d\mathbf{l}}{dl}.$$

The unit tangent along the curve is $d\mathbf{l}/dl = \boldsymbol{\tau}$, and, since the directional derivative of Φ is $d\Phi/dl = \nabla\Phi\cdot\boldsymbol{\tau}$, we have the result that

$$\nabla\Phi = \sum_{i=1}^{3} \frac{\partial \Phi}{\partial u_i}\, \nabla u_i$$

where ∇u_i may be referred to either the x_i or the u_i coordinate frame. Now ∇u_i has the direction of $\mathbf{e}_i$ and a magnitude of h_i^{-1}, so

$$\nabla\Phi = \sum_{i=1}^{3} \frac{1}{h_i} \frac{\partial \Phi}{\partial u_i}\, \mathbf{e}_i.$$

Components of a Vector in u_i Coordinates: The differential du_i is equal to

$$\nabla u_i \cdot \sum_{s=1}^{3} \mathbf{a}_s\, dx_s.$$

When only one of the dx_s is different from zero, we have a partial differential

$$\partial u_i = \nabla u_i \cdot \mathbf{a}_s\, \partial x_s$$

and this is the change in u_i when x_s changes by ∂x_s. The change in u_i is along the u_i coordinate curve; hence,

$$\partial u_i = |\nabla u_i|\partial x_s\, \cos\theta$$

or

$$\cos\theta = |\nabla u_i|^{-1} \frac{\partial u_i}{\partial x_s} = h_i \frac{\partial u_i}{\partial x_s}$$

where $\cos\theta$ is the direction cosine between the x_s axis and the u_i coordinate curve. In proving that the triple summation, in the discussion of the line elements in the previous section, vanished for $r \neq s$, we obtained as an intermediate step the result

$$\frac{1}{h_s^2} \frac{\partial x_k}{\partial u_s} = \frac{\partial u_s}{\partial x_k}.$$

Therefore, we have the following two equivalent expressions for the direction cosine between the unit vectors $\mathbf{a}_s$ and $\mathbf{e}_i$:

$$\cos\theta = h_i \frac{\partial u_i}{\partial x_s} = \frac{1}{h_i} \frac{\partial x_s}{\partial u_i}.$$

From the above results we may readily construct Table A.1 of direction cosines between the unit vectors $\mathbf{a}_i$ and $\mathbf{e}_s$.

TABLE A.1. TABLE OF DIRECTION COSINES

	$\mathbf{e}_1$	$\mathbf{e}_2$	$\mathbf{e}_3$
$\mathbf{a}_1$	$h_1\dfrac{\partial u_1}{\partial x_1}=\dfrac{1}{h_1}\dfrac{\partial x_1}{\partial u_1}$	$h_2\dfrac{\partial u_2}{\partial x_1}=\dfrac{1}{h_2}\dfrac{\partial x_1}{\partial u_2}$	$h_3\dfrac{\partial u_3}{\partial x_1}=\dfrac{1}{h_3}\dfrac{\partial x_1}{\partial u_3}$
$\mathbf{a}_2$	$h_1\dfrac{\partial u_1}{\partial x_2}=\dfrac{1}{h_1}\dfrac{\partial x_2}{\partial u_1}$	$h_2\dfrac{\partial u_2}{\partial x_2}=\dfrac{1}{h_2}\dfrac{\partial x_2}{\partial u_2}$	$h_3\dfrac{\partial u_3}{\partial x_2}=\dfrac{1}{h_3}\dfrac{\partial x_2}{\partial u_3}$
$\mathbf{a}_3$	$h_1\dfrac{\partial u_1}{\partial x_3}=\dfrac{1}{h_1}\dfrac{\partial x_3}{\partial u_1}$	$h_2\dfrac{\partial u_2}{\partial x_3}=\dfrac{1}{h_2}\dfrac{\partial x_3}{\partial u_2}$	$h_3\dfrac{\partial u_3}{\partial x_3}=\dfrac{1}{h_3}\dfrac{\partial x_3}{\partial u_3}$

In our curvilinear coordinate frame we can now express a vector $\mathbf{A}$ with rectangular components A_i as follows:

$$\mathbf{A} = \sum_{s=1}^{3}\left(A_1\frac{1}{h_s}\frac{\partial x_1}{\partial u_s}+A_2\frac{1}{h_s}\frac{\partial x_2}{\partial u_s}+A_3\frac{1}{h_s}\frac{\partial x_3}{\partial u_s}\right)\mathbf{e}_s$$

$$= \sum_{s=1}^{3}\left(A_1 h_s\frac{\partial u_s}{\partial x_1}+A_2 h_s\frac{\partial u_s}{\partial x_2}+A_3 h_s\frac{\partial u_s}{\partial x_3}\right)\mathbf{e}_s$$

$$= A_{u_1}\mathbf{e}_1 + A_{u_2}\mathbf{e}_2 + A_{u_3}\mathbf{e}_3.$$

These equations may be written in matrix form as

$$\begin{bmatrix}A_{u_1}\\A_{u_2}\\A_{u_3}\end{bmatrix}=\begin{bmatrix}\dfrac{1}{h_1}\dfrac{\partial x_1}{\partial u_1} & \dfrac{1}{h_1}\dfrac{\partial x_2}{\partial u_1} & \cdots \\ \dfrac{1}{h_2}\dfrac{\partial x_1}{\partial u_2} & \cdots & \cdots \\ & & \end{bmatrix}\begin{bmatrix}A_1\\A_2\\A_3\end{bmatrix}$$

$$=\begin{bmatrix}\dfrac{1}{h_1} & 0 & 0 \\ 0 & \dfrac{1}{h_2} & 0 \\ 0 & 0 & \dfrac{1}{h_3}\end{bmatrix}[T^*]\begin{bmatrix}A_1\\A_2\\A_3\end{bmatrix}$$

or as

$$\begin{bmatrix}A_{u_1}\\A_{u_2}\\A_{u_3}\end{bmatrix}=\begin{bmatrix}h_1\dfrac{\partial u_1}{\partial x_1} & h_1\dfrac{\partial u_1}{\partial x_2} & \cdots \\ h_2\dfrac{\partial u_2}{\partial x_1} & \cdots & \cdots \\ & & \end{bmatrix}\begin{bmatrix}A_1\\A_2\\A_3\end{bmatrix}$$

$$=\begin{bmatrix}h_1 & 0 & 0 \\ 0 & h_2 & 0 \\ 0 & 0 & h_3\end{bmatrix}[T]\begin{bmatrix}A_1\\A_2\\A_3\end{bmatrix}.$$

There are two expressions for the direction cosines and, consequently, two distinct but equivalent transformation laws for the vector components A_i. These transformation laws can be simplified so as to eliminate the diagonal matrices by redefining our vector components. There are two possibilities; for one we choose

$$\mathcal{A}_{ui}^{*} = h_i A_{u_i}.$$

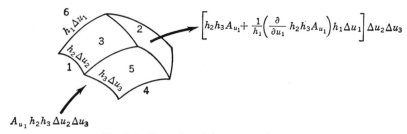

$$\left[h_2 h_3 A_{u_1} + \frac{1}{h_1}\left(\frac{\partial}{\partial u_1} h_2 h_3 A_{u_1}\right) h_1 \Delta u_1 \right] \Delta u_2 \Delta u_3$$

$$A_{u_1} h_2 h_3 \,\Delta u_2 \Delta u_3$$

Fig. A.1. Illustration of divergence of a vector.

We call the vector $\mathcal{C}^*$ the covariant vector corresponding to our real vector $\mathbf{A}$. The components $h_i A_{u_i}$ are its covariant components. The components of $\mathcal{C}^*$ transform according to the law

$$[\mathcal{C}^*_{u_i}] = [T^*][A_i].$$

We regard the matrix $[T^*]$ as our standard matrix, and the transformation effected by it as our standard transformation. For this reason $\mathcal{C}^*$ is called a covariant vector, the term covariant signifying having the same variation, i.e., having the same transformation law, as our standard.

The contravariant vector corresponding to the real vector $\mathbf{A}$ is defined as the vector $\mathcal{C}$ with contravariant components $h_i^{-1} A_{u_i}$. The contravariant components transform as follows upon change of coordinates:

$$[\mathcal{C}_{u_i}] = [T][A_i].$$

The term contravariant signifies variation or transformation unlike or different from our standard transformation.

The advantages of the above definitions are:

1. The transformation matrices are simpler.
2. The transformation laws are symmetrical in the x_i and u_i coordinates.
3. If the x_i are given as functions of u_i or vice versa, we can transform the covariant or contravariant components, respectively, in a straightforward manner, since we can readily evaluate the elements in the transformation matrix. After transforming, we convert back to our real vector by multiplying the covariant and contravariant components by h_i^{-1} and h_i, respectively.

If the x_i form an orthogonal curvilinear coordinate system as well, the above transformations are still valid for the covariant and contravariant components. The real vector in the x_i system is then obtained by multiplying by suitable scale factors. As an example, we have the following transformation for the differentials dx_i, $[du_i] = [T][dx_i]$, and, hence, the dx_i transform as the components of a contravariant vector. To convert to our real vector $d\mathbf{l} = \sum_{i=1}^{3} \mathbf{a}_i \, dx_i$, we multiply the components by h_i to get $d\mathbf{l} = \sum_{i=1}^{3} h_i \, du_i \mathbf{e}_i$, which is the line element in the u_i coordinate frame. As a second example, we consider the transformation of the gradient of a scalar function Φ. The gradient has components $\partial\Phi/\partial x_i$, and, if we regard these as the covariant components, then the covariant components in the u_i coordinate frame are given by

$$\left[\frac{\partial\Phi}{\partial u_i}\right] = [T^*]\left[\frac{\partial\Phi}{\partial x_i}\right].$$

The transformation law for the $\partial\Phi/\partial x_i$ is just the familiar rule for partial differentiation, and is one reason why we regard $[T^*]$ as our standard transformation matrix. The components of the real vector corresponding to the components of its covariant counterpart are given by $h_i^{-1} \partial\Phi/\partial u_i$, and these are the components of the gradient of Φ in the u_i coordinate frame.

Divergence and Curl in Orthogonal Curvilinear Coordinates: The divergence of a vector $\mathbf{A}$ in u_i coordinates is readily evaluated from the definition given in subsection b. Consider a small-volume element $\Delta u_1 \Delta u_2 \Delta u_3 / J$, as illustrated in Fig. A.1. The normal component of $\mathbf{A}$ at surface 1 is $A_{u_1} \mathbf{e}_1$, and the flux passing out through this surface is $- A_{u_1} h_2 h_3 \Delta u_2 \Delta u_3$. The flux through surface 2 is $A_{u_1} h_2 h_3 \Delta u_2 \Delta u_3 + [(1/h_1)(\partial/\partial u_1)(h_3 h_2 A_{u_1})] h_1 \Delta u_1 \Delta u_2 \Delta u_3$, and so the net outward flux from surfaces 1 and 2 is $(\partial/\partial u_1)(h_2 h_3 A_{u_1})\,\Delta u_1\,\Delta u_2\,\Delta u_3$. A cyclic interchange of subscripts gives the expressions for the net flux through surfaces 3 and 4 and that through surfaces 5 and 6. Combining all the terms gives

$$\nabla\cdot\mathbf{A} = \lim_{\Delta V\to 0} \frac{\displaystyle\oiint_{S} \mathbf{A}\cdot d\mathbf{S}}{\Delta V}$$

$$= \frac{1}{h_1 h_2 h_3}\left[\frac{\partial}{\partial u_1}(h_2 h_3 A_{u_1}) + \frac{\partial}{\partial u_2}(h_1 h_3 A_{u2}) + \frac{\partial}{\partial u_3}(h_1 h_2 A_{u3})\right].$$

The components of the curl of $\mathbf{A}$ may be evaluated in a similar fashion and, when expressed in determinant form, are given by

$$\nabla\times\mathbf{A} = \frac{1}{h_1 h_2 h_3}\begin{vmatrix} h_1\mathbf{e}_1 & h_2\mathbf{e}_2 & h_3\mathbf{e}_3 \\ \dfrac{\partial}{\partial u_1} & \dfrac{\partial}{\partial u_2} & \dfrac{\partial}{\partial u_3} \\ h_1 A_{u_1} & h_2 A_{u_2} & h_3 A_{u_3} \end{vmatrix}.$$

The Laplacian of a function is the divergence of the gradient and is written as $\nabla\cdot\nabla\Phi = \nabla^2\Phi$. In rectangular coordinates it is equal to

$$\nabla^2\Phi = \sum_{i=1}^{3}\frac{\partial^2\Phi}{\partial x_i^2}.$$

The form for the Laplacian in curvilinear coordinates is obtained by replacing the components A_{u_i} in the expression for $\nabla\cdot\mathbf{A}$ by the components of $\nabla\Phi$, that is, by $(1/h_i)(\partial\Phi/\partial u_i)$.

e. Properties of Scalar and Vector Fields

In the present subsection we shall derive some of the more basic properties common to all scalar and vector fields. Our frame of reference will be an xyz rectangular coordinate frame, but the results will hold in any coordinate frame.

Scalar Fields: A scalar function $\Phi(x, y, z)$ with continuous first-order partial derivatives defined at almost all points in a given space represents a scalar field. The points where Φ or its derivatives are discontinuous are the sources or sinks of the field. The gradient of Φ is a vector which measures the maximum rate of change of the field in space. The line integral of $\nabla\Phi$ between two points p_1 and p_2 connected by a curve C is $\int_{p_1}^{p_2}\nabla\Phi\cdot d\mathbf{l}$. The component of $\nabla\Phi$ along $d\mathbf{l}$ is just the directional derivative of Φ along the curve C, and, hence,

$$\int_{p_1}^{p_2}\nabla\Phi\cdot d\mathbf{l} = \int_{p_1}^{p_2}\frac{d\Phi}{dl}\,dl = \Phi(p_2) - \Phi(p_1)$$

independently of the path C followed. For Φ to represent a physical field, it must be single-valued, and, hence, the line integral around a closed contour vanishes; i.e.,

$$\oint_{C}\nabla\Phi\cdot d\mathbf{l} = 0.$$

Because of the above line-integral property, the scalar field Φ is called a conservative potential field. The constant-potential surfaces are the surfaces $\Phi = $ constant. The vector $\nabla\Phi$ is tangent to the system of lines which intersect the constant-potential surfaces orthogonally. When $\nabla\Phi$ has zero divergence, Φ is a solution of Laplace's equation $\nabla^2\Phi = 0$. If we let $u = v = \Phi$ in Green's first identity, we get

$$\iiint_{V}(\nabla\Phi)^2\,dV = \oiint_{S}\Phi\frac{\partial\Phi}{\partial n}\,dS$$

when $\nabla^2\Phi = 0$ in V. If Φ is constant on S, the surface integral vanishes, and, hence, $\nabla\Phi = 0$ also. Thus Φ is constant throughout V also. When Φ vanishes on S, it vanishes everywhere within V also, unless it has a singularity or a source inside S.

Vector Fields: A vector function $\mathbf{P}(x, y, z)$ represents a vector field. The points at which $\mathbf{P}$ or its derivatives are discontinuous correspond to the sources or sinks of the field or to idealized physical boundaries. When $\nabla\cdot\mathbf{P} = 0$, we say that $\mathbf{P}$ is a solenoidal or rotational vector field. Since $\nabla\cdot\nabla \times \mathbf{A} = 0$, we may derive $\mathbf{P}$ from the curl of a suitable vector potential function $\mathbf{A}$. All the lines of force in a divergenceless field close upon themselves; i.e., nowhere do they terminate in a source or sink. The relation $\mathbf{P} = \nabla \times \mathbf{A}$ determines $\mathbf{A}$ only to within an additive arbitrary vector $\nabla\Phi$, which is the gradient of a scalar function, since $\nabla \times \nabla\Phi = 0$. Thus, we may equally well derive $\mathbf{P}$ from the curl of a vector $\mathbf{A}'$, where $\mathbf{A}' = \mathbf{A} + \nabla\Phi$. The transformation to the new vector potential $\mathbf{A}'$ is called a gauge transformation.

When $\nabla \times \mathbf{P} = 0$, we call $\mathbf{P}$ an irrotational or lamellar field. From Stokes' theorem we have

$$\oint_C \mathbf{P}\cdot d\mathbf{l} = \iint_S \nabla \times \mathbf{P}\cdot d\mathbf{S} = 0$$

and, hence, we may derive $\mathbf{P}$ from the gradient of a scalar function as follows:

$$\mathbf{P} = -\nabla\psi$$

where ψ is called a scalar potential. It may readily be shown that, given a vector function $\mathbf{P}$ which is continuous with continuous derivatives in a region V and on its boundary S, any one of the following statements implies the rest:

1. The line integral $\int_1^2 \mathbf{P}\cdot d\mathbf{l}$ is independent of the path followed in V.
2. $\oint_C \mathbf{P}\cdot d\mathbf{l} = 0$ for every closed contour in V.
3. ψ exists such that $\mathbf{P} = -\nabla\psi$.
4. $\nabla \times \mathbf{P} = 0$ throughout V.

Helmholtz's theorem states that a general vector field will have both a solenoidal part and a lamellar part and may be derived from a vector and a scalar potential. Let the solenoidal part be $\mathbf{P}_s$ and the lamellar part be $\mathbf{P}_l$. The divergence of $\mathbf{P}$ does not vanish, and, hence, we put

$$\nabla\cdot\mathbf{P} = \nabla\cdot\mathbf{P}_l = \rho$$

where ρ is a scalar function constituting the source for $\mathbf{P}_l$. The source for $\mathbf{P}_s$ is $\mathbf{J}$ and gives the curl of $\mathbf{P}_s$, that is,

$$\nabla \times \mathbf{P} = \nabla \times \mathbf{P}_s = \mathbf{J}$$

where $\mathbf{J}$ is a vector source function. Since $\nabla\cdot\mathbf{P}_s = 0$, we may put $\mathbf{P}_s$ equal to the curl of a vector potential as follows:

$$\mathbf{P}_s = \nabla \times \mathbf{A}.$$

The lamellar part $\mathbf{P}_l$ may be put equal to $-\nabla\psi$, where ψ is a scalar potential. Substituting for $\mathbf{P}_s$ and $\mathbf{P}_l$ into the source equations gives

$$\nabla \times \nabla \times \mathbf{A} = \nabla\nabla\cdot\mathbf{A} - \nabla^2\mathbf{A} = \mathbf{J}$$

$$\nabla\cdot\nabla\psi = \nabla^2\psi = -\rho.$$

By a suitable gauge transformation, the divergence of $\mathbf{A}$ can be made zero, and so we may take, in general,

$$\nabla^2\mathbf{A} = -\mathbf{J}.$$

When the sources $\mathbf{J}$ and ρ are interrelated and are functions of time as well, the equations satisfied by the potentials are not as simple as those given above. A familiar example is the electromagnetic field.

The function $(4\pi R)^{-1}$, where $R^2 = (x - x_0)^2 + (y - y_0)^2 + (z - z_0)^2$, is a solution of Poisson's equation for a unit source located at (x_0, y_0, z_0), that is,

$$\nabla^2 \frac{1}{R} = -4\pi\delta(x - x_0)\delta(y - y_0)\delta(z - z_0)$$

where the product of the delta functions is used to represent a unit source. The delta function $\delta(x - x_0)$ is defined to be equal to zero at all points except $x = x_0$, where it becomes infinite in such a manner that

$$\int_{x-\alpha}^{x+\alpha} \delta(x - x_0)\, dx_0 = 1$$

where α is a small positive constant. If $\mathbf{B}(x_0)$ is an arbitrary vector function which is continuous at $x_0 = x$, we get the result

$$\int_{x_1}^{x_2} \mathbf{B}(x_0)\delta(x - x_0)\, dx_0 = \begin{cases} 0, & x \text{ not in the interval } x_1 - x_2 \\ \mathbf{B}(x), & x \text{ in the interval } x_1 - x_2. \end{cases}$$

A similar result holds for the product of the three delta functions $\delta(x - x_0)\delta(y - y_0)\delta(z - z_0)$. If the interval of integration over a given volume contains the point (x, y, z), then the integral of the product of an arbitrary continuous function times the three-dimensional delta function gives the value of the function at the point (x, y, z), that is,

$$\iiint_V \mathbf{B}(x_0, y_0, z_0)\delta(x - x_0)\delta(y - y_0)(z - z_0)\, dV_0 = \begin{cases} 0, & (x, y, z) \text{ not in } V \\ \mathbf{B}(x, y, z), & (x, y, z) \text{ in } V. \end{cases}$$

The exact analytical form of the delta function does not concern us since we will use only the above properties. It may be noted that the function $-\nabla^2(1/4\pi R)$ has the above properties. This function is symmetrical in the variables x, y, z and x_0, y_0, z_0, so that

$$\nabla_0^2 \frac{1}{R} = \nabla^2 \frac{1}{R}$$

where ∇_0 signifies differentiation with respect to the variables x_0, y_0, z_0.

From this property of the function $(4\pi R)^{-1}$ we may obtain the solutions for the potentials $\mathbf{A}$ and ψ by superposition. The results are

$$\mathbf{A} = \frac{1}{4\pi} \iiint_V \frac{\mathbf{J}(x_0, y_0, z_0)}{R}\, dV_0$$

$$\psi = \frac{1}{4\pi} \iiint_V \frac{\rho(x_0, \ldots)}{R}\, dV_0$$

which can be readily demonstrated by operating on $\mathbf{A}$ and ψ by ∇^2 and using the delta-function property of $-\nabla^2(1/4\pi R)$.

We will now examine the properties of our general vector field $\mathbf{P} = \mathbf{P}_s + \mathbf{P}_l$ in the interior of a volume V and on the surface S which completely surrounds V. Using the delta-function property of $(-1/4\pi)\nabla^2(1/R)$ permits us to write

$$\mathbf{P}(x, y, z) = -\frac{1}{4\pi} \nabla^2 \iiint_V \frac{P(x_0, y_0, z_0)}{R}\, dV_0.$$

The ∇^2 operator is equal to $\nabla\nabla\cdot\, - \nabla\times\nabla\times$, and so we obtain the result

$$\mathbf{P}(x,y,z) = -\frac{1}{4\pi}\nabla\nabla\cdot\iiint\limits_V \frac{\mathbf{P}(x_0,y_0,z_0)}{R}\,dV_0 + \nabla\times\nabla\times\frac{1}{4\pi}\iiint\limits_V \frac{\mathbf{P}(x_0,\ldots)}{R}\,dV_0$$

which will enable us to express $\mathbf{P}$ as the negative gradient of a scalar function plus the curl of a vector function.
  Consider the integral

$$\nabla\times\iiint\limits_V \frac{\mathbf{P}(x_0,\ldots)}{R}\,dV_0 = \iiint\limits_V \nabla\times\frac{\mathbf{P}(x_0,\ldots)}{R}\,dV_0.$$

Now

$$\nabla\times\frac{\mathbf{P}(x_0,\ldots)}{R} = \mathbf{P}(x_0,\ldots)\times\nabla_0\frac{1}{R} = -\nabla_0\times\frac{\mathbf{P}_0(x_0,\ldots)}{R} + \frac{1}{R}\nabla_0\times\mathbf{P}(x_0,\ldots).$$

The above integral becomes

$$\iiint\limits_V \frac{\nabla_0\times\mathbf{P}(x_0,\ldots)}{R}\,dV_0 - \iiint\limits_V \nabla_0\times\frac{\mathbf{P}}{R}\,dV_0 = \iiint\limits_V \frac{\nabla_0\times\mathbf{P}}{R}\,dV_0 + \oiint\limits_S \frac{\mathbf{P}\times\mathbf{n}}{R}\,dV_0.$$

The reduction of the second volume integral to a surface integral is readily accomplished by applying the divergence theorem to the vector function, $\mathbf{a}\times\mathbf{P}/R$, where $\mathbf{a}$ is an arbitrary constant vector. We get

$$\iiint\limits_V \left(\nabla_0\cdot\mathbf{a}\times\frac{\mathbf{P}}{R}\right)dV_0 = -\mathbf{a}\cdot\iiint\limits_V \nabla_0\times\frac{\mathbf{P}}{R}\,dV_0 = \oiint\limits_S \left(\mathbf{a}\times\frac{\mathbf{P}}{R}\right)\cdot\mathbf{n}\,dS_0 = \mathbf{a}\cdot\oiint\limits_S \frac{\mathbf{P}\times\mathbf{n}}{R}\,dS_0$$

and, since $\mathbf{a}$ is arbitrary, the required result follows at once by equating the two integrals that are dot-multiplied by $\mathbf{a}$.
  Next we consider the integral

$$\nabla\cdot\iiint\limits_V \frac{\mathbf{P}(x_0,\ldots)}{R}\,dV_0 = \iiint\limits_V \nabla\cdot\frac{\mathbf{P}(x_0,\ldots)}{R}\,dV_0$$

$$= -\iiint\limits_V \nabla_0\cdot\frac{\mathbf{P}}{R}\,dV_0 + \iiint\limits_V \frac{\nabla_0\cdot\mathbf{P}}{R}\,dV_0$$

$$= \iiint\limits_V \frac{\nabla_0\cdot\mathbf{P}}{R}\,dV_0 - \oiint\limits_S \frac{\mathbf{P}\cdot\mathbf{n}}{R}\,dS_0$$

since

$$\nabla\cdot\frac{\mathbf{P}}{R} = \mathbf{P}\cdot\nabla\frac{1}{R} = -\mathbf{P}\cdot\nabla_0\frac{1}{R} = \frac{1}{R}\nabla_0\cdot\mathbf{P} - \nabla_0\cdot\frac{\mathbf{P}}{R}.$$

Our results collected together show that the vector field $\mathbf{P}$ is given by the expression

$$\mathbf{P} = \nabla\times\left(\iiint\limits_V \frac{\nabla_0\times\mathbf{P}}{4\pi R}\,dV_0 + \oiint\limits_S \frac{\mathbf{P}\times\mathbf{n}}{4\pi R}\,dS_0\right) - \nabla\left(\iiint\limits_V \frac{\nabla_0\cdot\mathbf{P}}{4\pi R}\,dV_0 - \oiint\limits_S \frac{\mathbf{P}\cdot\mathbf{n}}{4\pi R}\,dS_0\right)$$

which is the mathematical statement of Helmholtz's theorem. We note that only if both $\nabla\times\mathbf{P}$ and $\mathbf{P}\times\mathbf{n}$ are equal to zero will the field in V be a pure lamellar field. Similarly, only if both $\nabla\cdot\mathbf{P}$ and $\mathbf{P}\cdot\mathbf{n}$ vanish will the field be a pure solenoidal field. This is not a surprising situation since, if the field outside S is to be zero, then either $\mathbf{P}\times\mathbf{n}$

and $\mathbf{P} \cdot \mathbf{n}$ vanish on S or else a surface distribution of sources must be placed on S to account for the discontinuous change in $\mathbf{P}$ as the surface S is traversed from the interior to the exterior of V. Thus, even though the volume sources $\mathbf{J} = \nabla \times \mathbf{P}$ and $\rho = \nabla \cdot \mathbf{P}$ vanish, the surface sources may not, and the field $\mathbf{P}$ in the interior of V will have both a lamellar and a solenoidal part. These results are of fundamental importance in the theory of electromagnetic cavities or resonators.

A.2. DYADIC ANALYSIS

When each component of a vector $\mathbf{A}$ is linearly related to all three components of a vector $\mathbf{B}$ as follows:

$$A_i = \sum_{j=1}^{3} C_{ij} B_j, \qquad i, j = x, y, z$$

then the relationship is conveniently represented by a matrix equation or by introducing the dyadic that corresponds to the matrix with elements C_{ij}. The dyadic is obtained by associating with each element C_{ij} the unit vectors $\mathbf{a}_i$, $\mathbf{a}_j$. The dyadic will be represented by boldface type with a bar above the quantity. The dyadic $\bar{\mathbf{C}}$ is given by

$$
\begin{aligned}
& \quad\; \mathbf{a}_x \mathbf{a}_x C_{xx} + \mathbf{a}_x \mathbf{a}_y C_{xy} + \mathbf{a}_x \mathbf{a}_z C_{xz} \\
\bar{\mathbf{C}} = {} & + \mathbf{a}_y \mathbf{a}_x C_{yx} + \mathbf{a}_y \mathbf{a}_y C_{yy} + \mathbf{a}_y \mathbf{a}_z C_{yz} \\
& + \mathbf{a}_z \mathbf{a}_x C_{zx} + \mathbf{a}_z \mathbf{a}_y C_{zy} + \mathbf{a}_z \mathbf{a}_z C_{zz}.
\end{aligned}
$$

The algebra of dyadics has a very close correspondence with matrix algebra. The dot product of two dyadics $\bar{\mathbf{C}}$ and $\bar{\mathbf{D}}$ is written $\bar{\mathbf{C}} \cdot \bar{\mathbf{D}}$, and is defined as the dyadic obtained by dotting the unit vectors that appear on either side of and adjacent to the dot. The result is seen to be the same as the matrix product of the two matrices that correspond to the dyadics $\bar{\mathbf{C}}$ and $\bar{\mathbf{D}}$. The product, in general, is not commutative, so that $\bar{\mathbf{C}} \cdot \bar{\mathbf{D}} \neq \bar{\mathbf{D}} \cdot \bar{\mathbf{C}}$ unless $\bar{\mathbf{C}}$, $\bar{\mathbf{D}}$ and $\bar{\mathbf{C}} \cdot \bar{\mathbf{D}}$ are symmetric. A dyadic is symmetric if its associated matrix is symmetric. The dot product of a dyadic with a vector is defined in a similar way. Thus, we have $\mathbf{A} = \bar{\mathbf{C}} \cdot \mathbf{B}$ for the system of equations introduced earlier.

The sum or difference of two dyadics is the dyadic with components that are the sum or difference of the components of the individual dyadics. Multiplication by a constant multiplies each component by this constant. Corresponding to the unit matrix is a unit dyadic called the idemfactor. The idemfactor is given by

$$\bar{\mathbf{I}} = \mathbf{a}_x \mathbf{a}_x + \mathbf{a}_y \mathbf{a}_y + \mathbf{a}_z \mathbf{a}_z$$

and multiplication of a vector or dyadic by $\bar{\mathbf{I}}$ leaves the vector or dyadic unchanged. The dyadic corresponding to the transposed matrix is called the conjugate dyadic and will be represented as $\bar{\mathbf{C}}_t$, where the subscript t indicates the transposed dyadic. The following results follow from the corresponding property for matrix multiplication:

$$\bar{\mathbf{C}} \cdot \bar{\mathbf{D}} = (\bar{\mathbf{D}}_t \cdot \bar{\mathbf{C}}_t)_t$$

$$\bar{\mathbf{C}} \cdot \mathbf{B} = \mathbf{B} \cdot \bar{\mathbf{C}}_t$$

$$\mathbf{A} \cdot (\bar{\mathbf{C}} \cdot \mathbf{B}) = (\mathbf{A} \cdot \bar{\mathbf{C}}) \cdot \mathbf{B}.$$

The vector, or cross, product of two dyadics is formed by crossing the unit vectors that appear on adjacent sides of the cross in the expression $\bar{\mathbf{C}} \times \bar{\mathbf{D}}$. A similar definition applies to the cross product between a vector and a dyadic. The following useful relation is also readily proved:

$$(\mathbf{A} \cdot \bar{\mathbf{C}}) \times \mathbf{B} = \mathbf{A} \cdot (\bar{\mathbf{C}} \times \mathbf{B}).$$

The dyadic $\bar{\mathbf{C}}$ may be written as a sum of three suitably defined vectors associated with the three unit vectors $\mathbf{a}_x$, $\mathbf{a}_y$, $\mathbf{a}_z$ as follows:

$$\bar{\mathbf{C}} = \mathbf{C}_1 \mathbf{a}_x + \mathbf{C}_2 \mathbf{a}_y + \mathbf{C}_3 \mathbf{a}_z$$

where the associated vectors are given by

$$\mathbf{C}_1 = C_{xx}\mathbf{a}_x + C_{yx}\mathbf{a}_y + C_{zx}\mathbf{a}_z$$

$$\mathbf{C}_2 = C_{xy}\mathbf{a}_x + C_{yy}\mathbf{a}_y + C_{zy}\mathbf{a}_z$$

$$\mathbf{C}_3 = C_{xz}\mathbf{a}_x + C_{yz}\mathbf{a}_y + C_{zz}\mathbf{a}_z.$$

The divergence of the dyadic $\bar{\mathbf{C}}$ is readily seen to be

$$\nabla \cdot \bar{\mathbf{C}} = (\nabla \cdot \mathbf{C}_1)\mathbf{a}_x + (\nabla \cdot \mathbf{C}_2)\mathbf{a}_y + (\nabla \cdot \mathbf{C}_3)\mathbf{a}_z$$

and is a vector. The curl of the dyadic $\bar{\mathbf{C}}$ is given by

$$\nabla \times \bar{\mathbf{C}} = (\nabla \times \mathbf{C}_1)\mathbf{a}_x + (\nabla \times \mathbf{C}_2)\mathbf{a}_y + (\nabla \times \mathbf{C}_3)\mathbf{a}_z.$$

Using the concept of associated vectors allows us to readily derive the equivalent of the divergence theorem and Stokes' theorem for dyadics. The results are

$$\iiint_V \nabla \cdot \bar{\mathbf{C}} \, dV = \oiint_S d\mathbf{S} \cdot \bar{\mathbf{C}} = \oiint_S \mathbf{n} \cdot \bar{\mathbf{C}} \, dS$$

$$\iint_S d\mathbf{S} \cdot \nabla \times \bar{\mathbf{C}} = \oint_C d\mathbf{l} \cdot \bar{\mathbf{C}}.$$

A dyadic may be referred to another coordinate system by finding the projection of the unit vectors $\mathbf{a}_x$, $\mathbf{a}_y$, $\mathbf{a}_z$ on the unit vectors in the second frame of reference. Of particular importance is a transformation to a coordinate frame in which a dyadic $\bar{\mathbf{C}}$ will have the simple diagonal form

$$\bar{\mathbf{C}} = C_{11}\mathbf{e}_1\mathbf{e}_1 + C_{22}\mathbf{e}_2\mathbf{e}_2 + C_{33}\mathbf{e}_3\mathbf{e}_3$$

where $\mathbf{e}_1$, $\mathbf{e}_2$, $\mathbf{e}_3$ are unit vectors along the preferred coordinate curves. This is referred to as a transformation to principal axes or normal coordinates. Let $\mathbf{n}$ be a unit vector along one of the principal axes, i.e., along one of the directions defined by $\mathbf{e}_1$, $\mathbf{e}_2$, or $\mathbf{e}_3$. The dot product of $\mathbf{n}$ with $\bar{\mathbf{C}}$ must then yield simply a vector in the direction of $\mathbf{n}$, say $\lambda\mathbf{n}$. Hence, we have

$$\bar{\mathbf{C}} \cdot \mathbf{n} = \lambda\mathbf{n} = \lambda\bar{\mathbf{I}} \cdot \mathbf{n}$$

or

$$(\bar{\mathbf{C}} - \lambda\bar{\mathbf{I}}) \cdot \mathbf{n} = 0.$$

This is a system of three homogeneous equations for which a solution for n_x, n_y, n_z can exist only if the determinant of the unknown variables vanishes. Therefore, we must have

$$\begin{vmatrix} C_{xx} - \lambda & C_{xy} & C_{xz} \\ C_{yx} & C_{yy} - \lambda & C_{yz} \\ C_{zx} & C_{zy} & C_{zz} - \lambda \end{vmatrix} = 0.$$

This equation is often called a secular equation. It is a cubic equation in λ, the solutions for λ being called the eigenvalues or latent roots. For each root of λ, a solution for n_x, n_y, n_z can be found, and this solution is called an eigenvector or modal column. When the dyadic $\bar{\mathbf{C}}$ is real and symmetric, all the eigenvalues are real, and the eigenvectors may be chosen real. When all the roots of λ are distinct, the eigenvectors are mutually orthogonal. If two or more roots are equal, the eigenvectors can be chosen so as to form an orthogonal set in an infinite number of ways.

A.3. MATRICES

In this section we will summarize some basic definitions as well as derive some useful properties of matrices. The basic rules of matrix algebra will not be discussed.

We will write $[A]$ for the matrix

$$
\begin{bmatrix}
a_{11} & a_{12} & \cdots & a_{1n} \\
a_{21} & a_{22} & \cdots & \cdots \\
\cdots\cdots\cdots\cdots\cdots\cdots \\
a_{n1} & \cdots & \cdots & a_{nn}
\end{bmatrix}
$$

whose elements are a_{ij}. The first subscript refers to the row and the second to the column. The determinant of $[A]$ will be written as $|A|$.

Definitions

Transpose Matrix: The matrix obtained by interchanging the rows and columns of $[A]$ is called the transpose matrix. It will be designated by $[A]_t$ and has the elements a_{ji} in the ith row and jth column.

Adjoint Matrix: The matrix composed of the elements A_{ji}, where A_{ji} are the cofactors of a_{ij} in the determinant $|A|$, is called the adjoint matrix and will be designated by $[A]_a$.

Inverse Matrix: The inverse matrix $[A]^{-1}$ is defined so that

$$[A]^{-1}[A] = [A][A]^{-1} = [I]$$

where $[I]$ is the unit matrix. It has elements $A_{ji}/|A|$ and hence is given by

$$[A]^{-1} = [A]_a/|A|.$$

Orthogonal Matrix: When $[A]_t = [A]^{-1}$, the matrix $[A]$ is an orthogonal matrix.

Reciprocal Matrix: The transpose of the inverse matrix is called the reciprocal matrix. It has elements $A_{ji}/|A|$ in the ith row and jth column.

Symmetrical Matrix: When $a_{ij} = a_{ji}$, the matrix is symmetrical.

Hermitian Matrix: When $a_{ij} = a_{ji}^*$, the matrix is a Hermitian matrix. The (*) denotes the complex conjugate value.

Unitary Matrix: When the matrix $[A]$ is equal to the inverse of the complex conjugate transposed matrix $([A^*]_t)^{-1}$, it is called a unitary matrix.

Matrix Eigenvalue Problem

The following equation is a typical matrix eigenvalue problem:

$$
\begin{bmatrix}
a_{11} & a_{12} & \cdots & a_{1n} \\
a_{21} & \cdots & \cdots & \cdots \\
\cdots\cdots\cdots\cdots\cdots \\
\cdots & \cdots & \cdots & a_{nn}
\end{bmatrix}
\begin{bmatrix} e_1 \\ e_2 \\ \vdots \\ e_n \end{bmatrix}
= \lambda
\begin{bmatrix}
p_{11} & p_{12} & \cdots & p_{1n} \\
p_{21} & \cdots & \cdots & \cdots \\
\cdots\cdots\cdots\cdots\cdots \\
p_{n1} & \cdots & \cdots & p_{nn}
\end{bmatrix}
\begin{bmatrix} e_1 \\ e_2 \\ \vdots \\ e_n \end{bmatrix}
$$

or $[A][E] = \lambda[P][E]$. The values of λ for which this equation has a solution are called the eigenvalues or latent roots. The above equation may be written as

$$([A] - \lambda[P])[E] = 0$$

and a solution exists only if the following determinant vanishes:

$$
\begin{vmatrix}
a_{11} - \lambda p_{11} & a_{12} - \lambda p_{12} & \cdots & a_{1n} - \lambda p_{1n} \\
\cdots\cdots\cdots\cdots\cdots\cdots\cdots\cdots\cdots\cdots\cdots\cdots\cdots\cdots \\
a_{n1} - \lambda p_{n1} & \cdots & \cdots & a_{nn} - \lambda p_{nn}
\end{vmatrix} = 0.
$$

The above characteristic equation is of the nth degree in λ and will have n roots, some of which may be equal or degenerate. For each value of λ, say λ_s, a solution for $[E]$, say $[E_s]$, may be found. Each solution is called an eigenvector or modal column. When $[A]$ and $[P]$ are real and symmetric or Hermitian, and $[P]$ is a positive or negative definite matrix, all the eigenvalues are real. For the sth solution, we have

$$[A][E_s] = \lambda_s[P][E_s]$$

$$[A^*][E_s^*] = \lambda_s^*[P^*][E_s^*].$$

The transpose of the second equation gives

$$[E_s^*]_t[A] = \lambda_s^*[E_s^*]_t[P]$$

when $[A]$ and $[P]$ are Hermitian. Postmultiply this latter equation by $[E_s,]$ premultiply the first by $[E_s^*]_t$, and subtract to get

$$(\lambda_s - \lambda_s^*)[E_s^*]_t[P][E_s] = 0.$$

Since $[E_s^*]_t[P][E_s]$ is not equal to zero, λ_s must equal λ_s^*, and, hence, λ_s is real. Thus all the eigenvalues are real since s is arbitrary. If $[P]$ is not a definite matrix complex eigenvalues can occur.

When $[A]$ and $[P]$ are real and symmetric or Hermitian, the eigenvectors form an orthogonal set with respect to the weighting factors P_{ij} when all the eigenvalues are distinct. The eigenvectors corresponding to degenerate roots are not uniquely determined. However, a basic set may be chosen in such a manner that the totality of eigenvectors still forms an orthogonal set with respect to the weighting factors P_{ij}.

We will prove the following orthogonality property for the case when $[A]$ and $[P]$ are symmetric and all the eigenvalues are distinct:

$$\sum_{i=1}^{n}\sum_{j=1}^{n} e_{is}e_{jr}P_{ij} = \delta_{sr}$$

where δ_{sr} is the Kronecker delta and equals unity for $r = s$, and zero otherwise. We begin with the equation

$$[A][E_s] = \lambda_s[P][E_s]$$

and its transpose for the rth solution,

$$[E_r]_t[A] = \lambda_r[E_r]_t[P].$$

Premultiplying the first by $[E_r]_t$, postmultiplying the second by $[E_s]$, and subtracting gives

$$(\lambda_s - \lambda_r)[E_r]_t[P][E_s] = 0.$$

Since $\lambda_s \neq \lambda_r$, we get

$$[E_r]_t[P][E_s] = 0$$

which when expanded gives the orthogonality condition stated above. Since the magnitude of each eigenvector is not determined, they may always be normalized so that

$$\sum_{i=1}^{n}\sum_{j=1}^{n} e_{is}e_{js}P_{ij} = 1.$$

Since $[E_r]_t[A][E_s] = \lambda_s[E_r]_t[P][E_s]$ the eigenvectors are also orthogonal with respect to $[A]$ as the weighting matrix. In many cases in practice, the matrix $[P]$ is simply the unit matrix $[I]$, and the orthogonality condition simplifies to $\sum_{i=1}^{n} e_{is}e_{ir} = \delta_{sr}$.

Cayley–Hamilton's Theorem

This theorem states that a square matrix $[A]$ satisfies its own characteristic equation, i.e.,

$$F([A]) = ([A] - \lambda_1[I])([A] - \lambda_2[I]) \cdots ([A] - \lambda_n[I]) = 0.$$

Postmultiplying by $[E_s]$ gives

$$([A] - \lambda_1[I]) \cdots ([A] - \lambda_{n-1}[I])[E_s](\lambda_s - \lambda_n) = ([A] - \lambda_1[I]) \cdots [E_s](\lambda_s - \lambda_s)(\lambda_s - \lambda_{s+1}) \cdots (\lambda_s - \lambda_n) = 0.$$

This result holds for all values of s. Any nonzero n-dimensional vector or column matrix $[B]$ may be expressed as a linear function of the $[E_s]$, say

$$[B] = \sum_{s=1}^{n} b_s[E_s].$$

Using the above result shows that $F([A])[B] = 0$, and, hence, $F([A]) = 0$, since $[B]$ is not equal to zero.

Sylvester's Theorem

Consider the equation

$$f([A]) = C_1([A] - \lambda_2[I])([A] - \lambda_3[I]) \cdots ([A] - \lambda_n[I])$$

$$+ C_2([A] - \lambda_1[I])([A] - \lambda_3[I]) \cdots ([A] - \lambda_n[I])$$

$$+ C_n([A] - \lambda_1[I]) \cdots ([A] - \lambda_{n-1}[I])$$

which defines an nth-degree polynomial function of $[A]$. Postmultiplying by $[E_s]$ gives

$$f([A])[E_s] = C_s(\lambda_s - \lambda_1) \cdots (\lambda_s - \lambda_{s-1})(\lambda_s - \lambda_{s+1}) \cdots (\lambda_s - \lambda_n)[E_s]$$

since all terms containing the eigenvalue λ_s vanish. Furthermore, it is seen that

$$f([A])[E_s] = f(\lambda_s)[E_s]$$

and, hence, the constants C_s are given by

$$C_s = \frac{f(\lambda_s)}{\displaystyle\prod_{\substack{i=1 \\ i \neq s}}^{n} (\lambda_s - \lambda_i)}$$

provided all the eigenvalues are distinct. Therefore $f([A])$ is given by

$$f([A]) = \sum_{s=1}^{n} f(\lambda_s) \left. \frac{\prod\limits_{i=1}^{n}([A] - \lambda_i[I])}{\prod\limits_{i=1}^{n}(\lambda_s - \lambda_i)} \right|_{i \neq s}$$

a result known as Sylvester's theorem. This theorem is useful for finding the mth power of a matrix. For a 2×2 matrix with eigenvalues λ_1 and λ_2, we have

$$[A]^m = \lambda_1^m \frac{[A] - \lambda_2[I]}{\lambda_1 - \lambda_2} + \lambda_2^m \frac{[A] - \lambda_1[I]}{\lambda_2 - \lambda_1}.$$

A.4. CALCULUS OF VARIATIONS

Consider the following integral in N-dimensional space:

$$\int_{V_n} F[x_1, x_2, \ldots, x_N; u(x_1, \ldots, x_N); v(x_1, \ldots, x_N); \dot{u}_1, \dot{u}_2, \ldots, \dot{u}_N; \dot{v}_1, \dot{v}_2, \ldots, \dot{v}_N] \, dx_1 \, dx_2 \cdots dx_N$$

where $\dot{u}_i = \partial u/\partial x_i$, $\dot{v}_i = \partial v/\partial x_i$. In the above, u and v are functions of the x_i only. We are given the function F and are required to find the two functions u and v so that the above integral will have a stationary value for arbitrary first-order variations in u and v that do not violate the boundary conditions; i.e., the values of u and v are specified on the boundary, and so the variations in u and v vanish at the ends of the ranges of integration. The functions F, u, and v are assumed continuous with at least piecewise-continuous partial derivatives. The function F is called the Lagrangian function. Let the first-order variations in u and v be $h(x_1, \ldots, x_N)$ and $g(x_1, \ldots, x_N)$, respectively, where h and g vanish on the boundary of V_N. The variations in $\dot{u}_i$ and $\dot{v}_i$ are $\partial h/\partial x_i$ and $\partial g/\partial x_i$, respectively. When u and v vary, the function F is a function of u, v, $\dot{u}_i$, and $\dot{v}_i$. The first-order change in F is obtained by expanding F in a Taylor series about the assumed correct functional forms of u and v; thus

$$F - F_0 = \delta F = \frac{\partial F}{\partial u} \, du + \frac{\partial F}{\partial v} \, dv + \sum_{i=1}^{N} \left(\frac{\partial F}{\partial \dot{u}_i} \, d\dot{u}_i + \frac{\partial F}{\partial \dot{v}_i} \, d\dot{v}_i \right)$$

$$= \frac{\partial F}{\partial u} h + \frac{\partial F}{\partial v} g + \sum_{i=1}^{N} \left(\frac{\partial F}{\partial \dot{u}_i} \frac{\partial h}{\partial x_i} + \frac{\partial F}{\partial \dot{v}_i} \frac{\partial g}{\partial x_i} \right)$$

when higher order terms are dropped. Hence we have

$$\delta \int_{V_N} F \, dx_1 \cdots dx_N = \int_{V_N} \delta F \, dx_1 \cdots dx_N$$

$$= \int_{V_N} \left[\frac{\partial F}{\partial u} h + \frac{\partial F}{\partial v} g + \sum_{i=1}^{N} \left(\frac{\partial F}{\partial \dot{u}_i} \frac{\partial h}{\partial x_i} + \frac{\partial F}{\partial \dot{v}_i} \frac{\partial g}{\partial x_i} \right) \right] dx_1 \cdots dx_N$$

where δ signifies the first variation. Consider the N-dimensional integral of a typical term like $(\partial F/\partial \dot{u}_i)(\partial h/\partial x_i)$. We have

$$\int_{V_{N-1}} \left(\int_{x_{i1}}^{x_{i2}} \frac{\partial F}{\partial \dot{u}_i} \frac{\partial h}{\partial x_i} \, dx_i \right) dx_1 \cdots dx_{i-1} \, dx_{i+1} \cdots dx_N$$

$$= \int_{V_{N-1}} \left(\left. \frac{\partial F}{\partial \dot{u}_i} h \right|_{x_{i1}}^{x_{i2}} - \int_{x_{i1}}^{x_{i2}} h \frac{\partial}{\partial x_i} \frac{\partial F}{\partial \dot{u}_i} \, dx_i \right) dx_1 \cdots dx_N$$

by an integration by parts. The points x_{i1} and x_{i2} lie on the boundary of V_N, and, since h vanishes on the boundary, the integrated term vanishes, and we are left with

$$-\int_{V_N} h \frac{\partial}{\partial x_i} \frac{\partial F}{\partial \dot{u}_i} \, dx_1 \cdots dx_N.$$

Using this result, the integral of δF over V_N becomes

$$\int_{V_N} \left[h \left(\frac{\partial F}{\partial u} - \sum_{i=1}^{N} \frac{\partial}{\partial x_i} \frac{\partial F}{\partial \dot{u}_i} \right) + g \left(\frac{\partial F}{\partial v} - \sum_{i=1}^{N} \frac{\partial}{\partial x_i} \frac{\partial F}{\partial \dot{v}_i} \right) \right] dx_1 \cdots dx_N.$$

In order for the original integral to be stationary, the first-order variation must vanish. Since g and h are arbitrary, we conclude that

$$\frac{\partial F}{\partial u} - \sum_{i=1}^{N} \frac{\partial}{\partial x_i} \frac{\partial F}{\partial \dot{u}_i} = 0 \quad \text{and} \quad \frac{\partial F}{\partial v} - \sum_{i=1}^{N} \frac{\partial}{\partial x_i} \frac{\partial F}{\partial \dot{v}_i} = 0.$$

These two partial differential equations determine the functional forms of u and v which make the integral of F over V_N stationary. These partial differential equations are called the Euler–Lagrange differential equations. If F was a function of any number of functions u, v, w, etc., we would find that each function u, v, w, etc., satisfied a partial differential equation of the above form.

Constraints

When the functions u and v are not independent but are related by an implicit functional form

$$G(u; v; \dot{u}_1, \ldots, \dot{u}_N; \dot{v}_1, \ldots, \dot{v}_N; x_1, \ldots, x_N) = 0$$

we call G an equation of constraint, since it constrains the functions u and v to satisfy a certain functional relationship. In our variational integral, the variations in u and v are now constrained in such a manner that $G = 0$ always. The integral of G over V_N is also identically zero, so that any multiple of G may be added to F without changing the value of the integral of F over V_N. Thus we consider the variation in the following integral:

$$\int_{V_N} (F + \lambda G) \, dx_1 \cdots dx_N.$$

The multiplier λ is, in general, a function of $x_1, \ldots, x_N$, and is called a Lagrange multiplier. Since the constraint function has been introduced into the integral, the variation may be conducted by treating u and v as independent functions. Thus, we obtain the following two partial differential equations, which, together with the original constraint function $G = 0$, allow us to find a solution for u, v, and λ:

$$\frac{\partial (F + \lambda G)}{\partial u} - \sum_{i=1}^{N} \frac{\partial}{\partial x_i} \frac{\partial (F + \lambda G)}{\partial \dot{u}_i} = 0$$

$$\frac{\partial (F + \lambda G)}{\partial v} - \sum_{i=1}^{N} \frac{\partial}{\partial x_i} \frac{\partial (F + \lambda G)}{\partial \dot{v}_i} = 0.$$

A.5. Infinite Products and the Gamma Function

Let $f(z)$ be an integral function, i.e., a function with no singularities in the finite complex z plane. The logarithmic derivative of this function is then a meromorphic function, i.e., a function whose only singularities are poles. Let

$f'(z)/f(z)$ have simple poles at $z = z_n$ with unit residues. The partial-fraction expansion is

$$\frac{f'(z)}{f(z)} = \frac{f'(z)}{f(z)}\bigg|_{z=0} + \sum_n \left(\frac{1}{z - z_n} + \frac{1}{z_n}\right).$$

Integrating with respect to z from 0 to z gives

$$\ln f(z)|_0^z = \frac{f'(0)}{f(0)}z + \sum_n \left(\ln \frac{z - z_n}{- z_n} + \frac{z}{z_n}\right)$$

or

$$f(z) = f(0)e^{[f'(0)/f(0)]z}\prod_n \left(1 - \frac{z}{z_n}\right)e^{z/z_n}$$

which is the infinite-product expansion of the integral function $f(z)$. When $f(z)$ is an even function $f_e(z)$ of z, $f_e'(0)$ equals zero, and we get

$$f_e(z) = f_e(0)\prod_n \left(1 - \frac{z^2}{z_n^2}\right).$$

Consider the function $\cos z$, which has zeros at $z = n\pi + \pi/2 = (n + \frac{1}{2})\pi$. Using the general formula gives the infinite-product expansion

$$\cos z = \prod_{n=0}^{\infty}\left[1 - \frac{z^2}{(n + \frac{1}{2})^2\pi^2}\right].$$

Similarly, the infinite-product expansion of $(\sin z)/z$ is found to be

$$\sin z = z\prod_{n=1}^{\infty}\left(1 - \frac{z^2}{n^2\pi^2}\right).$$

Gamma Function

The gamma function is defined by the integral

$$\Gamma(z) = \int_0^{\infty} e^{-u}u^{z-1}\,du.$$

The integral defines an analytic function of z for all z for which the real part is positive. An equivalent definition of the Γ function is the infinite-product representation

$$\Gamma(z) = \frac{e^{-\gamma z}}{z}\left[\prod_{n=1}^{\infty}\left(1 + \frac{z}{n}\right)e^{-z/n}\right]^{-1}$$

where $\gamma = 0.57722$ and is Euler's constant. The infinite-product representation defines $\Gamma(z)$ as an analytic function throughout the complex z plane, except in the vicinity of the negative-real axis, where $\Gamma(z)$ has simple poles. The following useful properties of the gamma function have been established:

1. $\Gamma(z + 1) = z\Gamma(z)$.
2. $\Gamma(n + 1) = n!$, n an integer.
3. $\Gamma(\frac{1}{2}) = \pi^{1/2}$.
4. $2^{2z-1}\Gamma(z)\Gamma(z + \frac{1}{2}) = \pi^{1/2}\Gamma(2z)$.
5. $\Gamma(z)\Gamma(1 - z) = [z\prod_{n=1}^{\infty}(1 - z^2/n^2)]^{-1} = \pi/\sin \pi z$.

6. As $z \rightarrow \infty$, the asymptotic value of $\Gamma(z)$ is $(2\pi)^{1/2}e^{z \ln z}z^{-1/2}e^{-z}$, valid for all z except in the region of the negative-real axis, where $\Gamma(z)$ has poles at $z = 0, -1, -2, \ldots$, etc. From property (5) an asymptotic expansion valid for z on the negative-real axis may be found.

7. For x finite and $|y|$ very large, $|\Gamma(x + jy)| \sim (2\pi)^{1/2}|y|^{x-1/2}e^{-\pi|y|/2}$.

Example 1

We wish to construct a function with simple poles at $w = \Gamma_n = [(n\pi/a)^2 - k_0^2]^{1/2}$ and find its asymptotic behavior for large w. n takes on the values $1, 2, 3, \ldots, \infty$.

Solution: The series $\sum_{n=1}^{\infty}(1/(w - \Gamma_n) + 1/(n\pi/a))$ is convergent since Γ_n approaches $n\pi/a$ for n large. Hence $e^{wa/n\pi}$ are suitable convergence factors to use in an infinite product. Let

$$f(w) = \prod_{n=1}^{\infty}\left(1 - \frac{w}{\Gamma_n}\right)e^{wa/n\pi}.$$

Then $1/f(w)$ has the required simple poles. The asymptotic behavior can be established by comparison with the gamma function. Consider

$$\frac{\displaystyle\prod_{n=1}^{\infty}(1 - wa/n\pi)e^{wa/n\pi}}{\displaystyle\prod_{n=1}^{\infty}(1 - w/\Gamma_n)e^{wa/n\pi}} \frac{1}{\displaystyle\prod_{n=1}^{\infty}(1 - wa/n\pi)e^{wa/n\pi}}$$

$$= \prod_{n=1}^{\infty}\frac{(1 - wa/n\pi)}{(1 - w/\Gamma_n)} \frac{\displaystyle\prod_{n=1}^{\infty}(1 + wa/n\pi)e^{-wa/n\pi}}{\displaystyle\prod_{n=1}^{\infty}(1 - w^2a^2/n^2\pi^2)}$$

$$= \prod_{n=1}^{\infty}\frac{(1 - wa/n\pi)}{(1 - w/\Gamma_n)} \frac{wa}{\sin wa} \frac{e^{-\gamma wa/\pi}}{(wa/\pi)\Gamma(wa/\pi)}.$$

Since $\Gamma_n \rightarrow n\pi/a$ the first product approaches the constant $\prod_{n=1}^{\infty}(\Gamma_n a/n\pi)$ as $w \rightarrow \infty$. Thus we find that

$$\frac{1}{f(w)} \sim K\frac{e^{-\gamma wa/\pi}}{\sin wa} \frac{\sqrt{wa}e^{-(wa/\pi)\ln wa/\pi}}{\sqrt{2\pi}e^{-wa/\pi}}$$

$$= \frac{K}{\sin wa}\frac{\sqrt{wa}}{\sqrt{2}}e^{-(wa/\pi)[\gamma - 1 + \ln(wa/\pi)]} \qquad \text{as } w \rightarrow \infty$$

where K is a suitable constant.

Example 2

We want to find the infinite-product expansion of the Bessel function $J_0(w)$ and the asymptotic behavior of the part that has zeros in the left half of the complex w plane.

Solution: Let $\pm w_n$ be the zeros of $J_0(w) = 0$. Since $J_0(w) = J_0(-w)$, $J_0'(0) = 0$; thus

$$J_0(w) = \prod_{n=1}^{\infty}(1 - w^2/w_n^2) = \prod_{n=1}^{\infty}(1 - w/w_n)e^{w/w_n}\prod_{n=1}^{\infty}(1 + w/w_n)e^{-w/w_n}$$

Other convergence factors can also be used. For large w we know that $J_0(w) \sim \sqrt{2/\pi w}\cos(w - \pi/4)$. Hence

w_n is asymptotic to $(2n-1)\pi/2 + \pi/4$, $n = 1, 2, 3, \ldots$; i.e., $w_n \sim (4n-1)\pi/4$. Hence we can also write

$$J_0(w) = \prod_{n=1}^{\infty} (1 - w/w_n) e^{4w/4n\pi} \prod_{n=1}^{\infty} (1 + w/w_n) e^{-4w/4n\pi}.$$

The part whose asymptotic behavior we want is

$$f(w) = \prod_{n=1}^{\infty} (1 + w/w_n) e^{-4w/4n\pi}$$

$$= \prod_{n=1}^{\infty} \frac{1 + w/w_n}{\left(1 + \dfrac{4w}{(4n-1)\pi}\right)} \prod_{n=1}^{\infty} \left(1 + \frac{4w}{(4n-1)\pi}\right) e^{-w/n\pi}.$$

The first product approaches a constant as $w \to \infty$. Hence we need to find the asymptotic behavior of

$$g(w) = \prod_{n=1}^{\infty} \left(1 + \frac{4w}{(4n-1)\pi}\right) e^{-w/n\pi}$$

which cannot be directly compared with the gamma function. We note that $1 + 4w/(4n-1)\pi = (n - 1/4 + w/\pi)/(n - 1/4) = (1 + (\hat{w} - 1/4)/n)/(1 - 1/4n)$ and also that $4w/4n\pi = (\hat{w} - 1/4)/n + 1/4n$ where $\hat{w} = w/\pi$. Hence

$$g(w) = \prod_{n=1}^{\infty} \frac{1 + \dfrac{\hat{w} - 1/4}{n}}{1 - \dfrac{1}{4n}} e^{-(\hat{w}-1/4)/n} e^{-1/4n}$$

$$= \frac{\displaystyle\prod_{n=1}^{\infty} \left(1 + \frac{\hat{w} - 1/4}{n}\right) e^{-(\hat{w}-1/4)/n}}{\displaystyle\prod_{n=1}^{\infty} \left(1 - \frac{1}{4n}\right) e^{1/4n}} = \frac{e^{-\gamma(\hat{w}-1/4)}}{\left(\hat{w} - \dfrac{1}{4}\right) \Gamma\left(\hat{w} - \dfrac{1}{4}\right) \displaystyle\prod_{n=1}^{\infty} \left(1 - \frac{1}{4n}\right) e^{1/4n}}$$

$$\sim K_1 \hat{w}^{-1/2} e^{-\hat{w}[\gamma - 1 + \ln \hat{w}] + (1/4)[\gamma - 1 + \ln \hat{w}]}$$

$$= K_2 \hat{w}^{-1/4} e^{-\hat{w}(\gamma - 1 + \ln \hat{w})}$$

since $e^{(1/4)\ln \hat{w}} = \hat{w}^{1/4}$. K_1 and K_2 are constants. If we had used simply $w_n \sim 4n\pi/4 = n\pi$ we would have gotten a $\hat{w}^{-1/2}$ instead of a $\hat{w}^{-1/4}$ factor which would not have been correct. A shift in the zeros of a function away from the integers n does affect the asymptotic behavior.

There is considerable latitude in the choice of exponential convergence factors that can be used. The requirement is that the series

$$S = \sum_{n=1}^{\infty} \left(\frac{1}{w - w_n} + \frac{1}{S_n}\right)$$

converges. If we choose $S_n = 1/n\pi$, then

$$S \sim \sum_{n=1}^{\infty} \left(\frac{1}{w - \dfrac{4n-1}{4}\pi} + \frac{1}{n\pi}\right)$$

$$= \sum_{n=1}^{\infty} \frac{w + \pi/4}{n\pi \left(w - \dfrac{4n-1}{4}\pi\right)}$$

which is a convergent series.

A.6. SUMMATION OF FOURIER SERIES

The solution of many waveguide problems involves Fourier series for which it is desirable to have summation formulas. There are several methods by which a wide variety of Fourier series may be summed in closed form or converted to more rapidly converging series. Some of the more useful methods will be considered here. The type of series to be considered is of the form

$$\sum_n \frac{e^{j\alpha n}}{P(n)}$$

where $P(n)$ is a rational function of the summation variable n. A number of the summation formulas that will be derived will be for the exponential function $e^{j\alpha n}$. The required summation formulas for the corresponding trigonometric series may be obtained by taking the real and imaginary parts separately.

Geometric-Series Method

The following geometric series is a useful one from which several summation formulas may be developed:

$$\sum_{n=1}^{\infty} r^n = \frac{r}{1-r}, \qquad |r| < 1.$$

Let $r = \lambda e^{j\alpha}$, where λ is a positive number less than unity. The sum of the following series follows at once:

$$\sum_{n=1}^{\infty} \lambda^n e^{j\alpha n} = \frac{\lambda e^{j\alpha}}{1 - \lambda e^{j\alpha}}.$$

Taking the limit as λ approaches unity, we get

$$\sum_{n=1}^{\infty} e^{j\alpha n} = \frac{e^{j\alpha}}{1 - e^{j\alpha}}.$$

Equating the real and imaginary parts gives

$$\sum_{n=1}^{\infty} \cos n\alpha = -\frac{1}{2} \qquad \sum_{n=1}^{\infty} \sin n\alpha = \frac{1}{2} \cot \frac{\alpha}{2}.$$

The above series may be differentiated with respect to α before taking the limit; this gives

$$\sum_{n=1}^{\infty} jn\lambda^n e^{j\alpha n} = \frac{j\lambda e^{j\alpha}}{(1 - \lambda e^{j\alpha})^2}.$$

Equating the real and imaginary parts and taking the limit as λ approaches unity gives

$$\sum_{n=1}^{\infty} n \cos n\alpha = \lim_{\lambda \to 1} \text{Re} \frac{\lambda e^{j\alpha}}{(1 - \lambda e^{j\alpha})^2} = -\frac{1}{4} \csc^2 \frac{\alpha}{2}$$

$$\sum_{n=1}^{\infty} n \sin n\alpha = 0.$$

In the above series it should be noted that, for $\lambda = 1$, the points $\alpha = 2m\pi$ are singular points. From the integral of the series $\sum_{n=1}^{\infty} e^{j\alpha n}$, further summation formulas may be obtained. For example, we have

$$j \int_{j\infty}^{\alpha} \sum_{n=1}^{\infty} e^{j\alpha n} \, d\alpha = \sum_{n=1}^{\infty} \frac{e^{j\alpha n}}{n} = j \int_{j\infty}^{\alpha} \frac{e^{j\alpha}}{1 - e^{j\alpha}} \, d\alpha = -\ln(1 - e^{j\alpha}).$$

To justify interchanging the order of integration and summation, we should include the factor λ, and take the limit as λ approaches 1; however, the result is the same as that given above.

Sometimes it is desirable to sum a series over only odd or even values of n. Such summations may be performed by replacing α by $\alpha + \pi$ and adding or subtracting series. To sum the series

$$\sum_{n=1,3,\ldots}^{\infty} \frac{e^{jn\alpha}}{n}$$

we consider

$$\sum_{n=1}^{\infty} \frac{e^{jn(\alpha+\pi)}}{n} = -\sum_{n=1,3,\ldots}^{\infty} \frac{e^{j\alpha n}}{n} + \sum_{n=2,4,\ldots}^{\infty} \frac{e^{j\alpha n}}{n} = -\ln(1 + e^{j\alpha}).$$

Subtracting this series from the series with $\alpha + \pi$ replaced by α gives

$$\sum_{n=1,3,\ldots}^{\infty} \frac{e^{j\alpha n}}{n} = \frac{1}{2} \ln \frac{1 + e^{j\alpha}}{1 - e^{j\alpha}} = \frac{1}{2} \ln \cot \frac{\alpha}{2} + j\frac{\pi}{4}.$$

Equating the real and imaginary parts gives

$$\sum_{n=1,3,\ldots}^{\infty} \frac{\cos n\alpha}{n} = -\frac{1}{2} \ln \tan \frac{\alpha}{2} \qquad \sum_{n=1,3,\ldots}^{\infty} \frac{\sin n\alpha}{n} = \frac{\pi}{4}.$$

Series involving products of trigonometrical terms may be summed by reducing them to the exponential type first and adding or subtracting series if necessary. Thus

$$\sum_{n=1,3,\ldots}^{\infty} \frac{\sin n\alpha \cos n\beta}{n}$$

may be written as

$$\frac{1}{2}\sum_{n=1,3,\ldots}^{\infty} \left[\frac{\sin n(\alpha+\beta)}{n} + \frac{\sin n(\alpha-\beta)}{n} \right] = \frac{1}{2} \operatorname{Im} \sum_{n=1,3,\ldots}^{\infty} \frac{e^{jn(\alpha+\beta)} + e^{jn(\alpha-\beta)}}{n} = \frac{\pi}{4}$$

if α and β are positive and real and $\beta < \alpha$ and $0 < \alpha + \beta < \pi$. These conditions follow from the properties of the logarithmic function since the sum is given by

$$\frac{1}{4} \operatorname{Im} \left(\ln j \cot \frac{\alpha+\beta}{2} + \ln j \cot \frac{\alpha-\beta}{2} \right).$$

The summation formulas given in part 2 of Table A.2 have been obtained by the above methods. To avoid an unduly large number of entries, only summation formulas for the exponential type of series are listed. Some of the series are divergent series and the sum is the limit as the convergence factor λ used above tends to unity.

Contour-Integration Method

The theory of contour integration provides a very powerful method for the summation of a variety of series, the restrictions on the series being that they be representable as a sum over both positive and negative integers, and also satisfy certain convergence criteria which will become clear after considering a few typical examples. In order for a series to be representable as a sum over both positive and negative integers, it is necessary that the series be an even function of the summation variable n. From this point on we will suppress the summation variable, it being understood that the sums are taken over the variable n.

TABLE A.2

Part 1: Maclaurin Series Expansions

$\ln \sin x$ $\qquad \ln x - \dfrac{x^2}{6} - \dfrac{x^4}{180} - \dfrac{x^6}{2{,}835} - \cdots, \qquad -\pi < x < \pi$

$\ln \cos x$ $\qquad -\dfrac{x^2}{2} - \dfrac{x^4}{12} - \dfrac{x^6}{45} - \dfrac{17x^8}{2{,}520} - \cdots, \qquad -\dfrac{\pi}{2} < x < \dfrac{\pi}{2}$

$\ln \tan x$ $\qquad \ln x + \dfrac{x^2}{3} + \dfrac{7x^4}{90} + \dfrac{62x^6}{2{,}835} + \cdots, \qquad -\dfrac{\pi}{2} < x < \dfrac{\pi}{2}$

$\sin x$ $\qquad x - \dfrac{x^3}{3!} + \dfrac{x^5}{5!} - \dfrac{x^7}{7!} + , \cdots, \qquad \text{all } x$

$\cos x$ $\qquad 1 - \dfrac{x^2}{2!} + \dfrac{x^4}{4!} - \dfrac{x^6}{6!} + \cdots, \qquad \text{all } x$

$\tan x$ $\qquad x + \dfrac{x^3}{3} + \dfrac{2x^5}{15} + \cdots, \qquad |x| < \dfrac{\pi}{2}$

$\cot x$ $\qquad \dfrac{1}{x} - \dfrac{x}{3} - \dfrac{x^3}{45} - \cdots, \qquad |x| < \pi$

$\sec x$ $\qquad 1 + \dfrac{x^2}{2!} + \dfrac{5x^4}{4!} + \cdots, \qquad |x| < \dfrac{\pi}{2}$

$\csc x$ $\qquad \dfrac{1}{x} + \dfrac{x}{3!} + \dfrac{7x^3}{3 \times 5!} + \cdots, \qquad |x| < \pi$

$\ln x$ $\qquad 2\left[\dfrac{x-1}{x+1} + \dfrac{1}{3}\left(\dfrac{x-1}{x+1}\right)^3 + \dfrac{1}{5}\left(\dfrac{x-1}{x+1}\right)^5 + \cdots \right], \qquad x >$

Part 2: Summations by Geometric Series

$\displaystyle\sum_1^\infty e^{jnx}$ $\qquad \dfrac{e^{jx}}{1 - e^{jx}} = -\dfrac{1}{2} + \dfrac{j}{2} \cot \dfrac{x}{2}, \qquad 0 < x < \pi$

$\displaystyle\sum_1^\infty n e^{jnx}$ $\qquad \dfrac{e^{jx}}{\left(1 - e^{jx}\right)^2} = -\dfrac{1}{4} \csc^2 \dfrac{x}{2}, \qquad 0 < x < \pi$

$\displaystyle\sum_1^\infty \dfrac{e^{jnx}}{n}$ $\qquad -\ln(1 - e^{jx}) = -\ln 2 \sin \dfrac{x}{2} + j\dfrac{\pi - x}{2}, \qquad 0 < x < 2\pi$

$\displaystyle\sum_1^\infty \dfrac{e^{jnx}}{n^2}$ $\qquad \dfrac{\pi^2}{6} - \dfrac{x}{4}(2\pi - x) - j\left(x \ln x - x - \dfrac{x^3}{72} - \dfrac{x^5}{14{,}400} - \cdots \right), \qquad 0 < x < 2\pi$

$\displaystyle\sum_1^\infty \dfrac{1}{n^2}$ $\qquad \dfrac{\pi^2}{6}$

$\displaystyle\sum_1^\infty \dfrac{e^{jnx}}{n^3}$ $\qquad j\left(\dfrac{\pi^2 x}{6} - \dfrac{\pi x^2}{4} + \dfrac{x^3}{12} \right) + \left(\dfrac{x^2}{2} \ln x - \dfrac{3x^2}{4} - \dfrac{x^4}{288} - \cdots \right) + \displaystyle\sum_1^\infty \dfrac{1}{n^3}, \qquad 0 < x < 2\pi$

$\displaystyle\sum_{1,3,\ldots}^\infty e^{jnx}$ $\qquad \dfrac{e^{jx}}{1 - e^{2jx}} = \dfrac{j}{2} \csc x, \qquad 0 < x < \pi$

$\displaystyle\sum_{1,3,\ldots}^\infty n e^{jnx}$ $\qquad -\dfrac{1}{2} \cot x \csc x, \qquad 0 < x < \pi$

$\displaystyle\sum_{1,3,\ldots}^\infty \dfrac{e^{jnx}}{n}$ $\qquad \dfrac{1}{2} \ln \dfrac{1 + e^{jx}}{1 - e^{jx}} = -\dfrac{1}{2} \ln \tan \dfrac{x}{2} + j\dfrac{\pi}{4}, \qquad 0 < x < \pi$

$\displaystyle\sum_{1,3,\ldots}^\infty \dfrac{e^{jnx}}{n^2}$ $\qquad \dfrac{\pi^2}{8} - \dfrac{\pi x}{4} - \dfrac{j}{2}\left(x \ln \dfrac{x}{2} - x + \dfrac{x^3}{36} + \dfrac{7x^5}{7{,}200} + \cdots \right), \qquad 0 < x < \pi$

TABLE A.2 (continued).

$$\sum_{1,3,\dots}^{\infty} \frac{1}{n^2} \qquad \frac{\pi^2}{8}$$

$$\sum_{1,3,\dots}^{\infty} \frac{e^{jnx}}{n^3} \qquad \sum_{1,3,\dots}^{\infty} \frac{1}{n^3} + \frac{1}{2}\left(\frac{x^2}{2}\ln\frac{x}{2} - \frac{3x^2}{4} + \frac{x^4}{144} + \cdots\right) + j\left(\frac{\pi^2 x}{8} - \frac{\pi x^2}{8}\right), \qquad 0 < x < \pi$$

$$\sum_{2,4,\dots}^{\infty} \frac{e^{jnx}}{P(n)} \qquad \sum_{1}^{\infty} \frac{e^{jnx}}{P(n)} - \sum_{1,3,\dots}^{\infty} \frac{e^{jnx}}{P(n)}$$

$$\sum_{1}^{\infty} (-1)^n e^{jnx} \qquad -\frac{1}{2} - \frac{j}{2}\tan\frac{x}{2}, \qquad 0 < x < \pi$$

$$\sum_{1}^{\infty} (-1)^n n e^{jnx} \qquad -\frac{1}{4}\sec^2\frac{x}{2}, \qquad 0 < x < \pi$$

$$\sum_{1}^{\infty} \frac{(-1)^n e^{jnx}}{n} \qquad -\ln 2\cos\frac{x}{2} - j\frac{x}{2}, \qquad -\pi < x < \pi$$

$$\sum_{1}^{\infty} \frac{(-1)^n e^{jnx}}{P(n)} \qquad \sum_{1}^{\infty} \frac{e^{jn(x+\pi)}}{P(n)}$$

$$\sum_{1,3,\dots}^{\infty} \frac{1}{n^3} \qquad \frac{1}{16}\left(\frac{3\pi^2}{2} + \pi^2\ln\frac{4}{\pi} - \frac{\pi^4}{288} - \frac{7\pi^6}{345,600} - \cdots\right) = 0.875\zeta(3) = 1.05175$$

where $\zeta(x)$ is the Riemann zeta function

Part 3: Summations by Contour Integration

$$\sum_{1}^{\infty} \frac{\cos nx}{n^2 - a^2} \qquad \frac{1}{2a^2} - \frac{\pi}{2a}\frac{\cos(x-\pi)a}{\sin \pi a}, \qquad 0 < x < 2\pi$$

$$\sum_{1}^{\infty} \frac{\sin nx}{n(n^2 - a^2)} \qquad \frac{(x-\pi)\sin \pi a + \pi\sin(\pi-x)a}{2a^2\sin \pi a}, \qquad 0 < x < 2\pi$$

$$\sum_{1}^{\infty} \frac{\cos nx}{n^2(n^2 - a^2)} \qquad \frac{\pi x}{2a^2} - \frac{\pi^2}{6a^2} - \frac{x^2}{4a^2} - \frac{\pi}{2}\frac{\cos(\pi-x)a}{a^3\sin \pi a} + \frac{1}{2a^4}, \qquad 0 < x < 2\pi$$

$$\sum_{1}^{\infty} \frac{\cos nx}{(n^2 - a^2)(n^2 - b^2)} \qquad -\frac{1}{2a^2 b^2} - \frac{\pi}{2}\left[\frac{\cos(\pi-x)a}{a(a^2 - b^2)\sin \pi a} + \frac{\cos(\pi-x)b}{b(b^2 - a^2)\sin \pi b}\right], \qquad 0 < x < 2\pi$$

$$\sum_{1}^{\infty} \frac{n\sin nx}{n^2 - a^2} \qquad \frac{\pi}{2}\frac{\sin(\pi-x)a}{\sin \pi a}, \qquad 0 < x < 2\pi$$

$$\sum_{1}^{\infty} \frac{n\sin nx}{(n^2 - a^2)(n^2 - b^2)} \qquad -\frac{\pi}{2}\left[\frac{\sin(\pi-x)a}{(b^2 - a^2)\sin \pi a} + \frac{\sin(\pi-x)b}{(a^2 - b^2)\sin \pi b}\right], \qquad 0 < x < 2\pi$$

$$\sum_{1,3,\dots}^{\infty} \frac{\cos nx}{n^2 - a^2} \qquad \frac{\pi\sin a(\pi/2 - x)}{4a\cos(\pi a/2)}, \qquad 0 < x < \pi$$

$$\sum_{1,3,\dots}^{\infty} \frac{\cos nx}{n^2(n^2 - a^2)} \qquad \frac{\pi}{4a^2}\left[\frac{\sin a(\pi/2 - x)}{a\cos(\pi a/2)} - x + \frac{\pi}{2}\right], \qquad 0 < x < \pi$$

$$\sum_{1,3,\dots}^{\infty} \frac{n\sin nx}{n^2 - a^2} \qquad \frac{\pi\cos a(\pi/2 - x)}{4\cos(\pi a/2)}, \qquad 0 < x < \pi$$

$$\sum_{1,3,\dots}^{\infty} \frac{\sin nx}{n(n^2 - a^2)} \qquad \frac{\pi\cos a(\pi/2 - x)}{4a^2\cos(\pi a/2)} - \frac{\pi}{4a^2}, \qquad 0 < x < \pi$$

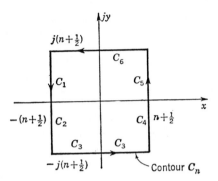

Fig. A.2. Contour used to sum series.

Consider the series $\sum_1^\infty ((\cos n\alpha)/P(n))$ where $P(n)$ is a rational even function of n. This series may be written as

$$\frac{1}{2} \sum_{-\infty}^{\infty} \frac{e^{j\alpha n}}{P(n)} - \frac{1}{2P(0)}.$$

Consider the following contour integral:

$$\oint_{C_n} \frac{e^{j\alpha z}\, dz}{P(z)(e^{2\pi jz} - 1)}$$

where the contour C_n consists of the square with vertices at $\pm (n + \frac{1}{2})$, $\pm j(n + \frac{1}{2})$, as in Fig. A.2, and n is chosen large enough so that all the zeros of $P(z)$ lie within C_n. The factor $e^{2\pi jz} - 1$ is introduced in the denominator of the integral to bring in poles at $\pm n$. If we can show that the integral over C_n vanishes as n approaches infinity, then, by Cauchy's integral formula, the sum of the residues of the integrand must equal zero, and we get

$$\sum_{-\infty}^{\infty} \frac{e^{j\alpha n}}{2\pi j P(n)} + \sum \left[\text{residues of } \frac{e^{j\alpha z}}{P(z)(e^{2\pi jz} - 1)} \text{ at zeros of } P(z) \right] = 0$$

since the residue at the poles of $[e^{2\pi jz} - 1]^{-1}$ is $1/2\pi j$. The above formula is for the case when none of the zeros of $P(z)$ are integers. If some of the zeros are integers, it is necessary to consider the residues at the double poles, triple poles, etc., of the integrand. These cases are easily treated in practice when they arise. If the summation is over only the odd or the even integers, then we introduce the factor $e^{j\pi z} + 1$ or $e^{j\pi z} - 1$, respectively, in place of $e^{2\pi jz} - 1$ in the denominator. When the terms alternate in sign, an additional exponential term such as $e^{\pm j\pi z}$ must be introduced in the numerator or denominator.

We will now show that the integral around C_n vanishes, provided $P(z)$ is of the order R^ϵ, $\epsilon > 1$, as $|z| = R \to \infty$, irrespective of the polar angle θ in the complex z plane. Since $P(z)$ is a rational function of z, there exists a positive constant M such that $|P(z)| \geq M$ everywhere on C_n for all n greater than some given finite n. On the segment C_1, we have

$$\left| \int_{C_1} \frac{e^{-j\alpha(n+1/2)} e^{-\alpha y}\, dz}{P(z)(e^{-j2\pi(n+1/2)} e^{-2\pi y} - 1)} \right| \leq \frac{1}{M} \int_{C_1} \frac{e^{-\alpha y}|dz|}{e^{-2\pi y} + 1} \leq \int_{C_1} \frac{e^{-\alpha y}}{M}|dz| \leq \frac{n + \frac{1}{2}}{M}.$$

A similar result holds for the segment C_5. On the segment C_2, we have

$$\left| \int_{C_2} \frac{e^{-j\alpha(n+1/2)} e^{-\alpha y}\, dz}{P(z)(e^{-j2\pi(n+1/2)} e^{-2\pi y} - 1)} \right| \leq \int_{C_2} \frac{e^{-\alpha y}|dz|}{M(e^{-2\pi y} + 1)} = \frac{1}{M} \int_{C_2} \frac{e^{(2\pi - \alpha)y}|dz|}{1 + e^{2\pi y}} \leq \frac{n + \frac{1}{2}}{M}.$$

since $y \leq 0$, provided $\alpha \leq 2\pi$. The same result also holds on the segment C_4. In a like manner, we may show that the integrals over C_3 and C_6 are both less than $2(n + \frac{1}{2})/M$. We thus find by adding the above results that the modulus of the integral over C_n is less than $(8n + 4)/M$. Hence, provided M increases as fast as n^ϵ, $\epsilon > 1$, the integral around C_n will vanish in the limit. It is, therefore, necessary that

$$\lim_{|z| \to \infty} |P(z)| = |z|^\epsilon, \qquad \epsilon > 1 \text{ for all values of } \theta.$$

If the summation is over only the odd or the even integers, then α is restricted to the range $\alpha < \pi$, in order that the integral shall converge in the lower half plane. If α is less than zero, then it is only necessary to change the sign of the exponential term in the denominator, and the integral around C_n may again be shown to vanish in the limit as n approaches infinity. The argument is the same as that used for positive α. Or, alternatively, the result follows at once by changing z to $-z$.

As an example, we will sum the series $\sum_1^\infty ((\cos n\alpha)/(n^2 - a^2))$. This series may be written as

$$\frac{1}{2} \sum_{-\infty}^{\infty} \frac{e^{jn\alpha}}{n^2 - a^2} + \frac{1}{2a^2}.$$

There are poles of order 1 at $n = \pm a$, with residues $e^{j\alpha a}/2a$ and $-e^{-j\alpha a}/2a$. By the previously given formula, we get

$$\frac{1}{2} \sum_{-\infty}^{\infty} \frac{e^{jn\alpha}}{n^2 - a^2} = \frac{-\pi j}{2a} \left(\frac{e^{j\alpha a}}{e^{2\pi ja} - 1} - \frac{e^{-j\alpha a}}{e^{-2\pi ja} - 1} \right) = \frac{-\pi}{2a} \frac{\cos(\alpha - \pi)a}{\sin \pi a}.$$

Thus the sum of our series is given by

$$\sum_1^\infty \frac{\cos n\alpha}{n^2 - a^2} = \frac{1}{2a^2} - \frac{\pi}{2a} \frac{\cos(\pi - \alpha)a}{\sin \pi a}.$$

When $\alpha = 0$, we get

$$\sum_1^\infty \frac{1}{n^2 - a^2} = \frac{1}{2a^2} - \frac{\pi}{2a} \cot \pi a$$

which is the well-known partial-fraction expansion of the cotangent function. A further result of interest is the limit of the above sum as a approaches zero. Thus, we find

$$\sum_1^\infty \frac{1}{n^2} = \lim_{a \to 0} \frac{1 - \pi a \cot \pi a}{2a^2} = \frac{1}{6} \pi^2.$$

Further summation formulas may be obtained by integration, differentiation, addition and subtraction of series, and so on. When differentiating a series, the convergence of the resulting series has to be established. Integration with respect to a parameter does not introduce any convergence difficulties. However, it is important to consider the constant term that arises upon integration. In practice, this is perhaps most readily taken into account by integrating between a fixed and a variable limit. For example,

$$j \sum_1^\infty \int_0^x \frac{e^{jnx}}{n} dx = \sum_1^\infty \frac{e^{jnx}}{n^2} - \sum_1^\infty \frac{1}{n^2} = -j \int_0^x \ln 2 \sin \frac{x}{2} dx$$

$$- \int_0^x \frac{\pi - x}{2} dx = -\frac{x}{4}(2\pi - x) - j \left(x \ln x - x - \frac{x^3}{72} \cdots \right)$$

where $\ln 2 \sin(x/2)$ has been expanded into a Maclaurin series in order to carry out the integration. This gives us the result

$$\sum_{1}^{\infty} \frac{1}{n^2} e^{jnx} = \frac{\pi^2}{6} - \frac{x}{4}(2\pi - x) - j\left(x \ln x - x - \frac{x^3}{72} \cdots\right).$$

The summation formulas given in part 3 of Table A.2 have been obtained by the above contour-integration method. A series such as $\sum_{1,3,\dots}^{\infty}((n \sin nx)/(n^2 - a^2))$ cannot be summed by this method, since the general term decreases only as n^{-1} for large n. However, this series is equal to $(d/dx)\sum_{1,3,\dots}^{\infty}((-\cos nx)/(n^2 - a^2))$. The convergence of the resulting series is the same as that of $\sum_{1,3,\dots}^{\infty}((\sin nx)/n)$, that is, for $|x| < \pi/2$.

There does not appear to be a similar procedure which may be used when the series to be summed is an odd function of n. This is quite a restriction, since many of the series that occur in practice are odd functions of n.

Integral Transform Methods

Integral transforms, such as the Laplace, Fourier, and Mellin transforms, can be used in a variety of ways to sum certain types of series in closed form. They also are found to be useful in many cases in converting relatively complicated series into simpler ones which are more easily summed, or in converting relatively slowly converging series into much more rapidly converging ones. We will consider the application of the Laplace transform to the summation of series first. The technique will be largely developed by means of particular examples.

Consider the general series $\sum_{0}^{\infty} f(\alpha n)$ where $f(x)$ is of such a form that its Laplace transform exists. We have by definition

$$F(p) = \int_{0}^{\infty} f(x) e^{-px} \, dx.$$

This integral determines $F(p)$ as an analytic function of the complex variable

$$p = u + jv$$

whose singularities all lie to the left of some value of $u = c$ in the p plane. The inversion integral gives

$$f(x) = \frac{1}{2\pi j} \int_{c-j\infty}^{c+j\infty} F(p) e^{px} \, dp.$$

If we replace x by αn and sum over n, we get

$$\sum_{0}^{\infty} f(\alpha n) = \frac{1}{2\pi j} \sum_{0}^{\infty} \int_{c-j\infty}^{c+j\infty} e^{p\alpha n} F(p) \, dp.$$

This is permissible since the inversion integral holds identically for all values of x, with the exception of certain values of x for which $f(x)$ may be discontinuous. In our case it is assumed that $f(x)$ is a continuous function of x. If, now, the inversion integral involving $F(p)$ is uniformly convergent, we may interchange the order of integration and summation to get

$$\sum_{0}^{\infty} f(\alpha n) = \frac{1}{2\pi j} \int_{c-j\infty}^{c+j\infty} F(p) \sum_{0}^{\infty} e^{p\alpha n} \, dp$$

$$= \frac{1}{2\pi j} \int_{c-j\infty}^{c+j\infty} \frac{F(p)}{1 - e^{p\alpha}} \, dp$$

provided $f(x)$ is of such a form that we may take $c < 0$ in order that $\sum_{0}^{\infty} e^{p\alpha n}$ will be a convergent series. This implies that $f(x)$ is asymptotic to $e^{-\epsilon x}$, $\epsilon > 0$, as x approaches infinity. In general, this condition is not satisfied, but, in practice, we can multiply $f(\alpha n)$ by $e^{-\epsilon n}$, sum the resulting series, and then take the limit as ϵ approaches zero. If we can evaluate the resulting integral, we have the sum of the series in closed form. Alternatively, we may expand the integral by the residue theorem. Let us choose for our contour the line $p = c$, $c < 0$, and the semicircle in the right half plane. Since $F(p)$ is analytic for all $u > c$, the only poles of the integrand occur at $e^{\alpha p} = 1$ or $p = \pm(2\pi nj/\alpha)$, $n = 0, 1, 2, \ldots$. The residues at these poles are $-(1/\alpha)F(\pm 2\pi nj/\alpha)$. Hence, provided the integral around the semicircle vanishes, we get

$$\sum_{0}^{\infty} f(\alpha n) = \frac{1}{\alpha} \sum_{0}^{\infty} F\left(\frac{2\pi jn}{\alpha}\right).$$

The change in sign is due to the fact that the contour is traversed in a clockwise sense. Sometimes it is possible to close the contour in the left half plane, and the sum of the series is then given in terms of the residues at the poles of $F(p)$.

As a first example, we will sum the series $\sum_{0}^{\infty} ne^{-na}$. The Laplace transform of xe^{-ax} is $(p + a)^{-2}$, and $\int_{c-j\infty}^{c+j\infty}(e^{px}/(p + a)^2)\,dp$ is uniformly convergent. Furthermore, $F(p)$ is analytic for all $\operatorname{Re} p > -a$. Therefore, we have

$$\sum_{1}^{\infty} ne^{-na} = \sum_{-\infty}^{\infty}(2n\pi j + a)^{-2} = \frac{-1}{4\pi^2}\sum_{-\infty}^{\infty}\left(n + \frac{a}{2\pi j}\right)^{-2} = -\frac{1}{4}\sinh^{-2}\frac{a}{2}$$

which is the same result as obtained in the first section by a different method. The above result follows when use is made of the partial-fraction expansion of $\csc^2(a/2)$; that is,

$$\sum_{-\infty}^{\infty}\frac{1}{(n + a)^2} = \pi^2 \csc^2 \pi a.$$

If we close the contour in the left half plane, the only pole of the integrand is a double pole at $p = -a$. The residue at this pole is

$$\frac{d}{dp}\left[(p + a)^2 \frac{1}{(1 - e^p)(p + a)^2}\right]\Bigg|_{p=-a} = \frac{e^{-a}}{(1 - e^{-a})^2} = \frac{1}{4}\sinh^{-2}\frac{a}{2}.$$

Hence, we see that closing the contour in the left half plane has the advantage of giving the sum in closed form directly.

We may extend the definition of the Laplace transform so that it is valid for negative values of x as follows:

$$\mathcal{L}f(x) = \int_{-\infty}^{0} f(x)e^{-px}\,dx, \qquad x < 0.$$

The inversion formula gives

$$f(x) = \frac{1}{2\pi j}\int_{c-j\infty}^{c+j\infty} e^{px} F(p)\,dp$$

where c now lies to the left of all the singularities of $F(p)$. For positive values of x, the contour may be closed in the left half plane, and, since no singularities are enclosed, the integral vanishes along with $f(x)$. For negative values of x, the contour may be closed in the right half plane, and the original function $f(x)$ is recovered.

It is now possible to consider a series such as $\sum_{-\infty}^{\infty} f(\alpha n)$. We will impose on $f(x)$ the condition that it be of integrable square over $-\infty \le x \le \infty$; that is, $\int_{-\infty}^{\infty}|f(x)|^2\,dx$ exists. The existence of the integral implies that $f(x)$ has no poles on the real axis. Let us break $f(x)$ up into two parts as follows:

$$f_{+}(x) = \begin{cases} f(x), & x \ge 0 \\ 0, & x < 0 \end{cases} \qquad f_{-}(x) = \begin{cases} 0, & x > 0 \\ f(x), & x < 0. \end{cases}$$

The corresponding Laplace transforms are

$$F_+(p) = \int_0^\infty e^{-px} f_+(x)\, dx \qquad F_-(p) = \int_{-\infty}^0 e^{-px} f_-(x)\, dx.$$

The inverse transforms are

$$f_+ = \frac{1}{2\pi j} \int_{c_+ - j\infty}^{c_+ + j\infty} e^{px} F_+ \, dp \qquad f_- = \frac{1}{2\pi j} \int_{c_- - j\infty}^{c_- + j\infty} e^{px} F_- \, dp.$$

From Parseval's theorem we have that $\int_{-j\infty}^{j\infty} |F(p)|^2\, dp$ exists if $f(x)$ is of integrable square. Consequently, $F(p)$ has no poles on the imaginary axis, and we may, therefore, take $c_+ < 0$ and $c_- > 0$ in the above inversion integrals. Under these conditions, we get

$$\sum_0^\infty f(\alpha n) = \frac{1}{2\pi j} \int_{c_+ - j\infty}^{c_+ + j\infty} F_+(p)(1 - e^{\alpha p})^{-1}\, dp = \frac{1}{\alpha} \sum_{-\infty}^\infty F_+ \left(\frac{2\pi n j}{\alpha} \right)$$

upon closing the contour in the right half plane. Similarly,

$$\sum_{-1}^{-\infty} f(\alpha n) = \frac{1}{\alpha} \sum_{-\infty}^\infty F_- \left(\frac{2\pi n j}{\alpha} \right)$$

upon closing the contour in the left half plane. Combining these results, we have, finally,

$$\sum_{-\infty}^\infty f(\alpha n) = \frac{1}{\alpha} \sum_{-\infty}^\infty F \left(\frac{2\pi n j}{\alpha} \right).$$

If the series on the right-hand side can be summed, or if it is more rapidly converging than the original series, some advantage will have been gained. As an example in the use of the above generalized Laplace transform, we will sum the series $\sum_{-\infty}^\infty (e^{j\alpha n} / (n^2 - a^2))$. The general term, considered as a function of x, has poles at $x = \pm a$. Provided a has a small imaginary part, this function is of integrable square over $-\infty \le x \le \infty$. The Laplace transforms are

$$F_+ = \int_0^\infty \frac{e^{j\alpha x - px}}{x^2 - a^2}\, dx \qquad F_- = \int_{-\infty}^0 \frac{e^{j\alpha x - px}}{x^2 - a^2}\, dx.$$

F_+ is analytic for all $\operatorname{Re} p \ge 0$, while F_- is analytic for all $\operatorname{Re} p \le 0$. Consequently, we may evaluate the transform for the particular value $\operatorname{Re} p = 0$. We have

$$F(p) = \int_{-\infty}^\infty \frac{e^{j\alpha x - px}}{x^2 - a^2}\, dx.$$

For $\alpha - v \ge 0$, we may replace x by the complex variable $z = x + jy$, and evaluate the integral by contour integration, closing the contour in the upper half plane. The only pole of the integrand in the upper half plane is at $z = a$. Hence,

$$F(p) = \frac{\pi j}{a} e^{(j\alpha - p)a}, \qquad p = jv, \; v \le \alpha.$$

For $\alpha - v \le 0$, the contour is closed in the lower half plane and gives

$$F(p) = \frac{\pi j}{a} e^{-(j\alpha - p)a}, \qquad p = jv, \; v \ge \alpha.$$

These results may be combined and give $F(p) = (\pi j/a)e^{j|\alpha - v|a}$ valid for Re $p = 0$ and all v. It may be easily shown that $F(p)$ is of integrable square over $-j\infty \leq p \leq j\infty$, and thus, we may apply the previously derived summation formula. That is,

$$\sum_{-\infty}^{\infty} \frac{e^{j\alpha n}}{n^2 - a^2} = \frac{\pi j}{a} \sum_{-\infty}^{\infty} e^{j|\alpha - 2\pi n|a}.$$

Restricting α to the range $0 \leq \alpha \leq 2\pi$, we have

$$\frac{\pi j}{a} \sum_{-\infty}^{\infty} e^{j|\alpha - 2n\pi|a} = \frac{\pi j}{a}\left[(e^{-j\alpha a} + e^{j\alpha a})\sum_{1}^{\infty} e^{j2\pi na} + e^{j\alpha a}\right]$$

$$= \frac{\pi j}{a}(j\cos\alpha a \cot \pi a + j\sin\alpha a) = -\frac{\pi}{a}\frac{\cos(\pi - \alpha)a}{\sin \pi a}.$$

The same result may be derived by the contour-integration method. In fact, for series of the type considered above, the contour-integration method is considerably more straightforward.

If we make the following changes in notation, $p = -j\omega$, $dp = -j\,d\omega$, then our generalized Laplace transforms become Fourier transforms, and the resulting formula is one form of the Poisson summation formula. We have

$$F(\omega) = \int_{-\infty}^{\infty} e^{j\omega t} f(t)\,dt = \mathfrak{F}f(t)$$

which is the Fourier transform of $f(t)$. The inversion integral becomes

$$f(t) = \frac{1}{2\pi}\int_{-\infty}^{\infty} e^{-j\omega t} F(\omega)\,d\omega.$$

By a procedure similar to that used for the Laplace transform, we obtain the following summation formula:

$$\sum_{-\infty}^{\infty} f(\alpha n) = \frac{1}{\alpha}\sum_{-\infty}^{\infty} F\left(\frac{2n\pi}{\alpha}\right)$$

where $F(\omega)$ is the Fourier transform of $f(t)$ as defined above.

The utility of the Poisson summation formula in converting a slowly convergent series into a rapidly convergent series may be appreciated from the following property of the Fourier transform. If $f(t)$ is a function which is localized near $t = 0$, then $F(\omega)$ is "spread out," or has an appreciable value for a large range of ω; for example, if $f(t) = \delta(t)$, then $F(\omega) = 1$. Conversely, if $f(t)$ is spread out along the t axis, then $F(\omega)$ is concentrated near $\omega = 0$. A slowly convergent series with a typical term $f(\alpha n)$ is thus converted to a rapidly converging series with a typical term $F(2n\pi/\alpha)$, since F will be small for large values of n when $f(\alpha n)$ decreases slowly with increasing n.

Many useful summation formulas may also be derived using Mellin transforms in an analogous way. The interested reader should consult the references at the end of this appendix for illustrations of the use of Mellin transforms in summing series.

Summation of Odd Series

In this section we shall consider a method by which many series that are odd functions of the summation variable n can be reduced to much more rapidly converging series. As an example to illustrate the method, consider the series $\sum_{1}^{\infty}((\sin n\alpha)/(n^2 - a^2))$, $0 \leq \alpha \leq \pi$, $a \gg 1$, and a not an integer. Since $a \gg 1$, the numerical evaluation of the above series must include terms up to $n = N$, where $N^2 \gg a^2$. For instance, if $a \approx 10$, at least 30 terms must be included. The oscillatory nature of the sine function further adds to the slow convergence of the series. If the above series is multiplied by $\tanh np$, it becomes an even function of n, and the resulting series is summable by the contour-integration method. We may write the above series as

$$\sum_{1}^{\infty} \frac{\sin n\alpha}{n^2 - a^2} = \sum_{1}^{\infty} \frac{\sin n\alpha \tanh np}{n^2 - a^2} + \sum_{1}^{\infty} \frac{\sin n\alpha}{n^2 - a^2}(1 - \tanh np)$$

where the second series on the right is rapidly converging, provided p is not too small, i.e., for $p = 1$; four terms are sufficient since $0.9993 < \tanh np < 1$ for $n > 4$. The first series on the right may be written as

$$\frac{1}{2j} \sum_{-\infty}^{\infty} \frac{e^{jn\alpha} \tanh np}{n^2 - a^2}.$$

This series is readily summed by the contour-integration method. The poles of the integrand are at $w = \pm a$ and at $w = \pm j[(2n + 1)/2p]\pi$, where $\tanh wp$ becomes infinite. Carrying through the algebra, we get

$$\sum_{1}^{\infty} \frac{\sin n\alpha}{n^2 - a^2} = \sum_{1}^{\infty} \frac{\sin n\alpha}{n^2 - a^2}(1 - \tanh np) + \frac{\pi}{2a} \tanh pa \frac{\sin(\pi - \alpha)a}{\sin \pi a}$$

$$-\frac{\pi}{p} \sum_{n=1,3,\dots}^{\infty} \frac{\sinh(n\pi/2p)(\pi - \alpha)}{(n^2\pi^2/4p^2 + a^2)[\sinh(n\pi/2p)\pi]}.$$

The choice of the value of p is governed by the values of the parameters α and a. When α is small, it is advantageous to take p small. The terms are asymptotic to

$$\frac{\pi}{p} e^{-n\pi\alpha/2p} \left(\frac{n^2\pi^2}{4p^2} + a^2 \right)^{-1}$$

and, hence, it is desirable to have $p = \pi\alpha$, so that the exponential term is less than 0.01 for $n > 9$. However, if α is very small, then it is desirable to take $p > \pi\alpha$, so that the correction series involving $1 - \tanh np$ will converge rapidly enough.

When $a = 1.5$, $\alpha = 1$, and $p = \pi$, it takes 25 terms of the original series to give three-figure accuracy for the sum. The same accuracy can be obtained from 1 term of the correction series plus 5 terms of the hyperbolic-sine series. The sum is -0.214, and we can see that the savings in computational labor is considerable.

In applied work many of the series that arise are associated with Green's functions. By constructing the relevant Green's function by Methods I and II of Chapter 2, a variety of summation formulas can be derived. Examples of summation formulas that can be obtained from the alternative representations of Green's functions are given in Chapter 2, in the Problems at the end of Chapters 2 and 5, in Sections 6.5, 7.1, 7.2, and 12.3, as well as in several other sections of the book.

A.7. FOURIER TRANSFORM IN THE COMPLEX DOMAIN

Contour Integration

Consider a function $f(x, y) = f(Z)$ of the complex variable $Z = x + jy$. f is analytic at all points at which it has a unique derivative. This means that $\lim_{\Delta Z \to 0}((f(Z + \Delta Z) - f(Z))/\Delta Z)$ must be independent of the direction of ΔZ in the complex plane (see Fig. A.3). If we choose $\Delta Z = \Delta x$ we get $df/dZ = \partial u/\partial x + j(\partial v/\partial x)$ where we have written $f = u(x, y) + jv(x, y)$. If we choose $\Delta Z = j\,\Delta y$ we get $df/dZ = (1/j)(\partial u/\partial y) + \partial v/\partial y$. For df/dZ to have a unique value we thus see that

$$\frac{\partial u}{\partial x} = \frac{\partial v}{\partial y} \qquad \frac{\partial u}{\partial y} = -\frac{\partial v}{\partial x}$$

which are called the Cauchy–Riemann equations. It may be shown that the Cauchy–Riemann equations are also sufficient to guarantee that $f(Z)$ be analytic.

Cauchy Integral Formula

Let $f(Z) = u + jv$ be analytic within and on a closed contour. Then we can show that $\oint_C f\,dZ = 0$ where C is arbitrary (see Fig. A.4(a)). To show the validity of this result consider a vector $\mathbf{A} = \mathbf{a}_x A_x(x, y) + \mathbf{a}_y A_y(x, y)$ and use Stokes' theorem to get $\oint_C \mathbf{A}\cdot d\mathbf{l} = \iint_S \nabla \times \mathbf{A}\cdot d\mathbf{S}$ or $\oint_C (A_x\,dx + A_y\,dy) = \iint_S (\partial A_y/\partial x - \partial A_x/\partial y)\,dx\,dy$. We have $\oint_C f\,dZ = \oint_C (u + jv)(dx + j\,dy) = \oint_C (u\,dx - v\,dy) + j\oint_C (v\,dx + u\,dy)$. In the first integral on the right let $u = A_x$, $v = -A_y$ and apply Stokes' theorem to obtain $\oint_C (u\,dx - v\,dy) = \iint_S (\partial(-v)/\partial x - \partial u/\partial y)\,dx\,dy = 0$ because $\partial v/\partial x = -\partial u/\partial y$ from the Cauchy–Riemann equations. To show that the second integral on the right vanishes let $v = A_x$ and $u = A_y$ and apply Stokes' theorem again.

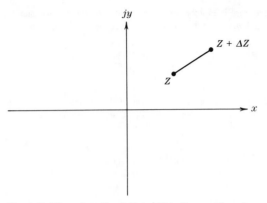

Fig. A.3. The points Z and $Z + \Delta Z$ in the complex plane.

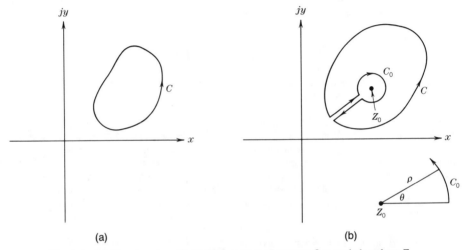

(a)   (b)

Fig. A.4. (a) Closed contour C. (b) Deformation of contour C to exclude pole at Z_0.

Consider now $f(Z) = g(Z)/(Z - Z_0)^n$ where g is analytic inside and on C and Z_0 is inside C. Consider a modified contour that includes a circle C_0 around the singular point Z_0 as in Fig. A.4(b). Then within and on $C + C_0$ f is analytic and hence

$$\oint_{C+C_0} f \, dZ = 0 = \oint_C f \, dZ + \oint_{C_0} f \, dZ$$

since the integral along the cut is traversed twice in opposite directions and vanishes. Thus we have

$$\oint_C f \, dZ = \oint_C \frac{g}{(Z - Z_0)^n} \, dZ = -\oint_{C_0} \frac{g}{(Z - Z_0)^n} \, dZ.$$

On C_0, $Z - Z_0 = \rho e^{j\theta}$, $dZ = -j\rho e^{j\theta} \, d\theta$; hence

$$\oint_C f \, dZ = \int_0^{2\pi} \frac{g(Z) j \rho e^{j\theta} \, d\theta}{(\rho e^{j\theta})^n}.$$

But g is analytic at Z_0 and we can choose ρ so small that $g(Z) \approx g(Z_0)$ everywhere on C_0. Thus

$$\oint_C f\,dZ = g(Z_0)j\rho^{-n+1}\int_0^{2\pi} e^{-j\theta(n-1)}\,d\theta = \begin{cases} 0, & n \neq 1 \\ 2\pi jg(Z_0), & n = 1. \end{cases}$$

Residue Theory

Consider a function f that has a pole of order n at Z_0. This means that near Z_0, f behaves as constant$/(Z - Z_0)^n$. The function $g = (Z - Z_0)^n f(Z)$ will be analytic at Z_0. A Taylor expansion about Z_0 gives $g(Z) = g(Z_0) + g'(Z_0)(Z - Z_0) + (1/2!)g''(Z_0)(Z - Z_0)^2 \cdots$. Hence we can write

$$f(Z) = \frac{g(Z_0)}{(Z - Z_0)^n} + \cdots + \frac{g^{n-1}(Z_0)}{(n-1)!(Z - Z_0)} + \frac{g^n(Z_0)}{n!} + \frac{g^{n+1}(Z_0)(Z - Z_0)}{(n+1)!} + \cdots .$$

The coefficient of $(Z - Z_0)^{-1}$ is $g^{n-1}(Z_0)/(n-1)!$ and is called the residue in the series expansion of $f(Z)$, the latter being called a Laurent series. We now see that if we form the contour integral of $f(Z)$ about any contour C enclosing Z_0 we get

$$\oint_C f(Z)\,dZ = 2\pi j \text{ (residue at } Z_0)$$

$$= 2\pi j \left. \frac{1}{(n-1)!}\frac{d^{n-1}}{dZ^{n-1}}(Z - Z_0)^n f(Z)\right|_{Z_0}$$

since only the term involving $(Z - Z_0)^{-1}$ gives a nonzero result. In general, if f has many pole singularities within C we get

$$\oint_C f\,dZ = 2\pi j \sum \text{ (residues of } f \text{ within } C).$$

The other type of singularity commonly encountered is the branch point singularity which arises in connection with multivalued functions. For example, $f = \sqrt{Z}$ has a branch point at $Z = 0$ since if we evaluate f on a small circular contour C enclosing $Z = 0$ we find that after going through an angle 2π that f becomes $-f$, i.e., $f(\rho e^{2j\pi}) = -f(\rho)$. The contour-integration theory given above must be modified when f has branch points.

Let $f = g(Z)/h(Z)$ where g is analytic at Z_0. If h has a zero of order one at Z_0 then f has a pole of order one. The residue at Z_0 is given by:

$$\text{residue} = \lim_{z \to Z_0} \frac{g(Z)}{h(Z)}(Z - Z_0).$$

By L'Hospitals' rule the limit is given by $\{d[g(Z - Z_0)]/dZ/(dh/dZ)\}$ at Z_0 or $[(Z - Z_0)g' + g]/h' = g(Z_0)/h'(Z_0)$ which is a convenient formula for the residue in this special case.

Jordan's Lemma

Jordan's lemma states that the integral over the semicircle shown in Fig. A.5

$$\lim_{\rho \to \infty} \int_C e^{jwt}F(w)\,dw = 0$$

provided the limit of $|F(w)| = K|w|^\epsilon$ as $|w|$ approaches infinity, where $\epsilon < 0$, $t > 0$. The proof is given below. On

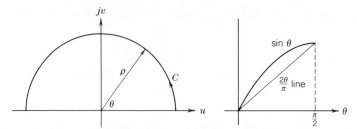

Fig. A.5. Illustration for Jordan's lemma.

the semicircle $w = \rho \cos \theta + j\rho \sin \theta$ and

$$\left| \lim_{\rho \to \infty} \int_C e^{jwt} F(w)\, dw \right| \leq \lim_{\rho \to \infty} \int_C |e^{jwt}|\, |F|\rho\, d\theta$$

$$= \lim_{\rho \to \infty} \int_0^{\pi} e^{-\rho t \sin \theta} K\rho^{1+\epsilon}\, d\theta$$

$$= \lim_{\rho \to \infty} K\rho^{1+\epsilon} 2 \int_0^{\pi/2} e^{-\rho t \sin \theta}\, d\theta$$

$$\leq \lim_{\rho \to \infty} K\rho^{1+\epsilon} 2 \int_0^{\pi/2} e^{-\rho t 2\theta/\pi}\, d\theta$$

since $\sin \theta \geq 2\theta/\pi$ for $0 \leq \theta \leq \pi/2$ as shown in Fig. A.5. We thus obtain

$$\lim_{\rho \to \infty} K\rho^{1+\epsilon} 2 \frac{e^{-\rho t} - 1}{-2\rho t/\pi} = 0 \qquad \text{for } \epsilon < 0.$$

This lemma is useful in establishing conditions under which an inverse Fourier transform can be evaluated using residue theory, which requires closing the contour by a semicircle in either the upper or lower half of the complex plane.

Fourier Transforms in the Complex Plane

Let $f(t)$ be of exponential order at $t = \pm \infty$, i.e.,

$$f(t) < e^{\alpha_+ t} \qquad \text{as } t \to +\infty$$
$$f(t) < e^{-\alpha_- t} \qquad \text{as } t \to -\infty.$$

Define

$$f_+(t) \equiv f(t), \quad t \geq 0 \qquad f_-(t) \equiv f(t), \quad t \leq 0$$
$$\equiv 0, \qquad t < 0 \qquad\qquad \equiv 0, \qquad t > 0.$$

Let

$$F_+(w) = \mathcal{F} f_+(t) = \int_{-\infty}^{\infty} e^{-jwt} f_+(t)\, dt = \int_0^{\infty} e^{-jwt} f_+(t)\, dt, \qquad w = \omega + j\sigma.$$

The integral is uniformly convergent for all w in the lower half plane $\sigma < -\alpha_+$ and defines an analytic function in the lower half plane; i.e., F_+ is analytic for $\sigma < -\alpha_+$. To recover f_+ we use the inversion formula

$$f_+(t) = \frac{1}{2\pi} \int_{C_+} F_+(w) e^{jwt}\, dw.$$

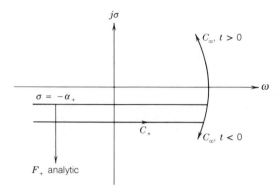

Fig. A.6. Inversion contour for F_+.

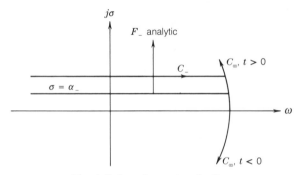

Fig. A.7. Inversion contour for F_-.

The contour C_+ must be chosen to be parallel with the w axis in the lower half plane $\sigma < -\alpha_+$ for the following reasons (see Fig. A.6): For $t < 0$, e^{jwt} becomes exponentially small in the lower half plane. Thus $\int_{C_\infty} F_+ e^{jwt}\, dw = 0$. Hence $\int_{C_+} F_+ e^{jwt}\, dw$ can be replaced by a closed contour integral $\oint_{C_+ + C_\infty} F_+ e^{jwt}\, dw$. But to obtain $f_+(t) \equiv 0$ for $t < 0$ we see that the above closed contour integral must vanish and hence C_+ must be chosen so that F_+ is analytic within $C_+ + C_\infty$. For $t > 0$ we can close the contour in the upper half plane to obtain $f_+ = (1/2\pi) \int_{C_+} F_+ e^{jwt}\, dw = (1/2\pi) \oint F_+ e^{jwt}\, dw = j\sum$ (residues of F_+ in upper half plane). The contour C_+ can be distorted in any arbitrary manner as long as it is not moved across a singularity of F_+.

The transform of f_- is handled in a similar way. Thus (see Fig. A.7)

$$F_-(w) = \int_{-\infty}^{0} f_-(t) e^{-jwt}\, dt, \qquad f_- = \frac{1}{2\pi} \int_{C_-} F_-(w) e^{jwt}\, dw.$$

If α_+ and α_- are such that F_+ and F_- have a common strip in which both are analytic, then we can choose $C_+ = C_- = C$ which is a common inversion contour as in Fig. A.8(a). Then $f = f_+ + f_- = (1/2\pi) \int_{C_+} F_+ e^{jwt}\, dw + (1/2\pi) \int_{C_-} F_- e^{jwt}\, dw = (1/2\pi) \int_C F e^{jwt}\, dw$ where $F = F_+ + F_-$.

Example

Let $f(t) = e^{-jk|t|}$; then

$$F_+ = \int_0^{\infty} e^{-jkt-jwt}\, dt = \frac{e^{-jkt-jwt}}{-j(k+w)}\Bigg|_0^{\infty} = \frac{1}{j(k+w)}, \qquad \sigma < 0$$

$$F_- = \int_{-\infty}^{0} e^{jkt-jwt}\, dt = \frac{e^{jkt-jwt}}{-j(w-k)}\Bigg|_{-\infty}^{0} = \frac{1}{-j(w-k)}, \qquad \sigma > 0.$$

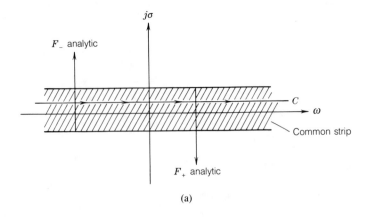

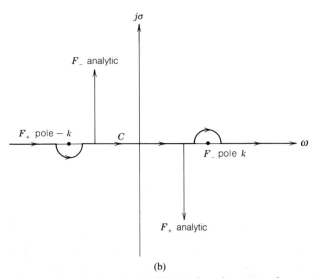

Fig. A.8. (a) Common inversion contour. (b) Common inversion contour for example discussed in text.

A common inversion contour C can be chosen as in Fig. A.8(b) by deforming C_+ and C_- without crossing the poles at $w = -k$ for F_+ and $w = k$ for F_-. Then we can write $f = (1/2\pi) \int_C e^{jwt} F(w)\,dw$ where $F = F_+ + F_- = (1/j)(1/(w+k) - 1/(w-k)) = -2k/(j(w^2 - k^2))$.

Final Value Theorem

This theorem gives the asymptotic behavior of the Fourier transform as $|w| \to \infty$ in terms of the behavior of $f(x)$ as $x \to 0$. Consider a function $f(x) \equiv 0$ for $x < 0$, $f(x) \neq 0$ for $x > 0$, and let $f(x) \sim Kx^\alpha$ as $x \to 0$. Also let this be the dominant singularity if $\alpha < 0$. We have

$$F(w) = \int_0^\infty e^{jwx} f(x)\,dx.$$

Let $wx = \lambda$, $dx = d\lambda/w$; then

$$F(w) = \int_0^\infty e^{j\lambda} f\left(\frac{\lambda}{w}\right) \frac{d\lambda}{w}.$$

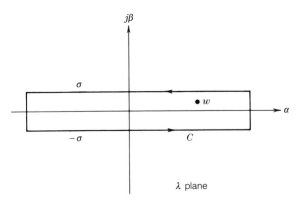

Fig. A.9. Contour for application of Cauchy's formula.

Now let $w \to \infty$ and use $f(\lambda/w) \sim K(\lambda/w)^\alpha$ to obtain

$$\lim_{w \to \infty} F(w) = \int_0^\infty e^{j\lambda} K \frac{\lambda}{w^{\alpha+1}} \, d\lambda = w^{-\alpha-1} \int_0^\infty K\lambda^\alpha e^{j\lambda} \, d\lambda.$$

Hence the asymptotic behavior of $F(w)$ is determined by that of $f(x)$ at $x = 0$. If $f(x) \sim x^{1/2}$ then $F(w) \sim w^{-3/2}$, while if $f(x) \sim x^{-1/2}$ then $F(w) \sim w^{-1/2}$.

A.8. WIENER–HOPF FACTORIZATION

In the solution of Wiener–Hopf integral equations we are often required to factorize a function $H(w)$ of the complex variable w into the sum form $H_+(w) + H_-(w)$ or into the product form $H_+(w)H_-(w)$ where $H_+(w)$ is analytic in the upper half plane and $H_-(w)$ is analytic in the lower half plane. When the factorization cannot be done by inspection it can often be carried out using Cauchy's formula.

Sum Factorization

Let $H(w)$ be analytic within and on the contour C shown in Fig. A.9. Then

$$H(w) = \frac{1}{2\pi j} \int_C \frac{H(\lambda)}{\lambda - w} \, d\lambda.$$

Now assume that $H(\lambda) \to 0$ as $\lambda \to \pm\infty$ for $-\sigma \le \beta \le \sigma$ so that as we extend C to $\pm\infty$ the contributions to the integral from the ends of the contour vanish. We will then obtain

$$H(w) = \frac{1}{2\pi j} \int_{-\infty-j\sigma}^{\infty-j\sigma} \frac{H(\lambda)}{\lambda - w} \, d\lambda - \frac{1}{2\pi j} \int_{-\infty+j\sigma}^{\infty+j\sigma} \frac{H(\lambda)}{\lambda - w} \, d\lambda$$

where in the second integral we have reversed the direction of integration. The first integral has its contour below the pole at $\lambda = w$ and is therefore analytic for all w with $\operatorname{Im} w > -\sigma$, i.e., in the upper half plane. It represents $H_+(w)$. Similarly, the second integral represents $H_-(w)$ since it is analytic in the lower half plane. Thus we have

$$H_+(w) = \frac{1}{2\pi j} \int_{-\infty-j\sigma}^{\infty-j\sigma} \frac{H(\lambda)}{\lambda - w} \, d\lambda$$

$$H_-(w) = \frac{1}{2\pi j} \int_{-\infty+j\sigma}^{\infty+j\sigma} \frac{H(\lambda)}{\lambda - w} \, d\lambda.$$

Example

Consider the function

$$H(w) = \frac{\sin wb}{(w^2 + a^2)}.$$

We need to associate the term e^{jwb} with H_+ since this decays in the upper half plane, while e^{-jwb} will grow in the upper half plane. We also need to include the pole at $w = -ja$ with H_+ and that at $w = ja$ with H_-. By using the factorization formulas derived above we get

$$H_+(w) = \frac{1}{2\pi j} \int_{-\infty-ja}^{\infty-ja} \frac{e^{j\lambda b} - e^{-j\lambda b}}{2j(\lambda^2 + a^2)(\lambda - w)} \, d\lambda.$$

The integral with the factor $e^{j\lambda b}$ can be closed in the upper half plane, while that with the factor $e^{-j\lambda b}$ can be closed in the lower half plane. The residue evaluation gives

$$H_+(w) = \frac{e^{-ab}}{2j(2ja)(ja - w)} + \frac{e^{jwb}}{2j(w^2 + a^2)} + \frac{e^{-ab}}{2j(2ja)(ja + w)}.$$

The second term approaches $e^{-ab}/2j(2ja)(w - ja)$ as $w \to ja$ and hence cancels the first term at the point $w = ja$. Thus $w = ja$ is not a pole. The only pole is that at $w = -ja$ and e^{jwb} decays in the upper half plane, so clearly $H_+(w)$ is analytic in the upper half plane. In a similar way we find that

$$H_-(w) = -\frac{e^{-jwb}}{2j(w^2 + a^2)} - \frac{e^{-ab}}{2j(2ja)(w + ja)} - \frac{e^{-ab}}{2j(2ja)(ja - w)}.$$

Product Factorization

If we want to factor $H(w)$ into the product form $H_+(w)H_-(w)$ we consider the function $\ln H(w) = \ln H_+(w) + \ln H_-(w)$ and use $\ln H(w)$ in Cauchy's formula. Thus we obtain

$$\ln H_+(w) = \frac{1}{2\pi j} \int_{-\infty-ja}^{\infty-ja} \frac{\ln H(\lambda)}{\lambda - w} \, d\lambda$$

$$\ln H_-(w) = -\frac{1}{2\pi j} \int_{-\infty+ja}^{\infty+ja} \frac{\ln H(\lambda)}{\lambda - w} \, d\lambda.$$

We now have

$$H_+(w) = e^{\ln H_+(w)}, \qquad H_-(w) = e^{\ln H_-(w)}.$$

A.9. ASYMPTOTIC EVALUATION OF INTEGRALS BY THE SADDLE-POINT METHOD

The saddle-point method or method of steepest descents is a generalization of Rayleigh's method of stationary phase for the asymptotic evaluation of certain types of integrals that commonly occur in diffraction and radiation problems. A typical integral is

$$I = \int_C F(\phi) e^{-jk_0 r \cos(\phi - \theta)} \, d\phi$$

where C is the contour shown in Fig. A.10. The method consists of deforming the contour C into a steepest-descent contour (SDC) along which $e^{-jk_0 r \cos(\phi - \theta)}$ will be exponentially decreasing at a maximum rate. The major contribution to I will then come from a short portion of the transformed contour along which F can be represented by a simple series expansion and such that the integral can be evaluated term by term. The first term is the dominant one.

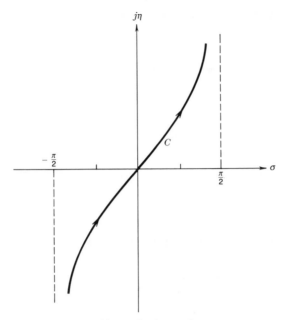

Fig. A.10. Contour of integration in complex $\phi = \sigma + j\eta$ plane.

We will consider two important cases: (I.) $F(\phi)$ has no singularity in the vicinity of the saddle point, and (II.) $F(\phi)$ has a simple pole near the saddle point.

Case I: F(ϕ) Is Analytic near the Saddle Point

Let $f(\phi) = -jk_0 r \cos(\phi - \theta)$. The saddle points are the values of ϕ for which $df/d\phi = 0$. We have $df/d\phi = jk_0 r \sin(\phi - \theta)$ so $\phi = \theta$ is a saddle point. A complex function such as f can have no maxima or minima and hence the stationary points are saddle points.

Near the saddle point the SDC passes through the saddle point at an angle $\pi/4$ relative to the σ axis. The SDC is specified by the condition $\text{Im}[f(\phi) - f(\theta)] = 0$. Thus $\text{Im}[-jk_0 r \cos(\sigma - \theta - j\eta) + jk_0 r] = jk_0 r[1 - \cos(\sigma - \theta) \cosh \eta] = 0$. Along the SDC

$$e^{f(\phi)} = e^{-jk_0 r \cos(\phi - \theta)} = e^{-jk_0 r - k_0 r \sin(\sigma - \theta) \sinh \eta}.$$

At this point it is convenient to change variables according to

$$w^2 = 2j[\cos(\phi - \theta) - 1] \qquad \text{or} \qquad w = 2e^{-j\pi/4} \sin \frac{(\phi - \theta)}{2}$$

i.e.,

$$w^2 = -\frac{2}{k_0 r}[f(\phi) - f(\theta)], \qquad dw = e^{-j\pi/4} \cos \frac{(\phi - \theta)}{2} d\phi$$

which gives

$$d\phi = \frac{e^{j\pi/4} dw}{\sqrt{1 - jw^2/4}}.$$

Our integral becomes

$$I = \int_C F(w) e^{-jk_0 r + j\pi/4} e^{-k_0 r w^2/2} (1 - jw^2/4)^{-1/2} dw$$

where C is the SDC consisting of the real axis in the $w = u + jv$ plane. We now expand $F(w)(1 - jw^2/4)^{-1/2}$ in a Taylor series about the saddle point $w = 0$. Thus

$$F(w)(1 - jw^2/4)^{-1/2} = \sum_{n=0}^{\infty} a_n w^n$$

where

$$a_n = \frac{1}{n!} \frac{d^n}{dw^n} [F(w)(1 - jw^2/4)^{-1/2}] \bigg|_{w=0}.$$

By using the result

$$\int_{-\infty}^{\infty} e^{-k_0 r w^2/2} w^{2n} \, dw = \left(\frac{2}{k_0 r}\right)^{n+1/2} \Gamma\left(n + \frac{1}{2}\right) = \frac{\sqrt{2\pi}\, 1 \cdot 3 \cdot 5 \cdot \cdots \cdot (2n-1)}{(k_0 r)^{n+1/2}}$$

and noting that terms involving odd powers of w integrate to zero, we get

$$I = e^{-jk_0 r + j\pi/4} \sum_{n=0}^{\infty} a_{2n} \left(\frac{2}{k_0 r}\right)^{n+1/2} \Gamma\left(n + \frac{1}{2}\right)$$

where Γ is the gamma function. In the above derivation it was assumed that the Taylor series expansion of $F(w)(1 - jw^2/4)^{-1/2}$ could be used for all w. Actually the function has branch points at $w = \pm 2\sqrt{-j}$ and $F(w)$ may also have singularities. However, for $k_0 r$ sufficiently large the error made in using this expansion can be shown to be exponentially small.

As an example we have

$$H_0^2(x) = \frac{1}{\pi} e^{-jx + j\pi/4} \int_{-\infty}^{\infty} e^{-xw^2/2} (1 - jw^2/4)^{-1/2} \, dw.$$

But

$$(1 - jw^2/4)^{-1/2} = 1 + \frac{jw^2}{8} - \frac{3}{8} \frac{w^4}{16} - j\frac{15w^6}{48 \cdot 64} + \cdots$$

and hence we obtain

$$H_0^2(x) \sim \sqrt{\frac{2}{\pi x}} e^{-jx + j\pi/4} \left[1 + \frac{j}{8x} - \frac{9}{128x^2} - j\frac{15^2}{48 \cdot 64 x^3} + \cdots \right]$$

which is the standard asymptotic expansion of the Hankel function.

Case II: $F(\phi)$ Has a Pole near the Saddle Point

In some problems $F(\phi)$ may have a pole very close to the saddle point; in addition, the parameters involved in the problem may be such that the pole can be made to coincide with the saddle point. In the w plane it is therefore of interest to consider the case when $F(w)$ has a pole at w_0 and w_0 lies very close to the saddle point at $w = 0$. If we write

$$\frac{F(w)}{\sqrt{1 - jw^2/4}} = \frac{F(w)}{\sqrt{1 - jw^2/4}} - \frac{R(w_0)}{\sqrt{1 - jw_0^2/4}} \frac{1}{w - w_0} + \frac{R(w_0)}{\sqrt{1 - jw_0^2/4}} \frac{1}{w - w_0}$$

where

$$R(w_0) = \lim_{w \to w_0} (w - w_0) F(w)$$

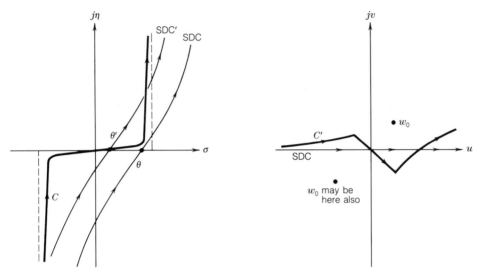

Fig. A.11. Illustration of a contour mapping from the ϕ plane into the w plane.

is the residue of $F(w)$ at the pole w_0, then

$$F_1(w) = \frac{F(w)}{\sqrt{1-jw^2/4}} - \frac{R(w_0)}{(w-w_0)\sqrt{1-jw_0^2/4}}$$

is analytic at w_0. The asymptotic evaluation of

$$I_1 = e^{-jk_0r+j\pi/4}\int_{-\infty}^{\infty}F_1(w)e^{-k_0rw^2/2}\,dw$$

may be carried out according to the method outlined under Case I. The original integral I is thus split into two parts, $I = I_1 + I_2$ where

$$I_2 = \frac{R(w_0)}{\sqrt{1-jw_0^2/4}}e^{-jk_0r+j\pi/4}\int_{-\infty}^{\infty}\frac{e^{-k_0rw^2/2}}{w-w_0}\,dw.$$

The integral

$$I_p = \int_{-\infty}^{\infty}\frac{e^{-k_0rw^2/2}}{w-w_0}\,dw$$

involved in I_2 can be evaluated exactly in terms of the error function or its complement. The original contour C and the SDC in the ϕ plane and the mapping of the SDC onto the w plane are illustrated in Fig. A.11. The contour C may or may not cross the SDC depending on the value of θ. In the ϕ plane we show a SDC labeled SDC' corresponding to the saddle point θ'. The contour C' in the w plane is then the contour C since the mapping $w = 2e^{-j\pi/4}\sin(\phi-\theta)/2$ depends on θ.

Initially we will assume that the point w_0 is not crossed in deforming C into the SDC. Then if $\mathrm{Im}\,w_0 > 0$ we have

$$I_p = j\pi e^{-k_0rw_0^2/2}\left[1+\mathrm{erf}\left(j\sqrt{\frac{k_0r}{2}}w_0\right)\right]$$

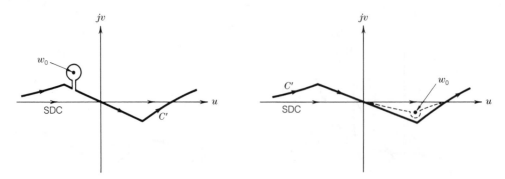

Fig. A.12. Deformation of contour C' into the steepest-descent contour showing the capture of a pole at w_0.

where the error function erf(x) is given by

$$\text{erf}(x) = \frac{2}{\sqrt{\pi}} \int_0^x e^{-t^2}\, dt.$$

For Im $w_0 < 0$ we have

$$I_p = j\pi e^{-k_0 r w_0^2/2}\left[-1 + \text{erf}\left(j\sqrt{\frac{k_0 r}{2}}\, w_0\right)\right]$$

$$= -j\pi e^{-k_0 r w_0^2/2}\,\text{erfc}\left(j\sqrt{\frac{k_0 r}{2}}\, w_0\right)$$

where

$$\text{erfc}(x) = 1 - \text{erf}(x) = \frac{2}{\sqrt{\pi}}\int_x^{\infty} e^{-t^2}\, dt$$

is the complement of the error function. When Im $w_0 = 0$ the Cauchy principal value is taken to give

$$I_p = j\pi e^{-k_0 r w_0^2/2}\,\text{erf}\left(j\sqrt{\frac{k_0 r}{2}}\, w_0\right).$$

Consider now what happens if w_0 is located such that C' crosses w_0 when it is deformed into the SDC. If Im $w_0 > 0$ initially, then when C' crosses w_0 we must add a term $-2j\pi e^{-k_0 r w_0^2/2}$ to I_p (see Fig. A.12 and note that the integral around w_0 is in a clockwise direction). If Im $w_0 < 0$, then we must add the negative of the above term to I_p since the integration around w_0 will now be in the positive sense. For the original integral along C we now have the following results:

(1)
$$I = I_1 + j\pi e^{-k_0 r w_0^2/2} A\left[1 + \text{erfc}\left(j\sqrt{\frac{k_0 r}{2}}\, w_0\right)\right]$$

when Im $w_0 > 0$ and the pole is not crossed in deforming C, or when Im $w_0 < 0$ and the pole is crossed in deforming C. The constant A is given by

$$A = \frac{R(w_0)}{\sqrt{1 - jw_0^2/4}}\, e^{-jk_0 r + j\pi/4}.$$

(2)
$$I = I_1 - j\pi A e^{-k_0 r w_0^2/2}\,\text{erfc}\left(j\sqrt{\frac{k_0 r}{2}}\, w_0\right),$$

Im $w_0 < 0$ and C not crossing the pole, or Im $w_0 > 0$ and C crossing the pole when it is deformed into the SDC.

$$(3) \qquad I = I_1 + j\pi A e^{-k_0 r w_0^2/2} \operatorname{erf}\left(j\sqrt{\frac{k_0 r}{2}} w_0\right), \qquad \operatorname{Im} w_0 = 0.$$

Note that we may express I_p in the form

$$I_p = \int_{-\infty}^{\infty} \frac{e^{-k_0 r w^2/2} - e^{-k_0 r w_0^2/2}}{w - w_0} \, dw + e^{-k_0 r w_0^2/2} \int_{-\infty}^{\infty} \frac{dw}{w - w_0}.$$

The first integral is analytic at w_0 and equal to

$$j\pi e^{-k_0 r w_0^2/2} \operatorname{erf}\left(j\sqrt{\frac{k_0 r}{2}} \, w_0\right)$$

while the second integral has the value

$$\int_{-\infty}^{\infty} \frac{dw}{w - w_0} = \begin{cases} j\pi, & \operatorname{Im} w_0 > 0 \\ 0, & \operatorname{Im} w_0 = 0 \\ -j\pi, & \operatorname{Im} w_0 < 0 \end{cases}$$

and accounts for the discontinuous behavior of I_p as a function of w_0.

Method of Stationary Phase

The method of stationary phase for finding the asymptotic values of integrals is closely related to the method of steepest descents. To explore this relationship consider the behavior of $f(\phi)$ in the vicinity of a saddle point as described in Section 11.8 in terms of the function $f_1 + jf_2 = f(\phi) - f(\theta)$. The contours $f_1 = \text{Const.}$ are orthogonal to the contours $f_2 = \text{Const.}$ because of the Cauchy–Riemann equations

$$\frac{\partial f_1}{\partial \sigma} = \frac{\partial f_2}{\partial \eta} \qquad \text{and} \qquad \frac{\partial f_1}{\partial \eta} = -\frac{\partial f_2}{\partial \sigma}.$$

The function f_1 is positive in the quadrants $-\pi/2 < \omega < 0$ and $\pi/2 < \omega < \pi$ and negative in the remaining quadrants. Also, f_1 increases in magnitude away from the saddle point and hence rises in value in the quadrants $-\pi/2$ to 0 and $\pi/2$ to π and falls in algebraic value in the other two quadrants as shown in Fig. 11.24.

There is a steepest-descent path through the saddle point along which f_1 increases negatively as rapidly as possible. Along this path the contours $f_1 = \text{const.}$ are cut at right angles so the path has the direction of

$$\nabla f_1 = \frac{\partial f_1}{\partial \sigma} + j\frac{\partial f_1}{\partial \eta}$$

and coincides with an $f_2 = \text{Const.}$ contour. If the contour C can be deformed into a steepest-descent contour it is clear that $e^{f(\phi)}$ decreases rapidly in value along this path.

Exactly the same reasoning can be applied to the function f_2. Thus the paths of steepest ascent and descent for f_2 occur along the lines given by

$$\nabla f_2 = \frac{\partial f_2}{\partial \sigma} + j\frac{\partial f_2}{\partial \eta}$$

which lie along the contours $f_1 = \text{Const.}$ The latter paths are located at an angle of $\pi/4$ with respect to the steepest-ascent or steepest-descent paths for f_1. Along a steepest-descent path for f_2 our integral would become

$$I = \int_{\text{SDC for } f_2} F(\phi) e^{-jf'' \rho^2/2} \, d\rho.$$

The argument can now be made that because of the rapid variation in phase of the integrand the major contribution to the integral comes from a small region near the saddle point. However, this argument can be replaced by a different one; that is, it can be shown that the contour for the above integral can be deformed into a steepest-descent contour for f_1 instead of f_2 and the value of the integral can be justified on the basis that it agrees with what is obtained by integrating along the SDC for f_1. Thus the method of stationary phase can be justified in terms of the more intuitively clear arguments used in the standard method of steepest descents by shifting the contour of integration. In deforming these contours any contributions from residues of poles swept across must be properly added on. Also the paths at infinity connecting the deformed contours must be considered. Usually these do not give a contribution. The contour yielding the maximum rate of change of phase is given exactly by

$$\text{Re}\,[f(\phi) - f(\theta)] = 0$$

for all values of ϕ.

A.10. SPECIAL FUNCTIONS

1. Bessel Functions

The two independent solutions of Bessel's differential equation are the Bessel and Neumann functions $J_\nu(x)$ and $Y_\nu(x)$. These have the series expansions

$$J_n(x) = \sum_{m=0}^{\infty} \frac{(-1)^m (x/2)^{n+2m}}{m!(n+m)!}$$

$$Y_n(x) = \frac{2}{\pi}\left(\gamma + \ln\frac{x}{2}\right)J_n(x) - \frac{1}{\pi}\sum_{m=0}^{n-1}\frac{(n-m-1)!}{m!}\left(\frac{2}{x}\right)^{n-2m}$$

$$-\frac{1}{\pi}\sum_{m=0}^{\infty}\frac{(-1)^m(x/2)^{n+2m}}{m!(n+m)!}\left(1+\frac{1}{2}+\frac{1}{3}+\cdots+\frac{1}{m}+1+\frac{1}{2}+\frac{1}{3}+\cdots+\frac{1}{n+m}\right)$$

when the order ν is an integer n. γ is Euler's constant and equals 0.5772.

Two related functions are the Hankel functions of the first and second kinds given by

$$H_n^1(x) = J_n(x) + jY_n(x)$$
$$H_n^2(x) = J_n(x) - jY_n(x).$$

When $x = jy$ we obtain the modified Bessel functions

$$I_n(x) = j^{-n}J_n(jy) = j^n J_n(-jy)$$

$$K_n(x) = \frac{\pi}{2}j^{n+1}H_n^1(jy).$$

Recurrence Relations:

$$\frac{2n}{x}Z_n(x) = Z_{n+1}(x) + Z_{n-1}(x)$$

where Z_n is any Bessel function or linear combination of Bessel functions.

$$\frac{2n}{y}I_n(y) = I_{n-1}(y) - I_{n+1}(y)$$

$$\frac{2n}{y}K_n(y) = K_{n+1}(y) - K_{n-1}(y)$$

Differentiation Formulas:

$$xZ'_n(x) = nZ_n(x) - xZ_{n+1}(x) = -nZ_n(x) + xZ_{n-1}(x)$$

$$Z'_0(x) = -Z_1(x)$$

$$yI'_n(y) = nI_n(y) + yI_{n+1}(y) = -nI_n(y) + yI_{n-1}(y)$$

$$yK'_n(y) = nK_n(y) - yK_{n+1}(y) = -nK_n(y) - yK_{n-1}(y)$$

$$I'_0(y) = I_1(y)$$

$$K'_0(y) = -K_1(y)$$

Small-Argument Approximations:

$$J_n(x) \sim \frac{1}{n!} \left(\frac{x}{2}\right)^n$$

$$Y_0(x) \sim \frac{2}{\pi} \left(\gamma + \ln \frac{x}{2}\right)$$

$$Y_n(x) \sim -\frac{(n-1)!}{\pi} \left(\frac{2}{x}\right)^n$$

$$I_n(y) \sim \frac{1}{n!} \left(\frac{y}{2}\right)^n$$

$$K_0(y) \sim -\left(\gamma + \ln \frac{y}{2}\right)$$

$$K_n(y) \sim \frac{(n-1)!}{2} \left(\frac{2}{y}\right)^n$$

Large-Argument Approximations:

$$J_n(x) \sim \sqrt{\frac{2}{\pi x}} \cos\left(x - \frac{\pi}{4} - \frac{n\pi}{2}\right)$$

$$Y_n(x) \sim \sqrt{\frac{2}{\pi x}} \sin\left(x - \frac{\pi}{4} - \frac{n\pi}{2}\right)$$

$$H_n^{1,2}(x) \sim \sqrt{\frac{2}{\pi x}} e^{\pm j(x - \pi/4 - n\pi/2)}$$

$$I_n(y) \sim \frac{e^y}{\sqrt{2\pi y}}$$

$$K_n(y) \sim \sqrt{\frac{\pi}{2y}} e^{-y}$$

Wronskian Relations:

$$J_n(x)Y'_n(x) - J'_n(x)Y_n(x) = \frac{2}{\pi x}$$

$$I_n(y)K'_n(y) - I'_n(y)K_n(y) = -\frac{1}{y}$$

All of the above formulas are also valid when n is a noninteger ν, in which case when $n!$ occurs in a formula it is replaced by $\Gamma(\nu + 1)$.

Analytic Continuation:

$$J_\nu(ze^{jm\pi}) = e^{j\nu m\pi} J_\nu(z)$$

$$H_\nu^1(ze^{j\pi}) = -e^{-j\nu\pi} H_\nu^2(z)$$

$$H_\nu^2(ze^{-j\pi}) = -e^{j\nu\pi} H_\nu^1(z)$$

$$J_{-n}(z) = (-1)^n J_n(z)$$

$$Y_{-n}(z) = (-1)^n Y_n(z)$$

$$H_{-\nu}^1(z) = e^{j\nu\pi} H_\nu^1(z)$$

$$H_{-\nu}^2(z) = e^{-j\nu\pi} H_\nu^2(z)$$

2. Spherical Bessel Functions

Let $Z_\nu(x)$ be any Bessel or modified Bessel function; then the corresponding spherical Bessel or modified Bessel function $z_\nu(x)$ is given by

$$z_\nu(x) = \sqrt{\frac{\pi}{2x}} Z_{\nu+1/2}(x).$$

Recurrence Relations and Differentiation Formulas:

$$z_{\nu-1}(x) + z_{\nu+1}(x) = \frac{2\nu+1}{x} z_\nu(x)$$

$$z_\nu'(x) = \frac{1}{2\nu+1}[\nu z_{\nu-1}(x) - (\nu+1)z_{\nu+1}(x)]$$

In the above z_ν is j_ν, y_ν, $h_\nu^{1,2}$, or any linear combination of these functions.

Series Expansions:

$$j_n(x) = \frac{1}{x}\left[P_{n+1/2}(x)\cos\left(x - \frac{n+1}{2}\pi\right) - Q_{n+1/2}(x)\sin\left(x - \frac{n+1}{2}\pi\right)\right]$$

$$y_n(x) = \frac{1}{x}\left[P_{n+1/2}(x)\sin\left(x - \frac{n+1}{2}\pi\right) + Q_{n+1/2}(x)\cos\left(x - \frac{n+1}{2}\pi\right)\right]$$

where

$$P_{n+1/2}(x) = 1 - \frac{n(n^2-1)(n+2)}{2^2 2! x^2} + \frac{n(n^2-1)(n^2-4)(n^2-9)(n+4)}{2^4 4! x^4} - \cdots$$

$$Q_{n+1/2}(x) = \frac{n(n+1)}{2 \cdot 1! x} - \frac{n(n^2-1)(n^2-4)(n+3)}{2^3 3! x^3} + \cdots.$$

Wronskian Relations:

$$j_\nu(x)y_\nu'(x) - j_\nu'(x)y_\nu(x) = \frac{1}{x^2}$$

$$j_\nu(x)[h_\nu^2(x)]' - j_\nu'(x)h_\nu^2(x) = -\frac{j}{x^2}$$

3. Legendre Functions

In the following formulas the argument is u. The first few polynomials are:

$$P_0 = 1 \qquad\qquad P_1^1 = (1-u^2)^{1/2} \qquad\qquad P_3^2 = 15u(1-u^2)^{1/2}$$

$$P_1 = u \qquad\qquad P_2^1 = 3u(1-u^2)^{1/2} \qquad\qquad P_3^3 = 15(1-u^2)^{1/2}$$

$$P_2 = \frac{3u^2-1}{2} \qquad P_2^2 = 3(1-u^2)$$

$$P_3 = \frac{5u^5-3u}{2} \qquad P_3^1 = \frac{3}{2}(5u^2-1)(1-u^2)^{1/2}$$

Special Values when n and m Are Integers:

$$P_n^m(u) = 0, \qquad m > n$$

$$P_{m+2s+1}^m(0) = 0$$

$$P_{m+2s}^m(0) = \frac{(-1)^s (2m + 2s)!}{2^{m+2s} s!(m+1)!}$$

Differentiation Formulas:

$$uP_n' - P_{n-1}' = nP_n$$

$$P_{n+1}' - P_{n-1}' = (2n+1)P_n$$

$$P_{n+1}' - uP_n' = (n+1)P_n$$

$$(u^2 - 1)P_n' = nuP_n - nP_{n-1}$$

$$P_n^m = (1 - u^2)^{m/2} \frac{d^m P_n(u)}{du^m} = \frac{(1 - u^2)^{m/2}}{2^n n!} \frac{d^{m+n}(u^2 - 1)^n}{du^{m+n}}$$

$$(1 - u^2)\frac{dP_n^m}{du} = (n+1)uP_n^m - (n-m+1)P_{n+1}^m = (n+m)P_{n-1}^m - nuP_n^m$$

Recurrence Formulas:

$$(n - m + 1)P_{n+1}^m = (2n+1)uP_n^m - (n+m)P_{n-1}^m$$

$$P_{n-1}^m = uP_n^m - (n-m+1)(1-u^2)^{1/2}P_n^{m-1}$$

$$P_{n+1}^m = uP_n^m + (n+m)(1-u^2)^{1/2}P_n^{m-1}$$

$$(1-u^2)^{1/2}P_n^{m+1} = (n+m+1)uP_n^m - (n-m+1)P_{n+1}^m$$

$$(1-u^2)^{1/2}P_n^{m+1} = 2muP_n^m - (n+m)(n-m+1)(1-u^2)^{1/2}P_n^{m-1}$$

$$(1-u^2)^{1/2}P_n^m = \frac{1}{2n+1}(P_{n+1}^{m+1} - P_{n-1}^{m+1})$$

$$\frac{m}{(1-u^2)^{1/2}}P_n^m = \frac{1}{2}u[(n-m+1)(n+m)P_n^{m-1} + P_n^{m+1}] + m(1-u^2)^{1/2}P_n^m$$

Integrals:

$$\int_{-1}^1 P_n^m(u)P_s^m(u)\, du = \frac{2}{2n+1}\frac{(n+m)!}{(n-m)!}\delta_{ns}$$

$$\int_{-1}^1 \frac{P_n^m(u)P_n^s(u)}{1-u^2}\, du = \frac{1}{m}\frac{(n+m)!}{(n-m)!}\delta_{ms}$$

A.11. VECTOR ANALYSIS FORMULAS

Rectangular Coordinates

$$u_1 = x \quad u_2 = y \quad u_3 = z$$

$$h_1 = 1 \quad h_2 = 1 \quad h_3 = 1$$

$$\nabla\Phi = \mathbf{a}_x \frac{\partial\Phi}{\partial x} + \mathbf{a}_y \frac{\partial\Phi}{\partial y} + \mathbf{a}_z \frac{\partial\Phi}{\partial z}$$

$$\nabla\cdot\mathbf{F} = \frac{\partial F_x}{\partial x} + \frac{\partial F_y}{\partial y} + \frac{\partial F_z}{\partial z}$$

$$\nabla \times \mathbf{F} = \mathbf{a}_x \left(\frac{\partial F_z}{\partial y} - \frac{\partial F_y}{\partial z} \right) + \mathbf{a}_y \left(\frac{\partial F_x}{\partial z} - \frac{\partial F_z}{\partial x} \right) + \mathbf{a}_z \left(\frac{\partial F_y}{\partial x} - \frac{\partial F_x}{\partial y} \right)$$

$$\nabla^2 \Phi = \frac{\partial^2 \Phi}{\partial x^2} + \frac{\partial^2 \Phi}{\partial y^2} + \frac{\partial^2 \Phi}{\partial z^2}$$

Cylindrical Coordinates

$$u_1 = r \quad u_2 = \phi \quad u_3 = z$$
$$h_1 = 1 \quad h_2 = r \quad h_3 = 1$$

$$\nabla \Phi = \mathbf{a}_r \frac{\partial \Phi}{\partial r} + \mathbf{a}_\phi \frac{1}{r} \frac{\partial \Phi}{\partial \phi} + \mathbf{a}_z \frac{\partial \Phi}{\partial z}$$

$$\nabla \cdot \mathbf{F} = \frac{1}{r} \frac{\partial}{\partial r} (rF_r) + \frac{1}{r} \frac{\partial F_\phi}{\partial \phi} + \frac{\partial F_z}{\partial z}$$

$$\nabla \times \mathbf{F} = \mathbf{a}_r \left(\frac{1}{r} \frac{\partial F_z}{\partial \phi} - \frac{\partial F_\phi}{\partial z} \right) + \mathbf{a}_\phi \left(\frac{\partial F_r}{\partial z} - \frac{\partial F_z}{\partial r} \right) + \mathbf{a}_z \left[\frac{1}{r} \frac{\partial (rF_\phi)}{\partial r} - \frac{1}{r} \frac{\partial F_r}{\partial \phi} \right]$$

$$\nabla^2 \Phi = \frac{1}{r} \frac{\partial}{\partial r} \left(r \frac{\partial \Phi}{\partial r} \right) + \frac{1}{r^2} \frac{\partial^2 \Phi}{\partial \phi^2} + \frac{\partial^2 \Phi}{\partial z^2}$$

Spherical Coordinates

$$u_1 = r \quad u_2 = \theta \quad u_3 = \phi$$
$$h_1 = 1 \quad h_2 = r \quad h_3 = r \sin \theta$$

$$\nabla \Phi = \mathbf{a}_r \frac{\partial \Phi}{\partial r} + \mathbf{a}_\theta \frac{1}{r} \frac{\partial \Phi}{\partial \theta} + \frac{\mathbf{a}_\phi}{r \sin \theta} \frac{\partial \Phi}{\partial \phi}$$

$$\nabla \cdot \mathbf{F} = \frac{1}{r^2} \frac{\partial}{\partial r} (r^2 F_r) + \frac{1}{r \sin \theta} \frac{\partial}{\partial \theta} (\sin \theta F_\theta) + \frac{1}{r \sin \theta} \frac{\partial F_\phi}{\partial \phi}$$

$$\nabla \times \mathbf{F} = \frac{\mathbf{a}_r}{r \sin \theta} \left[\frac{\partial}{\partial \theta} (F_\phi \sin \theta) - \frac{\partial F_\theta}{\partial \phi} \right] + \frac{\mathbf{a}_\theta}{r} \left[\frac{1}{\sin \theta} \frac{\partial F_r}{\partial \phi} - \frac{\partial}{\partial r} (rF_\phi) \right]$$
$$+ \frac{\mathbf{a}_\phi}{r} \left[\frac{\partial}{\partial r} (rF_\theta) - \frac{\partial F_r}{\partial \theta} \right]$$

$$\nabla^2 \Phi = \frac{1}{r^2} \frac{\partial}{\partial r} \left(r^2 \frac{\partial \Phi}{\partial r} \right) + \frac{1}{r^2 \sin \theta} \frac{\partial}{\partial \theta} \left(\sin \theta \frac{\partial \Phi}{\partial \theta} \right) + \frac{1}{r^2 \sin^2 \theta} \frac{\partial^2 \Phi}{\partial \phi^2}$$

Vector Identities

$$\nabla (\Phi + \psi) = \nabla \Phi + \nabla \psi$$

$$\nabla \cdot (\mathbf{A} + \mathbf{B}) = \nabla \cdot \mathbf{A} + \nabla \cdot \mathbf{B}$$

$$\nabla \times (\mathbf{A} + \mathbf{B}) = \nabla \times \mathbf{A} + \nabla \times \mathbf{B}$$

$$\nabla (\Phi \psi) = \Phi \nabla \psi + \psi \nabla \Phi$$

$$\nabla \cdot (\psi \mathbf{A}) = \mathbf{A} \cdot \nabla \psi + \psi \nabla \cdot \mathbf{A}$$

$$\nabla \cdot \mathbf{A} \times \mathbf{B} = \mathbf{B} \cdot \nabla \times \mathbf{A} - \mathbf{A} \cdot \nabla \times \mathbf{B}$$

$$\nabla \times (\Phi \mathbf{A}) = \nabla \Phi \times \mathbf{A} + \Phi \nabla \times \mathbf{A}$$

$$\nabla \times (\mathbf{A} \times \mathbf{B}) = \mathbf{A} \nabla \cdot \mathbf{B} - \mathbf{B} \nabla \cdot \mathbf{A} + (\mathbf{B} \cdot \nabla) \mathbf{A} - (\mathbf{A} \cdot \nabla) \mathbf{B}$$

$$\nabla (\mathbf{A} \cdot \mathbf{B}) = (\mathbf{A} \cdot \nabla) \mathbf{B} + (\mathbf{B} \cdot \nabla) \mathbf{A} + \mathbf{A} \times (\nabla \times \mathbf{B}) + \mathbf{B} \times (\nabla \times \mathbf{A})$$

$$\nabla \cdot \nabla \Phi = \nabla^2 \Phi$$

$$\nabla \cdot \nabla \times \mathbf{A} = 0$$

$$\nabla \times \nabla \Phi = 0$$

$$\nabla \times \nabla \times \mathbf{A} = \nabla(\nabla \cdot \mathbf{A}) - \nabla^2 \mathbf{A}$$

$$\iiint_V \nabla \Phi \, dV = \oiint_S \Phi \, d\mathbf{S}$$

$$\iiint_V \nabla \cdot \mathbf{A} \, dV = \oiint_S \mathbf{A} \cdot d\mathbf{S}$$

$$\iiint_V \nabla \times \mathbf{A} \, dV = \oiint_S \mathbf{n} \times \mathbf{A} \, dS$$

$$\iint_S \mathbf{n} \times \nabla \Phi \, dS = \oint_C \Phi \, d\mathbf{l}$$

$$\iint_S \nabla \times \mathbf{A} \cdot d\mathbf{S} = \oint_C \mathbf{A} \cdot d\mathbf{l}$$

REFERENCES AND BIBLIOGRAPHY

Vector and Dyadic Analysis

[A.1] L. Brand, *Vector and Tensor Analysis*. New York, NY: John Wiley & Sons, Inc., 1947.
[A.2] P. M. Morse and H. Feshbach, *Methods of Theoretical Physics*, part I. New York, NY: McGraw-Hill Book Company, Inc., 1953.
[A.3] B. Spain, *Tensor Calculus*, University Mathematical Texts. New York, NY: Interscience Publishers, Inc., 1953.

Matrices

[A.4] E. A. Guillemin, *The Mathematics of Circuit Analysis*. New York, NY: John Wiley & Sons, Inc., 1949.
[A.5] A. C. Aitken, *Determinants and Matrices*, University Mathematical Texts. New York, NY: Interscience Publishers, Inc., 1948.

Calculus of Variations

[A.6] R. Weinstock, *Calculus of Variations*. New York, NY: McGraw-Hill Book Company, Inc., 1952.
[A.7] R. Courant and D. Hilbert, *Methods of Mathematical Physics*, English ed. New York, NY: Interscience Publishers, Inc., 1953.

Infinite Products and Gamma Functions

[A.8] E. T. Copson, *Theory of Functions of a Complex Variable*. New York, NY: Oxford University Press, 1935.
[A.9] E. C. Titchmarsh, *Theory of Functions*, 2nd ed. New York, NY: Oxford University Press, 1939. (See also [A.2].)

Summation of Fourier Series

[A.10] L. A. Pipes, "The summation of Fourier series by operational methods," *J. Appl. Phys.*, vol. 21, pp. 298–301, Apr. 1950.
[A.11] A. D. Wheelon, "On the summation of infinite series in closed form," *J. Appl. Phys.*, vol. 25, pp. 113–118, Jan. 1954.
[A.12] G. G. Macfarlane, "The application of Mellin transforms to the summation of slowly convergent series," *Phil. Mag.*, vol. 40, pp. 188–197, Feb. 1949. See also [A.2] and [A.7, sect. 3.3 and probs. 26–28 at the end of sect. 13.96].)
[A.13] L. B. W. Jolley, *Summation of Series*. New York, NY: Dover Publications, 1961.

Name Index

Page numbers for entries that appear in a footnote are in italics

Subject Index

About the Author

Robert E. Collin (M'54-SM'60-F'72) was born in 1928 in Alberta, Canada. He received the B.Sc. degree in engineering physics from the University of Saskatchewan in 1951. He attended Imperial College in England for graduate work and obtained the Ph.D. degree in electrical engineering from the University of London in 1954.

From 1954 to 1958 he was a Scientific Officer at the Canadian Armament Research and Development Establishment where he worked on missile guidance antennas, radomes, and radar systems evaluation. He was also an adjunct professor at Laval University during this time. He immigrated to the United States in 1958 and became a citizen in 1964. He joined the Electrical Engineering Department at Case Institute of Technology (now Case Western Reserve University), Cleveland, OH, in 1958. During his tenure there he has served as Chairman of the Electrical Engineering and Applied Physics Department for five years and also as Interim Dean of Engineering for two years. He has been an Invited Professor at the Catholic University in Rio de Janeiro, Brazil; at Telebras Research Center, Campinas, Brazil; and the University of Beijing, Peoples Republic of China. He was also a Distinguished Visiting Professor at the Graduate School, Ohio State University, Columbus, during the 1982–83 academic year. He is the author/co-author of more than 100 technical papers. He is also the author of *Field Theory of Guided Waves, Foundations for Microwave Engineering, Antennas and Radiowave Propagation*, a co-author with R. Plonsey of *Principles and Applications of Electromagnetic Fields*, and a co-editor with F. Zucker and contributing author of *Antenna Theory, Parts I and II*.

Professor Collin is a member of the National Academy of Engineering and a member of the IEEE Antennas and Propagation Society, the IEEE Microwave Theory and Techniques Society, and the USA Commission B of URSI. He is currently a member of the Administrative Committee of the IEEE Antennas and Propagation Society. He has served on a variety of IEEE Committees for Awards, Publications, Technical Programs, and Standards. He is also currently an Associate Editor for the *IEEE Transactions on Antennas and Propagation* and on the Editorial Board for the *Electromagnetic Waves and Applications Journal*.